GREENLAND
Arctic Circle
ICELAND
NORWAY
FINLAND
SWEDEN
ESTONIA
LATVIA
LITHUANIA
RUSSIA
UNITED KINGDOM
DENMARK
NETH.
IRELAND
BELARUS
POLAND
GER.
BELG.
LUX.
CZECH REP.
SLOVAKIA
UKRAINE
MOLDOVA
ATLANTIC
OCEAN
SWITZ.
AUS.
HUNG.
FRANCE
SLOVENIA
CROATIA
ROM.
BOSNIA
MONTENEGRO
KOSOVO
BULG.
MACEDONIA
PORTUGAL
SPAIN
ITALY
ALBANIA
GREECE
TURKEY
GEORGIA
ARMENIA
AZERBAIJAN
KAZAKHSTAN
UZBEKISTAN
TURKMENISTAN
KYRGYZSTAN
TAJIKISTAN
MONGOLIA
CHINA
N. KOREA
S. KOREA
JAPAN
TUNISIA
CYPRUS
LEBANON
SYRIA
IRAQ
IRAN
AFGHANISTAN
PAKISTAN
NEPAL
BHUTAN
MOROCCO
ALGERIA
LIBYA
EGYPT
ISRAEL
JORDAN
KUWAIT
BAHRAIN
QATAR
U.A.E.
SAUDI ARABIA
OMAN
INDIA
BANGLADESH
MYANMAR
LAOS
TAIWAN
PACIFIC
OCEAN
Tropic of Cancer
WESTERN SAHARA
MAURITANIA
MALI
NIGER
CHAD
SUDAN
ERITREA
YEMEN
DJIBOUTI
THAILAND
VIETNAM
CAMBODIA
PHILIPPINES
SENEGAL
GAMBIA
GUINEA-BISSAU
GUINEA
BURKINA FASO
NIGERIA
CENTRAL AFRICAN REP.
SOUTH SUDAN
ETHIOPIA
SRI LANKA
SIERRA LEONE
LIBERIA
IVORY COAST
GHANA
TOGO
BENIN
CAMEROON
SOMALIA
INDIAN
OCEAN
MALDIVES
BRUNEI
MALAYSIA
SINGAPORE
EQUATORIAL GUINEA
GABON
CONGO
UGANDA
KENYA
RWANDA
BURUNDI
DRCONGO
TANZANIA
Equator
INDONESIA
PAPUA NEW GUINEA
EAST TIMOR
COMOROS
ATLANTIC
ANGOLA
ZAMBIA
MALAWI
MOCAMBIQUE
MADAGASCAR
ZIMBABWE
NAMIBIA
BOTSWANA
MAURITIUS
Tropic of Capricorn
AUSTRALIA
VANUATU
FIJI
NEW CALEDONIA
OCEAN
SWAZILAND
SOUTH AFRICA
LESOTHO
NEW ZEALAND
SOUTHERN
OCEAN
Antarctic Circle
60°
40°
20°
0°
100°
120°
140°
160°

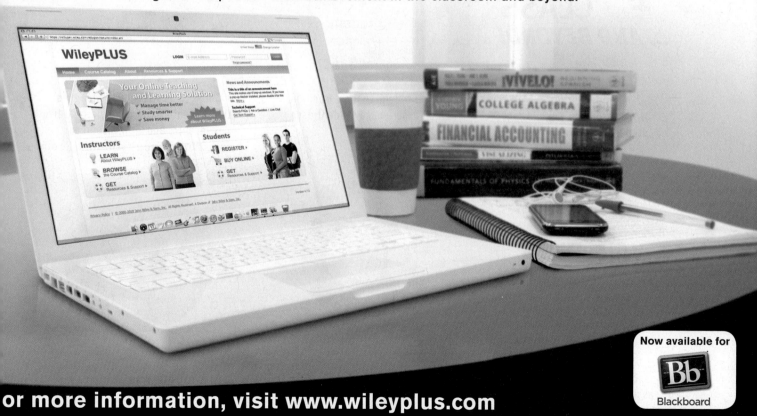

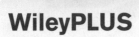

16th Edition

GEOGRAPHY

REALMS, REGIONS, AND CONCEPTS

H. J. de Blij
John A. Hannah Professor of Geography
Michigan State University

Peter O. Muller
Senior Professor, Department of Geography and Regional Studies
University of Miami

Jan Nijman
Professor of Urban Studies, University of Amsterdam
Professor Emeritus of Geography, University of Miami

WILEY

VICE PRESIDENT AND PUBLISHER	Petra Recter
EXECUTIVE EDITOR	Ryan Flahive
PRODUCT DESIGNER	Beth Tripmacher
PRODUCT DESIGNER	Howard Averback
ASSISTANT CONTENT EDITOR	Darnell Sessoms
EDITORIAL ASSISTANT	Julia Nollen
SENIOR CONTENT MANAGER	Micheline Frederick
SENIOR PRODUCTION EDITOR	Janet Foxman
SENIOR PHOTO EDITOR	Lisa Gee
MEDIA SPECIALIST	Anita Castro
CREATIVE DIRECTOR	Harry Nolan
SENIOR DESIGNER	Wendy Lai
SENIOR MARKETING MANAGER	Margaret Barrett
PRODUCTION SERVICES	Jeanine Furino/ Furino Production
PRODUCTION ASSISTANT	Helen Seachrist

Front cover photo: © Jan Nijman
Back cover photos: © H. J. de Blij

This book was set in 10.5/12 Adobe Garamond by Aptara, and printed and bound by Courier/Kendallville.

This book is printed on acid-free paper. ∞

Founded in 1807, John Wiley & Sons, Inc. has been a valued source of knowledge and understanding for more than 200 years, helping people around the world meet their needs and fulfill their aspirations. Our company is built on a foundation of principles that include responsibility to the communities we serve and where we live and work. In 2008, we launched a Corporate Citizenship Initiative, a global effort to address the environmental, social, economic, and ethical challenges we face in our business. Among the issues we are addressing are carbon impact, paper specifications and procurement, ethical conduct within our business and among our vendors, and community and charitable support. For more information, please visit our website: *www.wiley.com/go/citizenship*.

Library of Congress Cataloging-in-Publication Data

De Blij, Harm J.
Geography: realms, regions, and concepts/H. J. de Blij, Peter O. Muller, and Jan Nijman.—Sixteenth edition. pages cm.
Includes index.
ISBN 978-1-118-67395-9 (hardback)
1. Geography—Textbooks. I. Peter O. Muller. II. Jan Nijman. III. Title.
G128.D42 2013
910—dc23

2013024700

978-1-118-67395-9 (Main Book ISBN)
978-1-118-70760-9 (Binder-Ready Version ISBN)

Printed in the United States of America

10 9 8 7 6 5 4 3 2

PREFACE

For the past four decades, *Geography: Realms, Regions, and Concepts* has reported (and sometimes anticipated) trends in the discipline of Geography and developments in the world at large. In fifteen preceding editions, *Regions,* as the book has come to be called, has explained the contemporary world's **geographic realms** (the largest regional entities on the face of the Earth) and their natural environments and human dimensions. In the process, this book has become an introduction to Geography itself, the discipline that links the study of human societies and natural environments. We look at the ways people have organized their living space, adapted to changing social as well as environmental circumstances, and continue to confront forces beyond their control ranging from globalization to climate change. From old and still relevant concepts to new and untested ideas, *Regions* provides geographic perspective on our transforming world.

The book before you, therefore, is an information highway to geographic literacy. The first edition appeared in 1971, at a time when school geography in the United States (though not in Canada) was a subject in decline. It was a precursor of a dangerous isolationism in America, and geographers foresaw the looming cost of geographic illiteracy. Sure enough, the media during the 1980s began to report that polls, public surveys, tests, and other instruments were recording a lack of geographic knowledge at a time when our world was changing ever faster and becoming more competitive by the day. Various institutions, including the National Geographic Society, banks, airline companies, and a consortium of scholarly organizations, mobilized to confront an educational dilemma that had resulted substantially from a neglect of the very topics this book is about. Concern about geographic literacy and about the need for knowledge of the world around us continues to this day.

A useful discussion of such commonplace topics as globalization, and of popular misconceptions such as the "flat world" and the "death of distance," means that we must know the different parts of the world, the interrelated components that together make up the whole. This goes far beyond description or memorization. In each of the chapters of this book, we ask fundamental questions about the role of geography in the ongoing social, economic, and political developments in that particular realm. Geography helps to understand the dynamic differences among realms and regions and *why* they develop the way they do. Each chapter offers rich regional descriptions alongside critical analyses.

Knowing and understanding world geography is not just an academic exercise. You will find that much of what you encounter in this book is of immediate, practical value to you—as a citizen, a consumer, a traveler, a voter, a job-seeker. North America is a geographic realm with ever-intensifying global interests and involvements. Those interests and involvements require countless, often instantaneous decisions, whether they are concerned with international business, worldwide media, affairs of state, or disaster relief. Such decisions must be based on the best possible knowledge of the world beyond our continent. That knowledge can be gained by studying the layout of our world, its environments, societies, resources, policies, traditions, and other properties—in short, its regional geography.

REALMS, REGIONS . . . AND CONCEPTS

This book begins with an introductory chapter followed by 12 sets of chapters that each cover one of the world's major geographic realms and their constituent regions. The Introduction discusses the world as a whole, outlining the physical stage on which the human drama is being played out, providing environmental frameworks, demographic data, political background, and economic-geographical context. In the remaining chapters, we have divided the world into 12 major geographic realms and these are in turn subdivided into smaller regions. We explain why the realms and regions are delineated the way they are, we describe their prevailing geographic features, and we examine recent events around the world in their proper geographic context.

Along the way, we introduce more than 150 geographic concepts placed in their regional settings. Most of these concepts are indeed primarily geographical, but others are ideas about which, we believe, students of geography should have some knowledge. Although such concepts are listed on the opening page of every chapter, we have not, of course, enumerated every geographic notion used in that chapter. Many colleagues, we suspect, will want to make their own realm-concept associations, and as readers will readily perceive, the book's organization is quite flexible. It is possible, for example, to focus almost exclusively on substantive regional material or, alternatively, to concentrate mainly on conceptual issues.

The A-B Structure of the Regional Chapters

This sixteenth edition of *Regions* continues the formal separation of **A** and **B** chapters that was introduced in the fifteenth edition, an innovation that received substantial positive feedback from our users. As the table of contents shows, each of the world's geographic realms (with the exception of the Austral and Pacific Realms) is treated in a *pair* of chapters: **A** chapters discuss broad geographic features at the scale of the entire realm; **B** chapters provide a more detailed analysis at the scale of the realm's constituent regions and countries.

This structure increases flexibility for the user because it offers the option of customizing reading assignments. As anyone who has taught or taken a course in World Regional Geography knows, the subject is complex, the world is still a very large and highly diverse place, and semesters are short. In principle, each of the **A** or **B** chapters can be read in isolation. If users prefer to reduce reading assignments on one or more realms, they might concentrate on the **A** chapters for a main overview of the geography of the realm; the opposite approach focused on the **B** chapters is possible as well. Neither scale (realm or region) can be said to be more important than the other. They are most revealing, of course, in terms of their complementary nature: one informs the other—from the general to the specific, from the whole to its constituent parts, from the global to the local.

The second advantage of this **A-B** structure is more substantive and is anchored in current disciplinary trends. *Scale,* along with *region,* is one of the most crucial general concepts in Geography. Understanding geographic events almost always demands a multi-scalar perspective. Using the **A-B** structure, we consistently discuss processes at two different scales. First we employ the broader scale of the realm to appreciate, for example, matters of geopolitics or economic integration or core-periphery relationships—developments that unfold and project across an entire realm or major parts of it. Think, for example, about the realmwide ramifications of the Arab Spring movement. Subsequently, we "zoom in" on the regions nested within the realm to consider developments at a finer resolution. Think, for example, of Egypt's ongoing political challenges since the forced 2011 resignation of former President Mubarak. Typically, **B** chapters facilitate a closer look at the role of individual countries and national cultures (and, where necessary, their internal subdivisions).

The **A-B** structure should not be regarded as a rigid mechanism that separates geographic reality into two 'fixed' scales. Geography is a dynamic field; there is continuity in some ways, but change abounds. Geographic space is continuously reorganized and reshaped. The study of realms and regions is more based on "sliding" scales, from the level of the realm downward. **A** chapters typically deal with broader-scale issues; **B** chapters provide more close-up analysis. Thus international issues that do not always involve the entire realm may be discussed in either **A** or **B** chapters, or both (e.g., India-Pakistan relations within the South Asian realm).

Voice From the Region

Another new feature that was introduced in the fifteenth edition and that met with considerable approval from the readers is the boxed vignette entitled *Voice From the Region.* This feature provides a platform to let people from each geographic realm speak directly to the reader—a local voice, unfiltered, that presents a reflection on their corner of the world as seen through their own eyes. Located in each **B** chapter (plus Chapters 11 and 12), these vignettes, accompanied by the name and a photo of the individual providing the local perspective, focus on important current events ranging from Singapore's tenuous absorption of large numbers of Chinese immigrants to Egypt's stalled political reforms in the aftermath of its Arab Spring. Each *Voice* was solicited by the authors specifically for this purpose. It provides a useful tool for teachers to initiate class discussion about regional issues and perspectives, and it helps students identify more closely with developments and people across the spectrum of world regions.

NEWS ABOUT THE SIXTEENTH EDIITON

One of the most fascinating things about Geography is that it exudes a sense of permanence in certain respects (e.g., the natural environment), but in reality things change over time—sometimes quite slowly but at other times abruptly and dramatically. New editions of *Regions* have a reputation for comprehensive and detailed revision, and the sixteenth is a particularly substantial reprise. Occasionally the world changes so fundamentally that interpretations based on earlier understandings are overtaken by new developments—forcing us to rethink particular realms, regions, and countries, or even the global system that binds them together. Thus new editions of *Regions* not only contain myriad valuable updates but also reappraisals of the fundamental nature of the world's geography.

The world is undergoing some momentous changes in this decade: (1) widespread, longstanding pessimism concerning Subsaharan Africa's development prospects has rapidly given way to unprecedented hope and anticipation based on robust recent growth in this realm; (2) the European Union has become more disparate than ever—fiscally, economically, and politically—as rising core-periphery tensions weigh down the supranational federal framework; (3) the globally expanding influence and assertiveness of China is being met with increasingly mixed feelings in certain realms and regions; (4) ravenous demand for raw materials on the world market is altering landscapes in a wide range of commodity-rich countries from Mongolia to Australia to Chile; and (5) the *Arab Spring,* which has been destabilizing regional politics in North Africa and Southwest Asia, but has yet to deliver enduring democratic advances in any country. These developments, among others, have reverberated throughout the world in recent years and continue to transform their respective realms today. We will detail many of these changes, as well as the forces that drive them, in their geographic context as our global regional survey unfolds.

What to Look For

With a revision manuscript totaling tens of thousands of words, countless map modifications, dozens of original photos, and new *From the Field Notes* and other boxes, no summary can adequately encapsulate all that this new edition contains. One special note: we have highlighted the burgeoning outward growth of several of the world's megacities (e.g., São Paulo, Istanbul, Delhi-New Delhi, Tokyo, Moscow, Xian) and expanded the city maps in many of the *Great Cities of the World* boxes to reflect this metropolitan-scale urbanization.

Now let us turn to some of the high points by chapter. In the **Introduction**, besides various updates, your attention will be drawn to new sections on the use of geography in new media (smartphones, navigational mapping) as well as global population issues and the Earth's carrying capacity. Also look for several new photos and a new *Field Note* on Singapore as a *world-city*.

In **Chapters 1A** and **1B**, contemporary **Europe** is discussed within a core-periphery framework, an approach that could not be more pertinent given today's economic circumstances. The European Union is bigger and economically more disparate than ever and tensions between the crisis-ridden periphery and the core are straining the federalized, supranational framework. As Croatia becomes the EU's 28th member and Latvia joins the eurozone in 2014, Greece remains in a deep fiscal crisis and the United Kingdom is openly questioning the worth of continued EU membership. Germany finds itself increasingly at the center of European integration and power but there, too, voters are indicating that there are limits to their support for needy Union member-countries.

Chapters 2A and **2B**, which cover the **Russian Realm**, emphasize the Russian leadership's desire to revive the country's stature as a global power while internal calls for greater democracy are growing louder. There are continued difficulties with Muslim populations along the southern periphery in the Caucasus region, raising the haunting specter of more terrorism within Russia's borders. Climate change is likely to have a major impact on this gigantic country's northern, Arctic-fronting lowlands, yet Siberia and the Far East are rapidly losing population as their stay is no longer enforced by the government. In this new edition, we have designated eastern Ukraine as a new transition zone, given its large ethnic Russian population as well as its economic and political orientation to Russia.

Chapters 3A and **3B,** covering **North America**, present updated material on the U.S. housing crisis focusing on the geography of negative equity and foreclosures. Adding to the section on the realm's main industrial regions, there is new text on North American high-technology clusters with corresponding new map features. Also note the new map showing Miami's global city connections that reflect the city's role as a major business node in the Western Hemisphere.

In **Chapters 4A** and **4B** on **Middle America**, the emphasis continues on the realm's giant, Mexico, where the drug wars and accompanying violence pose a long-term challenge to national cohesion. Updated maps of the drug conflict display international trafficking, the latest carving up of Mexican turf by rival cartels, and the violent killings that have occurred. Elsewhere, look for new sections on rising, drug-related crime rates in Middle American countries (especially northern Central America) as well as new text on Haiti, plagued by the painfully slow recovery from its disastrous 2010 earthquake, and Cuba, where incremental economic reforms are gradually opening the country to tourism, trade, and investment.

Major portions of **Chapters 5A** and **5B** focus on the impact of the global commodities boom on **South America**. China's apparently insatiable appetite for raw materials, has resulted in the rapid expansion of the primary economic sectors (mining, agriculture) of such countries as Brazil, Chile, and Peru. This has changed the cultural and economic landscape, sometimes with serious environmental consequences, but also boosted government revenues while stimulating individual industries. Colombia continues to be profoundly affected by illicit cocaine production, exacerbated by the major 'export route' via western Venezuela that remained open following the 2013 death of Hugo Chavez. In 2014, the eyes of the world will be on Brazil when it hosts soccer's World Cup tournament, followed in 2016 by the Rio de Janeiro Olympic Games. This dynamic, fast-rising country intends to showcase its newfound modernity despite the unexpected mid-2013 outbreak of nationwide protests over the government's inattention to the needs of Brazil's enormous lower and middle classes.

Chapters 6A and **6B** document one of the biggest stories of the past few years: the emergence of **Subsaharan Africa** onto the global economic scene. The world's fastest growing national economies are located in this realm and they are reshaping the future prospects of a continent that has for so long been relegated to scenarios of doom. Africa's rapid growth, like South America's, is largely based on the exploitation and exporting of raw materials to China. Chinese investments in the primary sector, infrastructure, and construction are helping to fuel the ongoing transformation. Ironically, this realm's largest economy, South Africa, now finds itself lagging behind in terms of growth rates. There is a new map and accompanying table that cover differential economic growth and raw material dependency across the realm. A new *Regional Issue* box debates the pros and cons of foreign investments in agriculture (perhaps better known as 'land grabs').

By far, the most volatile realm of this decade is undoubtedly **North Africa/Southwest Asia**. **Chapters 7A** and **7B** contain new sections on the unfolding of the Arab Spring and how it has played out in different national contexts—supported by an important new map of the geography of national political regimes across the realm. Also look for new text on recent developments in Turkey, Iraq, Somalia, Egypt (with an updated *Voice from the Region* to underscore the continuation of simmering discontent), Mali, and, of course, Syria, whose catastrophic civil war had taken at least 100,000 lives as of mid-2013. The Arab Spring has so far failed to produce tangible democratic progress in any country, although the civil strife unleashed in Libya and Syria swiftly triggered major regional destabilization.

Chapters 8A and **8B** on **South Asia** now focus on: Pakistan's myriad challenges, especially in its unstable Afghanistan border zone; India's fragmented modernization, exhibited by its modest but highly developed IT industry juxtaposed against its vast underdeveloped, inefficient agricultural sector; the precarious India-Pakistan relationship that plays out at a different level among Hindus and Muslims inside

India; Sri Lanka's slow and arduous recovery from civil war; the developmental challenges of landlocked Nepal and Bhutan; and the Maldives' potentially fatal struggle with rising sea level as depicted on the book's covers and well as in the new *Voice from the Region*.

Chapters 9A and **9B** survey another incredibly dynamic realm, **East Asia**. If China's influence can now be felt all over the world, its dominance within its home realm is overwhelming. China's own economic geography is quickly changing as well, with the rapid westward expansion of the country's Coastal Core region into the Interior. In the meantime, the gap between rich and poor continues to widen and the Chinese government faces increasing demands from its workforce for higher wages. A new *Regional Issue* highlights the growing contentiousness of the One-Child-Only policy, and a new section describes how the unprecedented pace of Chinese government-led urbanization has raised questions about the possibility of an equally unprecedented real estate bubble—from the massive, still-empty developments on the outskirts of Tianjin to the entirely new city of Ordos in Inner Mongolia. Internationally, there is growing tension between China and Japan in the East China Sea, and between China and several Southeast Asian states in the South China Sea—while North Korea's belligerence reached new heights in early 2013. Look for an entirely rewritten section on Korea based on an extended 2012 visit of the senior author.

China's aggressive reach into **Southeast Asia**, and the tensions this provokes, are major themes in **Chapters 10A** and **10B**. There are new maps on conflicting claims in the South China Sea as well as on China's growing dominance of this realm's import and export trade (at the expense of Japan and United States). Also look for new text on the influx of affluent mainland-Chinese immigrants into Singapore and their challenged integration; on Indonesia's continued economic rise; and the sudden reopening of Myanmar to the realm and the rest of the world, ending decades of debilitating isolation and brutal repression.

Chapters 11 and 12, because of their limited size, remain single chapters without the **A/B** division. In **Chapter 11**, the **Austral Realm**, we continue our focus on three major themes: the environment (Australia has suffered both dreadful droughts and death-dealing fires, and is now confronted with difficult choices in balancing oil and gas exploitation against environmental protection); aboriginal peoples and their changing role in the modern, globalizing societies of Australia and New Zealand; and Australia's cultural diversification as immigration continues to transform its human geography.

Chapter 12, entitled **Pacific Realm and Polar Futures**, updates the three Pacific regions, and then turns to the growing impact of humans in (and designs on) both polar zones. In the far southerly latitudes, environmental change is facilitating the entry of thousands of tourists who cruise Antarctic waters and land on the Earth's southernmost shores. And in the northernmost latitudes, climate change has already resulted in the partial and temporary (August/September) opening of the long-frozen waterways of the Arctic Basin's Northeast Passage, along which a record 45 ships sailed through in 2012. Also look for an updated map of the Arctic showing the significant retreat of the polar icecap and for new text on recent technological advances that enable deep-sea mineral exploration in the waters of the Pacific and elsewhere.

This latest edition of *Regions* reflects our continuing commitment to bring to our readers our geographic perspectives on a fast-changing world and its dynamic regional components. Since the previous edition, the author team has spent much time traveling, researching, and lecturing in dozens of locations around the globe. The authors' international involvements are invaluable in the continuation, innovation, and strengthening of *Regions*, which after more than 40 years is still one of the most exciting projects in the domain of college geography textbooks. We trust that you will find this Sixteenth Edition as informative and challenging as you have our earlier ones.

DATA SOURCES

Numerous print and Internet sources were consulted during the updating of this book. For all matters geographical, of course, we consult *The Annals of the Association of American Geographers, The Professional Geographer, The Geographical Review, The Journal of Geography*, and many other academic journals published regularly in North America—plus an array of similar periodicals published in English-speaking countries from Scotland to New Zealand.

All quantitative information was updated to the year of publication and checked rigorously. Hundreds of other modifications were made, many in response to readers' and reviewers' comments. New spellings of place names continue, and we pride ourselves in being a reliable source for current and correct usage.

The statistical data that constitute Appendix B are derived from numerous sources. As users of such data are aware, considerable inconsistency marks the reportage by various agencies, and it is often necessary to make informed judgments about contradictory information. For example, some sources still do not reflect the rapidly declining rates of population increase or life expectancies in AIDS-stricken African countries. Others list demographic averages without accounting for differences between males and females in this regard.

In formulating Appendix B we have used among our sources the United Nations, the Population Reference Bureau, the World Bank, the Encyclopaedia Britannica *Books of the Year, The Economist* Intelligence Unit, the *Statesman's Year-Book*, and *The New York Times Almanac*. The urban population figures—which also entail major problems of reliability and comparability—are mainly drawn from the most recent database published by the United Nations' Population Division. For cities of less than 750,000, we developed our own estimates from a variety of sources. At any rate, the urban population figures used here are estimates for 2014 and they represent *metropolitan-area totals* unless otherwise specified.

ANCILLARIES

A broad spectrum of print and electronic ancillaries are available to support both instructors and students. To see a complete listing of these ancillaries, or to gain access to them upon adoption and purchase, please visit: http://www.wiley.com//college/sc/deblij/

ACKNOWLEDGMENTS

Over the 43 years since the publication of the First Edition of *Geography: Realms, Regions, and Concepts*, we have been fortunate to receive advice and assistance from literally thousands of people. One of the rewards associated with the publication of a book of this kind is the steady stream of correspondence and other feedback it generates. Geographers, economists, political scientists, education specialists, and others have written us, often with fascinating enclosures. We make it a point to respond personally to every such letter, and our editors have communicated with many of our correspondents as well. Moreover, we have considered every suggestion made and many who wrote or transmitted their reactions through other channels will see their recommendations in print in this edition.

Student Response

A major part of the correspondence we receive comes from student readers. We would like to take this opportunity to extend our deep appreciation to the several million students around the world who have studied from our books. In particular, we thank the students from more than 150 different colleges across the United States who took the time to send us their opinions. Students told us they found the maps and graphics attractive and functional. We have not only enhanced the map program with exhaustive updating but have added a number of new maps to this Sixteenth Edition as well as making significant changes in many others. Generally, students have told us that they found the pedagogical devices quite useful. We have kept the study aids the students cited as effective: a boxed list of each chapter's key concepts, ideas, and terms (numbered for quick reference in the text itself); a box that summarizes each realm's major geographic qualities; and an extensive and still-expanding Glossary.

Faculty Feedback

Several faculty colleagues from around the world assisted us with earlier editions, and their contributions continue to grace the pages of this book. Among them are:

James P. Allen, *California State University, Northridge*
Stephen S. Birdsall, *University of North Carolina*
J. Douglas Eyre, *University of North Carolina*
Fang Yong-Ming, *Shanghai, China*
Edward J. Fernald, *Florida State University*
Ray Henkel, *Arizona State University*
Richard C. Jones, *University of Texas at San Antonio*
Gil Latz, *Portland State University (Oregon)*
Ian MacLachlan, *University of Lethbridge (Alberta)*
Melinda S. Meade, *University of North Carolina*
Henry N. Michael, *late of Temple University (Pennsylvania)*
Clifton W. Pannell, *University of Georgia*
J. R. Victor Prescott, *University of Melbourne (Australia)*
John D. Stephens, *University of Washington*
Canute Vander Meer, *University of Vermont*

Faculty members from a large number of North American colleges and universities continue to supply us with vital feedback and much-appreciated advice. Our publishers arranged several feedback sessions, and we are most grateful to the following professors for showing us where the text could be strengthened and made more precise:

Martin Arford, *Saginaw Valley State University (Michigan)*
Donna Arkowski, *Pikes Peak Community College (Colorado)*
Greg Atkinson, *Tarleton State University (Texas)*
Christopher Badurek, *Appalachian State University*
Denis Bekaert, *Middle Tennessee State University*
Thomas L. Bell, *University of Tennessee*
Donald J. Berg, *South Dakota State University*
Jill (Alice) Black, *Missouri State University*
Kathleen Braden, *Seattle Pacific University*
David Cochran, *University of Southern Mississippi*
Joseph Cook, *Wayne County Community College (Michigan)*
Deborah Corcoran, *Southwest Missouri State University*
Marcelo Cruz, *University of Wisconsin at Green Bay*
William V. Davidson, *Louisiana State University*
Larry Scott Deaner, *Kansas State University*
Jeff de Grave, *University of Wisconsin, Eau Claire*
Jason Dittmer, *Georgia Southern University*
James Doerner, *University of Northern Colorado*
Steven Driever, *University of Missouri-Kansas City*
Elizabeth Dudley-Murphy, *University of Utah*
Dennis Ehrhardt, *University of Louisiana-Lafayette*
Bryant Evans, *Houston Community College*
William Flynn, *Oklahoma State University*
William Forbes, *Stephen F. Austin State University (Texas)*
Bill Foreman, *Oklahoma City Community College*
Eric Fournier, *Samford University (Alabama)*
Gary A. Fuller, *late of University of Hawai'i*
Randy Gabrys Alexson, *University of Wisconsin-Superior*
William Garbarino, *Community College of Allegheny County (Pennsylvania)*
Hari Garbharran, *Middle Tennessee State University*
Chad Garick, *Jones County Junior College*
Jon Goss, *University of Hawai'i*
Debra Graham, *Messiah College*
Richard Grant, *University of Miami*
Jason B. Greenberg, *Sullivan University*
Margaret M. Gripshover, *University of Tennessee*
Sara Harris, *Neosho County Community College (Kansas)*
John Havir, *Ashland Community & Technical College*
John Hickey, *Inver Hills Community College (Minnesota)*
Shirlena Huang, *National University of Singapore*
Ingrid Johnson, *Towson State University (Maryland)*
Kris Jones, *Saddleback College (California)*
Uwe Kackstaetter, *Front Range Community College, Westminster (Colorado)*
Cub Kahn, *Oregon State University*
J. Miguel Kanai, *University of Miami*
Thomas Karwoski, *Anne Arundel Community College (Maryland)*

Robert Kerr, *University of Central Oklahoma*
Eric Keys, *University of Florida*
Jack Kinworthy, *Concordia University-Nebraska*
Marti Klein, *Saddleback College (California)*
Christopher Laingen, *Kansas State University*
Heidi Lannon, *Santa Fe College (Florida)*
Unna Lassiter, *Stephen F. Austin State University (Texas)*
Richard Lisichenko, *Fort Hays State University (Kansas)*
Catherine Lockwood, *Chadron State College (Nebraska)*
George Lonberger, *Georgia Perimeter College*
Claude Major, *Stratford Career*
Steve Matchak, *Salem State College (Massachusetts)*
Chris Mayda, *Eastern Michigan University*
Trina Medley, *Oklahoma City Community College*
Dalton W. Miller, *Jr., Mississippi State University*
Ernando F. Minghine, *Wayne State University (Michigan)*
Veronica Mormino, *Harper College (Illinois)*
Tom Mueller, *California University (Pennsylvania)*
Irene Naesse, *Orange Coast College (California)*
Valiant C. Norman, *Lexington Community College (Kentucky)*
Richard Olmo, *University of Guam*
Pai Yung-Feng, *New York City*
J. L. Pasztor, *Delta College (Michigan)*
Iwona Petruczynik, *Mercyhurst College (Pennsylvania)*
Paul E. Phillips, *Fort Hays State University (Kansas)*
Rosann Poltrone, *Arapahoe Community College (Colorado)*
Jeff Popke, *East Carolina University*
David Privette, *Central Piedmont Community College (North Carolina)*
Joel Quam, *College of DuPage (Illinois)*
Rhonda Reagan, *Blinn College (Texas)*
Paul Rollinson, *Missouri State University*
Rinku Roy Chowdhury, *Indiana University*
A. L. Rydant, *Keene State College (New Hampshire)*
Justin Scheidt, *Ferris State University (Michigan)*
Kathleen Schroeder, *Appalachian State University*
Nancy Shirley, *Southern Connecticut State University*
Dmitrii Sidorov, *California State University, Long Beach*
Dean Sinclair, *Northwestern State University (Louisiana)*
Richard Slease, *Oakland, North Carolina*
Susan Slowey, *Blinn College (Texas)*
Dean B. Stone, *Scott Community College (Iowa)*
Jamie Strickland, *University of North Carolina at Charlotte*
Ruthine Tidwell, *Florida Community College at Jacksonville*
Irina Vakulenko, *Collin County Community College (Texas)*
Cathy Weidman, *Austin, Texas*
Kirk White, *York College of Pennsylvania*
Thomas Whitmore, *University of North Carolina*
Keith Yearman, *College of DuPage (Illinois)*
Laura Zeeman, *Red Rocks Community College (Colorado)*
Yu Zho, *Vassar College (New York)*

PERSONAL APPRECIATION

For assistance with the map of North American indigenous people, we are greatly indebted to Jack Weatherford, Professor of Anthropology at Macalester College (Minnesota), Henry T. Wright, Professor and Curator of Anthropology at the University of Michigan, and George E. Stuart, President of the Center for Maya Research (North Carolina). The map of Russia's federal regions could not have been compiled without the invaluable help of David B. Miller, then Senior Edit Cartographer at the National Geographic Society, and Leo Dillon of the Russia Desk of the U.S. Department of State. The map of Russian physiography was updated thanks to the suggestions of Mika Roinila of the State University of New York, College at New Paltz. Special thanks also go to Charles Pirtle, Professor of Geography at Georgetown University's School of Foreign Service for his advice on Chapter 4A; to Bilal Butt (post-doctoral fellow, University of Wisconsin-Madison) for his significant help with Chapter 6A; to Charles Fahrer of Georgia College and State University for his suggestions on Chapter 6B; and to C. Frederick Lohrengel II of Southern Utah University for his advice on Chapter 10A. Enormous gratitude goes to Beth Weisenborn, Virtual Course Coordinator for the Department of Geography at Michigan State University, and all the instructors of Geo 204 (World Regional Geography) for their invaluable suggestions to improve the book. And we are most grateful to Tanya de Blij, who holds a graduate degree in geography from Florida State University and whose computer and editing skills have helped keep our projects on track.

We also record our appreciation to those geographers who ensured the quality of this book's ancillary products. Elizabeth Muller Hames, D.O. (as well as M.A. in Geography, University of Miami) co-authored the original *Study Guide*. At the University of Miami's Department of Geography and Regional Studies, we are most grateful for the advice and support we received from faculty colleagues Richard Grant, Miguel Kanai, Shouraseni Sen Roy, Ira Sheskin, Justin Stoler, and Diana Ter-Ghazaryan as well as GIS Lab Manager Chris Hanson and our Office Manager Alexis Fernandez, assisted by Isabella Figueroa. Two of these colleagues (Richard and Diana) very kindly supplied us with photos for this edition, as did our friend Elena Grigorieva of the Institute for Complex Analysis of Regional Problems in Birobidzhan, Russia, who recently spent several months in the department as a visiting Fulbright scholar. We also take this opportunity to reconfirm our appreciation of the work done on the ancillaries for earlier editions by Eugene J. Palka, Professor of Geography at the United States Military Academy, West Point, NY, whose vision and enthusiasm contributed importantly to the achievement of our objectives.

We are privileged to work with a team of professionals at John Wiley & Sons that is unsurpassed in the college textbook publishing industry. As authors we are acutely aware of these talents on a daily basis during the crucial production stage, especially the outstanding coordination and leadership skills of Senior Production Editor Janet Foxman. Others who played a leading role in this process were Senior Photo Editor Lisa Gee, Senior Designer Wendy Lai, and Don Larson (ably assisted by Terry Bush and the late Ann Kennedy) of Mapping Specialists, Ltd. in Fitchburg, Wisconsin. For nearly 15 years, we have greatly appreciated the generosity and leadership of Executive Geosciences Editor Ryan Flahive, always the prime mover in launching and guiding this book, and who was so diligently assisted throughout the preparations for this latest edition by Julia Nollen. For this latest revision, we have especially benefitted from the managerial skills and savvy of our College Marketing Manager,

Margaret Barrett, whose boundless enthusiasm and energy perfectly complement the talents of the *Regions* team. We also thank Beth Tripmacher, Darnell Sessoms, Howard Averback, and Harry Nolan for their help and support. Beyond this immediate circle, we acknowledge the support and encouragement we continue to receive from many others at Wiley including Senior Content Manager Micheline Frederick and Publisher Petra Recter.

Most of all, we want to express our continuing appreciation and admiration for the superb efforts of Jeanine Furino of Furino Production, one of the most gifted production editors we have ever encountered. Her organizational skills, extraordinary knowledge, technical prowess, unerring professional instincts, and signal ability to work smoothly with every collaborator, truly make her an indispensable team leader in the challenging process of fashioning this attractive volume (in less than six months!) out of a mountain of text and graphic files, email attachments, sketches, design layouts, and multiple rounds of preliminary pages. We also are indebted to Jeanine for ushering us into the fast-changing digital era of book publishing—a painstaking, still-ongoing effort that began when we first worked together during the 1990s.

Finally, above all else, we express our gratitude to our families for yet again seeing us through the constantly challenging schedule of creating this latest edition of *Regions*.

H. J. de Blij
Sarasota, Florida

Peter O. Muller
Coral Gables, Florida

Jan Nijman
Coral Gables, Florida

June 26, 2013

To Malala Yousafzai

Shot in the head by the Taliban near her school in Pakistan, she survived unimaginable challenges and displayed boundless courage by ascending the global stage and exhorting her classmates and girls everywhere never to stop learning and always to seek the knowledge that gives women power.

CONTENTS

GEOGRAPHY

REALMS, REGIONS, AND CONCEPTS

WORLD REGIONAL GEOGRAPHY: Global Perspectives

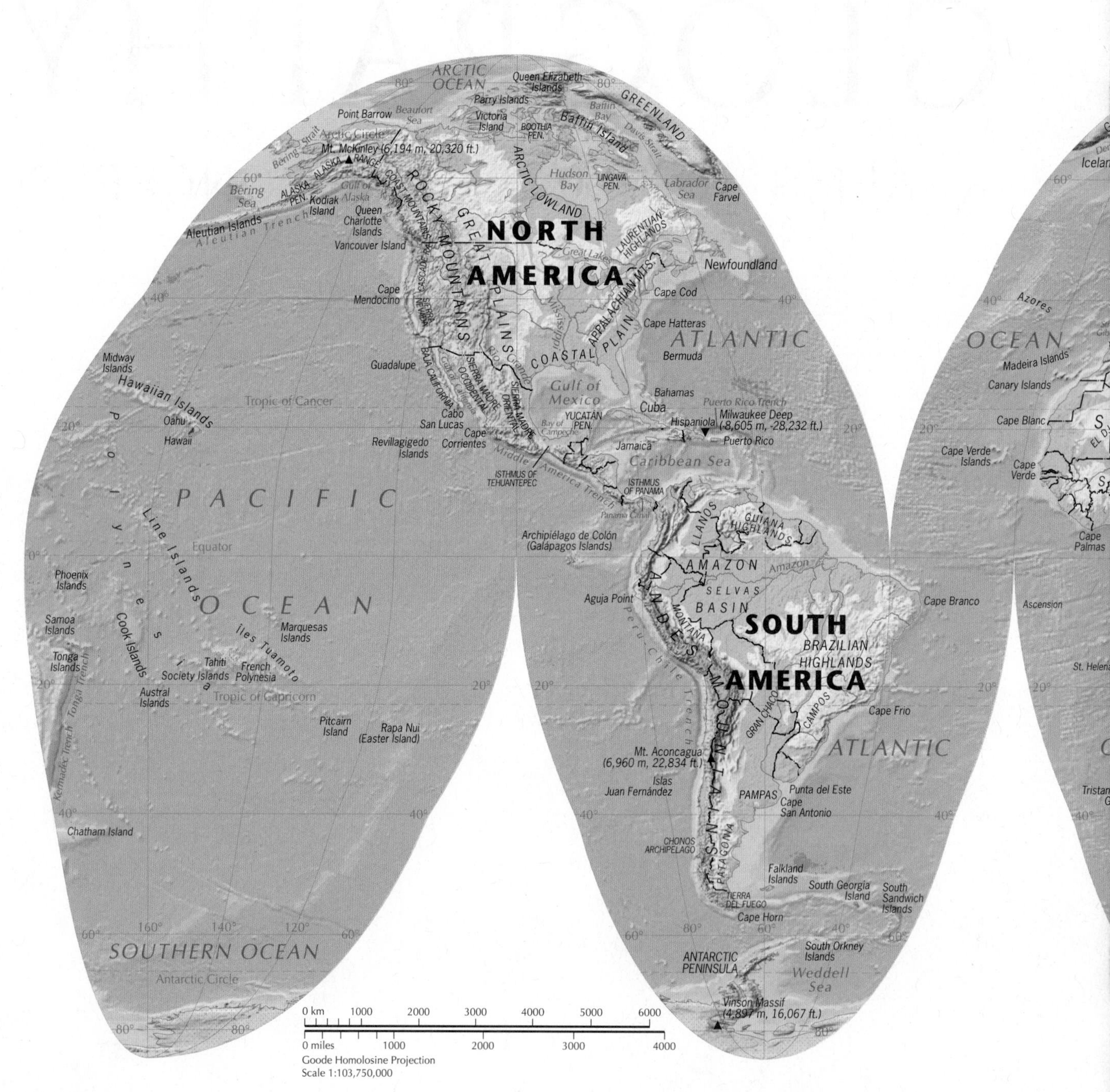

FIGURE G-1

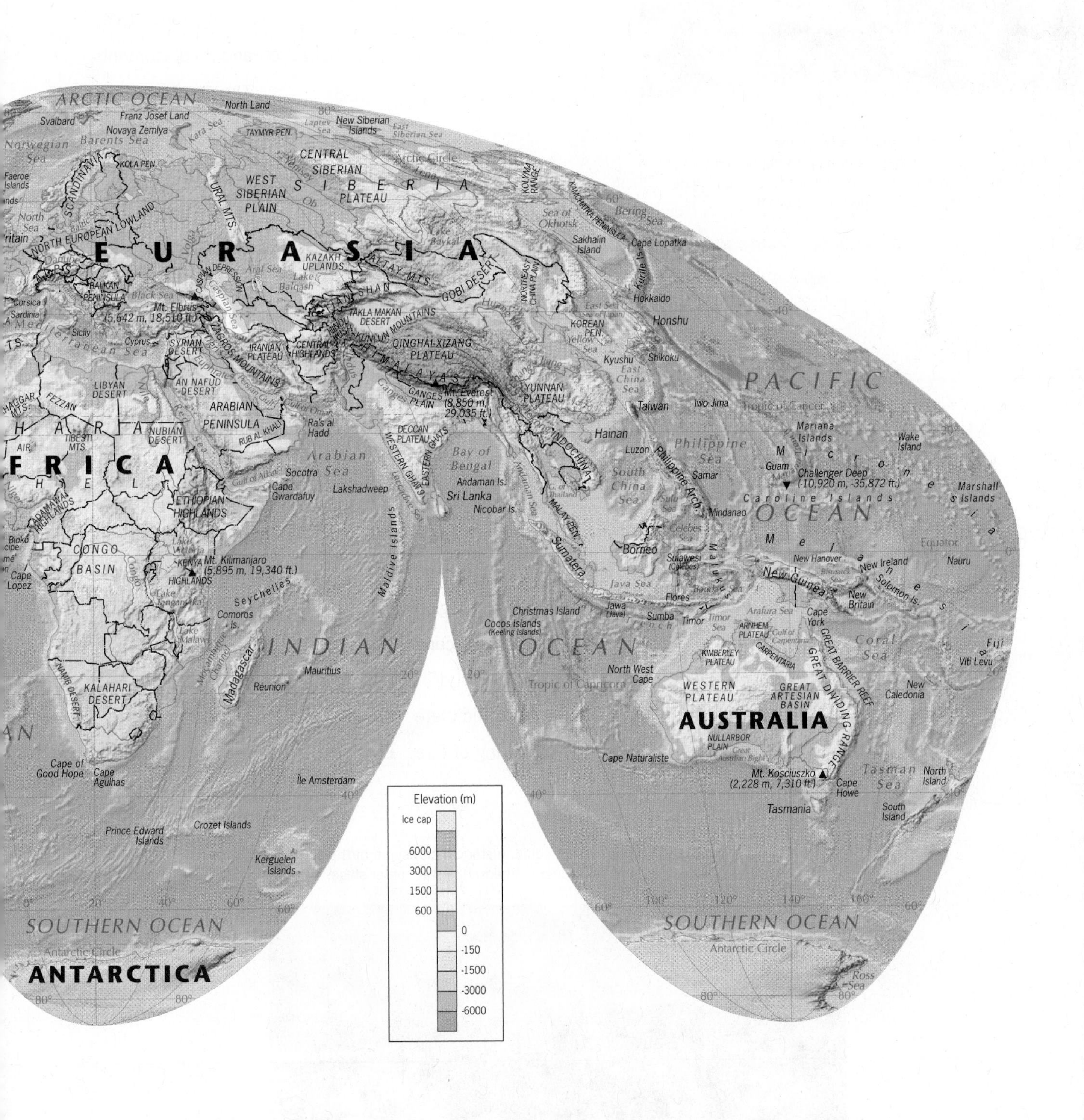

ARCTIC OCEAN
North Land
Svalbard
Franz Josef Land
Novaya Zemlya
Barents Sea
Kara Sea
TAYMYR PEN.
Laptev Sea
New Siberian Islands
East Siberian Sea
Norwegian Sea
KOLA PEN.
SCANDINAVIA
Faeroe Islands
CENTRAL SIBERIAN PLATEAU
WEST SIBERIAN PLAIN
S I B E R I A
Arctic Circle
Yenisey
Lena
Ob
URAL MTS.
KOLYMA RANGE
KAMCHATKA PENINSULA
Bering Sea
Sea of Okhotsk
Sakhalin Island
Cape Lopatka
North Sea
Baltic Sea
NORTH EUROPEAN LOWLAND
E U R A S I A
Volga
Danube
ALPS
Lake Baykal
KAZAKH UPLANDS
ALTAY MTS.
Aral Sea
Lake Balqash
CASPIAN DEPRESSION
BALKAN PENINSULA
Black Sea
Caspian Sea
Mt. Elbrus (5,642 m, 18,510 ft.)
TIAN SHAN
GOBI DESERT
NORTHEAST CHINA PLAIN
Kurile Is.
Hokkaido
Honshu
Corsica
Sardinia
Sicily
Mediterranean Sea
TAKLA MAKAN DESERT
KUNLUN MOUNTAINS
HINDU KUSH
Cyprus
SYRIAN DESERT
ZAGROS MOUNTAINS
IRANIAN PLATEAU
CENTRAL HIGHLANDS
QINGHAI-XIZANG PLATEAU
KOREAN PEN.
Yellow Sea
Kyushu
Shikoku
East China Sea
Euphrates
Persian Gulf
AN NAFUD DESERT
H I M A L A Y A S
Indus
Ganges
GANGES PLAIN
Mt. Everest (8,850 m, 29,035 ft.)
YUNNAN PLATEAU
Taiwan
Iwo Jima
PACIFIC
Tropic of Cancer
LIBYAN DESERT
HAGGAR MTS.
FEZZAN
ARABIAN PENINSULA
Gulf of Oman
Ra's al Hadd
RUB AL KHALI
NUBIAN DESERT
TIBESTI MTS.
AIR
Nile
Red Sea
DECCAN PLATEAU
WESTERN GHATS
EASTERN GHATS
Bay of Bengal
Hainan
INDOCHINA
Luzon
Philippine Sea
Mariana Islands
Wake Island
F R I C A
Arabian Sea
Gulf of Aden
Socotra
Cape Gwardafuy
Lakshadweep
Andaman Is.
Sri Lanka
Nicobar Is.
Andaman Sea
G. of Thailand
South China Sea
Samar
Philippine Arch.
Guam
Challenger Deep (-10,920 m, -35,872 ft.)
Marshall Islands
Caroline Islands
OCEAN
ETHIOPIAN HIGHLANDS
ADAMAWA HIGHLANDS
MALAY PEN.
Sulu Sea
Mindanao
Maldive Islands
Laccadive Sea
Bioko
CONGO BASIN
Congo
Lake Victoria
KENYA HIGHLANDS
Mt. Kilimanjaro (5,895 m, 19,340 ft.)
Sumatera
Borneo
Celebes Sea
Sulawesi (Celebes)
Maluku
Equator
Nauru
Cape Lopez
Lake Tanganyika
Seychelles
Java Sea
New Hanover
New Ireland
Bismarck Sea
New Guinea
Banda Sea
Solomon Is.
New Britain
Comoros Is.
Christmas Island
Jawa (Java)
Sumba
Flores
Timor
Timor Sea
Arafura Sea
Cape York
Lake Malawi
Cocos Islands (Keeling Islands)
ARNHEM PLATEAU
Gulf of Carpentaria
Coral Sea
Fiji
Viti Levu
Mozambique Channel
Madagascar
INDIAN
OCEAN
KIMBERLEY PLATEAU
CARPENTARIA
GREAT BARRIER REEF
GREAT DIVIDING RANGE
NAMIB DESERT
KALAHARI DESERT
Mauritius
Réunion
North West Cape
Tropic of Capricorn
WESTERN PLATEAU
GREAT ARTESIAN BASIN
New Caledonia
AUSTRALIA
NULLARBOR PLAIN
Great Australian Bight
Cape Naturaliste
Cape of Good Hope
Cape Agulhas
Île Amsterdam
Mt. Kosciuszko (2,228 m, 7,310 ft.)
Tasman Sea
North Island
Cape Howe
Tasmania
South Island
Prince Edward Islands
Crozet Islands
Kerguelen Islands
SOUTHERN OCEAN
Antarctic Circle
ANTARCTICA
Ross Sea
Elevation (m)
Ice cap
6000
3000
1500
600
0
-150
-1500
-3000
-6000

© Jan Nijman

Camel traders in Pushkar, a small town in India's Rajasthan State, relaxing at the end of a November day in 2011. Pushkar features the biggest annual camel fair on Earth, with tens of thousands of camels (and horses) changing hands during the five-day event.

IN THIS CHAPTER

- The power of maps
- The spatial order of the world
- Global climate change
- Dangerous places
- Globalization and its discontents
- The power of place

CONCEPTS, IDEAS, AND TERMS

Mental maps	1
Spatial perspective	2
Scale	3
Geographic realm	4
Transition zone	5
Regional concept	6
Absolute location	7
Relative location	8
Formal region	9
Spatial system	10
Hinterland	11
Functional region	12
Natural landscape	13
Continental drift	14
Tectonic plate	15
Pacific Ring of Fire	16
Climate	17
Ice age	18
Glaciation	19
Interglacial	20
Global climate change	21
Population distribution	22
Urbanization	23
Cultural landscape	24
State	25
Sovereignty	26
European state model	27
Development	28
Core area	29
Periphery	30
Globalization	31

A view of Florence, one of Europe's most iconic, historical cities and birthplace of the Renaissance. The famous Duomo (cathedral) lies at center stage.

© Jan Nijman

What are your expectations as you open this book? You have signed up for a course that will take you around the world to try to understand how it functions today. You will discover how interesting and unexpectedly challenging the discipline of geography is. We hope that this course, and this book, will open new vistas, bring new perspectives, and help you navigate our increasingly complex and often daunting world.

You could not have chosen a better time to be studying geography. The world is changing on many fronts, and so is the United States. Still the most formidable of all countries, the United States remains a great power capable of influencing nations and peoples, lives and livelihoods from pole to pole. That power confers on Americans the responsibility to learn as much as they can about those nations and livelihoods, so that the decisions of their government representatives are well-informed. But in this respect, the United States is no superpower. Geographic literacy is a measure of international comprehension and awareness, and Americans' geographic literacy ranks low among countries of consequence. That is not a good thing, neither for the United States nor for the rest of the world, because such geographic fogginess tends to afflict not only voters but also the representatives they elect, from the school board to Congress.

A WORLD ON MAPS

Just a casual glance at the pages that follow reveals a difference between this and other textbooks: there are almost as many maps as there are pages. Geography is more closely identified with maps than any other discipline, and we urge you to give as much (or more!) attention to the maps in this book as you do to the text. It is often said that a picture is worth a thousand words, and the same or more applies to maps. When we write "see Figure XX," we really mean it . . . and we hope that you will get into the habit. We humans are territorial creatures, and the boundaries that fence off our 200 or so countries reflect our divisive ways. Other, less visible borders—between religions, languages, rich, and poor—partition our planet as well. When political and cultural boundaries are at odds, there is nothing like a map to summarize the circumstances. Just look, for example, at the map of the African Transition Zone in Chapter 6B: this area's turbulence and challenges are steeped in geography.

Maps in Our Minds

All of us carry in our minds maps of what psychologists call our activity space: the apartment building or house we live in, the streets nearby, the way to school or workplace, the general layout of our hometown or city. You will know what lane to use when you turn into a shopping mall, or where to park at the movie theater. You can probably draw from memory a pretty good map of your hometown. These **mental maps [1]** allow you to navigate your activity space with efficiency, predictability, and safety. When you arrived as a first-year student on a college or university campus, a new mental map will have started forming. At first you needed a GPS, online, or hard-copy map to find your way around, but soon you dispensed with that because your mental map was sufficient. And it will continue to improve as your activity space expands.

If a well-formed mental map is useful for decisions in daily life, then an adequate mental map is surely indispensable when it comes to decision making in the wider world. You can give yourself an interesting test. Choose some part of the world, beyond North America, in which you have an interest or about which you have a strong opinion—for example, Israel, Iran, Pakistan, North Korea, or China. On a blank piece of paper, draw a map that reflects your impression of the regional layout there: the country, its neighbors, its internal divisions, major cities, seas (if any), and so forth. That is your mental map of the place. Put it away for future reference, and try it again at the end of this course. You will have proof of your improved mental-map inventory.

The Map Revolution

The maps in this book show larger and smaller parts of the world in various contexts. Some depict political configurations; others display ethnic, cultural, economic, or environmental features. ***Cartography*** (the making of maps) has undergone a dramatic technological revolution—a revolution that continues. Earth-orbiting satellites equipped with remote sensing technology (special on-board sensors and imaging instruments) transmit remotely sensed information to computers on the surface, recording the expansion of deserts, the shrinking of glaciers, the depletion of forests, the growth of cities, and myriad other geographic phenomena. Earthbound computers possess ever-expanding capabilities not only to organize this information but also to display it graphically. This allows geographers to develop a ***geographic information system (GIS)***, bringing information to a monitor's screen that would have taken months to assemble just a few decades ago.

There has also been a map revolution in the astounding proliferation of navigation systems in cars and on mobile phones. Smartphones allow the use of maps on the go, and many of us, in the developed world at least, have become dependent on them to traverse cities, to get to a store or restaurant, even to move around shopping malls. Google, the biggest company in this market, used to aim at cataloguing all of the world's information, but today it is also aiming to map the world in almost unimaginable detail. And the competition is now joined by Nokia and Apple. Whereas the maps on our smartphones allow us to move around more efficiently, the maps in this book are aimed at better *understanding* the world and its constituent parts.

Satellites—even spy satellites—cannot record everything that occurs on the Earth's surface. Sometimes the borders between ethnic groups or cultural sectors can be discerned by satellites—for example, in changing types of houses or religious shrines—but this kind of information tends to require on-the-ground verification through field research and reporting. No satellite view of Iraq could show you the distribution of Sunni and Shia Muslim adherents. Many of the boundaries you see on the maps in this book cannot be observed from space because long stretches are not even marked on the ground. So the maps you are about to "read" have their continued uses: they summarize complex situations and allow us to begin forming durable mental maps of the areas they represent.

There is one other point we should make that is especially important when it comes to world maps: never forget that the world is a sphere, and to project it onto a two-dimensional flat surface must necessarily entail some very significant distortions. Try peeling an orange and flattening the entire peel on a surface—you will have to tear it up and try to stretch it in places to get the job done. Take a look at Figure G-1 and note how the Atlantic Ocean and other segments of the planetary surface are interrupted. You can produce a map like this in many different ways, but you will always end up distorting things. When studying world maps, there is nothing like having a globe at hand to remind you of our three-dimensional reality.

GEOGRAPHY'S PERSPECTIVE

Geography is sometimes described as the most interdisciplinary of disciplines. That is a testimonial to geography's historic linkages to many other fields, ranging from geology to economics and from sociology to political science. And, as has been the case so often in the past, geography is in the lead on this point. Today, interdisciplinary studies and research are more prevalent than ever. The old barriers between disciplines are breaking down.

This is not to suggest that college and university departments are no longer relevant; they are just not as exclusive as they used to be. These days, you can learn some useful geography in economics departments and some good economics in geography departments. But each discipline still has its own particular way of looking at the world.

A Spatial Perspective

Most disciplines focus on one key theme: economics is about money; political science is about power; psychology

From the Field Notes . . .

"On the descent into Tibet's Lhasa Gongga Airport, I had a great view of the Yarlung Zangbo Valley, its braided stream channels gently flowing toward the distant east. The Yarlung Zangbo is the highest major river on Earth, running from the Tibetan Plateau into northeastern India where it joins the mighty Brahmaputra River that continues on to Bangladesh where it empties into the Indian Ocean. It was mid-October and the water levels were low. The landing strip of the airport can be seen in the center-right of the photo, on the south bank. The airport is quite far from Lhasa, the Tibetan capital, located about 62 kilometers (40 mi) to its southwest. Despite major road and tunnel construction, it is still more than an hour's drive. The airport had to be built away from the city and in this widest part of the valley because it allows the easiest landings and takeoffs in this especially rugged terrain. It lies at 3700 meters above sea level (12,100 ft), one of the highest airports in the world."

© Jan Nijman

www.conceptcaching.com

Concept Caching

is about the mind; biology is about life. Geography, then, is about space on the Earth's surface. More specifically, geographers are interested in the organization of ***terrestrial space***. Social space (cities, buildings, political boundaries, etc.) as well as natural space (climates, terrain, water bodies, etc.) are not randomly configured. Instead, there generally prevails a particular order, regularity, even predictability about the ways in which space is organized. Sometimes it is the deliberate work of human beings and sometimes it is the work of nature, but very often there are particular patterns. Geographers consider these spatial patterns and processes as not only interesting but also crucial to how we live and how we organize our societies. The **spatial perspective [2]** has defined geography from its beginning.

Environment and Society

There is another glue that binds geography and has done so for a very long time: an interest in the relationships between human societies and the natural (physical) environment. Geography lies at the intersection of the social and natural sciences and integrates perspectives from both, being the only discipline to do so explicitly. This perspective comes into play frequently: environmental change is in the news on a daily basis in the form of worldwide climate change, but this current surge of global warming is only the latest phase of endless atmospheric and ecological fluctuation. Geographers are involved in understanding current environmental issues not only by considering climate change in the context of the past, but also by looking carefully at the implications of global climate change for human societies.

More generally, think of this relationship between humans and their environment as a two-way street. On one hand, human beings have always had a transformative effect on their natural surroundings, from the burning of forests to the creation of settlements. On the other hand, humans have always been heavily dependent on the natural environment, their individual and collective behaviors very much a product of it. There are so many examples that it is hard to know where to begin or when to end: we eat what nature provides and traditional diets vary regionally; rivers allow us to navigate and connect with other peoples—or they serve as natural boundaries like the Rio Grande; wars are fought over access to water or seaports; landlocked countries seem to have different cultures from those of islands; and so on.

At times we are faced with the interrelationship between humans and their environment. For example, humans modify the environment through escalating carbon dioxide emissions (the so-called greenhouse effect) and are subsequently confronted with the need to adjust to rising sea levels. We will always be part of nature, no matter how far technology advances.

Spatial Patterns

Geographers, therefore, need to be conversant with the location and distribution of salient features on the Earth's surface. This includes the natural (physical) world, simplified in Figure G-1, as well as the human world, and our inquiry will view these in temporal (historical) as well as spatial perspective. We take a penetrating look at the overall geographic framework of the contemporary world, the still-changing outcome of thousands of years of human achievement and failure, movement and stagnation, stability and revolution, interaction and isolation. The spatial structure of cities, the layout of farms and fields, the networks of transportation, the configurations of rivers, the patterns of climate—all these form part of our investigation. As you will find, geography employs a comprehensive spatial vocabulary with meaningful terms such as area, distance, direction, clustering, proximity, accessibility, and many others we will encounter in the pages ahead. For geographers, some of these terms have more specific definitions than is generally assumed. There is a difference, for example, between *area* (surface) and *region*, between *boundary* and *frontier*, and between *place* and *location*. The vocabulary of geography holds some surprises, and what at first may seem to be simple ideas turn out to be complex concepts.

Scale and Scope

One prominent item in this vocabulary is the term **scale [3]**. Whenever a map is created, it represents all or part of the Earth's surface at a certain level of detail. Obviously, Figure G-1 displays a very low level of detail; it is little more than a general impression of the distribution of land and water as well as lower and higher elevations on our planet's surface. A limited number of prominent features such as the Himalayas and the Sahara are named, but not the Pyrenees Mountains or the Nile Delta. At the bottom of the map you can see that one inch at this scale must represent about 1650 miles of the real world, leaving the cartographer little scope to insert information.

A map such as Figure G-1 is called a *small-scale* map because the ratio between map distance and real-world distance, expressed as a fraction, is very small at 1:103,750,000. Increase that fraction (i.e., zoom in), and you can represent less territory—but also enhance the amount of detail the map can represent. In Figure G-2, note how the fraction increases from the smallest (1:103,000,000) to the largest (1:1,000,000). Montreal, Canada is just a dot on Map A but an urban area on Map D. Does this mean that world maps like Figure G-1 are less useful than larger-scale maps? It all depends on the purpose of the map. In this chapter, we often use world maps to show global distributions as we set the stage for the more detailed discussions to follow. In later chapters, the scale tends to become larger as we focus on smaller areas, even on individual countries and cities. But whenever you read a map, be aware of the scale because it is a guide to its utility.

The importance of the scale concept is not confined to maps. Scale plays a fundamental role in geographic research and in the ways we think about geographic problems: scale in terms of *level of analysis*. This is sometimes

EFFECT OF SCALE

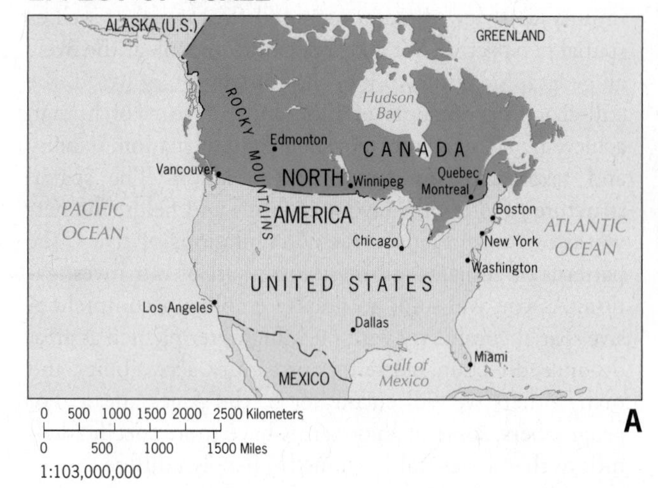

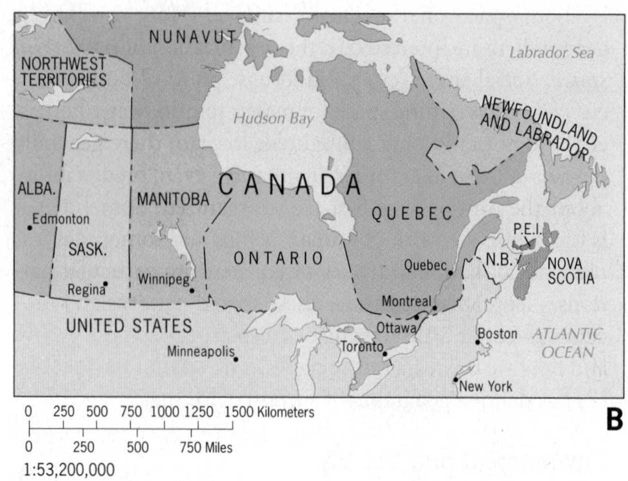

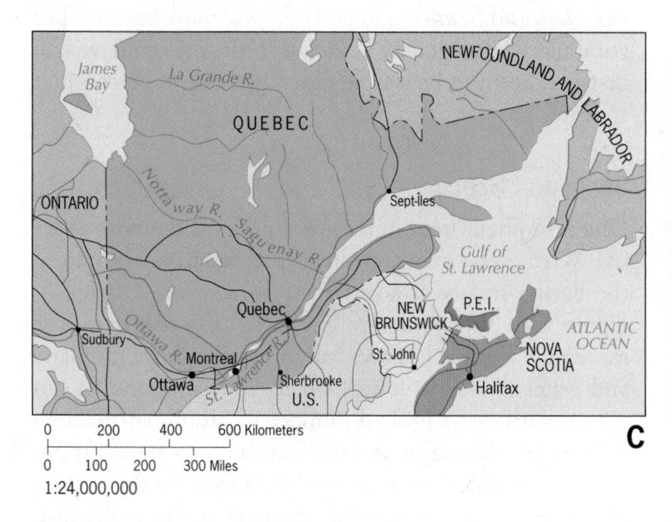

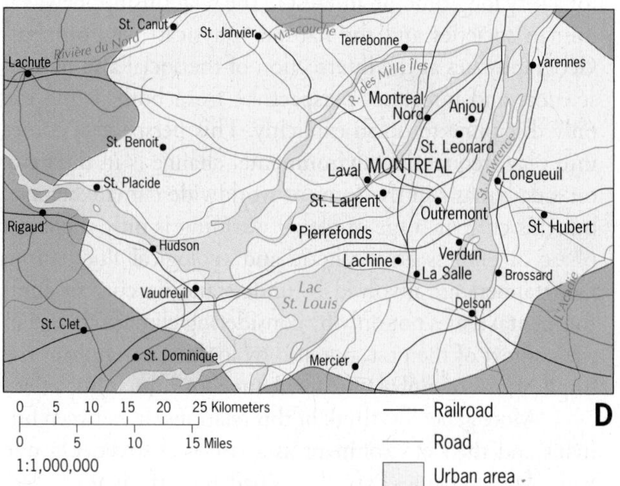

FIGURE G-2

referred to as *operational scale*, the scale at which social or natural processes operate or play out. For instance, if you want to investigate the geographic concentration of wealth in the United States, you can do so at a range of scales: within a neighborhood, a city, a county, a State,* or at the national level. You choose the scale that is the most appropriate for your purpose, but it is not always that straightforward. Suppose you had to study patterns of ethnic segregation: what do you think would be the most relevant scale(s)?

In this book, our main purpose is to understand the geography of the world at large and how it works, and so, inevitably, we must deal with large spatial entities. Our focus is on the world's realms and on the main regions within those realms, and in most cases we will have to forego analyses at a finer scale. For our purposes, it is the big picture that matters most.

*Throughout this book we will capitalize State when this term refers to an administrative subdivision of a country: for example, the U.S. State of Ohio or the Australian State of New South Wales. Since this term is also synonymous with country (e.g., the state of Brazil), we use the lower case when referring to such a national state.

WORLD GEOGRAPHIC REALMS

Ours is a globalized, interconnected world, a world of international trade and travel, migration and movement, tourism and television, financial flows and Internet traffic. It is a world that, in some contexts, has taken on the properties of a "global village"—but that village still has its neighborhoods. Their names are Europe, South America, Southeast Asia, and others familiar to us all. Like the neighborhoods of a city or town, these global neighborhoods may not have sharply defined borders, but their persistence, after tens of thousands of years of human dispersal, is beyond doubt. Geographers call such global neighborhoods **geographic realms [4]**. Each of these realms possesses a particular combination of environmental, cultural, and organizational properties.

These characteristic qualities are imprinted on the landscape, giving each realm its own traditional attributes and social settings. As we come to understand the human and environmental makeup of these geographic realms, we learn not only where they are located but also why they are located where they are (a key question in geography), how they are constituted, and what their future is likely to be in our fast-changing world. Figure G-3, therefore, forms the overall framework for our investigation in this book.

Criteria for Geographic Realms

The existence and identification of world geographic realms depends on a combination of factors. Our world offers a highly complex and variable environment of large and small continents, enormous oceans and countless waterways, innumerable islands, diverse habitats and cultures, and intricate political geographies. What constitutes a realm depends on the circumstances, but we can still identify three main sets of criteria:

- ***Physical and Human*** Geographic realms are based on sets of spatial criteria. They are the largest units into which the inhabited world can be divided. The criteria on which such a broad regionalization is based include both physical (that is, natural) and human (or social) yardsticks. On the one hand, South America is a geographic realm because physically it is a continent and culturally it is comprised of comparable societies. The realm called South Asia, on the other hand, lies on a Eurasian landmass shared by several other geographic realms; high mountains, wide deserts, and dense forests combine with a distinctive social fabric to create this well-defined realm centered on India.
- ***Functional*** Geographic realms are the result of the interaction of human societies and natural environments, a *functional* interaction revealed by farms, mines, fishing ports, transport routes, dams, bridges, villages, and countless other features that mark the landscape. According to this criterion, Antarctica is a continent but not a geographic realm.
- ***Historical*** Geographic realms must represent the most comprehensive and encompassing definition of the great clusters of humankind in the world today. China lies at the heart of such a cluster, as does India. Africa constitutes a geographic realm from the southern margin of the Sahara (an Arabic word for desert) to the Cape of Good Hope and from its Atlantic to its Indian Ocean shores.

Figure G-3 displays the 12 world geographic realms based on these criteria. As we will show in greater detail later, waters, deserts, and mountains as well as cultural and political shifts mark the borders of these realms. We shall discuss the positioning of these boundaries as we examine each realm.

Delineating Realms: Boundaries and Transition Zones

Oceans and seas are the most common natural boundaries of the world's realms, such as the South Atlantic to Subsaharan Africa's west or the North Atlantic to North America's east. But where two geographic realms meet, **transition zones [5]**, not sharp boundaries, often mark their contacts.

We need only remind ourselves of the border zone between the geographic realm in which most of us live, North America, and the adjacent realm of Middle America. The line in Figure G-3 coincides with the boundary between Mexico and the United States, crosses the Gulf of Mexico, and then separates Florida from Cuba and the Bahamas. But Hispanic influences are strong in North America north of this boundary, and the U.S. economic influence is strong south of it. The line, therefore, represents an ever-changing zone of regional interaction. Again, there are many ties between South Florida and the Bahamas, but the Bahamas resemble a Caribbean more than a North American society. Miami has so many Cuban and Cuban-American inhabitants that it is sometimes referred to as the second-largest Cuban city after Havana.

In Africa, the transition zone from Subsaharan to North Africa is so wide and well defined that we have put it on the world map; elsewhere, transition zones tend to be narrower and less easily represented. In the first half of this second decade of the twenty-first century, such countries as Belarus (between Europe and Russia) and Kazakhstan (between Russia and Muslim Southwest Asia) lie in inter-realm transition zones. Remember, over much (though not all) of their length, borders between realms are zones of regional change.

Transition zones are fascinating spaces: it is almost as if they rebel against a clear ordering of the world's geography. They remind us that the world is a restless and contested place with shifting boundaries and changing geographic fortunes. They challenge the geographer's mapping skills, and, above all, they underscore just how complex the study of geography is. As you will see, transition zones are often places of tension and/or conflict.

Geographic Realms: Dynamic Entities

Had we drawn Figure G-3 before Columbus made his voyages from 1492 (and assuming we had the relevant geographical knowledge), the map would have looked different: indigenous states and peoples would have determined the boundaries in the Americas; Australia and New Guinea would have constituted a single realm, and New Zealand would have been part of the Pacific Realm. The colonization, Europeanization, and Westernization of the world changed that map dramatically. Since World War II, the world map has been redrawn as a result of decolonization and the rise and then demise of the Cold War. That Cold War division between western and eastern Europe has now given way to far-reaching European integration across that geographic realm. Realms and regions are dynamic entities, and geography is always subject to change.

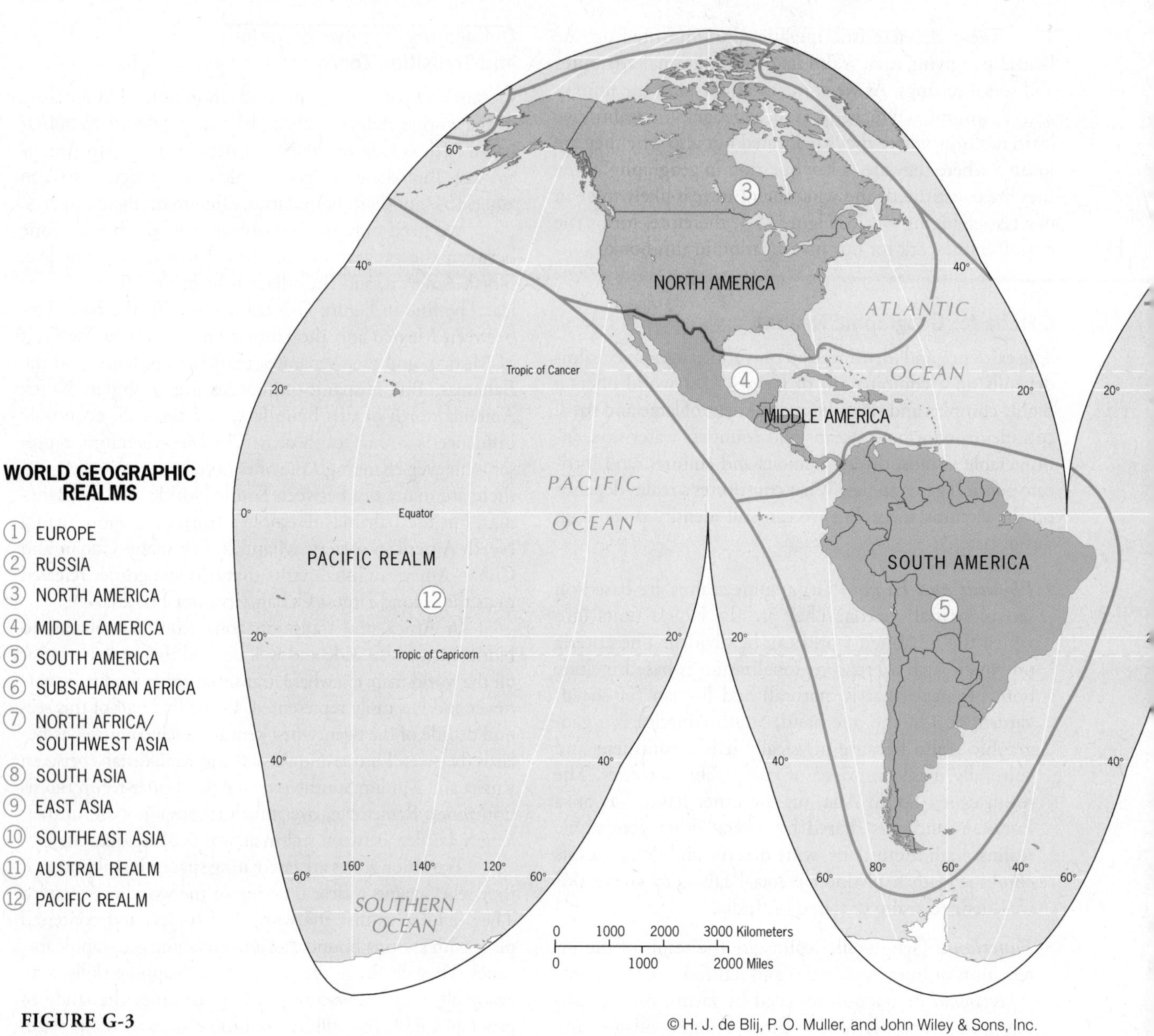

FIGURE G-3

Two Varieties of Realms

The world's geographic realms can be divided into two categories. The first are *monocentric* realms that are dominated by a single major political entity, in terms of territory and/or population. North America (United States), Middle America (Mexico), East Asia (China), South Asia (India), Russia, and the Austral Realm (Australia) are all monocentric realms. They are, in their entirety, heavily influenced by the presence of that one country. It is as if the realm is organized around them.

The second type of realm is *polycentric* in nature. In these, the appearance, functioning, and organization of the realm are dispersed among a number of more or less equally influential regions or countries. Europe, North Africa/Southwest Asia, Subsaharan Africa, and the Pacific Realm all fall into this category. Polycentric realms can be more volatile in some ways, their development determined by the sum of many different parts.

Two of the world's realms are a bit more difficult to categorize. Southeast Asia is a dynamic realm that contains almost a dozen countries, some of them regarded as emerging economies. Arguably, Indonesia is becoming the most influential power, but it may be premature to label this a monocentric realm. The other realm that seems to fall in-between is South America. Here it is Brazil that has the biggest population and increasingly the largest and most influential economy. South America, more emphatically than Southeast Asia, seems to be moving toward a monocentric reorganization of its realm.

Of course, some of the dominant powers in the monocentric realms influence events beyond their realm and demonstrate a truly global reach. The United States has dominated world events in an unprecedented manner since the

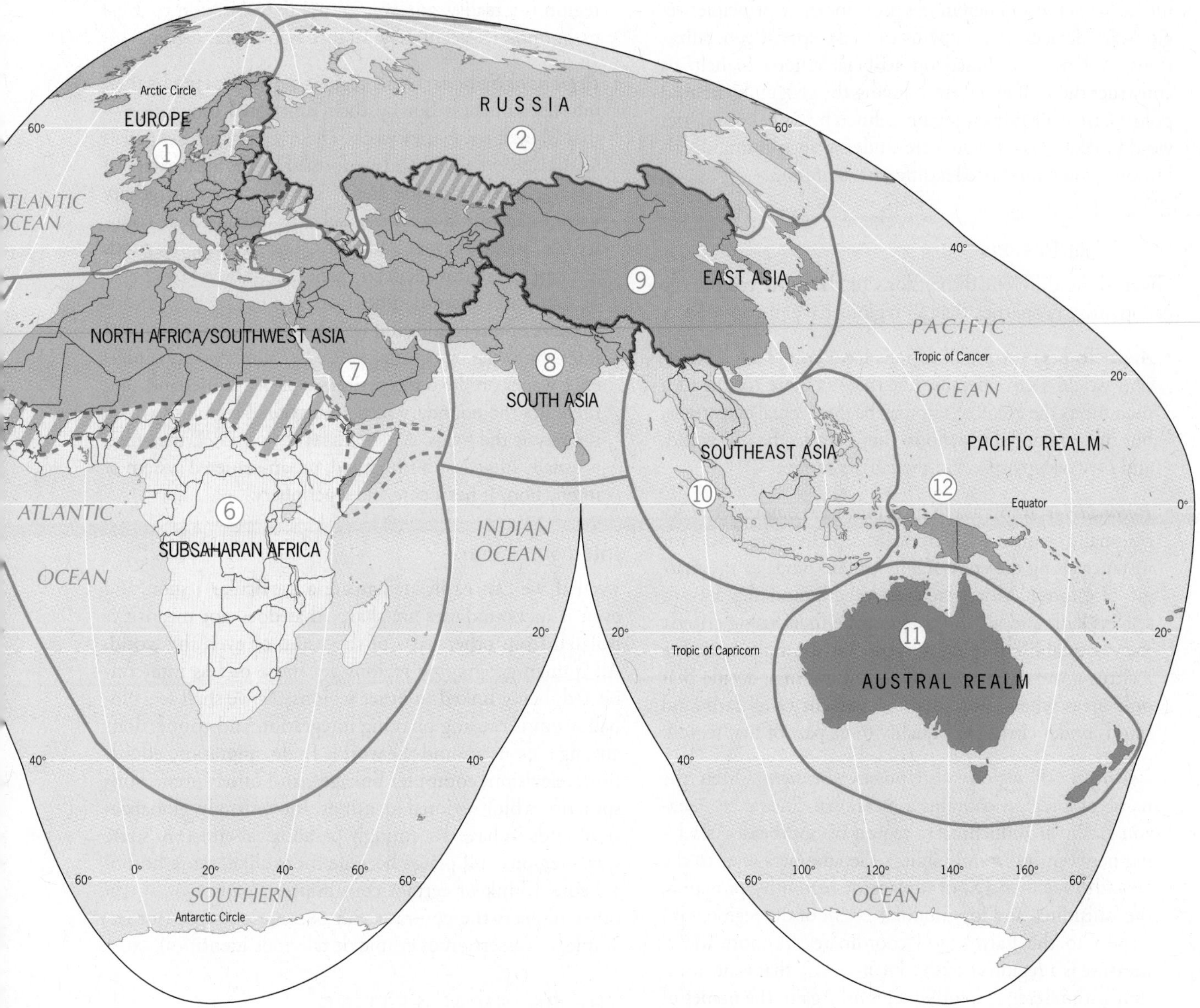

Second World War, but more recently it has had to make way for newly emergent powers. Nowadays, the rise of China is a hot topic of debate. We will see, for example, that China's role in South America has grown rapidly while that of the United States has waned. Our discussion of the various realms will give due consideration to the influence of global trends and outside powers.

REGIONS WITHIN REALMS

The compartmentalization of the world into geographic realms establishes a broad global framework, but for our purposes a more refined level of spatial classification is needed. This brings us to an important organizing concept in geography: the **regional concept [6]**. To establish regions within geographic realms, we need more specific criteria.

Let us use the North American realm to demonstrate the regional idea. When we refer to a part of the United States or Canada (e.g., the South, the Midwest, or the Prairie Provinces), we employ a regional concept—not scientifically but as part of everyday communication. We reveal our perception of local or distant space as well as our mental image of the region we are describing.

But what exactly is the Midwest? How would you draw this region on the North American map? Regions are easy to imagine and describe, but they can be difficult to outline on a map. One way to define the Midwest is to use the borders of States: certain States are part of this region, others are not. You could use agriculture as the principal criterion: the Midwest is where corn and/or soybeans occupy a certain percentage of the farmland. Look ahead to Figure 3B-6, where you will notice that a different name for this region is used—the Heartland—because of the differing (agricultural) criteria that define it. Each method results in a different delimitation; a Midwest based on States is different from a Midwest based on farm production or on industrial activity. Therein lies an important principle: regions are devices that allow us to make spatial generalizations, and they are based on artificial criteria to help us construct them. If you were studying the geography behind politics, then a Midwest region defined by State boundaries would make sense. If you were studying agricultural distributions, you would need a different definition.

Criteria for Regions

Given these different dimensions of the same region, we can identify properties that all regions have in common:

- ***Area*** To begin with, all regions have *area*. This observation would seem obvious, but there is more to this idea than meets the eye. Regions may be intellectual constructs, but they are not abstractions: they exist in the real world, and they occupy space on the Earth's surface.

- ***Boundaries*** It follows that regions have *boundaries*. Occasionally, nature itself draws sharp dividing lines, for instance along the crest of a mountain range or the margin of a forest. More often, regional boundaries are not self-evident, and we must determine them using criteria that we establish for that purpose. For example, to define a citrus-growing agricultural region, we may decide that only areas where more than 50 percent of all farmland stands under citrus trees qualify to be part of that region.

- ***Location*** All regions also possess *location*. Often the name of a region contains a locational clue, as in Amazon Basin or Indochina (a region of Southeast Asia lying between India and China). Geographers refer to the **absolute location [7]** of a place or region by providing the latitudinal and longitudinal extent of the region with respect to the Earth's grid coordinates. A more useful measure is a region's **relative location [8]**, that is, its location with reference to other regions. Again, the names of certain regions reveal aspects of their relative locations, as in *Mainland* Southeast Asia and *Equatorial* Africa.

- ***Homogeneity*** Many regions are marked by a certain *homogeneity* or sameness. Homogeneity may lie in a region's human (cultural) properties, its physical (natural) characteristics, or both. Siberia, a vast region of northeastern Russia, is marked by a sparse human population that resides in widely scattered, small settlements of similar form, frigid climates, extensive areas of permafrost (permanently frozen subsoil), and cold-adapted vegetation. This dominant uniformity makes it one of Russia's natural and cultural regions, extending from the Ural Mountains in the west to the Pacific Ocean in the east. When regions display a measurable and often visible internal homogeneity, they are called **formal regions [9]**. But not all formal regions are visibly uniform. For instance, a region may be delimited by the area in which, say, 90 percent or more of the people speak a particular language. This cannot be seen in the landscape, but the region is a reality, and we can use that criterion to draw its boundaries accurately. It, too, is a formal region.

- ***Regions as Systems*** Other regions are marked not by their internal sameness but by their functional integration—that is, by the way they work. These regions are defined as **spatial systems [10]** and are formed by the areal extent of the activities that define them. Take the case of a large city with its surrounding zone of suburbs, urban-fringe countryside, satellite towns, and farms. The city supplies goods and services to this encircling zone, and it buys farm products and other commodities from it. The city is the heart, the ***core*** of this region, and we call the surrounding zone of interaction the city's **hinterland [11]**. But the city's influence wanes on the outer periphery of that hinterland, and there lies the boundary of the functional region of which the city is the focus. A **functional region [12]**, therefore, is usually forged by a structured, urban-centered system of interaction. It has a core and a periphery.

Interconnections

Even if we can easily demarcate a particular region and even if its boundaries are sharp, that does not mean it is isolated from other parts of the realm or even the world. All human-geographic regions are more or less interconnected, being linked to other regions. As we shall see, globalization is causing ongoing integration and connections among regions around the world. Trade, migration, education, television, computer linkages, and other interactions sometimes blur regional identities. Interestingly, globalization tends to have a seemingly paradoxical effect: in some ways, regions and places become more alike, more homogeneous (think of certain consumption patterns), but in other respects the contrasts can become stronger (for example, a reassertion of ethnic or religious identities).

THE PHYSICAL SETTING

This book focuses on the geographic realms and regions produced by human activity over thousands of years. But we must not overlook the natural environments in which all this activity took place because we can still recognize the role of those environments in how people make their living. Certain areas of the world, for example, presented opportunities for plant and animal domestication that other areas did not. The people who happened to live in those favored areas learned to grow wheat, rice, or root crops and to domesticate oxen, goats, or llamas. We can still discern those early patterns of opportunity on the map in the twenty-first century. From such opportunities came adaptation and invention, and thereby arose villages, towns, cities, and states. But people living in other kinds of environments found it much harder to achieve this organization. The Americas, for instance, had no large animals that could be domesticated except llamas. This meant that societies created agricultural systems that did not involve ploughing as there were no draught animals. When Europeans introduced cows, horses,

From the Field Notes . . .

© H.J. de Blij

AP/Wide World Photos

"Flying over Iceland's volcanic topography (left photo) is to see our world in the making: this is some of the youngest rock on the planet, and even at rest you can sense its impermanence. Here nature shows us what mostly goes on deep below the surface along the mid-oceanic ridges, where tectonic plates pull apart and lava pours out of fissures and vents. When that happens on dry land, the results can be catastrophic. In the 1780s, an Icelandic volcano named Laki, in a series of eruptions, killed tens of thousands and caused a global ecological crisis. In 2010, the eruption of this far smaller volcano, Eyjafjallajökull (right photo), disrupted air travel for weeks across much of the Northern Hemisphere."

www.conceptcaching.com

and other livestock, this change completely revolutionized the environments and cultural systems of the Western Hemisphere. The modern map carries many such imprints of the past.

Natural (Physical) Landscapes

The landmasses of Planet Earth present a jumble of **natural landscapes [13]** ranging from rugged mountain chains to smooth coastal plains (Fig. G-1). Certain continents are readily linked with a dominant physical feature—for instance, North America and its Rocky Mountains, South America with its Andes and Amazon Basin, Europe with its Alps, Asia with its Himalaya Mountains and numerous river basins, and Africa with its Sahara and Congo Basin. Physical features have long influenced human activity and movement. Mountain ranges form barriers to movement but have also channeled the spread of agricultural and technological innovations. Large deserts similarly form barriers as do rivers, although rivers also permit accessibility and connectivity between people. River basins in Asia still contain several of the planet's largest population concentrations: the advantages of fertile soils and ample water supplies that first enabled clustered human settlement now sustain hundreds of millions in crowded South and East Asia.

As we study each of the world's geographic realms, we will find that physical landscapes continue to play significant roles in this modern world. That is one reason why the study of world regional geography is so important: it puts the human map in environmental as well as regional perspective.

Geology and Natural Hazards

Our planet may be 4.5 billion years old, but it is far from placid. As you read this chapter, Earth tremors are shaking the still-thin crust on which we live, volcanoes are erupting, storms are raging. Even the continents themselves are moving measurably, pulling apart in some areas, colliding in others. Hundreds of thousands of human lives are lost to natural calamities of this sort in almost every decade (over 350,000 in the 2010–2012 period alone), and such events have at times altered the course of history.

About a century ago a geographer named Alfred Wegener, a German scientist, used spatial analysis to explain something that is obvious even from a small-scale map like Figure G-1: the apparent jigsaw-like fit of the landmasses, especially across the South Atlantic Ocean. He concluded that the landmasses on the map are actually pieces of a supercontinent that existed hundreds of millions of years ago (he called it ***Pangaea***) that drifted away when, for some reason, that supercontinent broke up. His hypothesis of **continental drift [14]** set the stage for scientists in other disciplines to search for a mechanism that might make this possible, and much of the answer to that search proved to lie in the crust beneath the ocean surface. Today we know that

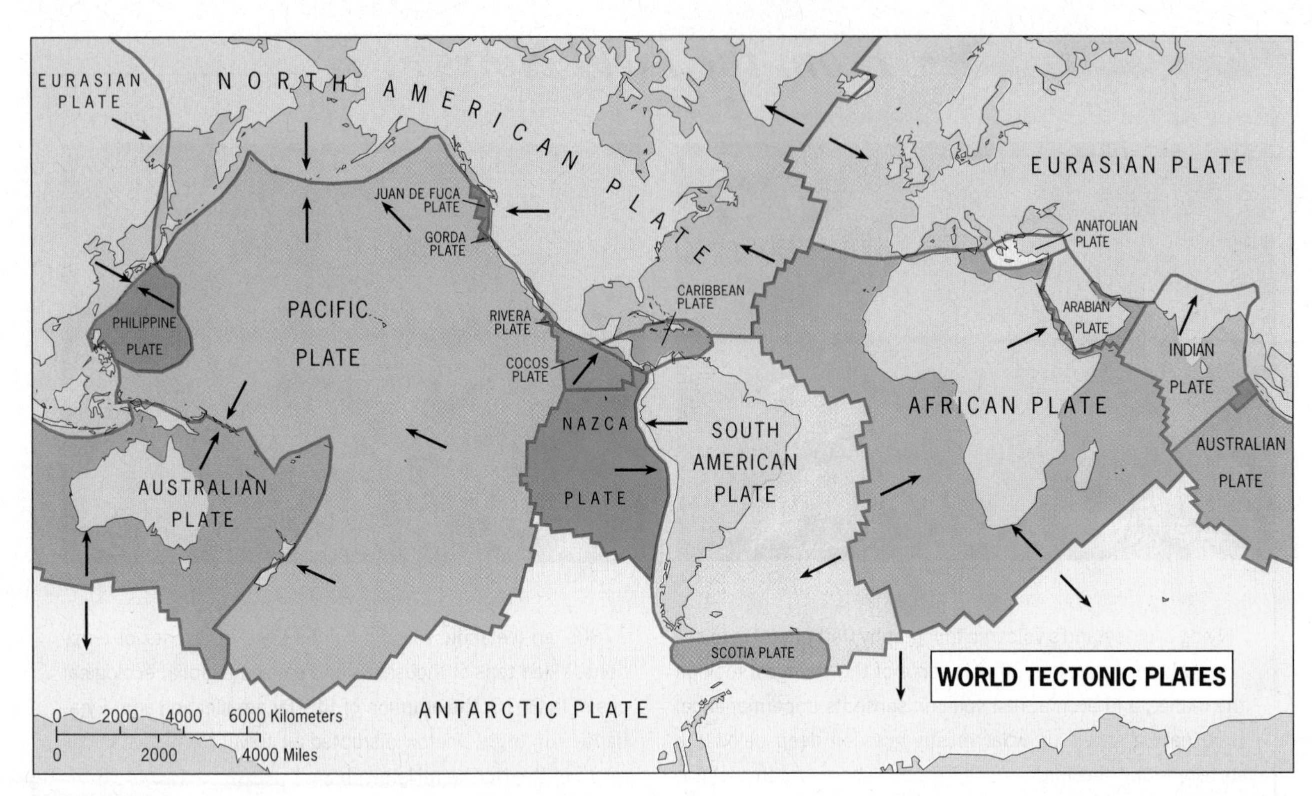

FIGURE G-4

the continents are "rafts" of relatively light rock that rest on slabs of heavier rock known as **tectonic plates [15]** (Fig. G-4) whose movement is propelled by giant circulation cells in the red-hot magma below (when this molten magma reaches the surface through volcanic vents, it is called lava).

Inevitably, moving tectonic plates collide. When they do, earthquakes and volcanic eruptions result, and the physical landscape is thrown into spectacular relief. Compare Figures G-4 and G-5, and you can see the outlines of the tectonic plates in the distribution of these hazards to human life. The 2010 earthquake adjacent to Port-au-Prince, Haiti measured 7.0 on the Richter scale. Although a shallow quake, its epicenter was located in a very densely populated area. Nearly 300,000 people died, a similar

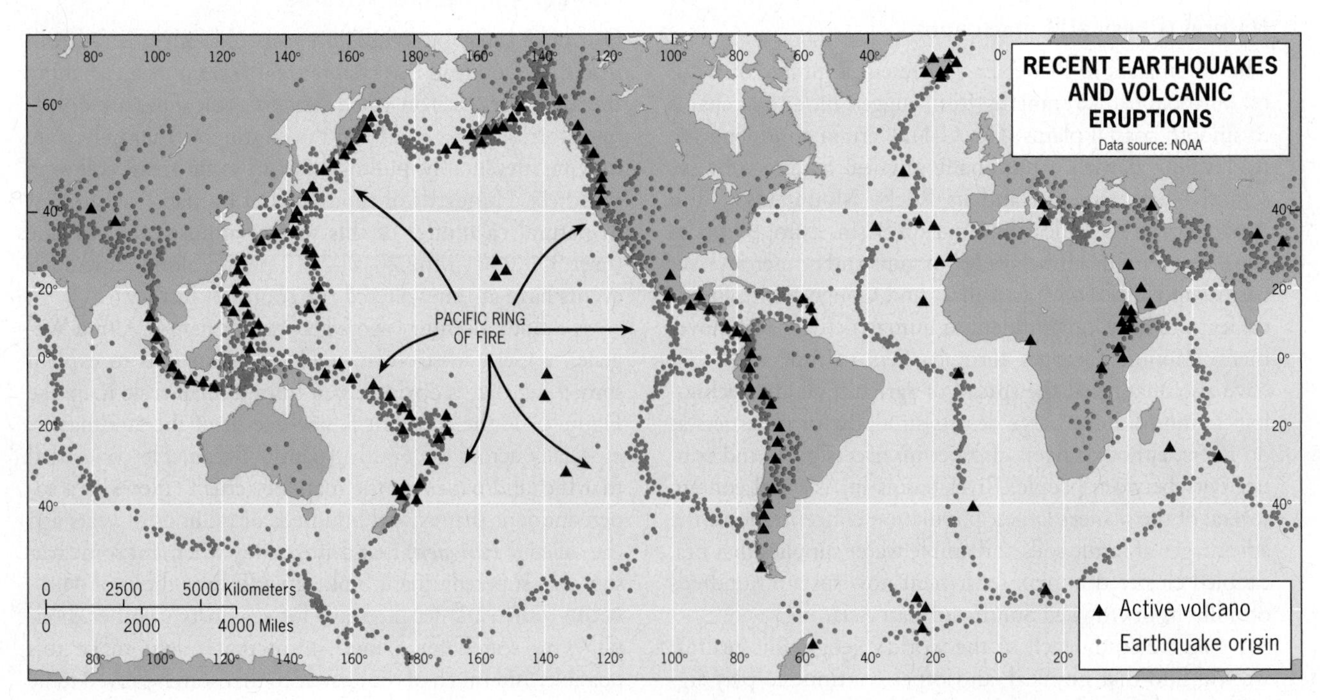

FIGURE G-5

number were injured, and 1.3 million were made homeless and destitute. The Earth's largest ocean is almost completely encircled by active volcanoes and earthquake epicenters. Appropriately, this is called the **Pacific Ring of Fire [16]**.

It is useful to compare Figure G-5 to Figure G-3 to see which of the world's geographic realms are most susceptible to the hazards inherent in crustal instability. Russia, Europe, Africa, and Australia are relatively safe; in other realms the risks are far greater in one sector than in others (western as opposed to eastern North and South America, for instance). As we shall discover, for certain parts of the world the activity mapped in Figure G-5 presents a clear and present danger. Some of the world's largest cities (e.g., Tokyo, Mexico City) lie in zones that are highly vulnerable to sudden disaster—as indeed occurred with Japan's huge 2011 earthquake and tsunamis not very far north of Tokyo.

Climate

The prevailing **climate [17]** constitutes a key factor in the geography of realms and regions (in fact, some regions are essentially defined by climate). But climates change: those dominating in certain regions today may not have prevailed there several thousand years ago. Thus any map of climate, including the maps in this chapter, is but a still-picture of our always-changing world.

Climatic conditions have swung back and forth for as long as the Earth has had an atmosphere. Periodically, an **ice age [18]** lasting tens of millions of years chills the planet and causes massive ecological change. One such ice age occurred while Pangaea was still in one piece, between 250 and 300 million years ago. Another started about 35 million years ago, and we are still experiencing it. The current epoch of this ice age, on average the coldest yet, is called the ***Pleistocene*** and has been going on for nearly 2 million years.

In our time of global warming this may come as a surprise, but we should remember that an ice age is not a period of unbroken, bitter cold. Rather, an ice age consists of surges of cold, during which glaciers expand and living space shrinks, separated by warmer phases when the ice recedes and life spreads poleward again. The cold phases are called **glaciations [19]**, and they tend to last longest, although milder spells create some temporary relief. The truly warm phases, when the ice recedes poleward and mountain glaciers melt away, are known as **interglacials [20]**. We are living in one of these interglacials today. It even has a geologic name: the ***Holocene***.

Imagine this: just 18,000 years ago, great icesheets had spread all the way south to the Ohio River Valley, covering most of the Midwest; this was the zenith of a glaciation that had lasted about 100,000 years, the ***Wisconsinan Glaciation*** (Fig. G-6). The Antarctic Icesheet was bigger

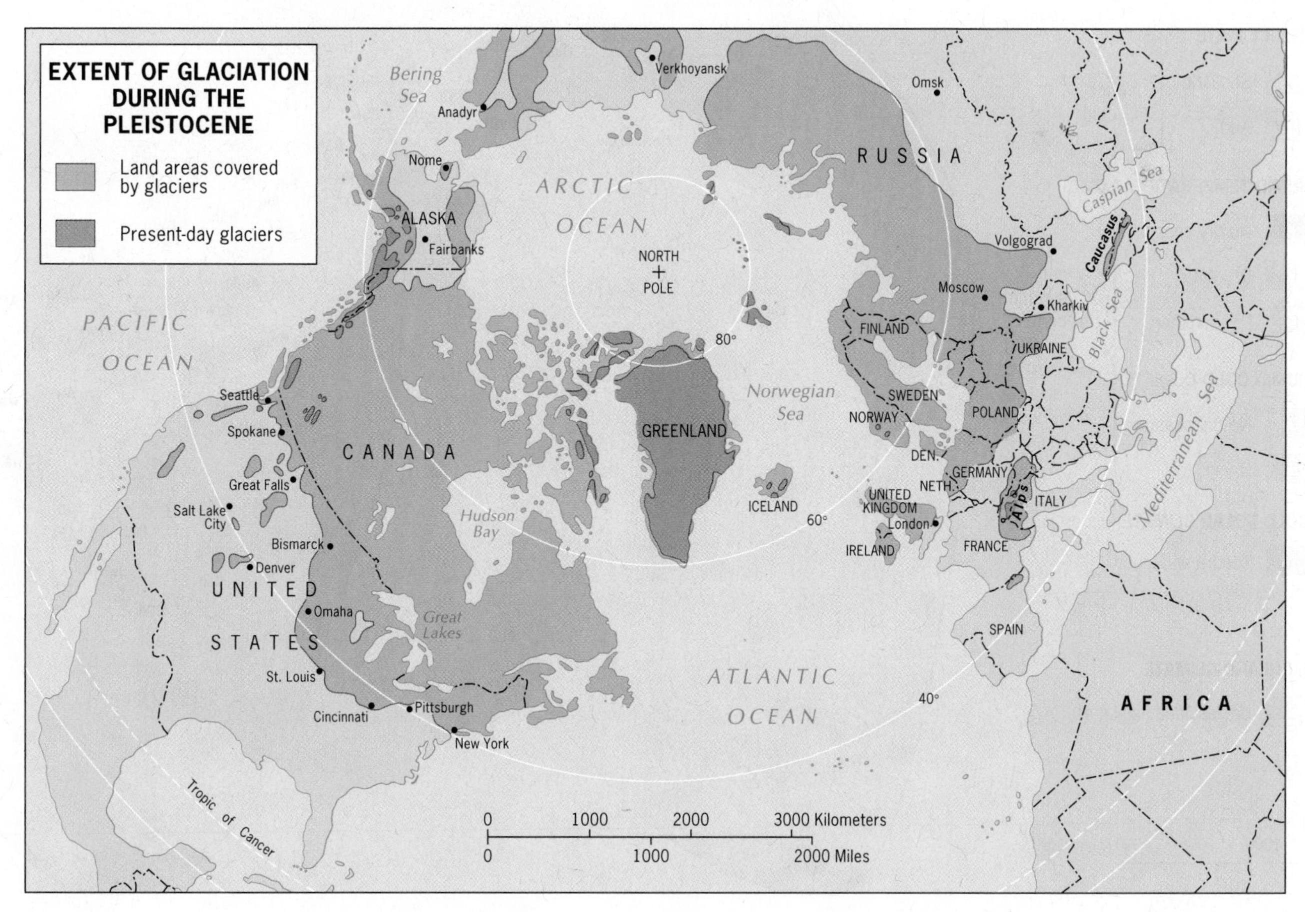

FIGURE G-6

than ever, and even in the tropics, great mountain glaciers pushed down valleys and onto plateaus. But then Holocene warming began, the continental and mountain glaciers receded, and ecological zones that had been squeezed between the advancing icesheets now spread north and south. In Europe particularly, where humans had arrived from Africa via Southwest Asia during one of the milder phases of the Wisconsinan Glaciation, living space expanded and human numbers grew.

Global Climate Change

Today we are living in an era of **global climate change [21]**, particularly natural global warming that has been accelerated by anthropogenic (human-source) causes. Since the Industrial Revolution, we have been emitting gases that have enhanced nature's ***greenhouse effect*** whereby the sun's radiation becomes trapped in the Earth's atmosphere. This is leading to a series of climate changes, especially the overall warming of the globe. One important international organization of experts, the Intergovernmental Panel on Climate Change (IPCC), predicts an increase of 2–3°C (3.6–5.4°F) overall for the globe, but with significant regional variability (e.g., more at higher latitudes, less at lower latitudes). Precipitation patterns are predicted to become more variable, particularly in regions where they are already seasonal.

This change in temperature may seem small but is expected to have significant impacts on global climate patterns, agricultural zones, and the quality of human lives. The full ramifications are not known, but scenarios

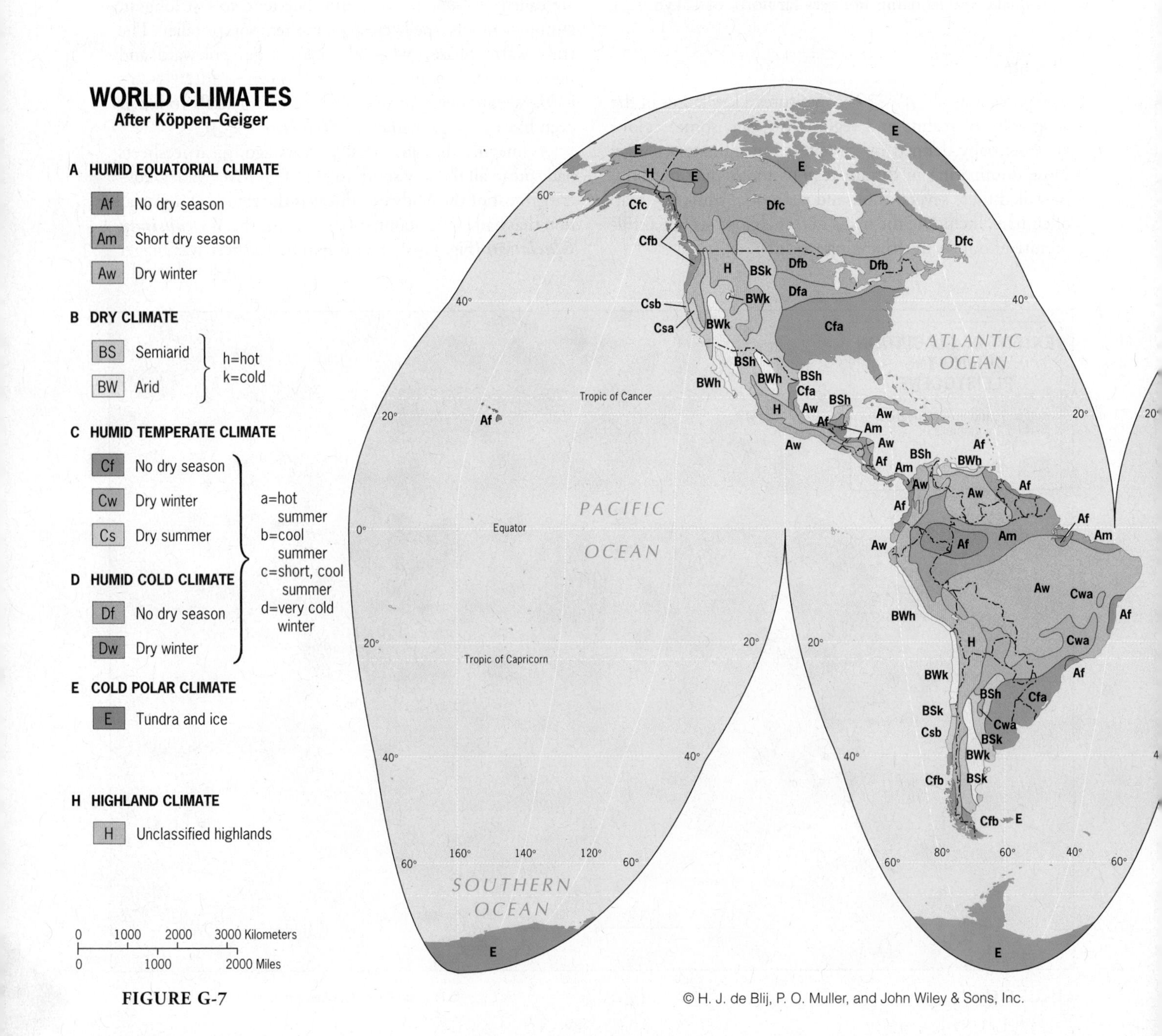

FIGURE G-7

are being modeled so that societies can confront the changes that are coming. Leaders of some countries are more skeptical than others, and some have already made greater adjustments than others. One of the most significant consequences of global climate change is that the icecap atop the Arctic Ocean is melting faster than even recent models predicted, with environmental and geopolitical implications. We pick up this issue in Chapters 2A and 12.

Climate Regions

We have just learned how variable climate can be, but in a human lifetime we see little evidence of this variability. We talk about the ***weather*** (the immediate state of the atmosphere) in a certain place at a given time, but as a technical term ***climate*** defines the aggregate, total record of weather conditions at a place or in a region over the entire period during which records have been kept.

Figure G-7 may appear very complicated, but this map is useful even at a glance. Devised long ago by Wladimir Köppen and subsequently modified by Rudolf Geiger, it represents climatic regions through a combination of colors and letter symbols. In the legend, note that the ***A*** climates (rose, orange, and peach) are equatorial and tropical; the ***B*** climates (tan, yellow) are dry; the ***C*** climates (shades of green) are temperate, that is, moderate and neither hot nor cold; the ***D*** climates (purple) are cold; the ***E*** climates (blue) are frigid; and the similar ***H*** climates (gray) prevail in highlands like the Tibetan Plateau and the upper reaches of the Andes.

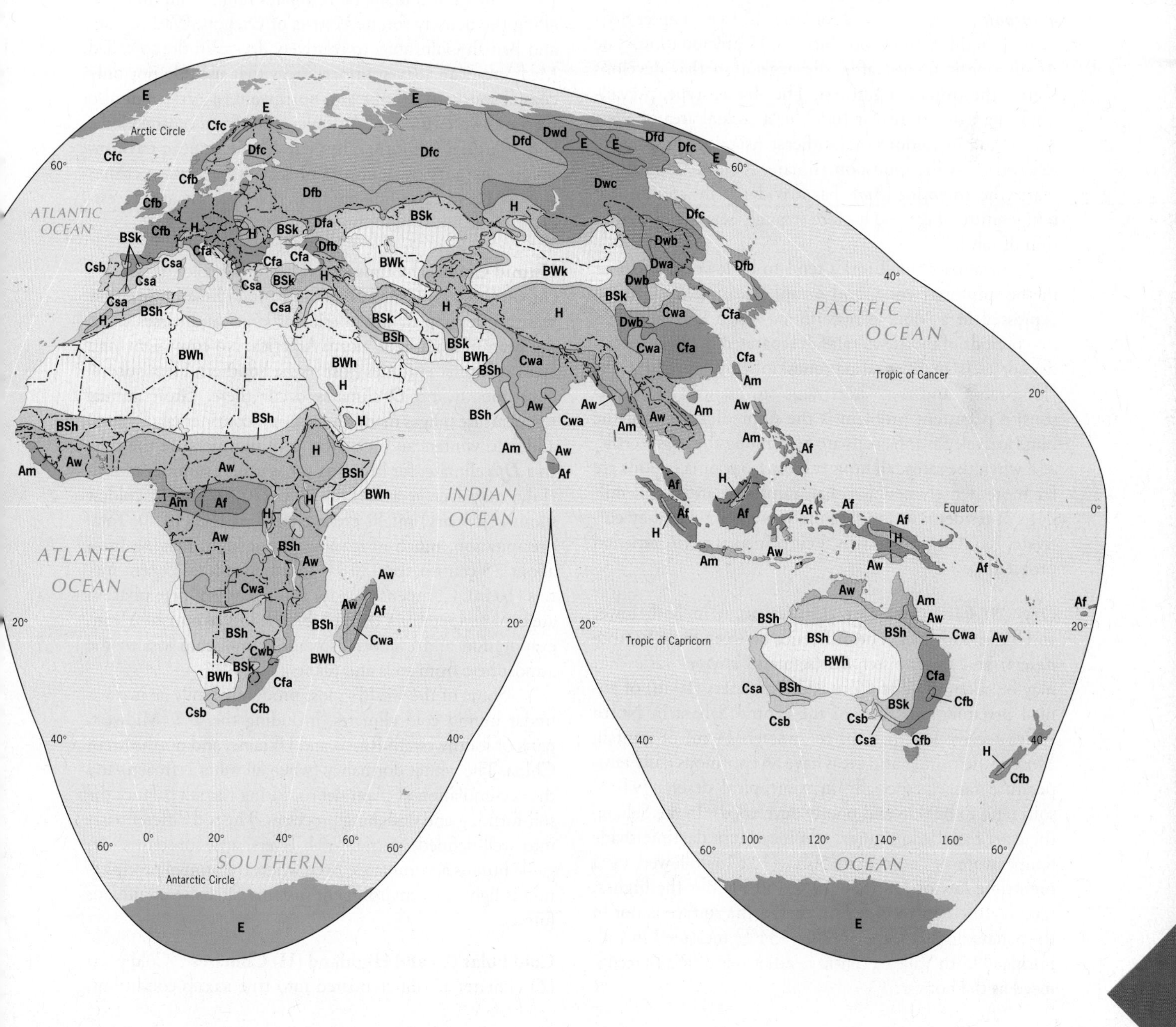

Figure G-7 merits your attention because familiarity with it will help you understand much of what follows in this book. The map has practical utility as well. Although it depicts climatic regions, daily weather in each color-coded region is relatively standard. If, for example, you are familiar with the weather in the large area mapped as ***Cfa*** in the southeastern United States, you will feel at home in Uruguay (South America), Kwazulu-Natal (South Africa), New South Wales (Australia), and Fujian Province (China). Let us look at the world's climatic regions in some detail.

Humid Equatorial (A) Climates The humid equatorial, or tropical, climates are characterized by high temperatures all year and by heavy precipitation. In the ***Af*** subtype, the rainfall arrives in substantial amounts every month; but in the ***Am*** areas, the arrival of the annual wet ***monsoon*** (the Arabic word for "season" [see Chapter 8A]) marks a sudden enormous increase in precipitation. The ***Af*** subtype is named after the vegetation that develops there—the tropical rainforest. The ***Am*** subtype, prevailing in part of peninsular India, in a coastal area of West Africa, and in sections of Southeast Asia, is appropriately referred to as the monsoon climate. A third tropical climate, the ***savanna*** (***Aw***), has a wider daily and annual temperature range and a more strongly seasonal distribution of rainfall.

Savanna rainfall totals tend to be lower than those in the rainforest zone, and savanna seasonality is often expressed in a "double maximum." Each year produces two periods of increased rainfall separated by pronounced dry spells. In many savanna zones, inhabitants refer to the "long rains" and the "short rains" to identify those seasons; a persistent problem is the unpredictability of the rain's arrival. Savanna soils are not among the most fertile, and when the rains fail hunger looms. Savanna regions are far more densely peopled than rainforest areas, and millions of residents of the savanna subsist on what they cultivate. Rainfall variability is their principal environmental problem.

Dry (B) Climates Dry climates occur in both lower and higher latitudes. The difference between the ***BW*** (true ***desert***) and the moister ***BS*** (semiarid ***steppe***) varies but may be taken to lie at about 25 centimeters (10 in) of annual precipitation. Parts of the central Sahara in North Africa receive less than 10 centimeters (4 in) of rainfall. Most of the world's arid areas have an enormous daily temperature range, especially in subtropical deserts (whose soils tend to be thin and poorly developed). In the Sahara, there are recorded instances of a maximum daytime shade temperature of more than 50°C (122°F) followed by a nighttime low of less than 10°C (50°F). But the highest temperature ever recorded on the Earth's surface is not in the Sahara: in 2013, the 56.7°C (134°F) measured in California's Death Valley a century earlier was officially recognized as the hottest.

Humid Temperate (C) Climates As the map shows, almost all these mid-latitude climate areas lie just beyond the Tropics of Cancer and Capricorn (23.5° North and South latitude, respectively). This is the prevailing climate in the southeastern United States from Kentucky to central Florida, on North America's west coast, in most of Europe and the Mediterranean, in southern Brazil and northern Argentina, in coastal South Africa, in eastern Australia, and in eastern China and southern Japan. None of these areas suffers climatic extremes or severity, but the winters can be cold, especially away from water bodies that moderate temperatures. These areas lie midway between the winterless equatorial climates and the summerless polar zones. Fertile and productive soils have developed under this regime, as we will note in our discussion of the North American and European realms.

The humid temperate climates range from moist, as along the densely forested coasts of Oregon, Washington, and British Columbia, to relatively dry, as in the so-called Mediterranean (dry-summer) areas that include not only coastal southern Europe and northwestern Africa but also the southwestern tips of Australia and Africa, central Chile, and Southern California. In these Mediterranean environments, the scrubby, moisture-preserving vegetation creates a natural landscape different from that of richly green western Europe.

Humid Cold (D) Climates The humid cold (or "snow") climates may be called the continental climates, for they seem to develop in the interior of large landmasses, as in the heart of Eurasia or North America. No equivalent land areas at similar latitudes exist in the Southern Hemisphere; consequently, no ***D*** climates occur there. Great annual temperature ranges mark these humid continental climates, and cold winters and relatively cool summers are the rule. In a ***Dfa*** climate, for instance, the warmest summer month (July) may average as high as 21°C (70°F), but the coldest month (January) might average only −11°C (12°F). Total precipitation, much of it snow, is not high, ranging from about 75 centimeters (30 in) to a steppe-like 25 centimeters (10 in). Compensating for this paucity of precipitation are cool temperatures that inhibit the loss of moisture from evaporation and evapotranspiration (moisture loss to the atmosphere from soils and plants).

Some of the world's most productive soils lie in areas under humid cold climates, including the U.S. Midwest, parts of southwestern Russia and Ukraine, and northeastern China. The winter dormancy (when all water is frozen) and the accumulation of plant debris during the fall balance the soil-forming and enriching processes. The soil differentiates into well-defined, nutrient-rich layers, and substantial organic humus accumulates. Even where the annual precipitation is light, this environment sustains extensive coniferous forests.

Cold Polar (E) and Highland (H) Climates Cold polar (***E***) climates are differentiated into true icecap conditions, where permanent ice and snow keep vegetation from gaining a foothold, and the tundra, which may have average temperatures above freezing up to four months of the year. Like rainforest, savanna, and steppe, the term ***tundra*** is vegetative as well as climatic, and the boundary between the ***D*** and ***E*** climates in Figure G-7 corresponds closely to that between the northern coniferous forests and the tundra.

Finally, the ***H*** climates—the unclassified highlands mapped in gray (Fig. G-7)—resemble the ***E*** climates. High elevations and the complex topography of major mountain systems often produce near-Arctic climates above the tree line, even in the lowest latitudes such as the equatorial section of the high Andes of South America.

Let us not forget an important qualification concerning Figure G-7: this is a still-picture of a changing scene, a single frame from an ongoing film. Climate continues to change, and only a few decades from now climatologists are likely to be modifying the climate maps to reflect new data. Who knows: we may even have to redraw those familiar coastlines. Environmental change is a never-ending challenge.

You will find larger-scale maps of climate in several of the regional chapters that follow, but it is useful to refer back to this Köppen-Geiger map whenever the historical or economic geography of a region or country is under discussion. The world climatic map reflects agricultural opportunities and limitations as well as climatic regimes, and as such helps explain some enduring patterns of human distribution on our planet. We turn next to this crucial topic.

REALMS OF POPULATION

Earlier we noted that population numbers by themselves do not define geographic realms or regions. Population distributions, and the functioning society that gives them common ground, are more significant criteria. That is why we can identify one geographic realm (the Austral) with less than 30 million people and another (South Asia) with more than 1.7 billion inhabitants. Neither population numbers nor territorial size alone can delimit a geographic realm. Nevertheless, the map of world population distribution shows some major clusters that are part of some specific realms (Fig. G-8).

Before we examine these clusters in some detail, remember that the world's human population now rounds off at 7.2 billion (see Appendix B for a detailed breakdown)—confined to the landmasses that constitute less than 30 percent of our planet's surface, much of which is arid desert, inhospitable mountain terrain, or frigid tundra. (Remember too that Fig. G-8 is another still-picture of an ever-changing scene: the rapid growth of humankind continues.) After thousands of years of slow growth, world population during the nineteenth and twentieth centuries grew at an increasing rate. That rate has recently been slowing down, even imploding in some parts of the world. But consider this: it took about 17 centuries following the birth of Christ for the world to add 250 million people to its numbers; now we are adding 250 million about every three years. While the *rate* of population growth has come down in some parts of the world, in absolute terms the global population continues to grow apace and is expected to reach 9.6 billion by 2050.

This raises the important question as to whether there are limits to the Earth's carrying capacity—will there be enough food to go around? That question has become more and more pressing over the past decade due to rapidly rising food prices resulting from increased demand in China and India, a dietary shift from grains to meat and vegetables, and the use of agricultural resources for the production of biofuels. The actual increase of population is only part of the problem; our growing appetite for certain

From the Field Notes . . .

"One early January morning in northern Vietnam, just outside the city of Ninh Binh, a girl came by on her bicycle, pulling an ox on a long rope. When I asked, through an interpreter, where she was taking the ox she replied that she had to take it to her uncle who still worked his land. The girl's own family was moving on to more urban lifestyles, with jobs in manufacturing or services. Vietnam has been changing and modernizing in recent years. Some sectors of the economy are growing vigorously and urbanization has proceeded apace. Sometimes people move to the cities, and sometimes the city comes to them (see the high-rise of the encroaching city in the left background). To the girl, it all seemed for the better. 'I like the city,' she said. She did not have much eye for the ox, but she proudly pointed to her new bicycle. To her, the future looked good."

© Jan Nijman

www.conceptcaching.com

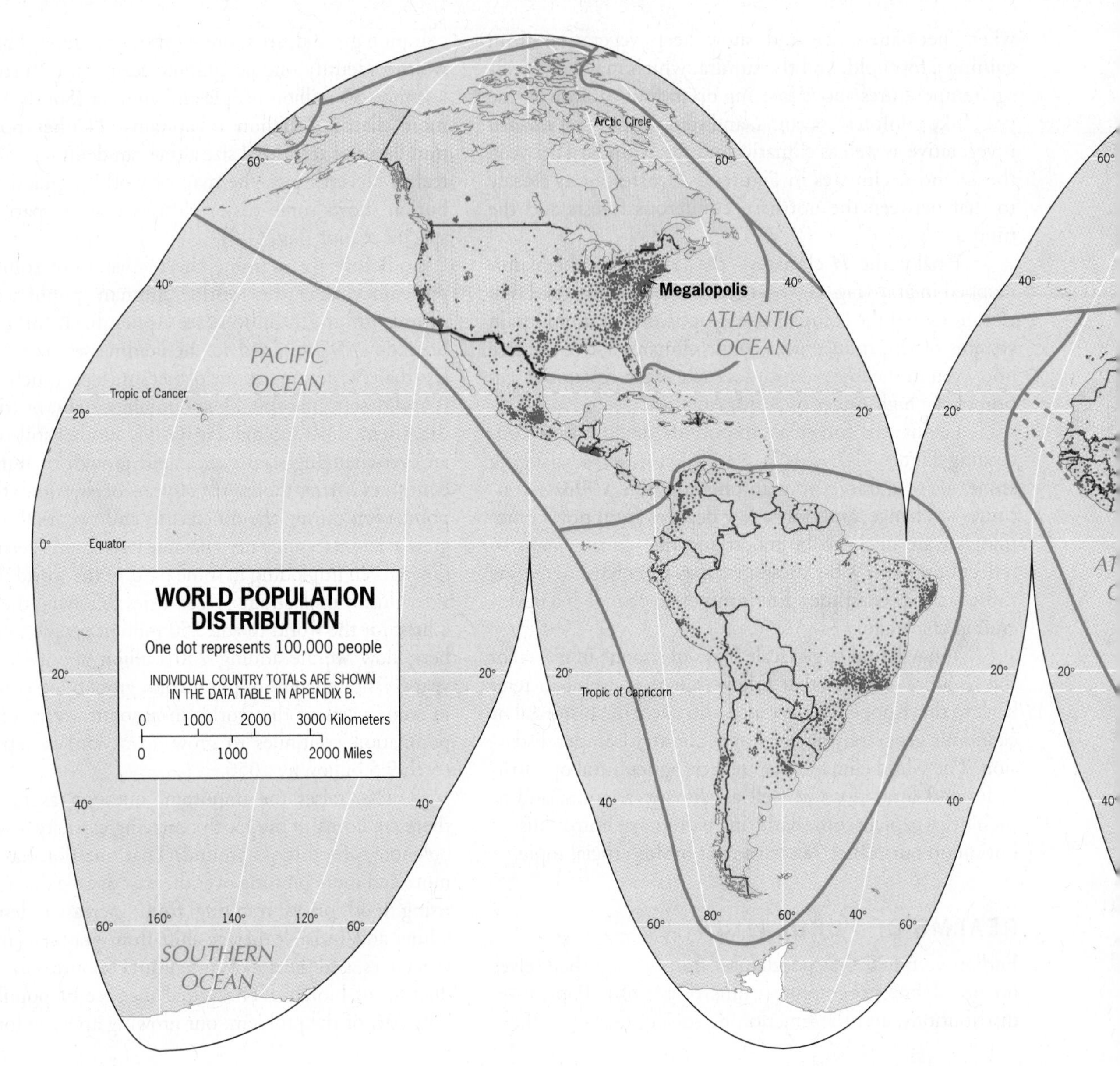

FIGURE G-8

products is another. And we are not just talking about food—think drinking water, fossil fuels, and minerals as well. Thus it seems inconceivable that 10 billion people by mid-century could be consuming the way we do today in the developed world.

Major Population Clusters

One way to present an overview of the location of people on the planet is to create a map of **population distribution [22]** (Fig. G-8). As you can see in the map's legend, each dot represents 100,000 people, and the clustering of large numbers of them in certain areas as well as the near-emptiness of others is readily evident. There is a technical difference between population distribution and ***population density***, which is another way of showing where people live. Density maps reveal the number of persons per unit area, requiring a different cartographic technique.

- ***South Asia*** The *South Asia* population cluster lies centered on India and includes its populous neighbors, Pakistan and Bangladesh. This huge agglomeration of humanity focuses on the wide plain of the Ganges River (**A** in Fig. G-8). South Asia recently became the world's largest population cluster, overtaking East Asia in 2010. A larger percentage of the people remain farmers here, although pressure on the land is greater, whereas agriculture is less efficient than in East Asia.

- ***East Asia*** Now surpassed by South Asia, the second-ranked *East Asia* population cluster lies centered on China and includes the Pacific-facing Asian coastal zone from the Korean Peninsula to Vietnam. Not long ago, we

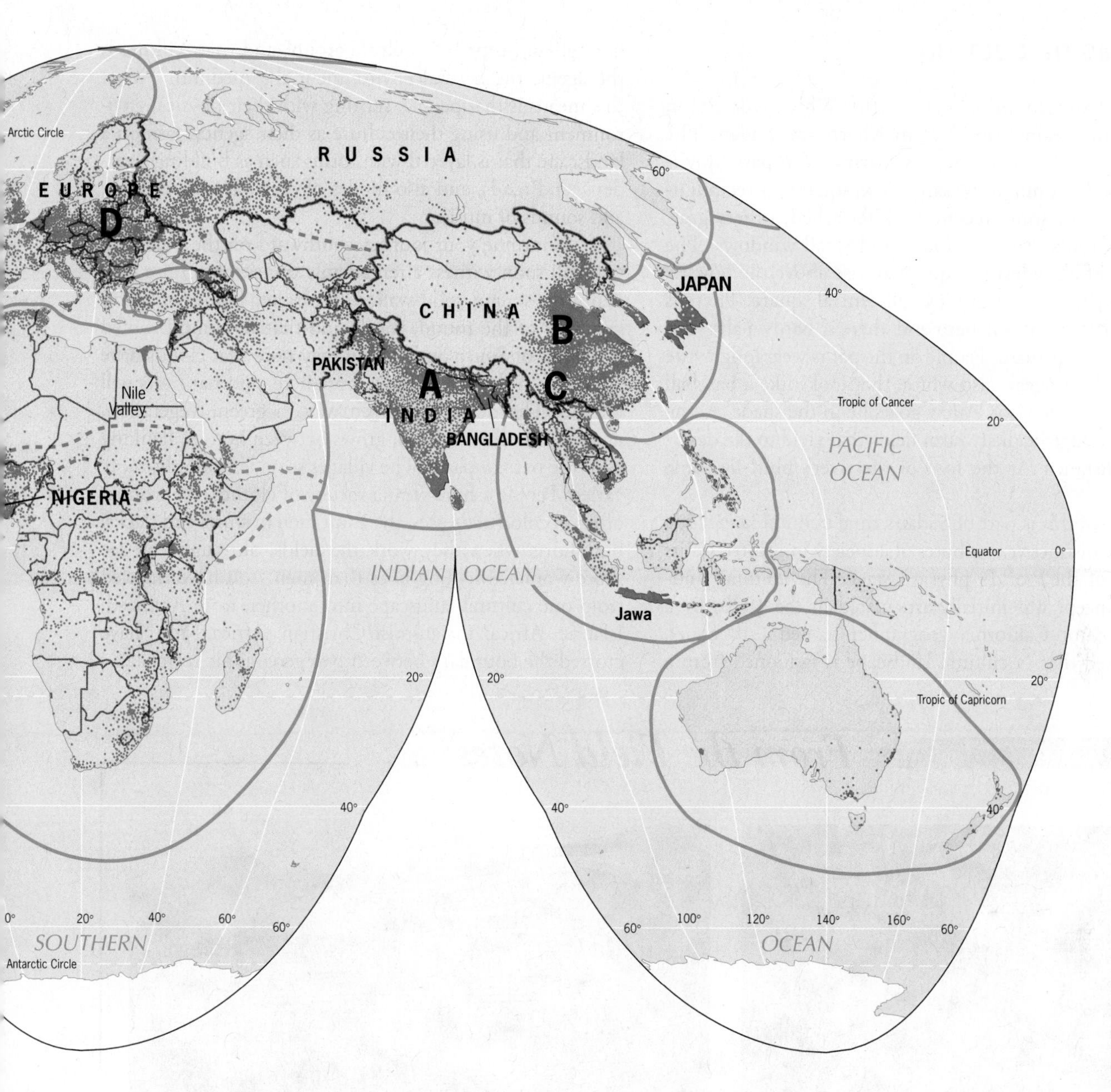

would have reported this as a dominantly rural, farming population, but rapid economic growth and associated urbanization have changed the picture. In China's interior river basins of the Huang (Yellow) and Chang/Yangzi (**B** and **C** on the map), and in the Sichuan Basin between these two letters, most of the people remain farmers. But the booming cities of coastal and increasingly interior China are attracting millions of new inhabitants, and in 2011 the Chinese urban population surpassed the 50-percent milestone.

- ***Europe*** The third-ranking population cluster, *Europe*, also lies on the Eurasian landmass but at the opposite end from China. The European cluster, including western Russia, counts more than 700 million inhabitants, which puts it in a class with the two larger Eurasian concentrations—but there the similarity ends. In Europe, the key to the linear, east-west orientation of the axis of population (**D** in Fig. G-8) is not a fertile river basin but a zone of raw materials for industry. Europe is among the world's most highly urbanized and industrialized realms, its human agglomeration sustained by factories and offices rather than paddies and pastures.

The three world population concentrations just discussed (South Asia, East Asia, and Europe) account for just about 4 billion of the world's 7.2 billion people. No other cluster comes close to these numbers. The next-ranking cluster, Eastern North America, is only about one-quarter the size of the smallest Eurasian concentrations. As in Europe, the population in this zone is concentrated in major metropolitan complexes; the rural areas are now relatively sparsely settled. Geographic realms and regions, therefore, display varying levels of **urbanization [23]**, the percentage of the total population residing in cities and towns. Some regions are urbanizing far more rapidly than others, a phenomenon we will explain as we examine each realm.

REALMS OF CULTURE

Imagine yourself in a boat on the White Nile River, headed upstream (south) from Khartoum, Sudan. The desert sky is blue, the heat is searing. You pass by villages that look much the same: low, square, or rectangular dwellings, some recently whitewashed, others gray, with flat roofs, wooden doors, and small windows. The minaret of a modest mosque may rise above the houses, and you get a glimpse of a small central square. There is very little vegetation; here and there a hardy palm tree stands in a courtyard. People on the paths wear long white robes and headgear, also white, that looks like a baseball cap without the visor. A few goats lie in the shade. Along the river's edge lie dusty farm fields that yield to the desert in the distance. At the foot of the river's bluff lie some canoes.

All of this is part of Sudan's rural **cultural landscape** [24], the distinctive attributes of a society imprinted on its portion of the world's physical stage. The cultural landscape concept was initially articulated in the 1920s by a University of California geographer named Carl Sauer, who stated that "a cultural landscape is fashioned from a natural landscape by a culture group" and that "culture is the agent; the natural environment the medium." What this means is that people, starting with their physical environment and using their culture as their agency, fashion a landscape that is layered with forms such as buildings, gardens, and roads, and also modes of dress, aromas of food, and sounds of music.

Continue your journey southward on the Nile, and you will soon witness a remarkable transition. Quite suddenly, the square, solid-walled, flat-roofed houses of Sudan give way to the round, wattle-and-thatch, conical-roofed dwellings of South Sudan. You may note that clouds have appeared in the sky: it rains more here, and flat roofs will not do. The desert has given way to green. Vegetation, natural as well as planted, grows between houses, flanking even the narrow paths. The villages seem less orderly, more varied. People ashore wear a variety of clothes, the women often in colorful dresses, the adult men in shirts and slacks, but shorts when they work the fields, although you see more women wielding hoes than men. You have traveled from one cultural landscape into another, from Arabized, Islamic Africa to animist/Christian Africa. You have crossed the boundary between two geographic realms.

From the Field Notes . . .

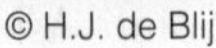

© H.J. de Blij

© H.J. de Blij

"The Atlantic-coast city of Bergen, Norway displayed the Norse cultural landscape more comprehensively, it seemed, than any other Norwegian city, even Oslo. The high-relief site of Bergen creates great vistas, but also long shadows; windows are large to let in maximum light. Red-tiled roofs are pitched steeply to enhance runoff and inhibit snow accumulation; streets are narrow and houses clustered, conserving warmth . . . The coastal village of Mengkabong on the Borneo coast of the South China Sea represents a cultural landscape seen all along the island's shores, a stilt village of the Bajau, a fishing people. Houses and canoes are built of wood as they have been for centuries. But we could see some evidence of modernization: windows filling wall openings, water piped in from a nearby well."

www.conceptcaching.com

Mustafa Ozer/AFP/Getty Images, Inc.

The *hajj* is the yearly pilgrimage of Muslims to the holy city of Mecca in Saudi Arabia. The pilgrimage is referred to as the fifth "pillar" of Islam, the obligation of every able-bodied Muslim to worship Allah in this holiest of sites at least once in their lifetime. This is the Grand Mosque of Mecca on November 17, 2010, as more than two million Muslim pilgrims launched into the final rituals of this largest religious pilgrimage in the world.

No geographic realm, not even the Austral Realm, has just one single cultural landscape, but cultural landscapes help define realms as well as regions. The cultural landscape of the high-rise North American city with its sprawling suburbs differs from that of urban South America; the organized terraced paddies of Southeast Asia are unlike anything to be found in the rural cultural landscape of neighboring Australia. Variations of cultural landscapes within geographic realms, such as between highly urbanized and dominantly rural (and more traditional) areas, help us define the world's regions.

The Geography of Language

Language is the essence of culture. People tend to feel passionately about their mother tongue, especially when they believe it is threatened in some way. In the United States today, the English Only movement reflects many people's fears that the primacy of English as the national language is under threat as a result of immigration. As we will see in later chapters, some governments try to suppress the languages (and thus the cultures) of minorities in mistaken attempts to enforce national unity, provoking violent reactions.

In fact, languages emerge, thrive, and die out over time, and linguists estimate that the number of lost languages is in the tens of thousands—a process that continues. One year from the day you read this, about 25 more languages will have become extinct, leaving no trace. Just in North America, more than 100 native languages were lost during the past half-century. Some major ones of the past, such as Sumerian and Etruscan, have left fragments in later languages. Others, like Sanskrit and Latin, live on in their modern successors. At present, about 6800 languages remain, half of them classified by linguists as endangered; some of the "hot spots" are the Amazon, Siberia, northern Australia, and the Andes. By the end of this century, the bulk of the world's population will be speaking just a few hundred languages, which means that many millions will no longer be able to speak their ancestral mother tongues.

Scholars have tried for many years to unravel the historic roots and branches of the "language tree," and their debates continue. Geographers trying to map the outcome of this research keep having to modify the pattern, so you should take Figure G-9 as a work in progress, not the final product. At minimum, there are some 15 so-called ***language families***, groups of languages with a shared but usually distant origin. The most widely distributed language family, the Indo-European (shown in yellow on the map), includes English, French, Spanish, Russian, Persian, and Hindi. This encompasses the languages of European colonizers that were carried and implanted worldwide, English most of all. Today, English serves as the national or official language of many countries and outposts, and remains the ***lingua franca*** (common second language) of government, commerce, and higher education in many multicultural societies (see Fig. G-9 inset map). In the postcolonial era, English became the chief medium of still another wave of ascendancy now in progress: globalization.

But even English may eventually go the way of Latin, morphing into versions you will hear (and learn to use) as you travel, forms of English that may, generations from now, be the successors that Italian and Spanish are to Latin. In Hong Kong, Chinese and English are producing a local "Chinglish" you may hear in the first taxi you enter. In Lagos, Nigeria, where most of the people

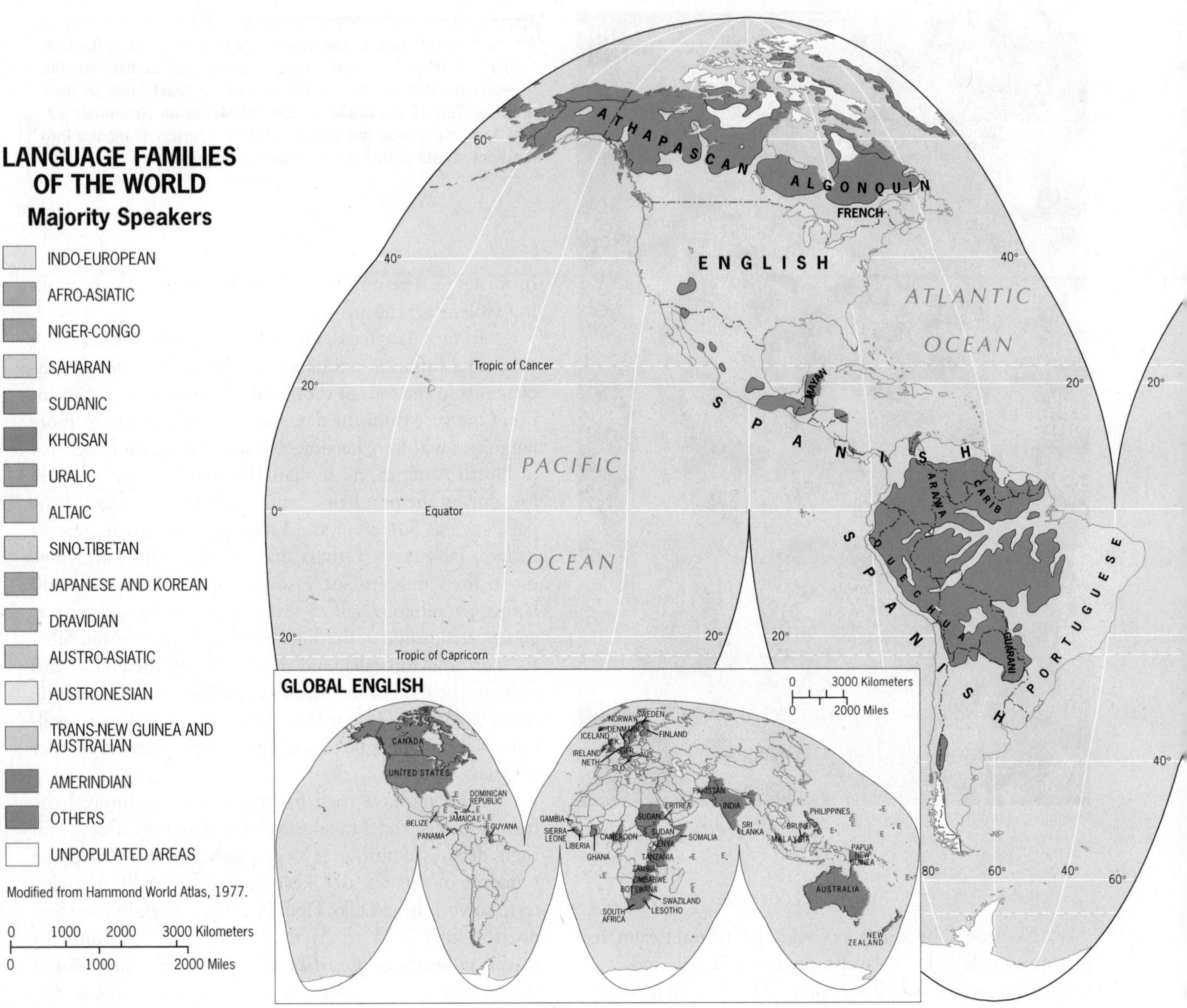

LANGUAGE FAMILIES OF THE WORLD
Majority Speakers

- INDO-EUROPEAN
- AFRO-ASIATIC
- NIGER-CONGO
- SAHARAN
- SUDANIC
- KHOISAN
- URALIC
- ALTAIC
- SINO-TIBETAN
- JAPANESE AND KOREAN
- DRAVIDIAN
- AUSTRO-ASIATIC
- AUSTRONESIAN
- TRANS-NEW GUINEA AND AUSTRALIAN
- AMERINDIAN
- OTHERS
- UNPOPULATED AREAS

Modified from Hammond World Atlas, 1977.

GLOBAL ENGLISH

FIGURE G-9

are culturally and ethnically Yoruba, a language called "Yorlish" is emerging. No map can keep up with the constant evolution of language.

Landscapes of Religion

Religion played a crucial part in the emergence of ancient civilizations and has shaped the course of world history. Hinduism, for example, was one of the earliest religions that helped shape an entire realm (South Asia). Later, Buddhism, Christianity, and Islam emerged as major belief systems, often splitting up into various branches stretching across realms and regions. Figure G-10 shows the current distribution of world religions. Our world has become a more complicated place in recent times, and its patterns of religion are increasingly diffuse and dynamic. But today, still, we find that geographic realms are often dominated by a single religion or family of religions: Christianity in Europe and the Americas, Islam in North Africa/Southwest Asia, Hinduism in South Asia, and Buddhism in mainland Southeast Asia. But the boundaries tend not to be very sharp and usually take the form of transition zones (e.g., between North and Middle America, or between North and Subsaharan Africa).

A WORLD OF STATES

Ours is a world of about 200 countries or **states [25]**. The political territorial organization of the world within a system of states hinges on the notion of **sovereignty [26]**. It is a concept from international law which means that the government of a state rules supreme within its borders.

Normally, states recognize each other's sovereignty, but this becomes a matter of contention at times of conflict and war.

In the tens of thousands of years of human history, the modern state is a relatively recent invention, and so is the international system of which it forms the cornerstone. The modern state emerged from other kinds of politico-territorial organization that have existed since the beginnings of complex civilizations. In the study of ancient history, scholars sometimes use the term polity or proto-state to indicate the difference. Ever since farm surpluses enabled the growth of large and prosperous towns, this was accompanied by the more sophisticated and centralized exercise of power and political organization. From these origins, the earliest states took shape.

Although ancient states such as the Greek city-states and the Roman Empire exhibited several qualities of modern states, it was not until the seventeenth century that European rulers and governments began to negotiate treaties that defined the state in international law. That is why the modern state is often described as based on the **European state model [27]**, with definitions of nationality and sovereignty. Often, the model assumed that state and nation were ideally conterminous, so that a ***nation-state*** would enclose an ethnically and culturally homogeneous people within a national boundary. That was never truly the case (even France, the "model of models," had its minorities), and today the ideal state is defined as a clearly and legally defined territory inhabited by a citizenry governed from a capital city by a representative government. As we shall discover in Chapters 1A and 1B, not even in Europe itself are all governments truly representative, but the European state model has, for better or worse, been adopted throughout the world.

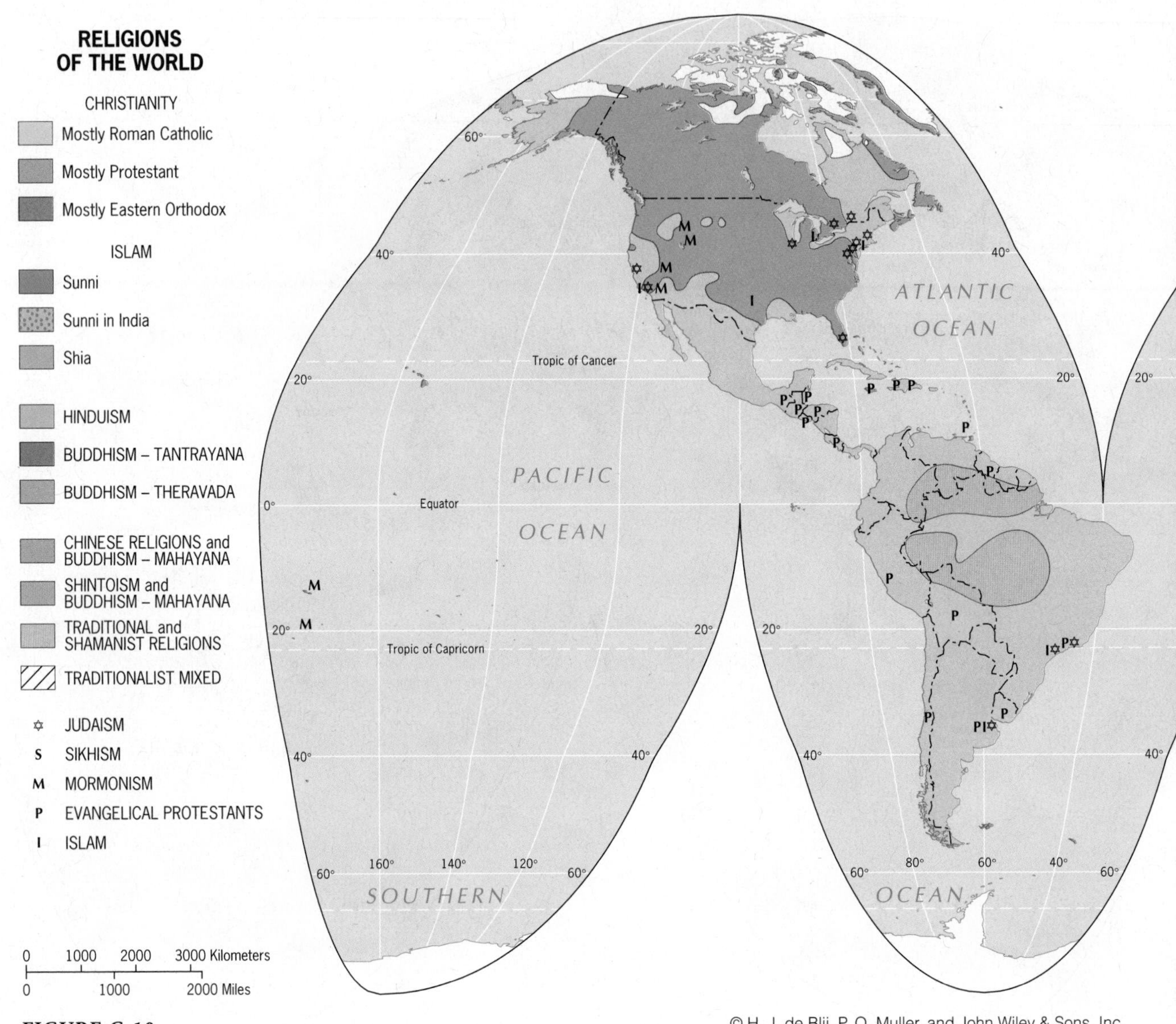

FIGURE G-10

So the modern state is a historical phenomenon, and there are also signs that it may not last forever. The state system today is challenged "from below" by ethnic minorities and regional secessionist movements (e.g., Tibetans in China; Scotland in the United Kingdom). And it is also challenged "from above" through increasingly powerful international organizations such as the European Union. Its member-states voluntarily transfer some of their power to "Brussels" (the EU's headquarters city) mainly because they think it will be to their economic advantage.

Even though many states find themselves negotiating these challenges through the decentralization of authority to regional governments or through the transfer of part of their authority to international bodies, it is important to keep in mind that they do so, almost always, with the capacity to retain their powers and to seize control at their discretion. For all our efforts to cooperate diplomatically (the United Nations), economically (the European Union), strategically (the North Atlantic Treaty Organization), and in other ways, it is the state and its government—not regions or realms—that holds the power and the authority to make decisions in the global arena. Decidedly, this is still a world of states.

Subdivisions of the State

Meanwhile, we are all too well aware that states contain subdivisions. Even the smallest states are partitioned in this manner. As all Americans—as well as Mexicans, Brazilians, and Australians—know, some larger states call their subdivisions *States*: the State of Virginia, the State of Chihuahua, the State of Bahia, the State of Victoria. (As pointed out earlier, a state denotes a sovereign country whereas a [capitalized] State signifies a subdivision.) The subdivisions of other states have alternate names: provinces (Canada),

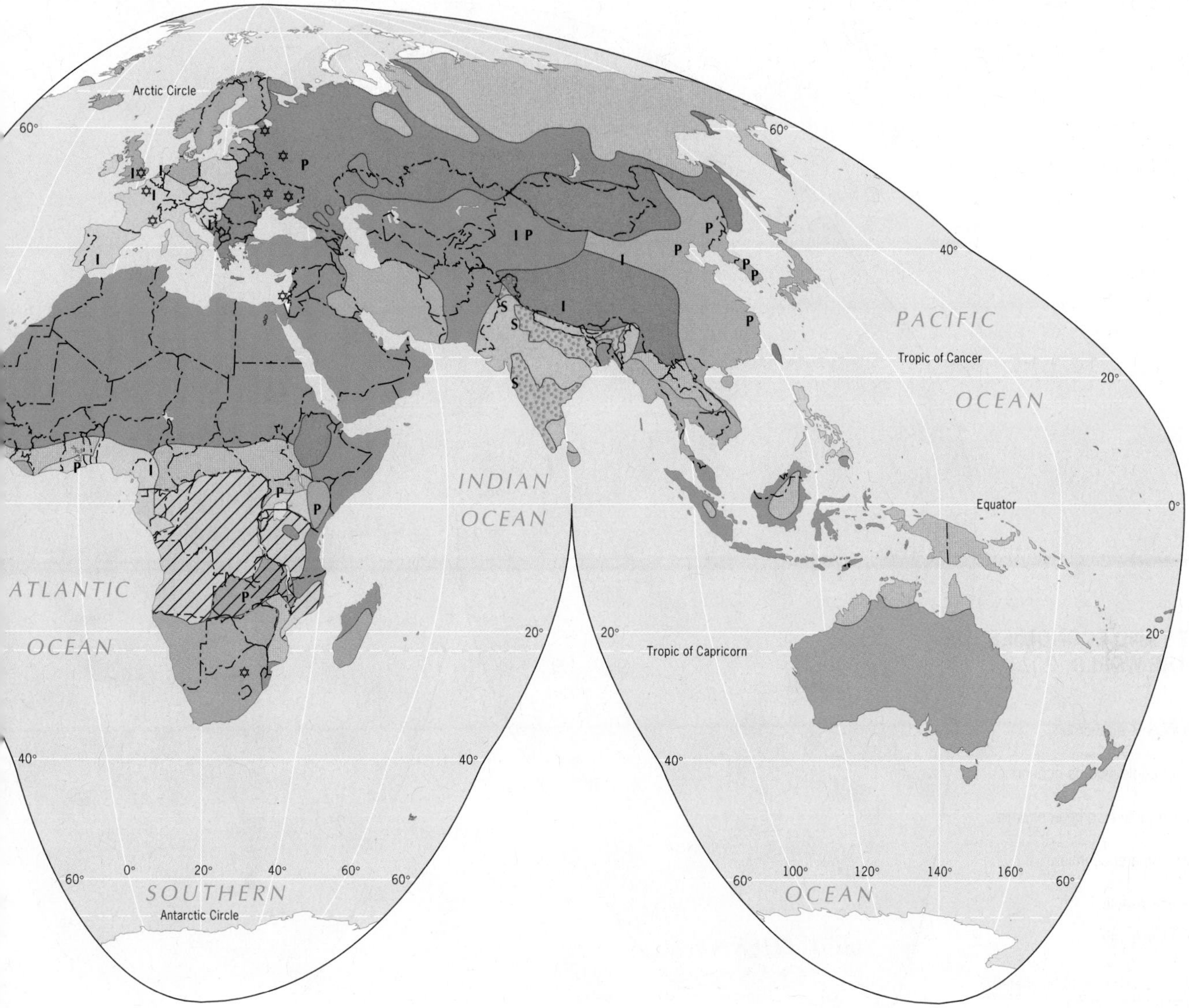

Adapted from E. H. Fouberg et al., *Human Geography*, 9e, based on several data sources.

regions (France), Autonomous Communities (Spain), Federal Districts (Russia), Divisions (Myanmar). And some of these subnational political units are becoming increasingly assertive, occasionally making their own decisions about their economic or social policies whether the central (state) government likes it or not. When that happens—in Quebec, in Catalonia, in Arizona—we should pay even closer attention to the map.

Our analysis of the world's regional geography requires data, and it is crucial to know the origin of these data. Unfortunately, we do not have a uniformly sized grid that we can superimpose over the globe: we must depend on the world's 190-plus countries to report vital information (think of this information as "state-istics"). Fortunately, all large and populous countries tend to also provide information on each of their subdivisions when they conduct their census. So at least some data are available at a finer scale.

Geopolitics and the State

As we shall observe in the pages that follow, states vary not only in terms of their dimensions, relative location, domestic resource base, productive capacity, and other physical and cultural properties, but also in terms of their influence in world affairs. Napoleon once remarked that "the politics of the state lies in its geography." There is no doubt that geography is vital to state affairs and to the relations among states within realms and regions. But the actual influence of geography is far from simple and difficult to measure. There can be many different aspects to a country's geography, and sometimes they are hard to separate from cultural or economic factors.

Take size, for example. Big countries tend to be more powerful than small countries, but this is hardly a perfect relationship, and economic prowess also counts for a great deal. Some relatively small countries carry a lot

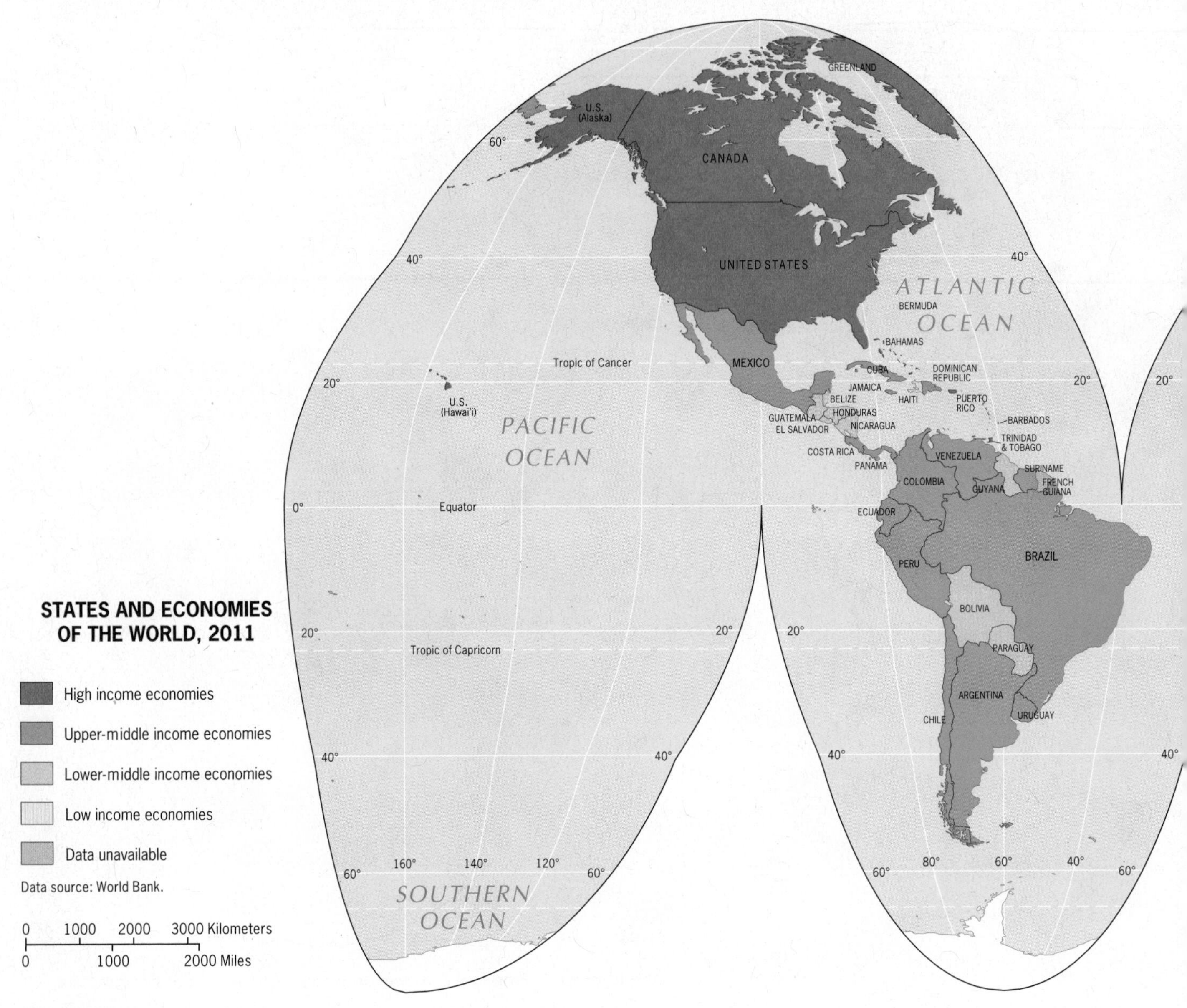

FIGURE G-11

of weight in international affairs, while other larger and more populous ones are much less influential. Think of Switzerland and its importance in the world of banking and finance; think of Israel and global geopolitics; think of Kuwait and oil.

When gauging the role of geography in international affairs we must also remember that while physical geography may be "permanent" in certain ways, the meaning ascribed to those geographic features can change over time. Think of the impact of transport and communications technology on relative distance, for example. And it is important that we consider the role of geography in historical context as well. For instance, the origins of the European Union must be understood in the context of World War II, particularly the role of Germany during and after the war.

States, Realms, and Regions

As Figures G-3 and G-11 suggest, geographic realms are mostly assemblages of states, and the borders between realms frequently coincide with the boundaries between countries—for example, between North America and Middle America along the U.S.-Mexico border. But a realm boundary can also cut across a state, as does the one between Subsaharan Africa and the Muslim-dominated realm of North Africa/Southwest Asia. Here the boundary takes on the properties of a wide transition zone, yet it still divides states such as Nigeria, Chad, and Ethiopia. The transformation of the margins of the former Soviet Union is creating similar cross-country transitions. Recently-independent states such as Belarus (between Europe and Russia) and Kazakhstan (between Russia and

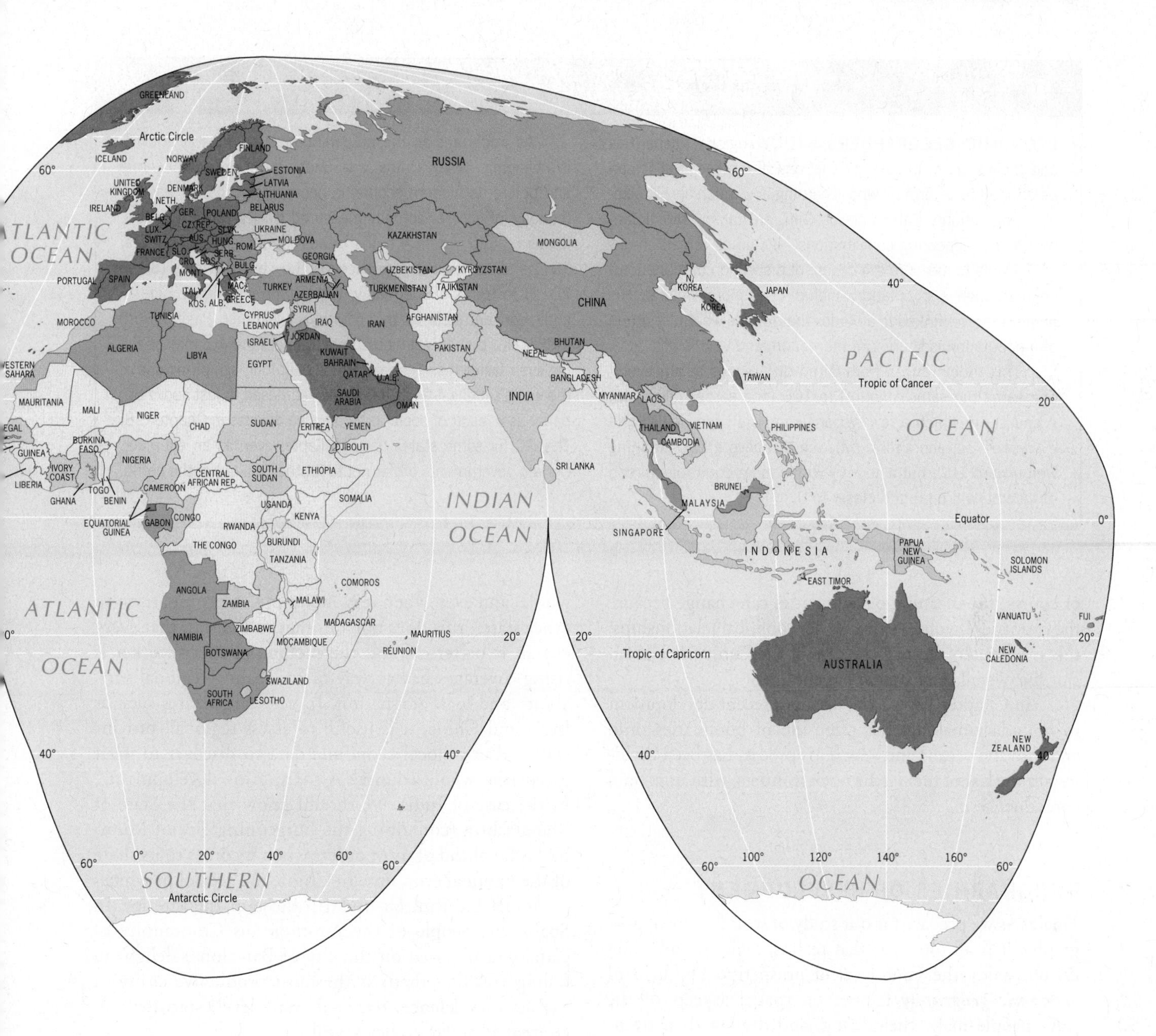

Muslim Southwest Asia) lie in transition zones of regional change.

Most often, however, geographic realms consist of groups of states whose boundaries also mark the limits of the realms. Look at Southeast Asia, for instance. Its northern border coincides with the political boundary that separates China (a realm practically unto itself) from Vietnam, Laos, and Myanmar (Burma). The boundary between Myanmar and Bangladesh (which is part of the South Asian realm) defines its western border. Here, the state boundary framework helps delimit geographic realms.

The global boundary framework is even more useful in delimiting regions within geographic realms. We shall discuss such divisions every time we introduce a Regions of the Realm (B) chapter, but an example is appropriate here. In the Middle American realm, we recognize four regions. Two of these lie on the mainland: Mexico, the giant of the realm, and Central America, which is constituted by the seven comparatively small states located between Mexico and the Panama-Colombia border (which also marks the boundary with the South American realm).

Political Geography

To our earlier criteria of physical geography, population distribution, and cultural geography, we now add ***political geography*** as a shaper of world-scale geographic regions. In doing so, we should be aware that the global boundary framework continues to change. Sometimes, new boundaries are created, as between Serbia and Kosovo when Kosovo declared its independence in 2008, or between (now-shrunken) Sudan and South Sudan in 2011. Occasionally, boundaries are eliminated, as was the case between former West and East Germany in 1990. And then,

The Gini Coefficient

ECONOMIC GEOGRAPHERS STUDY regional disparities and their causes, including variations of income. But it can be difficult to determine who is earning how much across sizeable populations. Enter Corrado Gini, an Italian statistician, who made pioneering contributions—including a mathematical formula to measure the degree of dispersion of a phenomenon through a population, including economic gains. His name is forever linked to an index that reveals what proportion of a population is sharing in the wealth, and who is not.

This index ranges from 0 (no differences at all; everyone earns the same amount) to 100 (one earner takes all). A country in which a few tycoons control all the wealth and everyone else labors for a pittance will have a "GC" leaning well toward 100; but a country with a more equitable spread of income will be much closer to 0.

As important as the actual number is the way the GC is changing. When China was under strict communist rule and before its modern economic boom began, its GC was low (nobody, of course, actually measured it). By 1993, however, it was reported to be 41; the newest figure—47 for the year 2012—shows that it now exceeds the United States (38 in 2012) in terms of inequality, with China's incomes increasingly concentrated in the country's wealthier Pacific Rim. India's GC, probably underestimated at around 40, may be rising even faster than China's. But Brazil, long exhibiting one of the world's highest GC's, has lately decreased to just above 50, partly as a result of social programs we discuss in Chapter 5B. The GC for some states is unavailable, even as an estimate. Certain governments prefer not to let the Gini out of the bottle.

of course, the meaning of boundaries can change because the countries involved agree on new political relationships; this is especially clear in the case of the European Union. But the overall state system has endured.

In Chapter 12, we will discuss a recent development in boundary-making: the extension of boundaries onto and into the oceans and seas. This process has been dividing up the last of the Earth's open frontiers, with uncertain consequences.

GEOGRAPHIES OF DEVELOPMENT

Finally, as we prepare for our study of world regional geography, it is all too clear that realms, regions, and states do not enjoy the same level of prosperity. The field of ***economic geography*** focuses on spatial aspects of the ways people make their living, and deals with patterns of production, distribution, and consumption of goods and services. As with all else in this world, these patterns reveal much variation. Individual states report the nature and value of their imports and exports, farm and factory output, and many other economic data to the United Nations and other international agencies. From such information, economic geographers can measure the comparative well-being of the world's countries (see the box titled "The Gini Coefficient"). The concept of **development [28]** is used to gauge a state's economic, social, and institutional growth.

Statistics: A Caution

The concept of development, as measured by data that reflect totals and averages for entire national populations, entails some pitfalls of which we should be aware from the start. When a state's economy is growing as a whole, and even when it is "booming" by comparison to other states, this does not automatically mean that every citizen is better off and the income of every worker is rising. Averages have a way of concealing regional variability and local stagnation. In very large states such as India and China, it is useful to assess regional, provincial, and even local economic data to discover to what extent the whole country is sharing in "development." In the case of India, we should know that the State of Maharashtra (containing the burgeoning city of Mumbai) is far ahead of most others when it comes to its share of the national economy. In China, the coastal provinces of the Pacific Rim far outstrip those of the interior. In Spain, the people of the Autonomous Community of Catalonia (focused on the city of Barcelona) delight in telling you that theirs is the most productive entity in the country. Hence, national (state-level) statistics can conceal as much as they reveal.

Development in Spatial Perspective

Various schemes to group the world's states into economic-geographic categories have come and gone, and others will probably arise in the future. For our purposes, the classification scheme used by the World Bank (one of the agencies that monitor economic conditions across the globe) is the most effective. It sorts countries into four categories based on the success of their economies: (1) high-income, (2) upper-middle-income, (3) lower-middle-income, and (4) low-income. These categories, when mapped, display interesting regional clustering (see Fig. G-11). Compare this map to our global framework (Fig. G-3), and you can see the role of economic geography in the layout of the world's geographic realms. Also evident are regional contrasts within realms—for instance, between Brazil and its western neighbors, between South Africa and most of

From the Field Notes . . .

"Thanks to a Brazilian intermediary I was allowed to enter and spend a day in two of Rio de Janeiro's hillslope *favelas*, an eight-hour walk through one into the other. Here live millions of the city's poor, in areas often ruled by drug lords and their gangs, with minimal or no public services, amid squalor and stench, in discomfort and danger. And yet life in the older *favelas* has become more comfortable as shacks are replaced by more permanent structures, electricity is sometimes available, water supply, however haphazard, is improved, and an informal economy brings goods and services to the residents. I stood in the doorway of a resident's single-room dwelling for this overview of an urban landscape in transition: satellite-television disks symbolize the change going on here. The often blue cisterns catch rainwater; walls are made of rough brick and roofs of corrugated iron or asbestos sheeting. No roads or automobile access, so people walk to the nearest road at the bottom of the hill. Locals told me of their hope that they will some day have legal rights to the space they occupy. The Brazilian government at times expresses support for these claims, but it is complicated. As the photo shows, people live quite literally on top of one another, and mapping the chaos will not be simple (but will be made possible with geographic information systems). This would allow the government to tax residents, but it would also allow residents to obtain loans based on the value of their *favela* properties, and bring millions of Brazilians into the formal economy. The hardships I saw on this excursion were often dreadful, but you could sense the hope for, and anticipation of, a better future."

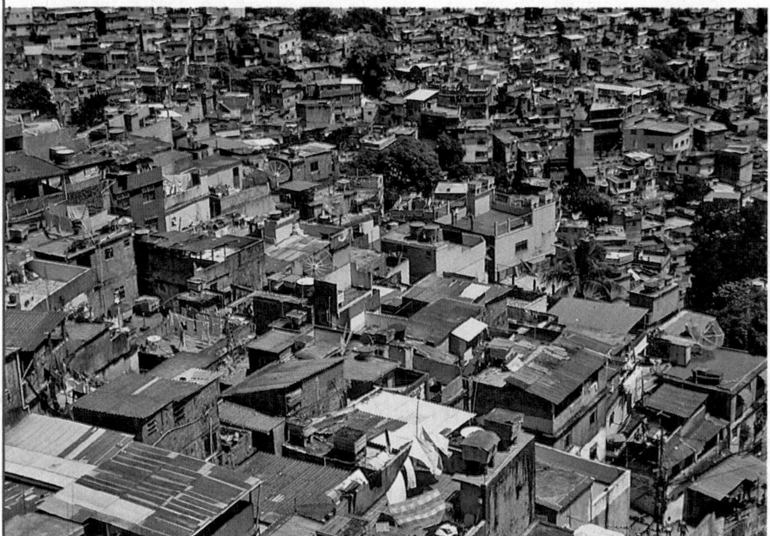

www.conceptcaching.com

the rest of Subsaharan Africa, and between west and east in Europe.

Economic geography is not the entire story, but along with factors of physical geography (such as climate), cultural geography (including resistance or receptivity to change and innovation), and political geography (history of colonialism, growth of democracy), it plays a powerful role in shaping our variable world.

A Core-Periphery World: Increasing Complexity

It has been obvious for a very long time that human success on the Earth's surface has focused on certain areas and bypassed others. The earliest cities and states of the Fertile Crescent, the empires of the Incas and the Aztecs, the dominance of ancient Rome, and many other hubs of activity tell the story of development and decay, of growth and collapse. In their heyday, such centers of authority, innovation, production, and expansion were the earliest **core areas [29]**, places of dominance whose inhabitants exerted their power over their surroundings near and far. Such core areas grew rich and, in many cases, endured for long periods because their occupants skillfully exploited those surroundings—controlling and taxing the local population, forcing workers to farm the land and mine the resources at their command. This created a **periphery [30]** that sustained the core for as long as the system endured, so that core-periphery interactions, one-sided though they were, created wealth for the former and enforced stability in the latter.

In modern times, the world can be said to revolve, economically speaking, around a global core and periphery—the world economy, after all, has become a single integrated spatial system. In the nineteenth century, this core more or less coincided with western Europe, controlling as it did vast areas of the world through its empires. In the twentieth century, the core expanded first to North America and then grew to include Japan, Australia, and New Zealand. Since the 1960s, Hong Kong, Singapore, Taiwan, South Korea, and some of the (oil-) rich Gulf States have become part of the core as well. And as the twenty-first century opened, the newest entrant stepped forward: Pacific-fronting China. The global core, therefore,

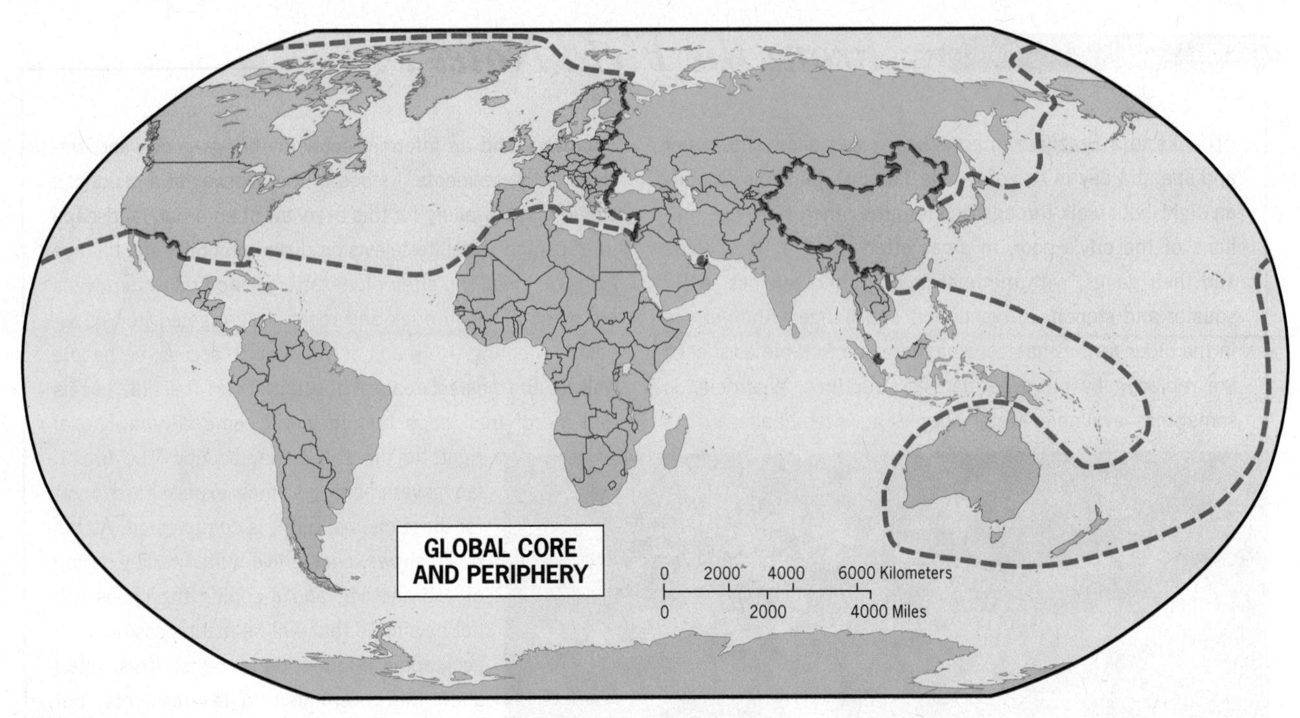

FIGURE G-12

continues to evolve. As an economic-geographic phenomenon, it is mapped in Figure G-12 above. But keep in mind that here the core is defined purely in economic terms: if we were to include such other criteria as representative Western-style government, for example, China would not qualify.

It is also important to realize that core-periphery relationships are not limited to the global scale. Countries themselves can and do exhibit such patterns as well. China again is a case in point: its coastal provinces form the core, while the interior and westernmost reaches of the country are part of China's periphery. Uneven development, therefore, exists at a range of scales, from the urban to the global. Most functional regions, in essence, are ***spatial networks*** comprising nodes of variable centrality and importance, and this usually translates into different levels of economic development. Except for a few special cases, all countries contain core areas. These national cores are often anchored by the country's capital and/or largest city: Paris (France), Tokyo (Japan), Buenos Aires (Argentina), and Bangkok (Thailand) are just a few notable examples. Larger countries may have more than one core area, such as Australia with its eastern and western coast cores and intervening periphery.

The world continues to exhibit major differences in productivity and well-being. One of the most intriguing economic-spatial outcomes of globalization is that a growing number of countries have accelerated their development—but this growth is often confined to specific city-regions whereas the rest of the country remains quite poor.

GLOBALIZATION

Globalization [31] is essentially a geographical process in which spatial relations—economic, cultural, political—shift to ever broader scales (now driven in no small part by recent rapid advances in communication and transport technologies). What this means is that what happens in one place has repercussions in places ever more distant, thereby integrating the entire world into an ever "smaller" global village. Globalization comes into our homes via television, computers, and smartphones: news today has never traveled faster, and sometimes even government leaders turn to the Internet on their personal electronic devices to get the latest reports on international events.

Globalization is not something entirely new. The second half of the nineteenth century, for instance, also witnessed major advances in the intensification of global interdependence. It was particularly affected by new technologies such as the steamship, the railway, and the telegraph, which subsequently were followed by the first motor vehicles and airplanes. With today's newest technologies, the world is becoming ever more interconnected. Thus geography and our knowledge of the world's realms and regions become increasingly important—because what happens elsewhere will have consequences wherever you are.

Global Challenges, Shared Interests

Globalization plays out in various spheres, from the environmental to the cultural to the economic. Today's most

From the Field Notes . . .

© Jan Nijman

"Singapore is an outstanding example of a city propelled into prominence by forces of globalization. Located strategically on one of the world's busiest shipping lanes (the Strait of Malacca) and with a good harbor, this city-state was bound to benefit from expanding world trade. It now boasts the busiest transshipment port in the world and the city has also become a shopping magnet for elites all across the Southeast Asian realm. During a visit to Singapore in the summer of 2012, I had spotted a newly built, futuristic-looking skyscraper across the bay from downtown and thought it would be worth a visit. It turned out to be the huge Marina Bay Sands Hotel (developed by the Sands Corporation of Las Vegas) that opened for business in 2010. The rooftop features a spectacular pool with a view of the city and its port down below. Like most 'world-cities', Singapore is a hub of both production and consumption."

www.conceptcaching.com

pressing environmental issue, no doubt, is global warming, a threat to the world at large. It is clear that we must confront this problem together, but it is far from easy to agree on strategies. Some countries are bigger polluters than others, some have more resources than others, and some are more developed than others. How to divide the burdens? At the Durban (South Africa) Conference on Climate Change in 2011, for the first time governments from around the world committed themselves to preparing a comprehensive global agreement to reduce greenhouse gas emissions. The good news was that the deal included developed and developing countries, as well as the participation of the United States, which had been reluctant to get involved in previous international efforts. The bad news is that the process will be an excruciatingly slow one: the target date for completing the agreement is 2015, and the actual reductions of emissions would not commence until 2020. On top of that, it remains to be seen if the agreement will be legally binding.

Culturally, too, the world is coming closer together, and this is most apparent in global migration flows. Such migration used to be uncommon because most people were rooted in their home environment, where they lived out their entire lives. When residential relocation did occur, it used to be one-way, with people migrating from one place to another and then staying put. But in the current globalization era, migration flows have intensified, in part because people now possess far greater knowledge about opportunities elsewhere. Moreover, it is now much easier to travel back and forth, which allows migrants to maintain close ties with their original home countries. Not surprisingly, as the number of highly mobile ***transnational migrants*** has increased, they have become instrumental in the spreading of cultures around the world. Examples include Algerians in Paris, Haitians in Montreal, Cubans in Miami, Mexicans in Los Angeles, Indians in Singapore, and Indonesians in Sydney.

But it is also important to keep in mind that people's mobility is often constrained, because some parts of this highly uneven world are so much better off than others. High-income countries are a magnet for migrants, but all too often they cannot get access. Millions of workers aspire to leave the periphery, which contains the world's poorest regions, to seek a better life somewhere in the core. Trying to get there, many of them die

© Scott Houston/© Corbis

The "Occupy Wall Street" demonstration in New York City on October 11, 2011. The protests that began here, in the form of deliberate illegal squatting and resistance to removal by the police, soon spread to cities across the United States and around the world. Under the banner of "We are the 99%," these demonstrators insist that the vast majority of people do not benefit from global finance and that the politicians are on the side of big business. In New York and elsewhere, these protests relied on a hard core of thousands of determined demonstrators and, while short-lived, made headlines around the world.

every year in the waters of the Mediterranean, the Caribbean, and the Atlantic. Others risk their lives at the barriers that encircle the global core as if it were a gated community—from the "security fence" between Mexico and the United States to the walls that guard Israel's safety to the razor wire that encircles Spain's outposts on North Africa's shore.

When the world economy entered a deep, extended downturn in 2008, unemployment skyrocketed and opportunities for migrants declined accordingly. In Europe, undocumented migration plunged 33 percent from 2008 to 2009 alone, and along the U.S.-Mexican border the number of interceptions fell by 23 percent during the same 12-month period. As the global economy recovers, immigration rates are expected to increase concomitantly.

Winners and Losers

If a geographic concept can arouse strong passions, globalization is it. To most economists, politicians, and businesspeople, this is the best of all possible worlds—the march of international capitalism, open markets, and free trade. In theory, globalization breaks down barriers to foreign trade, stimulates commerce, brings jobs to remote places, and promotes social, cultural, political, and other kinds of exchanges. High-tech workers in India are employed by computer firms based in California. Japanese cars are assembled in Thailand. American footwear is made in China. Fast-food restaurant chains spread standards of service and hygiene as well as familiar (and standardized) menus from Tokyo to Tel Aviv to Tijuana. If wages and standards of employment are lower in peripheral countries than in the global core, production will shift there and the gap will shrink. Everybody wins. Economic geographers can prove that global economic integration allows the overall economies of poorer countries to grow faster: compare their international trade to their national income, and you will find that the ***gross national income (GNI)**** of those that engage in more foreign trade (and thus are more "globalized") rises, while the GNI of those with less actually declines.

But there is another, more complicated issue. Although many countries, even lesser-developed ones that were able to latch onto globalization, have witnessed accelerated economic growth and rising per capita incomes, inequality within these countries has frequently increased just as fast. In other words, uneven development within countries has become more pronounced. As noted earlier, this is particularly obvious in China, the fastest-growing economy in the world over the past two decades: much of this growth took place in its Pacific coastal zone, not in the interior of the country, and income differentials became ever wider. And the same is true in India and most other ***emerging markets***. This is why a regional approach is so important to understanding what is going on in the world economy.

Globalization in the economic sphere is proceeding under the auspices of the World Trade Organization (WTO), of which the United States is the leading architect. To join, countries must agree to open their economies to foreign trade and investment. The WTO has 159 member-states (Russia being among the latest to join in 2012), all expecting benefits from their participation. But the leading global-core countries themselves do not always oblige when it comes to creating a "level playing field." The case of the Philippines is often cited: Filipino farmers found themselves competing against North American and European agricultural producers who receive subsidies to support production as well as the export of their products—and losing out. Meanwhile, low-priced, subsidized U.S. corn appeared on Filipino markets. As a result, the Philippine economy lost several hundred thousand farm jobs, wages went down, and WTO membership had the effect of severely damaging its agricultural sector. Not

*Gross national income (GNI) is the total income earned from all goods and services produced by the citizens of a country, within or outside of its borders, during a calendar year. *Per capita GNI* is a widely used indicator of the variation of spendable income around the globe and is reported for each country in the farthest-right column of the Data Table in Appendix B.

surprisingly, the notion of globalization is not popular among rural Filipinos.

Opposition to globalization is not confined to the periphery: in the United States and western Europe, WTO meetings have often been plagued by protests and demonstrations by those who believe that the global economy is "rigged" to benefit the few while most lose out. The global financial crisis that began in 2008 created a more specific, concrete target of such criticism: in 2011, the so-called "Occupy Wall Street" demonstrators in New York City triggered a global protest movement against corporate greed and corruption of the financial sector. Their slogan, "We are the 99%" (see photo), underscored their claim that the great majority of people in the world do not benefit from the workings of the global economy.

The Future

As with all significant transformations, the overall consequences of globalization are uncertain. Critics maintain that one of its most insidious outcomes is a steadily widening gap between rich and poor, a polarization of wealth that is likely to destabilize the world. Proponents argue that, as with the Industrial Revolution, it will take time for the benefits to spread—but that globalization's ultimate effects will be advantageous to all. Indeed, the world is functionally shrinking, and we will find evidence for that throughout this book. But the "global village" still retains its distinctive neighborhoods, and globalization has not erased their particular properties—in some cases even sharpening the contrasts. In the chapters that follow, we use the vehicle of geography to identify and investigate them.

REALMS AND REGIONS: THE STRUCTURE OF THIS BOOK

At the beginning of this chapter, we introduced a map of the great geographic realms of the world (Fig. G-3). We then addressed the task of dividing these realms into regions, and we used criteria ranging from physical geography to economic geography. The result is Figure G-14. On this map, note that we display not only the world geographic realms but also the regions into which they subdivide. The numbers in the legend reveal the order in which the realms and regions are discussed, starting with Europe (1) and ending with the Pacific Realm (12).

Before we launch our survey, here is a brief summary of the 12 geographic realms and their regional components:

Europe (1)

Territorially small, politically still fragmented but economically united, Europe has had a very turbulent history and has been disproportionately influential in global affairs. The regionalization of this realm is today best approached within a core/periphery framework. Generally speaking, Europe's core lies in its west, with a wide periphery curving across the realm's southern, eastern, and far northern domains.

Russia (2)

Territorially enormous and politically unified, Russia was the dominant force in the former Soviet Union that disintegrated in 1991. Undergoing a difficult transition from dictatorship to democracy and from communism to capitalism, Russia is geographically complex and continues to change. We define five regions: the Russian Core in the west, the Southeastern Frontier, Siberia, the Far East, and Transcaucasia.

North America (3)

Another realm in the global core, North America consists of the United States and Canada. We identify nine regions: the North American Core, the Maritime Northeast, French Canada, the South, the Southwest, the Pacific Hinge, the Western Frontier, the Continental Interior, and the Northern Frontier. Five of these regions extend across the U.S.-Canada border.

Middle America (4)

Nowhere in the world is the contrast between the global core and periphery as sharply demarcated as it is between North and Middle America. This small, fragmented realm divides into four regions: Mexico, Central America, and the Caribbean Basin's Greater and Lesser Antilles.

South America (5)

The continent of South America also defines a geographic realm in which Iberian (Spanish and Portuguese) influences dominate the cultural geography but indigenous imprints survive. We recognize four regions: the Caribbean North, composed of Caribbean-facing states; the Andean West, with its strong aboriginal influences; the Southern Cone; and Brazil, the realm's giant.

Subsaharan Africa (6)

Between the African Transition Zone in the north and South Africa's southernmost Cape lies Subsaharan Africa. The realm consists of five regions: Southern Africa, East Africa, Equatorial Africa, West Africa, and the African Transition Zone itself.

North Africa/Southwest Asia (7)

This vast geographic realm has several names, extending as it does from North Africa into Southwest and, indeed, Central Asia. It is a very complex and volatile realm, and much of

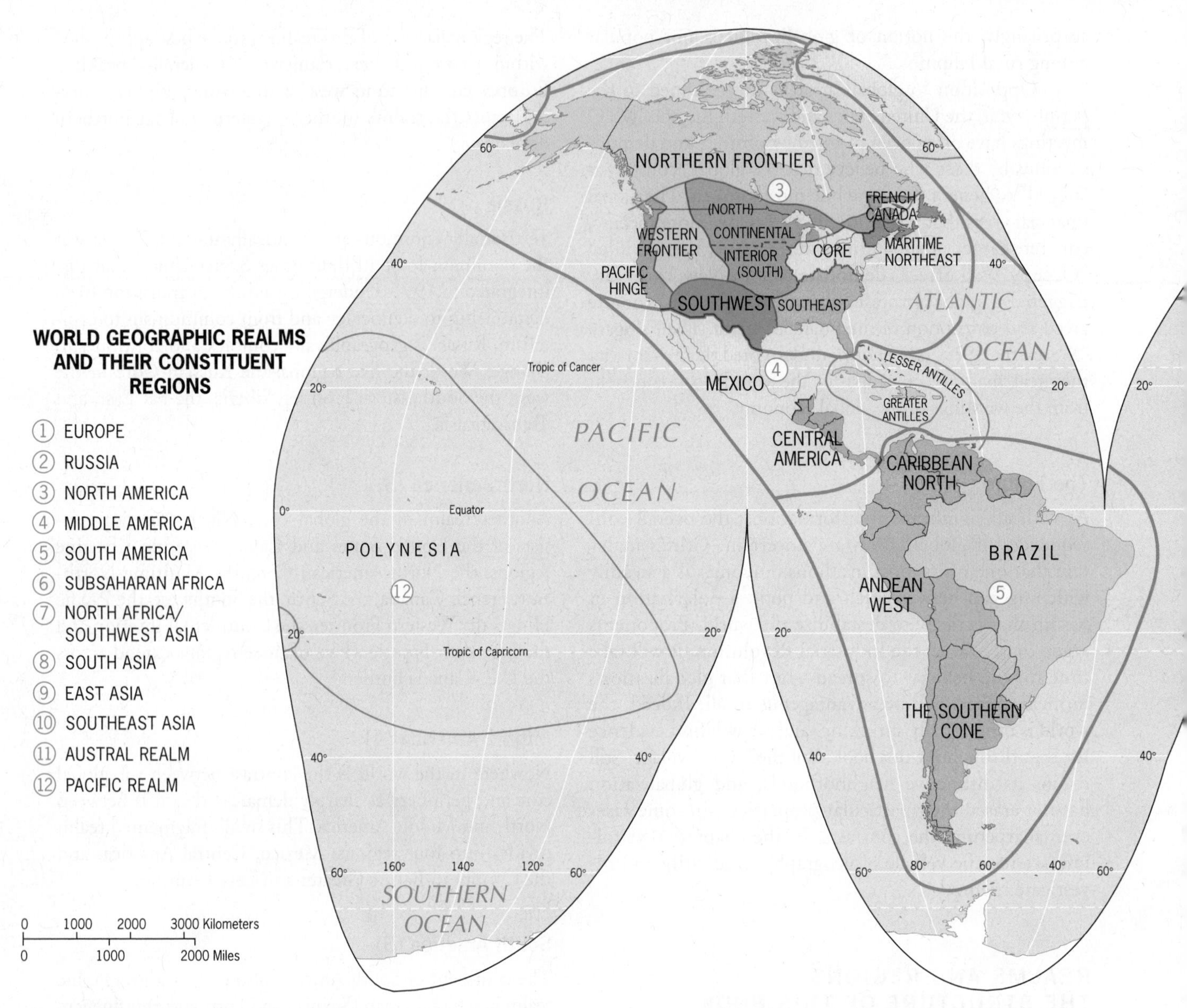

FIGURE G-13

this is related to its particular regional geographies. There are six regions: Egypt and the Lower Nile Basin, the Maghreb in North Africa, the Middle East, the Arabian Peninsula, and toward the east, the Empire States and Turkestan.

South Asia (8)

Physically, South Asia is one of the most clearly defined geographic realms, but has a complex cultural geography. It consists of five regions: India at the center, Pakistan to the west, Bangladesh to the east, the Mountainous North, and the Southern Islands that include Sri Lanka and the Maldives.

East Asia (9)

The vast East Asian geographic realm extends from the deserts of Central Asia to the tropical coasts of the South China Sea and from Japan to the Himalayan border with India. We identify seven regions: China's Coastal Core, Interior, and Western Periphery; Mongolia; the Korean Peninsula; Japan; and Taiwan.

Southeast Asia (10)

Southeast Asia is a varied and intriguing mosaic of natural landscapes, cultures, and economies. Influenced by India, China, Europe, and the United States, it includes dozens of religions and hundreds of languages plus economies reflecting both the global core and periphery. Physically, Southeast Asia consists of a broad peninsular mainland and an offshore arc consisting of thousands of islands. The two regions (Mainland and Insular) are based on this distinction.

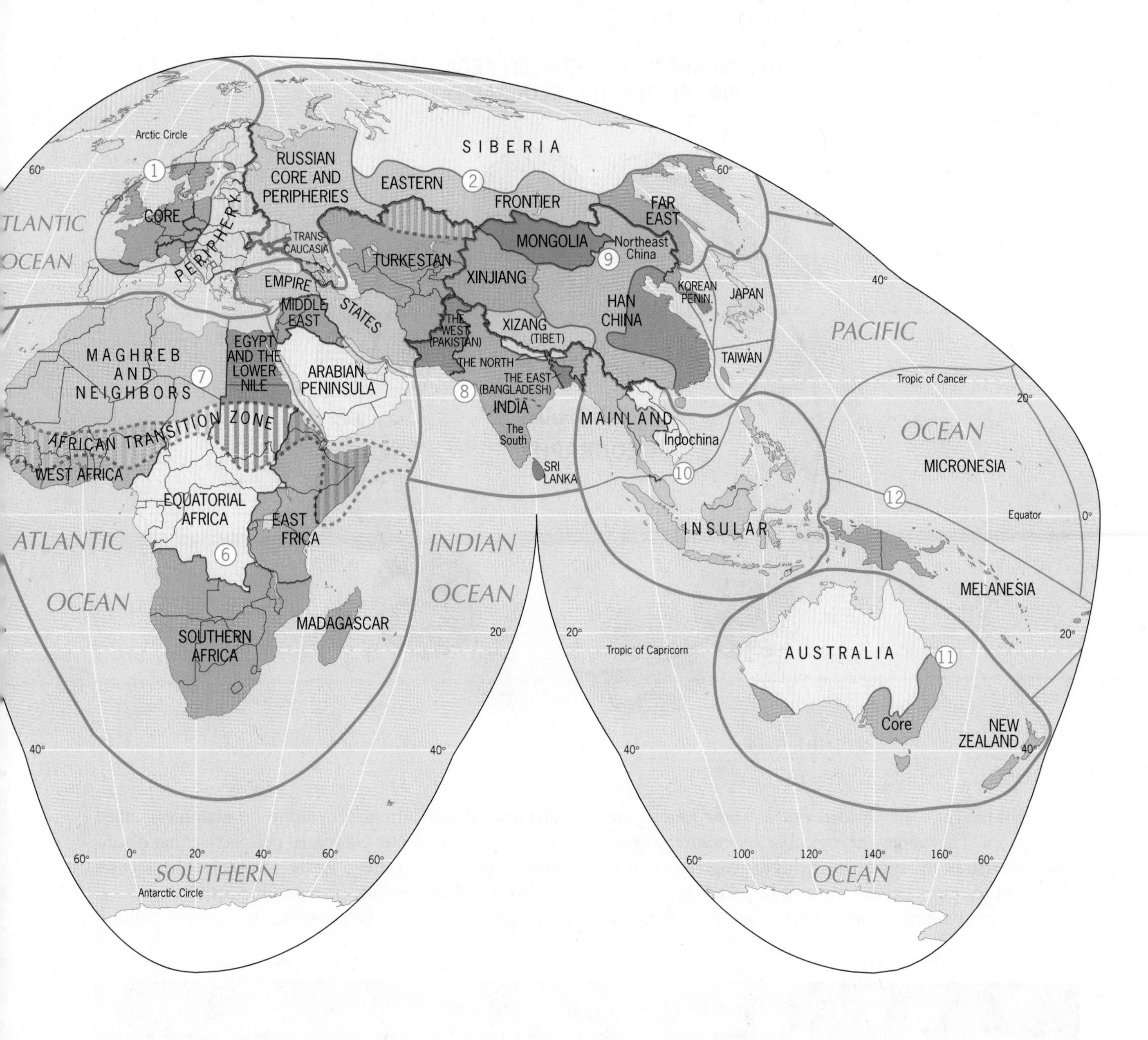

Austral Realm (11)

Australia and its neighbor New Zealand form the Austral geographic realm by virtue of continental dimensions, insular separation, and predominantly Western cultural heritage. The regions of this realm are defined by physical as well as cultural geography: in Australia, a highly urbanized, two-part core and a vast, desert-dominated interior; and in New Zealand, two main islands that exhibit considerable geographic contrast.

Pacific Realm (12)

The enormous Pacific Ocean, larger than all the landmasses combined, contains tens of thousands of islands large and small. Dominant cultural criteria warrant three regions: Melanesia, Micronesia, and Polynesia.

As this introductory chapter demonstrates, our world regional survey is no mere description of places and areas. We have combined the study of realms and regions with a look at geography's ideas and concepts—the notions, generalizations, and basic theories that make the discipline what it is. We continue this method in the chapters ahead so that we will become better acquainted with the world and with geography. By now you are aware that geography is a wide-ranging, multifaceted discipline. It is often described as a social science, but that is only half the story: in fact, geography straddles the divide between the social and the physical (natural) sciences. Many of the ideas and concepts you will encounter have to do with the multiple interactions between human societies and natural environments.

Regional geography allows us to view the world in an all-encompassing way. As we have seen, regional geography borrows information from many sources to create

THE RELATIONSHIP BETWEEN REGIONAL AND SYSTEMATIC GEOGRAPHY

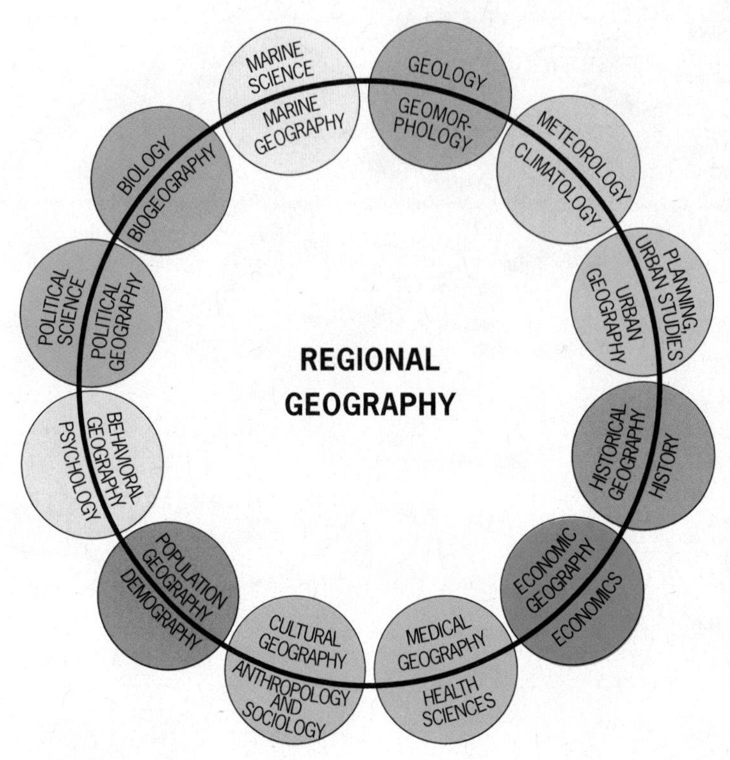

FIGURE G-14

an overall image of our divided world. Those sources are not random. They represent topical or ***systematic geography***. Research in the systematic fields of geography makes our world-scale generalizations possible. As Figure G-14 shows, these systematic fields relate closely to those of other disciplines. Cultural geography, for example, is allied with anthropology; it is the spatial perspective that distinguishes cultural geography. Economic geography focuses on the spatial dimensions of economic activity; political geography concentrates on the spatial imprints of political

What Do Geographers Do?

A SYSTEMATIC SPATIAL perspective and an interest in regional study are the unifying themes and enthusiasms of geography. Geography's practitioners include physical geographers, whose principal interests are the study of geomorphology (land surfaces), research on climate and weather, vegetation and soils, and the management of water and other natural resources. There also are geographers whose research and teaching concentrate on the ecological interrelationships between the physical and human worlds. They study the impact of humankind on our globe's natural environments and the influences of the environment (including such artificial contents as air and water pollution) on human individuals and societies.

Other geographers are regional specialists, who often focus their work for governments, planning agencies, and multinational corporations on a particular region of the world. Still other geographers—who now constitute the largest group of practitioners—are devoted to topical or systematic subfields such as urban geography, economic geography, and cultural geography (see Fig. G-14). They perform numerous tasks associated with the identification and resolution (through policy-making and planning) of spatial problems in their specialized areas. And, increasingly, there are geographers who combine their fascination for spatial questions with cutting-edge technical expertise.

Geographic information systems (GIS), digital mapping, remote sensing, geospatial data analysis, and geovisualization are among the myriad specializations listed by the 10,000-plus professional geographers of North America. On the book's website, you will find much information on the discipline, how one trains to become a geographer, and the many exciting (and rapidly expanding) career options that are open to the young professional.

behavior. Other systematic fields include historical, medical, behavioral, environmental, and urban geography. We will also draw on information from biogeography, marine geography, population geography, geomorphology, and climatology (as we did earlier in this chapter).

These systematic fields of geography are so named because their approach is global, not regional. Take the geographic study of cities, urban geography. Urbanization is a worldwide process, and urban geographers can identify certain human activities that all cities in the world exhibit in one form or another. But cities also display regional properties. The typical Japanese city is quite distinct from, say, the African city. Regional geography, therefore, borrows from the systematic field of urban geography, but it injects this regional perspective.

In the following chapters we call upon these systematic fields to give us a better understanding of the world's realms and regions. As a result, you will gain insights into the discipline of geography as well as the regions we investigate. As you will see, geography is vital to interpreting, comprehending, and coping with our rapidly transforming world (see box titled "What Do Geographers Do?" and the book's website for a detailed discussion of career opportunities).

POINTS TO PONDER

- Within the next few years or so China is expected to surpass the United States as the biggest national economy in the world.
- The global human population recently surpassed 7 billion and is predicted to reach 9.6 billion by 2050. Will there be enough food and water to go around?
- In this second decade of the twenty-first century, almost 1 billion people must survive on less than one U.S. dollar a day.
- Global warming is expected to cause a significant rise in sea level by the end of the twenty-first century, though estimates vary widely.
- The number of smartphones in use, worldwide, passed the 1 billion mark in 2012; the majority of these devices have map navigation applications.
- Globalization may cause the world to "shrink," but marked differences remain among geographic realms and regions.

GREENLAND (KALAALLIT NUNAAT) (DENMARK)
Jan Mayen (NORWAY)
Denmark Strait
Arctic Circle
Barents Sea
Hammerfest
Murmansk
KOLA PENINSULA
Reykjavík
ICELAND
Akureyri
Vík
Norwegian Sea
Narvik
Kiruna
Bodø
Rovaniemi
Kemi
Luleå
Oulu
FINLAND
NORWAY
SWEDEN
SCANDINAVIA
Gulf of Bothnia
Umeå
Kuopio
Joensuu
Vaasa
Trondheim
Östersund
Petrozavodsk
Faeroe Islands (DENMARK)
Sundsvall
Lake Ladoga
Tampere
Lahti
ATLANTIC OCEAN
Turku
Helsinki
ST. PETERSBURG
Shetland Islands (U.K.)
Gävle
Åland (FINLAND)
Gulf of Finland
Bergen
Tallinn
Uppsala
Oslo
Västerås
Stockholm
Hiiumaa
ESTONIA
Lake Peipus
Vänern
Tartu
Orkney Islands
Skien
Stavanger
Saaremaa
Pskov
Outer Hebrides
Norrköping
Gulf of Riga
Inverness
Kristiansand
Vättern
Visby
Gotland (SWEDEN)
SCOTTISH HIGHLANDS
Skagerrak
Riga
Inner Hebrides
Göteborg
Jönköping
LATVIA
Glasgow
Aalborg
Kattegat
Växjö
Liepaja
Edinburgh
North Sea
Öland
Londonderry
Aarhus
Baltic Sea
Daugavpils
Vitsyebsk
DENMARK
Klaipeda
LITHUANIA
Kaunas
Belfast
Newcastle
COPENHAGEN
Malmö
Vilnius
Isle of Man (U.K.)
Britain
UNITED KINGDOM
Odense
Bornholm (DENMARK)
Kaliningrad
RUSSIA
Ireland
IRELAND
Irish Sea
Gdansk
Limerick
Dublin
Leeds
BELARUS
Liverpool
Manchester
Frisian Islands
CAMBRIAN PLATEAU
HAMBURG
Szczecin
Bydgoszcz
Cork
St. George's Channel
Bremen
Elbe
Birmingham
Amsterdam
NETHERLANDS
Noteć
WARSAW
The Hague
NORTH EUROPEAN PLAIN
Poznan
Brest
Cardiff
LONDON
Rotterdam
Hannover
BERLIN
Vistula
POLAND
Bristol
Lodz
Celtic Sea
Plymouth
Strait of Dover
Antwerp
Essen
Oder
Leipzig
Land's End
Düsseldorf
Dresden
Wroclaw
Rivne
BRUSSELS
Isle of Wight
Cologne
GERMANY
Lille
Bonn
Katowice
Krakow
English Channel
BELGIUM
Frankfurt
PRAGUE
Lviv
Le Havre
Channel Islands (U.K.)
Rouen
LUXEMBOURG
Luxembourg
CZECH REPUBLIC
Dniester
Brest
Nuremberg
Brno
SLOVAKIA
Seine
PARIS
Marne
Stuttgart
Danube
Kosice
Chernivtsi
CARPATHIAN MTS.
Strasbourg
VIENNA
Bratislava
BUDAPEST
Tours
Loire
MUNICH
Salzburg
Debrecen
Nantes
FRANCE
Innsbruck
AUSTRIA
Oradea
Cluj
Zürich
LIECH.
HUNGARY
Bern
Graz
SWITZERLAND
ALPS
ROMANIA
Bay of Biscay
Geneva
SLOVENIA
Pécs
Ljubljana
Timisoara
TRANSYLVANIAN ALPS
Limoges
Lyon
Mt. Blanc (4,807 m, 15,771 ft.)
MILAN
Venice
Trieste
Zagreb
Bordeaux
MASSIF CENTRAL
Verona
Rijeka
CROATIA
BELGRADE
BUCHAREST
La Coruña
Turin
Po
Craiova
Gijón
Genoa
Bologna
BOSNIA
SERBIA
Danube
CANTABRIAN MOUNTAINS
Bilbao
Adriatic Sea
Sarajevo
Vigo
Pico de Aneto (3,404 m, 11,168 ft.)
Montpellier
Toulouse
Nice
Cannes
MONACO
SAN MARINO
DINARIC ALPS
Nis
BALKAN PENINSULA
Florence
Split
KOSOVO
Sofia
Porto
PYRENEES
Marseille
Ligurian Sea
Podgorica
Pristina
ANDORRA
Andorra la Vella
APENNINES
Elba
MONTENEGRO
Skopje
Plovdiv
PORTUGAL
Valladolid
Bastia
MACEDONIA
IBERIAN PENINSULA
Zaragoza
Corsica (FRANCE)
ALBANIA
MADRID
ROME
Tirane
Thessaloniki
BARCELONA
Bari
GREECE
Lisbon
Tagus
SPAIN
Olbia
Naples
Mount Olympus (2,917 m)
Balearic Sea
Minorca
Sassari
ITALY
PINDUS MTS.
Valencia
Majorca
Gulf of Taranto
Larisa
Guadiana
Palma de Majorca
Sardinia (ITALY)
Tyrrhenian Sea
Córdoba
Ibiza
Balearic Islands (SPAIN)
Seville
Guadalquivir
Alicante
Cagliari
Murcia
Palermo
Messina
Ionian Sea
Patrai

DEFINING the Realm

IN THIS CHAPTER

- A history of dominance in the modern world
- Has the EU reached its limits?
- The perennial question of Europe's eastern boundary
- Is Islam embedding itself in western Europe?
- Aging Europe
- The geography of the Euro crisis

CONCEPTS, IDEAS, AND TERMS

Physiography 1
Land hemisphere 2
City-state 3
Local functional specialization 4
Industrial Revolution 5
Nation-state 6
Nation 7
Indo-European language family 8
Complementarity 9
Transferability 10
Centrifugal forces 11
Centripetal forces 12
Supranationalism 13
Four Motors of Europe 14
Devolution 15

FIGURE 1A-1

Over the past five centuries, Europe and Europeans have influenced and changed the rest of the world more than any other realm or people has done. For good or bad, much of the world would look very different today if it had not been for Europe. The realm's empires spanned the globe and transformed societies far and near. European colonialism propelled an early wave of globalization. Millions of Europeans migrated from their homelands to the Old World as well as the New, changing (and sometimes nearly obliterating) traditional communities and creating new societies from Australia to North America. Colonial power and economic incentive combined to impel the movement of millions of imperial subjects from their ancestral homes to distant lands: Africans to the Americas, Indians to Africa, Chinese to Southeast Asia, Malays to South Africa's Cape, Native Americans from east to west. In agriculture, industry, politics, and other spheres, Europe generated revolutions—and then exported those revolutions across the world, thereby consolidating the European advantage.

But throughout much of that 500-year period of European hegemony, Europe also was a cauldron of conflict. Religious, territorial, and political disputes precipitated bitter wars that even spilled over into the colonies. And during the twentieth century, Europe twice plunged the world into war. The terrible, unprecedented toll of World War I (1914–1918) was not enough to stave off World War II (1939–1945), which again drew in the United States and ended with the first-ever use of nuclear weapons in Japan. In the aftermath of that war, Europe's weakened powers lost most of their colonial possessions and a new rivalry emerged: an ideological Cold War between the communist Soviet Union and the capitalist United States. This Cold War lowered an Iron Curtain across the heart of Europe, leaving most of the east under Soviet control and most of the west in the American camp. Western Europe proved resilient, overcoming the destruction of war and the loss of colonial power to regain economic strength. Meanwhile, the Soviet communist experiment failed at home and abroad, and in 1990 the last vestiges of the Iron Curtain were lifted. Since then, a massive effort has been underway to reintegrate and reunify Europe from the Atlantic coast to the Russian border, the key geographic story of this chapter.

DEFINING THE REALM

GEOGRAPHICAL FEATURES

As Figure 1A-1 shows, Europe is a realm of peninsulas and islands on the western margin of the world's largest landmass, Eurasia. It is a realm containing 600 million people and 40 countries, but it is territorially quite small. Yet despite its modest proportions it has had—and continues to have—a major impact on world affairs. For many centuries Europe has been a hearth of achievement, innovation, invention, and domination.

A view of Place de la Concorde, on the right (north) bank of the River Seine in Paris. This square was the main execution site of aristocrats and royalty during the French Revolution. Note the central 3300-year-old obelisk, which once stood at the entrance of Egypt's Luxor Temple, that was 'gifted' to France in 1829.

© Jan Nijman

Europe's Eastern Boundary

The European realm is bounded on the west, north, and south by Atlantic, Arctic, and Mediterranean waters, respectively. But where is Europe's eastern limit? Each episode in the historical geography of eastern Europe has left its particular legacy in the cultural landscape.

Twenty centuries ago, the Roman Empire ruled much of it (Romania is a cartographic reminder of

major geographic qualities of EUROPE

1. The European geographic realm lies on the western flank of the Eurasian landmass.
2. Though territorially small, Europe is heavily populated and fragmented into 40 states.
3. European natural environments are highly varied, and Europe's resource base is rich and diverse.
4. Europe's geographic diversity, cultural as well as physical, created strong local identities, specializations, and opportunities for trade and commerce.
5. The European Union (EU) is a historic and unique effort to achieve multinational economic integration and, to a lesser degree, political coordination.
6. Europe's relatively prosperous population is highly urbanized and rapidly aging.
7. Local demands for greater autonomy as well as cultural challenges posed by immigration are straining the European social fabric.
8. Despite Europe's momentous unification efforts, east-west contrasts still mark the realm's regional geography.
9. The ongoing "euro crisis" poses a threat to the EU and suggests that the single currency was introduced too soon for some members. It is possible that certain countries will either choose or be forced to exit the Union.

that period); for most of the second half of the twentieth century, the Soviet Empire controlled nearly all of it. In the intervening two millennia, Christian Orthodox church doctrines spread from the southeast, and Roman Catholicism advanced from the northwest. Turkish (Ottoman) Muslims invaded and created an empire that reached the environs of Vienna. By the time the Austro-Hungarian Empire ousted the Turks, millions of eastern Europeans had been converted to Islam. Albania and Kosovo today remain predominantly Muslim countries. Meanwhile, it is often said that western Europe's civilization had its cradle in ancient Greece, but that lies farther still to the southeast, beyond the former Yugoslavia.

Eastern Europe's tumultuous history, itself an expression of the absence of clear natural boundaries, played out on a physical stage of immense diversity, its landscapes ranging from open plains and wide river basins to strategic mountains and crucial corridors. Epic battles fought centuries ago remain fresh in the minds of many people living here today; pivotal past migrations are celebrated as though they happened yesterday. Nowhere in Europe is the cultural geography as complex as in the southeastern part of the realm. Illyrians, Slavs, Turks, Magyars, and other peoples converged here from near and far. Ethnic and cultural differences kept them in chronic conflict.

As we shall see, the geographic extent of Europe has always been debatable, and today it is a particularly contentious issue in terms of European Union expansion and with respect to relations with Russia. Europe's eastern boundary is a dynamic one, and it has changed with history. Some would say that there really is no clear boundary, that Europe's atmosphere, so to speak, just gets thinner toward the east. For now, our definition places Europe's eastern boundary between Russia and its numerous European neighbors to the west. This definition is based on several geographic factors including European-Russian contrasts in territorial dimensions, geopolitical developments, cultural properties, and history.

Climate and Resources

From the balmy shores of the Mediterranean Sea to the icy peaks of the Alps, and from the moist woodlands and moors of the Atlantic fringe to the semiarid prairies north of the Black Sea, Europe presents an almost infinite range of natural environments (Fig. 1A-2).

Europe's peoples have benefited from a large and varied store of raw materials. Whenever the opportunity or need arose, the realm proved to contain what was required. Early on, these requirements included cultivable soils, rich fishing waters, and wild animals that could be domesticated; in addition, extensive forests provided wood for houses and boats. Later, coal and mineral ores propelled industrialization. More recently, Europe proved to contain substantial deposits of oil and natural gas.

Landforms and Opportunities

In the Introduction chapter, we noted the importance of physical geography in the definition of geographic realms. The natural landscape with its array of landforms (such as mountains and plateaus) is a key element in the total physical geography—or **physiography [1]**—of any part of the terrestrial world. Other physiographic components include climate and the physical features that mark the natural landscape, such as vegetation, soils, and water bodies. Europe's area may be small, but its physical landscapes are varied and complex. Regionally, we identify four broad units: the Central Uplands, the southern

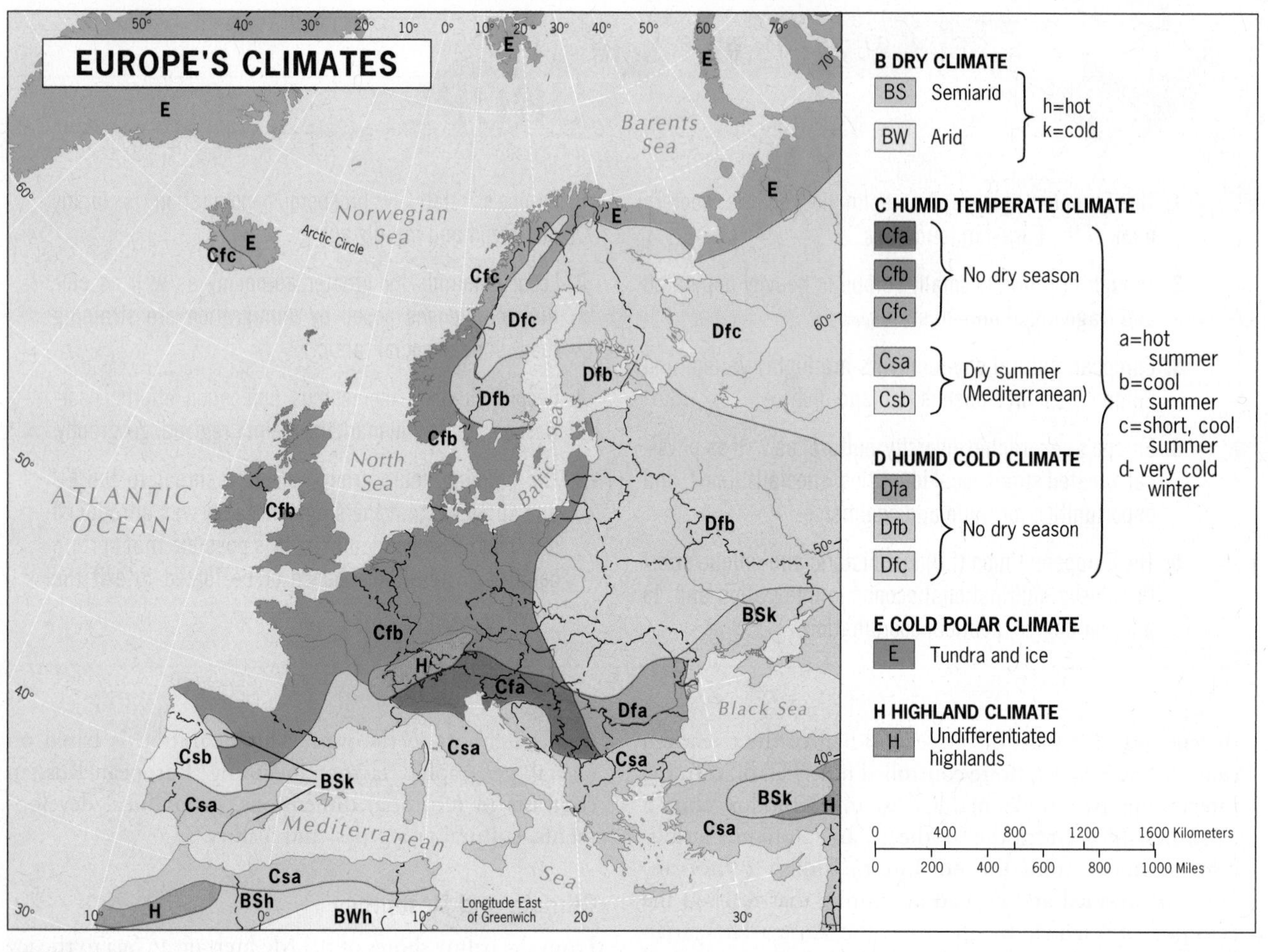

FIGURE 1A-2

From the Field Notes . . .

© Jan Nijman

"Driving across the Swiss Alps, I entered the mini-state of Liechtenstein and drove up to the town of Triesenberg to get a panoramic view of an upper stretch of the Rhine River, flowing from left to right. This is western Europe's leading river and one of its defining geographic features—as a means of transportation, sometimes as a barrier, and frequently as a natural boundary. Here it forms the mini-state's border with Switzerland, located on the far side. Although the river serves primarily as a political boundary, the physical and cultural landscapes on both sides of the Rhine have much in common."

www.conceptcaching.com

Alpine Mountains, the Western Uplands, and the North European Lowland (Fig. 1A-3).

The ***Central Uplands*** form the heart of Europe. It is a region of hills and low plateaus loaded with raw materials whose farm villages grew into towns and cities when the Industrial Revolution transformed this realm.

The ***Alpine Mountains***, a highland region named after the Alps, extend eastward from the Pyrenees on the French-Spanish border to the Balkan Mountains near the Black Sea, and include Italy's Appennines and the Carpathians of eastern Europe.

The ***Western Uplands***, geologically older, lower, and more stable than the Alpine Mountains, extend from Scandinavia through western Britain and Ireland to the heart of the Iberian Peninsula in Spain.

The ***North European Lowland*** stretches in a lengthy arc from southeastern Britain and central France across Germany and Denmark into Poland and Ukraine, from where it

FIGURE 1A-3

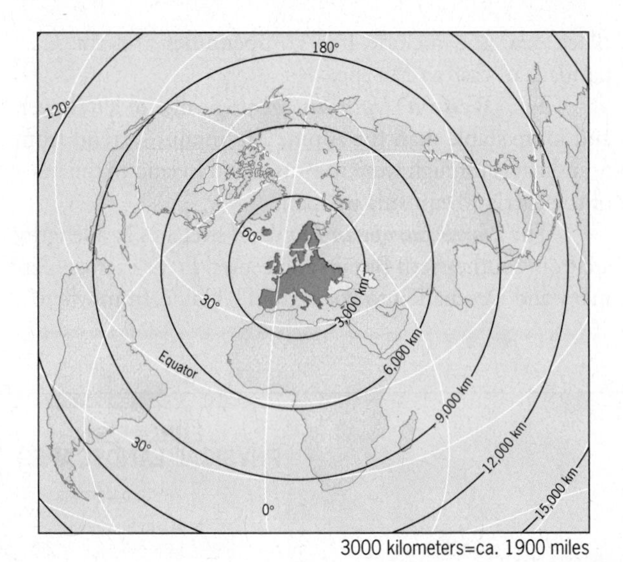

FIGURE 1A-4

© H. J. de Blij, P. O. Muller, and John Wiley & Sons, Inc.

continues well into Russia. Also known as the Great European Plain, this has been an avenue for human migration time after time, so that complicated cultural and economic mosaics developed here and together produced a jigsaw-like political map. As Figure 1A-3 shows, many of Europe's major rivers and connecting waterways serve this populous region, where a number of Europe's leading cities (London, Paris, Amsterdam, Copenhagen, Berlin, Warsaw) are located.

Locational Advantages

Europe also is endowed with some exceptional locational advantages. Its ***relative location***, at the crossroads of the **land hemisphere [2]**, creates maximum efficiency for contact with much of the rest of the world (Fig. 1A-4). A "peninsula of peninsulas," Europe is nowhere far from the ocean and its avenues of seaborne trade and conquest. Hundreds of kilometers of navigable rivers, augmented by an unmatched system of canals, open the interior of Europe to its neighboring seas and to the shipping lanes of the world. The Mediterranean and Baltic seas, in particular, were critical in the development of trade in early modern times, and in the emergence of Europe's early trading cities such as Venice (Italy) in the south and Lübeck (Germany) in the north.

And note the scale of the maps of Europe in this chapter. Europe is a realm of moderate distances and close proximities. Short distances and large cultural differences make for intense interaction, the constant circulation of goods and ideas. That has been the hallmark of Europe's geography for more than a millennium.

ANCIENT EUROPE

Modern Europe was peopled in the wake of the Pleistocene's most recent glacial retreat and global warming—a gradual warming that caused tundra to give way to deciduous forest and ice-filled valleys to turn into grassy vales. On Mediterranean shores, Europe witnessed the rise of its first great civilizations, on the islands and peninsulas of Greece and later in what is today Italy.

Ancient Greece and Imperial Rome

Ancient Greece lay exposed to influences radiating from the advanced civilizations of Mesopotamia and the Nile Valley, and in their fragmented habitat the Greeks laid the foundations of European civilization. Their achievements in political science, philosophy, the arts, and other spheres have endured for 25 centuries. One of the essential legacies of ancient Greece involved the formation of **city-states [3]** such as Athens and Sparta: relatively small territories comprised of cities and their hinterlands that were ruled by elected governments. This is where Western democracy had its origins. But the ancient Greeks never managed to unify their domain, and their persistent conflicts proved fatal when the Romans challenged them from the west. By 147 BC, the last of the sovereign Greek intercity leagues (alliances) had fallen to the Roman conquerors.

The center of civilization and power now shifted to Rome in present-day Italy. Borrowing from Greek culture, the Romans created an empire that stretched from Britain to the Persian Gulf and from the Black Sea to Egypt; they made the Mediterranean Sea a Roman lake carrying armies to distant shores and goods to imperial Rome. With an urban population that probably exceeded 1 million, Rome was the first metropolitan-scale urban center in Europe.

The Romans founded numerous other cities throughout their empire and linked them to the capital through a vast system of overland and water routes, facilitating political control and enabling economic growth in their provinces. It was an unparalleled ***infrastructure***, much of which long outlasted the empire itself.

Triumph and Collapse

Roman rule brought disparate, isolated peoples into the imperial political and economic sphere. By guiding (and often forcing) these groups to produce particular goods or materials, the Romans launched Europe down a road for which it would become famous: **local functional specialization [4]**. The workers on Elba, a Mediterranean island, mined iron ore. Those near Cartagena in Spain produced silver and lead. Certain farmers were taught irrigation to produce specialty crops. Others raised livestock for meat or wool. The *production of particular goods by particular people in particular places* became and remained a hallmark of the realm.

The Romans also spread their language across the empire, setting the stage for the emergence of the ***Romance languages***; they disseminated Christianity; and they established durable systems of education, administration, and commerce. But when their empire collapsed in the fifth century AD, disorder ensued and massive migrations soon brought Germanic and Slavic peoples to their present positions on the European stage. Capitalizing on Europe's weakness, the Arab-Berber Moors from North Africa, energized by Islam, conquered most of Iberia and penetrated France. Later the Ottoman Turks invaded eastern Europe and reached the outskirts of Vienna.

Visions of Our Land/The Image Bank/Getty Images, Inc.

The Rialto Bridge, one of Venice's famous landmarks, dates from the time that Venice was one of Europe's most powerful and richest city-states. The first bridge over the Grand Canal at this point dates back to the twelfth century; the present span was completed in 1591.

EARLY MODERN EUROPE

Europe's revival—its ***Renaissance***—did not begin until the fifteenth century. After a thousand years of feudal turmoil marking the "Dark" and "Middle" Ages, powerful monarchies began to lay the foundations of modern states. The discovery of continents and riches across the oceans opened a new era of ***mercantilism***, the competitive accumulation of wealth chiefly in the form of gold and silver. Best placed for this competition were the kingdoms of western Europe. Europe was on its way to colonial expansion and world domination.

Even as Europe's rising powers reached for world domination overseas, they fought with each other in Europe itself. Powerful monarchies and landowning ("landed") aristocracies had their status and privilege challenged by ever-wealthier merchants and businesspeople. Demands for political recognition grew; cities mushroomed with the development of industries; the markets for farm products burgeoned; and Europe's population, more or less stable at about 100 million since the sixteenth century, began to increase.

Early modern Europe also was the scene of a growing agricultural sector. As major ports and capital cities boomed, their expanding markets created widening economic opportunities for farmers. This led to revolutionary changes in land ownership and agricultural technology. Improved farming methods, better equipment, superior storage facilities, and more efficient transport to urban markets marked an agrarian revolution in the countryside. Moreover, the colonial merchants brought back new crops (the American potato soon became a European staple), causing market prices to rise and drawing ever more farmers into the economy.

The City-States of Early Modern Europe

From the fourteenth to the seventeenth centuries, growing agricultural production combined with expanding interregional trade to foster the emergence of a considerable number of city-states. These were cities dominated by merchant classes with powerful economic interests. The most important concentrations of city-states were on or near the Mediterranean and Baltic seas, locations that facilitated long-distance commerce. Well-known examples were Venice (for a time the richest place on Earth), Genoa, and Florence in present-day Italy; Lübeck and Hamburg in what is now northern Germany; and Bruges in today's Belgium. These city-states thrived in a Europe in which the modern nation-state had not yet appeared and whose political geography was fragmented at the finest of scales. But in economic terms the realm was being integrated at what were then unprecedented levels; along the shores of the Baltic Sea, the Mediterranean, and the North Sea; in the valleys of the Rhine and other major rivers; and overland.

MODERN HISTORICAL GEOGRAPHY

The Industrial Revolution

The term **Industrial Revolution [5]** suggests that an agrarian Europe was suddenly swept up in wholesale industrialization that changed the realm in a few decades. In reality, seventeenth- and eighteenth-century Europe had been industrializing in many spheres, long before the chain of events known as the Industrial Revolution began. From the textiles of England and Flanders to the iron farm implements of Saxony (in present-day Germany), from Scandinavian furniture to French linens, Europe had already entered a new era of *local functional specialization*. Nonetheless, by the end of the eighteenth century, industrial Europe took off as never before.

British Primacy

Britain was at the epicenter of this revolution. In the 1780s, the Scotsman James Watt and others devised a steam-driven engine, which was soon adapted to numerous industrial applications. At about the same time, coal (converted into carbon-rich coke) was recognized as a vastly superior substitute for charcoal in smelting iron—and, soon thereafter, far more durable steel. These momentous innovations had a rapid effect. The power loom revolutionized the weaving industry. Iron smelters, long dependent on Europe's dwindling forests for fuel, could now be concentrated near coalfields. Engines could move locomotives as well as power looms. Ocean shipping entered a new age.

Britain had an enormous advantage, for the Industrial Revolution occurred when British influence reigned worldwide and the significant innovations were achieved in Britain itself. The British controlled the flow of raw materials, they held a monopoly over products that were in global demand, and they alone possessed the skills necessary to make the machines that manufactured the products. Soon the fruits of the Industrial Revolution were being exported, and the modern industrial spatial organization of Europe began to take shape. In Britain, manufacturing regions developed near coalfields in the English Midlands, at Newcastle to the northeast, in southern Wales, and along Scotland's Clyde River around Glasgow.

Diffusion Onto the Continent

The Industrial Revolution diffused (spread) eastward from Britain onto the European mainland throughout the middle and late nineteenth century (Fig. 1A-5). Population skyrocketed, emigration mushroomed, and industrializing cities burst at the seams. European states already had acquired colonial empires before this revolution started; now colonialism gave Europe an unprecedented advantage in its dominance over the rest of the world.

In mainland Europe, a belt of major coalfields extends from west to east, roughly along the southern margins of the North European Lowland, due eastward from southern England across northern France and Belgium, Germany (the Ruhr), western Bohemia in the Czech Republic, Silesia in southern Poland, and the Donets Basin (Donbas) in eastern Ukraine. Iron ore is found in a broadly similar belt and together with coal provides the key raw material for the manufacturing of steel.

As in Britain, this cornerstone industry now spawned new concentrations of economic activity, growing steadily as millions migrated from the countryside to fill expanding employment opportunities. Densely populated and heavily urbanized, these emerging agglomerations became the backbone of Europe's world-scale population cluster (as shown in Fig. G-8).

Two centuries later, this east-west axis along the coalfield belt remains a major feature of Europe's population

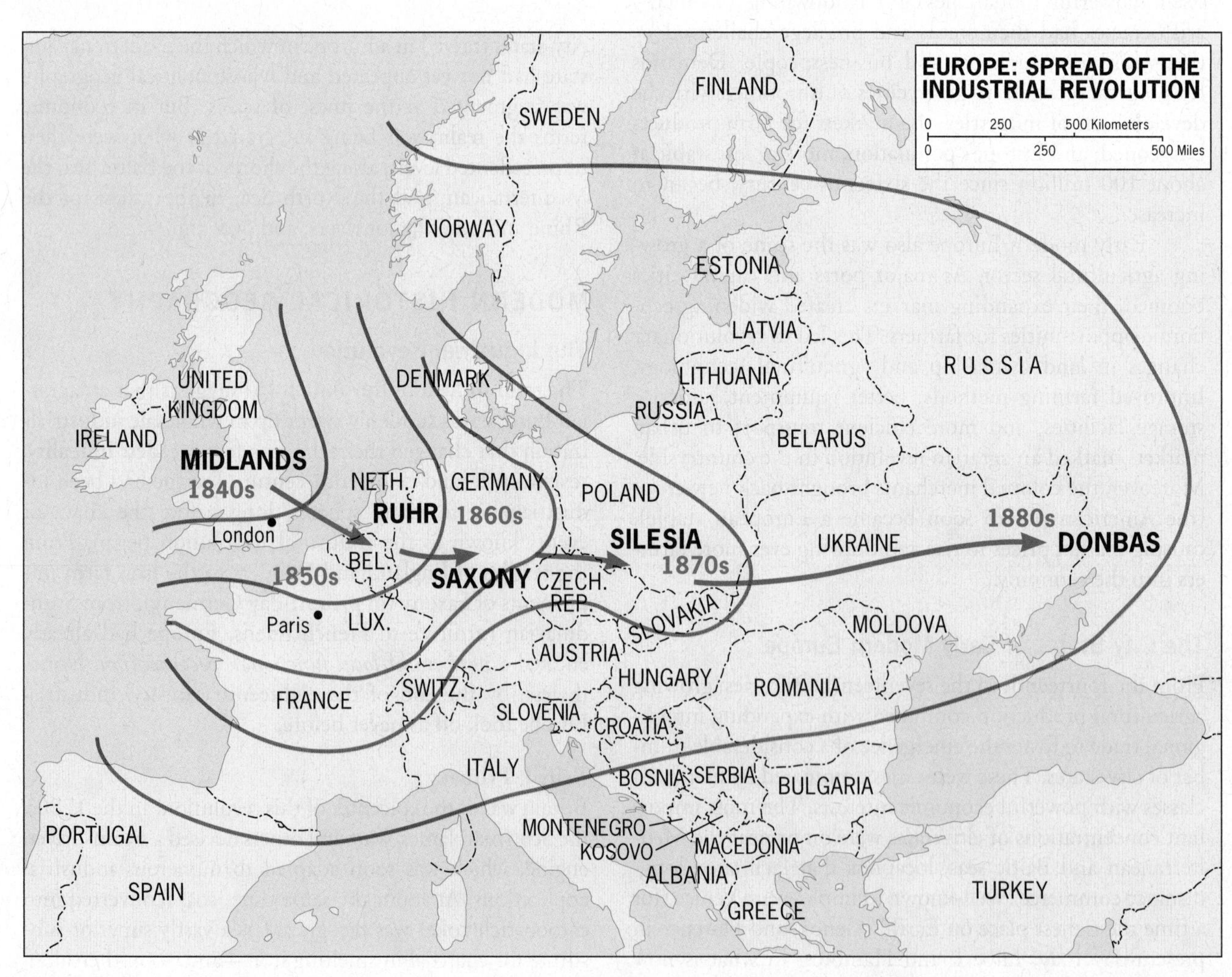

FIGURE 1A-5

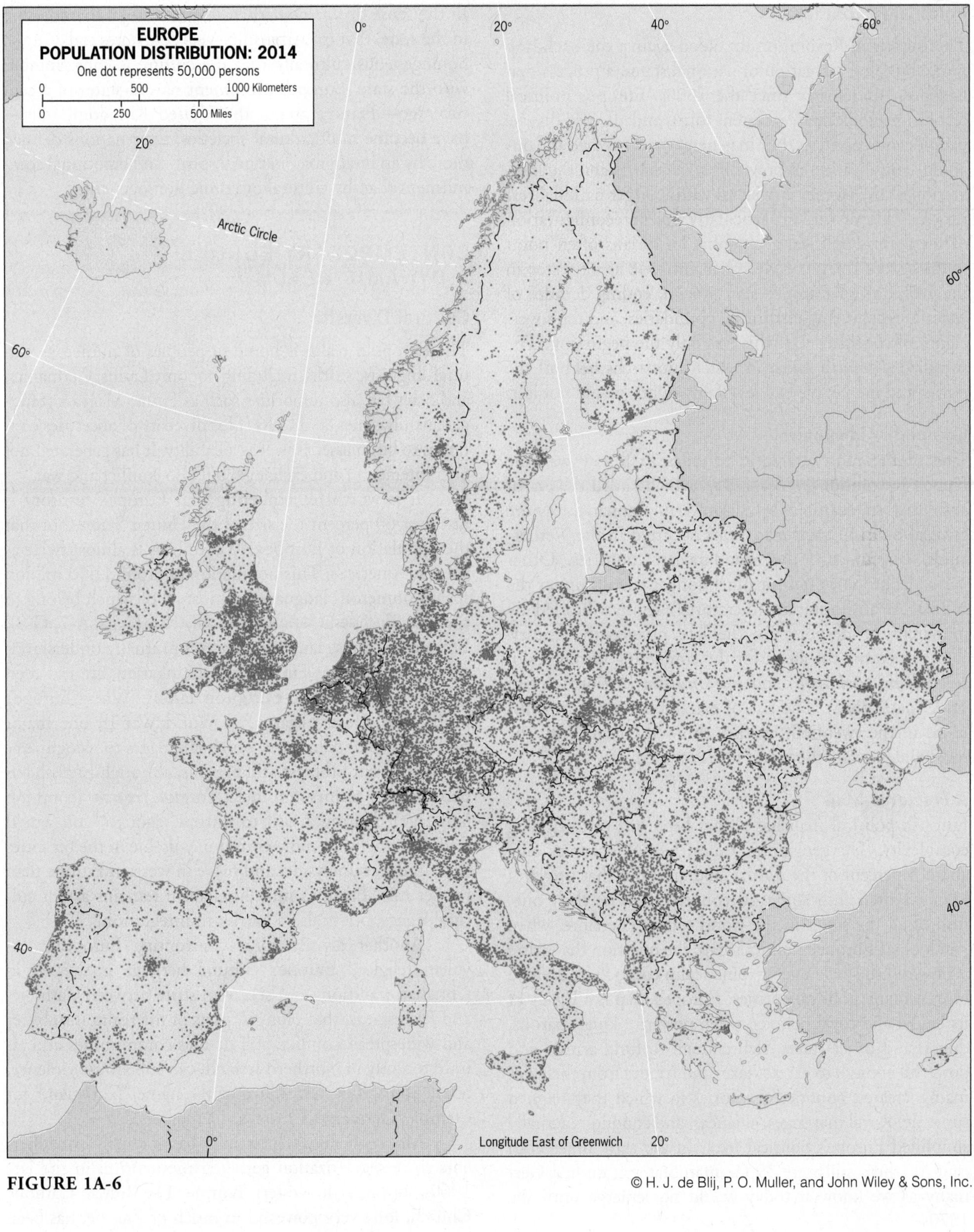

EUROPE
POPULATION DISTRIBUTION: 2014
One dot represents 50,000 persons
0 500 1000 Kilometers
0 250 500 Miles
Arctic Circle
Longitude East of Greenwich

FIGURE 1A-6

distribution map (Fig. 1A-6). It should also be noted that while industrialization spawned new cities, another set of manufacturing zones arose in and near many existing urban centers. London—already Europe's leading urban focus and Britain's richest domestic market—was typical of these developments. Many local industries were established here, taking advantage of the large supply of labor, the ready availability of capital, and the proximity of so great a number of potential buyers. Although the Industrial Revolution thrust other places into prominence, London did not lose its primacy: industries in and around the British capital multiplied.

Political Revolutions

The Industrial Revolution unfolded against the backdrop of the ongoing formation of national states, a process that had been under way since the 1600s. Europe's political revolutions took many different forms and affected diverse peoples and countries, but in general they involved the ongoing centralization of power by royal courts and the integration of the former city-states within larger national territories. This was often a particularly violent reconfiguration of the realm's political geography. Historians often point to the Peace (Treaty) of Westphalia in 1648 as a key step in the evolution of Europe's state system, ending decades of war and recognizing territories, boundaries, and the sovereignty of countries. But this was only the beginning. The so-called *absolutist states*, in which monarchs held all the power and the people had few if any rights, would not last.

Competing Ideologies

One of the most dramatic episodes to follow was the French Revolution (1789–1795), which ended the era of absolutist states ruled by all-powerful monarchs. More gradual political transformations occurred in the Netherlands, Britain, and the Scandinavian countries. Other parts of Europe would for much longer remain under the control of authoritarian (dictatorial) regimes headed by monarchs or despots. By the late nineteenth century, Europe became the arena for the competing ideologies of ***liberalism***, ***socialism***, and ***nationalism*** (national spirit, pride, patriotism). The rise of extreme nationalism (***fascism***) in the twentieth century threw the realm into the most violent wars the world had ever seen.

A Fractured Map

Europe's political map has always been one of intriguing complexity. As a geographic realm, Europe occupies only about 5 percent of the Earth's land area; but that tiny area is fragmented into some 40 countries—more than one-fifth of all the states in the world today. Therefore, when you look at Europe's political map, the question that arises is how did so small a geographic realm come to be divided into so many political entities? Europe's map is a legacy of its feudal and royal periods, when powerful kings, barons, dukes, and other rulers, rich enough to fund armies and powerful enough to extract taxes and tribute from their domains, created bounded territories in which they reigned supreme. Royal marriages, alliances, and conquests actually simplified Europe's political map. In the early nineteenth century, there still were 39 German States; a unified Germany as we know it today would not emerge until the 1870s.

State and Nation

Europe's political revolution produced a form of political-territorial organization known as the **nation-state [6]**, a territorial state embodied by its culturally distinctive population. The term **nation [7]** refers to a people with a single language, a common history, a similar ethnic background. In the sense of ***nationality***, it relates to legal membership in the state, that is, citizenship. Very few states today are so homogeneous culturally that the culture is conterminous with the state. Europe's prominent nation-states of a century ago—France, Spain, the United Kingdom, Italy—have become multicultural societies, their nations defined more by an intangible "national spirit" and emotional commitment than by cultural or ethnic homogeneity.

CONTEMPORARY EUROPE: A DYNAMIC REALM

Cultural Diversity

The European realm is home to peoples of numerous cultural-linguistic stocks, including not only Latins, Germanics, and Slavs but also minorities such as Finns, Magyars (Hungarians), Basques, and Celts. This diversity of ancestries continues to be an asset as well as a liability. It has generated not only interaction and exchange, but also conflict and war.

It is worth remembering that Europe's territory is just over 60 percent the size of the United States, but that the population of Europe's 40 countries is almost twice as large as America's. This population of around 600 million speaks numerous languages, almost all of which belong to the **Indo-European language family [8]** (Figs. 1A-7, G-9). But most of those languages are not mutually understandable; some, such as Finnish and Hungarian, are not even members of the Indo-European family. When Europe's unification efforts began after World War II, one major problem was to determine which languages to recognize as "official." That problem still prevails, although English has become the realm's unofficial ***lingua franca*** (common language). During a visit to Europe, though, you would find that English is more commonly usable in the big cities than in the countryside, and more in western Europe than farther east. Europe's multilingualism remains a rich cultural legacy, but is also a barrier to integration.

Another divisive force confronting Europeans involves religion. Europe's cultural heritage is steeped in Christian traditions, but sectarian strife between Catholics and Protestants, that plunged parts of the realm into bitter and widespread conflict, still divides communities and, as until recently in Northern Ireland, can still arouse violence. Some political parties still carry the name "Christian," for example, Germany's Christian Democrats.

More generally, Christianity has gradually lost adherents since secularization gathered momentum in the late 1960s, especially in western Europe. The Roman Catholic Church, long very powerful in much of Europe, has been losing its grip on society, and many church institutions—schools, universities, unions, political parties, charities, clubs—have been hollowed out. Moreover, many churches have closed down or have been converted into art galleries, public meeting halls, and even corporate offices.

Today, a new factor roils the religious landscape: the rise of Islam. In southeastern Europe, this takes the form

LANGUAGES OF EUROPE
0 200 400 600 Kilometers
0 100 200 300 Miles
Arctic Circle
ICELANDIC
FAEROESE
ATLANTIC OCEAN
SCOTS GAELIC
ENGLISH
IRISH GAELIC
North Sea
NORWEGIAN
SWEDISH
SAAMI (LAPP)
SAAMI
KARELIAN
FINNISH
SAMOYEDIC
KOMI
RUSSIAN
ESTONIAN
LATVIAN
LITHUANIAN
Baltic Sea
DANISH
FRYSIAN
DUTCH
WALLOON
GERMAN
ALSATIAN
POLISH
BELARUSSIAN
UKRAINIAN
CZECH
SLOVAK
HUNGARIAN
ROMANIAN
BRETON
FRENCH
GALICIAN
BASQUE
PROVENÇAL
SLOVENE
SERBO-CROATIAN
BULGARIAN
Black Sea
MACEDONIAN
ALBANIAN
GREEK
TURKISH
PORTUGUESE
SPANISH
CATALAN
CORSICAN-ITALIAN
SARDINIAN-ITALIAN
ITALIAN
Adriatic Sea
Mediterranean Sea
Longitude West of Greenwich
Longitude East of Greenwich
MAJOR INDO-EUROPEAN BRANCHES
GERMANIC GROUP
WESTERN GERMANIC
1 Dutch
2 German
3 Frysian
4 English
NORTHERN GERMANIC
5 Danish
6 Swedish
7 Norwegian
8 Icelandic
9 Faeroese
ROMANCE GROUP
10 Portuguese
11 Spanish
12 Catalan
13 Provençal
14 French
15 Italian
16 Rhaeto-Romansch
17 Romanian
18 Corsican-Italian
19 Sardinian-Italian
20 Walloon
SLAVIC GROUP
WEST SLAVONIC
21 Polish
22 Slovak
23 Czech
24 Lusatian
EAST SLAVONIC
25 Russian
26 Ukrainian
27 Belarussian
SOUTH SLAVONIC
28 Slovene
29 Serbo-Croatian
30 Macedonian
31 Bulgarian
OTHER INDO-EUROPEAN BRANCHES
CELTIC GROUP
BRITANNIC
32 Breton
33 Welsh
GAELISH
34 Irish Gaelic
35 Scots Gaelic
BALTIC GROUP
36 Latvian
37 Lithuanian
HELLENIC
38 Greek
THRACIAN/ILLYRIAN GROUP
39 Albanian
INDO-IRANIAN GROUP
40 Romani (dispersed)
URALIC LANGUAGE FAMILY
FINNO-UGRIC GROUP
41 Finnish
42 Karelian
43 Saami
44 Estonian
45 Hungarian
46 Komi
SAMOYEDIC GROUP
47 Samoyedic
ALTAIC LANGUAGE FAMILY
TURKIC GROUP
48 Turkish
OTHER LANGUAGES
BASQUE
49 Basque
Areas with significant concentrations of other languages (usually adjacent national languages)
Boundary between languages
After Murphy, 1998.

FIGURE 1A-7

of new Islamic assertiveness in an old Muslim bastion: the (Turkish) Ottoman Empire left behind millions of converts from Bosnia to Bulgaria among whom many are demanding greater political representation and power. In the west, this Islamic resurgence results from the relatively recent infusion of millions of Muslim immigrants from former colonies in North Africa and other parts of the far-flung Islamic world. Here, as mosques overflow with the faithful, churches stand nearly empty as a witness to secularism among Europeans.

For so small a realm, Europe's cultural geography is sharply varied. The popular image of Europe tends to be formed by British pageantry, the French wine country, or historic cities such as Venice or Amsterdam—but go beyond this core area, and you will find isolated Slavic communities in the mountains facing the Adriatic Sea, Muslim towns in poverty-mired Albania, Roma (Gypsy) villages in the interior of Romania, farmers using traditional methods unchanged for centuries in rural Poland. That map of 40 countries does not begin to reflect the diversity of European cultures.

Spatial Interaction

If not a single culture, then what does unify Europe? The answer lies in the realm's outstanding opportunities for productive interaction. The European realm is best understood as an enormous ***functional region***, an interdependent realm that is held together through highly developed, spatial economic and political networks. Modern Europe has seized on the realm's abundant geographic opportunities to create a huge, intensively used network of spatial interaction linking places, communities, and countries in countless ways. This interaction operates on the basis of two key principles.

First, regional **complementarity [9]** means that one area produces a surplus of a commodity required by another area. The mere existence of a particular resource or product is no guarantee of trade: it must be needed elsewhere. When two areas each require the other's products, we speak of *double complementarity*. Europe exhibits countless examples of this complementarity, from local communities to entire countries. Industrial Italy needs coal from western Europe; western Europe needs Italy's farm products.

Second, the ease with which a commodity can be transported by producer to consumer defines its **transferability [10]**. Distance and physical obstacles can raise the cost of a product to the point of unprofitability. But Europe is small, distances are short, and the Europeans have built the world's most efficient transport system of roads, railroads, and canals linking navigable rivers. Taken together, Europe's enormously diverse economic regions and its particularly efficient transportation infrastructure make for a highly interdependent economic realm.

From the Field Notes . . .

"If you were to be asked what city is shown here, would Paris spring to mind? Most images of the French capital show venerable landmarks such as the Eiffel Tower or Notre Dame Cathedral. I took this photo from the top of the famous Arc de Triomphe, looking toward another Paris, the northwestern business district just beyond the city line named *La Défense*. There, Paris escapes the height restrictions and architectural limitations of the historic center and displays an ultramodern face. Glass-box skyscrapers reflect the vibrant global metropolis this is, and there the landmark is the "Cube," a huge open structure admired as well as reviled (as was the Eiffel Tower in its time). *La Défense* is ingeniously incorporated into Paris's urban design, with a straight-line, broad avenue connecting it to the Arc de Triomphe and then connecting, via the *Champs Élysées* to the Place de la Concorde (see the first photo in this chapter). In terms of a grand geographic layout, Paris is hard to beat."

www.conceptcaching.com

Concept Caching

A Highly Urbanized Realm

About three of every four Europeans live in towns and cities, an average that is far exceeded in the west (see the Data Table in Appendix B) but not yet attained in much of the east. Large cities are production centers as well as marketplaces, and they also form the crucibles of their nations' cultures. Europe's major cities tend to have long histories and are compact, and in general the European cityscape looks quite different from its North American counterpart.

Seemingly haphazard inner-city street systems impede traffic; central cities may be picturesque, but they are also cramped. European city centers tend to be more vibrant today than those in the United States. They offer a mix of businesses, government functions, shopping facilities, educational and art institutions, and entertainment as well as housing for upper-income residents.

Wide residential sectors radiate outward from the central business district (CBD) across the rest of the central city, often inhabited by particular income groups. Beyond the central city lies a sizeable suburban ring, but even here residential densities are much higher than those in the United States because the European tradition is one of setting aside recreational spaces (in "greenbelts") and living in apartments rather than in detached single-family houses. There also is a greater reliance on public transportation, which further concentrates the suburban development pattern. That has allowed many nonresidential activities to suburbanize as well, and today ultramodern outlying business centers increasingly compete with the CBD in many parts of urban Europe (see photo).

A CHANGING POPULATION

Negative Natural Population Growth

There was a time when Europe's population was (in the terminology of population geographers) exploding, sending millions to the New World and the colonies and still growing at home. But today Europe's native population, unlike most of the rest of the world's, is actually shrinking. To keep a given population from declining, the (statistically) average woman must bear 2.1 children. For Europe as a whole that figure was 1.6 in 2012. Several countries recorded numbers at or below 1.3, including Poland, Portugal, Hungary, and Latvia. Such negative population growth poses serious challenges for any nation. When the population pyramid becomes top-heavy, the number of workers whose taxes pay for the social services of the aged goes down, leading to reduced pensions and dwindling funds for health care.

The Growing Multicultural Challenge

Meanwhile, immigration is partially offsetting Europe's population deficit. But it could be a mixed blessing, wherein demographic stability comes at the risk of social disruption. Millions of Turks, Turkish Kurds, Algerians, Moroccans, West Africans, Pakistanis, and West Indians are changing the social fabric of what once were homogeneous nation-states, especially France, Germany, the United Kingdom, Belgium, and the Netherlands. As noted earlier, one key dimension of this change is the spread of Islam in Europe. Muslim populations in eastern Europe (such as Albania's, Kosovo's, and Bosnia's) are indigenous communities converted during the period of Ottoman rule. The Muslim sectors of western European countries, on the other hand, represent more recent immigrations.

The majority of these immigrants are generally more religious than the Christian natives. They arrived in a Europe where native populations are stagnant or declining, where religious institutions are weakened, and where certain cultural norms are incompatible with Islamic traditions. Integration and assimilation of Muslim communities into

A 2009 referendum in Switzerland on new minaret construction produced some ugly symbolism (note the rocket-like depiction of minarets on the Swiss flag). The Swiss electorate approved a law that prohibits the construction of minarets (read: mosques). A majority of the Swiss seemed to feel that their culture was under threat from Islam and that this type of mosque architecture gave expression to an overly assertive minority—even though there were only *four* minarets in the entire country! No doubt, Switzerland's nearly 500,000 Muslims (just over 6 percent of the population, mainly Turks, and not known to be particularly zealous) must have felt uneasy about the vote.

the national fabric has been slow, their education and income levels considerably lower than average. In western European countries particularly, Islamic immigrants are highly concentrated in metropolitan areas. Thus in cities like Hamburg, London, and Brussels, the proportion of Muslim populations is considerably higher than the national average.

The social and political implications of Europe's cultural transformation are numerous and far-reaching. Long known for tolerance and openness, European societies are now trying to restrict immigration in various ways, and political parties with anti-immigrant platforms are gaining ground in several countries.

EUROPEAN UNIFICATION

Realms, regions, and countries can all be subject to dividing and unifying forces that cause them to become more or less cohesive and stable political units. Where the region or country is home to diverse populations with different political agendas, it may prove difficult to avoid divergence and territorial fragmentation. Political geographers use the term **centrifugal forces [11]** to identify and measure the strength of such division, which may result from religious, racial, linguistic, political, economic, or other regional factors.

Centrifugal forces are measured against **centripetal forces [12]**, the binding, unifying glue of the state or region. General satisfaction with the system of government and administration, legal institutions, and other functions of the state (notably including its treatment of minorities) can ensure stability at the state level. However, in the case of the former Yugoslavia, the centrifugal forces unleashed after the end of the Cold War exceeded the weak centripetal forces in that relatively young state, and it disintegrated with dreadful consequences.

Since World War II ended in 1945, Europe has witnessed a steady process of integration and unification at a much broader geographic scale. A growing majority of European states and their leaders recognize that closer association and regional coordination form the key to a more stable, prosperous, and secure future. A realmwide union has been in the making now for nearly three-quarters of a century. Despite relentless challenges and occasional setbacks, centripetal forces have thus far prevailed.

Background

At the end of World War II, much of Europe lay shattered, its cities and towns devastated, its infrastructure wrecked, its economies ravaged. If this was one of the world's most developed realms at the beginning of the twentieth century, its economic prowess had been almost completely destroyed by 1945. One of the primary motives for integration and collaboration among western European countries, therefore, was rapid economic recovery.

The United States played a leading role, initially, in spurring on such cooperation. In 1947, U.S. Secretary of State George C. Marshall proposed a European Recovery Program designed to help counter all this dislocation and to create stable political conditions in which democracy would survive. Over the next four years, the United States provided about $13 billion in assistance to Europe (more than $125 billion in today's money). The United States was driven by economic and political motives. It had become the largest producer in the world of manufactured goods and was eager to restore European markets. Politically, the ending of World War II witnessed accelerating tensions with the Soviet Union, which had taken control over the bulk of eastern Europe. In addition, communist parties seemed poised to dominate the political life of major western European countries. The United States was intent, through the Marshall Plan, to have a firm hand in western Europe and keep communist influences at bay. The Marshall Plan applied solely to 16 European countries, including defeated (West) Germany and Turkey.

Northwestern European countries themselves were also driven by political considerations. The two world wars had clearly shown the dangers of excessive nationalism and had laid bare the devastating problems that can arise from a lack of political cooperation and collaborative efforts (if the Allies had acted as one sooner, Hitler might have been stopped before things escalated into war in 1939). From the perspective of countries like France, the Netherlands, and Belgium, one of the key issues was to control Germany in the postwar years, and this could only be done through close political cooperation. Thus European integration was from the start both an economic and a political affair, and concerns about Germany played a major part.

As the economies recovered and Germany became firmly embedded in a pan-European structure, the motives for European unification received a different emphasis. Increasingly, the process has been driven by the need to facilitate an ever larger and more efficient open market that can compete globally with the United States, China, and Japan. Politically, the goals today are more and more about stabilizing a much larger and diverse Europe that now approaches the Russian frontier. It is not just about Germany anymore but rather about a much bigger and more complicated geopolitical zone that has to accommodate a large number of (still) sovereign states and at the same time maintain good relations with the Russian giant next door.

The Unification Process

The Marshall Plan did far more than stimulate European economies. It confirmed European leaders' conclusion that their countries needed a joint economic-administrative structure. Such a structure was needed not only to coordinate the financial assistance, but also to ease the flow of resources and products across Europe's mosaic of boundaries, to lower restrictive trade tariffs, and to seek ways to improve political cooperation.

SUPRANATIONALISM IN EUROPE

1944 Benelux Agreement signed.

1947 Marshall Plan created (effective 1948–1952).

1948 Organization for European Economic Cooperation (OEEC) established.

1949 Council of Europe created.

1951 European Coal and Steel Community (ECSC) Agreement signed (effective 1952).

1957 Treaty of Rome signed, establishing European Economic Community (EEC) (effective 1958), also known as the Common Market and "The Six." European Atomic Energy Community (EURATOM) Treaty signed (effective 1958).

1959 European Free Trade Association (EFTA) Treaty signed (effective 1960).

1961 United Kingdom, Ireland, Denmark, and Norway apply for EEC membership.

1963 France vetoes United Kingdom EEC membership; Ireland, Denmark, and Norway withdraw applications.

1965 EEC–ECSC–EURATOM Merger Treaty signed (effective 1967).

1967 European Community (EC) inaugurated.

1968 All customs duties removed for intra-EC trade; common external tariff established.

1973 United Kingdom, Denmark, and Ireland admitted as members of EC, creating "The Nine." Norway rejects membership in the EC by referendum.

1979 First general elections for a European Parliament held; new 410-member legislature convenes in Strasbourg. European Monetary System established.

1981 Greece admitted as member of EC, creating "The Ten."

1985 Greenland, acting independently of Denmark, withdraws from EC.

1986 Spain and Portugal admitted as members of EC, creating "The Twelve." Single European Act ratified, targeting a functioning European Union in the 1990s.

1987 Turkey and Morocco make first application to join EC. Morocco is rejected; Turkey is told that discussions will continue.

1990 Charter of Paris signed by 34 members of the Conference on Security and Cooperation in Europe (CSCE). Former East Germany, as part of newly reunified Germany, incorporated into EC.

1991 Maastricht meeting charts European Union (EU) course for the 1990s.

1993 Single European Market goes into effect. Modified European Union Treaty ratified, transforming EC into EU.

1995 Austria, Finland, and Sweden admitted into EU, creating "The Fifteen."

1999 European Monetary Union (EMU) goes into effect.

2002 The euro is introduced as historic national currencies disappear in 12 countries.

2003 First draft of a European Constitution is published to mixed reviews from member-states.

2004 Historic expansion of EU from 15 to 25 countries with the admission of Cyprus, the Czech Republic, Estonia, Hungary, Latvia, Lithuania, Malta, Poland, Slovakia, and Slovenia.

2005 Proposed EU Constitution is rejected by voters in France and the Netherlands.

2007 Romania and Bulgaria are admitted, bringing total EU membership to 27 countries. Slovenia adopts the euro.

2008 Cyprus and Malta adopt the euro.

2009 Slovakia adopts the euro.

2010 Financial crisis strikes heavily indebted Greece, requiring massive EU bailout. Later in the year, Ireland follows and raises fears of similar crises in Portugal, Spain, and Italy. The future of the EMU is clouded; the value of the euro declines after a long rise against the dollar.

2011 Estonia adopts the euro.

2012 Fiscal Compact agreement (effective 2013) signed by 20 member-states (and ratified by 16) stipulates their tight adherence to rules of fiscal responsibility. The agreement sparked intense debate in various countries and some had not ratified the agreement by mid-2013.

2013 Croatia admitted as the 28th EU member-state. Politicians in several member-states, including the United Kingdom and the Netherlands, demand a national referendum on leaving the EU.

2014 Latvia adopts the euro, increasing eurozone membership to 18 countries.

For all these needs, Europe's governments had some guidelines. While in exile in Britain, the leaders of three small countries—Belgium, the Netherlands, and Luxembourg—had been discussing an association of this kind even before the end of the war. There, in 1944, they formulated and signed the Benelux Agreement, intended to achieve total economic integration. When the Marshall Plan was launched, the Benelux precedent helped speed the creation of the Organization for European Economic Cooperation (OEEC), which was established to coordinate the investment of America's aid (see the box entitled "Supranationalism in Europe").

Soon the economic steps led to greater political cooperation as well. In 1949, the participating governments created the Council of Europe, the beginnings of what was to become a European Parliament meeting in Strasbourg, France. Europe was embarked on still another political revolution, the formation of a multinational union involving a growing number of European states. This is a classic example of **supranationalism [13]**, which geographers define as a voluntary association in economic, political, or cultural spheres of three or more independent states willing to yield some measure of sovereignty for their mutual benefit.

Under the Treaty of Rome, six countries joined to become the European Economic Community (EEC) in 1958, also called the "Common Market." In 1973 the United Kingdom, Ireland, and Denmark joined, and the renamed European Community (EC) now encompassed nine members. As Figure 1A-8 shows, membership reached 15 countries in 1995, after the organization had been renamed yet one more time to become the ***European Union (EU)***. Since then the number of member-states has climbed to 28, with Croatia the latest to join in mid-2013.

The EU's administrative, economic, and even political framework has become so advanced that the organization now has a headquarters with many of the trappings of a capital city. Early on, EU planners chose Brussels (already the national capital of Belgium, a member of Benelux) as the organization's center of governance. But in order to avoid giving Brussels too much prominence, they chose Strasbourg (located near the German border in the northeastern corner of France) as the seat of the European Parliament, whose elected membership represents all EU countries.

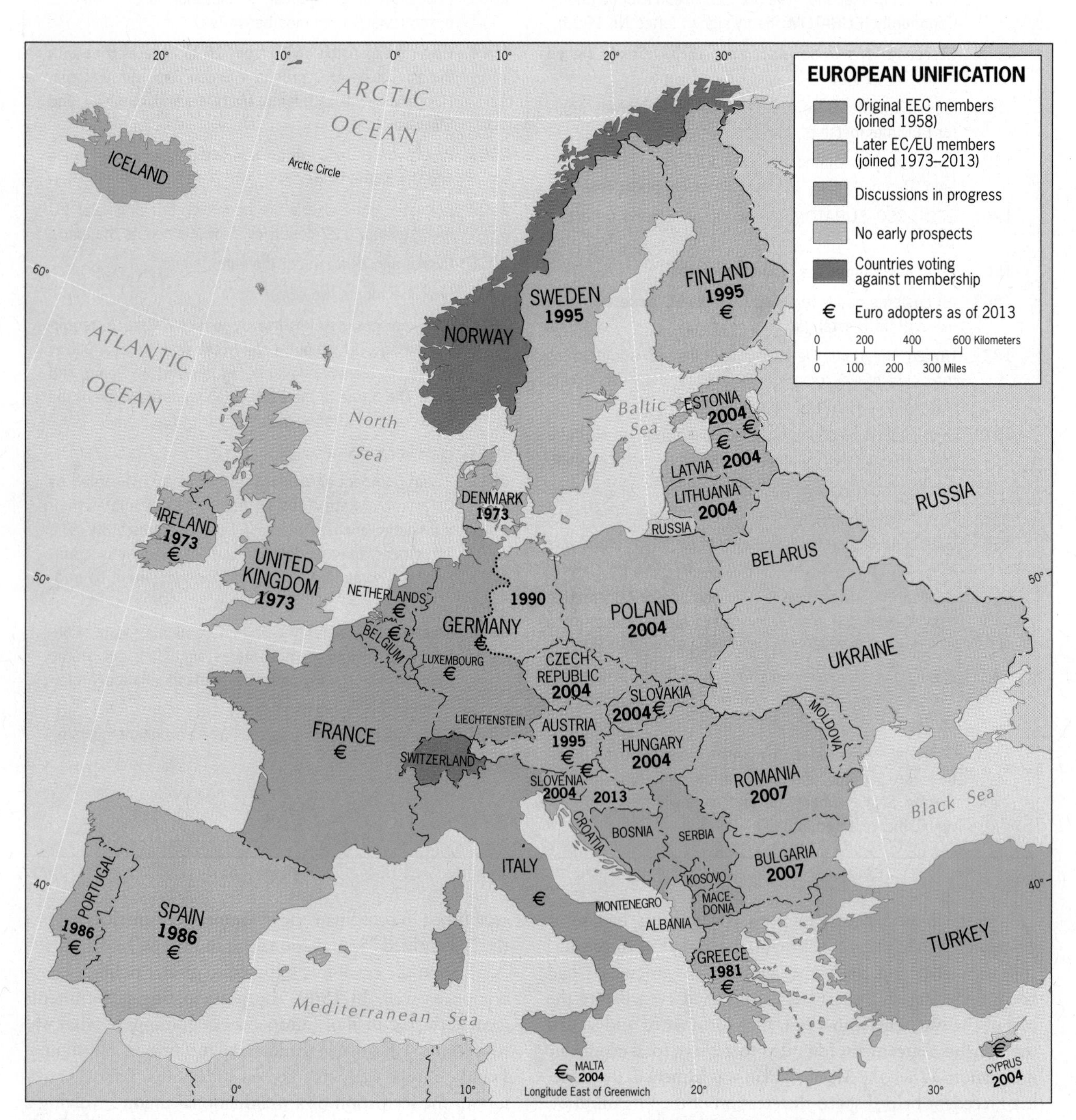

FIGURE 1A-8

CONSEQUENCES OF UNIFICATION

The European Union is not just a paper organization. It has a major impact on national economies, on the role of individual states, and on the daily lives of its member-countries' citizens.

One Market

EU directives are aimed at the creation of a single market for producers and consumers, businesses and workers. Corporations should be able to produce and sell anywhere within the Union without legal impediments, whereas workers should be able to move anywhere in the EU and find employment without legal restrictions. In order to make this happen and to keep things manageable, member-states have had to harmonize a wide range of national laws from taxation to the protection of the environment to educational standards. One major step came with the introduction of a single central bank (with considerable power over, for instance, interest rates) and a single currency, the ***euro***. The single currency was also meant to symbolize Europe's strengthening unity and to establish a counterweight to the once almighty American dollar. In 2002, twelve of the (then) 15 EU member-countries withdrew their currencies and began using the euro, with only the United Kingdom, Denmark, and Sweden staying out (Fig. 1A-8). More recently, Slovakia in 2009, Estonia in 2011, and Latvia became the eighteenth adopter—and member of the eurozone—in 2014. The single currency, hailed as a major triumph at its inception, has in reality become a major bone of contention as it implied a significant reduction in the members' freedom to formulate fiscal policies (more so than was initially realized). We will return to the evolving 'euro crisis' shortly.

A New Economic Geography

The establishment and continuing growth of the European Union have generated a new economic landscape that today not only transcends the old but is fundamentally reshaping the realm's regional geography. By investing heavily in new infrastructure and by smoothing the flows of money, labor, and products, European planners have dramatically reduced the divisive effects of their national boundaries. And by acknowledging demands for greater freedom of action by their provinces, States, departments, and other administrative units of their countries, European leaders unleashed a wave of economic energy that transformed some of these subnational units into powerful engines of growth (Fig. 1A-9). Four of these growth centers are especially noteworthy, to the point that geographers refer to them as the Four Motors of Europe [14]: (1) France's southeastern ***Rhône-Alpes Region***, centered on the country's second-largest city, Lyon; (2) ***Lombardy*** in north-central Italy, focused on the industrial metropolis of Milan; (3) ***Catalonia*** in northeastern Spain, anchored by the cultural and manufacturing center of Barcelona; and (4) ***Baden-Württemberg*** in southwestern Germany, headquartered by the high-tech city of Stuttgart.

Another change in regional economic geography involves agricultural activities and is closely related to deliberate policies designed in Brussels. Taxes tend to be high in Europe, and those collected in the richer member-states are used to subsidize growth and development in the less prosperous ones. This is one of the burdens of membership that is not universally popular in the EU, to say the least. But it has strengthened the economies of Portugal, Spain, and other national and regional economies to the betterment of the entire organization. Some countries also object to the terms and rules of the Common Agricultural Policy (CAP), which, according to some critics, supports farmers far too much and, according to others, far too little. (France in particular obstructs efforts to move the CAP closer to consensus, subsidizing its agricultural industry relentlessly while arguing that this protects its rural cultural heritage as well as its farmers.)

Despite economic integration and the harmonization of policies, Figure 1A-9 also shows that major differences persist among Europe's regions as well as within EU member-states. This is an important reminder that, as noted in the introductory chapter, uneven development is an unyielding phenomenon that plays out at various geographic scales.

Diminished State Power and New Regionalism

As states relinquish some of their power to Brussels and express agreement with the ideal of realmwide integration, some of the provinces and regions *within* states have seized the opportunity to assert their cultural identity and particular economic interests. Often, the local governments in these subregions simply bypass the governments in their national capitals, dealing not only with each other but even with foreign governments as their expanding business networks span the globe. In this they are imitated by other provinces, all seeking to foster their local economies and, in the process, strengthen their political position relative to the state.

Provinces, States, and other subnational political units on opposite sides of international boundaries can now also cooperate in pursuit of shared economic goals. Such cross-border cooperation creates a new economic map that seems to ignore the older political one, creating economically powerful regions that are more or less independent of surrounding national states. Note how European unification seems to simultaneously erode the power of states from above and from below: in order to retain membership, states are more or less compelled to concede major decision-making power to Brussels while at the same time the growing demands of subnational (and cross-national) regional authorities, from Catalonia to Lombardy, eat away at their territorial control at home.

Thus even as Europe's states have been working to join forces in the EU, many of those same states are confronting

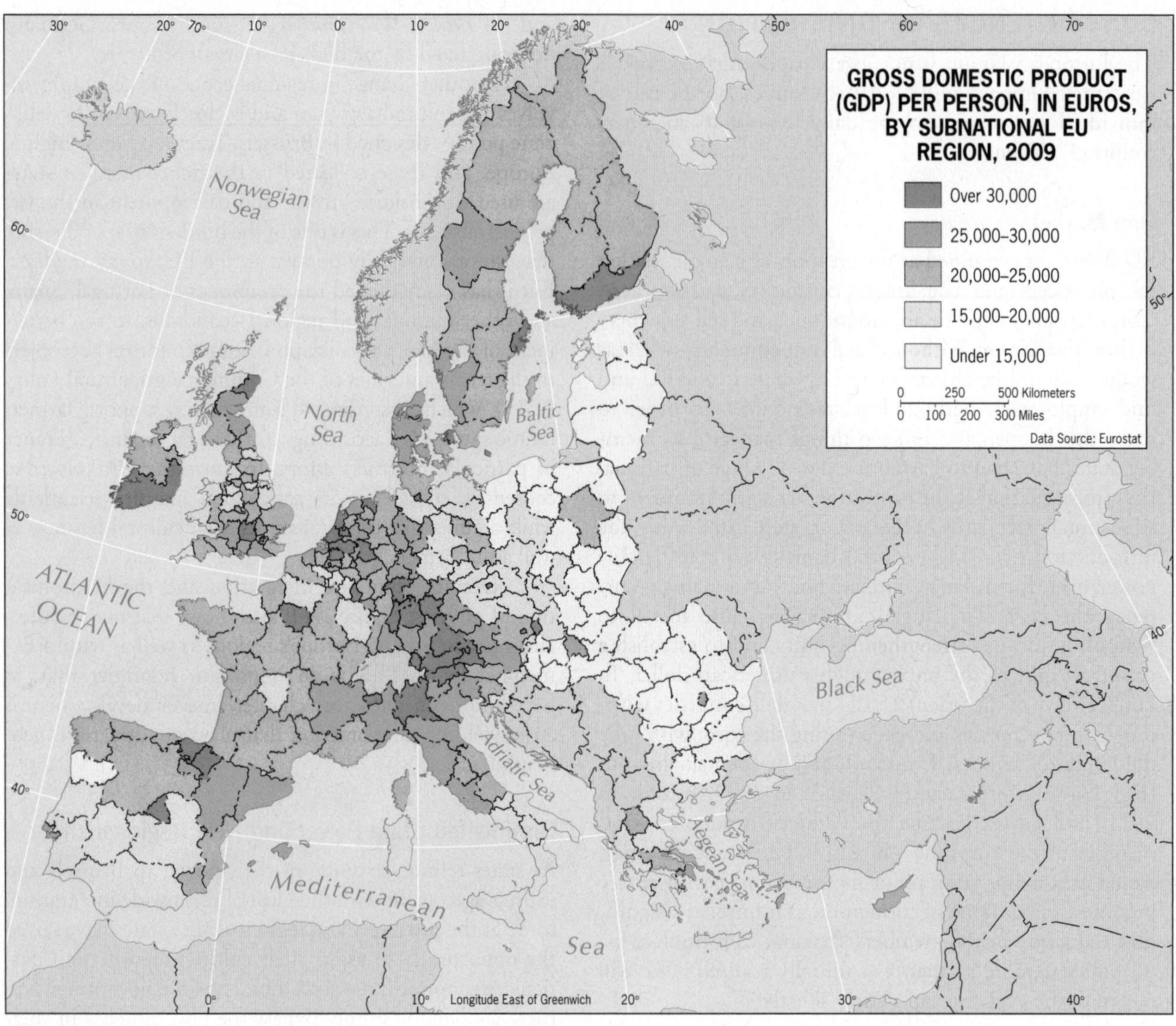

FIGURE 1A-9

severe centrifugal stresses. The term **devolution [15]** has come into use to describe the powerful centrifugal forces whereby regions or peoples within a state, through negotiation or active rebellion, demand and gain political strength and sometimes autonomy at the expense of the center. Most states exhibit some level of internal regionalism, but the process of devolution is set into motion when a key centripetal binding force—the nationally accepted idea of what a country stands for—erodes to the point that a regional drive for autonomy, or for outright secession, is launched. As Figure 1A-10 shows, numerous European countries are affected by devolution.

States respond to devolutionary pressures in various ways, ranging from accommodation to suppression. One way to deal with these centrifugal forces is to give historic regions (such as Scotland or Catalonia) certain rights and privileges formerly held exclusively by the national government. Another answer is for EU member-states to create new administrative divisions that will allow the state to meet regional demands, often in consultation with, or under pressure of, the ***European Commission*** (the name of the EU's central administration based in Brussels). We elaborate further in Chapter 1B, highlighting subnational autonomy movements faced by the governments of the United Kingdom, Spain, and Italy.

DEFINING EUROPE: DIFFICULT CHOICES

Widening or Deepening?

Expansion has always been an EU objective, and the subject never fails to arouse passionate debate. Will the incorporation of weaker economies undermine the strength of the whole? Should the ties and cooperation between existing members be deepened and solidified before other, less prepared countries are invited to join? Remember that member-states must adhere to strict economic policies and harmonize their political systems. This is much easier

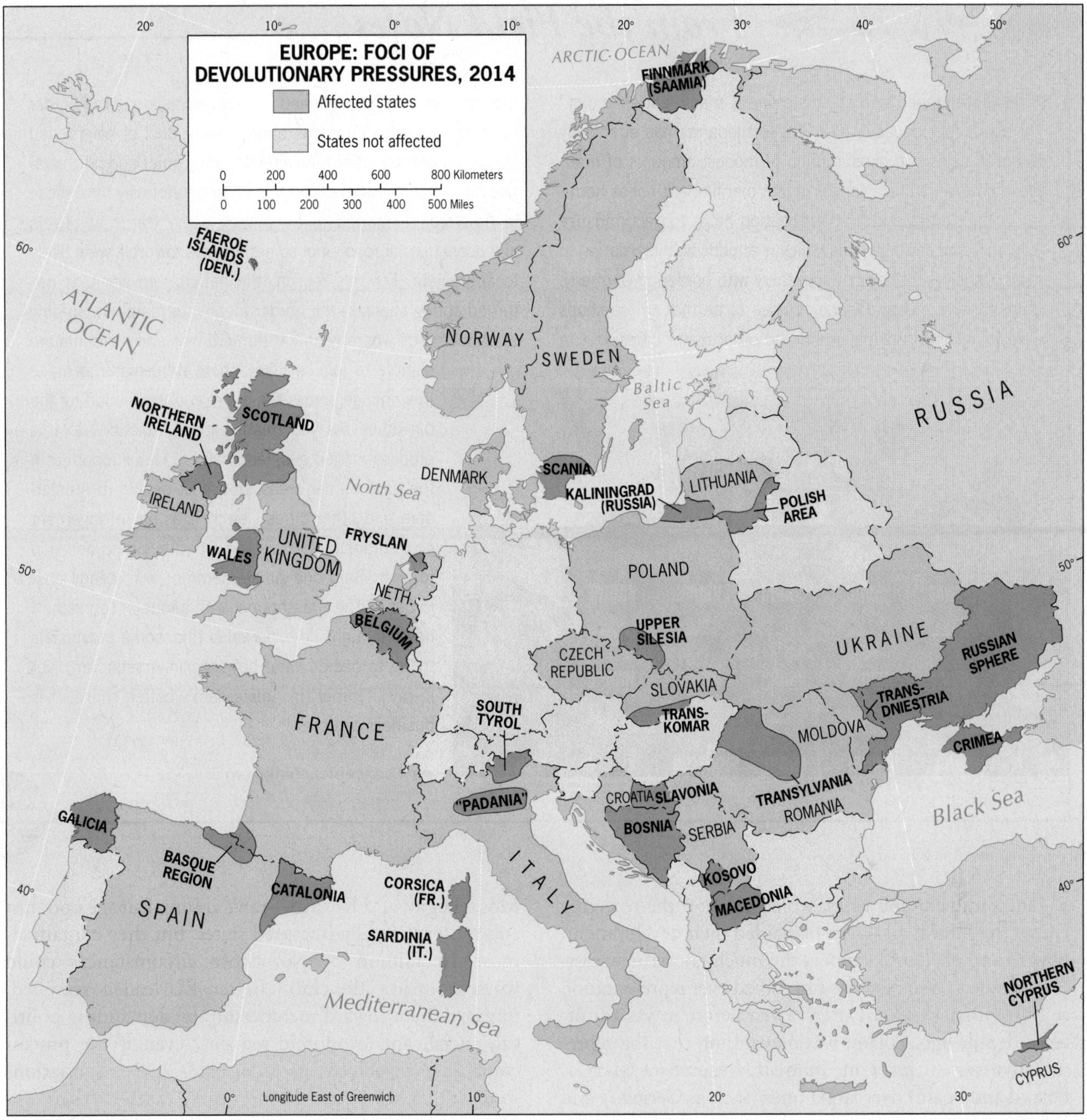

FIGURE 1A-10

for prosperous countries with longstanding democratic traditions than for poorer nations with a volatile political past.

Despite such misgivings, negotiations to expand the EU have long been in progress, and the gains of the past decade are mapped in Figure 1A-8. In 2004, a momentous milestone was reached: ten new members were added, creating a greater European Union with 25 member-states. Geographically, these ten fell into three groups: three Baltic states (Estonia, Latvia, and Lithuania); five contiguous states in eastern Europe, extending from Poland and the Czech Republic through Slovakia, Hungary, and Slovenia; and two Mediterranean island-states, the mini-state of Malta and the still-divided state of Cyprus. In 2007, both Romania and Bulgaria were incorporated, raising the number of members to 27 and extending the Union to the shores of the Black Sea. The newest and 28th member-state, Croatia, joined in 2013, the second component of former Yugoslavia (Slovenia was first) to be admitted to the EU.

Numerous structural implications arise from this expansion, affecting all EU countries. A common agricultural policy became even more difficult to achieve, given the poor condition of farming in most of the member-countries added since 2003. Also, some of the pre-2004

From the Field Notes . . .

"The Grasshopper, one of Amsterdam's many 'coffee shops.' For about four decades now, the Netherlands has effectively legalized sales and consumption of modest amounts of marijuana. The general consensus is that marijuana is not as harmful as hard drugs and that legalization helps to decriminalize its commerce and use. This policy is at odds with legislation in most of the rest of the EU, especially with bordering Germany, Belgium, and nearby France. It used to be that coffee shops were found only in Amsterdam and other major cities, but in recent years they have opened in small towns along the border to attract a foreign clientele (another byproduct of heightened cross-border interaction). As a result, once quiet and conservative rural villages (where lifestyles differ enormously from those in Amsterdam) started to experience heavy traffic of buyers and users from abroad—not something the townfolk were likely to appreciate. For this reason, the national government has turned coffee shops in the border areas into membership-only establishments whereby only Dutch citizens and residents are eligible to join. Initially, there were plans to make this change across the entire country, including the big cities. As you might expect, reactions to this proposal varied geographically: it was supported in the rural border areas, but most people in Amsterdam and other cities viewed it as an infringement on traditional liberties and a bad business decision to boot. Said one Amsterdammer and steady customer of The Grasshopper with a tone of sarcasm: 'I never thought I'd experience this: some provincials trying to decide what happens in Amsterdam. Not in my lifetime!' In Amsterdam's coffee shops, it is business as usual."

© Jan Nijman

www.conceptcaching.com

EU's less affluent countries, which were on the receiving end of the subsidy program that aided their development, now had to pay up to support the much poorer new eastern members. And disputes intensified over representation at EU's Brussels headquarters. Even before it was admitted a decade ago, Poland was demanding that the representative system favor medium-sized members (such as Poland and Spain) over larger ones (such as Germany and France).

The Remaining Outsiders

This momentous expansion has generated major geographic consequences for all of Europe, and not just the EU. As Figure 1A-8 shows, following the admission of Croatia in 2013, a number of countries and territories still remain outside the Union, and their prospect of joining seems a mixed bag. The first group includes the 5 remaining states that have emerged out of former Yugoslavia, plus Albania. In this cluster of western Balkan states, a majority are troubled politically and economically. Only Slovenia and Croatia have achieved full membership; the rest—Serbia, Bosnia and Herzegovina, Macedonia, Montenegro, and Kosovo—rank among Europe's poorest and most ethnically fractured states, but they contain almost 16 million people whose circumstances could worsen outside "the club." Better, EU leaders reasoned, to move them toward membership by demanding political, social, and economic reforms, even if the process could take several years. As of mid-2013, negotiations were underway with Montenegro, with Serbia and Macedonia still in the pre-negotiation stage after formally applying for membership. Elsewhere, ethnically and politically divided Bosnia as well as Albania have signed preliminary agreements with the EU, and their applications are expected in due course. Kosovo, however, remains in limbo following its incompletely-recognized declaration of independence from Serbia in 2008. Note, too, that this area of less-prepared countries just happens to be the most Islamic corner of Europe, which could be another impediment to accession.

The second group of outsiders is comprised of Ukraine and its neighbors. Four former Soviet republics in Europe's "far east" could some day join the EU (even Belarus, until recently the most disinterested, may now be reconsidering). The government of Ukraine has shown

For weeks in November 2012, protesters took to the streets of Athens to protest the Greek government's austerity measures to comply with EU bailout plans. This is part of the crowd of an estimated 40,000 people that turned out for the rally at Syntagma Square near Greece's Parliament, their banners screaming "NO".

interest, although its electorate is strongly (and regionally) divided between a pro-EU west and a pro-Russian east. Moldova views the EU as a potential supporter in a struggle against devolution along its eastern flank, and in 2006 a country not even on the map—Georgia, across the Black Sea from Ukraine—proclaimed its interest in EU membership for similar reasons: to find an ally against Russian intervention in its internal affairs.

Third and last, we should take special note of another important candidate for EU membership: Turkey. Some EU leaders would like to include a mainly "Muslim country" in what Islamic states sometimes call the "Christian club," but others are convinced that Turkey is just not European enough. Interestingly, the growing economic troubles inside the EU over the past few years contrast sharply against some eye-catching economic advances in Turkey—so it should come as no surprise that the Turks today are losing interest in joining the European Union.

Once Again: In Search of Europe's Eastern Boundary

In the past, "Eastern Europe" incorporated all of Europe east of Germany, Austria, and Italy; north of Greece; and south of Finland. The Soviet communist domination of Eastern Europe (1945–1990) behind an ideological and strategic "Iron Curtain" served to reinforce the division between "west" and "east." The collapse of the Soviet Empire in the early 1990s freed the European countries that had been under Soviet rule, and they swiftly turned their gaze from Moscow to the west. Meanwhile, the European Union had been expanding eastward, and membership in the EU became an overriding goal for the majority of the liberated eastern states. Today, with the Soviet occupation a rapidly fading memory and the European Union stretching from Ireland to the Black Sea, the boundary between western and eastern Europe has disappeared, at least in a political sense. Economically, significant contrasts remain between west and east (see the Data Table in Appendix B; compare, for instance, Bulgaria and France); but in this respect, too, things are evening out.

During the Cold War, Europeans, especially in the west, were preoccupied with the Iron Curtain (see the box titled "EU versus NATO"), and much less so with the question of the eastern boundary of Europe as a whole. After all, eastern Europe was drawn so tightly into the Soviet orbit politically, militarily, economically, and even culturally (the main foreign language taught in schools was Russian) that it seemed a moot point. But deep down, most of those eastern components remained quite European in spirit and tradition, and possessed historical ties that could not be easily erased. So now we are once again left to contemplate where, between Moscow and Vienna, Europe's eastern boundary lies or should be placed. More than anything else, it is the eastward expansion of the EU that is testing this question.

The Fiscal Crisis

The challenges to ongoing European integration are considerable, and they seem to be magnified during times of economic distress. Since 2009, the "euro crisis" has been an almost standard news item throughout the realm and is often reported in global news media as well. It was initially triggered by the financial crisis that erupted in the United States in 2008 and quickly reverberated around the globe, hitting Europe particularly hard. Economic growth rates plunged; unemployment rose; government deficits in certain cases skyrocketed; and politics in many countries became increasingly polarized.

But then the crisis took on its own form in Europe because it laid bare the limitations that accompanied the shift to a common currency and a common monetary policy. The evolving problems are not just about the euro

EU Versus NATO

FIGURE 1A-11

THE EMPHASIS PLACED nowadays on developments in the EU tends to overshadow another very important pan-European construct, the ***North Atlantic Treaty Organization (NATO)***. This alliance partially overlaps with the EU and includes many European countries (Fig. 1A-11). But NATO is an altogether different creature. First, it is primarily concerned with politics and security and not with economics: NATO's purpose is to provide military security to its members. Second, NATO is led by the United States, which provides the bulk of the alliance's military muscle, and also includes Canada. Third, NATO includes non-EU member Turkey, a critically important country because it is strategically located on Europe's southeastern edge and close to the Middle East (one can almost hear the Turks think "if we are good enough for NATO, why aren't we welcome in the EU?").

NATO was created in 1949 to provide a security umbrella for western Europe under U.S. leadership (in response, the Soviets and their satellites forged the Warsaw Pact). When the Cold War ended in 1991 with the disintegration of the Soviet Union, NATO became a more general security organization that would provide stability across Europe—and now opened its doors to membership for eastern European countries. This eastward expansion is not unlike that of the EU; the latest to join were Croatia and Albania in 2009, making for a current total of 28 member-states (Fig. 1A-11). Today, NATO is more concerned with the possibility of twenty-first-century types of threats to its members: terrorism, attacks by "rogue" states with weapons of mass destruction, and cyberwarfare (a very real and growing security danger with crucial global implications).

This also means that the alliance, while still mainly focused on Europe, is increasingly active on the global scene. For instance, one of the items atop its agenda is the possible nuclear missile threat from Iran. In the past few years, the largest military operation of the alliance (employing U.S. and European troops under a single command) has been far away from Europe, in Afghanistan. NATO intervention and air power also played a vital role in the support of Libyan rebels and subsequent defeat of the Qadhafi dictatorship in 2011.

Of course, both economic issues (EU) and security issues (NATO) are deeply political and are therefore often intertwined. Having separate organizations to deal with these two spheres adds another layer of complexity to the European realm. Economically, the United States and Europe are often perceived as competitors, and the Europeans expressly go their own way. But where security is concerned, the United States still functions as the guarantor of peace in the realm, with most Europeans happy to defer to the Americans who pick up the lion's share of the bill. The United States, in turn, views strategic involvement in Europe as fundamental to its own security interests.

itself but about the inability or unwillingness of individual countries to comply with fiscal standards (e.g., not running government deficits) and their inability to apply solutions on an individual country basis. Keep in mind that EU countries still have their own budgets, their own revenues and expenditures, and that the EU budget is still quite small (about 1 percent of the EU's GDP)—but the budget *rules* are in a number of crucial ways determined by Brussels.

The economic difficulties faced by several European national governments became more intractable because they had (only recently) introduced the single European currency and significantly constrained their policy options. Some of the EU's weaker member-states, such as Greece, accumulated enormous debts that violated terms in the agreements that established the eurozone. Moreover, having a single currency implied that countries could no longer independently pursue such monetary policies as devaluations (a common approach to promoting exports) or the lowering of interest rates (used to stimulate the economy). In other words, countries like Greece and Spain faced growing economic problems, but their hands were tied because of their dependence on the European Central Bank.

Greece accumulated such a massive government deficit that it required a bailout from Brussels in 2010 to avoid bankruptcy and chaos. A year later, it was Ireland's turn: the country that not long ago had been hailed as one of Europe's major success stories had to be rescued from imminent financial collapse. Portugal required a similar bailout in 2011, and Greece received a second one in 2012. By then, fiscal problems had also emerged in Italy (where the prime minister was forced to resign), in France (where austerity measures triggered widespread public protests), and even in the Netherlands (where a coalition government fell when it could not agree on severe budget cuts). Given the integrated financial system that the EU has become, the contagion of failing banks and faltering governments is a now a huge concern—as if Europe constituted a precarious row of dominoes. Not surprisingly, the future of the euro as a stable global currency has been thrown into uncertainty.

By mid-2012, fears of a breakup of the eurozone reached a peak when it seemed likely that Greece was about to exit. That country needed such enormous long-term bailouts that the richer EU member-states no longer appeared willing to provide economic support while the Greek government was unable to convince its constituents to accept the austerity measures that were part of the bailout agreement. The reason a breakup did not occur was that the consequences of a Greek exit would threaten the integrity of the EU as a whole. It would open the door to other countries leaving, an unraveling that could greatly undermine the very idea of a single European union—which is why the European Central Bank declared on July 26, 2012 that it would do "whatever it takes" to preserve the euro. This vow underscores the strong political will in much of Europe to stick together and bear the burden—but whether it endures will heavily depend on the economic direction Europe takes in the years immediately ahead.

Future Prospects of the EU

Europe's quest for enduring unity continues even as the fiscal crisis demonstrates that integration and unification is far more difficult to maintain during economic recessions. To be sure, unification of a realm of this size and complexity will always face recurring structural challenges. Even in good times, the sense of have and have-not, core and periphery, continues to pervade EU operations and negotiations. The European Union consists of a large number of countries of bewildering diversity, and for all to get along infinite patience, gigantic bureaucracies, and a penchant for compromise are essential.

The European Union now reaches deeply into eastern Europe; encompasses 28 members; has a common currency and a parliament; is developing a constitution; has facilitated political stability as well as economic progress for more than half a century; and is even considering expansion beyond the realm's borders. All this constitutes a tremendous achievement in this historically volatile and fractious part of the world. The EU now has a combined population of nearly 510 million that constitutes one of the world's richest markets; its member-states account for more than 40 percent of the world's exports.

Europe's leaders, by averting the 2012 breakup of the eurozone, have reinforced the political will to maintain the Union. Whether European integration stays on course will not only depend on the direction the realm's economy will take but also on the ability of Europe's political leadership to legitimize the cost of integration to their citizens. The problem of the so-called 'democratic deficit' is a serious one: from Germany to Greece, people often do not

Regional ISSUE — Should the Eurozone Be Maintained at All Costs?

A VIEW FROM GERMANY

"As a school teacher, I have followed the issues surrounding the euro crisis pretty closely. In a way it is amazing how my country's role has changed in the past couple of decades in the EU. In the old days, most Europeans used to be worried about us, especially after World War II. Now we are in the driver's seat, whether we like it or not. Germany has the strongest economy in Europe and when other countries are in trouble, from Greece to Portugal, everybody is looking to us for help.

I do think that European integration, overall, is a good thing and for Germany as well, because our economy has developed strong ties all over the continent and in recent years especially to the east. But I want to remind you how we got here: with very hard work and discipline. Our economy was in a shambles after the war and it took great efforts to rebuild it. Then, with reunification in the early 1990s, we had to reintegrate impoverished East Germany back into our society. If we are Europe's biggest economy and one of the most prosperous countries, it is because of hard work and dedication.

Now look at Greece: what a mess they have made of it there. And the same could be said of Portugal as well as Ireland, and it seems the same problems are surfacing in Italy and even France. Why did they let things get so far out of hand? Government finances are a disaster because their leaders just accumulated debt, spending their way into re-election time and again. They just have not been responsible and they have been living beyond their means for a long time. You know what? I am afraid this attitude was made possible because they think they can rely on the EU to bail them out.

If it is a one-time deal, we should make the sacrifice and help other countries by paying their debt for them. But it really worries me that this only makes them more dependent. When the European Central Bank comes up with these rescue plans (with our money!) they demand, rightly so, that these countries tidy up their finances and cut expenditures that they could not afford in the first place. But look at the angry mobs in Athens protesting these measures! What do these people want?

The EU is important and should be kept together, but not at any price. There are limits to our means, and we do have to look out for ourselves. We will not pay indefinitely for other people's lack of responsibility. Our next elections take place in the autumn of 2013 and I am going to listen closely to what our politicians have to say about this."

A VIEW FROM GREECE

"I am a student in Athens and I took part in the demonstrations in November of 2012 against the austerity measures imposed on us by the EU. Our government was required to enact three rounds (!) of budget cuts in order to appease the EU. They barely got enough votes in Parliament, but they passed new legislation that is just going to kill us financially. Can't they see that we are at the end, that there is just nothing more to give by the people who have to make ends meet every day?

I and most other Greeks believe in the idea of Europe and European integration. Of course we do: Greece is the cradle of European civilization. Greek philosophy and the notion of Western democracy come from us and we have shared it with the rest of the continent. We have been members of the EU for more than 30 years, long before other countries were let in, and we joined the eurozone at its outset. But some of the richer countries in the EU just cannot see what the impact is on less affluent member-states. Perhaps it was a mistake for Greece to join the euro and be subordinated to policies made in Germany or France.

As I see it, the fiscal crisis is the result of the rich members forcing the entire Union to abide by rules that the more vulnerable countries just cannot sustain. Like the 'law' that members are not allowed more than a 3 percent deficit on their annual budgets—that is a lot easier for rich countries than it is for we poorer ones. With a tax base like Germany's, it is easy enough. But if you are Greece or Portugal, it's an entirely different proposition. In the end, it is the more vulnerable people in countries like Greece that pay dearly for the lofty ideas coming from Brussels.

Did you know that the Greek economy has for the past five years shrunk by fully one-fifth? How do you expect us cover our necessary expenditures without running up debt? And now with these austerity packages, things will get even worse because everything from student loans to rent subsidies will decrease. And the economy is forced into a downward spiral because nobody can afford to buy anything anymore. Unemployment among Greek youths is now well above 50 percent and I can't see how there will be a job for me when I graduate from college.

When I hear comments from Brussels that we are being irresponsible or irrational, it makes me angry and I feel we should walk away from the EU altogether. Whatever happened to democracy? There are limits to our suffering. And I will not support another Greek government that dances to the tunes being played in Brussels."

Vote your opinion at www.wiley.com/go/deblijpolling

feel they are being heard and national governments need their votes to support the Union (see the *Regional Issue Box* "Should the Eurozone Be Maintained at all Costs?"). In early 2013, the British prime minister announced he was in favor of a referendum on continuing EU membership; but if the United Kingdom were to leave, even though it has always been a reluctant member, that would deal a serious blow to the Union. Debates also continue as to whether membership should be more flexible, wherein countries would not have to commit on all fronts in the same way. But others say that such a "Europe *à la carte*" would lead to endless negotiations and also undercut the entire notion of unity.

And yet, the EU has been very resilient and, thus far, has been able to avert any major reversals no matter how messy the process has been. Overall, it is truly remarkable what has been accomplished in little more than five decades. Some of the EU's leaders want more than an economic union: they envisage a United States of Europe, a political as well as an economic competitor for the United States. To others, such a "federalist" notion is an abomination not even to be mentioned (the British most of all are particularly wary of such an idea). Whatever the outcome, Europe is clearly undergoing still another of its revolutionary transformations, and when you study its evolving map you are looking at history in the making.

POINTS TO PONDER

- Germany has the strongest economy in Europe and now dominates much of the realm as well as EU policies, including efforts to deal with the fiscal crisis.
- Some western European countries accepted relatively large numbers of religious Muslim immigrants at a time when secularization among Europe's native population reached a peak.
- NATO's main purpose is to provide security to its member-states in Europe plus the United States and Canada; yet its most important military operations in recent years took place far outside the realm in Afghanistan and Libya.
- If Greece or any other fiscally troubled country were to leave the eurozone, a dangerous precedent would be set that could undermine the entire EU project.
- Unemployment rates among youths in Greece and Spain are well over 50 percent. What are the prospects of college graduates there?

EUROPE: MODERN CO
AND HISTORIC REGIO
European Core Bound
Western Europe
Britain and Ireland
Northern (Nordic) Eur
Mediterranean Europe
Eastern Europe
0 250 500 Kilom
0 100 200 300 Miles
ICELAND
Arctic Circle
Norwegian Sea
Faeroe Islands (Den.)
NORWAY
SWEDEN
FINLAND
ESTONIA
LATVIA
LITHUANIA
RUSSIA
BELARUS
UKRAINE
MOLDOVA
ROMANIA
BULGARIA
Black Sea
TURKEY
North Sea
DENMARK
Baltic Sea
IRELAND
Irish Sea
UNITED KINGDOM
Thames R.
ATLANTIC OCEAN
NETHERLANDS
BELGIUM
LUXEMBOURG
GERMANY
Elbe R.
Rhine R.
POLAND
Vistula R.
Western Dvina R.
Dnieper R.
Dniester R.
CZECH REP.
SLOVAKIA
HUNGARY
AUSTRIA
LIECHTENSTEIN
SWITZERLAND
Danube R.
Seine R.
Loire R.
FRANCE
Rhône R.
SLOVENIA
CROATIA
BOSNIA
SERBIA
KOSOVO
MONTENEGRO
MACE-DONIA
ALBANIA
GREECE
Aegean Sea
Adriatic Sea
Po R.
ITALY
PORTUGAL
SPAIN
Ebro R.
Tagus R.
Mediterranean Sea
MALTA
CYPRUS
MOROCCO
ALGERIA
TUNISIA
Longitude East of Greenwich

REGIONS

The Mainland Core

Britain and Ireland

The Discontinuous South

The Discontinuous North

The Eastern Periphery

IN THIS CHAPTER

- The EU: Croatia in, the UK out?
- The rebirth of east-central Europe
- Germany's persistent east-west divide
- The drive for Scottish independence
- Ethnicity and politics in the Balkans
- Turkey's new prominence

CONCEPTS, IDEAS, AND TERMS

Microstate 1
Site 2
Situation 3
Primate city 4
Estuary 5
Conurbation 6
Landlocked location 7
Hegemony 8
Metropolis 9
Central business district (CBD) 10
World-city 11
Shatter belt 12
Balkanization 13
Break-of-bulk 14
Entrepôt 15
Exclave 16
Irredentism 17

FIGURE 1B-1

© Jan Nijman

Where were these pictures taken?
Find out at www.wiley.com/college/deblij

Our objective in this and later regional (**B**) chapters is to become familiar with the regional and national frameworks of the realms under investigation, and to discover how various regions and individual states fit into the broader geographical mosaic. In these **B** chapters, therefore, we examine things at a finer scale. Even in this era of globalization, and especially within integrating Europe, the state still plays a key role in that process. The geography of these various regions and states, how they interact, and what their prospects may be in this changing world are among the key questions we will address.

REGIONS OF THE REALM

EUROPE'S REGIONAL COMPLEXITY

Europe presents us with a particular challenge because of its numerous countries and territories, and because of the continuing changes affecting its geography. Readers of this book who reside in North America tend to be accustomed to a relatively recent, familiar, and stable map of two (or three) countries: the United States and Canada, with Mexico to the south. In Europe, on the other hand, a long history of changing political geography continues to affect current developments. Just since 1990 no less than 15 new national names have appeared on the political map, some of them restorations of old entities such as Estonia and Lithuania, others new and less familiar (Moldova, Montenegro). And there may be more to come: Kosovo became nominally independent in 2008, and Belgium may yet fracture the way Czechoslovakia did more than two decades ago.

An additional contrast to North America lies in the protracted and tumultuous history of Europe. Many of Europe's cities were first built long before settlers reached the shores of the New World; a stroll through Venice, Bruges, or Prague takes us back many centuries. Because Europe's twenty-first-century cultures and political boundaries are the

© Jan Nijman

MAJOR CITIES OF THE REALM

Metropolitan Area	Population* (in millions)
Paris, France	11.0
London, UK	9.3
Madrid, Spain	7.1
Barcelona, Spain	5.9
Frankfurt, Germany	5.7
Milan, Italy	5.2
Berlin, Germany	3.5
Athens, Greece	3.5
Rome, Italy	3.3
Kiev, Ukraine	2.9
Stuttgart, Germany	2.4
Brussels, Belgium	2.0
Hamburg, Germany	1.8
Vienna, Austria	1.8
Budapest, Hungary	1.8
Warsaw, Poland	1.7
Lyon, France	1.6
Stockholm, Sweden	1.5
Prague, Czech Republic	1.3
Dublin, Ireland	1.2
Amsterdam, Netherlands	1.1

*Based on 2014 estimates.

result of deeply rooted historical processes that were conditioned by geography, we should keep that in mind when we examine the realm's current spatial framework.

And Europe displays something else North America does not: **microstates [1]** that do not have the attributes of "complete" states but are on the map as tiny, yet separate entities nonetheless, such as Monaco, Andorra, San Marino, and Liechtenstein. Add to these the Russian exclave (outlier) of Kaliningrad on the Baltic Sea and the British dependency of Gibraltar at the outlet of the Mediterranean Sea, and Europe seems to display a bewildering political mosaic indeed.

Traditional Formal Regions, Modern Spatial Network

In the old days, Europe divided rather easily into Western, Northern (Nordic), Mediterranean (Southern), and Eastern regions, and countries were grouped accordingly. Some of this traditional historical geography lingers in cultural landscapes and still has relevance. But in the era of European unification, a more dominant ***core-periphery*** framework has been superimposed on that regional scheme, defined in terms of how central—or peripheral—countries are to the workings of the EU. Put differently, Europe can be understood as a set of formal, more or less homogeneous, cultural regions; but at the same time, the entire realm increasingly functions as a tightly interdependent economic and political regional system, a spatial network that revolves around the EU.

The original Common Market of 1957 still anchors what has become a ***core area*** for all of the European realm. Both Britain and Ireland form part of this core, but the British—whose EU membership was delayed by a French veto and who never adopted the euro—still are not the full-fledged members the Germans, the French, and the Belgians are. It is therefore appropriate to consider the United Kingdom and Ireland as a geographically distinct, offshore component of the European Core.

Take a close look at the map in Figure 1B-1 and note that today the core area of the EU does not coincide with national borders—the core is primarily defined on the basis of regional economic performance, not political geography. For example, Northern Italy (but not the South) or Southern Sweden (but not the North) are part of this core. You see how complicated and challenging the work of geographers can be when trying to understand the world's spatial organization. And the European realm illustrates that challenge like no other. But this chapter will bring clarity to what at first glance may appear to be a chaotic regional landscape.

In the discussion that follows, we will focus first on countries such as Germany and France that lie entirely within the Core region. Next, we discuss countries with significant regions inside as well as outside the Core, such as Spain, Italy, and Sweden. Lastly, our attention focuses on Europe's Periphery.

THE MAINLAND CORE

Not counting the microstate of Liechtenstein, eight states form the Mainland Core of Europe: dominant Germany and France, the three Benelux countries, the two landlocked mountain states of Switzerland and Austria, and the Czech Republic (Fig. 1B-2). This is the European region sometimes still referred to as "Western Europe," and its most populous country (Germany) also contains Europe's largest economy.

Germany

Twice during the twentieth century Germany plunged Europe and the world into war, until, in 1945, the defeated and devastated German state was divided into two parts, West and East (see the red-line delimitation in Fig. 1B-3), the latter of which was forced to surrender some important industrial areas to Poland.

West and East Germany then set forth on widely different economic and political trajectories. Soviet rule in East Germany was established on the Soviet-communist model, and in return for the extreme hardships the USSR had suffered at German hands during the war, it was harshly punitive. The American-led authority in

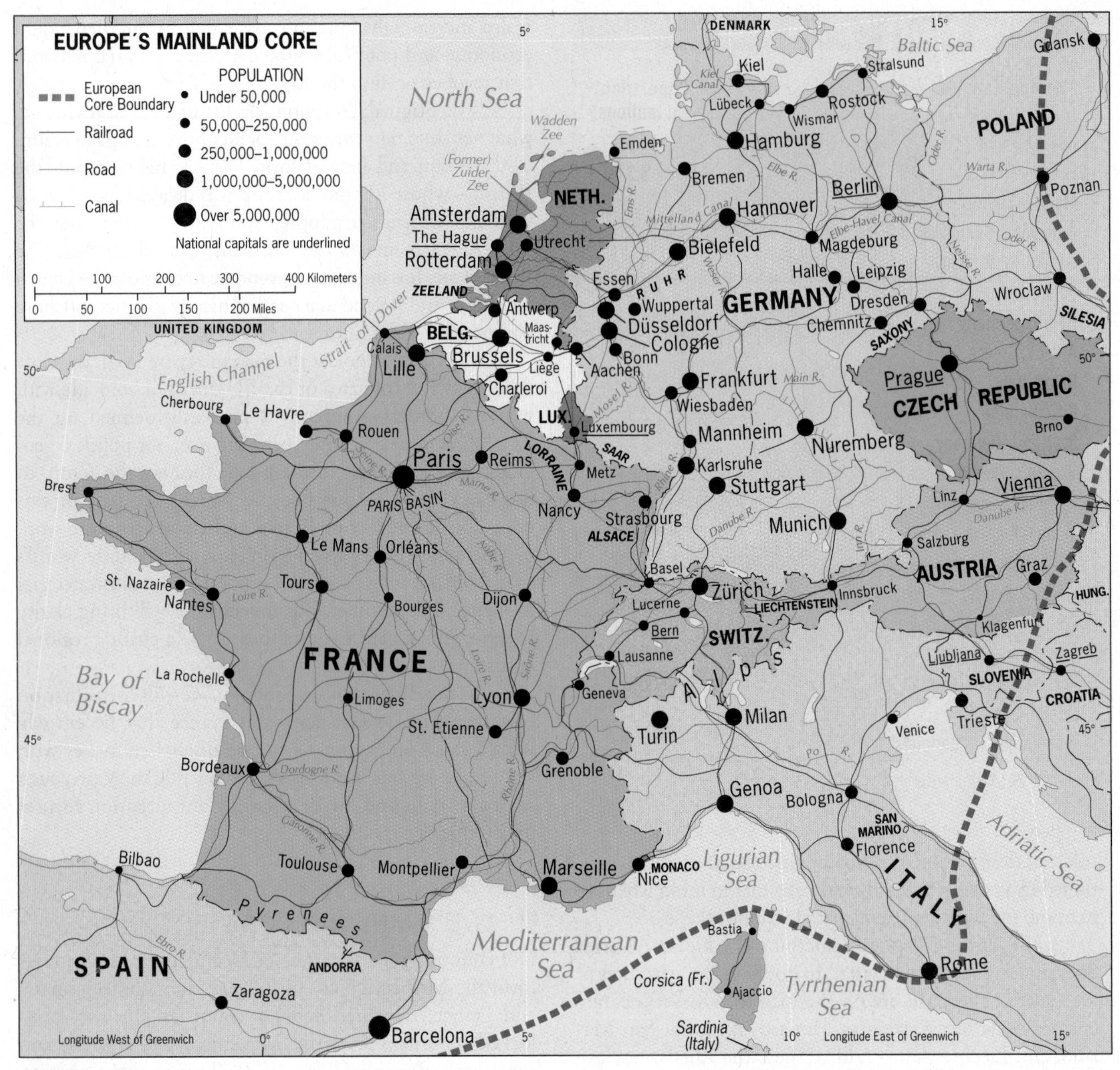

FIGURE 1B-2

© H. J. de Blij, P. O. Muller, and John Wiley & Sons, Inc.

West Germany was less strict and aimed more at rehabilitation. When the Marshall Plan was instituted, West Germany was included, and its economy recovered rapidly. Meanwhile, West Germany was reorganized politically into a modern federal state based on democratic foundations.

West Germany's economy soon thrived. Between 1949 and 1964 its gross national income (GNI) tripled while industrial output rose 60 percent. Simultaneously, West Germany's political leaders participated enthusiastically in the negotiations that led to the six-member Common Market in 1957. Geography worked in West Germany's favor: it had common borders with all but one of the initial member-states. Its transport infrastructure, rapidly rebuilt, was second to none in the realm. More than compensating for its loss of Saxony and Silesia in the east were the expanding Ruhr (in the hinterland of the Dutch port of Rotterdam) and the newly emerging industrial complexes centered on Hamburg in the north, Frankfurt (the leading financial hub as well) in the center, and Stuttgart in the south. West Germany exported huge quantities of iron, steel, motor vehicles, machinery, textiles, and farm products. Until just a few years ago, Germany had the largest export economy by value in the world; today it is surpassed by China, but Germany is still second and outranks the third-place United States.

In 1990, West Germany had a population of about 62 million and East Germany 17 million. Communist misrule in the East had yielded outdated factories, crumbling infrastructures, polluted environments, drab cities,

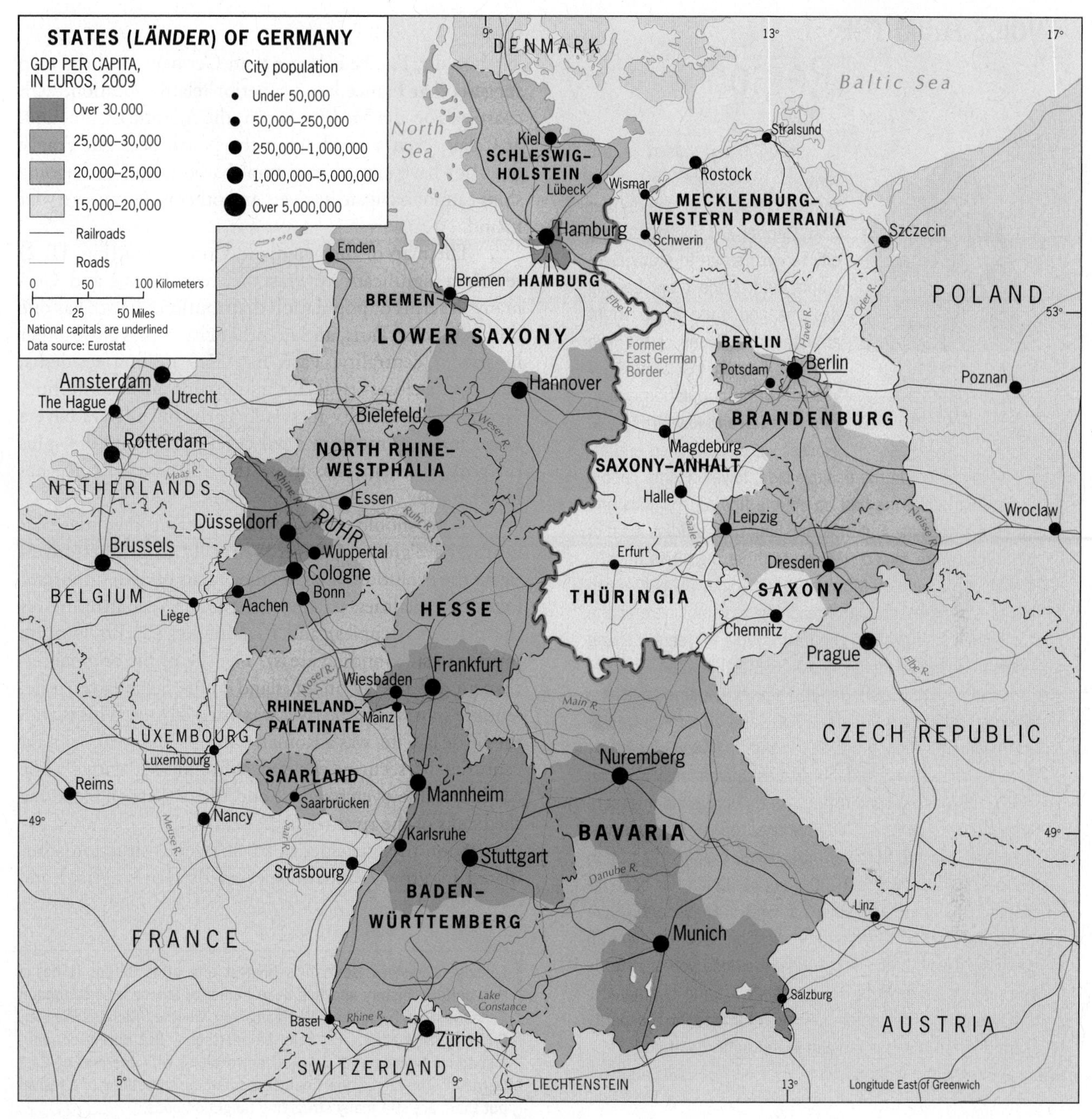

FIGURE 1B-3

inefficient farming, and inadequate legal and other institutions. Reunification that year was more a rescue than a merger, and the cost to West Germany was gargantuan. Today, unemployment rates in the East are twice as high as in the West. Regional disparity will continue to plague Germany for some time to come.

Upon reunification, East Germany was reorganized into six new States to fit the federal system of West Germany. Today's reunified Federal Republic of Germany consists of 16 States (or *Länder*). Figure 1B-3 makes a key point: regional disparity in terms of gross domestic product (GDP) per person remains a serious problem between the former East and West. Note that more than half of former East Germany's six States are still in the lowest income category, while most of the ten former West German States rank in the two higher income categories.

Another challenge facing Germany, along with other western European countries, is the integration of sizeable immigrant minorities. Germany has more than 15 million immigrants, about 19 percent of the total population of 79.8 million. They are culturally different and economically worse off. About 40 percent of immigrant households live below the official poverty line, and immigrants are far

VOICE FROM THE Region

© Franziska Bader

Franziska Bader, 29 years of Age, Leipzig, Germany

"I was born and raised in the town of Gera, about 50 miles south of Leipzig in the former East Germany. I moved to the city ten years ago to study Geography and, since then, Leipzig has changed quite a lot. Right after the [1989] fall of the Iron Curtain, Leipzig lost almost 20 percent of its population due to outmigration and low birth rates. Houses were abandoned and often dilapidated. On the positive side, lower-income people and students like me had no trouble finding spacious and affordable housing. But after finishing their studies most of my friends were forced to move to other big cities (mostly in the south, such as Munich) to get a job. I was lucky to find one as a research and teaching assistant here at the University of Leipzig. In the past couple of years, Leipzig seems to be making a comeback, with national newspapers even calling it the *New Berlin*. The city became attractive to more and more students, artists, and cultural entrepreneurs, generally young people looking for housing in empty factories and other affordable spaces. Increasingly, you can see beautifully renovated houses, thriving shopping centers, and even groups of international tourists. The government has invested in major infrastructure projects to try to lure people back. Since 2000, the population has been increasing again and now stands at 550,000, close to where it was right before German reunification. But the city still has a long way to go. In one neighborhood you see newly built luxury apartments while others continue to stagnate. It may be some time before Leipzig returns to the prosperity of times past, but it is moving in the right direction."

more likely to be unemployed. And Germany, too, has suffered from the recession of recent years, if only because it has been dragged down by the EU as a whole. The German economy has barely grown since 2010, all the while making major contributions to the bailout of fiscally troubled Greece, Ireland, and Portugal.

Nonetheless, Germany traversed the global recession better than any other European country, and retains its dominance over Europe. However, warning signs in mid-2013 showed that even Germany is not immune to Europe's economic difficulties. Still, over the past several years, Germany has developed close ties with eastern Europe and, from the Balkans to the Baltic Sea, is widely perceived to be the realm's economic leader. Today, German trade with fast-growing Poland exceeds its trade with Russia.

France

Territorially, France is larger than Germany, and the map suggests that France has a superior relative location, with coastlines on the Mediterranean, the Atlantic Ocean, and, at Calais, even a window on the North Sea. But France does not have any good natural harbors, and oceangoing ships cannot navigate its rivers and other waterways very far inland.

The map of the Mainland Core region (Fig. 1B-2) reveals a significant contrast between France and Germany in terms of population distribution. France has one dominant city, Paris, and no other city comes close either in size or centrality: Paris has 11.0 million residents, whereas its closest rival, Lyon, contains only 1.6 million. Germany has no city to match Paris, but it does have a wider range of medium-sized cities: in the 100,000-plus category, Germany has more than 80 while France has less than 40.

Why should Paris, without major raw materials nearby, have grown so large? Whenever geographers investigate the evolution of a city, they focus on two important locational qualities: its **site [2]** (the physical attributes of the place it occupies) and its **situation [3]** (its location relative to surrounding areas). The site of the original settlement at Paris lay on an island in the Seine River, a defensible spot where the river was often crossed. This island, the Île de la Cité, was a Roman outpost 2000 years ago; for centuries its security ensured continuity. Eventually the island became overcrowded, and the city expanded along the banks of the river (Fig. 1B-4A).

Soon the settlement's advantageous situation stimulated its growth and prosperity. Its fertile agricultural

Leipzig is eastern Germany's biggest city after Berlin. It has a rich cultural history and has been home to, among others, Johann Sebastian Bach, Johann Wolfgang von Goethe, Richard Wagner, Friedrich Nietzsche, and Felix Mendelssohn. But during communist times Leipzig was dragged down along with the rest of East Germany. The city is now finally experiencing something of a revival but there are still many struggling neighborhoods.

© Jan Nijman

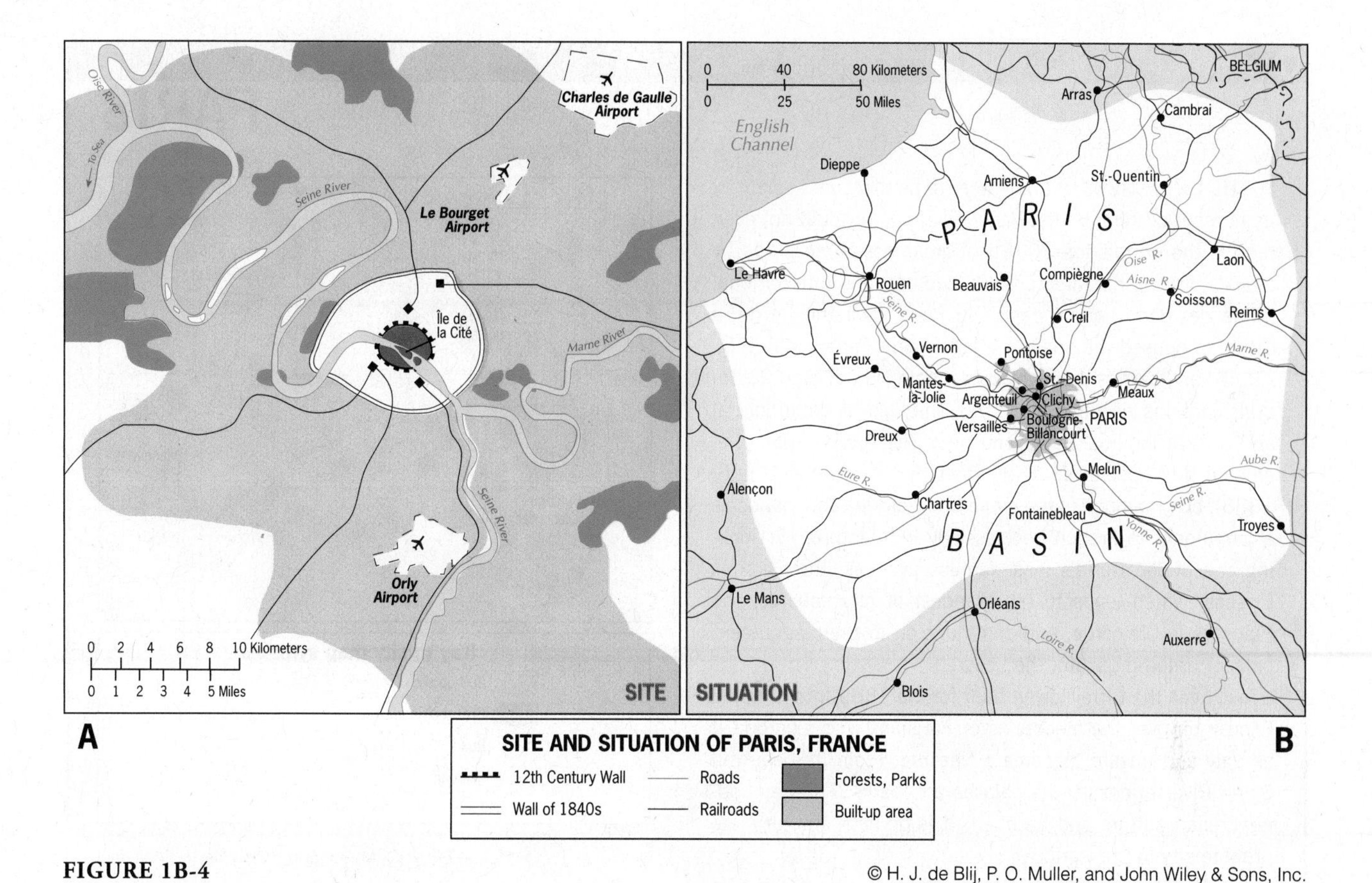

FIGURE 1B-4

© H. J. de Blij, P. O. Muller, and John Wiley & Sons, Inc.

hinterland thrived, and, as an enlarging market, Paris's focality increased steadily. The Seine River is joined near Paris by several navigable tributaries (the Oise, Marne, and Yonne). When canals extended these waterways even farther, Paris was linked to the Loire Valley, the Rhône-Saône Basin, Lorraine (an industrial area in the northeast), and the northern border zone with Belgium. When Napoleon reorganized France and built a radial system of roads—followed later by railroads—that focused on Paris from all parts of the country, the city's *primacy* was assured (Fig. 1B-4B). A **primate city [4]** is one that is disproportionately large compared to all others in the urban system and exceptionally expressive of the nation's culture. Paris is the quintessential example, not only personifying France but also serving as its capital.

France early on developed the strong tradition of a highly centralized state, in which the central government in Paris maintained tight control over 96 *départements*. It remained so for nearly two centuries, but today France is decentralizing. A new layer of governance, consisting of 26 larger *régions*, has been inserted at the level between Paris and the *départements* both to accommodate devolutionary forces as well as seize the opportunities of regional and local growth throughout the country (Fig. 1B-5).

France has one of the world's most productive and diversified economies. Northern French agriculture remains vigorous and varied, exploiting the country's wide range of soils and microclimates and enjoying state subsidies and protections. In the southeast, Lyon, France's second city and headquarters of the *région* named Rhône-Alpes, has become a focus for growth industries and multinational firms. This *région* is evolving into a self-standing economic powerhouse that is becoming a driving force in the European economy. Indeed, it is one of the *Four Motors of Europe* (discussed in Chapter 1A) with its own international business connections to countries as far away as China and Chile. France's economic geography is marked by new high-technology industries, and it is a leading producer of high-speed trains, aircraft, fiber-optic communications systems, and space-related technologies. It also is the world leader in nuclear power, which currently supplies about 75 percent of its electricity and thereby reduces dependence on foreign oil imports; but rising public opposition today makes it unlikely that France will build many more nuclear plants.

France is also one of those European states with a rapidly aging population, posing some major challenges for the years ahead. Natural population growth is negative and must be compensated for through immigration. However, most immigrants hail from poor Islamic countries with very different cultures, and their integration and acceptance in French society has been a painful process. The graying of the French population means that in the future more and more elderly people must be cared for with the incomes earned by a shrinking productive workforce.

The French economy has already displayed evidence that it might encounter the same kinds of problems that

AMONG THE REALM'S GREAT CITIES . . .

PARIS

IF THE GREATNESS of a city were to be measured solely by its number of inhabitants, Paris (11 million) would not even rank in the world's top 20. But if greatness is measured by a city's historic heritage, cultural content, and international influence, Paris has no peer. Old Paris, near the Île de la Cité that housed the original village where Paris began (Fig. 1B-4A) and carries the eight-century-old Notre Dame Cathedral, contains an unparalleled assemblage of architectural and artistic landmarks old and new. The Arc de Triomphe, erected by Napoleon in 1806 (though not completed until 1836), commemorates the emperor's victories and stands as a monument to French neoclassical architecture, overlooking one of the world's most famous avenues, the Champs Elysées, which leads to the grandest of city squares, the Place de la Concorde, and continues on to the magnificent palace-turned-museum, the Louvre.

Even the Eiffel Tower, built for the 1889 International Exposition over the objections of Parisians who regarded it as ugly and unsafe, became a treasure. From its beautiful Seine River bridges to its palaces and parks, Paris embodies French culture and tradition. It is perhaps the ultimate primate city in the world.

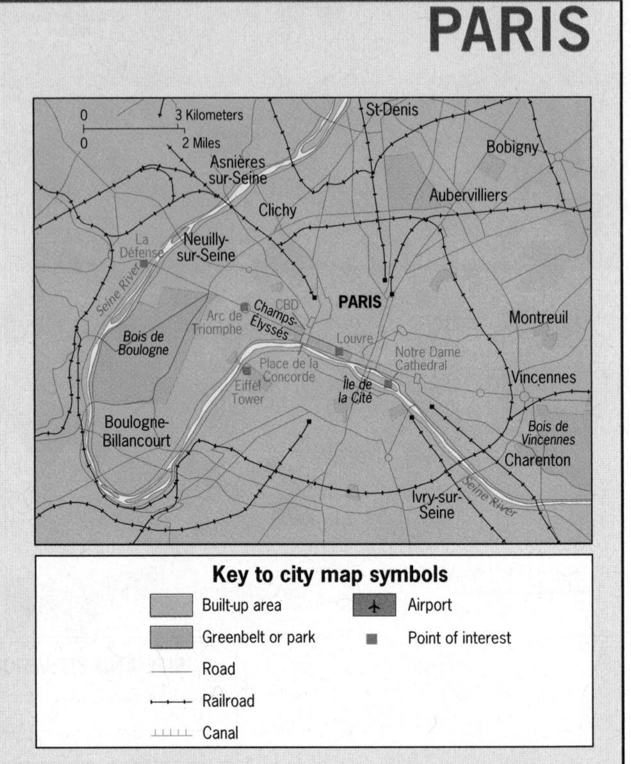

As the capital of a globe-girdling empire, Paris was the hearth from which radiated the cultural forces of Francophone assimilation, transforming much of North, West, and Equatorial Africa, Madagascar, Indochina, and many smaller colonies into societies on the French model. Distant cities such as Dakar, Abidjan, Brazzaville, and Saigon acquired a Parisian atmosphere. France, meanwhile, spent heavily to keep Paris, especially Old Paris, well maintained—not just as a relic of history, but as a functioning, vibrant center, an example to which other cities can aspire.

Today, Old Paris is ringed by a new and different Paris. Stand atop the Arc de Triomphe and turn around 180° from the Champs Elysées, and the tree-lined avenue gives way to the vista of *La Défense*, an ultramodern high-rise complex that is one of Europe's leading business districts (see photo in Chapter 1A). But from the top of the Eiffel Tower you can see as far as 80 kilometers (50 mi) and discern parts of a Paris visitors rarely experience: grimy, aging industrial quarters, and poor, crowded neighborhoods where discontent and unemployment fester—and where immigrants cluster in a world apart from the splendor of the old city.

have bedeviled countries like Greece and Spain—which would raise major concerns because France's economy is much bigger and more important to the EU as a whole. Government debt is high and rising; the public sector is very large (nearly twice the size as Germany's), draining expenditures while youth unemployment hovers around 25 percent. The French electorate did not take kindly to austerity measures, and in 2012 they elected a new president from the Socialist Party. By mid-2013, the French economy showed weaknesses similar to those afflicting Greece, Portugal, and Italy, a troubling prospect indeed.

Benelux

Three small countries are crowded into the northwestern corner of the Mainland Core—Belgium, the Netherlands, and Luxembourg—and are collectively referred to by their first syllables (***Be-Ne-Lux***). Their total population, 28.4 million, reminds us that this is one of the most densely peopled corners of our planet. Not coincidentally, the *Low Countries*, as they are also sometimes called, are situated at the **estuary [5]** of the great Rhine and Scheldt rivers and have access to the seas that has, throughout modern history, been the envy of the Germans and the French. They are a highly productive trio: both the Netherlands and Belgium rank among the top 20 economies of the world, and tiny Luxembourg has the world's highest per capita gross national income. The seafaring Dutch had a thriving agricultural economy and amassed a rich colonial empire; the Belgians forged ahead during the Industrial Revolution; and Luxembourg came into its own after the Second World War as

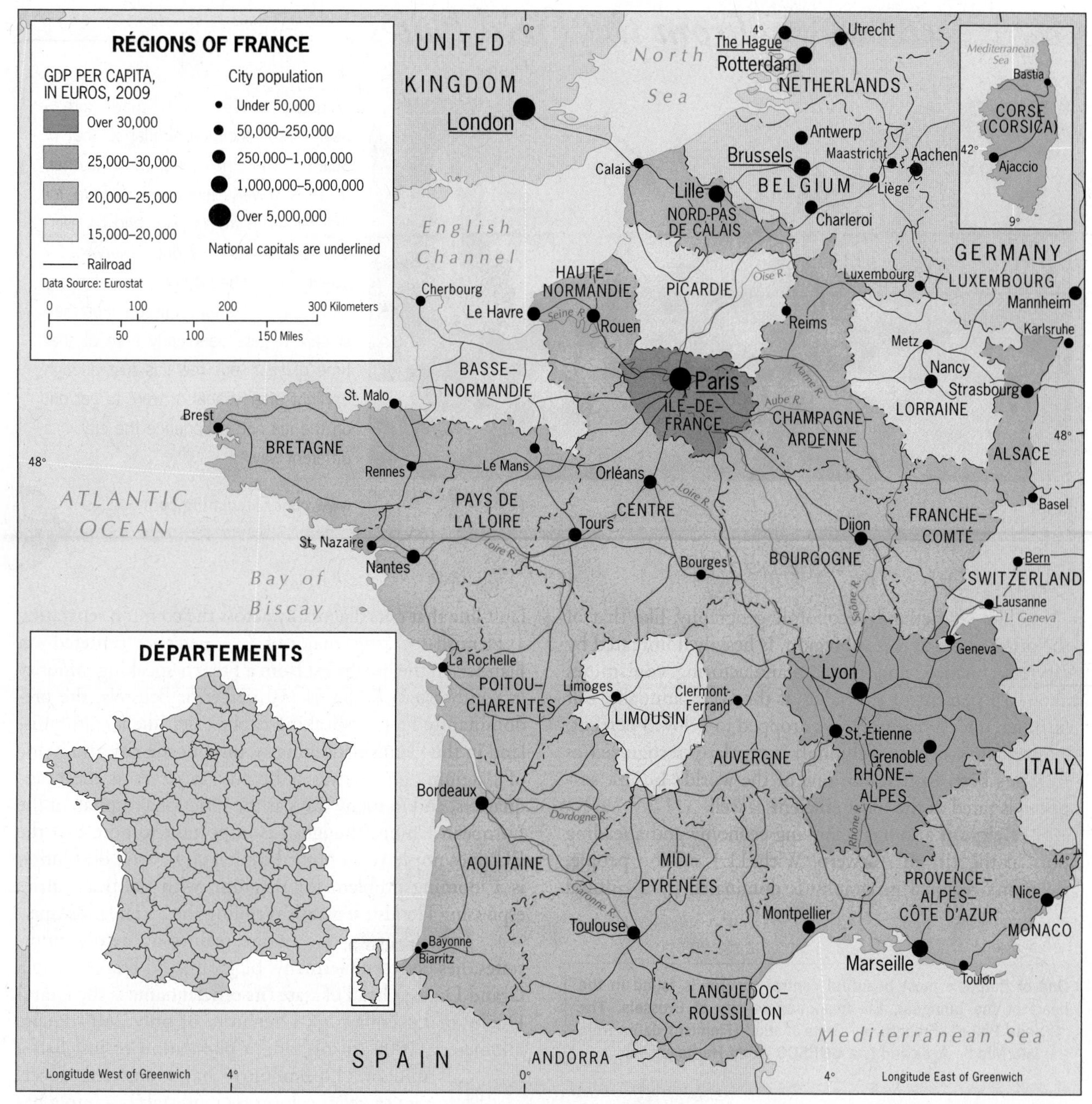

FIGURE 1B-5

a financial center for the evolving European Union and, indeed, the world.

The Netherlands, one of Europe's oldest democracies and a constitutional monarchy today, has for centuries been expanding its territory—not by warring with its neighbors but by wresting land from the sea. Its greatest project so far, the draining and reclaiming of almost the entire Zuider Zee (Southern Sea), started in 1932 and continues. About a quarter of the country lies below sea level, land the Dutch refer to as ***polders***. In the far southwestern province of Zeeland, islands have been connected by dikes and bridges in a project known as the Delta Works, an ingenious system that protects against flooding and storm surges yet leaves the coastal ecosystem intact. Dutch water management became famous for it and the country's engineering firms are employed around the world.

The regional geography of this highly urbanized country (16.8 million) is noted for the ***Randstad***, a roughly triangular urban core area anchored by Amsterdam, the constitutional capital; Rotterdam, Europe's largest port; and The Hague, the seat of government. This **conurbation [6]**, as geographers call large urban areas when two or more cities merge spatially, now forms a ring-shaped complex that surrounds a still-rural center (in Dutch, *rand* means edge or margin; *stad* means city).

From the Field Notes . . .

© Jan Nijman

"Amsterdam has a latitude and a near-coastal location similar to that of Seattle—and it has comparable weather. It takes a very serious cold wave for the canals to freeze over, and on average it happens only once every six or seven years. When it does, as in January 2012, people cannot wait to put on their skates. It gets dark early here at that time of year, but there is just enough daylight after school or work to get out on the ice and experience the city in a different way."

www.conceptcaching.com

Concept Caching

The Netherlands' economic geography, like that of the other states in this subregion, is heavily dominated by services, finance, and trade; manufacturing contributes about 15 percent of the value of the GDP annually, but farming, once a mainstay, has dropped to below 3 percent. Amsterdam's airport, Schiphol, is regularly recognized as Europe's best; Rotterdam, one of the world's busiest seaports, is rated as the most efficient as well.

Belgium also has a thriving economy and a leading port in the city of Antwerp. With 11.1 million people, Belgium's regional geography is dominated by a cultural fault line that cuts diagonally across the country, separating a Flemish-speaking majority (59 percent) centered on Flanders in the northwest from a French-speaking minority in southeastern Wallonia (40 percent). Brussels, the predominantly French-speaking capital, lies like a cultural island in the Flemish-speaking sector; but the city also is one of Belgium's greatest assets because it serves as the headquarters, and in many ways as the functional capital, of the European Union (though it is important to note that the EU does not have an official capital city). Still, devolution is a looming problem for Belgium, with political parties espousing Flemish separatism roiling the social landscape.

Luxembourg, one of the realm's numerous ministates, lies between Germany, Belgium, and France, with a Grand Duke as head of state (its official name is the Grand Duchy of Luxembourg), a territory of only 2600 square kilometers (1000 sq mi), and a population of just half a million. Luxembourg has translated sovereignty, relative location, and stability into a haven for financial, service, and information-technology industries. In the late 2000s, there were more than 160 banks in this tiny country and nearly 14,000 holding companies (corporations that hold controlling stock in other businesses in Europe and worldwide). With its unmatched per capita income (U.S. $61,240), Luxembourg is in some ways the greatest beneficiary of the advent of the European Union. And in no country in Europe is support for the EU stronger than it is here.

One of Europe's most beautiful central squares is found in the heart of the bilingual, EU headquarters city of Brussels. The "Grande Place" (French) or "Grote Markt" (Flemish) dates from the late Middle Ages and is a UNESCO World Heritage site.

© Jan Nijman

The Alpine States

Switzerland, Austria, and the ***microstate*** of Liechtenstein on their border share an absence of coasts and the mountainous terrain of the Alps and little else (Fig. 1B-2). Austria speaks one language; the Swiss speak German in their north, French in the west, Italian in the southeast, and even a bit of Rhaeto-Romansch in the remote central highlands (Fig. 1A-7). Austria has a large primate city; multicultural Switzerland does not. Austria has a substantial range of domestic raw materials; Switzerland does not. Austria is twice the size of Switzerland and has a larger population, but far more trade moves across the Swiss Alps between western and Mediterranean Europe than crosses Austria.

Switzerland, not Austria, is in most ways the leading state in the Alpine subregion of the Mainland Core. Mountainous terrain and **landlocked location [7]** can constitute crucial barriers to economic development but, as is so often the case, geography poses constraints in one way while offering opportunities in another. That is why Switzerland is such an important lesson in the complexities of human geography. Through the skillful maximization of their limited opportunities (including the transfer needs of their neighbors), the Swiss have transformed their seemingly restrictive environment into a prosperous state. They deftly utilized the waters cascading from their mountains to generate hydroelectric power to develop highly specialized industries. Swiss farmers perfected ways to optimize the productivity of mountain pastures and valley soils. Swiss leaders converted their country's isolation into stability, security, and neutrality, making it a world banking behemoth, a global magnet for money. Zürich, in the German-speaking sector, is the financial center; Geneva, in the French-speaking sector, is one of the world's most internationalized cities. The Swiss feel that they do not need to join the EU—and they have not done so to this day.

Austria, which joined the EU in 1995, is a remnant of the old Austro-Hungarian Empire and has a historical geography that is far more reminiscent of unstable eastern Europe than that of Switzerland. In fact, even Austria's physical geography seems to demand that the country look eastward: it is at its widest, lowest, and most productive in the east, where the Danube links it to Hungary, its old ally in the anti-Muslim wars of the past.

Vienna, by far the Alpine subregion's largest city, also lies on the country's eastern perimeter. One of the world's most expressive primate cities with magnificent architecture and monumental art, Vienna today is the Mainland Core's easternmost major city, but Vienna's relative location changed dramatically with EU enlargement over the past decade. Vienna found itself in a far more centralized position when the EU border shifted eastward but many Austrians had their doubts about their neighbors to the east becoming EU members of potentially equal standing. In Austria public support for the European Union has fallen to low levels.

The Czech Republic

The Czech Republic, product of the 1993 Czech-Slovak "velvet divorce," centers on the historic province of Bohemia, the mountain-encircled national core area that focuses on the capital, Prague. This is a classic primate city as well, its cultural landscape faithful to Czech traditions, and it is also an important industrial center. The surrounding mountains contain many valleys with small towns that specialize, Swiss-style, in fabricating high-quality goods. In the old Eastern Europe, even during the communist period, the Czechs always were leaders in technology and engineering; their products could be found on markets in foreign countries near and far.

Bohemia always was cosmopolitan and Western in its exposure, outlook, development, and linkages; Prague lies in the upper basin of the Elbe River, its traditional outlet through northern Germany to the North Sea. Today, the Czech Republic (10.5 million) is reclaiming its position at the center of European action.

THE CORE OFFSHORE: BRITAIN AND IRELAND

As Figure 1B-1 shows, two countries form the maritime component of the European Core Area: the United Kingdom and Ireland. These countries lie on two major islands, surrounded by a constellation of tiny ones. The larger island, a mere 34 kilometers (21 mi) off the mainland at the closest point, is ***Britain***; its smaller neighbor to the west is ***Ireland*** (Fig. 1B-6).

The names attached to these islands and the countries they encompass are the source of some confusion, even irritation. They are sometimes still referred to as the British Isles, a usage that should have terminated when British dominance over most of Ireland ended almost a century ago. The state that occupies Britain and the northeastern corner of Ireland is officially called the United Kingdom—abbreviated UK. But this country often is referred to simply as Britain, and its people are known as the British. The state of Ireland officially is known as the Republic of Ireland (*Eire* in Irish Gaelic), but it does not include the entire island of Ireland.

During the long British occupation of Ireland, which is overwhelmingly Roman Catholic, many Protestants from northern Britain settled in northeastern Ireland. In 1921, when British domination ended, the Irish were set free—except in that corner in the north, where London kept control to protect the area's Protestant settlers. That is why the country to this day is officially known as the United Kingdom of Great Britain and Northern Ireland.

Northern Ireland was home not only to Protestants from Britain, but also to a substantial population of Irish Catholics who found themselves on the wrong side of the border when Ireland was liberated. Ever since, conflict has intermittently engulfed Northern Ireland and spilled over into Britain.

Although all of Britain lies in the United Kingdom, political divisions exist here as well. England is the largest of these units, the center of power from which the rest of the region was originally brought under unified control. The English conquered Wales in the Middle Ages, and

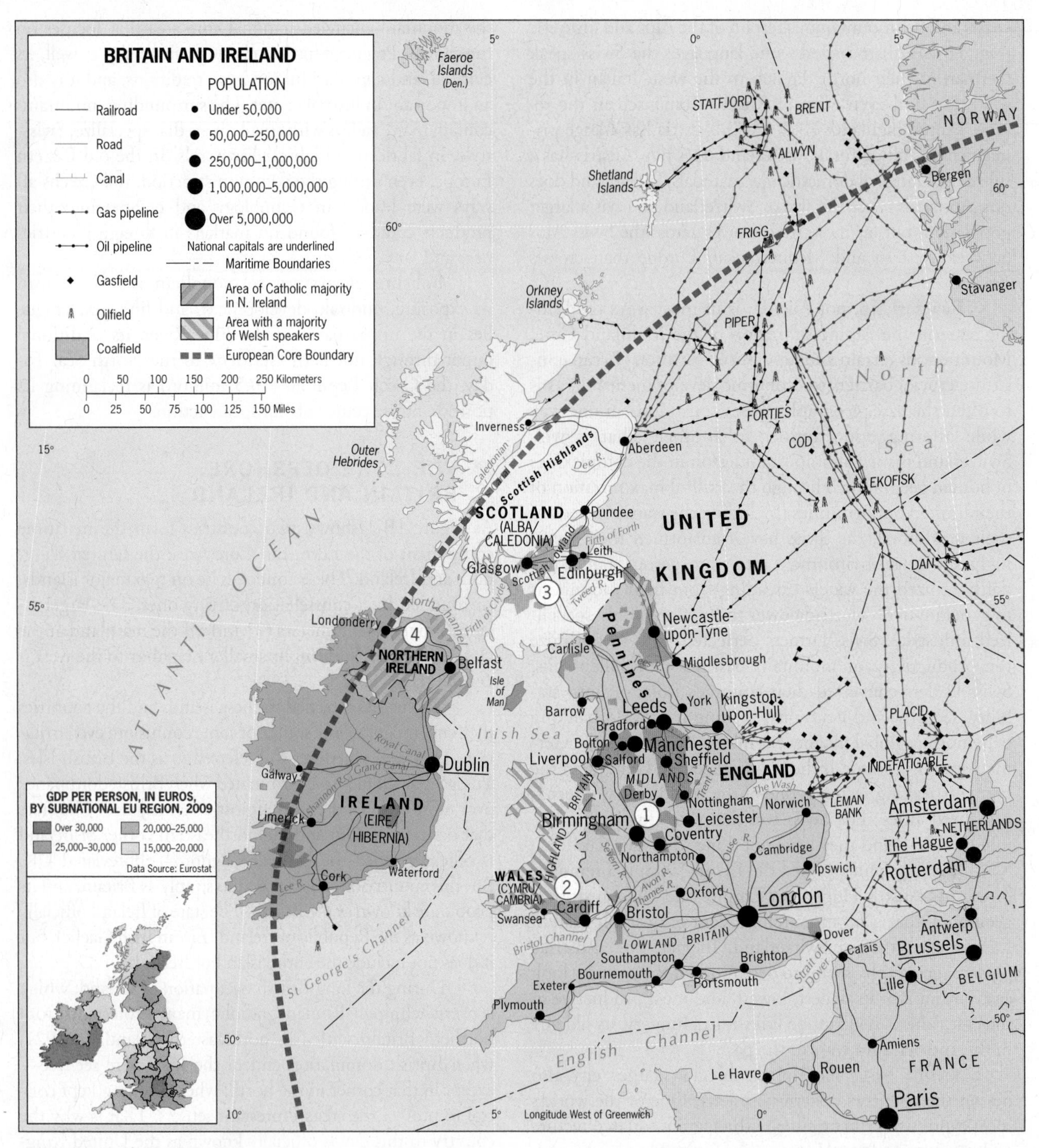

FIGURE 1B-6

Scotland's link to England, cemented when a Scottish king ascended the English throne in 1603, was ratified by the Act of Union of 1707. Thus England, Wales, Scotland, and Northern Ireland became the United Kingdom.

Having united the Welsh, Scots, and Irish, the British set out to forge what would become the world's largest colonial empire. An era of mercantilism (competitive accumulation of wealth among countries) and domestic manufacturing (the latter based on water power from streams flowing off the Pennines, Britain's highland backbone) foreshadowed the momentous Industrial Revolution that transformed Britain—and much of the world.

But British **hegemony [8]** (all-out dominance) came to an end in the early twentieth century, challenged by the rise of other powers such as Russia, Germany, Japan, and the United States. World War II marked the end of the age of colonial empires and the beginning of U.S. global dominance—now with Britain as a "junior partner." For Britain, change after 1945 was profound: not only was it forced to let go of most of its overseas territories (though many

AMONG THE REALM'S GREAT CITIES . . .

LONDON

ALONG THE BANKS of the Thames, this city of more than 9 million inhabitants displays the heritage of state and empire: the Tower and the Tower Bridge, the Victoria Embankment, and the Houses of Parliament (officially known as the Palace of Westminster). But in recent times this historic landscape has become studded with ultramodern office buildings, museums, and entertainment parks such as the EYE, the giant ferris wheel that provides a panoramic view of this giant **metropolis [9]** (central city plus its suburban ring). A bit further downriver, the economic heart of the city (and of the UK) beats 24/7 in what is known as the *City of London*, an incredibly dense concentration of office activity surrounding the Bank of England that houses London's enormous finance sector and related producer services (accounting, advertising, consulting, insurance, etc.). London, together with New York, is one of the most pre-eminent world-cities, in which business decisions are made, on a daily basis, that have consequences all across the globe.

This is also one of the world's most cosmopolitan cities, with countless nationalities and immigrant ethnic communities. But London has also become hugely expensive and the gulf between its rich and poor has steadily widened. Lately, a substantial number of lower- and middle-income households have had no choice but to move to the suburbs or even farther away because life in this city simply became unaffordable.

The urban layout of the London region reveals much about the internal spatial structure of the European metropolis in general. Such a ***metropolitan area*** remains focused on the large city at its center, especially the downtown **central business district (CBD) [10]**, which is the oldest commercial hub of the urban agglomeration. Cities with a preindustrial past are often located on a river (or harbor), expanding from their historical nucleus in a more or less concentric manner—as the case of London illustrates. It should also be noted that an encircling greenbelt was set aside during the 1930s for recreation, horticultural farming, and other nonresidential, noncommercial uses. Although London's growth eroded parts of this greenbelt, in places leaving only "green wedges," the design preserved critically needed open space in and around the city, channeling suburbanization outward into the urban fringe beyond 40 kilometers (25 mi) from the center.

Compared to large North American cities, London is highly compact. This allows for effective public transportation but it also requires careful urban planning. Consider this: Heathrow is the world busiest international airport but it has only two runways. It has long been operating at full capacity, but London's ever-growing air traffic now requires additional landing facilities. There are four other airports in the metropolis—Gatwick, Stansted, Luton, and City—but they are insufficient. So now the government is planning yet another sizeable airport to the east of London near Gravesend on the south bank of the Thames.

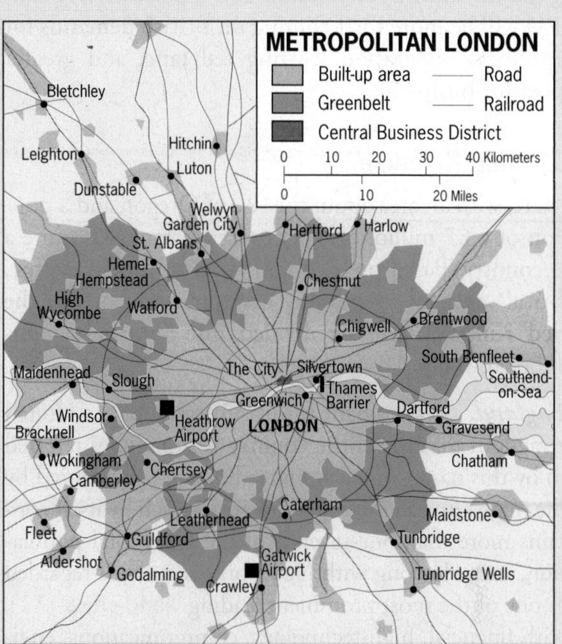

From the national government's point of view, London poses a dilemma of sorts: it is absolutely vital to the British economy but as a highest-echelon global city it is in certain ways better connected to the rest of the world than it is to, say, the outer reaches of the UK. One result is the still-widening gap in prosperity between southeastern England and the rest of the country. Yet London's needs are always top-priority: for example, it receives almost 90 percent of UK government funding for major transportation projects in order to keep this premier world-city moving.

remained part of the largely symbolic British Commonwealth), but it also had to deal with the rapid resurgence of the European mainland. Always ambivalent about the EC and EU, and with its first membership application vetoed by the French in 1963, Britain (admitted in 1973) has worked to restrain moves toward tighter integration. When most member-states in 2002 adopted the new euro in favor of their national currencies, the British kept their pound sterling and delayed their participation in the European Monetary Union (EMU)—a decision that they did not regret when the 'euro crisis' developed in the late 2000s.

Indeed, British criticism of the European Union has only increased in recent years, so much so that the UK's prime minister in 2013 announced his support for a referendum to discontinue EU membership. Importantly, the date for that referendum would not be until 2017, with the opportunity to renegotiate the terms of UK membership and the general direction of the EU. In other words, the prime minister may be gambling that, with the threat of a referendum and a possible UK exit, the EU will be more forthcoming on British demands for less political integration, cutting red tape, and greater support for business.

The United Kingdom

The UK, with an area about the size of Oregon and a population of 63.7 million, is by European standards quite a large country. Based on a combination of physiographic, historical, cultural, economic, and political criteria, the United Kingdom can be divided into four subregions (numbered in Fig. 1B-6):

1. ***England.*** So dominant is this subregion of the United Kingdom that the entire country is sometimes referred to by this name. Small wonder: England is anchored by the huge London metropolitan area, which by itself contains more than one-seventh of the UK's total population. Indeed, along with New York, London is regarded as one of the most prominent leading **world-cities [11]**, with financial, high-technology, communications, engineering, and related industries reflecting the momentum of its long-term growth and agglomeration. Within England, there is a notable difference between the crowded, fast-paced, and globally connected southeast centered on London, and the far less prosperous north and west (see inset map in Fig. 1B-6); the latter were the hearth of the Industrial Revolution, but have now lost their dynamism and portions of their populations in the age of deindustrialization. Inequality in England has also been on the rise between rich and poor within regions and cities. This became painfully clear when the country was rocked by violent riots in London and many other cities during the summer of 2011. These riots were triggered by the police shooting of a biracial man in London, but the deeper causes were generally said to lie within the growing socioeconomic polarization among England's classes.
2. ***Wales.*** This nearly rectangular, rugged territory was a refuge for ancient Celtic peoples, and in its western counties more than half the inhabitants still speak Welsh. Because of the high-quality coal reserves in its southern tier, Wales too was engulfed by the Industrial Revolution, and Cardiff, the capital, was once the world's leading coal exporter. But the fortunes of Wales also declined, and many Welsh emigrated. Among the 3 million who remained, the flame of Welsh nationalism survived, and in 1997 voters approved the establishment of a Welsh Assembly to administer public services in Wales, a first devolutionary step. But in many respects, economically and culturally, Wales is closely wedded to England and independence is not in the cards. In the Welsh Assembly, the nationalist party has only 11 of the 60 seats. The situation in Scotland, however, is quite different.
3. ***Scotland.*** Nearly twice as large as the Netherlands but with a population of just 5.3 million, Scotland is a major component of the United Kingdom. It is also relatively removed from Britain's thriving core in the southeast. Most of the population resides in the Scottish Lowlands, anchored by Edinburgh, the capital, in the east and Glasgow in the west. Attracted there by the labor demands of the Industrial Revolution (coal and iron reserves lay in the area), the Scots developed a world-class shipbuilding industry. Subsequent decline and obsolescence were followed by high-tech development, notably in the hinterland of Glasgow, and Scottish participation in the exploitation of oil and gas reserves beneath the North Sea, which transformed the eastern ports of Aberdeen and Leith (Edinburgh). But many Scots feel that they are disadvantaged within the UK and should play a major role in the EU. Therefore, when the British government placed the option of a Scottish Parliament before the voters in 1997, 74 percent approved. Unlike the situation in Wales, the Scottish nationalist party has a comfortable majority in their parliament. And it may not end there: a scheduled referendum in 2014 for full independence might deliver just that, although much will depend on the exact terms of "independence." The breakup of the UK seems too much for many people to even contemplate, but it could happen.
4. ***Northern Ireland.*** A declining majority of the overall population of 1.8 million, now just under 42 percent, are Protestants and trace their ancestry to Scotland or England; a growing minority, currently very close behind at 41 percent, are Roman Catholics and share their religion with the Irish Republic on the other side of the border. Although Figure 1B-6 suggests that there are majority areas of Protestants and Catholics in Northern Ireland, no such clear separation exists; mostly they live in clusters throughout the territory, including the walled-off neighborhoods that mark the major cities of Belfast and Londonderry (see photo). Partition is no solution to a conflict that has raged for more than four decades at a cost of thousands of lives; Catholics accuse London as well as the local Protestant-dominated administration of discrimination, whereas Protestants accuse Catholics of seeking union with the Republic of Ireland. The Northern Ireland Assembly, to which powers are supposed to be devolved from London, had to be suspended from 2002 to 2007 before the parties could be brought together at the table again. The prospects for greater autonomy are clouded.

Republic of Ireland

The Republic of Ireland fought itself free from British colonial rule just three generations ago. Its cool, moist climate had earlier led to the adoption of the American potato as the staple crop, but excessive rain and a blight in the late 1840s, coupled with colonial mismanagement, caused famine and

"The Troubles" between Protestants and Catholics, pro- and anti-British factions that have torn Northern Ireland apart for decades at a cost of more than 3000 lives, are etched in the cultural landscape. The so-called Peace Wall across West Belfast, shown here separating Catholic and Protestant neighborhoods, is a tragic monument to the failure of accommodation and compromise, a physical manifestation of the emotional divide that still runs deep. A steady stream of sectarian-based attacks continue across Belfast (2012 was a particularly violent year). Political negotiations and agreements have in a way become even more difficult since the original Irish Republican Army (IRA) splintered during the 1990s into numerous factions with different perspectives and strategies. What an analyst said to one of the authors back in 2005 still is the view across the cityscape you see here: "Good fences make good neighbors . . . the Peace Wall will have to stay up for a few more decades yet."

cost over a million lives. Another 2 million Irish left for North America and other shores.

Hard-won independence in 1921 did not bring real economic prosperity until the final decades of the twentieth century, when Ireland became known as the ***Celtic Tiger*** (likening it to the miraculous, Hong Kong-style economic development of the so-called *Asian Tigers*). Participation in the EU, adoption of the euro, business-friendly tax policies, comparatively low wages, an English-speaking workforce, and an advantageous relative location for some time combined to produce the highest rate of economic growth in the entire European Union. In the 1990s, this booming, service-based economy, accompanied by burgeoning cities and towns, skyrocketing real estate prices, mushrooming industrial parks, and bustling traffic, transformed a country long known for emigration into a magnet for industrial workers, producing new social challenges for a closely knit, long-isolated society. Among these immigrants were thousands of people of Irish descent returning from foreign places to take jobs at home, workers from elsewhere in the European Union (including large numbers of Poles), and job-seekers from African and Caribbean countries. For some time, Ireland had the highest economic growth rate in the EU, and its capital, Dublin, became an international business node almost overnight.

By 2007, however, Ireland's economic boom began to fade, and a year later things proceeded to fall apart: the real estate market stagnated, service industries found more favorable conditions in eastern Europe, and unemployment accelerated. Soon it became painfully clear that the country's financial troubles were far greater than initially thought: several banks failed after overextending credit (part of the global financial crisis that took hold in the United States in 2008), and the government itself had amassed such huge deficits that an EU bailout was required in 2011. In the Irish case, there is hope that the bailout was a one-time event, and by 2013 the economy showed signs of recovery. But keep in mind that Ireland relies heavily on foreign investment, so its fiscal health depends heavily on the fortunes of the EU and U.S. economies.

THE DISCONTINUOUS SOUTH

As Figures 1B-1 and 1B-7 reveal, the northern sectors of two major southern states (Italy and Spain) form parts of the European Core. Portugal, on the western flank of the Iberian Peninsula, remains outside the European Core Area, far less urbanized, much more agrarian, and not strongly integrated into it. The fourth national entity in this southern domain is the island mini-state of Malta, south of Sicily, an historically important crossroads with a population of approximately 400,000 and a booming tourist industry.

Italy

Centrally located within Mediterranean Europe, most populous of the realm's southern states, best connected to the European Core, and economically most advanced is Italy (population: 60.8 million), a charter member of Europe's Common Market.

Administratively, Italy is organized into 20 internal regions, many with historic roots dating back centuries (Fig. 1B-8). Several of these regions have become powerful economic entities centered on major cities, such as Lombardy (Milan) and Piedmont (Turin); others are historic hearths of Italian culture, most prominently Tuscany (Florence) and Veneto (Venice). These regions in the

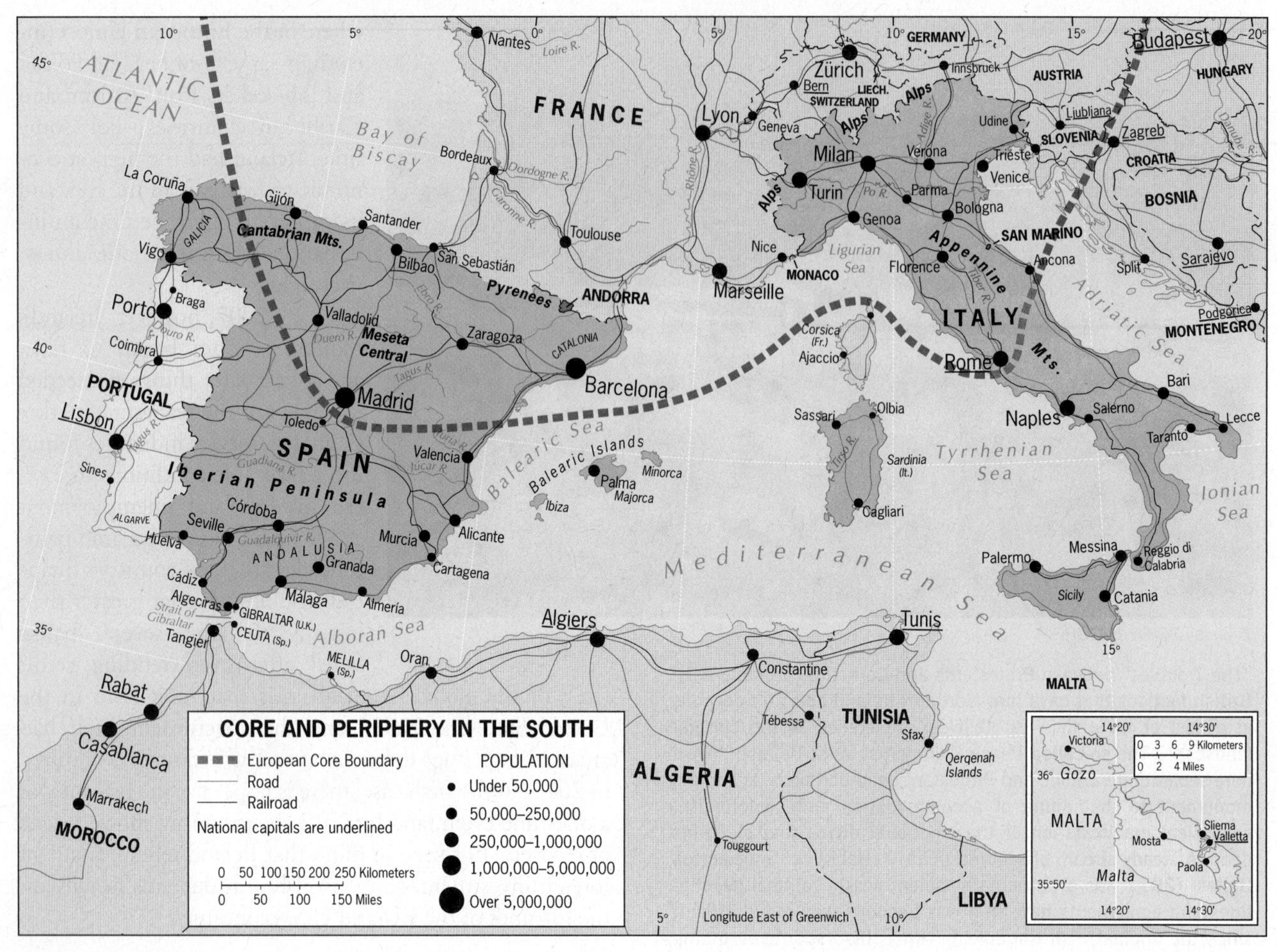

FIGURE 1B-7

northern half of Italy stand in strong social, economic, and political contrast to such southern regions as Calabria (the "toe" of the Italian "boot") and Italy's two major Mediterranean islands, Sicily and Sardinia. Not surprisingly, Italy is often described as two countries—a progressive north and a stagnant south (known as the ***Mezzogiorno***). The urbanized, industrialized north is part of Europe's Core; the low-income south typifies the Periphery.

North and south are bound by the ancient headquarters, Rome, which lies astride the narrow transition zone between Italy's contrasting halves. This clear manifestation of Europe's Core-Periphery contrast is referred to in Italy as the ***Ancona Line***, named after the city on the Adriatic coast where it reaches the other side of the peninsula (Fig. 1B-8, blue line). Whereas Rome remains Italy's capital and cultural focus, the functional core area of the country has shifted northward into Lombardy in the basin of the Po River. Here lies southern Europe's leading manufacturing complex. The Milan–Turin–Genoa triangle exports appliances, instruments, automobiles, ships, and many specialized products. Meanwhile, the Po Basin, lying on the margins of southern Europe's dominant Mediterranean climatic regime (with its hot, dry summers), enjoys a more even pattern of rainfall distribution throughout the year, making it a productive agricultural zone as well.

Metropolitan Milan embodies the new, modern Italy. Not only is Milan (5.2 million) Italy's largest city and leading manufacturing center—making Lombardy one of Europe's Four Motors—but it also is the country's financial and service-industry headquarters. Today the Milan area, a cornerstone of the European Core, contains just 7 percent of Italy's population but accounts for more than one-third of the entire country's national income. Overall, however, Italy's economic prospects are now so troubled that a crisis may be unavoidable.

Spain, Portugal, and Malta

At the western end of southern Europe lies the broad Iberian Peninsula, separated from France and western Europe by the rugged mountain wall of the Pyrenees and from North Africa by the narrow Strait of Gibraltar. Spain (population: 46.4 million) occupies most of this compact Mediterranean landmass, and peripheral Portugal lies in its southwestern corner. Both countries benefited enormously from their admission to the European Union in 1986, an advantage that soon withered.

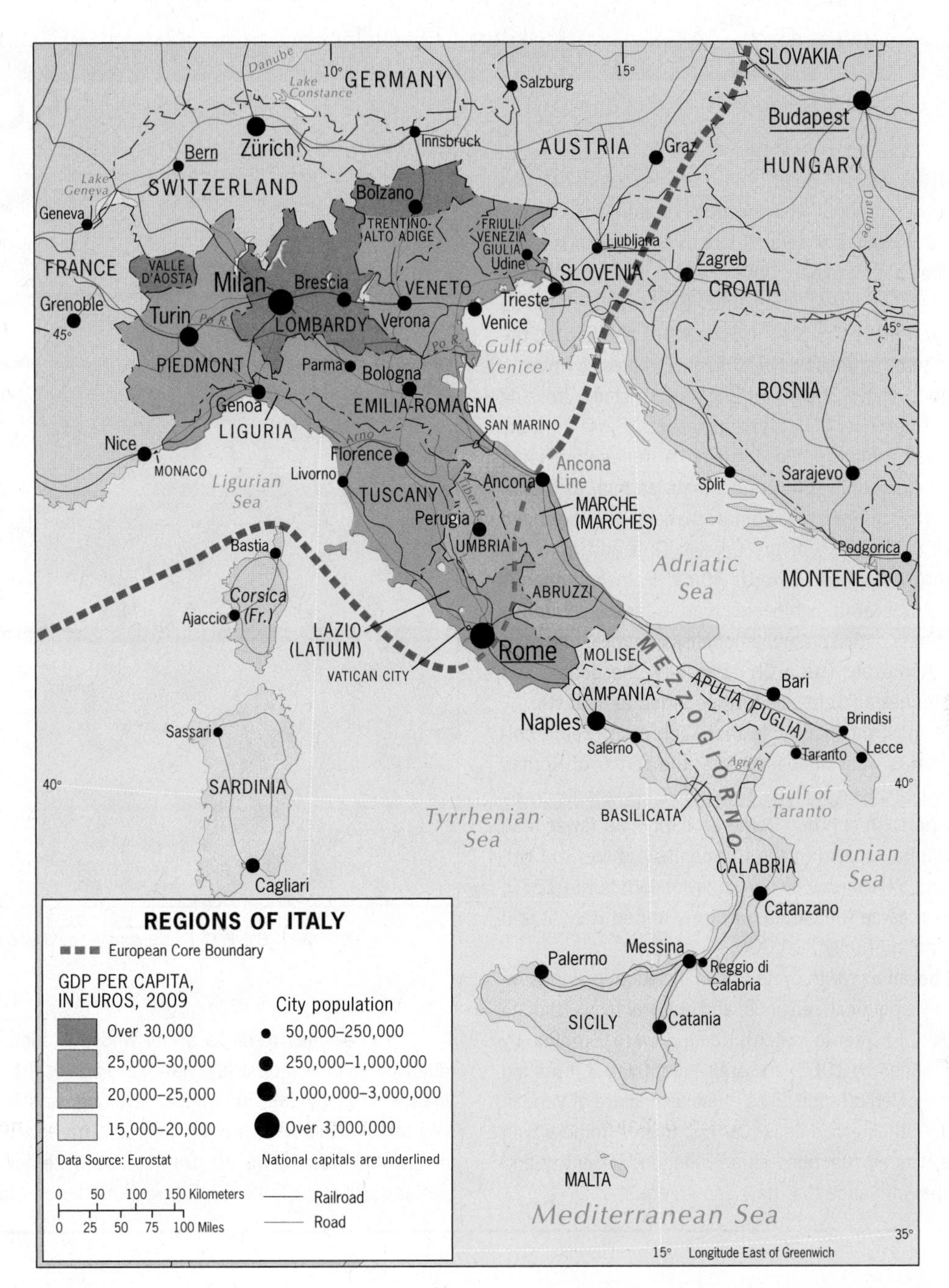

FIGURE 1B-8

Spain followed the leads of Germany and France and decentralized its administrative structure in response to devolutionary pressures. These pressures were especially strong in Catalonia, the Basque Country, and Galicia, and in response the Madrid government created so-called *Autonomous Communities (ACs)* for all 17 of its regions (Fig. 1B-9). Every AC has its own parliament and administration that control planning, public works, cultural affairs, education, environmental policy, and even, to some extent, international commerce. Each AC can negotiate its own degree of autonomy with the central government in the capital. Some Spanish observers feel that devolution has gone too far and that a federal system (such as Germany's) would have been preferable, but now there is no turning back.

Even so, the AC framework has not defused the secessionist-minded movements. Relations between Madrid and the Basque Country remain tense and in Catalonia the urge for independence is strengthening as the Spanish economy struggles while Catalonia continues to do quite well. Centered on the prosperous, productive coastal city of Barcelona, the AC named Catalonia is Spain's leading industrial area and has become one of Europe's Four Motors. Catalonia is endowed with its own distinctive language and culture that find vivid expression in Barcelona's urban landscape. In recent years, Catalonia—with 6 percent of Spain's territory and 17 percent of its population—has annually produced 25 percent of all Spanish exports and nearly 40 percent of its industrial exports.

AMONG THE REALM'S GREAT CITIES . . .

ROME

FROM A HIGH vantage point, Rome seems to consist of an endless sea of tiled roofs, above which rise numerous white, ochre, and gray domes of various sizes; in the distance, the urban perimeter is marked by high-rises fading into the urban haze. This historic city lives amid its past as perhaps no other as busy traffic encircles the Colosseum, the Forum, the Pantheon, and other legacies of Europe's greatest empire.

Founded about 3000 years ago at an island crossing point on the Tiber River about 25 kilometers (15 mi) from the sea, Rome had a high, defensible site. A millennium later, with a population some scholars estimate as high as 1 million, it was the capital of a Roman domain that extended from Britain to the head of the Persian Gulf and from the shores of the Black Sea to North Africa. Rome's emperors endowed the city with magnificent, marble-faced, columned public buildings, baths, stadiums, obelisks, arches, and statuary; when Rome became a Christian city, the domes and spires of churches and chapels further added to its luster.

It is almost inconceivable that such a city could collapse, but that is what happened after the center of Roman power shifted eastward to Constantinople (now Istanbul). By the end of the sixth century, Rome probably had fewer than 50,000 inhabitants, and in the thirteenth, a mere 30,000. Papal rule and a Renaissance revival lay ahead, but in 1870, when Rome became the capital of newly united Italy, it still had a population of only 200,000.

Now began a growth cycle that eclipsed all previous records. As Italy's political, religious, and cultural focus (though not an industrial center to match), Rome grew to 1 million by 1930, to 2 million by 1960, and subsequently to 3.3 million where it has leveled off today. The religious enclave of Vatican City, Roman Catholicism's headquarters, makes Rome a twin capital; the Vatican functions as an independent entity and exerts a global influence that Italy cannot equal.

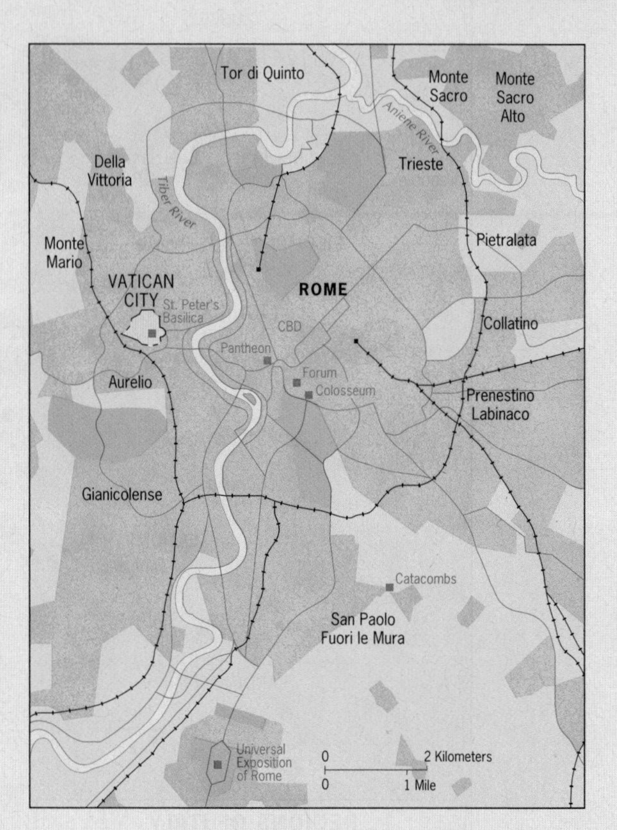

Rome today remains a city whose economy is dominated by service industries; national and local government, finance and banking, insurance, retailing, and tourism employ three-quarters of the labor force. The new city sprawls far beyond the old, walled, traffic-choked center where the Roman past and the Italian future come face to face.

Relations between Catalonia and Castile (where Madrid is located) are also a product of history. During the Spanish Civil War (1936–1939), the military dictator Francisco Franco brought all of Spain under his command by violently crushing the resistance in several corners of the country, especially in independent-minded Catalonia. From 1939 to 1975, Franco's dictatorial rule was bitterly resented in Barcelona. Even though things have greatly improved since Spain converted to democracy after his death in 1975, there has always remained a sense of animosity that is nowadays perhaps best expressed (and more happily so) in "*el clasico*"—the highly anticipated annual soccer match between the country's leading teams, Barcelona and Real Madrid. But secession remains a serious issue: on September 11, 2012, the important Catalan holiday of National Day, well over a million protesters (more than one-seventh of the AC's population) marched in the streets of Barcelona to demand full independence.

As Figure 1B-9 shows, Spain's capital and largest city, Madrid, lies near the geographic center of the state. It also lies within an economic-geographic transition zone. In terms of personal annual income, the Spanish north is far more affluent than the south, a direct result of the country's distribution of resources, climate (the south suffers from drought and inferior soils), and overall development opportunities. The most prosperous ACs, apart from Madrid, are located to the north along the Pyrenees, and they include Catalonia and Basque Country.

Economically, Spain is going through a rough patch. Economic growth has declined since 2008, and fears persist that the country's banking and financial sectors may be following those of Greece, Ireland, and Portugal. Unemployment in

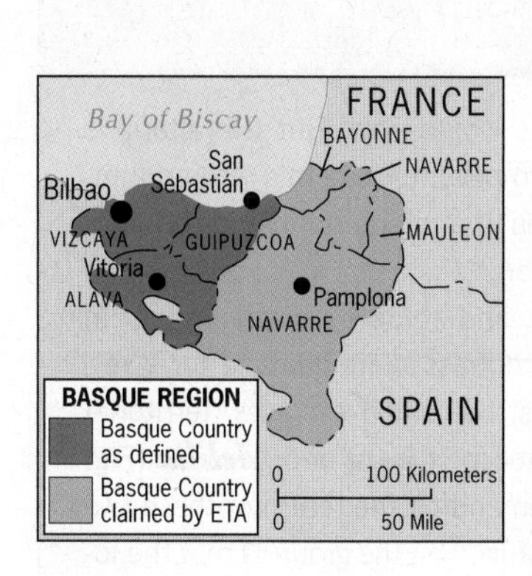

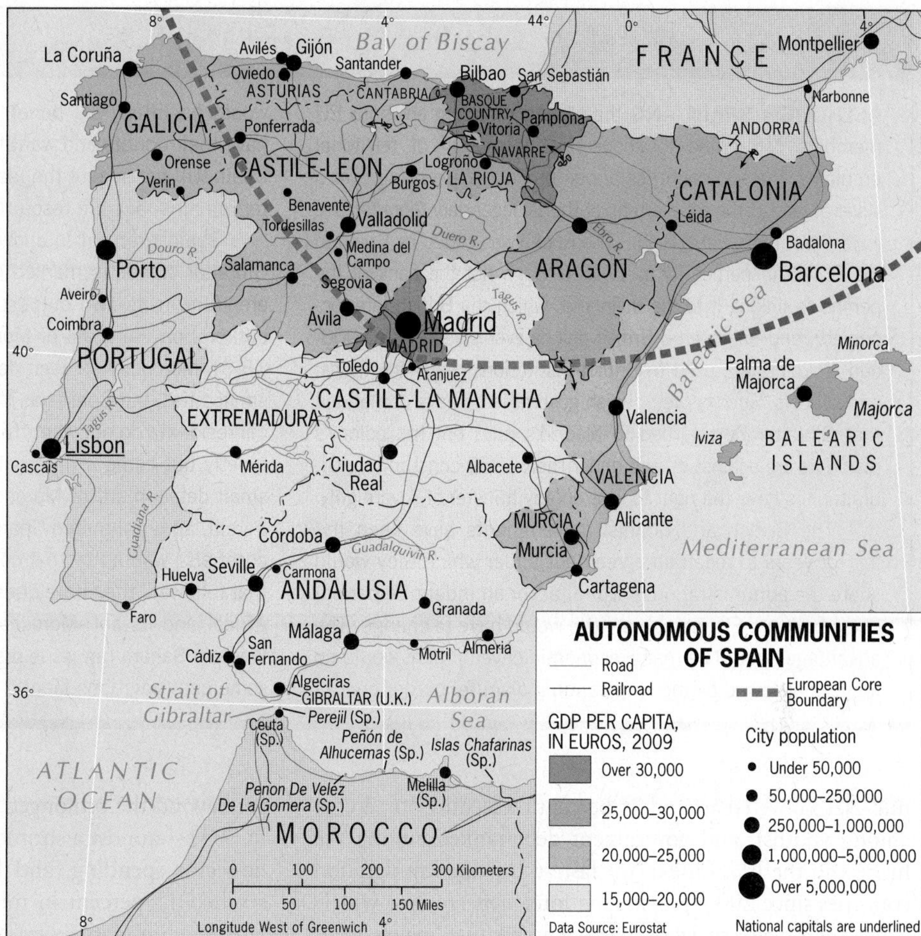

FIGURE 1B-9

From the Field Notes . . .

"Catalonian nationalism is visible both obviously and subtly in Barcelona's urban landscape. Walking toward the Catalonian Parliament, I noticed that the flags of Spain (left) and Catalonia (right) flew from slightly diverging flagpoles above the entrance to the historic building. Is there a message here?"

www.conceptcaching.com

Gibraltar, Ceuta, and Melilla

ALTHOUGH SPAIN AND the United Kingdom are both EU members committed to cooperative resolution of territorial problems, the two countries are embroiled in a dispute over a sliver of land at the southern tip of Iberia: legendary ***Gibraltar*** (see lower inset map, Fig. 1B-9). "The Rock" was ceded by Spain to the British in perpetuity in 1713 (though not the neck of the peninsula linking it to the mainland, which the British later occupied), and has been a British colony ever since. Its 30,000-odd residents are used to British institutions, legal traditions, and schools. Successive Spanish governments have demanded that Gibraltar be returned to Madrid's rule, but the colony's residents are against it. And under their 1969 constitution, Gibraltarians have the right to vote on any transfer of sovereignty.

The British and Spanish governments have been trying for years to reach an agreement under which they would share the administration of Gibraltar for an indefinite period; since both are EU members, little would have to change. The advantages would be many: economic development would no longer be slowed by the longstanding dispute, border checks would be lifted, EU benefits would flow. But Gibraltarians have their doubts and want to put the issue to a referendum. Spain will not accept the idea of a referendum, and the matter remains far from resolution.

Spain's refusal to allow, and abide by, a referendum in Gibraltar stands in marked contrast to its demand for a referendum in its own outposts, two small *exclaves* (territorial outliers) on the coast of Morocco, ***Ceuta*** and ***Melilla*** (Fig. 1B-9). Morocco has been demanding the return of these two small cities, but Spain has refused on the grounds that the local residents do not want this. The matter has long simmered quietly but came to international attention in 2002 when a small detachment of Moroccan soldiers seized the island of Perejil, an uninhabited Spanish possession off the Moroccan coast also wanted by that country. In the diplomatic standoff that followed, the entire question of Spain's holdings in North Africa (and its anti-Moroccan stance in the larger matter of Western Sahara [an issue discussed in Chapter 7B]) exposed some contradictions Madrid had preferred to conceal.

mid-2013 hovered around 25 percent (exceeding 50 percent among youths), and government debt ranked among the highest in the EU. One of the fastest-growing immigration countries since 2000, job-seeking foreigners are now widely viewed as a major social problem. And because Spain's economy accounts for a much larger proportion of the EU than Greece's or Ireland's, a Spanish default could be far more consequential. Whether Spain's financial sector will hold in order to enable the country to turn the economic corner once the extended recession gives way to renewed growth, remains to be seen.

Much the same can be said of **Portugal** (10.6 million), which occupies the southwestern corner of the Iberian Peninsula. Because EU agricultural policies entail transfers from rich to poor areas in the Union and because the EU funds major infrastructural projects in hitherto isolated areas, Portugal at first benefited substantially from its 1986 admission to the EU—but today it remains far behind Europe's leading economies. Unlike Spain, which has major population clusters on its interior plateau as well as its coastal lowlands, the Portuguese are concentrated along and near the Atlantic coast. Lisbon and the second city, Porto, are coastal cities. The country has some excellent natural harbors, including Lisbon and Sines (Fig. 1B-7), with considerable potential to attract a larger share of container shipping (especially for transferring onto freight trains bound for Spain), but these ports need to become more efficient and competitive. The best farmlands lie in the moister western and northern zones of the country; but farms here are small and inefficient, and even though Portugal remains dominantly rural it must import as much as half of its food. Most recently, Portugal's shaky finances and its struggle to get beyond the prolonged recession required an EU bailout in 2011—amidst a shrinking economy, major cuts in public-sector spending, and high unemployment levels (that reached 15 percent in mid-2013). The economic outlook remains dim, and expectations are that the Portuguese government will require a second EU bailout to make ends meet.

Southernmost Europe also contains the mini-state of **Malta**, located in the central Mediterranean Sea just south of Sicily. Malta is a small archipelago (island chain) of three inhabited and two uninhabited islands with a population of just over 400,000 (Fig. 1B-7, inset map). An ancient crossroads and culturally rich with Phoenician, Arab, Italian, and British infusions, Malta became a British dependency and has long served the UK's shipping and its military. It suffered terribly during World War II bombings, but despite limited natural resources Malta recovered strongly during the postwar era. Today it has a booming tourist industry and a relatively high standard of living, and was one of the ten new member-states to join the European Union as part of the historic expansion of 2004.

Greece and Cyprus

If the label "discontinuous geography" applies anywhere, it is the area between Austria and the Aegean Sea, covering the western Balkans (former Yugoslavia) and Greece. This is an area encompassing various (and conflicting) religions and linguistic groups; prosperous as well as very poor nations; lands with volatile and violent histories; EU members and nonmembers (some without any prospect of gaining entry anytime soon); and, furthest removed from the Core is Greece, the birthplace of European civilization and a longtime EU member and

Western ally. We turn first to Greece and the Greek-connected island-state of Cyprus, and then to the Balkans.

Seemingly dangling from the southern end of eastern Europe, **Greece** was the wellspring of one of the ancient world's greatest civilizations, its scientists and philosophers still cited to this day, its famous tragedies still staged in the amphitheaters built more than 2000 years ago. Its familiar peninsulas and islands are bounded by Turkey to the east and by Bulgaria, Macedonia, and Albania to the north (Fig. 1B-10). Note that some of Greece's islands in the Aegean Sea lie on the very doorstep of Turkey; in addition, Greeks represent the majority of the population of Cyprus in the distant northeast corner of the Mediterranean Sea.

A member of the EU since 1981, Greece has endured a turbulent modern history, with alternating leftist and fascist

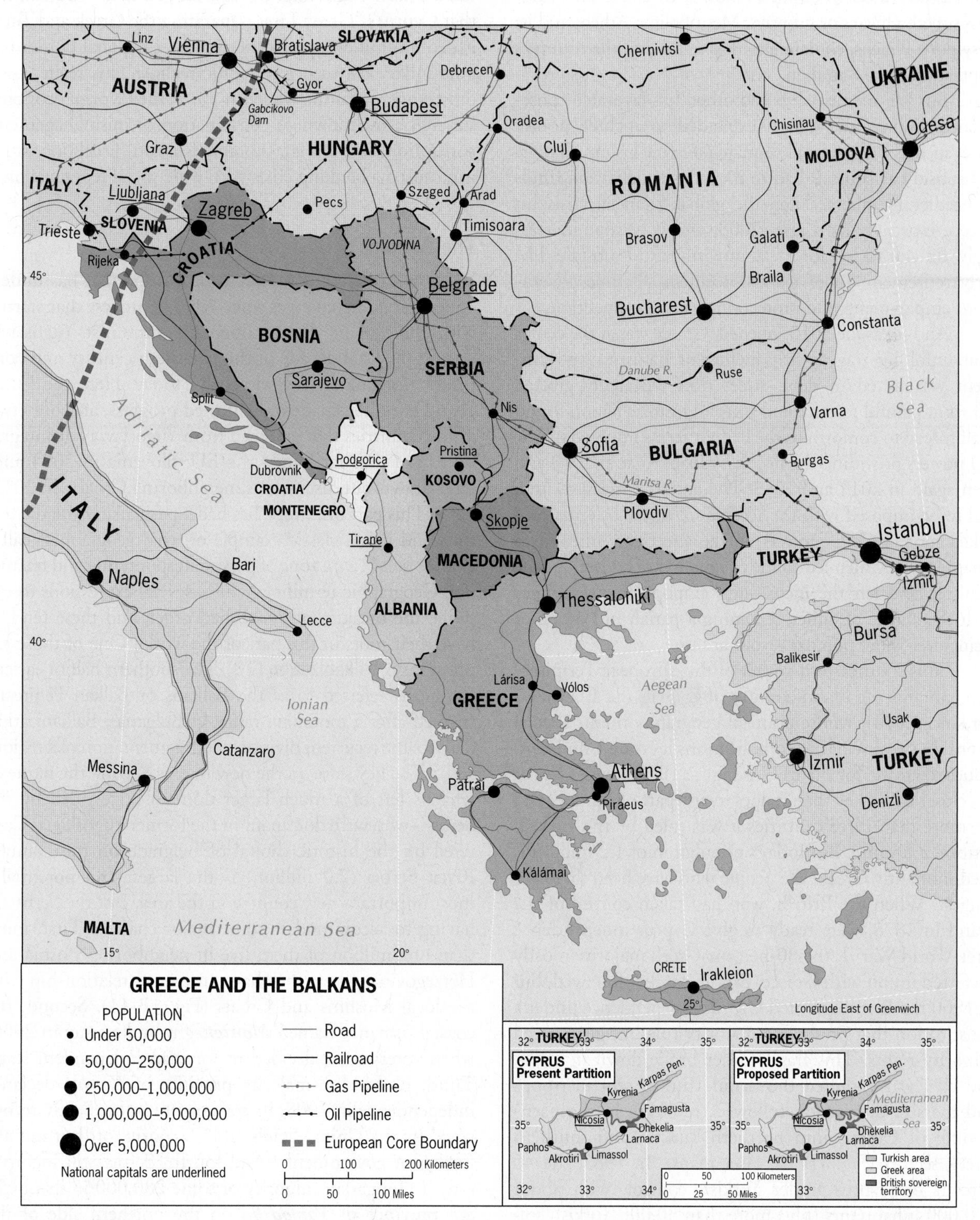

FIGURE 1B-10

dictatorships giving way to more democratic government in the 1980s and economic upheavals marking the EU period. Economic stagnation in the 1980s was followed by such progress that Greece came to be called the "locomotive of the Balkans," an EU beacon in its remote corner of the Union. Metropolitan Athens burgeoned, benefiting from EU assistance, the hosting of the 2004 Olympics, and membership in the EMU. Arriving by air, you find yourself at one of the world's most modern airports followed by a ride on world-class superhighways or subways. Metropolitan Athens and its busy port, Piraeus, contain almost one-third of the country's population of 10.8 million.

But EU membership had other, less favorable consequences. When the Union expanded to include poorer states such as Bulgaria and Romania, Greece lost much of its EU subsidy as Brussels had to divert its equalization funds to needier members. Then the global financial crisis hit Greece especially hard after 2008, and it became increasingly difficult to pay for social provisions and subsidies like early retirement, loose tax enforcement, lifetime government employment, and protection against competition.

And so, as the 2010s opened, Greece was at the center of an unfolding tragedy as its inefficient, hamstrung government was mired in debt, street riots repeatedly greeted efforts at painful reform, and a proud historic nation found itself forced to conform to the bailout terms imposed by the EU powers providing the necessary loans, first in 2010 and then again in 2012 and 2013. The austerity measures that had to be imposed with the acceptance of the bailouts provoked more violent protests on the streets of Athens (see Chapter 1A), which led to the resignation of a Greek prime minister, and left the increasingly unpopular government with the choice of either becoming a pariah in Europe or facing ever wider popular revolts at home.

This southeastern corner of the European Periphery also contains the island country of **Cyprus** (an EU member since 2004), whose political geography merits special attention because of the complications it created, and continues to create, for the EU. As Figure 1B-1 shows, Cyprus lies closer to Turkey than it does to any part of Europe, and for more than three centuries it was ruled by the Turkish Ottoman Empire. But today's population of 1.2 million is predominantly Greek, the people who have been there the longest. When the British, who had taken control of the island in 1878, were ready to give Cyprus independence after World War II, the 80-percent-Greek majority mostly preferred union with Greece. Ethnic conflict followed, but in 1960 the British granted Cyprus independence under a constitution that prescribed majority rule but guaranteed minority rights. This fragile order broke down in 1974, and civil war engulfed the island. Turkey sent in troops and massive dislocation followed, resulting in the partitioning of Cyprus into northern Turkish and southern Greek sectors (Fig. 1B-10, inset maps). In 1983, the 40 percent of Cyprus under Turkish control, with about 100,000 inhabitants (plus more than 30,000 Turkish soldiers), declared itself the independent ***Turkish Republic of Northern Cyprus***. Only Turkey recognizes this mini-state (which today contains a population of about 300,000); meanwhile, the international community recognizes the government on the Greek side as legitimate.

Things got even more complicated when only the Greek side of the island was admitted to the EU in 2004. Resentment was high on both the Turkish side and in Turkey itself—just as discussions on Turkey's own admission to the EU were beginning. It was—and remains—a reminder that Cyprus's "Green Line" separating the Greek and Turkish communities constitutes not just a regional border but a boundary between geographic realms. This reality took on stronger meaning when, in 2013, the Cypriot economy suffered a meltdown. A combination of mismanagement, corruption, and foreign (notably Russian) fiscal opportunism caused a banking crisis with wide social repercussions.

The Balkans

We now turn to a component of Europe that has undergone wrenching changes since 1990, a process that started with the violent dismantling of communist Yugoslavia during the final decade of the twentieth century and continues (fortunately less violently) today. Here the European Union has made only limited progress, and only two of the countries that emerged from Yugoslavia's disintegration have thus far joined the EU: Slovenia in 2004 and, after drawn-out discussions, neighboring Croatia in 2013.

This part of Europe has had a particularly volatile history, and it is a classic example of what geographers call a **shatter belt [12]**, a zone of persistent splintering and fracturing. Geographic terminology uses several expressions to describe the breakup of established order, and these tend to have their roots in this part of the world. One of those expressions is **balkanization [13]**. The southern half of eastern Europe is referred to as the Balkans or Balkan Peninsula (named after a mountain range in Bulgaria). Balkanization denotes the recurrent division and fragmentation of a region.

The key state on the new map is **Serbia**, the name of what is left of a much larger domain once ruled by the Serbs—who were dominant in the former Yugoslavia. Centered on the historic capital of Belgrade on the Danube River, Serbia (7.0 million) is the largest and potentially most important new country in the area. But the Serbs are having to accommodate some major changes. First, more than one million of them live in neighboring Bosnia and Herzegovina, where they have an uneasy relationship with the local Muslims and Croats (Fig. 1B-11). Second, the coastal province named ***Montenegro*** broke away in 2006, when voters there decided to form an independent state. Third, its Muslim-majority province of ***Kosovo*** declared independence in 2008, its sovereignty immediately recognized by the United States and a majority of (but not all) European governments. And fourth, Serbia still incorporates a Hungarian minority of some 280,000 in its northern province of ***Vojvodina*** on the northern side of the Danube at a time when Hungary has already joined the

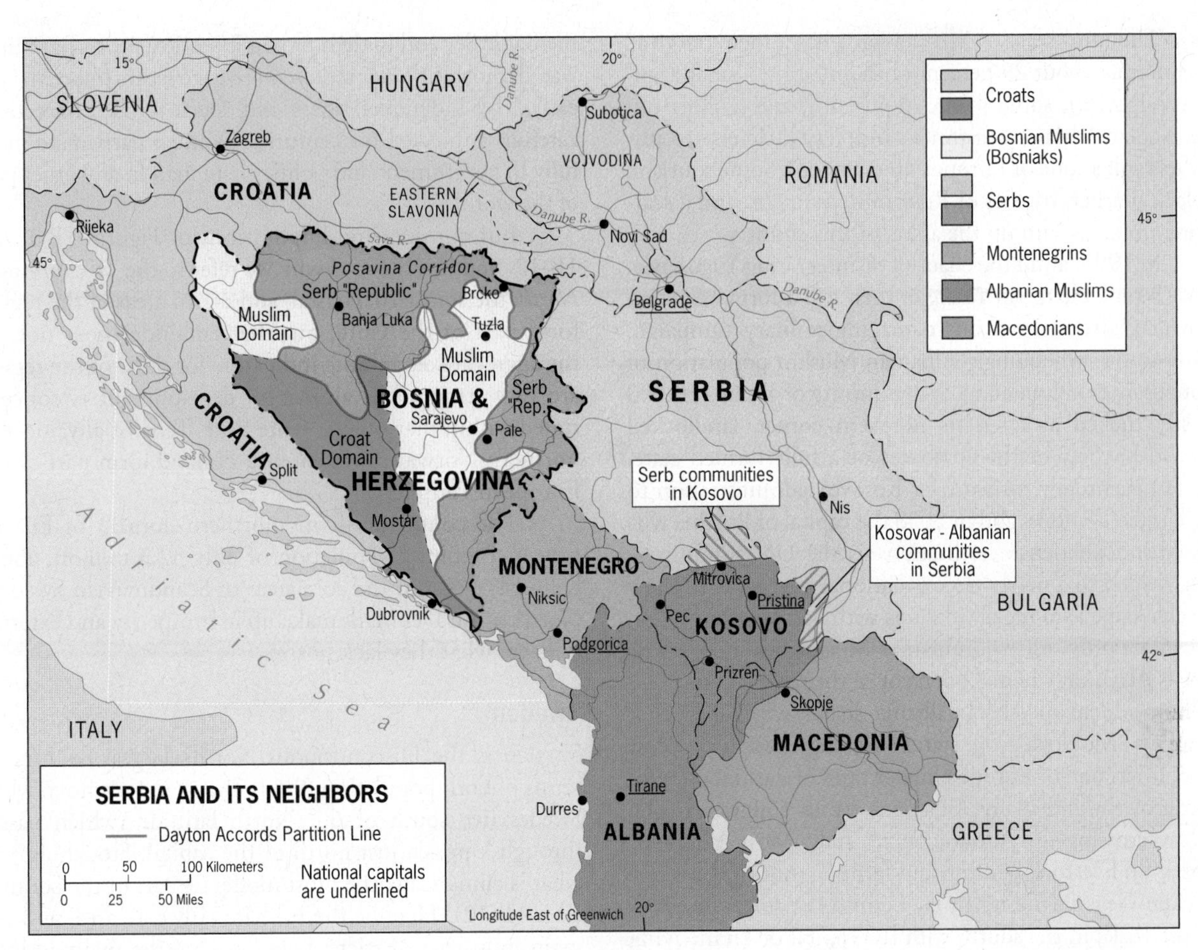

FIGURE 1B-11

EU. Clearly, the Balkans represent a virtual kaleidoscope of ethnic identities that severely complicates any kind of territorial arrangement (see Fig. 1B-11), but in early 2013 Serbia and Kosovo signed a compromise agreement that may allow both to pursue EU membership.

EU expansion into this southeastern sector in 2013 incorporated **Croatia**, which had previously joined NATO in 2009 (see Chapter 1A). But here, too, in this crescent-shaped country with prongs along the Hungarian border and the Adriatic coast, there are significant challenges. About 91 percent of Croatia's 4.3 million citizens are ethnic Croats, but the country's Serb minority has faced discrimination and has declined from 12 percent of the population to less than 5. The EU took a dim view of human rights issues in Croatia, but these improved as EU membership was being negotiated. Meanwhile, about 600,000 Croats live outside Croatia in neighboring Bosnia, where their relationships with Muslims and Serbs are not always cooperative.

When former Yugoslavia collapsed, **Bosnia and Herzegovina** (*Bosnia* in our shorthand form) was the cauldron of calamity. No ethnic group was overwhelmingly dominant here, and this multicultural, effectively landlocked triangle of territory, lying between the Serbian stronghold to the east and the Croatian republic to the west and north, fell victim to disastrous conflict among Serbs, Croats, and ***Bosniaks*** (now the official name for Bosnia's Muslims, who constitute about 41 percent of the population of 3.8 million). As many as 250,000 people perished in concentration camps associated with ***ethnic cleansing*** practices; in 1995, a U.S. diplomatic effort resulted in a truce that partitioned the country as shown in Figure 1B-11. This remains one rough corner of Europe, and fears of renewed conflict linger.

The southernmost "republic" of former Yugoslavia was **Macedonia**, which emerged from the collapse as a state with a mere 2 million inhabitants, of whom two-thirds are Macedonian Slavs. As the map shows, Macedonia adjoins Muslim Albania and Kosovo, and its northwestern corner is home to the 37 percent of the Macedonians who are nominally Muslims. The remainder of this culturally diverse population are Turks, Serbs, and Roma (discussed later in this chapter under Slovakia). Macedonia is one of Europe's poorest countries, landlocked and powerless. Even its very name caused it problems: Greece, Macedonia's neighbor to the south, argued that this name was Greek property and therefore would not recognize it. Next, Macedonia faced an autonomy movement among its Albanian citizens, requiring allocation of scarce resources in the effort to hold the fledgling state together. Macedonians cling to the hope that eventual EU admission will bring it subsidies and better economic times.

The mini-state of **Montenegro** has a mere 600,000 inhabitants, about 29 percent of them Serbs, a small capital (Podgorica), some scenic mountains, and a short but spectacular Adriatic coastline—and very little else to justify its rank as one of Europe's 40 countries. Some tourism, a black market, trickles of Russian investment, and a scattering of farms sum up the assets of this country.

In 1999, amid the chaos of disintegrating Yugoslavia, NATO took charge in **Kosovo**, then still a formal Serbian province, after a brief but damaging military campaign. With an overwhelmingly Albanian-Muslim population of more than 2 million and a Serb minority of about 150,000 mostly tucked away in its northern corner, landlocked Kosovo has few of the attributes of a fully-formed state. NATO eventually turned over Kosovo's administration to the United Nations, and in 2008 the capital of Pristina witnessed independence celebrations as the UN administration yielded to a newly elected national government. Now the challenge is to repair relations with Serbia.

The only other dominantly Muslim country in Europe is **Albania**, where 82 percent of the population of 2.8 million adhere to Islam. Albania shares with one other country—Moldova—the status of being Europe's poorest state. It has one of Europe's higher rates of natural population growth, and many Albanians try to emigrate to the EU by crossing the Adriatic Sea to Italy. Most Albanians subsist on livestock herding and farming, but the poverty-stricken Gegs in the north lag behind the somewhat better-off Tosks in the south, with the capital of Tirane lying close to this cultural divide. For all of Europe's globalization, Albania would represent the symptoms of the global periphery anywhere in the world.

THE DISCONTINUOUS NORTH

It is a long way from the Balkans to the basin of the Baltic Sea—in terms of distance, physiography, and culture. The north's remoteness, isolation, and environmental severity, especially at higher latitudes, have had positive binding effects for part of this domain and seem to have fostered similar cultures. The Scandinavian Peninsula lay removed from most of the wars of mainland Europe and developed in relatively tranquil circumstances (even though Norway was overrun by Nazi Germany during World War II). The three major languages (Swedish, Norwegian, and Danish) are mutually intelligible, and in terms of religion there is overwhelming adherence to the same Lutheran church in the three Scandinavian countries as well as Iceland, Finland, Estonia, and Latvia. Lithuania and Latvia have distinct languages, and in Lithuania the Roman Catholic religion prevails—but all these northern states have a shared history on the shores of the Baltic Sea and in their proximity to Russia. In Scandinavia, democratic and representative governments emerged early, and individual rights and social welfare have been carefully protected for centuries. Women participate more fully in government and politics here than in any other part of the world.

But consider the implications of Figures 1B-1 and 1B-12, and it is obvious why we refer to the *Discontinuous North*: the southern, coastal, and urban areas of the region form part of the Core, but the remainder does not. In the aggregate, economic indicators for most of the region are very strong, but almost all development is concentrated within that smaller Core zone. Territorially, most of northern Norway, Sweden, and Finland form part of the European Periphery.

The countries of this northern domain of Europe have a combined population of only 32.3 million, about half of Italy's. But the core areas of Scandinavian Sweden, Norway, and Denmark make up in prosperity and external linkages what they lack in size.

Sweden

Sweden is the Discontinuous North's largest country in terms of both population (9.5 million) and territory. Most Swedes live south of 60° North latitude (which passes through Uppsala just north of the capital, Stockholm) in what is climatically the most moderate part of the country (Fig. 1B-12). Here lie the primate city, core area, and the main industrial districts; here, too, are the main agricultural areas that benefit from the lower relief, better soils, and milder climate.

Outer Stockholm's prosperous suburban landscape along the shore of Lilla Värtan, the narrow strait that separates the Swedish capital from the adjacent island of Lidingö. Stockholm, which today is home to 1.5 million people, has a long history that dates back to the thirteenth century. Southeastern Sweden is interlaced with myriad waterways, and Stockholm is no exception. Its layout is quite unusual in that this metropolis has been built on a cluster of no less than 14 islands.

© Jan Nijman

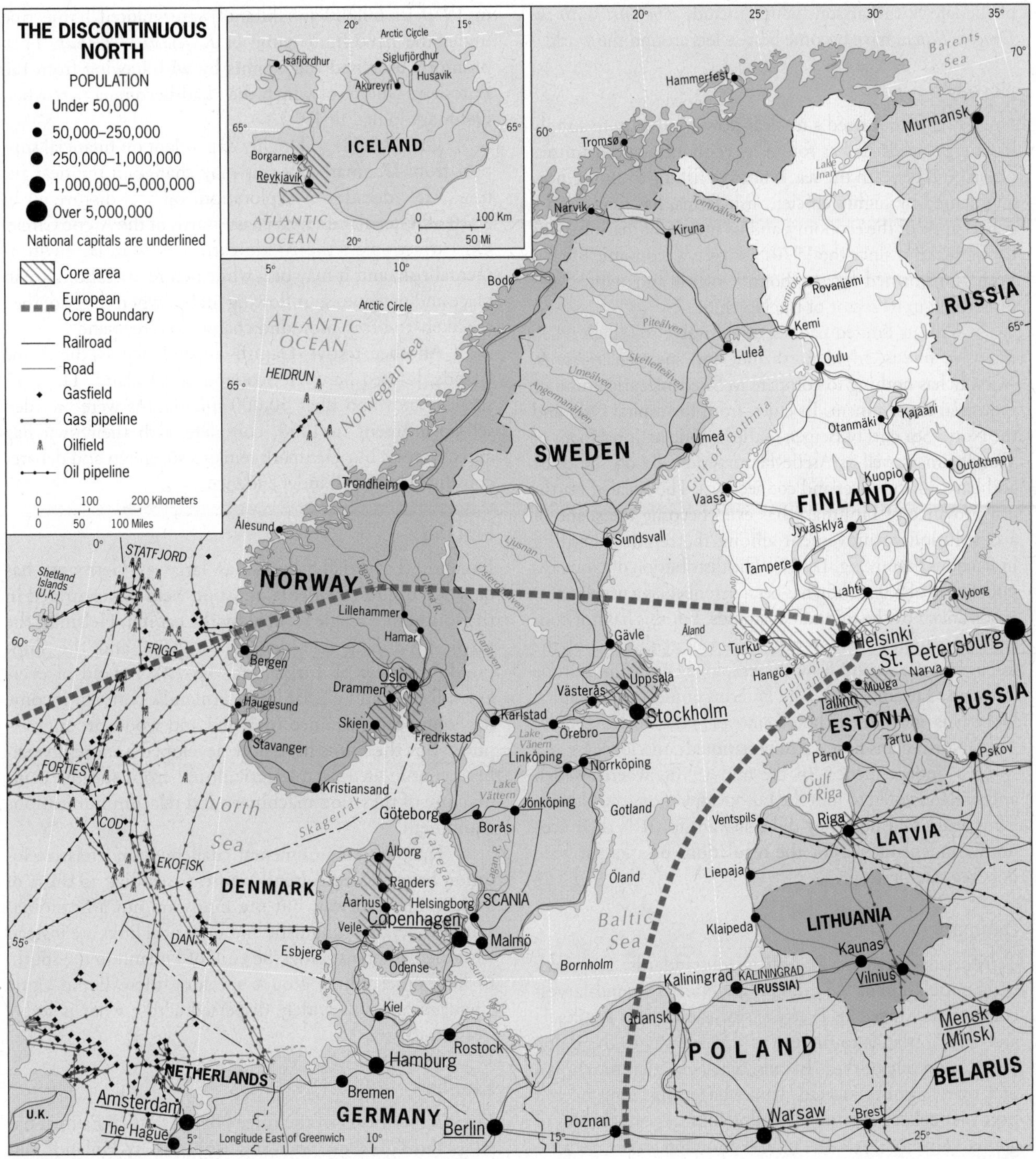

FIGURE 1B-12

Sweden long exported raw or semi-finished materials to industrial countries, but today the Swedes make finished products themselves, including motor vehicles, electronics, stainless steel, furniture, and glassware. Much of this production is based on local resources, including a major iron ore reserve at Kiruna in the far north (there is a steel mill at Luleå). Swedish manufacturing, in contrast to that of several western European countries, is based in dozens of small and medium-sized towns specializing in particular products. Energy-poor Sweden was a pioneer in the development of nuclear power, but a national debate over the risks involved has reversed that course. Sweden joined the EU in 1995, but opted out of the eurozone in 2002.

This is northern Europe's biggest and most influential country. Historically, Sweden has dominated both Norway and Denmark. Culturally, Sweden has time and again made its presence known on the global scene, from popular music to films to works of fiction. Most recently, the crime novels of the late Stieg Larsson (which include *The Girl With the Dragon Tattoo*) have become best-sellers around the world.

Norway

Norway does not need a nuclear power industry to supply its energy needs. It has found its economic opportunities on, in, and beneath the sea. Norway's fishing industry, now augmented by highly efficient fish farms, has long been a cornerstone of the economy, and its merchant marine spans the world. But since the 1970s, Norway's economic life has been transformed by the bounty of oil and natural gas discovered in its sector of the North Sea.

With its limited patches of cultivable soil, high relief, extensive forests, frigid northland, and spectacular ***fjords***, Norway has nothing to compare to Sweden's agricultural or industrial development. Its cities, from the capital Oslo and the North Sea port of Bergen to the historic national focus of Trondheim as well as Arctic Hammerfest, lie on the coast and have difficult overland connections. The isolated northern province of Finnmark has even become the scene of a devolutionary movement among the reindeer-herding indigenous Saami (Fig. 1A-11). The distribution of Norway's population of 5.0 million has been described as a necklace, its beads linked by the thinnest of strands. But this has not constrained national development. Norway at the end of the 2000s had the second-lowest unemployment rate in Europe (after tiny Luxembourg). And in terms of income per capita, Norway is one of the richest countries in the world.

Norwegians have a strong national consciousness and spirit of independence. In the mid-1990s, when Sweden and Finland voted to join the European Union, the Norwegians again said no. They did not want to trade their economic independence for the regulations of a larger, even possibly safer, Europe.

Denmark

Territorially small by Scandinavian standards, Denmark has a population of 5.6 million, the North's second-largest country after Sweden. It consists of the Jutland Peninsula and several islands to the east at the gateway to the Baltic Sea; it is on one of these islands, Sjaelland, that the capital of Copenhagen is located. Copenhagen has long been a port that collects, stores, and transships large quantities of goods. This **break-of-bulk [14]** function exists because many oceangoing vessels cannot enter the shallow Baltic Sea, making the city an ***entrepôt* [15]** where transfer facilities and activities prevail. The completion of the Øresund bridge-tunnel link to southern Sweden in 2000 further enhanced Copenhagen's situation (Fig. 1B-12).

Denmark remains a kingdom, and in centuries past Danish influence extended far beyond its present confines. Remnants of that period continue to challenge Denmark's governance. Greenland came under Danish rule following union with Norway (1380) and continued as a Danish possession when that union ended in 1814. In 1953, Greenland's status changed from colony to province, and in 1979 its 60,000 inhabitants were granted home rule under the new, Inuit name of ***Kalaallit Nunaat***. They promptly exercised their rights by withdrawing from the European Union, of which they had become a part when Denmark joined in 1973.

Greenland is still highly dependent on financial support from Denmark but this may change in the near future. After decades of exploration, oil was discovered in 2010 off Greenland's west coast, north of the Arctic Circle. Oil extraction in the Arctic carries substantial environmental risk, and it may be a while before operations are in place and revenues start flowing in; but when they do, they are likely to bring profound change to Greenland.

Another restive Danish dependency is the Faroe Islands, located between Scotland and Iceland. These 17 small islands and their 50,000 inhabitants were awarded self-government in 1948, complete with their own flag and currency, but Denmark remains sovereign and debates continue about full independence.

Finland

Finland, territorially almost as large as Germany, has only 5.4 million residents, most of them concentrated in the southern triangle formed by the capital, Helsinki, the textile-producing complex of Tampere, and the shipbuilding center of Turku (Fig. 1B-12). A land of evergreen forests and glacial lakes, Finland's national income has long been sustained by wood and wood-product exports. But the Finns have now developed a more diversified economy with staple agricultural crops and the manufacture of precision machinery and telecommunications equipment.

Finland's environmental challenges and relative location have created cultural landscapes similar to those of Sweden and Norway, but the Finns are not a Scandinavian people; their linguistic and historical links are instead with the Estonians across the Gulf of Finland to the south. In fact, other ethnic groups speaking these Finno-Ugric languages are also widely dispersed across what is today western Russia.

Estonia

The northernmost of the three "Baltic states," Estonia has longstanding ethnic and linguistic ties to Finland. But during the period of Soviet control from 1940 to 1991, Estonia's demographic structure transformed drastically: today about 27 percent of its 1.3 million inhabitants are Russians, most of whom came there as colonizers.

After a difficult period of adjustment, Estonia today is forging well ahead of its Baltic neighbors, Latvia and Lithuania (see Appendix B). Busy traffic links Tallinn, the capital, with its counterpart in Finland, Helsinki, and a free-trade zone at Muuga Harbor facilitates commerce with Russia. But more important for this country's future was its entry into the European Union in 2004. Since then, Estonia has gained attention as a result of its economic

From the Field Notes . . .

© H.J. de Blij

© H.J. de Blij

"The most cheerful corner of Nuuk, the capital of Greenland, is enlivened by the bright colors of the homes built on what was the original site of settlement here on the Davis Strait. Nuuk offers a few surprises: a supermarket featuring vegetables grown in Greenland, a nine-hole golf course being expanded to 18 holes, and a vigorous debate over the prospect of independence. With a population of only about 60,000, considerable autonomy from former colonial power Denmark, and substantial investment from Copenhagen, the citizens of *Kalaallit Nunaat* are nonetheless divided on the question of their future. Climate change, the prospect of oil reserves to be found offshore, and freedom to fish (including whales) like the Norwegians do, are factors seen by many indigenous Greenlanders as potential rewards of independence. Danish residents and those of Danish (but local) ancestry tend to see it differently. In the 2009 election, the party representing indigenous interests did better than expected, and 'KN' seemed to be on course toward sovereignty."

www.conceptcaching.com

advances, particularly its creative entrepreneurship in the high-tech and software industries (Skype, for instance, was originally an Estonian company). Having entered the eurozone in 2011, Estonia's economy has remained relatively stable, and this country may soon be joining the ranks of the European Core.

Latvia

As Figure 1B-12 reminds us, the boundary of the European Core that traverses Europe's North includes southern Norway and Sweden but excludes Estonia and the other two Baltic states. The latter have experienced economic improvement in recent years, but are not yet on a par with Estonia, let alone Scandinavia. Latvia, the middle Baltic state centered on the port and capital city of Riga, was tightly integrated into the Soviet system throughout Moscow's long domination. As a consequence, today only 63 percent of the population of 2.0 million is Latvian, and about 27 percent is Russian.

Following independence, ethnic tensions arose between these Baltic and Slavic sectors, but the prospect of EU membership mandated an end to discriminatory practices. Latvians concentrated instead on the economy (which had been left in dreadful shape), and by 2004 the country qualified for admission. Consider this: 25 years ago, virtually all of Latvia's trade was with the Soviet Union. Today its principal trading partners are Germany, the United Kingdom, and Sweden. Russia figures in only one category: Latvia's import of oil and gas.

Nevertheless, Latvia was hammered by the fiscal crisis, and by 2012 its economy had contracted by one-fifth as unemployment approached 20 percent. Even though Latvia also required an EU bailout in 2012 to keep government finances afloat, plans remained in place for the country to join the eurozone in 2014.

Lithuania

This southernmost Baltic state of 3.2 million has a residual Russian minority of only about 5 percent, but relations with this gigantic neighbor are tense—despite a continuing dependence on Russia as a trading partner. One reason for this strained relationship has to do with neighboring ***Kaliningrad***, Russia's **exclave [16]** facing the Baltic Sea (Fig. 1B-12). When Kaliningrad became a Russian territory after World War II (Russia still regards this exclave as an important outpost), Lithuania was left with only about 80 kilometers (50 mi) of Baltic coastline and a small port that was not even connected by rail to the interior capital of Vilnius. Significantly, Lithuania joined NATO in 2004 and a year later it called for the demilitarization of Kaliningrad as a matter of national security. Russia did not oblige.

Economically, things at first went well, initially, spurred by growing foreign investment and profits from the oil refinery at Mazeikiai. Lithuania's economy in 2003 had the highest growth rate in Europe and it was in a mood of optimism that it joined the EU in 2004 (without adopting the euro). Here as in so many other countries, the fiscal crisis at the end of the decade hit the economy hard and austerity measures had to be introduced. There seemed to be light at the end of the tunnel in mid-2013, and there was talk of joining the eurozone by 2017. Whether enthusiasm for the euro continues both here and in Latvia, remains to be seen.

Iceland

Iceland, the volcanic, glacier-studded island in the frigid waters of the North Atlantic just south of the Arctic Circle, is the eighth country of the Discontinuous North. Inhabited by people with Scandinavian ancestries (population: 330,000), Iceland and its small neighboring archipelago, the Westermann Islands, are of special scientific interest because they lie astride the Mid-Atlantic Ridge. Here, the Eurasian and North American tectonic plates of the Earth's crust are diverging (see Fig. G-4), new land can be seen forming, and spectacular volcanic eruptions are periodically on display (as occurred memorably in 2010 when the Eyjafjallajökull volcano brought transatlantic and domestic air traffic across northern Europe to a halt for several days).

Iceland's population is almost completely urban, and the capital, Reykjavik, contains about half the country's inhabitants. The nation's economic geography is traditionally oriented to the surrounding waters, whose seafood harvests have given Iceland a high standard of living. In the 1990s, Iceland embarked on economic liberalization policies and its financial industries and banks grew rapidly. For a while, Iceland was referred to as the *Nordic Tiger*; but then, not unlike Ireland, which experienced a similar boom, the economy went into a nosedive with the onset of global recession in 2008. Iceland's government, closely involved with some of the troubled banks, had to be bailed out by the International Monetary Fund. There were plans to join the EU in 2012, but the application process was suspended in early 2013 as an unexpectedly rapid economic recovery took hold, giving Icelanders pause as to whether joining the European trading bloc was their best option.

THE EASTERN PERIPHERY

Europe's Eastern Periphery is a fascinating region, not least because it is so difficult to discern exactly where its boundaries lie. It is a region that in some ways becomes less European toward the east and less Slavic going westward; one that throughout history has found itself in the path of the territorial ambitions of major powers on either side; and with a changeable political geography to match. Some historians have morbidly dubbed the area extending from the Baltic to the Black Sea "Bloodlands," the arena in which Hitler's Germany and Stalin's Soviet Union did their most murderous work during the second quarter of the twentieth century. By far, more civilians and military personnel died here than in any other part of Europe. After World War II, the countries of eastern Europe were, forcibly for the most part, incorporated into the Soviet orbit as satellites, their various national identities suppressed along with basic freedoms and their economies subjugated to Moscow's interests and commands.

Europe has changed fundamentally in the quarter-century since the end of the Cold War. The old Eastern Europe, its western edge sharply defined by the Iron Curtain, no longer exists, and in some ways we have seen the rebirth of the older cultural region known as Central Europe or, in its prevailing German language, *Mitteleuropa*. Many of these changes were consolidated as NATO and the EU expanded eastward. As we saw, the Czech Republic has already been integrated into Europe's Core, and Slovenia is on track to become the first component of former Yugoslavia to merit similar inclusion. By 2013, a relatively large number of states in the Eastern Periphery had already joined the Union, and others are knocking on the door. Undoubtedly, the transformation of this sector of the European realm will continue to advance steadily as this decade unfolds.

East-Central Europe

As Figures 1B-1 and 1B-13 show, the four states in this group are all immediately adjacent to or already partially integrated into the European Core. All have four joined NATO as well: Poland and Hungary in 1999, and Slovakia and Slovenia in 2004. The largest and most populous of these countries is **Poland**. With 38.2 million people and lying between two historic enemies, Poland has borders that have shifted time and again; but its current status

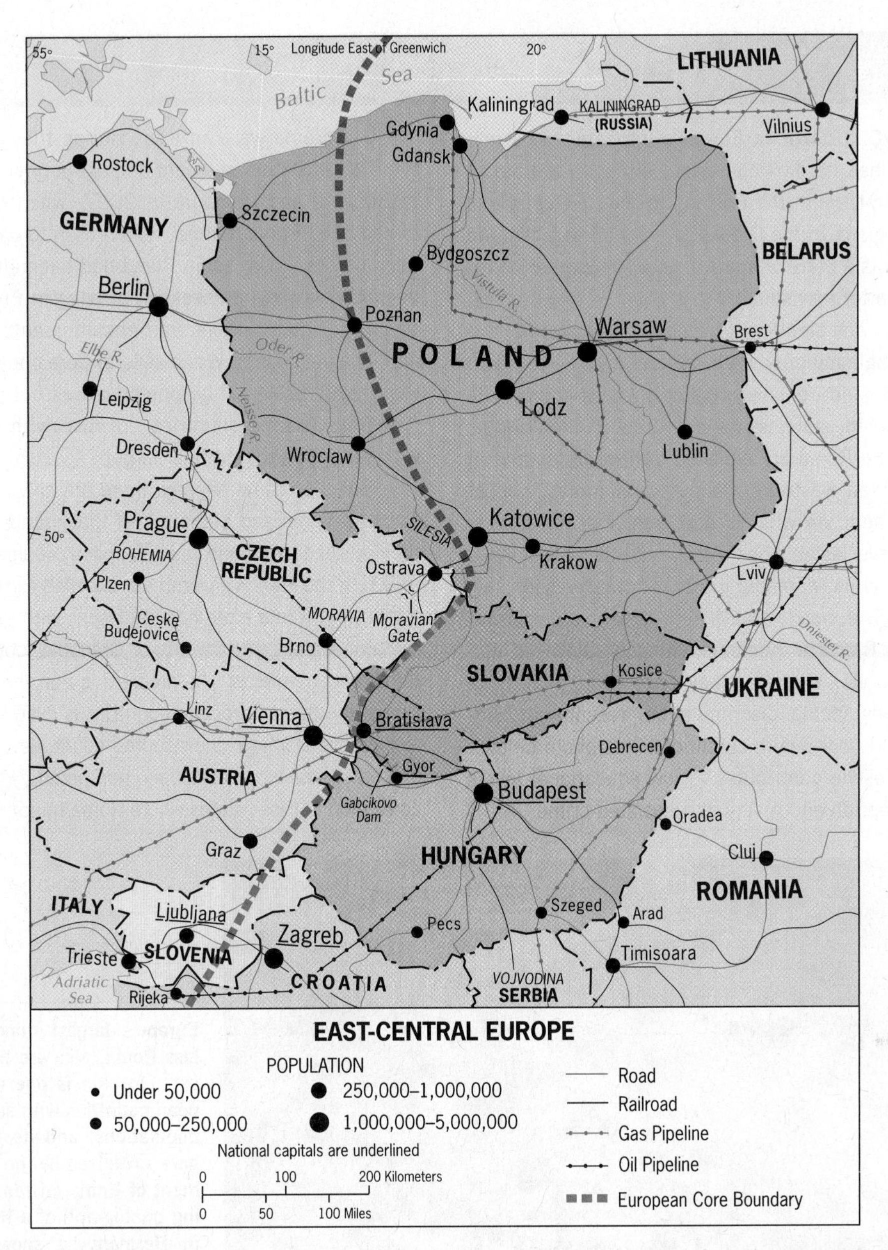

FIGURE 1B-13

appears to be more durable than in the past. As the map shows, the historic and once-central capital of Warsaw now lies closer to Russia than to Germany, but the country today looks westward, not to the east. Throughout the Soviet-communist period, Silesia in the southwest was Poland's industrial heartland, and Katowice, Wroclaw, and Krakow grew into major manufacturing centers amid some of the world's worst environmental degradation. The Soviets invested far less in agriculture, collectivizing farms without modernizing technologies and leaving post-Soviet farming in abysmal condition. All this made governing Poland difficult, but the prospect of EU membership (with the promise of EU subsidies) motivated the government to get its house in order.

After entering the EU in 2004, Poland saw hundreds of thousands of workers depart for jobs in the European Core, but many have now returned and the economy is growing robustly. Today, Poland is in the best shape it has been in for a very long time, and its economic relations with Germany are closer than ever.

Government also was the problem in neighboring, landlocked **Slovakia**, where during the communist period its people were far more pro-Soviet than the Czechs next door (with whom they were united within the old Czechoslovakia before it broke up in 1993). Many observers of the 2004 EU expansion wondered whether misgoverned Slovakia should be admitted at all because its capital city, Bratislava, had become synonymous with corruption and inefficiency. Slovakia's Hungarian minority, comprising about 10 percent of the population of 5.4 million and concentrated in the south along the Danube River, was at odds with the Slovak regime. Moreover, additional

Europe's Stateless Nation

EVERY ETHNIC GROUP in Europe's intricate social mosaic, it seems, has its own homeland, either in a state or a subnational unit. From the Frisians to the Corsicans and from the Norwegians to the Greeks, everyone has a historic base, whether it is a State or a province or some other political entity, no matter how small.

But there is one significant exception: the ***Roma***, Europe's largest single minority. Perhaps as many as 14 million Roma—formerly (and derisively) called *gypsies*—constitute the realm's troubled, stateless nation. Although their origins are uncertain, the Roma are believed to have originated in India and migrated westward along several routes, one of which carried them via what is now Iran and Turkey into Europe. They never established a territorial base; they have retained their mobile, nomadic lifestyle to the present day; and they mostly live, nearly always in poverty, in a discontinuous arc across Bulgaria, Romania, Hungary, Slovakia, and the Czech Republic. Wherever they have gone, they have found themselves facing discrimination, resentment, unemployment, and poor health conditions (see photo below). Their mobile lifestyle contributes to low educational levels and to the perception and reality of associated crime.

The Roma were an issue during the debate preceding the EU's 2004 enlargement and again prior to the accession of Bulgaria and Romania in 2007, when many Europeans talked of a "Roma deluge" as a reason to oppose expansion. Even before 2004, some Roma had been entering the European Core as asylum-seekers, arriving in English towns and setting up their wagons and encampments in public parks and village squares. When several Core countries toughened their asylum rules, their objective was not only to stem the flow from Islamic countries but also within the EU itself, with the Roma the primary target.

But they came anyway: there are now close to a million Roma in Spain and perhaps half that number in France, and their presence has been cause for highly charged public debate. In 2010, the French government expelled a group of more than a thousand Roma after violent clashes with police in the town of Saint Aignan and the city of Grenoble, both sites of illegal Roma encampments. The problem is that the situation for the Roma in eastern European countries is even worse. EU efforts to help the Roma in their source countries through subsidies and other assistance have been hampered by the high levels of corruption in those states where Roma minorities are largest.

Tomasz Tomaszewski/ngs/Getty Images, Inc.

Europe's largest minority, the stateless Roma, also are the realm's poorest. Slovakia is one of several European countries with substantial Roma populations, and its government has been criticized by the EU for its treatment of Roma citizens. This depressing photograph of a Roma settlement in Hermanovce shows a cluster of makeshift dwellings virtually encircled by a moat bridged only by a walkway. The village of Hermanovce may itself not be very prosperous, but that moat represents a social chasm between comparative comfort and inescapable deprivation.

concerns were raised over reports of mistreatment of the smaller Roma (Gypsy) minority (see the box titled "Europe's Stateless Nation"). But again the promise of EU membership led to sufficient reforms, and after 2004 the economy perked up. Following another downturn in 2007, a more enlightened administration took over, and in 2008 Slovakia surprised many by meeting the terms of admission to the EMU, adopting the euro in 2009 before the Poles or the Hungarians could.

Many economic geographers anticipated a bright future for also-landlocked **Hungary** following the collapse of the Soviet Empire and the dismantling of the Iron Curtain. The Hungarians (Magyars) moved into the middle Danube Basin more than a thousand years ago from an Asian source; they have neither Slavic nor Germanic roots. They converted their fertile lowland into a thriving nation-state and also created an imperial power that held sway over an area far larger than present-day Hungary.

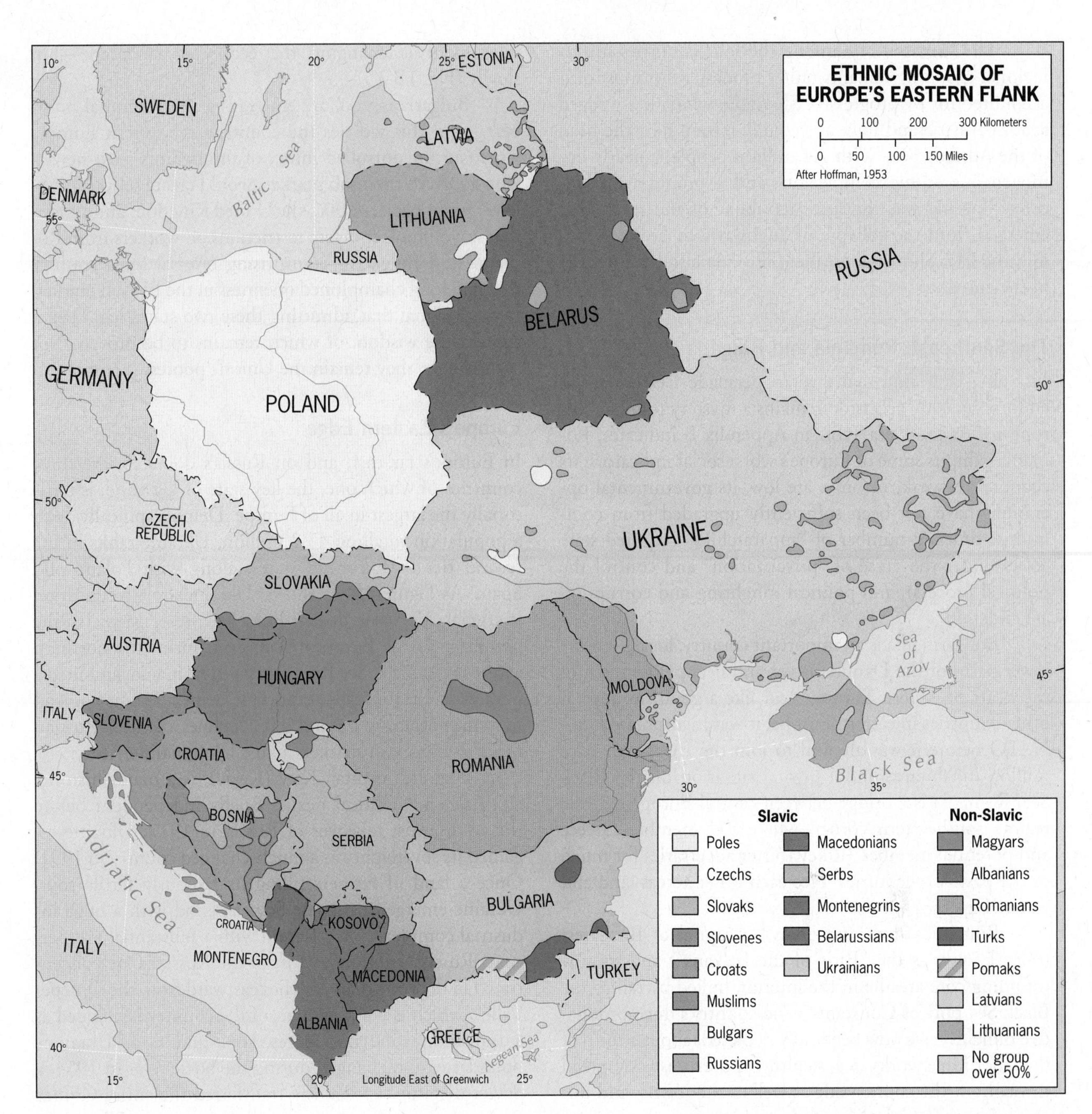

FIGURE 1B-14

Ethnic Magyar remnants of this Greater Hungary can still be found in parts of Romania, Serbia, and Slovakia (Fig. 1B-14), and the government in the twin-cities capital astride the Danube River, Budapest, has a history of irredentism toward these external minorities.

The concept of **irredentism [17]**—a government's open support for fellow ethnic or cultural cohorts in neighboring or more distant countries—derives from a nineteenth-century campaign by Italy to incorporate the territory inhabited by an Italian-speaking minority of Austria, calling it ***Italia Irredenta*** or "Unredeemed Italy." One advantage of joining the EU, obviously, was that much of the reason for Hungarian irredentism disappeared. If and when Serbia joins the EU, it may disappear altogether.

With a population of 9.8 million, a distinctive culture, and a considerable and diversified resource base, Hungary should have good prospects, and its economic potential was a strong factor in its 2004 admission into the EU. But developments soon spiraled downward with the onset of global recession. The Hungarian government came into conflict with Brussels about the violation of EU monetary policies and the state of Hungary's finances (racking up debt) while local political parties accused the government of a dictatorial turn. Hungary's challenges are both political and economic.

The success story among the four countries in this regional grouping remains rather modest, with most of it accounted for by progressive **Slovenia**, which lies wedged against Austria and Italy in the hilly terrain near the head of the Adriatic Sea. With 2.1 million people, a nearly homogeneous ethnic complexion, and a productive economy, Slovenia was the first "republic" of the seven that emerged from the collapse of Yugoslavia to be invited to join the EU. Shortly thereafter, Slovenia also became part of the eurozone.

The Southeast: Romania and Bulgaria

Just how **Romania** managed to persuade EU leaders to endorse its 2007 accession remains a mystery to many Europeans. As the Data Table in Appendix B indicates, Romania exhibits some of Europe's worst social indicators: its economy is weak, incomes are low, its governmental operations have not been sufficiently upgraded from communist times (a number of "apparatchiks" acquired state assets under the guise of "privatization" and control the political process), and political infighting and corruption are endemic.

But Romania is an important country, located in the lower basin of the Danube River and occupying much of the heart of eastern Europe. And like a good number of other countries in eastern Europe, it was incorporated into NATO before it was allowed to join the EU. With 21.2 million inhabitants and a pivotal situation on the Black Sea, Romania is a bridge between central Europe and the realm's southeastern corner, where EU member Greece and potential member Turkey (bitter adversaries for much of the past two centuries) face each other across land and water (Fig. 1B-15).

Romania's drab and decaying capital of Bucharest (once known as the "Paris of the Balkans") and its surrounding core area lie in the interior, linked by rail to the Black Sea port of Constanta. The country's once-productive oilfields have now been fully depleted, about a third of the labor force works in agriculture, poverty is widespread throughout the countryside as well as the towns, and unemployment is high. Many talented Romanians continue to leave the country in search of opportunities elsewhere.

Across the Danube lies Romania's southern neighbor, **Bulgaria**. The rugged Balkan Mountains form Bulgaria's physiographic backbone, separating the Danube and Maritsa basins. As the map shows, Bulgaria has five neighbors, several of which are in political turmoil. Bulgaria became a member of the European Union, which required judicial and other social reforms. But today it is still the Union's poorest country, and its prospects are not particularly enticing.

Bulgaria has a Black Sea coast and an outlet, the port of Varna, but the main advantage it derives from its coastal location is the tourist trade its beaches generate. The capital, Sofia, lies not on the coast but near the opposite border with Serbia, and in its core-area hinterland some foreign investment is changing the economic landscape—but slowly (Fig. 1B-15).

Bulgarians, too, are emigrating in substantial numbers—and this worries the countries of western Europe, where an uncontrolled influx of immigrants with newly-won access to their job markets would create serious problems. As far back as 2006, the United Kingdom announced that it would place severe restrictions on workers from Romania and Bulgaria—a surprising reversal for a country that had long championed openness in the EU job market. It was a signal that admitting these two states had been a stretch, the wisdom of which remains to be proven. Not surprisingly, they remain the Union's poorest members.

Europe's Eastern Edge

In Europe's far east, and on Russia's doorstep, are three countries of which one, the key state of **Ukraine**, is territorially the largest in all of Europe. Demographically, with a population totalling 45.2 million, Ukraine ranks in the second tier of European states, along with Poland and Spain. As Figure 1B-15 shows, Ukraine's relative location is crucial. Not only does it link the core of Russia to the periphery of the European Union: it forms the northern shore of the Black Sea from Russia to Romania, and incorporates the strategically important Crimea Peninsula. And most importantly, oil and gas pipelines connect Russian fields to European markets across Ukrainian territory.

Ukraine's capital, Kiev (Kyyiv), is a major historic, cultural, and political nexus. Briefly independent before the communist takeover of Russia in 1917, Ukraine regained its sovereignty as a much-changed country in 1991. Once a land of farmers tilling its famously fertile soils, Ukraine emerged from the Soviet period with a huge industrial complex in its east and with a substantial (18 percent) Russian minority. Most of the Russians are concentrated in the zone that stretches eastward from the Dnieper Valley, which is why that part of the country is mapped as a transition zone in Figures 1B-1 and G-3. Ukraine's boundaries also changed during the Soviet era. In 1954, a Soviet dictator capriciously transferred the entire Crimea Peninsula, including its Russian inhabitants, to Ukraine as a reward for its productivity.

The Dnieper River forms a valuable geographic reference to comprehend Ukraine's spatial division that creates the transition zone (Fig. 1B-15). To its west lies agrarian, rural, mainly Roman Catholic Ukraine; in its great southern bend and eastward lies industrial, urban, Russified (and Russian Orthodox) Ukraine. Soviet planners transformed eastern Ukraine into a communist Ruhr based on abundant local coal and iron ore, making the Donets Basin (*Donbas* for short) a key industrial complex. Meanwhile, the Russian Soviet Republic supplied Ukraine with oil and gas.

As the map shows, Ukraine, with the exception of its eastern, urban-concentrated Russian minority (still 18 percent today), possesses an ethnically homogeneous population by eastern European standards. Ukraine is a critically

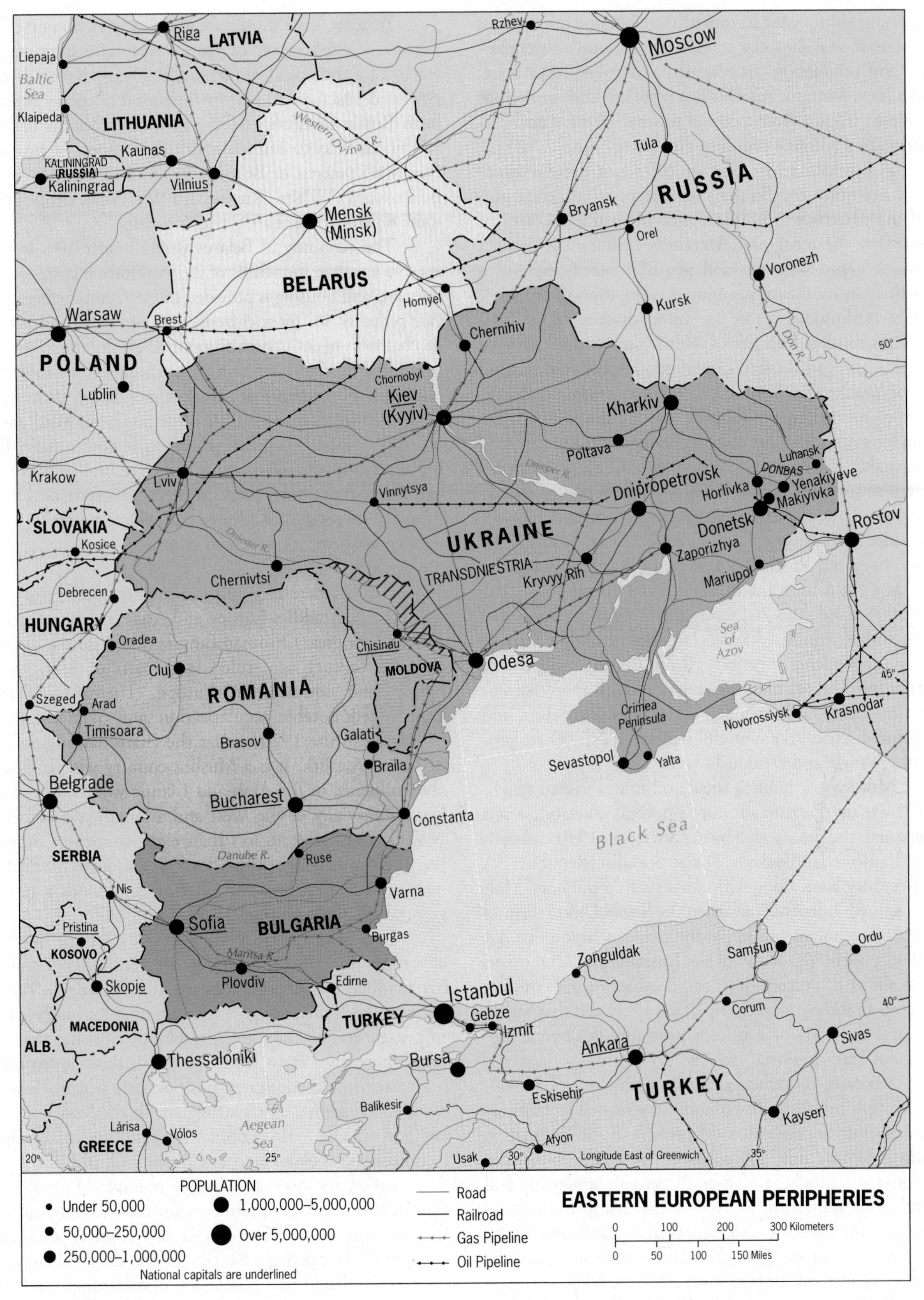

Riga
LATVIA
Rzhev
Moscow
Liepaja
Baltic Sea
Klaipeda
LITHUANIA
Western Dvina R.
Kaunas
KALININGRAD (RUSSIA)
Kaliningrad
Vilnius
Tula
RUSSIA
Bryansk
Orel
Mensk (Minsk)
BELARUS
Voronezh
Homyel
Kursk
Warsaw
Brest
Chernihiv
Don R.
POLAND
Lublin
Chornobyl
Kiev (Kyyiv)
Kharkiv
Poltava
Krakow
Lviv
Dnieper R.
Luhansk
DONBAS
Vinnytsya
Dnipropetrovsk
Horlivka
Yenakiyeve
Makiyivka
SLOVAKIA
Kosice
UKRAINE
Donetsk
Rostov
Dniester R.
Zaporizhya
TRANSDNIESTRIA
Chernivtsi
Kryvyy Rih
Mariupol
Debrecen
HUNGARY
Sea of Azov
Oradea
Chisinau
MOLDOVA
Odesa
Cluj
ROMANIA
Szeged
Arad
Crimea Peninsula
Krasnodar
Novorossiysk
Timisoara
Brasov
Galati
Braila
Sevastopol
Yalta
Belgrade
Bucharest
Constanta
Black Sea
SERBIA
Danube R.
Ruse
Varna
Nis
Pristina
Sofia
BULGARIA
Burgas
KOSOVO
Maritsa R.
Zonguldak
Samsun
Ordu
Skopje
Plovdiv
Edirne
Istanbul
Corum
MACEDONIA
TURKEY
Gebze
Izmit
Sivas
ALB.
Thessaloniki
Bursa
Ankara
Eskisehir
TURKEY
Balikesir
Kayseri
Lárisa
Vólos
Aegean Sea
GREECE
Afyon
20°
25°
Usak
30°
Longitude East of Greenwich
35°
50°
45°
40°
POPULATION
Under 50,000
50,000–250,000
250,000–1,000,000
1,000,000–5,000,000
Over 5,000,000
National capitals are underlined
Road
Railroad
Gas Pipeline
Oil Pipeline
EASTERN EUROPEAN PERIPHERIES
0 100 200 300 Kilometers
0 50 100 150 Miles

FIGURE 1B-15

important country for Europe's future because it has access to international shipping lanes, a large resource base, massive farm production, an educated and skilled labor force, and a large domestic market. But it suffers from numerous problems ranging from political mismanagement and corruption to a faltering economy and rising crime.

A presidential election in 2004 first drew international attention to Ukraine's unique political geography, and the pattern was neatly replicated in the elections of 2010. For the most part, Ukraine's electorate is divided between a pro-Western (and pro-EU) northwest and a pro-Russian southeast, resulting in a situation reminiscent of Czechoslovakia's before its "velvet divorce." The 2010 elections rendered a split of 46/49 percent for the pro-Western and pro-Russian candidates, respectively, putting Victor Yanukovych in office. In Ukraine's case, the transit of much needed energy supplies through the east creates a complication. Ukraine's own dependence on Russian supplies makes it vulnerable to Moscow's decisions on prices as well as supplies. Some of Ukraine's leaders urge speedy integration into the European sphere; others tend to prefer a middle road between Europe and Russia. A 2012 law that recognized Russian as a second official language was seen as a tilt toward the east and away from Europe. On the other hand, the 2013 mega-deal with the Shell Oil Company to exploit Ukraine's large gas reserves will make it less dependent on energy from Russia and provide greater freedom to strengthen relations with the West. For the time being, EU membership prospects are dim, and ongoing discussions about affiliating with NATO are proceeding slowly and cautiously.

Moldova, Ukraine's small and impoverished neighbor (by many measures Europe's poorest country), was a Romanian province seized by the Soviets in 1940 and converted into a landlocked "Soviet Socialist Republic." A half-century later, along with other such "republics," Moldova gained independence when the Soviet Union disintegrated. Romanians remain in the majority among its 4.1 million people, but many of the Russians and Ukrainians (each about 13 percent) have migrated across the Dniester River to an industrialized strip of land between that waterway and the Ukrainian border, proclaiming there a "Republic of Transdniestria" (see Fig. 1B-15).

Separatist movements constitute only one of Moldova's multiple problems. Its economy, dominated by farming, is in decline; an estimated 40 percent of the population works at jobs outside its borders because unemployment in Moldova is typically as high as 30 percent; smuggling and illegal arms trafficking are rife. These many misfortunes translate into weakness, and Russia's malign influence in the form of support for Transdniestria's separatists keeps the country in turmoil. In fact, since 2007 the Moldovan government has been pressured by Russia to recognize the regime of breakaway Transdniestria as legitimate. The country also is highly dependent on Russian energy supplies, so these attempts at coercion cannot be totally disregarded.

Belarus is the third state in this eastern margin of the European Periphery and in many ways it is the most peripheral of all. Landlocked, autocratic, and heavy-handedly misgoverned, sustained in part by the transit of energy supplies from Russia to Europe's Core countries, Belarus has few functional links to Europe and little prospect of progress. Nearly 85 percent of Belarus's 9.4 million inhabitants are Belarussians ("White" Russians), a West Slavic people; only some 8 percent are (East Slavic) Russians.

The economy of Belarus is reminiscent of a bygone era. No less than four-fifths of the workforce is employed by the state, and housing is provided through employment. Private property, for most citizens, does not exist, and neither do chambers of commerce or unions. All education is tightly controlled by the state: high school students are told they must join the government's Youth Union if they want to attend university, but a university education is not worth much unless you want to prepare for a career in government. This is a society stuck in old Soviet ways in a realm that continues to undergo profound change. In effect, Belarus trades energy subsidies for serving as a stable and willing buffer to European encroachment on Russia's doorstep.

Finally, there is **Turkey**, a mighty country of 76.7 million that straddles Europe and Asia. Turkey is the successor of the great Ottoman Empire that, before it disintegrated a century ago, ruled large parts of Asia, North Africa, and southeastern Europe. Thereafter, Turkey experienced notable secularization and Westernization, beginning in the 1920s under the charismatic leadership of Kemal Atatürk. It is a Muslim country with considerable influence in the Arab and Islamic worlds, but also a key military ally of the West and a longtime member of NATO. Bordering no less than eight countries, controlling the straits between the Black Sea and the Mediterranean, and facing three seas, Turkey commands a pivotal geostrategic position (Fig. 1B-1).

This country also plays a key role in any Western efforts to do something about the terrible civil war in Syria, which borders southeastern Turkey. Indeed, Turkey itself has been the strongest voice arguing for intervention to put an end to the especially bloody uprising that has raged there since early 2011. But the Turkish government finds itself in a complicated position here because it is already involved in a costly struggle with the Kurds of eastern Turkey. The Syrian conflict in its full regional geopolitical context is elaborated in Chapters 7A and 7B.

Turkey has been an associate member of the EU for four decades, and its application for full membership was first submitted in 1987. But is this country European enough? Or is it perhaps too big (second in population size only to Germany) and potentially destabilizing? Or is the Turkish case less urgent than others because it is already firmly embedded within NATO and related security structures led by the United States? At any rate, full Turkish membership in the EU remains on hold. The primary reasons, it is said, are the continuing poor treatment of Kurds in eastern Turkey and the apparent inability of the Turkish government to conform to the Union's economic standards—issues that are not likely to be resolved in the immediate future. Ironically, in recent years Turkey's economy has performed extremely well while its stature in the Islamic world has grown (see Chapters 7A and 7B). Whereas an economically troubled Europe is still unable to make up its mind about Turkey, the Turks themselves are steadily losing interest in the EU altogether.

With its 600 million inhabitants in 40 countries, including some of the world's highest-income economies, a politically stable and economically integrated Europe would be a superpower in any new world order. However, Europe's political geography is anything but stable as devolutionary forces and cultural stresses continue to trouble the realm even as EU expansion proceeds. The global economic crisis of the past few years has underscored the enormous difficulty of managing such a large and diverse realm as a single economic system. Europeans have not yet found a way to give voice to collective viewpoints in the world, or to generate collective action in times of crisis. Importantly, even after the latest expansion, the EU incorporates just over two-thirds of the realm's national economies, and additional enlargement will become increasingly difficult for economic, political, and cultural reasons.

POINTS TO PONDER

- Surveys show that a small European state, the Netherlands, has the world's most globalized economy.
- Scotland will hold a referendum on independence from the UK in 2014; the UK government wants to hold a referendum on continuing membership in the European Union.
- Brussels is the main seat of the European Union, yet it is the capital of a country that threatens to come apart.
- In northern European countries, the south tends to be more developed than the north, whereas in southern European countries the north tends to be more developed than the south.
- Fiscally troubled national governments in the eurozone that accept conditions for a bailout from the EU must impose austerity measures on their population and thereby risk losing the next election.
- By admitting such countries as Romania, Bulgaria, and Croatia while keeping Turkey at arm's length, the EU has made some debatable choices.

Elevation (m)
3000
1500
600
300
150
0
below sea level
0
-150
-1500
-3000
Roads
Railroads
0 km 200 400 600 800 1000 1200
0 miles 200 400 600 800
Lambert Azimuthal Equal-Area Projection
Scale 1:20,000,000
UNITED KINGDOM
NETHERLANDS
Amsterdam
North Sea
HAMBURG
GERMANY
BERLIN
COPENHAGEN
DENMARK
NORWAY
SWEDEN
SCANDINAVIA
Stockholm
Gulf of Bothnia
FINLAND
Helsinki
Baltic Sea
Kaliningrad
RUSSIA
POLAND
WARSAW
LITHUANIA
LATVIA
Riga
Gulf of Riga
ESTONIA
Tallinn
Gulf of Finland
Lake Peipus
Lake Ladoga
ST. PETERSBURG
Pskov
Novgorod
Vilnius
BELARUS
MENSK (MINSK)
DNIEPER LOWLAND
UKRAINE
(KYIV) KIEV
MOLDOVA
Chisinau
ODESA
CRIMEAN PEN.
Sevastopol
KHARKIV
DNIPROPETROVSK
DONETSK
Taganrog
Sea of Azov
Black Sea
Novorossiysk
Krasnodar
ROSTOV
Shakhty
Sochi
Armavir
Maykop
Stavropol
Cherkessk
Nalchik
Mt. Elbrus (5,633 m, 18,481 ft.)
CAUCASUS MOUNTAINS
Elista
GEORGIA
TBILISI
ARMENIA
YEREVAN
AZERBAIJAN
BAKI
AZER.
TURKEY
IRAQ
TABRIZ
Vladikavkaz
Nazran
Grozny
Makhachkala
Astrakhan
CASPIAN DEPRESSION
Caspian Sea
VOLGOGRAD
Volzhsky
Don
Kursk
Orel
Bryansk
Smolensk
Kaluga
Tula
Belgorod
Staryy Oskol
Voronezh
Lipetsk
Tambov
Ryazan
MOSCOW
Tver
Dubna
Zelenograd
Vladimir
Ivanovo
Yaroslavl
Rybinsk
Cherepovets
Vologda
Kostroma
Dzerzhinsk
NIZHNIY NOVGOROD
Petrozavodsk
Lake Onega
EUROPEAN LOWLAND
Northern Dvina
Volga
Cheboksary
Yoshkar-Ola
Ulyanovsk
Saransk
Penza
Saratov
Engels
Balakovo
Tolyatti
SAMARA
KAZAN
Nizhnekamsk
Naberezhnye Chelny
Izhevsk
Kama
Vyatka
Kudymkar
Syktyvkar
PERM
YEKATERINBURG
Nizhny-Tagil
Sterlitamak
Ufa
Magnitogorsk
Orenburg
Orsk
Ural
URAL MOUNTAINS
CHELYABINSK
Kurgan
Tyumen
Murmansk
Lovozero
KOLA PENINSULA
Arctic Circle
White Sea
Severodvinsk
Onega
Arkhangelsk
Barents Sea
Kolguyev I.
Novaya Zemlya
Vaygach I.
Kara Sea
Naryan Mar
Pechora
PECHORA BASIN
Gornyatskiy
Vorkuta
Mt. Narodnaya (1,895 m, 6,214 ft.)
Salekhard
YAMAL PENINSULA
GYDA PENINSULA
Gulf of Ob
Novyy Port
Belyy I.
Dickson
Dudinka
Norilsk
Yenisey
Urengoy
Sergino
Ob
WEST SIBERIAN PLAIN
Khanty Mansiysk
Surgut
Nizhnevartovsk
Irtysh
RUSSIA
Kolpashevo
Lesosibirsk
Petropavl
OMSK
NOVOSIBIRSK
Tomsk
Krasnoyarsk
Kemerovo
Prokopyevsk
Novokuznetsk
Barnaul
Biysk
Gorno Altaysk
Rubtsovsk
Semey
Mt. Belukha (4,506 m, 14,783 ft.)
ALTAY
Astana
KAZAKH UPLANDS
KAZAKHSTAN
Lake Balqash
SARYESIK-ATYRAU DESERT
MOYYNQUM
USTYURT PLATEAU
Aral Sea
UZBEKISTAN
QIZILQUM
TASHKENT
Bishkek
TURKMENISTAN
Garabogaz Bay
Ashgabat
ELBURZ MOUNTAINS
TEHRAN
IRAN
QOM
KARAJ
ZAGROS
JUNGGAR BASIN
Svalbard (NORWAY)
Franz Josef Land
Alexandra Land
George Land
Graham Bell I.
Wilczek Land
Ushakov
October Re
ARCTIC

OCEAN
Bering Strait
UNITED STATES
St. Lawrence I.
Chukchi Sea
CHUKCHI PENINSULA
Wrangel I.
Gulf of Anadyr
St. Matthew I.
CHUKCHI RANGE
East Siberian Sea
Pevek
Anadyr
Mt. Ledyanaya (2,562 m, 8,406 ft.)
KORYAK RANGE
Bering Sea
Aleutian Islands
UNITED STATES
New Siberian Islands
Novaya Sibir I.
Kotelnyy I.
Malyy Lyakhovskiy I.
Bolshoy Lyakhovskiy I.
Ambarchik
Laptev Sea
Gulf of Yana
KOLYMA RANGE
Karaginskiy I.
Begichev I.
Nordvik
Tiksi
Kolyma
Omsukchan
Shelikhov Gulf
Palana
Commander Islands (RUSSIA)
NORTH SIBERIAN LOWLAND
CHERSKIY RANGE
Ust Nera
Magadan
KAMCHATKA PENINSULA
Mt. Klyuchevskaya (4,750 m, 15,584 ft.)
Petropavlovsk-Kamchatskiy
VERKHOYANSK RANGE
Mt. Mus Khaya (2,959 m, 9,708 ft.)
Lena
CENTRAL
Okhotsk
Bolsheretsk
Cape Lopatka
Sea of Okhotsk
YAKUTSK BASIN
Namtsy
Yakutsk
Ust Maya
Paramushir I.
SIBERIAN
Tura
Vilyuy
Suntar
Mirnyy
Lena
Aldan
DZHUGDZHUR RANGE
Kurile Islands
Shantar Islands
Uda Bay
Nogliki
Sakhalin Island
Lensk
PLATEAU
Lazarev
STANOVOY RANGE
Tatar Strait
Urup
Angara
Amur
Vanino
Yuzhno Sakhalinsk
Novikovo
Etorofu
Ust Ilimsk
Mt. Skalistyy (2,467 m, 8,094 ft.)
Skovorodino
Komsomolsk
Mt. Tardoki-Yani (2,090 m, 6,857 ft.)
Kunashiri
Shikotan
Habomai
Ust Kut
La Perouse Strait
Wakkanai
Tayshet
Bratsk
Amur
Blagoveshchensk
Birobidzhan
Khabarovsk
Kushiro
Lake Baykal
YABLONOVYY RANGE
Heihe
SIKHOTE-ALIN RANGE
SAPPORO
Hokkaido
Ust Ordynskiy
Chita
GREATER KHINGAN RANGE
Dalnegorsk
PACIFIC OCEAN
Angarsk
Irkutsk
Ulan-Ude
Aginskoye
Songhua
Lake Khanka
Aomori
Mt. Munku Sardyk
HARBIN
Vladivostok
Nakhodka
SENDAI
NORTHEAST CHINA PLAIN
JILIN
Ulaanbaatar
CHANGCHUN
East Sea (Sea of Japan)
TOKYO
KAWASAKI
YOKOHAMA
Mt. Fuji (3,776 m, 12,388 ft.)
MONGOLIA
MONGOLIAN PLATEAU
SHENYANG
NORTH KOREA
NAGOYA
KYOTO
CHINA

IN THIS CHAPTER

- Geopolitics of the "heartland"
- Global warming in the Arctic
- From Czars to Soviets to 21st-century Russians
- Post-Soviet Russia and the Near Abroad
- Russia's natural riches

CONCEPTS, IDEAS, AND TERMS

Heartland 1
Continentality 2
Permafrost 3
Tundra 4
Taiga 5
Steppe 6
Forward capital 7
Federation 8
Russification 9
Command economy 10
Satellite state 11
Near Abroad 12

FIGURE 2A-1

Every geographic realm has a dominant, distinguishing feature: Europe's jigsaw-puzzle map of 40 countries, South America's familiar triangle, Subsaharan Africa's straddling of the equator, Southeast Asia's thousands of islands and peninsulas. But the Russian geographic realm has what is perhaps the most distinctive hallmark of all—its gigantic territorial size. Almost (but not quite) all of this realm is constituted by the state that dominates it, which is why we call this the Russian realm. By itself, Russia is nearly twice as large as Canada, the world's second-largest country. Russia is three times as big as neighboring Europe. Stretching east-west from the Pacific Ocean to the Baltic Sea, Russia has *nine* time zones. When Russians are getting ready for bed in Moscow, they're having breakfast in Vladivostok.

DEFINING THE REALM

As Figure 2A-1 reveals and a globe shows even better, Russia's entire northern coast, from its border with Norway in the west to the Bering Strait in the east, faces the Arctic Ocean, with all the climatic consequences you would expect. And for all its huge dimensions, most of the Russian realm lies at high latitudes, shut out by mountains and deserts from the warmer and moister parts of Eurasia. That has been a geographic problem for Russian rulers and governments for as long as Russia has existed as a state. Russia's ports on the Pacific Ocean in the east lie about as far from where most Russians live as they could be. And in the west, where Russia's core area is situated, every maritime outlet is restricted in some way. The Arctic ports operate on seasonal access. The Baltic Sea ports lie far from the open ocean. The Black Sea ports (some leased from Ukraine) require navigation through narrow straits across Turkish territory simply to reach the eastern Mediterranean Sea.

The Russian realm is vast, but its human numbers are modest. As the Data Table in Appendix B reports, this entire realm's population of barely 160 million is now smaller than

major geographic qualities of RUSSIA

1. Russia is the largest territorial state in the world. Its area is nearly twice as large as that of the next-ranking country (Canada).
2. Russia is the northernmost large and populous country in the world; much of it is cold and/or dry. Extensive rugged mountain zones separate Russia from warmer subtropical air, and the country lies open to Arctic air masses.
3. Russia is by far the "widest" realm on Earth, stretching from west to east for some 10,000 kilometers (over 6000 mi) and spanning nine time zones.
4. Russia was one of the world's major colonial powers. Under the czars, the Russians forged the world's biggest contiguous empire; the Soviet rulers who succeeded the czars expanded that empire.
5. For so large a territory, Russia's shrunken population of just under 143 million is comparatively small. The population remains heavily concentrated in the westernmost one-fifth of the country.
6. Development in Russia is concentrated west of the Ural Mountains; here are the major cities, leading industrial regions, densest transport networks, and most productive farming areas. National integration and economic development east of the Urals extend mainly along a narrow corridor that stretches from the southern Urals region to the southern Far East around Vladivostok.
7. Russia is a multicultural state with a complex domestic political geography and administrative structure.
8. Its huge territorial size notwithstanding, Russia faces land encirclement within Eurasia; it has few good and suitably located ports.
9. Regions long part of the Russian and Soviet empires are still realigning themselves in the post-communist era. Eastern Europe and the heavily Muslim Southwest Asia realm are encroaching on Russia's imperial borders.
10. The failure of the Soviet communist system left Russia in economic disarray. It is often considered to be one the world's major emerging economies, but its growth is fragmentary and precarious. Russia is overly dependent on exports of oil and natural gas.

Courtesy of Dr. Elena Grigorieva, ICARP FEB RAS, Birobidzhan, Russia

The mid-winter landscape beyond Siberia in far eastern Russia. This is the World War II Memorial in the city of Birobidzhan, one of the more than 70,000 such monuments the Soviets erected all across the USSR in the postwar era. No country suffered greater losses (26.6 million) during that "Great Patriotic War" of the early 1940s, a human catastrophe that will forever scar Russian history.

those of such individual countries as Nigeria and Pakistan. Moreover, the realm's population has declined substantially since 1990 before leveling off at the beginning of this decade. For a combination of reasons we will examine later, the 1991 collapse of the Soviet Union created political, economic, and social conditions that continue to produce negative demographic effects today. This issue has come to the forefront in Russia at a time when the country's leaders want to restore Russia's power and prestige in the wider world. To accomplish that restoration, Russia will need to strengthen its social institutions, diversify its economy, and restore the confidence lost when the Soviet system disintegrated. The Russian geographic realm as defined by our framework consists of four political entities: the giant Russian state itself plus three small countries in Transcaucasia, the mountainous area between the Black Sea and the Caspian Sea. Georgia, Armenia, and Azerbaijan are small in size, and their combined population is just under 11 percent of the realm's total. They are located in a historically turbulent zone of conflict between Russian and non-Russian peoples.

Take a look at the Russian realm in Figure G-3, and you will note that many of Russia's neighbors, even those countries that formed part of the former Soviet Union, are now absorbed into other realms. The Baltic states, for example, have become an integral part of Europe as members of the European Union. The Central Asian countries once under Moscow's sway, such as Kazakhstan and Uzbekistan, now form a discrete region in an adjoining realm in which Islam is resurgent. But the three Transcaucasian states are not collectively European (although Georgia has European aspirations), or Islamic (only in Azerbaijan does Islam have a strong foothold), or Russian. What these three states have in common is Russia's still-powerful influence in one form or another. Georgia endures Russian political, economic, and even military intervention; Azerbaijan seeks ways to avoid having to depend on Russian transit for its energy exports; and Armenia continues to view Russia as its most dependable ally. For these reasons, elaborated in the following discussion, we map these states as part of the Russian geographic realm.

Even though our first task is to look in some detail at the giant physical stage on which the Russian geographic realm is built, we should note that the human geography of this realm is not neatly defined by sharp boundaries. Look again at Figure G-3, and you will see that the margins of the Russian realm in two prominent places (and elsewhere not shown at that map's small scale) are marked by ***transition zones***. There, geographic features of this realm spill over into neighboring realms, sometimes creating social and political problems for the adjacent countries affected. But first, let us examine the physical landscapes, climate, and ecologies of the Russian realm.

PHYSICAL GEOGRAPHY OF THE RUSSIAN REALM

Physiographic Regions

The first feature you notice when looking at the physiographic map of the Russian realm is a prominent north-south trending mountain range that extends from the Arctic Ocean southward to Kazakhstan (Figs. 2A-1 and 2A-2), dividing Russia into two parts: the Russian Plain to the west and Siberia to the east. This range, the Ural Mountains, is sometimes designated as the "real" eastern boundary of Europe, but as we noted in Chapters 1A and 1B, Russia is not Europe. Cultural life to the east of the Urals is pretty much the same as it is to the west, and there is no geographic justification for putting the city of Samara (see Fig. 2A-1) in Europe but Chelyabinsk, on the other side of the Urals, outside of it. Neither is European; both are Russian.

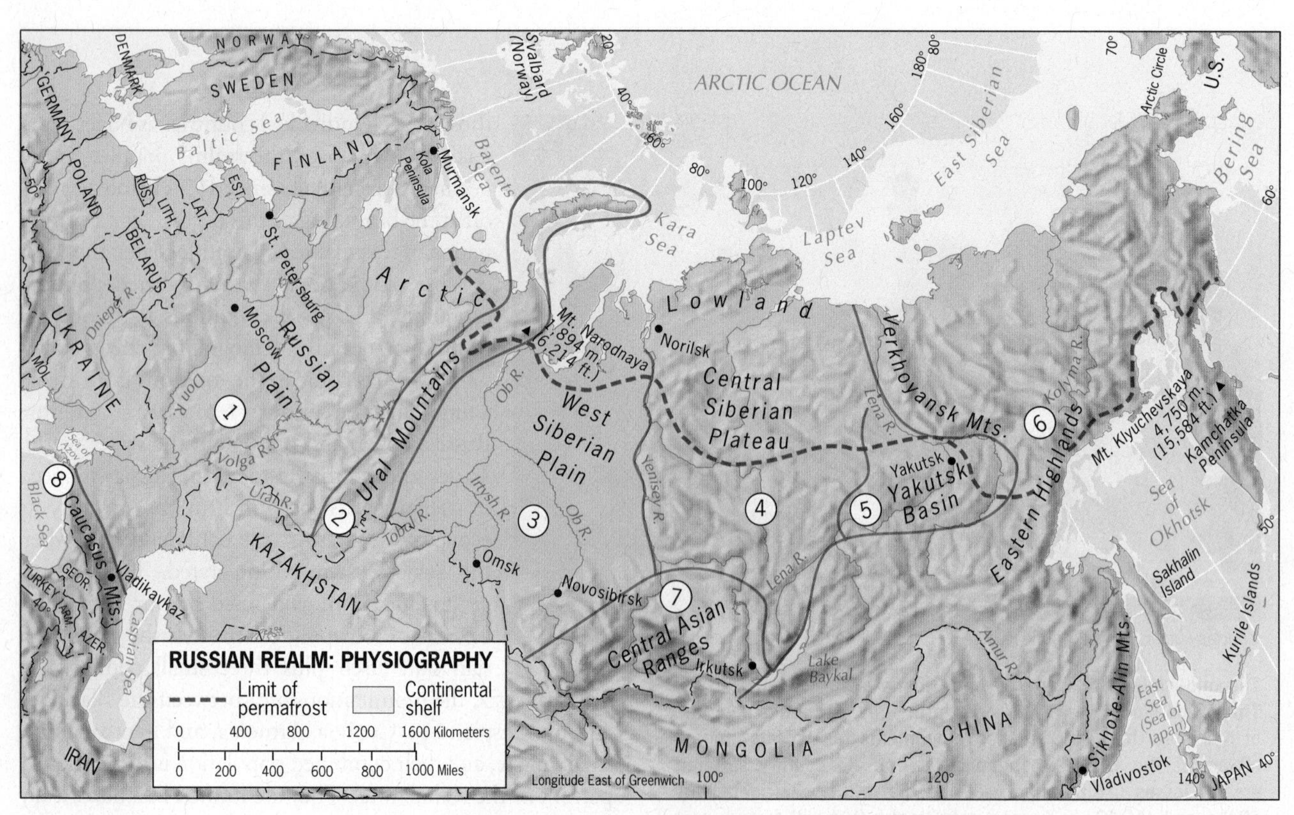

FIGURE 2A-2

The Russian Plain

The Russian Plain, which lies west of the Ural Mountains, is the eastward continuation of the North European Lowland, and here lies Russia's ***core area***. Travel northward from the centrally located capital of Moscow, and the countryside soon changes to coniferous (needleleaf) forests like those of Canada. Centuries ago, those forests afforded protection to the founders of the Russian state when invaders on horseback from eastern Asia swept into this region. The Volga River, Russia's Mississippi, flows southward in a wide arc across the plain from the forested north—but unlike the Mississippi, the Volga drains into a landlocked lake, the Caspian "Sea." Travel southward from Moscow today, and the land is draped in grain fields and pastures. Eastward, the Ural Mountains form a range not tall enough to create a major barrier to transportation, but it is prominent because it separates two vast expanses of low ***relief*** (elevation range).

In geopolitical texts, the western Russian Plain has also been dubbed the Eurasian **heartland [1]** because it lies deep within the world's greatest landmass and because it has had a major influence throughout history on the shaping of societies to the west (Europe), southeast (Central Asia), and far east (China, Japan, Korea, and even across the Bering Strait into North America). But this exceptional geographic position not only provides opportunities for expansion; it also implies potential vulnerability and a sense of encirclement. This apparent contradiction has been a feature of the realm for centuries.

The village of Tara on the Irtysh River in the Omsk Region of the Siberian Federal District displays little or no evidence of modernization. Czarist-era wooden houses and wobbly sheds are seen here on a sunny summer day. But even though this is one of Russia's southernmost administrative units (see Fig. 2B-2), it lies exposed to the harsh Siberian winter; when the river freezes over, you can walk to the inhabited meander knoll seen in the near distance. The short growing season allows cultivation of spring wheat and some other early-ripening crops, and small farm settlements like this dot the flat landscape.

Siberia

The West Siberian Plain, on the far side of the Urals, is often described as the world's largest unbroken lowland, and here the rivers flow north, not south. Over the last 1600 kilometers (1000 mi) of its course to the Arctic Ocean, the Ob River falls less than 90 meters (300 ft) in huge meanders flanked by forests until the trees give way in the near-Arctic cold to mosses and lichens. Follow an imaginary trail eastward, and in just about the middle of Russia the West Siberian Plain gives way to the higher relief of the Central Siberian Plateau, one of the most sparsely populated areas in the habitable world. Where the West Siberian Plain meets the Central Siberian Plateau, the Yenisey River, also flowing northward into the Arctic Ocean, marks the change of landscape.

Continuing our eastward transect, we next encounter the Yakutsk Basin, drained by the Lena River and forming the last vestige of moderate ***topography*** (surface configuration). Now we approach the realm's mountainous eastern perimeter (look again at Fig. 2A-1), where the few roads and railroads must wend their way through tunnels and valleys with hairpin turns. We map this jumble of ranges collectively as the Eastern Highlands, but this is one of the most spectacular, diverse, and still-remote regions on Earth.

Kamchatka and Sakhalin

In its farthest eastern reaches, the Russian realm makes contact with the Pacific Ring of Fire (see Fig. G-5). There are no active volcanoes in Siberia and earthquakes are rare, but the Kamchatka Peninsula has plenty of both. This is one of the most volatile segments of the Pacific Ring of Fire, with more than 60 volcanoes, many of them active or dormant, forming the spine of the peninsula. Nothing here resembles Siberia (the climate is moderated by the Pacific waters offshore; the vegetation is mixed, not coniferous). It is one of the most difficult places on Earth to live, virtually disconnected from the rest of Russia (and entirely so overland).

Also on the Pacific fringe of the Russian realm is an island named Sakhalin, where earthquakes rather than volcanic activity are the leading environmental hazard. Long a battleground between Russians and Japanese, Sakhalin finally fell to the Russians at the end of World War II. In subsequent decades the island and its maritime environs proved to contain substantial reserves of oil and natural gas, making it a valuable asset in Russia's energy-based export economy.

The Southern Perimeter

A final look at Figures 2A-1 and 2A-2 reminds us that mountainous topography encircles much more than just the eastern edge of the Russian realm. Some of the highest relief in Eurasia prevails in the southern interior, where the Eastern Highlands meet the Central Asian Ranges; here lies Lake Baykal in a tectonic trough that is more than 1500 meters (5000 ft) deep—the deepest lake of its kind in the world. And between the Caspian Sea and the Black Sea, the Caucasus Mountains, geologically an extension of Europe's Alps, rise like a wall between Russia and the lands beyond—a multi-tiered barrier over which Russians have fought with their neighbors for centuries.

When you approach the forbidding Caucasus ranges from the low relief of the Russian Plain to the north, they seem impenetrable, and you realize why so few routes cross them, even today. It is also clear why Russia's armies had such difficulty pushing across the Caucasus into the areas beyond. At present, these mountains shelter insurgents defying Russia's authority in such territories as Chechnya and Ingushetiya. And the zone beyond is no easy target: to the south of the Caucasus, high relief dominates all the way into neighboring Turkey and Iran.

Harsh Environments

The historical geography of Russia is the story of Slavic expansion from its populous western heartland across interior Eurasia to the east, and into the mountains and deserts of the south. This eastward march was hampered not only by vast distances but also by harsh natural conditions. As the northernmost populous country on Earth, Russia has virtually no natural barriers against the onslaught of Arctic air. Moscow lies farther north than Edmonton, Canada, and St. Petersburg lies at latitude 60° North—the latitude of the southern tip of Greenland. Winters are long, dark, and bitterly cold in most of Russia; summers are short and growing seasons limited. Many a Siberian frontier outpost was doomed by cold, snow, and hunger.

It is therefore useful to view Russia's past, present, and future in the context of its climate. Even today, in this age of global warming, Russia still suffers from severe cold and associated drought. Precipitation, even in western Russia, ranges from modest to minimal because the warm, moist air carried across Europe from the North Atlantic Ocean loses much of its warmth and moisture by the time it reaches Russia. Figure 2A-3 reveals the consequences.

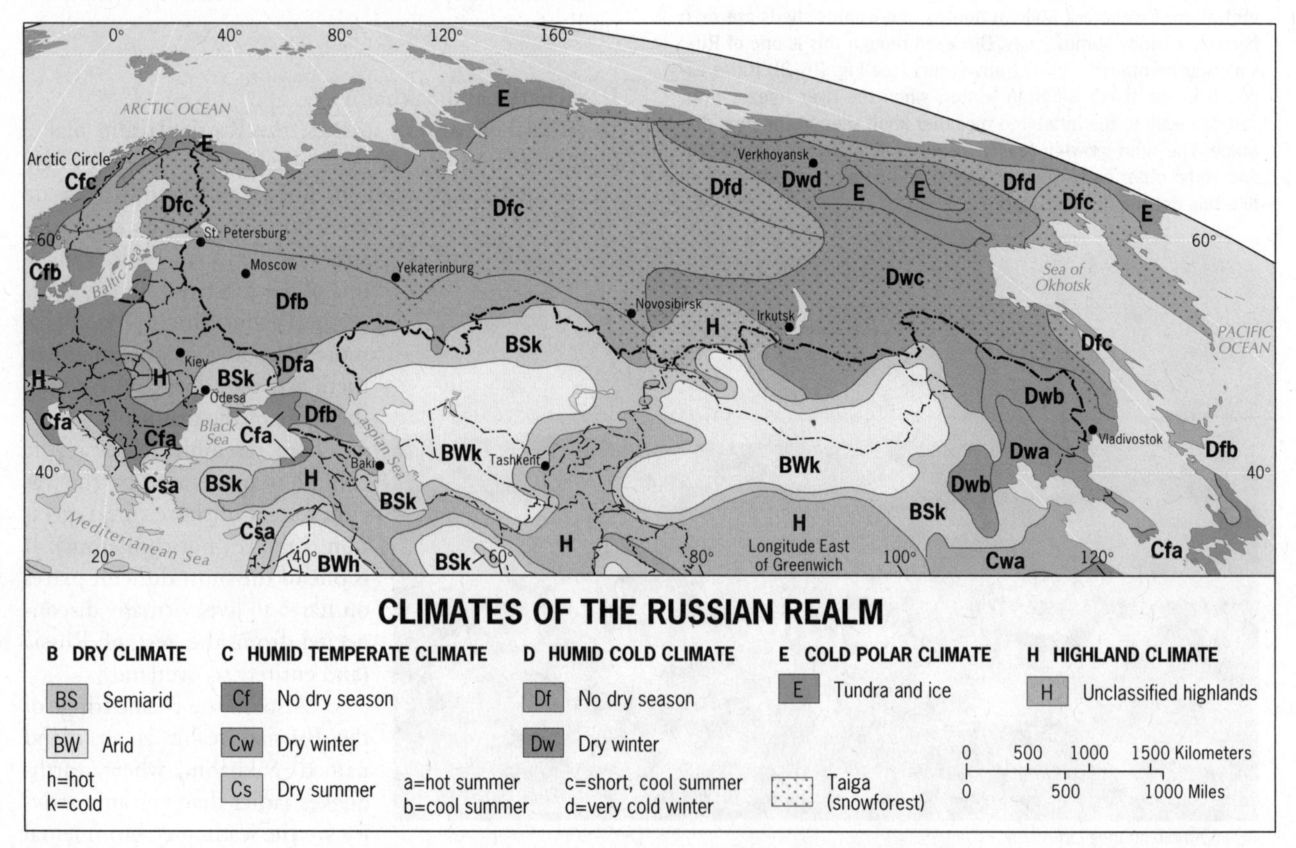

FIGURE 2A-3

DEA/C. SAPPA/Getty Images

As Figure 2A-3 shows, Siberia is cold and, in Russia's far northeast, dry as well. In the north lies the tundra, treeless and windswept, and beyond is the ice of the Arctic. But where somewhat more moderate conditions prevail (moderate being a relative concept in northwestern Russia and Siberia), coniferous forests known as *taiga* cover the countryside. Also called boreal (cold-temperate) forests, these evergreen, needleleaf pines and firs create a dense and vast high-latitude girdle of vegetation across northern Eurasia as well as northern North America. This view from the air shows how tightly packed the trees are; they are slow-growing, but long-lived. Most of the world's taiga forest, among the largest surviving stands of primary forest on the planet, remains protected by distance from the threat of exploitation—but lumbering is nevertheless taking its toll. Recent studies indicate that climate change in high latitudes is enabling the forest to expand northward at a faster rate than it diminishes due to lumbering activities elsewhere, a rare case of good news relating to global warming.

Russia's climatic **continentality [2]** (inland climatic environment remote from moderating and moistening maritime influences) is expressed by its prevailing ***Dfb*** and ***Dfc*** conditions. Compare the Russian map to that of North America (Fig. G-7), and you note that, except for a small corner off the Black Sea, Russia's climatic conditions resemble those of the Upper Midwest of the United States and interior Canada. Along its entire northern edge, Russia has a zone of ***E*** climates, the most frigid on the planet. In these Arctic latitudes originate the polar air masses that dominate its environments.

The Russian realm's harsh northern climates affect people, animals, plants—and even the soil. In Figure 2A-2 you can find a direct consequence of what you see in Figure 2A-3: a dashed blue line that starts at the shore of the Barents Sea, crosses Siberia eastward, and reaches the coast of the Pacific Ocean north of the neck of the Kamchatka Peninsula. North of this line, water in the ground is permanently frozen, creating an even more formidable obstacle to settlement and infrastructure than the severe weather alone. It is referred to as **permafrost [3]**, and it affects other high-latitude environments as well (Alaska, for example). Where the permafrost ends, seasonal temperature changes cause alternate thawing and freezing, with destructive impacts on buildings, roads, railroad tracks, and pipelines.

Examine Figure 2A-3 carefully, and you will see two ecological terms: *tundra* and *taiga*. **Tundra [4]**, as the map suggests, refers to both climate and vegetation. The blue area mapped as ***E*** marks the coldest and ice-affected environmental zone in Russia and elsewhere in the high Arctic: this is frigid, treeless, windswept, low-elevation terrain where bare ground and rock prevail and mosses, lichens, patches of low grass, and a few hardy shrubs are all that grows. **Taiga [5]**, the stippled area on the map, a Russian word meaning "snowforest" (also called boreal forest), extends across vast reaches of Eurasia as well as northern North America and is dominated by coniferous (as in pine cone) trees. As the map shows, the taiga extends southward into relatively moderate environs, and there it becomes a mixed forest of coniferous and deciduous trees. Although the taiga prevails from northern Scandinavia to the Russian Far East, this is the vegetative landscape most often associated with Siberia: endless expanses of rolling countryside draped in dense stands of evergreen pine trees (see photo).

Climates and Peoples

Climate and weather (there is a distinction: ***climate*** refers to a long-term average, whereas ***weather*** describes a set of immediate atmospheric conditions at a given place and time) have always challenged the peoples of this realm. The high winds that drive the bitter Arctic cold southward deep into the landmass, the blizzards of Siberia, the temperature extremes, rainfall variability, and short and undependable growing seasons even in the more moderate western parts of this realm have always made farming difficult. That was true during the time of the ruling czars, when the threat of famine never receded. It was also a constant problem during the Soviet-communist period, when Moscow was the capital of an empire far larger than this realm and the non-Russian sectors of that empire (such as Ukraine and parts of Central Asia) produced food staples Russia needed. Despite a massive effort to restructure agriculture throughout the Soviet Empire by means of collectivization and irrigation projects, the Russians often had to import grain.

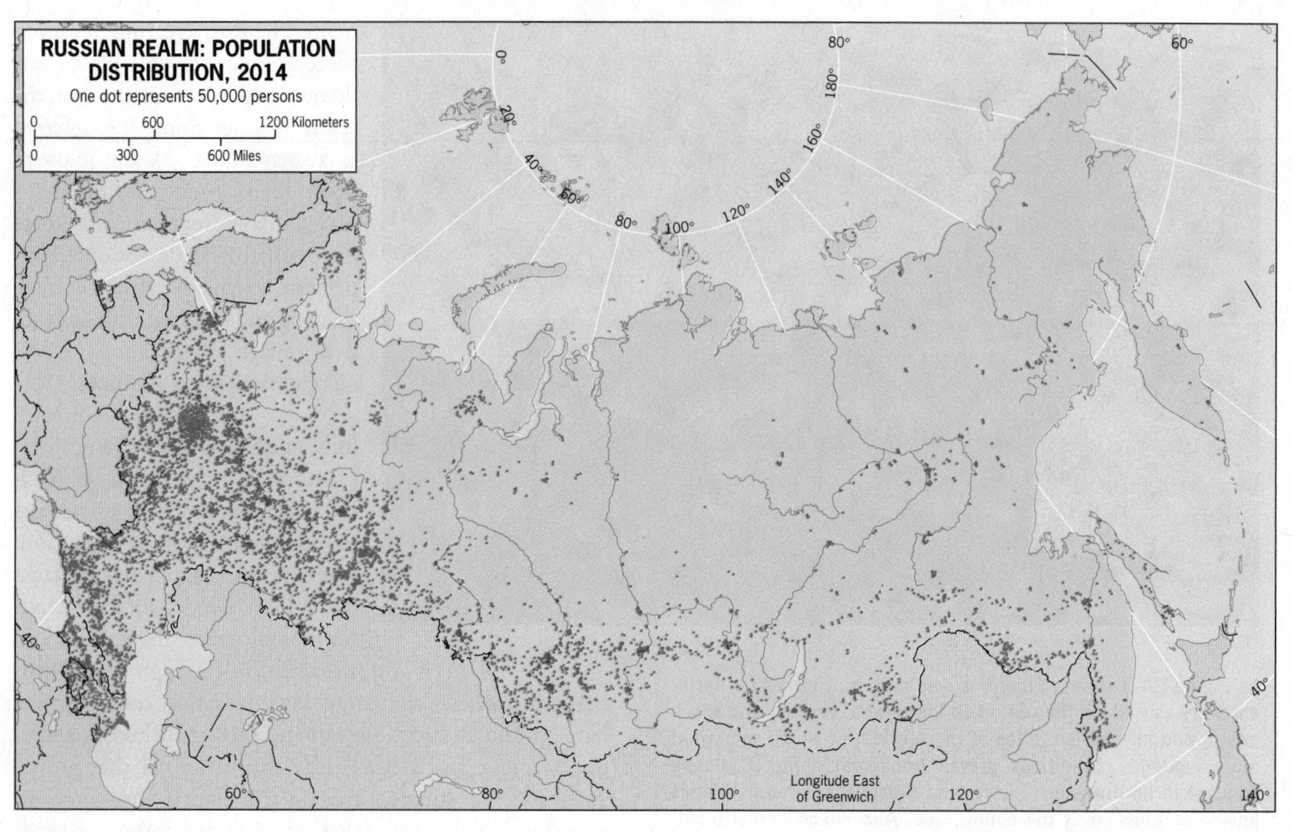

FIGURE 2A-4

As we noted in the Introduction, and as we will observe time and again throughout this book, humankind's long-term dependence on agriculture remains etched on the population map, even as our planet becomes ever more urbanized and its economy more globalized. By studying the climates of the Russian realm, we can begin to understand what the map of population distribution communicates (Fig. 2A-4). The overwhelming majority of the realm's 160 million inhabitants remain concentrated in the west and southwest, where environmental conditions were least difficult at a time when farming was the mainstay of most of the people. To the east, the population is sparser and tends to cluster along the southern margin of the realm, becoming even more thinly distributed east of Lake Baykal. If you consider that three-quarters of this realm's population today lives in cities and towns, it is not difficult to imagine just how empty vast stretches of countryside must be—especially in frigid northern latitudes.

Climate Change and Arctic Prospects

As every map of this high-latitude realm shows, its northern coast lies entirely on the polar side of the Arctic Circle. Such a lengthy coastline on the Arctic Ocean does not exactly constitute an advantage: most of the Arctic Ocean is frozen much of the year, and only some warmth from the North Atlantic Drift ocean current keeps the western ports of Murmansk and Arkhangelsk open a bit longer. Ports like this (and even St. Petersburg on the Gulf of Finland, an arm of the Baltic Sea) would never have developed the way they did if Russia had better access to the world ocean; but as we noted, this has been one of the country's historic impediments.

Now it looks as though nature may give Russia a helping hand. If, as most climatologists anticipate, global warming causes long-term melting of large portions of the Arctic Ocean's ice cover, that body of water may come to play a new and different role in this realm's future. Milder atmospheric conditions may shrink the area of permafrost mapped in Figure 2A-2; moister air masses may improve agriculture on the Russian Plain; warmer water may keep Arctic ports open year-round; and the so-called Northern Sea Route (also known as the Northeast Passage) might even open up between the Bering Strait and the North Sea (see Fig. 12-5), shortening some international sea routes by thousands of kilometers and ushering in a new era of trade.

That, at least, is how many Russians like to envision global warming—as a potential advantage nature has long denied them. And indeed, the Arctic provides much evidence for such warming: the melting of parts of the Greenland Ice Sheet, the shrinking of the average extent of permanent Arctic Ocean ice, and the reduced incidence of icebergs. Russian economic planners look at the shallow waters offshore along the Arctic Ocean coastline (see the white- and light-blue areas in Fig. 2A-1) and hope that oil and gas reserves beneath those waters will come within reach of exploitation.

These developments may even have the effect of expanding the Russian geographic realm far into the Arctic. As we note when we discuss the polar areas in greater detail in Chapter 12, states with coastlines in the Arctic are likely to demand certain exclusive rights not only over waters offshore but also the resources on and beneath the ocean floor. It is therefore in their interest to extend those rights as far from shore as possible, even hundreds of (nautical) miles outward. In 2007, the Russian government even placed a metal Russian flag at the North Pole on the seafloor, under the permanent ice of the Arctic Ocean, symbolizing its intentions (see final photo in Chapter 12). The map of the northern perimeter of the Russian realm is changing—but not always for the better.

Ecologies at Risk

If, as computer models predict, global warming continues and perhaps intensifies in the polar latitudes, these ecologically sensitive environments are sure to be severely affected. Animals as well as humans have long adapted to prevailing climatic conditions, and such adaptations would be significantly disrupted by environmental change. The intricate web of relationships among species and their environments on the one hand, and among species themselves on the other, can be quickly damaged by temperature change. The polar bear is just one prominent example: it depends on ample floating sea ice to hunt as well as raise cubs. When sea ice contracts and wider stretches of open water force polar bears to swim greater distances during the cub-rearing season, fewer infants make it to adulthood. If ice-free Arctic summers are indeed in the offing, the polar bear could become extinct in your lifetime. Rapid ecosystem change would also endanger seal, bird, fish, and other Arctic wildlife populations.

Such changes affect human populations as well. For example, Inuit (formerly called Eskimo) communities still pursuing traditional lives in parts of the Arctic domain (though not in the Russian realm) are similarly adapted to the harsher environments of the past. Their traditions, already under pressure from political and economic forces resulting from their incorporation into modern states, would be further threatened by environmental changes that are likely to alter, or even destroy, the ways of life they developed over thousands of years.

And if a new era of oil and natural gas exploration and exploitation is indeed about to open, already-fragile offshore environments face even greater peril. As technologies of recovery put ever more of these reserves within reach, oil platforms, drills, pumps, and pipelines will make their appearance—along with risks of oil spills, pollution, disturbance, and damage of the kind we have seen all over the lower-latitude world. Clearly, the forces of globalization are finally penetrating a part of the world long protected from it by distance and nature.

RUSSIA'S NATURAL RICHES

Given the enormous territorial dimensions of the Russian realm, it is not surprising that its natural resources are vast and varied. The Russians today are excited about the energy reserves that are to be found and exploited beneath Arctic waters, but this realm already is a major producer and exporter of oil and natural gas from a growing number of fields dispersed across Russia from the North Caucasus in the west to the island of Sakhalin in the east and from the Arctic shore of Siberia in the north to the Caspian Basin in the south (Fig. 2A-5).

And petroleum and natural gas are not the only underground riches of this realm. Deposits of so many minerals lie within Russian territory that virtually all of the raw materials required by modern industry are present. Nor are oil and natural gas the only energy reserves. Major coalfields are found east as well as west of the Urals, and in Siberia as well as in more southerly latitudes (the string of mines producing high-quality coal along the Trans-Siberian Railroad corridor were vital in Russia's industrial development and its successful war against Nazi Germany). Large deposits of iron ore, too, lie widely scattered across this enormous realm, from the so-called Kursk Magnetic Anomaly near the border with Ukraine to the remotest corners of Siberia's Arctic north. And when it comes to other metallic ores, this realm has it all—or almost all—from gold to lead and from aluminum to zinc. Moreover, one of the world's largest concentrations of nonferrous (non-iron) metals lies in and around the Ural Mountains, where Russia's metallurgical industries emerged.

And there is likely to be much more. Vast reaches of this far-flung realm have not yet been fully explored, especially in the far northeast. Given such material assets, you would imagine that this realm has always been in the forefront of industrial production and economic diversification. But, as we shall also see, the availability of raw materials is only one part of such development. You are likely to buy Chinese and Japanese products on an almost daily basis—but you are likely to own very few if any items of Russian manufacture. That is part of the story we relate in Chapter 2B.

Environment—in the form of climate, relief, vegetation, and other elements—has much to do with the way Russia evolved as the dominant power over this vast realm. To better understand the map, let us briefly trace the historical geography of Russia, its defenders, and rulers who eventually pushed Russian influence far beyond the Russian Plain—even into western North America. It may come as a surprise to many Americans that Russians not only colonized Alaska, but even reached central California and built a fort not very far north of present-day San Francisco. So let us focus first on place and time in Russia's evolution.

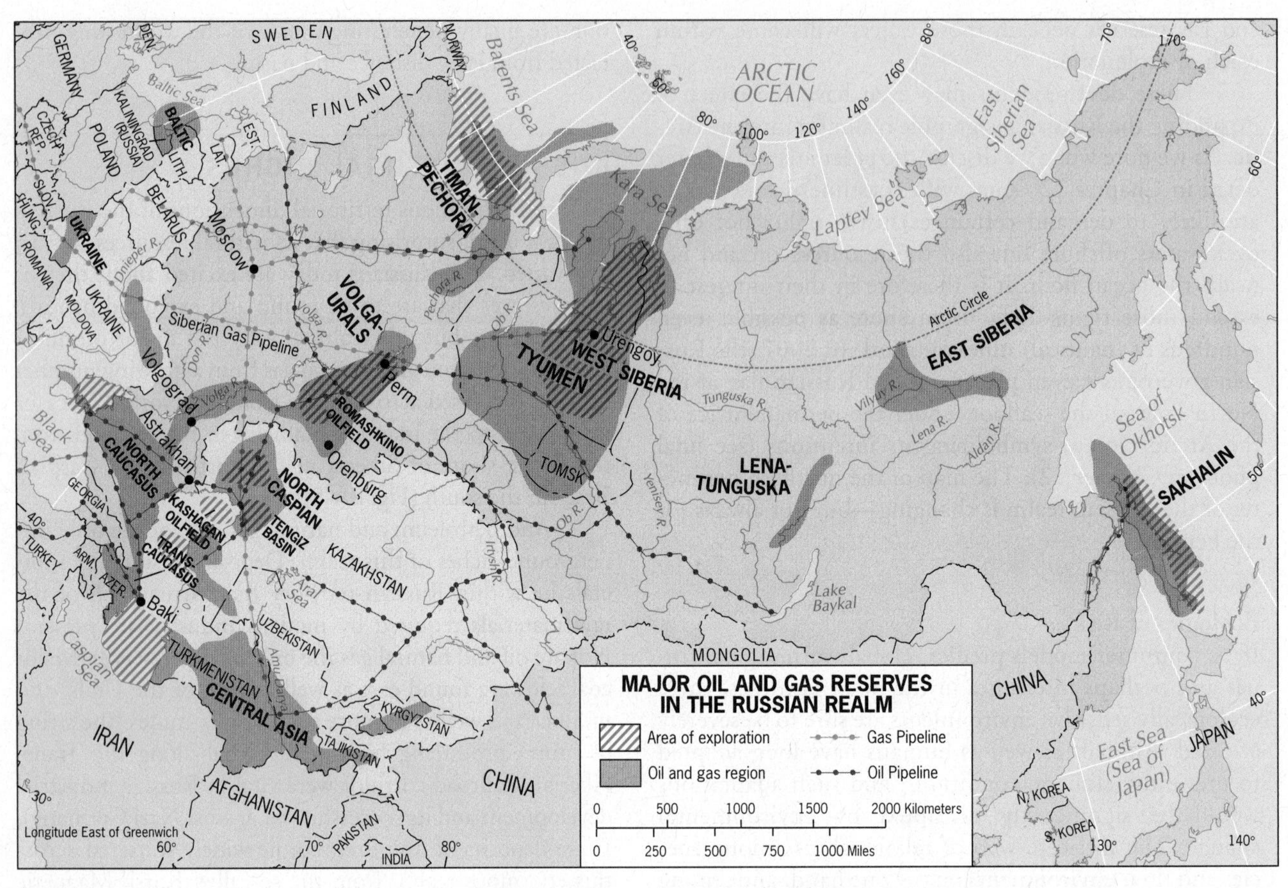

FIGURE 2A-5 © H. J. de Blij, P. O. Muller, and John Wiley & Sons, Inc.

RUSSIAN ROOTS

A thousand years ago, Eurasian peoples of many ethnic sources and cultural backgrounds were migrating across the plains south of the taiga (coniferous snowforest) in search of new and secure homelands; Scythians, Sarmatians, Goths, Huns, and others came, fought, settled, survived, absorbed neighbors, or were driven off. Eventually the Slavs, comparative latecomers to this turbulent stage, established settlements in the area of present-day Ukraine, north of the Black Sea and in the southwestern corner of the physiographic region we have defined as the Russian Plain (Fig. 2A-2).

Here the Slavic peoples found fertile soils, a relatively moderate climate, and a physical landscape with many advantages and opportunities: the Dnieper River and its tributaries, stands of forest, relatively low relief, and easy contact among settlements. The Slavs used the name ***Rus*** to designate such settlements, and the largest and most successful of these early Russes was the one located where the capital of modern Ukraine, Kiev (Kyyiv), lies today. This is one reason why many "Russians" today cannot imagine an independent, Europe-oriented Ukraine: this, after all, is their historic heartland. And from this southern base, the Slavs expanded their domain across the Russian Plain, first establishing their northern headquarters at Novgorod on Lake Ilmen (see Fig. 2A-1). This northern Rus was well positioned to benefit from the trade between the Hanseatic ports on the Baltic Sea and the trading centers on the Black and Mediterranean seas. During the eleventh and twelfth centuries, the Kievan Rus and the Novgorod Rus combined to form a large and prosperous state astride both the northern forest and southern **steppe [6]** (semiarid grassland).

The Mongol Invasion

Prosperity attracts attention, and knowledge about the Russes spread far and wide. In the distant east, north of China, another successful state had been building: the empire of the Mongol peoples under Genghis Khan. Also under the sway of this legendary ruler was a group of nomadic, Turkic-speaking peoples known as the Tatars. Together, Mongol-Tatar armies rode westward on horseback into the domain of the Russes to challenge the power of the Slavs.

Geography had much to do with their early success: on the open steppes of the southern Russian Plain, the Russes lay exposed to the fast-charging Mongol forces, and by the middle of the thirteenth century, the Kievan Rus had fallen. Slavic refugees were fleeing into the northern forests, where they reorganized to face their enemies in newly built Russes. The forest environment was their ally: Mongol tactics were effective for the open plain, but not in the woods. What ensued was not a victory for either side,

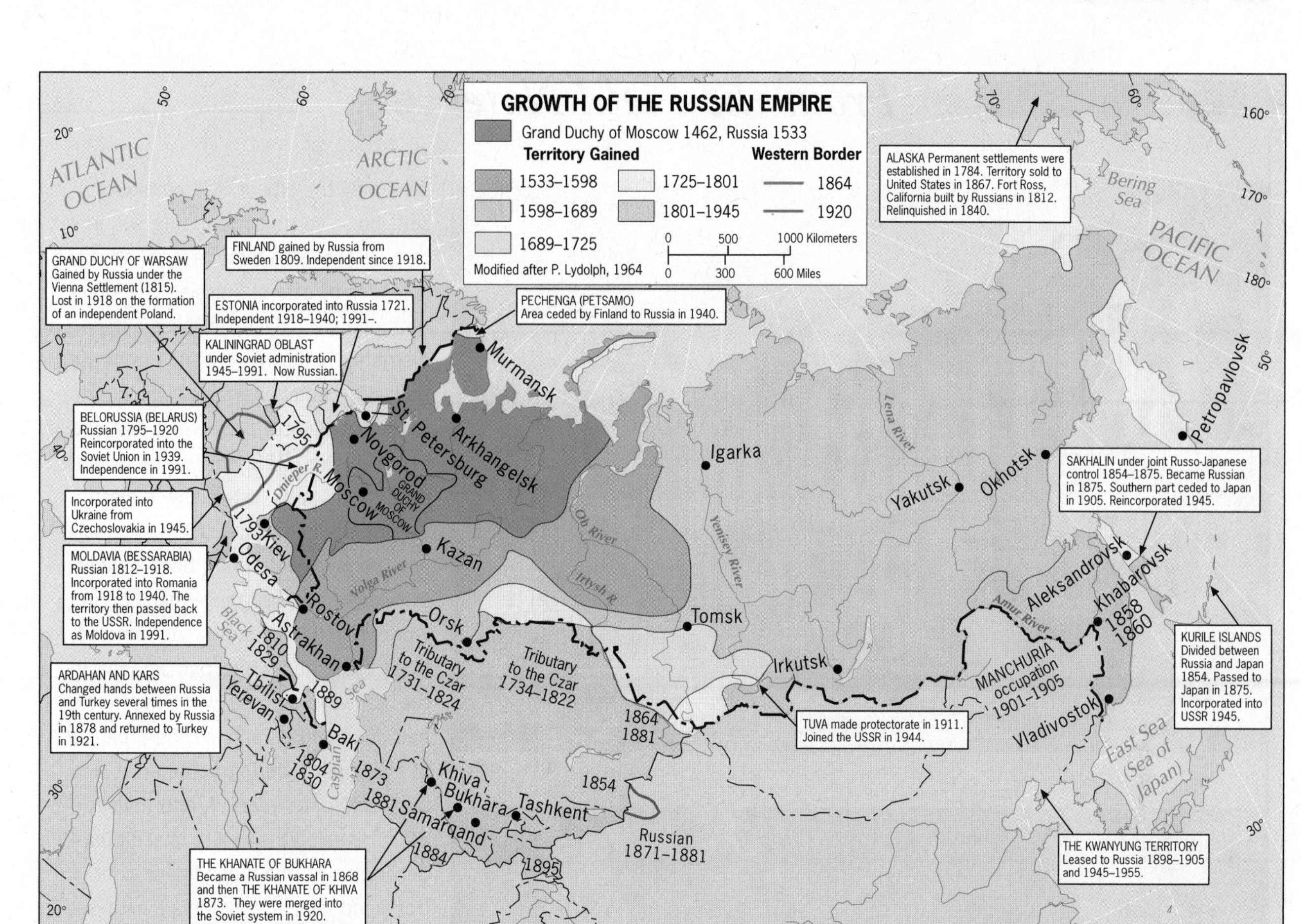
GROWTH OF THE RUSSIAN EMPIRE
Grand Duchy of Moscow 1462, Russia 1533
Territory Gained
1533–1598
1598–1689
1689–1725
1725–1801
1801–1945
Western Border
1864
1920
0 500 1000 Kilometers
0 300 600 Miles
Modified after P. Lydolph, 1964
ATLANTIC OCEAN
ARCTIC OCEAN
Bering Sea
PACIFIC OCEAN
GRAND DUCHY OF WARSAW Gained by Russia under the Vienna Settlement (1815). Lost in 1918 on the formation of an independent Poland.
FINLAND gained by Russia from Sweden 1809. Independent since 1918.
ESTONIA incorporated into Russia 1721. Independent 1918–1940; 1991–.
KALININGRAD OBLAST under Soviet administration 1945–1991. Now Russian.
PECHENGA (PETSAMO) Area ceded by Finland to Russia in 1940.
ALASKA Permanent settlements were established in 1784. Territory sold to United States in 1867. Fort Ross, California built by Russians in 1812. Relinquished in 1840.
BELORUSSIA (BELARUS) Russian 1795–1920 Reincorporated into the Soviet Union in 1939. Independence in 1991.
Incorporated into Ukraine from Czechoslovakia in 1945.
MOLDAVIA (BESSARABIA) Russian 1812–1918. Incorporated into Romania from 1918 to 1940. The territory then passed back to the USSR. Independence as Moldova in 1991.
ARDAHAN AND KARS Changed hands between Russia and Turkey several times in the 19th century. Annexed by Russia in 1878 and returned to Turkey in 1921.
THE KHANATE OF BUKHARA Became a Russian vassal in 1868 and then THE KHANATE OF KHIVA 1873. They were merged into the Soviet system in 1920.
SAKHALIN under joint Russo-Japanese control 1854–1875. Became Russian in 1875. Southern part ceded to Japan in 1905. Reincorporated 1945.
KURILE ISLANDS Divided between Russia and Japan 1854. Passed to Japan in 1875. Incorporated into USSR 1945.
TUVA made protectorate in 1911. Joined the USSR in 1944.
THE KWANYUNG TERRITORY Leased to Russia 1898–1905 and 1945–1955.
MANCHURIA occupation 1901–1905
Murmansk
St. Petersburg
Arkhangelsk
Novgorod
GRAND DUCHY OF MOSCOW
Moscow
Kazan
1795
Dnieper R.
1793 Kiev
Odesa
Rostov
Astrakhan
Black Sea
1810 1829
Tbilisi
Yerevan
1889
Baki
1804 1830
Caspian Sea
Volga River
Orsk
Tributary to the Czar 1731–1824
Tributary to the Czar 1734–1822
1873
1881 Samarqand
Khiva
Bukhara
Tashkent
1884
1895
1854
1864 1881
Russian 1871–1881
Ob River
Irtysh R.
Yenisey River
Lena River
Tomsk
Igarka
Irkutsk
Yakutsk
Okhotsk
Petropavlovsk
Aleksandrovsk
Khabarovsk
1858 1860
Amur River
Vladivostok
East Sea (Sea of Japan)
Longitude East of Greenwich

FIGURE 2A-6

but a standoff: the Tatars threatened and tried to lay siege, but they could not win outright. So the leaders of the forest-based Russes paid tribute to the Mongol-Tatar invaders, and in exchange they were left alone.

One of these Russes was Moscow, deep in the forest on the Moscow River and destined to become the capital of a vast empire. On a defensible site and in a remote situation, Moscow's leaders began to establish trade links with even safer Novgorod, and the city grew and thrived. When the Mongols, worried about Moscow's growing power and influence, attacked and were successfully repulsed, Moscow emerged as the leader among Russes, its destiny assured. The Mongol-Tatar campaign had essentially failed.

Before we go on, we should take note of a crucial geographic development. Whereas the Mongols and Tatars fought together, their failure to conquer the Slavic-Russian heartland had both cultural and geographic consequences that have endured into the present. Although most of the Mongols withdrew following their unsuccessful fourteenth-century assault on Moscow, many Tatars remained on what, at the time, was the periphery around the Slavic/Russian core in such places as the Volga River Basin, the Crimean Peninsula, and other smaller pockets. There they were converted in the wave of Islam that was rolling into the Black Sea Basin and beyond, creating a new kind of division between Christian Slavs and their old-time adversaries.

Grand Duchy of Muscovy

During the fourteenth century, the Grand Duchy of Muscovy rose to preeminence under the rule of princes or dukes. They extended Moscow's trade links from the Baltic to Black Sea shores and forged valuable religious ties with the leadership of the Eastern Orthodox Church in Constantinople. Then, from around 1450, there began a period lasting more than three centuries during which powerful Russian rulers etched their imprints on the map of Russia, Eurasia, and, by extension, the world (Fig. 2A-6).

By the sixteenth century, during the reign of Ivan IV, better known as the infamous *Ivan the Terrible*, the Grand Duchy of Muscovy had been transformed into a major military power and an imperial state. He expanded Moscow's empire by conquering the Islamic regions to the south and gaining control over present-day Estonia and Latvia on the Baltic Sea. It was a time of almost continuous warfare, and Ivan the Terrible acquired his reputation from the reign of terror he unleashed in his pursuit of military discipline, centralized administrative control, and retaliation against many members of the nobility.

From the Field Notes . . .

© H.J. de Blij

© H.J. de Blij

"Not only the city of St. Petersburg itself, but also its surrounding suburbs display the architectural and artistic splendor of czarist Russia. The czars built opulent palaces in these outlying districts (then some distance from the built-up center), among which the Catherine Palace, begun in 1717 and completed in 1723 followed by several expansions, was especially majestic. During my first visit in 1994, the palace, parts of which had been deliberately destroyed by the Germans during World War II, was still being restored; large black and white photographs in the hallways showed what the Nazis had done and chronicled the progress of the repairs during communist and postcommunist years. A return visit in 2000 revealed the wealth of sculptural decoration on the magnificent exterior (top) and the interior detail of a set of rooms called the 'golden suite,' of which the ballroom (bottom) exemplifies eighteenth-century Russian Baroque at its height."

www.conceptcaching.com

BUILDING THE RUSSIAN EMPIRE

Czarist Russia

When Peter the Great became czar (he ruled from 1682 to 1725), Moscow already lay at the center of a great empire—great, at least, in terms of the territories it controlled. The Islamic threat had been ended with the defeat of the Tatars. The influence of the Russian Orthodox Church was represented by its distinctive religious architecture and powerful bishops. Peter consolidated Russia's gains and endeavored to make a modern, European-style state out of his loosely knit country. He built St. Petersburg as a **forward capital [7]** on the doorstep of Swedish-held Finland, fortified it with major military installations, and made it Russia's leading port.

Peter the Great, an extraordinary leader, was in several ways the founder of modern Russia. In his desire to remake Russia—to pull it from the forests of the interior to the seas of the west, to open it to outside influences, and to relocate its population—he left nothing to chance. Prominent merchant families were forced to move from other cities to St. Petersburg. Ships and wagons entering the city had

Russians in North America

THE FIRST WHITE settlers in Alaska were Russians, not western Europeans, and they came across Siberia and the Bering Strait, not across the Atlantic and North America. Russian hunters of the sea otter, valued for its high-priced pelt, established their first Alaskan settlement at Kodiak Island in 1784. Moving south along the North American coast, the Russians founded a chain of additional villages and forts to protect their tenuous holdings until they reached as far as the area just north of San Francisco Bay, where they built Fort Ross in 1812.

But these Russian settlements were isolated and vulnerable. European fur traders began to put pressure on their Russian competitors, and St. Petersburg soon found the distant settlements a burden and a risk. In any case, American, British, and Canadian hunters were decimating the sea-otter population, and profits declined. When U.S. Secretary of State William Seward offered to purchase Russia's North American holdings in 1867, St. Petersburg quickly agreed—for $7.2 million (more than $100 million in today's dollars). Thus Alaska, including its lengthy southward coastal extension, became U.S. territory and, in 1959, the forty-ninth State. Although Seward was ridiculed for his decision—Alaska was called "Seward's Folly" and "Seward's Icebox"—he was vindicated when gold was discovered there in the 1890s. The twentieth century further underscored the wisdom of Seward's action, strategically as well as economically. At Prudhoe Bay off Alaska's northern Arctic slope, large oil reserves were tapped in the 1970s and are still being exploited. And like Siberia, Alaska probably contains yet unknown riches. From the point of view of the United States, besides the 24 dollars allegedly paid to acquire Manhattan Island, this $7.2 million transaction is often regarded as the most lucrative real estate deal in world history.

to bring building stones as an entry toll. The czar himself, aware that to become a major power Russia had to be strong at sea as well as on land, went to the Netherlands to work as a laborer in the famed Dutch shipyards to learn the most efficient method for building ships. Meanwhile, the czar's forces continued to conquer people and territory: Estonia was incorporated in 1721, widening Russia's window to the west, and in Siberia major expansion soon occurred south of the city of Tomsk (Fig. 2A-6).

Under Czarina Catherine the Great, who ruled from 1760 to 1796, Russia's empire in the Black Sea Basin grew at the expense of the Ottoman Turks. The Crimea Peninsula, the port city of Odesa (Odessa), and the entire northern coastal zone of the Black Sea fell under Russian control. Also during this period, the Russians made a fateful move: they penetrated the corridor between the Black and Caspian seas, whose spine is the mountainous Caucasus with its dozens of ethnic and cultural groups, many of which were Islamized. The cities of Tbilisi (now in Georgia), Baki (Baku) in Azerbaijan, and Yerevan (Armenia) were captured. Eventually, the Russian push toward an Indian Ocean outlet was thwarted by the British, who held sway in Persia (modern Iran), and also by the Turks.

Meanwhile, Russian colonists (descendants of Cossacks who originally hailed from Ukraine and southwestern Russia) had advanced far to the east, settling along the southeastern frontier, crossing the Bering Strait, and entering Alaska in 1784 (see the box titled "Russians in North America"). Catherine had made Russia a huge colonial power, but within a century the Russians gave up on their North American outposts. The sea-otter pelts that had attracted the early pioneers were running out, European and white American hunters were cutting into the profits, and indigenous American resistance was growing. When the United States offered to purchase Russia's Alaskan holdings in 1867, the Russian government swiftly agreed—setting Alaska on a new course that culminated in its joining the U.S. as the forty-ninth State in 1959.

Nineteenth-Century Expansion

Although Russia had withdrawn from North America, Russian expansionism during the nineteenth century continued in Eurasia. While extending their empire southward, the Russians also took on the Poles, old enemies to the west, and succeeded in taking most of what is today the Polish state, including the capital of Warsaw. To the northwest, Russia took over Finland from the Swedes in 1809.

During most of the nineteenth century, however, the Russians were preoccupied with Central Asia—the region between the Caspian Sea and western China—where Tashkent and Samarqand (Samarkand) came under St. Petersburg's control (Fig. 2A-6). The Russians here were still bothered by raids of nomadic horsemen, and they sought to establish their authority over the Central Asian drylands as far as the edges of the high mountains that lay to the south. Thus Russia gained many Muslim subjects, because this was Islamic Asia they were penetrating. Under czarist rule, though, these people retained some autonomy.

Much farther to the east, a combination of Japanese expansionism and a decline of Chinese influence led Russia to annex from China a number of provinces north and southeast of the Amur River. Soon thereafter, in 1860, the Russians founded the port of Vladivostok on the Pacific. Now began the series of events that were to lead to the first involuntary halt in the Russian drive for territory. As

Figure 2A-1 shows, the most direct route from western Russia to the port of Vladivostok lay across northeastern China, the territory then still called Manchuria. The Russians had begun construction of the Trans-Siberian Railroad in 1892, and they wanted China to permit the tracks to cross Manchuria. But the Chinese resisted. Then, taking advantage of the Boxer Rebellion in China in 1900 (see Chapter 9A), Russian forces occupied Manchuria so that railway construction might proceed.

That move, however, threatened Japanese interests in this area, and Japan confronted Russia in the Russo-Japanese War of 1904–1905. Not only were the Russians beaten and forced out of Manchuria: Japan even took possession of the southern portion of Sakhalin Island, which they named Karafuto and retained until 1945.

Thus Russia, like Britain, France, and other European powers, expanded through ***imperialism***. Whereas the other European powers expanded overseas, Russian influence traveled overland into Central Asia, Siberia, China, and the Pacific coastlands of the Far East. What emerged was not the greatest empire but the largest territorially contiguous empire in the world.

The czars embarked on their imperial conquests in part because of Russia's relative location: Russia always lacked warm-water ports. Had the early-twentieth-century Revolution not intervened, their southward push might have reached the Persian Gulf or even the Mediterranean Sea. Czar Peter the Great envisaged a Russia open to trading with the entire world; he developed St. Petersburg on the Baltic Sea into Russia's leading port. But in truth, Russia's historical geography is one of remoteness from the mainstreams of change and progress, as well as one of self-imposed isolation.

A Multinational Empire

Centuries of Russian expansionism did not confine itself to empty land or unclaimed frontiers. The Russian state became an imperial power that annexed and incorporated numerous nationalities and cultures. This was done by employing force of arms, by overthrowing uncooperative rulers, by annexing territory, and by stoking the fires of ethnic conflict. By the time the ruthless Russian regime had begun to confront revolution among its own citizens, czarist Russia controlled peoples representing more than 100 nationalities. The winners in the revolutionary struggle that ensued—the communists who forged the Soviet Union—did not liberate these subjugated peoples. Rather, they changed the empire's framework, binding the peoples colonized by the czars into a new system that would in theory give them autonomy and identity. In practice, it doomed those peoples to bondage and, in certain cases, extinction. The Soviet Union, which arose in the wake of the communist revolution of 1917, constituted another effort to incorporate all these peoples and regions into this vast multinational state, notwithstanding the mantra-like socialist condemnations of imperialism.

THE SOVIET UNION

Communism found fertile ground in the Russia of the 1910s and 1920s. In those days Russia was infamous for the wretched serfdom of its peasants, the cruel exploitation of its workers, the excesses of its nobility, and the ostentatious palaces and riches of the czars. Ripples from the European Industrial Revolution introduced a new age of misery for those laboring in factories. There were workers' strikes and ugly retributions, but when the czars finally tried to better the lot of the poor, it was too little too late.

There was no democracy, and the people had no way to express or channel their grievances. Europe's democratic revolution had passed Russia by, and its economic revolution touched the czars' domain only slightly. Most Russians, as well as tens of millions of non-Russians under the czars' control, faced exploitation, corruption, starvation, and harsh subjugation. When the people began to rebel in 1905, there was no hint of what lay in store; even after the full-scale Revolution of 1917, Russia's political future hung in the balance.

The Political Framework

Russia's great expansion had brought many nationalities under czarist control; now the revolutionary government sought to organize this heterogeneous ethnic mosaic into a smoothly functioning state. The czars had conquered, but they had done little to bring Russian culture to the peoples they ruled. In 1917, when the old order was overthrown, the Russians themselves constituted only about one-half of the population of the entire empire.

In the wake of the Revolution, the first response of many of Russia's subject peoples was to proclaim independent republics, as occurred in Ukraine, Georgia, Armenia, Azerbaijan, and even Central Asia. But Vladimir Lenin, the communist leader and chief architect of the new political system, had no intention of permitting the Soviet state to break up. In 1923, when his blueprint for the new Soviet Union went into effect, the last of these briefly independent units was fully absorbed into the sphere of the Moscow regime.

The political framework for the Soviet Union was based on the ethnic identities of its many incorporated peoples. Given the size and cultural complexity of the empire, it was impossible to allocate territory of equal political standing to all the nationalities (the communists controlled the destinies of well over 100 peoples, both large nations and small isolated groups). It was decided to divide the vast realm into 15 *Soviet Socialist Republics* (*SSRs*), each of which was delimited to correspond broadly to one of the major nationalities (Fig. 2A-7). As was noted, Russians constituted only about half of the developing Soviet Union's population, but they also were (and still are) the most widely dispersed ethnic group in the realm (see Fig. 2A-8). The Russian Republic, therefore, was by far the largest designated SSR, encompassing just over three-quarters of the total Soviet territory.

FIGURE 2A-7

Within the SSRs, smaller minorities were assigned political units of lesser rank. These were called Autonomous Soviet Socialist Republics (ASSRs), which in effect were republics within republics; other areas were designated as Autonomous Regions or other nationality-based units. It was a complicated, cumbersome, often poorly designed framework, but in 1924 it was launched officially under the banner of the ***Union of Soviet Socialist Republics (USSR)***.

A Phantom Federation

The Soviet planners called their system a **federation [8]**. Federalism involves the sharing of power between a country's central government and its political subdivisions (provinces, States, or, in the Soviet case, "Socialist Republics"). Study the map of the former Soviet Union (Fig. 2A-7) and an interesting geographic corollary emerges: every one of the 15 SSRs had a boundary with a non-Soviet neighbor. This seemed to give geographic substance to the notion that any Republic was free to leave the USSR if it so desired. Reality, of course, was quite different, and Moscow's absolute control over the SSRs made the Soviet Union a federation in theory only.

In practice, it was very difficult to accommodate the shifting multinational mosaic of the Soviet realm. The republics quarreled among themselves over boundaries and territory. Demographic change, forced migrations, war, and economic factors soon made much of the layout of the 1920s obsolete. Moreover, the communist planners made it Soviet policy to relocate entire peoples from their homelands in order to better fit the grand design, and to reward or punish—sometimes capriciously. The overall effect, however, was to move minority peoples eastward and to replace them with Russians. This **Russification [9]** of the Soviet Empire produced substantial ethnic Russian minorities in all the non-Russian republics.

The centerpiece of the tightly controlled Soviet "federation" was the Russian Republic. With half of the vast state's population, the capital city, the realm's core area, and 76 percent of the Soviet Union's territory, Russia was the empire's nucleus. In other republics, "Soviet" often was simply equated with "Russian"—it was the reality with which the lesser republics lived. Russians came to the other republics to teach (Russian was taught in the colonial schools), to organize (and frequently dominate) the local Communist Party, and to implement Moscow's economic decisions. This was colonialism, but somehow the communist disguise—how could socialists, as the communists called themselves, be colonialists?—and the contiguous spatial nature of the empire made it appear to

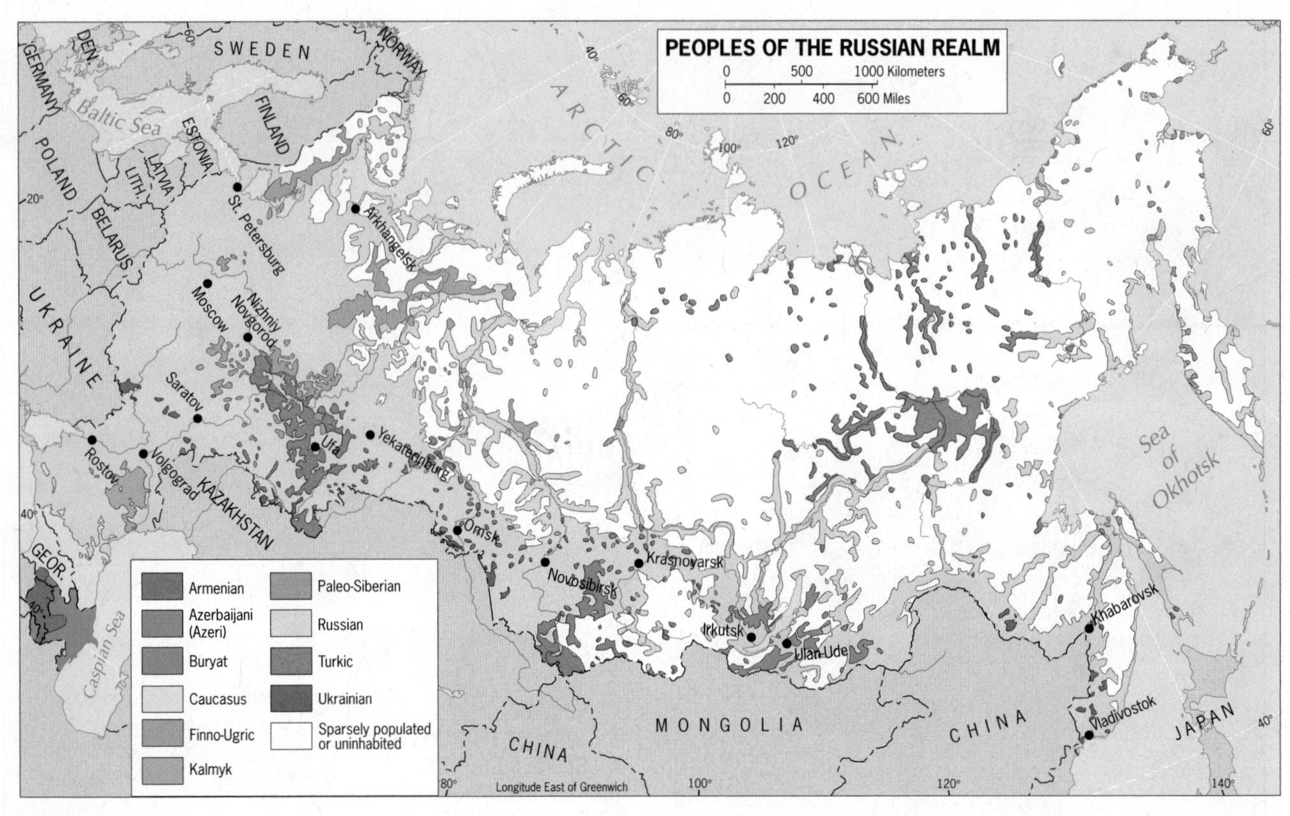

FIGURE 2A-8

the rest of the world as something else. Indeed, on the world stage the Soviet Union became a champion of oppressed peoples, a force in the decolonization process. It was an astonishing contradiction that would, in time, be fully exposed.

The Soviet Economic Framework

The geopolitical changes that resulted from the establishment of the Soviet Union were accompanied by a gigantic economic experiment: the conversion of the empire from a czarist autocracy with a capitalist veneer to communism. From the early 1920s onward, the country's economy would now be *centrally planned*—the communist leadership in Moscow would make all decisions regarding economic planning and development. Soviet planners had two principal objectives: (1) to accelerate industrialization, and (2) to ***collectivize*** agriculture. For the first time ever on such a scale, an entire country was organized to work toward national goals prescribed by a central government.

The Soviet planners believed that agriculture could be made more productive by organizing it into huge state-run enterprises. The holdings of large landowners were expropriated, private farms were taken away from their farmers, and the land was consolidated into collective farms. Initially, all such land was meant to be part of a ***sovkhoz***, literally a grain-and-meat factory in which agricultural efficiency, through maximum mechanization and minimum labor requirements, would be at its peak.

Soviet agriculture never attained such productivity and those who obstructed the communists' grand design suffered a dreadful fate. It has been estimated that between 30 and 60 million people lost their lives from imposed starvation, constant political purges, Siberian exile, and forced relocation. The Soviet grand experiment amounted to an incalculable human tragedy, but the secretive character of Soviet officialdom made it possible to hide everything from the world.

The USSR practiced a **command economy [10]**, in which state planners assigned the production of particular manufactures to particular places, often disregarding the rules of economic geography. For example, the manufacture of railroad cars might be assigned (as indeed it was) to a factory in Latvia. No other factory anywhere else would be permitted to produce this equipment—even if supplies of raw materials would make it cheaper to build them near, say, Volgograd 2000 kilometers (1250 mi) away. Yet, despite an expanded and improved transportation network (see Fig. 2A-9), such practices made manufacturing in the USSR extremely expensive, and the absence of competition made managers complacent and workers far less productive than they might have been.

Of course, the Soviet planners never imagined that their experiment would fail and that a market-driven economy would replace their command economy. When that happened, the transition was predictably difficult;

indeed, it is far from over and continues to severely stress the now more democratic, post-communist state.

THE NEW RUSSIA

On December 25, 1991, the inevitable occurred: the Soviet Union disintegrated. The centrally planned economy went into structural failure, the arms race with the United States during the Cold War had drained resources, and efforts at Russification had only fueled the drive for independence among so many of the peoples that had been forcibly incorporated into the Union. The last Soviet president, Mikhail Gorbachev, resigned, and the Soviet hammer-and-sickle flag flying atop the Kremlin was lowered for the last time and immediately replaced by the white, red, and blue Russian tricolor. A new and turbulent era began—but Soviet institutions and systems will long cast their shadows over transforming Russia.

When the Soviet system failed and the Soviet Socialist Republics became independent states in 1991, Russia was left without the empire that had taken centuries to build and consolidate—and that contained crucial agricultural and mineral resources. No longer did Moscow control the farms of Ukraine and the oil and natural gas reserves of Central Asia. But look at Figures 2A-1 and especially 2B-2 and you will see that, even without its European and Central Asian colonies, Russia remains an empire. Russia lost the "republics" on its periphery, but Moscow still rules over a domain that extends from the borders of Finland to North Korea. Inside that domain Russians are in the overwhelming majority, but many subjugated nationalities, from Tatars to Yakuts, continue to inhabit ancestral homelands. Accommodating these many indigenous peoples is still one of the challenges facing the Russian Federation today.

A Complex Cultural Mosaic

Although Russia's dominance of this geographic realm justifies our naming it as such, this is a culturally and ethnically diverse part of the world whose traditions and customs spill over into neighboring realms even as neighbors have come to live here. Ethnic Russians still form the great majority, but sizeable parts of this realm are home to non-Russian peoples—and not just along the borders. As Figure 2A-8 shows, the Russian realm contains Finnish, Turkic, Armenian, and dozens of other "nationalities," but the scale of this map cannot reflect the complexity of the overall ethnic mosaic. Here is an interesting example: in the southwest, facing the Caspian Sea, lies a small "republic" that is part of modern Russia, named Dagestan (which achieved notoriety in 2013 in conjunction with the Boston Marathon bombings). It is half the size of Maine, has a population of 3 million, and contains nearly 30 ethnic "nationalities" speaking their own languages, most of which are variants of languages spoken in the Caucasus, Turkey, and Iran.

As the map shows, the Slavic peoples known collectively as the Russians not only form the majority of the population but also are the most widely dispersed. Although the Russian Plain is the core area of the Russian state and was its historic hearth, Russian settlement extends from the shores of the Arctic Sea to the Black Sea coast and from St. Petersburg on the Gulf of Finland to Vladivostok on the Sea of Japan. Discontinuous nuclei and ribbons of Russian settlement are scattered across Siberia, but as we noted in Figure 2A-4, the population, Russian and non-Russian, tends to concentrate in the southern sector of the realm.

Russians are not the only Slavic peoples. More than one thousand years ago, when many ethnic groups struggled to establish themselves in Europe, the early Slavs, whose original home may have been on the North European Plain in the area north of the Carpathians, achieved stability and security in a homeland that expanded steadily, not only eastward into what is today Russia, but also westward and southwestward into the valley of the Danube and beyond. As we noted in Chapters 1A and 1B, the Serbs, Croats, Slovaks, and Czechs are Slavic peoples, as are the Poles, Ukrainians, and Belarussians, among others (see Fig. 1B-14). Today you can discern evidence of this distant past in the Slavic languages these descendants speak (see Fig. 1A-7), which still have much in common despite centuries of divergence.

Even as the early Russian state consolidated, non-Russian peoples incubated in and beyond the Caucasus Mountains in the corridor between the Black and Caspian seas. Georgians, Armenians, Azeris, and many others created a jigsaw mosaic of nationalities in this region (the aforementioned Dagestan is only a tiny part of it). Then, after about AD 1400, Islam propagated by the Ottoman Empire pushed northward here, and by 1500 Islam's vanguards were challenging Slavic peoples along the northern shore of the Black Sea. Remnants of such advances and invasions still remain on the cultural map (Fig. 2A-8): look at the crescent of Turkic peoples (shaded red) extending from near the city of Nizhniy Novgorod to the border of Kazakhstan, and you see part of the evidence. Today, Russia has a far higher percentage of Muslims in its population than western European countries do; although the data are not precise, the consensus is that roughly 15 percent of Russia's inhabitants are Muslims.

Long before Islam appeared on the scene, Slavic peoples from one end of their domain to the other had accepted the teachings of the Eastern Orthodox Church, and in the Russian realm the Russian church became Eastern Orthodoxy's dominant institution. It stayed that way until the triumph of the communist revolutionaries, who in 1917 put an end to rule by the Russian czars and began the disestablishment of the church. For more than seven decades, atheism was official policy—until the communist system imploded. Since the early 1990s, the Russian Orthodox Church has made a vigorous comeback, attended by nationalist and ethnic propaganda.

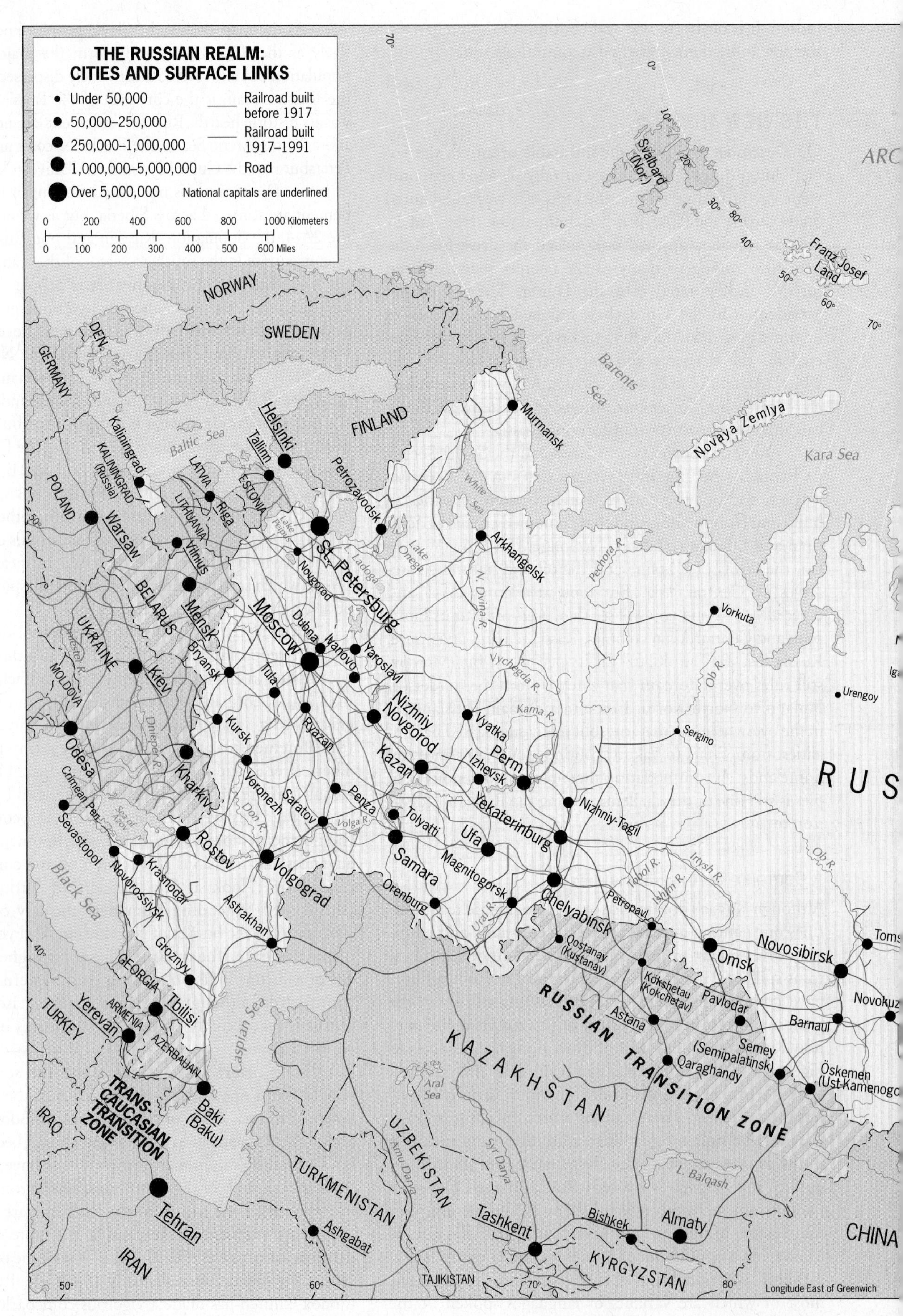

THE RUSSIAN REALM: CITIES AND SURFACE LINKS
Under 50,000
50,000–250,000
250,000–1,000,000
1,000,000–5,000,000
Over 5,000,000
Railroad built before 1917
Railroad built 1917–1991
Road
National capitals are underlined
0 200 400 600 800 1000 Kilometers
0 100 200 300 400 500 600 Miles
NORWAY
SWEDEN
FINLAND
DEN.
GERMANY
POLAND
KALININGRAD (Russia)
Kaliningrad
Baltic Sea
LATVIA
LITHUANIA
ESTONIA
Helsinki
Tallinn
Riga
Vilnius
Warsaw
Lake Peipus
Novgorod
St. Petersburg
L. Ladoga
Lake Onega
Petrozavodsk
Murmansk
White Sea
Arkhangelsk
Barents Sea
Svalbard (Nor.)
Franz Josef Land
Novaya Zemlya
Kara Sea
Pechora R.
N. Dvina R.
Vorkuta
BELARUS
Mensk
UKRAINE
MOLDOVA
Dniester R.
Dnieper R.
Kiev
Bryansk
Moscow
Dubna
Ivanovo
Yaroslavl
Tula
Kursk
Ryazan
Vychegda R.
Urengoy
Kama R.
Vyatka
Sergino
Nizhniy Novgorod
Kazan
Izhevsk
Perm
Nizhniy-Tagil
Yekaterinburg
RUS
Odesa
Kharkiv
Voronezh
Don R.
Saratov
Penza
Volga R.
Tolyatti
Samara
Ufa
Magnitogorsk
Orenburg
Ural R.
Chelyabinsk
Tobol R.
Ishim R.
Irtysh R.
Ob R.
Petropavl
Crimean Pen.
Sea of Azov
Sevastopol
Rostov
Volgograd
Black Sea
Novorossiysk
Krasnodar
Astrakhan
Groznyy
GEORGIA
Tbilisi
TURKEY
ARMENIA
Yerevan
AZERBAIJAN
Caspian Sea
Baki (Baku)
TRANS-CAUCASIAN TRANSITION ZONE
IRAQ
Tehran
IRAN
Qostanay (Kustanay)
Omsk
Novosibirsk
Toms
Kökshetau (Kokchetav)
Pavlodar
Astana
Barnaul
Novokuz
RUSSIAN TRANSITION ZONE
KAZAKHSTAN
Semey (Semipalatinsk)
Qaraghandy
Öskemen (Ust-Kamenogo
Aral Sea
UZBEKISTAN
Amu Darya
Syr Darya
TURKMENISTAN
Ashgabat
Tashkent
Bishkek
Almaty
L. Balqash
KYRGYZSTAN
TAJIKISTAN
CHINA
50°
60°
70°
80°
Longitude East of Greenwich

FIGURE 2A-9

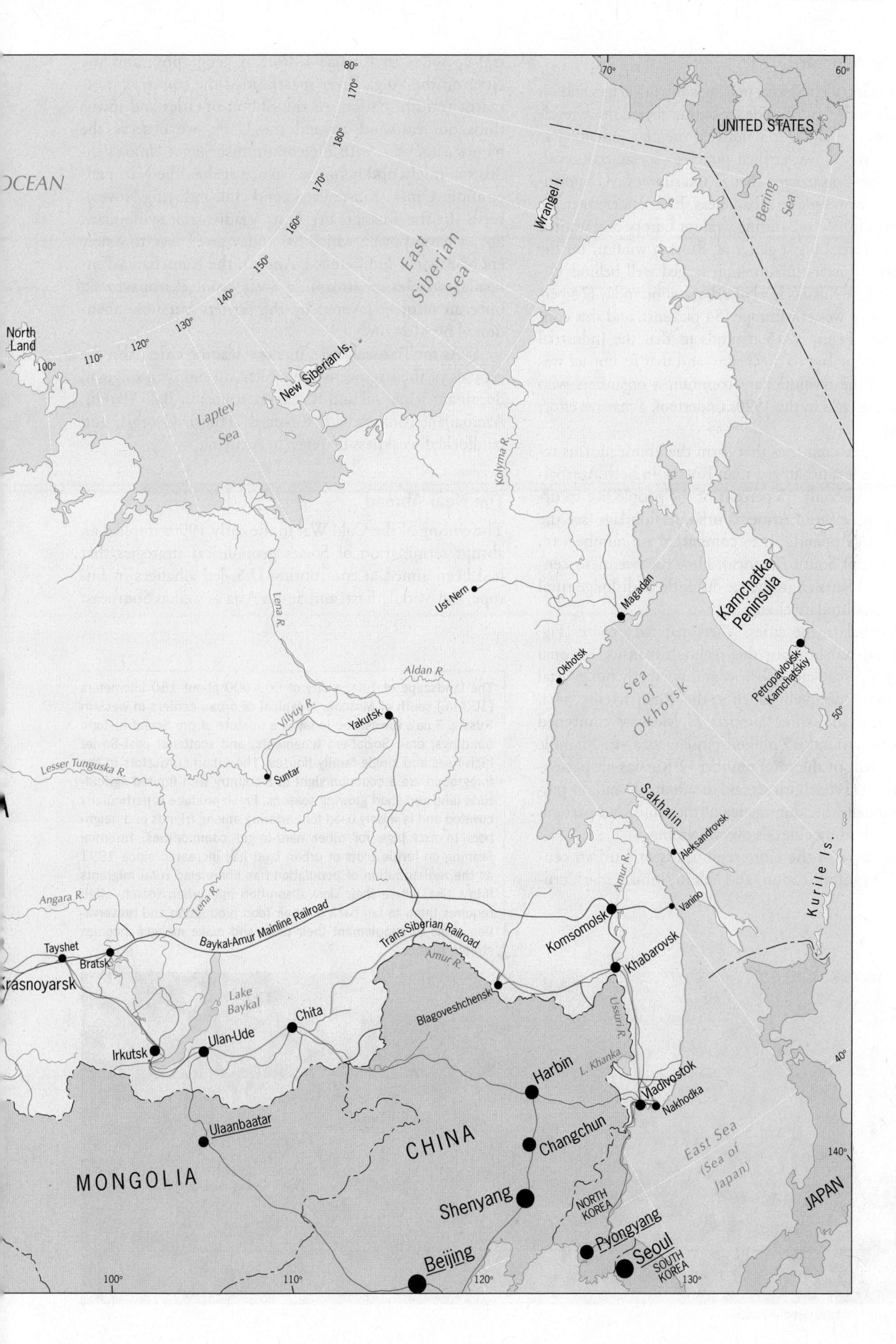

OCEAN
UNITED STATES
Bering Sea
Wrangel I.
East Siberian Sea
North Land
New Siberian Is.
Laptev Sea
Kolyma R.
Kamchatka Peninsula
Ust Nern
Magadan
Lena R.
Okhotsk
Petropavlovsk-Kamchatskiy
Aldan R.
Sea of Okhotsk
Yakutsk
Vilyuy R.
Lesser Tunguska R.
Suntar
Sakhalin
Aleksandrovsk
Kurile Is.
Amur R.
Angara R.
Vanino
Komsomolsk
Baykal-Amur Mainline Railroad
Trans-Siberian Railroad
Tayshet
Bratsk
Khabarovsk
Krasnoyarsk
Lake Baykal
Blagoveshchensk
Ussuri R.
Chita
Ulan-Ude
Irkutsk
Harbin
L. Khanka
Vladivostok
Nakhodka
Ulaanbaatar
CHINA
Changchun
East Sea (Sea of Japan)
MONGOLIA
Shenyang
NORTH KOREA
JAPAN
Pyongyang
Beijing
Seoul
SOUTH KOREA

Cities Near and Far

When the czars ruled Russia, people living in villages and on homesteads in the countryside far outnumbered those residing in towns and cities. It is not that the czars intended to keep it that way, however their policies—social, economic, and otherwise—obstructed change and suppressed opportunities that draw people to urban areas. Even under Peter the Great, who admired urbanizing western Europe and wanted to make St. Petersburg a glittering Russian window on the wider world, Russia's urbanization lagged well behind Europe's. Compare Russia's level of urbanization today (74 percent) to that of western Europe (84 percent), and this contrast endures. Figure 1A-5 reminds us that the Industrial Revolution came late to this realm, and that its impact was slowed until the revolutionary communist organizers who succeeded the czars in the 1920s undertook a massive effort to catch up.

The three countries that form the Transcaucasus region are even less urbanized than Russia: in both Azerbaijan and Georgia only 53 percent of the people live in urban areas, and even in Armenia urbanization lags (see the Data Table in Appendix B to compare these numbers to, say, East Asia or South America). Here the twentieth-century period of Russian influence did little to dislodge rural traditions or stimulate change.

Nonetheless, the cities, when mapped by size (Fig. 2A-9), reveal much about this realm's expansive regional geography. Primate-city Moscow and northerly but coastal St. Petersburg decidedly anchor the Russian core area. Consider this: the three airports of Moscow combined (which serve around 65 million passengers a year) handle more than half of the total number of Russia's air passengers. With St. Petersburg second at about 11 million passengers annually, the dominance of the Core, and particularly its two leading cities, is overwhelming.

Elsewhere in the Core region, historic urban centers like Novgorod, Kazan, and Yekaterinburg mark crucial episodes in Russia's historical geography; and the cities on the Volga River spearheaded the country's post-czarist transformation. As the ribbon of cities and towns thins out eastward beyond the Urals, we observe the names associated with the communist Soviet Union's industrial might and heroic resistance against the Nazi penetration: Omsk, Krasnoyarsk, and chill-inducing Novosibirsk. In the Russian Far East, Vladivostok symbolized Soviet naval power, which has today given way to a new era of Russian indifference. And on the Kamchatka Peninsula, people are streaming away from Petropavlovsk, once an outpost favored by the Soviets but now abandoned by Moscow.

As for Transcaucasia, its three leading cities, also the capitals of their respective republics, all signify a singular, dominant issue: oil and its export routes in Baki (Baku), Azerbaijan; conflict with Russia in Tbilisi, Georgia; and landlocked weakness in Yerevan, Armenia.

The Near Abroad

The ending of the Cold War in the early 1990s implied an abrupt termination of Soviet geopolitical strategies that had been aimed at confronting U.S.-led alliances in Europe, the Middle East, and South Asia as well as Southeast

The landscape of Tula, a city of 505,000 about 150 kilometers (100 mi) south of Moscow, is typical of urban centers in western Russia. Tula's townscape today is a mixture of pre-Soviet historic buildings, drab Soviet-era tenements, and scattered post-Soviet high-rises and single-family houses. The urban cultivators in the foreground are a common sight in a country with limited agricultural land and short growing seasons. Fresh produce is particularly coveted and is widely used for bartering among friends and neighbors in exchange for other hard-to-get commodities. Informal farming on fertile plots of urban land has increased since 1991 as the redistribution of population has channeled rural migrants into cities, where their slow absorption into urban society often requires them to fall back on their food production and preservation skills to supplement their diets and make meager incomes stretch further.

Mauro Galligani/Contrasto/Redux Pictures

What Are Russia's Rights in the Near Abroad?

IN FAVOR OF STRONG RUSSIAN INFLUENCE

"I am delighted that [former] President Medvedev showed the world that we Russians will protect ourselves and our allies in the Near Abroad wherever it's necessary. Those Georgians who think they can ride roughshod over our friends the Ossetians have learned a lesson, and I hope they heed it. As a former officer in the Soviet Army, I have no doubt that nations have to show their strength or they'll get trampled on. I'm still sorry we left Afghanistan in 1989 the way we did. And even if the USSR is no more, we Russians earned the right not to be bullied by upstart leaders like that dreadful Saakashvili in Georgia or that supposedly democratically-elected Yushchenko in Ukraine. Have those people forgotten what we did for them and their countries? We created their national identities, established their republics, built their economies, put them on the map. We built their roads, railroads, bridges, and airports; we taught them Russian and nourished their cultures. We saved the Armenians from the Turks, the Estonians from the Germans, the Mongolians from the Chinese. If the Americans hadn't armed the Afghan extremists, we wouldn't have the Taliban threatening the place today.

"So what thanks do we get? Saakashvili starts a war in South Ossetia and kills Russian soldiers and Ossetian villagers. Chechen terrorists kill Russians in the Caucasus, and all we hear about is that we mistreat minorities. The Latvians insult and offend us at every turn, treating its Russian-speaking residents as second-class citizens. The Ukrainians are cozying up to the European Union even though the millions of Russians in the country oppose it. Now the Americans want to put so-called missile defense systems in Poland and the Czech Republic, and I'm pleased that our president played our Kaliningrad card. I've had it with the high-handed treatment we're getting around our perimeter, and it's time to put a stop to it.

"You can look at the Near Abroad in two ways: as the ring of countries that encircles us or as the Slavic outposts that persist from Kazakhstan to Kosovo. Whichever way you define it, we Russians reserve the right to take care of our own, and we will not stand by as our kinspeople are mistreated, or our homeland threatened, by those who think we are weak and impotent. I hope the Ukrainians and the Moldovans and the Albanians and others who might miscalculate learn a lesson from what happened in Georgia. If we have to, we will take appropriate action, and that includes military action. Our military forces suffered during the Yeltsin years, but those days are over. We have the money to rebuild our forces and we're doing so. We will not be pushed around, and it would be a grave mistake for the rest of the world to see us as weak or lacking resolve. The Soviet Union may have collapsed, but Russia will always be a force to reckon with."

OPPOSED TO STRONG RUSSIAN INFLUENCE

"Talk about the Near Abroad is all the rage in Russia these days. Whether it's the television news, the magazines and newspapers, or talk radio, the Near Abroad is the topic that gets everyone riled up. In truth, this is nothing new—it started even before the USSR broke up two decades ago. Not only were Russians marooned in the former republics, but their friends and allies among the locals had a tough time as well. Those who tried to help save the situation Gorbachev had created were seen as traitors to their people, and in places like Lithuania and Georgia it got pretty ugly.

"As a history teacher, I try to put things in long-term perspective. I don't care what state or nation it is, when you've got the power you tend to abuse it, and even if you don't, you're accused of doing so by association. That certainly was true of the western European colonial powers. Look at France and its 'Near Abroad' in its former colony of Algeria. How many tens of thousands of Frenchmen were killed in the run-up to Algeria's independence, and how many afterward? We Russians feel that we weren't in a similar situation, because our Soviet republics really were not colonies. But millions of Russians did go to the republics to govern, to build infrastructure, to teach, or just to work for the fatherland, and you can be sure that many locals saw these 'foreigners' as power-hungry outsiders. In the Muslim societies we never managed to persuade the people of the irrelevance of religion. In other republics, Latvia for example, there had been sympathies for Nazi Germany and we were viewed as occupiers. Get this: the Lithuanians not long ago launched a (U.S.) $34 billion claim against us for what they call Russia's '50-year occupation.'

"Make no mistake: I don't like it when I hear or read about the mistreatment of Russians or our allies in countries of the Near Abroad. Trying to force Russians to abandon their citizenship and compelling them to pass difficult language tests as a condition for local citizenship is no way to get over the past. But I feel that nothing is gained by the kind of violence that recently took place in South Ossetia and Abkhazia. Some reports say that Georgia's president started the whole thing by unleashing attacks on Russian protective forces; others tell a different story. But the fact is that Russian forces were within Georgia's international borders even before the conflict took its deadly turn, and that was asking for trouble. I have no doubt that the Georgian regime was guilty of discrimination against South Ossetians (and Abkhazians as well), but we're talking here about only tens of thousands of people in a tiny enclave—was it worth risking a wider war over what is essentially a minority-treatment issue?

"I don't think so, but let me tell you—I'm in a small minority here. The small shooting war in Georgia has galvanized Russians. I hope it's not the beginning of a new era in the Near Abroad."

Vote your opinion at www.wiley.com/go/deblijpolling

Asia. Indeed, Soviet foreign policy and power had been projected beyond the Soviet frontier to other continents, from Africa to Middle America. The collapse of the Soviet Union meant that the new Russia, shorn of **satellite states [11]** in eastern Europe and abandoned by all of the other former Soviet republics, had to completely rethink its position in the world and most immediately in Eurasia. And it had to do so in the knowledge that a total makeover of its economy was the top priority.

The Russian government in 1992 introduced the term **Near Abroad [12]** for the newly formed countries that surround Russia today. These are for the most part the other former Soviet republics, from the Baltic states to Kazakhstan. These states may now be located "abroad," but seen from Moscow they are so "near"—and together constitute such a pronounced encirclement of the country—that the Russian government feels entitled to take a special interest in them (compare Figs. 2A-7 and 2A-9). What this really means is that Russia will not allow any of these bordering countries to develop a threat of any sort and that it reserves the right to intervene if necessary. It is not that hard to imagine this sense of territorial encirclement in Russia, particularly with the eastward expansion of the EU (now practically on Russia's western doorstep), Islamic fundamentalism in the south, and the swift rise and growing assertiveness of China in the east.

And then there is the issue of Russian minorities who now find themselves living abroad. Although many Russian nationals have returned from the former SSRs, millions of others stayed on. They lived in areas where Russians were in the majority (for example, in eastern Ukraine and in northern Kazakhstan) and where post-Soviet life was not all that different from what it was during Moscow's rule. But elsewhere, Russians who remained in the former SSRs found themselves mistreated in various ways, and they often appealed to Moscow for help. Thus the concept of a Russian "Near Abroad" found its way into Russian discourse. Beyond Russia's borders but of concern to its leaders and people, the "Near Abroad" became geographic shorthand for a Russian sphere of influence, whereby Moscow stood ready to help its fellow ethnics in case of trouble. On the map, the most vivid example of this is the so-called Russian Transition Zone in northern Kazakhstan (Fig. 2A-9), which was tightly integrated into the adjacent Russian sphere during Soviet times. (A larger-scale map would show similar but smaller Russian clusters in Estonia, Latvia, Moldova, and elsewhere.) And Ukraine, which occupies part of the western transition zone, is of the utmost importance as well (as noted in Chapter 1B), with its eastern half heavily dominated by Russia and home to a large Russian population.

A REALM IN FLUX

Almost 25 years ago, the Soviet Empire collapsed and disintegrated. What remains today is the massive centerpiece of that empire, the Russian Federation. Two-plus decades may seem like a long time, but the current situation still bears the marks of the Soviet era and this very much remains a realm in flux. Soviet times were repressive and economically stagnant, whereas the new Russia has offered greater freedom and business opportunities. Nonetheless, the benefits of this new era have mostly been limited to the favored few, and income inequality continues to widen. By 2011, the masses were increasingly showing their discontent, most notably in anti-regime street protests in Moscow and other cities. They are frustrated with the widespread corruption as well as a post-communist economy that is not a true free market (it is sometimes derisively referred to as mafia-capitalism) and does not provide any of the securities of the old Soviet system. Indeed, a recent survey among Russians showed that half of the population agreed that it was "a great misfortune that the Soviet Union no longer exists."

Demographically, economically, administratively, politically, and geostrategically, Russia is likely to continue to adjust for many years to come. Ideally, relations with

At a soccer match between Russia and Poland during the 2012 Euro Tournament, Russian fans 'greeted' the Polish team with a gigantic banner that left little doubt about their convictions—the game was played in Warsaw, Poland, and those fans knew very well that it was televised around the globe, along with their message. There is plenty of breeding ground in Russia for aggressive foreign policy claims in the Near Abroad and beyond. The game ended in a 1-1 tie.

© Photo ITAR-TASS/Valery Sharifulin/Corbis

the Near Abroad remain cordial and peaceful, the Russian Federation itself maintains cohesion, and overall economic well-being improves. Clearly, this depends on more than decision-making in Moscow. The geography of the Russian realm is such that it finds itself at the heart of the Eurasian continent, the Russian Federation itself bordering no less than 15 countries as well as four other realms. Perhaps, in such circumstances, long-term stability is an illusion. To be sure, it is a formidable challenge to manage a country of this enormous size and diversity. In Chapter 2B, we discuss the evolving administrative (spatial) structure of the new Russia and take a closer look at the regions that constitute this unique realm.

POINTS TO PONDER

- High-speed train travel, urgently needed in this far-flung realm, has been very slow to develop. In 2010 the first such route opened between the capital, Moscow, and the leading port and second city, St. Petersburg.
- All of the major transnational empires that were created in the nineteenth century came to an end in the wake of World War I or World War II—except the Russian Empire, which continued in a different guise after 1917 until it finally imploded in 1991.
- Global warming can have major disruptive effects on the ways that people interact with their habitat. But Russians consider themselves to be potential beneficiaries as Siberia's harsh environments moderate and the Arctic Ocean's ice recedes, probably opening up new sea lanes and improving access to major energy deposits buried beneath the seafloor.
- The Russian Federation borders 15 countries and four other geographic realms. Compare that to the conterminous United States, which borders only two countries and two other realms.

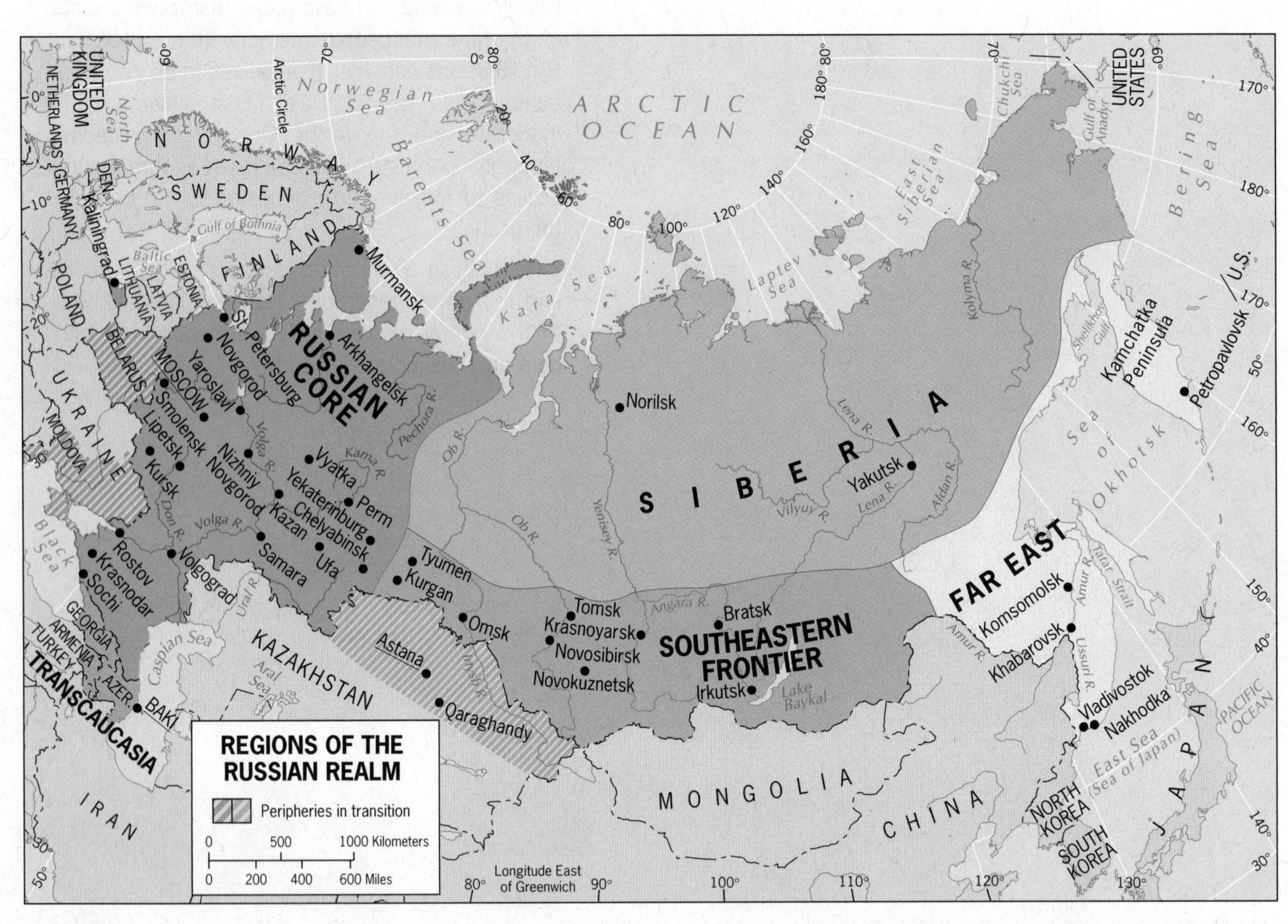
REGIONS OF THE RUSSIAN REALM
Peripheries in transition
RUSSIAN CORE
SIBERIA
SOUTHEASTERN FRONTIER
FAR EAST
TRANSCAUCASIA
ARCTIC OCEAN
Norwegian Sea
Barents Sea
Kara Sea
Laptev Sea
East Siberian Sea
Chukchi Sea
Bering Sea
Sea of Okhotsk
East Sea (Sea of Japan)
PACIFIC OCEAN
Caspian Sea
Aral Sea
Black Sea
Baltic Sea
North Sea
Gulf of Bothnia
Gulf of Anadyr
Shelikhov Gulf
Tatar Strait
Lake Baykal
Kamchatka Peninsula
Arctic Circle
NORWAY
SWEDEN
FINLAND
ESTONIA
LATVIA
LITHUANIA
POLAND
BELARUS
UKRAINE
MOLDOVA
GEORGIA
ARMENIA
AZER.
TURKEY
IRAN
KAZAKHSTAN
MONGOLIA
CHINA
NORTH KOREA
SOUTH KOREA
JAPAN
UNITED KINGDOM
NETHERLANDS
GERMANY
DEN.
UNITED STATES
U.S.
Kaliningrad
Murmansk
St. Petersburg
Arkhangelsk
Novgorod
Yaroslavl
MOSCOW
Smolensk
Lipetsk
Kursk
Nizhniy Novgorod
Vyatka
Perm
Yekaterinburg
Chelyabinsk
Kazan
Ufa
Samara
Volgograd
Rostov
Krasnodar
Sochi
Baki
Tyumen
Kurgan
Omsk
Astana
Qaraghandy
Tomsk
Krasnoyarsk
Novosibirsk
Novokuznetsk
Bratsk
Irkutsk
Norilsk
Yakutsk
Komsomolsk
Khabarovsk
Vladivostok
Nakhodka
Petropavlovsk
Volga R.
Kama R.
Pechora R.
Don R.
Ural R.
Ob R.
Irtysh R.
Yenisey R.
Angara R.
Lena R.
Vilyuy R.
Aldan R.
Kolyma R.
Amur R.
Ussuri R.
0 500 1000 Kilometers
0 200 400 600 Miles
Longitude East of Greenwich

FIGURE 2B-1

REGIONS

The Russian Core

The Southeastern Frontier

Siberia

The Russian Far East

Transcaucasia

IN THIS CHAPTER

- Running a country with nine time zones
- Moscow: From Soviet capital to global city
- Where Russia meets China, Japan, and the Korean Peninsula
- The cold beauty of Siberia
- Trouble in Transcaucasia
- Hosting the 2014 Sochi Winter Olypics

CONCEPTS, IDEAS, AND TERMS

Distance decay	1
Unitary state system	2
Federal system	3
Population implosion	4
BRICs	5
Centrality	6
Double complementarity	7

Connie Coleman/Getty Images, Inc.

Where were these pictures taken?
Find out at www.wiley.com/college/deblij

No other single state so thoroughly dominates its geographic realm as does Russia. It is important to keep in mind the many changes that have taken place in recent decades along the edges of this realm. After the USSR imploded in 1991 and all of the other former Soviet Republics had seceded from the Union, Russia was left all by itself—a gigantic country of unmatched proportions, but one that had lost control of major bordering regions in Europe and Southwest Asia, from Estonia to Kazakhstan. Today we refer to the Russian realm as comprising the country of Russia as well as the three comparatively tiny Transcaucasian countries of Georgia, Azerbaijan, and Armenia. To be sure, Russia's historic influences continue to spill over into now-independent states all around its borders from the Baltic to the Far East, but most of these countries are now drawn more firmly into the orbits of other realms. However, such is not the case with Transcaucasia, which remains very tightly embedded within the Russian sphere.

REGIONS OF THE REALM

Russia itself is enormous and diverse physiographically, ethnically, as well as economically, and could be subdivided in a variety of ways and at different scales. Let us try to keep things simple and delimit what we might call first-order regions, recognizing that each of this realm's enormous regions contains numerous subdivisions. Four of the five regions we discuss lie entirely inside Russia (Fig. 2B-1):

1) The Russian Core
2) The Southeastern Frontier
3) Siberia
4) The Russian Far East

And only one lies outside the Russian state:

5) Transcaucasia

Jochem D Wijnands/Getty Images, Inc.

POST-SOVIET RUSSIA

When the USSR disintegrated at the end of 1991, Russia's former empire devolved into 14 independent states, and Russia itself was a changed country. Russians now made up about 83 percent of the population of just under 150 million, a far higher proportion than in the days of the Soviet Union. But numerous minority peoples remained under Moscow's new flag, and millions of Russians found themselves under new governments in the former Soviet Republics.

Russia's Changing Political Geography

The new Russian government also faced an old Soviet problem: the sheer size of the country, its vast distances, and the remoteness of many of its Regions. Geographers refer to the principle of **distance decay [1]** in order to explain how increasing distances between places tend to reduce interactions among them. Because Russia is the world's largest country, distance is a significant factor in the relationships between the capital and outlying areas. Furthermore, Moscow lies in the far west of this gigantic country, half a world away from the shores of the Pacific. Not surprisingly, one of the most obstreperous Regions has been remote Primorskiy, which contains Vladivostok (Fig. 2B-1). Still another problem is the stupendous size variation among the country's myriad administrative units, as can be seen in Figure 2B-2. Whereas the (territorially) smallest divisions are concentrated in the Russian core area, the largest lie far to the east, where Sakha is nearly a thousand times as large as Ingushetiya. On the other hand, the populations of the enormous eastern divisions are tiny compared to those of the smaller ones in the Russian Core. Such diversity spells administrative difficulty.

There is significant social disharmony between Moscow and many of the subnational political entities. Almost wherever one goes in Russia today, Moscow is disliked and frequently berated by angry locals. The capital is seen as the privileged playground for those who have benefited most from the post-Soviet transition—bureaucrats and hangers-on whose economic policies have driven down standards of living, whose greed and corruption have hurt the economy, whose heavy-handed actions in Chechnya have been an unmitigated disaster, who have allowed foreigners (prominently the United States) to erode Russian power and prestige, who fail to pay the wages of workers toiling in industries still owned by the state, and who do not represent the Russian people. At the same time, though, many have migrated to Moscow and other cities in the search for better livelihoods, higher paid jobs, or business opportunities.

And the dissatisfaction of many people with their lack of economic progress since the dismantling of the Soviet Union provides a pretext for leading politicians to turn to authoritarian ways and promise restoration of Russian prestige in the world. This apparent turn away from democracy has other groups worried (see photo), those who have been hoping for a democratic and free Russia after so many decades of Soviet oppression.

AP/Wide World Photos

A portion of the huge crowd on the streets of downtown Moscow in February 2012, braving the −22°C (−4°F) cold to challenge the bid of Vladimir Putin to reclaim the presidency of the Russian Federation. The efforts of these several thousand protesters notwithstanding, Putin coasted to victory a month later proclaiming that he had received 63% of the vote.

The Soviet Legacy

Soviet planners had created a highly complex administrative structure for their "Russian Soviet Federative Socialist Republic," and Russia's post-communist leaders have been forced to utilize this framework to make their country function. Figure 2B-2 reflects this complex Russian administrative system that dates back to Soviet times. There are now 83 entities in all: 2 Autonomous Federal Cities (Moscow and St. Petersburg), 21 Republics, 46 Provinces (Oblasts), 9 Territories (Krays), 4 Autonomous Regions (Okrugs), and 1 Autonomous Oblast. These administrative units operate with varying degrees of power and autonomy from the federal government. The 21 Republics, established and recognized in order to accommodate substantial ethnic minorities, include a cluster in the far south, another lying to the

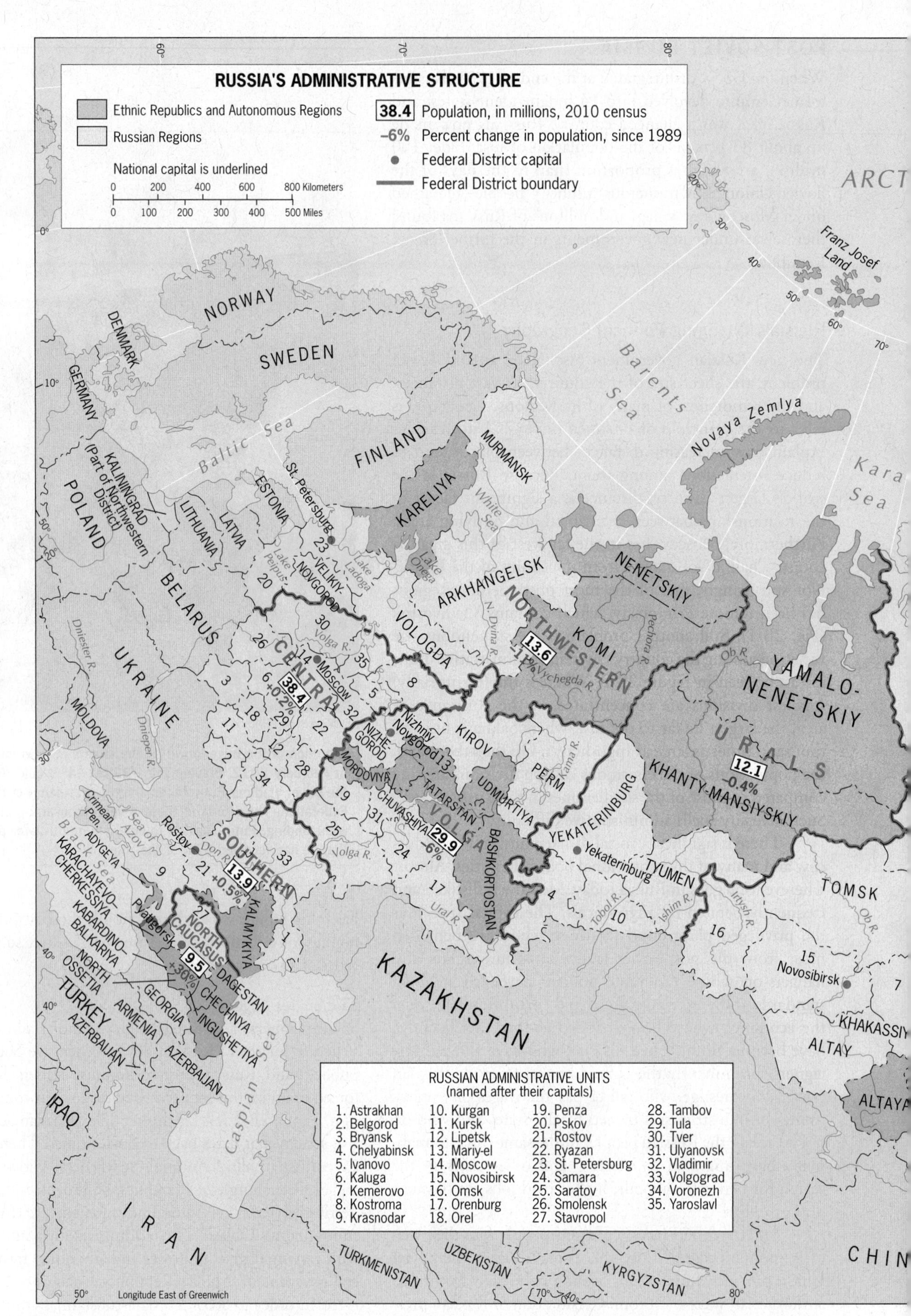

RUSSIA'S ADMINISTRATIVE STRUCTURE
Ethnic Republics and Autonomous Regions
Russian Regions
National capital is underlined
38.4 Population, in millions, 2010 census
–6% Percent change in population, since 1989
Federal District capital
Federal District boundary
0 200 400 600 800 Kilometers
0 100 200 300 400 500 Miles
NORWAY
SWEDEN
FINLAND
DENMARK
GERMANY
POLAND
KALININGRAD (Part of Northwestern District)
LITHUANIA
LATVIA
ESTONIA
BELARUS
UKRAINE
MOLDOVA
KAZAKHSTAN
TURKEY
GEORGIA
ARMENIA
AZERBAIJAN
IRAQ
IRAN
TURKMENISTAN
UZBEKISTAN
KYRGYZSTAN
Baltic Sea
Barents Sea
Kara Sea
White Sea
Black Sea
Sea of Azov
Caspian Sea
Novaya Zemlya
Franz Josef Land
St. Petersburg
Moscow
Nizhny Novgorod
Rostov
Pyatigorsk
Yekaterinburg
Novosibirsk
MURMANSK
KARELIYA
ARKHANGELSK
NENETSKIY
KOMI
VOLOGDA
VELIKIY NOVGOROD
KIROV
PERM
UDMURTIYA
TATARSTAN
CHUVASHIYA
MORDOVIYA
BASHKORTOSTAN
YEKATERINBURG
TYUMEN
YAMALO-NENETSKIY
KHANTY-MANSIYSKIY
TOMSK
KHAKASSIY
ALTAY
ALTAYA
ADYGEYA
KARACHAYEVO-CHERKESSIYA
KABARDINO-BALKARIYA
NORTH OSSETIA
CHECHNYA
INGUSHETIYA
DAGESTAN
KALMYKIYA
NORTHWESTERN
CENTRAL
VOLGA
SOUTHERN
NORTH CAUCASUS
URALS
13.6
-11%
38.4
+0.2%
29.9
-6%
13.9
+0.5%
9.5
+30%
12.1
-0.4%
RUSSIAN ADMINISTRATIVE UNITS (named after their capitals)
1. Astrakhan
2. Belgorod
3. Bryansk
4. Chelyabinsk
5. Ivanovo
6. Kaluga
7. Kemerovo
8. Kostroma
9. Krasnodar
10. Kurgan
11. Kursk
12. Lipetsk
13. Mariy-el
14. Moscow
15. Novosibirsk
16. Omsk
17. Orenburg
18. Orel
19. Penza
20. Pskov
21. Rostov
22. Ryazan
23. St. Petersburg
24. Samara
25. Saratov
26. Smolensk
27. Stavropol
28. Tambov
29. Tula
30. Tver
31. Ulyanovsk
32. Vladimir
33. Volgograd
34. Voronezh
35. Yaroslavl
Longitude East of Greenwich

FIGURE 2B-2

OCEAN
UNITED STATES
Wrangel I.
East Siberian Sea
Bering Sea
CHUKOTSKIY (CHUCKCHI)
North Land
New Siberian Is.
Laptev Sea
Kolyma R.
MAGADAN
KAMCHATKA
SAKHA (YAKUTIYA)
FAR EASTERN
6.3
–21%
Lena R.
Vilyuy R.
Sea of Okhotsk
KHABAROVSK
KRASNOYARSK
Lesser Tunguska R.
Aldan R.
SIBERIAN
19.3
–9%
SAKHALIN
Kurile Is.
Angara R.
Lena R.
AMUR OBLAST
Amur R.
YEVREYSKAYA
Khabarovsk
IRKUTSK
BURYATIYA
CHITA
L. Baykal
Ussuri R.
PRIMORSKIY KRAY
TYVA
L. Khanka
East Sea (Sea of Japan)
CHINA
MONGOLIA
JAPAN
NORTH KOREA
SOUTH KOREA

east of Moscow, and a third which lies in the Mongolia border zone.

When Russia's post–1991 government took over, it faced complicated administrative problems. Not only were some entities reluctant to sign the Russian Federation Treaty, but despite the ranking implied by the list above, certain Regions were more equal than others—that is, some were used to privilege under the old system, and their local leaders expected that to continue. Meanwhile, a multinational, multicultural state that had long been accustomed to authoritarian rule and government control over virtually everything—from factory production to everyday life—now had to be governed in a new way. Democratization of the political system, the transition to a market economy, privatization (the sale of state-owned industries), and other far-reaching changes had to be handled quickly or the country risked chaos.

MAJOR CITIES OF THE REALM

City	Population* (in millions)
Moscow, Russia	12.1
St. Petersburg, Russia	4.9
Baki (Baku), Azerbaijan	2.3
Novosibirsk, Russia	1.5
Yekaterinburg, Russia	1.4
Nizhniy Novgorod, Russia	1.2
Kazan, Russia	1.2
Chelyabinsk, Russia	1.2
Tbilisi, Georgia	1.1
Yerevan, Armenia	1.1
Volgograd, Russia	1.0
Vladivostok, Russia	0.6

*Based on 2014 estimates.

Unitary versus Federal Options

Russia's leaders recognized that their options were limited. They could continue to hold as much power as possible at the center, making decisions in Moscow that would apply to all the Republics as well as the lesser subdivisions of the state. Such a **unitary state system [2]**, with its centralized government and administration, marked authoritarian kingdoms of the past and serves totalitarian dictatorships of today. Or they could share power with the subnational units, allowing elected local leaders to come to Moscow to represent the interests of their people. That is the federal system Russia chose as the only way to accommodate the country's economic and cultural diversity.

In a **federal system [3]**, the national government usually is responsible for matters such as defense, foreign policy, and international trade. The provinces, States, and other subdivisions retain authority over affairs ranging from education to transportation. A federal system does not create unity out of diversity, but it does allow diverse components of the state to coexist, their common interests represented by the national government and their regional interests by their local administrations. Certain countries owe their survival as coherent states to their federal frameworks; India and Australia are cases in point.

But to maintain a generally acceptable balance of power between the center and the subnational units is difficult. Disputes concerning "States' rights" even continue to roil the U.S. political scene more than 225 years after the Constitution was adopted. Early on, the Russian government decided to end, for all practical purposes, the hierarchical regional system the Soviets had established. The Republics retained their special status, but all the others—okrugs, krays, and so forth—were redesignated as Regions. This was intended to address the favoritism of Soviet times as well as streamline the system in general.

Russia's New Federal Structure

In 1992, most of Russia's internal "republics," autonomous regions, oblasts, and krays (all of them components of the administrative hierarchy) signed a document known as the Russian Federation Treaty, which committed them to cooperate in the new federal system. At first a few units refused to sign, including Tatarstan, scene of Ivan the Terrible's brutal conquest more than four centuries earlier, and a republic in the remote Caucasus periphery, Chechnya-Ingushetiya, where Muslim rebels had launched a campaign for independence (Fig. 2B-2).

Chechnya-Ingushetiya subsequently split into two separate republics, whose names at the time were spelled Chechenya and Ingushetia. Eventually, only Chechnya refused to sign the Russian Federation Treaty; Russian military intervention followed and triggered a prolonged and violent conflict, with disastrous consequences for Chechnya's people and infrastructure (the capital, Groznyy, was completely destroyed). We will discuss Russia's contentious southern periphery in more detail later in this chapter.

In 2000, the new Putin administration moved to diminish the influence of the Regions by creating a new geographic framework that combined the 83 Regions, Republics, and other entities into a set of eight new administrative units—not to enhance their influence in Moscow, but to increase Moscow's authority over them. As Figure 2B-2 shows, each of these new Federal Districts has a capital, elevating cities such as Rostov and Novosibirsk to a status secondary only to Moscow. In a related move, the Russian president proposed, and the parliament approved, that Regional governors would be appointed rather than elected—thereby concentrating still more power in Moscow. Russia's federal system now appears to be moving in a unitary direction.

A Shrinking Population

Some of these administrative problems are compounded by an alarming trend: the Russian population has been shrinking steadily since 1990. And, as the boxed numbers in Figure 2B-2 shows, certain parts of the country have witnessed a veritable **population implosion [4]**. When the Soviet Union disintegrated in 1991, Russia's population totaled about 149 million. Today, nearly a quarter-century later, this number is down to around 143 million—despite the in-migration of several million ethnic Russians from the former Soviet SSRs beyond Russia's borders. Since the end of communist rule, Russia has witnessed over 15 million more deaths than births, with severe social dislocation the primary cause.

Undoubtedly, the transition away from Soviet rule is one reason: uncertainty tends to cause families to have fewer children, and abortion is widespread in Russia. But the birth rate has stabilized at around 13 per thousand; it is Russia's death rate that has skyrocketed, now exceeding more than 14 per thousand. According to the most recent data series through 2012, this produces an annual loss of population that exceeds 0.1 percent or about 150,000 per year.

The heart of downtown Moscow, with the triangular Kremlin (right of center) fronting on adjacent Red Square.

Russian males are the ones most affected. Male life expectancy dropped from 71 in 1991 to 63 in 2012 (over the same period, the rate for females dipped only slightly to 75—a life expectancy almost 20 percent higher). Males are more likely to be afflicted by alcoholism and related diseases, by AIDS (significantly underreported in Russia, according to international agencies), by heavy smoking, and by suicide, accidents, and murder. On average, a Russian male is nine times more likely to die a violent or an accidental death than his European counterpart. Shockingly, fewer than half of today's male Russian teenagers will survive to age 60.

In 2011 and 2012, Russia seemed to be bucking the trend: for the first time in 20 years the overall population expanded again, for two years straight, albeit minimally so. It is too early to tell if this turnaround will persist and become more significant; indeed, it could well prove to be only a momentary reprieve. At any rate, the situation appears grim at the regional scale. Look again at the percentages of population decline by Federal District (Fig. 2B-2): clearly, out-migration plays an important role too. People are no longer forced to stay put in these challenging environments by a dictatorial regime, and they are drawn to the far more prosperous, opportunity-rich western parts of Russia. Understandably, the emptying out of villages has been occurring all across Russia: more than 20,000 have turned into "ghost villages"—fully 15 percent of all such settlements in the country two decades ago.

What is the answer? Russia's leaders hope that improvements in the social circumstances of the average Russian will reduce the overall rate of population loss. Campaigns against alcoholism and careless lifestyles are under way. Immigration is another option, and should Russia relax its rules, hundreds of thousands of Koreans and Chinese would eagerly move into the Russian Far East to counter the outflow of Russians there. But Moscow is not interested in the least to see its Pacific Rim transformed into an extension of East Asia. Thus few of the ghost villages are likely to be reborn.

A Volatile Economy

Since the beginning of this century, Russia's economy has been turned 180 degrees. Private property, upstart companies, trade, foreign investment, and stock exchanges, all the things we consider vital components of the free market, have found their way into Russia. The city of Moscow has changed beyond recognition with a new high-rise skyline and unprecedented linkages to the global economy. After some major (and for many people painful) adjustments were made, Russia's GDP has grown vigorously as the price of shares on the Moscow stock exchange skyrocketed. In everyday parlance, this performance has catapulted Russia into membership in the exclusive club of the **BRICs [5]**, the world's four biggest emerging markets (Brazil, Russia, India, China). Today, Russia's *nouveaux riche* are known for their conspicuous consumption during shopping trips to Paris, Milan, and New York, for sending their children to high-cachet Swiss and British boarding schools, and even for purchasing premier soccer-team franchises in Europe (as some billionaires have done).

But to many economists Russia's rise seems precarious: much of it is based exclusively on exports of oil and gas; incoming foreign investment is quite limited compared to other emerging markets; the get-rich-quick schemes of a handful of highly successful businessmen (women are rarely part of this group) do not benefit the masses; and the gargantuan state apparatus left by the Soviets is responsible for widespread corruption and the close connections to

AMONG THE REALM'S GREAT CITIES . . .

MOSCOW

IN THE VASTNESS of Russia's territory, Moscow, capital of the Federation, seems to lie far from the center, close to its western margin. But in terms of Russia's population distribution, Moscow's centrality is second to none among the country's cities (see Fig. 2A-4). The Russian population as a whole has been shrinking but Moscow is bursting at the seams. The metropolis lies at the heart of Russia's primary core area and at the focus of its circulatory systems.

On the banks of the meandering Moscow River, Moscow's skyline of onion-shaped church domes and modern buildings rises from the forested flat Russian Plain like a giant oasis in a verdant setting. Archeological evidence points to ancient settlement of the site, but Moscow enters recorded history only in the middle of the twelfth century. Forest and river provided defenses against Tatar raids, and when a Muscovy force defeated a Tatar army in the late fourteenth century, Moscow's primacy was assured. A huge brick Kremlin (citadel; fortress), with walls 2 kilometers (1.2 mi) in length and with 18 towers, was built to ensure the city's security. From this base Ivan the Terrible expanded Muscovy's domain and laid the foundations for Russia's vast empire.

The triangular Kremlin and the enormous square in front of it (Red Square of Soviet times), flanked by St. Basil's Cathedral and overlooking the Moscow River, is the center of old Moscow and is still the heart of this metropolitan complex of 12.1 million. From here, avenues and streets radiate in all directions to the Garden Ring and beyond. Until the 1970s, the Moscow Circular Motorway enclosed most of the built-up area, but today the metropolis sprawls far beyond this beltway.

In the post-communist era, the center of Moscow has undergone a complete makeover as Soviet-dictated urbanism made way for the emergence of a fast-paced, international business complex—one that increasingly resembles those of other global cities. Once the free market, however imperfectly, shaped up, Moscow's CBD became the preferred location for foreign investors, multinational corporations, upscale shopping, and elite housing. It quickly became one of the most expensive residential areas in the world, its exclusiveness even more striking in the heart of a metropolis that is still mostly dominated by drab, austere Soviet architecture.

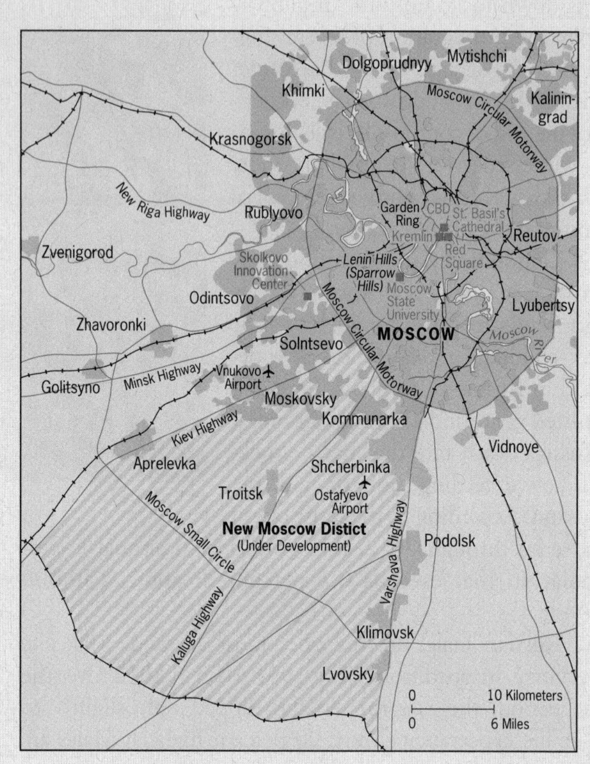

Moscow has grown so fast and congestion has been so overwhelming that the government has embarked on an enormous urban expansion plan. It began by incorporating a huge swath of land southwest of the capital into Moscow City proper that will the double the size of the metropolitan area, in the process swallowing up extensive tracts of forest as well as numerous municipalities. A consortium of design and planning companies is now working on the plan for *New Moscow*—a project unprecedented in scale and ambition that is to include a new federal government complex, a highest-order financial district, a world-class high-technology cluster, and a wide range of housing and amenities for nearly two million new residents. The full implementation of the plan will not occur until 2030 at the earliest.

organized crime. The value of the "bribes market" is estimated to be about 20 percent of the national economy. Tellingly, unlike the other BRICs, the Russian economy was severely impacted by the global financial crisis that began in the late 2000s.

Seventy years of Soviet-style socialism have left their mark. If statistics on the Russian economy over the past decade suggest robust growth, there is also ample evidence that most of the people aren't seeing the fruits. And that only makes it harder to preserve this federation, which is not helped by its especially complicated spatial administrative structure that is riddled with corruption and intransigence. Not unexpectedly, Russians are increasingly frustrated with the economic and political situation. As noted earlier, the (rigged) parliamentary elections of December 2011, which handed victory to President Putin's United Russia Party, triggered unprecedented demonstrations on Moscow's streets. Nonetheless, little stood in the way of Putin's March 2012 return to the presidency for another six years—and his previously engineered

VOICE FROM THE Region

CITY LIFE IN MOSCOW

© Mitia Kolossov

Mitia Kolossov, Moscow

"I was born and raised in Moscow, not far from the campus of Moscow State University. The main university building is 236 meters [775 ft] high; I could see the tower from my room in the house where I grew up. My parents and grandparents are also from Moscow. That is pretty rare because the capital has always been a city of migrants. Moscow has changed incredibly in the last 15 years. For me, the most important thing is that it is a city with so many opportunities: for instance, Moscow now has more than 200 theaters, dozens of art galleries, and its restaurants represent all possible world cuisines. Unlike the Soviet past, the city is now brightly illuminated and looks nicer, more romantic at night than during the day. But it has also become much busier and hectic everywhere, with endless and often unpredictable traffic jams and very limited parking spaces. Some years ago I graduated from Moscow State University and am presently in the advertising business. For a large number of my clients, both international and local companies, this city became the main center of control and decision-making. About 95 percent of Russian advertising activity is located in Moscow, and you cannot enter the market from anywhere else. Sometimes I dream about living in a smaller town, being tired of the monster-city tempo. But Moscow is where the action is. And it is home."

© Mitia Kolossov

Moscow State University, Main Building

constitutional change allows him to run for reelection for yet another six-year term (his fourth) after that.

It is now time to turn our attention to the realm's regional geography. As was noted at the beginning of this chapter, so vast is Russia, so varied its physiography, and so diverse its cultural landscape that regionalization requires a broad perspective and a high level of generalization. Let us begin by outlining the five-region framework that is mapped in Figure 2B-1: (1) the ***Russian Core***, everything lying west of the Urals; (2) the ***Southeastern Frontier***, the eastward extension of the Core into the heart of Eurasia; (3) ***Siberia***, the frigid territory sprawling across the country's immense northeastern quadrant; (4) the ***Far East***, Russia's elongated window on the Pacific Ocean; and (5) ***Transcaucasia***, whose three small states lie just outside Russia in the corridor between the Black and Caspian seas. As we shall see, each of these first-order regions contains major subregions.

THE RUSSIAN CORE

The heartland of a state is its ***core area***. Here much of the population is concentrated, and here lie its biggest cities, leading industries, densest transportation networks, most intensively cultivated farmlands, and other key components of the country. Core areas of long standing strongly reflect the imprints of both culture and history. The Russian Core, broadly defined, extends from the western border of the Russian realm to the Ural Mountains in the east (Fig. 2B-1). This is the Russia of Moscow and St. Petersburg, of the Volga River and its industrial cities.

Moscow is the megacity hub of a wider area comprising some 50 million inhabitants (more than one-third of the country's entire population), most of them concentrated in cities that, during the communist period, specialized in assigned industries such as Nizhniy Novgorod (the "Soviet Detroit" where automobiles were made) and Ivanovo (known for its textiles). But today, Russia is less a diversified manufacturing country than a commodity producer and exporter (mainly oil and natural gas). The tall steel-and-glass towers you see rising above the central business districts for the most part belong to state-controlled energy companies such as Gazprom. And the cars being built are likely to display foreign names such as Fiat or Volkswagen.

Moscow itself has grown rapidly in recent years and today has a population of just over 12 million, of which more than 2 million are nonregistered migrants. It is the economic and business opportunities that continue to draw people in, but steadily worsening congestion and an increasingly harsh social climate mean that many dream of leaving the city as soon as they can afford to. The Moscow metro system now carries up to nine million passengers daily, and new route extensions and metro stations are built almost every year. To handle all this overcrowding, the government in 2011 announced plans for a massive, 20-year-long expansion to the southwest (see box on Moscow).

St. Petersburg (the former Leningrad) remains Russia's second city, with a population of 4.9 million. In the late seventeenth and early eighteenth century, Czar Peter the Great and his architects transformed this city at the head of the Baltic Sea's Gulf of Finland into the Venice of the north. St. Petersburg's palaces, churches, waterfront façades, bridges, and monuments still give the city a European look unlike that of any other city in Russia. It was proclaimed the capital of Russia in 1712. Under the czars, St. Petersburg was the focus of Russian political and cultural life, and Moscow served as a distant second city. The Imperial Winter Palace and the adjoining Hermitage (one of the biggest art museums in the world) at the heart of the city are among a host of surviving architectural treasures.

Today, after a long interval of neglect during the communist days of Leningrad, St. Petersburg's ports on the Gulf

A new research facility at the Skolkovo Innovation Center, 25 kilometers (15 mi) southwest of central Moscow. This budding high-tech business complex has attracted wide international interest and aspires to become nothing less than Russia's Silicon Valley.

of Finland are again important to Russia's trade with the rest of the world, and the city is now a major destination for many cruise ships. But St. Petersburg possesses none of Moscow's locational advantages, at least not with respect to the domestic market. It lies well outside the Central Industrial Region (Fig. 2B-3) near the northwestern corner of the country, 650 kilometers (400 mi) from the capital. Neither is it better off than Moscow in terms of resources: fuels, metals, and foodstuffs must all be brought in, mostly from far away. With Moscow these days primarily defined in terms of global finance and producer services, St. Petersburg's niche involves trade, culture, and tourism.

Many realms around the world have core areas that are quite distinct from their peripheries, but it is important to appreciate the extreme dominance of Russia's Core region. An urbanized region that contains a third of the entire national population in a country with nine time zones is truly astonishing. And the economic and social disparities seem to be increasing by the day. Moscow's middle class is expanding and its changing consumer preferences are today on view at nearly 90 American-style shopping malls across the city, all built since the 1990s. Meanwhile, in Russia's outlying areas, many continue to live at near-subsistence levels. Moreover, as the core keeps growing, the periphery is emptying out.

Central Industrial Region

At the heart of the Russian Core lies the Central Industrial Region (Fig. 2B-3). The precise definition of this subregion varies, because all regional definitions are subject to debate. Some geographers prefer to call this the Moscow Region, thereby emphasizing that for over 400 kilometers (250 mi) in all directions from the capital, everything is oriented toward this historic focus of the state. As Figure 2B-3 (even better, Fig. 2A-9) underscores, Moscow has maintained its decisive **centrality [6]**: roads and railroads converge in all directions—from Ukraine in the south; from Minsk (Belarus) and the rest of eastern Europe in the west; from St. Petersburg and the Baltic coast in the northwest; from Nizhniy Novgorod and the Urals in the east; from the cities and waterways of the Volga Basin in the southeast (a canal links Moscow to the Volga, the country's single most important navigable river); and even to the subarctic northern periphery that faces the Barents Sea, where the strategic naval port of Murmansk and the lumber-exporting outpost of Arkhangelsk lie.

Povolzhye: The Volga Region

Another major subregion within the Russian Core is the ***Povolzhye***, the Russian name for an area that extends along the middle and lower valley of the Volga River as well as its tributary, the Kama, above the city of Kazan. It would be appropriate to call this the Volga Region, for this greatest of Russia's rivers is its lifeline and most of the Povolzhye's cities lie on its banks (Fig. 2B-3). In the 1950s, a canal was completed to link the lower Volga with the lower Don River (and thereby directly access the Black Sea). The city of Volgograd is centrally located in this region, on the west bank of the Volga. It was named Stalingrad from 1925 to 1961, and that name is associated with a game-changing battle in World War II. In early 2013, the city government decided to temporarily restore the name Stalingrad at the time of the annual war commemorations. This was widely regarded as a way to rehabilitate Stalin, part of a contentious turn toward state-sponsored patriotism and nostalgia for the Soviet days.

The Volga corridor was an important historic route in old Russia, but for a long time neighboring regions overshadowed it. The Moscow area as well as Ukraine were far ahead in industry and agriculture. The Industrial Revolution that came late in the nineteenth century to the Central Industrial Region did not have much effect in the Povolzhye. Its major function remained the transit of foodstuffs and raw materials to and from other regions.

The elongated Povolzhye (the *vol* refers to the Volga) forms the eastern flank of the Central Industrial Region, its cities and farms sustained by Russia's greatest river. For a long time this subregion was bypassed in the Soviet industrialization drive, but during World War II it

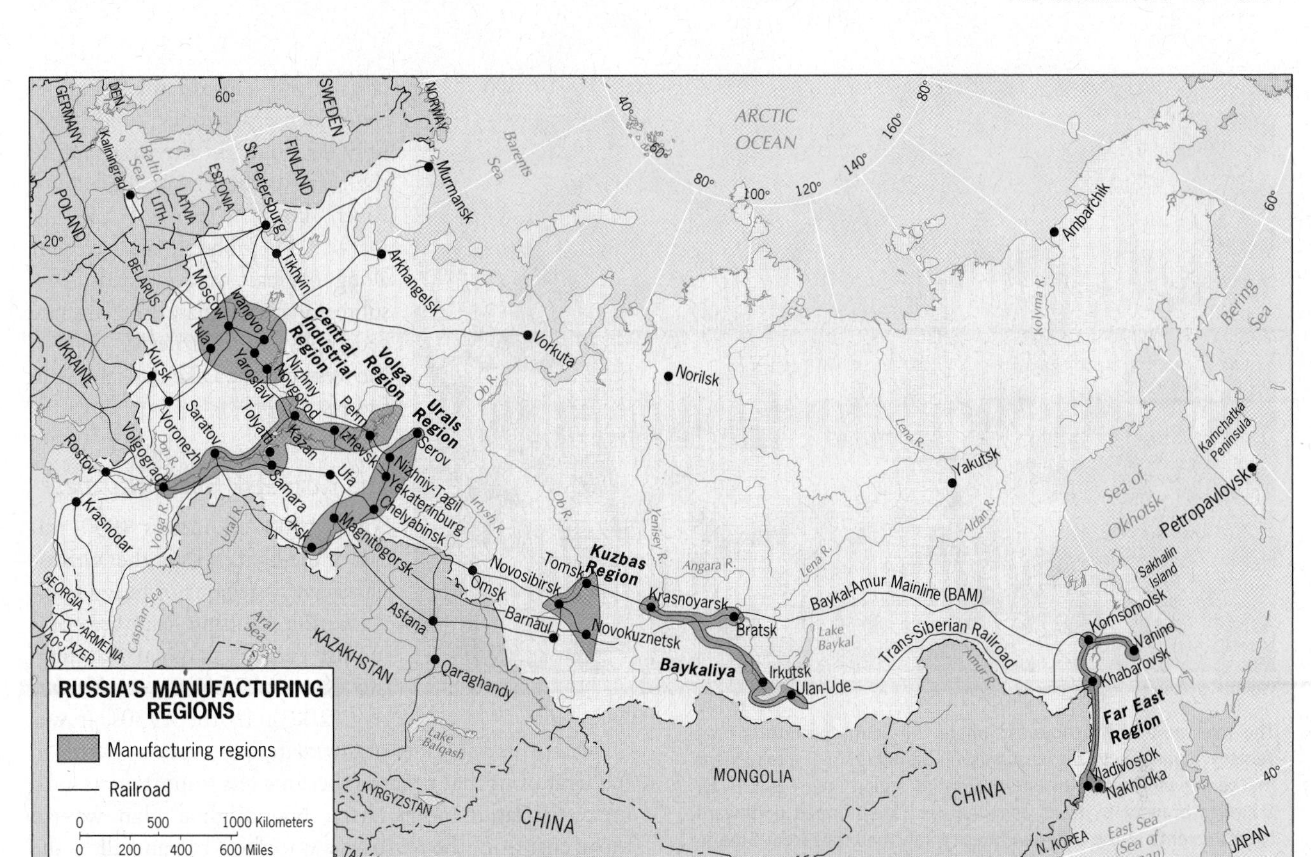

FIGURE 2B-3

suddenly took on much greater importance because its cities lay relatively remote from the German armies invading from the west. Accelerated industrial development put the Volga Valley's cities on the map as never before. Then, in the 1950s, the crucial Volga-Don Canal was opened, creating new linkages and opportunities. Next, the full dimensions of the oil- and gasfields in and near the Povolzhye became clear (see Fig. 2A-5), and for some time these reserves were the largest in the entire Soviet Union. And finally, the Volga Region's northwestward linkages, via the Moscow and Mariinsk canals to the Baltic Sea, were improved and augmented by upgraded rail and road connections. Again, the industrial composition of the Volga's riverside cities continues to change, but if the Central Industrial Region is Russia's heart, the Povolzhye is its key artery.

Russian President Putin (left) and Prime Minister Medvedev on the ski slopes of the southern city of Sochi on the Black Sea, site of the 2014 Winter Olympics. Getting to host the games was a major coup for the Russian government, which will surely take advantage of the global media exposure to advance Russia's image in the world. But some observers describe Sochi as a "political tinderbox" because of its proximity to the turbulent Caucasus (as Fig. 2B-5 shows, the city is close to the Georgian border). Under Czar Alexander II in the 1860s, the Circassians native to this area were forcibly removed to make way for Russians. Today's Russian leadership acts increasingly autocratic, and relations between Putin and Medvedev are souring.

The Urals Region

The Ural Mountains form the eastern edge of the Russian Core. They are not particularly high; in the north they consist of a single range, but southward they broaden into

Wolfgang Kaehler/SuperStock

The river port at Krasnoyarsk, where the south-to-north-flowing Yenisey River meets the east-west Trans-Siberian Railroad near the center of the Southeastern Frontier region (see Fig. 2B-1). Supplies brought by train from the west are shipped northward to settlements in the Siberian interior; raw materials from Siberia arrive by boat and barge for dispatch to Russia's factories and markets. Large storage facilities stand across the river.

a hilly zone. Nowhere are they an obstacle to east-west surface transportation. An enormous storehouse of metallic mineral resources located in and near the Urals has made this area a natural place for industrial development. Today, the Urals Region, well connected to the nearby Volga and Central Industrial Regions, extends from Serov in the north to Orsk in the south (Fig. 2B-3).

As Figure 2B-3 reveals, the Soviet legacy still survives on Russia's economic landscape, despite the country's conversion from a domestic manufacturing system to what some observers now call a "petro-ruble state." Not only the Central Industrial Region but also the essential Volga corridor, the Urals manufacturing complex, the industrial nodes of the Southeastern Frontier, and even the outliers of the Far East still mark the manufacturing map, even if many factories have fallen silent and people are leaving. Urban dwellers tend to find new ways to make their living, and such cities as Volgograd, Chelyabinsk, Novosibirsk, Irkutsk, and Vladivostok will never be empty.

THE SOUTHEASTERN FRONTIER

From the southeastern flanks of the Ural Mountains to the headwaters of the Amur River, and from the latitude of Tyumen to the northern zone of neighboring Kazakhstan, lies Russia's vast Southeastern Frontier region, product of a gigantic Soviet experiment in the eastward extension of the Russian Core (Fig. 2B-1). As the maps of cities and transportation (Fig. 2A-9) as well as population distribution (Fig. 2A-4) suggest, this region is more densely peopled and more fully developed in the west than in the east. To the east of the Yenisey River, linear settlement prevails, marked by ribbons and clusters along the east-west railroads. Two subregions dominate the human geography: the Kuznetsk Basin in the west and the Lake Baykal area in the east.

The Kuznetsk Basin (*Kuzbas*)

Some 1500 kilometers (950 mi) east of the Urals lies another of Russia's primary regions of heavy manufacturing resulting from the communist period's national planning: the Kuznetsk Basin, or ***Kuzbas*** (Fig. 2B-3). In the 1930s, it was opened up as a supplier of raw materials (especially coal) to the Urals, but that function became less important as local industrialization accelerated. The original plan was to move coal from the Kuzbas west to the Urals and allow the returning trains to carry iron ore east to the coalfields—a classic case of **double complementarity [7]**. However, good quality iron ore deposits were subsequently discovered close to the Kuznetsk Basin itself. As the new resource-based Kuzbas industries expanded, so did its urban centers. The leading city, located just outside the Kuzbas, is Novosibirsk, which stands at the intersection of the Trans-Siberian Railroad and the Ob River as the symbol of Russian enterprise in the vast eastern interior. To the northeast lies Tomsk, one of the oldest Russian towns east of the Urals, founded in the seventeenth century and now caught up in the modern development of the Kuzbas. Southeast of Novosibirsk lies Novokuznetsk, a city that produces steel for this subregion's machine and metal-working plants as well as aluminum products from Urals bauxite.

The Lake Baykal Area (*Baykaliya*)

To the east of the Kuzbas, development becomes more insular, and distance becomes a stronger adversary. North of the central segment of the Mongolian border zone and eastward around Lake Baykal, larger and smaller settlements cluster along the two railroads to the Pacific coast (Fig. 2B-3). West of the lake, these rail corridors lie in the headwater zone of the Yenisey River and its tributaries. A number of dams and hydroelectric projects serve the valley of the Angara River, particularly the city of Bratsk. Mining, lumbering, and some farming sustain life here, but isolation dominates it. The city of Irkutsk, near the southern end of Lake Baykal, is the principal service center for the enormous Siberian region to the north as well as for a lengthy east-west stretch of southeastern Russia.

Beyond Lake Baykal, the Southeastern Frontier really lives up to its name: this is southern Russia's most rugged, remote, and forbidding country. Settlements are rare, with many being mere camps. The Buryat Republic (see Fig. 2B-2) is part of this zone; the territory bordering it to the east was taken from China by the czars and may well become an issue in the future. Where the Russian-Chinese boundary turns southeastward along the Amur River, the region named the Southeastern Frontier ends and Russia's Far East begins.

SIBERIA

Before we assess the potential of Russia's Pacific Rim, we should remember that the ribbons of settlement just discussed hug the southern perimeter of this gigantic country, avoiding the huge Siberian region to the north (Fig. 2B-1). Siberia extends from the Ural Mountains to the Kamchatka Peninsula—a vast, frigid, forbidding land. Larger than the conterminous United States but inhabited by only an estimated 20 million people, Siberia quintessentially symbolizes the Russian environmental plight: vast distances, cold temperatures worsened by strong Arctic winds, difficult terrain, poor soils, and limited options for survival.

But Siberia also possesses resources. From the days of the first Russian explorers and Cossack adventurers, Siberia's riches have beckoned. Gold, diamonds, and many other precious minerals were found. Later, metallic ores including iron and bauxite were discovered. Still more recently, the Siberian interior proved to contain massive deposits of oil and natural gas (Fig. 2A-5) that now contribute significantly to Russia's energy supply and exports.

As the physiographic map (Fig. 2A-2) reveals, major rivers—the Ob, Yenisey, and Lena—flow gently northward across Siberia and the Arctic Lowland into the Arctic Ocean. Hydroelectric power development in the basins of these rivers has generated electricity used to extract and refine local ores, and run the lumber mills that have been set up to exploit the vast Siberian forests.

The human geography of Siberia is fragmented, and most of the region is virtually uninhabited (Fig. 2A-8). Ribbons of Russian settlement have developed; the Yenisey River, for instance, can be traced on this map of Soviet peoples (a series of small settlements north of Krasnoyarsk), and the upper Lena Valley is similarly fringed by ethnic Russian settlement. Yet hundreds of kilometers of empty territory separate these narrow ribbons and other islands of habitation.

Siberia, Russia's freezer, is stocked with goods that may become mainstays of future national development. Already, precious metals and mineral fuels are bolstering the Russian economy. In time, we may expect Siberian resources to play a growing role in the economic development of the Southeastern Frontier and the Russian Far East as well. One step in that process was already taken during Soviet times: the completion of the BAM (Baykal-Amur Mainline) Railroad in the 1980s. This route, lying north of and parallel to the old Trans-Siberian Railroad, extends 3500 kilometers (2200 mi) eastward from near the important center of Krasnoyarsk directly to the Far East city of Komsomolsk (Fig. 2B-3). In the post-Soviet decades, the BAM Railroad has been beset by equipment breakdowns and workers' strikes. Nonetheless, it is a key element of the infrastructure that will serve the economic growth of easternmost Russia in the twenty-first century.

Oymyakon is known as the coldest town on Earth. It is located north of the city of Yakutsk in eastern Siberia. The average daily temperature during winter hovers around −46° Celsius (−51° Fahrenheit), and the record low recorded here in 1933 was −68°C (−90°F). A few hundred people live here, and there is even a school—which does not open if the temperature drops below −52°C (−2°F). Cellphones would not work here, even if there was satellite coverage, and visitors are advised not to wear glasses—they freeze to your face.

THE RUSSIAN FAR EAST

When the Putin government in Moscow in 2000 superimposed its framework of Federal Districts over Russia's matrix of Soviet-era regions (Fig. 2B-2), it gave the name *Far East* to the largest one (in terms of territory) of all. In one stroke, the Far East incorporated four residual ethnic republics and six Russian Regions. But many Russian citizens have a different view of what constitutes the geographic region they call the Far East. Justifiably, they see most of the northern zone of the official Far East as a continuation of Siberia. To them, the real Far East is constituted by the area beyond the Southeastern Frontier to the Pacific coast, the island of Sakhalin, the Kamchatka Peninsula, and a narrow stretch of land along

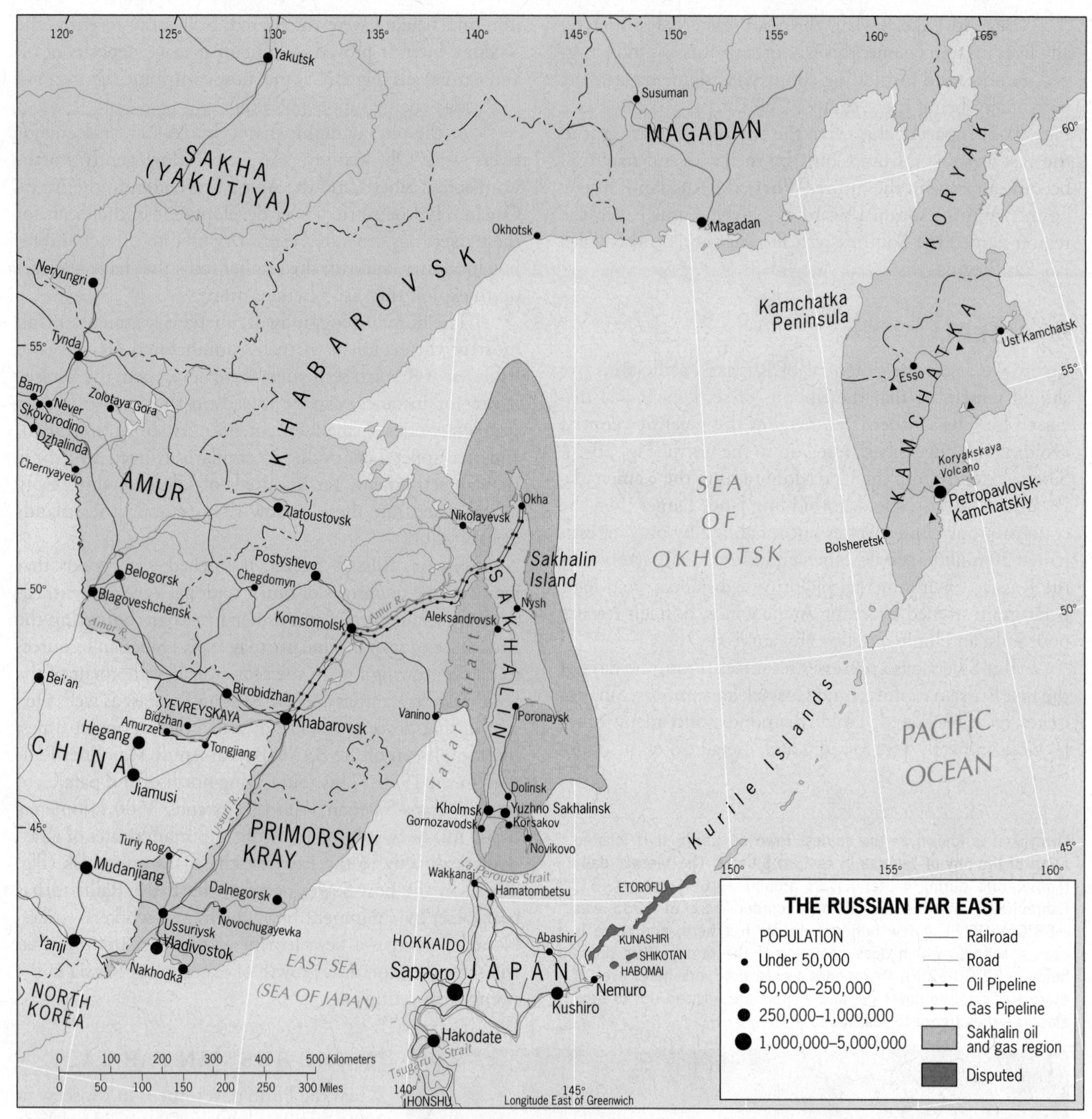

FIGURE 2B-4

the northern shore of the Sea of Okhotsk (Figs. 2B-1 and 2B-4).

In Soviet times, citizens who agreed to move to this most distant region to work in factories or for the administration were rewarded with special privileges to compensate for their hardship. Make no mistake: the Pacific Ocean's proximity tends to temper natural environments here, but life throughout the Russian Far East is tough. Moreover, in the post-communist era, Moscow's favors were withdrawn, so that hundreds of thousands of inhabitants under the new democracy voted with their feet and simply departed. Take the largest urban center—Petropavlovsk—on the Kamchatka Peninsula (see photos in *From the Field Notes*). In 1989, shortly before the Soviet Union collapsed, it had a population of 269,000; by 2002, it had dropped to 198,000; today, it contains just over 170,000 residents. And Petropavlovsk also has the distinction of being the largest place on Earth without an overland link to the outside world: there simply is no road off the huge peninsula.

Almost everywhere in the Far East, people now feel abandoned by Moscow. The once-great naval base of Vladivostok (the city's name means "We Own the East") lies dormant. Factory closures are widespread. And when the locals tried something new—clandestinely importing used Japanese cars—the government sent its federal police

Renaud Visage/Photographer's Choice/Getty Images

The Siberian tiger, the largest member of the cat family, is native to this environment. It needs a vast and undisturbed habitat, its coat protects against the cold, and its partially white fur serves as camouflage. But as a result of human encroachment and poaching, even in these remote parts, there are only a few hundred left and these animals have become a protected species. The Siberian tiger can now only be found in the forests and mountain ranges of Russia's Far East, and sightings are rare.

to close down the business. Meanwhile, the nearby container port at Nakhodka suffers repeated breakdowns. Cross-border commerce with China is minimal. Trade with Japan is inconsequential. The Trans-Siberian and BAM railroads that were to stimulate cross-country traffic and trade cannot operate efficiently without massive subsidies—available during Soviet times but no longer in the twenty-first century's market economy.

The Far East does have significant reserves of what underpins the Russian economy—oil and natural gas in and around the island of Sakhalin. And, as Figure 2B-4 reminds us, the southern sector of the Russian Far East lies between other reserves and potential markets: China next door and Japan across a narrow sea. Already, both the Japanese and the Chinese are bidding for Russia's energy supplies, and undoubtedly there will be further construction of pipelines and other energy-related infrastructure. But what this region needs is a diversified economy of the kind that is reviving China's Northeast across the Amur and Ussuri rivers. Such development, however, is currently not in sight.

From the Field Notes . . .

© H.J. de Blij

© H.J. de Blij

"The Russian Far East is losing people at the fastest rate among all Russian Federal Districts, and my field trip to Petropavlovsk-Kamchatsky on the Kamchatka Peninsula only reinforced what I had learned in Vladivostok some years ago. During Soviet times, Russians willing to move to the Far East were given special privileges (an older man told me that 'being as far away from Moscow as possible was reward enough'), but today locals here feel abandoned, even repressed. In 1989, Petropavlovsk had about 270,000 residents; today that total is estimated to be fully 100,000 fewer. It is not just the bleak environment—the ubiquitous ash and soot from dozens of active volcanoes on Kamchatka, covering recent snowfalls and creating a pervasive black mud—it is also a loss of purpose that people feel here. There was a time when the Soviet windows on the Pacific were valuable and worthy; now the people scramble to make a living, and when they find a way, for example by selling used cars bought cheaply in Japan and brought to the city by boat, Moscow sends law enforcement to halt the illegal trade. Petropavlovsk has no surface links to the rest of the world, and a 2009 springtime view of the city at the foot of the volcano Koryakskaya (top) and a street scene near the World War II museum (bottom) suggest why the Far East is depopulating."

www.conceptcaching.com

THE SOUTHERN PERIPHERY

Now, having followed Russia's eastward expansion and organization, let us take another look at the realm's regional map (Fig. 2B-1) to focus on the most complicated region of all—the area where Russia's leaders, communists as well as czars, pushed the expanding state's influence across and beyond the Caucasus Mountains. On this map, it looks as though the Russian Core extends to the international boundary (which coincides, more or less, with the crestline of this towering range). But things are not that simple. When you travel southward from Volgograd or Rostov, the landscape—physical as well as cultural—does not change; the Russian Plain's gently rolling countryside with its villages and farms is very much like that south of Moscow. Taking the train south from Astrakhan in the Volga River Delta, you will soon see the Caucasus wall looming in the distance, but you may not realize that you have entered a different world until you get off near its base at Groznyy, the capital of Chechnya, one of Russia's 21 ethnic Republics (Fig. 2B-5). Just observing the mix of peoples makes you aware that you have left the heartland and reached Russia's periphery.

Figure 2B-5 is important enough to spend some serious time studying it. Here the communist leaders who succeeded the czars confronted multiple challenges. First, inside the Russian state itself, there were many minority peoples (such as the Chechens of Chechnya) who had fought the czars' armies and who now faced new rulers. Second, there were peoples who had sided with the czars and who wanted to cooperate with the Soviet regime, such as the Ossetians. To recognize the aspirations of these peoples, some (but not all) of whom were Muslim, the communist administration designated ethnic "republics" for them. On our map, these are colored pink to distinguish them from the rest of Russia, which is shown in tan.

Russia's Internal Periphery

From Buddhist-infused Kalmykiya on the perimeter of the Caspian Sea westward to mainly Orthodox Christian Adygeya near the Black Sea, the Internal Periphery of southern Russia is defined by eight ethnic Republics, each with its own identity (Fig. 2B-5).

Physiographically as well as culturally divided Chechnya, as the inset map suggests, lies in a transition zone between plain and mountains, the mountains serving as refuge for those opposed to Moscow's rule. When the post-Soviet Russian government asked all Regions and Republics within Russia to sign its proposed Russian Federation Treaty, Chechnya's leaders refused, believing that they now had the opportunity to finally end Russian control. Its western neighbor, Ingushetiya, also got caught up in anti-Russian activism, and the ensuing drawn-out conflict took a terrible toll in adjacent North Ossetia where, in 2004, more than 350 students, teachers, and parents died in a terrorist attack on a school. Still farther west, things remained calmer, and no problems at all arose in generally pro-Moscow Adygeya.

This tier of ethnic republics along Russia's southern border displays all the properties of a disadvantaged periphery: not only do they share a history of subjugation by the powerful core to the north, but they also lag behind the rest of Russia by almost any measure of social progress. In terms of health, education, income, opportunity, and other such indices, the peoples of the Internal Periphery, whether they are pro-Russian or not, have a long way to go to achieve Russian living standards. And there is much simmering resentment of Russian rule, past and present, in such republics as Chechnya and Ingushetiya.

In the case of Chechnya, such anger has its most recent roots in the Second World War, when Soviet leaders accused the Chechens of sympathizing, and even collaborating, with the Nazi invaders. Josef Stalin, then the Soviet dictator, ordered the entire Chechen population loaded on trains and exiled to the desert of Kazakhstan across the Caspian Sea. Tens of thousands died along the way or after arriving there, and even though Stalin's successor, Nikita Khrushchev, in 1957 allowed the survivors to return to their homeland, the Chechens never forgave their tormentors. Then in 1992, just after the Soviet Union disbanded, they saw their opportunity, refusing to sign the Russian Federation Treaty that would make their republic an integral component of the new Russian state, and mounting a campaign for independence that led to a bitter and costly war that destabilized a number of neighboring republics as well. In fact, Chechen terrorists eventually carried this campaign to the very heart of Moscow, killing hundreds of civilians there and changing Russia's political landscape. When presidential candidate Vladimir Putin promised victory, the Russians elected him overwhelmingly in 2000. Massive military intervention did overpower the Chechens and the war came to an end in 2007; Chechnya now has a pro-Russian regime that is receiving massive economic support from Moscow. However, dissent among the Chechens continues, and the Internal Periphery as a whole remains restive and unstable.

■ TRANSCAUCASIA: RUSSIA'S EXTERNAL PERIPHERY

Russian (and later Soviet) manipulation of the ethnic map did not terminate at the boundary we see in Figure 2B-5, but it did take a different turn. Beyond the Russian border, the Soviet Empire extended its power over three adjoining Transcaucasian entities where the Russians had for centuries exerted influence and waged war. These three units were **Georgia** (mapped in yellow in Fig. 2B-5), facing the Black Sea; **Azerbaijan** (green), bordering on the Caspian Sea; and **Armenia** (lavender), landlocked in the middle. During Soviet times, Georgia was a loyal partner (Josef Stalin was a Georgian by birth). Christian Armenia appreciated membership in the Soviet Union because of the security it provided next door to hostile Islamic Turkey. And Islamic Azerbaijan, whose ethnic majority comprises a

SOUTHERN RUSSIA:
INTERNAL AND EXTERNAL PERIPHERIES
POPULATION
Under 50,000
50,000–250,000
250,000–1,000,000
1,000,000–5,000,000
Over 5,000,000
National capitals are underlined
Railroad
Road
Canal
Oil pipeline
Proposed oil pipeline
0 50 100 150 200 250 300 350 Kilometers
0 50 100 150 200 Miles
RUSSIA
KAZAKHSTAN
KALMYKIYA
Volgograd
Astrakhan
Elista
Rostov
Mariupol
UKR.
Sea of Azov
Tsimlyansk Res.
Don River
Volga River
Tikhoretsk
Kropotkin
Krasnodar
Novorossiysk
Armavir
Stavropol
Budennovsk
Maykop
ADYGEYA
Mineral'nyye Vody
Cherkessk
Pyatigorsk
KARACHAYEVO-CHERKESSIYA
KABARDINO-BALKARIYA
Nalchik
INGUSHETIYA
Groznyy
CHECHNYA
Makhachkala
Vladikavkaz
NORTH OSSETIA
Nazran
Sochi
ABKHAZIA
Sokhumi
CAUCASUS MOUNTAINS
Pankisi Gorge
SOUTH OSSETIA
Tskhinval
Gori
DAGESTAN
Caspian Sea
Black Sea
Samtredia
Kutaisi
Supsa
Rioni R.
Batumi
AJARIA
Tbilisi
Rustavi
GEORGIA
Borcka
TURKEY
Trabzon
Ceyhan
Lake Mingacevir
Kura R.
AZERBAIJAN
Sumgait
Baki (Baku)
Gänjä
Gyumri
ARMENIA
L. Sevan
Yerevan
NAGORNO-KARABAKH
Xankändi (Stepanakert)
Aras R.
NAXCIVAN AZER.
Armenian controlled
AZERBAIJAN PROVINCE
Agri
Lake Van
Van
Marand
Tabriz
Ardabil
IRAN
Lake Urmia
Longitude East of Greenwich
CHECHNYA
Kargalinskaja
Russian North (Plains)
Iscerskaja
Selkovskaja
Terek R.
Cervlënnaja
Groznyy
Gudermes
Argun
Sunzha R.
Shali
Urban-Industrial Middle Zone
Vedeno
Chechen South (Mountains)
Shatoy
Argun R.
NORTH OSSETIA
INGUSHETIYA
DAGESTAN
CAUCASUS MTS.
GEORGIA
0 20 40 Kilometers
0 10 20 30 Miles

FIGURE 2B-5

© Diana K. Ter-Ghazaryan

Yerevan, Armenia's capital, is one of the world's oldest continually inhabited urban settlements, always ready to adapt to changed regional circumstances. No other city of the old USSR has undergone as dramatic a transformation of its landscape. New construction was everywhere in mid-2013, with this growing high-rise framing Dalma Gardens Mall. As the cars out front attest, this is the entrance to the country's first upscale shopping center. Turn around in the parking lot, and your view is dominated by Mount Ararat across the nearby Turkish border, reputedly where Noah's Ark landed.

people known as the *Azeris* (who have close kinship with Iranian Azeris directly across their southern boundary with Iran), was a valued member of the USSR because of its rich oil reserves.

As if this were not enough, note in Figure 2B-5 that Azerbaijan has a large exclave on the Iranian border (Naxcivan), separated from the main part of the country by Armenian territory—and that the Armenians hold another exclave named Nagorno-Karabakh inside what would appear to be Azerbaijan's territory (which is home to more than 130,000 Christian Armenian citizens who were placed under Islamic Azerbaijan's jurisdiction by Soviet mandate).

Soviet rule kept the lid on potential conflict in this historically turbulent part of Transcaucasia, and after the collapse of the USSR Russia sought to retain its influence and maintain stability in this vital sector of the Near Abroad. The Russians wanted to keep Azerbaijan's oil flowing to and through Russia (see the pipeline that leads from Baki [Baku] via Chechnya's Groznyy to the Black Sea terminal at Novorossiysk in Fig. 2B-5); they also wanted to keep Georgia in the Slavic fold, and they looked for ways to mitigate the long-running clash between Azerbaijan and Armenia over the enclave of Nagorno-Karabakh.

From the Field Notes . . .

"Upon first entering Sukhum, the capital of the 'de facto' state of Abkhazia on the Black Sea coast (see Fig. 2B-5), I visited the former government house—now an eerie and unintended war memorial. It was the scene of an intense battle in September 1993 when the Georgian government unsuccessfully tried to suppress Abkhazian separatists. The building bears the marks of bullets, bombs, and fires, and is deliberately kept in this state as a reminder of the war. The Abkhazians repelled the Georgians and insist on their own spelling of the city's name (Georgians use Sokhumi), and note that it is the Abkhazian "national" flag flying on the top of the burnt-out building. The legacies of war in the Caucasus are dramatically evident in the landscape with the omnipresence of ruined buildings, war memorials, empty houses of displaced persons, refugee camps, military checkpoints, minefields, abandoned farms, ruined roads, and destroyed statues such as the one of Lenin that used to stand on the now empty base in front of the building. Like many other structures (over half by some estimates) in Sukhum, this one remains as it was when the fighting ended, but its vast size and high-rise prominence in a city of small, low buildings makes it visible from everywhere in town and a daily reminder of the war's brutality."

© John O'Loughlin

www.conceptcaching.com

But things have not gone well here for Moscow. Spurred by Western investments, Azerbaijan in 2006 began exporting oil via a new pipeline across Georgia and Turkey to the latter's Mediterranean port of Ceyhan, providing it with a more favorable economic as well as political alternative to Russian transit. Armenia and Azerbaijan have persisted in their sometimes-violent quarrel over Nagorno-Karabakh. And things really fell apart in Georgia, where the government pursued pro-European policies and allegedly mistreated its pro-Russian Ossetian minority, and where Russia openly supported devolutionary initiatives in the northwestern corner of the country, a small ethnic entity known as Abkhazia (Fig. 2B-5; see *From the Field Notes* box).

Soon after the USSR's disintegration in 1991, the disagreement between Russia and Georgia escalated into serious discord. The Russians closed their border with Georgia, shutting Georgian farmers out of their crucial market. Then the Russians started issuing Russian passports to Abkhazian citizens of Georgia. Next, the Russians signaled even stronger support for their Ossetian allies, and in a still-disputed sequence of events during 2008 Russian forces entered South Ossetia and engaged the Georgian military in a brief but costly conflict. After a cease-fire was brokered by international action, Moscow took the extraordinary step of recognizing the two pro-Russian autonomous regions of Georgia—Abkhazia and South Ossetia—as independent states.

These actions sent shudders through Europe and much of the world. If Russia could intervene militarily in this part of its Near Abroad, where else might Russian forces be sent to enforce Russia's will? Georgia, as the map shows, has a lengthy border with Russia, but so does Ukraine, where Russia could claim to have even stronger interests, including leased military bases. Unlike Azerbaijan, predominantly agricultural Georgia has no energy card to play—although it did agree to allow the new pipeline from Azerbaijan to Turkey to traverse its territory, thereby angering the Russians even further. And the Georgian government has courted the European Union and even NATO, seeking the security Georgians have lacked for centuries. The Azeris of Azerbaijan, whose prevailing religion is Shi'ite Islam like that of neighboring Iran, are caught in the global web (and political geography) of energy demand and supply, but most ordinary people have benefited little from their country's petroleum wealth. The landlocked 3.3 million Armenians live in a world of unsettled scores: with the Turks over their decimation by Ottoman Muslims, with the Azeris over their exclave of Nagorno-Karabakh, and with the international community over the proper recognition of their plight. Transcaucasia may be the smallest region of the Russian realm, but it is also the second-most populous and undoubtedly the most volatile.

AN UNCERTAIN FUTURE

It seems that Russia is paying an increasingly high price for its efforts to control the Southern Periphery. Terrorist attacks struck Moscow in both 2010 and 2011, killing dozens of people on subway trains, at airports, and at railroad stations. It could happen again and there are fears terrorists might strike during the 2014 Winter Olympics here, in and around Sochi. Mostly, the perpetrators are assumed to be violent extremists from the northern Caucasus.

Although Russia is now one of the BRICs, this stature is based mainly on its massive output of raw materials and the accelerated economic growth that took place during the commodity boom of recent years. But the benefits of growth have been distributed most unevenly and tens of millions of Russians today suffer deprivation. They have lost their secure (if meager) livelihoods from the Soviet past and many are without decent jobs and face ever rising costs of living. Small wonder that some long for the old days of communism. At the same time, the Russian leadership has a penchant for autocratic rule that is hard to reconcile with democracy. If Russia's population is divided along ethnic lines, it is also fragmented economically and politically.

Even after the dismantling of the Soviet Empire, keeping a country together the size of Russia is a colossal challenge. Russia has not yet found a balance, not in its so-called managed democracy, not in dealing with free markets, not in a partnership with the West, not in accommodating minorities, not in forging better relations with the former Soviet republics that surround it. Russia is rigid, but that doesn't mean it is stable. Nonetheless, this is a huge and very powerful country that will continue to throw its weight around on the Eurasian continent.

POINTS TO PONDER

- Is Russia too large to be ruled effectively as a single state? Might there be some "ideal" size for the proper functioning of a democracy?
- One in seven Russian villages has been abandoned and is now a ghost town.
- The 2014 Winter Olympics at Sochi provide the Russian leadership with an unparalleled public relations opportunity, but they also pose a significant security risk.
- According to a recent poll marking the twentieth anniversary of the Soviet Union's demise, 70 percent of Russians agreed that the changes since 1991 had not benefited ordinary people.
- In terms of economic development, Russia is by far the most unpredictable of the four BRICs.

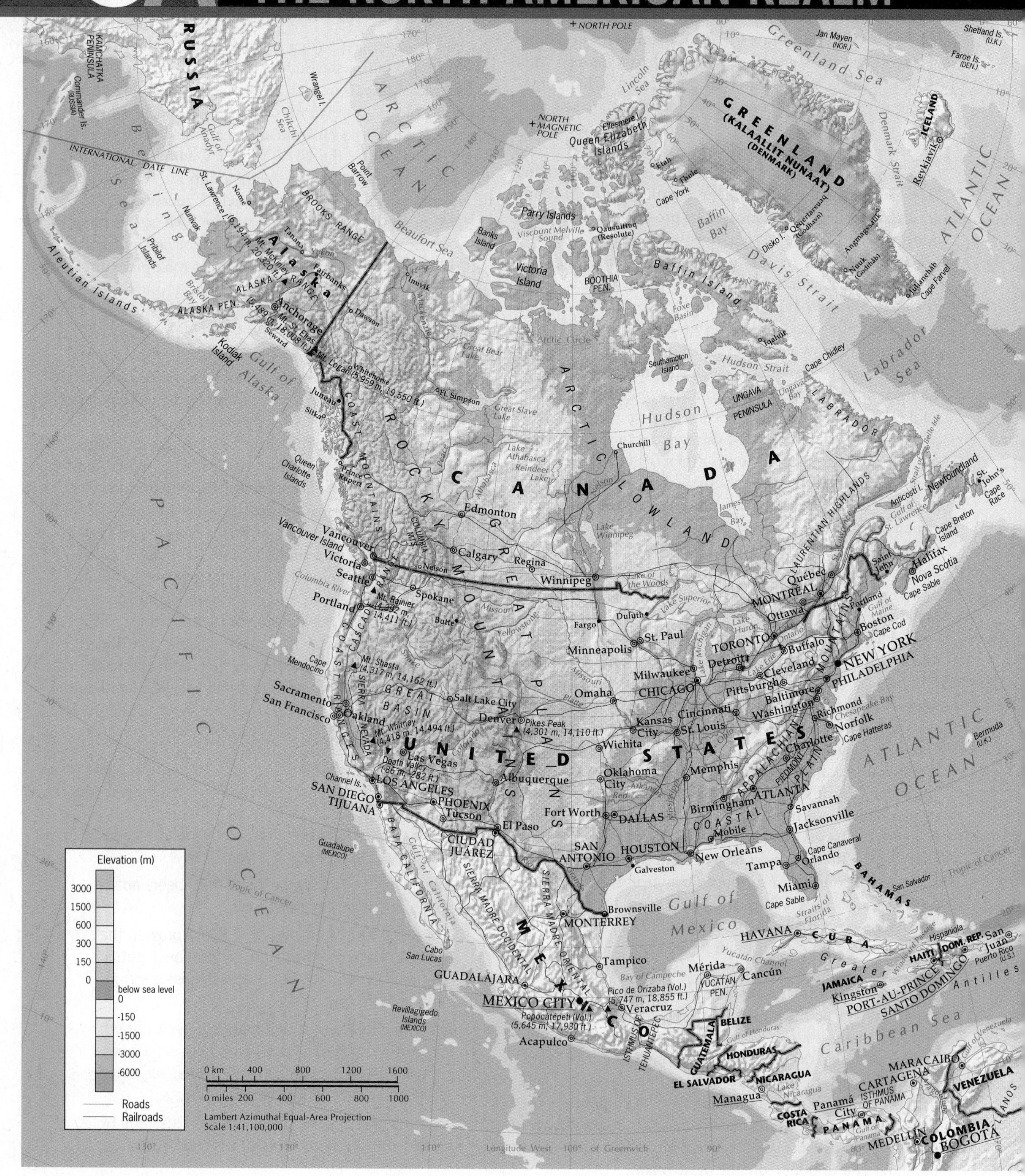
NORTH POLE
RUSSIA
KAMCHATKA PENINSULA
Commander Is. (RUSSIA)
Wrangel I.
ARCTIC OCEAN
Chukchi Sea
Gulf of Anadyr
NORTH MAGNETIC POLE
Lincoln Sea
Ellesmere I.
Queen Elizabeth Islands
Greenland Sea
Jan Mayen (NOR.)
Shetland Is. (U.K.)
Faroe Is. (DEN.)
GREENLAND (KALAALLIT NUNAAT) (DENMARK)
ICELAND
Reykjavík
Denmark Strait
Etah
Thule
Cape York
INTERNATIONAL DATE LINE
Bering Sea
St. Lawrence I.
Nome
Point Barrow
Nunivak
Pribilof Islands
BROOKS RANGE
Alaska
Mt. McKinley (6,194 m, 20,320 ft.)
Tanana
Yukon
Fairbanks
ALASKA RANGE
Beaufort Sea
Banks Island
Parry Islands
Viscount Melville Sound
Qausuittuq (Resolute)
Baffin Bay
Disko I.
Qeqertarsuaq (Godhavn)
Angmagssalik
Nuuk (Godthåb)
Julianehåb
Cape Farvel
ATLANTIC OCEAN
Davis Strait
Baffin Island
Victoria Island
BOOTHIA PEN.
Inuvik
Mackenzie
Aleutian Islands
Bristol Bay
ALASKA PEN.
Anchorage
Mt. St. Elias (5,489 m, 18,008 ft.)
Seward
Dawson
Kodiak Island
Gulf of Alaska
Mt. Logan (5,959 m, 19,550 ft.)
Whitehorse
Great Bear Lake
Arctic Circle
Foxe Basin
Iqaluit
Southampton Island
Hudson Strait
Cape Chidley
Labrador Sea
Juneau
Sitka
Ft. Simpson
Great Slave Lake
ARCTIC
Hudson Bay
UNGAVA PENINSULA
Ungava Bay
LABRADOR
COAST MOUNTAINS
ROCKY
Peace
Lake Athabasca
Reindeer Lake
Athabasca
Churchill
CANADA
Queen Charlotte Islands
Prince Rupert
Nelson
LOWLAND
James Bay
Strait of Belle Isle
Newfoundland
St. John's
Cape Race
Anticosti I.
Gulf of St. Lawrence
Cape Breton Island
Edmonton
COLUMBIA MTS.
GREAT
Lake Winnipeg
LAURENTIAN HIGHLANDS
PACIFIC OCEAN
Vancouver
Vancouver Island
Victoria
Calgary
Regina
Nelson
Winnipeg
Lake of the Woods
Québec
Saint John
Halifax
Nova Scotia
Cape Sable
Seattle
Columbia River
Mt. Rainier (4,392 m, 14,411 ft.)
Spokane
Missouri
Lake Superior
MONTRÉAL
Ottawa
Portland
Gulf of Maine
Portland
Boston
Cape Cod
CASCADE RANGE
COAST RANGES
Butte
Yellowstone
Duluth
Fargo
St. Paul
Minneapolis
Lake Huron
Lake Michigan
Lake Ontario
TORONTO
Buffalo
Detroit
Lake Erie
Cape Mendocino
Mt. Shasta (4,317 m, 14,162 ft.)
Snake
Milwaukee
Cleveland
NEW YORK
PHILADELPHIA
MOUNTAINS
Sacramento
San Francisco
Oakland
SIERRA NEVADA
GREAT BASIN
Salt Lake City
Platte
Omaha
CHICAGO
Pittsburgh
Baltimore
Washington
Chesapeake Bay
Richmond
Norfolk
Cape Hatteras
Mt. Whitney (4,418 m, 14,494 ft.)
Denver
Pikes Peak (4,301 m, 14,110 ft.)
Kansas City
Cincinnati
St. Louis
Ohio
Bermuda (U.K.)
UNITED STATES
Las Vegas
Death Valley (-86 m, -282 ft.)
Colorado
Wichita
PLAINS
APPALACHIAN
Charlotte
PIEDMONT
Channel Is.
LOS ANGELES
Albuquerque
Oklahoma City
Arkansas
Memphis
SAN DIEGO
TIJUANA
PHOENIX
Tucson
Red
Fort Worth
DALLAS
Mississippi
Birmingham
ATLANTA
COASTAL PLAIN
Savannah
Jacksonville
BAJA CALIFORNIA
El Paso
Guadalupe (MEXICO)
CIUDAD JUÁREZ
SAN ANTONIO
HOUSTON
Mobile
New Orleans
Galveston
Cape Canaveral
Orlando
Tampa
Gulf of California
SIERRA MADRE OCCIDENTAL
SIERRA MADRE ORIENTAL
Rio Grande
Tropic of Cancer
Miami
Cape Sable
Straits of Florida
BAHAMAS
San Salvador
Brownsville
Gulf of Mexico
MONTERREY
HAVANA
CUBA
Hispaniola
HAITI
DOM. REP.
San Juan
Puerto Rico (U.S.)
Cabo San Lucas
MEXICO
Tampico
Bay of Campeche
Mérida
Cancún
Yucatán Channel
YUCATÁN PEN.
Greater Antilles
Windward Passage
JAMAICA
Kingston
PORT-AU-PRINCE
SANTO DOMINGO
GUADALAJARA
Pico de Orizaba (Vol.) (5,747 m, 18,855 ft.)
MEXICO CITY
Veracruz
Popocatépetl (Vol.) (5,645 m, 17,930 ft.)
Revillagigedo Islands (MEXICO)
Acapulco
ISTHMUS OF TEHUANTEPEC
BELIZE
Gulf of Honduras
GUATEMALA
HONDURAS
Caribbean Sea
EL SALVADOR
NICARAGUA
Lake Nicaragua
Managua
COSTA RICA
Panamá City
ISTHMUS OF PANAMA
PANAMA
Gulf of Panama
MARACAIBO
Gulf of Venezuela
CARTAGENA
VENEZUELA
LLANOS
MEDELLÍN
COLOMBIA
BOGOTÁ
Elevation (m)
3000
1500
600
300
150
0
below sea level
0
-150
-1500
-3000
-6000
Roads
Railroads
0 km 400 800 1200 1600
0 miles 200 400 600 800 1000
Lambert Azimuthal Equal-Area Projection
Scale 1:41,100,000
Longitude West 100° of Greenwich

FIGURE 3A-1

IN THIS CHAPTER

- The geography of home foreclosures
- North America's high-tech regions
- The American creed
- Bilingualism in Canada
- Immigration and the future of multiculturalism

CONCEPTS, IDEAS, AND TERMS

Cultural diversity	1
Physiographic regions	2
Rain shadow effect	3
Fossil fuel	4
Urban system	5
American Manufacturing Belt	6
Outer city	7
Deindustrialization	8
Gentrification	9
Sunbelt	10
Migration	11
Melting pot	12

We turn now to the Western Hemisphere, the two great interconnected continents that separate the Atlantic and Pacific oceans and extend, very nearly, from pole to pole, flanked by numerous islands large and small, indented by gulfs and bays of historic and economic import, and endowed with an enormous range of natural resources. Two continents—North and South—form the Americas, but three geographic realms blanket them (North, Middle, and South). In the context of physical (natural) geography, North America from Canada's Ellesmere Island in the far north to Panama in the south is a continent. In terms of modern human geography, the northern continent is divided into the North American and Middle American realms along a transition zone marked by a political as well as physical boundary between the United States and Mexico (Fig. 3A-1). In Texas, from the Gulf of Mexico to El Paso/Ciudad Juárez, the Rio Grande forms this border. From El Paso westward to the Pacific Ocean, straight-line boundaries, reinforced by fences and walls, separate the North from the Middle. Here, global core and global periphery meet, sometimes contentiously. We begin in North America.

DEFINING THE REALM

North America is constituted by two of the world's most highly advanced countries by virtually every measure of social and economic development. Blessed by an almost unlimited range of natural resources and bonded by trade as well as culture, Canada and the United States are locked in a mutually productive embrace that is reflected by the economic statistics: in an average year of the recent past, about two-thirds of Canadian exports went to the United States, and just over half of Canada's imports came from its southern neighbor. For the United States, Canada is not quite as dominant, but it is still its leading export market and the second biggest importer.

This realm is also defined on the basis of a range of broad cultural traits, from urban landscapes and a penchant for mobility to religious beliefs, language, and political persuasions. Both countries rank among the world's most highly urbanized: nothing symbolizes the North American city as strongly as the skyscrapered panoramas of New York, Chicago, and Toronto—or the vast, beltway-connected suburban expanses of Los Angeles, Washington, and Houston. North Americans are also the most mobile people on Earth. Each year about one out of every eight individuals changes his or her residence, a proportion that has recently declined but still leads the world.

Despite noteworthy multilingualism, in most areas English is both countries' *lingua franca*. And notwithstanding all the religions followed by minorities, the great majority of both Canadians and Americans are Christians. Both states are stable democracies, and both have federal systems of government.

Canadians and Americans also share a passion for the same sports. Baseball and a modified version of American football are popular sports in Canada; more than 20 years ago, the Toronto Blue Jays became the first team outside the United States to win baseball's World Series. Canada's national sport, ice hockey, now has more NHL teams in the U.S. than in Canada, although curling, a slower-paced Canadian pastime, has not exactly captured the American imagination.

No metropolis escaped the U.S. housing crisis of the late 2000s: a foreclosed home on the upscale outskirts of Washington, D.C., 2012.

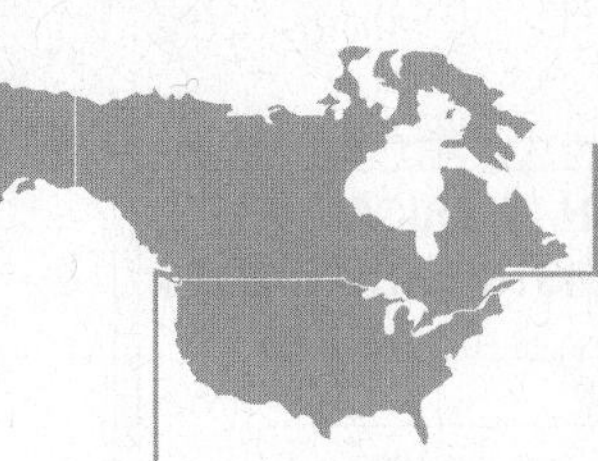

major geographic qualities of NORTH AMERICA

1. North America encompasses two of the world's biggest states territorially (Canada is the second-largest in size; the United States is third).
2. Both Canada and the United States are federal states, but their systems differ. Canada's is adapted from the British parliamentary system and is divided into ten provinces and three territories. The United States separates its executive and legislative branches of government, and it consists of 50 States, the Commonwealth of Puerto Rico, and a number of island territories under U.S. jurisdiction in the Caribbean Sea and the Pacific Ocean.
3. Both Canada and the United States are plural societies. Canada's pluralism is most strongly expressed in regional bilingualism. In the United States, divisions occur largely along ethnic, racial, and income lines.
4. A substantial number of Quebec's French-speaking citizens support a movement that seeks independence for this Canadian province. The movement's high-water mark was reached in the 1995 referendum in which (minority) non-French-speakers were the difference in the narrow defeat of separation. Prospects for a breakup of Canada have been diminishing since 2000 but have not disappeared.
5. North America's population, not large by international standards, is one of the world's most highly urbanized and mobile. Largely propelled by a continuing wave of immigration, the realm's population total is expected to grow by more than one-third over the next half-century.
6. By world standards, this is a rich realm where high incomes and high rates of consumption prevail. North America possesses a highly diversified resource base, but nonrenewable fuel and mineral deposits are consumed prodigiously.
7. North America has long been home to one of the world's great manufacturing complexes that generated a dense, mature urban system. Over the past few decades, North America has come to rely increasingly on a highly advanced information economy and high-technology industries.
8. The two countries heavily depend on each other for supplies of critical raw materials (e.g., Canada is a leading source of U.S. energy imports) and have long been each other's primary trading partners. Today, the North American Free Trade Agreement (NAFTA), which also includes Mexico, is linking all three economies ever more tightly as the remaining barriers to international trade and investment flow are dismantled.
9. Continued immigration and high transnational mobility make for an exceptionally diverse multicultural realm.

Driving from Michigan across southern Ontario to upstate New York does not involve a sharp contrast in cultural landscapes. Americans visiting Canadian cities (and vice versa) find themselves in mostly familiar settings. Canadian cities have fewer impoverished neighborhoods, no ethnic ghettos (although low-income ethnic districts do exist), lower crime rates, and, in general, better public transportation; but rush hour in Toronto very much resembles rush hour in Chicago. The boundary between the two states is porous, even after the events of 9/11. Americans living near the border shop for more affordable medicines in Canada; Canadians who can afford it seek medical treatment in the United States.

None of this is to suggest that Canadians and Americans don't have their differences—they do. Most of these differences may seem subtle to people from outside the realm, but they are important nonetheless, particularly so to Canadians who at times feel dominated by Americans whose political or cultural values they don't always share. You can even argue over names: if this entire hemisphere is called the Americas, aren't Canadians, Mexicans, and Brazilians as well as others Americans too?

POPULATION CLUSTERS

Although Canada and the United States share many historical, cultural, and economic qualities, they also differ in significant ways, diversifying this realm. The United States, somewhat smaller territorially than Canada, occupies the heart of the North American continent and, as a result, encompasses a greater environmental range. The U.S. population is dispersed across most of the country, forming major concentrations along both the (north-south trending) Atlantic and Pacific coasts (Fig. 3A-2). The overwhelming majority of Canadians, on the other hand, live in an interrupted east-west corridor that extends across southern Canada, mostly within 300 kilometers (200 mi) of the U.S. border. The United States, again unlike Canada, is a

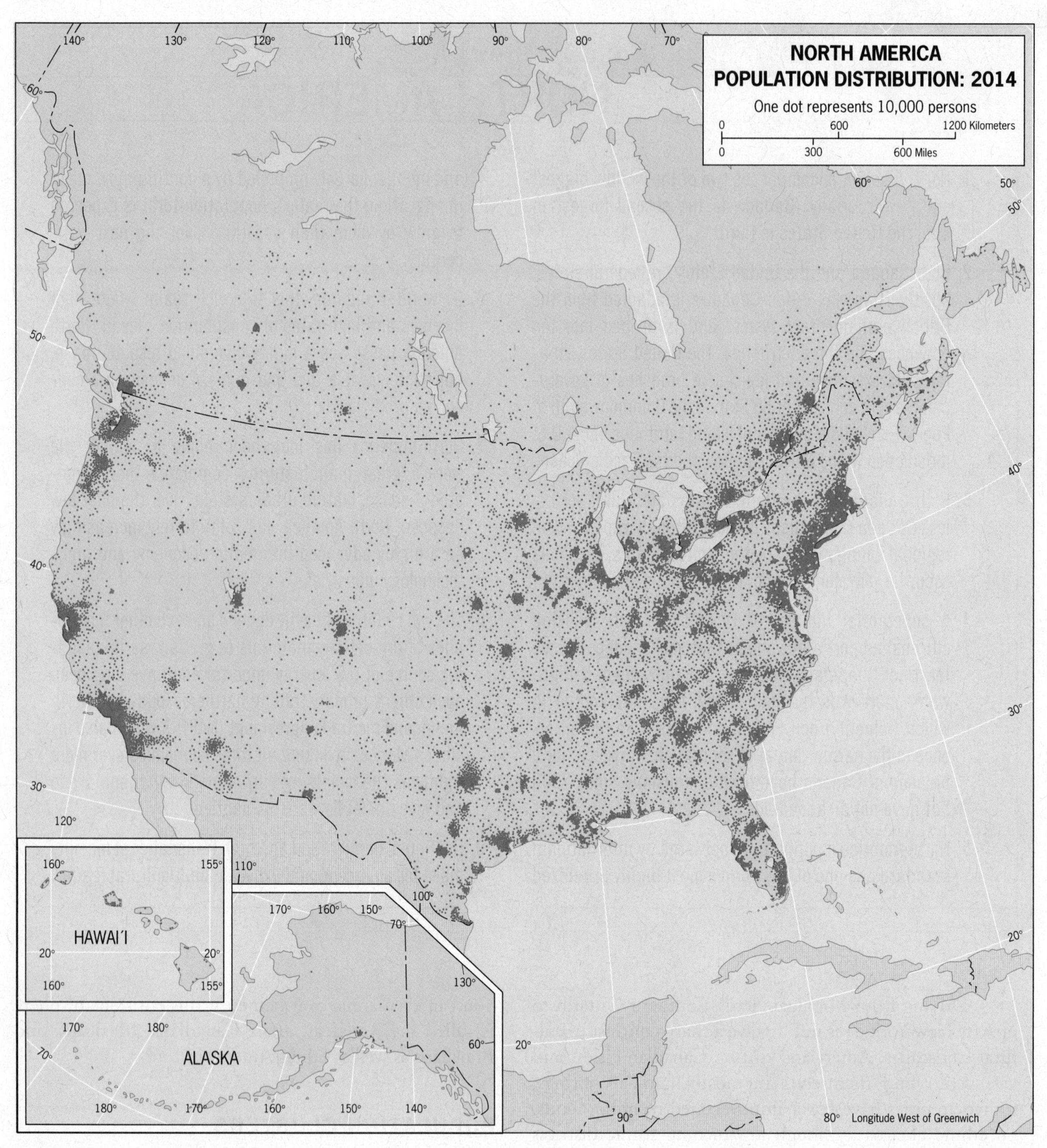

FIGURE 3A-2 © H. J. de Blij, P. O. Muller, and John Wiley & Sons, Inc.

fragmented country in that the broad peninsula of Alaska is part of it (offshore Hawai'i, however, belongs in the Pacific Realm).

Figure 3A-2 reveals that the great majority of both the U.S. and Canadian population resides to the east of a line drawn down the middle of the realm, reflecting the historic core-area development in the east and the later and still-continuing shift to the west, and, in the United States, to the south. Certainly this map shows the urban concentration of North America's population: you can easily identify cities such as Toronto, Chicago, Denver, Dallas-Fort Worth, San Francisco, and Vancouver. Just under 80 percent of the realm's population is concentrated in cities and towns, a higher proportion even than Europe's.

As the Data Table in Appendix B indicates, the population of the United States, which reached 317 million in 2014, is growing at a rate 25 percent higher than Canada's, and the 400-million mark may be reached as soon as 2050. This is an unusually high rate

Vito Palmisano/Photographer's Choice/Getty Images, Inc.

The central business districts (CBDs) of North America's large cities have distinctive skylines, often featuring architectural icons like the Empire State Building. Indeed, you should have little trouble identifying this CBD: the black profile of Hancock Tower on the shoreline of Lake Michigan just to the right of center is one of the signature structures of downtown Chicago, America's third-largest city. In fact, the only skyscraper more famous than Hancock Tower is Willis Tower, the Western Hemisphere's second-tallest building, from whose 110-story-high Skydeck this photo was taken. Willis Tower??? Absolutely, although it is still far more widely known by its former name of Sears Tower (the original 1973 tenant that relocated its corporate headquarters to suburban Hoffman Estates a quarter-century ago). Today, according to its website, Willis Tower has reinvented itself as one of the world's greenest buildings, in the vanguard of a movement that marks Chicago as the U.S. leader in working toward a more sustainable urban future.

of growth for a high-income economy, resulting from a combination of natural increase (which leads today) and substantial immigration (expected to lead after 2030).

Although Canada's overall growth rate is significantly lower than that of the United States, immigration contributes proportionately even more to this increase than in the United States. With just over 35 million residents, Canada, like the United States, has been relatively open to legal immigration, and as a result both societies exhibit a high degree of **cultural diversity [1]** in ancestral and traditional backgrounds. Indeed, Canada recognizes two official languages, English and French (the United States does not even designate English as such); and by virtue of its membership in the British Commonwealth, East and South Asians form a larger sector of Canada's population than Asians and Pacific Islanders do in the United States. On the other hand, cultural diversity in the United States reflects large Hispanic (16.7 percent), African American (13.1 percent), and Asian (5 percent) minorities as well as a wide range of other ethnic backgrounds.

Robust urbanization, substantial immigration, and cultural diversity are defining properties of the human geography of the North American realm. But first we need to become familiar with the physical stage on which the human drama is unfolding.

NORTH AMERICA'S PHYSICAL GEOGRAPHY

Physiographic Regions

One of the distinguishing properties of the North American landmass, all of which lies on the North American tectonic plate, is its remarkable variegation into regional physical landscapes. So well defined are many of these landscape regions that we use their names in everyday parlance—for example, when we say that we flew over the Rocky Mountains or drove across the Great Plains or hiked in the Appalachians. These landscapes are called **physiographic regions [2]**, and nowhere else in the hemisphere is their diversity greater than to the north of the Rio Grande (Fig. 3A-3).

In the Far West, the Pacific Mountains extend all the way from Southern California through coastal Canada to Alaska. In the western interior, the Rocky Mountains form a continental backbone from central Alaska to New Mexico. Around the Great Lakes, the low-relief landscapes of the Interior Lowlands and the Great Plains to the west are shared by Canada and the United States, and the international boundary even divides the Great Lakes. In the east, the Appalachian Mountains (as the cross-sectional inset shows, no match for the Rockies) form a corridor of ridges, valleys, and plateaus that represent a familiar topography from Alabama and Georgia to Nova Scotia and Newfoundland. If there is a major physiographic province that belongs to only one of the realm's two countries, it is the Canadian Shield, scoured bare by the Pleistocene glaciers that deposited their pulverized rocks as fertile soil in the U.S. Midwest, that is, in the Interior Lowlands.

Climate

This diversity of landscape is matched by a variety of climates. Take another look at Figure G-7, and you can see the lineaments of landscape mirrored in the contrasts of climate. North America may not have it all—there are no areas of true tropical environment in the North American realm to speak of except at the southern tip of Florida—but does exhibit a great deal of variation. North America contains moist coastal zones and arid interiors, well-watered plains, and even bone-dry deserts. On the world map, ***Cf*** and ***Df*** climates are especially good for commercial farming; note how large North America's share of these environments is.

The map leaves no doubt: the farther north you go, the colder it gets, and even though coastal areas derive moderation from warmer offshore waters, which is why there is a Vancouver and a Halifax, the rigors of *continentality* set in not far from the coast. Hot summers, frigid winters, and limited precipitation make high-latitude continental interiors

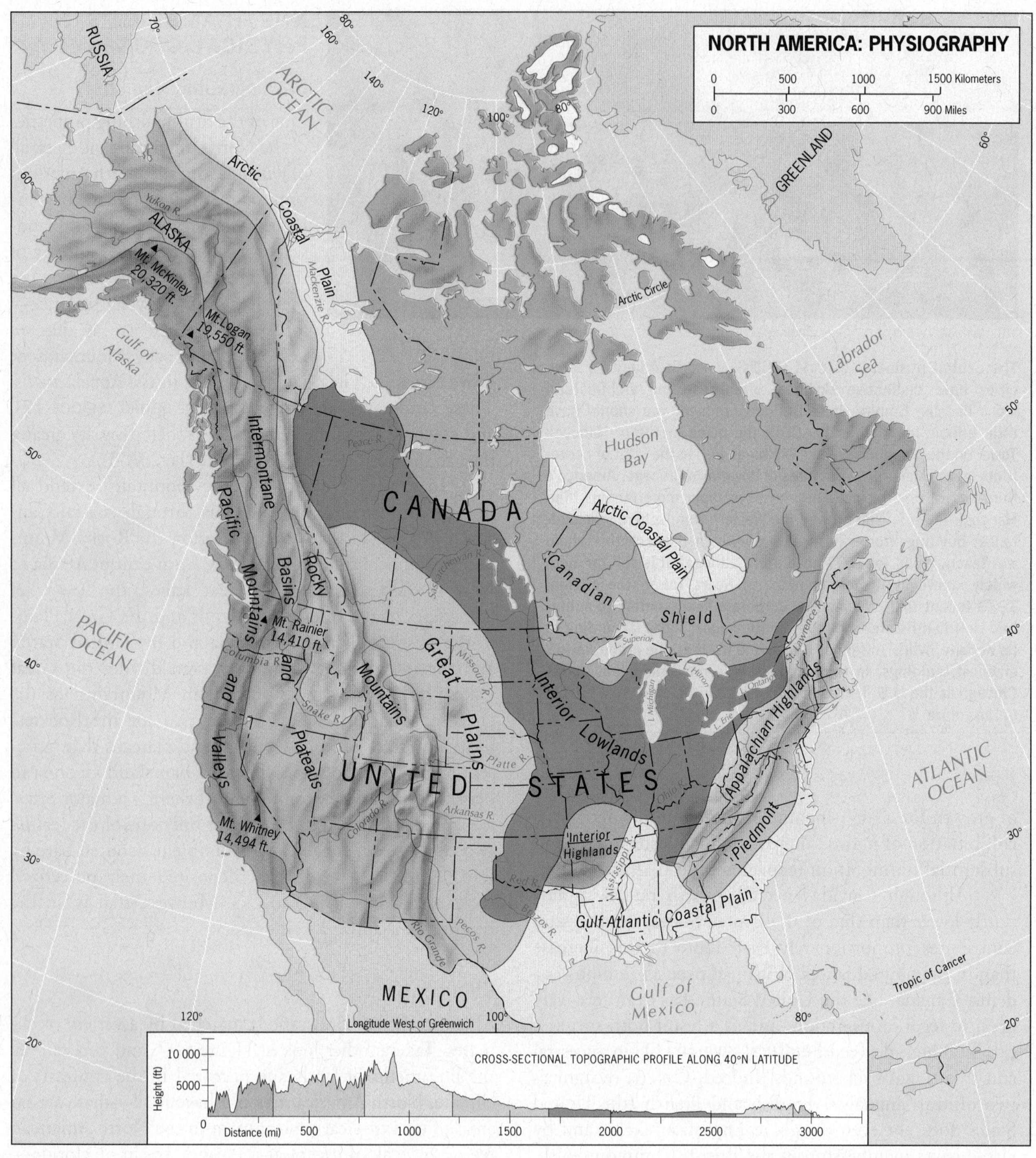

FIGURE 3A-3 © H. J. de Blij, P. O. Muller, and John Wiley & Sons, Inc.

difficult places to make it on the land. Figure G-7 has much to do with the southerly concentration of Canadian population and with the lower densities in the interior throughout the realm.

In the west, especially in the United States, you can see what the Pacific Mountains do to areas inland. Moisture-laden air arrives from the ocean, the mountain wall forces the air upward, cools it, condenses the moisture in it, and produces rain—the rain for which Seattle and Portland as well as other cities of the Pacific Northwest are (in) famous. By the time the air crosses the mountains and descends on the landward side, most of the moisture has been drawn from it, and the forests of the ocean side give way to scrub and brush. This **rain shadow effect [3]** extends all the way across the Great Plains; North America does not turn moist again until the Gulf of Mexico sends humid tropical air northward into the eastern interior via the Mississippi-Missouri River Basin.

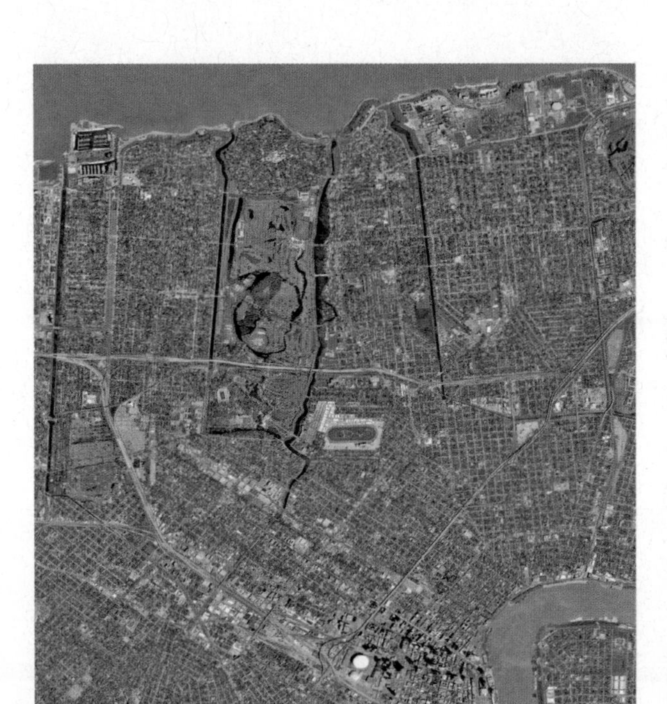

Digital Globe/Eurimage/Science Source

On August 29, 2005, Category-4 Hurricane Katrina cut a 230,000-square-kilometer (90,000-sq-mi) path of destruction along the Louisiana and Mississippi Gulf Coast after making landfall just west of the mouth of the Mississippi River. Most devastated was the city of New Orleans, 80 percent of which lay underwater in the immediate aftermath of the storm. The pair of aerial photos above shows central New Orleans under normal conditions in the spring of 2004 (left) and two days after the tropical cyclone struck (right). Katrina was by far the most costly natural disaster in U.S. history, and its toll was truly staggering: more than 1800 lives lost, at least 200,000 homes destroyed, nearly a million people displaced, and more than $25 billion in insured property damage. Despite $180 billion in federal aid during the years following the disaster, rebuilding has been patchy and painfully slow. Today, the city's population is still more than 20 percent smaller than before the storm. Tens of thousands of residential properties remain unoccupied, most of them still damaged. The proportion of the population living in poverty stood at 27 percent in 2012, and the number of homeless people in that year was estimated at 6,700, three times the national average. The Big Easy continues to struggle toward an acceptable level of restoration, yet it remains highly vulnerable to hurricanes in the future.

In a very general way, therefore, and not including the coastal strips along the Pacific, nature divides North America into an arid west and a humid east. Again the population map reveals more than a hint of this: draw a line approximately from Lake Winnipeg to the mouth of the Rio Grande, and look at the contrast between the (comparatively) humid east and drier west when it comes to population density. Water is a large part of this story.

Between the Rocky Mountains and the Appalachians, North America lies open to air masses from the frigid north and tropical south. In winter, southward-plunging polar fronts send frosty, bone-dry air masses deep into the heart of the realm, turning even places like Memphis and Atlanta into iceboxes; in summer, hot and humid tropical air surges northward from the Gulf of Mexico, giving Chicago and Toronto a taste of the tropics. Such air masses clash in low-pressure systems along weather fronts loaded with lightning, thunder, and, frequently, dangerously destructive tornadoes. And the summer heat brings an additional threat to the Gulf-Atlantic Coastal Plain (Fig. 3A-3): hurricanes capable of inflicting catastrophic devastation on low-lying areas (see photo pair above). These tropical cyclones also prune natural vegetation, replenish underground water reservoirs, fill natural lakes, and flush coastal channels.

Great Lakes and Great Rivers

Two great drainage systems lie between the Rockies and the Appalachians: (1) the five Great Lakes that drain into the St. Lawrence River and the Atlantic Ocean, and (2) the mighty Mississippi-Missouri system that carries water from a vast interior watershed to the Gulf of Mexico, where the Mississippi forms one of the world's major ***deltas***. Both natural systems have been modified by human engineering. In the case of the St. Lawrence Seaway, a series of locks and canals has created a direct shipping route, via the Great Lakes and their outlet, from the Midwest to the Atlantic. The Mississippi and Missouri rivers have been fortified by artificial levees that, while failing to contain the worst of flooding, have enabled farmers to cultivate the most fertile of American soils.

NATIVE AMERICANS AND EUROPEAN SETTLEMENT

When the first Europeans set foot on North American soil, the continent was inhabited by millions of people whose ancestors had reached the Americas from Asia via Alaska, and possibly also across the Pacific, more than 14,000 years before (and perhaps as long as 30,000 years ago). In search of Asia, the Europeans misnamed them Indians. These Native Americans or ***First Nations***—as they are now called, respectively, in the United States and Canada—were organized into hundreds of nations with a rich mosaic of languages and a great diversity of cultures (Fig. 3A-4). Canada's First Nations also include Métis (of mixed native and European descent) and the Inuit (formerly called Eskimos) of the far north.

The eastern nations were the first to bear the brunt of the European invasion. By the end of the eighteenth century, ruthless and land-hungry settlers had driven most of the native peoples living along the Atlantic and Gulf coasts from their homes and lands, initiating a westward push that was to devastate indigenous society. Today, there are only about 3 million remaining Native Americans in the United States and about one-third that number in Canada. In the U.S. they are left with only about 4 percent

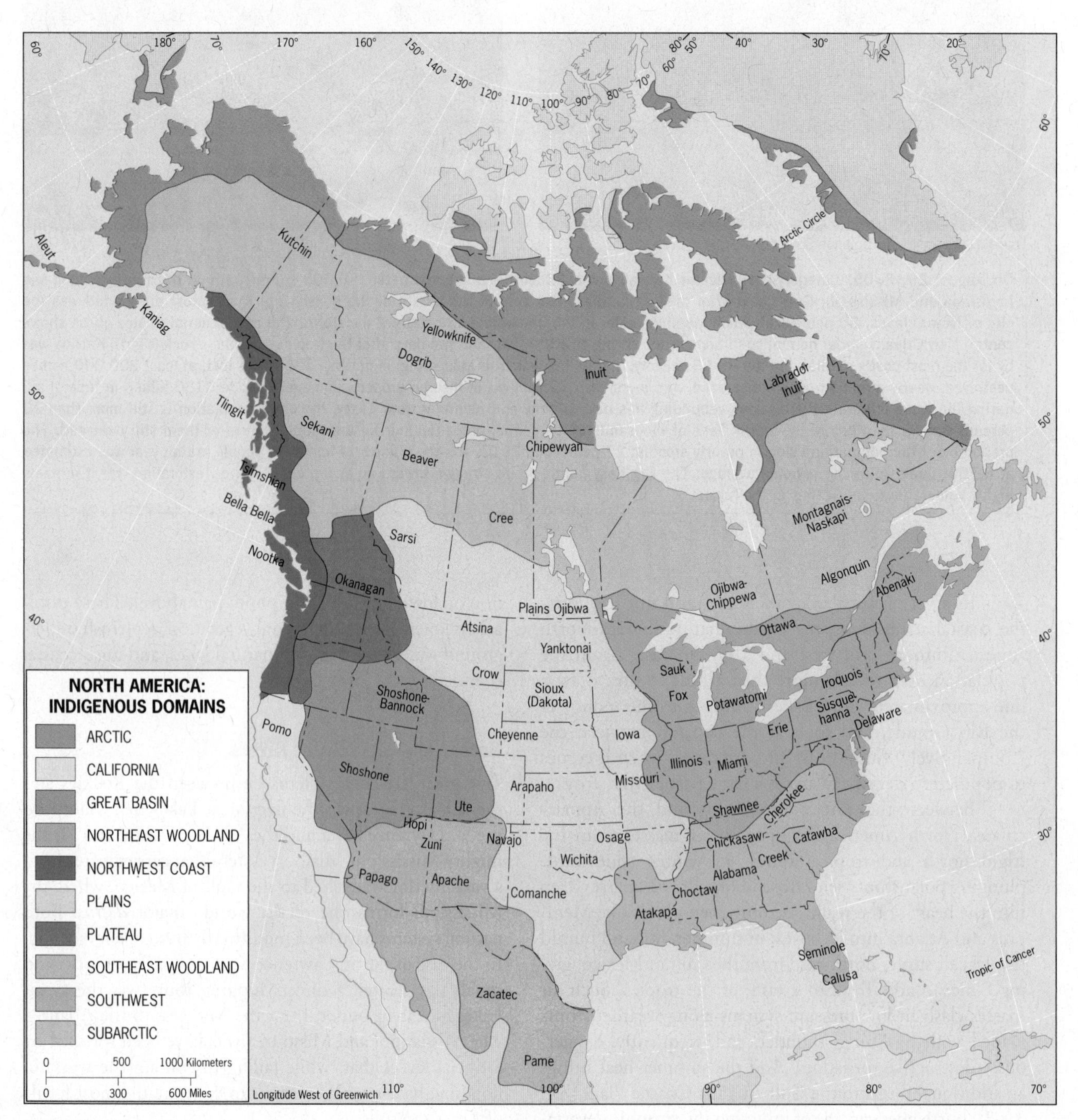

FIGURE 3A-4

of the national territory in the form of mostly impoverished reservations. Aboriginal peoples today are better off territorially in Canada where they hold titles to large tracts of land, especially in the northern sectors of British Columbia and Quebec (but keep in mind that northern Canada is relatively empty to begin with, and that environmental conditions up north can be quite harsh).

The current population geography of North America is rooted in the colonial era of the seventeenth and eighteenth centuries that was dominated by Britain and France. The French sought mainly to organize a lucrative fur-trading network, while the British established settlements along the coast of what is now the eastern seaboard of the United States (the oldest, Jamestown, was founded in Virginia more than four centuries ago).

The British colonies soon became differentiated in their local economies, a diversity that was to endure and later shape much of North America's cultural geography. The northern colony of New England (Massachusetts Bay and surroundings) specialized in commerce; the southern Chesapeake Bay colony (Virginia and Maryland) emphasized the plantation farming of tobacco; the Middle Atlantic area lying in between (southeastern New York, New Jersey, eastern Pennsylvania) formed the base for a number of smaller, independent-farmer colonies.

These neighboring colonies thrived and yearned to expand, but the British government responded by closing the inland frontier and tightening economic controls. This policy provoked general opposition in the colonies, followed by the revolutionary challenge that was to lead to independence and the formation of the United States of America by the late 1780s. The acquisition of more distant western territories followed in short order (Fig. 3A-5).

CULTURAL FOUNDATIONS

The modern American creed, if one can be identified, has been characterized by urban geographer Brian Berry as exhibiting an adventurous drive, a liking for things new, an ability to move, a sense of individualism, an aggressive pursuit of goals and ambition, a need for societal acceptance, and a firm sense of destiny. These qualities are not unique to modern U.S. culture, of course, but in combination they seem to have created a particular and pervasive mindset that is reflected in many ways in this geographic realm. Generally speaking, these tendencies tend to translate into an intense pursuit of educational and other goals and high aspirations for upward socioeconomic mobility. Thus the *American Dream*, whether a genuine prospect or an ideological fabrication, is very much an expression of this set of cultural values.

Facilitating these aspirations is ***language***. None of these goals would be within reach for so many were it not for the use of English throughout most of the realm. In Europe, by contrast, language inhibits the mobility so routinely practiced by Americans: workers moving from Poland to Ireland found themselves at a disadvantage in competition with immigrants from Ghana or Sri Lanka. In North America, a worker from Arkansas would not even consider language to be an issue when applying for a job in Vancouver.

As we shall see, the near-universality of English in North America is diminishing, but English continues to be the dominant medium of interaction. Interestingly, English in the United States is undergoing a change that is affecting it worldwide: in areas where it is the second language it is blending with the local tongue, producing hybrids just as is happening in Nigeria ("Yorlish"), Singapore ("Singlish"), and the Philippines ("Taglish"). English, therefore, may become the Latin of the twenty-first century: Latin also produced blends that eventually consolidated into Italian, French, Spanish, and the other Romance languages.

Also reflecting the cultural values cited above is the role of ***religion***, which sets American (more so than Canadian) culture apart from much of the rest of the Western world. The overwhelming majority of North Americans express a belief in God, and a large majority regularly attend church, in contrast to much of Europe. Three out of four people in the United States say they belong to a religion, and four out of ten say they attend a religious service once a week (more than in Iran!). Religious observance is a virtual litmus test for political leaders; no other developed country prints "In God We Trust" on its currency.

Figure 3A-6 shows the mosaic of Christian faiths that blankets the realm, but a map at this scale can only suggest the broadest outlines of what is a much more intricate pattern. Protestant denominations are estimated to number in the tens of thousands in North America, and no map could show them all. Southern (and other) Baptists form the majority across the U.S. Southeast from Texas to Virginia, Lutherans in the Upper Midwest and northern Great Plains, Methodists in a belt across the Lower Midwest, and Mormons in the interior West centered on Utah. Roman Catholicism prevails in most of Canada as well as the U.S. Northeast and Southwest, where ethnic Irish and Italian adherents form majorities in the Northeast and Hispanics in the Southwest. Behind these patterns lie histories of proselytism, migration, conflict, and competition, but tolerance of diverse religious (even nonreligious) views and practices is a hallmark of this realm.

On the other hand, U.S. culture does have its complexities, and one example will suffice. This is in many respects the freest country in the world, and it certainly prides itself as such. Yet the United States also has the largest prison population of the developed Western world: about 2.4 million or 1 for every 100 adults. That is 8 times as many as in Canada, 9 times as many as in Germany, and 12 times as many as in Japan. It is also 3 times the rate in Iran and even exceeds the number in Russia by a wide margin. It wasn't always like this: the proportion of the population behind bars has quadrupled since 1970, mainly as a result of (overly?) harsh laws and sentencing. This may have produced a safer country, but the cost is enormous—financially (at around $25,000 per year per inmate) as well as socially, because incarceration affects so many people (not just the inmates but also their relatives, dependents, and social networks).

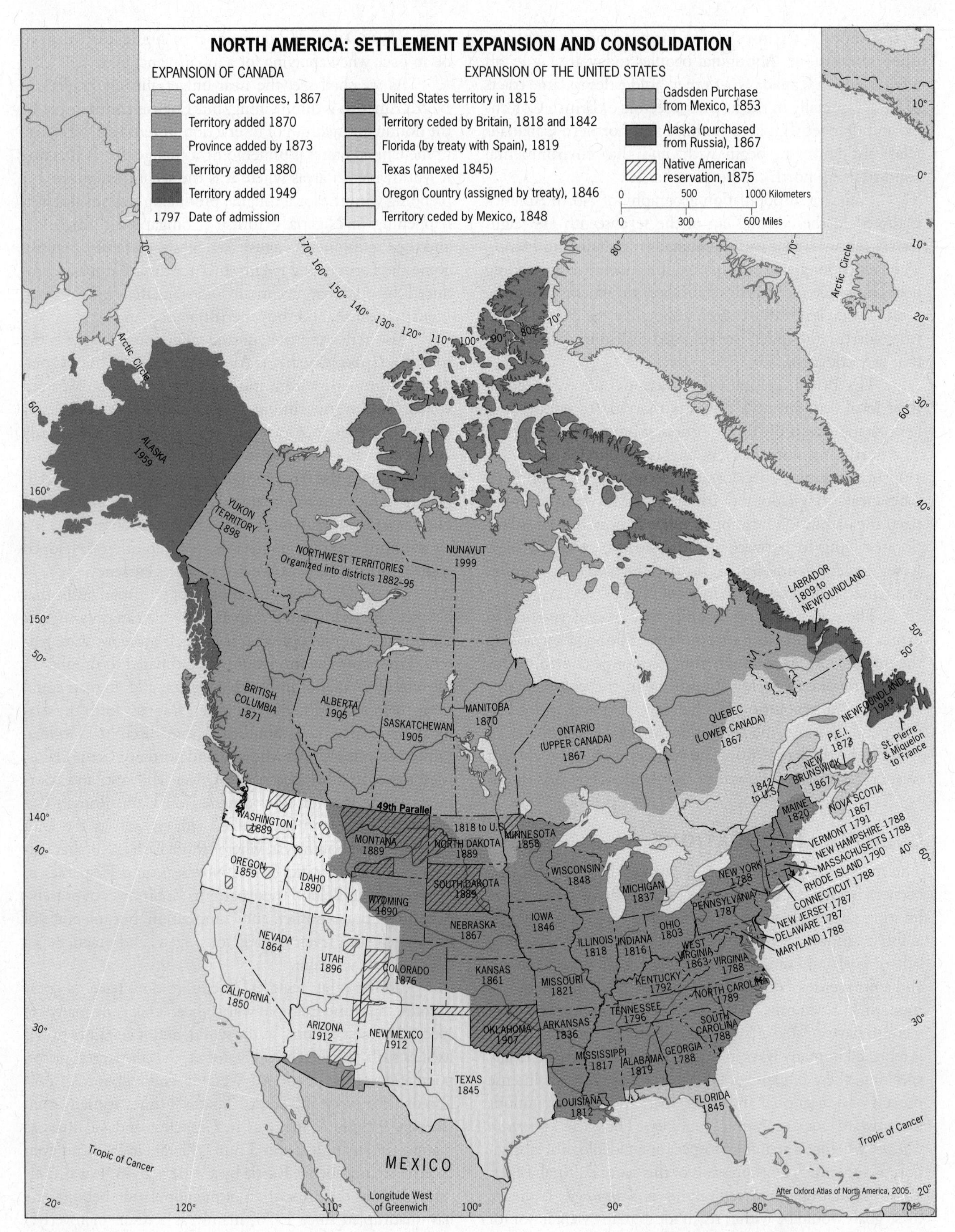
NORTH AMERICA: SETTLEMENT EXPANSION AND CONSOLIDATION
EXPANSION OF CANADA
Canadian provinces, 1867
Territory added 1870
Province added by 1873
Territory added 1880
Territory added 1949
1797 Date of admission
EXPANSION OF THE UNITED STATES
United States territory in 1815
Territory ceded by Britain, 1818 and 1842
Florida (by treaty with Spain), 1819
Texas (annexed 1845)
Oregon Country (assigned by treaty), 1846
Territory ceded by Mexico, 1848
Gadsden Purchase from Mexico, 1853
Alaska (purchased from Russia), 1867
Native American reservation, 1875
0 500 1000 Kilometers
0 300 600 Miles
ALASKA 1959
YUKON TERRITORY 1898
NORTHWEST TERRITORIES Organized into districts 1882–95
NUNAVUT 1999
BRITISH COLUMBIA 1871
ALBERTA 1905
SASKATCHEWAN 1905
MANITOBA 1870
ONTARIO (UPPER CANADA) 1867
QUEBEC (LOWER CANADA) 1867
LABRADOR 1809 to NEWFOUNDLAND
NEWFOUNDLAND 1949
P.E.I. 1873
NEW BRUNSWICK 1867
NOVA SCOTIA 1867
St. Pierre & Miquelon to France
1842 to U.S.
49th Parallel
1818 to U.S.
WASHINGTON 1889
OREGON 1859
IDAHO 1890
MONTANA 1889
NORTH DAKOTA 1889
SOUTH DAKOTA 1889
WYOMING 1890
MINNESOTA 1858
WISCONSIN 1848
MICHIGAN 1837
MAINE 1820
VERMONT 1791
NEW HAMPSHIRE 1788
MASSACHUSETTS 1788
RHODE ISLAND 1790
CONNECTICUT 1788
NEW YORK 1788
PENNSYLVANIA 1787
NEW JERSEY 1787
DELAWARE 1787
MARYLAND 1788
NEVADA 1864
UTAH 1896
COLORADO 1876
NEBRASKA 1867
IOWA 1846
ILLINOIS 1818
INDIANA 1816
OHIO 1803
WEST VIRGINIA 1863
VIRGINIA 1788
KENTUCKY 1792
CALIFORNIA 1850
KANSAS 1861
MISSOURI 1821
NORTH CAROLINA 1789
TENNESSEE 1796
SOUTH CAROLINA 1788
ARIZONA 1912
NEW MEXICO 1912
OKLAHOMA 1907
ARKANSAS 1836
MISSISSIPPI 1817
ALABAMA 1819
GEORGIA 1788
TEXAS 1845
LOUISIANA 1812
FLORIDA 1845
MEXICO
Arctic Circle
Tropic of Cancer
Longitude West of Greenwich
After Oxford Atlas of North America, 2005.

FIGURE 3A-5

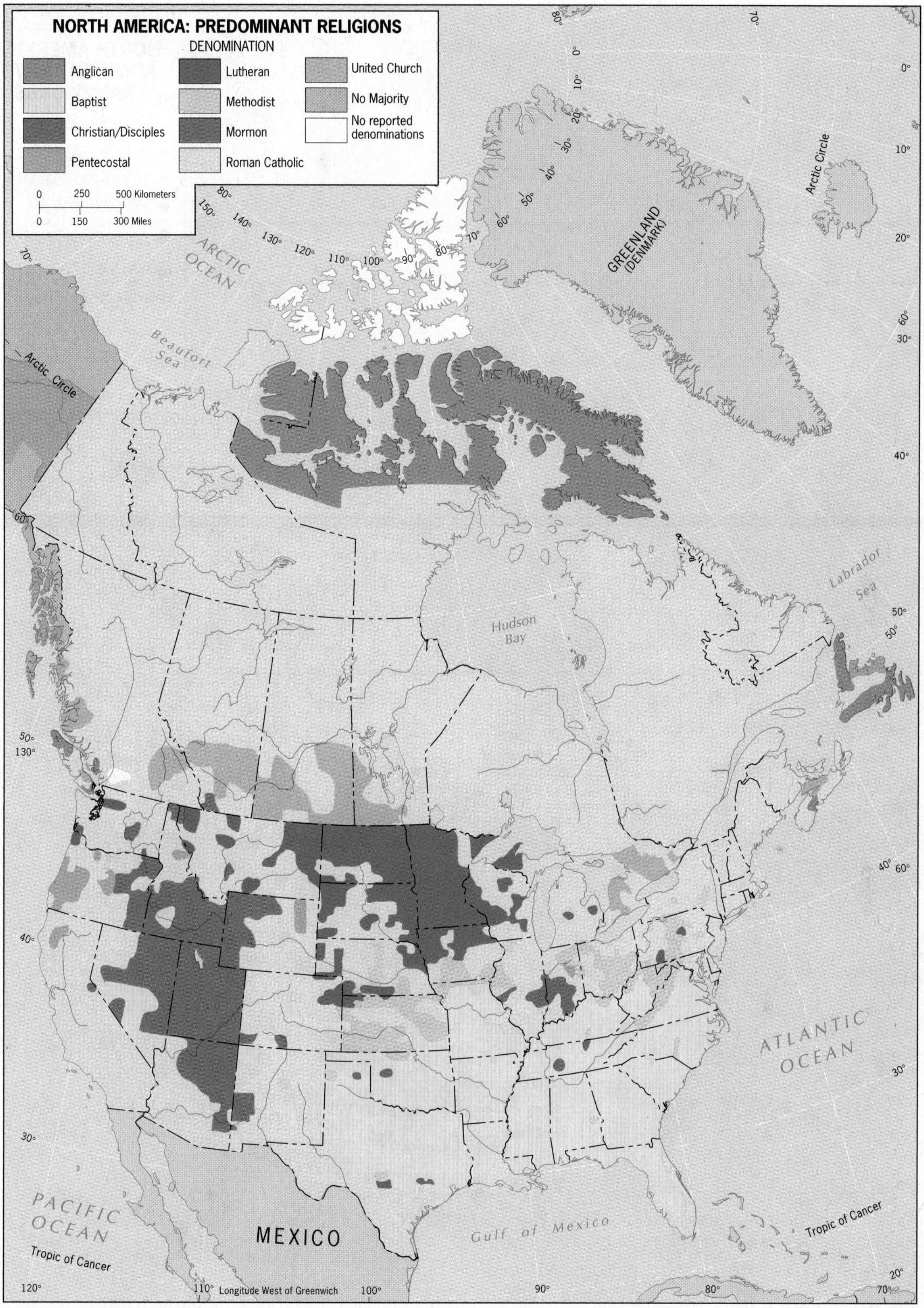

FIGURE 3A-6

THE FEDERAL MAP OF NORTH AMERICA

The two states that constitute North America may have arrived at their administrative frameworks with different motives and at different rates of speed, but the result is unmistakably similar: their internal political geographies are dominated by straight-line boundaries of administrative convenience (Fig. 3A-7). In comparatively few places, physical features such as rivers or the crests of mountain ranges mark internal boundaries, but by far the greater length of

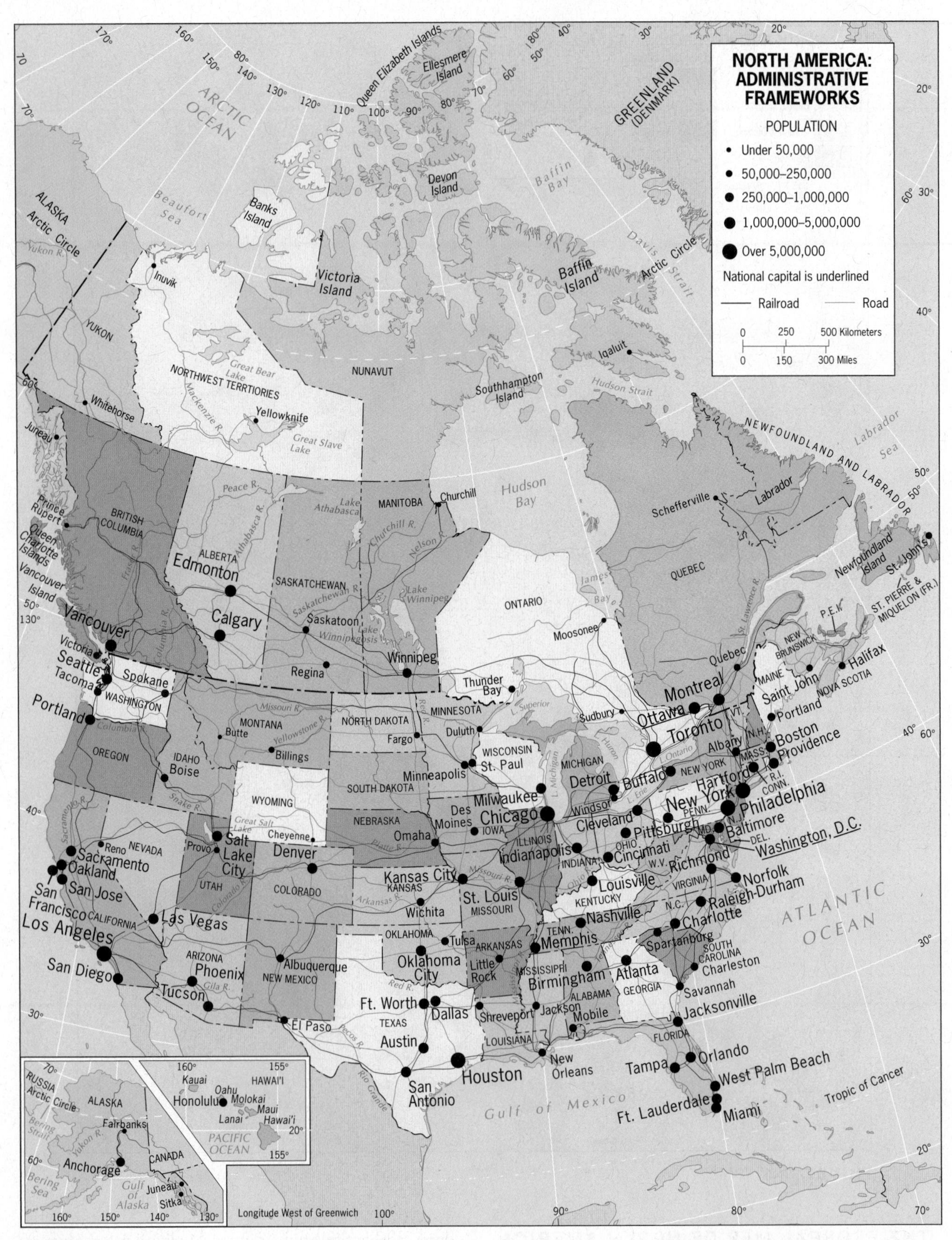
NORTH AMERICA: ADMINISTRATIVE FRAMEWORKS
POPULATION
Under 50,000
50,000–250,000
250,000–1,000,000
1,000,000–5,000,000
Over 5,000,000
National capital is underlined
Railroad
Road
0 250 500 Kilometers
0 150 300 Miles
ARCTIC OCEAN
Beaufort Sea
Baffin Bay
Davis Strait
Hudson Strait
Hudson Bay
James Bay
Labrador Sea
ATLANTIC OCEAN
Gulf of Mexico
Tropic of Cancer
Arctic Circle
GREENLAND (DENMARK)
Queen Elizabeth Islands
Ellesmere Island
Devon Island
Banks Island
Victoria Island
Baffin Island
Southhampton Island
Iqaluit
ALASKA
YUKON
NORTHWEST TERRITORIES
NUNAVUT
Inuvik
Whitehorse
Yellowknife
Great Bear Lake
Great Slave Lake
Juneau
Prince Rupert
Queen Charlotte Islands
Vancouver Island
BRITISH COLUMBIA
ALBERTA
SASKATCHEWAN
MANITOBA
ONTARIO
QUEBEC
NEWFOUNDLAND AND LABRADOR
Labrador
Newfoundland Island
NEW BRUNSWICK
NOVA SCOTIA
P.E.I.
ST. PIERRE & MIQUELON (FR.)
Edmonton
Calgary
Vancouver
Victoria
Saskatoon
Regina
Winnipeg
Churchill
Lake Athabasca
Lake Winnipeg
Thunder Bay
Moosonee
Schefferville
Sudbury
Ottawa
Toronto
Montreal
Quebec
Saint John
Halifax
St. John's
Seattle
Tacoma
Spokane
Portland
WASHINGTON
OREGON
IDAHO
Boise
MONTANA
Butte
Billings
WYOMING
Cheyenne
NORTH DAKOTA
Fargo
SOUTH DAKOTA
MINNESOTA
Duluth
Minneapolis
St. Paul
WISCONSIN
Milwaukee
MICHIGAN
Detroit
Chicago
IOWA
Des Moines
NEBRASKA
Omaha
NEVADA
Reno
UTAH
Salt Lake City
Provo
COLORADO
Denver
KANSAS
Kansas City
Wichita
MISSOURI
St. Louis
ILLINOIS
INDIANA
Indianapolis
OHIO
Cleveland
Cincinnati
Columbus
Louisville
KENTUCKY
Nashville
TENN.
Memphis
CALIFORNIA
Sacramento
Oakland
San Francisco
San Jose
Los Angeles
San Diego
Las Vegas
ARIZONA
Phoenix
Tucson
NEW MEXICO
Albuquerque
El Paso
OKLAHOMA
Oklahoma City
Tulsa
TEXAS
Ft. Worth
Dallas
Austin
San Antonio
Houston
ARKANSAS
Little Rock
LOUISIANA
Shreveport
New Orleans
MISSISSIPPI
Jackson
ALABAMA
Birmingham
Mobile
GEORGIA
Atlanta
Savannah
FLORIDA
Jacksonville
Orlando
Tampa
West Palm Beach
Ft. Lauderdale
Miami
SOUTH CAROLINA
Charleston
Spartanburg
N.C.
Charlotte
Raleigh-Durham
VIRGINIA
Richmond
Norfolk
W.V.
Pittsburgh
PENN.
Philadelphia
Baltimore
MD.
DEL.
Washington, D.C.
N.J.
NEW YORK
New York
Buffalo
Albany
Hartford
CONN.
R.I.
Providence
MASS.
Boston
VT.
N.H.
MAINE
Portland
RUSSIA
ALASKA
Fairbanks
Anchorage
Juneau
Sitka
Bering Sea
Gulf of Alaska
CANADA
HAWAI'I
Kauai
Oahu
Honolulu
Molokai
Maui
Lanai
Hawai'i
PACIFIC OCEAN
Longitude West of Greenwich

FIGURE 3A-7

boundaries is ruler-straight. Even most of the international boundary between Canada and the United States west of the Great Lakes coincides with a parallel of latitude, 49°N.

The reasons are all too obvious: the framework was laid out before, in some areas long before, significant white settlement occurred. There was something arbitrary about these delineations, but by delimiting the internal administrative units early and clearly, governments precluded later disputes over territory and resources. In any case, neither Canada nor the United States was to become a unitary (centralized) state: both countries are ***federal states***. The Canadians call their primary subdivisions provinces, whereas in the United States they are, as the country's name implies, States.

© Jim Havey/Alamy

Wind "farms" are spreading to many more places in North America as the clean-energy campaign to generate electricity from sources other than fossil fuels accelerates. This cluster of modern wind turbines on a cattle ranch in southern Wyoming (the livestock provide a sense of scale) exploits a flat-floored windstream corridor between areas of higher relief.

It is no matter of trivial pursuit to take a careful look at some of the provinces and States that compartmentalize this realm, because some are of greater consequence than others. We have already noted the special significance of mainly French-cultured Quebec, which still occupies a unique position in federal Canada. In many ways the key Canadian province is Ontario, containing the country's largest and most globalized city (Toronto) as well as the capital (Ottawa) on its eastern river border with Quebec. In the United States, key States in the political and economic geography of the federation (indeed the four most populated, in descending order) are California facing the Pacific, Texas bordering Mexico, Florida pointing toward the Caribbean and South America beyond, and New York with its window on the Atlantic.

The rectangular layout of the realm seems to symbolize North America's modernity and its stability, whereas the federal structure itself (with considerable autonomy at the State or provincial level) reflects the culture of freedom and independence. If federalism works within these high-income economies, shouldn't the rest of the world emulate it? Federalism does have its assets, but it also carries some significant liabilities.

THE DISTRIBUTION OF NATURAL RESOURCES

One of these liabilities has to do with the variable allotment of natural resources among the States and provinces. In Canada, Alberta is favored with massive energy reserves (and more may be in the offing), but its provincial government is not always eager to see the federal administration take that wealth to assist less affluent provinces. In the United States, when oil prices are high Texas and Oklahoma benefit while California's budget suffers. So it is important to compare the map of natural resources (Fig. 3A-8) with the one of States and provinces.

Water certainly is a natural resource, and North America as a realm is comparatively well supplied with it despite concerns over long-term prospects in the American Southwest and the Great Plains, where some States depend on sources in other States. Another concern focuses on lowering water tables in some of North America's most crucial aquifers (underground water reserves) in which a combination of overuse and decreasing replenishment suggests that supply problems may arise.

North America is endowed with abundant reserves of ***minerals*** that are mainly found in three zones: the Canadian Shield north of the Great Lakes, the Appalachian Mountains in the east, and the mountain ranges of the west. The Canadian Shield contains substantial iron ore, nickel, copper, gold, uranium, and diamonds. The Appalachians yield lead, zinc, and iron ore. And the western mountain zone has significant deposits of copper, lead, zinc, molybdenum, uranium, silver, and gold.

In terms of **fossil fuel [4]** energy resources (oil, natural gas, coal), North America is also quite well endowed, though the realm's voracious demand, the highest in the world, cannot be met by domestic supplies alone and necessitates an enormous need for imports. Figure 3A-8 displays the distribution of oil, natural gas, and coal.

The leading ***oil***-production areas lie (1) along and offshore from the Gulf Coast, where the floor of the Gulf of Mexico is yielding a growing share of the output; (2) in the Midcontinent District, from western Texas to eastern Kansas; and (3) along Alaska's North Slope facing the Arctic Ocean. An important development is taking place in Canada's northeastern Alberta, where oil is being drawn from deposits of *tar sands* in the vicinity of the boomtown of Fort McMurray. The process is expensive and can reward

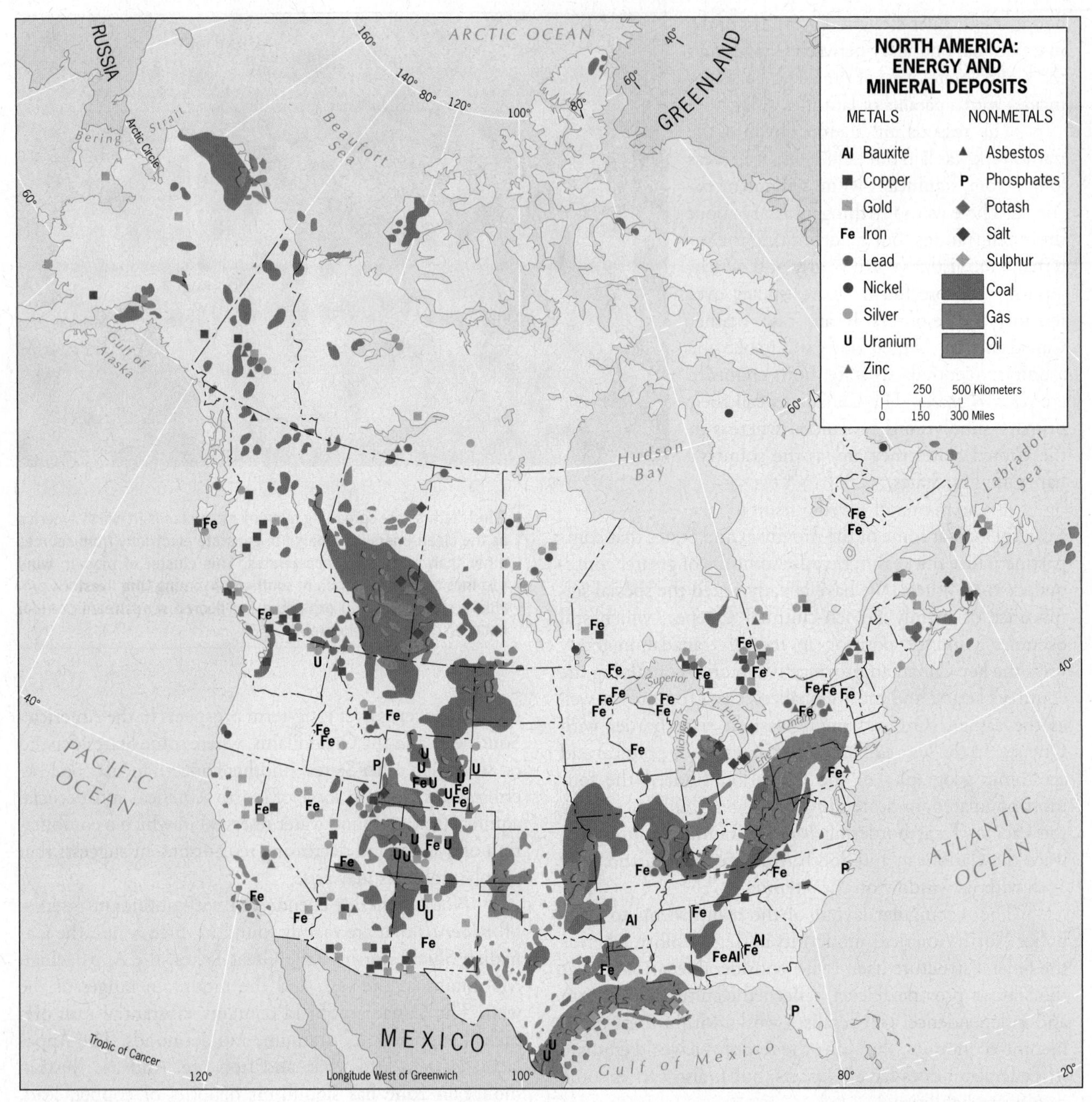

FIGURE 3A-8

investors only when the price of oil is comparatively high, but the reserves of oil estimated to be contained here may exceed those of Saudi Arabia (we will have more to say about the impact of Alberta's tar sands in Chapter 3B).

The distribution of ***natural gas*** reserves resembles that of oilfields because petroleum and natural gas tend to be found in similar geological formations (i.e., the floors of ancient shallow seas). What the map cannot reveal is the volume of production, in which this realm leads the world (Russia and Iran lead in proven reserves). That output has risen dramatically since 2000 as natural gas has become the fuel of choice for electricity generation in North America. Driving this expansion has been the widespread application of hydraulic fracturing (*fracking*) technology. By injecting pressurized fluids into deeply buried shale rocks to create fractures, vast quantities of trapped gas can be extracted, resulting in surging supplies and nosediving prices.

The ***coal*** reserves of North America, perhaps the largest on the planet, are found in Appalachia, beneath the Great Plains of the United States as well as Canada, and in the southern Midwest among other places. These reserves guarantee an adequate supply for centuries to come, although coal has become a less desirable fuel due its polluting emissions that contribute to global warming.

Given the relative scarcity of oil and the geopolitical disadvantages of having to import nearly half of what is needed in the United States (ca. 40 percent in 2012), you might expect nuclear energy to be more ubiquitous, especially since it is relatively cheap to produce. There are just over 100 nuclear power plants in the United States providing about one-fifth of all electrical energy (Canada has 19 plants that produce 15 percent of the country's electricity). But ever since the notorious leak and near-calamity at Three Mile Island (Pennsylvania) in 1979, the fear of nuclear accidents is pervasive. Since then, not a single new nuclear plant has been approved by the government.

URBANIZATION AND THE SPATIAL ECONOMY

Industrial Cities

When the Industrial Revolution crossed the Atlantic and touched down in America in the 1870s, it took hold so successfully and advanced so robustly that only 50 years later North America was surpassing Europe as the world's mightiest industrial complex. Industrialization progressed in tandem with urbanization as manufacturing plants in need of large labor supplies were built in and near cities. This, in turn, propelled rural-urban migration (including the migration of African Americans from the South to northern cities during the first half of the twentieth century) and the growth of individual cities. At a broader scale, a *system* of new cities emerged that specialized in the collection and processing of raw materials and the distribution of manufactured products. This **urban system [5]** was interconnected through a steadily expanding network of railroads, themselves a product of the Industrial Revolution.

As new technologies and innovations emerged, and specializations such as Detroit's automobile industry strengthened, an **American Manufacturing Belt [6]** (that extended into southern Ontario) evolved into the foundation of a North American Core (Fig. 3A-9). This core area, on the way to becoming the world's most productive and important, contained the majority of the realm's industrial activity and leading cities, including New York, Chicago, Toronto, and Pittsburgh (the "steel city").

In the course of intertwined urbanization and industrialization, the spatial economy of North America underwent profound changes. The ***primary*** sector, involving the extraction of raw materials from nature (agriculture, mining, fishing), was rapidly mechanized, and the workforce in this sector shrank considerably. Employment in the ***secondary*** sector, using the input of raw materials and manufacturing them into finished products, grew rapidly. The ***tertiary*** sector, entailing all kinds of services to support production and consumption (banking, retail, transport) expanded as well. Both the secondary and tertiary sectors were overwhelmingly concentrated in cities.

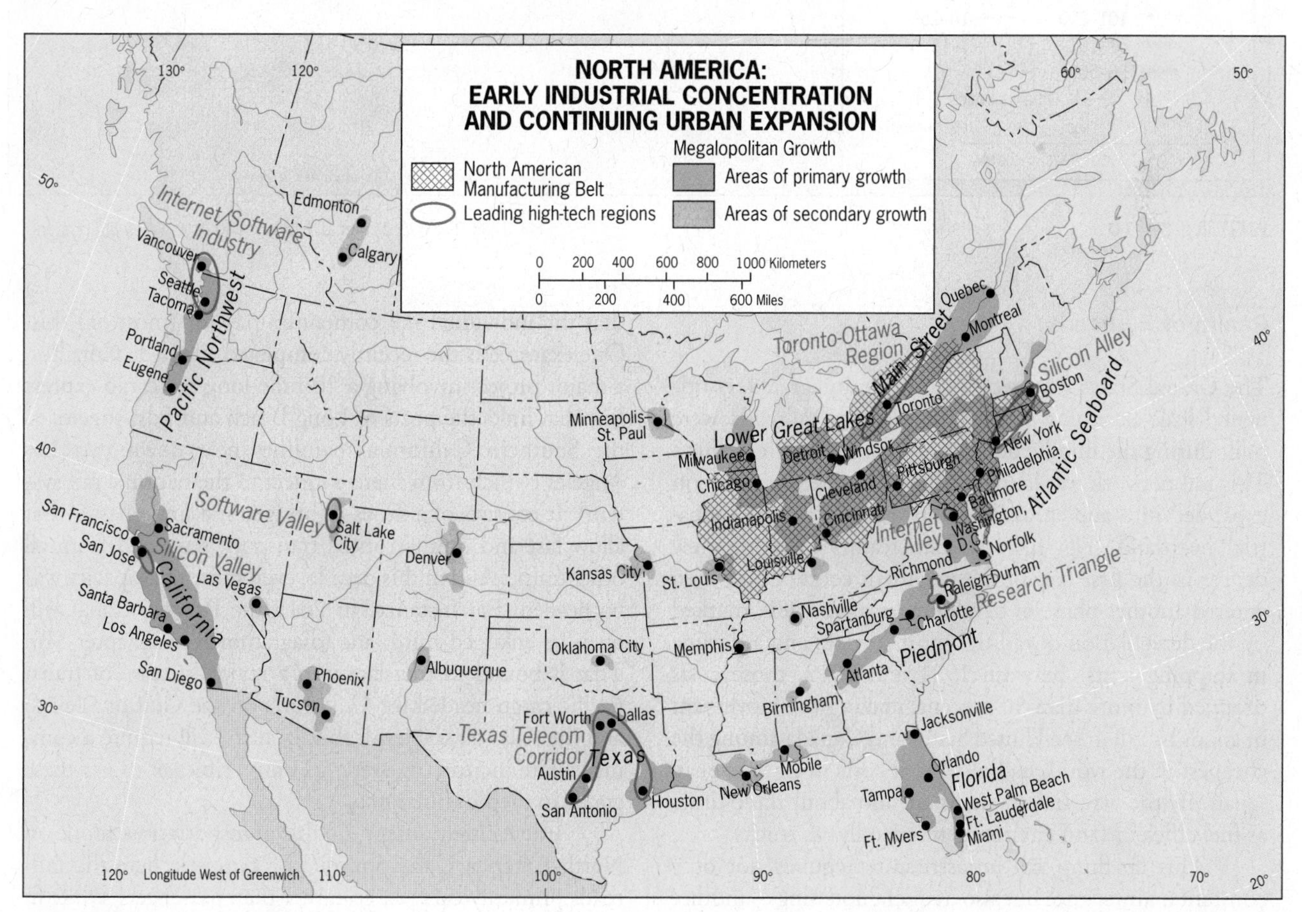

FIGURE 3A-9

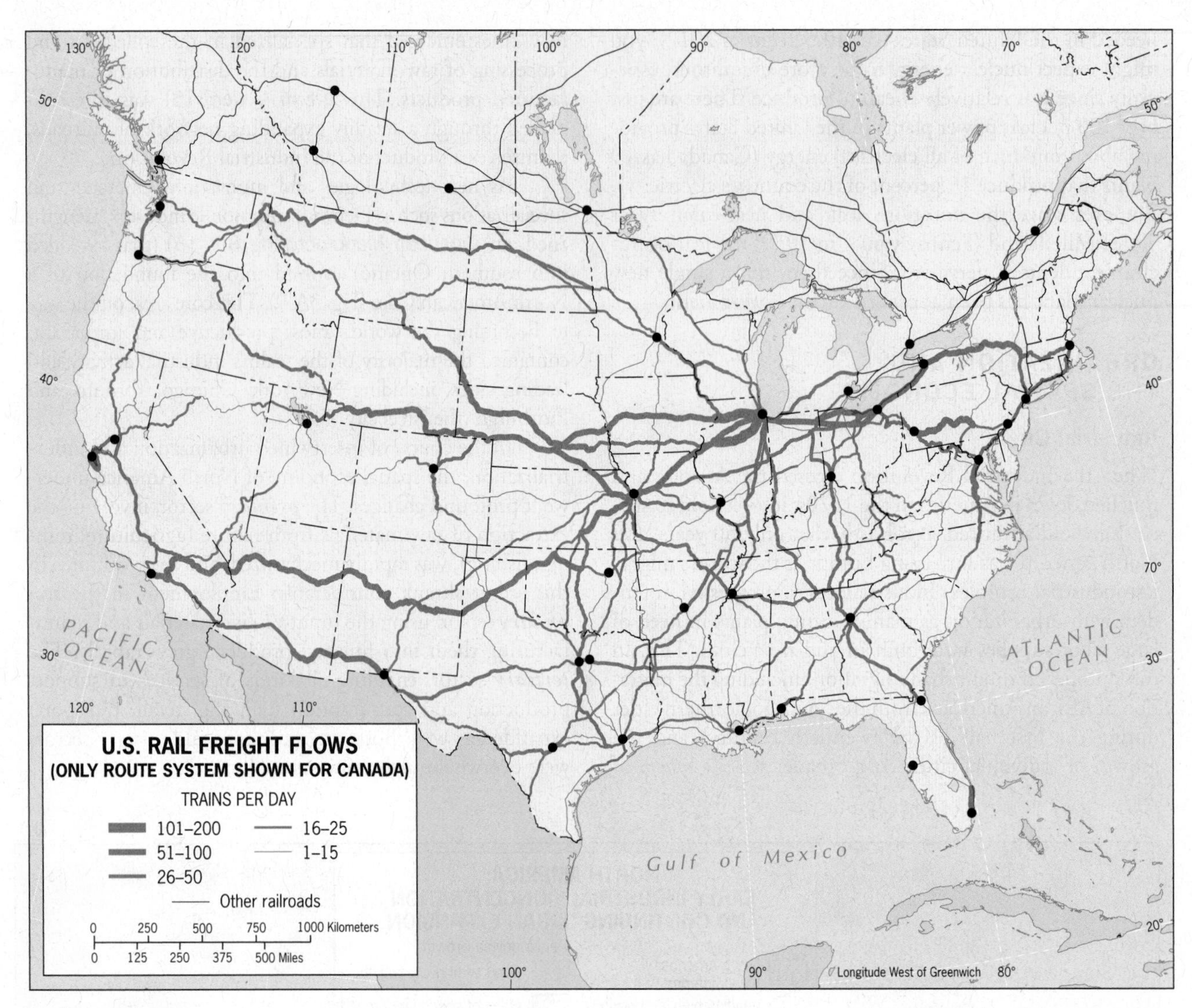

FIGURE 3A-10

Realm of Railroads

The United States first became a single, integrated, continental-scale economy when long-distance railroads were built during the middle decades of the nineteenth century. This rail network was launched in the Northeast and soon expanded into and across the Midwest, the realm's industrial heartland (Fig. 3A-10). The realm's network grew denser in the first half of the twentieth century and then entered another phase of development after 1980, marked by the deregulation of rail transport services and a decline in shipping costs. Between 1981 and 2012, those costs dropped by more than 50 percent, making the movement of goods by rail in the United States and Canada among the cheapest in the world, half of what it costs in Europe or in Japan. By the way, freight trains are also about three times as fuel efficient (and environment-friendly) as trucks.

This far-flung rail infrastructure requires not only constant maintenance but also frequent updating—another advantage of the privatized North American rail system is that the individual rail companies pay for (most of) this. One example is the recently completed Alameda Corridor, a major project involving a 20-mile-long rail cargo express line that links the ports of Long Beach and Los Angeles to the Southern California mainlines near downtown Los Angeles (which from there connect to the national rail system). It consists of a series of bridges and underpasses that allow fast and uninterrupted transport, separated from all other traffic. And in this decade, even more rail capacity will be needed. For instance, in 2015 the Panama Canal will open its enlarged third lane (diagrammed in Chapter 4B). That is bound to translate into a major increase of trans-Pacific cargo headed for U.S. ports on the Gulf of Mexico and the Atlantic seaboard; this, in turn, will require a commensurate increase in overland transfer by rail to get these goods to their destinations.

Interestingly, usage of this transportation mode in North America is the opposite of Europe's: here the railroads primarily carry cargo rather than passengers, while in the European realm it is the other way around. Rail freight in North America accounts for about 42 percent of the entire freight market (that also includes trucking, shipping, and air transport), which ranks it among the highest in the world's wealthy countries. Although the largest single cargo share is claimed by coal (38 percent and dropping), the fastest growing category of cargo is what is called "intermodal"—truck-trailer containers stacked onto railroad flatcars at ports and other break-of-bulk points.

Deindustrialization and Suburbanization

Cities continued to evolve, along with the nature of the economy and with the development of new transportation and communication technologies. The mass introduction of automobiles and the accompanying construction of a large-scale highway system, especially after World War II, had two major effects on cities. First, it resulted in much higher mobility and interconnectedness among cities. Second, it propelled a process of *suburbanization*—the transformation from compact city to widely dispersed metropolis through the evolution of residential suburbia into a complete **outer city [7]** with its own businesses and industries, sports and entertainment complexes, and myriad other amenities. As the newly urbanized suburbs increasingly captured major economic activities, many central cities saw their comparative status diminish.

Importantly, this transformation coincided with the **deindustrialization [8]** that hit American cities after 1970: the loss of manufacturing (and jobs) due to automation and the relocation of production to countries with lower wages. While the suburban outer cities became the destination of intrametropolitan migrants and of economic activity and employment in the service sector, many once-thriving and predominant CBDs were now all but reduced to serving the less affluent populations that increasingly dominated the central city's close-in neighborhoods. The fate of Detroit is often cited as exemplifying this sequence of events, but to a certain degree all major U.S. cities suffered.

The Information Economy and City Regions

Since the 1980s, a number of North American cities (but not all) have bounced back from the destructive effects of deindustrialization and unemployment. The ***information economy***—embodied in the *quaternary* sector—had arrived. Many northern cities had lost population and underwent "hollowing out" for nearly two decades, but the tide turned again. Employment in the tertiary sector started to grow rapidly, especially in high-technology, producer services (such as consulting, advertising, and accounting), finance, research and development, and the like. Indeed, with the onset of the digital era, most of these information-rich services evolved into quaternary activities. The geography of the information economy is quite different from that of the old manufacturing industries and is still not fully understood. Some of it concentrates in established CBDs like Midtown Manhattan; some of it locates around important hubs of the Internet infrastructure to benefit from maximum bandwidth (e.g., North Dallas); and still other activities cluster in suburban areas containing large pools of highly skilled workers such as

From the Field Notes . . .

"Monitoring the urbanization of U.S. suburbs for the past four decades has brought us to Tyson's Corner, Virginia on many a field trip and data-gathering foray. It is now hard to recall from this recent view that less than 50 years ago this place was merely a near-rural crossroads. But as nearby Washington, D.C. steadily decentralized, 'Tyson's' capitalized on its unparalleled regional accessibility (its Capital Beltway location at the intersection with the radial Dulles Airport Toll Road) to attract a seemingly endless parade of high-level retail facilities, office complexes, and a plethora of supporting commercial services. Today, this suburban downtown ranks among the largest business districts in all of North America. But it also exemplifies urban sprawl, and in 2009 its developers, tenants, and county government formed a coalition to transform Tyson's into a true city characterized by smarter, greener development."

www.conceptcaching.com

Silicon Valley or North Carolina's Research Triangle. The largest and best known high-technology clusters in North America are mapped in Figure 3A-9.

Northern California's ***Silicon Valley***—the world's leading center for computer research and development, and the headquarters of the United States' microprocessor industry—exquisitely illustrates the locational dynamics of this newest sector of the spatial economy. Proximity to Stanford, a world-class research university; adjacent to cosmopolitan San Francisco; the major local concentration of highly educated and skilled workers; a strong business culture; ample local investment capital; high-quality housing; and a scenic area with good weather—all combining to make Silicon Valley a prototype for similar developments elsewhere, and not just in North America. Under different names (*technopolis* is used in Brazil, France, and Japan; *science park* in China, Taiwan, and South Korea), such ultramodern, campus-like complexes symbolize the digital economy just as the smoke-belching factory did the industrial age of the past.

Polycentric Cities

Fly into any large American metropolitan region, and you will see the high-rises of suburban downtowns encircling the old central-city CBD, some of them boasting their own impressive skylines (see photo in *From the Field Notes*). In Canada, where such deconcentration has had more to do with lower land values outside the city centers, this spatial pattern has become common as well, although most Canadian cities remain more compact than the far-flung U.S. metropolis.

Thus the overall structure of the modern North American metropolis is polycentric and resembles a pepperoni pizza (Fig. 3A-11) in its general form. The traditional CBD still tends to be situated at the center, much of its former cross-traffic diverted by beltways; but the outer city's CBD-scale nodes are both ultramodern and thriving. Efforts to attract businesses and higher-income residents back to the old CBD sometimes involve the construction of multiple-use high-rises that often displace low-income residents, resulting in conflicts and lawsuits. The revitalization and upgrading—or **gentrification [9]**—of crumbling downtown-area neighborhoods increases real estate values as well as taxes, which tends to drive lower-income, long-time residents from their homes. The revival of old CBDs and/or part of their immediate surroundings is happening in numerous urban areas, from Manhattan's Harlem to downtown Seattle; but the growth and development of large interconnected metropolitan regions marked by multiple urban centers is certain to continue.

Effects of the Great Recession

The financial crisis that erupted in the United States in 2008, and soon assumed global proportions, was concentrated in the banking sector; it had much to do with excessive mortgage lending and borrowing, and with the increasingly nontransparent trade in mortgages. Although

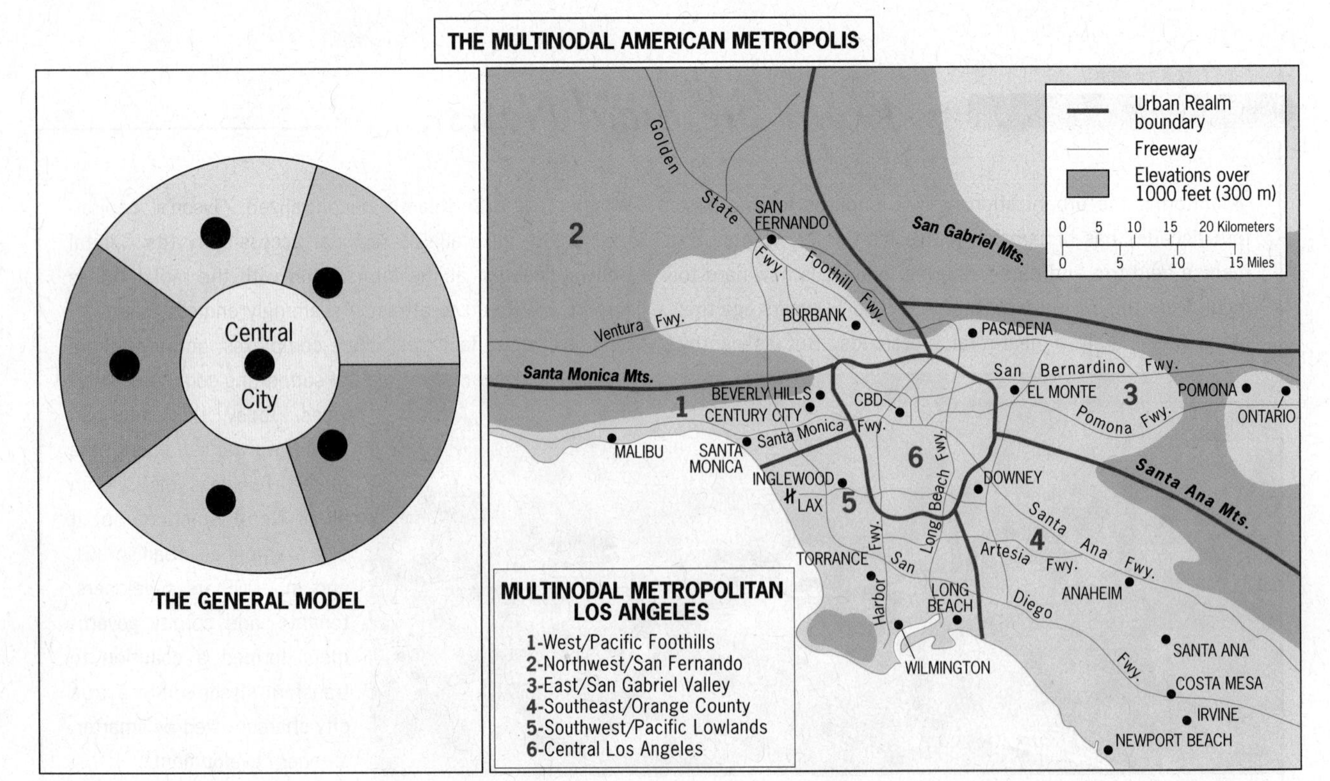

FIGURE 3A-11

many banks got into serious trouble for having accumulated what were clinically labeled "toxic assets," millions of American households fell victim as their mortgages became ever more difficult to afford. The crisis was in large part a housing (or mortgage), crisis and one of the lessons learned from it was how essential this sector has become to the twenty-first-century economy.

The resulting major economic downturn—quickly dubbed the Great Recession—took on unusually clear geographical dimensions precisely because it was so closely connected to housing (which can be regarded as capital fixed in place). Thus the crisis affected some places more severely than others, and the impact was most devastating where housing values had risen unrealistically in the years preceding 2008. In those real estate markets, there had seemed no limit to the escalation of home prices. Buyers were betting on continued appreciation, and many bought for speculative purposes with the intention of selling in the near future for a tidy profit. The banks propelled everything along with easy-to-obtain home loans, sometimes with variable interest rates they knew would prove fatal to buyers at a later stage. When the bubble finally burst, housing prices nosedived and myriad homeowners went under water as their home now became worth less than the value of their mortgage. This situation is referred to as ***negative equity*** (equity being the difference between market value and mortgage balance), a problem because it renders homeowners less affluent. In these circumstances, they cannot afford to sell because to pay off the bank would result in a significant loss.

The scale of this housing crisis was enormous, and its duration has been especially painful. The first peak was reached in early 2009, when a staggering 28 percent of all mortgaged residential properties in the United States—almost 14 million homes—were affected. But worse was yet to come. After a dip in 2010, a second peak was reached in early 2011 (31 percent/just over 15 million homes). The level has fluctuated since then, and the newest data at press time (late 2012) indicated a return to 28 percent, possibly the beginning of a downward trend in response to improving national economic conditions.

Figure 3A-12 shows that negative-equity conditions vary considerably by State. Hardest hit were Nevada, Florida, Arizona, Georgia, and Michigan (in that order)—where the housing bubble had been at its greatest during the 2000s. It should also be noted that the crisis was particularly severe in what were until recently fast-growing metropolitan areas. In Nevada, worst off were Las Vegas and its suburbs as well as Reno; in Florida, topping the list were Miami, Orlando, and Tampa; and in Arizona, most affected were Phoenix and especially its suburb of Glendale. Michigan's high ranking, however, reflected a different situation because this State's economic problems were deeply rooted in the job losses associated with the long-term decline of the automobile industry. This is yet another reminder that

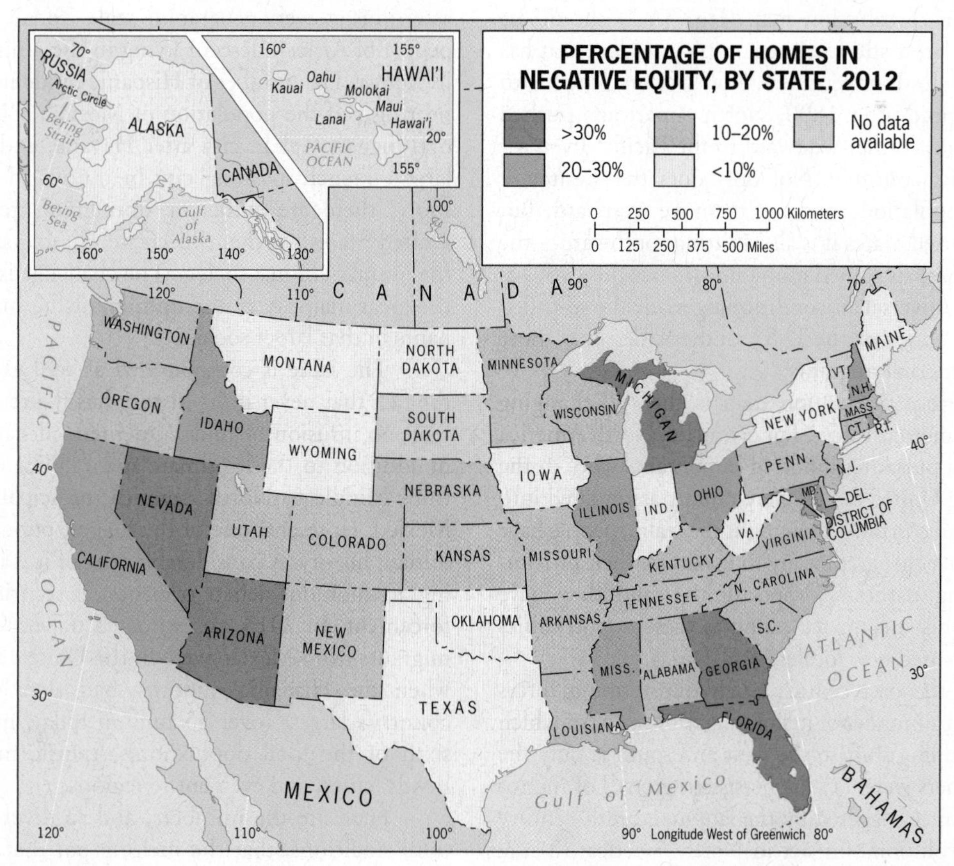

FIGURE 3A-12

deindustrialization in the American Manufacturing Belt has forged a widening economic gap between Rustbelt cities like Detroit and such booming Sunbelt counterparts as Atlanta and Dallas-Fort Worth.

As the U.S. economy spiraled downward during the prolonged Great Recession, unemployment rose, stock markets experienced major jolts, and more and more people had difficulty maintaining homes financed by negative-equity-plagued mortgages. This swiftly led to an alarming rise in foreclosures, led by the States with the highest negative-equity percentages. Nevada has topped all the others since 2008, and even in 2012 an astonishing 1 out of 56 mortgaged properties in that State were foreclosed.

The effects of the economic crisis on Canada's housing market were less severe. Canadian banks had been more conservative than their U.S. counterparts in their mortgage lending, and the housing sector remained relatively healthy. Nonetheless, the crisis quickly became global in scope and Canada was not spared some of its harmful effects.

THE MAKING OF A MULTICULTURAL REALM

The Virtues of Mobility and Immigration

Given the mobility of North America's population, the immigrant streams that continue to diversify it, the economic changes affecting it, and the forces acting upon it, the population distribution map (Fig. 3A-2) should be viewed as the latest still in a motion picture, one that has been unreeling for four centuries. Slowly at first, then with accelerating speed after 1800, North Americans pushed their settlement frontier westward to the Pacific. Even today, such shifts continue. Not only does the "center of gravity" of population continue to move westward, but within the United States it is also shifting southward—the latter a drift that has gained momentum since the 1960s as the advent of universal air conditioning made the so-called **Sunbelt [10]** States of the U.S. southern tier ever more attractive to internal migrants.

The current population map is the still-changing product of numerous forces. For centuries, North America has attracted a pulsating influx of immigrants who, in the faster-growing United States, were rapidly assimilated into the societal mainstream. Throughout the realm, people have sorted themselves geographically to maximize their proximity to evolving economic opportunities, and they have shown little resistance to relocating as these opportunities successively favored new locales.

During the past century, such transforming forces have generated a number of major **migrations [11]** of which the still-continuing shift to the west and south is only the latest. Five others were: (1) the persistent growth of metropolitan areas, first triggered by the late-nineteenth-century Industrial Revolution's impact in North America; (2) the large-scale movement of African Americans from the rural South to the urban North during the latter stages of the industrial era; (3) the shift of tens of millions of urban residents from central cities to suburbs and subsequently to exurbs even farther away from the urban core; (4) the return migration of millions of African Americans from the deindustrializing North back to the growing opportunities in the South (in metropolises such as Atlanta and Charlotte); and (5) the strong and steady influx of immigrants from outside North America including, in recent times, Mexicans, Cubans, and other Latinos; South Asians from India and Pakistan; and East and Southeast Asians from Hong Kong, Vietnam, and the Philippines.

Several of the migration streams affecting North America rank among the largest in world history, creating the pair of plural societies that characterize this realm (see the box titled "The Migration Process"). North America functions, in some ways, as a global magnet for human resources, and its persistent prosperity hinges at least in part on wave after wave of ambitious immigrants, be they engineers from China, physicians from India, nurses from Jamaica, or farm workers from Mexico.

The Challenge of Multiculturalism

But growing diversity, especially at a time of globalization and ***transnationalism***, comes with challenges. The sheer dimensions of immigration into the United States, from virtually all corners of the world, creates an increasingly complex ethnic and cultural mosaic that tests **melting pot [12]** assertions at every geographic scale. Today there are more people of African descent living in the United States than in Kenya. The number of Hispanic residents in America is nearing half the population of Mexico. Miami is the second-biggest Cuban city after Havana, and Montreal the largest French-speaking city in the world after Paris. In short, there are sufficient immigrant numbers in the United States for them to create durable societies within the overall national society. The challenge is to ensure that the great majority of the immigrants become full participants in that larger society.

The issue is compounded as well as politicized by the fact that never in its history has the realm received so large an infusion of undocumented (illegal) immigration in addition to the legitimate stream. The ongoing influx from Middle to North America, principally from or via Mexico, represents one of the biggest population shifts in human history. A considerable part of it is illegal, producing acrimonious debate over border security and law enforcement. In 2013, an estimated 6.7 million illegal immigrants from Mexico were in the United States at a time when the Hispanic minority had already become the country's largest (over 55 million today, more than one-sixth of the total population)—transforming neighborhoods, cities, and even entire regions.

Such are the numbers, and so diverse are the cultural traditions, that the melting pot that America once was is today morphing into something else. Spatially, the

The Migration Process

BOTH CANADA AND the United States are products of international migration. Europeans first crossed the Atlantic with the intent of establishing permanent colonies in the early 1600s, and from these colonies evolved the two countries of North America. The Europeanization of North America doomed the realm's indigenous societies, but this was only one of many areas around the world where local cultures and foreign invaders came face to face. Between 1835 and 1935, perhaps as many as 75 million Europeans departed for distant shores—most of them bound for the Americas (Fig. 3A-13). Some sought religious freedom, others escaped poverty and famine, still others simply hoped for a better life. A comparative few were transported against their will to penal colonies, and then there was of course forced movement related to slavery (mapped in Chapter 6A).

Studies of the ***migration decision*** show that migration flows vary in size with (1) the perceived difference between home (source) and destination; (2) the effectiveness of information flow, that is, the news about the destination that emigrants send back to those left behind waiting to decide; and (3) the distance between source and destination (shorter moves attract many more migrants than longer ones).

Every migration stream produces a counter-stream of returning migrants who cannot adjust, are unsuccessful, or are otherwise persuaded or compelled to return home. Migration studies also conclude that several discrete factors are at work in the process. ***Push factors*** motivate people to move away from an undesirable locale that may be afflicted by famine, armed conflict, religious persecution, or some other adversity. ***Pull factors*** attract them to destinations perceived to hold a promise of security, opportunity, or another desired goal.

To the early (and later) European immigrants, North America was a new frontier, a place to escape persecution and acquire a piece of land. Opportunities were reported to be unlimited. That perception of opportunity has never changed. Immigration continues to significantly shape the human-geographic complexion of the United States as well as Canada. Today's immigrants account for 43 percent of the annual population growth of the United States (and will surpass 50 percent by 2030). Never in its history, however, has the U.S. received so large an undocumented (illegal) immigration flow in addition to the legitimate stream, and the ongoing influx from Middle to North America is one of the largest population shifts in human history.

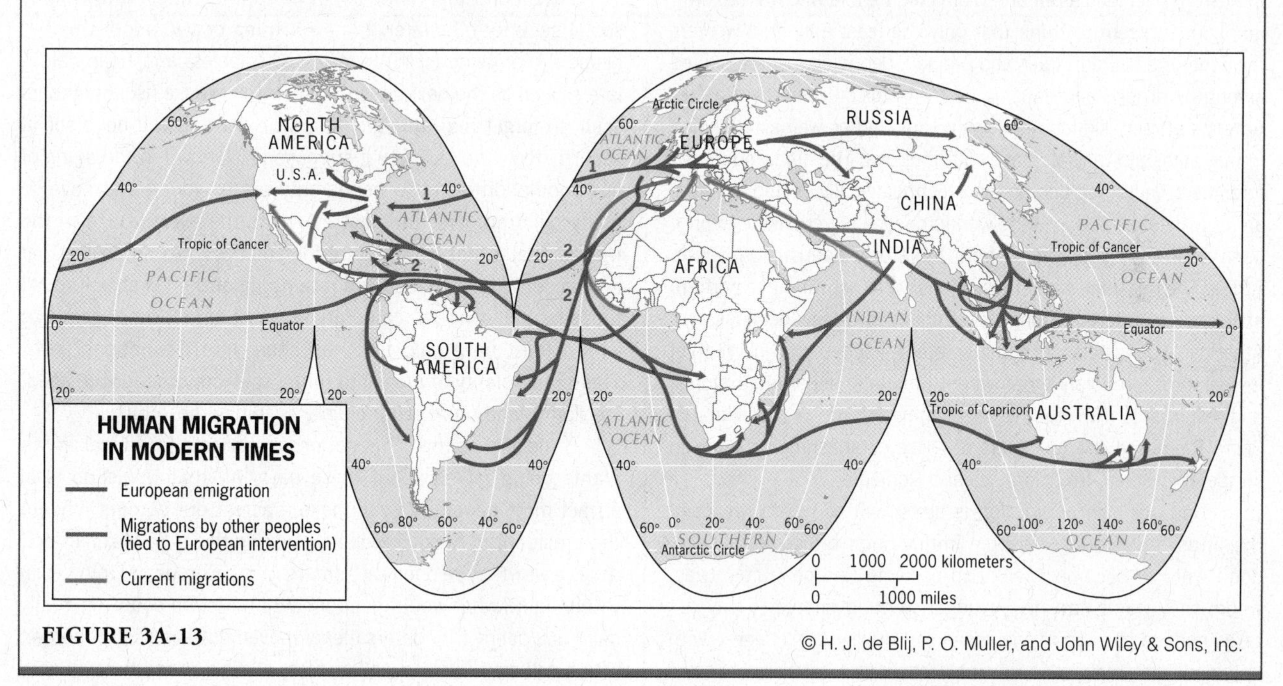

FIGURE 3A-13

United States is now completing its transformation into a ***mosaic culture***, an increasingly heterogeneous complex of separate, more or less uniform "tiles" whose residents spend less time than ever interacting and "melting." This applies not only to new immigrant groups but also to existing communities, underscored by the proliferation of walled-off, gated housing complexes in metropolitan areas across the realm. There is a serious downside to all this, as Bill Bishop writes in his landmark book, *The Big Sort: Why the Clustering of Like-Minded America Is Tearing Us Apart.* As is the case in the world at large, balkanization fueled by people wanting to interact only with those closely resembling themselves leads to misunderstanding of others, miscalculations when it comes to decisions and policies, and widening fissures between communities. This could even threaten the very survival of the democratic values that have underpinned the evolution of American society.

Regional ISSUE Immigration

IMMIGRATION BRINGS BENEFITS—THE MORE THE MERRIER!

"The United States and Canada are nations of immigrants. What would have happened if our forebears had closed the door to America after they arrived and stopped the Irish, the Italians, the eastern Europeans, and so many other nationalities from entering this country? Now we're arguing over Latinos, Asians, Russians, Muslims, you name it. Fact is, newcomers have always been viewed negatively by most of those who came before them. When Irish Catholics began arriving in the 1830s, the Protestants already here accused them of assigning their loyalty to some Italian pope rather than to their new country, but Irish Catholics soon proved to be pretty good Americans. Sound familiar? Muslims can be very good Americans too. It just takes time, longer for some immigrant groups than others. But don't you see that America's immigrants have always been the engine of growth? They become part of the world's most dynamic economy and make it more dynamic still.

"My ancestors came from Holland in the 1800s, and the head of the family was an architect from Rotterdam. I work here in western Michigan as an urban planner. People who want to limit immigration seem to think that only the least educated workers flood into the United States and Canada, depriving the less-skilled among us of jobs and causing hardship for citizens. But in fact America attracts skilled and highly educated as well as unskilled immigrants, and they all make contributions. The highly educated foreigners, including doctors and technologically skilled workers, are quickly absorbed into the workforce; you're very likely to have been treated by a physician from India or a dentist from South Africa. The unskilled workers take jobs we're not willing to perform at the wages offered. Things have changed! A few decades ago, American youngsters on summer break flooded the job market in search of temporary employment in hotels, department stores, and restaurants. Now they're vacationing in Europe or trekking in Costa Rica, and the managers of those establishments bring in temporary workers from Jamaica and Romania.

"And our own population is aging, which is why we need the infusion of younger people immigration brings with it. We don't want to become like Japan or some European countries, where they won't have the younger working people to pay the taxes needed to support the social security system. I agree with opponents of immigration on only one point: what we need is legal immigration, so that the new arrivals will get housed and schooled, and illegal immigration must be curbed. Otherwise, we need more, not fewer, immigrants."

LIMIT IMMIGRATION NOW!

"The percentage of recent immigrants in the U.S. population is the highest it has been in 70 years, and in Canada in 60 years. America is adding the population of San Diego every year, over and above the natural increase, and not counting illegal immigration. This can't go on. By 2020, more than one-sixth of the U.S. population will consist of recent immigrants. At least one-third of them will not have a high school diploma. They will need housing, education, medical treatment, and other social services that put a huge strain on the budgets of the States they enter. The jobs they're looking for often aren't there, and then they start displacing working Americans by accepting lower wages. It's easy for the elite to pontificate about how great immigration is for the American melting pot, but they're not the ones affected on a daily basis. Immigration is a problem for the working people. We see company jobs disappearing across the border to Mexico, and at the same time we have Mexicans arriving here by the hundreds of thousands, legally and illegally, and more jobs are taken away.

"And don't talk to me about how immigration now will pay social security bills later. I know a thing or two about this because I'm an accountant here in Los Angeles, and I can calculate as well as the next guy in Washington. Those fiscal planners seem to forget that immigrants grow older and will need social security too. And as for that supposed slowdown in the aging of our population because immigrants are so young and have so many children, over the past 20 years the average age in the United States has dropped by four months. So much for that nonsense. What's needed is a revamping of the tax structure, so those fat cats who rob corporations and then let them go under will at least have paid their fair share into the national kitty. There'll be plenty of money to fund social services for the aged. We don't need unskilled immigrants to pay those bills.

"And I'm against this notion of amnesty for illegal immigrants being talked about these days. All that would do is to attract more people to try to make it across our borders. I heard the president of Mexico propose opening the U.S.-Mexican border the way they're opening borders in the European Union. Can you imagine what would happen? What our two countries really need is a policy that deters illegal movement across that border, which will save lives as well as jobs, and a system that will confine immigration to legal channels. These days, that's not just a social or economic issue; it's a security matter as well."

Vote your opinion at www.wiley.com/go/deblijpolling

In Canada, the melting-pot notion was put to the test during the 1990s when East Asian immigrants from Hong Kong arrived during and following the 1997 takeover of the British crown colony by Beijing's communist government. Affluent Chinese families arriving in Vancouver proceeded to purchase and renovate (or frequently replace) traditional homes in long-stable Vancouver neighborhoods, arousing the ire of numerous locals. But in general, Canadian views on immigration differ from those in the United States. Canada has long faced critical labor shortages, especially in its western provinces including energy-booming Alberta and Asian-trade-burgeoning British Columbia. In need not only of professionals but also truck drivers and dock workers, Canada tries to balance its legal immigration process to keep it compatible with the country's employment as well as demographic needs. Across the realm, multicultural tendencies now vary on a regional basis depending on the subtle interplay of economics, existing cultural patterns, immigration, and local politics.

POINTS TO PONDER

- Whereas more than 70 percent of Canada's population lives within 200 kilometers (125 mi) of the northern U.S. border, only about 12 percent of the Mexican population lives that close to the southern U.S. border.
- There are only two states in the world that share the same country code for international telephone traffic: the United States and Canada.
- The United States does not have an official language. Should it?
- The North American rail freight system is unique among the world's realms in terms of its transport capacity and low cost.
- The U.S. and Canada are still each other's biggest overall trading partners, despite the rapid economic rise of China.

NORTH AMERICAN REGIONS:
CORE AND PERIPHERIES
0 500 1000 1500 Kilometers
0 300 600 900 Miles
ARCTIC OCEAN
PACIFIC OCEAN
ATLANTIC OCEAN
Hudson Bay
Gulf of Mexico
Arctic Circle
Tropic of Cancer
120° Longitude West of Greenwich 110°
NORTHERN FRONTIER
WESTERN FRONTIER
PACIFIC HINGE
CONTINENTAL INTERIOR
(North)
(South)
SOUTHWEST
SOUTH
CORE
FRENCH CANADA
MARITIME NORTHEAST
1
2
3
4
5
6
7
8
9
Anchorage
Fort McMurray
Edmonton
Calgary
Vancouver
Seattle
Portland
San Francisco
Los Angeles
San Diego
Las Vegas
Phoenix
Salt Lake City
Denver
Winnipeg
Minneapolis
Omaha
Kansas City
Dallas
Houston
San Antonio
Chicago
Milwaukee
Detroit
St. Louis
Memphis
New Orleans
Atlanta
Charlotte
Jacksonville
Tampa
Miami
Norfolk
Washington, D.C.
Philadelphia
New York
Boston
Toronto
Montréal
Halifax

FIGURE 3B-1

REGIONS

North American Core

Maritime Northeast

French Canada

Southeast

Southwest

Pacific Hinge

Western Frontier

Continental Interior

Northern Frontier

IN THIS CHAPTER

- U.S.-Canada cross-border linkages
- The Hispanicization of the Southwest
- Miami: South of the South
- China's impact on the Pacific Hinge
- The natural riches of the Northern Frontier
- The Alaskan frontier and U.S. geopolitics

CONCEPTS, IDEAS, AND TERMS

First Nations 1

Devolution 2

Cross-border linkages 3

Secession 4

Megalopolis 5

American Manufacturing Belt 6

Technopole 7

Racial profiling 8

Pacific Rim 9

Tar sands 10

Boreal forest 11

© Jan Nijman

Where were these pictures taken?
Find out at www.wiley.com/college/deblij

Geology, climate, culture, and history have combined to create an interesting matrix of regions in North America (Fig. 3B-1). Mountain ranges, coastal zones, deserts, low-lying plains, and a range of climatic environments all demand different kinds of human adaptation. Successive waves of immigrants brought with them a variety of cultures; sometimes the differences were subtle, but sometimes they were quite stark. The emergence of two large countries, the United States and Canada, provided a pair of major political and institutional territorial structures within which regional differences have been accommodated. As Figure 3B-1 shows, a majority of North America's regions traverse both countries regardless of the international boundary. But some other regional developments are best understood within the national context of each country. As much as Canada and the United States have in common, there is ample reason to focus the first part of this chapter on each individually. In the discussion that follows, therefore, we begin by highlighting the regional geographies of two concordant neighbors—in the knowledge that Canadians tend to know more about the United States than Americans know about Canada. After that, we will examine the nine evolving regions of the realm as a whole.

REGIONS OF THE REALM

REGIONALISM IN CANADA: DIVISIVE FORCES

Canada's Spatial Structure

Territorially the second-largest state in the world after Russia, Canada is administratively divided into only 13 subnational entities—10 provinces and 3 territories (the United States has 50). The provinces—where 99.7 percent

© Jan Nijman

MAJOR CITIES OF THE REALM

City	Population* (in millions)
New York	21.1
Los Angeles	13.2
Chicago	9.6
Dallas-Ft. Worth	7.0
San Francisco	6.6
Houston	6.3
Washington, D.C.	6.2
Philadelphia	6.1
Miami-Ft. Lauderdale	5.9
Toronto, Canada	5.8
Atlanta	5.7
Boston	4.7
Phoenix	4.5
Detroit	4.3
Montreal, Canada	4.0
Seattle	3.8
San Diego	3.3
Denver	3.1
Vancouver, Canada	2.6
Ottawa, Canada	1.3

*Based on 2014 estimates.

of all Canadians live—range in dimensions from tiny, Delaware-sized Prince Edward Island to Quebec, nearly twice the size of Texas (Fig. 3B-2).

As in the United States, the smallest provinces lie in the northeast. Canadians call these four the Atlantic Provinces (or simply Atlantic Canada): Prince Edward Island, Nova Scotia, New Brunswick, and the mainland-island province named Newfoundland and Labrador. Gigantic, mainly ***Francophone*** (French-speaking) Quebec and populous, heavily urbanized Ontario, both flanking Hudson Bay, form the heart of the country. Most of western Canada (which is what Canadians call everything west of Ontario) is organized into the three Prairie Provinces: Manitoba, Saskatchewan, and Alberta. In the far west, beyond the Canadian Rockies and extending to and along the Pacific Ocean, lies the tenth province, British Columbia.

Of the three territories in Canada's Arctic North, the Yukon is the smallest—but nothing is small here in Arctic Canada. With the Northwest Territories and recently created Nunavut, the three Territories cover almost 40 percent of Canada's total area. The population, however, is extremely small: only about 120,000 people inhabit this vast, frigid frontier zone.

The case of Nunavut merits special mention. Created in 1999, this newest Territory is the outcome of a major aboriginal land claim agreement between the *Inuit* people (formerly called Eskimos) and the federal government, and encompasses all of Canada's eastern Arctic as far north as Ellesmere Island (Fig. 3B-2), an area far larger than any other province or territory. With around one-fifth of Canada's total area, Nunavut—which means "our land"—contains only about 35,000 residents, of whom 80 percent are Inuit.

Canada's population is clustered in a discontinuous ribbon, roughly 300 kilometers (200 mi) wide, nestled against the U.S. border and along ocean shores, whereas most of the country's energy reserves lie far to the north of this corridor. As such, the map of Canada's human settlement (Figs. 3A-2 and 3B-2) bears out this generalization and emphasizes the significance of environment. People choose to live in the more agreeable areas to the south, but territorial control of abundant northern resources is highly important. The population pattern creates cross-border affinities with major American cities in several places (Toronto-Buffalo, Windsor-Detroit, Vancouver-Seattle), although very little of this occurs in the three Prairie Provinces of the Canadian West, which for the most part adjoin sparsely settled North Dakota and Montana.

Cultural Contrasts

Canada's capital, Ottawa, is symbolically located astride the Ottawa River, which marks the boundary between English-speaking Ontario and French-speaking Quebec. Though comparatively small, Canada's population is markedly divided by culture and tradition, and this division has a pronounced regional expression. Linguistically, 68 percent of Canada's citizens speak English as their mother tongue; 13 percent speak French only; 17 percent of the population is bilingual in English and French; and the remaining 2 percent are speakers of other languages. The spatial concentration of more than 85 percent of the country's Francophones in Quebec accentuates this division (Fig. 3B-2). Quebec's population is more than 80 percent French-Canadian, and this province remains the historic, traditional, and emotional focus of French culture in Canada.

Over the past half-century a strong nationalist movement has emerged in Quebec, and at times it has demanded outright separation from the rest of Canada. This ethnolinguistic division continues to be the litmus test for Canada's federal system, and the issue continues to pose a latent threat to the country's future national unity. The historical roots of the matter are related to French-British competition since the seventeenth century in which Britain gained the upper hand. To this day, the British monarch remains the official head of state (as Canada remains part of the British Commonwealth), represented by the governor general. It is one of the issues that riles Quebec nationalists.

There does seem to be an overall decline in support for Quebec's independence. This is partly due to a number of laws that reassure the primacy of Québécois culture in the province (a 2006 federal parliamentary resolution effectively recognizes the Québécois as a *distinct nation*). In addition, there appears to be a growing realization among the French-speaking people that continued integration in

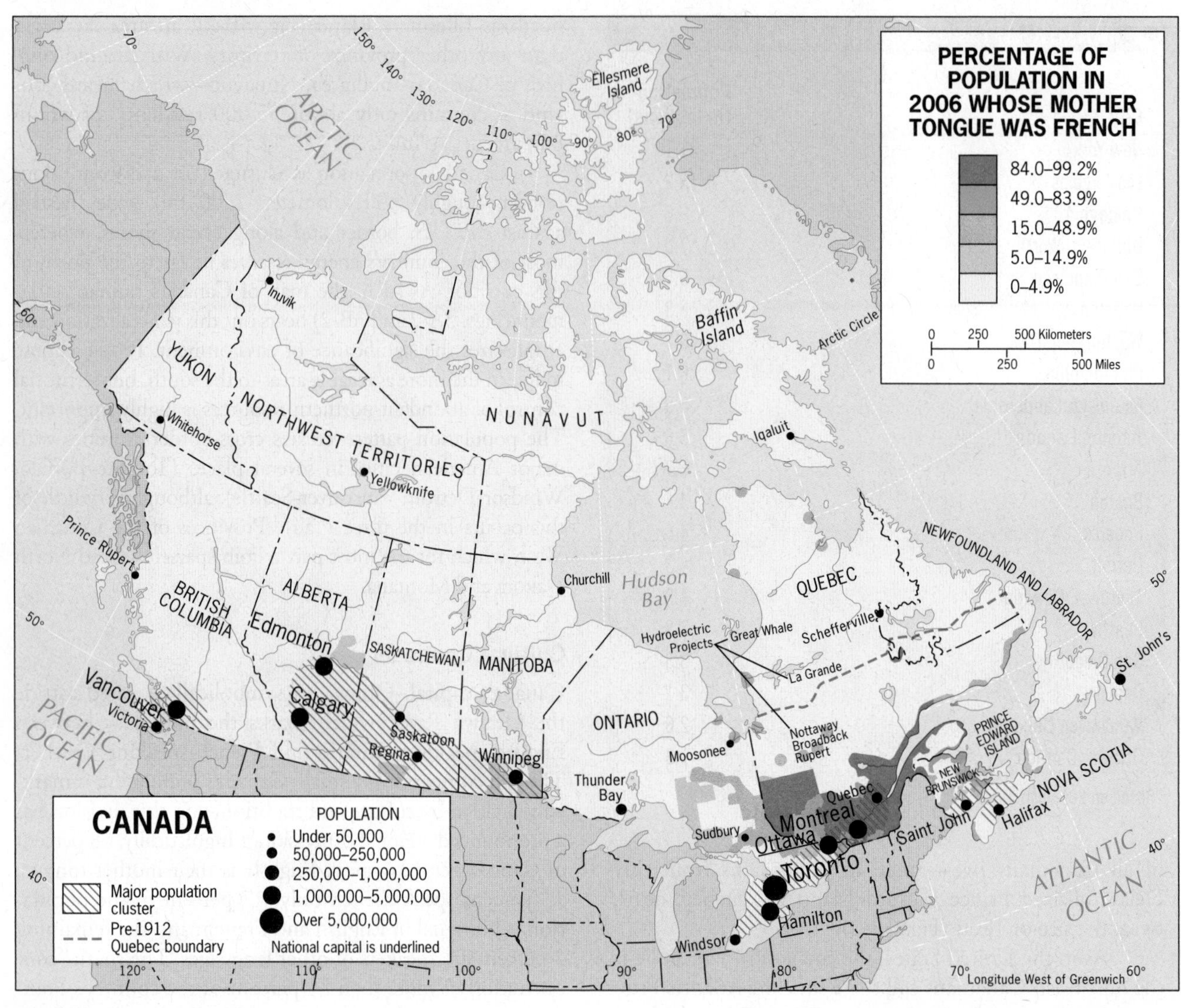

FIGURE 3B-2

Canada combined with a high degree of autonomy is the best possible option. But the struggle for a maximum degree of autonomy continues, along with the accompanying political and ethnic tensions.

The Ascendancy of Indigenous Peoples

On the wider Canadian scene, the culture-based Quebec struggle has stirred ethnic consciousness among the country's 1.2 million native peoples—dominated by **First Nations [1]**, but also including Métis and Inuit. They, too, have received a sympathetic hearing in Ottawa, and in recent years breakthroughs have been achieved with the creation of Nunavut as well as local treaties for limited self-government in northern British Columbia.

A foremost concern among these peoples is that their aboriginal rights be protected by the federal government against the provinces of which they are a part. This is especially true for the largest First Nation of Quebec, the Cree, whose historic domain covers the northern half of the province. As Figure 3B-2 shows, the territory of the Cree (north of the red dashed line) is no unproductive wilderness: it contains the James Bay Hydroelectric Project, a massive scheme of dikes and dams that has transformed much of northwestern Quebec and generates hydroelectric power for a huge market within and outside the province. In 2002, after the federal courts supported its attempts to block construction of the project, the Cree negotiated a groundbreaking treaty with the Quebec government whereby this First Nation dropped its opposition in return for a portion of the income earned from electricity sales. The Cree also secured the right to control their own economic and community development.

Centrifugal Forces

The events of the past four decades have had a profound impact on Canada's national political geography, and not only in Quebec. Leaders of the western provinces objected to federal concessions to Quebec, insisting that the equal treatment of all ten provinces is a basic principle that precludes the designation of special status for any single one. Recent

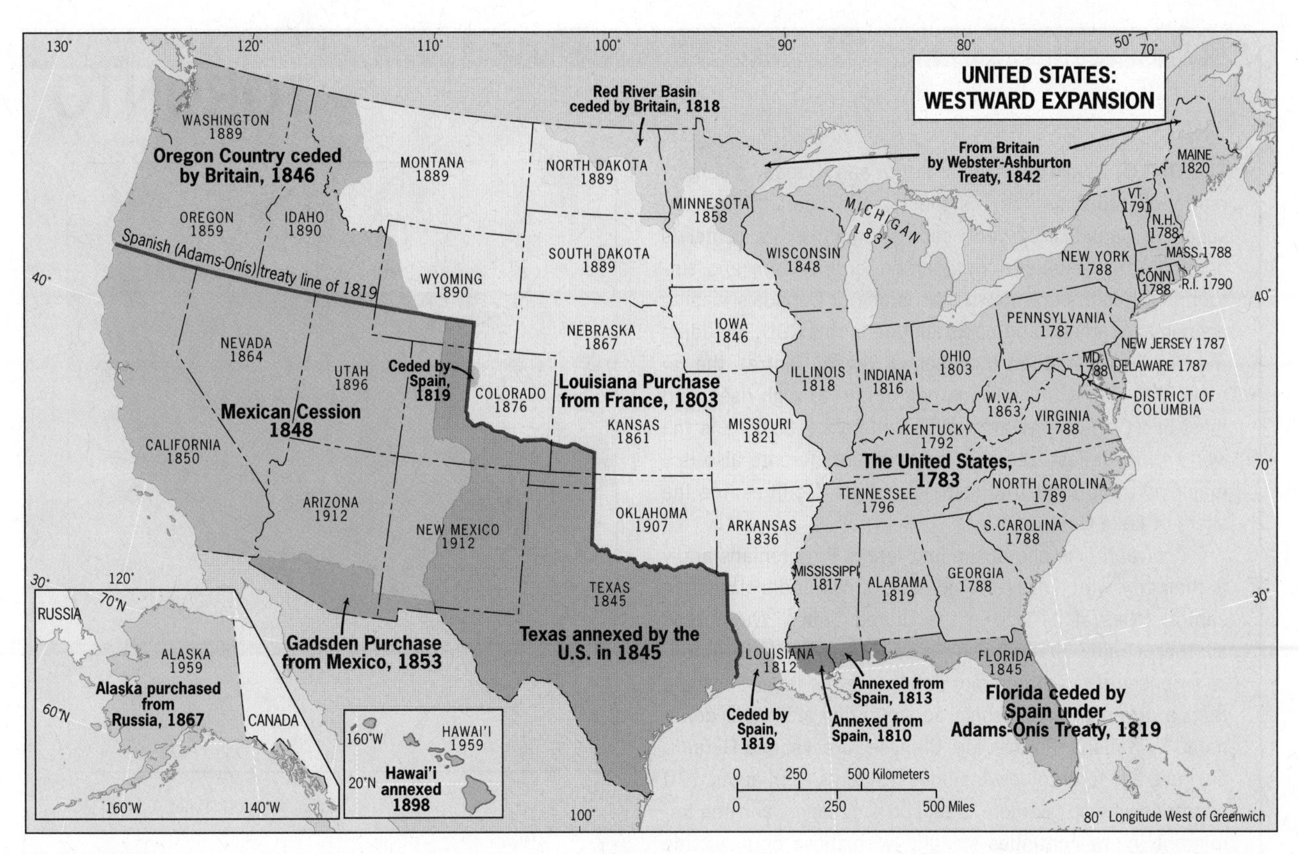

FIGURE 3B-3

federal elections and opinion surveys have clearly revealed the emerging fault lines that surround Quebec and set off the western provinces and British Columbia from the rest of the country. Thus, with national unity always a challenge in this comparatively young federation, Canada confronts forces of **devolution [2]** that threaten to weaken the state.

In addition, the continuing economic integration of Canada and the United States (galvanized by the North American Free Trade Agreement of 1994) tends to pull the southern parts of Canada's most populous provinces into an ever closer relationship with its U.S. neighbor. Local **cross-border linkages [3]** are already strongly developed and are likely to intensify in the future: the Atlantic Provinces with neighboring New England; Quebec with New York State; Ontario with Michigan and adjacent Midwestern States; the Prairie Provinces with the Upper Midwest; and British Columbia with the (U.S.) Pacific Northwest. Such functional reorientations constitute yet another set of potentially powerful devolutionary challenges confronting the Ottawa government.

REGIONALISM AND ETHNICITY IN THE UNITED STATES

Nothing in the United States compares with the devolutionary challenges Canada has faced throughout its evolution as a modern state. The United States has about four times as many subnational political units as Canada, but none since the Civil War (1861–1865) has mounted a serious campaign for **secession [4]** of the kind Quebec has. And in only one (Hawai'i) have indigenous people demanded recognition of a part of the State as sovereign territory.

In Canada, indigenous peoples were able to hold on to sparsely populated areas in the north. In the United States, on the other hand, the relentless push westward by European settlers and the U.S. government in effect dispossessed—and decimated—the original inhabitants (Fig. 3B-3). In the end, they were confined to some 300 reservations located mainly in the western half of the country (Fig. 3B-4, lower-right map). There are only about 3 million Native Americans left today, with the Navajo and Cherokee being the largest surviving nations. Whereas in Canada the First Nations hold significant territorial power, in the United States they are in a far weaker position to challenge the federal government.

Within U.S. society, much more so than in Canada, the social fabric and ethnic relations are determined by past and present waves of immigration. In certain cases, immigration has helped forge a particular regional geography. This was especially true in the case of the hundreds of thousands of Africans who were forcibly brought to the United States in bondage from the seventeenth to the mid-nineteenth centuries and put to work as slaves on plantations in the American South.

Slavery was abolished in 1865, but racial segregation persisted in a variety of legal forms until the civil rights

AMONG THE REALM'S GREAT CITIES . . .

TORONTO

TORONTO, CAPITAL OF Ontario and Canada's largest metropolis (5.8 million), is the historic heart of English-speaking Canada. The landscape of much of its center is dominated by exquisite Victorian-era architecture and surrounds a healthy downtown that is one of Canada's leading economic centers. Landmarks abound in this CBD, including the Sky Dome stadium (now called Rogers Centre), the famous City Hall with its facing pair of curved high-rises, and mast-like CN Tower, which at 553 meters (1815 ft) is the world's fifth-tallest freestanding structure. Toronto also is a major port and industrial complex, whose facilities line the shore of Lake Ontario.

Livability is one of the first labels Torontonians apply to their city, which has retained more of its middle class than central cities of its size in the United States. *Diversity* is another leading characteristic because this is North America's richest urban ethnic mosaic. Among the largest of more than a dozen thriving ethnic communities are those dominated by Italians, Portuguese, Chinese, Greeks, and Ukrainians; overall, Toronto now includes residents from about 170 countries who speak more than 100 languages; and the immigrant inflow continues steadily, with those born outside Canada constituting fully half of the city's population (in all the world only two other million-plus cities exhibit a higher percentage).

Originally, the city's advantageous location on the northwestern shore of Lake Ontario was a function of its centrality in what has always been Canada's most populated region, its access to regional waterways, and relative proximity to the U.S. core area. Subsequently, it was the advent of the railroad during the late-nineteenth-century industrialization era that solidified Toronto's position as Canada's leading city. Today, the Greater Toronto area produces about one-fifth of Canada's entire economic output and is home to nearly half of the country's corporate headquarters.

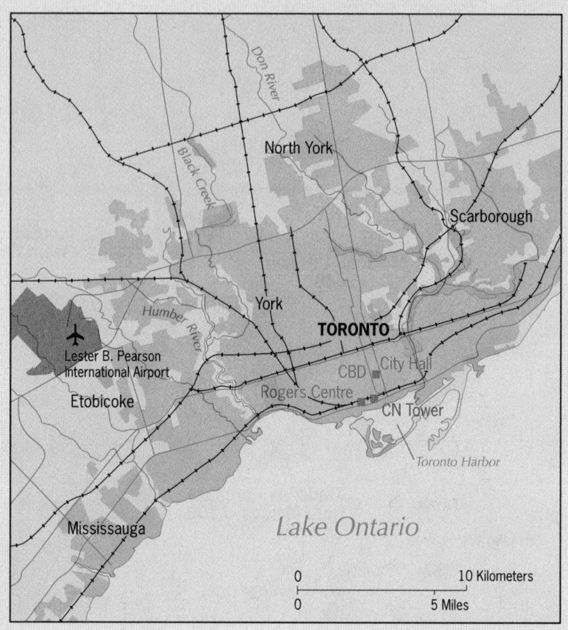

Toronto has functioned successfully in recent decades, thanks to a metropolitan government structure that fostered central city–suburban cooperation. But that relationship is increasingly stressed as the outer city gains a critical mass of population, economic activity, and political clout. Externally, the city has become the centerpiece of the Main Street conurbation (see Fig. 3A-9), which extends for 1000 kilometers (650 mi) from Windsor-Detroit in the southwest through Toronto and Montreal to Quebec City in the northeast.

movement peaked in the 1960s. In violation of the Constitution, a wide range of State and local Jim Crow laws were enacted to codify the repression of African Americans. This historical legacy is still visible today in the geography of race in the United States, with blacks regionally concentrated in the Southeast, from eastern Texas to Maryland (Fig. 3B-4, upper-right map). We noted in Chapter 3A that substantial numbers of blacks migrated north to work in the industrial centers of the American Manufacturing Belt during the second quarter of the twentieth century; major African American population clusters still exist in those northern cities, but they are not clearly apparent at the scale of the national map.

Race continues to be a sensitive issue in the United States, more so than in Canada, because of this terrible history and because until recently a large number of African Americans were effectively denied equal opportunities. The 2008 election of the first black president was a major historical event but even now, following Barack Obama's re-election to a second term, it is still not clear if his presidency has served as a catalyst of progressive change in terms of race relations.

Another highly visible regional expression of ethnic migration concerns the presence of Hispanics in the southwestern United States (Fig. 3B-4, upper-left map). This migration, too, should be understood in historical and geographical contexts. In these parts of the country, many Hispanic place names reflect Mexico's ouster from a vast area acquired in 1848 by the United States at the end of the Mexican War through the Treaty of Guadalupe Hidalgo. This treaty established the Mexican-American border at the Rio Grande and assigned to the United States no less than 1.3 million square kilometers (525,000 sq mi) of Mexican territory, including present-day California, Arizona, Nevada, New

DISTRIBUTION OF MAJOR RACIAL/ETHNIC GROUPS, CONTERMINOUS UNITED STATES, 2010

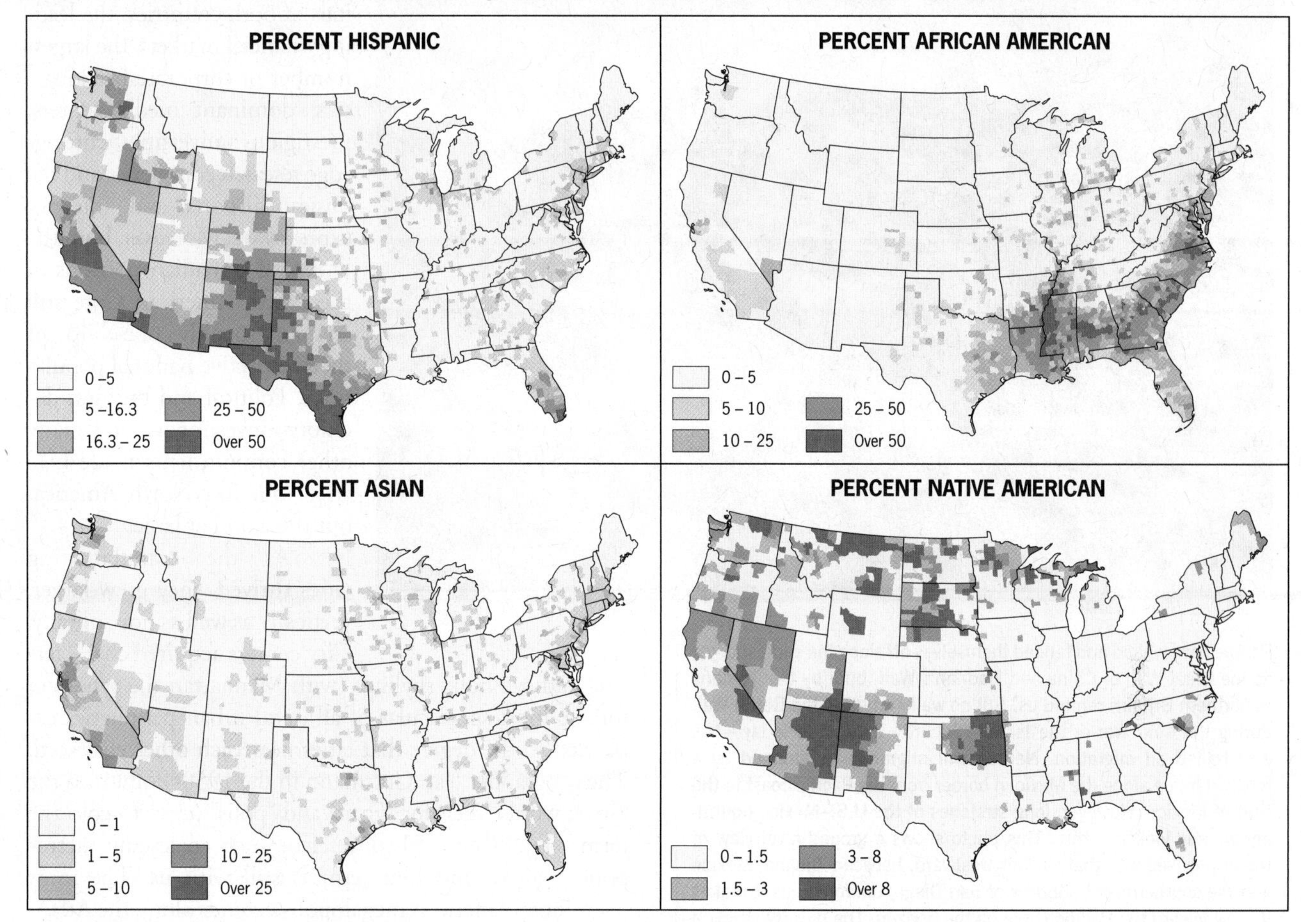

FIGURE 3B-4

Mexico, Texas (which had already declared independence from Mexico), southern Colorado, and Utah (Fig. 3B-3). In Mexico, this disastrous event (which precipitated that country's civil war) is much more vivid in the national consciousness and memory than it is in the United States. Certain Mexican nationalists even proclaim that much of the current migration across the U.S.-Mexican border, illegal though it may be (see photo), is a matter of resettlement of unjustly lost lands.

Since 1980, the number of Hispanics in the United States has more than tripled and now exceeds 55 million, outnumbering the African American minority and changing the ethnic-cultural map of America. Hispanics are still strongly clustered in the Southwest, California, and Florida (see Fig. 3B-4, upper-left map). In fact, along virtually the entire multi-State tier north of the Mexican border, America has become a cultural transition zone (think of the expression "Tex-Mex"), where immigration, legal as well as illegal, has become a contentious issue. Language, not a point of controversy for the African American community, is an emotional matter for some Anglophone Americans who fear erosion of the status of English in their communities and in the country as a whole. Indeed, Hispanics are now dispersing into many parts of the United States—including rural areas in the Southeast and Midwest as well as urban communities from Massachusetts to Washington State. Even some small and dwindling towns in the Midwest and on the Great Plains, in States like Kansas, are now being revived due to the influx of Hispanics.

Finally, the Asian immigrant population, whose arrival rate has also risen in recent decades, is geographically the most agglomerated of the main ethnic minorities (Fig. 3B-4, lower-left map). Chinese remain the most numerous among the 19-plus million Americans who identify themselves as Asian, but this diverse minority also includes Japanese who have been in the United States for generations, Filipinos who began arriving after World War II, Koreans who have migrated since the 1950s, Vietnamese who were resettled in the 1970s following the Indochina War, more gradual arrivals of Pacific Islanders including Hawaiians, and considerable numbers of South Asian Indians in the past few decades. Whatever their origins, the great majority have remained in California, where many have achieved notable economic success and upward mobility. This regional geography of ethnicity within the national context of the United States was largely shaped by migration histories as well as the processes of voluntary or involuntary inclusion and exclusion.

© Jim West/Alamy Limited

People have walled and fenced themselves off since the earliest days, as the Great Wall of China and Hadrian's Wall (built by the Romans in northern Britain) remind us. But no wall—whether the Berlin Wall during the Cold War or the Israeli "Security Barrier" of today—has ever halted all migration. Neither will migration be stopped by a fortified fence along the Mexican border from the Pacific coast to the Gulf of Mexico. However, long stretches of the U.S.–Mexico boundary already look like this. This photo shows a ground-level view of the improvised wall that slashes westward, between Tijuana, Mexico and the southernmost suburbs of San Diego, California, as it makes its way toward the Pacific shore on the horizon. The poverty-stricken landscape of this eastern part of Tijuana extends almost to the rusty barrier itself, while development on the U.S. side has kept its distance from this well-patrolled segment of the border.

REGIONS OF NORTH AMERICA

Now let us consider the various regions within the North American realm on the basis of physical, cultural, and historical geography. Ample justification exists for recognizing the two countries of the North American realm as discrete regions, based on several contrasts we have discussed. There is, however, another way to conceptualize North America's regional geography: by combining functional and formal principles. The result, Figure 3B-1, is by no means the only solution because the drawing of a regional boundary, though well-informed, usually remains arbitrary to a certain degree. But for our general purpose these boundaries are the most pertinent and the following regional distinctions will be helpful in understanding the overall fabric of the realm. As we note in the following regional survey based on our map, there is more to the framework of North America than ethnicity and politics.

THE NORTH AMERICAN CORE (1)

The historic Core region of the North American realm integrates what have long been the most prominent parts of the United States and Canada (Fig. 3B-1). They contain the largest cities and federal capitals of both countries, the leading financial markets, the largest number of corporate headquarters, dominant media centers, prestigious universities, cutting-edge research complexes, and the busiest airports and intercity expressways. Moreover, both the U.S. and Canadian portions of the North American Core still contain roughly one-third of their respective national populations. Political and business decisions, investments, and many other commitments made here affect not just North America, but the world at large.

As the manufacturing cities thrived, they grew larger, vertically as well as horizontally. City centers acquired characteristic and symbolic skylines (with Manhattan the synonym for the most spectacular of all), and urban peripheries expanded to the degree that cities near each other coalesced. The geographer Jean Gottmann in the 1950s combined the Greek words "megalo" (great) and "polis" (city) to coin the term **megalopolis [5]** to describe such coalescing metropolitan areas, and that term is still with us. The most prominent American megalopolis extends along the Atlantic seaboard from north of Boston to south of Washington, D.C. Sometimes also referred to as "Bosnywash," this was the economic anchor of the North American core area: the seat of the U.S. government, the nucleus of business and finance, the hearth of culture, and the trans-Atlantic trading interface between much of the realm and Europe. In Figure 3A-9, note that Canada's predominant megalopolis is its most highly urbanized zone extending from Windsor (adjacent to Detroit) through Toronto to Montreal and Quebec City; urban geographers call this Windsor-Quebec axis ***Main Street***, and it also forms part of the realm's core area.

Nonetheless, the dominance of the historic Core has been declining. As was true of the **American Manufacturing Belt [6]**—the region with which the Core coincides—it once commanded unmatched supremacy. But deindustrialization and the subsequent emergence of the information economy have eroded that dominance as competitors to the south and west continue to siphon away some of its key functions. As already noted, the State of Michigan is experiencing long-term economic decline as jobs in the automobile industry have steadily disappeared over the past few decades. In fact, household incomes in Michigan fell by more than 21 percent just between 2000 and 2009, the most precipitous drop in the nation. Second and third on this unenviable list were two other States of the old Manufacturing Belt: Indiana and Ohio.

AMONG THE REALM'S GREAT CITIES . . .

NEW YORK

NEW YORK IS much more than the largest city of the North American realm. It is one of the most famous places on Earth; it is the hemisphere's gateway to Europe and the rest of the Old World; it is a tourist mecca; it is the seat of the United Nations; it is one of the globe's most important financial centers—in many ways its most prominent world-city.

New York City consists of five boroughs, centered by the island of Manhattan, which contains the CBD south of 59th Street (the southern border of Central Park). Here is a skyscrapered landscape unequaled anywhere, studded with more fabled landmarks, streets, squares, and commercial facilities than any other cities except London and Paris. This is also the cultural and media capital of the United States, which means that the city's influence constantly radiates across the planet thanks to New York-based television networks, newspapers and magazines, book publishers, fashion and design leaders, and artistic and new media trendsetters.

At the metropolitan scale, New York forms the center of a vast urban region—250 square kilometers (150 sq mi) in size, containing a population of 21.1 million—which sprawls across parts of three States in the heart of Megalopolis. That outer city has become a giant in its own right with its population of 13 million, massive business complexes, and flourishing suburban downtowns.

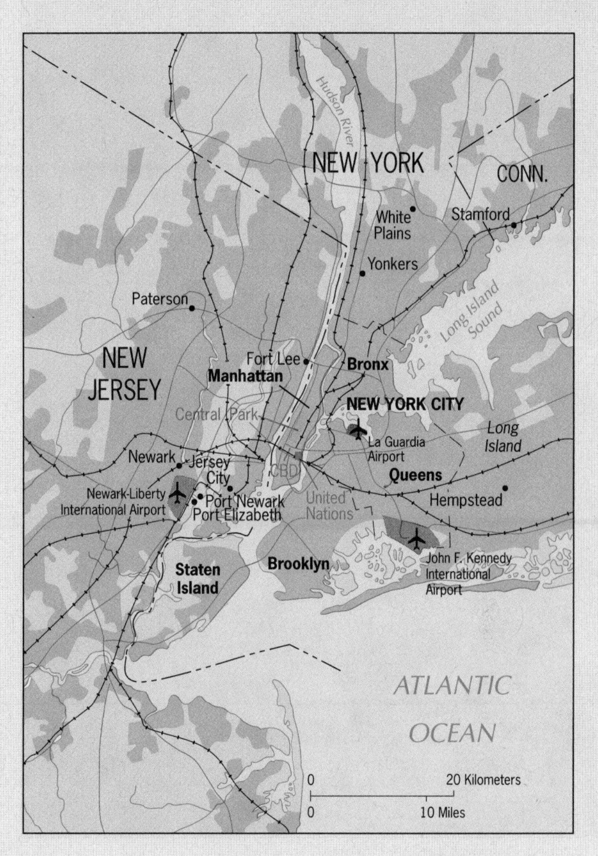

New York's prominence has certainly been put to the test over the past half-century. It was hit hard by waves of deindustrialization that swept across North America during the 1960s and 70s, so much so that bankruptcy was declared in 1975. The population began to shrink while inner city decay advanced. But New York took off again, beginning in the mid-eighties, led by its expanding economic sectors in high- technology, information, and producer services such as accounting, finance, insurance, advertising, and consulting.

Then there was 9/11—the horrendous act of terrorism that profoundly shook New York and the rest of the nation. It changed the Big Apple's skyline forever and left a permanent mark on the city's psyche. It also disrupted and displaced a great deal of economic activity, especially in finance and producer services, but these effects now seem to have been temporary. More recently, the 2008 fiscal crisis that triggered the Great Recession took a major toll on Wall Street (where, arguably, that crisis originated in the first place)—as well as on 'Main Street.'

Ever resilient New York today appears to be prosperous enough, despite widening internal inequalities, its steadily rising cost of living, and all the challenges that come with managing a metropolis of this magnitude. The population overall is growing, crime rates are lower than they have been in decades, and even such aging ghettoes as Harlem are being gentrified and redeveloped.

THE MARITIME NORTHEAST (2)

When you travel north from Boston into New Hampshire, Vermont, and Maine, you move across one of North America's historic culture hearths whose identity has remained strong for four centuries. Even though Massachusetts, Connecticut, and Rhode Island—traditional New England States—share these qualities, they have become part of the Core, so that on a functional as well as formal basis, the Maritime Northeast extends from the northern border of Massachusetts to Newfoundland, thereby incorporating all four Canadian Atlantic Provinces (Fig. 3B-1).

Difficult environments, a maritime orientation, limited resources, and a persistent rural character have combined to slow economic development here. Primary industries—fishing, logging, some farming—are mainstays, although recreation and tourism have boosted the regional economy in recent decades. Upper New England experiences the spillover effect of the Boston area's prosperity to some degree, but not enough to ensure steady and sustained economic growth. Atlantic Canada has endured economic hard times as well in the recent past. For example, overfishing has depleted fish stocks. Alternate opportunities focus on this subregion's spectacular scenery

© Clarence Holmes/Demotix/Corbis

The skyline of Lower Manhattan, with the Statue of Liberty on its island in the foreground, on the evening of September 11, 2012. The parallel beams of the Tribute in Light, representing the fallen Twin Towers, shine into the night sky to conclude the annual ceremonies that memorialize the terrorist attacks of 9/11 (2001).

and the tourism it attracts. Economic prospects have also been boosted in Canada's poorest province, Newfoundland and Labrador, by the discovery of significant offshore oil reserves. Production is now well underway, and in the 2010s additional energy deposits have been discovered both on and around Newfoundland Island.

FRENCH CANADA (3)

Francophone Canada constitutes the inhabited southern portion of Quebec, focused on the St. Lawrence Valley from near Montreal northeastward to the river's mouth (Fig. 3B-1). It also includes the French-speaking Acadians who reside in neighboring New Brunswick (Fig. 3B-2). The French cultural imprint on the region's cities and towns is matched in the rural areas by the narrow, rectangular *long lots* perpendicular to the river, also of French origin.

The economy of French Canada was historically less industrial and more rural than the adjacent Core region (and Toronto far outranks Montreal here). Nevertheless, Montreal's economic role has advanced since 2000, with information technology, telecommunications, and biopharmaceuticals all growing in importance. But Quebec's economic prospects tend to be overshadowed by the region's political uncertainties, and bursts of nationalism (such as laws prohibiting shops from advertising in English) have at times eroded business confidence. When the provincial government sought to address the declining birth rate resulting from accelerating urbanization by encouraging immigration, tolerance for multiculturalism in Quebec proved to be so low that a special commission held hearings to investigate the matter. Parochialism has cost Quebec dearly and sets this region off from the rest of the Canada that virtually encircles it.

The Acadians in the neighboring province of New Brunswick, Canada's largest cluster of French-speakers outside Quebec, take a different view. Not only do they reject the notion of independence for themselves, but they also actively promote all efforts to keep Quebec within the Canadian federation. Accommodation with the Anglophone majority and acceptance of multiculturalism in New Brunswick set an example for the rest of French Canada.

Yet in some ways, French Canada is a very rich part of North America. Northern Quebec produces so much

Small fishing ports still dot the New England coast: Chatham, Massachusetts, at the elbow of Cape Cod.

© H. J. de Blij

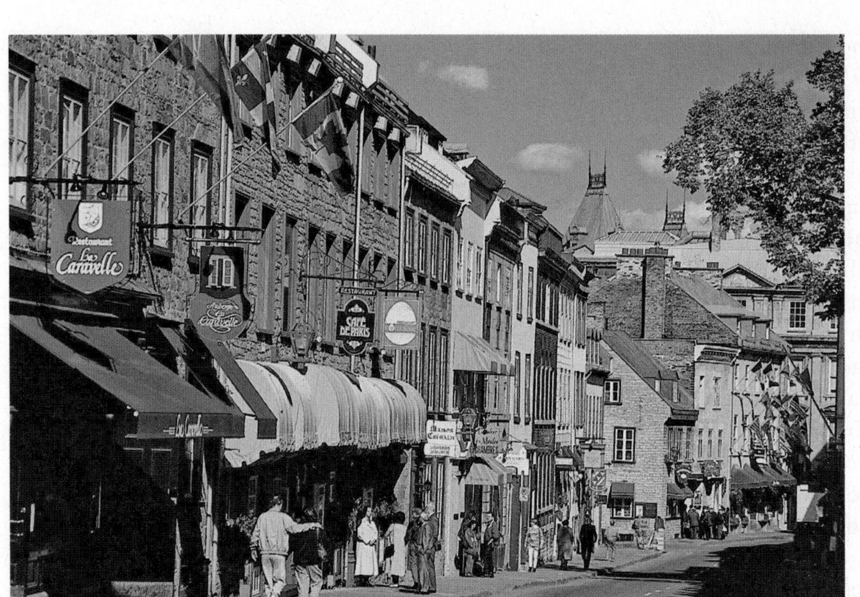

An unmistakably French cultural landscape in North America: the Rue Saint-Louis in Quebec City, the capital of Quebec Province. Not an English sign in sight in this center of French/Québécois culture. The green spires in the background belong to Le Chateau Frontenac, a grand hotel built on the site of the old Fort St. Louis (the Count of Frontenac was a prominent governor of what was then known as New France). French-speaking, Roman Catholic Quebec lies at the heart of French Canada.

hydroelectricity that it exports the surplus to Ontario and to the northeastern United States. Most of the power is generated through the James Bay Project, a series of dams and power stations on the eastern side of Hudson Bay along the La Grande River (Fig. 3B-2). Quebec is so well off in this regard that the recent discovery of natural gas in the St. Lawrence River Basin was met with tempered enthusiasm. Many Québécois feel that they have no need for gas; unlike hydroelectric power, it is a nonrenewable energy resource that would boost carbon emissions and heighten the risk of widespread water pollution. Not surprisingly, these views are disputed by energy-industry interests, and it remains to be seen if drilling can begin in the foreseeable future.

THE SOUTHEAST (4)

For more than a century following the U.S. Civil War (1861–1865), the Southeast (Fig. 3B-1) remained in economic stagnation and cultural isolation from the rest of the country. During the 1970s, however, things changed so fast that a New South was born almost overnight. The Sunbelt migration stream drove people and enterprises into the long-dormant cities. Core-region companies looking for headquarters or subsidiary offices found Atlanta, Charlotte, Tampa, Miami-Fort Lauderdale, and other urban areas both economical and attractive, swiftly turning them into boomtowns. Racial segregation had been dismantled in the wake of the civil rights movement. A new social order was matched by new facilities ranging from airports (Atlanta's quickly became one of the world's busiest) to theme parks open to all. Soon the nation was watching Atlanta-based CNN on television, vacationing at Orlando's Walt Disney World, and monitoring space flights that originated at Cape Canaveral.

The geography of development in the South, however, is rather uneven. While many cities and certain agricultural areas have benefited, others have not; the South contains some of the realm's poorest rural areas, and the gap between rich and poor here is wider than in any other region in the realm. Northern Virginia's Washington suburbs, central North Carolina's Research Triangle, eastern Tennessee's Oak Ridge complex, and metropolitan Atlanta's corporate campuses are locales that typify the New South; but Appalachia and rural Mississippi still represent the Old South, where depressed farming areas and stagnant small industries restrict both incomes and change.

Climatically, the Southeast is notably warmer than the Core region to the north and more humid than the Southwest. It is not for nothing that this is where the tobacco, cotton, and sugar plantations thrived. But the summer heat brings a serious threat to the Gulf-Atlantic Coastal Plain of the South: hurricanes capable of inflicting catastrophic devastation on low-lying areas. These tropical cyclones are fueled by the rapidly rising air over the warm waters of the Gulf of Mexico and the Caribbean that reach peak temperatures in late summer. Coastal Florida and southern Louisiana have been particularly hard hit during the short span of years since the turn of the century.

A major disaster of another kind struck the Gulf Coast in 2010: an explosion at a British Petroleum rig about 100 kilometers (60 mi) southeast of the Mississippi Delta caused the worst oil spill in U.S. history. It took BP five months to seal the resulting seafloor leak, but not before an estimated 4.9 million barrels of crude oil spewed into the Gulf, with much of it washing ashore from western Louisiana as far east as Florida's Panhandle. The damage inflicted on wildlife, fisheries, and local industries (including tourism) was considerable, and this environmental disaster fully exposed the enormous risks of deep-sea oil extraction. As a result, prospects for expanding offshore drilling elsewhere in U.S. waters were dealt a huge setback.

There is another, relatively small part of this region that warrants special attention. South Florida used to be a particularly remote and nearly unpopulated corner of the United States. Before the twentieth century, it was too far from the northeastern Core of the country (and even from closer urban centers like Atlanta) to warrant development; the climate and mosquitoes posed a formidable challenge to prospective settlers; and there were no natural resources to exploit. Indeed, it was not until the 1960s that this subregion of the Southeast became fully integrated into the national economy and urban system—thanks to the widespread introduction of air conditioning and affordable air travel, not to mention the mass immigration of Cubans fleeing Fidel Castro's communist regime.

In short order, burgeoning South Florida became part of an international region that encompassed the Caribbean Basin and much of Middle and South America. Located south of America's "Deep South," today it continues to evolve into an interface between North America and the rest of the hemisphere. South Florida's largest urban center, Miami, emerged in the 1980s as an important *world-city* (Fig. 3B-5). This new role was certainly a result of the growth of its large, well-educated Hispanic population, but that is not the entire story. Take another look at the world map in Figure G-1 and note Miami's longitude: it is not only located at the southern end of North America's Atlantic seaboard, but also lies on the same meridian as Guayaquil, Ecuador, the westernmost major city in South America. Now compare Miami's relative location to that of Los Angeles (three time zones to the west) and it becomes clear why Miami is far better connected to countries such as Brazil, Argentina, and Colombia—while L.A.'s southern linkages are almost entirely with Mexico. In fact, Los Angeles is much more heavily oriented to the Pacific Rim, whereas Miami is the realm's most important world-city focused on the Americas.

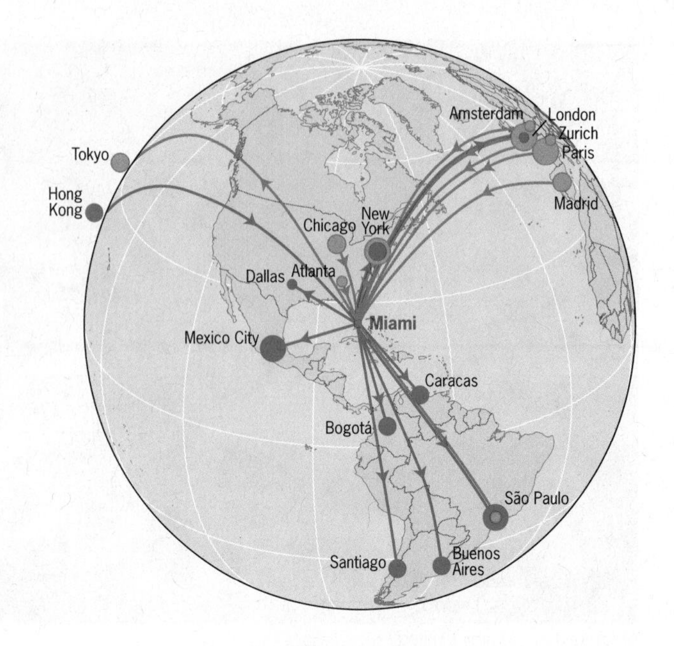

FIGURE 3B-5 © H. J. de Blij, P. O. Muller, J. Nijman, and John Wiley & Sons, Inc.

From the Field Notes . . .

"Domino Park lies at the center of Little Havana, perhaps the most iconic neighborhood in Greater Miami. After Fidel Castro came to power in 1959, thousands of Cubans fled the island and resettled here. Today there are more than 900,000 Cubans in the metropolitan area, the biggest concentration of any single ethnic group in the United States. In fact, Miami would qualify as the second-largest city in Cuba. Domino Park is situated on Southwest 8th Street, better known locally as 'Calle Ocho.' It is only a short walk from the Bay of Pigs Monument that commemorates the fallen soldiers of the failed invasion of Cuba by exiles in 1961. Taking a stroll along Calle Ocho, you can stop at a cafe for a *cortadito*, watch cigars being made by hand, or get fitted for a guayabera shirt. Domino Park recently opened to women (!), but it is mostly older men who come here to play games, talk politics, and meet with friends—just as they would in Cuba."

© Jan Nijman

www.conceptcaching.com

THE SOUTHWEST (5)

The Southwest has taken on its own regional identity in recent years, and today it is firmly established in the North American consciousness for several reasons. One is environmental: here, steppes and deserts dominate. Another is cultural: Anglo, Hispanic, and Native American cultures coexist here, sometimes amicably, at other times uneasily.

As Figure 3B-1 shows, by our definition the Southwest begins in East Texas and extends all the way to the eastern edge of Southern California. Texas leads this region in most respects. Its economy, once heavily dependent on oil and natural gas, has been restructured and diversified so that today the Dallas-Forth Worth–Houston–San Antonio triangle has become one of the world's most productive **technopoles [7]**—state-of-the-art, high-technology industrial complexes—including Austin at the heart of it. This is also a hub of international trade and the northern anchor of a NAFTA-generated transnational growth corridor that extends into Mexico as far as Monterrey.

New Mexico, in the middle of the Southwest region, is the least developed economically and ranks low on many U.S. social indices, but its environmental and cultural attractions benefit Albuquerque and Santa Fe, with Los Alamos a famous name in research-and-development circles. In the west, Arizona's technologically transformed desert now harbors two large coalescing metropolises, Phoenix and Tucson. Growth continues, but rising concerns about climate change add to the uncertainties of maintaining water supplies already stretched to the limit.

The huge State of Texas, larger than France, is surpassed in areal size only by Alaska and is also the second most populous U.S. State after California (27 and 39 million, respectively). It is a most affluent State, with much of its wealth still emanating from its energy sector: Texas claims about a quarter of all U.S. oil reserves and nearly a third of its natural gas deposits. Today, the Lone Star State has become a leader in the generation of wind energy as well, with wind turbines increasingly dotting the sparsely settled landscape of West Texas.

Texas and the other southwestern States border Mexico, and all are home to substantial Mexican populations, both legal and illegal. In Arizona, the Hispanic population has nearly tripled since 1990 as the majority of non-Hispanic whites dwindled from 72 to 56 percent. Among the elderly almost 80 percent are non-Hispanic whites, but the State's 18-and-younger age cohort is now 45 percent Hispanic. In 2010, Arizona passed a widely debated (and criticized) law making it mandatory for people to carry proof of legal residency, failure of which can result in detention and presumably deportation. Much of the debate centered on the apparent inevitability of **racial profiling [8]** this would involve (i.e., who will be asked to produce such documentation?). Despite legal challenges, the mood among Arizonans seemed clear: 81 percent of registered voters were in favor of the law.

VOICE FROM THE Region

© Cinthia Carvajal

Cinthia Carvajal, student in Anthropology and Sustainability, Arizona State University

ILLEGAL IMMIGRATION IN ARIZONA

"Senate Bill 1070 was signed into law in Arizona in 2010. This law was created to reduce illegal immigration into the State, where crossing the international border without documents has occurred for years. As a Hispanic immigrant living in Arizona, I can see how the law has affected friends, family, and other people around me. Basically, it has caused polarization in Arizona at a scale not seen before. Conversations about immigration and the new law are not pleasant. The law has divided the State, has divided its classrooms, and has divided people who once coexisted in peaceful and respectful ways. Those who support the law are convinced it is the best solution and has no effects on documented and law-abiding immigrants—that it just targets those who are undocumented. Others oppose the law; they are people of many races and ethnicities, but most are documented immigrants or children of immigrants whose background can be seen in their faces or heard in their accents. People like me. The ambiguity of the law has caused many to leave the State. Those of us who stay behind live every day with that little doubt in our minds: who is going to be the next person to question my 'legality'?"

THE PACIFIC HINGE (6)

The Southwest meets the Far West near metropolitan San Diego on California's border with Mexico, and from there the region we call the Pacific Hinge extends all the way north into Canada where Vancouver forms the northern anchor in southwestern British Columbia (Fig. 3B-1). We include almost all of California in this region, but (for environmental as well as economic reasons) only the western portions of Oregon and Washington State. This is a particularly important part of North America and increasingly a counterweight to the historic Core in the U.S. Northeast.

The Pacific Hinge is a major economic region not just in this realm but in the global economy too. It includes such leading cities as Los Angeles, San Francisco, San Diego, Portland, and Seattle as well as America's most populous State: in fact, California's economy ranks among the world's ten largest *by country*. This region also encompasses one of the realm's most productive agricultural areas in California's Central Valley, magnificent scenery, and a culturally diverse population drawn by its long-term economic growth and pleasant living conditions.

Our regional definition is mainly based on economic considerations and intensifying trade connections across the Pacific Ocean. This part of North America is deeply involved in the development of countries in eastern Asia. In the postwar era, Japan's success had a salutary impact here, but what has since taken place in China, South Korea, Taiwan, Singapore, and other Asian-Pacific economies has created unprecedented opportunities. The term **Pacific Rim [9]** has come into use to describe the discontinuous regions surrounding the great Pacific Ocean that have experienced spectacular economic growth and progress over the past four decades: not only coastal China and various parts of East and Southeast Asia, but also Australia, South America's Chile, and the western shores of Canada and the conterminous United States. That is why in this chapter we speak of the *Pacific Hinge* of North America, now the key interface between the realm and the Rim.

This development is expressed in the growing importance of the region's ports. Since 1995, the seaports of Los Angeles and nearby Long Beach have seen their container traffic double; and despite the global recession, a recent report predicted another doubling of traffic by 2030. Moreover, the region's airports (especially Los Angeles and San Francisco) annually handle millions of Asia-bound passengers as well as a steadily growing number of cargo shipments.

The Pacific Rim is therefore a classic example of a functional region, with economic activity in the form of capital flows, raw-material movements, and trading linkages that generate urbanization, industrialization, and labor migration. As part of this process, human landscapes from Sydney to Santiago are being transformed within a 32,000-kilometer (20,000-mi) corridor that lines the globe's largest body of water.

Greater Los Angeles, the region's largest and most dynamic activity complex, is profiled in the *Among the Realm's Great Cities* box. Farther north, the San Francisco Bay Area has seen rather more orderly expansion, with the Silicon Valley technopole the key component of its high-technology success story. In what Americans refer to as the Pacific Northwest, the Seattle-Tacoma area originally benefited from cheap hydroelectricity generated by Columbia River dam projects, attracting aluminum producers and aircraft manufacturers (most notably Boeing). Meanwhile, the technopole centered on suburban Redmond (Microsoft's headquarters) heralds the Seattle area's transformation into a prototype metropolis of the new information economy.

Vancouver, located at the northern edge of the Pacific Hinge, has the locational advantage of being closer, on air and sea routes, to East Asia than any other large North American city. Ethnic Chinese here constitute 20 percent of the metropolitan population, and the total number of resident Asians now exceeds 50 percent.

From the Field Notes . . .

"On this street corner in central Los Angeles, four different cultures converge. Los Angeles is one of the most ethnically diverse metropolitan areas in the United States and in the world. More than a third of the population was born abroad, and more than half speak a language other than (or in addition to) English. The largest non-native ethnic group are Hispanics (mainly Mexicans and Central Americans), but there also are many Asians, especially Chinese and Koreans. Cesar E. Chavez, after whom this avenue is named, was a charismatic Mexican American farm worker turned union organizer and became nationally known as a leading civil rights activist. His birthday, March 31, is a State holiday in both California and Texas—and an optional holiday in Arizona (!). Chavez Avenue is a major east-west thoroughfare through downtown Los Angeles, and even a short stroll along it offers a sampling of global cultures.

© Jan Nijman

www.conceptcaching.com

AMONG THE REALM'S GREAT CITIES . . .

LOS ANGELES

IT IS HARD to think of another city that has been in more headlines in the past several years than Los Angeles. Much of this publicity has been negative. But despite its earthquakes, mudslides, wildfires, pollution episodes, showcase trials, and riots, L.A.'s glamorous image has hardly been dented. This is compelling testimony to the city's resilience and vibrancy, and to its unchallenged position as the Western world's entertainment capital.

The plane-window view during the descent into Los Angeles International Airport, almost always across the heart of the metropolis, gives a good feel for the immensity of this urban landscape. It not only fills the huge natural amphitheater known as the Los Angeles Basin, but it also oozes into adjoining coastal strips, mountain-fringed valleys and foothills, and even the margins of the Mojave Desert more than 90 kilometers (60 mi) inland. This quintessential spread city, of course, could only have materialized in the automobile age. In fact, most of it has been built rapidly since 1920, propelled by the swift postwar expansion of a high-speed freeway network unsurpassed anywhere in metropolitan America. In the process, Greater Los Angeles became so widely dispersed that today it has reorganized within a sprawling, multinodal geographic framework (see Fig. 3A-11).

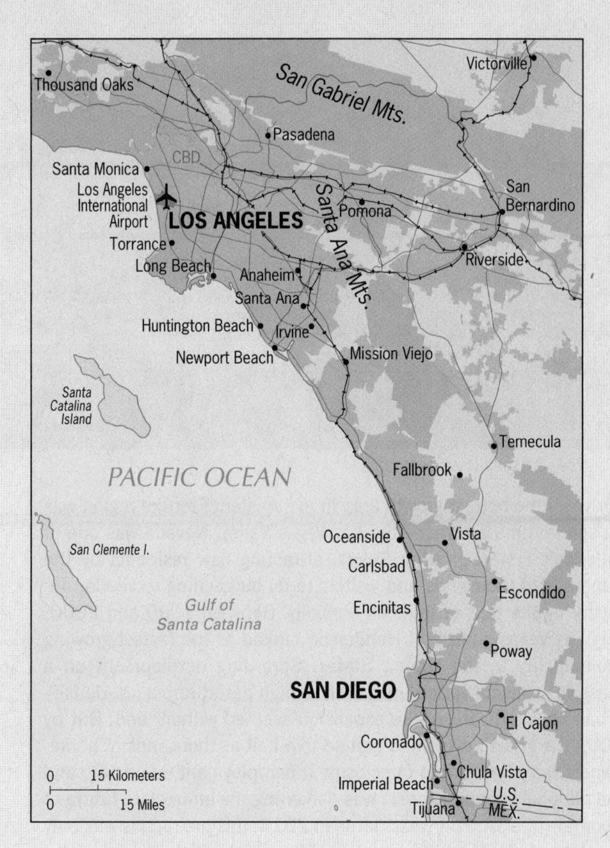

The metropolis as a whole is home to 13.2 million Southern Californians, constitutes North America's second-largest urban agglomeration, and forms the southern anchor of the huge California megalopolis that parallels the Pacific coast from San Francisco southward to the Mexican border. It also is the Pacific Hinge's leading trade, manufacturing, and financial center. In the global arena, Los Angeles is the eastern Pacific Rim's dominant city and the origin of the greatest number of transoceanic flights and sailings to Asia.

THE WESTERN FRONTIER (7)

Where the forests of the Pacific Hinge yield to the scrub of the rain shadow on the inland slopes of the mountain wall that parallels the west coast, and the terrain turns into intermontane ("between the mountains") basins and plateaus, lies a region stretching eastward from the Sierra Nevada and Cascades to the Rockies, encompassing segments of southern Alberta and British Columbia, eastern Washington State and Oregon, all of Nevada, Utah, and Idaho plus the western components of Montana, Wyoming, and Colorado. As Figure 3B-1 shows, Edmonton, Calgary, and Denver are situated along the inland edge of this aptly named Western Frontier region, and Salt Lake City, anchoring the Software Valley technopole and symbolizing the new high-tech era, lies at the heart of it.

Remoteness, dryness, and sparse population typified this region for a very long time. It became the redoubt of the Mormon faith and a place known for boom-and-bust cycles of mining, logging, and, where possible, livestock raising. However, in recent decades advances in communications and transportation technologies, sunny climates, wide open spaces, lower costs of living, and growing job opportunities have combined to form effective pull factors for myriad outsiders. Every time an earthquake struck in California, eastward migration got a boost. And during the most recent ten-year period, millions of Californians emigrated from the overcrowded, overpriced, and recession-plagued Pacific coast to the inland high desert.

Development in the Western Frontier centers on its widely dispersed urban areas, which are making this the realm's fastest growing region (albeit from a low base). Here thousands of new high-technology manufacturing and specialized service jobs propelled a two-decade-long influx, slowed but not reversed by the Great Recession. Yet this growth has not come without problems as locals see land values rise, prevailing conservative ideas challenged, and traditional lifestyles endangered.

Ethan Miller/Getty Images, Inc.

For years the hottest growth area in the Western Frontier region was its southernmost corner—the Las Vegas Valley. Nevada was one of America's fastest-growing States, attracting new residents by the hundreds of thousands and visitors to its burgeoning recreation industry by the tens of millions annually. Between 1990 and 2000, this Las Vegas suburb of Henderson ranked as the fastest-growing municipality in the United States. Sprawling development on a massive scale converted this patch of high desert into a low-density urban checkerboard whose expansion seemed without end. But by 2009 the growth cycle had ground to a halt as thousands of homeowners were forced into foreclosure, unemployment was rising, and the national economic crisis was darkening the immediate future of Boomtown, USA. Four years later, in 2013, this metropolis was only slowly emerging from the depths of the crisis.

One of the fastest-growing metropolitan areas in North America, Las Vegas, lies on the border between the Western Frontier and Southwest regions. Far more than a gambling and entertainment magnet that draws some 40 million visitors annually, Las Vegas attracts immigrants because of its history of job creation, relatively cheap land and low-cost housing, moderate taxes, and sunny if seasonally hot weather. No doubt, proximity to Southern California has had much to do with its explosive growth (which has leveled off since 2008—see the photo above), but Las Vegas is also known far and wide as the ultimate frontier city—where (almost) anything goes.

■ THE CONTINENTAL INTERIOR (8)

The Western Frontier meets the Continental Interior region along a line marked by cities that influence both regions, from Denver in the south to Edmonton in the north. This vast North American heartland extends from interior Canada to the borders of the Southeast and Southwest (Fig. 3B-1). The region contains many noteworthy cities, including Kansas City, Omaha, Minneapolis, and Winnipeg; nonetheless, agriculture is the dominant story here. As the map of farm resource regions (Fig. 3B-6) shows, this is North America's breadbasket. This is the land of the beef- and pork-producing Corn Belt—*Meat Belt* would be a more accurate label—which extends into the western half of the Core region (and, thanks to global warming, has spawned a new outlier in the southernmost wheatlands of the Canadian Prairie Provinces); of the mighty soybean, America's and the world's most rapidly expanding crop of the past half-century; and of spring wheat in the Dakotas and Canada's Prairie Provinces as well as winter wheat in Kansas. Indeed, the cities and towns of this region share histories of food processing, packing, and marketing; flour milling; and soybean, sunflower, and canola oil production. This is also the scene of the struggle for survival of the family farm as the unrelenting incursion of large-scale corporate farming threatens a longstanding way of life.

This may be farmland, but it is not immune from what might seem to be the nonagricultural realities of the contemporary world, not even the energy issue. Corn continues to be grown mainly as feed for meat-producing hogs and beef cattle, but a sizeable proportion of the crop is now dedicated to making ethanol, the gasoline substitute (or additive). Whereas farmers hope that rising prices will revitalize the Corn Belt, others worry that accelerating corn prices will increasingly strain meat-loving consumers. In any case, the amount of corn needed to make a real dent in U.S. gasoline consumption is many times greater than the entire annual crop. In 2011, when more than half of the corn grown in the Midwest was converted into biofuels, ethanol accounted for about 10 percent of national gasoline consumption while raising alarms about possible food prices.

Even a cursory look at demographic statistics makes it clear why many States in the Continental Interior are losing population or gaining far fewer than the national average. The Great Plains, the western component of this region, is especially hard hit: younger people and those better off are leaving, whereas older and less affluent residents stay behind. Villages and small towns continue to die, and the notion of abandoning certain areas to return them to their natural state is being seriously discussed.

But things are not the same everywhere. The Continental Interior may not have as prominent a presence on the national scene as New York or California, and the region is not as wealthy, but coal-rich Wyoming and oil-booming North Dakota had the highest increases in household income in the United States during the first decade of this century (at 8.0 and 7.4 percent, respectively—while the national figure was *negative* 7.1 percent). Moreover, unemployment rates here rank among the lowest in the

U.S. FARM RESOURCE REGIONS

1 BASIN AND RANGE
Largest share of nonfamily farms, smallest share of U.S. cropland.
4 percent of farms, 4 percent of value of production, 4 percent of cropland.
Cattle, wheat, and sorghum farms.

2 EASTERN UPLANDS
Most small farms of any region.
15 percent of farms, 5 percent of production value, and 6 percent of cropland.
Part-time cattle, tobacco, and poultry farms.

3 FRUITFUL RIM
Largest share of large and very large family and nonfamily farms.
10 percent of farms, 22 percent of production value, 8 percent of cropland.
Fruit, vegetable, nursery, and cotton farms.

4 HEARTLAND
Most farms (22 percent), highest value of production (23 percent), and most cropland (27 percent).
Cash grain and cattle farms.

5 MISSISSIPPI PORTAL
Higher proportions of both small and larger farms than elsewhere.
5 percent of farms, 4 percent of value, 5 percent of cropland.
Cotton, rice, poultry, and hog farms.

6 NORTHERN CRESCENT
Most populous region.
15 percent of farms, 15 percent of value of production, 9 percent of cropland.
Dairy, general crop, and cash grain farms.

7 NORTHERN GREAT PLAINS
Largest farms and smallest population.
5 percent of farms, 6 percent of production value, 17 percent of cropland.
Wheat, cattle, and sheep farms.

8 PRAIRIE GATEWAY
Second in wheat, oat, barley, rice, and cotton production.
13 percent of farms, 12 percent of production value, 17 percent of cropland.
Cattle, wheat, sorghum, cotton, and rice farms.

9 SOUTHERN SEABOARD
Mix of small and larger farms.
11 percent of farms, 9 percent of production value, 6 percent of cropland.
Part-time cattle, general field crop, and poultry farms.

Modified after USDA, ERS, Agric. Info. Bull. No. 760.

FIGURE 3B-6

country. With an economy largely driven by agriculture and mining, and with oil and gas exploration a leading growth industry, the effects of the Great Recession were minimal in this region.

THE NORTHERN FRONTIER (9)

Figure 3B-1 leaves no doubt as to the dimensions of this final North American region: it is by far the largest of the realm, covering just about 90 percent of Canada and all of the biggest U.S. State, Alaska. Not only does this region include the northern parts of seven of Canada's provinces, but it also comprises the Yukon and Northwest Territories as well as recently established Nunavut. The very sparsely populated Northern Frontier remains a land of isolated settlement based on the exploitation of newly discovered resources. The Canadian Shield, underlying most of the eastern half of the region, is a rich storehouse of mineral resources, including metallic ores such as nickel, uranium, copper, gold, silver, lead, and zinc. The Yukon and Northwest Territories have proven especially bountiful, with the mining of gold and especially diamonds (Canada currently ranks third in the world); at the opposite, eastern edge of the Shield, near Voisey's Bay on the central coast of Labrador, the largest body of high-grade nickel ore ever discovered is now open for extration.

The province of Alberta is experiencing one of the most spectacular economic booms in the realm's history. Recent estimates indicate that Alberta's oil reserves, including the **tar sands [10]** around Fort McMurray (see Fig. 3B-7 and the chapter's final photo), rank third in the world behind only those of Venezuela and Saudi Arabia. Alberta has become Canada's fastest-growing province, and Calgary its fastest-expanding major city. The boom makes Calgary a magnet for business and highly skilled workers; the city's skyline is in spectacular transformation, comparable to what has recently occurred on the Arabian Peninsula. The big question is how long these good times will last. Productive locations in the Northern Frontier constitute a far-flung network of mines, oil and gas installations, pulp mills, and hydropower stations that have spawned hundreds of small settlements as well as thousands of kilometers of interconnecting transportation and communications lines. Inevitably, these activities have infringed on the lands of indigenous peoples without the preparation of treaties or agreements, which has led to recent negotiations between the government and leaders of First Nations over resource development in the Northern Frontier. But essentially this

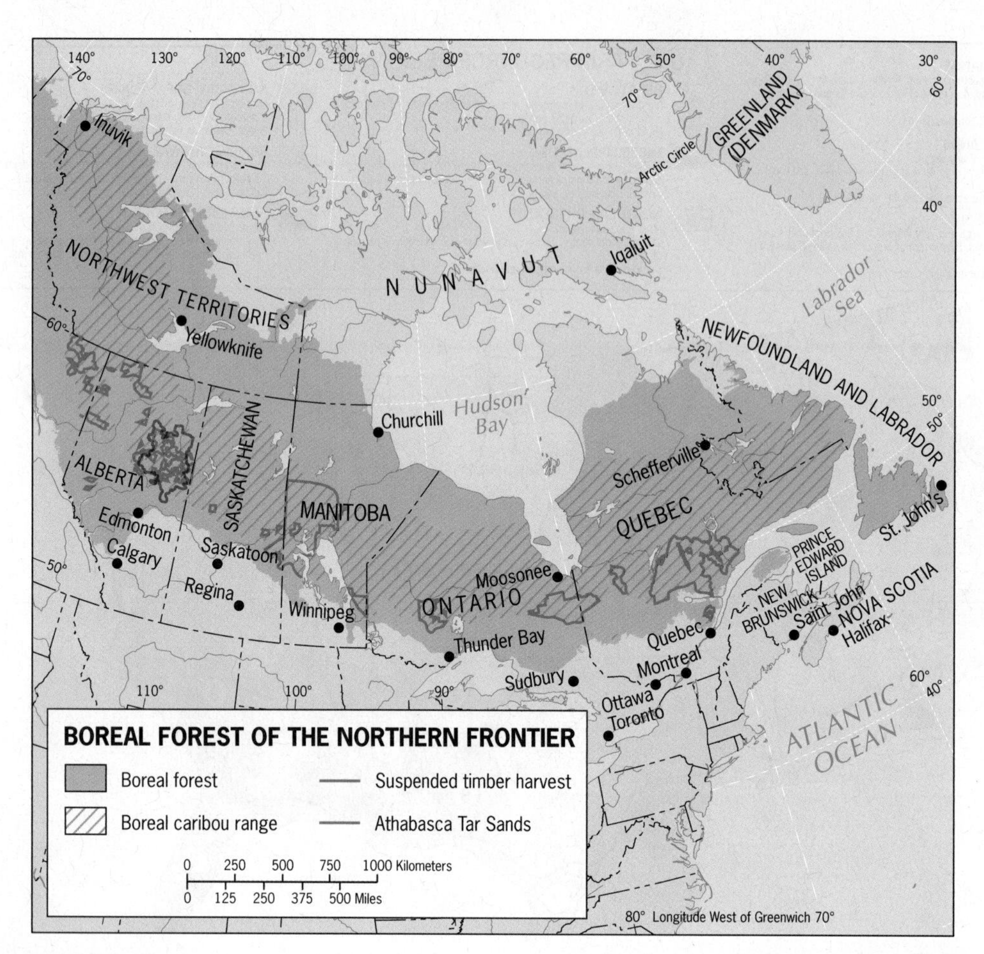

FIGURE 3B-7

vast region remains a frontier in the truest sense of that term.

Alaska's regional geography differs from the rest of the Northern Frontier in that the State contains several urban settlements and an incipient core area in the coastal zone around Anchorage. The population of Alaska, which has just passed the 750,000 milestone, accounts for more than one-third of the region's total. Here also lies the only metropolis of any size, Anchorage (population 300,000). The State's internal communications (air and surface) are also better developed.

Another difference lies in Alaska's North Slope oil exploitation, one of the hemisphere's key energy sources since the 1300-kilometer (800-mi) Trans-Alaska Pipeline began operating in 1977. The North Slope is the lowland north of the Brooks Range that faces the Arctic Ocean. Dwindling supplies at the main reserve at Prudhoe Bay are now compelling producers to turn to huge additional reserves in Alaska's northland, but opposition from preservationist organizations has slowed their plans. The issue remains in contention—as is another project to construct a natural gas pipeline to parallel the oil pipeline to carry gas, a byproduct of oil drilling, to the growing market in the Lower-48.

Climate change is likely to affect the Northern Frontier in as-yet-unforeseeable ways. As the map shows, this region extends all the way north to the Arctic Ocean and therefore includes waters that, in the event of longer-term sustained global warming and associated contraction of the Arctic ice cap, will open, possibly year-round, near-shore waterways that would dramatically alter global shipping routes and intercontinental distances (a topic discussed in Chapter 12). The question of ownership of these waters—and the submerged seafloor beneath them, coming within reach of exploitation—will be a matter of significant international contention in the decades ahead.

In 2010, Canada solidified its reputation as an environmentally conscious nation through the signing of an unprecedented agreement. The Canadian Boreal Forest Agreement involves 21 logging firms and 9 environmental organizations as well as the Canadian government. As shown in Figure 3B-7, it imposes a moratorium on logging and any other development within a huge zone of 290,000 square kilometers (112,000 sq mi). Greenpeace, which originated in Canada, has called it the all-time biggest agreement to preserve nature. The word "boreal" is derived from the Latin word for north; **boreal forests [11]**—identical to the Russian *taiga* (snowforest) discussed in Chapter 2A—

© Jim Wark/AirPhoto

As the price of a barrel of oil (and a gallon of gasoline) rises, it becomes profitable to derive oil from sources other than liquid reserves. The Canadian province of Alberta contains vast deposits of "oil sands" in which the petroleum is mixed with sand, requiring a relatively expensive and complicated process to extract it. The quantity of oil locked in these Athabasca Tar Sands is estimated to constitute one of the world's largest reserves, and this huge open-pit mining project is under way around the town of Fort McMurray to recover it. Given the enormous size of the area to be mined, exploitation is not without its environmental critics. Those who assume that Canada will sell this oil to the United States should be aware that efforts are underway to overcome the obstacles blocking the construction of pipelines across the mountains to Canada's Pacific coast. If successful, these would open up Asian markets, led by energy-voracious China; thus the rising cost of oil has political as well as economic ramifications.

are dominated by dense stands of coniferous (cone-bearing) needleleaf trees such as spruce, fir, and pine. The boreal forests of the Northern Frontier also constitute the habitat of such endangered wildlife species as the caribou, lynx, American black bear, and wolverines.

POINTS TO PONDER

- California's ports are growing rapidly thanks to Pacific Rim trade, but how will they be affected by the newly widened Panama Canal beginning in 2015?
- More than 750,000 foreign students annually attend U.S. colleges and universities; in academic year 2011–2012, 39 percent of them came from just two countries—China and India.
- In 2012, 57 percent of all mortgaged homes in the State of Nevada had market values that were less than the balance of the mortgage.
- Counting all sources—onshore as well as offshore, proven oil and gas reserves, and particularly Alberta's tar-sands deposits—Canada now ranks among the world's top three countries in actual and potential fossil-fuel energy resources.

THE MIDDLE AMERICAN REALM
UNITED STATES
MEXICO
GUATEMALA
EL SALVADOR
PACIFIC OCEAN
Gulf of Mexico
Bay of Campeche
Gulf of California (Sea of Cortés)
Gulf of Tehuantepec
BAJA CALIFORNIA
SONORAN DESERT
ROCKY MOUNTAINS
EDWARDS PLATEAU
COASTAL
SIERRA MADRE OCCIDENTAL
SIERRA MADRE ORIENTAL
SIERRA MADRE DEL SUR
ISTHMUS OF TEHUANTEPEC
PETÉN
YUCATAN
Tropic of Cancer
Channel Is.
Salton Sea
Colorado
Rio Grande
Arkansas
Mississippi
SAN DIEGO
TIJUANA
Mexicali
Ensenada
PHOENIX
Tucson
Nogales
Las Cruces
El Paso
CIUDAD JUÁREZ
Amarillo
Lubbock
Oklahoma City
Little Rock
Fort Worth
DALLAS
Austin
SAN ANTONIO
HOUSTON
Galveston
Beaumont
Baton Rouge
Corpus Christi
Del Rio
Ciudad Acuña
Nuevo Laredo
Laredo
Monclova
McAllen
Reynosa
Brownsville
Matamoros
Saltillo
Guadalupe
MONTERREY
Torreón
Chihuahua
Hermosillo
Tiburón Island
Los Mochis
Culiacán
Durango
Mazatlán
La Paz
Cabo San Lucas
False Cape
Cape San Quintin
Guadalupe (MEXICO)
Archangel Island
Cedros Island
Eugenia Point
Ciudad Victoria
Zacatecas
San Luis Potosí
Aguascalientes
Tampico
Tepic
Marias Islands
GUADALAJARA
LEÓN
Guanajuato
Cape Corrientes
Puerto Vallarta
Querétaro
Poza Rica
Pachuca
MEXICO CITY
Morelia
Colima
Toluca
PUEBLA
Xalapa
Veracruz
Popocatépetl (Vol.) (5,465 m, 17,930 ft.)
Pico de Orizaba (Vol.) (5,747 m, 18,855 ft.)
Revillagigedo Islands (MEXICO)
Tuxtepec
Coatzacoalcos
Chilpancingo
Oaxaca
Acapulco
Salina Cruz
Campeche
Ciudad del Carmen
Villahermosa
Ciudad Pemex
Monclova
Tuxtla Gutiérrez
Tapachula
Quetzaltenango
Guatemala City
Santa Ana
San Salvador
Archipiélago de Colón (Galápagos Islands) (ECUADOR)
Pinta I.
Marchena
120°
110°
100°
30°
20°
10°
0°
Elevation (m)
3000
1500
600
300
150
0
below sea level
0
-150
-1500
-3000
-6000
Roads
Railroads
0 km 200 400 600 800
0 miles 100 200 300 400 500
Albers Equal-Area Projection
Scale 1:15,000,000

Chattanooga
Charlotte
Columbia
Atlanta
Birmingham
Montgomery
P L A I N
Charleston
Savannah
Wilmington
Cape Fear
Cape Lookout
Tallahassee
Jacksonville
Orlando
Cape Canaveral
Tampa
St. Petersburg
Sarasota
Lake Okeechobee
Fort Myers
West Palm Beach
Fort Lauderdale
Miami
The Everglades
Florida Keys
Straits of Florida
Hamilton
Bermuda Islands (U.K.)
ATLANTIC OCEAN
Tropic of Cancer
Grand Bahama Island
Freeport
Abaco Island
New Providence
Nassau
Eleuthera Island
Andros Island
Cat Island
San Salvador
Great Exuma
Long Island
Crooked Island
Acklins Island
Mayaguana Island
BAHAMAS
Turks & Caicos Islands (U.K.)
Cockburn Town
Great Inagua Island (BAHAMAS)
HAVANA
Matanzas
Pinar del Río
Santa Clara
Cienfuegos
Gulf of Batabanó
Ciego de Ávila
Camagüey
Las Tunas
Holguín
Bayamo
Santiago de Cuba
Guantánamo
Naval Station (U.S.)
CUBA
Cabo de San Antonio
Yucatán Channel
Isle of Youth
Cozumel Island
Cayman Islands (U.K.)
George Town
Greater Antilles
Windward Passage
DOMINICAN REPUBLIC
Milwaukee Deep (-8,605 m, -28,232 ft.)
Cap-Haïtien
Santiago de los Caballeros
Gonaïves
HAITI
Hispaniola
PORT-AU-PRINCE
Carrefour
Les Cayes
San Pedro de Macorís
SANTO DOMINGO
Mona Passage
Mona Island (U.S.)
Puerto Rico (U.S.)
San Juan
Ponce
U.S. Virgin Islands (U.S.)
British Virgin Islands (U.K.)
St. Croix (U.S.)
Charlotte Amalie
Anguilla (U.K.)
St. Martin (SINT MAARTEN & FRANCE)
St.-Barthélemy (FRANCE)
Barbuda
ANTIGUA & BARBUDA
St. John's
Antigua
Basseterre
ST. KITTS & NEVIS
Montserrat (U.K.)
Grande-Terre
GUADELOUPE (FRANCE)
Pointe-à-Pitre
Basse-Terre
DOMINICA
Roseau
Fort-de-France
MARTINIQUE (FRANCE)
Castries
ST. LUCIA
BARBADOS
Bridgetown
Kingstown
ST. VINCENT & THE GRENADINES
St. George's
GRENADA
Leeward Islands
Windward Is.
Lesser Antilles
Montego Bay
JAMAICA
Kingston
Caribbean Sea
Bay Islands (HONDURAS)
Gulf of Honduras
La Ceiba
San Pedro Sula
HONDURAS
Tegucigalpa
Caratasca Lagoon
Mosquito Island (NICARAGUA)
Puerto Cabezas
MOSQUITO COAST
Providence Island (COLOMBIA)
Corn Islands (NICARAGUA)
San Andrés Island (COLOMBIA)
NICARAGUA
León
Lake Managua
Managua
Granada
Bluefields
Lake Nicaragua
San Juan
NICOYA PENINSULA
COSTA RICA
Alajuela
San José
Limón
ISTHMUS OF PANAMA
Panama Canal
Gulf of Mosquitos
Colón
Panama City
Gulf of Urabá
OSA PENINSULA
David
PANAMA
Yaviza
DARIÉN
Turbo
Puerto Armuelles
Burica Point
Gulf of Chiriquí
Chitré
AZUERO PEN.
Gulf of Panama
Mala Point
Coiba Island
Mariato Point
Cocos Island (COSTA RICA)
Malpelo Island (COLOMBIA)
ARUBA
Oranjestad
Willemstad
CURAÇAO
Kralendijk
Bonaire (NETH.)
Islas Los Roques (VENEZUELA)
Isla Blanquilla (VENEZUELA)
TRINIDAD & TOBAGO
Tobago
Port-of-Spain
Trinidad
San Fernando
Gulf of Paria
GUAJIRA PENINSULA
Gulf of Venezuela
Isla de Margarita
Isla la Tortuga
Santa Marta
BARRANQUILLA
CARTAGENA
MARACAIBO
Lake Maracaibo
MARACAY
CARACAS
VALENCIA
Barcelona
COAST RANGE
Barquisimeto
CORDILLERA DE MÉRIDA
Ciudad Guayana
Orinoco
LLANOS
VENEZUELA
GUYANA
GUIANA HIGHLANDS
MEDELLÍN
Magdalena
CORDILLERA OCCIDENTAL
CORDILLERA CENTRAL
CORDILLERA ORIENTAL
Cauca
BOGOTÁ
Buenaventura
CALI
COLOMBIA
Negro
Casiquiare
Vaupés
Uaupés
Equator
Branco
BRAZIL
BASIN
Tumaco
Esmeraldas
Galera Point
QUITO
Caquetá
80°
70°
60°
30°
20°
10°
0°

IN THIS CHAPTER

- Building a wall between two realms
- Ciudad Juárez: The curse of cocaine
- Drug-related violence invades northern Central America
- The natural vulnerability of Hispaniola and other Caribbean islands
- Is small beautiful? The predicament of tiny island-nations

FIGURE 4A-1 © H. J. de Blij, P. O. Muller, and John Wiley & Sons, Inc.

CONCEPTS, IDEAS, AND TERMS

Indigenous 1
NAFTA 2
Maquiladoras 3
Land bridge 4
Archipelago 5
Hurricane Alley 6
Altitudinal zonation 7
Tierra caliente 8
Tierra templada 9
Tierra fría 10
Tierra helada 11
Tierra nevada 12
Tropical deforestation 13
Culture hearth 14
Mestizo 15
Hacienda 16
Plantation 17
Connectivity 18
Small-island developing economies 19

Look at a world map, and it is obvious that the Americas comprise two landmasses: North America extending from Alaska to Panama and South America from Colombia to Argentina. But here we are reminded that continents and geographic realms do not necessarily coincide. In Chapters 3A and 3B we discussed a North American realm whose southern boundary is the U.S.-Mexican border and the Gulf of Mexico. Between North America and South America lies the small but important geographic realm known as Middle America. Consisting of a mainland corridor and myriad Caribbean islands, Middle America is a highly fragmented realm.

DEFINING THE REALM

From Figure 4A-1 it is clear that Middle America is much wider than it is long. The distance from Baja California to Barbados is about 6000 kilometers (3800 mi), but from the latitude of Tijuana to Panama City is only half that distance. In terms of total area, Middle America is the second-smallest of the world's geographic realms. As the map shows, the dominant state of this realm is Mexico, larger than all its other countries and territories combined.

Middle America may be a small geographic realm by global standards, but comparatively it is densely peopled. Its population passed the 200-million milestone in 2011, more than half of it residing in Mexico alone, and the rate of natural increase (at 1.5 percent) remains above the world average. Figure 4A-2 shows Mexico's populous interior core area quite clearly, but note that the population in Guatemala and Nicaragua tends to cluster toward the Pacific rather than the

major geographic qualities of MIDDLE AMERICA

1. Middle America is a relatively small realm consisting of the mainland countries from Mexico to Panama and all the islands of the Caribbean Basin to the east.
2. Middle America's mainland constitutes a crucial barrier between Atlantic and Pacific waters. In physiographic terms, this is a land bridge connecting the continental landmasses of North and South America.
3. Middle America is a realm of intense cultural and political fragmentation. The presence of many small, insular, and remote countries poses major challenges to economic development.
4. Middle America's cultural geography is complex. Various African and European influences dominate the Caribbean, whereas Spanish and indigenous traditions survive on the mainland.
5. The realm contains the Americas' least-developed territories and a substantial number of so-called small-island developing economies.
6. In terms of area, population, and economic potential, Mexico leads the realm.
7. Mexico and Panama are uniquely connected beyond the realm, the former because of its border with the United States and the latter because of the Panama Canal; the rest of the realm is relatively isolated.

Caribbean coast. Among the islands, this map reveals how crowded Hispaniola (containing Haiti and the Dominican Republic) is, but at this scale we cannot clearly discern the pattern on the smaller islands of the eastern Caribbean and the Bahamas, some of which are also densely populated.

What Middle America lacks in size it makes up in physiographic and cultural diversity. This is a realm of soaring volcanoes and spectacular shorelines, of tropical forests and barren deserts, of windswept plateaus and scenic islands. It holds the architectural and technological legacies of ancient **indigenous [1]** civilizations. Today it is a mosaic of immigrant cultures from Africa, Europe, and elsewhere, richly reflected in music and the visual arts. Material poverty, however, is endemic: island Haiti is the poorest country in the Americas; Nicaragua, on the mainland, is almost as badly off. As we will discover, a combination of factors has produced a distinctive but challenged realm between North and South America.

The setting of central Mexico City, which lies at the heart of the world's largest urban population agglomeration (estimated to have surpassed 30 million in 2013).

© David R. Frazier/Danita Delimont

GEOGRAPHICAL FEATURES

Sometimes you will see Middle and South America referred to in combination as "Latin" America, alluding to their prevailing Spanish-Portuguese heritage. This is an imperfect regional designation, just as "Anglo America," a term once commonly used for North America, was also improper. Such culturally based terminologies reflect historic power and dominance, and they tend to make outsiders out of those people they do not represent.

In North America, the term *Anglo* (as a geographic appellation) was offensive to many Native Americans, African Americans, Hispanics, Quebecers, and others. In Middle (and South) America, millions of people of indigenous-American, African, Asian, and European ancestries do not fit under the "Latin" rubric. You will not find the cultural landscape particularly "Latin" in the Bahamas, Barbados, Jamaica, Belize, or lengthy stretches of Guatemala and Mexico. So let us adopt the geographic neutrality of North, Middle, and South.

But is Middle America sufficiently different from either North or South America to merit distinction as a realm? Certainly, in this age of globalization and migration, many border areas are becoming transition zones, as is happening along the U.S.-Mexican

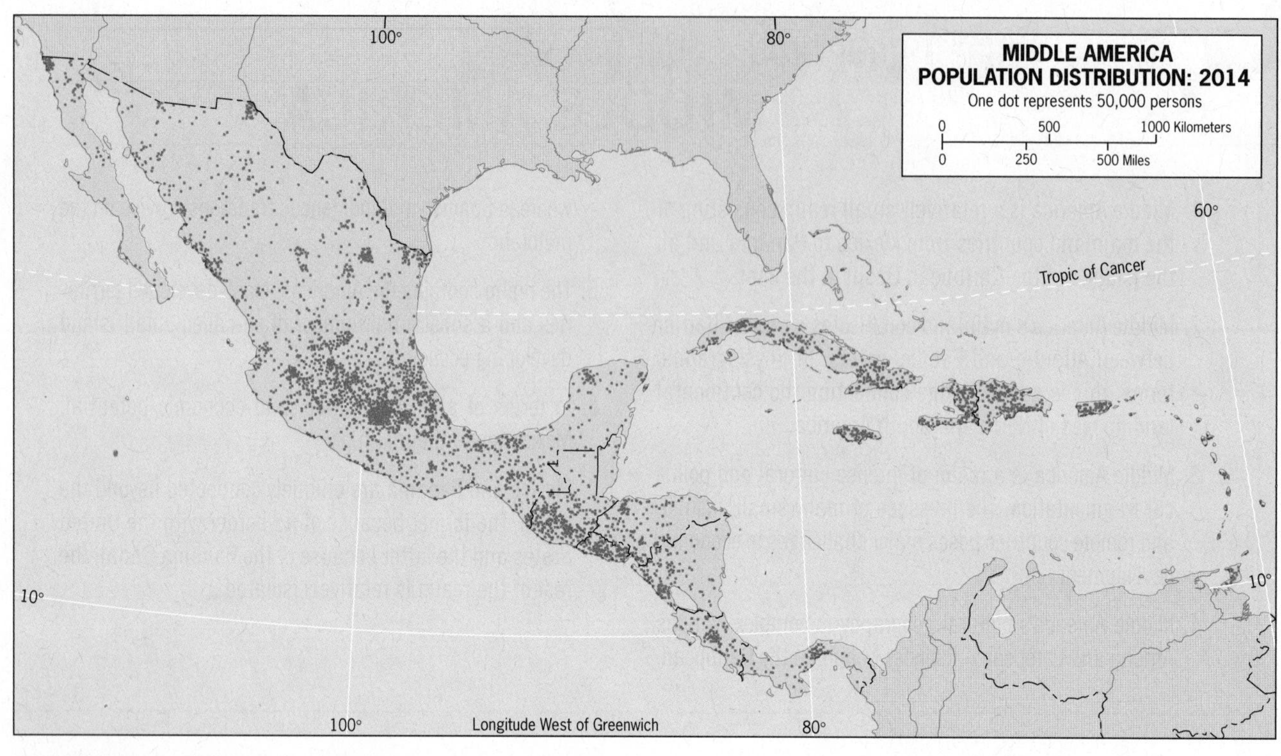

FIGURE 4A-2

© H. J. de Blij, P. O. Muller, and John Wiley & Sons, Inc.

boundary. Yet consider this: North America encompasses just a pair of states, and the entire continent of South America only 12 (plus France's dependency on the northeast coast). But far smaller Middle America, as we define it, incorporates more than three-dozen political entities, including several dependencies (or quasi-dependencies) of the Netherlands, the United Kingdom, and France as well as a few constituent territories of the United States. Therefore, unlike South America, Middle America is a multilingual patchwork of independent states, territories in political transition, and residual colonial dependencies, with strong continuing ties to the United States and non-Iberian Europe. Middle America is defined in large measure by its vivid multicultural geographiy.

The Realm's Northern Land Boundary

The 3169-kilometer (1969-mi) land border between North America and Middle America is the longest in the world separating a rich realm from a poor one. The U.S.-Mexican boundary crosses half the continent from the Pacific to the Gulf, but Mexican cultural influences penetrate deeply into the southwestern States and American impacts reach far into Mexico. To Mexicans the border is a reminder of territory lost to the United States in historic conflicts; to Americans it is a symbol of economic contrasts and illegal immigration. Along the Mexican side, the effects of **NAFTA [2]** (the North American Free Trade Agreement between Canada, the United States, and Mexico that went into effect in 1994) have transformed Mexico's economic geography.

The implementation of NAFTA led to an economic boom as Mexico became part of a free-trade zone and

Automobiles wait at the border to enter the United States from Mexico at the Tijuana-San Ysidro port of entry just south of San Diego—the busiest land border crossing in the world. Here, in 2011, some 50,000 northbound vehicles and 25,000 pedestrians entered the U.S. every day. A major expansion of this California facility, located at the southern terminus of Interstate-5, is underway to increase its capacity and reduce vehicle waiting times. The number of Mexico-bound inspection lanes doubled from 11 to 22 in 2012; the U.S.-bound lanes will similarly expand from 34 to 63 by 2016.

© David R. Frazier Photolibrary, Inc./Alamy

market encompassing almost 475 million people. A major beneficiary was the strip of land that ran along the boundary with the United States. The resulting boom changed urban landscapes all along the emerging border zone—but it could not, of course, close the massive economic gap between the two sides.

Under NAFTA, factories based in Mexico could assemble imported, duty-free raw materials and components into finished products, which were then exported back into the U.S. market. Logically, most of these factories, called **maquiladoras [3]**, are located as close to the U.S. border as possible. Thus manufacturing employment in the cities and towns along that border, from Tijuana on the Pacific to Matamoros at the mouth of the Rio Grande, expanded rapidly. By 2001, after only seven years of NAFTA's existence, there were some 3000 factories with more than 1.2 million workers in the border zone and in northern Yucatán (where Mérida was part of the process). Since then, these numbers have roughly stayed the same.

Mexico's growing economic linkages with its mighty northern neighbor have had a positive overall effect on the national economy, but they have increased regional inequalities and also made Mexico more vulnerable to economic crises in the United States. In 2008, Mexico entered a deep slump as exports to the recession-plagued U.S. declined sharply; by 2012, a slow recovery began to take hold. Mexico now provides about 12 percent of all U.S. imports annually, the third-largest source of foreign trade after China (18 percent) and Canada (14 percent). China has been increasing its share more rapidly, even though it faces tariffs that NAFTA-member Mexico does not. This puts pressure on Mexico to improve its competitiveness in order to build an even closer economic relationship with the United States.

A Troubled Border Zone

Despite the recent economic development of the borderlands, the boundary itself has become increasingly troubled, with many in the United States crusading to reinforce it to deter illegal immigration and the smuggling of drugs. Although the flow of illegal crossings from Mexico into the U.S. has slowed, during the first eight years of this century American authorities annually apprehended up to a million would-be, illegal migrants, many of them repeat offenders. Some were so desperate to make it across that they risked their lives: the number of deaths in the border zone since 2000 has varied between 200 and 500 yearly—a painful reminder of the enormous gap in living standards and life opportunities between the two realms.

As the Great Recession took hold in the U.S. and reduced the demand for labor, illegal crossings and border apprehensions declined commensurately. Nonetheless, even if this traffic flow swings upward again as economic recovery proceeds, it appears that migrants in Middle America increasingly prefer to remain *within* their realm rather than head north. In Mexico, urbanization is proceeding apace, from 66 percent in 1980 to 77 percent in 2014, and it is the mid-sized and smaller cities with lower costs of living that are growing fastest today. If Mexico is able to develop and modernize its economy in a more regionally balanced manner (a topic discussed in greater detail in Chapter 4B), this internal population shift may well be sustainable in the years ahead.

Another huge concern in the border zone is the burgeoning cocaine trade, and to a lesser extent the smuggling of marijuana. Most cocaine is produced in northwestern South America, and almost all of what is destined for the U.S. market passes through Mexico (see Fig. 4B-6). Much of the marijuana consumed within the U.S. is grown in western Mexico and finds its way across the border as well. The contraband is mostly carried in trucks or cars, often disguised as regular cargo in myriad ways (e.g., stuffed in teddy bears, inside hollowed out furniture, or simply in boxes labeled anything from candy to engine parts). It is the sheer volume of this unlawful cross-border traffic that makes it virtually impossible for authorities to conduct sufficient

The boundary between Mexico and the United States displays some stunning cross-border contrasts. In this scene, the crowded Mexican town of San Luis Colorado ends at a tall metal fence, where the Arizona desert takes over. This photo, therefore, looks eastward; south is to the right. Like other Mexican towns pressed against the border, San Luis Rio Colorado (locals simply call it San Luis) grew rapidly and chaotically in response to the opening of several maquiladoras, assembly plants making products for export to the United States.

FIGURE 4A-3 © H. J. de Blij, P. O. Muller, J. Nijman, and John Wiley & Sons, Inc.

inspections; and it is the infrastructure of the cities and industrial areas on the Mexican side of the border that facilitates the operation of the narcotics trade.

The city of Ciudad Juárez, probably the largest node in the hemispheric drug trade, is ideally located to handle this clandestine illicit activity. It is situated directly on the international boundary and forms a single built-up urban area with El Paso, Texas on the U.S. side—though the two are separated by a continuous artificial barrier running northwest-southeast across the entire metropolis (Fig. 4A-3). This map also shows the four ports of entry, which together account for about 25 million border crossings every year. Juárez has grown quite rapidly in the (post–1993) NAFTA era, and today exhibits numerous manufacturing districts as well as a sprawling mosaic of neighborhoods, ranging from clusters of upper-income ***colonias*** to the teeming slums that house the poorest of the new migrants from the south. With a plentiful supply of poor people willing to take the risk of driving or carrying drugs across the border, Juarez also has dozens of maquiladoras that can be bribed to store, conceal, and transport the contraband north. With Mexican authorities having little or no control over the area, it is hardly surprising that local law enforcement personnel are often implicated in the drug business. Once the prohibited substances cross over to El Paso, Interstate-10 is the designated east-west expressway in the United States. As the power of Mexican organized crime syndicates has grown enormously since 2000, the narcotics trade has triggered a concomitant, frightening increase in violence throughout northern Mexico (these drug wars are detailed in Chapter 4B).

In 2006, the U.S. government, responding both to homeland security concerns (potential terrorists are thought to enter the country via Mexico) and the tidal wave of cocaine smuggling, initiated the construction of a fortified fence along the entire length of the Mexican border. The very idea seems at odds with the ideals of closer economic and political cooperation, and many security experts questioned its efficacy. At any rate, in 2011, with about one-third of the border (ca. 1000 kilometers [650 mi]) hardened by fences and walls, the project to build a "virtual fence" to fill the remaining gaps had become such a boondoggle that it was canceled. However, formidable physical barriers are now concentrated around all the main crossing points, and it remains unclear how the longer-term economic health of the border zone they divide will be affected.

The Regions

As Figure 4A-1 suggests, Middle America can be divided into four distinct regions. Dominant ***Mexico*** occupies the largest part of the collective territory, with much of its northern border defined by the Rio Bravo (Rio Grande) from El Paso to the Gulf of Mexico. To the southeast, Mexico yields to the region named ***Central America***, consisting of the seven republics of Guatemala, Belize, Honduras, El Salvador, Nicaragua, Costa Rica, and Panama. (Sometimes the entire mainland portion of the realm is incorrectly referred to as Central America, but only these seven countries constitute this region.) The ***Greater Antilles*** is the regional

In 2010, groups of students in Mexico City staged this protest against the drug trafficking violence in Ciudad Juárez. An estimated 2738 people were murdered that year in Juárez, making it the most violent city on Earth. Since then, a declining murder rate has seen it fall into second place as a surge of drug-related violence during 2012 bestowed upon the Honduran city of San Pedro Sula the dubious distinction of being the "murder capital of the world."

© ALEX CRUZ/epa/Corbis

name that refers to the four large islands in the northern sector of the Caribbean Sea: Cuba, Jamaica, Hispaniola, and Puerto Rico; two countries, Haiti and the Dominican Republic, share the island of Hispaniola. And the ***Lesser Antilles*** form an extensive crescent of smaller islands from the Virgin Islands off Puerto Rico to Aruba near the northwestern coast of Venezuela; they also include the Bahamas island chain north of the Greater Antilles. In the Caribbean islands large and small, moist winds sweep in from the east, watering windward (wind-facing) coasts while leaving leeward (wind-protected) areas dry. On the map, the "Leeward" and "Windward" Islands actually are not "dry" and "wet," these terms having navigational rather than environmental implications.

PHYSICAL GEOGRAPHY

A Land Bridge

The funnel-shaped mainland, a 4800-kilometer (3000-mi) connection between North and South America, is wide enough in the north to contain two major mountain chains and a vast interior plateau, but narrows to a slim 65-kilometer (40-mi) ribbon of land in Panama. Here this strip of land—or ***isthmus***—bends eastward so that Panama's orientation is east-west. Thus mainland Middle America is what physical geographers call a **land bridge [4]**, an isthmian link between continents.

If you examine a globe, you can see some other present and former land bridges: Egypt's Sinai Peninsula between Asia and Africa, the (now-broken) Bering land bridge between northeasternmost Asia and Alaska, and the shallow waters between New Guinea and Australia. Such land bridges, though temporary features in geologic time, have played crucial roles in the dispersal of animals and humans across the planet. But even though mainland Middle America forms a land bridge, its internal fragmentation has always inhibited movement. Mountain ranges, swampy coastlands, and dense rainforests make contact and interaction difficult.

Island Chains

As shown in Figure 4A-1, the approximately 7000 islands of the Caribbean Sea stretch in a lengthy arc from Cuba and the Bahamas eastward and then southward to Trinidad, with numerous outliers outside (such as Barbados) and inside (e.g., the Cayman Islands) the main chain. As we noted above, the four large islands—Cuba, Hispaniola (containing Haiti and the Dominican Republic), Puerto Rico, and Jamaica—are called the Greater Antilles, and all the remaining smaller islands constitute the Lesser Antilles. The entire Antillean **archipelago [5]** (island chain) consists of the crests and tops of mountain chains that rise from the floor of the Caribbean, the result of collisions between the Caribbean Plate and its neighbors (Fig. G-4). Some of these crests are relatively stable, but elsewhere they contain active volcanoes, and almost everywhere in this realm earthquakes are an ever-present danger—in the islands as well as on the mainland (Fig. G-5). Add to this the realm's seasonal exposure to Atlantic/Caribbean hurricanes, and it amounts to some of the highest-risk real estate on Earth.

Dangerous Landscapes

The danger from below is dramatically illustrated in the Haitian, western half of Hispaniola, which is laced with geologic fault lines associated with the nearby boundary that separates the North American and Caribbean plates (refer to Fig. 4B-10, inset map). On January 12, 2010, Haiti was hit by a massive earthquake; the epicenter of that 7.0 temblor was located just outside Port-au-Prince and virtually destroyed this teeming, impoverished capital city of over 2 million (see photo). At least 300,000 people died, and within a week more than a million had fled to the countryside. It was the worst natural disaster in Haiti's history, but certainly not the first.

The environmental hazard from above comes in the form of hurricanes, powerful tropical cyclones that annually

This shocking street scene in central Port-au-Prince is typical of the widespread devastation resulting from the 2010 earthquake that struck the Haitian capital. The human casualty toll was enormous, thousands of buildings were made uninhabitable, debris clogged the streets, and the economy was wrecked. Survivors depended almost exclusively on assistance from foreign governments and international relief agencies. But life somehow found a way to go on, as these resourceful people proved. All too soon, though, Haiti's plight disappeared from the headlines even as the human drama continued. The hemisphere's poorest country and weakest economy continues to face a colossal struggle to rebuild, its infrastructure still in shambles. In 2013, more than 350,000 people still lived in the hundreds of dirty and unsafe encampments where they sought refuge three years earlier.

threaten the Caribbean Basin and its surrounding coastlines. The eastern half of Middle America is one of the most hurricane-prone areas in the world. One of the key conditions for the formation of hurricanes is very warm ocean water, because it further heats the hot moist air rising above it to "fuel" the evolving storm. The prolonged Atlantic/Caribbean hurricane season extends from June 1 to December 1, with the greatest number of tropical cyclones occurring in August and September when seawater reaches its highest temperatures. Most storms travel in a westerly direction from their low-latitude spawning ground off the coast of West Africa, steered across the Atlantic by the trade winds to reach the Caribbean Sea. Once there, many of these cyclones follow similar routes within **Hurricane Alley [6]** whose wide axis lies along all of the Greater Antilles and then broadens to include southern Florida, Mexico's Yucatán Peninsula, and all of the Gulf of Mexico. On average, every season sees the development of four to eight major hurricanes and rarely does a year go by without a destructive landfall on at least one of these densely populated areas.

Altitudinal Zonation of Environments

Continental Middle America and the western margin of South America are areas of high relief and strong environmental contrasts. Even though settlers have always favored temperate intermontane basins and valleys, people also cluster in hot tropical lowlands as well as high plateaus just below the snow line in South America's Andes Mountains. In each of these zones, distinct local climates, soils, vegetation, crops, domestic animals, and modes of life prevail. Such **altitudinal zones [7]** (diagrammed in Fig. 4A-4) are known by specific names as if they were regions with distinguishing properties—as in reality they are.

The lowest of these vertical zones, from sea level to 750 meters (2500 ft), is known as the ***tierra caliente*** **[8]**, the "hot land" of the coastal plains and low-lying interior basins where tropical agriculture predominates. Above this zone lie the tropical highlands containing Middle and South America's largest population clusters, the ***tierra templada*** **[9]** of temperate land reaching up to about 1800 meters (6000 ft). Temperatures here are cooler; prominent among the commercial crops is coffee, while corn (maize) and wheat are the staple grains. Still higher, from about 1800 to 3600 meters (6000 to nearly 12,000 ft), is the ***tierra fría*** **[10]**, the cold country of the higher Andes where hardy crops such as potatoes and barley are mainstays. Above the tree line, which marks the upper limit of the *tierra fría*, lies the ***tierra helada*** **[11]**; this fourth altitudinal zone, extending from about 3600 to 4500 meters (12,000 to 15,000 ft), is so cold and barren that it can support only the grazing of sheep and other hardy livestock. The highest zone of all is the ***tierra nevada*** **[12]**, a zone of permanent snow and ice associated with the loftiest Andean peaks. As we will see, the varied human geography of mainland Middle and western South America closely reflects these diverse environments.

Tropical Deforestation

Before the Europeans arrived, two-thirds of continental Middle America (at lower altitudes) was covered by tropical rainforests. It is estimated that at present only about 10 percent of this vegetation remains. Between 2000 and 2010, Central America lost almost 12 percent of its woodlands, a deforestation rate close to ten times the global average. El Salvador today has lost virtually all of its forests, and most of the six other republics in its region will soon approach that stage. As for bedeviled Haiti, its ravaged woodlands have already reached the stage of complete denudation (as revealed in the aerial photo in Chapter 4B).

The causes of **tropical deforestation [13]** are related to the persistent economic and demographic problems of

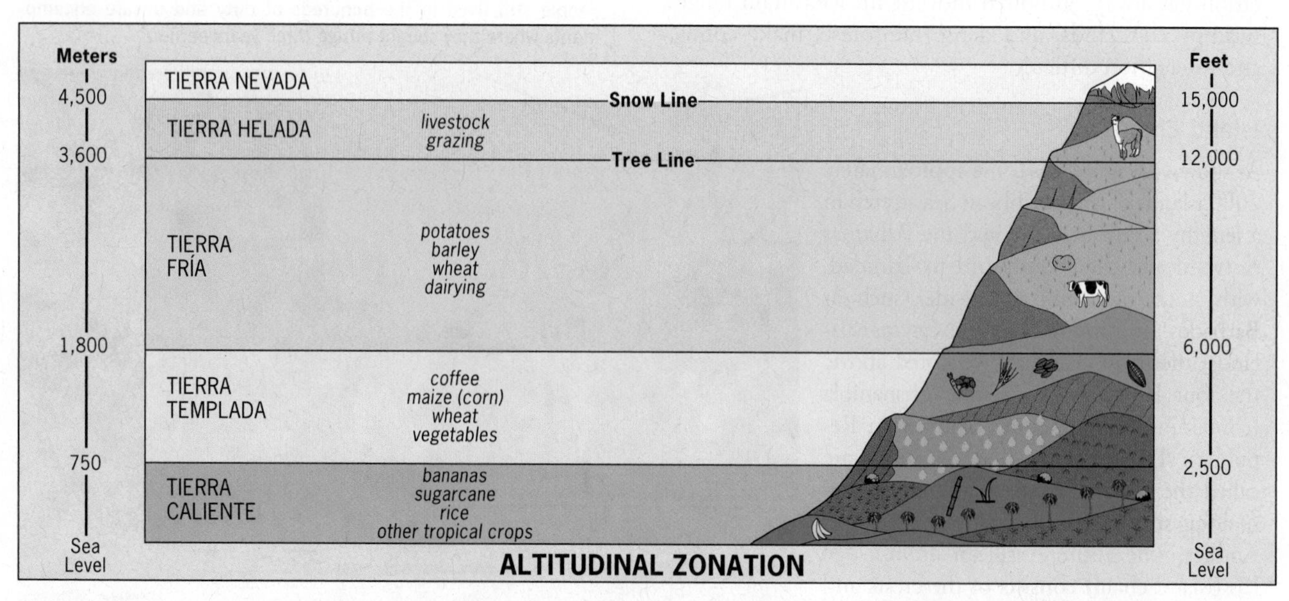

FIGURE 4A-4

© Humberto Olarte Cupas/Alamy

Despite scattered attempts to reverse this landscape scourge, deforestation continues to afflict Central America. Its worst effects plague the steeper slopes of interior highlands, as here in western Panama. In the wake of recent deforestation, the land near the top of this hill has already begun to erode, because in the absence of binding tree roots the copious tropical rains are making short work of the unprotected topsoil.

disadvantaged countries. In Central America, the leading cause has been the need to clear rural lands for cattle pasture as many countries, especially Costa Rica, became meat producers and exporters. Because tropical soils are so nutrient-poor, newly deforested areas are able to function as pastures for only a few years at most. These fields are then abandoned for other freshly cut lands and quickly become a ravaged landscape (see photo). Without the protection of tree roots, local soil erosion and flooding immediately become problems, affecting still-productive areas nearby. A second cause of deforestation is the rapid logging of tropical woodlands as the timber industry increasingly turns from the exhausted forests of the midlatitudes to harvest the rich tree resources of the equatorial zones, responding to accelerating global demands for housing, paper, and furniture. The third major contributing factor is related to the region's population explosion: as more and more peasants are required to extract a subsistence from inferior lands, they have no choice but to cut down the remaining forest for both firewood and additional crop-raising space, and their intrusion prevents the trees from regenerating.

CULTURAL GEOGRAPHY

Mesoamerican Legacy

Mainland Middle America was the scene of the emergence of a major ancient civilization. Here lay one of the world's true **culture hearths [14]**, a source area from which new ideas radiated outward and whose population could expand and make significant material as well as intellectual progress. Agricultural specialization, urbanization, trade, and transportation networks developed, and writing, science, art, and other spheres of achievement saw major advances. Anthropologists refer to the Middle American culture hearth as ***Mesoamerica***, which extended southeast from the vicinity of present-day Mexico City to central Nicaragua. Its development is particularly remarkable because it occurred in highly different geographic environments, each presenting obstacles that had to be overcome in order to unify and integrate large territories. First, in the low-lying tropical plains of what is now northern Guatemala, Belize, and Mexico's Yucatán Peninsula, and perhaps simultaneously in Guatemala's highlands to the south, the Maya civilization arose more than 3000 years ago. Later, far to the northwest on the high plateau in central Mexico, the Aztecs founded a major civilization centered on the largest city ever to exist in pre-Columbian times.

The Lowland Maya

The Maya civilization is the only major culture hearth in the world that arose in the lowland tropics. Its great cities, with their stone pyramids and massive temples, still yield archeological information today. Maya culture reached its zenith from the third to the tenth centuries AD. The Maya civilization, anchored by a network of city-states, unified an area larger than any of the present-day Middle American countries except Mexico. Its population probably totaled between 2 and 3 million; certain Maya languages are still used in the area to this day. The Maya city-states were marked by dynastic rule that functioned alongside a powerful religious hierarchy, and the great cities that now lie in ruins were primarily ceremonial centers. We also know that this culture produced skilled artists and scientists, and the Maya achieved a great deal in agriculture and trade as well.

The Highland Aztecs

In what is today the intermontane highland zone of Mexico, significant cultural developments were also taking place. Here, just north of present-day Mexico City, lay Teotihuacán, the first true urban center in the Western Hemisphere, which prospered for nearly seven centuries after its founding around the beginning of the Christian era.

The Aztec state, the pinnacle of organization and power in pre-Columbian Middle America, is thought to

From the Field Notes . . .

"We spent Monday and Tuesday upriver at Lamanai, a huge, still mostly overgrown Maya site deep in the forest of Belize. On Wednesday we drove from Belize City to Altun Ha, which represents a very different picture. Settled around 200 BC, Altun Ha flourished as a Classic Period center between AD 300 and 900, when it was a thriving trade and redistribution center for the Caribbean merchant canoe traffic and served as an entrepôt for the interior land trails, some of them leading all the way to Teotihuacán. Altun Ha has an area of about 6.5 square kilometers (2.5 sq mi), with the main structures, one of which is shown here, arranged around two plazas at its core. I climbed to the top of this one to get a perspective, and sat down to have my sandwich lunch, imagining what this place must have looked like as a bustling trade and ceremonial center when the Roman Empire still thrived, but a more urgent matter intruded. A five-inch tarantula emerged from a wide crack in the sunbaked platform, and I noticed it only when it was about two feet away, apparently attracted by the crumbs and a small piece of salami. A somewhat hurried departure put an end to my historical-geographical ruminations."

© H.J. de Blij

www.conceptcaching.com

have originated in the early fourteenth century with the founding of a settlement on an island in a lake that lay in the *Valley of Mexico* (the area surrounding what is now Mexico City). This urban complex, a functioning city as well as a ceremonial center, named Tenochtitlán, was shortly to become the greatest city in the Americas and the capital of a large powerful state.

The Aztecs produced a wide range of impressive accomplishments, although they were better borrowers and refiners than they were innovators. They developed irrigation systems, and they constructed elaborate walls to terrace slopes where soil erosion threatened. Indeed, the greatest contributions of Mesoamerica's indigenous peoples surely came from the agricultural sphere and included the domestication of corn (maize), the sweet potato, cacao beans (the raw material of chocolate), and tobacco.

Spanish Conquest

Spain's defeat of the Aztecs in the early sixteenth century opened the door to Spanish penetration and supremacy. The Spaniards were ruthless colonizers but not more so than other European powers that subjugated other cultures. The Spaniards first enslaved the indigenous people and were determined to destroy the strength of their society. But biology accomplished what ruthlessness could not have achieved in so short a time: diseases introduced by the Spaniards and the slaves they imported from Africa killed millions of indigenous people.

Middle America's cultural landscape was drastically modified. Unlike the indigenous peoples, who had utilized stone as their main building material, the Spaniards employed great quantities of wood and used charcoal for heating, cooking, and smelting metal. The onslaught on the forests was immediate, and rings of deforestation swiftly expanded around the colonizers' towns. The Spaniards also introduced large numbers of cattle and sheep, and people and livestock now had to compete for available food (requiring the opening of vast areas of marginal land that further disrupted the region's food-producing balance). Moreover, the Spaniards introduced their own crops (notably wheat) and farming equipment, and soon large wheatfields began to encroach upon the small plots of corn that the indigenous people cultivated.

The Spaniards' most far-reaching cultural changes derived from their traditions as town dwellers. The indigenous people were moved off their land into nucleated villages and towns that the Spaniards established and laid out. In these settlements, the Spaniards could exercise the kind of rule and administration to which they were accustomed (Fig. 4A-5). The internal focus of each Spanish town was the central ***plaza*** or market square, around which both the local church and government buildings were located. The surrounding street pattern was deliberately laid out in gridiron form, so that any insurrections could be contained by having a small military force seal off the affected blocks and then root out the troublemakers. Each town was located near what was thought to be good agricultural land

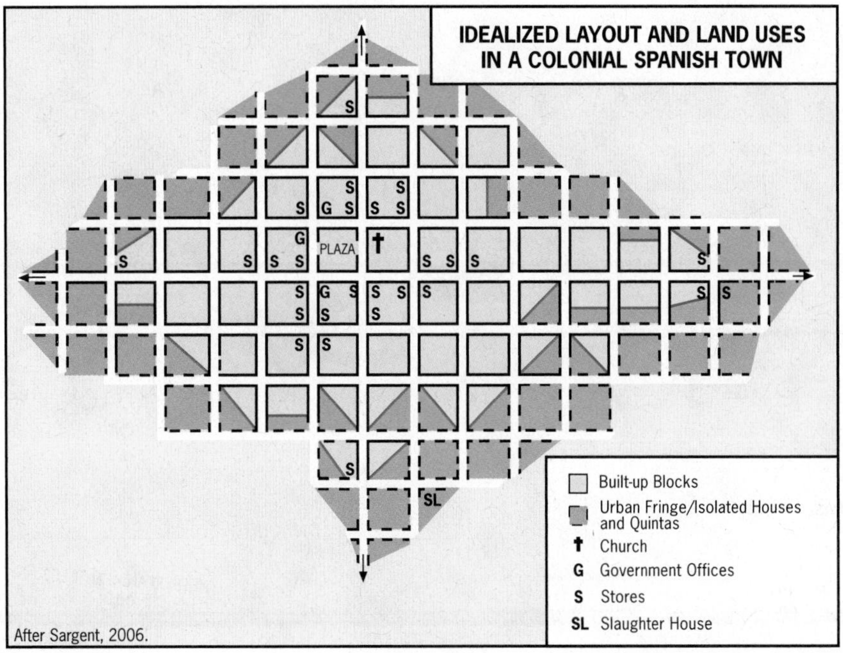

IDEALIZED LAYOUT AND LAND USES IN A COLONIAL SPANISH TOWN

Built-up Blocks
Urban Fringe/Isolated Houses and Quintas
Church
G Government Offices
S Stores
SL Slaughter House

After Sargent, 2006.

FIGURE 4A-5

(which often was not so good), so that the indigenous people could venture out each day and work in the fields. Packed tightly into these towns and villages, they came face to face with Spanish culture. Here they (forcibly) learned the Europeans' Roman Catholic religion and Spanish language, and they paid their taxes and tribute to a new master. Many of Middle America's leading cities still bear this Spanish imprint.

Collision of Cultures

But Middle America is not Spain. The cultural fabric of Middle America reflects the collision of indigenous, Spanish, and other European influences. Indeed, in more remote areas in southeastern Mexico and interior Guatemala, the nucleated indigenous village survived and to this day native languages prevail over Spanish (see Fig. 4B-4).

In Middle America outside Mexico, only Panama, with its twin attractions of interoceanic transit and gold deposits, became an early focus of Spanish activity. From there, following the Pacific-fronting side of the isthmus, Spanish influence radiated northwestward through Central America and into Mexico. The major arena of international competition in Middle America, however, lay not on the Pacific side but on the islands and coasts of the Caribbean Sea. Here the British gained a foothold on the mainland, controlling a narrow coastal strip that extended southeast from Yucatán to what is now Costa Rica. As the colonial-era map (Fig. 4A-6) shows, in the Caribbean the Spaniards faced not only the British but also the French and the Dutch, all interested in the lucrative sugar trade, all searching for instant wealth, and all seeking to expand their empires.

Much later, after centuries of European colonial rivalry in the Caribbean Basin, the United States entered the picture and made its influence felt in the coastal areas of the mainland, not through colonial conquest but through the introduction of widespread, large-scale, banana plantation agriculture. The effects of these plantations were as far-reaching as the impact of colonialism on the Caribbean islands. Because the diseases the Europeans had introduced were most rampant in these hot humid lowlands (as well as the Caribbean islands to the east), the indigenous population that survived was too small to provide a sufficient workforce. This labor shortage was quickly remedied through the trans-Atlantic slave trade from Africa that transformed the population composition of the Caribbean Basin.

The cultural variety of the Caribbean Basin is especially striking, and it is hardly an arena of exclusive Hispanic cultural heritage. For example, Cuba's southern neighbor, Jamaica (population 2.7 million, mostly of African ancestry), has a legacy of British involvement, while to the east in Haiti (10.7 million, overwhelmingly of African ancestry) the strongest imprints have been African and French. The Lesser Antilles also exhibit great cultural diversity. There are the (once Danish) U.S. Virgin Islands; French Guadeloupe and Martinique; a group of British-influenced islands, including Barbados, St. Lucia, and Trinidad and Tobago; and the Dutch St. Maarten (shared with the French) as well as the now-autonomous islands of the former Netherlands Antilles—Aruba, Curaçao, and Bonaire—off the northwestern Venezuelan coast.

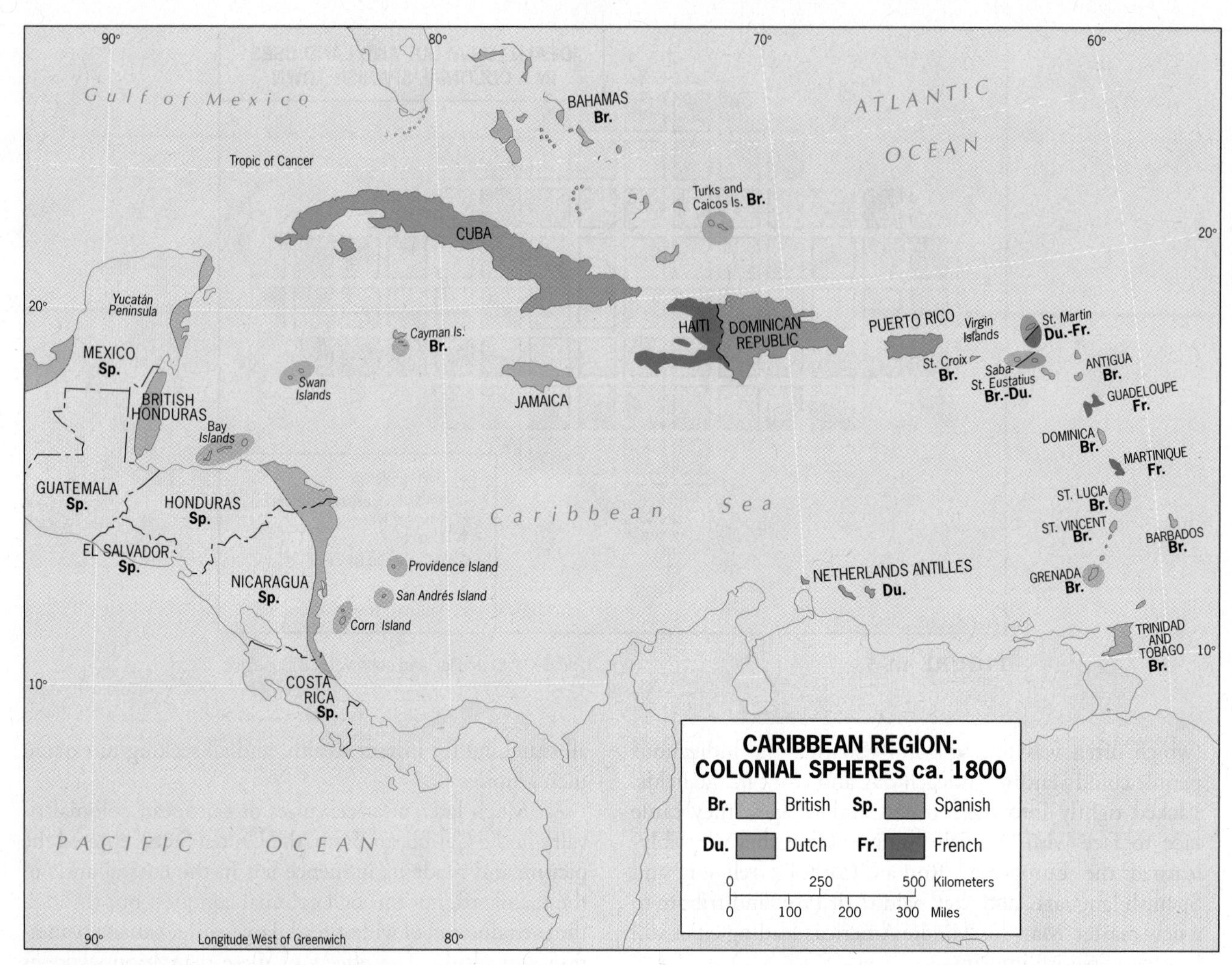

FIGURE 4A-6

POLITICAL AND ECONOMIC FRAGMENTATION

Independence

Independence movements stirred Middle America at an early stage. On the mainland, insurrections against Spanish authority (beginning in 1810) achieved independence for Mexico by 1821 and for the Central American republics by the end of the 1820s, resulting in the creation of eight different countries. The United States, concerned over European designs in the realm, proclaimed the Monroe Doctrine in 1823 to deter any European power from reasserting its authority in the newly independent republics or from further expanding its existing domains.

By the end of the nineteenth century, the United States itself had become a major force in Middle America. The Spanish-American War of 1898 made Cuba independent and placed Puerto Rico under the U.S. flag; soon thereafter, the Americans were in Panama constructing the Panama Canal. Meanwhile, with U.S. corporations propelling a boom based on massive banana plantations, the Central American republics had become colonies of the United States in all but name.

Independence came to the Caribbean Basin in fits and starts. Afro-Caribbean Jamaica as well as Trinidad and Tobago, where the British had brought in a large South Asian population, attained full sovereignty from the United Kingdom in 1962; other British islands (among them Barbados, St. Vincent, and Dominica) became independent later on. France, however, retains Martinique and Guadeloupe as *Overseas Départements* of the French Republic, and the Dutch islands are at various stages of autonomy. No less than 33 states are found on the political map of the Caribbean Basin today.

Regional Contrasts

There are some striking contrasts, socially and economically, between the Middle American uplands on the one hand and the Caribbean coasts and islands on the other (Fig. 4A-7). These were conceptualized by cultural geographer John Augelli into the ***Mainland-Rimland framework***. The Euro-Indigenous ***Mainland***, from Mexico southeast to Panama, is dominated by European (Spanish) as well as indigenous influences and also includes **mestizo [15]** sectors where the two ancestries mixed. As

From the Field Notes . . .

© Jan Nijman

"Driving around the small Caribbean island of Bonaire and coming face to face with this surreal landscape, I first thought I had experienced a mirage. These glistening white, perfectly cone-shaped hills are actually salt piles. This small (formerly Dutch) island off the coast of Venezuela possesses the perfect geographic conditions for salt production: it has a series of salt water inlets, it is very hot and dry, and lies in the zone of persistent trade winds. Remember that salt, nowadays taken for granted in every household, has long been one of the world's most precious spices, and was widely used for the preservation of meat and fish. The Dutch began large-scale production of salt in the 1620s; today this local industry is in the hands of the Antilles International Salt Company and continues to be an important source of Bonaire's foreign revenues."

www.conceptcaching.com

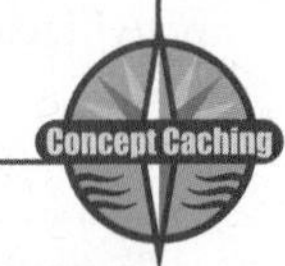

Figure 4A-7 shows, the Mainland is subdivided into several areas based on the strength of the indigenous legacy. The Mainland's Caribbean coastal strip and all of the Basin's islands to the east constitute the ***Rimland***, which for the most part possesses a very different cultural heritage, based on a fusion of European and African influences.

Supplementing these contrasts are regional differences in outlook and orientation. The Caribbean Rimland was an area of sugar and banana plantations, of high accessibility, of seaward exposure, and of maximum cultural contact and mixture. The Middle American Mainland, being farther removed from these contacts, was an area of greater isolation. The Rimland was the domain of the great *plantation*, and its commercial economy was therefore susceptible to fluctuating world markets and tied to overseas investment capital. The Mainland was dominated by the *hacienda*, which was far more self-sufficient and less dependent on external markets.

This contrast between plantation and hacienda land tenure in itself constitutes strong evidence for the Rimland-Mainland dichotomy. The **hacienda [16]** was a Spanish institution, whereas the modern plantation was the concept of northwestern Europeans. On their haciendas, Spanish landowners possessed a domain whose productivity they might never push to its limits: the very possession of such a vast estate brought with it social prestige and a comfortable lifestyle. Native workers lived on the land—which may once have been their land—and had plots where they could grow their own subsistence crops. All this is written as though it is mostly in the past, but the legacy of the hacienda system, with its inefficient use of land and labor, still exists throughout mainland Middle America.

The **plantation [17]**, in contrast, is all about efficiency and profit. Foreign ownership and investment is the norm, as is production for export. Most plantations grow only a single crop, be it sugar, bananas, or coffee. Much of

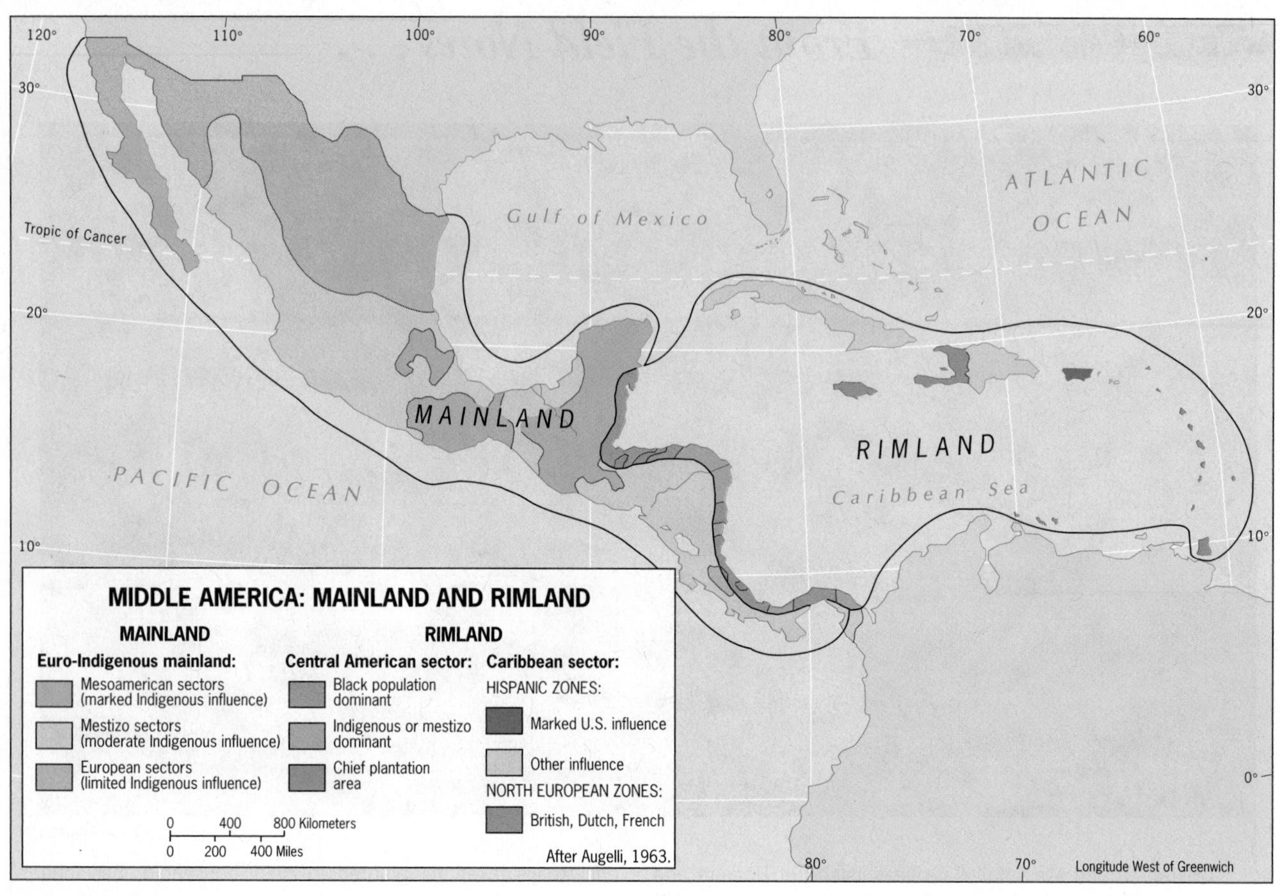

FIGURE 4A-7 © H. J. de Blij, P. O. Muller, and John Wiley & Sons, Inc.

the labor is seasonal—needed in large numbers mainly during the harvest period and such labor has been imported because of the scarcity of indigenous workers. With its "factory-in-the-field" operations (left photo), the plantation is far more efficient in its use of land and labor than the hacienda. Profit and wealth, rather than social prestige, are the dominant motives for the plantation's establishment and operation.

Contrasting land uses in the Middle American Rimland and Mainland give rise to some very different rural cultural landscapes. Huge stretches of the realm's best land continue to be controlled by (often absentee) landowners whose haciendas yield export or luxury crops, or foreign corporations that raise fruits for transport and sale on their home markets. The banana plantation shown here (left) lies near the Caribbean coast of Costa Rica. The vast fields of banana plants stand in stark contrast to the lone peasant who ekes out a bare subsistence from small cultivable plots of land, often in high-relief countryside where grazing some goats or other livestock is the only way to use most of the land.

John Coletti/Getty Images, Inc.

J. Gerard Sidaner/Science Source

Connections Matter

The unusual layout of the Middle American realm offers an opportunity to speculate about the role of spatial factors in (economic) development. Geographic fragmentation and magnitude undoubtedly play a part, and so does accessibility versus remoteness in the form of **connectivity [18]**. This is true for both the Caribbean and mainland components of the realm. As noted earlier, Mexico's northern margin is strongly connected to the United States, and this provides an impulse for the entire Mexican economy even if it tends to accentuate regional differences (discussed in Chapter 4B). Among Central American countries, there is no question that Panama is by far the best connected to the world economy; indeed, the Panama Canal provides an enormous range of global linkages in trade, finance, producer services, and other sectors of international business.

Now take a look at Figure 4A-8 and note how connectivity correlates with economic development. Of the eight mainland countries, Mexico and Panama have the highest GDP per capita; moreover, as we move away from the realm boundaries with North and South America toward the increasingly remote central interior of the Mexico-Panama land corridor, GDP per capita declines substantially. Nicaragua and Honduras, tucked away in the middle of that narrowing land bridge, are the most distant from the bigger and more vibrant economies of the neighboring realms and decidedly lack the expanding external connections of Panama or even the historical connections of Belize (to the UK, its former colonial ruler) or Costa Rica (to the United States, triggered well over a century ago by the banana trade).

© Liane Cary/Age Fotostock America, Inc.

Hundreds of thousands of North American retirees, and many affluent purchasers of second homes, have sought the sun and low-cost residential opportunities of Middle America, converting some areas into virtual exclaves of the North. From Mexico to Panama, waterfront real estate is among the attractions for these permanent and seasonal migrants, as here in Cabo San Lucas at the southern tip of Mexico's Baja California peninsula. This is yet another example of international connections that matter.

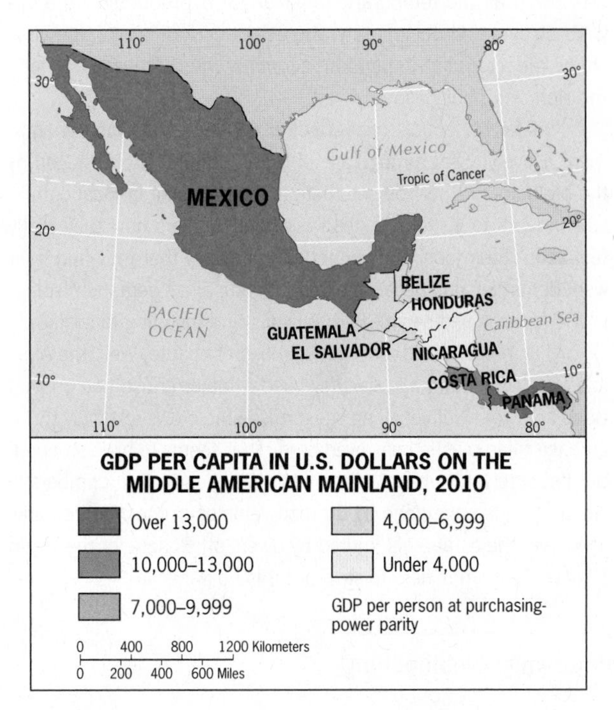

FIGURE 4A-8 © H. J. de Blij, P. O. Muller, J. Nijman, and John Wiley & Sons, Inc.

Ironically, the isolation of parts of the land bridge has in recent years made it a preferred route for narcotrafficking, particularly cocaine from Colombia on its way to Mexico where powerful drug cartels control its shipment into the United States. The inaccessible interior stretches and the many waterways surrounding the land bridge combine to thwart the ability of fiscally challenged governments to adequately patrol the region and enforce laws. The rise of drug trafficking also plays a huge role in the alarming increase of organized criminal activity and violence, especially in Honduras, El Salvador, and Guatemala. In 2011, the head of the U.S. Southern Military

Regional ISSUE The Role of the Tourist Industry in Middle American Economies

IN SUPPORT OF THE TOURIST INDUSTRY

"As the general manager of a small hotel in St. Maarten, on the Dutch side of the island, I can tell you that without tourists, we would be in deep trouble economically. Mass tourism, plain and simple, has come to the rescue in the Caribbean and in other countries of Middle America. Look at the numbers. Here it's the only industry that is growing, and it already is the largest worldwide industry. For some of the smaller countries of the Caribbean, this is not just the leading industry but the only one producing external revenues. Whether it's hotel patrons or cruise-ship passengers, tourists spend money, create jobs, fill airplanes that give us a link to the outside world, require infrastructure that's good not just for them but for locals too. We've got better roads, better telephone service, more items in our stores. All this comes from tourism. I employ 24 people, most of whom would be looking for nonexistent jobs if it weren't for the tourist industry.

"And it isn't just us here in the Caribbean. Look at Belize. I just read that tourism there, based on their coral reefs, Mayan ruins, and inland waterways, brought in U.S. $100 million last year, in a country with a total population of only about 300,000! That sure beats sugar and bananas. In Jamaica, they tell me, one in every four workers has a job in tourism. And the truth is, there's still plenty of room for the tourist industry to expand in Middle America. Those Americans and Europeans can close off their markets against our products, but they can't stop their citizens from getting away from their awful weather by coming to this tropical paradise.

"Here's another good thing about tourism. It's a clean industry. It digs no mine shafts, doesn't pollute the atmosphere, doesn't cause diseases, doesn't poison villagers, isn't subject to graft and corruption the way some other industries are.

"Last but not least, tourism is educational. Travel heightens knowledge and awareness. There's always a minority of tourists who just come to lie on the beach or spend all their time in some cruise-ship bar, but most of the travelers we see in my hotel are interested in the place they're visiting. They want to know why this island is divided between the Dutch and the French, they ask about coral reefs and volcanoes, and some even want to practice their French on the other side of the border (don't worry, no formalities, just drive across and start talking). Tourism's the best thing that happened to this part of the world, and other parts too, and I hope we'll never see a slowdown."

CRITICAL OF THE TOURIST INDUSTRY

"You won't get much support for tourism from some of us teaching at this college in Puerto Rico, no matter how important some economists say tourism is for the Caribbean. Yes, tourism is an important source of income for some countries, like Kenya with its wildlife and Nepal with its mountains, but for many countries that income from tourism does not constitute a real and fundamental benefit to the local economies. Much of it may in fact result from the diversion to tourist consumption of scarce commodities such as food, clean water, and electricity. More of it has to be reinvested in the construction of airport, cruise-port, overland transport, and other tourist-serving amenities. And as for items in demand by tourists, have you noticed that places with many tourists are also places where prices are high?

"Sure, our government people like tourism. Some of them have a stake in those gleaming hotels where they can share the pleasures of the wealthy. But what those glass-enclosed towers represent is globalization, powerful multinational corporations colluding with the government to limit the opportunities of local entrepreneurs. Planeloads and busloads of tourists come through on prearranged (and prepaid) tour promotions that isolate those visitors from local society.

"Picture this: luxury liners sailing past poverty-stricken villages, luxury hotels towering over muddy slums, restaurants serving caviar when, down the street, children suffer from malnutrition. If the tourist industry offered real prospects for economic progress in poorer countries, such circumstances might be viewed as the temporary, unfortunate byproducts of the upward struggle. Unfortunately, the evidence indicates otherwise. Name me a tourism-dependent economy where the gap between the rich and poor has narrowed.

"As for the educational effect of tourism, spare me the argument. Have you sat through any of those 'culture' shows staged by the big hotels? What you see there is the debasing of local culture as it adapts to visitors' tastes. Ask hotel workers how they really feel about their jobs, and you'll hear many say that they find their work dehumanizing because expatriate managers demand displays of friendliness and servitude that locals find insulting to sustain.

"I've heard it said that tourism doesn't pollute. Well, the Alaskans certainly don't agree—they sued a major cruise line on that issue and won. Not very long ago, cruise-ship crews routinely threw garbage-filled plastic bags overboard. That seems to have stopped, but I'm sure you've heard of the trash left by mountain-climbers in Nepal, the damage done by off-road vehicles in the wildlife parks of Kenya, the coral reefs injured by divers off Bonaire or the Virgin Islands. Tourism is here to stay, but it is no panacea."

Vote your opinion at www.wiley.com/go/deblijpolling

Command warned that this part of Central America had become the "deadliest zone in the world." Not surprisingly, national murder rates in El Salvador and Honduras for that year turned out to be the highest and second-highest on Earth, respectively, with fourth-ranked Guatemala close behind (by 2012, Honduras had moved up to first place).

Is Small Beautiful?

As noted, the Middle American realm is exceptional in the modest size of its territorial extent and population. But the number of constituent countries is considerable, and they tend to be quite small. Caribbean islands often invoke images of beautiful scenery, tropical cocktails, and shiny blue waters, yet their economic realities are almost invariably harsh.

The limited land areas of the Lesser Antilles each average well below half a million people, and that, combined with insularity, remoteness, and low connectivity, poses formidable challenges that are common to **small-island developing economies [19]**. First, natural resources are frequently limited, which requires a heavy reliance on imports made more expensive by added transport costs. Second, the cost of government is relatively high on a per capita basis: even the smallest population will require services such as schools, hospitals, and waste disposal. Third, these specialized services must often be brought in from elsewhere. And fourth, local production cannot really benefit from economies of scale, which means that local producers can be put out of business by cheaper imports, thereby driving up unemployment.

Given the Caribbean Basin's limited economic options, does the tourist industry offer better opportunities? Opinions on this question are divided. The resort areas, scenic treasures, and historic locales of Caribbean America attract between 20 and 25 million visitors annually, with about half of them traveling on Florida-based cruise ships. Certainly, Caribbean tourism is a prospective moneymaker for many islands. In Jamaica alone, this industry accounts for about one-fifth of the gross domestic product and employs more than one-fourth of the labor force.

But Caribbean tourism also has serious drawbacks. The invasion of poor communities by affluent tourists contributes to rising local resentment, which is further fueled by the glaring contrasts of shiny new hotels towering over substandard housing and luxury liners gliding past poverty-stricken villages. At the same time, tourism can have the effect of debasing local culture, which often is adapted to suit the visitors' tastes at hotel-staged "culture" shows. In addition, the cruise industry tends to monopolize revenues (accommodations, meals, entertainment) with relatively few dollars flowing into the local economy. Finally, even though tourism does generate income in the Caribbean, the intervention of island governments and multinational corporations can remove opportunities from local entrepreneurs in favor of large operators and major resorts.

The Push for Regional Integration

Another challenge for the Middle American realm is to foster greater economic integration. Many of the countries on the mainland as well as in the Caribbean are poorly connected within the realm and are heavily dependent on major outside countries, particularly the United States. For the large majority of these countries, the United States is their primary trading partner. Consider this: of all the trade involving Middle American countries, less than 10 percent occurs within the realm. And less than 1 percent takes place between the Caribbean Basin and the Middle American mainland.

Over the years, efforts have been made to advance economic integration and convert this realm into more of a ***functional region***. The mainland saw the creation of the Central American Common Market in 1960, but it became moribund within a decade because of the 1969 war between El Salvador and Honduras that was followed by wider intraregional conflict throughout the 1970s and 1980s (some with U.S. involvement). The organization was resuscitated in the 1990s, but since 2005 it has been strongly overshadowed by ***CAFTA***, the Central American Free Trade Agreement with the United States. CAFTA seems a mixed blessing: it may increase access to U.S. markets and lead to cheaper imports, but it may also galvanize the dominant position of the United States at the cost of greater intraregional integration.

Within the Caribbean Basin itself, ***CARICOM*** (Caribbean Community) was established in 1989 and today consists of 15 full members, including nearby Guyana and Suriname in South America. To a certain extent, CARICOM follows the model of the European Union and in 2009 even introduced a common passport. But economic change has been painfully slow, a reminder that the geography of the Middle American realm poses many formidable challenges. Clearly, these cannot be easily overcome even with the best political and economic intentions.

POINTS TO PONDER

- The 3169-kilometer (1969-mi) land border between North America and Middle America is the longest in the world that separates a rich realm from a poor one.
- The United States is the single most important economic partner of just about every country and territory in this realm—is that a positive or a negative for Middle America?
- "Poor Mexico . . . so far from God and so close to the United States"—former Mexican President Porfirio Diaz.
- Most of the small Caribbean island-nations contain less than half a million people—is that enough to sustain a viable economy?

4B MIDDLE AMERICAN REGIONS

115° 110° 105° 100° 95° 90° 85° 80° 75° 70°

UNITED STATES
Los Angeles
San Diego
Tijuana
Phoenix
Tucson
El Paso
Ciudad Juarez
Hermosillo
Chihuahua
Dallas
Houston
San Antonio
Nashville
Memphis
Atlanta
Charlotte
Norfolk
Charleston
Jacksonville
New Orleans
Orlando
Tampa
Ft. Myers
W. Palm Beach
Ft. Lauderdale
Miami
BAHAMAS
TURKS CAICOS
Gulf of Mexico
Baja California
Gulf of California (Sea of Cortés)
Sierra Madre Occidental
Plateau of Mexico
Sierra Madre Oriental
La Paz
Tropic of Cancer
Durango
Torreón
Sabinas Basin
Nuevo Laredo
Monterrey
Matamoros
MEXICO
Mazatlán
San Luis Potosí
Tampico
León
Guadalajara
Popocatépetl 5,452 m. (17,887 ft.)
Mexico City
Puebla
Cuernavaca
Veracruz
Balsas Basin
Pan-American Highway
Southern Highlands
Oaxaca
Acapulco
Bay of Campeche
Mérida
Cancún
Yucatán Peninsula
Villahermosa
Tuxtla Gutiérrez
Isthmus of Tehuantepec
Belize City
Belmopan
BELIZE
GUATEMALA
San Pedro Sula
Quezaltenango
Guatemala City
HONDURAS
Tegucigalpa
San Salvador
EL SALVADOR
Volcanic Axis
CENTRAL AMERICA
NICARAGUA
León
Managua
San José
Limón
COSTA RICA
PANAMA
Colón
Panama City
Pan-American Highway
Havana
CUBA
Camagüey
Guantánamo
Santiago
HAITI
Puerto Pl.
Port-au-Prince
JAMAICA
Montego Bay
Port Antonio
Kingston
GREATER ANTILLES
Caribbean
PACIFIC OCEAN
Santa Marta
Barranquilla
Cartagena
Maracaibo
Medellín
COLOMBIA
Bogotá
30° 25° 20° 15° 10° 5°
Longitude West of Greenwich
110° 105° 100° 95° 90° 85° 80° 75°

FIGURE 4B-1

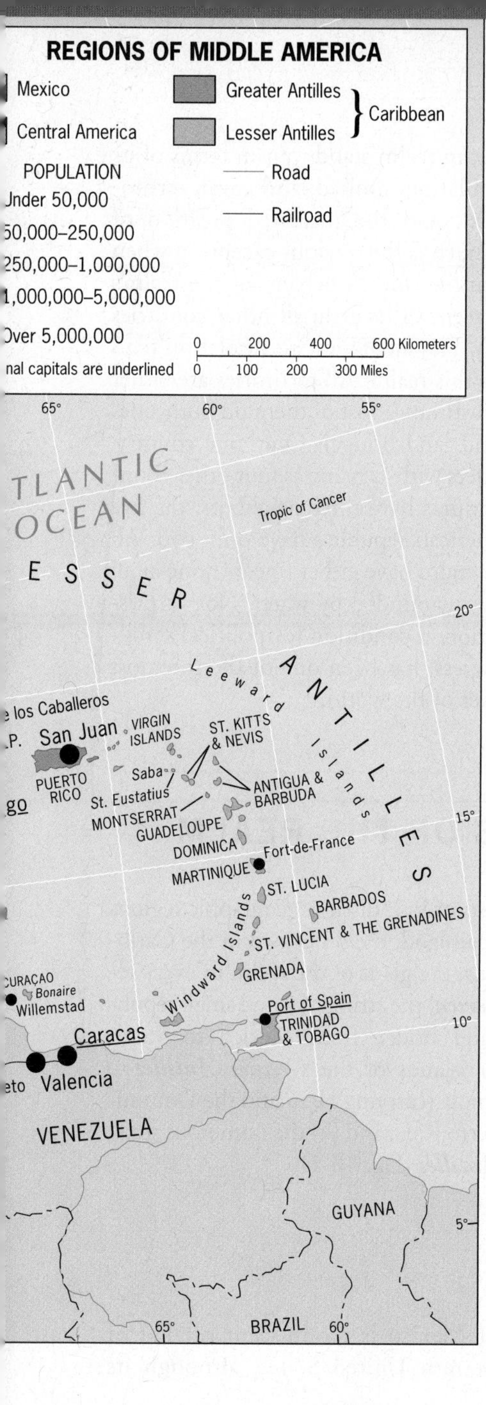

REGIONS

Mexico

Central America

Caribbean Basin

Greater Antilles

Lesser Antilles

IN THIS CHAPTER

- Mexico's shifting drug wars
- Indigenous peoples demand recognition and rights
- Panama Canal expansion fuels boom in Panama City
- Aftermath of Haiti's 2010 killer earthquake
- The debate over Puerto Rico's status

CONCEPTS, IDEAS, AND TERMS

Acculturation 1

Transculturation 2

Ejidos 3

Maquiladoras 4

Failed state 5

Dry canal 6

Biodiversity hot spot 7

Tierra templada 8

Tierra caliente 9

Offshore banking 10

Social stratification 11

Mulatto 12

Small-island developing economies 13

© Jan Nijman

Where were these pictures taken
Find out at www.wiley.com/college/deblij

The Middle American realm stands out in terms of its overall small population, limited land cover, territorial fragmentation, and the insulated position of many its countries. Mexico is the obvious exception when it comes to size, accounting for 58 percent of the realm's population and 72 percent of its area; all other countries vary from small to tiny. Fragmentation and isolation tend to go hand in hand in this realm. All territories are either islands or have long coastlines (most of them on both sides of the narrow mainland land bridge). Only one country, Guatemala, shares borders with as many as four other countries; Mexico and Honduras have three neighbors; the five remaining Central American republics have only two. All other territories in the realm have either one or none at all as they are completely surrounded by water. Not surprisingly, regional integration, a condition for political stability and economic progress, has been one of the foremost challenges in this corner of the world.

REGIONS OF THE REALM

Middle America consists of four distinct geographic regions (the first two on the mainland; the other two in the Caribbean Basin): (1) ***Mexico***, the giant of the realm in every respect; (2) ***Central America***, the string of seven small republics occupying the land bridge from Mexico to South America; (3) the four islands of the ***Greater Antilles***—Cuba, Jamaica, Hispaniola (carrying Haiti and the Dominican Republic), and Puerto Rico; and (4) the numerous small islands of the ***Lesser Antilles*** (Fig. 4B-1).

MEXICO

Physiography

The physiography of Mexico is reminiscent of that of the conterminous western United States, although its

© Jan Nijman

environments are more tropical. Figure 4B-2 shows several prominent features: the elongated peninsula of Baja (Lower) California in the northwest, the far eastern Yucatán Peninsula, and the Isthmus of Tehuantepec in the southeast where the Mexican landmass tapers to its narrowest extent. Here in the southeast, Mexico most resembles Central America physiographically; a mountainous backbone forms the isthmus, curves southeastward into Guatemala, and extends northwestward toward Mexico City. Shortly before reaching the capital, this mountain range divides into two chains, the Sierra Madre Occidental in the west and the Sierra Madre Oriental in the east (Figs. 4B-1, 4B-2). These diverging ranges frame the funnel-shaped Mexican heartland, the center of which consists of the rugged, extensive Plateau of Mexico (the important Valley of Mexico lies near its southeastern end). As Figure G-7 reveals, Mexico's climates are characterized by dryness, particularly in the broad, mountain-flanked north. Most of the better-watered areas lie in the southern half of the country where a number of major population concentrations have developed (see Fig. 4A-2).

Regional Diversity

Physiographic, demographic, economic, historical, and cultural criteria combine to reveal a regionally diverse Mexico extending from the lengthy ridge of Baja California to the tropical lowlands of the Yucatán Peninsula, and

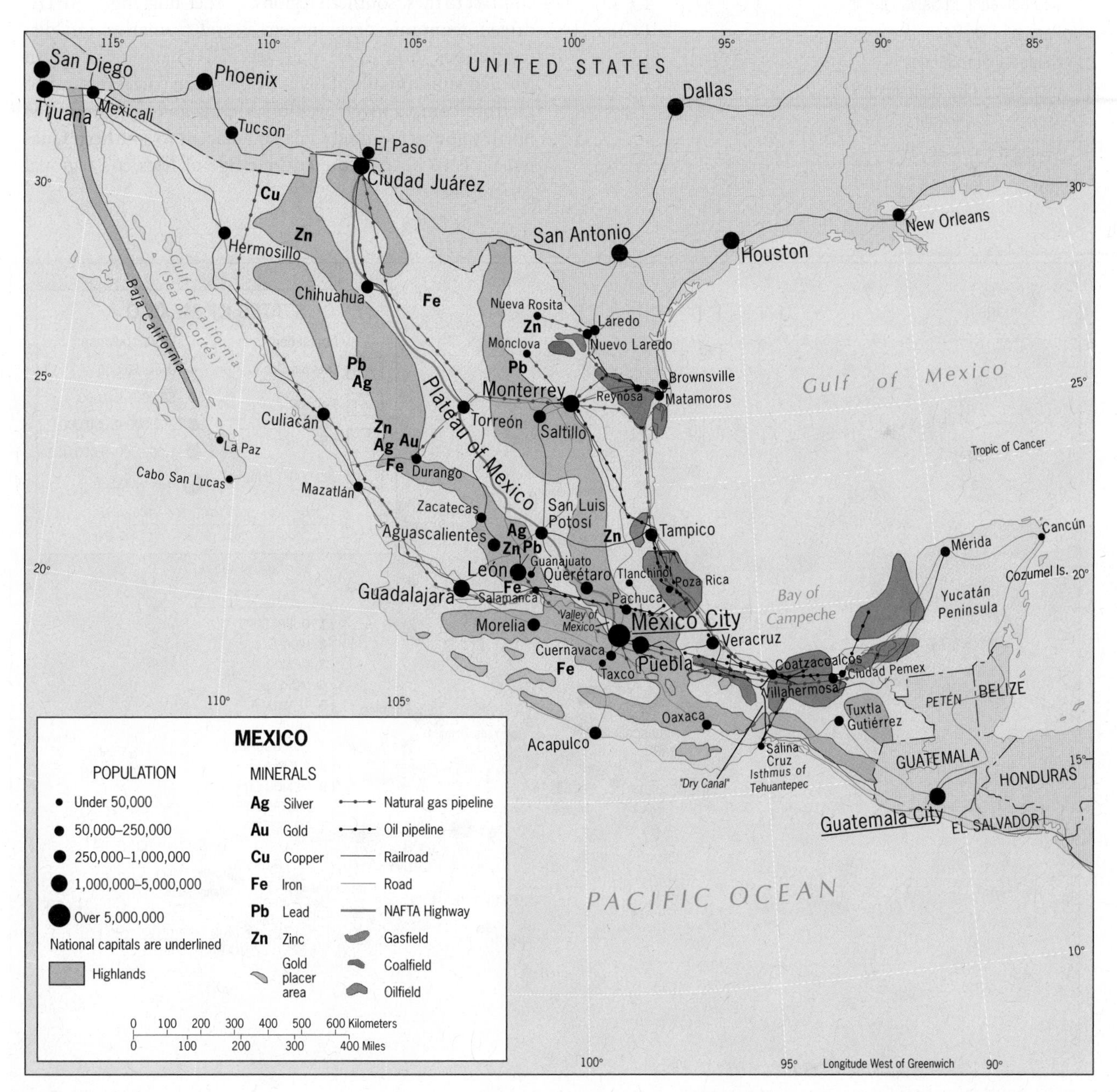

FIGURE 4B-2 © H. J. de Blij, P. O. Muller, and John Wiley & Sons, Inc.

MAJOR CITIES OF THE REALM

Metropolitan Area	Population* (in millions)
Mexico City, Mexico	30.3
Guadalajara, Mexico	4.8
Monterrey, Mexico	4.6
San Juan, Puerto Rico	2.5
Puebla, Mexico	2.5
Port-au-Prince, Haiti	2.4
Santo Domingo, Dominican Rep.	2.3
Havana, Cuba	2.1
Tijuana, Mexico	2.0
San José, Costa Rica	1.7
San Salvador, El Salvador	1.7
Panama City, Panama	1.6
Ciudad Juárez, Mexico	1.3
Guatemala City, Guatemala	1.3
Tegucigalpa, Honduras	1.2
Managua, Nicaragua	1.0

*Based on 2014 estimates.

from the economic dynamism of the U.S. borderland to the indigeous traditionalism of far southeastern Chiapas (Fig. 4B-2). The country's core area, anchored by Mexico City and extending westward to Guadalajara, lies within the transition zone between the more Hispanic-mestizo north and the dominantly indigenous-mestizo south. East of the core area lies the Gulf Coast, once dominated by major irrigation projects and sprawling livestock-raising schemes but now the mainland center of Mexico's petroleum industry. The region south of the core is dominated by the rugged, Pacific-fronting Southern Highlands, where coastal Acapulco's luxurious resorts stand in stark contrast to the indigenous villages and communal-farm settlements of the interior.

The dry, far-flung north stands in particularly sharp contrast to these southern regions. Border-hugging, NAFTA-driven economic development is still formative and discontinuous, but, as we shall see, it is changing northern Mexico substantially. This is also true in Yucatán, and to go from comparatively well-off Mérida and Cancún in the north to poverty-mired Chiapas adjacent to southern Guatemala is to observe the entire range of Mexico's regional geography.

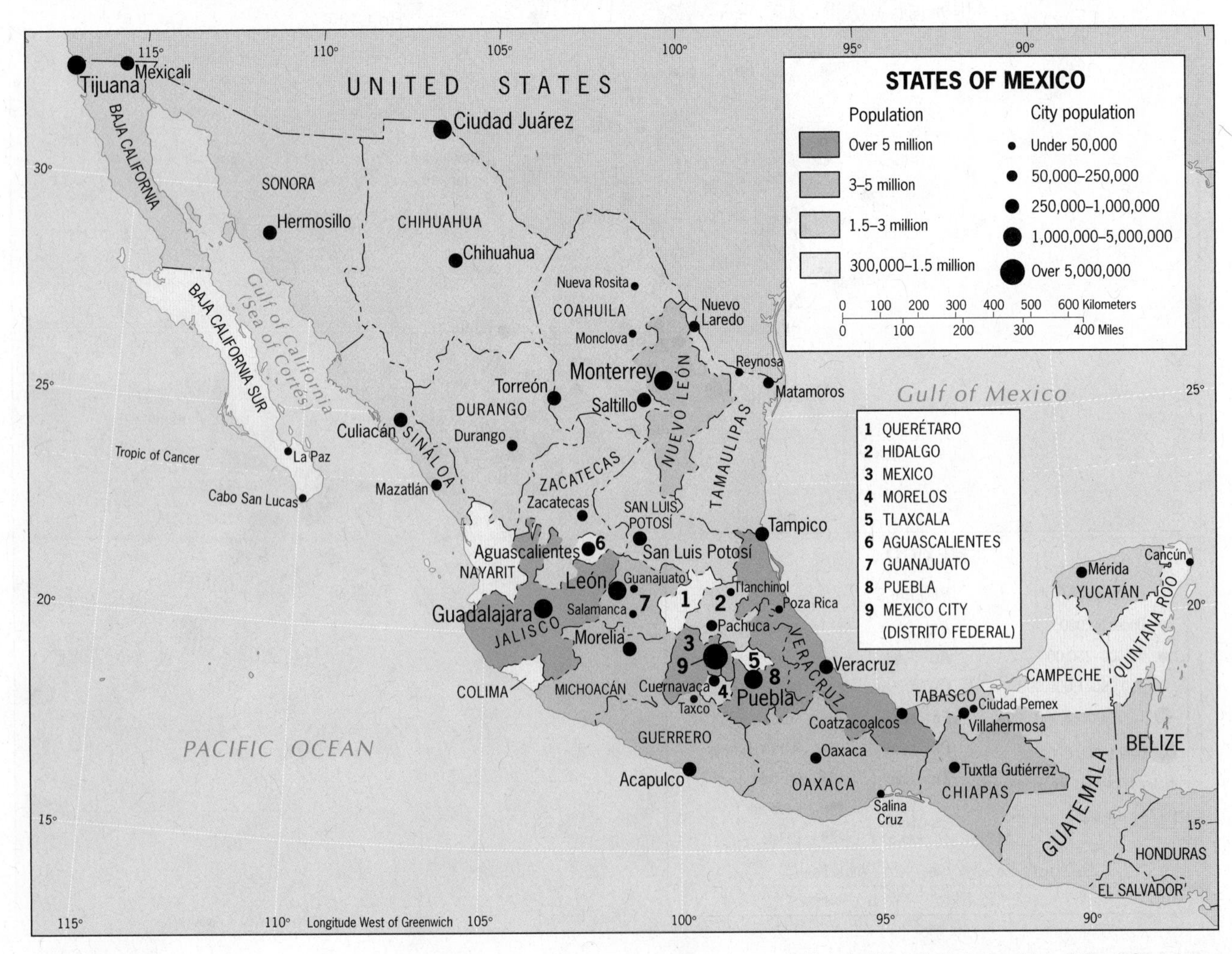

FIGURE 4B-3

AMONG THE REALM'S GREAT CITIES . . .

MEXICO CITY

MIDDLE AMERICA HAS only one great metropolis: Mexico City. With 30.3 million inhabitants, Mexico City is home to just over one-fourth of Mexico's population and grows by more than 300,000 every year. Even more significantly, Mexico City has surpassed Tokyo, Japan (current population: 26.6 million) to become the world's largest urban agglomeration.

Lakes and canals marked this site when the Aztecs built their city of Tenochitlán here seven centuries ago. The conquering Spaniards made it their headquarters, and following independence the Mexicans made it their capital. Centrally positioned and well connected to the rest of the country, Mexico City, hub of the national core area, became the quintessential primate city.

Vivid social contrasts mark the cityscape. Historic plazas, magnificent palaces, churches, villas, superb museums, ultramodern skyscrapers, and luxury shops fill the city center. Beyond lies a zone of comfortable middle-class and struggling, but stable, working-class neighborhoods. Outside this belt, however, lies a ring containing more than 500 established slum areas and countless, even poorer *ciudades perdidas*—the "lost cities" where newly arrived peasants live in miserable poverty and squalor. (These squatter settlements contain one-third of the metropolitan area's population.) Mexico City's more affluent residents have also been plagued by problems in recent times as the country's social and political order came close to unraveling. Rampant crime remains a serious concern, much of it associated with corrupt police.

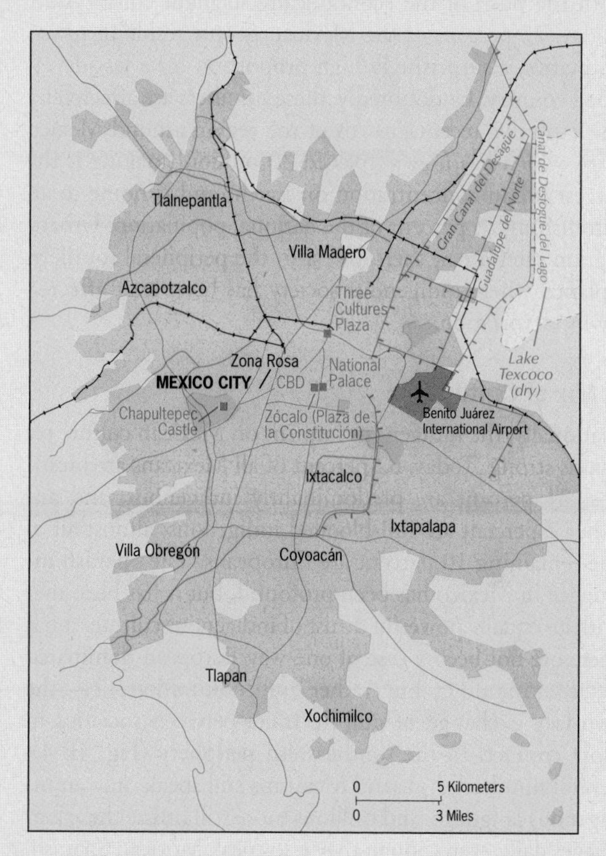

Environmental crises parallel the social problems. Local surface waters have long since dried up, and groundwater supplies are approaching depletion; to meet demand, the metropolis must now import much of its water by pipeline from across the mountains (with almost half of that supply lost through leakages in the city's crumbling water pipe network). Air pollution here is among the world's worst as nearly 5 million motor vehicles and tens of thousands of factories churn out smog that in Mexico City's thin, high-altitude air sometimes reaches 100 times the acceptably safe level. And add to all this a set of geologic hazards: severe land subsidence as underground water reservoirs are overdrawn; the ever-present threat of earthquakes that can wreak havoc on the city's unstable surface (the last big one occurred in 1985); and even the risk of volcanic activity, as nearby Mount Popocatépetl occasionally shows signs of ending centuries of dormancy.

In spite of it all, the great city continues to beckon, and every year perhaps 100,000 of the desperate and the dislocated arrive with hope—and little else.

Population Patterns

Mexico's population expanded rapidly throughout the closing decades of the twentieth century, doubling in just 28 years; but demographers have recently noted a sharp drop in fertility, and they are predicting that Mexico's population (currently at 120 million) will cease growing altogether by about 2050. That would have enormous implications for the country's economy and for the United States as well since it will reduce cross-border migration in the future.

The distribution of Mexico's population relative to the country's 31 internal States is shown in Figure 4B-3 (and even more precisely in Fig. 4A-2). The largest concentration, containing the core area and more than half the Mexican people, extends across the densely populated "waist" of the country from Veracruz State on the eastern Gulf Coast to Jalisco State on the Pacific. The center of this corridor is dominated by the most populous State, Mexico (**3** on the map), at whose heart lies the Federal District of Mexico City (**9**). In the dry and rugged terrain to the north of this central corridor lie Mexico's least-populated States. Southern Mexico also exhibits a sparsely peopled periphery in the hot and humid lowlands of the Yucatán, but to the southwest most of the highlands of the continental spine contain sizeable populations.

Another major feature of Mexico's population map is urbanization, driven by the pull of the cities (with their perceived opportunities for upward mobility) in tandem with the push of the economically stagnant countryside. Today, 77 percent of the Mexican people reside in towns and cities, a surprisingly high proportion for a less-developed country. Undoubtedly, these numbers are affected by the explosive recent growth of the region around Mexico City, which has now surpassed 30 million (making it the largest urban concentration on Earth) and is home to an astonishing 25 percent of the national population. Urbanization rates are at their lowest in the peripheral southern uplands where indigenous society has been least affected by modernization.

A Mix of Cultures

Nationally, the indigenous imprint on Mexican culture remains strong. Today, 62 percent of all Mexicans are mestizos, 21 percent are predominantly indigenous, and another 7 percent are full-blooded indigenous; almost all of the remaining 10 percent are Europeans. The Spanish influence in Mexico has been profound, but it has been met with an equally powerful thrust of indigenous culture. It has therefore not been a case of one-way European-dominated **acculturation [1]** but rather **transculturation [2]**—the two-way exchange of culture traits between societies in close contact. In the southeastern periphery (Fig. 4B-4), several hundred thousand Mexicans still speak only an indigenous language, and millions more still utilize these languages daily even though they also speak Mexican Spanish. The latter has been strongly shaped by indigenous influences, as have Mexican modes of dress, foods and cuisine, artistic and architectural styles, and folkways. This fusion of heritages, which makes Mexico unique, is the end product of an upheaval that began to reshape the country just over a century ago.

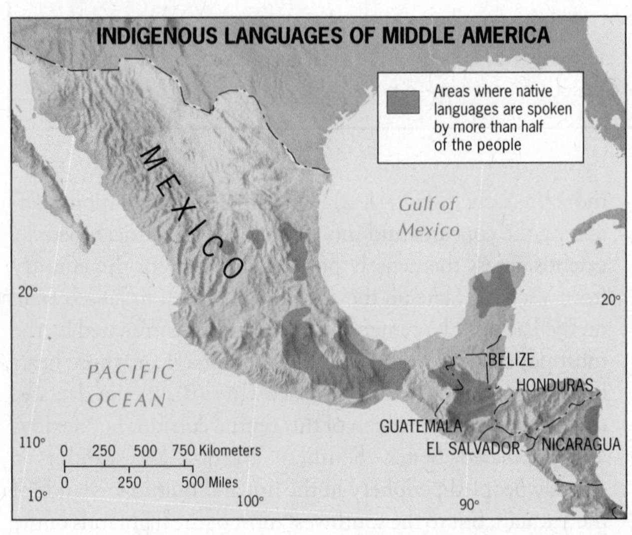

FIGURE 4B-4 © H. J. de Blij, P. O. Muller, and John Wiley & Sons, Inc.

Agriculture: Fragmented Modernization

Modern Mexico was forged in a revolution that began in 1910 and set into motion events that are still unfolding today. At its heart, this revolution was about the redistribution of land. As late as 1900, only about 8000 haciendas (large, traditional, family-owned farms) blanketed virtually all of Mexico's better farmland. About 95 percent of all rural families owned no land whatsoever and toiled as *peones* (landless, constantly indebted serfs) on the haciendas. The triumphant revolution produced a new constitution in 1917 that launched a program of expropriation and parceling out of the haciendas to rural communities.

Since 1917, more than half the cultivated land in Mexico has been redistributed, mostly to peasant communities consisting of 20 families or more. On such farmlands, known as ***ejidos* [3]**, the government holds title to the land, but the rights to its use are parceled out to village communities and then to individuals for cultivation. This system of land management is an indigenous legacy, and not surprisingly most *ejidos* lie in central and southern Mexico where nativist social and agricultural traditions are strongest. About half of Mexico's land continues to be held in such "social landholdings." However, the reforms did not lead to increased productivity. Some of the land is excessively fragmented, causing both inferior crop yields and widespread rural poverty. In the 1990s, the Mexican government attempted to privatize *ejidos*, hoping to promote consolidation and increase productivity, but that effort did not succeed. To date, less than 10 percent of the *ejidos* have been privatized.

On the other hand, larger-scale commercial agriculture has diversified during the past three decades and made major gains with respect to both domestic and export markets. The country's arid northern tier has led the way as major irrigation projects have been built on streams flowing down from the interior highlands. Along the booming northwestern coast of the mainland, which lies within a day's drive of Southern California, large-scale production of fruits, vegetables, cut flowers, and cotton have become the cornerstones of an increasingly profitable export trade to the U.S. and elsewhere.

Shifting Economic Geographies

The **maquiladoras [4]** in the country's northern borderland—foreign-owned factories that assemble imported raw materials and components and then export finished manufactures back to the United States (see Chapter 4A)—account for about one-fifth of Mexico's industrial jobs and approximately half of its total exports. Many companies are headquartered in cities just north of the border, such as San Diego and El Paso, and this is where design, marketing, and other strategic work takes place. But most of the workforce is located south of the border at the assembly plants (which is why the cities on the Mexican side have grown so much faster than their U.S. counterparts). Think electronics, machinery, garments, construction materials, automobile parts, and much more. Although the impact of NAFTA has been enormous, it should be noted that Mexican employees work long hours for low wages with meager fringe benefits; moreover, most reside in the overcrowded clusters of rudimentary shacks and slum settlements that encircle the burgeoning urban centers of the border zone. Belatedly, wages have increased, with workers now earning between U.S. $2 and $6 an hour, depending on skills and tasks performed.

There is no job security, however. During the past few years, this borderland region received two serious blows. First, the Great Recession in the United States slowed the demand for exports, which immediately curtailed production, profits, growth, and employment all across Mexico's north. And second, in what could be more of a structural threat, hundreds of American and other foreign manufacturers who had moved their plants to northern Mexico in the post–1990 period decided to relocate once again—to East and Southeast Asia, where wages were lower than those paid by the maquiladoras. Although factories assembling bulky products such as motor vehicles and major home appliances were still better off situated directly across the U.S. border, many others producing lighter and smaller goods such as electronic equipment and cameras were quick to relocate to China, Vietnam, and other Asia-Pacific countries. As a result, thousands of Mexican workers found themselves unemployed.

Mexico's only way to reverse this trend is through the growth of higher-paying jobs in more advanced sectors—particularly electronic goods—that are less likely to be lost to other parts of the disadvantaged world. Expansion of education and training in high-technology and management fields is a critical first step. The northeastern city of Monterrey in the relatively high-income State of Nuevo Léon has become a successful model in this effort, nurturing both an international business community and a complex of up-to-date industrial facilities that have attracted leading multinational corporations. Here indeed lies hope for Mexico's future, and one frequently hears reports these days that rising labor costs in China may induce certain manufacturers to return to northern Mexico, allowing them to more directly control flexible (so-called 'on-time') production at a close distance. Since the restless global economy constantly requires the world's regions to adjust in order to keep pace, Mexico's prospects may well brighten in coming years.

States of Contrast

Countries with strong regional disparities face serious challenges that can be difficult to overcome and may deteriorate over time. Mexico's southernmost States—Chiapas, Oaxaca, and Guerrero, all bordering the Pacific Ocean—are by far the poorest. The States bordering the United States in the north, including Nuevo Léon, Chihuahua, and Baja California, have the highest incomes. Using rural poverty as a measure, only about 10 percent of people in the countryside in the north are in the poorest category, but nearly 50 percent in the south.

Mexico's north-south divide is especially noticeable in the economic data mapped in Figure 4B-5. In general terms, annual per capita income in the northern States exceeds U.S. $10,000; but in the southern States it falls below $5000. Moreover, Mexico's infrastructure, already inadequate, serves the south far less well than the north. Whatever the index—literacy, electricity use, water availability—the south lags by a wide margin.

This is a serious problem because the impoverished, heavily indigenous population of the south not only trails far behind the rest of Mexico in overall development, but is also the least well educated, the least productive agriculturally, and the most isolated part of the country. Since the early 1990s, a radical group of Maya peasant farmers in Chiapas State calling themselves the Zapatista National Liberation Army (ZNLA) has engaged in guerrilla warfare, demanding better treatment for Mexico's 33 million indigenous citizens. Despite substantial public support, their 20-year struggle has not yielded significant results. In fact, NAFTA has only widened the gap between north and south. For upwardly mobile Mexicans in the north, the cause of the Zapatistas is a distant one indeed.

These still-widening contrasts were thrown into sharp relief in 2006 and again in 2012, when Mexico's presidential elections were contested by three candidates broadly representing the conservative elite, the working classes, and the middle classes. When the ballots were counted, there was a clear spatial divide in the voting patterns, which reflected the varying economic fortunes across the country. Generally, most of the votes in the north went to the conservative and centrist candidates while the poorer southern States supported the leftist candidate (see inset map in Fig. 4B-5). It was the centrist candidate, Enrique Peña Nieto, who narrowly won the presidency in 2012—providing the latest evidence of Mexico's sharp politico-geographical cleavages.

The Drug Wars

Since the late 2000s, Mexico has been plagued by yet another obstacle in its struggle to achieve sustained economic progress. This problem has received worldwide attention as the cocaine-producing drug cartels centered in Colombia established new bases in U.S.-border cities in northern Mexico and launched a vicious war for supremacy. They responded in part to the success of the antidrug campaign in Colombia, but the cartels also saw new opportunities beckoning in Mexican territory adjacent to their main market in the United States. So serious was the situation by 2010 that the drug barons and their allies were killing law-enforcement officers by the hundreds as well as competitors (and often innocent bystanders) by the thousands, intimidating government and threatening all who stood in their way. Fueling this rampage was the steady stream of guns, purchased easily in the United States with the flood

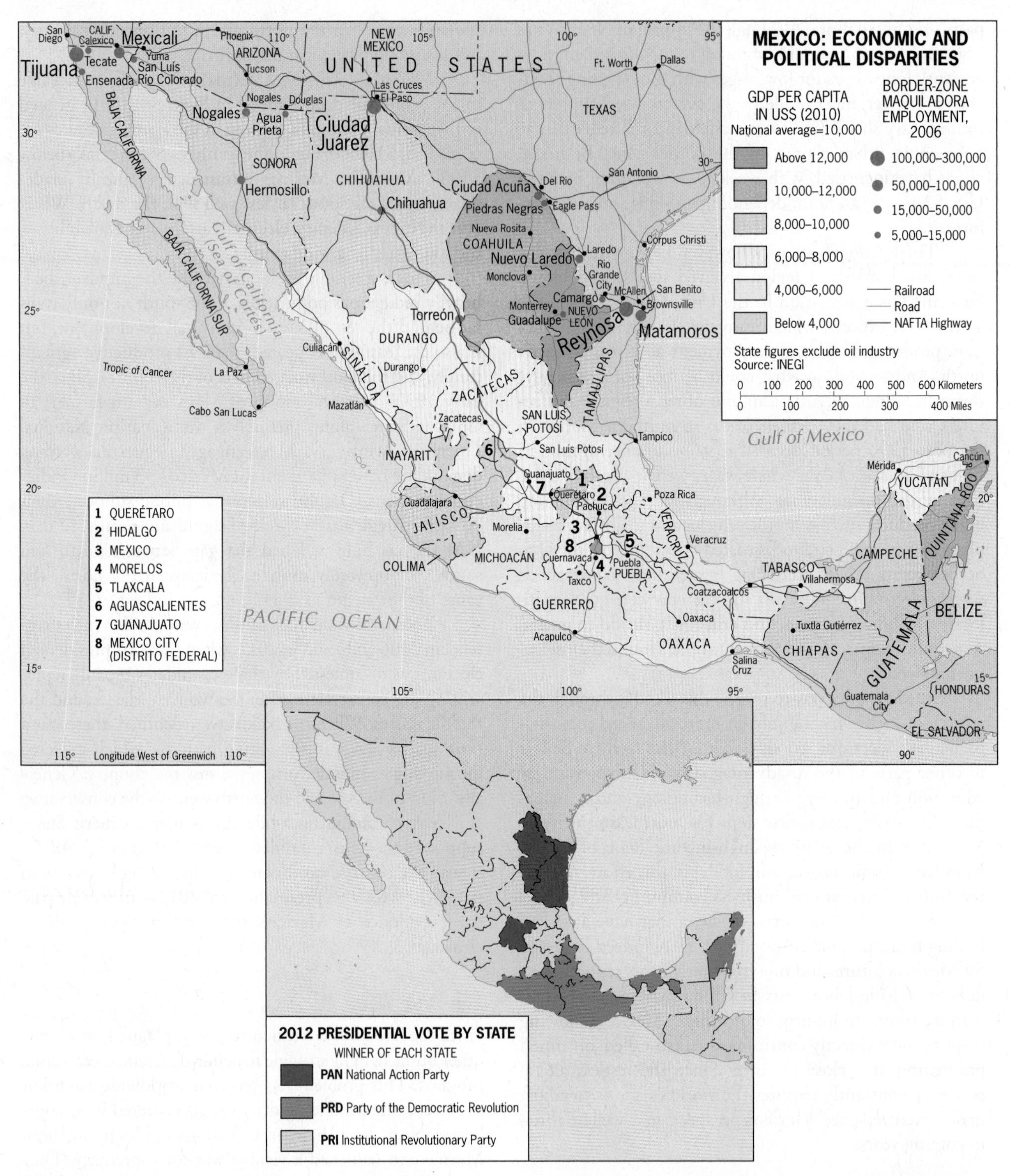

FIGURE 4B-5

of drug money generated by that illicit trade, which flowed across the border and further empowered the cartels.

Figure 4B-6 provides an overview of the geography of this drug war. Cocaine is produced in Colombia, Bolivia, and Peru (see Chapter 5A), and just about all of it enters the United States through Mexico. Bolivian and Peruvian cocaine, constituting roughly half of Mexico's "imports," is shipped by sea and illegally enters the country in the Pacific Coast States of Guerrero, Michoacán, Colima, and Sinaloa. The other half originates in Colombia and enters either overland via the Middle American land bridge through Guatemala, or by boat across the Caribbean and Gulf of Mexico to be smuggled into the eastern States of Yucatán, Quintana Róo, and Veracruz. The illicit powder is then repackaged

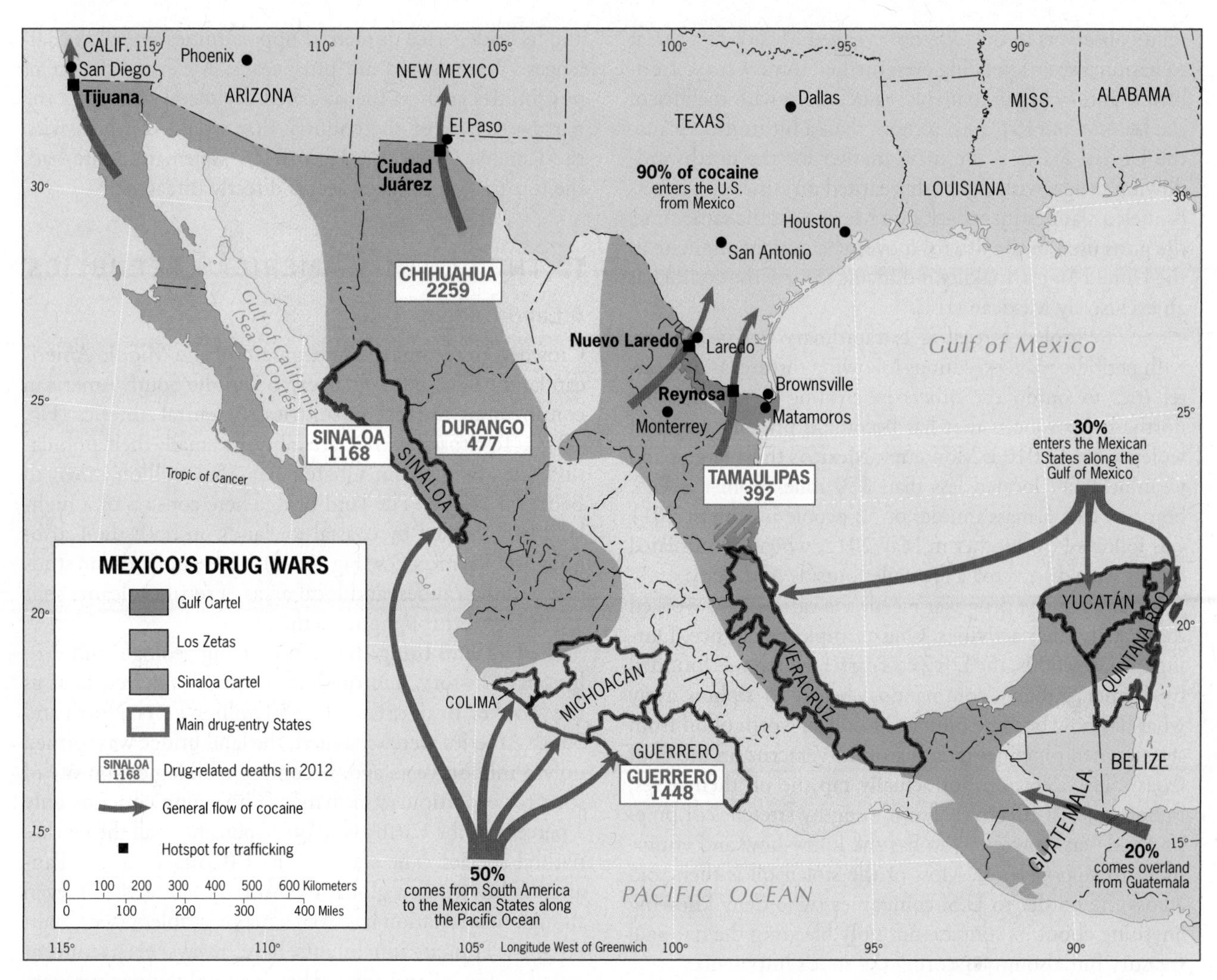

FIGURE 4B-6

(and/or crystallized to form crack) in smaller quantities for "retailing" purposes prior to being smuggled into the United States, mainly through the border cities of Reynosa, Nuevo Laredo, Ciudad Juárez, and Tijuana (Fig. 4B-6).

Interestingly, when Mexican criminal organizations were coöpted by the Colombian cartels, they demanded to be paid in cocaine and this allowed the Mexicans to start up their own distribution networks—it also meant that a good amount of the stuff 'stuck around' and ended up on the streets of Mexican cities, resulting in heightened drug addiction and rising crime. Several major Mexican cartels sprang up and are involved in this drug-trafficking operation, and competition is literally murderous. Territorial control is key in this "business": each cartel dominates its own turf, with various components being heavily contested from time to time. The cartels themselves are not particularly stable, with countless mergers and splits occurring since 2007. The two leading rivals are Los Zetas and the Sinaloa Cartel, which control huge swaths of eastern and western Mexico, respectively. A few years ago, the Sinaloa formed an alliance with two smaller cartels, La Familia and the Gulf, creating what is occasionally referred to as The Federation—which is engaged in a violent conflict for supremacy with Los Zetas. More recently, the Sinaloa Cartel extended its grip on western Mexico with takeovers of the smaller but strategically located Beltran-Leyva (along the coast of Sonora State) and the Juárez Cartel that used to control some key "export valves" in Ciudad Juárez. Next came the absorption of the even smaller but important Tijuana cartel, which used to command the well-known westernmost border city of the same name—but here, too, the Sinaloa is now in control.

Much of the worst violence occurs where control is contested, especially around the most crucial ports of entry and along primary transport routes. During 2010, nearly half of Mexico's 15,273 drug-related killings occurred in the States of Sinaloa (cocaine entry; marijuana production) and Chihuahua (which contains Ciudad Juárez, the leading U.S. port of entry). In early 2013, the murder total since the war erupted surpassed 60,000. By then, some of the violence was shifting to smaller Mexican provinces and cities as key control points like Juarez had more or less been 'consolidated' by either the Sinaloa or the Zetas.

As the conflict intensified and the Mexican government clearly began to lose control of parts of its territory, many observers in the U.S. now viewed Mexico as a threat to national security; going even further, some were so troubled that they began to invoke associations with the idea of the **failed state [5]**. This is more than a bit ironic because the United States is the main market for the northward-flowing drugs, with the profits earned on American streets funneled clandestinely back into Mexico. At the same time, the guns used in the drug war overwhelmingly originate in the United States, making it difficult to view the conflict as an exclusively Mexican crisis.

The violence itself is extraordinary and gruesome, with periodic mass executions following torture, as one cartel tries to outdo the others in creating terror. Another northern metropolis that has become a primary target of violence since 2010 is Monterrey, Mexico's third-largest and wealthiest city located less than 150 miles from the U.S. border. Here, a mass murder of 52 people in August 2011 was followed by another in May 2012, when 49 mutilated bodies were discovered a few miles outside of the city.

Even as the drug war raged, the cartels have worked to diversify their activities. One example will suffice. During the late 2000s, the Los Zetas cartel, which was founded by former military commandos, began to siphon away what has now become billions of dollars worth of oil from state-owned pipelines in and around Veracruz on the Gulf Coast. The Zetas do not actually tap the oil themselves, but are said to effectively "own" lengthy stretches of pipeline and "tax" anyone who has the know-how and equipment to siphon it off. Most of the stolen oil is then sold across the border to U.S. companies (who deny knowing anything about its source), not only bleeding the national treasury but also underscoring the state's impotence.

Mexico's Future

The future of this country depends on the ability of the government to meet an array of formidable challenges. In the short term, it must end the violence that has destabilized so much of Mexico. In the longer term, it needs to narrow the gap between rich and poor to reduce massive—and still widening—regional inequalities. And from the standpoint of economic geography, the government has to spearhead the effort to spread the positive effects of NAFTA from north to south, especially through upgrading infrastructure, increasing investment in education, and implementing antipoverty programs that work.

The future of Mexico is inextricably bound up with the United States, in good times and bad. The good refers to these countries' economic interaction: Mexico's economy has benefited enormously by developing closer economic ties with North America, and the United States continues to be the country's most important trading partner. Given this foundation, Mexico's longer-term problems may seem less acute—but it would be a mistake for its federal government to lose sight of these problems, no matter how urgent the pursuit of the drug cartels. Clearly, Mexico is in the midst of a tumultuous transformation, and its leaders face numerous opportunities as well as challenges. When times are prosperous, Mexicans dream of possibilities such as the so-called **dry canal [6]** across the narrowest part of the country that would compete with the Panama Canal (see Fig. 4B-2). When times are bad, the integrity of the state seems directly threatened.

THE CENTRAL AMERICAN REPUBLICS

A Land Bridge

Crowded onto the narrow segment of the Middle American land bridge between Mexico and the South American continent are the seven countries of Central America (Fig. 4B-7). Territorially, they are all quite small; their population sizes range from Guatemala's 15.7 million down to Belize's 325,000. The land bridge here consists of a highland belt flanked by coastal lowlands on both the Caribbean and Pacific sides (Fig. 4B-8). These uplands are studded with volcanoes, and local areas of fertile volcanic soils are scattered throughout them.

The land bridge has a fascinating geologic and evolutionary history, and one famous study referred to it as the Monkey Bridge. For some 50 million years, North and South America were separated; the land bridge was formed only 3 million years ago, becoming a biologic highway of sorts for evolutionary exchange. The region contains only 1 percent of the Earth's land but 7 percent of all the world's natural species. The southern part (Costa Rica and Panama) is known as a global **biodiversity hot spot [7]**, even though deforestation has been a major problem (see Chapter 4A). Human inhabitants have always been concentrated in the upland zone, where tropical temperatures are moderated by elevation and rainfall is sufficient to support a variety of crops.

Central America is not a large region, but because of its physiography it contains many isolated, comparatively inaccessible locales. The population tends to concentrate in the much cooler uplands—***tierra templada* [8]**—and population densities are generally greater toward the Pacific than toward the Caribbean side (see Fig. 4A-2). The most significant exception is El Salvador, whose political boundaries confine its inhabitants mostly to its tropical coastal lowlands—***tierra caliente* [9]**—moderated here by the somewhat cooler waters of the Pacific.

Central America, as noted earlier, actually begins within Mexico, in Chiapas and in Yucatán, and the region's republics face many of the same problems as the least-developed parts of Mexico. Population pressure is one of them. A population explosion began in the mid-twentieth century, increasing the region's inhabitants from 9 million to over 45 million by 2014. Today, the region also continues its slow emergence from a grim period of turmoil that lasted through most of the 1980s and 1990s. In fact, devastating inequities, repressive governments, external interference, and the frequent unleashing of armed forces have destabilized Central America for much of its modern history.

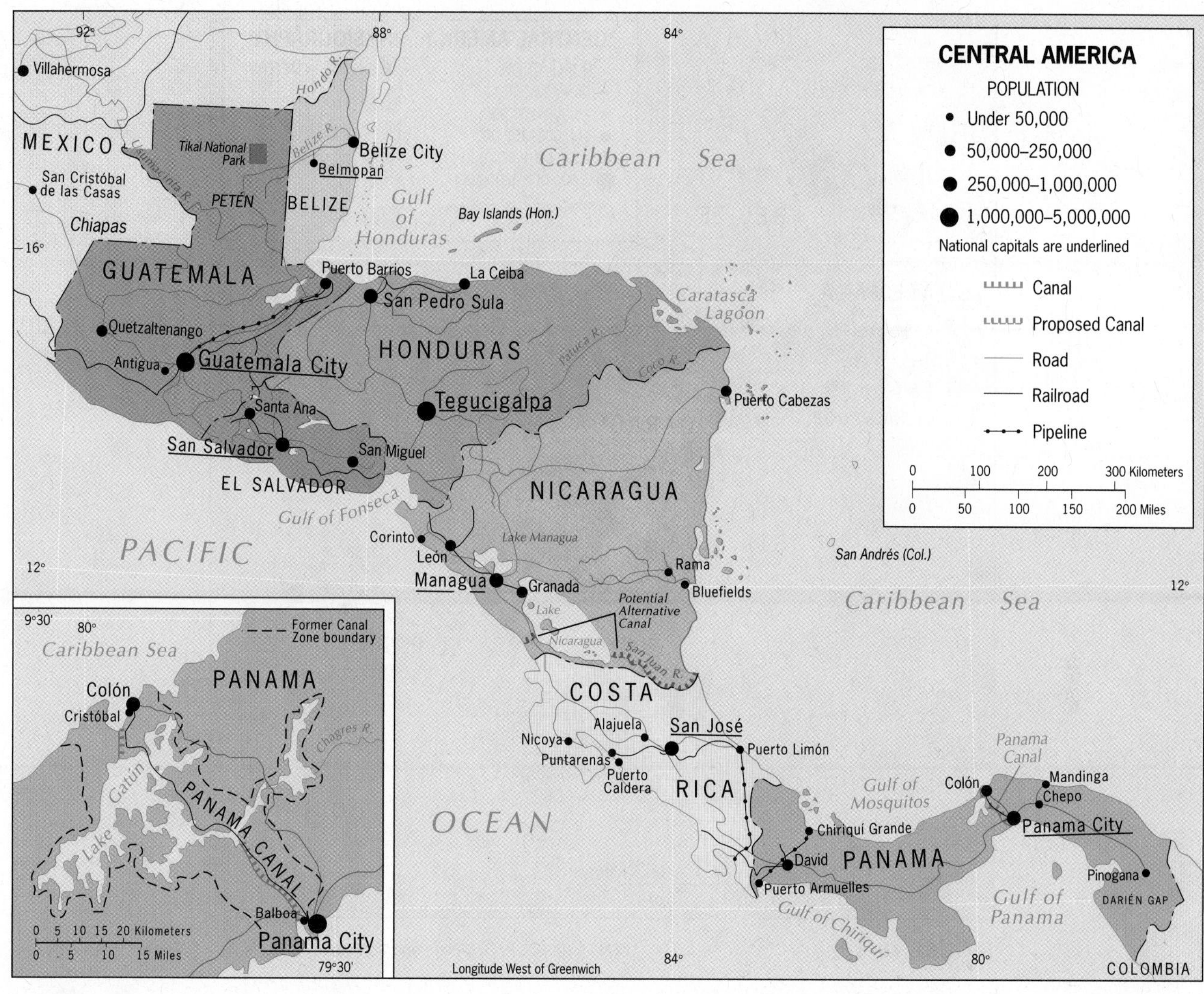

FIGURE 4B-7

Guatemala

The westernmost of Central America's republics, Guatemala has more land neighbors than any other. Straight-line boundaries lying across the tropical forest mark much of the border with Mexico, creating the box-like region of Petén between Chiapas State on the west and Belize on the east; also to the east lie Honduras and El Salvador (Fig. 4B-7). This heart of the ancient Maya Empire, which remains strongly permeated by indigenous culture and traditions, has only a small window on the Caribbean but a longer Pacific coastline. Guatemala was still part of Mexico when the Mexicans threw off the Spanish yoke, and though independent from Spain after 1821, it did not become a separate republic until 1838. Mestizos, not the indigenous majority, secured the country's independence. Most populous of the seven republics with 15.7 million inhabitants (mestizos are in the majority with 57 percent, indigenous 43 percent), Guatemala has seen a great deal of conflict. Repressive regimes brokered deals with U.S. and other foreign economic interests that stimulated development, but at a high social cost. Over the past half-century, military regimes have dominated political life. The deepening split between the wretchedly poor indigenous populations and the better-off mestizos, who here call themselves ***ladinos***, generated a civil war that started in 1960 and claimed more than 200,000 lives as well as 50,000 "disappearances" before it ended in 1996. An overwhelming number of the victims were of Maya descent; the mestizos continue to control the government, army, and land-tenure system.

Guatemala's tragedy is that its economic geography demonstrates considerable potential, but has long been shackled by unrelenting internal conflicts and the widening of one of the hemisphere's biggest gaps between rich and poor. In tandem, they keep the income of at least 60 percent of the population below the poverty line (some estimates range as high as 75 percent). The country's mineral wealth includes nickel in the highlands and oil in the lower-lying north. Agriculturally, soils are fertile and moisture is ample over highland areas large enough to produce a wide range of crops, including excellent coffee.

Since 2010, a dangerous new problem has threatened the country as the drug wars increasingly spill across

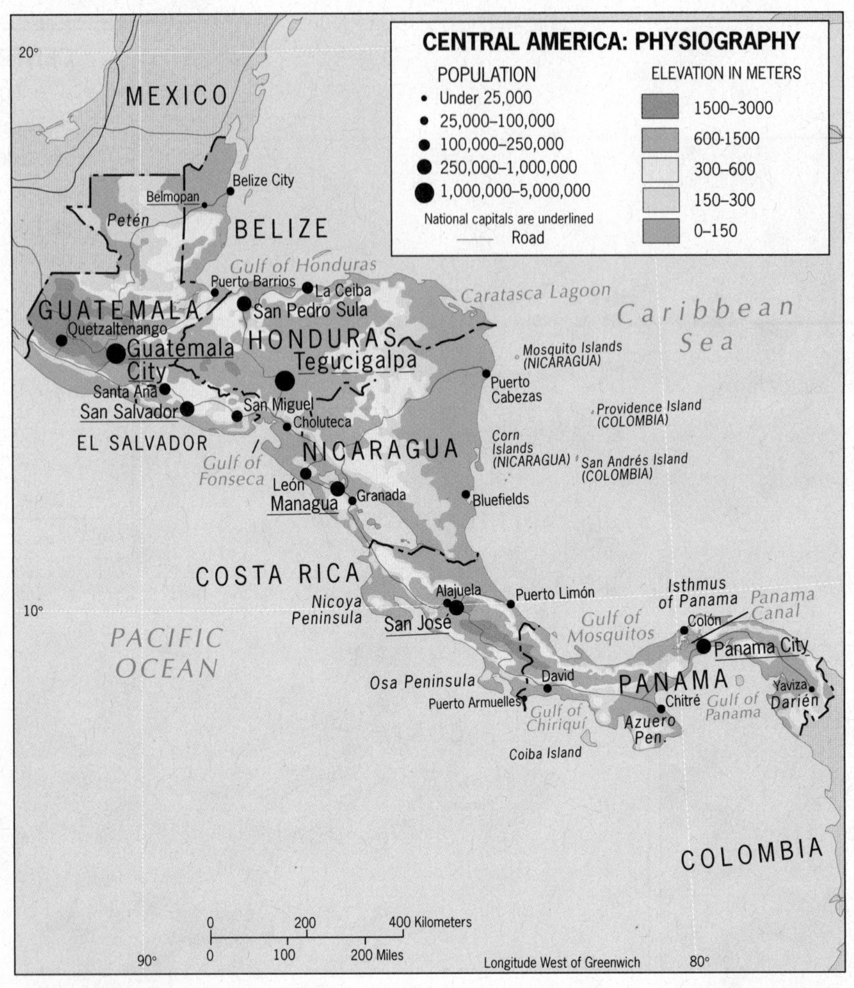

FIGURE 4B-8

the Mexican border. Figure 4B-6 shows that both Los Zetas and the Sinaloa Cartel have staked out turfs inside southwestern Guatemala, the staging area for the 20 percent of Colombian cocaine that penetrates Mexico overland—a threat both to Mexico and Guatemala. If this activity continues to intensify, it will raise a huge new challenge for the Mexican government: already stretched thin fighting the drug-related conflicts in its northern border zone, the country's beleaguered security forces will also need to more heavily police the opposite, 929-kilometer (577-mi) border with Guatemala.

Belize

Strictly speaking, Belize is not a Central American republic in the same tradition as the other six. Until 1981, this country, a wedge of land between northern Guatemala, Mexico's Yucatán Peninsula, and the Caribbean Sea (Fig. 4B-7), was a dependency of the United Kingdom known as British Honduras. Slightly larger than Massachusetts and containing a minuscule population of about 325,000 (many of African descent), Belize is much more reminiscent of a Caribbean island than of a continental Middle American state. Today, all that is changing as the demographic complexion of Belize is being reshaped. Thousands of residents of African descent have emigrated (many went to the United States) and were replaced by tens of thousands of Spanish-speaking immigrants. Most of the latter have been land seekers and refugees from strife in nearby Guatemala, El Salvador, and Honduras, and their proportion of the Belizean population has risen from 33 in 1980 to more than 50 percent today. And with the newcomers now in the majority, Belize's cultural geography is becoming increasingly Hispanicized; in fact, by 2012 Spanish speakers outnumbered those who use English or creole as a first language.

The Belizean transformation extends to the economic sphere as well. No longer just an exporter of sugar and bananas, Belize today is producing new commercial crops, and its seafood-processing and clothing industries have become major revenue earners. Also important is tourism, which annually lures more than 200,000 vacationers to the country's Mayan ruins, resorts, and newly legalized casinos; a growing specialty is ***ecotourism***, based on the natural attractions of the country's near-pristine environment, including its magnificent offshore coral reefs.

Belize is also known as a center for **offshore banking [10]** a financial haven for foreign companies and individuals who want to avoid paying taxes in their home countries.

Honduras

Honduras is another small Middle American state that regularly the suffers the wrath of destructive hurricanes—particularly the devastating impact of Hurricane Mitch in 1998. With 8.8 million inhabitants, almost 90 percent mestizo, disadvantaged Honduras more than 15 years later has still not fully restored what was already the third-poorest economy in the Americas (after Haiti and Nicaragua). Agriculture, livestock, forestry, and limited mining form the mainstays of the economy, with the now-familiar Central American products—apparel, bananas, coffee, and shellfish—earning most of the external income.

Honduras, directly in contrast with Guatemala, has a long Caribbean coastline and a relatively small window on the Pacific (Fig. 4B-7). The country also occupies a critical place in the political geography of Central America, flanked as it is by Nicaragua, El Salvador, and Guatemala—all continuing to grapple with the aftermath of years of internal conflict and, more recently, natural disaster. Comparable in natural beauty and biodiversity to Costa Rica (see photo), Honduras hopes to exploit its potential for ecotourism but is constrained by a defective infrastructure and the lack of funding for new facilities.

Today its problems are worsening as Honduras has become a key station in the trafficking of cocaine along the land bridge from Colombia to Mexico. The escalation of drug-gang violence since 2010 gave this country the dubious distinction of having the highest murder rate in the world, eliminating any immediate efforts to promote tourism. Combined with its chaotic and unstable political situation, this country is increasingly regarded as being the closest in the realm to failed-state status.

Nonetheless, Honduras has unveiled an innovative development strategy, and the constitution has been amended to allow for the creation of quasi-independent "special development regions." The idea is to attract foreign investment into these protected zones, which are to be independent in their economic, fiscal, and budgetary policies. This initiative follows the charter-cities model that has proven successful in Hong Kong—but critics fear that it may only clear the way for organized crime to take full control of these new economic enclaves.

El Salvador

This is Central America's smallest country territorially, smaller even than Belize, but with a population 20 times as large (6.5 million) it is the most densely peopled. El Salvador adjoins the Pacific in a narrow coastal plain backed by a chain of volcanic mountains, behind which lies the country's heartland. Unlike neighboring Guatemala, El Salvador

From the Field Notes . . .

"Invited by the Honduran government to survey the country's ecotourism potential, I toured the beautiful countryside with a National Geographic Society delegation. There is no doubt that Honduras is exceptionally endowed by nature and that there is considerable potential for a vibrant tourist industry. But this is a poor country, and the government has great difficulty making the required investments—in infrastructure, for example. We flew in these small propeller planes to get from the capital, Tegucigalpa, to the Mayan ruins of Copán. There was no airport nearby and overland travel is arduous and time-consuming. Ecotourism by its very nature is small-scale and the revenues tend to be relatively modest, so it is a major challenge to raise the substantial funds needed for costly infrastructural improvements."

© Jan Nijman

www.conceptcaching.com

has a far more homogeneous population (86 percent mestizo and just 1 percent indigenous). Yet ethnic homogeneity has not translated into social or economic equality or even opportunity. Whereas other Central American countries were called banana republics, El Salvador was a coffee republic, and the coffee was produced on the huge landholdings of a few landowners and on the backs of a subjugated peasant labor force.

From 1980 to 1992, El Salvador was torn by a devastating civil war that was exacerbated by outside arms supplies from the United States (supporting the government) and Nicaragua (aiding the Marxist rebel forces). But ever since the negotiated end to that war, efforts have been under way to prevent a recurrence because El Salvador is having difficulty overcoming its legacy of searing inequality. The civil war did have one positive outcome: affluent citizens who left the country and did well in the United States and elsewhere send substantial funds back home; these ***remittances*** now provide the largest single source of foreign revenues. This has helped stimulate such industries as apparel and footwear manufacturing as well as food processing. But a major stumbling block to revitalization of the agricultural sector has again been land reform, and in this regard El Salvador's future still hangs in the balance.

Another serious issue looms as well. Here, as in certain other Central American countries, many of the former military and paramilitary troops appear to have gone 'freelance' and now play a role in organized crime, increasingly connected to the cocaine business and the Mexican cartels. Ominously, in 2011 El Salvador was reported to have had the second-highest murder rate in the world, surpassed only by neighboring Honduras.

Nicaragua

This country is best approached by re-examining the map (Fig. 4B-7), which underscores Nicaragua's position tucked away in the heart of Central America. The Pacific coast follows a southeasterly direction here, but the Caribbean coast is oriented north-south so that Nicaragua forms a triangle of land, with its lakeside capital, Managua, situated in a valley on the mountainous, earthquake-prone, Pacific side (the country's core area has always been located here). The Caribbean side, where the uplands yield to a wide coastal plain of tropical rainforest, savanna, and swampland, has for centuries been home to indigenous peoples such as the Miskito, who have lived remote from the focus of national life.

Until the 1990s, Nicaragua had a checkered history of political instability, conflict, and economic backwardness. The strife ended in 1990 when the first democratic government in decades was voted into office. But economic progress has been meager for Nicaragua's population of 6.2 million, and the economy has long been ranked as continental Middle America's poorest.

The country's options are limited: agriculture dominates the economy and manufacturing is weak. There is a growing reliance on remittances from Nicaraguans who have emigrated (now ca. 15 percent of the economy) as well as on foreign aid. For decades there has been talk of an interoceanic canal to rival the waterway across Panama, even though the land bridge here is three times wider. Nonetheless, in 2013 the government commissioned a Chinese company to plan and build a canal capable of handling ships twice the maximum size of Panama's.

Costa Rica

If there is one country that underscores Middle America's variety and diversity it is Costa Rica—because it differs significantly from its neighbors and from the norms of Central America as well. Bordered by two volatile countries (Nicaragua to the north and Panama to the east), Costa Rica is a nation with an old democratic tradition and, in this boiling cauldron, no standing army since 1949! Although the country's Hispanic imprint is similar to that found elsewhere on the mainland, its early independence, its good fortune to lie remote from regional strife, and its leisurely pace of settlement allowed Costa Rica the luxury of concentrating on its economic development. Perhaps most important, internal political stability has predominated over much of the nearly 200 years since its independence from Spain.

Like its neighbors, Costa Rica (2014 population: 4.6 million) is divided into environmental zones that parallel its coastlines. The most densely settled is the central highland zone, lying in the cooler *tierra templada*, whose heartland is the Valle Central (Central Valley), a fertile basin that contains Costa Rica's main coffee-growing area and the leading population cluster focused on the cosmopolitan capital of San José (Fig. 4B-7).

The long-term development of Costa Rica's economy has given this country the region's highest standard of living, literacy rate, and life expectancy (though even here, one-fifth of the population is trapped in poverty). Agriculture continues to dominate (with bananas, coffee, tropical fruits, and seafood the leading exports), and tourism—especially ecotourism—has expanded steadily. Despite the deforestation of more than 80 percent of its original woodland cover, Costa Rica is widely known for its superb scenery and for its (belated) efforts to protect what is left of its diverse tropical flora and fauna.

Panama

Panama owes its existence to the idea of a canal connecting the Atlantic and Pacific oceans to avoid the lengthy circumnavigation of South America. The Panama Canal (see the inset map in Fig. 4B-7) was opened in 1914, a symbol of U.S. power and influence in Middle America. The Canal Zone was held by the United States under a treaty that granted it "all the rights, powers, and authority" in the area "as if it were the sovereign of the territory." In the 1970s,

From the Field Notes . . .

"The Panama Canal remains an engineering marvel 90 years after it opened in August 1914. The parallel lock chambers each are 1000 feet long and 110 feet wide, permitting vessels as large as the Queen Elizabeth II to cross the isthmus. Ships are raised by a series of locks to Gatún Lake, 85 feet above sea level. We watched as tugs helped guide the QEII into the Gatún Locks, a series of three locks leading to Gatún Lake, on the Atlantic side. A container ship behind the QEII is sailing up the dredged channel leading from the Limón Bay entrance. The lock gates are 65 feet wide and 7 feet thick, and range in height from 47 to 82 feet. The motors that move them are recessed in the walls of the lock chambers. Once inside the locks, the ships are pulled by powerful locomotives called mules that ride on rails that ascend and descend the system. It was still early morning, and a major fire, probably a forest fire, was burning near the city of Colón, where land clearing was in progress. This was the beginning of one of the most fascinating days ever."

© H.J. de Blij

www.conceptcaching.com

as the canal was transferring more than 14,000 ships per year (that number has held fairly steady—14,500 transits were logged in 2012—but the total cargo tonnage has risen significantly) and generating hundreds of millions of dollars in tolls, Panama sought to terminate U.S. control in the Canal Zone. Delicate negotiations were launched. In 1977, an agreement was reached on a staged withdrawal by the United States from the territory, first from the Canal Zone and then from the Panama Canal itself (a process completed at the end of 1999).

Panama today reflects some of the usual geographic features of the Central American republics. Its population of 3.7 million is about 70 percent mestizo and also contains substantial indigenous, European, and black minorities. Spanish is the official language, but English is also widely used. Ribbon-like and oriented east-west, Panama's topography is mountainous and hilly. Eastern Panama, especially Darien Province adjoining Colombia, is densely covered by tropical rainforest, and here is the only remaining gap in the intercontinental Pan American Highway (Fig. 4B-7). Most of the rural population lives in the uplands west of the canal; there, Panama produces bananas, coffee, sugarcane, rice, and, along its narrow coastal lawlands, shrimps and other seafood. Much of the urban population is concentrated in the vicinity of the artificial waterway, anchored by the cities at each end of the canal.

Near the northern, Caribbean end of the Panama Canal lies the city of Colón, site of the Colón Free Zone, a huge trading entrepôt designed to transfer and distribute goods on their way to South America. It is augmented by the Manzanillo International Terminal, an ultramodern port facility that in 2011 transshipped nearly 10,000 containers a day. Near the southern, Pacific end lies Panama City, often likened to Miami because of its waterfront location and high-rise-dominated skyline. The capital is the financial center that handles the revenues generated by the canal, but its skyscrapered profile also reflects the proximity of Colombia's illicit drug industry and associated money-laundering and corruption. The city's modern image presents a stark contrast to the (rural) poverty that afflicts one-third of Panama's population.

The 2015 completion of the Panama Canal's third lane will surely boost interoceanic traffic and increase business opportunities in Panama. U.S. ports on the Gulf of Mexico and the Atlantic seaboard will also get busier as they handle this heightened volume of trade. The ports of the New York–New Jersey metropolitan region have already begun to expand their capacity, and leading retailers such as Walmart (a voracious importer of Chinese goods)

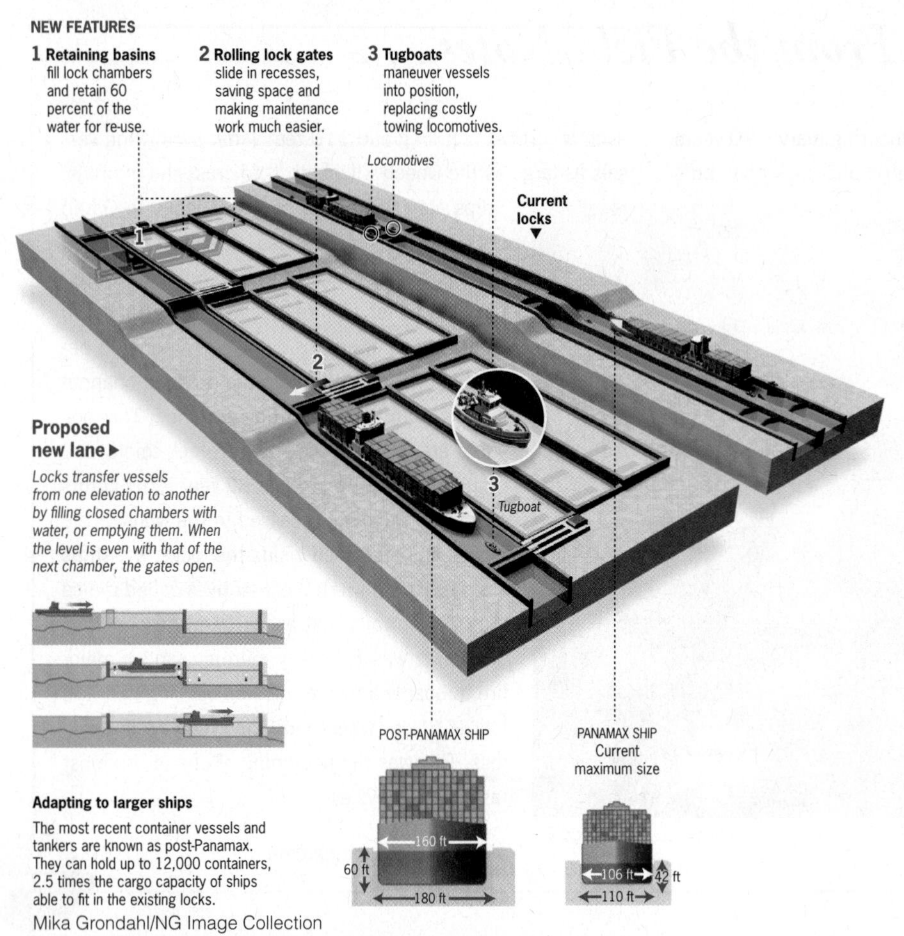

NEW FEATURES
1 Retaining basins fill lock chambers and retain 60 percent of the water for re-use.
2 Rolling lock gates slide in recesses, saving space and making maintenance work much easier.
3 Tugboats maneuver vessels into position, replacing costly towing locomotives.
Locomotives
Current locks
Proposed new lane
Locks transfer vessels from one elevation to another by filling closed chambers with water, or emptying them. When the level is even with that of the next chamber, the gates open.
Tugboat
POST-PANAMAX SHIP
PANAMAX SHIP
Current maximum size
Adapting to larger ships
The most recent container vessels and tankers are known as post-Panamax. They can hold up to 12,000 containers, 2.5 times the cargo capacity of ships able to fit in the existing locks.
160 ft
60 ft
180 ft
106 ft
42 ft
110 ft
Mika Grondahl/NG Image Collection

Diagram showing the expansion and upgrading of the Panama Canal's lock system in order to accommodate ships that currently exceed the waterway's ("Panamax") capacity. The target date for completing the new third lane was August 15, 2014, the Canal's one-hundredth anniversary, but it has already been pushed back to 2015 because of construction delays.

are building additional storage facilities at the port of Houston and other key coastal distribution points.

Nevertheless, China will be the biggest beneficiary because its rapid recent development has made it the world's number-one exporter. Even during the post–2007 global recession, which primarily affected the United States and Europe, the Chinese economy maintained its growth. And China's imports passing through an expanded Panama Canal are expected to be considerably cheaper, further enhancing its economic strength and standing on the international scene. Most of all, this is an outstanding example of just how tightly interconnected the world economy has become and how the relative location of one small country in Middle America is of the utmost global significance.

THE CARIBBEAN BASIN

Fragmentation and Insularity

As Figure 4B-1 reveals, the Caribbean Basin, Middle America's island region, consists of a broad arc of numerous islands extending from the western tip of Cuba to the southern coast of Trinidad. The four larger islands or Greater Antilles (Cuba, Hispaniola, Jamaica, and Puerto Rico) are clustered in the western segment of this arc. The smaller islands, or Lesser Antilles, of the eastern segment extend southward within a crescent-shaped chain from the Virgin Islands to Trinidad and Tobago. (Breaking this tectonic-plate-related regularity are the Bahamas as well as the Turks and Caicos, located north of the Greater Antilles, and numerous other islands too small to appear on a map at the scale of Fig. 4B-1.)

On these myriad islands, whose combined land area accounts for only 9 percent of Middle America, lie 33 states and numerous other political entities. (Europe's colonial flags have not totally disappeared from this region, and the U.S. flag flies over both Puerto Rico and the Virgin Islands.) The populations of these states and territories, however, constitute 21 percent of the entire geographic realm, making this the most densely peopled part of the Western Hemisphere.

The island-nations of the Caribbean Basin are generally (very) small, and their territories are separated, often at considerable distances from other islands. These fragmented geographic conditions create a set of circumstances that has proven to be quite challenging. Economic opportunities are few, most things are relatively expensive, and due to limited interaction with the outside world the island societies tend to be static.

Ethnicity and Class

Social stratification [11] in most Caribbean islands is rigid and social mobility is limited. Class structures tend to be closely associated with ethnicity, and as such the region still carries imprints of colonial times. The historical geography of Cuba, the Dominican Republic, and Puerto Rico is suffused with Hispanic culture; Haiti and Jamaica carry stronger African legacies. But the reality of this ethnic diversity is that European lineages still hold the advantage. Hispanics tend to be in the best positions in the Greater Antilles; people who have mixed European-African ancestries, and who are described as **mulatto [12]**, rank next. The largest part of this social pyramid is also the most underprivileged: the Afro-Caribbean majority. In virtually all Caribbean societies, the minorities hold disproportionate power and exert overriding influence. In Haiti, the mulatto minority accounts for less than 5 percent of the population but has long held most of the power. In the neighboring Dominican Republic, the pyramid of power puts Hispanics (16 percent) at the top, the mixed sector (73 percent) in the middle, and the Afro-Caribbean minority (11 percent) at the bottom. In the Caribbean social mosaic, historical advantage has a way of perpetuating itself.

The composition of the population of the islands is further complicated by the presence of Asians from both China and India. During the nineteenth century, the emancipation of slaves and subsequent local labor shortages brought some far-reaching solutions. More than 100,000 Chinese emigrated to Cuba as indentured laborers; and Jamaica, Guadeloupe, and especially Trinidad saw nearly 250,000 South Asians arrive for similar purposes. To the African-modified forms of English and French heard in the Caribbean Basin, therefore, can be added several Asian languages. The ethnic and cultural diversity of the societies of Caribbean Middle America seems endless.

THE GREATER ANTILLES

The four islands of the Greater Antilles (whose populations constitute 90 percent of the Caribbean Basin's total) contain five political entities: Cuba, Haiti, the Dominican Republic, Jamaica, and Puerto Rico (Fig. 4B-1). Haiti and the Dominican Republic share the island of Hispaniola.

Cuba

The largest Caribbean island-state in terms of both territory (111,000 square kilometers/43,000 sq mi) and population (11.3 million), Cuba lies only 145 kilometers (90 mi) from the southernmost island of the Florida Keys. (Fig. 4B-9). Havana, the now-dilapidated capital, lies almost directly south of outermost Key West on the northwestern coast of the elongated island.

Cuba was a Spanish possession until 1898 when, with U.S. help in the ten-week-long Spanish-American War, it achieved independence. Fifty years later, an American-backed dictator was fully in control, and by the 1950s Havana had become an American playground. The island was ripe for revolution, and in 1959 Fidel Castro's insurgents gained control, thereby converting Cuba into a communist dictatorship and a client-state of the Sovier Union. Nearly a million Cubans fled the island for the United States, and Miami swiftly became the second-largest Cuban city after Havana.

In the fall of 1962, the world was on the brink of nuclear war when the United States called on the Soviet Union to remove its newly installed nuclear missiles from Cuba (aimed at the United States) or face reprisals. In the end, the Soviets conceded and in return Washington promised not to attack Cuba. The United States also continued to retain the Guantánamo Bay Naval Base located near the southeastern tip of the island (Fig. 4B-9), which was perpetually leased to the United States by the Cuban government in 1903. Since 2002, the Guantánamo base has acquired notoriety as a detention camp for 9/11-related terrorists as

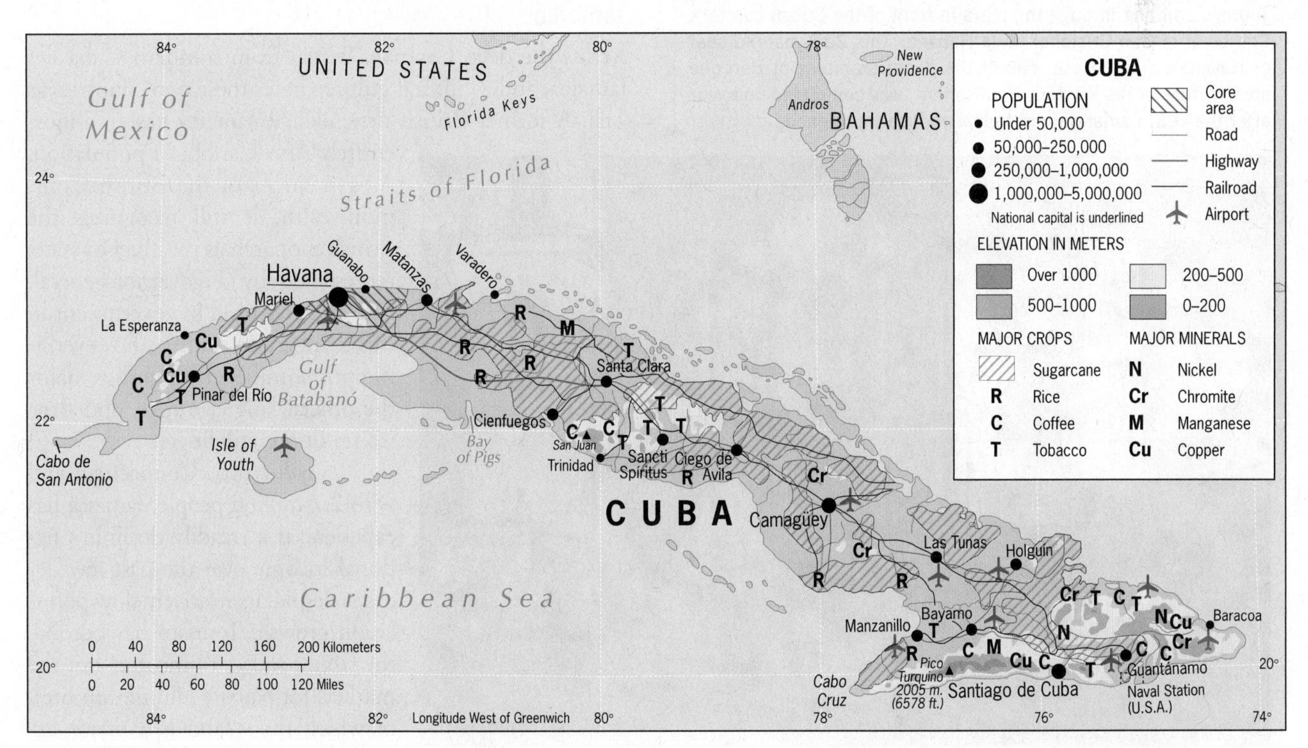

FIGURE 4B-9

well as prisoners of war from both the Iraq and Afghanistan conflicts.

Castro's rule survived the 1991 collapse of the Soviet Empire despite the loss of subsidies and sugar markets on which it had long relied. As the map indicates, sugar was Cuba's economic mainstay for many years; the plantations, once the property of rich landowners, extend all across the island. But sugarcane is losing its position as the leading Cuban foreign exchange earner. Mills have been closing down, and cane fields are being cleared for other crops as well as for pastures. However, Cuba has additional economic opportunities, especially in its highlands.

There are three mountainous zones, of which the southeastern Sierra Maestra is the highest and most extensive. These highlands create considerable environmental diversity as reflected by extensive timber-producing tropical forests and varied soils on which crops ranging from tobacco to coffee to rice to subtropical and tropical fruits are grown. Rice and beans are the staples, but Cuba is unable to meet its dietary needs and so must import food. The central and western savannas support livestock-raising. Even though Cuba has only limited mineral reserves, its nickel deposits are extensive and have been mined for more than a century.

Early in the twenty-first century, Cuba found a crucial new supporter in Venezuela's leader, Hugo Chávez. Cuba has no domestic petroleum reserves, but Venezuela is oil-rich and, since 2003, has been providing all of the fuel that Cuba needs. In return, Castro sent 30,000 health workers and other professionals to Venezuela, where they aid the poor. In 2011, it was estimated that the Chávez government subsidized Cuba to the yearly tune of about U.S. $3.5 billion. The Chávez regime came to a premature end with his death in 2013; the consequences for Cuba were still unknown at press time, but are likely to be serious in the long run.

Poverty, crumbling infrastructure, crowded slums, and rampant unemployment mark the Cuban cultural landscape, but the Castro regime still seems to have support among the general population. In 2006, Fidel Castro became too ill to lead the country and turned the reins over to his brother Raúl. Since then, a series of (modest) liberalization measures have slowly taken effect. In 2011, hundreds of thousands of government workers were laid off, forcing them to find employment in the private sector, part of a plan to privatize one-third of the public sector by 2015. These changes were undoubtedly necessitated by the persistently deplorable condition of the Cuban economy. One of the few potentially bright spots is the construction of a new deepwater container port at Mariel (the notorious refugee port of the early 1980s) 25 miles west of Havana. This facility is being built by a Brazilian consortium and will be operated by Singapore Ports, underscoring the global interest in Cuba's future possibilities in the changed environment of Caribbean shipping after the enlarged Panama Canal opens for business.

In the United States, many hold the view that Cuba could be the shining star of the Caribbean, its people free, its tourist economy booming, its products flowing to nearby North American markets. But a great deal will need to change for that vision to become reality.

Tourists ambling through the plaza in front of the Cuban capital's Cathedral of San Cristóbal de la Habana. This 237-year-old seat of Havana's archdiocese, one of the finest examples of Baroque architecture in the Western Hemisphere, was completed one year after the U.S. Declaration of Independence was signed.

© Jerry Ginsberg/Danita Delimont

Jamaica

Across the deep Cayman Trench from southern Cuba lies Jamaica, and a cultural gulf separates these two countries as well. A former British dependency, Jamaica has an almost entirely Afro-Caribbean population. As a member of the British Commonwealth, it still recognizes the British monarch as the chief of state, represented by a governor-general. The effective head of government in this democratic country, however, is the prime minister. English remains the official language here, and British traditions still linger.

Smaller than Connecticut and with 2.7 million people, Jamaica has experienced a steadily declining national income over the past few decades despite its relatively slow population growth. Tourism has become the largest source of income, but the markets for bauxite (aluminum ore), of which this island is a major exporter, have dwindled. And like other

Caribbean countries, Jamaica has difficulty in making money from its sugar exports. Jamaican farmers also produce crops ranging from bananas to tobacco, but the country faces the disadvantages on world markets common to those in the global periphery. Meanwhile, Jamaica must import all of its oil and much of its food because the densely populated coastal flatlands suffer from overuse and shrinking harvests.

The capital, Kingston, lies on the southeastern coast and reflects Jamaica's economic struggle. Almost none of the hundreds of thousands of tourists who visit the country's beaches, explore its Cockpit Country of awesome limestone towers and caverns, or populate the many cruise ships calling at Montego Bay or other points along the north coast get even a glimpse of what life is like for the ordinary Jamaican.

Haiti

Already the poorest state in the Western Hemisphere for several decades, Haiti in the fall of 2008 was staggered by no less than four tropical cyclones that struck within a few weeks of each other. The storms and the floods they unleashed killed more than 800 people and dislocated some 800,000. Then, barely 15 months later, the killer earthquake of January 12, 2010 caused devastation on an unimaginable scale (see photo in Chapter 4A). The capital, Port-au-Prince, lay in ruins, most of the country's infrastructure collapsed, and the Haitian government became effectively invisible. Schools, hospitals, and railroads as well as the seaport and the airport all came to a standstill. Millions became unemployed overnight. But all of that paled in comparison to the scale of human suffering. In the months following the earthquake, the estimated death toll reached and then surpassed 300,000; at least another 300,000 were injured. About 1.5 million homeless people initially fled the capital for the countryside, partly out of fear of aftershocks.

It was a tragedy of colossal proportions, its horrendous images broadcast around the world on television and the Internet. Large-scale international emergency aid arrived, but it became clear that, in the long run, much of the affected area would have to be totally rebuilt. In the catastrophe's aftermath, millions of Haitians continue to subsist in the most wretched circumstances.

The challenge of recovery remains a gargantuan undertaking because nature has not provided this country of 10.7 million with any breaks. Haiti lies along a particularly dangerous tectonic fault zone where the North American and Caribbean plates meet (Fig. 4B-10, inset map). The quake of 2010, geologists say, had been in the making for centuries, and more of them could easily follow in the

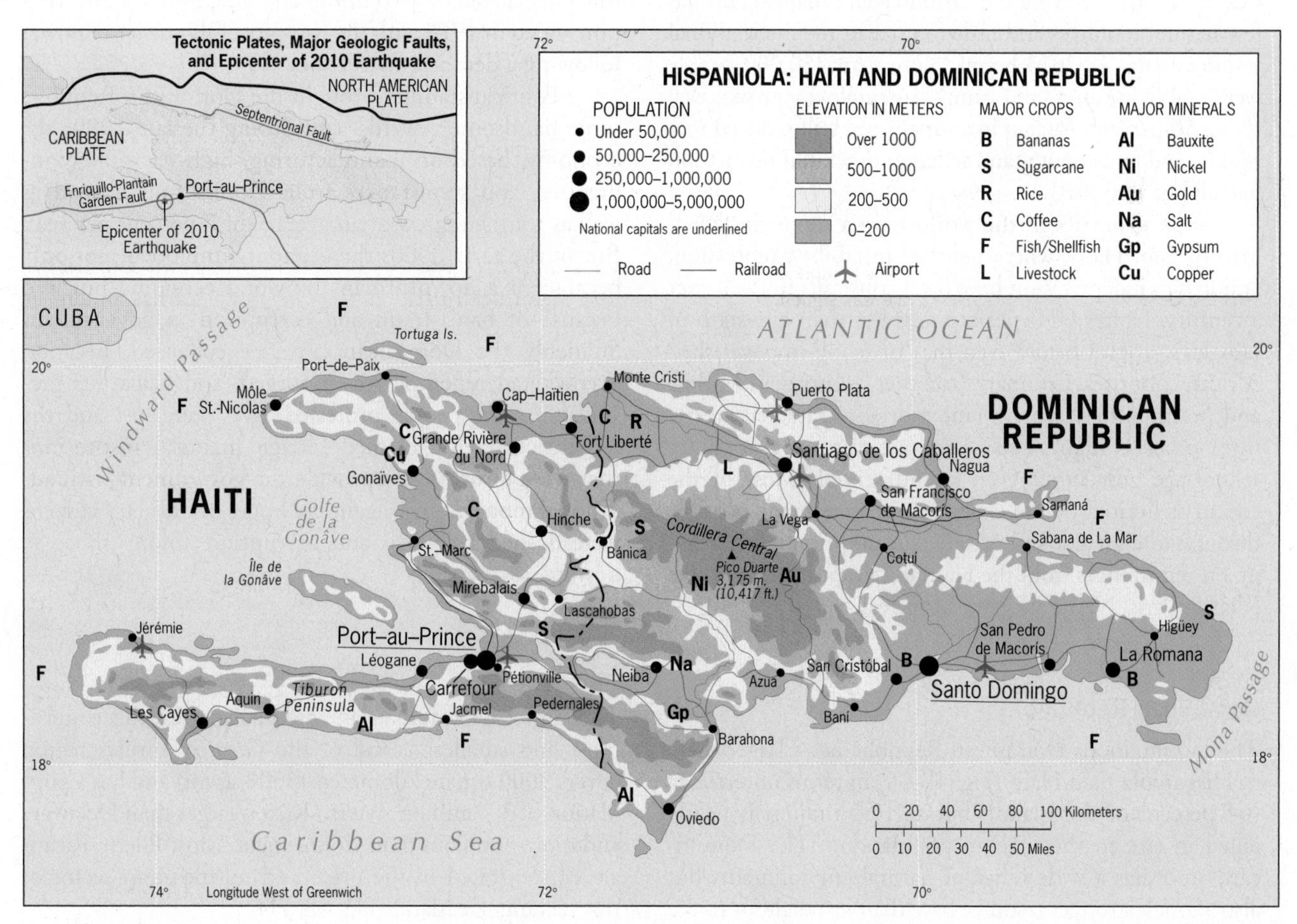

FIGURE 4B-10

From the Field Notes . . .

© H.J. de Blij

"Fly along the political boundary between Haiti and the Dominican Republic, and you see long stretches of the border marked by a stark contrast in vegetation: denudation prevails to the west in Haiti while the forest survives on the Dominican (eastern) side. Overpopulation, widespread appalling poverty, lack of governmental control, and mismanagement on the Haitian side combine to create one of the region's starkest spatial contrasts."

www.conceptcaching.com

future. This country also lies astride the axis of Hurricane Alley, and even a single year without a major storm is something to be grateful for (fortunately including the three hurricane seasons following the earthquake). Finally, Haiti has few natural resources and little to offer in the international trading arena. As 2013 began, more than 350,000 people were still living in tented camps, unemployment was well above 50 percent, foreign investment was still reduced to a trickle, and talk about grand schemes to rebuild this nation had all but vanished.

Few countries in the world have a more checkered history than Haiti, where political instability, repression, and material deprivation have been constants for well over a century. Haiti's GNI per capita is less than one-sixth of Jamaica's, a level below even that of many impoverished African countries. Compared to other countries in Middle and South America, its infant mortality rate was nearly three times as high. A shocking one-third of elementary school-age girls never even attend school. Most of the country's limited public expenditures were made possible through foreign aid, and private consumption relied heavily on remittances from the Haitian Diaspora in cities like Miami, Paris, New York, and Montreal.

Dominican Republic

The mountainous Dominican Republic has a larger share of Hispaniola than Haiti (Fig. 4B-10) in terms of territory (64 percent of the island), but at 10.4 million is nearly equal in size to the Haitian population. "The Dominican," too, has a wide range of natural environments but also a much stronger resource base than its neighbor to the west. Nickel, gold, and silver have long been exported along with sugar, tobacco, coffee, and cocoa, but tourism is the leading industry. An extended period of dictatorial rule punctuated by revolutions and U.S. military intervention ended in 1978 with the first peaceful transfer of power following a democratic election.

Political stability brought the Dominican Republic some handsome rewards, and during the late 1990s the economy, based on manufacturing, high-technology industries, and remittances from Dominicans abroad as well as tourism, grew at an average of 7 percent per year. But in the early 2000s the economy imploded, not only because of a downturn in the world economy but also because of bank fraud and corruption in government. Suddenly the Dominican currency collapsed, inflation skyrocketed, unemployment surged, and blackouts prevailed. As the people protested, lives were lost and the self-enriched elite blamed foreign financial institutions that were unwilling to provide the government with additional money. Once again the hopes of ordinary citizens were dashed by greed and corruption among those in power.

Puerto Rico

The largest U.S. domain in Middle America, this easternmost and smallest island of the Greater Antilles region covers 9000 square kilometers (3500 sq mi) and has a population of 3.7 million. Puerto Rico is larger than Delaware and more populous than Connecticut. Most Puerto Ricans are concentrated in the urbanized northeastern sector of this rectangular island (Fig. 4B-11).

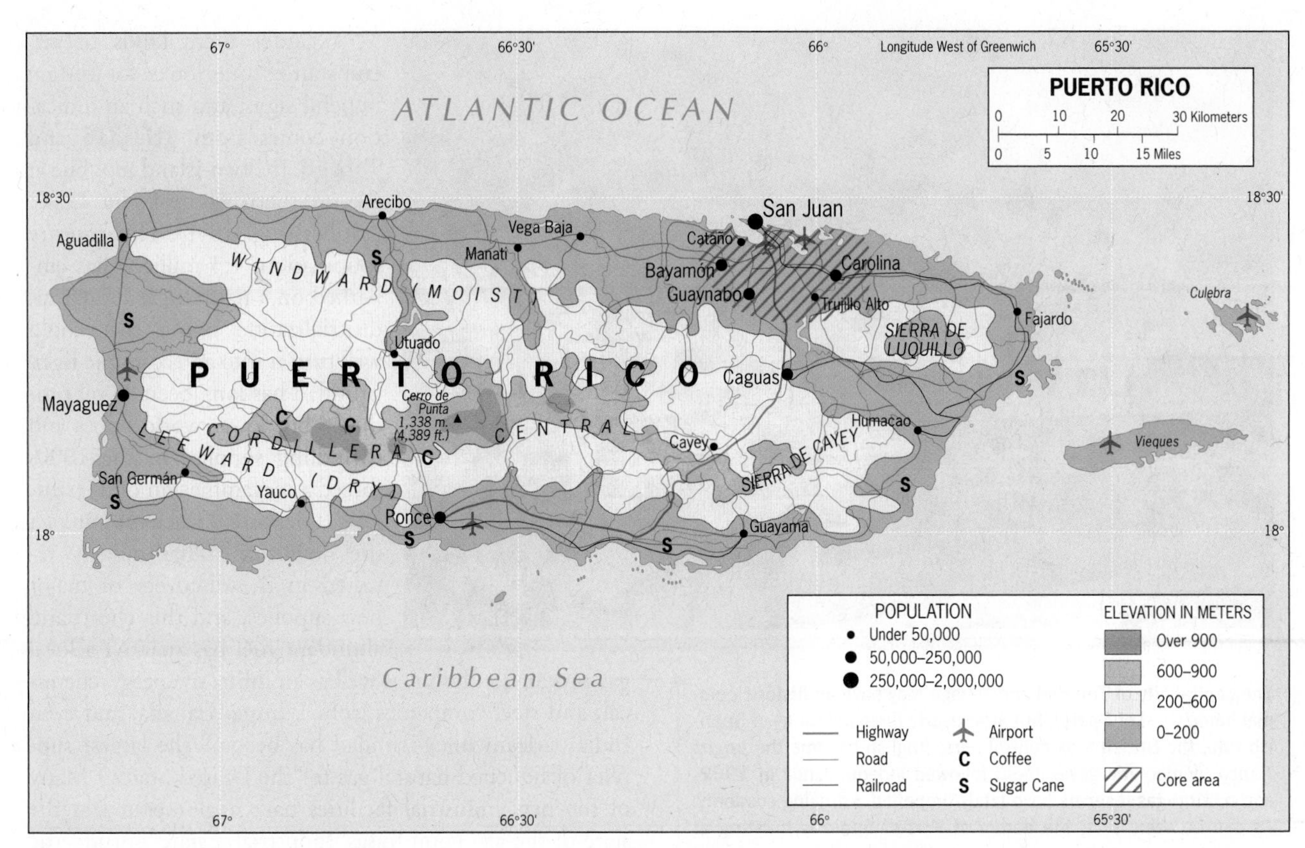

FIGURE 4B-11

Puerto Rico fell to the United States well over a century ago during the Spanish-American War of 1898. Since the Puerto Ricans had been struggling for some time to free themselves from Spanish imperial rule, this transfer of power was, in their view, only a change from one colonial master to another. As a result, the first half-century of U.S. administration was difficult, and it was not until 1948 that Puerto Ricans were permitted to elect their own governor. When the island's voters approved the creation of a Commonwealth in a 1952 referendum, Washington, D.C. and San Juan, the seats of government involved, entered into a complicated arrangement. Puerto Ricans are U.S. citizens but pay no federal taxes on their local incomes. The Puerto Rican Federal Relations Act governs the island under the terms of its own constitution and awards it considerable autonomy. Puerto Rico also receives a sizeable annual subsidy from Washington, averaging approximately U.S. $4.2 billion during the past few years.

Despite these apparent advantages in the poverty-plagued Caribbean, Puerto Rico has not thrived under U.S. administration. Long dependent on a single-crop economy (sugar), the island based its industrial development during the 1950s and 1960s on its comparatively cheap labor, tax breaks for corporations, political stability, and special access to the U.S. market. Consequently, pharmaceuticals, electronic equipment, and apparel top today's list of exports, not sugar or bananas. But this industrialization failed to stem a tide of emigration that carried more than one million Puerto Ricans to the New York City area alone. The same wage advantages that favored corporations kept many Puerto Ricans poor or unemployed. The level of unemployment on the island in 2013 stood at more than 14 percent, the highest of any U.S. subnational entity. Most residents receive federal support; another 21 percent work in the public sector, that is, in government. The private sector remains severely underdeveloped.

Most Puerto Ricans are now fed up with their lack of economic progress. In the 2012 U.S. presidential election, which included a non-binding referendum on Statehood, 61 percent voted in favor of becoming the 51st State. Discussions about implementation (which has political implications for the deeply divided U.S. Congress) as well as the process and conditions of accession are likely to continue well into the second half of this decade.

THE LESSER ANTILLES

As Figure 4B-1 shows, the Greater Antilles are flanked by two clusters of islands: the extensive Bahamas/Turks and Caicos archipelago to the north and the Lesser Antilles to

Boutin/Sipa Press

The capital city of Trinidad and Tobago may have an historic colonial name (Port of Spain), but after nearly three centuries of Spanish rule, the British took control here. English became the *lingua franca*, democratic government followed independence in 1962, and natural-gas reserves have lately propelled a thriving economy. As can be seen here, the harbor of Port of Spain is bursting at the seams as a car carrier delivers automobiles from Japan, and containers are stacked high on the docks.

the southeast. **The Bahamas**, the former British colony that is now the closest Caribbean neighbor to the United States, alone consists of nearly 3000 coral islands—most of them rocky, barren, and uninhabited—but approximately 700 carry vegetation, of which roughly 30 are inhabited. Centrally positioned New Providence Island houses most of the country's 400,000-plus inhabitants and also contains the capital, Nassau, a leading tourist attraction.

The Lesser Antilles are grouped geographically into the Leeward Islands and the Windward Islands, a (climatologically incorrect) reference to the prevailing airflows in this tropical zone. The Leeward Islands extend from the U.S. Virgin Islands to the French dependencies of Guadeloupe and Martinique, and the Windward Islands from St. Lucia to the Dutch-affiliated islands of Aruba and Curaçao off the northwestern Venezuelan coast.

It would be impractical to detail the individual geographic characteristics of each island of the Lesser Antilles, but we should note that nearly all share the insularity and environmental risks of this region—major earthquakes, volcanic eruptions, and hurricanes; that they confront, to varying degrees, similar socioeconomic problems in the form of limited domestic resources, overpopulation, soil deterioration, land fragmentation, and market limitations; and that tourism has become the leading industry for many.

Under these kinds of circumstances one looks for certain hopeful signs, and such an indication comes from **Trinidad and Tobago**, the two-island republic at the southern end of the Lesser Antilles (Fig. 4B-1). This country (population: 1.3 million) has embarked on a natural-gas-driven industrialization boom that could well turn it into an economic tiger. Trinidad has long been an oil producer, but lower world prices and dwindling supplies in the 1990s forced a reexamination of its natural gas deposits to help counteract the downturn. That quickly resulted in the discovery of major new supplies, and this cheap and abundant fuel has sparked a local gas-production boom as well as an influx of energy, chemical, and steel companies from Europe, Canada, and even India. (Meanwhile, Trinidad has become the largest supplier of liquefied natural gas for the United States.) Many of the new industrial facilities have agglomerated at the state-of-the-art Point Lisas Industrial Estate outside the capital, Port of Spain, and they have propelled Trinidad to become the world's leading exporter of ammonia and methanol. Natural gas also is an efficient fuel for the manufacturing of metals, and steelmakers as well as aluminum refiners have been attracted to locate here. With Trinidad lying just off the Venezuelan coast of South America, it is also a sea-lane crossroads that is strongly connected to the vast, near-coastal supplies of iron ore and bauxite that are mined in nearby countries, most importantly the Brazilian Amazon.

But Trinidad is an exception. It is a reflection of the predicament of **small-island developing economies [13]** that, after the initial wave of decolonization in the 1950s and 1960s, several territories decided that they were probably better off affiliating with the European country that had ruled them. Thus, for example, Guadeloupe remained with France and the Cayman Islands are still British. The former Netherlands Antilles passed up full independence to continue their affiliation with the Dutch but did so individually and with considerable autonomy: **Curaçao, Aruba**, and **St. Maarten** are now "countries" within the Kingdom of the Netherlands, whereas the other Caribbean islands of the former Dutch Empire have acquired the status of overseas municipalities. Although economic logic seems to dictate continued association with the former colonial powers, local politics and pride demand some minimal degree of autonomy and avoidance of paternalistic European interference. The combined result is often a complicated legal framework that shapes the contemporary political status of the islands.

VOICE FROM THE Region

Megan Elizabeth Chong, Curaçao

THE PULL OF CURAÇAO'S ETHNIC MOSAIC

"Growing up on the island of Curaçao, I have had the opportunity to interact with people of various races, ethnicities, religions, and nationalities. As is true for most Caribbean islands, immigrants have come from many different parts of the world. This ethnic mix makes for the unique culture of Curaçao. Personally, I have an American mother of German origin and an Antillean father of Chinese, indigenous, and Spanish descent. It would be difficult for me to find someone with identical origins, but that goes for many people living here. Since the island is so small, only 60 kilometers (38 mi) long and 15 kilometers (9 mi) wide, most people know each other even if they have very different backgrounds. I think this has helped me to develop a deeper understanding and appreciation for others. On the other hand, the opportunities for work and studies are quite limited on small islands like Curaçao. Each year, many graduating high school students move abroad to attend college. The majority end up in the Netherlands, with which we have historical ties and where students from Curaçao get the same subsidies as Dutch nationals. Some choose to study in the United States, but a good number stay on the island to take classes at the local university or look for work. This year, I will be attending Maastricht University in the Netherlands to major in psychology. Many of the students who leave Curaçao to study abroad do not return after their studies—or at least not right away—as they normally begin a career elsewhere. I will probably do the same, having connections in Europe as well as the United States; but I know I will always be drawn back to Curaçao, the island I am proud to call home."

POINTS TO PONDER

- Northern Mexico highlights the volatility of globalization: the inception of NAFTA in the mid-1990s led to foreign investment and the proliferation of maquiladoras, but a decade later many companies had departed for Asian-Pacific destinations that offered even lower wages.
- Costa Rica has known political stability for more than 190 years, and it has been a continuous democracy for over six decades. What could other Central American countries learn from this example?
- The geography of a small Middle American country, Panama, significantly impacts the global economy and especially its two biggest players, the United States and China.
- Honduras now seeks to solve its massive problems by offering parts of its territory to foreign investors and allowing them to become more or less independent of the Honduran government—a stroke of genius or an act of desperation?

THE SOUTH AMERICAN REALM

Elevation (m)

3000
1500
600
300
150
0
below sea level
0
-150
-1500
-3000
-6000

Roads
Railroads

0 km 400 800 1200 1600
0 miles 200 400 600 800 1000

Lambert Azimuthal Equal-Area Projection
Scale 1:30,000,000

IN THIS CHAPTER

- South America's commodities boom
- The growing assertiveness of indigenous peoples
- Inequality and violence: A hallmark of South America?
- Brazil on the move
- The Chinese are coming . . .

CONCEPTS, IDEAS, AND TERMS

Unity of place 1
Indigenous 2
Altiplano 3
Land alienation 4
Liberation theology 5
Cultural pluralism 6
Commercial agriculture 7
Subsistence agriculture 8
Uneven development 9
Rural-to-urban migration 10
Megacity 11
"Latin" American City model 12
Informal sector 13
Barrio (favela) 14
Dependencia theory 15

FIGURE 5A-1

Of all the continents, South America has the most familiar shape—a gigantic triangle connected by mainland Middle America's tenuous land bridge to its neighbor in the north. South America also lies not only south but mostly east of its northern counterpart. Lima, the capital of Peru—one of the continent's westernmost cities—lies farther east than Miami, Florida. Thus South America juts out much more prominently into the Atlantic Ocean toward southern Europe and Africa than does North America. But lying so far eastward means that South America's western flank faces a much wider Pacific Ocean, with the distance from Peru to Australia nearly twice that from California to Japan.

DEFINING THE REALM

As if to reaffirm South America's northward and eastward orientation, the western margins of the continent are rimmed by one of the world's longest and highest mountain ranges, the Andes, a giant wall that extends unbroken from Tierra del Fuego near the continent's southern tip in Chile to northeastern Venezuela in the far north (Fig. 5A-1). The other major physiographic feature of South America dominates its central north—the Amazon Basin; this vast humid-tropical amphitheater is drained by the mighty Amazon River, which is fed by several major tributaries. Much of the remainder of the continent can be classified as plateau, with the most important components being the Brazilian Highlands that cover most of Brazil southeast of the Amazon Basin, the Guiana Highlands located north of the lower Amazon Basin, and the cold Patagonian Plateau that blankets the southern third of Argentina. Figure 5A-1 also reveals two other noteworthy river basins beyond Amazonia: the Paraná-Paraguay Basin of south-central South America, and the Orinoco Basin in the far north that drains interior Colombia and Venezuela.

Symbol of the ongoing development thrust into South America's interior, the new Ponte Rio Negro, the first major bridge in the Amazon Basin and Brazil's longest, links the city of Manaus to its satellite, Iranduba.

© Ian Trower/Robert Harding Imagery/Corbis

PHYSIOGRAPHY

Explorers' Continent

It was here in northern South America that the great German explorer and scientist Alexander von Humboldt, one of the founders of the modern discipline of geography, embarked on his legendary expeditions in the early nineteenth century. After landing on the coast of Venezuela and trekking across the continent's northern interior, the 30-year-old Humboldt was struck by the area's biodiversity, its majestic natural beauty, and the adaptive abilities of the human populations. He discovered and named many species of flora and fauna, traversed tropical grasslands and jungles, met with indigenous peoples, crossed dangerous rivers, and reached the summit of the highest mountain in the Americas climbed by Europeans at the time (Ecuador's Chimborazo). He compiled large numbers of maps and was one of the first scientists to note how the coastlines of eastern South America and western Africa fitted together like pieces of a jigsaw puzzle, speculating about the continents' geologic movements.

Humboldt was most important in the rise of the modern discipline of geography because of his many discoveries as well as his views on the **unity of place [1]**: that in a particular locale or region intricate connections exist between climate, geology, biology, and human cultures. As such, he laid the foundation for modern geography as an *integrative discipline* with a spatial perspective. It took him four decades after his arrival in South America to produce his magnum opus that articulated this holistic perspective, the highly ambitious and appropriately titled *Cosmos* (published in five volumes in 1845).

Almost three centuries earlier, the opposite end of South America was the scene of a crucial stage in the first circumnavigation of the globe and expedition led by Ferdinand Magellan. Once it became clear that Columbus

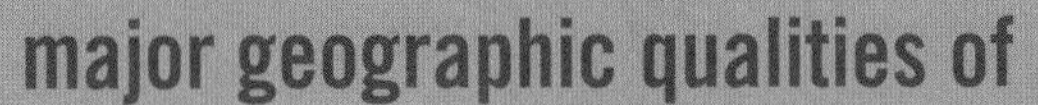

major geographic qualities of SOUTH AMERICA

1. South America's physiography is dominated by the Andes Mountains in the west and the Amazon Basin in the central north. Much of the remainder is plateau country.
2. Almost half of the realm's area and just under half of its total population are concentrated in one country—Brazil.
3. South America's population remains concentrated along the continent's periphery. Most of the interior is sparsely peopled, but sections of it are now undergoing significant development.
4. Interconnections among the states of the realm are improving rapidly. Economic integration has become a growing force, but is still at an early stage.
5. Regional economic contrasts and disparities, both in the realm as a whole and within individual countries, remain strong.
6. Cultural pluralism prevails in almost all of the realm's countries and is often expressed regionally.
7. Rapid urban growth continues to mark much of the South American realm, and urbanization overall is today on a par with the levels of the United States and western Europe.
8. This realm contains abundant natural riches, and it has benefited in recent years from increased global demand for raw materials.

had stumbled upon America, not India, Spanish and Portuguese explorers continued their efforts to discover a westward passage from the Atlantic to the Pacific. This took them to the far south along what is now the Argentinean coast. Magellan and his crew spent some five months in Patagonia, named by Magellan for "big-foot people" (official, but never-verified, reports of the voyage told of the crew's mysterious encounters with 8-foot giants). With the onset of spring in 1520, Magellan's five ships set sail for the treacherous waters of what he named the *Estrecho de Todos los Santos*—now called the Strait of Magellan (Fig. 5A-1). Of the five ships in Magellan's fleet, only three survived the daring, 600-kilometer (375-mile)-long passage; the other two crashed on the rocks in the icy waters of the Southern Ocean.

Myriad Climates and Habitats

If the Russian realm is the widest in east-west extent, the South American realm is the longest measured from north to south. Within South America, no other country is more emblematic of this elongated geography than Chile, averaging only about 150 kilometers (90 mi) in width but 4000 kilometers (2500 mi) in length. As a consequence of this latitudinal span, fully one-tenth of our planet's circumference, the realm contains an enormous variety of climates and vegetation. Combine this with substantial variation in elevation from west to east, and it is clear why South America has such an impressive range of natural habitats.

Take another look at the map showing the global distribution of climates (Fig. G-7) and note the variation in climatic types, particularly in the realm's northwest and southern half. Travel northeast from Lima, Peru for about 600 kilometers (375 mi) and you encounter no less than four different climate zones: arid, highland, and two varieties of humid tropical. A transect of similar length from Santiago, Chile eastward across the continent to Buenos Aires, Argentina will take you through five climate zones: interior highland, arid, and semiarid environments bracketed by a different humid temperate climate along each coast. Vegetation in South America varies accordingly, from lush tropical rainforests to rocky and barren snow-covered mountaintops to grasslands, fertile as well as parched. This natural diversity also makes for considerable cultural differences, as we shall soon see.

STATES ANCIENT AND MODERN

Thousands of years before the first European invaders appeared on the shores of South America, peoples now referred to as **indigenous [2]** had migrated into the continent via North and Middle America and founded societies in coastal valleys, in river basins, on plateaus, and in mountainous locales. These societies achieved different and remarkable adaptations to their diverse natural environments, and by about 1000 years ago, a number of regional cultures thrived in the elongated valleys between mountain ranges of the Andes from present-day Colombia southward to Bolivia and Chile. These high-altitude valleys, called ***altiplanos* [3]**, provided fertile soils, reliable water supplies, building materials, and natural protection to their inhabitants.

The Inca State

One of these *altiplanos*, at Cuzco in what is now Peru, became the core area of South America's greatest indigenous empire, that of the ***Inca***. The Inca were expert builders whose stone structures (among which Machu Picchu near Cuzco is the most famous), roads, and bridges helped to

From the Field Notes . . .

"Traveling on Peru's Cuzco *altiplano*, it was easy to get a sense of the advantages of this highland environment to the Incas of the past. The soils are fertile, temperatures are moderate, and the surrounding snowcapped ranges of the Andes feed the Urubamba River and many other streams. The Urubamba Range in the background, partially hidden behind the clouds, rises up to nearly 6000 meters (19,000 ft) and the city of Cuzco itself lies at 3350 meters (11,000 ft). For centuries, this area has produced abundant corn (seen in the foreground), potatoes, and other *tierra fría* crops. But the retreating glaciers in the higher elevations are now a cause of concern. Global warming may jeopardize what has been nature's gift to this *altiplano* for many centuries: a reliable water supply."

© Jan Nijman

www.conceptcaching.com

unify their vast empire; they also proved themselves to be efficient administrators, successful farmers and herders, and skilled manufacturers; their scholars studied the heavens, and physicians even experimented with brain surgery. Great military strategists, the Inca integrated the peoples they vanquished into a stable and well-functioning state, an amazing accomplishment given the high-relief terrain they had to contend with (see photo).

As a minority ruling elite in their far-flung empire, the Inca were at the pinnacle in their rigidly class-structured, highly centralized society. So centralized and authoritarian was their state that a takeover at the top was enough to gain immediate power over all of it—as a small army of Spanish invaders discovered in the 1530s. The European invasion brought a quick end to thousands of years of indigenous cultural development and changed the map forever.

The Iberian Invaders

The modern map of South America started to take shape when the Iberian colonizers began to understand the location and economies of the indigenous societies. The Inca, like Mexico's Maya and Aztec peoples, had accumulated gold and silver at their headquarters, possessed productive farmlands, and constituted a ready labor force. Not long after the defeat of the Aztecs in 1521, Francisco Pizarro sailed southward along the continent's northwestern coast, learned of the existence of the Inca Empire, and withdrew to Spain to organize its overthrow. He returned to the Peruvian coast in 1531 with 183 men and two dozen horses, and the events that followed are well known. In 1533, his party rode victorious into Cuzco.

At first, the Spaniards kept the Incan imperial structure intact by permitting the crowning of an emperor who was under their control. But soon the breakdown of the old order began. The new order that gradually emerged in western South America placed the indigenous peoples in serfdom to the Spaniards. Great haciendas were formed by **land alienation [4]** (the takeover of indigenously held land by foreigners), taxes were instituted, and a forced-labor system was introduced to maximize the profits of exploitation.

Lima, the west-coast headquarters of the Spanish conquerors, soon became one of the richest cities in the world, its wealth based on the exploitation of vast Andean silver deposits. The city also served as the capital of the Viceroyalty of Peru, as the Spanish authorities quickly integrated the new possession into their colonial empire (Fig. 5A-2). Subsequently, when Colombia and Venezuela came under Spanish control and, later, when Spanish settlement expanded into what is now Argentina and Uruguay, two additional viceroyalties were added to the map: New Granada and La Plata.

Meanwhile, another vanguard of the Iberian invasion was penetrating the east-central part of the continent, the coastlands of present-day Brazil. This area had become a Portuguese sphere of influence because Spain and Portugal had signed a treaty in 1494 to recognize a north-south line 370 leagues west of the Cape Verde Islands as the boundary between their New World spheres of influence. This border ran approximately along the meridian of 50°W longitude, thereby cutting off a substantial triangle of eastern South America for Portugal's exploitation (Fig. 5A-2). But a brief look at the political boundaries of South America (Fig. 5A-1) shows that this treaty did not limit Portuguese colonial territory to lands east of the 50th meridian. Instead, Brazil's boundaries were bent far inland to include almost the entire Amazon Basin, and the country came to be only slightly smaller in territorial size than all the other South American countries combined. This westward thrust was the result of

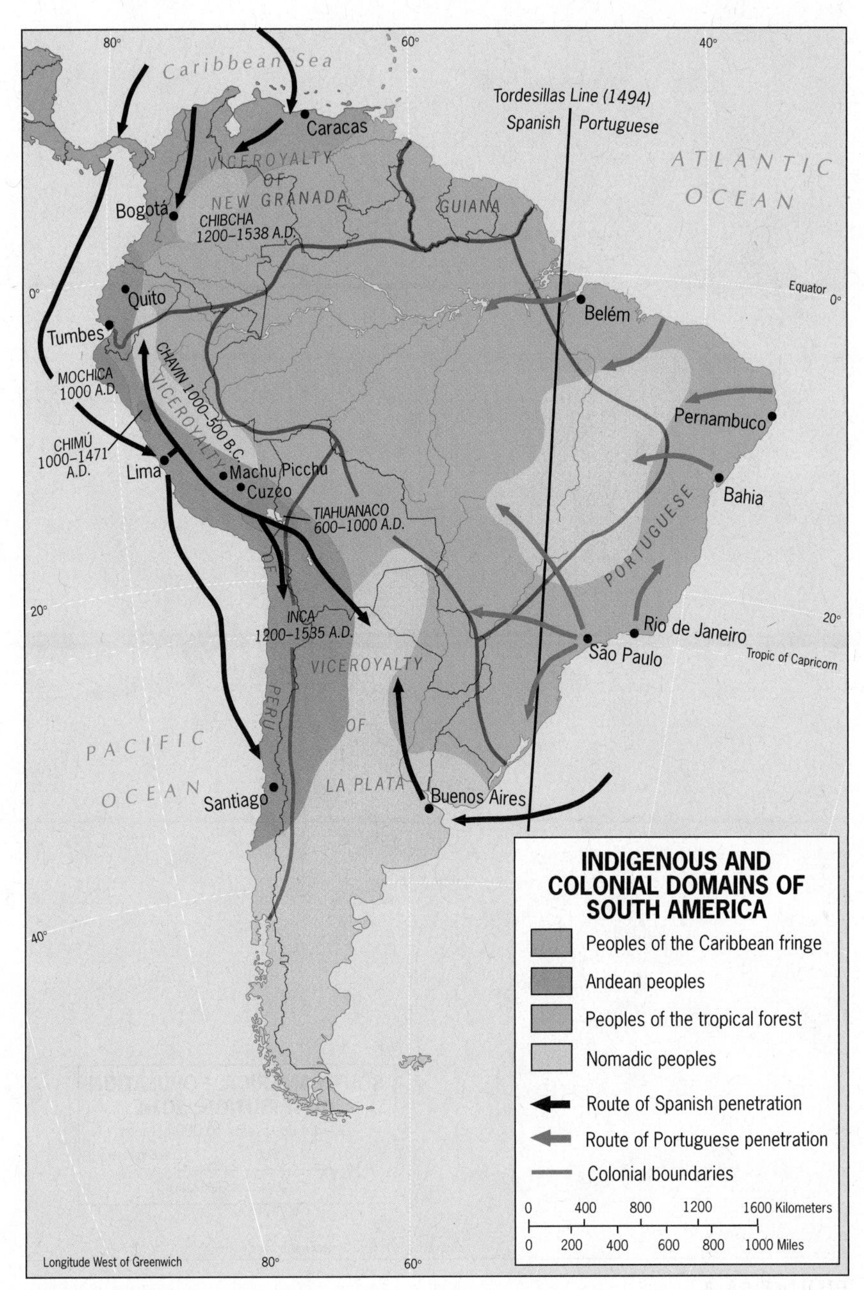

FIGURE 5A-2

Portuguese and Brazilian penetration, particularly by the ***Paulistas***, the settlers of São Paulo who needed indigenous slave labor to run their plantations.

Independence and Isolation

Despite their adjacent location on the same continent, their common language and cultural heritage, and their shared national problems, the countries that arose out of South America's Spanish viceroyalties (together with Brazil) until quite recently existed in a considerable degree of isolation from one another. Distance as well as physiographic barriers reinforced this separation, and the realm's major population agglomerations still adjoin the coast, mainly the eastern and northern coasts (Fig. 5A-3). The viceroyalties existed primarily to extract riches and fill Spanish coffers. In Iberia, there was little interest in developing the American lands for their own sake. Only after those who had made Spanish and Portuguese America their new home and who had a stake there rebelled against Iberian authority did things begin to change, and then very slowly. Thus South America was saddled with the values, economic outlook, and social attitudes of eighteenth-century Iberia—not the best tradition from which to begin the task of forging modern nation-states.

Certain isolating factors had their effect even during the wars for independence. Spanish military strength was always concentrated at Lima, and those territories that lay farthest from their center of power—Argentina as well as

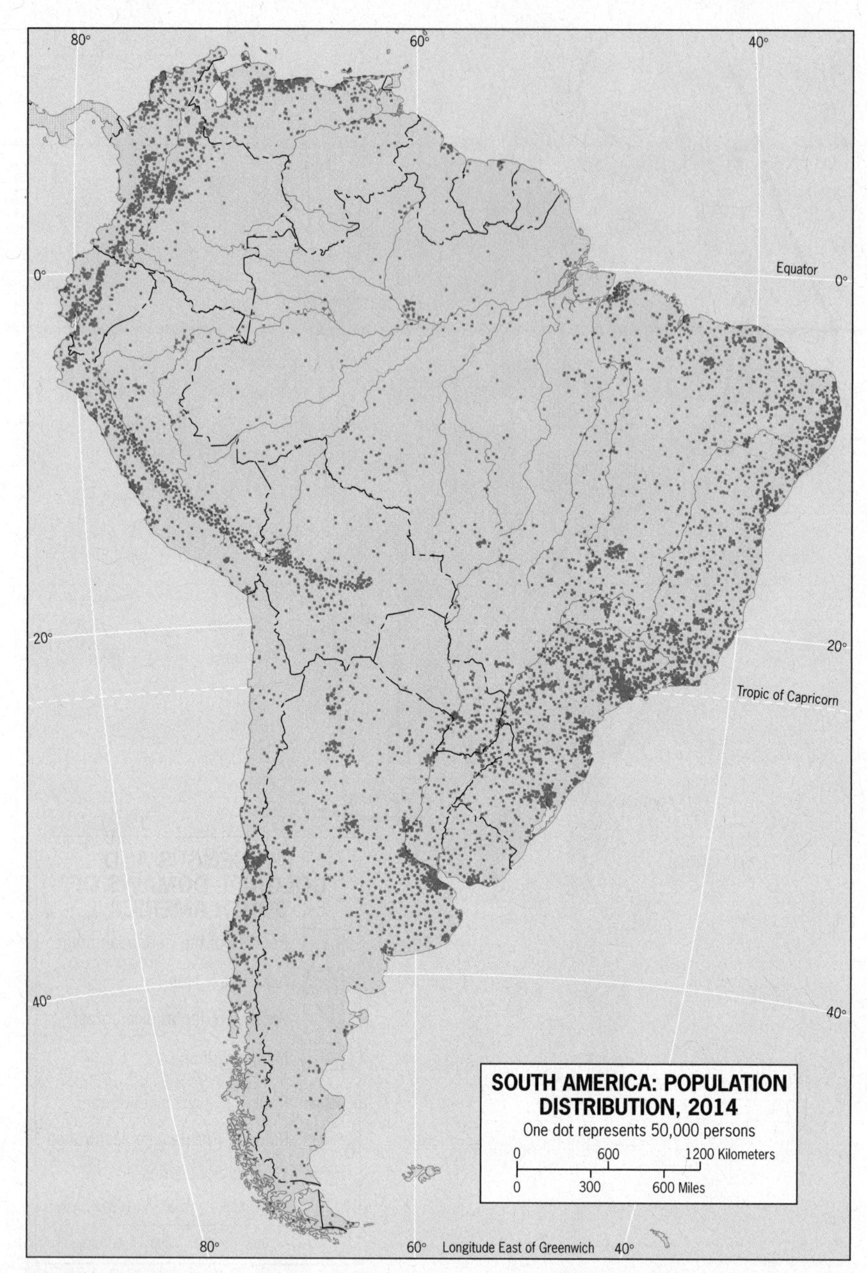

FIGURE 5A-3

Chile—were the first to gain independence from Spain (in 1816 and 1818, respectively). In the north, Simón Bolívar led the burgeoning independence movement, and in 1824 two decisive military defeats there spelled the end of Spanish power in South America.

This joint struggle, however, did not produce unity because no fewer than nine countries emerged from the three former viceroyalties. It is not difficult to understand why this fragmentation took place. With the Andes intervening between Argentina and Chile and the Atacama Desert between Chile and Peru, overland distances seemed even greater than they really were, and these obstacles to contact proved quite effective. Hence, from their outset the new countries of South America began to grow apart amid friction and even wars. Only within the past two decades have the countries of this realm finally begun to recognize the mutual advantages of increasing cooperation and to make lasting efforts to steer their relationships in this direction.

THE CULTURAL MOSAIC

When we speak of the interaction of South American countries, it is important to keep in mind just who does the interacting. The fragmentation of colonial South America into ten individual republics, and the subsequent postures of each of these states, was the work of a small minority that constituted the landholding, upper-class elite. Therefore, in every country a vast majority—be they indigenous people in Peru or those of African descent in Brazil—could only watch as their European masters struggled with one another for supremacy.

From the Field Notes . . .

© H.J. de Blij

"Near the waterfront in Belém lie the now-deteriorating, once-elegant streets of the colonial city built at the mouth of the Amazon. Narrow, cobblestoned, flanked by tiled frontages and arched entrances, this area evinces the time of Dutch and Portuguese hegemony here. Mapping the functions and services here, we recorded the enormous diversity of activities ranging from carpentry shops to storefront restaurants and from bakeries to clothing stores. Dilapidated sidewalks were crowded with shoppers, workers, and people looking for jobs (some newly arrived, attracted by perceived employment opportunities in this growing city of 2.2 million). The diversity of population in this and other tropical South American cities reflects the varied background of the region's peoples and the wide hinterland from which these urban magnets have drawn their inhabitants."

www.conceptcaching.com

The Population Map—Then and Now

If we were able to reconstruct a map of South America's population before the arrival of the Europeans (a "pre-Columbian" map, as it would be called), it would look quite different from the current one in Figure 5A-3. Indigenous societies inhabited not only the Andes and adjacent lowlands but also riverbanks in the Amazon Basin, where settlements numbering in the thousands subsisted on fishing and farming. They did not shy away from harsh environments such as those of the island of Tierra del Fuego in the far south, where the fires they kept going against the bitter cold led the Europeans to name the place "land of fire."

Today the map looks quite different. Many of the indigenous societies succumbed to the European invaders, not just through warfare but also because of the diseases the Iberian conquerors brought with them. Geographers estimate that 90 percent of native Amazonians died within a few years of contact, and the peoples of Tierra del Fuego also are no longer there to build their fires. From one end of South America to the other, the European arrival spelled disaster.

Spanish and Portuguese colonists penetrated the interior of South America, but the great majority of the settlers stayed on or near the coast, a pattern still visible today. Almost all of the realm's major cities have coastal or near-coastal locations, and the current population distribution map gives you the impression of a continent yet to be penetrated and inhabited. But look carefully at Figure 5A-3, and you will see a swath of population located well inland from the settlements along the northwest coast, most clearly in Peru but also extending northward into Ecuador and southward into Bolivia. That is the legacy of the Inca Empire and its incorporated peoples, surviving in their mountainous redoubt and still numbering in the tens of millions.

Indigenous Reawakening

Today, South America's long-downtrodden indigenous peoples are staging a social, political, and economic reawakening. In several of the realm's countries, where their numbers are large enough to translate into political strength, they have begun to realize their potential. The indigenous peoples in Peru constitute 45 percent of the national population, and in Bolivia they are in the majority at 55 percent.

Newly empowered indigenous political leaders are emerging and bringing the plight of the realm's aboriginal peoples to both local and international attention. South America's indigenous peoples were conquered, decimated by foreign diseases, robbed of their best lands, subjected to involuntary labor, denied the right to grow their traditional crops, socially discriminated against, and swindled out of their fair share of the revenues from resources in their traditional domains. They may still be the poorest of the realm's poor, but they are increasingly asserting themselves. For some South American states, the consequences of this movement will be far-reaching.

The indigenous reawakening is in part related to changing religious practices in South America. Officially, just over 80 percent of the population is Roman Catholic, and traditionally South Americans tend to be viewed as devout followers of the Vatican. But recent surveys show that many do not attend church regularly and that more than half of all Catholics describe themselves as believers of the doctrine, rather than adherents of the Church. Since the late nineteenth century, the Catholic Church has often been criticized for its conservative position on social issues and for siding with the establishment. During the 1950s, a powerful movement known as **Liberation theology [5]** emerged in South America and subsequently gained followers around the world. The movement was a blend of Christian religion and socialist philosophy that interpreted the teachings of Christ as a quest to liberate the impoverished masses from oppression. Although the Church has lately been trying to make amends, it could not avoid losing popular support, especially among indigenous peoples. In recent years, Protestant-evangelical faiths have also been gaining many new adherents, largely at the expense of Catholicism.

African Descendants

As Figure 5A-2 shows, the Spaniards initially got very much the better of the territorial partitioning of South America—not just in land quality but also in the size of the aboriginal labor force. When the Portuguese began to develop their territory, they turned to the same lucrative activity that their Spanish rivals had pursued in the Caribbean—the plantation cultivation of sugar for the European market. And they, too, found their workforce in the same source region, as millions of Africans (nearly half of all who came to the Americas) were brought in bondage to the tropical Brazilian coast north of Rio de Janeiro (see Fig. 6A-6). Not surprisingly, Brazil now has South America's largest black population, which is still heavily concentrated in the country's poverty-mired northeastern States. With Brazilians of direct or mixed African ancestry today accounting for just over half of the population of 198 million, the Africans decidedly constitute a major immigration of foreign peoples into South America.

FIGURE 5A-4 © H. J. de Blij, P. O. Muller, and John Wiley & Sons, Inc.

Ethnic Landscapes

The cultural landscape of South America, similar to that of Middle America, is a layered one. Indigenous inhabitants cultivated and crafted diverse landscapes throughout the continent, some producing greater impacts than others. When the Europeans arrived, the cultural transformation that resulted from depopulation severely impacted the environment. Native peoples now became minorities in their own lands, and Europeans introduced crops, animals, and ideas about land ownership and land use that irreversibly changed South America. They also brought in Africans from various parts of Subsaharan Africa. Europeans from non-Iberian Europe also started immigrating to South America, especially during the first half of the twentieth century. Japanese settlers arrived in Brazil and Peru during the same era. All of these elements have contributed to the present-day ethnic complexion of this realm. Figure 5A-4 shows the distinct concentrations of indigenous and African cultural dominance, as well as areas where these groups are largely absent and people of European ancestry predominate.

Of course, ethnic origins are not always so straightforward, and patterns can change as a result of internal migrations and ethnic mixing. In recent years, research on individual DNA and genetics has taught us a lot about the regional and group origins of people. On the map, Argentina is indicated as having predominantly European ancestry. More specifically, recent research shows that, for the average Argentine, nearly 80 percent of his or her genetic structure is European, 18 percent aboriginal, and 2 percent African. If nobody had any mixed ancestors, these percentages would translate perfectly into European, indigenous, and African shares of the population—but of course that is not the case. Many people do not have perfect knowledge about their ethnic ancestry, and while some may be strictly of one single origin, many others will have some degree of mixed ancestry. In the aggregate, however, there is no doubt that the population of cone-shaped southern South America is predominantly of European origin. South America, therefore, is a realm marked by **cultural pluralism [6]**, where indigenous peoples of various cultures, Europeans from Iberia and elsewhere, Africans mainly from western tropical Africa, and even some Asians from Japan, India, and Indonesia cluster in adjacent areas but generally do not mix. The bottom line is a cultural mosaic of almost endless variety.

ECONOMIC GEOGRAPHY

Agricultural Land Use and Deforestation

The internal divisions of this cultural kaleidoscope are further reflected in the realm's economic landscape. In South America, larger-scale **commercial [7]** or market (for-profit) and smaller-scale **subsistence agriculture [8]** (primarily for household use) exist side by side to a greater degree than anywhere else in the world (Fig. 5A-5). The geography of plantations and other commercial agricultural systems was initially tied to the distribution of landholders of European background, whereas subsistence farming (such as highland mixed subsistence-market, agroforestry, and shifting cultivation) is historically associated with the spatial patterns of indigenous peoples as well as populations of African and Asian descent.

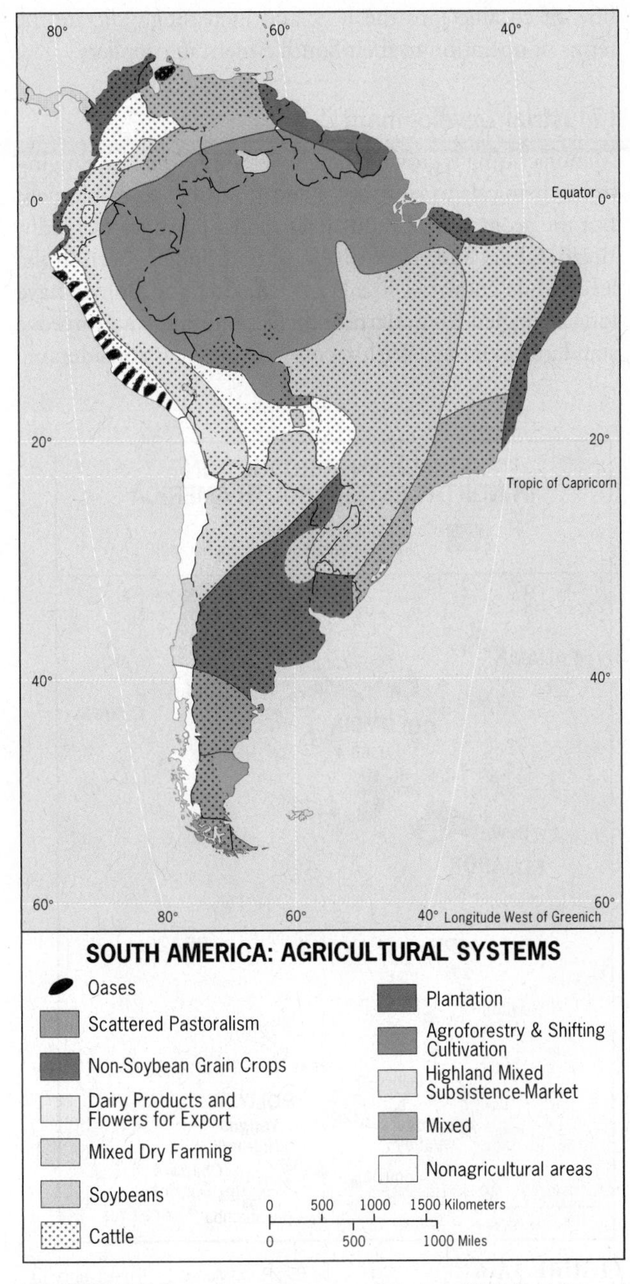

FIGURE 5A-5 © H. J. de Blij, P. O. Muller, and John Wiley & Sons, Inc.

At a broader level of generalization, agricultural systems and land use in South America vary in close relationship to physiography, as one would expect. Forestry and agroforestry prevail in the Amazon Basin; ranching dominates in the grasslands of the Southern Cone; and different forms of agriculture, with or without irrigation, are found in an extensive zone from northeastern Brazil to northern Argentina, as well as scattered among pockets of moderate elevation in the Andean highlands. Land uses throughout South America are changing rapidly today, mainly due to ongoing deforestation and the introduction or expansion of new crops. The fastest-growing crop is soybeans, the cultivation of which now dominates much of east-central Brazil and spills over into adjacent areas of Paraguay, Uruguay, and Argentina (Fig. 5A-5).

Deforestation is a particularly acute problem in northern Brazil. In the past, deforestation was attributed mainly to small-scale landholders, colonists who had made their way into the Amazon rainforest to eke out a new living. The usual pattern of settlement often went like this. As main and branch highways were cut through the forest, settlers followed (frequently in response to government-sponsored land occupation schemes), moving out laterally to clear land for farming (see final photo in Chapter 5B). Subsistence crops were planted, but within a year or two, weed infestation and declining soil fertility made these plots unproductive. As soil fertility continued to decline, the settlers planted grasses and then sold their land to cattle ranchers. The peasant farmers then moved on to newly opened nearby areas, cleared more land for planting, and the cycle repeated itself.

Unfortunately, this activity not only entrenches low-grade land uses across widespread areas, but also entails the burning and clearing of vast stands of tropical woodland (in fact, since the 1980s, an area of rainforest about the size of Ohio has been lost *annually* in northern Brazil). To make matters even worse, the deforestation crisis continues to intensify as ***agro-industrial*** operations engaged in large-scale production for export markets increasingly penetrate Amazonia and transform its land cover.

As South American economies modernize, agriculture remains highly important. Almost one-fifth of this realm's workforce is still employed in the primary sector that includes farming, cattle raising, and fishing—a much greater share than in North America. The South American contribution to global trade in grains, soybeans, coffee, orange juice, sugar, and many other crops is significant and growing. And in addition to all these successful agricultural endeavors, illegal farming thrives in the form of narcotics production, particularly cocaine.

The Geography of Cocaine

All of the cocaine that enters the United States comes from South America, mainly Colombia, Peru, and Bolivia. Within these three countries, cocaine annually brings in billions of (U.S.) dollars and "employs" tens of thousands of workers, constituting a powerful industry. The first of the three stages of cocaine production involves the extraction of coca paste from the coca plant, a raw-material-oriented activity that is located near the areas where the plant is grown. The main zone of coca-plant cultivation is along the eastern slopes of the Andes and in adjacent tropical lowlands in Bolivia, Peru, and Colombia. Today, five areas dominate in the growing of coca leaves for narcotic production: Bolivia's Chaparé district in the marginal Amazon Lowlands northeast of the city of Cochabamba; the Yungas Highlands north of the Bolivian capital, La Paz; north-central Peru's Huallaga and neighboring valleys; south-central Peru's Apurimac Valley, southeast of Huancayo; and the areas around the guerrilla-controlled territories of southern Colombia (Fig. 5A-6).

Coca leaves harvested in the source areas of the Andes and nearby interior lowlands make their way to local collection centers, located at the convergence of rivers and trails, where coca paste is extracted and prepared. The second stage of production involves the refining of that coca paste (about 40 percent pure cocaine) into cocaine hydrochloride (more than 90 percent pure), a lethal concentrate that is diluted with substances such as sugar or flour before being sold on the streets to consumers. Cocaine refining requires sophisticated chemicals, carefully controlled processes, and a labor force skilled in their supervision. Most of this processing is dominated by Colombia and takes place in the rebel-held territory of the lowland central-south and east, beyond the reach of the Bogotá government (see Fig. 5B-3). Interior Colombia also possesses the geographic advantage of intermediate location, lying between the source areas to the south (as well as locally) and the U.S. market to the north.

The final stage of production entails the distribution of cocaine to the marketplace, which depends on an efficient, clandestine transportation network that leads into the United States. Private planes operating out of remote airstrips were the preferred "exporting" method until about a decade ago, and for some time Miami was at the center of cocaine distribution in the United States. But aggressive U.S. measures along the coasts of Florida as well as the Gulf of Mexico have effectively closed down trans-Caribbean flight paths and caused a shift. Much of the cocaine now travels overland to be smuggled by sea through northwestern South America's Pacific and Caribbean seaports. As noted in Chapter 4A, its main destination is Mexico (often via countries in northern Central America), which today supplies almost all the cocaine that enters the United States.

Since 2010, Venezuela has become an additional important link in northbound trafficking, and today about one-quarter of all U.S.-bound cocaine passes through that country. Colombia's FARC and other insurgent groups, originally based in the country's central south, have expanded to establish themselves in the Venezuelan border zone because the savanna on the other side—flat, remote, and beyond the effective control of distant Caracas—is ideal for operating airstrips (Fig. 5A-6). Indeed, they can be fashioned in a matter of hours by dragging a log behind a pick up truck to smooth the ground, important whenever Venezuelan forces bomb the runways. Thus cocaine is increasingly smuggled across the border, particularly into Venezuela's Apure State from which small aircraft transport it to various cartel strongholds in Mexico for distribution in the United States (see Fig. 4B-6).

The rerouting of drugs through Mexico has been aided by the increased northward flow of goods into the United States under NAFTA—providing ever wider opportunities for smuggling via the millions of trucks that cross the border each year. It has also triggered the rapid rise of organized crime as several Mexican drug gangs evolved into sophisticated international cartels that now control the inflow of cocaine into the U.S. and increasingly dictate the terms of operation to their South American suppliers.

Industrial Development

Manufacturing is growing rapidly in South America, ranging from chemicals to electronics and from textiles to biofuels. But the geography of industrial production varies markedly. Brazil, Chile, and Argentina have traditionally been in the lead, while countries like Peru, Ecuador, and Bolivia have long struggled to modernize their economies and improve standards of living. Within countries, there is considerable

FIGURE 5A-6 © H. J. de Blij, P. O. Muller, J. Nijman, and John Wiley & Sons, Inc.

uneven development [9] as well. Most manufacturing is concentrated in and around major urban centers, leaving vast empty spaces in the interior of the realm.

Brazil, in particular, is increasingly drawing attention for its momentous growth and the rising sophistication of a number of economic sectors. In the past, typical Brazilian exports were foods and footwear. Today, the leading exports include oil (thanks to the fortuitous discovery of huge offshore reserves in the past few years), steel, and state-of-the-art Embraer aircraft. For almost a decade now, Brazil has been included as one of the four biggest emerging markets in the world, along with Russia, India, and China—the so-called ***BRICs***.

Despite recent impressive economic growth across much of the realm, it is clear that the longstanding isolation of its individual countries remains a serious obstacle. Nonetheless, most governments agree that tighter integration, political as well as in terms of transportation infrastructure, would greatly enhance economic opportunities.

Economic Integration

The separatism that has long characterized international relations in this realm is finally giving way as South American countries discover the benefits of forging new partnerships with one another. But integration is only beginning and the realm continues to be fragmented: only about 25 percent of its exports stay within South America (a very low share when compared to Europe, where intrarealm exports total close to 70 percent). Nonetheless, things are moving in the right direction. With mutually advantageous trade the catalyst, a new continentwide spirit of cooperation is blossoming at every level. Periodic flare-ups of boundary disputes now rarely escalate into open conflict. Cross-border rail, road, and pipeline projects, stalled for years, are multiplying steadily. For example in southern South America, five formerly contentious nations are developing the *hidróvia* (water highway), a system of river locks that is opening most of the Paraná-Paraguay Basin to barge transport; similar proposals have been advanced to connect this basin to the Amazon River system. And importantly, investments today flow more freely than ever from one country to another, particularly in the agricultural sector.

Recognizing that free trade may well solve several of the realm's economic-geographic problems, governments are now pursuing multiple avenues of economic supranationalism. In 2014, South America's republics were affiliating with the following major trading blocs:

- ***Mercosur/l*** (***Mercosur*** in Spanish; ***Mercosul*** in Portuguese)**:** Launched in 1995 by countries of the Southern Cone and Brazil, this Common Market established a free-trade zone and customs union linking Brazil, Argentina, Uruguay, and Paraguay. Venezuela joined in 2012 and Bolivia will fully accede by 2014. Meanwhile, founding member Paraguay was suspended in 2012 following the overthrow of its elected leader. With Colombia, Chile, Ecuador, and Peru participating as associate members, this organization aspires to become the dominant free-trade organization for all of South America.
- ***Andean Community:*** First formed as the Andean Pact in 1969 and then restarted in 1995 as a customs union with common tariffs for imports, this bloc consists of Colombia, Peru, Ecuador, and Bolivia. Venezuela was a member until it withdrew in 2006.
- ***Union of South American Nations (UNASUR):*** Founded in 2008 in the Brazilian capital of Brasília, the 12 independent countries of South America signed a treaty to create a union envisioned to be similar to the European Union (see Chapter 1A), whose proclaimed goals are a continental parliament, a coordinated defense effort, a single passport for all its citizens, and greater cooperation on infrastructure development. However, significant disagreement among member-states concerning specifics puts the completion of these efforts many years into the

An aerial image of one of South America's most spectacular raw-material sites: northern Brazil's Carajás Iron Ore Mine, located where the southeastern rim of the Amazon Basin meets the Brazilian Highlands in the center of Pará State. This is be the world's richest as well as largest proven deposit of iron ore, whose open-pit mines steadily deepen and sprawl outward as minerals are extracted from the surface—one stripped-away, environmentally degrading layer at a time. In 2012, 320 million metric tons were hauled away via a dedicated, 850-kilometer (535-mi) railroad to the Atlantic port of São Luis (see Fig. 5B-9).

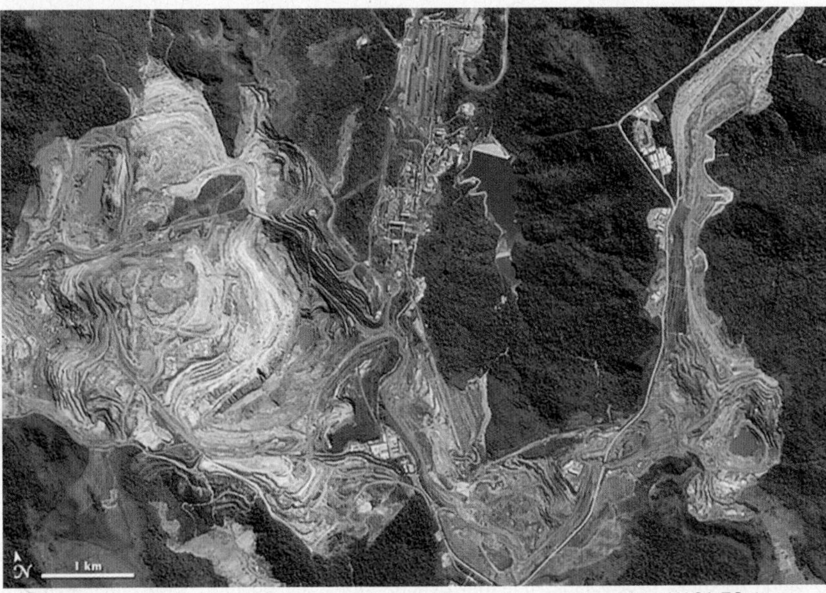

NASA image created by Jesse Allen, using EO-1 ALI data provided courtesy of the NASA EO-1 team

future. UNASUR was preceded by the South American Community of Nations.

- ***Pacific Alliance (PA):*** Inaugurated in 2012, this newest bloc's founding members were Mexico, Colombia, Peru, and Chile. They swiftly formed a free-trade area among themselves and announced their commitment to further economic integration "with a clear orientation to Asia." In 2013, while the Mercosur/l countries squabbled with one another, the PA was advancing steadily with a collective economic growth rate almost twice as high. Candidates for new membership were lining up, led by Costa Rica, Panama, and Guatemala; other countries, led by Canada and the United States, expressed interest in acquiring *observer* status.
- ***Free Trade Area of the Americas (FTAA):*** The United States and other NAFTA proponents have tried to move this hemispheric free-trade idea forward, but it has been resisted by peasants and workers in South America, and formally opposed by Mercosur/l. As long as the terms of trade remain set by the North, the Southern partners will be reluctant to participate in this initiative.

The Commodities Boom

Since shortly after the turn of this century, the South American economy in the aggregate has grown robustly by about 5 percent annually. Much of this growth has occurred in Brazil, Chile, and Peru, but other parts of the continent also fared well. A leading reason has been the steadily rising demand for the realm's abundant raw materials in the global marketplace, in no small part resulting from the explosive growth (and voracious raw-material appetites) of China and India. These two Asian giants have been gobbling up Brazilian soybeans, Chilean copper, Peruvian silver, Venezuelan oil, and so much more. Over the past few years, commodity prices have skyrocketed and South America has reaped the benefits. Thus much of the realm was virtually unaffected by the global recession that arose during the late 2000s, which enabled South America to strengthen its position in the global economy. Even though growth has leveled off since 2011, partially in response to slowing Asian demand, in absolute terms the exporting of raw materials remains at a high level.

Heavy reliance on the production of raw materials (slightly more than 50 percent of the realm's exports today) is not always beneficial. As the saying goes, what goes up must come down: world commodity prices—and therefore the foreign revenues of countries overly tied to them—can be notoriously unstable. A second cautionary note is that high demands for commodities can drive up the value of a producing country's currency, which can limit sales of other exports because the latter become more expensive for foreign trading partners. Third, natural resource management, because of complex ownership issues and exploitation rights, invites corruption and complacency (note, for instance, that discoveries of major oil reserves do not often result in a society's advancement). Finally, and most importantly, many resources (especially minerals and energy deposits) are finite, and it is important for producing countries to plan for a future without them. That, of course, is a major challenge to political leaders who tend to think no further than the next election. But if governments engage in wise resource management and development strategies, the commodities boom can only contribute to their nation's overall economic growth.

From the Field Notes . . .

© H.J. de Blij

© H.J. de Blij

"Two unusual perspectives of Rio de Janeiro form a reminder that here the wealthy live near the water in luxury high-rises, such as these overlooking Ipanema Beach—while the poor have million-dollar views from their hillslope *favelas*, such as Rocinho."

www.conceptcaching.com

URBANIZATION

Rural-Urban Migration

As in most other realms, South Americans are leaving the land and migrating to towns and cities. South America's modern urbanization process got an early start and has been intensifying since 1950 as the growth of urban settlements averaged 5 percent a year. Meanwhile, rural areas have grown by less than 2 percent annually over those same six-plus decades. As a result, the realm's urban population has climbed to its current level of 82 percent, ranking it with those of western Europe and the United States. These numbers underscore not only the dimensions but also the durability of the **rural-to-urban migration [10]** from the countryside to the cities.

In South America, as in Middle America, Africa, and Asia, people are attracted to the cities and driven from the poverty of the rural areas. Both ***push*** and ***pull*** factors are at work. Rural land reform has been very slow in coming, and for this and other reasons every year tens of thousands of farmers simply give up and leave, seeing little or no possibility for economic advancement. The urban centers lure them because they are perceived to provide opportunity—the chance to earn a regular wage; visions of education for their children, better medical care, upward social mobility, and the excitement of life in a big city draw hordes to places such as São Paulo and Lima.

But the actual move can be traumatic. Cities in developing countries are surrounded and often invaded by squalid slums, and this is where the urban immigrant most often finds a first—and sometimes permanent—abode in a makeshift shack without even the most basic amenities and sanitary facilities (see photo). Unemployment remains persistently high, often exceeding 25 percent of the available labor force. But still the people come, hopeful for a better life, the overcrowding in the shantytowns worsens, and the threat of regional-scale disease and other disasters rises.

Regional Patterns

The generalized spatial pattern of South America's urban transformation is displayed in Figure 5A-7, which shows a ***cartogram*** of the continent's population. Here we see not only the realm's countries in population space relative to each other, but also the proportionate sizes of individual large cities within their total national populations.

Regionally, the Southern Cone (countries colored green) is the most highly urbanized. Today in Argentina, Chile, and Uruguay, almost all of the population resides in cities. Ranking after them in urbanization is Brazil (tan). The next highest group of countries (beige) borders the Caribbean Sea in the north. Not surprisingly, the Andean countries (brown) constitute the realm's least urbanized zone. Figure 5A-7 tells us a great deal about the relative positions of major metropolises in their countries. Three of them—Brazil's São Paulo and Rio de Janeiro, and Argentina's Buenos Aires—rank among the world's **megacities [11]** (cities whose populations exceed 10 million). But today even the population of the Amazon Basin is more than 70 percent urbanized.

The "Latin" American City Model

The urban experience in the South and Middle American realms varies because of diverse historical, cultural, and economic influences. Nevertheless, there are a number of common threads that have prompted geographers to search for useful generalizations. One is the model of the intraurban spatial structure of the **"Latin" American City model [12]** proposed by Ernst Griffin and Larry Ford (Fig. 5A-8).

The idea behind a ***model*** is to create an idealized representation of reality, displaying as many key real-world elements as possible. In the case of South America's cities, the

A 2012 satellite image of Manaus, the fastest-growing large Brazilian metropolis, now with a population of 2 million. Located at the confluence of the Rio Negro and the coffee-colored, sediment-rich Rio Solimões (the local Brazilian name for the Amazon River), this used to be one of Brazil's most remote places. But today it has become a leading hub in the ongoing colonization and urbanization of the Amazon Basin—symbolized by its Ponte Rio Negro Bridge (thin white line at the central left edge of the image), which opened in 2011 (see chapter-opening photo).

© NASA/Corbis

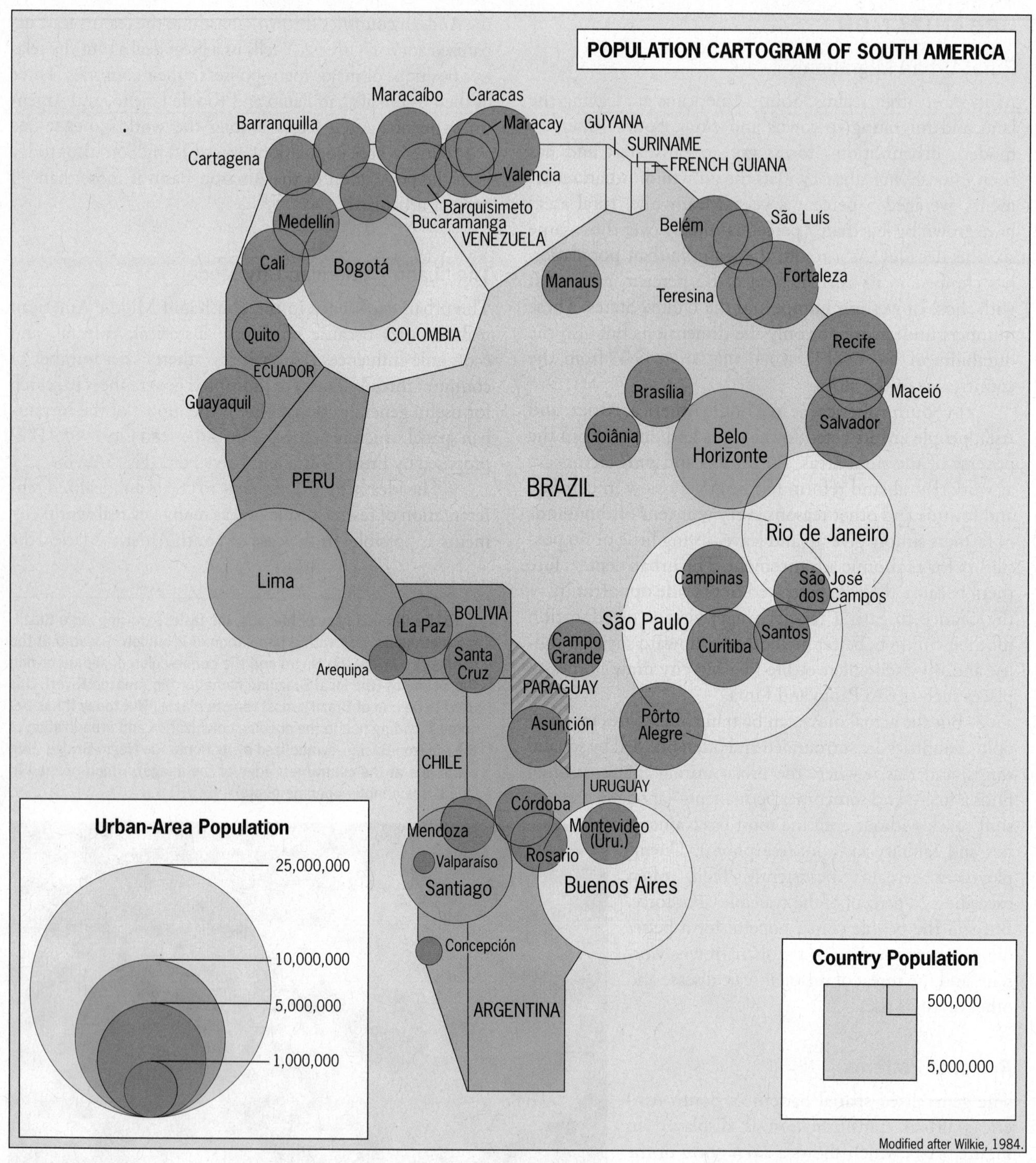

FIGURE 5A-7

basic spatial framework of city structure, which blends traditional elements of South and Middle American culture with modernization forces now reshaping the urban scene, is a composite of radial sectors and concentric zones. Anchoring the model is the ***central business district (CBD)***, the primary business, employment, and entertainment focus of the surrounding metropolis. The CBD contains many modern high-rise buildings but also mirrors its colonial beginnings. As shown in Figure 4A-5, by colonial law Spanish colonizers were required to lay out their cities around a central square, or plaza, dominated by a church and government buildings. Santiago's *Plaza de Armas*, Bogotá's *Plaza Bolívar*, and Buenos Aires's *Plaza de Mayo* are classic examples (see photo in *From the Field Notes*). The plaza was the hub of the city, which later outgrew its old core as new commercial districts formed nearby; but to this day the plaza remains a ceremonial center and an important link with the past.

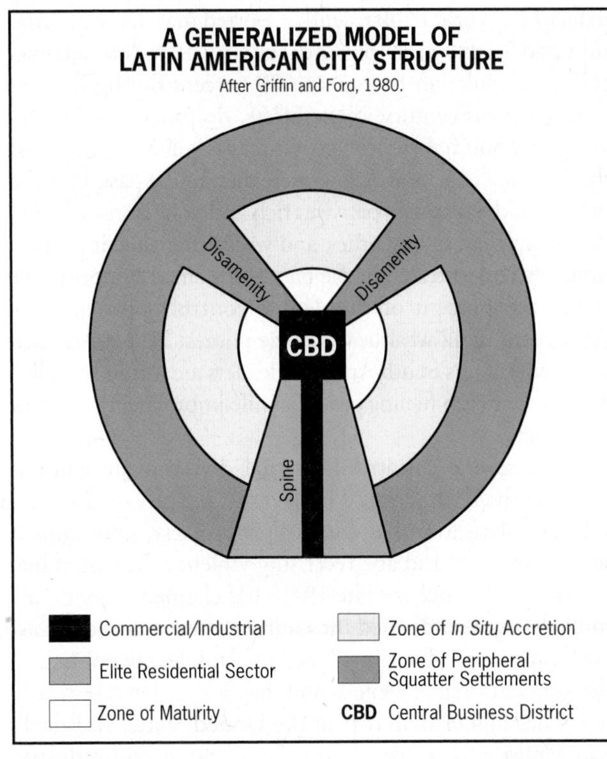

FIGURE 5A-8 © H. J. de Blij, P. O. Muller, and John Wiley & Sons, Inc.

Radiating outward from the urban core along the city's most prestigious axis is the commercial ***spine***, which is adjoined by the ***elite residential sector*** (shown in green in Fig. 5A-8). This widening corridor is essentially an extension of the CBD, featuring offices, retail facilities, and housing for the upper and upper-middle classes.

The three remaining concentric zones are home to the less fortunate residents of the city, with income level and housing quality decreasing as the distance from the CBD increases. The ***zone of maturity*** in the inner city contains housing for the middle class, who invest sufficiently to keep their aging dwellings from deteriorating. The adjacent ***zone of* in situ *accretion*** is one of much more modest housing interspersed with unkempt areas, representing a transition from inner-ring affluence to outer-ring poverty and taking on slum characteristics.

The outermost ***zone of peripheral squatter settlements*** is home to the millions of comparatively poor and unskilled workers who have recently migrated to the city. Here many newcomers earn their first cash income by becoming part of the **informal sector [13]**, in which workers are undocumented and money transactions are beyond the control of government. The settlements consist mostly of self-help housing, vast shantytowns known as ***barrios* [14]** in Spanish-speaking South America and *favelas* in Brazil. Some of their entrepreneurial inhabitants succeed more

From the Field Notes . . .

"Cuzco is the historic capital of the Inca Empire, located in the Andes of southeastern Peru. The city center is a UNESCO-designated World Heritage Site and attracts well over a million tourists each year. I drove out of the valley up the hillside to get a good view of the layout of the center, so typical of Spanish colonial cities. The *Plaza de Armas* with its parks and benches is surrounded by the Cathedral of Santo Domingo and other religious and municipal buildings, and adjacent areas around them are used for commerce and retailing. Located farther away from the *Plaza* are newer residential areas. This historic Inca capital was sacked in the early sixteenth century by Pizarro, and only some Inca ruins remain in the center; outside the city lies the impressive Sacsayhuaman fortress, with most of its walls still intact. This is not a very large city—it is too remote for that—but the population has tripled over the past quarter-century to more than 500,000."

© Jan Nijman **www.conceptcaching.com**

than others, transforming parts of these shantytowns into beehives of activity that can propel resourceful workers toward a middle-class existence.

A final structural element of many South American cities forms an inward, narrowing sectoral extension of the zone of peripheral squatter settlements and is known as the ***zone of disamenity***. It consists of undesirable land along highways, rail corridors, riverbanks, and other low-lying areas; people here are so poor that they are forced to live in the open. Thus the realm's cities present enormous contrasts between poverty and wealth, squalor and comfort—harsh contrasts all too frequently observed in the urban landscape.

FUTURE PROSPECTS

The Need for Stability

As this decade opened, a number of South American countries were celebrating two centuries of independence. What mattered most, for nationalistic purposes, was the date when independence was declared, even if it took additional years for Spain to recognize that new political status. The wars with Spain lasted from 1808 (not coincidentally, soon after Spain had been weakened by Napoleon's invasion of Iberia) to 1838. Brazil's independence from Portugal was achieved during that same period, in 1822.

But the first two centuries of freedom have not seen the evolution of the kind of stable, mature political climates and institutions that one might have expected. South American countries have been in frequent political turmoil. Dictatorial regimes ruled from one end of the realm to the other; unstable governments fell with damaging frequency. Widespread poverty, harsh regional disparities, poor internal surface connections, limited international contact, and economic stagnation prevailed.

Today all that is being cast aside as South America enters a new era of dramatic transition. Well into the second decade of the twenty-first century, formal democracy seems to have taken hold almost everywhere. Long-isolated countries are becoming more interconnected through new transportation routes and trade agreements. New settlement frontiers are being opened. Energy resources, some long exploited and others newly discovered, are boosting national economies as world prices have risen. Foreign states and corporations have appeared on the scene to buy commodities and invest in infrastructure. The pace of globalization is accelerating from Bogotá to Buenos Aires.

Yet progress still varies enormously across the realm (as detailed in Chapter 5B), and these exciting developments need to be viewed against a backdrop of persistent problems of deep inequality and persistent class divisions, ethnic tension, and political malfeasance.

Problems of Inequality and Violence

An important indicator of South America's economic growth since 2000 has been the steady expansion of its middle class: in fact, the World Bank recently reported that this social stratum had increased its presence within the realm's income-earning population from 20 to 29 percent during the first decade of this century. Nonetheless, deep divisions persist, and many South American governments will need to address these inequalities more aggressively than in the past. By most measures, the disparity between rich and poor is still wider in this realm than in any other, and wealth remains disproportionately concentrated in the hands of a small minority (the richest 20 percent of the people control approximately 70 percent of all wealth, while the poorest 20 percent own 2 percent). Thus South America's leaders are forced to walk a fine line between fueling growth while improving the lives of the masses.

In several countries, internal divisions are complicated by the resurgence of indigenous peoples, and the resulting polarization has harmed civil society, undermined social cohesion, and abetted rising violence. In Colombia, internal strife since the late 1940s has claimed more than a million lives as it divided the country, drew countless peasants into the production of cocaine, and deeply politicized the population. Venezuela now has a murder rate nearly ten times higher than that of the United States. In Brazil's São Paulo and other metropolises, gated communities for higher-income residents have now become standard, armed with state-of-the-art security systems to wall off crimes committed by the poor; and in Rio de Janeiro relentless, grinding poverty in its teeming slums has made them ever more dangerous communities.

The Shadow of the United States

The United States has long played a key role in this realm, beginning with the Monroe Doctrine's 1823 assertion that European colonial powers had no rights to South America. During the Cold War (1945–1990), the United States became politically involved in a number of countries, mainly to keep Soviet influence out of the Western Hemisphere (as in Chile during the early 1970s). This did not always enhance its local standing; in fact, anti-Americanism never seems far from the surface in South America, based heavily on past U.S. support for right-wing dictatorships and continuing perceptions of imperialistic behavior.

To be sure, relations with the United States have never been smooth and were always asymmetrical. For instance, South America today attracts only about 4 percent of all U.S. foreign trade. But the United States remains the biggest trading partner for the realm as a whole, accounting for almost one-fifth of the exports and imports of all South American countries. Thus it was hardly a coincidence that ***dependencia* theory [15]** originated here in South America during the 1960s. It was a new way of thinking about economic development and underdevelopment that explained the persistent poverty of some countries in terms of their unequal relations with the world's more affluent countries. Whatever the current validity of *dependencia* theory, the asymmetry remains, and U.S. foreign policy,

Photo by Fabio Rossi/Globo/Getty Images, Inc.

Nothing symbolizes China's new economic drive in South America more than its growing interest in its burgeoning fellow BRIC. In 2011, the two countries concluded a (U.S.) $7 billion deal in which Chinese investors helped finance a major expansion of Brazil's already-booming soybean production. Billions of additional dollars of Chinese investment contributed to the new *Superporto do Açu* that opened for business in late 2012. The photo shows a Chinese delegation visiting the construction site in 2010. Açu is located about 400 kilometers (250 mi) northeast of Rio de Janeiro, and its massive 3-kilometer (2-mi)-long concrete pier jutting out into the Atlantic contains ten berths capable of accommodating the largest ships afloat (which are also Chinese). Brazilian investors have called this facility the "Highway to China," and today it claims to be the world's largest port for handling bulk cargoes.

as viewed from South America, is not always credible. Interestingly, America's special interests in the realm are now under increased pressure from another major power that would have been furthest from President Monroe's mind back in 1823: China.

China Calling

By 2010, China had displaced the United States as the leading trading partner of Brazil and Chile, and in 2011 Peru joined that list. Argentina and Colombia may soon be next because in both China is already a close second. These remarkable developments have yet to fully register in the United States; meanwhile, China is well on its way to becoming the most important world power in the South American realm.

Nearly invisible just ten years ago, China has made its presence felt across South America in a variety of ways that include establishing new embassies and consulates, buying up companies, partnering joint ventures, financing infrastructure projects and development assistance, and sending as well as inviting high-level trade delegations. Their motives are clear: the Chinese need key raw materials such as oil, copper, and a plethora of minerals to fuel the enormous economic growth of their country (see photo). At the same time, they are seeking to expand markets for Chinese exports, and the steadily expanding middle classes of Brazil and other countries make enticing targets.

Whether it is the United States or China, South Americans will need to carefully consider how and where to focus their economic attention. Within the realm itself, there is still much to be done in terms of building political accord and economic integration. Undoubtedly, those Chinese investments mentioned above will help to build the bridges so badly needed to bring this realm more closely together.

POINTS TO PONDER

- The Amazonian rainforest is mainly located within Brazil but sometimes is referred to as "the lungs of the Earth." Should its preservation be a global responsibility?
- Is it a coincidence that both Liberation theology and *dependencia* theory originated in this realm?
- The map of South America's religions is changing. Brazil is still nominally the world's largest Roman Catholic country, but the number of adherents is declining and now accounts for barely 60 percent of the national population (down from 90 percent as recently as 1970).
- Should China's rapidly growing involvement in South America be a cause for concern in the United States?

5B SOUTH AMERICAN REGIONS

FIGURE 5B-1

The Caribbean North

The Andean West

The Southern Cone

Brazil

IN THIS CHAPTER

- Oil and political change in Venezuela
- The cocaine curse
- Resurgence of indigenous peoples
- Brazil's unmatched natural resource base

CONCEPTS, IDEAS, AND TERMS

Insurgent state 1

Failed state 2

Altiplano 3

Triple Frontier 4

Elongation 5

Buffer state 6

BRIC 7

El Niño 8

Forward capital 9

Cerrado 10

Growth pole concept 11

Steve Allen/Photodisc/Getty Images, Inc.

Where were these pictures taken?
Find out at www.wiley.com/college/deblij

On the basis of physiographic, cultural, and political criteria, South America can be divided into four rather clearly defined regions (Fig. 5B-1):

REGIONS OF THE REALM

1. ***The Caribbean North***, located almost entirely north of the equator, consists of five entities that display a combination of Caribbean and South American features: Colombia, Venezuela, and those that represent three historic colonial footholds by Britain (Guyana), the Netherlands (Suriname), and France (French Guiana).
2. ***The Andean West*** is formed by four republics that share a strong indigenous cultural heritage as well as powerful influences resulting from their Andean physiography: Ecuador, Peru, Bolivia, and, transitionally, Paraguay.
3. ***The Southern Cone***, for the most part located south of the Tropic of Capricorn (latitude 23½°S), includes three countries: Argentina, Chile, and Uruguay (all with strong European imprints and little remaining indigenous influence) plus aspects of Paraguay. The far northern part of this region that lies east of the Andes is occupied by the Gran Chaco, a sparsely populated, hot, semiarid, grassy lowland centered on western Paraguay.
4. ***Brazil*** occupies an enormous part of interior and eastern South America. In the Amazon Basin of the north, its own interior overlaps with the continent's "green heart"—often referred to the world's "lungs"—the largest tropical rainforest on Earth. As South America's giant, accounting for just about half the realm's territory as well as population, it is swiftly developing into the Western Hemisphere's second superpower. In Brazil the dominant Iberian influence is Portuguese, not Spanish, and here

Ian Trower/Robert Harding World Imagery/Getty Images, Inc.

MAJOR CITIES OF THE REALM

City	Population* (in millions)
São Paulo, Brazil	26.4
Buenos Aires, Argentina	14.1
Rio de Janeiro, Brazil	12.3
Lima, Peru	9.6
Bogotá, Colombia	9.5
Santiago, Chile	6.3
Belo Horizonte, Brazil	5.8
Brasília, Brazil	4.1
Caracas, Venezuela	3.5
Guayaquil, Ecuador	2.4
Asunción, Paraguay	2.4
Manaus, Brazil	2.0
Santa Cruz, Bolivia	1.9
Quito, Ecuador	1.7
Montevideo, Uruguay	1.7

*Based on 2014 estimates.

Africans, not indigenous peoples, form a significant component of demography and culture. And Brazil is in some ways a bridge between the Americas and Africa, which is why we discuss it last in this chapter, just before turning to Subsaharan Africa in Chapter 6A.

Each of these four regions shares some important commonalities in terms of physical environment, ethnic origins, cultural milieu, and international outlook. But there also are important differences, especially involving economic performance, democratic functioning, and social stability.

THE CARIBBEAN NORTH

The countries of South America's northern tier have something in common besides their coastal location: each has a coastal tropical-plantation zone based on the Caribbean colonial model. Especially in the three Guianas, early European plantation development encompassed the forced immigration of African laborers and eventually the absorption of this element into the population matrix. Far fewer Africans were brought to South America's northern shores than to Brazil's Atlantic coasts, and tens of thousands of South Asians also arrived as contract laborers and stayed as settlers, so the overall situation here is not comparable to Brazil's. Moreover, it is also distinctly different from that of the rest of South America.

To this day, Guyana, Suriname, and French Guiana still display the coastal orientation and plantation dependency with which the colonial period endowed them, although the logging of their tropical forests is penetrating and ravaging the interior. In Colombia and Venezuela, however, farming, ranching, and mining drew the population inland, overtaking the coastal-plantation economy and creating diversified economies (Fig. 5B-2).

Colombia

Imagine a country more than twice the size of France but with only three-quarters of the French population, with an environmental geography so varied that it can produce crops ranging from the temperate to the tropical, and possessing world-class oil reserves as well as many other natural resources. This country is situated in the crucial northwestern corner of South America, with 3200 kilometers (2000 mi) of coastline on both Atlantic (Caribbean) and Pacific waters, closer than any of its neighbors to the markets of the north and sharing a border with giant Brazil to the southeast. Its nation uses a single language and adheres to one dominant religion. Wouldn't such a country thrive?

The answer, unfortunately, is no. Colombia has a history of strife and violence, its politics unstable, its economy damaged, and its overall future clouded. Colombia's cultural uniformity did not produce social cohesion. Its spectacular, scenic physical geography also divides its population of 48.6 million into clusters not sufficiently interconnected to foster integration; even today, this enormous country has less than 1600 kilometers (1000 mi) of four-lane highways. Its proximity to U.S. markets is a curse as well as a blessing: at the root of Colombia's latest surge of internal conflict lies its role as one of the world's leading producers of illicit drugs.

History of Conflict

Colombia's recent disorder is not its first. In the past, civil wars between conservatives and liberals (based on Roman Catholic religious issues) developed into conflicts pitting rich against poor, elites against workers. In Colombia today, people still refer to the last of these wars as *La Violencia*, a decade of strife beginning in 1948 during which as many as 200,000 people died. In the 1970s, disaster struck again. In remote parts of the country, groups opposed to the political power structure began a campaign of terrorism, damaging the developing infrastructure and destroying confidence in the future. Simultaneously, the U.S. market for narcotics expanded rapidly, and many Colombians got involved in the drug trade. Powerful and wealthy drug cartels formed in major cities such as Medellín and Cali, with networks that influenced all facets of Colombian life from the peasantry to the politicians. In all, over the past half-century, it is estimated that one million people died violently as the fabric of Colombian society unraveled.

People and Resources

As Figure 5B-2 reveals, Colombia's physiography is mountainous in its Andean west and north and comparatively flat in its interior. Colombia's scattered population (see Fig. 5A-3) tends to cluster in the west and north, where the resources and the agricultural opportunities (including the coffee for which Colombia is famous) lie.

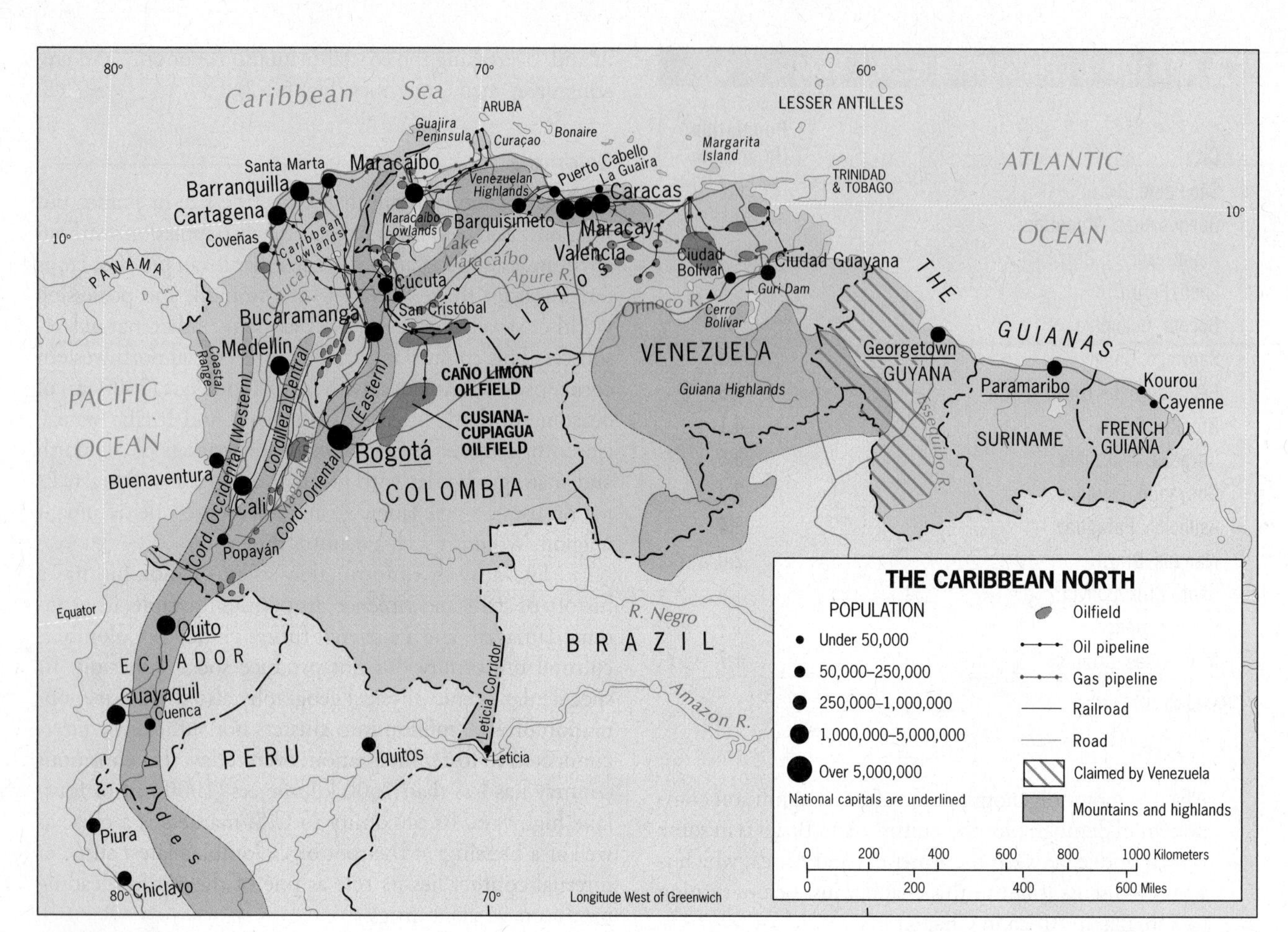

FIGURE 5B-2

Colombia's population clusters remain poorly interconnected. Several lie on the Caribbean coast, centered on Barranquilla, Cartagena, and Santa Marta, old colonial entry points. Others are anchored by major cities such as Medellín and Cali. What is especially interesting about Figure 5B-2 in this context is how little development has taken place on the country's lengthy Pacific coast, where the port city of Buenaventura, across the mountains from Cali, is the only place of any size. The map suggests that Colombia's Pacific Rim era, already in full force in Chile, Peru, and Ecuador, has yet to arrive.

The map also shows that Colombia and neighboring Venezuela share the oil and gas reserves in and around the "Lake" Maracaibo area ("Lake" because this is really a gulf with a narrow opening to the sea), but Venezuela has the bigger portion. Recent discoveries, however, have boosted Colombia's production along the base of the easternmost Andean cordillera, and our map shows a growing system of pipelines from the interior to the coast. Meanwhile, the vast and remote southeastern interior proved fertile ground for that other big Colombian money-maker—drugs.

Cocaine's Curse

The rise of the narcotics industry, fueled by outside (especially U.S. but also Brazilian and European) demand and coupled with its legacy of violence, crippled the state for decades and threatened its very survival. Drug cartels based in the cities controlled vast networks of producers and exporters; they infiltrated the political system, corrupted the army and police, and waged wars with each other that cost tens of thousands of lives and destroyed much of Colombia's social order. The drug cartels also organized their own armed forces to combat attempts by the Colombian government to control the illegal narcotics economy. Meanwhile, the owners of large haciendas in the countryside hired private security guards to protect their properties, banding together to expand these units into what quickly became, in effect, private armies. Colombia was in chaos as narco-terrorists committed appalling acts of violence in the cities, rebel forces and drug-financed armies of the political "left" fought against paramilitaries of the political "right" in the countryside, and Colombia's legitimate economy, from coffee-growing to tourism, suffered grievously. To further weaken the national government, the rebels even took to bombing oil pipelines.

Threats of the Insurgent State

What happened in Colombia beginning in the 1970s and then escalating during the subsequent four decades was not unique in the world—states have succumbed to chaos in the past—but this was an especially clear-cut case of a process long studied and modeled by political geographers.

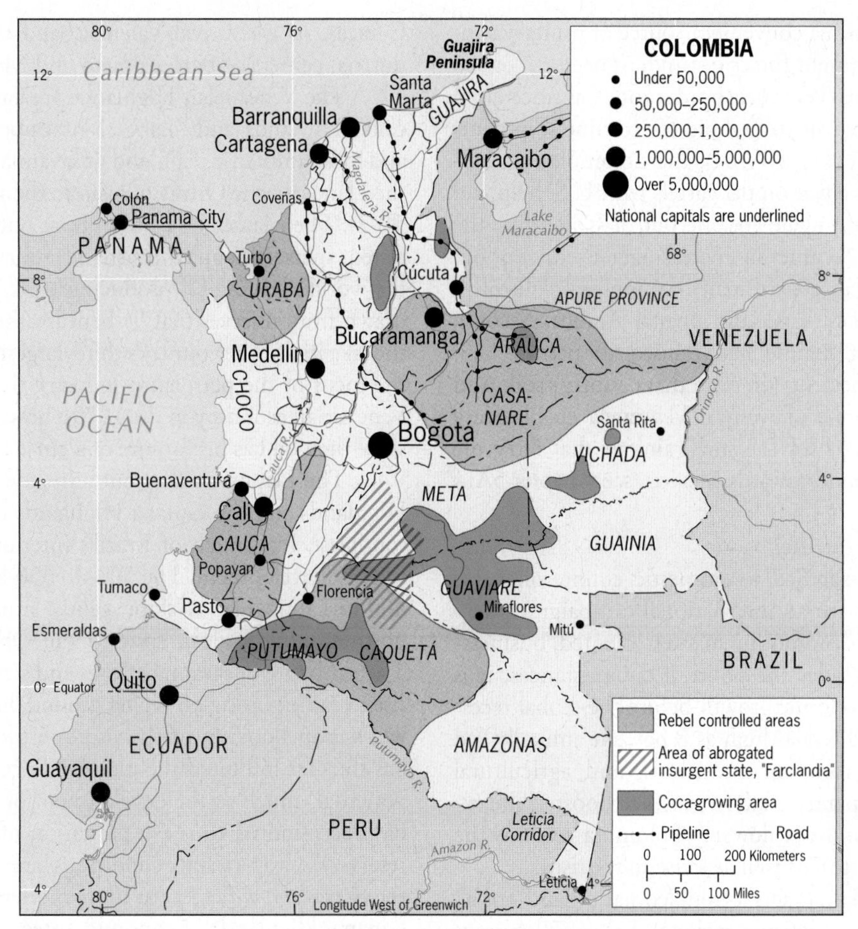

FIGURE 5B-3 © H. J. de Blij, P. O. Muller, J. Nijman, and John Wiley & Sons, Inc.

By the turn of this century, certain parts of Colombia were beyond the control of its government and armed forces. There, insurgents of various stripes created their own domains, successfully resisting interference and demanding to be left alone to pursue their goals, illegitimate though they might be. Leaders of some of these insurgent domains even sent emissaries to Bogotá, the capital, to negotiate their terms for "independence."

We use the concept of an **insurgent state [1]** to indicate the effective control of part of the state's territory by rebels, complete with boundaries, a core area as well as a capital, a local government, and schools and other social services that substitute for those the formal state may have previously provided. Colombia during the 1990s and early 2000s contained several entities that were taking on the properties of insurgent states, and one of these, the red-striped zone shown in Figure 5B-3, even acquired a name—"Farclandia"—after the initials of the Revolutionary Armed Forces of Colombia (FARC), one of the most brutal and by far the most powerful among Colombia's insurgent groups. At the end of 1999, FARC had the Colombian government on the ropes, forcing it to demilitarize its area south of Bogotá (about the size of Switzerland) and announcing plans for a second insurgent state centered on the remote southeastern town of Mitú. It appeared to be just a matter of time before the national government in Bogotá would lose control, the status quo would turn into disintegration, and Colombia would devolve into a **failed state [2]** (a country whose institutions have collapsed and in which anarchy prevails).

For the past decade, Colombia has mounted a twin campaign to defeat the rebels militarily and to persuade them through legal means to give up their arms. The government did succeed in transforming Colombia from a nearly failed state to a more stable, safer, and prosperous country. Securing assistance from the United States, it was able to increase the pressure on armed rebels as well as coca growers, invaded "Farclandia," and scored a number of noteworthy successes in killing and arresting leaders as well as freeing hostages held by the rebels. Colombia's nascent insurgent states weakened, but rebel troops still control important areas and, as along the Venezuelan border, they have in places been able to expand their presence.

If rebel groups like FARC initially pursued a purely political or ideological agenda, their widespread guerrilla activities soon brought them into contact with the cocaine cartels, which had their own reasons for wanting to exercise territorial control. In effect, this led to close, if murky, ties between the cartels and the rebel armies: the insurgents had found themselves a source of funding while the cartels linked up with a convenient source of military support against government forces in times of need.

The Colombia-Venezuela border zone has since 2010 emerged as a key transit stronghold for cocaine shipments to the United States (Fig. 5B-3). As the Colombian government stepped up the war on the cartels with U.S. help and made it harder to smuggle cocaine out of Colombia, the Venezuelan route became an increasingly useful option. First, the Colombian government has minimal control in the border provinces, especially pivotal Arauca. Second, Venezuela's Chávez regime had refused to permit U.S. troops to expand the drug war onto that country's soil. And third, the flat savanna of western Venezuela enables easy takeoffs and landings for the small aircraft that ferry the cocaine northwestward towards Mexico (see Chapter 5A).

A New Economic Future?

The government matched its domestic counteroffensive against the rebels with an international campaign to help revive the economy, promoting market-oriented, business-friendly policies. Under the political circumstances, it is remarkable that economic growth before the global recession began in 2008 was as high as 8 percent annually, led by the revenues from energy, metals, and agricultural products; the comparative lull in violence boosted the exporting of coffee and cut flowers (Colombia remains the leading U.S. importer) as well as a rise in tourism.

But Colombia remains a troubled nation and not all of the problems are of its own making. At least 70 percent of the exported cocaine winds up in the United States; almost all of the remainder goes to Brazil and/or Europe. Colombians maintain that the cocaine keeps flowing mainly because Americans continue to buy so much of it—and thereby indirectly contribute to Colombia's ongoing domestic troubles.

And cocaine is not alone among the very difficult challenges Colombia continues to face. Economic development is hampered by extreme inequality that has only worsened over time, particularly in the countryside. In 1954, 3 percent of landowners controlled 55 percent of the land; in 2005, a half-century later, less than half of 1 percent of the landowners held a whopping 63 percent of all agricultural lands. If the government wants to restore belief in the political system among the rural poor (many of whom have turned to employment in the cocaine industry and/or have been engaged by the rebels), land reform is a vital necessity.

Venezuela

A long boundary separates Colombia from Venezuela, its neighbor to the northeast. Much of what is important in Venezuela is concentrated in the northern and western parts of the country, where the Venezuelan Highlands form the eastern spur of the northern end of the Andes system. Most of Venezuela's 30.7 million people are concentrated in these uplands, which include the capital of Caracas, its early rival Valencia, and the commercial/industrial centers of Barquisimeto and Maracay.

The Venezuelan Highlands are flanked by the Maracaibo Lowlands and "Lake" Maracaibo to the northwest and by an immense plainland of savanna country, known as the Llanos, in the Orinoco Basin to the south and east (Fig. 5B-2). The Maracaibo Lowlands, at one time a disease-infested, sparsely peopled coastland, today constitute one of the world's leading oil-producing areas; much of the oil is drawn from reserves that lie beneath the shallow waters of the lake itself. The country's third-largest city, Maracaibo, is the focus of the petroleum industry that transformed the Venezuelan economy in the 1970s; however, as we shall see, since then oil has been more of a curse than a blessing.

The Llanos on the southern side of the Venezuelan Highlands and the Guiana Highlands in the country's far southeast, like much of Brazil's interior, are in a nascent stage of development. The 300- to 650-kilometer- (180- to 400-mi) long Llanos slope gently from the base of the northernmost Andean spur to the Orinoco's floodplain. Their mixture of savanna grasses and scrub woodland gives rise to cattle grazing on higher ground, however widespread wet-season flooding of the more fertile lower-lying areas has thus far inhibited the plainland's commercial farming potential (much of the development of the Llanos to date has been limited to the exploitation of its substantial oil reserves). Crop-raising conditions are more favorable in the *tierra templada* areas at more moderate altitudes in the Guiana Highlands. Economic integration of this even more remote interior zone with the heartland of Venezuela has been spearheaded by the discovery of rich iron ores on the northern side of the Guiana Highlands southwest of Ciudad Guayana. Local railroads now connect with the Orinoco, and from there ores are shipped directly by sea to foreign markets.

Oil and Politics

Despite these opportunities, Venezuela since the late 1990s has been in upheaval as longstanding economic and social problems finally intensified to the point where the electorate decided to push the country into a new era of radical political change. For nearly two decades since the euphoric 1970s, oil had not bettered the lives of most Venezuelans. A primary reason was that the government unwisely acquired the habit of living off oil profits, forcing the country to suffer the consequences of the prolonged global oil depression that began in the early 1980s. Venezuela found itself heavily burdened by a huge foreign debt it had incurred through borrowing against future oil revenues that were not materializing fast enough. By the mid-1990s, the government was required to sharply devalue the currency, and a political crisis ensued that resulted in a severe recession and widespread social unrest.

With more and more Venezuelans enraged at the way their oil-rich country was approaching bankruptcy without making progress toward the more equitable distribution of the national wealth, voters resoundingly turned in an extreme direction in the 1998 presidential election. Expressing their disgust with Venezuela's ruling elite of both political parties, they elected Hugo Chávez, a former colonel who in 1992 had led a failed military coup. Reaffirming their decision in 2000, Venezuelans gave nearly 60 percent of their votes to Chávez, apparently providing him with a mandate to act as strongman on behalf of the urban poor and the increasingly penurious middle class.

Venezuela's Autocratic Turn

Chávez, clinging to power tenaciously, did indeed pursue such a course after entering office in 1999, sweeping aside Congress and the Supreme Court, supervising the rewriting of the Venezuelan constitution in his own image, and proclaiming himself the leader of a "peaceful leftist revolution" that would transform the country. Even though he professed that social equality ranked atop his agenda, Chávez stirred up racial divisions by actively promoting mestizos (about 70 percent of the population) over people of European background (18 percent).

In the international arena, Chávez sparked controversy at every turn: infuriating the government of neighboring Colombia by expressing his "neutrality" in its confrontation with cocaine-producing insurgent forces; interfering with another neighbor, Guyana, by aggressively reviving a century-old territorial claim to the western zone of that country (the striped area in Fig. 5B-2); embracing Cuba's communist regime; and proclaiming his solidarity with Iran's leadership. Chávez had positioned himself as the champion of Venezuela's (and South America's) poor, trumpeting his "Bolivarian Revolution" as a local alternative to what he described as the insidious encroachment of U.S. imperialism. And yet, while alienating many other South American governments, Venezuela did receive support from Brazil and Argentina to formally join Mercosur in 2012, which in some ways should bring Venezuela back into the realm's fold.

Hugo Chávez's second (and final) presidential term was scheduled to end in 2012, but in 2009 he organized and won a referendum that abolished term limits. Vilified by much of the Western world and most of Venezuela's entrepreneurial citizens (many of whom had fled the country), Chávez had a powerful message that continues to resonate among South America's impoverished masses—itself testimony to the persistent inequality that characterizes this realm. Although Chávez was reelected to a third term in late 2012, terminal illness prevented his swearing-in, and his death in March 2013 plunged the country into a period of political uncertainty. Clearly, what happens next in Venezuela matters far beyond its borders because of its enormous oil reserves—which according to a number of the latest estimates may be the largest of any country on Earth.

The "Three Guianas"

Three small entities form the eastern flank of the realm's northern region: Guyana, Suriname, and French Guiana. They are good reminders of why the name "Latin" America is inappropriate: the first is a legacy of British colonialism and employs English as its official language; the second is a remnant of Dutch influence where Dutch is still official among its polyglot of tongues; and the third is still a dependency—of France. None has a population exceeding 1 million, and all three exhibit social characteristics and cultural landscapes far more representative of the Caribbean Basin than South America. Here British, Dutch, and French colonial powers acquired possessions and established plantations, brought in African and Asian workers, and created economies similar to those that mark a number of the Caribbean islands.

Guyana, with 840,000 people, still has more inhabitants than Suriname and French Guiana combined. When Guyana became independent in 1966, its British rulers left behind an ethnically and culturally divided population in which people of South Asian (Indian) ancestry now make up about 44 percent and those of African background (including African-European ancestry) 30 percent. This makes for contentious politics, given the religious mix, which is approximately 57 percent Christian, 28 percent Hindu, and 7 percent Muslim. Guyana remains dominantly rural, and plantation crops continue to figure prominently among exports (gold from the interior is the most valuable single product).

Oil may become a factor in the economy, though, because a recently discovered reserve that lies offshore from Suriname extends westward beneath Guyana's waters. Still, Guyana is among the realm's poorest and least urbanized countries, and it is strongly affected by the region's narcotics industry. Its thinly populated interior, beyond the reach of antidrug campaigns, has become a staging area for drug distribution to Brazil, North America, and Europe. A recent official report suggests that drug money amounts to as much as one-fifth of Guyana's total economy.

Venezuela's territorial claim against Guyana may be in abeyance, but the United Nations in 2007 settled a dispute with Suriname over a potentially oil-rich maritime zone in Guyana's favor. Some geologists believe that this offshore basin might hold more oil than Europe's North Sea, which would—if properly managed—be able to transform Guyana's economy. Exploratory drilling began in 2009 and the search continues today.

Suriname actually progressed more rapidly than Guyana did after it was granted independence in 1975, but persistent political instability soon ensued. The Dutch colonists brought South Asians, Indonesians, Africans, and even some Chinese to their colony, making for a notably fractious nation. More than 100,000 residents—about one-quarter of the entire population—emigrated to the Netherlands, and were it not for support from its former colonial ruler, Suriname would have collapsed. Nonetheless, its rice farms give Suriname self-sufficiency, even allowing for some exporting, and plantation crops continue to rank among the exports. Suriname's leading income producer, however, is its bauxite (aluminum ore) mined in a zone across the middle of the country. Offshore oil finds (exploration began to accelerate in 2012) could provide important revenues in the years ahead.

Suriname, most of whose population of 550,000 resides along the northern coastal strip, is one of South America's poorest countries, and its opportunities continue to be limited. Along with Guyana, Suriname is involved in an environmental controversy centering on its luxuriant tropical rainforests. Timber companies based in Asia's Pacific Rim offer lucrative rewards for the right to cut down the magnificent hardwood trees, but conservationists are trying to slow this destruction by buying up concessions before they can be opened to logging.

Suriname's cultural geography is enlivened by its many languages. Dutch is the official tongue, but other than by officials and in schools it is not heard much. A mixture of Dutch and English, Sranan Tongo, serves as a kind of common language, but you can also hear indigenous, Hindi, Chinese, Indonesian, and even some French Creole in the streets.

French Guiana, the easternmost outpost of the Caribbean North region, is a dependency—mainland South America's only one. This territory is an anomaly in other ways as well. Consider this: it is almost as large as South Korea (population: just under 50 million) with a population of only 220,000. Its status is that of an Overseas Départment of France, and its official language is French. Nearly half the population resides in the immediate vicinity of the coastal capital, Cayenne.

In 2013, there still was no prospect of independence for this decidedly underdeveloped relic of the former French Empire. Gold remains the most valuable export, and the small fishing industry sends some exports to France. But what really matters here in French Guiana is the European Space Agency's launch complex at Kourou on the coast, which accounts for more than half of the territory's entire economic activity. From plantation farming to spaceport . . . but with minimal involvement, or benefits, for most ordinary people.

▪▪ THE ANDEAN WEST

The second regional grouping of South American states—the Andean West (Fig. 5B-4)—is dominated physiographically by the great Andes mountain chain and historically by indigenous peoples. This region encompasses Peru, Ecuador, Bolivia, and transitional Paraguay, the last with one foot in the West and the other in the South (Fig. 5B-1). Bolivia and Paraguay also constitute South America's only two landlocked countries. For several decades the three main countries of this region (Peru, Ecuador, and Bolivia) have been members of the Andean Community, a trading bloc that also includes Colombia.

Spanish conquerors overpowered the indigenous nations, but they did not reduce them to small minorities as happened to so many indigenous peoples in other parts of the world. Today, roughly 45 percent of the residents of Peru, the region's most populous country, are indigenous; in Bolivia, they are in the majority at 55 percent. About 15 percent of Ecuador's population identifies itself as indigenous, and in Paraguay the ethnic mix, not regionally clustered as in the other three countries, is overwhelmingly weighted toward indigenous ancestry.

As the Data Table in Appendix B reports, this is South America's poorest region economically, with lower incomes, higher numbers of subsistence farmers, and fewer opportunities for job-seekers. For a very long time, the urbane lives of the land-owning elite have been worlds apart from the hard-scrabble existence of the landless peonage (the word *peon* is an old Spanish term for an indebted day laborer).

But today, this region, like the realm as a whole, is stirring, and oil and natural gas are part of the story. In Bolivia, the first elected president of indigenous ancestry is trying to gain control over an energy industry not used to his aggressive tactics. In Ecuador, a populist leftist became president in 2007 and was twice reelected on promises to divert more of the country's oil revenues toward domestic needs and improving the lot of the poor. In Peru, where an energy era is dawning as new reserves are discovered, the government is under pressure to protect Amazonian peoples and environments, limit foreign involvement, and put domestic needs and rights first.

Peru

Peru straddles the Andean spine for more than 1600 kilometers (1000 mi) and is the largest of the region's four republics in both territory and population (31.0 million). Physiographically and culturally, Peru divides into three parts (Fig. 5B-4). Lima and its port, Callao, lie at the center of the ***desert coast*** subregion, and it is symptomatic of the cultural division prevailing in Peru that for nearly 500 years the capital city has been positioned on the periphery, not in a central location in a basin of the Andes. From an economic point of view, however, the Spaniards' choice of a headquarters on the Pacific coast proved to be sound, for the coastal strip has become commercially the most productive part of the country. A thriving fishing industry contributes significantly to the export trade; so do the products of irrigated agriculture in some 40 oases dispersed across the arid coastal plain, which include fruits such as citrus and olives, and vegetables such as asparagus (a big moneymaker) and lettuce.

The Andean or ***Sierra*** subregion occupies just about one-third of the country and is the ancestral home of the largest component in the total population, the speakers of Quechua, the *lingua franca* that emerged during the Inca Empire. Their physical survival during the harsh Spanish colonial regime was made possible by their adaptation to the high-altitude environments they inhabited, but their social fabric was ripped apart by communalization, forced cropping, religious persecution, and involuntary migration to towns and haciendas where many became serfs. Although indigenous peoples constitute almost half of the population

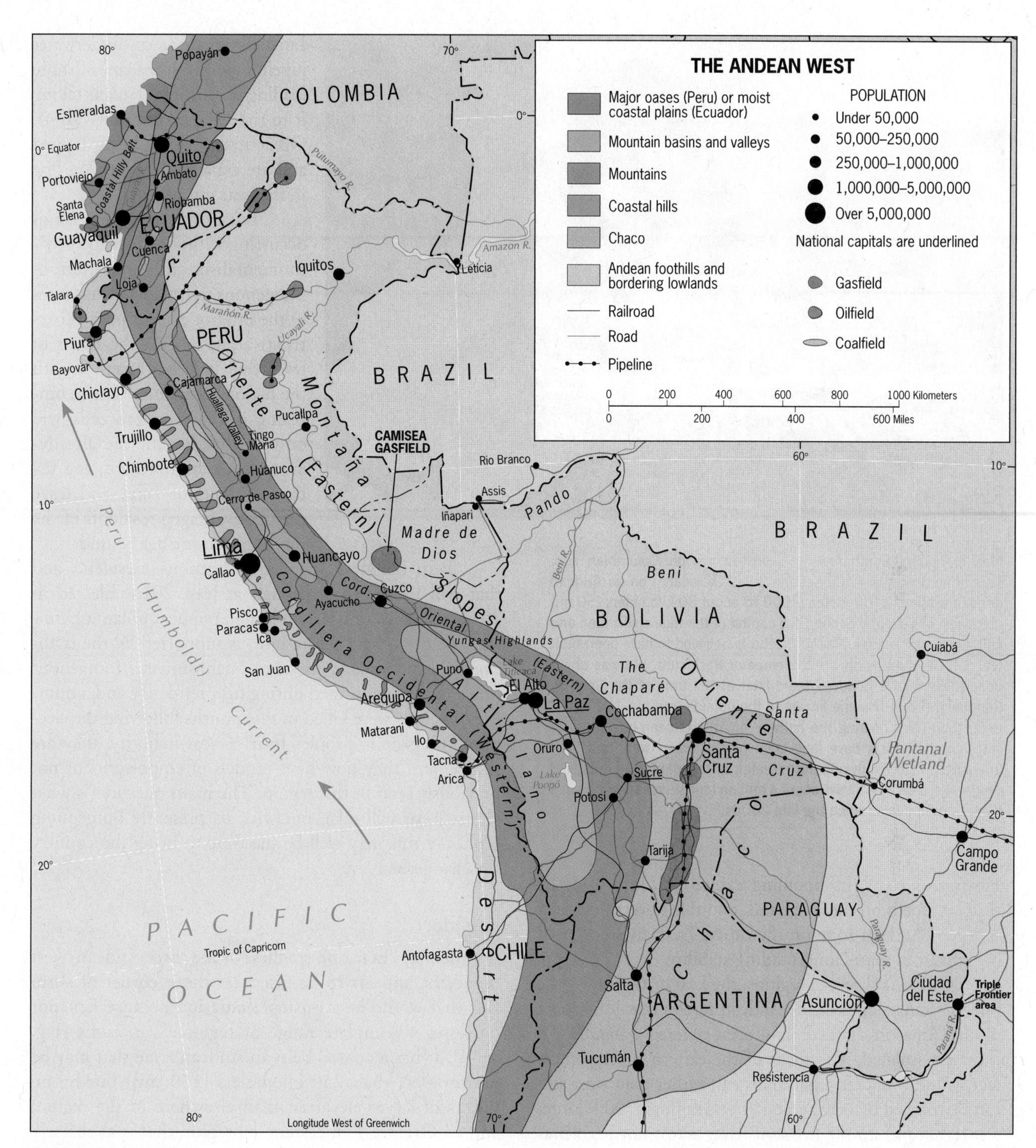

FIGURE 5B-4

of Peru, their political influence remains slight. Nor is the Andean subregion a major factor in Peru's commercial economy—except, of course, for its enormous mineral storehouse, which yields copper, zinc, and lead from mining centers, the largest of which is Cerro de Pasco. Peru is also the world's sixth-largest producer of gold, the price of which has been soaring in recent years, contributing to this country's steadily growing economy. The goldmines are located in the Andes north of Lima.

In the high valleys and intermontane basins, the indigenous population is concentrated either in isolated villages, around which people practice a precarious subsistence agriculture, or in the more favorably located and fertile areas where they are tenants, peons on white- or mestizo-owned haciendas. It was a very different story during the time of the Incas, when the indigenous social structure was still unharmed and the Andean Sierra provided bountiful harvests. These days, there is a painful contrast between the

© Jan Nijman

Machu Picchu, South America's best-known pre-Columbian site. This fifteenth-century "City of the Incas" is located on an Andean ridge (elevation: 2400 meters/7900 ft) about 80 kilometers (50 mi) northwest of Cuzco, the long-time capital of the Inca civilization and still a major city today. Machu Picchu is believed to have been built (ca. 1450) as an estate for the emperor Pachacuti, but was abandoned less than a hundred years later at the time of the Spanish Conquest. It first became known to the outside world in 1911, and restoration (now nearing the halfway point) has been ongoing ever since. Archeologists have long been fascinated as well, and their intensive on-site fieldwork and research efforts continue. Improved overland access now allows about a million tourists to visit each year, a one-third increase since the late 2000s.

prosperous coast, its booming capital of Lima, and the thriving north on the one hand, and this poverty-plagued southern Andean zone on the other. That division could be a threat to Peru's longstanding stability.

Of Peru's three subregions, the East or ***Oriente***—the inland-facing slopes of the Andean ranges that lead down to the Amazon-drained, rainforest-covered *montaña*—is the most isolated. The focus of the eastern subregion, in fact, is Iquitos, a city that looks east rather than west and can be reached by oceangoing vessels sailing 3700 kilometers (2300 mi) up the Amazon River across northern Brazil. Iquitos grew rapidly during the Amazon's wild-rubber boom of just over a century ago, and then declined; now it is finally growing again and reflects Peruvian plans to open up the eastern interior.

Today, the Oriente subregion, and perhaps Peru as a whole, appears on the threshold of a new era due to major new discoveries of oil and gas reserves in the Oriente. Indeed, while still one of the poorer countries in this realm, Peru has experienced higher economic growth rates than any other between 2005 and 2013. Already, pipelines transport natural gas from the Camisea reserve (north of Cuzco) to a conversion plant on the Paracas Peninsula south of Lima. From there, an underwater pipeline sends it to an offshore loading platform for tankers taking it to the U.S. market (Fig. 5B-4). Other reserves now being tapped and are expected to come on line in the near future.

But this development is not occurring without adversity. Environmentalists and supporters of indigenous rights are raising issues in the interior even as political activists argue against the terms of trade that Peru has accepted with the big oil companies. Their complaint is that the proceeds will only further benefit the already-favored coastal, northern, and urban residents of Peru, and leave the disadvantaged residents of the interior even further behind.

Peru has never had an indigenous president, and that will be the case through at least 2016. Elected in 2011, the country's current leader, populist Ollanta Humala, like the others before him, is supported by the traditional Hispanic (and mestizo) establishment. Indigenous protests have continued during his presidency and a number of people were killed in the months following the election. Indigenous peoples form a near-majority, they are restive, and they now have models of empowerment not previously seen in this region. The main question is what course Peru will take, but with its presently burgeoning economy this may well be the time to bring the country closer together.

Ecuador

On the map, Ecuador, smallest of the three Andean West republics, appears to be just a northern corner of Peru. But that would be a misrepresentation because Ecuador possesses a complete range of regional variations (Fig. 5B-4). It has a coastal belt; an Andean zone that may be narrow (less than 250 kilometers [150 mi]) but by no means of lower elevation than elsewhere in the region; and an Oriente—an eastern subregion that is as sparsely settled and as economically marginalized as that of Peru. As with Peru, just about half of Ecuador's population (which totals 15.4 million) is concentrated in the Andean intermontane basins and valleys, and the most productive subregion is the coastal strip. Here, however, the similarities end.

Ecuador's Pacific coastal zone consists of a belt of hills interrupted by lowlands, of which the most important lies in the south between the hills and the Andes, drained by the Guayas River. Guayaquil—the country's largest city, main port, and leading commercial center—forms the focus of this subregion. Unlike Peru's coastal

From the Field Notes . . .

© H.J. de Blij

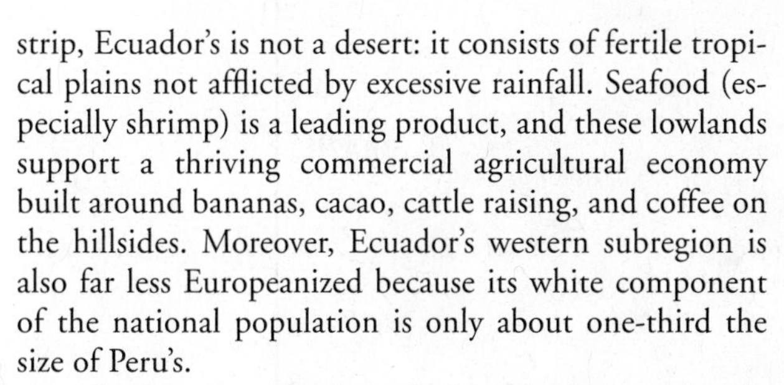

© H.J. de Blij

"I can't remember being hotter anyplace on Earth, not in Kinshasa, not in Singapore . . . not only are you near the equator here in steamy Guayaquil, but the city lies in a swampy, riverine lowland too far from the Pacific to benefit from any cooling breezes and too far from the Andes foothills to enjoy the benefits of suburban elevation. But Guayaquil is not the disease-ridden backwater it used to be. Its port is modern, its city-center waterfront on the Guayas River has been renovated, its international airport is the hub of a commercial center, and it has grown into a metropolis of 2.4 million. Ecuador's oil revenues have made much of this possible, but from the White Hill (a hill that serves as a cemetery, with the most elaborate vaults near its base and the poorest at the feet of the giant statue of Jesus that tops it) you can see that globalization has not quite arrived here. High-rise development remains limited; international banks, hotels, and other businesses remain comparatively few; and the "middle zone" encircling the city shows little evidence of prosperity (top). Beating the heat and glare is an everyday priority: whole streets have been covered by makeshift tarpaulins and more permanent awnings to protect shoppers (bottom). Talk to the locals, though, and you find that there is another daily concern: the people 'up there in the mountains who rule this country always put us in second place.' Take the 45-minute flight from Guayaquil to cool and comfortable Quito, the capital, and you're in another world, and you quickly forget Guayaquil's problems. That's just what the locals here say the politicians do."

www.conceptcaching.com

Concept Caching

strip, Ecuador's is not a desert: it consists of fertile tropical plains not afflicted by excessive rainfall. Seafood (especially shrimp) is a leading product, and these lowlands support a thriving commercial agricultural economy built around bananas, cacao, cattle raising, and coffee on the hillsides. Moreover, Ecuador's western subregion is also far less Europeanized because its white component of the national population is only about one-third the size of Peru's.

A greater proportion of whites is engaged in administration and hacienda ownership in the central Andean zone, where most of the Ecuadorians who are indigenous also reside—and, not surprisingly, where land-tenure reform is an explosive issue. The differing interests of the Guayaquil-dominated coastal lowland and the Andean-highland subregion focused on the capital (Quito) have long fostered a deep regional cleavage between the two. This schism has intensified in recent years, and autonomy and other devolutionary remedies are now being openly discussed in the coastlands (see *From the Field Notes*).

In the rainforests of the Oriente subregion, oil production is expanding as a result of the discovery of additional reserves. Some analysts are predicting that interior Ecuador as well as Peru will prove to contain energy resources comparable to those of Colombia and Bolivia, and that an "oil era" will soon transform their economies. As it is, oil already tops Ecuador's export list, and the industry's infrastructure is being modernized. This is essential because considerable ecological damage has already taken place: the trans-Andean pipeline to the seaport of Esmeraldas, constructed in 1972, was the source of numerous oil spills, and toxic waste was dumped along its route in a series of dreadful environmental disasters. A second, more modern pipeline went into operation in 2003; but like Peru, Ecuador faced growing opposition from environmentalists and activists who, in 2005, shut it down for a week. With revenues from oil and gas exports exceeding 50 percent of all exports, Ecuador's leaders are hearing an increasingly familiar refrain from its people: demand more from the oil companies and give these funds to those who need them most. But it is not as simple as that: oil and gas exploration and exploitation require huge investments, and foreign companies can afford to spend what the government's own state company, Petro Ecuador, cannot. That leads to difficult choices because the state needs income to cover its obligations. Clearly, energy riches are a double-edged sword.

Bolivia

Nowhere are the problems typical of the Andean West more acute than in landlocked, volatile Bolivia. Before reading further, take a careful look at Bolivia's regional geography in both Figures 5B-4 and 5B-5. Bolivia is bounded by remote peripheries of both Brazil and Argentina, mountainous Andean highlands as well as intermontane ***altiplanos*** **[3]** (high-elevation basins and valleys) of Peru, and coveted coastal zones of northern Chile. As the maps show, the Andes in this zone broaden into a vast mountainous complex some 700 kilometers (450 mi) wide.

On the boundary between Peru and Bolivia, freshwater Lake Titicaca lies at 3700 meters (12,500 ft) above sea level and helps make the adjacent *Altiplano* (see map) liveable by ameliorating the coldness in its vicinity, where the snow line lies just above the plateau surface. On the lake's surrounding cultivable land, potatoes and grains have been raised for centuries dating back to pre-Inca times, and the Titicaca Basin still supports a major cluster of indigenous (Aymara) subsistence farmers. This portion of the *Altiplano* is the heart of modern Bolivia and also contains the capital city, La Paz.

The European/Indigenous Divide

The Bolivian state is the product of the Hispanic impact, and the country's indigenous peoples (who still make up about 55 percent of the national population of 11.2 million) no more escaped the loss of their land than did their Peruvian or Ecuadorian counterparts. What made the richest Europeans in Bolivia wealthy, however, was not land but minerals. The town of Potosí in the eastern cordillera became a legend for the immense deposits of silver in its vicinity; tin, zinc, copper, lead, and several ferroalloys were also discovered there. Indigenous workers were forced to work in the mines under the most appalling conditions.

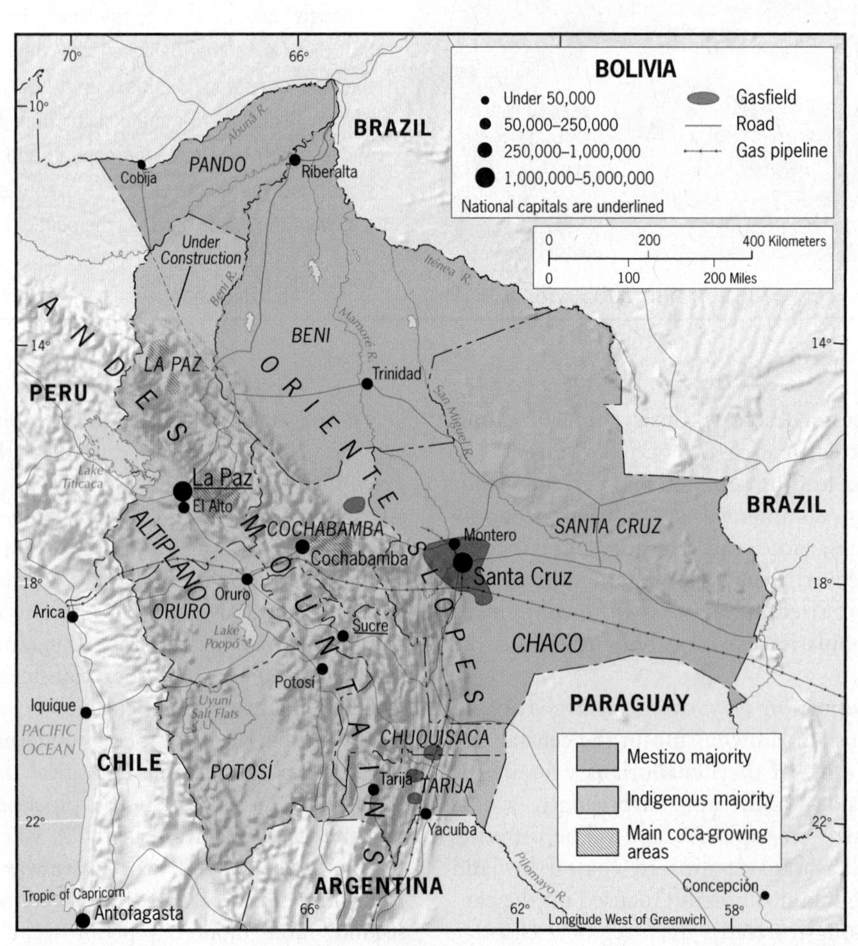

FIGURE 5B-5

Aizar Raldes/AFP/Getty Images, Inc.

Rodrigo Buendia/AFP/Getty Images, Inc.

Bolivia's political geography is as divided as its physical geography, the power base of the indigenous population lying in the altiplanos and mountains, and that of the mestizo minority centered in the interior lowlands. These photos, taken a day apart in Santa Cruz, where mestizo strength is concentrated, show the crucial difference of opinion: a majority of mestizos want autonomy for their eastern provinces (left); virtually all indigenous people protest the prospect of such autonomy, seen as a prelude to secession and independence, as fatally fracturing their country.

Today, natural gas and oil, exported to Argentina and Brazil, are leading sources of foreign revenues, and zinc has replaced tin as the leading export metal. But Bolivia's economic prospects will always be impeded by the loss of its outlet to the Pacific Ocean during its war with Chile in the 1880s, despite its transit rights and dedicated port facilities at Antofagasta.

More critical than its economic limitations or its landlocked situation is Bolivia's social predicament. The government's history of mistreatment of indigenous peoples and harsh exploitation of its labor force hangs heavily over a society where about two-thirds of the people, most of them indigenous, live in dire poverty. In recent years, however, this underrepresented majority has been making a growing impact on national affairs. In 2003, violent opposition to a government plan to export natural gas to the United States via a new pipeline to the Chilean coast led to chaos and the government's resignation. In 2005, Bolivian voters elected their first president of indigenous (Aymara) ancestry, Evo Morales, whose agenda included the nationalization of the country's natural gas resources.

Departments and Capitals

Landlocked, physiographically bisected, culturally bifurcated, and economically divided, Bolivia is a severely challenged state. Moreover, the country's prospects are worsened by its political geography: look again at Figure 5B-5 and you can see that Bolivia's nine provinces are regionally divided between indigenous-majority *departments* (as subnational units are called here) in the west and those with mestizo majorities in the east. The capital city, La Paz, lies on the indigenous-majority Andean *Altiplano*, but many mestizo Bolivians do not recognize it as such: historically, the functions of central government have been divided between La Paz (the administrative headquarters) and Sucre (which lost most of its government branches in 1899 during a civil war but retained the Supreme Court, thereafter calling itself the "constitutional capital"). And extremist mestizo Bolivians even suggest that the southeastern city of Santa Cruz (now Bolivia's biggest) should become the "compromise capital" of their country.

The Santa Cruz *Department*, like the others in the Oriente, stands in sharp contrast to those of the Andes in the west. Here the hacienda system (see Chapter 4A) persists almost unchanged from colonial times, its profitable agriculture supporting a wealthy aristocracy. Now this eastern zone of Bolivia has proved to contain abundant energy resources as well, adding to its economic advantage. So there is talk—abetted by frequent public demonstrations—in support of autonomy, even secession here (see the photo pair), but neither the haciendas nor the energy industry could operate without the indigenous labor force.

When Evo Morales was elected and began trying to gain control of the energy industry while making moves to alter the division of national wealth, the struggle between east and west took a new turn. When energy prices were high, there was sufficient money to spend on the social programs President Morales had promised to expand. But when, in 2008, the price of natural gas declined (and remained at a relatively low level through early 2013), the specter of state failure rose again. The combination of rising expectations among indigenous citizens and devolutionary hopes among the minority mestizos continues to cast an ominous cloud over Bolivia's future.

Paraguay

Paraguay, Bolivia's landlocked neighbor to the southeast, is one of those transitional countries lying in between regions, exhibiting properties of each (see Fig. 5B-1). Certainly, Paraguay (population: 7.0 million) is not an "Andean" country: it has no highlands of consequence. Its well-watered eastern

Martin Bernetti/AFP/Getty Images, Inc.

These buckets of salt brine sitting in the moonscape of southwestern Bolivia's Uyuni Salt Flats represent one of the country's most promising new development opportunities. Once the liquid evaporates, the residue consists mainly of lithium. This lightest of all metals is used to manufacture the lithium-ion batteries that power the portable electronic devices that dominate our daily lives, especially smartphones, tablet computers, and laptops. Larger versions of these batteries are also used to power hybrid motor vehicles, and Bolivia contains as much as half of the world's lithium reserves—enough to support close to 5 billion electric cars. The facility under construction here in this remote part of the *Altiplano* is a pilot lithium extraction plant that opened in 2011. The Bolivian government is the major player in this effort and is determined to keep lithium production a domestic enterprise—despite considerable pressure by foreign companies and investors to become involved. Another fascinating angle is that much of the land of the Salt Flats is the property of indigenous groups, who are seeking a share in the income to be earned from mineral exploitation. Their claims are consistent with the country's 2010 constitution, which enables indigenous communities to have control over resources in their territory—and potentially shape the future course of development.

plains give way to the dry scrub of the Gran Chaco in the northwest. Nor does it have clear, spatially entrenched ethnic divisions between indigenous, mestizo, and other peoples as do Bolivia and Peru. But aboriginal ancestry dominates the ethnic complexion of Paraguay, and continuing protests by landless peasants are similar to those taking place elsewhere in the Andean West. Moreover, the native language, Guaraní, is so widely spoken that this is one of the world's most thoroughly bilingual societies.

The poet Augusto Roa Bastor has described his country as "an island surrounded by land." This relative isolation has surely contributed to the unusual dominance of Guaraní, which is spoken by about 90 percent of the Paraguayans even though only 1.5 percent of the population is indigenous. Because the Spanish colonial rulers never reached far into the interior of the country, Christian missionaries were required to learn this language to connect with the locals. Today, Guaraní is officially on an equal footing with Spanish and is taught in every school—a remarkable exception to the worldwide trend of disappearing indigenous languages.

In every other sphere, however, Paraguay's longstanding isolation is eroding. Nowhere is this more evident than in the decreasingly remote Gran Chaco, which is now rapidly opening up as a burgeoning new cattle-ranching frontier. Deforestation is occurring at a prodigious rate, with satellite land-cover analyses confirming the razing of a woodland area nearly equal in size to the State of Delaware during 2010–2011 alone. This is a particular nightmare for indigenous peoples like the Ayoreo, whose traditional livelihoods are threatened with imminent destruction (the word for bulldozer in their language means "attacker of the world"). The disappearance of the Chaco has also raised alarms among biologists and conservationists, because so much of it remains unexplored and is certain to contain myriad unknown plant and animal species.

Looking south, there is little in Paraguay's economic geography to compare to Argentina, Uruguay, or Chile, as Appendix B confirms. In a sense, Paraguay is a bridge between the Andean West and the Southern Cone—but not yet a heavily traveled one.

And Paraguay is transitional in yet another way: more than 300,000 Brazilians have crossed the border to settle in eastern Paraguay, where they have created a thriving commercial agricultural economy that produces soybeans, livestock, and other farm products exported to or through Brazil. Brazil, of course, is the giant in the Mercosur/l free-trade zone, but Paraguay is in a geographically difficult position, often complaining that Brazil is not interested in living up to its regional-trade obligations and creates unacceptable difficulties for Paraguayan exporters. Meanwhile, politicians raise fears that Brazilian immigration is creating a steadily expanding foreign enclave within Paraguay, where people speak Portuguese (including in the local schools), the Brazilian rather than the Paraguayan flag flies over public buildings, and a Brazilian cultural landscape is beginning to evolve. Like Bolivia, Paraguay pays a heavy price for its landlocked weakness.

Paraguay's low GNI resembles that of countries of the Andean West, not the more advantaged Southern Cone region. This is also one of South America's least urbanized states, and poverty dominates the countryside as well as the slums encircling the capital, Asunción, and other towns (Fig. 5B-4). In 2011, about 35 percent of the population lived at or below the official poverty level. Moreover, research suggests that 1 percent of the population owns about 77 percent of the land, which may be a record for inequality in the South American realm.

Another, and quintessentially geographical, problem stemming from Paraguay's long-term weakness and misrule lies in the southeast, where the borders of Brazil, Argentina, and Paraguay converge in a chaotic scene of smuggling, money-laundering, political intrigue, and even terrorist activity, centered on the town of Ciudad del Este. Locals call this the **Triple Frontier [4]** (Fig. 5B-4). This area is home to a large Middle Eastern community, identified by the U.S. government as a source of funding for Islamic militant groups—an allegation that was rejected by the governments of Argentina, Brazil, and Paraguay. Nevertheless, Paraguay's state system decidedly needs strengthening, and not just for domestic reasons.

In 2008, Paraguay ended its history of ruthless dictatorial rule with the election of a radical, pro-Guaraní priest, Fernando Lugo, who during his days as a missionary supported hacienda invasions by landless peasants and promised land reform and other remedies for the poor. Even though money for such remedies is always hard to come by, Lugo did score a major success in 2009 when he negotiated better terms with Brazil for the sale of Paraguay's share of the electricity generated by Itaipu Dam located on the Paraná River between the two countries. Furthermore, Brazil aided Paraguay in building a state-of-the-art transmission line from Itaipu to Asunción, which in 2012 led to the start of a crucial expansion of Paraguay's electrical-power grid. Nevertheless, his political opponents engineered Lugo's swift impeachment on dubious grounds in mid-2012. Outraged neighbors viewed his ouster as a *coup d'état* and, led by Brazil and Argentina, quickly responded by suspending Paraguay from Mercosur/l (whose economic ties it urgently needs); in mid-2013, the country remained in supranational limbo.

THE SOUTHERN CONE

This region of South America, consisting of Argentina, Chile, and Uruguay, acquired its name from its tapered, ice-cream-cone shape (Fig. 5B-6). Since 1995, the countries of the Southern Cone have been drawing closer together in an economic union named ***Mercosur*** (the Spanish acronym for *Mercado Común del Sur*), the hemisphere's second-largest trading bloc after NAFTA. Despite setbacks and disputes, this organization has expanded and today encompasses Argentina, Brazil, Uruguay, Venezuela, and Bolivia (full accession in 2014)—as well as currently suspended Paraguay. Chile, Peru, Ecuador, and Colombia participate as associate members. (Note that in Portuguese-speaking Brazil, this bloc is known as ***Mercosul*** [*Mercado Comum do Sul*]—which is why we use the appellation *Mercosur/l* throughout this chapter).

Argentina

The largest Southern Cone country by far is Argentina, whose territorial size ranks second only to Brazil in this geographic realm; its population of 41.7 million ranks third overall after Brazil and Colombia. Although Argentina exhibits a great deal of physical-environmental variety within its boundaries, the overriding majority of the Argentines are concentrated in the physiographic subregion known as the ***Pampa*** (a word meaning "plain"). Figure 5A-3 underscores the degree of clustering of Argentina's inhabitants on the land and in the cities of the Pampa. It also shows the relative emptiness of the other six subregions (shown in Fig. 5B-6): the scrub-forested ***Chaco*** in the northwest; the mountainous ***Andes*** in the west, along whose crestline lies the boundary with Chile; the arid plateaus of ***Patagonia*** south of the Rio Colorado; and the undulating transitional terrain of intermediate ***Cuyo, Entre Rios*** (also known as "Mesopotamia" because it lies between the Paraná and Uruguay rivers), and the ***North***.

From the Field Notes . . .

"At the heart of Buenos Aires lies the *Plaza de San Martín*, flanked by impressive buildings but, unusual for such squares in Iberian America, carpeted with extensive lawns shaded by century-old trees. Getting a perspective of the plaza was difficult until I realized that you could get to the top of the 'English Tower' across the avenue you see in the foreground. From there, one can observe the prominent location occupied by the monument to the approximately 700 Argentinian military casualties of the Falklands War of 1982 (center), where an eternal flame behind a brass map of the islands symbolizes Argentina's undiminished determination to wrest these islands from British control."

 www.conceptcaching.com

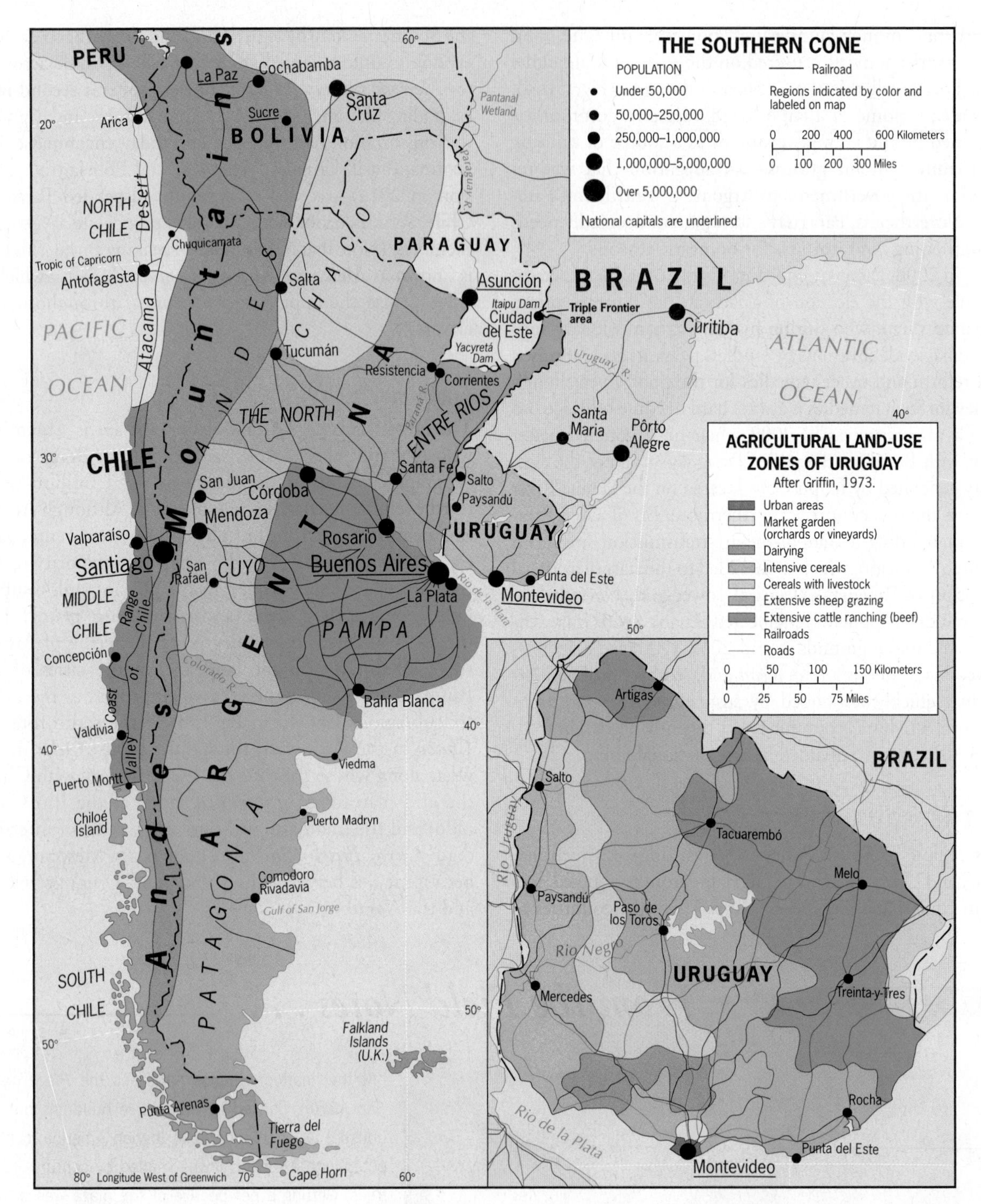

FIGURE 5B-6

The Argentine Pampa is the product of the past 150 years. During the second half of the nineteenth century, when the great grasslands of the world were being opened up (including those of the interior United States [Great Plains], Russia, and Australia), the economy of the long-dormant Pampa began to emerge. The food needs of industrializing Europe grew by leaps and bounds, and the advances of the Industrial Revolution—railroads, more efficient ocean transport, refrigerated ships, and agricultural machinery—helped to make large-scale commercial meat and grain production in the Pampa not only feasible but also highly profitable. Large haciendas were laid out and farmed by tenant workers; railroads were built and radiated ever farther outward from the booming capital of Buenos Aires, soon bringing the entire Pampa into production.

A Culture Urban and Urbane

Argentina once was one of the richest countries in the world. Its historic affluence is still reflected in its architecturally splendid cities whose plazas and avenues are flanked by ornate public buildings and private mansions. This is true not only of the capital, Buenos Aires, at the head of the Rio de la Plata estuary—it also applies to interior cities such as Mendoza and Córdoba. The cultural imprint is dominantly Spanish, but the cultural landscape was diversified by a massive influx of Italians and smaller but influential numbers of British, French, German, and Lebanese immigrants.

Argentina has long been one of the realm's most urbanized countries: 91 percent of its population is concentrated in cities and towns, a higher percentage even than western Europe or the United States. Fully one-third of all Argentinians live in metropolitan Buenos Aires, by far the leading industrial complex where processing Pampa products dominates. Córdoba has taken its place as the next-largest city and industrial center, and has been selected by foreign automobile manufacturers as the auto-assembly center for the growing Mercosur/l market. But what concentrates the urban populations is the processing of products from the sprawling, sparsely peopled interior: Tucumán (sugar), Mendoza (wines), Santa Fe (forest products), and Salta (livestock). Argentina's product range is enormous. There is even a petroleum reserve near Comodoro Rivadávia on the central coast of Patagonia.

Economic Volatility

Despite all these riches, Argentina's economic history is one of boom and bust. Possessing a vast territory with diverse natural resources, adequate infrastructure, and above-average international linkages, Argentina should still be one of the world's wealthiest countries, as it once was. But political infighting and economic mismanagement have combined to shackle a vibrant and varied economy.

AMONG THE REALM'S GREAT CITIES . . .

BUENOS AIRES

ITS NAME MEANS "fair winds," which first attracted European mariners to the site of Buenos Aires alongside the broad estuary of the muddy Rio de la Plata. The shipping function has remained paramount, and to this day the city's residents are known throughout Argentina as the *porteños* (the "port dwellers"). Modern Buenos Aires was built on the back of the nearby Pampa's grain and beef industry. It is often likened to Chicago and the Corn Belt in the United States because both cities have thrived as interfaces between their immensely productive agricultural hinterlands and the rest of the world.

Buenos Aires (14.1 million) is yet another classic South American primate metropolis, housing over one-third of all Argentines, serving as the capital since 1880, and functioning as the country's economic core. Moreover, Buenos Aires is a cultural center of global standing, a monument-studded city that contains the world's widest street (*Avenida 9 de Julio*).

During the half-century between 1890 and 1940, the city was known as the "Paris of the South" for its architecture, fashion leadership, book publishing, and performing arts activities (it still has the world's biggest opera house, the newly-renovated *Teatro Colón*). With the recent restoration of democracy, Buenos Aires is now trying to recapture its golden years. Besides reviving these cultural functions, the city has added a new one: the leading base of the hemisphere's motion picture and television industry for Spanish-speaking audiences.

During the past few years, the city has been showing signs of distress as inflation ran at nearly 25 percent annually and the cost of living skyrocketed. Economic growth was slowed to about 2 percent, unemployment rose, and the homeless became a common sight along the city's elegant boulevards. There really was no good economic reason for this downturn, and most residents point to the leadership, the politicians, as the culprits. But this is not the first economic crisis faced by the city in recent times, and resilient Buenos Aires is likely to return to its grander ways in due course.

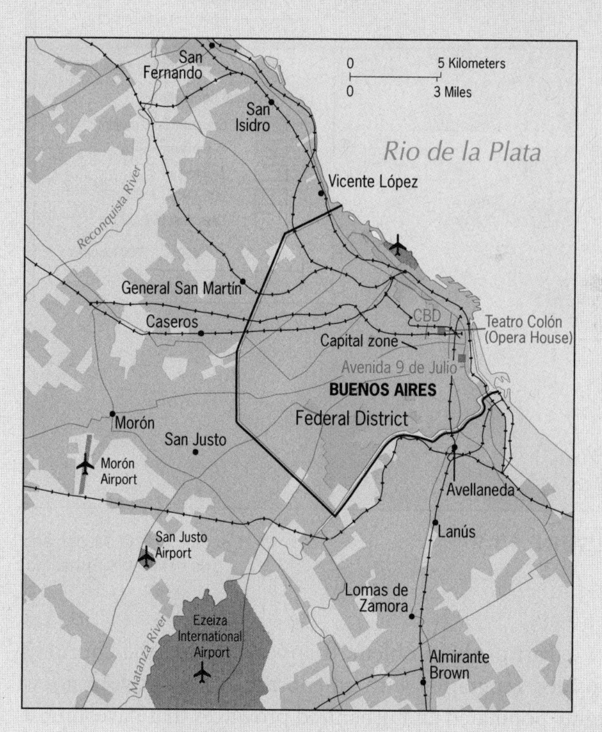

FIGURE 5B-7 © H. J. de Blij, P. O. Muller, and John Wiley & Sons, Inc.

Part of the problem, it appears, lies in the country's lopsided geography with an enormous gap between a few highly populated and urbanized provinces that have almost all the political power and a large number of provinces with very small populations (Fig. 5B-7). Buenos Aires Province has 15.7 million people (out of a total 42 million) while Tierra del Fuego has barely more than 125,000. A dozen other provinces contain less than 750,000 inhabitants, so that the larger ones in addition to dominant Buenos Aires are also disproportionately influential in domestic politics, especially Córdoba and Santa Fe.

Another seemingly never-ending problem for Argentina has been corrupt politics and associated mismanagement, going back at least as far as the mid-twentieth century. The darkest episode was the "Dirty War" of 1976–1983, in which a repressive military junta caused more than 10,000 (and perhaps as many as 30,000) Argentinians to disappear without a trace. Then in 1982, this ruthless military clique decided to launch an invasion of the British-held Falkland Islands (which the Argentines call the Malvinas) off the far southern Patagonian coast, resulting in a crushing defeat for Argentina (Fig. 5B-6). By the time civilian government replaced the discredited junta, inflation and national debt were soaring. Economic revival during the 1990s was followed by another severe downturn that exposed the flaws in Argentina's fiscal system, including scandalously inefficient tax collection and unconditional federal handouts to politically powerful provinces.

The current administration, led by the country's first (and now reelected) female president, Cristina Fernandez de Kirchner (whose late husband preceded her in this office), has thus far failed to make progress. Indeed, corruption and mismanagement reached so high a level in early 2012 that the authoritative international weekly, *The Economist,* decided it would no longer include statistics provided by the Argentine government; pointedly, the newsmagazine also announced that it did not want to be part of the government's "deliberate attempt to deceive voters and swindle investors"—a reflection of the country's sorry condition today.

Argentina was a net exporter of oil until 2003, but it must now rely on imports to serve its consumption needs. In response, the government in 2012 decided to nationalize the biggest oil company, YPF—widely considered to be a disastrous move that merely served populist political needs. In the longer term, it is feared that nationalization will only further erode the company's efficiency and competitiveness. And to add to these problems, the old Falklands dispute erupted once again on the conflict's 30th anniversary in 2012, just as the British began preparations for oil-drilling in the potentially rich seabed surrounding those far-off South Atlantic islands (Fig. 5B-6). Once again, Buenos Aires registered strong protests and aggressively restated its claims that the Malvinas must be returned to Argentina.

Chile

For over 4000 kilometers (2500 mi) between the crestline of the Andes and the coastline of the Pacific lies the narrow strip of land that is the Republic of Chile (Fig. 5B-8). On average just 150 kilometers (90 mi) wide (and only rarely over 250 kilometers or 150 miles in width), Chile is the world's quintessential example of what **elongation [5]** means to the functioning of a state. Accentuated by its north-south orientation, this severe territorial attenuation not only results in Chile extending across numerous environmental zones; it has also contributed to the country's external political, internal administrative, and general economic challenges.

FIGURE 5B-8 © H. J. de Blij, P. O. Muller, and John Wiley & Sons, Inc.

Nevertheless, throughout most of their modern history, the Chileans have made the best of this potentially disruptive centrifugal force: from the beginning, the sea has constituted an avenue of longitudinal communication; the Andes Mountains continue to form a barrier to prevent encroachment from the east; and when confrontations loomed at the far ends of the country, Chile proved to be quite capable of coping with its northern rivals, Bolivia and Peru, as well as Argentina in the far south.

Three Subregions

As Figures 5B-6 and 5B-8 indicate, Chile is a three-subregion country. About 90 percent of its 17.7 million people are concentrated in what is called Middle Chile, where Santiago, the capital and largest city, and Valparaíso, the chief port, are located. North of Middle Chile lies the Atacama Desert, wider, drier, and colder than the coastal desert of Peru. To the south of Middle Chile, the coast is punctuated by a plethora of fjords and islands, the topography is mountainous, and the climate—wet and cool near the Pacific—soon turns drier and much colder against the Andean interior. South of the latitude of Chiloé Island, there are few permanent overland routes and very few settlements.

Not surprisingly, the land in Middle Chile is the most fertile and valuable, with hardly any agriculture to be found either in the North or South (Fig. 5A-5). The three subregions are also apparent on the realm's cultural map, with Europeans dominating the Middle, mestizos prevailing in the North, and indigenous groups forming the majority in the South (Fig. 5A-4).

Still, prior to the 1990s, the arid Atacama region in the North accounted for more than half of Chile's foreign revenues. It contains the world's largest exploitable deposits of nitrates, which was the country's economic mainstay before the discovery of methods of synthetic nitrate production a century ago. Subsequently, copper became the dominant export (Chile again possesses the world's largest reserves, which in 2012 accounted for more than half of all export revenues). It is mined in several places, but the main concentration lies on the eastern margin of the Atacama near Chuquicamata (see photo), not far from the port of Antofagasta.

Political and Economic Success

Following the withdrawal of its vicious military dictatorship in 1990, Chile embarked on a highly successful program of free-market economic reform that brought stable growth, lowered inflation as well as unemployment, reduced poverty, and attracted massive foreign investment. The last is of particular significance because these new international connections enabled the export-led Chilean economy to diversify and develop in some badly needed new directions. Copper remains at the top of the export list, but many other mining ventures have been launched. In the agricultural sphere, fruit and vegetable production for export has soared because Chile's harvests coincide with the winter farming lull in the affluent countries of the Northern Hemisphere. Industrial

Peter McBride/Aurora Photos, Inc.

A view into the gaping maw of northern Chile's Chuquicamata copper mine, the world's largest and second-deepest open-pit mine. For most of the century since mining began here in 1910, 'Chuqui' was the single leading source of copper on Earth. But the increasingly negative environmental impact of its moonscape of slag heaps and toxic materials has now forced the relocation of the town of Chuquicamata and the 2013 closure of the surface mine. Nonetheless, huge quantities of copper will continue to be extracted from this still-rich deposit after a major new underground mine is built below this site: it will begin operating in 2018—with plans to keep enlarging it until at least 2060.

expansion is taking place as well, though at a more leisurely pace, and new manufactures have included an array of goods that range from basic chemicals to computer software.

Chile's increasingly globalized economy has also propelled the country into a prominent role on the international economic scene. The United States, long Chile's leading trade partner, now ranks behind the Asian Pacific Rim, where China consumes the bulk of Chile's exports followed by Japan and South Korea. (Argentina remains Chile's leading source of imports, mostly energy, a potential problem given Argentina's rising domestic needs.) Chile's regional commerce is growing as well, and it has affiliated with Mercosur/l as an associate member.

One of Chile's noteworthy initiatives in this decade is its 'Start-Up Chile' project, nicknamed *Chilecon Valley*. Since 2010, the government has granted startup funding of up to (U.S.) $40,000 for roughly 500 new cutting-edge information technology companies from around the world, and it has provided visas for their personnel to live and work in Chile. The aim is to stimulate and nurture entrepreneurship and innovation in Chile. At the same time, much has been done to increase access to higher education: in 2012, 1.1 million students were enrolled in the country's universities, up from about 200,000 only ten years earlier. Tellingly, 70 percent of today's students are the first in their families to go to college.

Chile's post-1990 governments, regardless of the party in power, seem to have discovered a winning formula: a pragmatic approach that circumvents ideological extremism, exhibits transparency in decision-making, and maintains a desirable balance between sustained economic growth and poverty alleviation. If things can stay on track, Chile's goal of becoming the first South American country to join the ranks of the world's most developed economies may well be rapidly approaching.

Uruguay

Uruguay, in comparison with Argentina or Chile, is compact, small, and rather densely populated. This **buffer state [6]** of the early independence era, lying between (then) potentially hostile Argentina and Brazil, has evolved into a fairly prosperous agricultural country—in effect a smaller-scale Pampa, though possessing less favorable soils and topography. Montevideo, the coastal capital, contains half of the country's population of 3.4 million; from here, railroads and roads radiate outward into the productive agricultural hinterland (Fig. 5B-6). In the immediate vicinity of Montevideo lies Uruguay's major farming zone, producing vegetables and fruits for the metropolis as well as wheat and fodder crops; most of the rest of the country is used for grazing cattle and sheep, with beef products, wool and textile manufactures, and hides dominating the export trade (Fig. 5B-6, inset map). Tourism is another major economic activity as Argentines, Brazilians, and other visitors flock to the Atlantic beaches at Punta del Este and other thriving resort towns.

About the size of Florida but with less than one-fifth the population, Uruguay offers significant agricultural promise. Nonetheless, its government seeks to diversify the economy, and one of its plans—the construction of two large cellulose (paper) factories on the Uruguayan side of the Uruguay River where it forms the border with Argentina—has caused a quarrel between the two Mercosur/l neighbors that has exposed some serious rifts. Montevideo is the administrative headquarters of Mercosur/l, but, like Paraguay, Uruguay has long felt neglected and even obstructed by its much larger neighbors. When Argentines across the river began demonstrating, barricaded a bridge, and stopped taking vacations on Uruguay's beaches, they severely impacted the Uruguayan economy.

The Argentines, who in any case dislike major foreign investments (the factories were going to be built by Finnish and Spanish companies), argued that the paper mills would accelerate deforestation, cause river pollution, create acid rain, and damage the farming, fishing, and tourist industries. When the presidents of the two countries met to negotiate a solution and Uruguay's leader agreed to halt construction for a "study period," Uruguayan public opinion showed strong opposition to the compromise. This issue reveals just how suddenly nationalist feelings can overwhelm the need for international cooperation. It also shows that Mercosur/l partners are far from united on economic matters and that the collaboration implied by membership in the organization is still a distant reality.

BRAZIL: GIANT OF SOUTH AMERICA

The next time you board an airplane, don't be surprised if the aircraft was built in Brazil. When you go to the supermarket to buy provisions, take a look at their sources. Chances are some of them will come from Brazil. When you listen to your car radio, some of the best music you hear may have originated in Brazil. The emergence of Brazil as the regional superpower of South America and an economic superpower in the world at large is going to be one of the main stories of the first quarter of the twenty-first century. Along with Russia, India, and China, Brazil is designated as one of the four biggest emerging markets in the world, the **BRICs [7]** (an acronym based on the first letter of this quartet of countries). In 2013, Brazil surpassed the UK to become the world's sixth-largest economy.

Why is Brazil so upward-bound today? First, after a long period of dictatorial rule by a minority elite that used the military to stay in power, Brazil embraced democratic government in 1989 and has not looked back since. The era of military coups and repeated crushing of civil liberties is over. Second, with its vast storehouse of natural resources, Brazil has benefited enormously from increased demands for commodities in the world market, mainly from China and India. In some ways, Brazil's political and economic turnaround runs parallel to that of Chile, but because of its sheer size Brazil matters far more to the rest of the realm and, indeed, the world.

By every measure, Brazil is South America's giant. It is so large that it has common boundaries with all the realm's other countries except Ecuador and Chile (Fig. 5B-1). Its tropical and subtropical environments range from the equatorial rainforest of the Amazon Basin to the humid temperate climate of the far south. Territorially as well as in terms of population, Brazil ranks fifth in the world, and on both counts it represents just about half of this entire realm. And by 2016, the Brazilian economy, with its ultramodern industrial base, will also be fifth in size, trailing only the U.S., China, Japan, and Germany.

Population and Culture

Brazil's population grew rapidly during the world's twentieth-century population explosion. But over the past three decades, the rate of natural increase has slowed from nearly 3.0 percent to 1.0 percent today, and the average number of children born to a Brazilian woman has been more than halved from 4.5 in 1975 to 1.9 in 2012. It is a demographic trend consistent with Brazil's overall modernization since 1980.

Brazil's population of 198.2 million is as diverse as that of the United States. In a pattern quite familiar in the Americas, the indigenous inhabitants of the country were decimated following the European invasion (approximately 800,000 now survive, about two-thirds of them deep within the Amazonian interior). Africans came in very large numbers as well, and they currently total more than 15 million.

Brazil's culture is infused with African themes, a quality that has marked it from the very beginning. Three centuries ago, the Afro-Brazilian sculptor and architect affectionately known as Aleijadinho was Brazil's most famous artist. The world-renowned composer Heitor Villa-Lobos used numerous Afro-Brazilian folk themes in his music. So many Africans were brought in bondage to the city and hinterland of Salvador in Bahia State (Fig. 5B-9) from what is today Benin (formerly named Dahomey) in West Africa that Bahia has become a veritable outpost of African culture. Indeed, Brazil can be said to have the second-largest black (African) population in the world, after Nigeria.

Significantly, however, there was also much racial mixing, and the 2010 census reported that 97 million Brazilians (50.7 percent of the total population) have combined European, African, and minor indigenous ancestries. White Brazilians of European origin—the descendants of immigrants from Portugal, Italy, Germany, and eastern Europe—are no longer in the majority: between 2000 and 2010 their share of the population dropped from 53.7 percent to 47.7 percent.

Yet another significant, although small, minority began arriving in Brazil in 1908: the Japanese, who today are concentrated in the States of São Paulo and Paraná. The more than 1 million Japanese-Brazilians form the largest ethnic Japanese community outside Japan, and in their multicultural environment they have risen to the top ranks of Brazilian society as business leaders, urban professionals, commercial farmers—and even as politicians in the city of São Paulo. Committed to their Brazilian homeland as they are, the Japanese community also retains its contacts with Japan, resulting in many a trade connection.

Brazilian society, to a much greater degree than is true elsewhere in the Americas, has made progress in dealing with its racial divisions. Yet blacks remain the least advantaged among the country's leading population groups, and community leaders continue to complain about discrimination. But ethnic mixing in Brazil is so pervasive that hardly any group is unaffected, and official census statistics concerning "blacks" and "Europeans" are rather pointless.

What the Brazilians do have is a true national culture, expressed in a traditional adherence to the Roman Catholic faith (its adherents now constitute only about 65 percent of the population [down from 90 percent as recently as the 1970s], a share that steadily continues to erode under Protestant-evangelical and secular pressures); in the universal use of a modified form of Portuguese as the common language

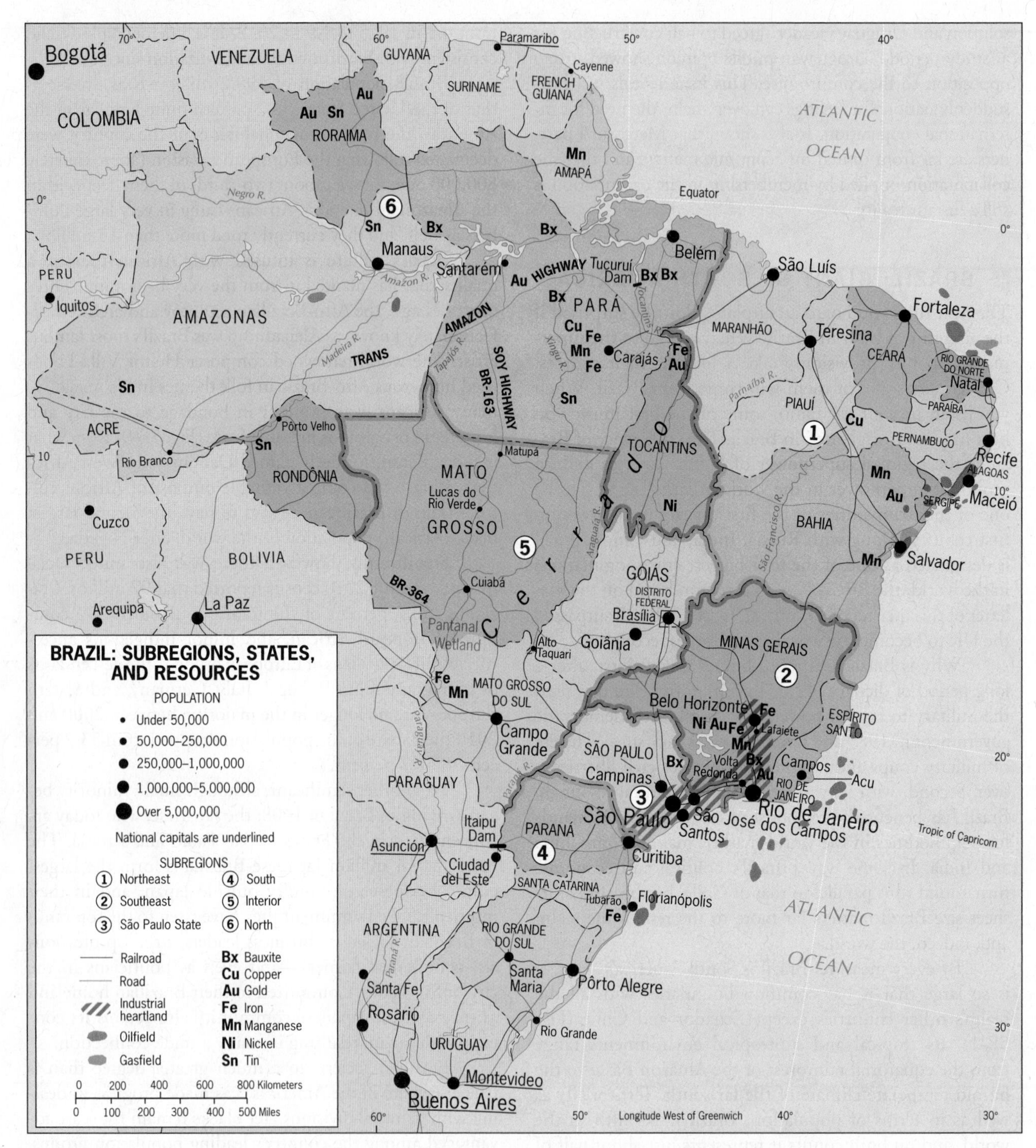

FIGURE 5B-9

("Brazilian"); and in a set of lifestyles in which soccer, "beach culture," distinctive music and dance, and an intensifying national consciousness and pride are fundamental ingredients.

Inequality and Poverty

For all its accomplishments in multiculturalism, Brazil remains a country of stark, appalling social inequalities (Fig. 5B-10). Although such inequality is hard to measure precisely, South America is frequently cited as the geographic realm exhibiting the world's sharpest division between affluence and poverty. And within South America, Brazil was traditionally reputed to have the widest gap of all.

But since 2002, the Brazilian government has been implementing a set of policies aimed at bringing relief to the poor while maintaining robust economic growth. It has enacted land reform and increased access to education for Brazil's masses. It has also started a subsidy program that has had

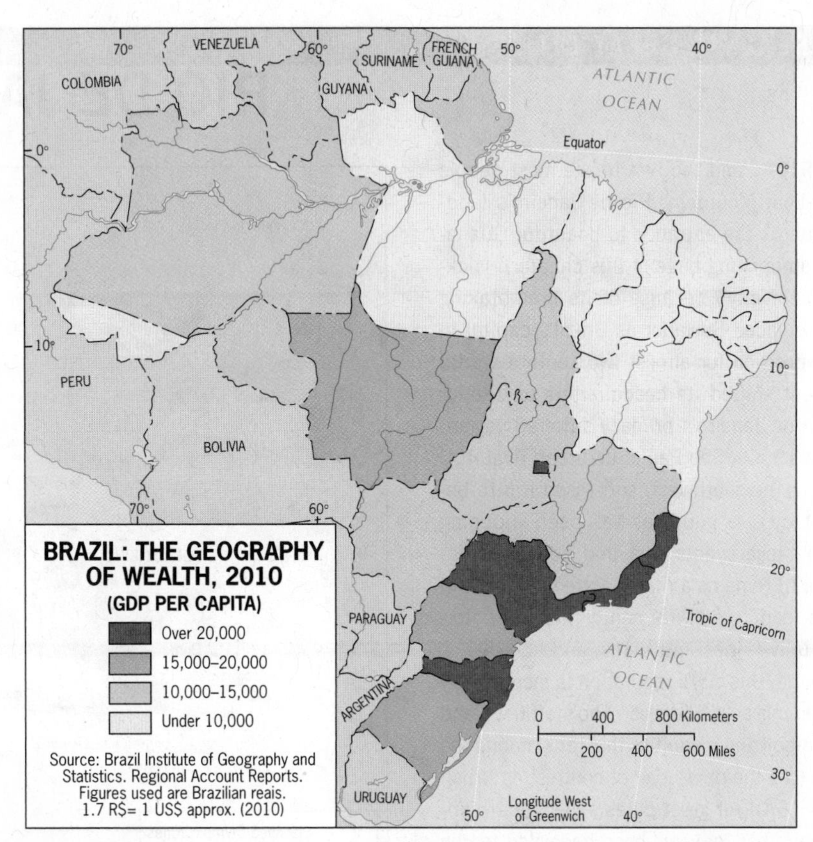

FIGURE 5B-10 © H. J. de Blij, P. O. Muller, J. Nijman, and John Wiley & Sons, Inc.

significant results in the poorest States. This *Bolsa Familia* (Family Fund) plan, instituted under former President Cardoso in the 1990s, provides families with small payments of cash to keep their children in school and ensure their vaccinations against diseases that especially afflict the poor. In just a few years this program has become so successful that it now serves as a model for antipoverty campaigns in many other parts of the world. At the same time, the percentage of people officially living in poverty in Brazil dropped from roughly 30 percent in 2000 to around 21 percent in 2012.

Regional inequalities are being targeted as well by government policies. Major new legislation was passed in 2012 that would allow revenues from oil to be shared by all of Brazil's States (previously this income was mainly channeled to the three oil-producing States of Rio de Janeiro, Espirito Santo, and São Paulo). In addition, overall federal spending is increasingly being steered to poorer States—particularly the Northeast—in order to accelerate economic growth.

Development Prospects

Brazil is richly endowed with mineral resources, including enormous iron and aluminum ore reserves, extensive tin and manganese deposits, and highly promising oil- and gasfields (Fig. 5B-9). Other significant energy developments involve massive new hydroelectric facilities and the successful integration of sugarcane-based anhydrous ethanol with gasoline to allow Brazil's motor vehicles to utilize this "gasohol" instead of costlier imported petroleum (all Brazilian cars today are configured to use this biofuel, whose ethanol content must be at least 18 percent). Besides these natural endowments, Brazil has more arable land than any other country in the world. Brazilian soils sustain a bountiful agricultural output that makes the country the world's leading exporter of coffee, orange juice, sugar, tobacco, ethanol, beef, and chicken. Its government has set the goal of Brazil becoming the number-one food exporter in the world as soon as 2015, displacing the United States. Commercial agriculture, in fact, is now the fastest growing economic sector, driven by mechanization and the opening of a booming new farming frontier in the fertile grasslands of southwestern Brazil (discussed later in conjunction with the Interior subregion).

Industrialization has driven Brazil's rise as a global economic power. Much of the momentum for this continuing development was unleashed in the early 1990s after the government opened the country's long-protected industries to international competition and foreign investment. These new policies proved to be highly effective because productivity has risen by at least one-third since 1990 as Brazilian manufacturers attained world-class quality. More than a decade ago, revenues from industrial exports surpassed those from agriculture. Heightened commerce with Argentina made that other key member of Mercosur/l one of Brazil's leading trade partners.

AMONG THE REALM'S GREAT CITIES . . .

RIO DE JANEIRO

SAY "SOUTH AMERICA" and the first image most people conjure up is Sugar Loaf Mountain, Rio de Janeiro's landmark sentinel that guards the entrance to beautiful Guanabara Bay (see photo on opening page of this chapter). Nicknamed the "magnificent city" because of its breathtaking natural setting, Rio replaced Salvador as Brazil's capital in 1763 and held that position for almost two centuries until the federal government shifted its headquarters to interior Brasília in 1960. Rio de Janeiro's primacy suffered yet another blow in the late 1950s: São Paulo, its urban rival 400 kilometers (250 mi) to the southwest, surpassed Rio to become Brazil's largest city—a gap that has been widening ever since. Although these events triggered economic decline, Rio (12.3 million) remains a major *entrepôt*, air-travel and tourist hub, and leading cultural center with its entertainment industries, universities, museums, and libraries.

On the darker side, this city's reputation is increasingly tarnished by the widening abyss between Rio's affluent and poor populations—symbolizing inequities that rank among the world's most extreme (see the photo pair of contrasting urban landscapes in Chapter 5A). All great cities experience problems, and Rio de Janeiro has for years been bedeviled by the drug use and crime waves emanating from its most desperate hillside *favelas* (slums) that continue to grow explosively.

Rio's planners recently launched a wide-ranging project (known as "Rio-City") to improve urban life for all residents. This ambitious scheme, it is said, aims to reshape nearly two dozen of the aging city's neighborhoods, introduce an ultramodern crosstown expressway to relieve nightmarish traffic congestion, and—most importantly—bring electrical power, paved streets, and a sewage-disposal network to the beleaguered *favelas*.

Much of this was related to Rio's hosting of the Summer Olympics in 2016 and soccer's World Cup tournament in 2014, which motivated local and federal authorities to do all they could to showcase their city in the "happy-go-lucky" image of Brazilian soccer. But in the early summer of 2013, hundreds of thousands people took to the streets of Rio to vent their frustration with government plans to raise transit fares, protest runaway corruption, demand better schools and other public services, and even to protest what many considered wildly disproportionate investments in new soccer stadiums that would sit unused after the World Cup. Facing a race against time to prepare the country for the events of 2014, this unrest was the last thing Brazil's government needed.

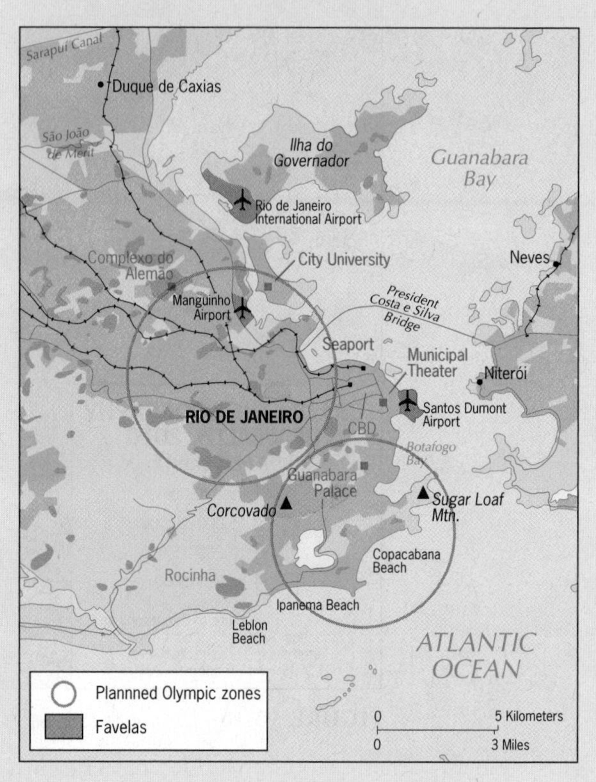

On the global stage, Brazil has become a formidable presence in other ways. The country's enormous and easily accessible iron ore deposits, the relatively low wages of its workers, and the mechanized efficiency of its steelmakers enable Brazil to produce that commodity at half the cost of steel made in the United States (which caused American steel producers to demand protectionist measures). The biggest consumer of iron ore and steel, as well as many other raw materials, is China. In 2012, the Açu Superport, 400 kilometers (250 mi) northeast of Rio de Janeiro, opened for business as the largest bulk cargo port on Earth, accommodating the biggest ships that sail the oceans—those, too, are Chinese (see final photo in Chapter 5A).

A Highly Promising Oil Future

Compared to the other BRICs, Brazil stands out in terms of the diversity and accessibility of its natural resources. It has been paying for government programs with revenues mainly derived from commodities—iron ore, soybeans, coffee, orange juice, beef, sugar—that in recent years have commanded high prices on international markets.

Brazil can now add oil—lots of it—to its portfolio. In the past, a leading concern for Brazil was energy and its rising costs. Recessions were deepened by high oil and gas prices, and other than some modest domestic reserves and those in neighboring Bolivia there were few nearby sources to acquire what was needed. But in 2009 the state-owned (but publicly traded) petroleum company, Petrobras, confirmed the discovery of an enormous oil reserve, quite possibly among the world's largest, capable of yielding an estimated 2 billion barrels. It was proof, said then-President Lula with a broad smile on his face, that "God is Brazilian."

By 2010, the country was already self-sufficient in oil. Production is expected to double by 2020, and before then Brazil will become a major exporter, thereby adding significantly to its foreign revenues. Moreover, the newfound reserves lie off the coast in a cluster of oilfields—most fortuitously located near Rio de Janeiro and São Paulo (see Fig. 5B-9)—and are being explored further as you read this. Although these oil deposits are massive and reserve estimates are constantly being revised upward as new discoveries are made, they are buried quite far beneath the ocean floor. Thus exploitation will be costly and require substantial foreign investment in addition to leading-edge drilling methods, but is likely to take Brazil's technology sector to an ever more advanced level.

Brazil's Subregions

Brazil is a federal republic consisting of 26 States and the federal district of the capital, Brasília (Fig. 5B-9). As in the United States, the smallest States lie in the northeast and the larger ones farther west; their populations range from about 475,000 in the northernmost, peripheral Amazon State of Roraima to more than 42 million in burgeoning São Paulo State. Although Brazil is about as large as the 48 contiguous United States, it does not exhibit a clear physiographic regionalism. Even the Amazon Basin, which covers just about 60 percent of the country, is not entirely a plain: between the tributaries of the great river lie low but extensive tablelands. Given this physiographic ambiguity, the six Brazilian subregions discussed next exhibit no absolute or even generally accepted boundaries. In Figure 5B-9, those boundaries have been drawn to coincide with the borders of States, making identifications easier.

The **Northeast** was Brazil's source area, its culture hearth. The plantation economy took root here at an early date, attracting Portuguese planters, who soon began to import the country's largest contingent of African slaves to work in the sugar fields. But the ample rainfall occurring along the coast soon gives way to drier and more variable patterns in the interior, which is home to about half of the region's 50-plus million people. This drier inland backcountry—called the ***sertão***—is not only seriously overpopulated but also contains some of the worst poverty to be found anywhere in the Americas. The Northeast produces less than one-sixth of Brazil's gross domestic product, but its inhabitants constitute more than one-fourth of the national population. Given this staggering imbalance, it is not surprising that this subregion contains half of the country's poor, a literacy rate 20 percent below Brazil's mean, and an infant mortality rate twice the national average.

Much of the Northeast's misery is rooted in its unequal system of land tenure. Farms must be at least 100 hectares (250 acres) to be profitable in the hard-scrabble *sertão*, a size that only large landowners are able to afford. Moreover, the Northeast is tormented by a monumental environmental problem: the cyclical recurrence of devastating droughts at least partly attributable to ***El Niño*** **[8]** (periodic events of sea-surface warming off the continent's northwestern coast that skew regional weather patterns).

Brazil's great contradiction today, the Northeast is finally receiving greater attention from the central government, largely in the form of federally funded (think oil money) infrastructure projects and incentive-driven investment. In cities such as Recife and Salvador, hordes of peasants driven from the land constantly arrive to expand the surrounding shantytowns. As yet, few of the generalizations about emerging Brazil apply here, but there are some bright spots. A petrochemical complex has been built near Salvador, creating thousands of jobs and luring foreign investors. Irrigation projects have nurtured a number of productive new commercial agricultural ventures. Tourism is booming along the entire Northeast coast, whose thriving beachside resorts attract tens of thousands of vacationing Europeans (flying times are 8 hours or less). Recife has now spawned a budding software industry and a major medical complex. And Fortaleza is the center of new clothing and shoe industries that have already put the city on the global economic map.

The **Southeast** has been modern Brazil's ***core area***, with its major cities and leading population clusters. Gold first drew many thousands of settlers, and other mineral finds also contributed to the influx—with Rio de Janeiro itself serving as the terminus of the "Gold Trail" and then as the long-time capital of Brazil until 1960. "Rio" became the cultural capital as well, the country's most international center, *entrepôt*, and tourist hub. The third quarter of the twentieth century brought another mineral age to the Southeast, based on the iron ores around Lafaiete carried to the nearby steelmaking complex at Volta Redonda (Fig. 5B-9). The surrounding State of Minas Gerais (the name means "General Mines") formed the base from which industrial diversification in the Southeast has steadily mushroomed. The burgeoning metallurgical center of Belo Horizonte paved the way and is now the endpoint of a rapidly developing, ultramodern manufacturing corridor that stretches 500 kilometers (300 mi) southwest to metropolitan São Paulo (Fig. 5B-9, striped zone).

São Paulo State is both the leading industrial producer and the primary focus of ongoing Brazilian development. This economic-geographic powerhouse accounts for nearly half of the country's GDP, with a booming economy that today matches Argentina's in overall size. Not surprisingly, this subregion is growing phenomenally (it contains more than 20 percent of Brazil's population) as a magnet for migrants, especially from the Northeast.

The wealth of São Paulo State was built on its coffee plantations (known as ***fazendas***), and Brazil is still the world's leading producer. But coffee today has been eclipsed by other farm commodities. One of them is orange juice concentrate (here, too, Brazil leads the world). São Paulo

São Paulo may not have an imposing skyline to match New York or a skyscraper to challenge the Sears (now Willis) Tower of Chicago, but what this Brazilian megacity does have is mass. São Paulo is more than just another city—it is a vast conurbation of numerous cities and towns containing more than 26 million inhabitants who constitute the third-largest human agglomeration on Earth. This view shows part of the razor-sharp edge of the CBD, a concrete jungle with little architectural distinction but home to a vibrant urban culture. In the foreground, juxtaposed against the affluence of downtown, the teeming inner city reflects the opposite end of the social spectrum—a chaotic jumble of *favelas* that mainly house the rural-urban migrants, especially from Brazil's hard-pressed Northeast.

State now produces more than double the annual output of Florida, thanks to a climate all but devoid of winter freezes, ultramodern processing plants, and a fleet of specially equipped tankers that ship the concentrate to foreign markets. Another leading pursuit is soybeans, in which Brazil today ranks second among the world's producers.

Matching this agricultural prowess is the State's industrial strength. The revenues derived from the coffee plantations provided the necessary investment capital, ores from Minas Gerais supplied the vital raw materials, São Paulo City's outport of Santos facilitated access to the ocean, and immigration flows from Europe, Japan, and other parts of Brazil contributed the increasingly skilled labor force. As the capacity of the expanding domestic market grew, the advantages of central location and large-scale agglomeration nailed down São Paulo's primacy. This also resulted in metropolitan São Paulo becoming the country's—and South America's—leading industrial complex and megacity. But with this came massive problems of overcrowding, pollution, and congestion (see *Great Cities* box).

The **South** consists of three States, whose combined population exceeds 27 million: Paraná, Santa Catarina, and Rio Grande do Sul (Fig. 5B-9). Southernmost Brazil's exceptional agricultural potential has long attracted sizeable numbers of European immigrants. They introduced their advanced farming methods to several areas in this part of the country. Portuguese rice farmers clustered in the valleys of Rio Grande do Sul, where tobacco production has now propelled Brazil to become the world's number-one exporter. The Germans, specialists in raising grain and cattle, occupied the somewhat higher areas to the north and in Santa Catarina. The Italians selected the highest slopes, where they established thriving vineyards. All of these fertile lands proved highly productive, and with growing markets in the mushrooming urban areas to the north, this tri-State subregion became Brazil's most affluent corner.

With Brazil's South firmly rooted in the European/commercial agricultural sphere (Figs. 5A-4, 5A-5), European-style standards of living accompany the diverse Old World heritage that is reflected in the urban centers and countryside (where German and Italian are spoken alongside Portuguese). This has led to hostility against non-European Brazilians, and a number of communities actively discourage poor job-seeking migrants from the north by offering to pay return bus fares or even blocking their household-goods-laden vehicles. Moreover, extremist groups have proliferated and continue to openly espouse the secession of the South from the rest of Brazil.

Economic development within the South is not limited to the agricultural sector. Coal from Santa Catarina and Rio Grande do Sul is shipped north to the steel plants of Minas Gerais. Local manufacturing is growing as well, especially in Pôrto Alegre and Tubarão. During the 1990s, a major center of the computer software industry was established in Florianópolis, the island-city as well as State capital just off Santa Catarina's coast. Known as *Tecnópolis*, this budding technopole continues to grow by capitalizing on its seaside amenities, skilled labor force, superior air-travel and global communications linkages, and government and private-sector initiatives to support new companies.

The **Interior** subregion—constituted by the States of Goiás, Mato Grosso, and Mato Grosso do Sul—is also known as the ***Central-West***. This is the subregion that Brazil's developers long sought to make a part of the country's productive heartland, and in 1960 the new capital of Brasília was deliberately situated on its margins (Fig. 5B-9).

By locating the new capital city in the wilderness 650 kilometers (400 mi) inland from its predecessor, Rio de Janeiro, the nation's leaders dramatically signaled the opening of Brazil's development thrust toward the west. Brasília is

AMONG THE REALM'S GREAT CITIES . . .

SÃO PAULO

SÃO PAULO, WHICH lies on a plateau 50 kilometers (30 mi) inland from its Atlantic outport of Santos, at first appears to possess no obvious locational advantages. Yet here on this site we find the third-largest metropolis on Earth, whose population has multiplied so uncontrollably that São Paulo has more than doubled in size (from 11 to just over 26 million) during the past three decades.

Founded in 1554 as a Jesuit mission, the initial choice of location was based on access to the relatively large native population groups of the interior that were targets of conversion to Catholicism—hence the naming after St. Paul. The mission was also situated on the Tietê River, a convenient means of transport that originates near the coast but flows inland toward the northwest interior. In the seventeenth century, São Paulo became the home base of the so-called *bandeirantes* (explorers, gold prospectors, and slave traders), an unruly lot for whom the relatively remote location, far from the rule of law, was an advantage. Important gold mines were soon discovered in neighboring Minas Gerais State, turning São Paulo into a busy gateway that was further enhanced as the fertile lands to the west attracted a growing number of settlers.

Modern São Paulo, however, was built on the nineteenth-century coffee boom. Ever since, it has grown steadily as both an agricultural processing center (soybeans, orange juice concentrate, and sugar besides coffee) and a manufacturing complex (today accounting for about half of all of Brazil's industrial jobs). Along the way, it also evolved into Brazil's primary focus of commercial and financial activity. Twenty-first-century São Paulo's bustling, high-rise CBD (see photo) is the very symbol of urban South America, and attracts the realm's largest flow of foreign investment as well as the trade-related activities that befit the city's emergence as the business capital of Mercosur/l.

Nonetheless, even for this metropolitan industrial giant of the Southern Hemisphere, the increasingly global tide of postindustrialism is rolling in and São Paulo is learning to adapt. To avoid becoming a Detroit-style Rustbelt, the aging automobile-dominated manufacturing zone on the central city's southern fringes is today attracting new industries. Internet companies have flocked here, as well as to the nearby city of Campinas, in such numbers that they now form the country's largest high-tech cluster—increasingly referred to as "Silicon Valley of Brazil."

Elsewhere in São Paulo's vast urban constellation—whose suburbs now sprawl outward up to 100 kilometers (60 mi) from the CBD—many additional opportunities are being exploited. In the outer northeastern sector, new research facilities as well as computer and telecommunications-equipment factories are springing up. And to the west of central São Paulo, lining the ring road that follows the

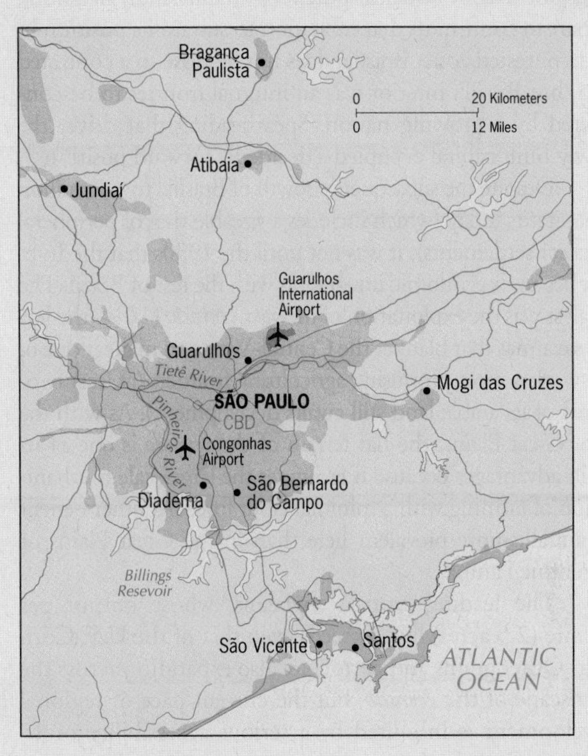

Pinheiros River, is South America's largest suburban office complex replete with a skyline of ultramodern high-rises.

These advances notwithstanding, the colossal recent growth of this megacity has come at a price. Whereas incomes are about 25 percent higher than in Rio, so is the cost of living. And there are massive problems of overcrowding, pollution, and congestion. Traffic jams here are among the world's worst, with more than twice as many gridlocked motor vehicles on any given day than in Manhattan. The recent expansion of the metro—the last of only four such transit lines—was not completed until 2010; it has somewhat eased commuting, but overcrowding is immense with a staggering 5.2 million passengers riding the trains every work day. The urban region has expanded in all directions and is now relatively well connected by major highways. One of the world's most spectacular expressways leads to Santos, with enormous bridges, tunnels, and crisscrossing north- and southbound lanes—mostly amidst breathtaking scenery.

But there is also another São Paulo, this one plagued by grinding poverty on a scale with Mexico City's (see foreground of photo). This is the most pressing urban crisis of all as the ever-expanding belt of shantytowns tightens its grip on much of the metropolis. With its rapid growth rate expected to persist through the foreseeable future, can anything prevent Greater São Paulo from surpassing the 30 million milestone less than 10 years from today?

noteworthy in another regard because it represents what political geographers call a **forward capital [9]**. A state will sometimes relocate its capital to a sensitive area, perhaps near a peripheral zone under dispute with an unfriendly neighbor, in part to confirm its determination to sustain its position in that contested zone. Brasília does not lie close to a contested area, but Brazil's interior was an internal frontier to be conquered by a growing nation. Spearheading that drive, the newly-built capital occupied a decidedly forward position.

Despite the subsequent growth of Brasília to 4.1 million inhabitants today (which includes a sizeable ring of peripheral squatter settlements), it was not until the 1990s that the Interior began its economic integration with the rest of Brazil. The catalyst was the exploitation of the vast ***cerrado* [10]**—the fertile savannas that blanket the Central-West and make it one of the world's most promising agricultural frontiers (at least two-thirds of its arable land still awaits development). As with the U.S. Great Plains, the flat terrain of the *cerrado* is one of its main advantages because it facilitates the large-scale mechanization of farming with a minimal labor force. Another benefit is rainfall, more prevalent here than in the Great Plains or Argentine Pampa.

The leading crop is soybeans, whose output per hectare (2.5 acres) here exceeds even that of the U.S. Corn Belt. Other grains and cotton are also expanding across the farmscape of the *cerrado*, but the current pace of regional development is inhibited by a serious accessibility problem—forcing the Interior's products to travel along poor roads and intermittent railroads to reach the markets and ports of the Atlantic seaboard. Today several projects are finally underway to alleviate these bottlenecks, including the privately financed Ferronorte railway that links Santos to the southeastern corner of Mato Grosso State, and the much-improved, so-called ***Soy Highway*** leading northward to the Amazon River port city of Santarém.

The **North** is Brazil's territorially largest and most rapidly developing subregion, which consists of the seven States of the Amazon Basin (Fig. 5B-9). This was the scene of the great rubber boom of a bit more than a century ago, when the wild rubber trees in the *selvas* (tropical rainforests) produced huge profits and the central Amazon city of Manaus enjoyed a brief period of wealth and splendor. But the rubber boom ended in 1910, and for most of the seven decades that followed, Amazonia was a stagnant hinterland lying remote from the centers of Brazilian settlement. All that changed quite dramatically during the 1980s as new development began to stir throughout this awakening subregion, which currently is the scene of the world's largest migration into virgin territory. More than 200,000 new settlers arrive each year. The North of Brazil is also well known for its high rates of ***deforestation*** (an issue discussed in Chapter 4A). Removing the rainforest results directly from logging operations, but more of it is now a matter of clearing space for land occupation by settlers as well as the expansion of large-scale agribusiness.

From 1995 to 2005, an area of 20,000 square kilometers (7700 sq mi) was cleared away every year. But the worst may be over (despite a surge in 2012–13). In 2008, the Brazilian government pledged to achieve an 80-percent reduction in deforestation by 2020, and to terminate the practice by 2030. At least part of its thinking is that Brazil has enormous potential as a "green economy" with its abundance of land, water, and sunshine. Today, about 40 percent of the country's energy is already obtained from renewable sources (versus less than 12 percent in the United States), and it is becoming easier to envision Brazil as a leading environmental power.

Development projects abound today in the Amazonian North. One of the most durable is the ***Grande Carajás***

VOICE FROM THE Region

© Karina Felicio

Karina Felicio, geography student, Roraima, Brazil

LIVING IN ONE OF BRAZIL'S FAR CORNERS

"I have lived all my life in the town of Boa Vista, capital of Roraima which is the northernmost State of Brazil. Roraima has dense, lowland tropical rainforest in the south and beautiful highland rainforest in the north. There are less than 500,000 people in the entire State and we are extremely isolated—almost 3000-kilometers (1850 mi) from São Paulo, separated by the Amazon Basin. Most of our border is with Guyana and Venezuela. To go anywhere in Brazil from here, you have to first drive to Manaus; there is one road that goes there, the BR-174. Before that road was built, we were much more isolated still. Manaus is on the Rio Negro and from there you can travel the Amazon by boat, or you can fly to the Southeast. I went to São Paulo once and it's a different world from Roraima. The drive to Manaus normally takes 12 hours but on the way back the bus broke down and I remember having to stand there for 6 hours in 97-degree heat, waiting for it to be repaired. Because of the distances, the products we buy are more expensive, from food to clothing. Lately, many people like me cross the border to nearby Venezuelan towns where things are cheaper. Despite the distances and some of the hardships, I love living here: I belong with the indigenous culture, the natural environment is beautiful, and my friends and family are here. In addition, the Brazilian government now invests a lot in remote border areas like Roraima and induces migrants from other parts of the country to settle here. Despite the small population, we have a good public university in Boa Vista where I study geography, an important subject in Brazil. I might move somewhere else for a career opportunity, the Southeast maybe, but I already know that I will miss Roraima."

NRSC/Science Source

This is what the Amazon's equatorial rainforest looks like from an orbiting satellite after the human onslaught in preparation for settlement. The colors on this Landsat image emphasize the destruction of the trees, with the dark green of the natural forest contrasted against the pale green and pinks of the leveled forest. The linear branching pattern of deforestation is the preferred approach here in Rondônia State's Highway BR-364 corridor. But farming is not likely to succeed for very long, and much of the cleared land is likely to be abandoned. Then the onslaught would resume to clear additional land—a cycle the Brazilian government has committed itself to end by 2030.

Project in central Pará State, a huge multifaceted scheme centered on the world's largest known deposit of iron ore in the hills around Carajás (Fig. 5B-9; aerial photo Chapter 5A). Besides a vast mining complex, other new construction here includes the Tucuruí Dam on the nearby Tocantins River and an 850-kilometer (535-mi) railroad to the Atlantic port of São Luis. This ambitious development project further emphasizes the exploitation of additional minerals, cattle raising, crop farming, and forestry. What is taking place here is a manifestation of the **growth-pole concept [11]**. A growth pole is a location where a set of activities, given a start, will expand and generate widening ripples of development in the surrounding area. According to this scenario, the stimulated hinterland could one day cover one-sixth of all Amazonia.

Understandably, tens of thousands of settlers have descended on this part of the Amazon Basin. Those seeking business opportunities have been in the vanguard, but they have been followed by masses of lower-income laborers and peasant farmers in search of jobs and land ownership. The initial stage of this colossal enterprise has boosted the fortunes of many urban centers, particularly Manaus northwest of Carajás. Here, a thriving industrial complex (specializing in the production of electronic goods) has emerged within the free-trade zone adjoining the city thanks to the outstanding air-freight facilities and operations at Manaus's ultramodern airport. But many problems have also arisen as the tide of pioneers rolled across central Amazonia. One of the most tragic involved the Yanomami people, whose homeland in Roraima State was overrun by thousands of claim-stakers (in search of newly discovered gold), who triggered violent confrontations that ravaged the fragile aboriginal way of life.

Another important, pathbreaking development scheme, known as the ***Polonoroeste Plan***, is located about 1600 kilometers (1000 mi) to the southwest of Grande Carajás concentrated in the 2400-kilometer (1500-mi)-long Highway BR-364 corridor that parallels the Bolivian border and interconnects the western Brazilian towns of Cuiabá, Pôrto Velho, and Rio Branco (Fig. 5B-9). Even though the government had planned for the penetration of western Amazonia to proceed via the east-west Trans-Amazon Highway, the migrants of the 1980s and 1990s preferred to follow BR-364 and settle within the Basin's southwestern rim zone, mostly in Rondônia State (see photo). Agriculture has been the dominant activity here, but in the quest for land, bitter conflicts continue to break out between peasants and landholders as the Brazilian government grapples with the volatile issue of land reform.

Brazil is the cornerstone of South America, the dominant economic force in Mercosur/l, the only dimensional counterweight to the United States in the Western Hemisphere, a maturing democracy, and an emerging global giant as an industrial and agricultural powerhouse. Its rapidly intensifying relationship with China, in particular, is a clear sign that South America no longer remains in the shadow of the United States. The future of the South American realm depends in large part upon Brazil's stability and social as well as economic progress.

POINTS TO PONDER

- Brazil is well on its way to displacing the United States as the world's biggest food exporter; it aims to do so by 2015.
- Much of South America's economic growth in the past ten years can be attributed to China's demand for raw materials.
- Rising cost of living becomes a focal point for social protest when a government seeks to attract world-class events such as the Olympic Games. As Brazil found out, with millions pouring into new infrastructure to accommodate what locals regard as the jet-set elite, bus fares and food prices rise, creating an atmosphere that can turn violent.

THE SUBSAHARAN AFRICAN REALM

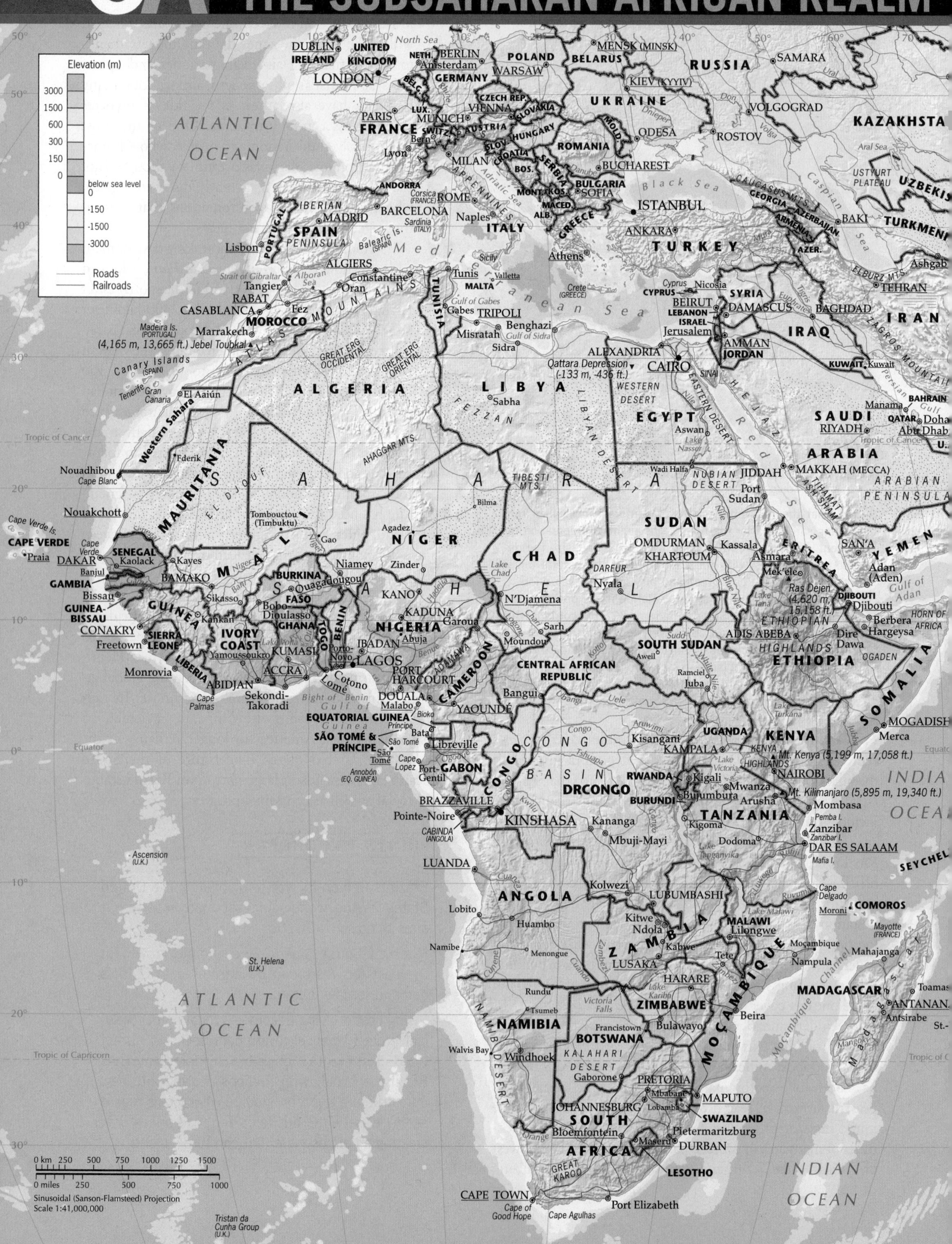

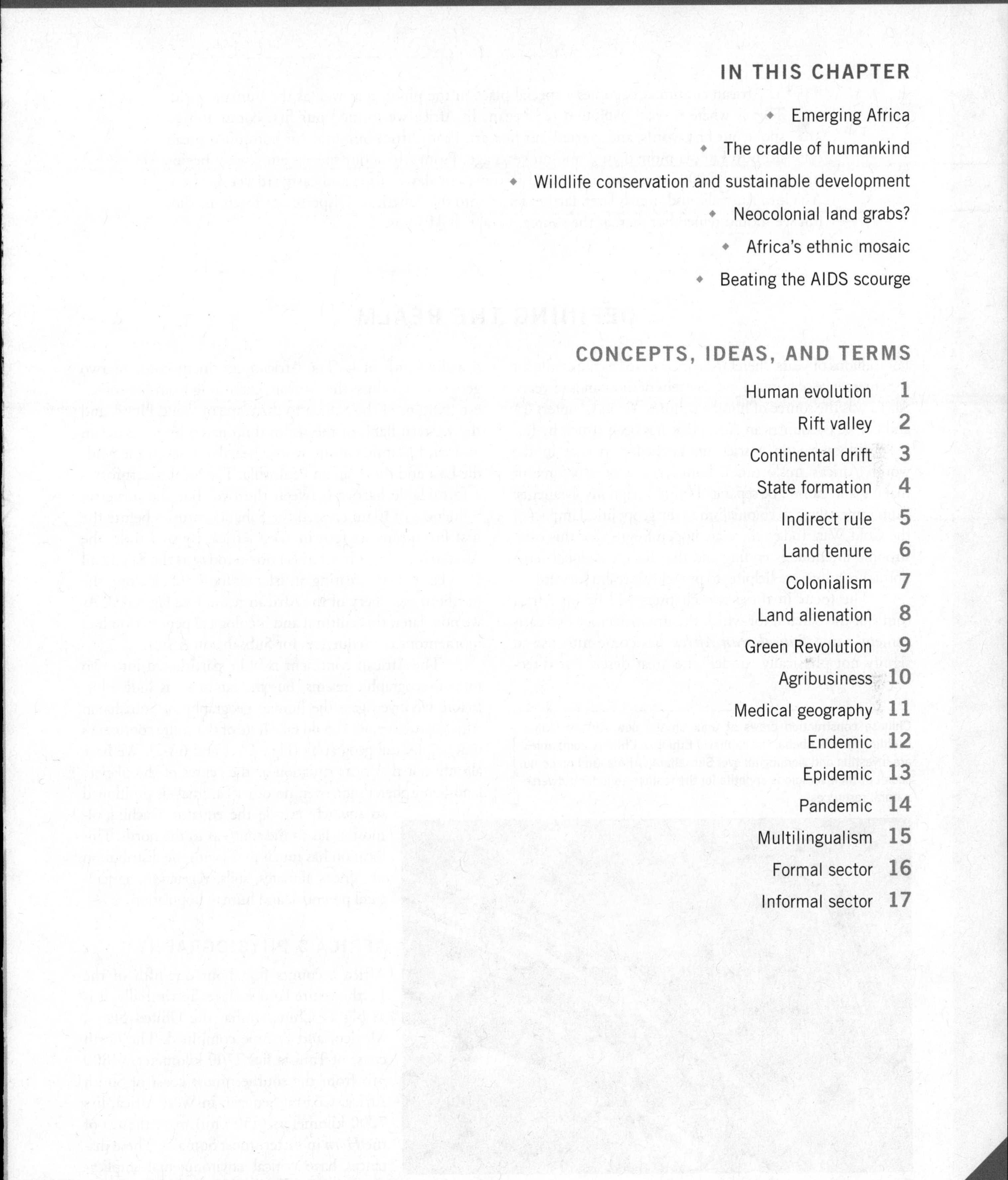

IN THIS CHAPTER

- Emerging Africa
- The cradle of humankind
- Wildlife conservation and sustainable development
- Neocolonial land grabs?
- Africa's ethnic mosaic
- Beating the AIDS scourge

CONCEPTS, IDEAS, AND TERMS

Human evolution 1
Rift valley 2
Continental drift 3
State formation 4
Indirect rule 5
Land tenure 6
Colonialism 7
Land alienation 8
Green Revolution 9
Agribusiness 10
Medical geography 11
Endemic 12
Epidemic 13
Pandemic 14
Multilingualism 15
Formal sector 16
Informal sector 17

FIGURE 6A-1

The African continent occupies a special place in the physical as well as the human world. This is where **human evolution [1]** began. In Africa we formed our first communities, spoke our first words, and created our first art. From Africa our ancestor hominins spread outward into Eurasia more than 2 million years ago. From Africa our species emigrated, beginning perhaps 95,000 years ago, northward into present-day Europe and eastward via southern Asia into Australia and, much later, farther afield into the Americas. Disperse our forebears did, but we should remember that, at the source, we are all Africans.

DEFINING THE REALM

For millions of years, therefore, Africa served as the cradle for the emergence of humankind. For tens of thousands of years, Africa was the source of human cultures. Yet in Chapters 6A and 6B we encounter an Africa that has been struck by human dislocation on a scale unmatched anywhere in the world. Africa's misfortunes, however, are of more recent making and cannot be separated from foreign involvements: from imperialism to colonialism to the geopolitical impact of the Cold War. Today, there are hopeful signs that this catastrophic interlude is ending and that for once global (economic) relations are helping to propel this realm forward.

The focus in these two chapters will be on Africa south of the Sahara, for which the unsatisfactory but convenient name ***Subsaharan Africa*** has come into use to signify not physically "under" the great desert but directionally south of it. The African continent contains two geographic realms: the African, extending from the southern margins of the Sahara to the Cape of Good Hope, and the western flank of the realm dominated by the Muslim faith and Islamic culture whose heartland lies in the Middle East and the Arabian Peninsula. The great desert forms a formidable barrier between the two, but the powerful influences of Islam crossed the Sahara centuries before the first Europeans set foot in West Africa. By that time, the African kingdoms in what is known today as the Sahel had been converted, creating an Islamic foothold all along the northern periphery of the African realm (see Fig. G-3). As we note later, this cultural and ideological penetration had momentous consequences for Subsaharan Africa.

The African continent may be partitioned into two human-geographic realms, but the landmass is indivisible. Before we investigate the human geography of Subsaharan Africa, therefore, we should take note of the entire continent's unique physical geography (Figs. 6A-1 and 6A-2). We have already noted Africa's situation at the center of the planet's land hemisphere; moreover, no other landmass is positioned so squarely astride the equator, reaching almost as far to the south as to the north. This location has much to do with the distribution of Africa's climates, soils, vegetation, agricultural potential, and human population.

Chinese construction crews at work on the new African Union Building in Adis Abeba, the capital of Ethiopia. Chinese companies are investing and working all over Subsaharan Africa, and make no secret of their voracious appetite for the realm's abundant, diverse natural resources.

AFRICA'S PHYSIOGRAPHY

Africa accounts for about one-fifth of the Earth's entire land surface. Territorially, it is as big as China, India, the United States, Mexico, and Europe combined. The north coast of Tunisia lies 7700 kilometers (4800 mi) from the southernmost coast of South Africa. Coastal Senegal, in West Africa, lies 7200 kilometers (4500 mi) from the tip of the *Horn* in easternmost Somalia. These distances have critical environmental implications. Much of Africa is far from maritime

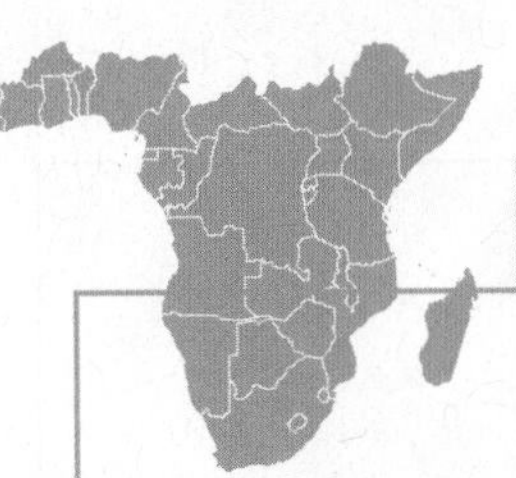

major geographic qualities of SUBSAHARAN AFRICA

1. Physiographically, Africa is a huge plateau continent without a major "spinal" mountain range but with a set of Great Lakes, several major river basins, variable rainfall, soils of generally inferior fertility, and mainly savanna and steppe vegetation.
2. Hundreds of distinct ethnic groups make up Subsaharan Africa's culturally rich and varied population. They far outnumber the states in this realm, and only rarely do state and ethnic boundaries coincide.
3. Most of Subsaharan Africa's peoples depend on farming for their livelihood.
4. The realm is rich in raw materials vital to industrialized countries, but many economies continue to rely on primary activities—the extraction of resources—and not the greater income-generating activities of manufacturing and assembly.
5. The realm is famous for its wildlife, but many species are threatened. Combining bioconservation and sustainable development is a major challenge when the outside world covets elephant tusks and rhinoceros horns.
6. Severe dislocation still affects a number of Subsaharan African countries, from the Sudans to Zimbabwe. This realm has by far the largest refugee population in the world today.
7. Foreign interest in the realm's natural resources, from commodities to agricultural land, has been expanding rapidly in recent years.
8. Subsaharan Africa has recently experienced swift economic growth, with a half-dozen countries among the fastest growing in the world today.

sources of moisture. In addition, as Figure G-7 shows, large parts of the landmass lie in latitudes where global atmospheric circulation systems produce arid conditions. The Sahara in the north and the Kalahari in the south form parts of these globe-girdling desert zones. Water supply is one of Africa's foremost problems.

Rifts and Rivers

Africa's topography reveals several properties that are not replicated on other landmasses. Alone among the continents, Africa does not have a mountain backbone; neither the northwestern Atlas nor the far southern Cape Ranges are in the same league as the Andes or Himalayas. Where Africa does have high mountains, as in Ethiopia and South Africa, these are deeply eroded plateaus or, as in East Africa, high snowcapped volcanoes. Furthermore, Africa is one of only two continents containing a cluster of Great Lakes, and the only one whose lakes result from powerful tectonic forces in the Earth's crust. These lakes (with the exception of Lake Victoria) lie in deep trenches called **rift valleys [2]**, which form when huge parallel fractures or faults appear in the Earth's crust and the strips of crust between them sink, or are pushed down, to form great, steep-sided, linear valleys. In Figure 6A-2 these rift valleys, which stretch more than 9600 kilometers (6000 mi) from the Red Sea to Swaziland, are marked by red lines.

Africa's rivers, too, are unusual: their upper courses frequently bear landward, seemingly unrelated to the coast toward which they eventually flow. Several rivers, such as

Farms along the fertile, terraced wall of Kenya's Eastern Rift Valley—a key corridor in the evolution of our human species.

© H.J. de Blij

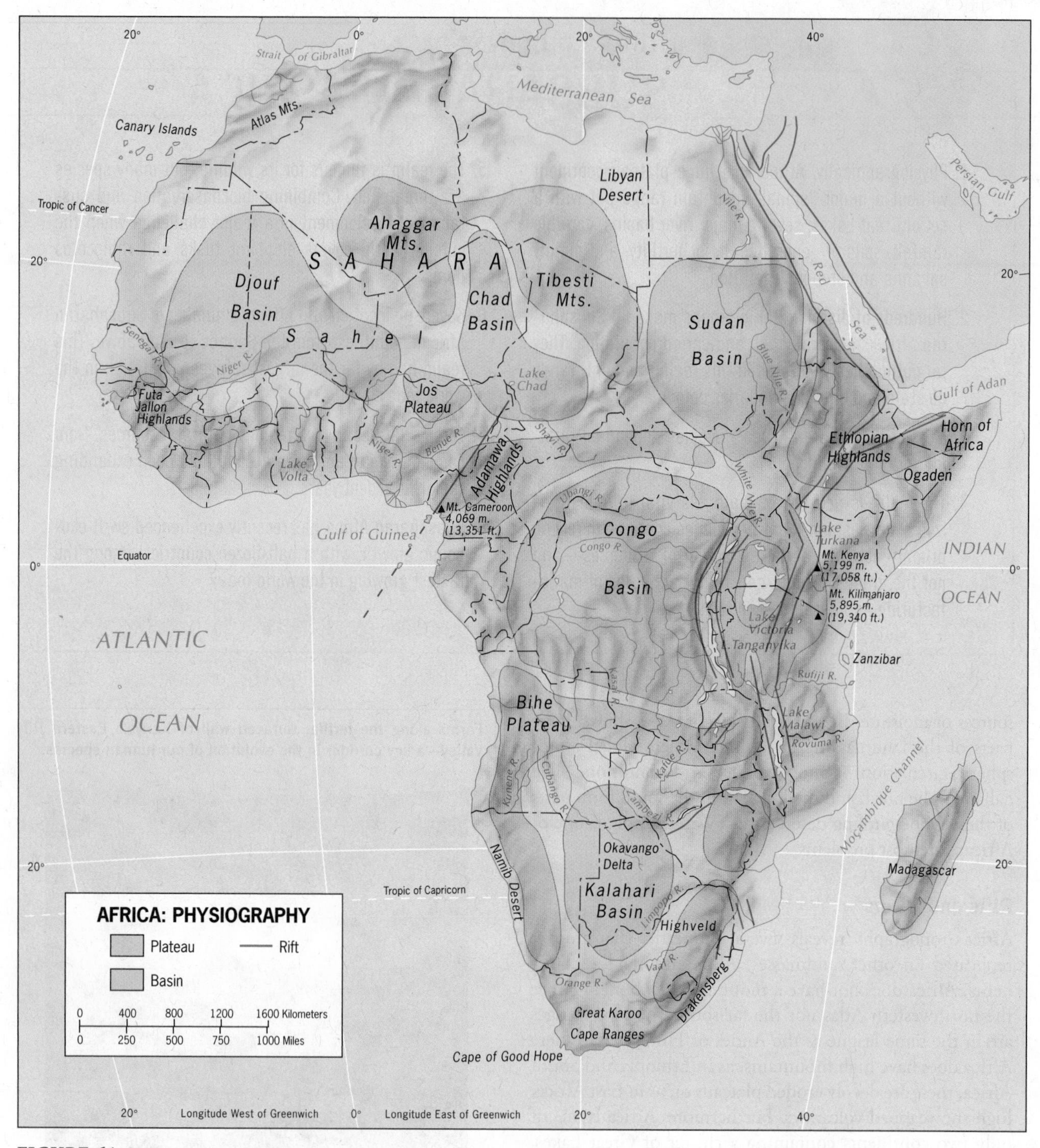

FIGURE 6A-2

the Nile and the Niger, have inland as well as coastal deltas. Major waterfalls, notably Victoria Falls on the Zambezi, or lengthy systems of cataracts, separate the upper from the lower river courses.

Finally, Africa may be described as the plateau continent. Except for some comparatively limited coastal plains, almost the entire continent lies above 300 meters (1000 ft) in elevation, and fully half of it lies over 800 meters (2500 ft) high. As Figure 6A-2 indicates, the plateau surface has sagged under the weight of accumulating sediments into a half-dozen major basins (three of them in the Sahara). The margins of Africa's plateau are marked by escarpments, often steep and step-like. Most notable among these is the Great Escarpment of southeastern South Africa, marking the eastern edge of the Drakensberg Mountains.

Continental Drift and Plate Tectonics

Africa's remarkable and unusual physiography was a key piece of evidence that geographer Alfred Wegener used to

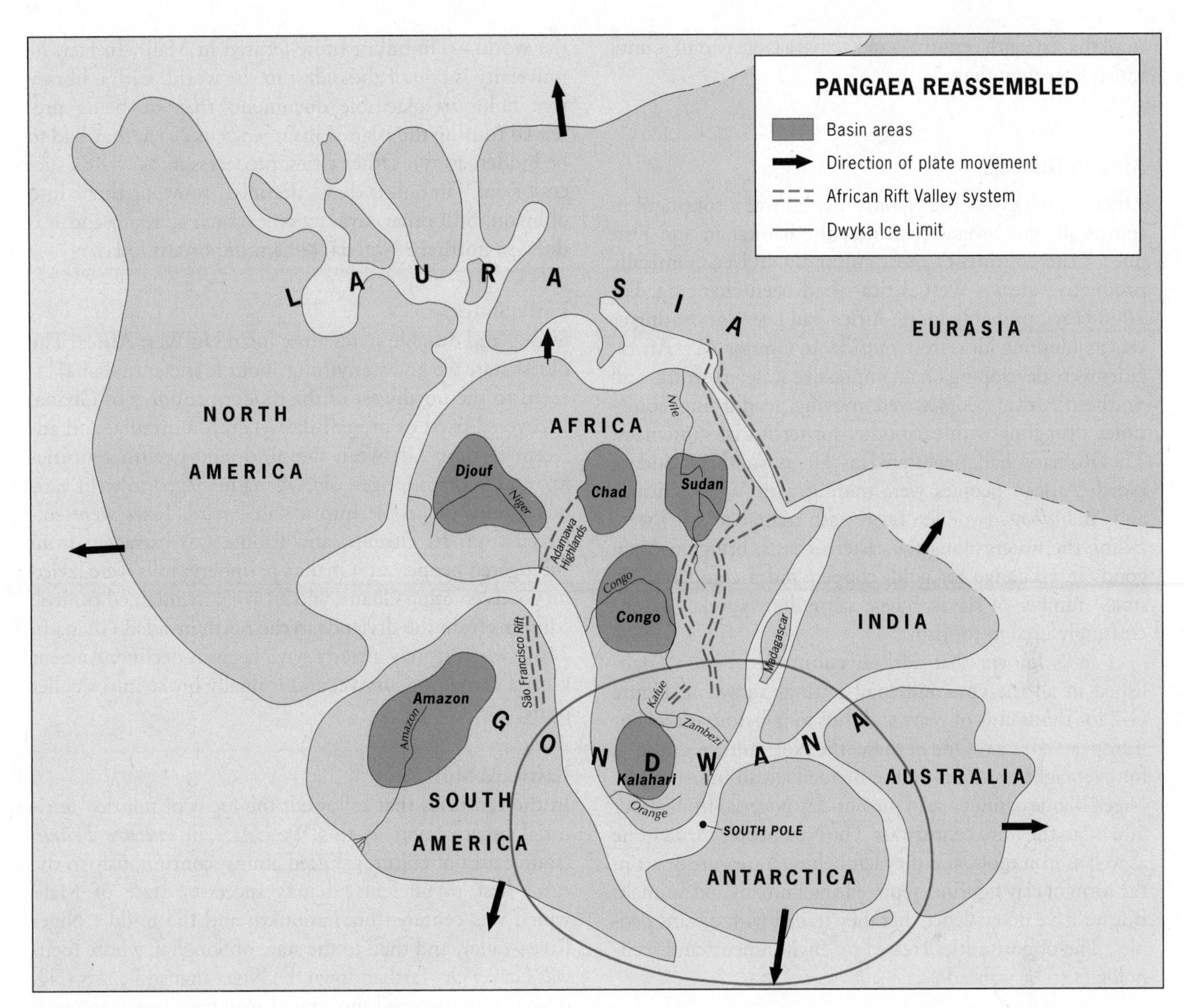

FIGURE 6A-3

construct his hypothesis of **continental drift [3]**. All of the present-day continents, Wegener reasoned, lay assembled into one giant landmass called ***Pangaea*** not very long ago (220 million years) in geologic time. The southern part of this supercontinent was *Gondwana*, of which Africa formed the core (Fig. 6A-3). When, roughly about 200 million years ago, tectonic forces began to split Pangaea apart, Africa (and the other landmasses) acquired its present configurations. That process, now known as ***plate tectonics***, continues and is marked by earthquakes and volcanic eruptions. By the time it started, however, Africa's land surface had begun to acquire some of the features that mark it today—and make it unique. The rift valleys, for example, demarcate the zones where plate movement continues—hence the linear shape of the Red Sea, where the Arabian Plate is separating from the African Plate (Fig. G-4). And yes, the rift valleys of East Africa probably mark the further fragmentation of the African Plate (some geophysicists have already referred to a Somali Plate, which they predict will separate, Madagascar-like, from the rest of Africa).

Africa's ring of escarpments, its rifts, its river systems, its interior basins, and its lack of significant mountains all relate to the continent's central position within Pangaea—all pieces of the puzzle that led to an explanation based on plate tectonics theory.

AFRICA'S HISTORICAL GEOGRAPHY

Africa is the cradle of humankind. Archeological research has chronicled 7 million years of transition from Australopithecenes to hominins to *homo sapiens*. It is therefore ironic that we know comparatively little about Subsaharan Africa from 5000 to 500 years ago—that is, before the onset of European colonialism. This is partly due to the colonial period itself, during which African history was neglected, numerous African traditions and artifacts were destroyed, and many misconceptions about African cultures and institutions became entrenched. It is also a result of the absence of a written history over most of Africa south of the Sahara until the sixteenth century—and across a large part of it until much later than that.

African Genesis

Africa on the eve of the colonial period was a continent in transition. For several centuries, the habitat in and near one of the continent's most culturally and economically productive areas—West Africa—had been changing. For 2000 years, probably more, Africa had been innovating as well as adopting ideas from outside. In interior West Africa, cities were developing on an impressive scale; in central and southern Africa, peoples were moving, readjusting, sometimes struggling with each other for territorial supremacy. The Romans had penetrated as far as southern Sudan, North African peoples were trading with West Africans, and Arab *dhows* (wooden boats with triangular sails) were sailing the waters along the eastern coasts, bringing Asian goods in exchange for gold, copper, and a comparatively small number of slaves. These same dhows today are increasingly used in tourism.

It is known that African cultures had been established in all the environmental settings shown in Figure G-7 for thousands of years and thus long before Islamic or European contact. One of these, the Nok culture, endured for over eight centuries on the Benue Plateau (north of the Niger-Benue confluence in present-day Nigeria) from about 500 BC to the third century AD. The Nok people made stone as well as iron tools, and they left behind a treasure of art in the form of clay figurines representing humans and animals. But we have no evidence that they traded with distant peoples. The opportunities created by environments and technologies still lay ahead.

Early Trade

West Africa, over a north-south span of a few hundred kilometers, displayed an enormous contrast in environments, economic opportunities, modes of life, and products. The peoples of the tropical forest produced and needed goods that were different from the products and requirements of the peoples of the dry, distant north. For example, salt is a prized commodity in the forest, where humidity precludes its formation, but it is plentiful in the desert and semiarid steppe. This enabled the desert peoples to sell salt to the forest peoples in exchange for ivory, spices, and dried foods. Thus there evolved a degree of ***regional complementarity*** between the peoples of the forest and those of the drylands. And the savanna peoples—those located in between—found themselves in a position to channel and handle the trade (which is always economically profitable).

The markets in which these goods were exchanged prospered and grew, and urban centers arose in the savanna belt of West Africa. One of these old cities, now an epitome of isolation, was once a thriving center of commerce and learning and one of the leading urban places in the world—Timbuktu (now located in Mali). In fact, its university is one of the oldest in the world, with a library that holds irreplaceable documents that are being preserved (during the Islamic insurgency of 2013, they had to be hidden away). Other cities, predecessors as well as successors of Timbuktu, have declined, some of them into oblivion. Still other savanna cities, such as Kano and Kaduna in northern Nigeria, remain important today.

Early States

Strong and durable states arose inland in West Africa. The oldest state we know anything about is ancient Ghana, located to the northwest of the modern country of Ghana. It covered parts of present-day Mali, Mauritania, and adjacent territory. Between the ninth and twelfth centuries AD, and perhaps longer, old Ghana managed to weld various groups of people into a stable state. Taxes were collected from its citizens, and tribute was extracted from subjugated peoples on Ghana's periphery; tolls were levied on goods entering Ghana, and an army maintained control. Muslims from the drylands to the north invaded Ghana in 1067, when it may already have been in decline. Ancient Ghana could not survive, and it finally broke into smaller units.

Eastward Shift

In the centuries that followed, the focus of politico-territorial organization in this West African ***culture hearth*** (source area of culture) shifted almost continuously to the east—first to ancient Ghana's successor state of Mali, which was centered on Timbuktu and the middle Niger River Valley, and then to the state of Songhai, whose focus was Gao, a city farther down the Niger that still exists (Fig. 6A-4). This eastward movement may have been the result of the growing influence and power of Islam. Traditional animist religions prevailed in ancient Ghana, but Mali and its successor states sent massive, gold-laden pilgrimages to Mecca along the savanna corridor south of the Sahara, passing through present-day Khartoum and Cairo. Of the tens of thousands who participated in these pilgrimages to Islam's holiest city (located in western Arabia), some remained behind. Today, many Sudanese trace their ancestry to the West African savanna kingdoms.

Beyond the West

West Africa's savanna zone undoubtedly experienced momentous cultural, technological, and economic developments, but other parts of Africa made progress as well. Early states emerged in present-day Sudan, Eritrea, and Ethiopia. Influenced by key innovations from the Egyptian culture hearth to the north, these kingdoms were stable and durable: the oldest, Kush, lasted 23 centuries (Fig. 6A-4). The Kushites built elaborate irrigation systems, forged iron tools, and created impressive structures as the ruins of their long-term capital and industrial center, Meroe, reveal. Nubia, to the southeast of Kush, became Christianized until

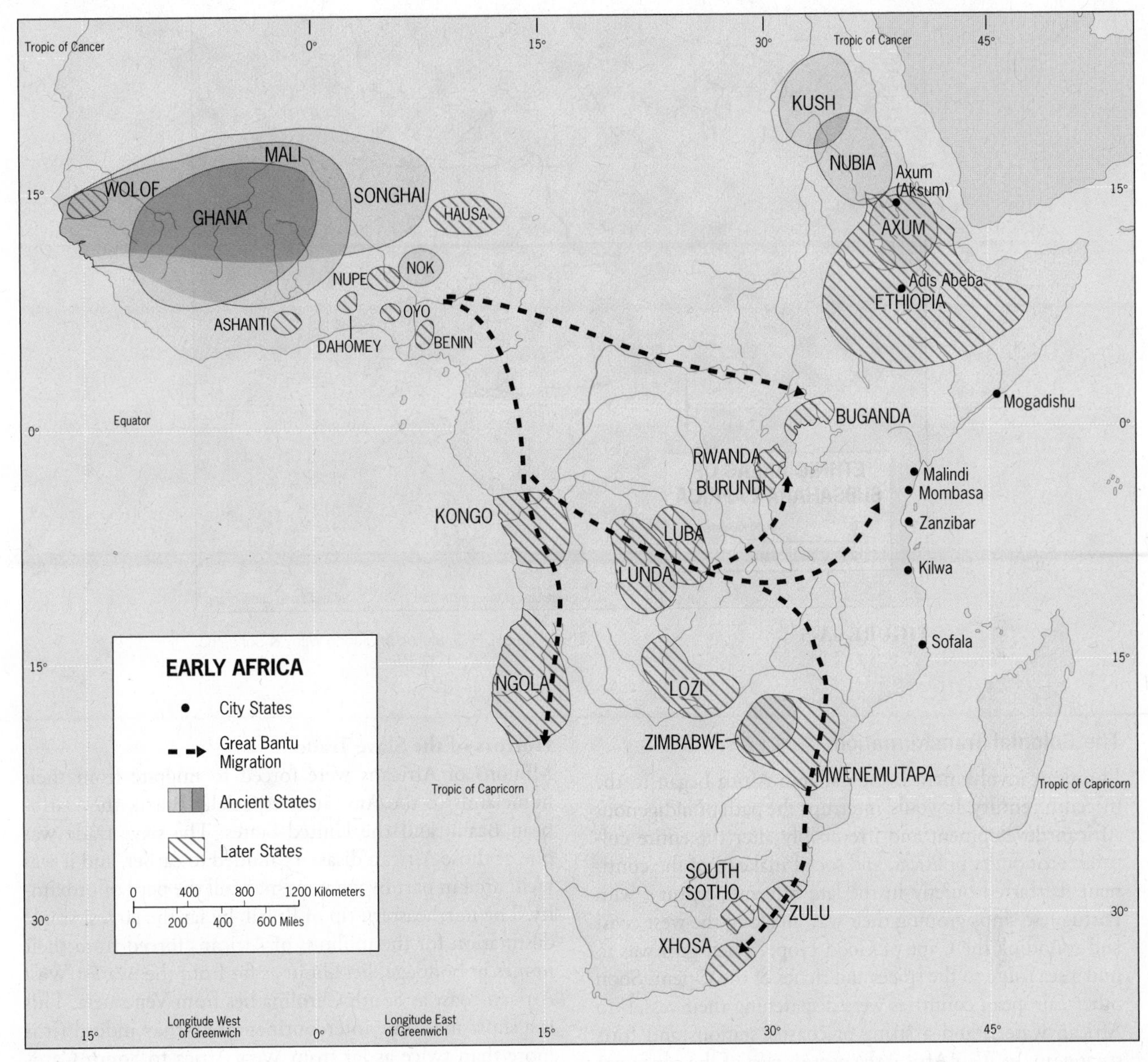

FIGURE 6A-4

the Muslim wave overtook it in the eighth century. And Axum was the richest market in northeastern Africa, a powerful kingdom that controlled Red Sea trade and endured for six centuries. Axum, too, was a Christian state facing the Islamic surge, but Axum's rulers deflected the Muslim advance and gave rise to the Christian dynasty that eventually shaped modern Ethiopia.

The process of **state formation [4]** spread throughout Africa and was still in progress when the first European contacts occurred in the late fifteenth century. Large and effectively organized states developed on the equatorial west coast (notably Kongo) and on the southern plateau from the southern part of the Congo River Basin southeastward to Zimbabwe. East Africa had several city-states, including Mogadishu, Kilwa, Mombasa, and Sofala.

Bantu Migration

A crucial event commencing about 5000 years ago affected virtually all of Equatorial, West, and Southern Africa: the Great Bantu Migration from present-day Nigeria and Cameroon southward and eastward across the continent. This epic advance appears to have occurred in waves that populated the Great Lakes area and ultimately penetrated South Africa, where it resulted in the formation of the powerful Zulu Empire in the nineteenth century (Fig. 6A-4).

All this reminds us that, before European colonization, Africa was a realm of rich and varied cultures, diverse lifestyles, technological progress, and external trade. It was, however, also a highly fragmented realm, its cultural mosaic (Fig. 6A-5) spelling weakness when European intervention came to change the social and political map forever.

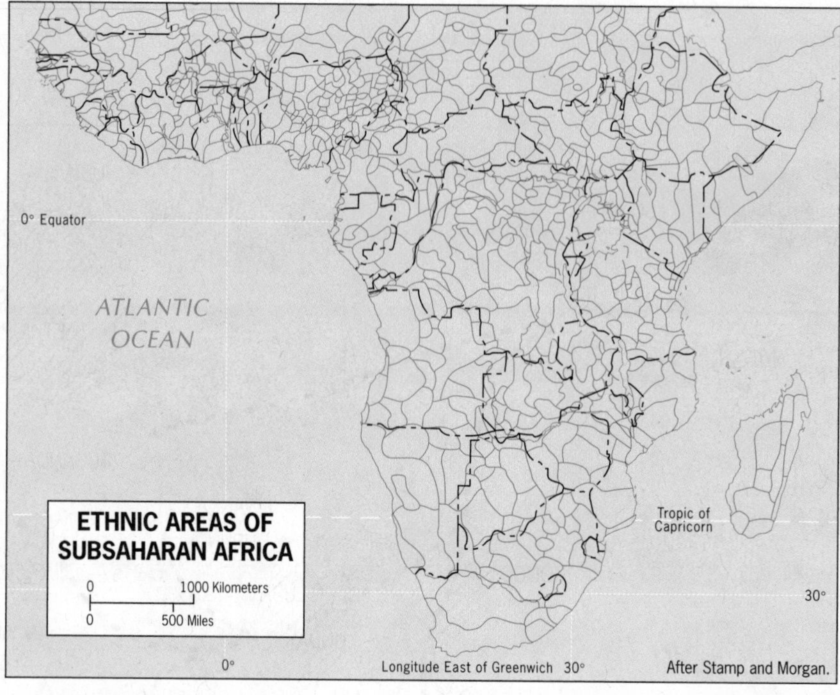

FIGURE 6A-5

The Colonial Transformation

European involvement in Subsaharan Africa began in the fifteenth century. It would interrupt the path of indigenous African development and irreversibly alter the entire cultural, economic, political, and social makeup of the continent. It started quietly in the late fifteenth century, with Portuguese ships groping their way all along the west coast and rounding the Cape of Good Hope. Their goal was to find a sea route to the spices and riches of the Orient. Soon other European countries were dispatching their vessels to African waters, and a string of coastal stations and forts sprang up. In West Africa, the nearest part of the continent to European spheres in Middle and South America, the initial impact was strongest. At their coastal control points, the Europeans traded with African intermediaries for the slaves who were destined to work New World plantations, for the gold that had been flowing northward across the desert, and for highly prized ivory and spices.

Coastward Reorientation

Suddenly, the centers of activity lay not with the inland cities of the savanna belt but with the foreign stations on the Atlantic coast. As the interior declined, the coastal peoples thrived. Small forest states gained unprecedented wealth, transferring and selling slaves captured in the interior to the European traders on the coast. Dahomey (now called Benin) and Benin (now part of neighboring Nigeria) were states built on the slave trade. When slavery was eventually abolished in Europe, those who had inherited the power and riches it had brought vigorously opposed abolition on both continents.

Horrors of the Slave Trade

Millions of Africans were forced to migrate from their homelands to the Americas, especially Brazil, the Caribbean Basin, and the United States. The slave trade was one of those African disasters alluded to earlier, and it was facilitated in part by what we may call the peril of proximity. The northeastern tip of Brazil, by far the largest single destination for the millions of Africans forced from their homes in bondage, lies about as far from the nearest West African coast as South Carolina lies from Venezuela. This is a short maritime intercontinental journey indeed (it is more than twice as far from West Africa to South Carolina). That proximity facilitated the forced migration of millions of West Africans to Brazil, which in turn contributed to the emergence of an African cultural diaspora in Brazil that is without equal in the New World (see Chapter 5B).

Although slavery was not new to West Africa, the *kind* of slave raiding and trading the Europeans introduced certainly was. In the interior of Africa and within city-states, kings, chiefs, and prominent families traditionally took a few slaves, but the status of those slaves was unlike anything that lay in store for those who were shipped across the Atlantic. In fact, large-scale slave trading had been introduced in East Africa long before the Europeans brought it to West Africa. African intermediaries from the coast raided the interior for able-bodied men and women and marched them in chains to the Arab markets on the coast (the island of Zanzibar was one such slave trading market). There, packed in specially built dhows, they were carried off to Arabia, Persia, and India. When the European slave

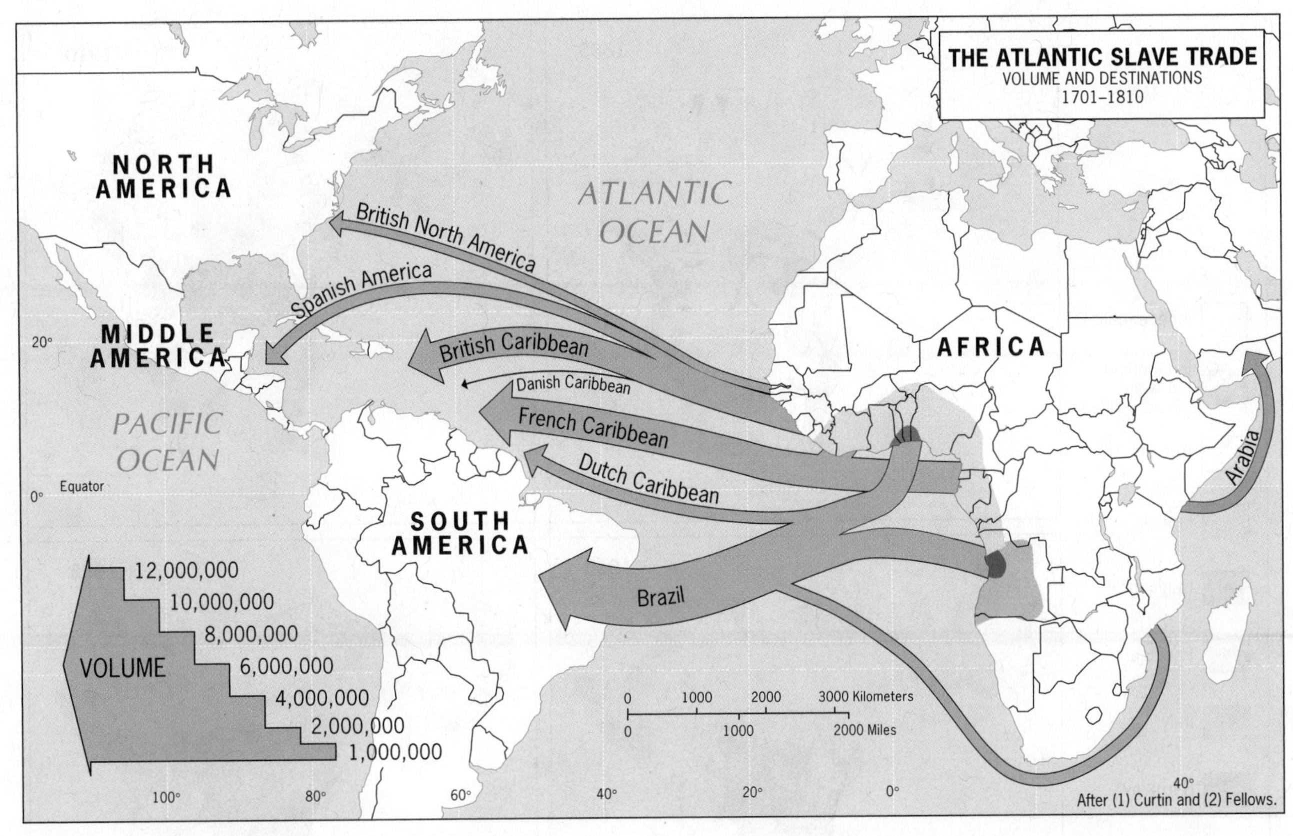

FIGURE 6A-6

trade took hold in West Africa, however, its volume was far greater. Europeans, Arabs, and collaborating Africans ravaged the continent, forcing perhaps as many as 30 million persons away from their homelands in captivity (Fig. 6A-6). Families were destroyed, as were entire communities and cultures; those who survived their exile suffered unfathomable misery.

The European presence on the West African coast completely reoriented its trade routes, for it initiated the decline of the interior savanna states discussed earlier and strengthened the coastal forest states. Moreover, the Europeans' insatiable demand for slaves ravaged the population of the interior. But it did not lead to any major European thrust toward the interior or produce colonies overnight. The African intermediaries were well organized and strong, and they held off their European competitors, not just for decades but for centuries. Although the Europeans first appeared in the fifteenth century, they did not carve up West Africa until nearly 400 years later, and they did not conquer many other areas until after the beginning of the twentieth century.

Colonization

In the second half of the nineteenth century, whether or not they had control, the European powers finally laid claim to virtually all of Africa. Colonial competition was intense, and spheres of influence began to overlap. It was time for negotiation among the powerful, and in 1884 a conference was convened in Berlin to carve up the African map (see box titled "The Berlin Conference"). The major colonial contestants were Britain, France, Portugal, Belgium, and Germany itself. On maps spread across a large table, representatives from these powers drew boundaries, exchanged real estate, and forged a new map that would become a liability in Africa decades later. As Figure 6A-7 indicates, when the three-month conference was in progress, most of Africa remained under traditional African rule. Not until after 1900 did the colonial powers manage to control all the areas they had marked off and acquired on their new maps.

It is important to examine Figure 6A-7 carefully because the colonial powers governed their new dependencies in very different ways, and their contrasting legacies are still in evidence to this day in the countries their colonies spawned. Some colonial powers were democracies at home (Great Britain and France); others were dictatorships (Portugal and Spain). The British established a system of **indirect rule [5]** over much of their domain, leaving indigenous power structures in place and making local rulers representatives of the British Crown. This was unthinkable in the Portuguese colonies, where harsh, direct control predominated. The French sought to create culturally assimilated elites that would represent French ideals in the colonies.

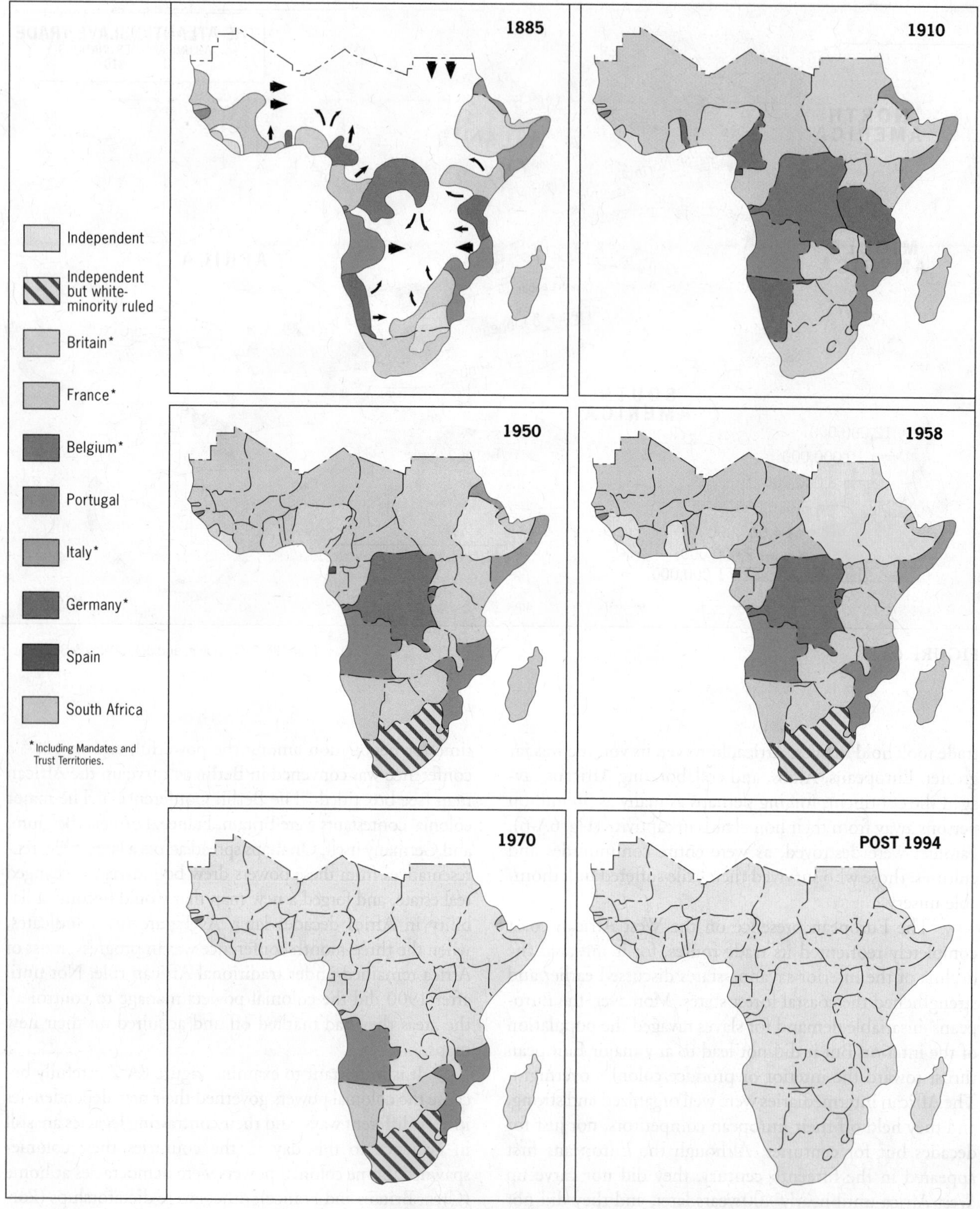

COLONIZATION AND DECOLONIZATION SINCE 1885

FIGURE 6A-7

The Berlin Conference

IN NOVEMBER 1884, the imperial chancellor and architect of the German Empire, Otto von Bismarck, convened a conference of 14 states (including the United States, which had no African claims) to settle the political partitioning of Africa. Bismarck wanted not only to expand German spheres of influence in Africa but also to play off Germany's colonial rivals against one another to the Germans' advantage. The major colonial contestants in Africa were: (1) the British, who held beachheads along the West, South, and East African coasts; (2) the French, whose main sphere of activity was in the area of the Senegal River as well as north and northwest of the Congo Basin; (3) the Portuguese, who now desired to extend their coastal stations in Angola and Moçambique deep into the interior; (4) King Leopold II of Belgium, who was amassing a personal domain in the Congo Basin; and (5) Germany itself, active in areas where the designs of other colonial powers might be obstructed, as in Togo (between British holdings), Cameroon (a wedge into French spheres), South West Africa (taken from under British noses in a swift strategic move), and East Africa (where German Tanganyika broke the British design to assemble a solid block of territory stretching northward from the Cape to Cairo).

When the conference convened in Berlin, more than 80 percent of Africa was still under traditional African rule. Nonetheless, the colonial powers' representatives drew their boundary lines across the entire map. These lines were drawn through known as well as unknown areas, pieces of territory were haggled over, boundaries were erased and redrawn, and African real estate was exchanged among European governments. In the process, African peoples were divided, unified regions were ripped apart, hostile societies were thrown together, hinterlands were disrupted, and migration routes were closed off (for an idea, see Fig. 6A-5). Not all of this was felt immediately, of course, but these were some of the effects when the colonial powers began to consolidate their holdings and the boundaries on paper became barriers on the African landscape (Fig. 6A-7).

The Berlin Conference was Africa's undoing in more ways than one. The colonial powers superimposed their domains on the African continent. By the time Africa regained its independence after the late 1950s, the realm had acquired a legacy of political fragmentation that could neither be eliminated nor made to operate satisfactorily. The African politico-geographical map is therefore a permanent liability that resulted from three months of ignorant, greedy acquisitiveness during a period when Europe's search for minerals and markets had become insatiable.

King Leopold II of Belgium sought to claim Congo Free State as his own personal fiefdom. After financing the expeditions that staked Belgium's claim in Berlin, he embarked on a campaign of exploitation and genocide of such ruthlessness that, after the impact of the slave trade, Leopold's reign of terror was Africa's most severe demographic disaster, with an estimated 10 million murdered (chronicled unforgettably in Joseph Conrad's classic novella, *Heart of Darkness*). By 1909, the Belgian government had taken over following Leopold's death and began to mirror Belgium's own internal divisions: corporations, government administrators, and the Roman Catholic Church each pursued their sometimes competing interests. But no one thought to change the name of the colonial capital: it remained Leopoldville until the Belgian Congo achieved independence in 1960.

Colonialism did transform Africa, but in its post-Berlin form it lasted less than a century. In Ghana (then named the Gold Coast), for example, the Ashanti (Asante) Kingdom was still fighting the British in the early years of the twentieth century; by 1957, Ghana was independent again. These days, most of Subsaharan Africa has already marked half a century of independence, and, in retrospect, the colonial period should be seen as an interlude rather than a paramount chapter in modern African history—although it did leave a deep and lasting geographic imprint, especially on the realm's cultural landscape.

POSTCOLONIAL AFRICA

In Subsaharan Africa, more than in any other realm, there is a huge discrepancy between the geography of territorial states (imposed by the former colonial powers) and the geography of national and ethnic identities (look again at Fig. 6A-5). The political map of Subsaharan Africa has 45 states but no nation-states (apart from some microstates and ministates in the islands and the far south). Centrifugal forces are powerful, cultural pluralism prevails, and outside interventions during the Cold War, when communist and anticommunist foreigners took sides in local civil wars, worsened conflict within many African states.

Clearly, the failings of democracy across the realm in past decades cannot be understood without considering this precarious legacy. Corruption and incompetence have seemed rampant, but at a deeper level it was often the (political exploitation of) ethnic fragmentation and tribal tensions that impeded or undermined enlightened governance.

Colonialism's economic legacy was not much better. In Africa, capitals, core areas, port cities, and transport systems were laid out to maximize colonial profit and facilitate the exploitation of minerals and soils; the colonial mosaic inhibited interregional communications except where cooperation enhanced efficiency. Colonial Zambia and Zimbabwe, for instance (then respectively named Northern and Southern Rhodesia), were landlocked and needed outlets, so railroads were built to Portuguese-owned ports. But

From the Field Notes . . .

© H.J. de Blij

"Looking down on this enormous railroad complex northeast of Johannesburg, South Africa, we were reminded of the fact that almost an entire continent was turned into a wellspring of raw materials carried from interior to coast and shipped to Europe and other parts of the world. This complex lies near Witbank in the eastern Rand, a huge inventory of freight trains ready to transport ores from the plateau to Durban and Maputo. But at least South Africa acquired a true transportation network in the process, ensuring regional interconnections; in most African countries, railroads serve almost entirely to link resources to coastal outlets."

www.conceptcaching.com

such routes did little to create internal African linkages. The modern map reveals the results: in West Africa you can travel from the coast into the interior of all the coastal states along railways or adequate roads. But no modern routeways were ever built to connect these coastal neighbors to each other.

A Realm in Need of Infrastructure

In Chapters 6A and 6B you will encounter numerous references to African countries with fast-growing economies, rising revenues, and newly opened resource reserves. But you will read much less about intra-African trade and transportation, diversifying economies, and the kind of regional complementarity that made the old West African states so stable and successful. Africa needs infrastructure: a network of high-speed highways, a system of railroads, and a speedier way to traverse its few connecting rivers. Take a boat up the Congo River from Kinshasa, and get ready for about a week of slow going, stops, transfers, delays, and lost perishables. Try taking the surface route from Cairo via the Nile to Khartoum, and you will surely fly the next time.

That is one thing for visitors and tourists, but another for businesses. According to World Bank estimates, only 13 percent of Subsaharan African trade is internal, the lowest percentage of intra-realm trade in the world. Try to import goods from abroad or export your products overseas, and the cost of transport can add 50 percent or more to the retail price. About the only way to manufacture something in northern Nigeria and sell it in Senegal is to truck it via potholed roads to Port Harcourt or Lagos, load it onto a boat, ship it to Dakar, and then hope that your customer is in that port and not somewhere in the interior. No decent road links the eleven countries along the coast from Nigeria to Senegal, so forget about selling things along the way.

Africa needs north-south and east-west roads as well as railroads. It also needs more efficient border posts; fewer export and import restrictions, tariffs, and tolls to help farmers sell to their neighbors; and less corruption. On the roads, too much traffic crawls along at slow speeds, easily victimized at roadside checkpoints where bribes have to be paid. There is nothing like a 100-kph (63-mph) truck barreling along to market on a four-lane, hard-surfaced highway to discourage this.

NATURAL ENVIRONMENTS

Only the southernmost tip of Subsaharan Africa lies outside the tropics. Although African elevations are comparatively high, they are not high enough to ward off the heat that comes with a tropical location except in especially favored locales such as the Kenya Highlands and parts of Ethiopia. And, as we already noted, Africa's bulky shape means that much of the continent lies far from maritime moisture sources. Variable weather and frequent droughts, therefore, are among Africa's unrelenting environmental challenges.

It is useful at this point to refer back to Figure G-7. As that map shows, Africa's climatic regions are distributed almost symmetrically about the equator, though more so in the center of the landmass than in the east, where elevation changes the picture. The hot, rainy climate of the

Congo Basin merges gradually, both northward and southward, into climates with distinctly dry winter seasons. Winter, however, is marked more by drought than by cold. In parts of the area mapped ***Aw*** (tropical savanna), the annual seasonal cycle produces two rainy seasons, often referred to locally as the long rains and the short rains, separated by two winter dry periods. As you go farther north and south away from the moist Congo Basin, the dry season(s) grow longer and the rainfall diminishes and becomes less and less dependable.

Wildlife Under Threat

Africa's shrinking rainforests and vast savannas form the world's last refuges for wildlife ranging from primates to wildebeests. Gorillas and chimpanzees survive in dwindling numbers in threatened forest habitats, while millions of herbivores range in great herds across the savanna plains where people compete with them for space. European colonizers, who introduced hunting as a "sport" (a practice that did not exist within African cultural traditions) and who brought their capacities for mass destruction to animals as well as people in Africa, helped clear vast areas of wildlife and pushed species to near-extinction.

Later, colonial rulers and subsequent national governments laid out game reserves and other kinds of conservation areas, but these were not sufficiently large or well enough connected to allow herd animals to adhere to their seasonal and annual migration routes. The same climatic variability that affects farmers also affects wildlife, and when the rangelands wither, the animals seek better pastures. When the fences of a game reserve wall them off, they cannot survive. When there are no fences, the wildlife invades neighboring farmlands and destroys crops, and the farmers retaliate. After thousands of years of equilibrium, the competition between humans and animals in Africa has taken a new turn.

But it is not solely population pressure that is threatening African wildlife. The horn of the rhinoceros became a valued property for Arabs who fashion them into dagger handles and, in powdered form, a purported aphrodisiac for wealthy East Asians able to afford this as a luxury. As a result, the northern white rhinoceros is now nearing extinction; the number of all species of rhino has fallen from several hundred thousand to only about 20,000 today. A terrible tragedy is befalling the African elephant as the value of its ivory tusks has skyrocketed; the Chinese illicit market is likely to exterminate this species within the next decade. Note that it is external demand, not internal circumstance, that is hastening this calamity. Meanwhile, competition for land has more to do with the decline of Africa's lion population (from around 400,000 half a

From the Field Notes . . .

"Flying a small propeller plane over East Africa's semiarid grasslands, we had a good view of a Maasai village near the border between Kenya and Tanzania. The Maasai are a pastoral people whose main livelihood is herding cattle, sheep, and goats. I could clearly see the fenced 'kraals' where they keep their animals at night (during the day they are moved around) and the small dwellings that surround them. The more animals a man (!) has, the higher his status. But this way of life is under pressure: many have decided to settle in towns or cities because they offer more and better opportunities. And this area is also the northern extension of Tanzania's famous Serengeti National Park, and here wildlife preservation comes into direct conflict with the Maasai way of life. 'It is really impossible,' said the park guide on the 10-seater plane with us, pointing down at the settlement. 'Of course they want to protect their cattle, but they are killing lions and other animals that should be protected. And we must tell them all the time that they have to stay out of certain areas because the cattle compete for food with the wild animals.' Nodding agreeably, I thought to myself: yes, but isn't this their land? As with so many other places around the world where we want to protect nature, we must involve the local people. That is much easier said than done."

© Jan Nijman

www.conceptcaching.com

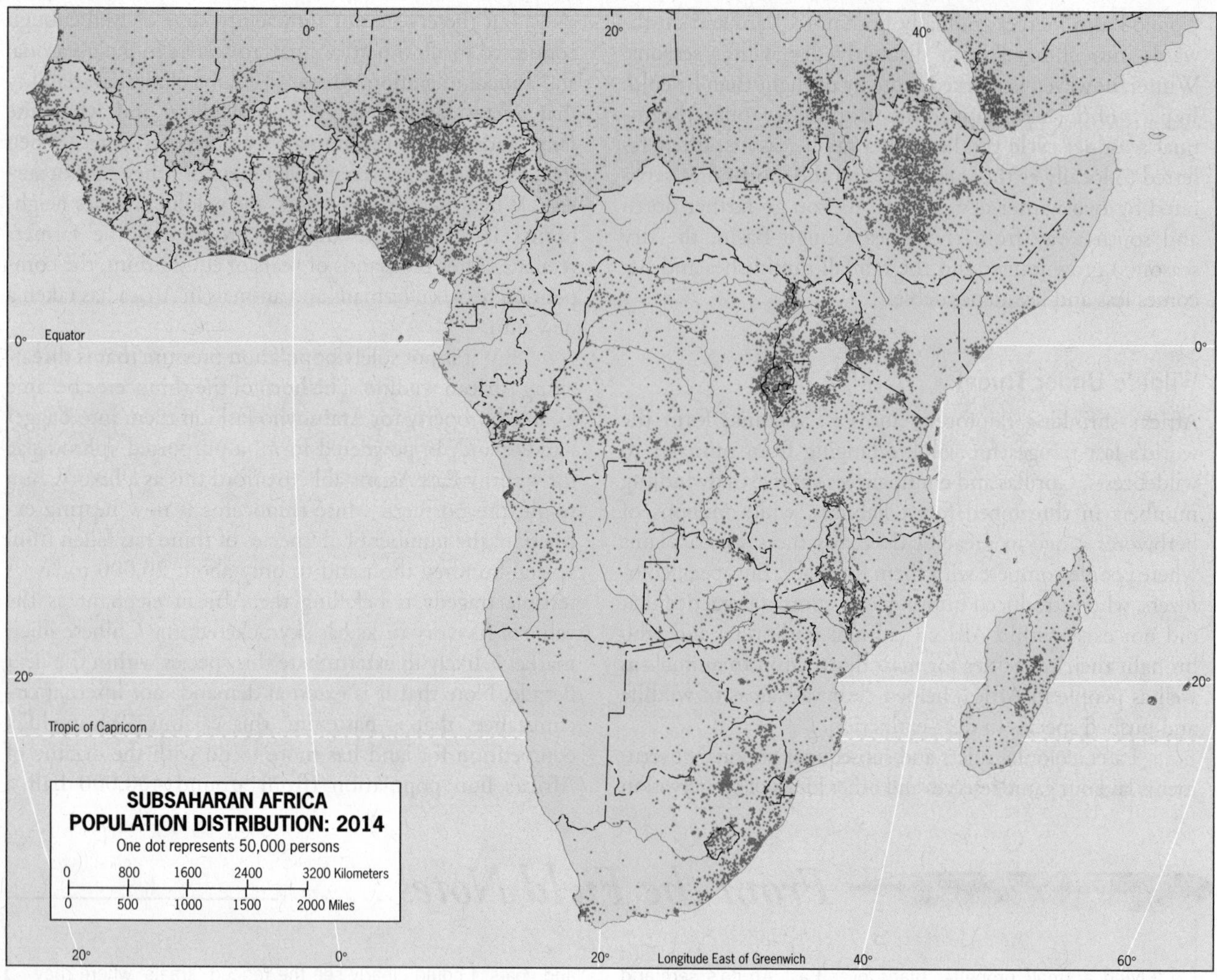

FIGURE 6A-8

century ago to barely more than 20,000 today) as well as the threat to other species. It is in some ways the end of the Cenozoic era, the last gasp of a fauna that emerged after the demise of the dinosaurs, survived ice ages and glaciations, and now falls victim to humankind's relentless rise.

People, Farmlands, and Environments

It would seem that there could be space for wildlife as well as humans in Subsaharan Africa. Figure 6A-8 does not present a picture of a densely-peopled realm: although there are major clusters of population in West Africa (where Nigeria is the realm's most populous country), East Africa (encircling Lake Victoria), and the African Horn (where Ethiopia's highlands sustain a large concentration), most of the rest of the realm seems relatively sparsely populated. Our Data Table in Appendix B bears out this impression: all the countries of Subsaharan Africa *combined* have a population only slightly above two-thirds of China's alone.

But much of this population continues to depend on farming for their livelihood, and we have already seen that African environments are difficult for millions of farmers. Not everywhere, of course: Africa has its areas of good soil, ample water, and robust farm productivity. However, these areas are not extensive. The major population concentration in the Great Lakes region reflects the volcanic soils of Mount Kilimanjaro as well as those of the Western Rift Valley zone. The Ethiopian Highlands were once known as the "breadbasket of northeastern Africa" and retain their capability for greater productivity than is presently the case. Higher-latitude, cooler, and moister areas of South Africa are exceptionally productive in a wide range of farm produce. And parts of West Africa sustain intensive farming. Taken together, however, these highly productive areas do not add up to the huge, fertile, alluvial basins of China or India—or even the (North African) Nile Valley and Delta. The great majority of African farmers face daunting challenges that include: (1) climatic variability, (2) the economic policies of national governments, and (3) the difficulties African

© AP/Wide World Photos

The cell phone revolution has made a dramatic impact on farmers in many parts of the developing world, especially in Subsaharan Africa where distances can be far, land lines absent, and market information scarce. Between 2005 and 2012, the number of cell phone subscriptions increased sixfold to encompass nearly 60 percent of the population. Now farmers and market women, such as the Kenyan woman pictured here, can stay in touch to better gauge when it is best to market crops and to better manage logistics.

farmers face in reaching world markets. Look again at Figure G-7, and you can see the relatively short distances involved as the moist tropics of the equatorial Congo Basin give way to the deserts of the north and south. Moreover, as the annual rainfall total declines, its variability increases, so that farmers in the drier zones cannot count on sufficient moisture in any given year.

Moreover, the economic policies of national governments frequently disadvantage farmers as the prices of their products are kept artificially low to please urban (and politically powerful) consumers. And in this era of globalization, African farmers often get a raw deal on world markets. For all the talk of free trade in the wealthy global core, many governments favor their local farmers over those of Africa (as well as other parts of the global periphery), from rice subsidies in Japan to "market support" in France.

AFRICANS AND THEIR LAND

Although rainfall is a critical physical criterion for farming, a number of political and economic factors are also influential. Among them are land tenure; commercial (for-profit) versus subsistence (household-level) agriculture; type of farming system (rotational, shifting cultivation, intercropping, etc.); crop prices; government policies promoting the planting of one crop over another; indigenous agricultural knowledge of particular crops; and the degree of technology adoption and mechanization.

With nearly two-thirds of the realm's livelihoods dependent on farming, the issue of land tenure is crucial. **Land tenure [6]** refers to the way people own, occupy, and use land. African traditions of land tenure are different from those found in Europe or the Americas. In most of Subsaharan Africa, communities, not individuals, customarily hold land. Occupants of the land have temporary, custodial rights to it and cannot sell it. Land may be held by large (extended) families, a village community, or even a traditional chief who holds the land in trust for the people. His subjects may house themselves on it and farm it, but in return they must follow his rules.

Land Ownership Versus Land Alienation

At the onset of **colonialism [7]**, colonial administrators intended to control the most fertile areas of the occupied colonies through eviction of indigenous peoples. In some cases this was done by physical force (mostly through military conquest), and in other areas by coercion. Prior to colonialism, many traditional African livelihood systems adhered to sustainable land management practices such as leaving the land fallow or rotational grazing. Viewing this land as unused, colonial planners initiated a process of **land alienation [8]** not unlike what happened in Middle and South America. Many of the most fertile and productive areas were placed under the direct control of colonial settlers and governments. As time went by, these lands were bought and sold and became private legal property. At the end of the colonial era, several newly independent African countries initiated programs whereby land would revert to traditional forms of ownership and management. However, the legacy of colonialism has been difficult to overcome, and some African governments have adopted policies on land management that continue to marginalize small farmers.

Rapid population growth, such as Africa has experienced since the end of the colonial era, makes access to land even more complicated. Traditional systems of land use, which involve subsistence farming in various forms ranging from shifting cultivation to pastoralism, work best when the population is fairly stable and tenure is communal. Land must be left fallow to recover from cultivation, and pastures must be kept free of livestock so that the grasses can revive. The African population explosion that began in the mid-twentieth century set into motion a cycle of land overuse. When soils cannot rest and pastures are overgrazed, the land becomes degraded and yields decline.

Subsistence Farming

Although there is commercial farming in parts of Africa, most African farmers remain subsistence farmers who grow grain crops (maize [corn], millet, sorghum) in drier areas and root crops (yams, manioc [cassava], sweet potatoes) where moisture is adequate. Others herd livestock, mostly cattle and goats, as they counter environmental and climatic variability. Farmers and pastoralists alike have been unable to secure access to regular markets and stable prices for their products because of government policies that often promote one particular type of export-oriented crop (for example, peanuts in Gambia; tea and coffee in Kenya; cacao [cocoa] in Ivory Coast) over other crops that do not fetch high prices on the world market.

Even though African farmers learned to adapt to these problems, farm yields in Africa have remained modest. Governments, in response to World Bank policies, paid greater attention to high-profile industrial projects and as a result have neglected agriculture. The decline of African agriculture has been disproportionally hard on Africa's women who, according to current estimates, produce 75 percent of all local food in Subsaharan Africa. Development policies frequently pay too little attention to this situation and its related household dynamics, thereby dooming many well-intentioned projects.

The **Green Revolution [9]** the development of more productive, drought-tolerant, pest-resistant, higher-yielding types of grain has had less impact in Africa than elsewhere (see box titled "A Green Revolution for Africa?"). Where people depend mainly on rice and wheat, the Green Revolution pushed back the prospect of increased hunger. But not enough research was done on this realm's dominant crops, mostly tubers, so the advances of the Green Revolution barely touched Africa. And, in any case, the Green Revolution is hardly an unqualified remedy: the poorest farmers, who need help the most, can least afford the more expensive, higher-yielding seeds or pesticides that are so often trequired.

Foreign Agribusiness in Africa

Low agricultural productivity resulting from inefficiency, coupled with governments' need for more revenues, have converged in some African countries to result in the selling or long-term leasing of huge tracts of public land to the big-time investors of multinational agro-industrial corporations. The idea is that if you do not have the means to develop your land, then large-scale **agribusiness [10]** is the best option to increase yields and simultaneously fill government coffers. Foreign investors, in turn, are motivated by steadily rising global food prices, brightening prospects for biofuels, and growing shortages of arable land in their own

A Green Revolution for Africa?

THE TERM *Green Revolution* refers to the development of higher-yielding varieties of grains through genetic manipulation. Since the 1970s, it has narrowed the worldwide gap between population and food production, and it was especially influential in regions where people depend mainly on rice and wheat for their staples. The Green Revolution has had less impact in Subsaharan Africa, however. In part, this relates to the realm's high rate of population growth (which is substantially higher than that of India or China). Other reasons have to do with Africa's staples: rice and wheat support only a small part of this realm's population. Corn (maize) supports many more, along with sorghum, millet, and other grains. In moister areas, root crops, such as the yam and cassava as well as the plantain (similar to the banana) supply most calories. These crops were not priorities in Green Revolution research.

Lately, there have been a few signs of improvement. Scientists have worked toward two goals: first, to develop strains of corn and other crops that would be more resistant to Africa's virulent crop diseases, and second, to increase the productivity of those strains. But these efforts faced serious problems. An average African hectare (2.5 acres) planted with corn, for instance, yields only about half a ton of corn, whereas the global average is 1.3 tons. When a virus-resistant variety was developed and distributed throughout Subsaharan Africa in the 1980s, yields rose significantly where farmers could afford to buy the new strain as well as the fertilizers it required. Nigeria saw a near-doubling of its production before it leveled off. Hardier varieties of root crops also raised yields in some areas.

Clearly, Africa needs more than this to reverse the trend toward food deficiency because food production has simply not kept pace with population growth. Lack of capital, inefficient farming methods, inadequate equipment, soil exhaustion, male dominance, apathy, and devastating droughts also contributed to this decline—not to mention affordability problems with more expensive higher-yielding seeds and pesticides. And the realm's seemingly endless series of civil conflicts have also reduced farm output. The Green Revolution may narrow the gap between production and need, but the battle for food sufficiency and security in Subsaharan Africa is far from won.

Regional ISSUE Neocolonial Land Grabs?

THESE LAND DEALS ARE A SOLUTION TO OUR PROBLEMS

"When the dust of all this drama settles around so-called 'land grabs', it should be clear to anybody that our government has come up with a brilliant solution to the country's problems. As an accountant here in the capital of Adis Abeba, I know a few things about costs and benefits, and in these land deals Ethiopia comes out as a winner.

"We mustn't fool ourselves: decade after decade this country has been hit by famines and food shortages. It was not just people going hungry; hundreds of thousands have died as a result. Surely one reason for this has been that we have failed to produce enough food. The landholdings are too small to be efficient and most farmers don't have the means or the knowledge to improve yields.

"The way forward for us and several other countries in Africa is to use our abundant land and natural resources to our advantage, to allow foreign investors to come in and exploit these resources for us. It's not like we are giving it away. Not at all! In most cases it is not a permanent sale but a lease, and the agreements contain provisions that they must sell a minimum share of the harvest in Ethiopia so all of it cannot be exported. And they must hire Ethiopian workers.

"Food in this country will become cheaper, you will see, simply because there will be more of it. Foreign agri-businesses know their stuff. Government income will rise from these deals, and these revenues can be invested in education and infrastructure. That, in turn, will create more employment opportunities. I can understand the fears of people about these land deals but they are misplaced and old-fashioned. In the global economy today, you must be willing to deal with the outside world, use the resources you've got to your advantage. That is what we are doing."

AFRICA IS BEING ROBBED, ONCE AGAIN

"It is hard to believe this is really happening, but it's true and I've seen it with my own eyes here in Ethiopia. Families who have been farming for many years are forced off the land so the government can lease it to foreign investors. It is a disaster because we Ethiopians no longer control our own land and what we grow on it. I own a shop here in a small town in western part of the country and I have seen prices go up. Many of my customers are rural people who used to grow their own food but now they must pay in cash. The government promised them all kinds of things when they terminated the leases on these small farms, yet little has come of it.

"Where is a country going when it sells its land to outsiders? That is my question to you. How can you ever believe that big foreign companies will keep the interests of the Ethiopian people in mind? Of course they don't! Why do you think they come here in the first place? They grow just one or two crops to export back to their own country. Quite a few are said to be rich Arabs from the Gulf states, where fertile land is scarce. Many just grow rice, a food we Ethiopians don't eat much of. I even heard a story of one company growing sorghum for their camels in Arabia!

"And don't believe the argument that these lands were not being cultivated. Do you really think we Africans would let good land sit idle? We have millions of small farmers looking for every bit of fertile soil they can find on which to earn a living. Worst of all is when the outsiders buy up the land and don't do anything with it. Foreign investors get such good deals from our government that they just sit on the land as a speculative investment. They believe Africa is rising and that the land will increase in value, so they wait until they can sell it for a profit in the future. In the meantime, we can't touch our own natural resources. It is a scandalous situation."

Vote your opinion at www.wiley.com/go/deblijpolling

countries. For many, acquiring land in Africa is a long-term proposition based on expectations that natural resources will become ever more scarce in the future as demand increases while supply declines.

In 2009 alone, 44 million hectares (110 million acres) were involved in such land transactions—an area the size of California and West Virginia combined. Most of this land was in Sudan, Ethiopia, and Moçambique. Some of the investors were African, but many others were foreign (always a contentious issue). When the government of Madagascar in that same year declared its intention to sell half of all arable public lands to a South Korean corporation, protests ensued and quickly escalated to a level that forced the government to cancel the transaction and resign.

The impact of these gigantic land deals is mixed. In some cases, output on the large commercial farms far exceeds previous levels of production. But in many others, investors seem to have had mainly speculative intentions and much farmland remains idle. In either situation, significant numbers of small farmers are uprooted and their villages destroyed. Understandably, there is now growing mistrust among grass-roots populations, and the success of these ventures increasingly depends on how they are handled by the government, especially regarding assistance provided to locals most directly affected. To date, the record of countries involved in such schemes has been highly variable: Sudan has performed miserably, but others (including Moçambique) have shown considerable economic growth.

ENVIRONMENT AND HEALTH

The study of human health in spatial context is the field of **medical geography [11]**, and medical geographers employ modern methods of analysis (including geographic information systems) to track disease outbreaks, identify their sources, detect their carriers, and prevent their recurrence. Alliances between medical personnel, epidemiological researchers, and geographers have already yielded significant results. Doctors know how a disease ravages the body; geographers know how climatic conditions such as wind direction or variations in river flow can affect the distribution and effectiveness of disease carriers. This collaboration helps protect vulnerable populations.

Tropical Africa is the source area of many serious illnesses and has thereby become the focus of much research in medical geography. Investigators look at the carriers (*vectors*) of infectious diseases, the environmental conditions that give rise to them, and also the cultural and social geography of disease dispersion and transmission. Comparing medical, environmental, and social/cultural maps can integrate crucial evidence to help combat the scourge.

In Africa today, hundreds of millions of people carry one or more maladies, often without knowing exactly what ails them. A disease that infects many people (the *hosts*) in a kind of equilibrium, without causing rapid and widespread deaths, is said to be **endemic [12]** to that population. People affected may not die suddenly or dramatically, but their quality of life and productive capacity are hindered as their overall health is weakened and can deteriorate rapidly when a more acute illness strikes. In tropical Africa, hepatitis, venereal diseases, and hookworm are among many public-health threats in this category.

Epidemics and Pandemics

When a disease outbreak has local or regional dimensions, it is called an **epidemic [13]**. It may claim thousands, even tens of thousands, of lives, but it remains confined to a certain area, perhaps one defined by the range of its vector. In tropical Africa, trypanosomiasis, the disease known as sleeping sickness and vectored by the tsetse fly, has regional dimensions (Fig. 6A-9). The extensive herds of savanna wildlife form the *reservoir* of this disease, and the tsetse fly transmits it both to livestock and people. It is endemic to wildlife, but it also kills cattle, so Africa's herders try to keep their animals in tsetse-free zones. African sleeping sickness appears to have originated in a West African source area during the fifteenth century, and from there it spread throughout much of tropical Africa. Its

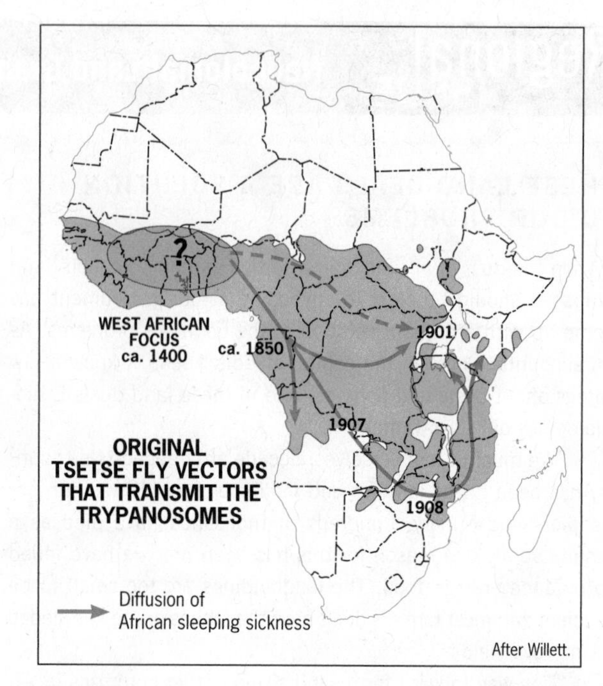

FIGURE 6A-9 © H. J. de Blij, P. O. Muller, and John Wiley & Sons, Inc.

epidemic range was limited by that of the tsetse fly: where there are no tsetse flies, there is no sleeping sickness. More than anything else, the tsetse fly has kept substantial parts of Subsaharan Africa's savannas free of livestock and open to wildlife.

When a disease spreads worldwide, it is described as a **pandemic [14]**. Africa's and the world's most deadly vectored disease is malaria, transmitted by a mosquito and the killer of at least 1 million people each year. Eradication campaigns against the mosquito vector have had some success, but always the carrier has come back with renewed vigor. At present, as many as 300 million people are affected by malaria globally. Although current efforts at combatting the disease through increased use of mosquito nets appears to be having good results, most of the million-plus annual deaths are African children under the age of five. Indeed, the short life expectancies for the realm's tropical countries listed in Appendix B partly reflect this mortality from malarial infection.

Despite these 'natural' challenges associated with the environment, there has been remarkable progress in battling various diseases in recent years. This is especially evident in the sharp decline of infant mortality rates across the realm, averaging 4 percent annually over the past decade. Yet even though one in ten African children still does not reach the age of five, these rates are now declining faster than any other realm has ever experienced. Three countries—Kenya, Senegal, and Rwanda—have seen their rates drop dramatically by more than 8 percent for several years in a row. This indicates that the spreading of infectious diseases is steadily coming under control—through vaccinations, education, and especially the widespread introduction of 'insecticide-treated bednets' (a simple, cheap, yet highly effective innovation that has found its way into even the most remote villages all across the realm).

The Battle Against AIDS

Malaria remains Africa's deadliest disease to this day, but since 1980 AIDS has dominated the medical news from this part of the world. AIDS first erupted in Subsaharan Africa and quickly became a global pandemic; no geographic realm has been spared.

AIDS stands for Acquired Immune Deficiency Syndrome, the body's failure to protect itself against a virus. That virus, for want of a better name at the time researchers were trying to identify it, is called the Human Immunodeficiency Virus (HIV). Thus the disease is properly called HIV/AIDS. Since it was first recognized in the early 1980s, more than 70 million people worldwide have contracted HIV/AIDS and about half that number have died of it, as many as 80 percent of them Africans.

By the early 1990s, HIV/AIDS had spread most virulently in Equatorial and East Africa, and medical geographers referred to an "AIDS Belt" from DRCongo* to Kenya. A decade later, however, the worst-afflicted countries lay in Southern Africa. In 2012, 23.5 million HIV-positive people were living in Subsaharan Africa, 69 percent of the world's total. The largest HIV-infected national populations were in South Africa (about 6 million) and Nigeria (almost 3 million); ranked next were Moçambique, Tanzania, and Zimbabwe, with each recording about 1.5 million victims. In certain other countries, such as DRCongo, the outbreak of the disease is known to be severe: there, at least 1.3 million people are believed to be infected, but reliable estimates have not been available in recent years. When standardized for population size, the dire situation of several countries becomes even clearer: around 25 percent of the population aged 15–49 are HIV-positive in Lesotho, Swaziland, and Botswana; 18 percent in neighboring South Africa; and close to 14 percent in Zimbabwe, Zambia, and Namibia.

More than 60 percent of all those infected are women, reflecting cultural and social circumstances. Overall, no part of tropical Africa has been spared. Life expectancies plummeted. Children by the millions were orphaned. Companies lost workers and were unable to replace them. National economies contracted. Associated costs—benefits, treatment, medicines—skyrocketed.

Why is Africa suffering so disproportionately? First, HIV/AIDS originated in tropical African forest margins and rapidly spread through all segments of society. Second, the social stigma associated with HIV/AIDS, which is sexually transmitted, makes acknowledging and treating it especially problematic. Third, life-prolonging medications are expensive and particularly difficult to provide in remote rural areas. Fourth, governmental leadership during the AIDS crisis has varied from highly aggressive and effective (as in Uganda, where political and medical officials cooperated in a massive campaign to distribute free condoms and advocate their use) to catastrophically negligent, as in South Africa where government ministers for a time misled the public and unnecessarily delayed the mass medical intervention that was critically needed.

The situation remained serious in 2013, but some important progress in the battle against HIV/AIDS was being made. A new South African government continues to aggressively address the AIDS crisis. Public-health campaigns elsewhere are having a beneficial effect. Lower-cost, generic anti-HIV medicines are today becoming more widely available through international help. The number of new infections across the realm in 2012 dropped to 1.7 million, down from 2.4 million in 2001—a decline mainly due to better education and the growing use of condoms. Those who carry the disease also live longer as result of improved treatment and the wider availability of medication—even though many still do not have access to it. Most encouraging, more than 6 million Africans were receiving antiretroviral therapy in 2012, up from only 100,000 in 2003.

CULTURAL PATTERNS

We tend to think of Africa in terms of its prominent countries and well-known cities, its development problems and political dilemmas, but Africans themselves have another perspective. The colonial period created states and capitals, introduced foreign languages to serve as the *linguae francae*, and brought railways and roads. The colonizers stimulated labor movements to the mines they opened, and they disrupted many other migrations that had been part of African life for many centuries. But they did not change the ways of life of most of the people. Despite accelerating urbanization, 63 percent of the realm's population still live in, and work near, Africa's hundreds of thousands of villages. They speak one of well over a thousand languages in use in this realm. The villagers' concerns are local; their focus is on subsistence, health, and safety. They worry that the conflicts over regional power and/or political ideology will engulf them, as has happened to tens of millions in Liberia, Sierra Leone, Ethiopia, Rwanda, DRCongo, and Angola since the 1970s. Africa's (numerically) largest cultural groups form major nations, such as the Yoruba of Nigeria and the Zulu of South Africa. Africa's smallest groups of people number just a few thousand. As a geographic realm, Subsaharan Africa has the most complex and fragmented cultural mosaic on Earth (see Fig. 6A-5).

*Two countries in Africa have the same short-form name, *Congo*. In this book, we use ***DRCongo*** for the much larger Democratic Republic of the Congo, and ***Congo*** for the smaller Republic of the Congo just to its northwest.

African Languages

Africa's linguistic geography is a key component of that cultural intricacy. Most of Subsaharan Africa's more than one thousand languages do not have a written tradition, making classification and mapping difficult. Scholars have attempted to delimit an African language map, and Figure 6A-10 is a composite of their efforts. One feature is common to all language maps of this continent: the geographic realm begins approximately where northern Africa's Afro-Asiatic language family (mapped in yellow in Fig. 6A-10) gives way, although the pattern is sharper in West Africa than to the east.

In Subsaharan Africa, the dominant language family is the Niger-Congo family, of which the Kordofanian subfamily is a small, historic, far northeastern outlier (Fig. 6A-10), and the Niger-Congo languages carry the other subfamily's name. This subfamily (mapped in purple) extends across the realm from West to East and Southern Africa. The Bantu language forms the largest branch of this subfamily, but Niger-Congo languages in West Africa, such as Yoruba and Akan, also have millions of speakers. Another important language family is the Nilo-Saharan family (mapped in orange), extending from Maasai in Kenya northwest to Teda in Chad. No other language families are of similar extent or importance: the Khoisan family, of ancient origins, now survives among the dwindling Khoi and San peoples of the Kalahari in the southwest; the small white minority in South Africa speak Indo-European languages; and Malay-Polynesian languages prevail in Madagascar, which was peopled from Southeast Asia before Africans reached it.

The Most Widely Used Languages

About 40 African languages are spoken by 1 million people or more, and a half-dozen by 10 million or more: Hausa (43 million), Yoruba (22 million), Ibo, Swahili, Lingala, and Zulu. Although English and French have become important *linguae francae* in multilingual countries

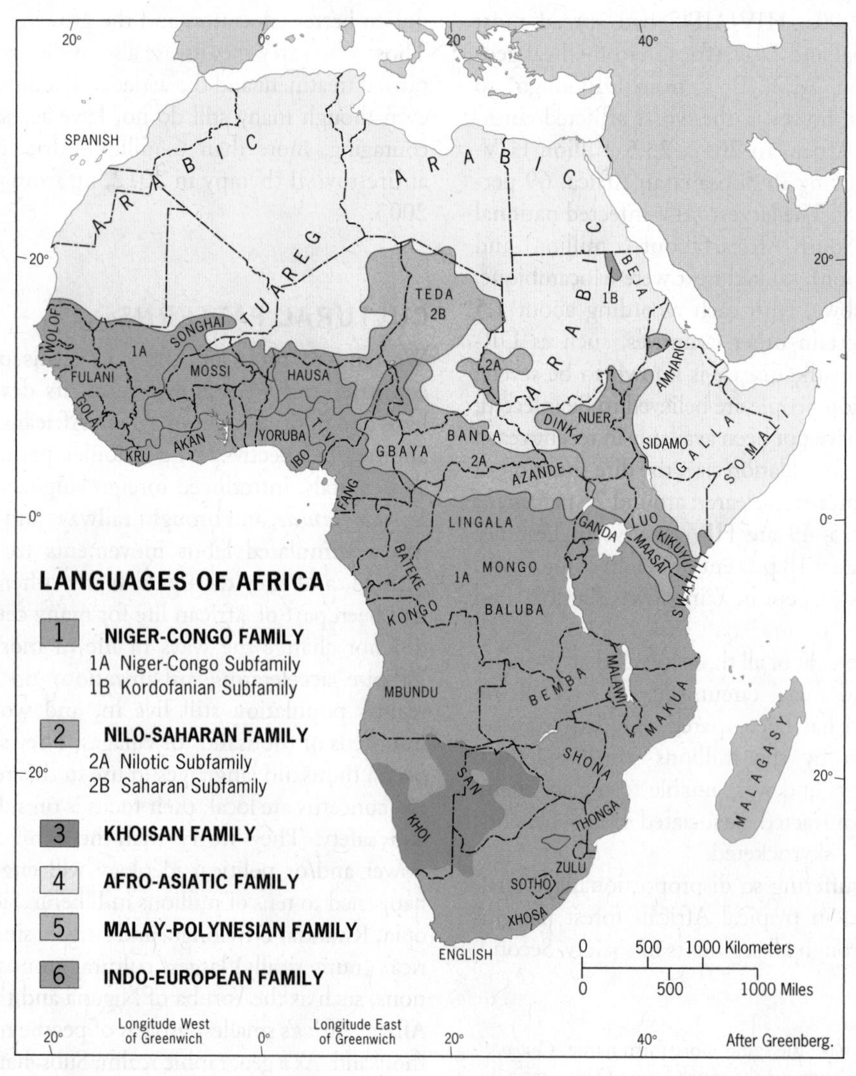

FIGURE 6A-10

© Pascal Maitre/Cosmos/Aurora Photos, Inc.

The faithful kneel during Friday prayers outside a mosque in Kano, northern Nigeria. The survival of Nigeria as a unified state is an African success story; the Nigerians have overcome strong centrifugal forces in a multiethnic country that is dominantly Muslim in the north, Christian-animist in the south. In the 1990s, some Muslim clerics began calling for an Islamic Republic in Nigeria, and after the death of the dictator Abacha and the election of a non-Muslim president, the Islamic drive intensified. A number of Nigeria's northern States adopted Sharia (strict Islamic) law, which led to destructive riots between the majority Muslims and minority Christians who felt threatened by this turn of events—a situation that continues to deteriorate (elaborated in Chapter 6B). Can Nigeria avoid the fate of Sudan?

such as Nigeria and Ivory Coast (where officials even insist on spelling the name of their country *Côte d'Ivoire* in the Francophone manner), African languages also serve this purpose. Hausa is a common language across the West African savanna; Swahili is widely used in East Africa. And pidgin languages—mixtures of African and European tongues—are found along West Africa's coast; millions of Pidgin English (called *Wes Kos*) speakers use this medium in Nigeria and Ghana. However, not all African languages are going to survive: Nigeria alone has 17 endangered (near-extinct) languages, and in the continent at large as many as 63 languages are on the verge of extinction.

Language and Culture

Multilingualism [15] can be a powerful centrifugal force in society, and African governments have tried with varying success to establish national alongside local languages. Nigeria, for instance, made English its official language because none of its 500-odd languages, not even Hausa, had sufficient internal interregional use. But using a European, colonial language as an official medium invites criticism, and Nigeria remains divided on this issue. On the other hand, making a dominant local language official would give rise to negative reactions from ethnic minorities. Language remains a potent force in Africa's cultural life.

Religion in Africa

Africans had their own belief systems long before Christians and Muslims arrived to convert them. And for all of Subsaharan Africa's cultural diversity, its people had a consistent view of their place in nature. Spiritual forces, according to African tradition, are manifest everywhere in the natural environment (a religious world view known as ***animism***), not in a supreme deity that exists in some remote place. Thus gods and spirits affect people's daily lives, witnessing every move, rewarding the virtuous and punishing (through injury or crop failure, for example) those who misbehave. Ancestral spirits can inflict misfortune on the living. They are everywhere: in the forest, rivers, and mountains.

As with land tenure, the religious views of Africans clashed fundamentally with those held by the colonizers. Monotheistic ***Christianity*** first penetrated Africa in the northeast when Nubia and Axum were converted, and Ethiopia has been a Coptic Christian stronghold since the fourth century AD. But the Christian churches' real invasion did not commence until the onset of colonialism after the turn of the sixteenth century. Christianity's various denominations made inroads in different locales: Roman Catholicism in much of Equatorial Africa, mainly at the behest of the Belgians; the Anglican Church in British colonies; and Presbyterians and others elsewhere. And these days, as in South America, Evangelical churches continue to rapidly gain adherents.

Some of those churches are now far more conservative than their counterparts in Europe and North America, even those appealing to the most conservative congregations in the United States. A split in the U.S. Episcopal (Anglican) Church regarding the treatment of gays and lesbians has some congregations aligning themselves under the Bishop of Uganda. But almost everywhere, Christianity's penetration led to a blending of traditional and Christian beliefs, so that much of Subsaharan Africa is nominally, though not exclusively, Christian (see Fig. G-10). Go to a church in Gabon or Uganda or Zambia, and you may hear drums instead of church bells, sing African music rather than hymns, and see African carvings alongside the usual statuary.

Islam had a rather different arrival and impact. Today, about 32 percent of the realm's population is Muslim. By the time of the colonial invasion, Islam had advanced out of Arabia, across the Sahara, and part-way down the coasts of Africa (particularly along the Indian Ocean). Muslim clerics converted the rulers of African states and commanded them to convert their subjects. They Islamized the savanna states and penetrated into present-day northern Nigeria, Ghana, and Ivory Coast. They encircled and isolated Ethiopia's Coptic Christians and Islamized the Somali people in Africa's Horn. They established beachheads on the coast of Kenya and took over offshore Zanzibar. Arabizing Islam and European Christianity competed for African minds, but Islam proved to be a far more pervasive force. The tension between Islam and Christianity continues in the twenty-first century (see photo), and is one of the defining regional characteristics of the African Transition Zone elaborated Chapter 6B.

URBANIZATION AND SOCIAL CHANGE

As can be discerned in Appendix B, Subsaharan Africa remains the least urbanized world realm, but people are moving to its towns and cities at an accelerating pace. Today, 37 percent of the Subsaharan African population resides in urban settlements. This means that over 325 million people (more than the entire current population of the United States) now live in towns and cities, of which many were founded and developed by the colonial powers. But the infrastructure of the cities has been unable to keep up with the tide of incoming migrants.

The biggest African cities became centers of embryonic national core areas, and many of them incorporated government headquarters. This **formal sector [16]** of the city used to be the dominant one, with governmental control and regulations affecting civil service, business, industry, and all of their workers; these districts were endowed with the largest and most modern buildings. Today, however, African cities look different. From a distance, the skyline resembles that of a modern urban center. But in the streets, on the sidewalks right below the shop windows, there are hawkers, basket weavers, jewelry sellers, garment makers, wood carvers—a second economy, most of it beyond governmental control. This **informal sector [17]** now dominates most African cities. It is peopled by the rural immigrants, who also work as servants, apprentices, construction workers, and in countless other menial jobs. Millions of these migrants cannot find formal employment and are condemned to living in squalid circumstances in the squatter-settlement tracts around (as well as inside) nearly all of Africa's cities.

A typical urban landscape in Subsaharan Africa. This aerial view of the city of Arusha (population: ca. 550,000) in northern Tanzania shows a signature feature of the settlement layout of urban Africa: an irregular pattern of low-rise, low-density development. It reflects a general lack of planning, complex land ownership issues, and piecemeal as well as patchwork investment in construction.

EMERGING AFRICA

Despite colonial legacies, formidable environmental challenges, and a long history of serious adversity, Africa in the early twenty-first century is taking a turn for the better. Although conditions remain harsh in many places the achievements of the past decade are quite encouraging, even stunning in some cases. We have already noted some of these successes, including a substantial decline in the infant mortality rate and progress toward containing malaria and HIV/AIDS. As a more general indicator of health and well-being, life expectancy overall has increased by about 6 years since 2000. And there is more.

Annual economic annual growth in the realm over the past decade has averaged 5-6 percent and for certain countries it has been higher than that (Fig. 6A-11). As a whole, this realm outperformed all other realms except East Asia, and as of mid-2013 it exhibited the highest growth rate in the world; moreover, among the ten fastest growing national economies since 2005, no fewer than six were located here. Notwithstanding Subsaharan Africa's low level of development in absolute terms, it has been moving upward with remarkable speed. And with higher national incomes as well as an expanding tax base, government spending and investments have increased accordingly, especially in the spheres of public health and education (secondary school enrollment has grown by almost 50 percent during the last ten years).

However, as Figure 6A-11 shows, economic performance across this realm has been highly uneven. The fastest growing countries tend to located on the coast, away from the troubled African Transition Zone across the north, and are generously endowed with natural resources. The poorest performers are first and foremost the victims of war or political mismanagement: South Sudan's economy shrank by about a third in just two years, from 2010 to 2012, and Zimbabwe had an average negative annual growth of nearly −3 percent.

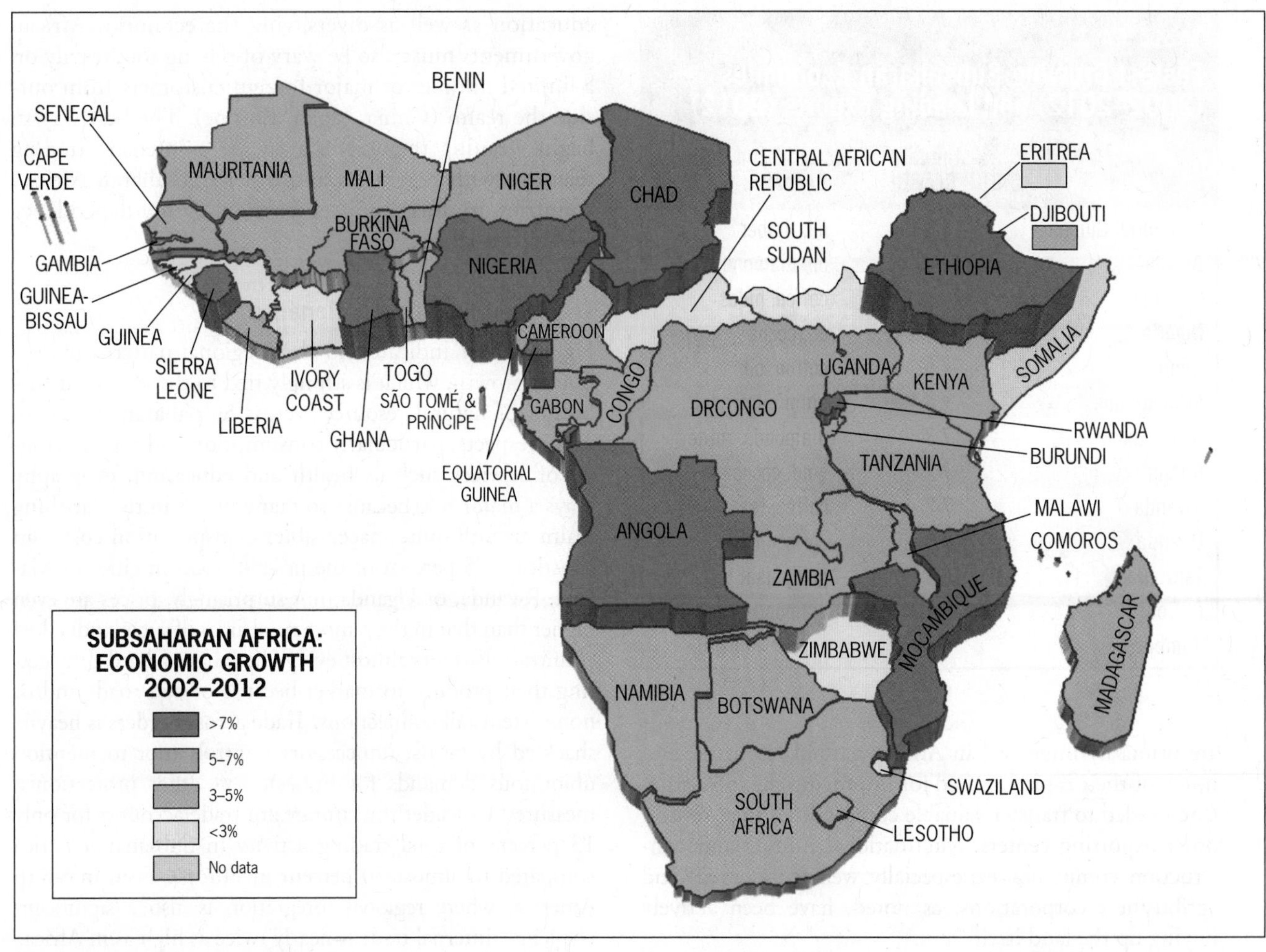

FIGURE 6A-11 © H. J. de Blij, P. O. Muller, J. Nijman, and John Wiley & Sons, Inc.

Middle-class youths at a cultural festival in Kampala, Uganda, during the summer of 2012. Demands for education, cultural goods, computer and Internet use, and access to leading entertainment media are surging, and upscale consumption is on the rise in many of Subsaharan Africa's cities.

© Yannick Tylie/Corbis

But, overall, the economy in most of the realm shows more vibrancy than ever and, with the recent subsidence of wars and violent conflict, democracy is taking hold as well. A number of countries that used to be notorious for perennial strife, including Angola, Sierra Leone, and Chad, have turned a corner and stabilized. Democratic advances, in the form of open elections and a free press, have been especially notable in parts of Southern and West Africa. Political dislocation, repression, and dysfunction do persist in various countries, of which Zimbabwe is the most glaring example—but the Mugabe regime in Harare is fast becoming an anomaly.

Vulnerable Growth

Subsaharan Africa's fate in modern times has been tied to foreign intervention and outside interests. That is still true today as the realm is being drawn into the global economy. Foreign direct investment between 2002 and 2012 more than tripled from U.S. $11 billion to $35 billion. Outside investors

TABLE 6A-1

Fastest Growing Subsaharan Economies (2002–2012) and Leading Exports (2012)

	% Annual GDP Growth	1st and 2nd Biggest Export Products
Equatorial Guinea	10.8	oil, timber
Angola	10.6	oil, diamonds
Ethiopia	8.9	coffee, hides
Nigeria	7.5	oil, cocoa
Chad	7.5	cotton, oil
Moçambique	7.4	aluminum, coal
Sierra Leone	7.3	diamonds, rutile
Ghana	7.2	gold, cocoa
Rwanda	7.2	coffee, tea
Uganda	7.0	coffee, fish
Tanzania	6.9	gold, sisal
DRCongo	6.2	diamonds, copper
Zambia	6.1	copper, minerals

are primarily interested in Africa's natural resources, and much of their capital is used for improving the infrastructure needed to transfer valuable commodities to ports and other exporting centers. Multinational mining and construction companies are especially well represented, and agribusiness corporations, as noted, have been actively buying up the land itself.

Whereas foreign investments are needed and can make vital contributions to a developing economy, the concern is that they provide too little in terms of manufacturing or even processing of raw materials, the kind of economic activity that adds value to a stage of production. Table 6A-1 lists the fastest growing economies of Subsaharan Africa together with their two leading export products in terms of monetary value. In every case, these exports are raw materials, commodities that have fared well in the global marketplace in recent years, particularly because of massive demand from China.

Although Subsaharan Africa has been successfully riding the global commodity boom, there are drawbacks and heightening risks when a realm's economy is so one-sided. First, many resources are finite, and at some point mineral and energy reserves are going to be depleted. Second, heavy reliance on commodity exports means heavy dependence on world market prices that can fluctuate suddenly and substantially. Third, even if international commodity prices are high, the terms of trade tend to be unbalanced if an exporting country must import high-value-added manufactured goods such as cars and computers. And finally, raw-material exporting makes only a modest contribution to the employment economy because it involves minimal domestic processing.

The bottom line here is that governments must focus on using their revenues, while they last, to invest in education as well as diversifying the economy. African governments must also be wary of relying too heavily on a limited number of major foreign customers from outside the realm (China, Japan, Europe). The best way to begin weaning themselves away is to intensify trading relations with neighbors and other Subsaharan African countries to forge greater economic interdependence within this realm.

The Need for Supranationalism

Figure 6A-11 indicates no clear regional patterns of economic growth, which is strongly tied to the uneven distribution of natural resources across Subsaharan Africa. In other respects, particularly consumption and the availability of services such as health and education, geography plays a major role because so many places in this far-flung realm are still quite inaccessible. Transportation costs can constitute 75 percent of the price of food in cities in Malawi, Rwanda, or Uganda; not surprisingly, prices are even higher than that in the remote rural areas of such landlocked countries. Farmers almost everywhere have a hard time getting their produce to market because of bad roads and/or nonexistent rail connections. Trade across borders is heavily shackled by tariffs, unnecessary controls (not to mention ubiquitous demands for bribes), and other protectionist measures. Consider this: intrarealm trade accounts for only 13 percent of total trading activity in Subsaharan Africa compared to almost 70 percent in Europe. Even in South America, where regional integration is also a significant challenge, internal trade is nearly twice as high as in Africa.

To overcome such disadvantages, African states must improve international cooperation, continentwide as well as regionally. The Organization of African Unity (OAU) was established for this purpose in 1963 and was superseded by the African Union (AU) in 2001. In 1975, the Economic Community of West African States (ECOWAS) was established by 15 countries to promote trade, transportation, industry, and social affairs in that region. And in the early 1990s, important steps were taken when 12 countries joined in forming the Southern African Development Community (SADC) and when the Common Market for Eastern and Southern Africa (COMESA) was created. Hence, the institutional structures for integration exist (indeed, there is a need to streamline the various initiatives) and exports across the realm are today at an all-time high. Now it is a matter of summoning the considerable political will to create lasting regional cooperation and realmwide interdependence.

If there is one geographic realm that is subject to stereotypical judgment, it is Subsaharan Africa. All too often, this one is portrayed as a single, monolithic, economic basket case, where dysfunctional government goes hand in hand with economic underperformance, frequent famines, and aimless violence. In the 2010s, that characterization couldn't be further off the mark. Subsaharan Africa not only exhibits enormous geographic, cultural, and economic diversity but also harbors the world's newest emerging market. Economic forecasts tell of another decade lying ahead in which growth and development here is likely to outstrip that of all other realms. In the early 1960s, when much of Africa became politically independent, there was short-lived talk about the 'wind of change' bringing self-determination to these new countries. Instead, adversity was the overwhelming fate of so many of the realm's inhabitants. Now, after an especially traumatic half-century, the 'wind of change' is once again gathering force, prospects are brightening, and Subsaharan Africa is evolving into a realm of hope.

POINTS TO PONDER

- Shipping a new car from China to Tanzania costs less than hauling it from Tanzania to neighboring Uganda.
- Subsaharan Africa is home to 69 percent of all HIV/AIDS cases in the world.
- The realm contains about four-dozen countries; the number of languages spoken within it is estimated to be between 1300 and 2000.
- China's deepening impact on Subsaharan Africa is expanding steadily through activities ranging from trade to investment to education. Brazil, too, is increasingly making its presence felt here.
- At the end of the first two decades of this century, Subsaharan Africa is likely to have recorded the fastest economic growth among all the world geographic realms.

SUBSAHARAN AFRICAN REGIONS

PORTUGAL
SPAIN
ITALY
GREECE
TURKEY
CYPRUS
SYRIA
LEBANON
ISRAEL
JORDAN
IRAQ
IRA
KUWAIT
BAHRAIN
QA
SAUDIA ARABIA
YEMEN
Mediterranean Sea
Red Sea
MOROCCO
TUNISIA
ALGERIA
LIBYA
EGYPT
WESTERN SAHARA
MAURITANIA
NIGER
MALI
CHAD
SUDAN
ERITREA
DJIBOUTI
SENEGAL
GAMBIA
GUINEA
GUINEA-BISSAU
SIERRA LEONE
LIBERIA
IVORY COAST
BURKINA FASO
GHANA
TOGO
BENIN
NIGERIA
Niger R.
Nile R.
Lake Chad
CAMEROON
CENTRAL AFRICAN REPUBLIC
SOUTH SUDAN
ETHIOPIA
SOMALIA
EQUATORIAL GUINEA
GABON
CONGO
Congo R.
DRCONGO
UGANDA
KENYA
RWANDA
BURUNDI
Lake Victoria
Lake Tanganyika
TANZANIA
CABINDA (Angola)
ATLANTIC OCEAN
INDIAN OCEAN
Equator
ANGOLA
ZAMBIA
Zambezi R.
MALAWI
MOCAMBIQUE
MADAGASCAR
ZIMBABWE
NAMIBIA
BOTSWANA
Tropic of Capricorn
SWAZILAND
Vaal R.
Orange R.
LESOTHO
SOUTH AFRICA
15°
30°
0°
45°

REGIONS OF SUBSAHARAN AFRICA

- Southern Africa
- East Africa
- Equatorial Africa
- West Africa
- African Transition Zone
- Approximate southern boundary of Islam

0 600 1200 Kilometers
0 300 600 Miles

REGIONS

Southern Africa

East Africa

Equatorial Africa

West Africa

African Transition Zone

IN THIS CHAPTER

- South Africa stumbling
- Angola's continuing rise
- Unrest in Nigeria's north
- China in Ethiopia
- Somalia's political patchwork

CONCEPTS, IDEAS, AND TERMS

Apartheid	1
Exclave	2
Landlocked state	3
Sharia law	4
Enclave	5
Periodic market	6
Islamic Front	7
Choke point	8
Failed state	9

FIGURE 6B-1

© Jan Nijman

Where were these pictures taken?
Find out at www.wiley.com/college/deblij

On the face of it, Africa seems to be so massive, compact, and continuous that any attempt to justify a contemporary regional breakdown is doomed to fail. No deeply penetrating bays or seas create peninsular fragments as they do in Europe. No major islands (other than Madagascar) provide the broad regional contrasts we see in Middle America. Nor does Africa really taper southward to the peninsular proportions of South America. And Africa is not cut by an Andean or a Himalayan mountain barrier. Given Africa's colonial fragmentation and cultural mosaic, is regionalization possible? Indeed it is.

REGIONS OF THE REALM

Maps of environmental distributions, ethnic patterns, cultural landscapes, historic culture hearths, and other spatial data yield a four-region structure complicated by a fifth, overlapping zone as shown in Figure 6B-1. Beginning in the south, we identify the following regions:

1. ***Southern Africa***, extending from the southern tip of the continent to the northern borders of Angola, Zambia, Malawi, and Moçambique. Ten countries constitute this region, which extends beyond the tropics and whose giant is South Africa.
2. ***East Africa***, where equatorial natural environments are moderated by elevation and where plateaus, lakes, and mountains, some carrying permanent snow, exemplify the countryside. Six countries, including the highland part of Ethiopia, comprise this region. The island-state of Madagascar, marked by Southeast Asian influences, is neither Southern nor East African, but is included here because of its relative location.
3. ***Equatorial Africa***, much of it defined by the basin of the Congo River, where elevations are lower than in East Africa, temperatures are higher and moisture more

© Jan Nijman

MAJOR CITIES OF THE REALM

City	Population* (in millions)
Lagos, Nigeria	12.7
Kinshasa, DRCongo	9.9
Abidjan, Ivory Coast	4.8
Johannesburg, South Africa	4.0
Nairobi, Kenya	3.8
Cape Town, South Africa	3.8
Dakar, Senegal	3.4
Dar es Salaam, Tanzania	3.4
Ibadan, Nigeria	3.3
Durban, South Africa	3.2
Adis Abeba, Ethiopia	3.2
Accra, Ghana	2.9
Lusaka, Zambia	2.1
Harare, Zimbabwe	1.6
Mombasa, Kenya	1.1

*Based on 2014 estimates

abundant, and where most of Africa's surviving tropical rainforests remain. Among the nine countries that make up this region, which now includes newly formed South Sudan, DRCongo* dominates territorially and demographically.

4. ***West Africa***, which includes the countries of the western coast and those on the margins of the Sahara in the interior, a populous region anchored in the southeast by the realm's demographic giant, Nigeria. Fifteen countries form this crucial African region.

5. ***The African Transition Zone***, the complicating factor on the regional map of Africa. In Figure 6B-1, note that this striped zone of increasing Islamic influence completely dominates some countries (e.g., Somalia in the east and Senegal in the west) while cutting across others, in the process creating Islamized northern areas and non-Islamic southern zones (e.g., Nigeria, Chad, Ivory Coast). Note, too, that an elongated strip of Islamic dominance extends southward from the African Horn, lining the Indian Ocean coastlands of both Kenya and Tanzania.

SOUTHERN AFRICA

Southern Africa, as a geographic region, consists of all the countries and territories lying south of Equatorial Africa's DRCongo and East Africa's Tanzania (Fig. 6B-2). Thus defined, the region extends from Angola and Moçambique (on the Atlantic and Indian Ocean coasts, respectively) to South Africa and includes a half-dozen landlocked states. Also marking the northern limit of the region are Zambia and Malawi. Zambia is nearly cut in half by a long land extension from DRCongo, and Malawi penetrates deeply into Moçambique.

Africa's Resource-Rich Region

Southern Africa constitutes a geographic region in both physiographic as well as human terms. Its northern zone marks the southern limit of the Congo Basin in a broad upland that stretches across Angola and into Zambia (the tan corridor extending eastward from the Bihe Plateau in Fig. 6A-2). Lake Malawi is the southernmost of the East African rift-valley lakes; Southern Africa has none of East Africa's volcanic and earthquake activity. Most of this region is plateau country, and the Great Escarpment is prominent here. There are two consequential river systems: the Zambezi (which forms the border between Zambia and Zimbabwe) and the Orange-Vaal (South African rivers that combine to demarcate southern Namibia from South Africa).

Southern Africa has long been the continent's richest region materially. A great zone of mineral deposits extends through the heart of the region from Zambia's Copperbelt through Zimbabwe's Great Dyke and South Africa's Bushveld Basin and Witwatersrand to the goldfields and diamond mines of Free State and Northern Cape provinces in the heart of South Africa. Ever since these minerals began to be exploited in colonial times, many migrant laborers have come to work in the mines. Major on- and offshore oil reserves boost the economy of Angola. The region's agricultural diversity matches its mineral wealth, especially so in South Africa: vineyards drape the slopes of South Africa's Cape Ranges and the country's relatively high latitudes and its range of altitudes create environments for apple orchards, citrus groves, banana plantations, pineapple farms, and many other crops.

Despite this considerable wealth and potential, not all of the ten countries of Southern Africa have prospered and some are on divergent tracks (see also Fig. 6A-11). As Figure G-11 shows, Angola has moved into the World Bank's upper-middle-income group but three other countries remain mired in the low-income category (Malawi, Moçambique, and Zimbabwe); and even though South Africa, Lesotho, Swaziland, Zambia, Namibia, and Botswana fall into the two middle-income ranks, the (desert-dominated) last two rank among the realm's three most sparsely populated countries. Zimbabwe remained the basket case of this region—and of the realm at large—in 2013. South Africa has long been the realm's most important country by many measures, at least in terms of size and dominance. But South Africa's economy is suffering from a combination of ailments.

South Africa

The Republic of South Africa (RSA) is the giant of Southern Africa, its economy still the largest of this entire realm.

---

*Two countries in Africa have the same short-form name, *Congo.* In this book, we use ***DRCongo*** for the much larger Democratic Republic of the Congo, and ***Congo*** for the smaller Republic of Congo just to its northwest.

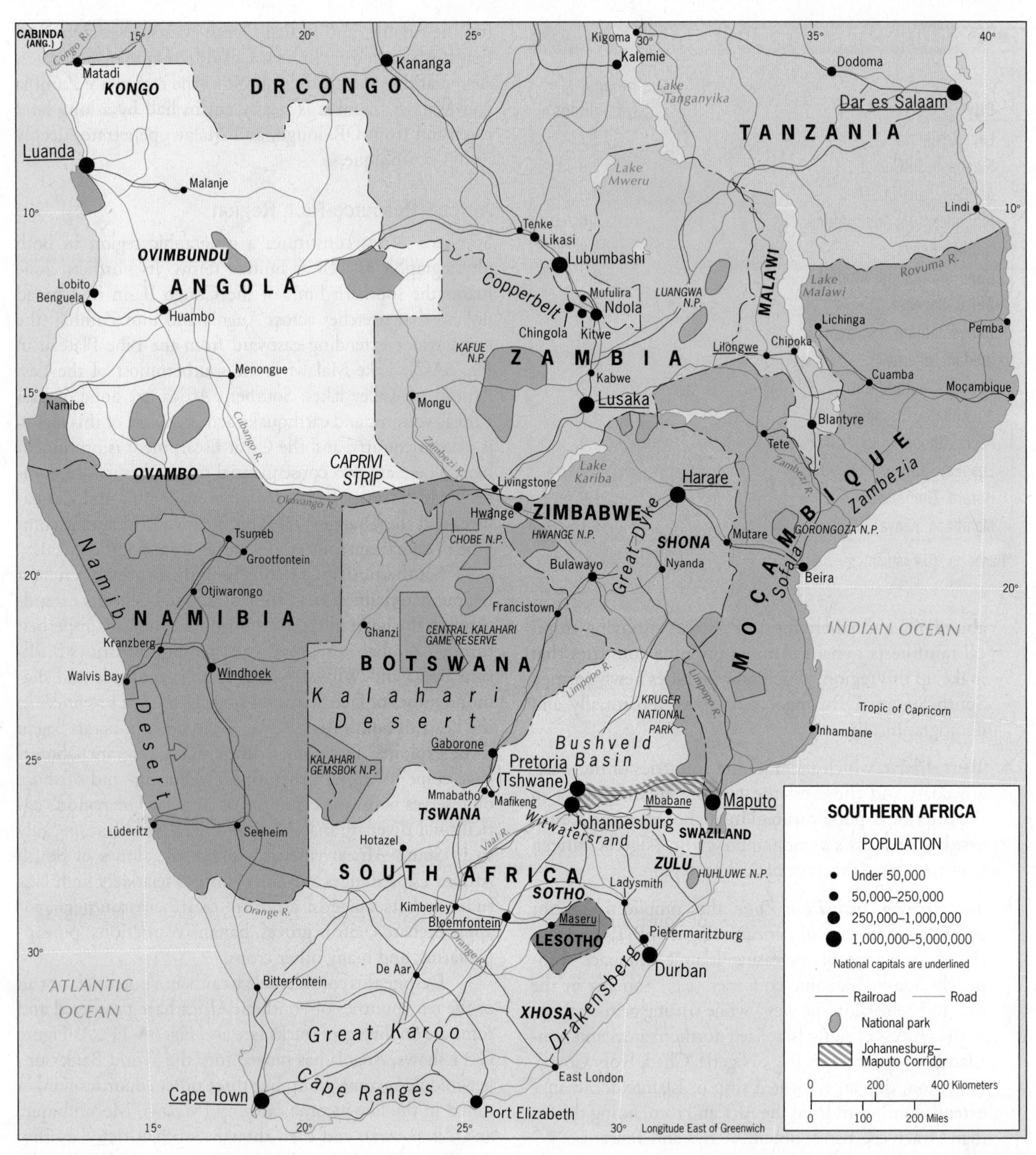

FIGURE 6B-2

Its historical geography differs somewhat from much of the rest of Subsaharan Africa. The country's lands were fought over by various African nations before the Europeans arrived and the colonial "scramble for Africa" took place. Peoples migrated southward—first the Khoisan-speakers and then the Bantu peoples—into the South African *cul-de-sac*. The Zulu and Xhosa nations fought over lands at about the time the first European settlers arrived.

South Africa is one of the most strategic places on Earth, the gateway from the Atlantic to the Indian Ocean, a key source of provisions on the route to Asia's riches. The Dutch East India Company founded Cape Town as early as 1652, and the Hollanders and their descendants, known as ***Boers***, have been a part of the South African cultural mosaic ever since. The British took over about 150 years later, and both colonial powers vied for control throughout South Africa's early history as a state. The British came to dominate the Cape, while the Boers trekked into the South African interior and, on the high plateau they called the *highveld*, founded their own republics. A war ensued, but by 1910 the Boers and the British had negotiated a power-sharing arrangement, although the Boers eventually achieved dominance that lasted from 1948 to 1994. Having long since shed their European ties, they came to call themselves ***Afrikaners***, their word for Africans.

For more than 40 years, between about 1950 and the mid-1990s, multicultural South Africa was in the grip of the world's most notorious racist policy, **Apartheid [1]**—the word itself means "apartness", but in practice it involves strict racial segregation and severe discrimination. Out of the concept of Apartheid grew a notion, promulgated by the white (European) minority then in control of the state, known as ***separate development.*** This would apply Apartheid to the entire country, carving it up into racially-based entities whose inhabitants would be citizens of those ethnic domains, but not of South Africa as a whole. Predictably, such racist social engineering aroused strong opposition within South Africa and beyond, leading to worldwide condemnation and international sanctions.

© H.J. de Blij

South Africa's Cape Town, whose magnificent site is dominated by Table Mountain, was founded by the Dutch in 1652 as a waystation for empire building in Southeast Asia's East Indies.

South Africa seemed headed for a violent revolution, but disaster was averted by what was, at the time, an almost inconceivable turn of events. A leader of the white-minority government that for decades had ruthlessly pursued its Apartheid policies, and a revered leader of the multicultural majority who had for 28 years languished in an island prison not far from Cape Town, struck an accord that, in effect, created a new South Africa virtually overnight. Nelson Mandela walked out of prison a free man in February 1990, and in the Republic of South Africa's first democratic election in 1994 Mandela became the country's president on a platform representing the African National Congress (ANC), the multiracial anti-Apartheid movement that had fought the policy for many years. The ANC's old foes, the architects of Apartheid, were seated as part of the loyal opposition in a parliament that was now indeed a rainbow assembly. Before 1994 South Africa had been divided into four provinces, and this was replaced with a new federal arrangement consisting of nine provinces (see Fig. 6B-3, including the inset).

The Ethnic Mosaic

In addition to the various indigenous African nations and the Europeans who settled in South Africa came peoples from Asia. The Dutch brought thousands of Southeast Asians to the Cape to serve as domestics and laborers. The British imported laborers from their South Asian colonies to work on sugarcane plantations, adding further cultural diversity to this multiethnic state. Moreover, a substantial population of mixed ancestry, clustered at the Cape, constitutes today's so-called *Coloured* sector of the country's citizenry. In the process, South Africa became Africa's most pluralistic and heterogeneous society. Even so, Africans now outnumber non-native Africans by about 5 to 1 (see Table 6B-1).

TABLE 6B-1
Demographic Data for South Africa

Population Groups	2014 Estimated Population (in millions)
African nations	**41.4**
Zulu	10.0
Xhosa	7.5
Tswana	3.3
Sotho (N and S)	3.3
Others (6)	17.3
Mixed (Coloureds)	**4.7**
African/White	4.5
Malayan	0.2
Whites	**4.6**
Afrikaners	2.9
English-speakers	1.6
Others	0.1
South Asian	**1.3**
Muslims	0.7
Hindus	0.6
TOTAL	**52.0**

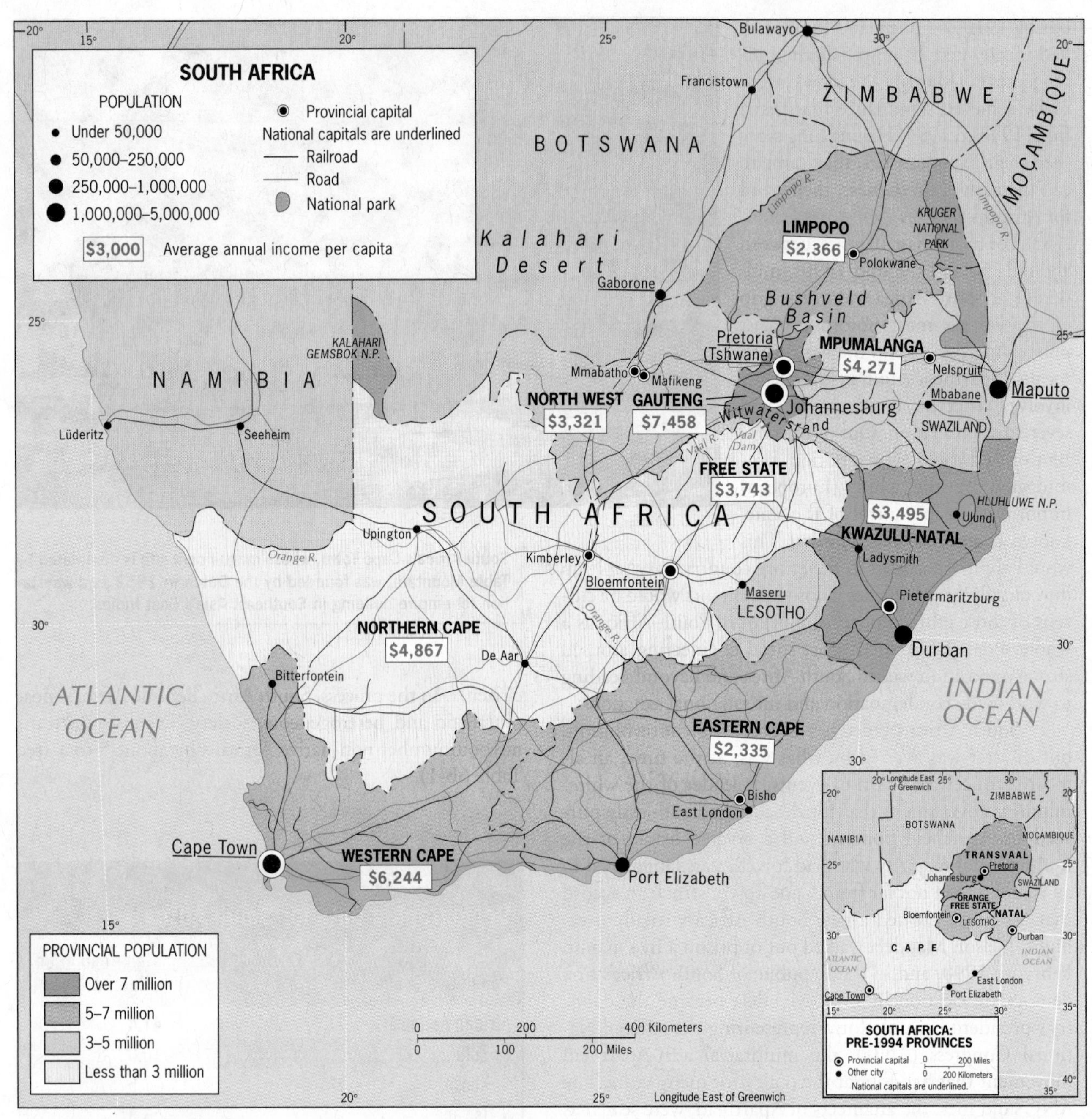

FIGURE 6B-3

Although heterogeneity marks the spatial demography of South Africa, regionalism pervades the human mosaic. The Zulu nation still is largely concentrated in today's Kwazulu-Natal Province (Fig. 6B-3). The Xhosa still cluster in the Eastern Cape, from the city of East London to the Kwazulu-Natal border and below the Great Escarpment. The Tswana still occupy ancestral lands along the border with Botswana. Cape Town remains the heartland of the Coloured population; Durban still has the strongest South Asian imprint. Travel through South Africa, and you will recognize the remarkable diversity of rural cultural landscapes as they change from Xhosa to Ndebele to Venda.

South Africa's Economic Geography

South Africa's economy is the largest in the realm and it has a major influence in its region. The country stretches from the warm subtropics in its far north to Antarctic-chilled waters in the south. With a land area in excess of 1.2 million square kilometers (470,000 sq mi) and a heterogeneous population of 52 million, South Africa contains the bulk of the region's minerals, most of its good farmlands, its largest cities, best harbors, most productive factories, and best developed transport networks. Mineral exports from Zambia and Zimbabwe move through South African ports. Workers from as far away as Malawi and as close as Lesotho work in South Africa's mines, factories, and fields.

Ever since diamonds were discovered at Kimberley in the 1860s, South Africa has been synonymous with minerals. Rail lines were laid from the coast to the diamond complex even as fortune seekers, capitalists, and tens of thousands of African workers, many from as far afield as Moçambique, streamed to the site. Subsequently, prospectors discovered what was long to be the world's greatest goldfield on a ridge called the Witwatersrand (Fig. 6B-3). There, Johannesburg soon became the gold capital of the world. During the twentieth century, additional goldfields were discovered in Orange Free State. Coal and iron ore were found in abundance, which gave rise to a major iron and steel industry. Other metallic minerals, which included chromium and platinum, yielded large revenues on world markets. Asbestos, manganese, copper, nickel, antimony, and tin were mined and sold; a thriving metallurgical industry developed in South Africa itself. During much of the twentieth century, capital flowed into the country, white immigration expanded, farms and ranches were laid out, and overseas markets multiplied.

Along the way, South Africa's cities grew rapidly. Johannesburg was no longer just a mining complex: it became an industrial agglomeration and a financial center as well. The old Boer capital, Pretoria, just a short distance north of the Witwatersrand, became the country's administrative center during Apartheid. In Orange Free State, substantial industrial growth (including oil-from-coal technology) matched the expansion of mining. As the core area centered on Johannesburg developed megalopolitan characteristics, most of the coastal cities also expanded. Durban's port served not only the

AMONG THE REALM'S GREAT CITIES . . .

JOHANNESBURG

SUBSAHARAN AFRICA IS urbanizing swiftly, but it still has only one true conurbation, and South Africa's Johannesburg lies at the heart of it. Little more than a century ago, Johannesburg was a small (though rapidly growing) mining town based on the newly discovered gold reserves of the Witwatersrand. In its early history, this future national core area was dominated by the ancestors of Dutch settlers and the still-remaining names on the map are a reminder of that era.

Today Johannesburg forms the focus of a conurbation (that anchors a province named *Gauteng*) of almost 8 million, extending from Pretoria in the north to Vereeniging in the south, and from Springs in the east to Krugersdorp in the west. The population of metropolitan Johannesburg itself passed the 3 million mark in 2003 (it now stands at 4.0 million), thereby overtaking Cape Town to become South Africa's largest urban agglomeration.

Johannesburg's skyline is the most impressive in all of Subsaharan Africa, a forest of skyscrapers reflecting the wealth generated here over the past century (see left photo). Look southward from a high vantage point, and you see the huge mounds of yellowish-white slag from the mines of the "Rand," the so-called mine dumps, partly overgrown today, interspersed with suburbs and townships. In a general way, Johannesburg developed as a white city in the north and a black city in the south. Well-known Soweto (SOuth WEstern TOwnships) lies to the southwest (see right photo). Houghton and other spacious, upper-class suburbs, formerly exclusively white residential areas, lie to the north.

Johannesburg has neither the scenery of Cape Town nor the climate and beaches of Durban. The city lies at an elevation of 1750 meters above sea level (5750 ft—more than a mile high), and its thin air often is polluted from smog created by motor vehicles, factories, mine-dump dust, and countless cooking fires in the townships and shantytowns that ring much of the metropolis.

Over the past century, the Johannesburg area produced nearly one-half of all the world's gold by value. Today Johannesburg lies at the heart of an industrial, commercial, and financial complex that forms South Africa's economic core.

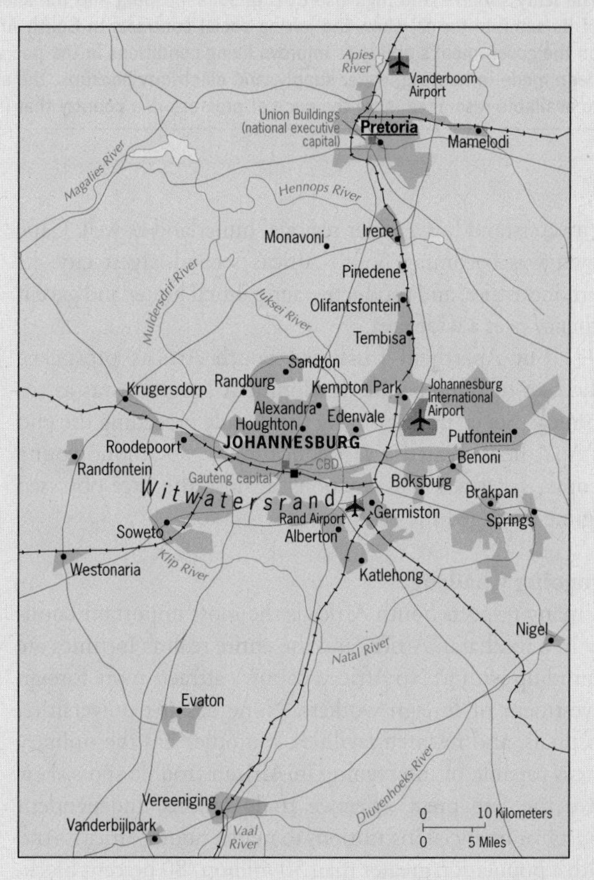

© H. J. de Blij, P. O. Muller, and John Wiley & Sons, Inc.

Martin Harvey/Gallo Images/Getty Images

© Sergio Pitamitz/Robert Harding World Imagery/Corbis

The leafy suburbs and high-rise CBD of Johannesburg and the shantytown housing of an especially poor section of crowded Soweto, one of its satellite towns, show the jarring social contrasts in South Africa's ever-more-unequal society. Future political stability will depend on the government's ability to improve living conditions in the poorest areas, rural as well as urban. Although significant advancement has been made in housing, water supply, and electricity hookups, the enormity of the problem, and constraints ranging from political realities to available resources, slow the pace of progress in a country that is in a race against time.

Witwatersrand but a wider regional hinterland as well. Cape Town was becoming South Africa's second-largest city; its port, industries, and productive agricultural hinterland gave it primacy over a wide area.

But Apartheid frustrated South Africa's prospects. The cost of the separate development program was astronomical. Social unrest during the decade preceding the end of Apartheid created a vast educational gap among young people. International sanctions against the race-obsessed regime damaged the economy.

Ongoing Challenges

In many respects South Africa is the most important country in Subsaharan Africa, and the entire realm's fortunes are bound up with it. No African country attracts more foreign investment or foreign workers. None has the universities, hospitals, and research facilities. No other has the military forces capable of intervening in African trouble spots. Few have the free press, effective trade unions, independent courts, or financial institutions to match South Africa's. And with a population greater than 50 million (80 percent black, 9 percent Coloured, just under 9 percent white, 2.5 percent Asian), South Africa has a large, multiracial, middle class.

Meanwhile, the political system began to falter. In South Africa's first (and subsequent) free and democratic elections, the multiracial "rainbow" African National Congress (ANC) Party, heir to the anti-Apartheid resistance movement, was the winner. As economic opportunities opened for all, a black middle class quickly arose but whites continued to enjoy the benefits of a system that could not be changed in short order. At the same time, many millions of people living in South Africa's periphery saw little change—except that they seemed poorer than ever compared to their fellow citizens in the cities and on white-owned farmlands. South Africa's economy boomed, but mainly because the old exploitive system continued under a new, now multiracial elite.

This troubled many in the ANC, and a split developed between those who were vested in the system and who wanted stability with slow and affordable change, and those who wanted faster, even revolutionary transformation that would upend the status quo, accelerate the redistribution of wealth, and empower the poor. Two decades after the country's first democratic election, South Africa's principal division is no longer racial: it is economic (in 2012, the RSA recorded the world's worst inequality with a Gini coefficient [see box in Introduction chapter] of 0.63). And in the ANC, the revolutionary wing is strengthening and the pendulum may swing from the haves to the have-nots.

South Africa's economy, for a decade or so after the momentous transition of 1994, thrived on its old assets—profitable commodities like gold, platinum, and diamonds—because local labor was cheap and world prices were high. Investors once deterred by Apartheid now rushed in, and black entrepreneurs prospered. But things were changing. Global competition was eroding South Africa's advantages: in 1970 South Africa produced nearly 80 percent of the world's gold, but today that has dwindled to barely 7 percent. And mine workers were demanding fairer wages, mounting protests and strikes; in 2012, an ugly incident at the Marikana platinum mine (about 100 km [60 mi] northwest of Johannesburg) cost 38 striking miners their lives when police along with the mining company's security personnel opened fire on the protesting workers. In the rural areas, meanwhile, hundreds of white landowners were killed by invaders taking advantage of their isolation. Not surprisingly, tens of thousands of skilled people have left South Africa, in doubt over the country's future.

Such developments erode confidence and deter investment, and while South Africa still has Africa's largest and most diversified economy, its dominance is declining. The Republic suffers from massive unemployment which, among the African majority, is estimated to be as high as 50 percent. The public schools are failing and there is not enough money to reverse the tide: South Africa now ranks 132nd out of a total of 144 countries on elementary school performance, an ominous harbinger for the future. Urgently needed land reform is slowed by the lack of available funds. Millions of houses have been built in the post-Apartheid era and utilities provided, but far fewer than needed. Many in the left wing of the ANC blame South Africa's political leaders of the post-Mandela period: Mandela retired in 1999 and was succeeded by Thabo Mbeki, a businessman who was in the stability/slow-transition camp. Then in 2007 Jacob Zuma, a populist, polygamist, and father of 16 children by wives and others assumed the presidency hoping to reconcile the diverging ANC factions, staving off a challenge by a young firebrand ready to restart the revolution. Clearly, South Africa's political ground is now shifting.

South Africa and the World

South Africa has many of the geographic endowments for a strong economy, but it will take time, stability, and competent management to achieve its potential. Meanwhile, South Africa's leaders seek ways to raise the country's profile in a globalizing, competitive world and, for example, aspire to join the exclusive BRICs (dynamic Brazil-Russia-India-China). In what some observers regarded as a publicity stunt to advance this cause, the government in 2013 maneuvered to host a BRIC summit meeting by inviting the group to convene in the port city of Durban. Its goal was not just to raise South Africa's standing, but to convert the BRICs into the BRICS—with South Africa added via the capital "S"—a continuing compaign you can monitor when you surf the Web.

But South Africa's economy is not only much smaller, but also much weaker than those of the four BRICs. Indeed, the South African economy lags well behind those of such other potential members of this group as Turkey, Indonesia, and Mexico. South Africa does not have an oil bonanza; it still depends heavily upon commodity sales on fickle world markets; it is losing its regional advantage in education and skills; it has critical infrastructure needs; and what it lacks now is persuasive and capable leadership that can do for the economy what Mandela did for the society. Such leadership is not in sight.

The Middle Tier

Between South Africa's northern border and the region's northern limit lie two groups of states: those possessing borders that adjoin South Africa and those beyond. Five countries form the Middle Tier, and all border on South Africa: Zimbabwe, Namibia, Botswana, and the two ministates of Lesotho and Swaziland (Fig. 6B-2). As the map reveals, four of these five are landlocked. Diamond-exporting (and upper-middle-income) **Botswana** occupies the heart of the Kalahari Desert and surrounding steppe; despite its lucrative diamonds, the majority of its 1.9 million inhabitants are subsistence farmers. In 2012, Botswana continued to be the most severely AIDS-afflicted country in all of tropical Africa. **Lesotho** and **Swaziland**, both traditional kingdoms, depend very heavily on remittances from their workers in South African mines, fields, and factories.

Southern Africa's youngest independent state, **Namibia** (2.5 million), is a former German colony with a territorial peculiarity: the so-called Caprivi Strip linking it to the right bank of the Zambezi River (Fig. 6B-2), another consequence of colonial partitioning. Administered by South Africa from 1919 to 1990, Namibia is named after one of the world's driest deserts (the Namib, which lines its coast). This state is about as large as Texas and Oklahoma combined, but only its far north receives enough moisture to enable subsistence farming, which is why most of the people live close to the Angolan border. Mining in the Tsumeb area and ranching in the vast steppe country of the south form the leading commercial activities. The capital, Windhoek, is centrally situated opposite Walvis Bay, the main port. German influence still lingers in what used to be called South West Africa, as does an Afrikaner presence from the Apartheid period. Although orderly land reform is under way, unresolved issues remain, and much of the population still lives in poverty.

The Tragedy of Zimbabwe

Botswana's other neighbor, Zimbabwe, may be said to lie at the heart of Southern Africa, between the Zambezi River in the north and the Limpopo River in the south, between the escarpment to the east and the desert to the west. Landlocked but endowed with good farmlands, cool uplands, a wide range of mineral resources, and varied natural environments, Zimbabwe at independence had one of Southern Africa's most vibrant economies and seemed to have a bright future. Zimbabwe's core area is defined by the mineral-rich Great Dyke that extends across the heart of the country from the vicinity of its capital, Harare, in the north to the second city, Bulawayo, in the south. Gold, copper, asbestos, chromium, and platinum are among its natural endowments. Its farms are capable of producing tobacco, tea, sugar, cotton, and even cut flowers in addition to staples for the local market.

But these are not good times for Zimbabwe. During the colonial period, the tiny minority of whites that controlled what was then called Southern Rhodesia (after Cecil Rhodes, the British capitalist of diamond fame and scholarship honors), took the most productive farmlands and organized and ran the agricultural economy. Following independence, their descendants continued to hold huge estates, only parts of which were being farmed. This enormous inequality inhibited broad-based development. What was needed was a comprehensive land reform program and, indeed, democratic rule.

In the wake of their successful joint campaign to end white-minority rule, the two nations that constitute most of Zimbabwe's population, the Shona (82 percent) and the Ndebele (14 percent), engaged in bitter ethnic conflict. Proper legal reforms did not take effect, and the rule of President Mugabe turned from ineffective to disastrous. As he encouraged squatters to invade white farms (where a number of owners were killed and many others fled), the agricultural economy began to collapse. Corruption in the Mugabe government resulted in the transfer of land, not to needy squatters but to friends of top officials. Foreign investment in other Zimbabwean enterprises, including the mining industry, dried up. By the mid-1990s, Zimbabwe was in economic free fall.

Mugabe also turned against the informal sector of the economy, which was all most Zimbabweans had left when jobs on farms and in factories and mines disappeared. He ordered his henchmen to destroy the dwellings and shacks of some 700,000 "informal" urban slum dwellers (many of whom were said to support the political opposition), leaving them in the streets without shelter or livelihood. As conditions worsened, people by the hundreds of thousands streamed out of Zimbabwe. Today, an estimated 3.2 million of the country's 13.1 million people are displaced, and about 80 percent are jobless.

Heroic Zimbabweans and persistent outsiders have put pressure on Mugabe to force him to permit some form of representative government. In 2008, this dictator was actually defeated in legislative elections that should have produced a runoff between Mugabe and his challenger, Morgan Tsvangirai; but the regime managed to avoid this, forcing Tsvangirai to accept an ineffective power-sharing arrangement. After difficult negotiations, a new constitution was approved by referendum in early 2013, clearing the way for another "free election." But only a few weeks later, media reports showed Mugabe's thugs roughing up campaigners in the dilapidated streets of Harare. The human catastrophe of Zimbabwe stands in sharp contrast to what is being achieved elsewhere in the region.

The Northern Tier

In the four countries that extend across the Northern Tier of Southern Africa—Angola, Zambia, Malawi, and Moçambique—change is everywhere. **Angola** (22.3 million), formerly a Portuguese colony, together with its **exclave** [2] (separated outlier) of ***Cabinda*** had a thriving (colonial) economy based on a wide range of mineral and agricultural exports at the time of independence in 1975. But the Portuguese had done a miserable job of preparing the country for self-rule, and soon Angola fell victim to the Cold War, with northern peoples opting to follow a communist course and southerners falling under the sway of a rebel movement backed by South Africa and the United States. The results included devastated infrastructure, idle farms, looting of diamonds, hundreds of thousands of casualties, and millions of landmines that continue to kill and maim.

Over the past decade, Angola's economy has revived, spectacularly so, and statistically it became the fastest growing national economy in the world! In 2012, the World Bank moved Angola up from the category of "lower middle income" to that of "upper middle income" country. Almost all of this growth comes from oil, much of which is exported to China, and Angola now vies for first place with Nigeria in Subsaharan Africa in oil production. Yielding about U.S. $5 billion per year, the return of stability is attracting investors to begin rebuilding this ruined country. But the dependence on oil is too great: this fuel accounts for almost half of Angola's GDP, more than 90 percent of its exports, and around 80 percent of government revenues—so that those "upper middle income" statistics hardly reflect better economic circumstanes for the average Angolan.

On the opposite coast, the other major former Portuguese dependency, **Moçambique** (25.0 million), experienced a comparable trajectory. Upon independence Moçambique also chose a Marxist course with unfortunate economic and political consequences. Here, too, a rebel movement supported by South Africa caused civil conflict, created famines, and generated a stream of more than a million refugees toward Malawi. Railroad and port facilities lay idle, and Moçambique at one time was ranked by the United Nations as the world's poorest country.

Moçambique has neither oil nor as much agricultural land as Angola. But it does have considerable bauxite (aluminum ore) and coal deposits, and it has a major advantage in terms of relative location. In recent years, port traffic has been revived and the country is working with South Africa on a joint Maputo Development Corridor (Fig. 6B-2). In 2011, construction began on a new bridge over the Zambezi River (only the country's fourth) at the interior city of Tete; it will open by 2015, and is expected to expedite the rising flow of road transport to and from the neighboring landlocked states.

Moçambique's economy has grown rapidly over the past decade, but here again we have an African economy that overly depends on a single resource, in this case the bauxite that accounts for one-third of all exports. And, like Angola, the country has performed poorly in distributing the benefits of growth. Late 2010 saw riots on the streets of the capital city, Maputo, to protest higher food prices. Among Moçambique's masses, poverty still reigns, but the 2012 discovery of the Rovuma offshore natural gas reserve has raised the country's prospects.

Landlocked **Zambia** (14.5 million), another product of British colonialism, shares the mineral riches of the Copperbelt with its northern neighbor, DRCongo's Katanga Province. Not only have commodity prices on which Zambia depends fluctuated wildly, but Zambia's outlet ports—Lobito in Angola as well as Beira in Moçambique—and the railways leading there were rendered inoperative by Cold War conflicts. Most recently, China has taken an interest in Zambia's minerals, and the Chinese are investing in railroad repairs as well as the expansion of mining operations. Zambia's annual economic growth has exceeded 6 percent over the past decade.

Neighboring **Malawi** (16.8 million), like Zambia, has been able to sustain democracy for nearly two decades now. Malawi has an almost totally agricultural economic base and is also unremittingly challenged by environmental degradation. This country's dependence on corn as its food staple, its variable climate, and its severely fragmented land-use pattern combine to create cycles of boom and bust from which Malawi has yet to escape. Still, the economy has grown steadily at nearly 6 percent per year since the turn of this century.

EAST AFRICA

To the east of the chain of Great Lakes that marks the eastern border of DRCongo (Lakes Albert, Edward, Kivu, and Tanganyika), the land rises from the Congo Basin to the East African Plateau. Hills and valleys, fertile soils, and copious rains mark this transition in Rwanda and Burundi. Eastward the rainforest disappears and the open savanna cloaks the countryside. Great volcanoes tower above a rift-valley-dissected highland. At the heart of the region lies Lake Victoria. Farther north the surface rises above 3300 meters (10,000 ft), and so deep are the trenches cut by faults and rivers there that the land was called, appropriately, Abyssinia (now Ethiopia). Five countries, in addition to the highland component of Ethiopia, comprise this East African region: Kenya, Tanzania, Uganda, Rwanda, and Burundi (Fig. 6B-4). Here the Bantu peoples who make up most of the population met Nilotic peoples from the north.

Kenya

Kenya (45.4 million) is neither the largest nor the most populous country in East Africa, but over the past half-century it has been the dominant state in the region. Its skyscrapered capital at the heart of its core area, Nairobi, home to 3.8 million, is the region's largest city and hub for many activities; its chief port, Mombasa, is East Africa's busiest.

After independence, Kenya chose the capitalist path of development, aligning itself with Western interests. Without major known mineral deposits, Kenya depended on exports of coffee and tea as well as other food products, and on a tourist industry based on its magnificent landscapes. Tourism became its largest single earner of foreign exchange, and Kenya prospered, apparently proving the wisdom of its economic planners.

But serious problems soon emerged. Kenya during the 1980s had the highest rate of population increase in the world, and population pressure on farmlands and on the fringes of the wildlife reserves mounted. Poaching became widespread, and tourism declined. During the late 1990s, El Niño-triggered rains affected parts of Kenya, causing landslides and washing away large segments of the crucial highway that connects Nairobi and Mombasa. This disaster was followed by a severe drought lasting several years, bringing widespread famine to the interior. Meanwhile, government corruption siphoned off funds that should have been invested. An elected president, Daniel arap Moi, turned into a dictator who severely damaged Kenya's prospects for stability. The AIDS epidemic brought another major setback. And then Kenya sustained the impact of terrorist attacks in Nairobi and Mombasa.

On the surface, Kenya today seems relatively prosperous and exhibits fairly healthy economic statistics. Nairobi's CBD breathes an atmosphere of modernity and worldliness, and its largest slum, Kibera, is world-renowned for its vibrant informal economy. But the country has utterly failed at building a unified nation out of its 41 (!)

Kenya has become one of the world's major exporters of flowers, but not without controversy. Here, workers are picking roses in a huge, 2-hectare (5-acre) greenhouse, part of an industry that currently concentrates around Lake Naivasha in the Eastern Rift Valley. European companies have set up these enterprises, buying huge swaths of land and attracting workers far and wide. Locals see wildlife and waters threatened, but the government appreciates the revenues from an industry that now accounts for more than 20 percent of Kenya's agricultural export income. This is a revealing core-periphery issue: companies growing and exporting the flowers to European markets use pesticides not tolerated in Europe itself, so that the ecological damage prevented by regulations in the global core falls on the periphery, where rules are less stringent and the environment suffers. The burgeoning flower industry around Lake Naivasha also has created social problems as thousands of workers, attracted by rumored job opportunities, have converged on the area and find themselves living in especially squalid conditions.

Marta Nascimento/REA/Redux

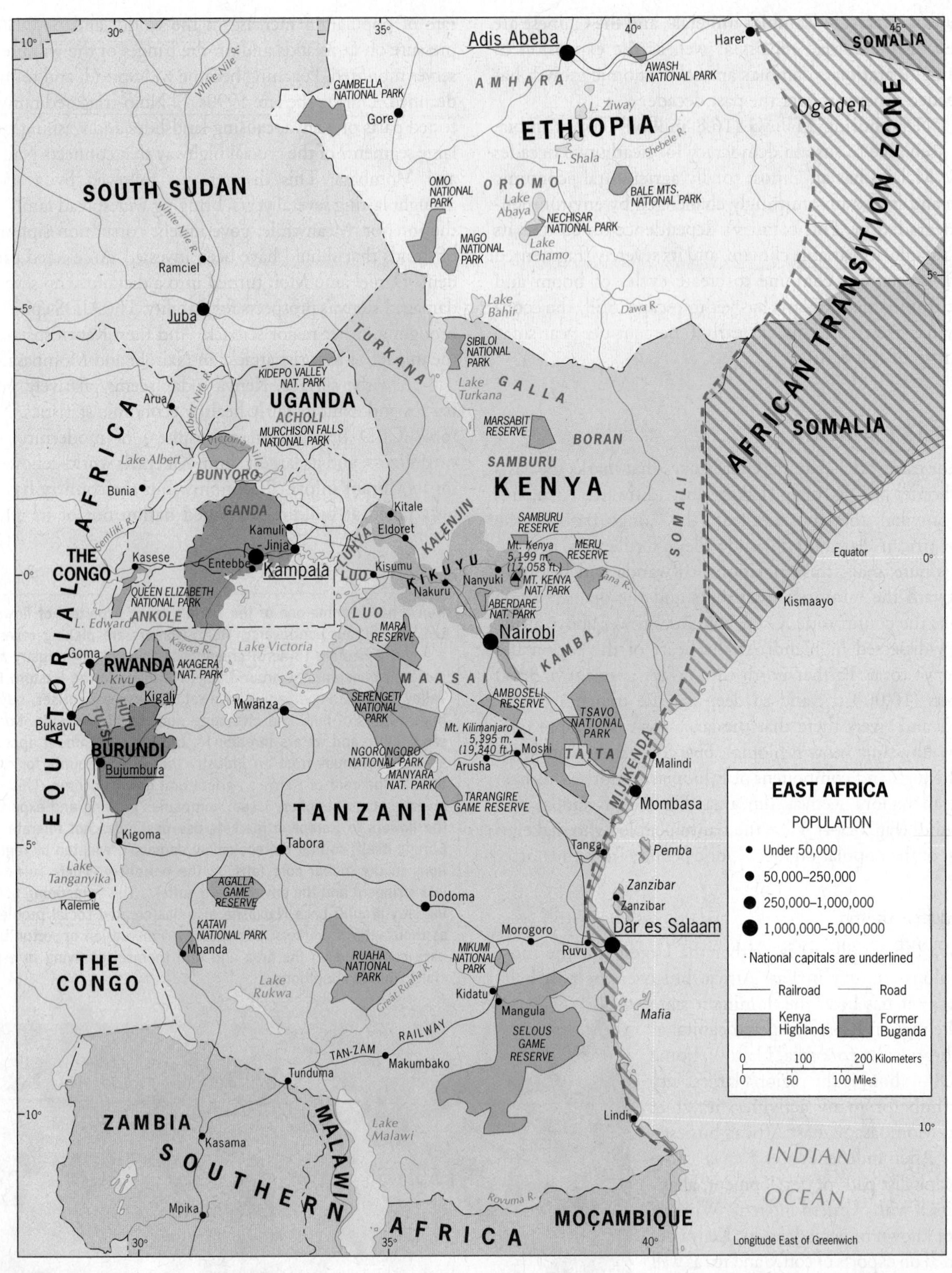

FIGURE 6B-4

constituent ethnic groups. Geography, history, and politics have placed the Kikuyu—who, according to the latest census, account for only 17 percent of the population—in a position of dominance. This is contested by several other groups (see Fig. 6B-4). The Luhya, Kalenjin, Luo, and Kamba together constitute almost 50 percent of the population, and in the country's periphery there are major peoples such as the Maasai, Turkana, Boran, and Galla.

Kenya is notorious for the corruption of its elected officials. The country's average annual per-capita income is

AMONG THE REALM'S GREAT CITIES . . .

NAIROBI

NAIROBI IS THE quintessential colonial legacy: there was no African settlement on this site when, in 1899, the railroad the British were building from the port of Mombasa to the shores of Lake Victoria reached it. However, it possessed something even more important: water. The fresh stream that crossed the railway line was known to local Maasai cattle herders as Enkare Nairobi (Cold Water). The railroad was extended farther into the interior, but Nairobi grew, and Indian traders set up shop. The British established their administrative headquarters here. Not surprisingly, when Kenya became independent in 1963, Nairobi was chosen to be the national capital.

Nairobi owes its primacy to its governmental functions, which ensured its priority through the colonial and independence periods, and to its favorable situation. To the north and northwest lie the Kenya Highlands, the country's leading agricultural area and the historic base of the largest nation in Kenya, the Kikuyu. Beyond the rift valley to the west lie the productive lands of the Luo in the Lake Victoria Basin. To the east, elevations descend rapidly from Nairobi's 1600 meters (5000 ft), so that highland environments make a swift transition to tropical savanna; to the north, increasing aridity produces semiarid steppe.

A moderate climate, a modern city center, several major visitor attractions (including Nairobi National Park, on the city's doorstep), and an ultramodern airport have boosted

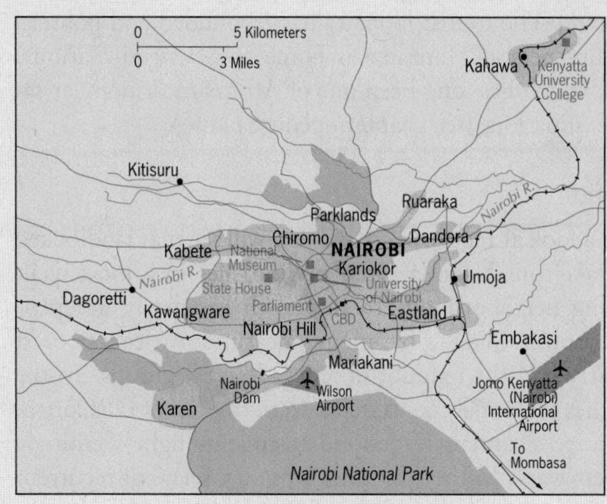

Nairobi's fortunes as a major tourist destination, though wildlife destruction, security concerns, and political conditions have damaged the industry in recent years.

Nairobi is Kenya's principal commercial, industrial, and educational center. But its growth (to 3.8 million today) has come at a price: its modern central business district stands in stark contrast to the squalor in the shantytowns that house the countless migrants its perceived opportunities attract.

now about U.S. $1600; yet every member of the Kenyan Parliament earns $156,000 tax-free. A recent poll indicated that the typical urban Kenyan pays about 16 bribes per month for anything from accessing running water to avoiding traffic tickets.

The highly contested outcome of Kenya's 2008 presidential election—the opposition felt it was cheated by allegedly fraudulent results that kept the Kikuyu president in office—uncorked what had been longstanding tensions, and violence erupted that left 1300 people dead and hundreds of thousands displaced. An important step forward came with a new constitution in 2010 that more effectively guaranteed a separation of powers and a multiparty political system. New elections were held in March 2013 without a reprise of the violence of 2008. The presidency was narrowly won by Uhuru Kenyatta, a politician deeply implicated in the violence following the 2008 election. In the outside world, his election was greeted with skepticism; inside Kenya, his main challenge is to unite this painfully divided state.

Tanzania

The name of this country is a hybrid derived from ***Tan***ganyika plus ***Zan***zibar. This is the biggest and most populous East African state with just over 50 million people. Tanzania has been described as a country without a core because its clusters of population and zones of productive capacity lie dispersed—mostly on its margins along the Indian Ocean coast, near the shores of Lake Victoria in the northwest, next to Lake Tanganyika in the far west, and facing Lake Malawi in the interior south. Although the capital, Dar es Salaam, is located on the east coast, some government agencies are based in interior Dodoma, sometimes called Tanzania's "co-capital." This in sharp contrast to Kenya, which has a well-defined core area in the Kenya Highlands centered on Nairobi in the heart of the country. Moreover, Tanzania consists of more than 100 peoples, with no ethnic group large enough to dominate the state; 30 percent of the population, mainly on the coast, are Muslims.

After achieving full independence in 1964, Tanzania embarked on a socialist path toward development, including an extensive but poorly planned farm collectivization program. The tourist industry declined sharply, and Tanzania became one of the world's poorest countries. But Tanzania did achieve remarkable political stability and a degree of democracy. Since 1990, the government has changed course to improve living standards, but the AIDS crisis, problems with the Zanzibar merger (which continue to this day), and involvement in the troubles of neighboring countries, including Rwanda, made things difficult. Yet today, Tanzania's prospects have improved. It is a leading gold producer and the economy has grown at nearly 7 percent per year for a decade. The tourist industry has rebounded and political stability persists. Tanzania is home to Serengeti National Park, which has long been one of Africa's most popular safari destinations (see chapter-opening photo).

Uganda

Take a look at Figure 6B-4 and you will see that landlocked but lakefront Uganda (38.0 million) lies in a most challenging neighborhood. To the north lies newly independent South Sudan, by many measures Africa's poorest country and the redoubt of renegade rebel groups. To the west lies the heavily contested eastern frontier of DRCongo, where government forces and insurgents fight seemingly endless wars. To the south lies Rwanda, scene of recurrent mass genocides dating back as far as the 1950s. All this violence has not left Uganda untouched, so that its hopes have long been pinned on comparatively stable Kenya, its outlet to the ocean, and on the buffer of Lake Victoria.

Not that Uganda itself was a haven of peace and stability: when the British colonialists arrived, Uganda was the domain of East Africa's most important indigenous state, the Kingdom of Buganda, peopled by the Ganda (Fig. 6B-4; also see Fig. 6A-4). The British established their headquarters near the Ganda capital of Kampala and used the kingdom's dominance over its neighbors to control Uganda through indirect rule. Thus the Ganda became the dominant nation in modern, multicultural Uganda, and when the British left they bequeathed Uganda a complicated federal system designed to perpetuate Ganda supremacy.

That system soon failed, bringing to power one of Africa's most brutal dictators, Idi Amin. Uganda had a strong economy based on coffee, cotton, and other farm exports as well as copper mining; its sizeable Asian minority of about 75,000 dominated local commerce. Amin ousted all the Asians, exterminated his opponents, and destroyed the economy. After that, the AIDS epidemic struck Uganda severely. Recovery was slow after Amin's expulsion in 1980, and subsequently Uganda progressed toward more representative government only to witness another case of growing autocracy. A massive official campaign made Uganda a leader in the struggle against AIDS, but ugly homophobia has tarnished its reputation as well.

As the map indicates, Uganda is a **landlocked state** [3] and depends on Kenya for a distant outlet to the ocean. This entails additional transport costs, affecting both exports (mainly coffee) and imports. Still, Uganda's economy shows healthy growth; should its external periphery be stabilized, a bright future may yet be in prospect.

Rwanda and Burundi

Rwanda and Burundi would seem to occupy Tanzania's northwest corner, and indeed they were part of the German colonial domain forcibly assembled before World War I. But during that war Belgian forces attacked the Germans from their Congo bases and were awarded these territories when the conflict ended in 1918. Local peoples were powerless to prevent this imperial land-grab.

Rwanda and Burundi (each with 11.3 million people), Africa's most densely populated countries, are physiographically part of East Africa, but their cultural geography is linked to the north and west. Here, Tutsi pastoralists from the north subjugated Hutu farmers, who had themselves made serfs of the local Twa (pygmy) population, setting up a conflict that was originally ethnic but became cultural. Certain Hutu were able to advance in the Tutsi-dominated society, becoming to some extent converted to Tutsi ways, leaving subsistence farming behind, and rising within the social hierarchy. These so-called moderate Hutu were—and are—frequently targeted by other Hutus, who resent their position in society. This longstanding discord, exacerbated by colonial policies, repeatedly devastated both countries and in 1994 resulted in the horrific Rwanda genocide. Unfortunately, this conflict continues to simmer and has spilled over into DRCongo. As many as 4 million people have perished as Hutu, Tutsi, Ugandan, and Congolese rebel forces fought for control over the eastern margins of DRCongo. Rwanda's official economic growth figures (7 percent annually over the past decade) provide the impression of a healthy economy, but the country's main products are coffee and tea, and an estimated two-thirds of the population lives in dire poverty.

Ethiopia

The highland zone of Ethiopia also forms part of East Africa. Adis Abeba, the historic capital, was the headquarters of a Coptic Christian, Amharic empire that held its own against the colonial intrusion except for a brief period from 1935 to 1941, when the Italians defeated it. Indeed, the Ethiopians, based in their mountain fortress (Adis Abeba lies more than 2400 meters/8000 ft above sea level), became colonizers themselves, taking control of much of the Islamic part of the African Horn to the east.

This country's natural outlets are toward the Gulf of Adan and the Red Sea, but in 1993 its government was forced to yield independence to Eritrea (part of the African Transition Zone) and Ethiopia became effectively landlocked. A bitter border war (1998–2000) between the two countries followed, and cost some 100,000 lives. Physiographically and culturally, however, Ethiopia is part of East Africa, and the Oromo (35 percent) and Amhara (27 percent) peoples are neither Arabized nor Muslim: they are Africans. Figure 6B-4 shows that functional linkages between Ethiopia and the rest of East Africa remain weak—but this is likely to improve in the future.

During the past decade, Ethiopia has made considerable economic progress and ranked as the fifth-fastest-growing economy in the world (but starting from a low base). Much of this growth has been attributed to new governmental initiatives, such as selling enormous tracts of publicly-owned arable land to foreign agribusinesses and courting Chinese foreign investment. By 2012, Chinese investment in Ethiopian infrastructure projects (roads, bridges, hydroelectric facilities) had risen to U.S. $2.5 billion (see opening photo, Chapter 6A). These investments have contributed to fast economic growth at around 9 percent per year since 2002.

Ethiopia's diverse environments, spectacular scenery, and archeological fame sustain an economy that depends primarily on farm products (coffee, tea, and spices) and tourism. But no less than 85 percent of Ethiopia's 91.2 million citizens are subsistence farmers, and the country, once a regional breadbasket, now requires food imports from relief agencies.

Not all of Ethiopia is properly considered part of East Africa: its northern parts lie in the African Transition Zone and as such they face challenges of a completely different regional order. We will return to these issues later in the chapter.

Madagascar

Only 400 kilometers (250 mi) off Africa's central east coast lies the world's fourth-largest island, Madagascar (Fig. 6B-5). But Madagascar is not part of either East Africa or Southern Africa. About 2000 years ago, the first human settlers arrived here—not from Africa but from distant Southeast Asia. A powerful Malay kingdom of the Merina came to flourish in the highlands, and its language, Malagasy, became the indigenous tongue of the entire island (see Fig. 6A-10).

The Malay immigrants later brought Africans to their island, but today the Merina (26 percent) and the Betsimisaraka (15 percent) remain the largest of about 20 ethnic groups in the population of 23.1 million. Like mainland Africa, Madagascar experienced colonial invasion and competition. Portuguese, British, and French colonizers appeared after 1500, but the Merina were well organized and resisted colonizers conquest. Eventually, Madagascar became part of France's empire, and French became the *lingua franca* of the educated elite.

Because of its Southeast Asian imprint, Madagascar's staple food is rice, not corn. It has some minerals, including chromite, iron ore, and bauxite, but the economy is weak, damaged by long-term political turmoil and burdened by rapid population growth. The infrastructure has crumbled; the "main road" from the capital to the nearest port (Fig. 6B-5) is now a potholed 250 kilometers (150 mi) that takes 10 hours for a truck to navigate.

Meanwhile, Madagascar's unique flora and fauna retreated before the human onslaught. Madagascar's long-term isolation kept evolution here so distinct that the island is a discrete zoogeographic realm. Primates living on the island are found nowhere else; 33 varieties of lemurs are unique to Madagascar. Many species of birds, amphibians, and reptiles are also exclusive to this island. Their home, the rainforest, covered 168,000 square kilometers (65,000 sq mi) in

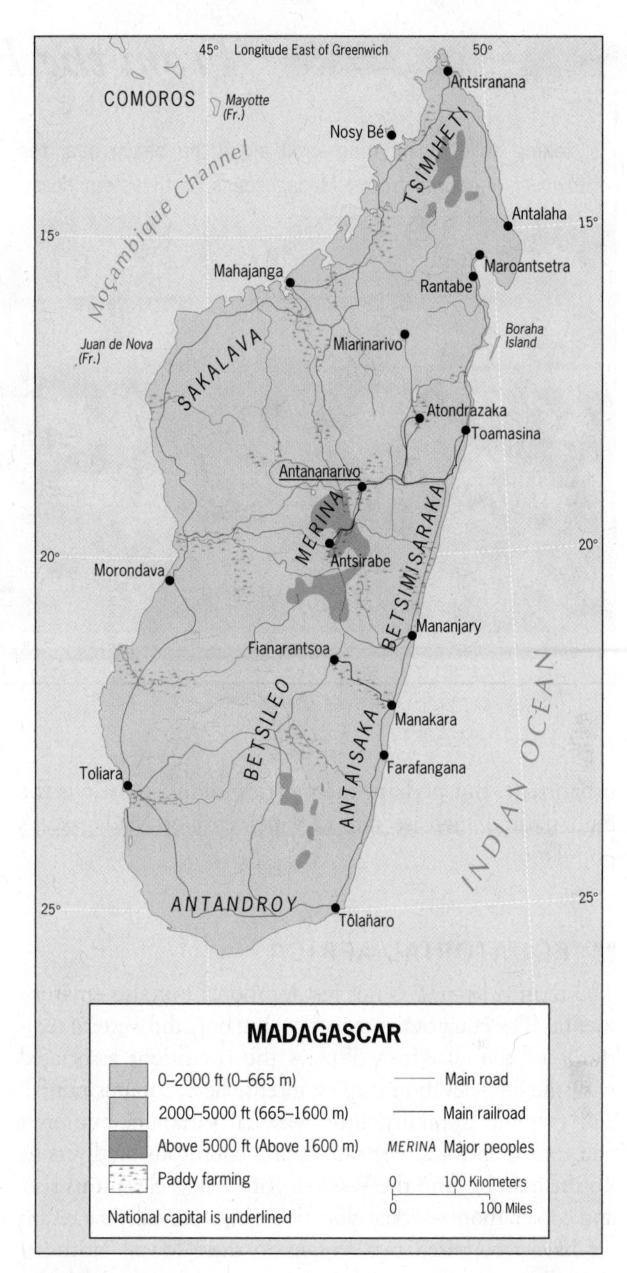

FIGURE 6B-5 © H. J. de Blij, P. O. Muller, and John Wiley & Sons, Inc.

1950, but today less than one-third of it is left. Logging, introduced by the colonists, damaged it; slash-and-burn agriculture continues to destroy it; and severe droughts since 1980 have intensified the impact. Obviously, Madagascar should be a global conservation priority, but funds are limited and the needs are enormous. Malnutrition and poverty are powerful forces when survival is at stake for villages and families.

Madagascar's cultural landscape retains its Southeast Asian imprints, in the towns as well as the paddies. The capital, Antananarivo, is the country's primate city, its architecture and atmosphere combining traces of Asia and Africa. Poverty dominates the townscape here, too, and there is little to attract in-migrants (Madagascar is only 31 percent

From the Field Notes . . .

"Taking an early morning stroll along the beach near the town of Maroantsetra on Madagascar's northeastern coast, I stopped to observe a group of fishermen at work. It was a very basic way of fishing, with minimal equipment. Using a couple of small rowboats, they dropped their net about 20 meters (65 ft) from shore and then pulled it in toward the beach. It was a meager haul. Most of this catch, they said, was intended for consumption by the men's families and the rest would be sold informally. More than 70 percent of the people in this country live below the national poverty level, and in rural and coastal areas the percentage is even higher. These economic problems stand in sharp contrast to the country's natural beauty—or what remains of it after years of severe environmental degradation."

© Jan Nijman

www.conceptcaching.com

urbanized). But perhaps the most ominous statistic is the high natural increase rate (2.9 percent) of Madagascar's population.

EQUATORIAL AFRICA

The term *equatorial* is not just locational but also environmental. The equator bisects Africa, but only the western two-thirds of central Africa displays the conditions associated with the low-elevation tropics: intense heat, copious rainfall and extreme humidity, little seasonal variation, rainforest and monsoon-forest vegetation, and enormous biodiversity. To the east, beyond the Western Rift Valley, elevations rise, and cooler, more seasonal climatic regimes prevail. As a result, we have recognized two regions in these lowest latitudes: (just-discussed) East Africa to the east, and Equatorial Africa to the west.

Equatorial Africa is physiographically dominated by the huge, bowl-shaped Congo Basin. In the far northwest, the Adamawa Highlands separate this region from West Africa; rising elevations and climatic change mark its southern limits (see the ***Cwa*** boundary in Fig. G-7). Its political geography consists of nine states, of which DRCongo (formerly Zaïre) is by far the largest in both territory and population (Fig. 6B-6).

Five of the other eight states—Gabon, Cameroon, Congo (Republic), Equatorial Guinea, and São Tomé and Príncipe—all have coastlines on the Atlantic Ocean. The Central African Republic, the southern part of Chad that belongs to this region, and newly independent South Sudan are all landlocked. In many ways, vast and complex Equatorial Africa is the most troubled region in the entire realm.

DRCongo

As the map shows, DRCongo—known officially as *The Democratic Republic of the Congo*—has only a tiny window (37 kilometers/23 mi) facing the Atlantic Ocean, barely enough to accommodate the wide mouth of the Congo River. Oceangoing ships can reach the port of Matadi, inland from which falls and rapids make it necessary to move goods by road or rail to the megacity capital, Kinshasa. Lack of navigability also necessitates transshipment between Kisangani and Ubundu, and at Kindu. Follow the railroad south from Kindu, and you reach another narrow corridor of territory at the city of Lubumbashi in the southeastern corner of the country. That vital part of Katanga Province contains most of DRCongo's major mineral resources, including copper and cobalt.

With a territory not much smaller than the United States east of the Mississippi, a population of nearly 75 million, a rich and varied mineral base, and much serviceable agricultural land, DRCongo would seem to have all the ingredients needed to lead this region and, indeed, all of Subsaharan Africa. But powerful centrifugal forces, arising from its physiography and cultural geography, pull this country apart. The immense forested heart of basin-shaped DRCongo creates communication and transportation barriers between east and west, north and south. Many of the country's productive areas lie along its periphery, separated by considerable distances. These areas tend to look across the border, to one or more of DRCongo's nine neighbors, for outlets, markets, and often ethnic kinship as well.

DRCongo's civil wars of the 1990s started in one such neighbor, Rwanda, and spilled over into what was then still known as Zaïre. Rwanda has for centuries been the scene of conflict between sedentary Hutu farmers and invading Tutsi

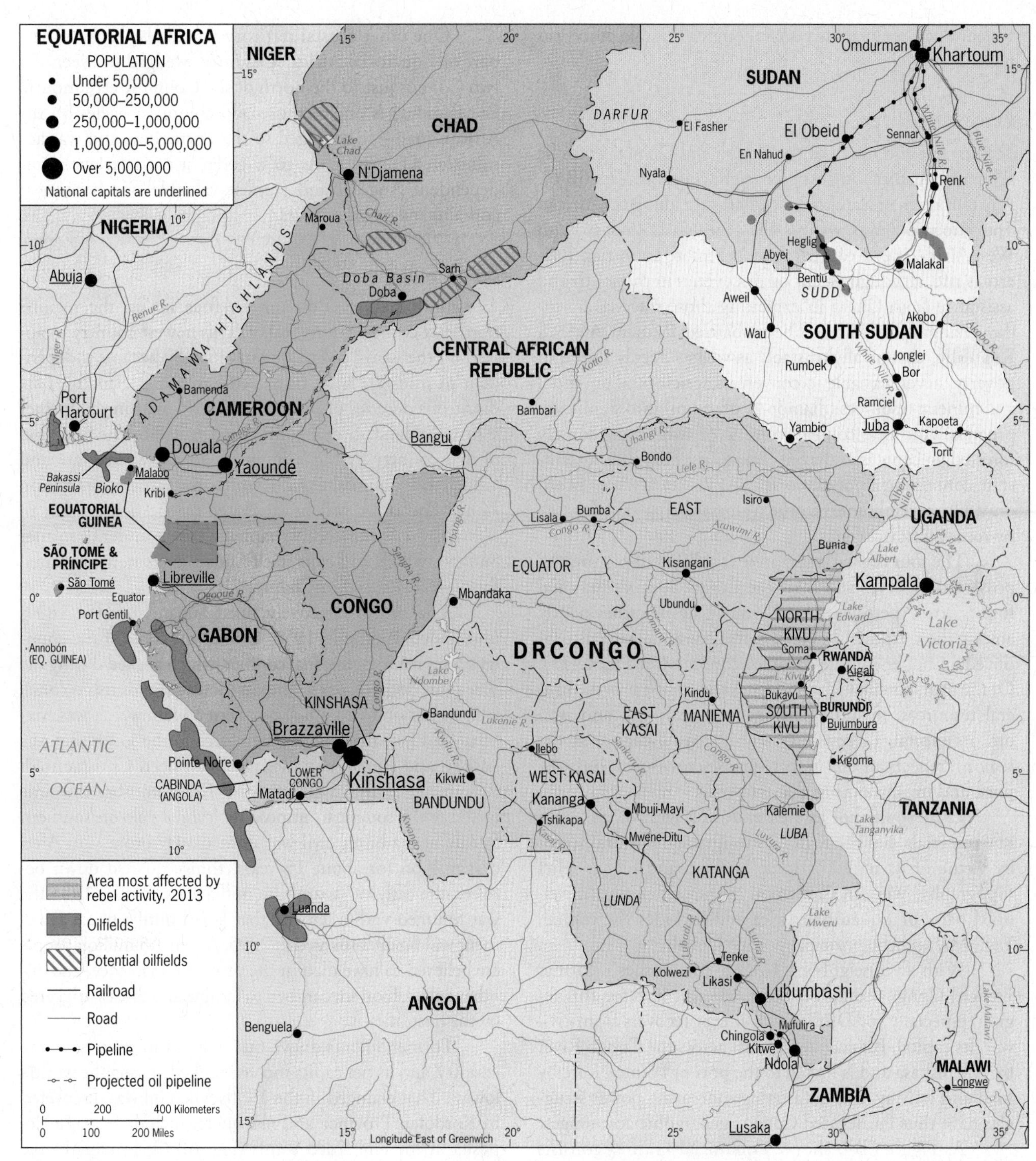

FIGURE 6B-6

pastoralists. Colonial borders and practices worsened the situation, and after independence in the early 1960s a series of terrible crises followed. In the mid-1990s, another of these crises generated one of the largest refugee streams ever seen in the world, and that conflict engulfed eastern (and later northern and western) DRCongo. The death toll will never be known, but estimates range upward from 5 million, a calamity that was still not enough to propel the international community into concerted peacemaking action.

By 2004, a combination of power transfer in Kinshasa, negotiation among various rebel groups and the African states involved in the conflict, UN assistance, and general exhaustion produced a semblance of stability in most of the eastern margins of DRCongo (Fig. 6B-6). But the Kinshasa-based government proved unable to completely pacify its eastern periphery (especially North and South Kivu Provinces), and yet another series of rebel attacks, allegedly encouraged by Rwanda, occurred. In mid-2013, the level of violence had again abated even though a durable peace was not in sight.

Across the River

To the west and north of the Congo and Ubangi rivers lie seven of Equatorial Africa's other eight countries (Fig. 6B-6). A pair of them are landlocked. **Chad**, straddling the African Transition Zone as well as the regional boundary with West Africa, is one of Africa's most remote countries. Poverty is rife, although recent oil discoveries in the south and assistance from China in exploiting those reserves are today changing its status. The embattled **Central African Republic**, chronically unstable as well as deeply mired in poverty, never was able to convert its agricultural potential and mineral resources (diamonds, uranium) into significant progress. And one country consists of two small, densely forested volcanic islands: **São Tomé and Príncipe**, a ministate containing a population of just 240,000 whose economy, like several others in the realm, is being transformed by recent oil discoveries.

The four coastal states present a different picture. All possess oil reserves and share the Congo Basin's equatorial forests; thus petroleum and timber figure prominently among their exports. In **Gabon**, this combination has produced an upper-middle-income economy (see Fig. G-11). Of the four, coastal Gabon also has the largest proven mineral resources, including manganese, uranium, and iron ore. Its capital, Libreville (the only coastal capital in the region), reflects all this in its high-rise downtown, bustling port, and mushrooming shantytowns.

Cameroon, not as well endowed with oil or other raw materials, has the region's strongest agricultural sector by virtue of its higher-latitude location and higher-relief topography. Western Cameroon is one of the more developed parts of Equatorial Africa and includes the capital, Yaoundé, and the principal port of Douala.

With five neighbors, **Congo** (the "other" [former French] Congo) could be a major transit hub for this region, especially for DRCongo if it ever recovers from civil war. Its capital, Brazzaville, lies just across the Congo River from Kinshasa and is linked to the port of Pointe Noire by road and railway. But devastating, unceasing power struggles have thus far negated Congo's geographic advantages.

As Figure 6B-6 shows, **Equatorial Guinea** consists of a rectangle of mainland territory and the island of Bioko, where the capital of Malabo is located. A former Spanish colony that remained one of the realm's least-developed territories, Equatorial Guinea has now been so dramatically affected by increased oil production and exporting that it has become Africa's only high-income economy (Fig. G-11) with GDP growth rates exceeding 10 percent for the last decade. However, as in so many other oil-rich countries, this bounty has had minimal impact on the well-being of most of the people, but ostentatious displays of wealth by members of the ruling elite have elicited international condemnation.

One other coastal territory would also seem to be a part of Equatorial Africa: ***Cabinda***, wedged between the two Congos just to the north of the Congo River's mouth. But Cabinda is one of those odd colonial legacies on the African map—it belonged to the Portuguese and was administered as part of Angola. Today it is an exclave of independent Angola, and a most valuable one because it contains major oil reserves.

South Sudan

The ninth state of Equatorial Africa lies in the region's northeast corner: South Sudan. This newest country to appear on the world political map officially became independent in mid-2011. Its birth came in the aftermath of six disastrous decades of postcolonial strife within the former state of Sudan, a brutal conflict magnified by the bisection of that country by the religious divide between Islam and Christianity-animism that signifies the African Transition Zone. The latter will be discussed as the final region of Subsaharan Africa in this chapter; the remainder of former Sudan—which still calls itself Sudan in its new truncated form—is discussed in Chapter 7B.

The British effectively ruled Sudan from the 1890s until independence in 1956. One of the roots of the country's long-running internal conflict, which peaked during the past three decades, lies in the decision of the British colonial administration to combine northern Sudan, which was Arabized and Islamized, with a large area to the south that was African and where many villagers had been Christianized. As soon as the British departed, the Khartoum-based regime in the north sought to impose its Islamic rule on southern Sudan, and a bitter civil war immediately broke out. After dragging on for about 15 years, things quieted down between the early1970s and the mid-1980s. But by 1985 the war resumed with a ferocity that lasted until a peace agreement was finally brokered in 2005. About 1.5 million people are believed to have died in the protracted violence, and another half million succumbed to famine and disease triggered by the hostilities.

Former Sudan's drawn out civil war impoverished the country, and its per capita income ranked among the world's lowest. That changed in the 1990s when oil was discovered in Kordofan Province and elsewhere, including major deposits in the embattled south (Fig. 6B-6). Leaders in the south saw oil as a ticket to self-sufficiency, and the north felt that it too would walk away with control of several oilfields.

The 2005 peace agreement that ended the war stipulated a division of the oil revenues between north and south, and also laid the groundwork for a referendum in the south on secession and its reconstitution as an independent country. That referendum was finally held in the southern provinces in early 2011, with 99 percent of their population voting for independence, which was implemented on July 1 of that year.

Nevertheless, the new country continues to exhibit a landscape of grinding poverty. Pre-2011 governance there had been so deficient that the size of South Sudan's population is not even known—for 2014, the best estimate is just about 10 million. A great majority of the people are either subsistence farmers or cattle herders (stuck in a traditional livelihood in which individual and family status depends on the number of one's livestock). Most people are illiterate, and child mortality rates rank among the highest in Subsaharan Africa.

Sudan and South Sudan have had a contentious relationship ever since the secession of the latter. Long stretches of their joint boundary remain at issue. But in 2013 the two neighbors, aware of the cost their disputes, came to an *ad hoc* agreement that allowed oil exports to flow again.

WEST AFRICA

West Africa occupies most of Africa's Bulge, extending south from the margins of the Sahara to the Gulf of Guinea coast and from Lake Chad west to Senegal (Fig. 6B-7). Politically, the broadest definition of this region includes all those states that lie to the south of Western Sahara, Algeria, and Libya, and to the west of Chad (itself sometimes included) and Cameroon. Within West Africa, a rough division is sometimes made between the large, mostly steppe-and-desert states that extend across the southern Sahara (Chad could also be included here) and the smaller, better-watered coastal states.

Apart from once-Portuguese Guinea-Bissau and long-independent Liberia, West Africa encompasses four former British and nine former French dependencies. The British-influenced countries (Nigeria, Ghana, Sierra Leone, and Gambia) lie separated from one another, whereas Francophone West Africa is contiguous. As Figure 6B-7 shows, political boundaries extend perpendicularly from the coast into the interior, so that from Mauritania to Nigeria, the West African habitat is parceled out among parallel, coast-oriented states. Across these boundaries, especially across those between former British and former

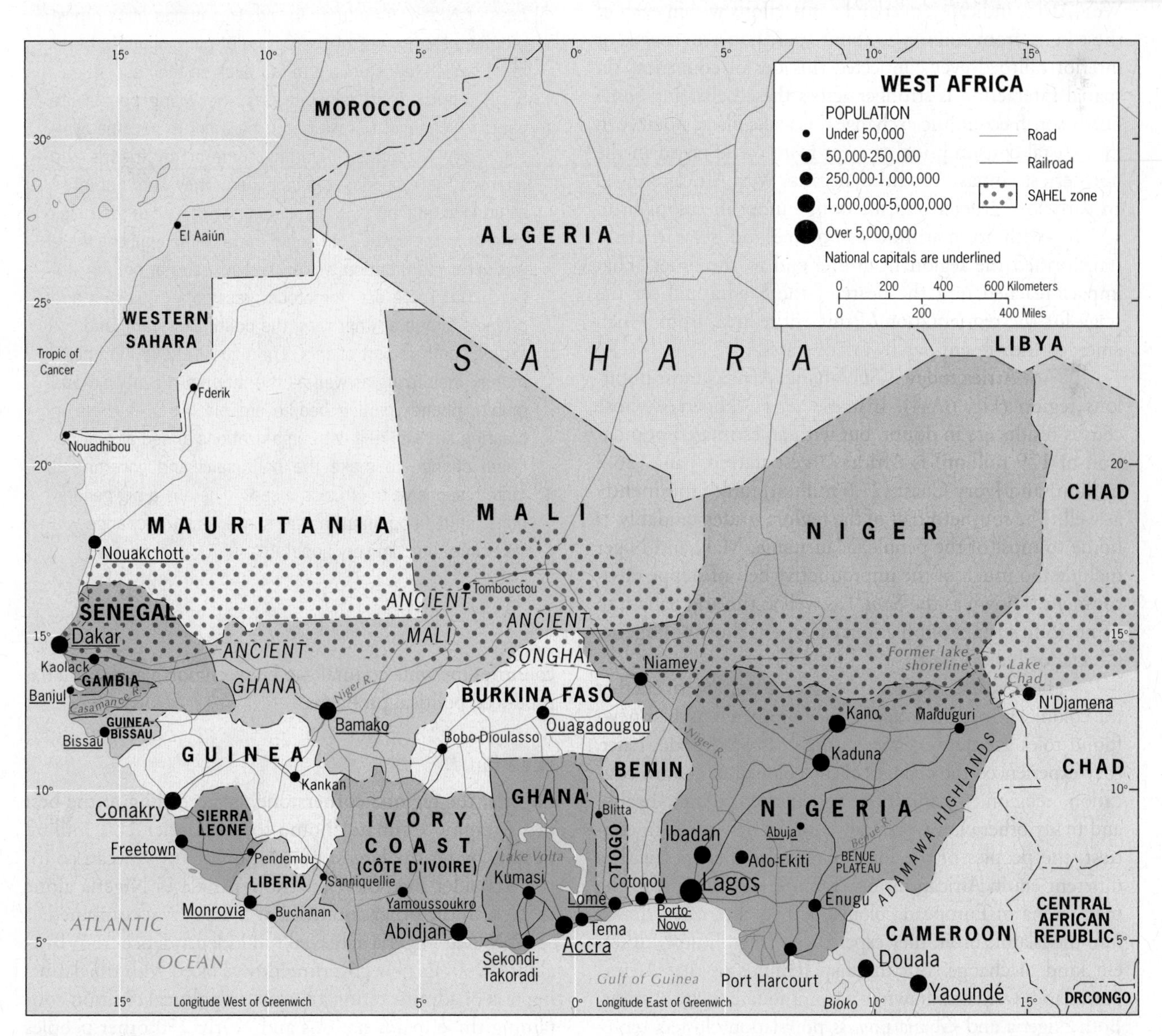

FIGURE 6B-7

French territories, there is only limited interaction. For example, in terms of value, Nigeria's trade with Britain is about 100 times as great as its trade with nearby Ghana. The countries of West Africa are not interdependent economically, and their incomes are largely derived from the sale of their products on the non-African international market.

In light of these cross-currents of subdivision within West Africa, why are we justified in speaking of a single West African region? First, because this part of the realm exhibits remarkable cultural and historical momentum. The colonial interlude failed to extinguish West African vitality, expressed not only by the old states and empires of the savanna and the cities of the forest, but also by the vigor and entrepreneurship, the achievements in sculpture, music, and dance, of peoples from Senegal to Nigeria's southeastern Iboland. Second, because West Africa contains a set of parallel east-west environmental belts, very clearly reflected in Figure G-7, whose role in the development of the region is pervasive. As the transport-route pattern on the map of West Africa indicates, overland connections within each of these belts, from country to country, are poor; no coastal or interior railroad ever connected this tier of countries. Yet spatial interaction is stronger across these belts, and some north-south economic exchange does take place, notably in the coastal consumption of meat from cattle raised in the northern savannas. And third, because West Africa received an early and crucial imprint from European colonialism, which—with its maritime commerce and slave trade—transformed the region from one end to the other. That impact reached into the heart of the Sahara and set the stage for the reorientation of this entire area, from which emerged the present patchwork of states.

West Africa today is Subsaharan Africa's most populous region (Fig. 6A-4). In these terms, Nigeria (whose census results are in doubt, but with an estimated population of 179 million) is Africa's largest state; Ghana (26.7 million) and Ivory Coast (21.6 million) rank prominently as well. The southern half of the region, understandably, is home to most of the people. Mauritania, Mali, and Niger include too much of the unproductive belt of steppe environment (known as the *Sahel*) as well as the arid Sahara to its north to sustain populations as large as those of Nigeria, Ghana, or Ivory Coast.

The peoples along the coast reflect the modern era that the colonial powers introduced: they prospered in their newfound roles as intermediaries in the coastward trade. Later, they experienced the changes of the colonial period; in education, religion, urbanization, agriculture, politics, health, and many other endeavors, they adopted new ways. In contrast, the peoples of the interior retained their ties with a different era in African history. Distant and aloof from the main arena of European colonial activity and often drawn into the Islamic orbit, they experienced a significantly different kind of change. But the map reminds us that Africa's boundaries were not drawn to accommodate such contrasts. Both Nigeria and Ghana possess population clusters representing the interior as well as the coastal peoples, and in both countries the wide cultural gap between north and south has produced political problems.

VOICE FROM THE Region

© Victoria Okoye

Victoria Okoye, Nigeria

YOUNG PEOPLE AND POLITICS IN NIGERIA

"My family is from Nigeria and I've lived in Abuja and Lagos. Democracy in Nigeria is slowly improving, but many young people still don't see the political system as accountable or as a tool to improve conditions in the country. Instead, politics remains an avenue for personal gain and those in government seem disconnected from the everyday struggles of average citizens. Problems are often attributed to Nigeria's ethnic diversity, but this diversity is part of the richness of Nigeria's culture. To me, this is one of the most prominent examples of the failure of politics and government, which leads people to fall back on their social, cultural, familial, and religious ties. We young people are eager for change, but we find it hard not to become cynical. Many Nigerians, including some of my friends and relatives, have become disillusioned. They were not interested in voting in the 2011 elections because of their lack of faith in the system. In some parts of the country, there was voter intimidation, violence, and fraud at polling stations. But there are numerous grassroots initiatives and projects aimed at improving the political system, and engaging youth is part of this. Through on-the-ground community mobilizing as well as the Internet, Facebook, and mobile phones, young people are playing a key role in creating and working with organizations aimed at real political change to make the politicians and government more responsive to citizens' needs. There is a big generational shift happening in that sense. Lasting change will take time, but I'm very hopeful."

Nigeria

Nigeria, the region's cornerstone, is (according to the best demographic estimate) home to just under 180 million people, by far the largest population of any African country. Consider this: there are more people in Nigeria alone than in all the other countries of West Africa combined.

When Nigeria achieved full independence from Britain in 1960, its new government was faced with the daunting task of administering a European political creation containing three major nations and nearly 250 other peoples ranging from several million to a few thousand in number.

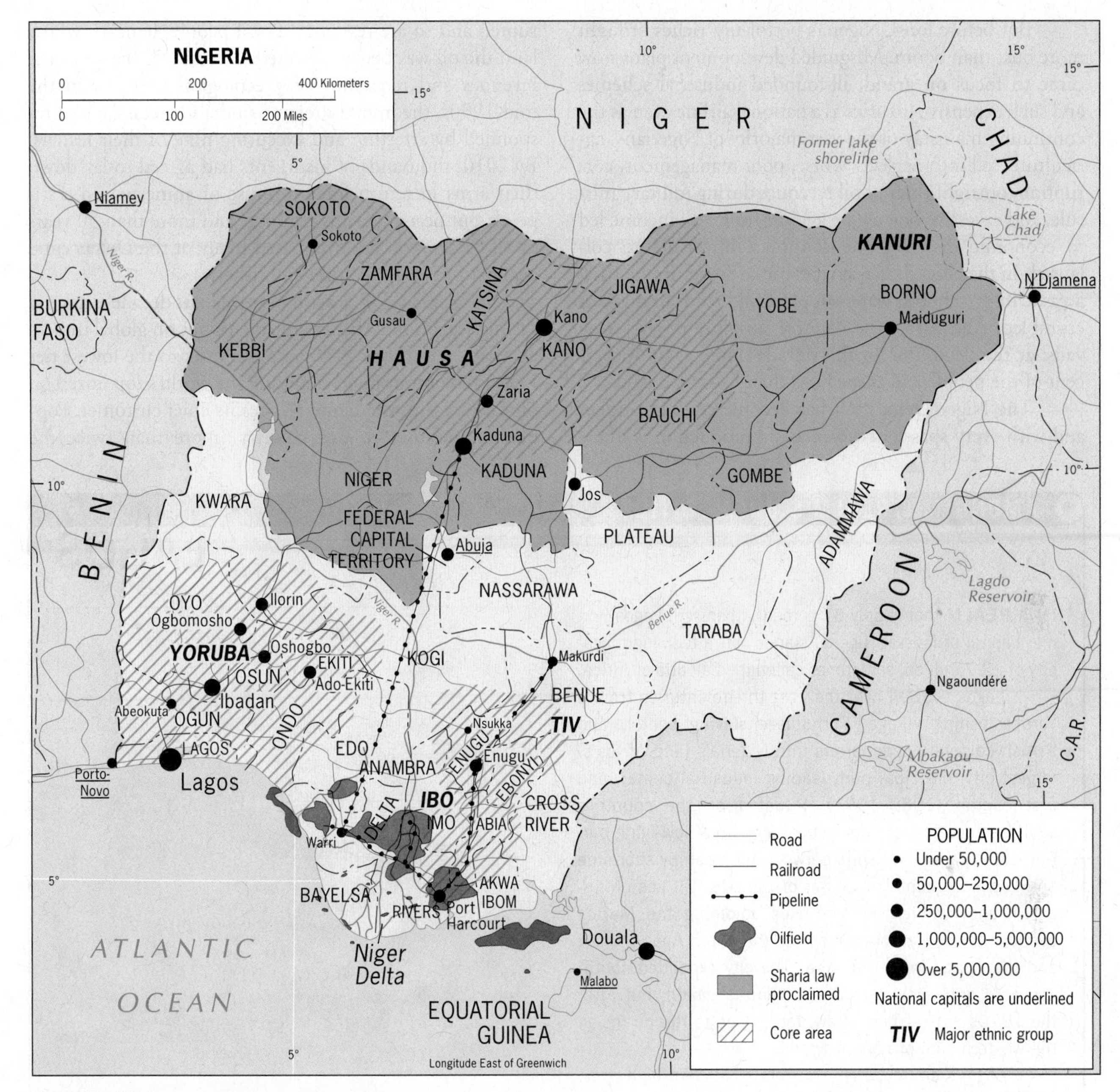

FIGURE 6B-8

For reasons obvious from the map (Fig. 6B-8), Britain's colonial imprint was always stronger in the two southern subregions than in the north. Christianity became the dominant faith in the south, and southerners, especially the Yoruba, took a lead role in the transition from colony to independent state. The choice of Lagos, the dominant port of the Yoruba-dominated southwest, as the capital of a federal Nigeria (and not one of the cities in the more populous north) reflected British aspirations for the country's future. A three-region federation, two of which lay in the south, would ensure the primacy of the non-Islamic part of the state. But this framework did not last long. In 1967, the Ibo-dominated Eastern Region declared its own independence as the Republic of Biafra, leading to a three-year civil war at a cost of about a million lives. Since then, Nigeria's federal system has been modified repeatedly; today there are 36 States, and the capital has been moved from Lagos to more centrally located Abuja (Fig. 6B-8).

Fateful Oil

Large oilfields were discovered beneath the Niger River Delta during the 1950s, when Nigeria's agricultural sector was producing most of its exports (peanuts, palm oil, cocoa, cotton) and farming still had priority in national and State development plans. Soon, revenues from oil production dwarfed all other sources, bringing the country a brief period of prosperity and promise.

But before long, Nigeria's petroleum riches brought more bust than boom. Misguided development plans now came to focus on grand, ill-founded industrial schemes and such expensive luxuries as a national airline even as the continuing mainstay of the vast majority of Nigerians, agriculture, fell into neglect. Worse, poor management, corruption, outright theft of oil revenues during military misrule, and excessive borrowing against future oil income led to economic disaster. The country's infrastructure collapsed. In the cities, basic services broke down. In the rural areas, clinics, schools, water supplies, and roads to markets crumbled. And across the country, absolute poverty prevails: at the beginning of this decade, more than 60 percent of the people still earned less than U.S. $1 a day.

The Niger Delta itself has become conflict-ridden, and with every spasm of violence, oil production is interrupted and so are revenues. Local people, beneath whose land the oil was being exploited, demanded a share of the revenues and reparations for ecological damage; in the mid-1990s, the military regime under General Abacha responded by arresting and executing nine of their leaders. By 2010, thousands of insurgents had agreed to lay down their arms in return for a promise of amnesty and a stipend. But peace remains tenuous, and more than 50 years of oil production has decimated many of the Delta's rural occupations in agriculture and fisheries.

Nigeria's GDP growth over the past decade is among the highest on the African continent; but on global indices of national well-being, Nigeria has sunk to the lowest tier even as its output ranks it among the world's top dozen oil producers, with the United States its chief customer. Population increases in this country (more than twice the

AMONG THE REALM'S GREAT CITIES . . .

LAGOS

IN A REALM that is only 37 percent urbanized, Lagos, former capital of federal Nigeria, stands out: a teeming megacity of 12.7 million, sometimes called the Calcutta of Africa.

Lagos evolved over the past three centuries from a Yoruba fishing village, Portuguese slaving center, and British colonial headquarters into Nigeria's (and Africa's) largest city, principal port, leading industrial center, and first capital (1960–1991). Situated on the country's southwestern coast, it consists of a group of low-lying barrier islands and sand spits between the swampy shoreline and Lagos Lagoon. The center of the city still lies on Lagos Island, where the high-rises adjoining the Marina overlook Lagos Harbor and, across the water, Apapa Wharf and the Apapa industrial area. The city expanded southeastward onto Ikoyi Island and Victoria Island, but after the 1970s most urban sprawl took place to the north, on the western side of Lagos Lagoon.

Lagos's cityscape is a mixture of modern high-rises, dilapidated residential areas, and squalid slums. From the top of a high-rise one sees a seemingly endless vista of rusting corrugated roofs, the houses built of cement or mud in irregular blocks separated by narrow alleys. On the outskirts lie the shantytowns of the least fortunate, where shelters are made of plywood and cardboard and lack even the most basic facilities. Slum demolitions are a regular occurrence, often to make way for new construction. The government has proclaimed its ambition to make Lagos into Africa's premier business center.

Traffic in Lagos is as bad as it gets anywhere, with daily commutes of three or more hours the rule rather than the exception. During rush hour (which is almost always) cars move at about 5 kilometers (3 mi) an hour. The only beneficiary of this chaos is the *okada*, the motor-bike taxi that in recent years has taken the city by storm. These small

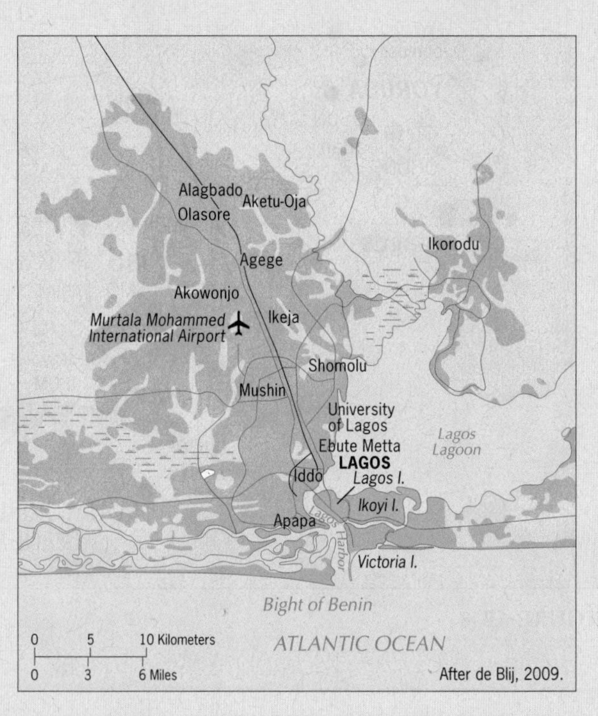

vehicles sometimes carry entire families, weaving swiftly through traffic and violating every rule in the book—and proliferated so rapidly that they became a problem and are now banned from certain roads and local areas.

By world standards, Lagos ranks among the most severely polluted, congested, and disorderly cities. Mismanagement and official corruption are endemic. Laws, rules, and regulations, from zoning to traffic, are flouted. The international airport is notorious for its inadequate security and for extortion by immigration and customs officers. In many ways, Lagos remains a metropolis out of control.

world average) are disturbingly high and this tends to offset economic growth.

Islam Ascendant

In 1999, Nigeria's hopes were raised when, for the first time since 1983, a democratically elected president was sworn into office. But soon its problems (which now also included a deepening AIDS crisis) magnified when northern States, beginning with Zamfara, proclaimed ***Sharia* law [4]** (strict Islamic) **law [4]**. When Kaduna State followed suit, riots between Christians and Muslims devastated that State's venerable capital city of Kaduna. There, and in 11 other northern States (Fig. 6B-8), the imposition of *Sharia* law led to the departure of thousands of Christians, intensifying the cultural fault line that threatens the cohesion of the country. That schism continues to deepen (see photo), and the kind of Islamic revivalism now taking hold in the north raises the prospect that Nigeria, West Africa's cornerstone and one of this realm's most important states, may succumb to devolutionary forces arising from its location astride the African Transition Zone. For Subsaharan Africa, that would be a calamity.

Nigeria—especially the north—is increasingly unsafe for locals as well as foreigners. A string of kidnappings and killings in early 2013 has put the expatriate community and foreign corporations on alert. The radical Islamic organization responsible for most of these acts of terrorism is ***Boko Haram*** (which means "Western education is sinful"), a jihadist group that originated in the far northeastern State of Borno (Fig. 6B-8), an impoverished, underrepresented corner of Nigeria where repressive government policies have triggered growing militancy. Boko Haram's "religious cleansing" activities were initially confined to Borno, but lately operations have expanded westward to the key city of Kano and even the federal capital, Abuja, to its south. Support for Islamic militants is drawn from young northern Nigerian men, half of whom are unemployed, who have lost hope in the politicians and in their own future.

Development Prospects

Nigeria's rapid economic growth over the past decade has exceeded that of South Africa, and some observers suggest that Nigeria is poised to overtake the realm's leader. But growing exports of crude oil and an improving economy do not make for a leadership change just yet. Nigeria's numbers are in part a reflection of its demography: Nigeria has more than three times the population of South Africa. Nigeria's commodities are in demand; South Africa's world markets are static or declining. But consider this: South Africa has a well-developed, modern road and railroad network (respectively, 754,000 kilometers [470,000 mi] and 12,500 kilometers [7770 mi] in length); the comparable data for Nigeria are 12,500 and 2200 kilometers [1365 mi]. Whether it is level of urbanization, literacy, agricultural diversity, manufacturing capacity, higher education, or infrastructure, South Africa will stay in the lead for some time to come. Certainly Nigeria has the potential to take its place as Africa's economic leader, but catching up from a low base is not the same as overtaking in total. For all its own shortcomings, South Africa remains the realm's undisputed pacesetter.

Coast and Interior

Nigeria is one of 17 states (counting Chad and offshore Cape Verde) that form the region of West Africa. Four of these countries, comprising a huge territory under steppe and desert environments but containing relatively small populations (Fig. 6B-7), are landlocked: Burkina Faso, Mali, Niger, and Chad (all are discussed further in Chapter 7B). Figure G-7 clearly shows the prevailing arid conditions in these four interior states, and Figure 6A-8 reveals the concentration of population within the steppe zone and along the ribbon of water provided by the Niger River—a lifeline dependency most crucial in Mali and Niger (both are also challenged by troubled, remote northern frontiers deep in the Sahara). Scattered oases form the remaining settlements and anchor regional trade as well. And it should also be noted that even the coastal states do not escape the aridity, especially the marginal lands of their interiors.

A man walks past the wreckage at police headquarters in Kano, Nigeria following a suicide bomber's attack in January 2012. More than 150 people were killed within days in a series of coordinated attacks by the jihadist Islamist group, Boko Haram, in this largest city of northern Nigeria. A shadowy organization with a Taliban-like agenda, Boko Haram is linked to al-Qaeda. It seeks to impose strict Sharia law (already in practice throughout the north, as Fig. 6B-8 indicates) on Nigerians all across the country, using any means it deems necessary.

© AP/Wide World Photos

West Africa's countries all share the effects of the environmental zonation depicted in Figure G-7, but they also exhibit distinct regional geographies. Typical of the coastal zone is **Benin**, Nigeria's neighbor with a population of 9.9 million, which has a growing cultural and economic link with the Brazil's northeastern Bahía State, where many of its people were taken in bondage and where elements of West African culture survive. Benin's geography is comparable to that of neighboring **Togo** (population: 6.3 million): both are markedly elongated narrow political units with savannas in the north and humid tropical lowlands near the coast, but without usable rivers. On the other hand, **Burkina Faso** is representative of the much drier interior: impoverished and landlocked, but with undeveloped reserves of gold, its commercial economy relies on the exporting of cotton. Importantly, with Muslims constituting 59 percent of Burkina Faso's 18.6 million residents, this country—together with Mali (93 percent), Niger (99 percent), and transitional Chad (56 percent)—lies firmly within the Islamic orbit of northern West Africa; among other things, this translates into a fast-growing population well above twice the world average, a demographic trend also exhibited by Mali, Niger, and Chad.

Among the coastal states, **Ghana**, known in colonial times as the Gold Coast, was the first modern West African state to achieve independence, with a democratic government and a sound commercial economy based on cocoa exports. A pair of grandiose post-independence schemes can be seen on the map: the port of Tema (near the capital, Accra), intended to serve a vast West African hinterland, and Lake Volta, which resulted from the region's largest dam project. Both democracy and the economy soon failed, but during the 1990s a military regime was replaced by a stable representative government and recovery began. In 2007, Ghana celebrated 50 years of independence, its democracy maturing, its economy forging ahead, corruption persistent but declining, and its international stature rising. The same year also brought news of a major discovery of oil reserves off Ghana's coast and caused a surge of optimism. This came only a few months after its government had received a nearly (U.S.) $550 million aid package from the Millennium Challenge Corporation in the United States to expand commercial agriculture, further improve infrastructure, and combat poverty—a grant made in recognition of Ghana's achievements that featured a democratic, closely contested, and peaceful presidential election. Later elections and a peaceful transfer of power after the 2012 death of President John Atta Mills have solidified Ghana's reputation as an African trailblazer.

Ivory Coast (officially *Côte d'Ivoire*) has had a turbulent history. Following independence in 1960, it translated the next three decades of autocratic but stable rule into economic progress, yet excesses by the country's president-for-life eventually cost it dearly. One of those excesses involved the late-1980s construction of a Roman Catholic basilica to rival St. Peter's in Rome in the president's home village, Yamoussoukro, also designated to replace Abidjan as the country's new capital. By the turn of this century, the political succession had become badly entangled in a north-south, Muslim-Christian schism that degenerated into two rounds of civil war between 2002 and 2011. Conflict has now largely subsided, but the country remains deeply troubled as it tries to rebuild its shattered society and cocoa-based commercial economy.

Democratic **Senegal** on the extreme west coast demonstrates what stability can facilitate: without oil, diamonds, or other valuable sources of income and with an overwhelmingly subsistence-farming population, Senegal nevertheless managed to achieve some of the region's highest GNI levels. More than 95 percent Muslim and dominated by the Wolof, the ethnic group concentrated in and around the capital of Dakar, and with continuing close ties to France, Senegal has even been capable of overcoming both a failed effort to unify with its English-speaking **enclave [5]**, **Gambia**, and a secession movement in its southwestern Casamance district. In 2010, Senegal marked both 50 years of independence and four decades of democracy, and in 2012 a runoff election ousted a president who tried to secure a third term.

Other parts of West Africa have been afflicted by civil war and horror. **Liberia**, founded in 1822 by freed American slaves who returned to Africa with the help of

© Richard Grant

Even in the poorer countries of the world, you see something that has become a phenomenon of globalization: gated communities. Widening wealth differences, security concerns, and real estate markets in societies formerly characterized by traditional forms of land ownership combined to produce this new element in the cultural landscape. This is the entrance to Golden Gate, the first private gated-community development in Accra, Ghana, started in 1993 as a joint venture between a Texas-based construction company and a Ghanaian industrial partner.

From the Field Notes . . .

© H.J. de Blij

"Wherever you go in West Africa, you are struck by the bright colors, especially in the markets. And Freetown, Sierra Leone on this Friday seems to be one huge market, the streets lined with stalls, young and old crowding intersections in huge numbers. Most of the goods were for local trade; compared to what you see in East Africa, where tourism plays a bigger role, these markets still have a strong domestic atmosphere. Music plays everywhere, in shops, out in the open; and the whole city seems suffused with the aromas of spicy foods."

www.conceptcaching.com

American colonization societies (today, their descendants make up about 2.5 percent of the population), was governed by Americo-Liberians for six generations and sold rubber and iron ore abroad. A military coup in 1980 ended that era, and full-scale civil war beginning in 1989 embroiled virtually every ethnic group in the country. About a quarter of a million people, one-tenth of the population, perished; hundreds of thousands of others fled the country. But Liberia turned a corner in 2006 with the election of Ellen Johnson Sirleaf as Africa's first female president. She succeeded in stabilizing the political system and in restoring democracy. Sirleaf subsequently received the Nobel Peace Prize in 2011 and later that year was reelected.

Sierra Leone followed a similar route from self-governing British Commonwealth member to republic to one-party state and military dictatorship. In the 1990s, a rebel movement, financed by diamond sales, inflicted dreadful punishment on the local population, but a remarkable turnaround made possible by British intervention occurred early in the twenty-first century as free elections were held. In 2010, the United Nations lifted all sanctions against the country, which has now stabilized and is rebuilding. Diamonds are still the leading export, and Sierra Leone has been among Africa's fastest growing economies in recent years. The same cannot yet be said of neighboring Guinea, where dictatorial mismanagement and a violent power struggles combined to ruin economic opportunities in both agriculture and mining.

As Figure 6B-7 shows, **Guinea** borders all three of the troubled countries just discussed, and it has involved itself in the affairs of all of them—receiving in return a stream of refugees in its border zones. Guinea, with a population of 12.1 million, has been under dictatorial rule ever since its founding president, Sekou Touré, turned from father of the nation into usurper of the people. As Figure G-11 shows, Guinea is one of Africa's lowest-income states, but representative government and less corrupt administration could have made it a far better place. Guinea has significant gold deposits, may possess as much as one-third of the world's bauxite (aluminum ore) reserves, and can produce far more coffee and cotton than it does; offshore lie productive fishing grounds. As the map shows, this is no ministate like its coastal neighbor **Guinea-Bissau**, the former Portuguese colony. Guinea's capital, Conakry, has a substantial hinterland reaching well into West Africa's interior, giving the country a wide range of environments.

We should remember that the many conflicts we noted above are often geographically concentrated in a few places and they are not representative of vast and populous West Africa. Tens of millions of farmers and herders who manage to cope with fast-changing environments in this zone between desert and ocean live remote from the news-making conflicts along the well-watered coast. Time-honored systems continue to serve: for instance, the local village markets that drive the traditional economy. Visit the countryside, and you will find that some village markets are not open every day but only every third or fourth day or in some other rotation adjusted to factors of product supply, distance, and population distribution. Such a system ensures that all villages of a certain size get to participate in the exchange network. These **periodic markets [6]** represent one of the many traditions enduring here, even as the cities beckon the farmers and burst at the seams. This region's great challenges are economic survival and nation-building through political stability under some of the most challenging circumstances in the world.

Saharan Shadows

The huge western bulge of the African landmass is environmentally dominated by the Sahara, the great desert whose dust is spread widely by *Harmattan* "winter" winds across the entire region and far out to sea. Under nature's regime, the Sahara expands and contracts over time, its semiarid southern margin or ***Sahel*** (an Arabic word meaning "shore"—as if the great desert were a vast sea) shifting north and south. Figure 6B-7 shows the Sahel in an average position, but people and their animals are increasingly affecting its configuration through the process of ***desertification*** (human-induced desert expansion). Herds of livestock quickly occupy the desert-adjoining steppes when they turn temporarily green, but their grazing in numbers far beyond the carrying capacity of this land removes vast areas of this grass as they yank it out by the roots and trample it into the fragile soil layer, making the now-dusty surface highly vulnerable to wind erosion. Inevitably, drought soon returns, the great desert advances, and the people again retreat into the moister savannas to the south, where they invade farmlands and confront sedentary inhabitants. It is an ancient cycle intensified by unrelenting population growth and political partitioning, and it challenges virtually every state in this key zone of Africa.

THE AFRICAN TRANSITION ZONE

From Mauritania and Senegal in the far west to the Horn of Africa in the east, a semiarid belt of land, about 500 kilometers (300 mi) wide, lies between the bone-dry Sahara to the north and the more humid savanna zone to the south. Here in the ***Sahel***—mapped in Figure 6B-7—steppe (***BS***) climates separate desert (***BW***) from tropical savanna (***Aw***) climates (see Fig. G-7). Life in this stressed belt of steppelands is quite unpredictable. During the nine-month dry season, from October to June, there is virtually no rain and temperatures rise to 50°C (122°F). When the rains finally come, they are erratic and some areas get a little while others are flooded. Some years the harvest is good, sometimes it is almost nonexistent. According to a recent UNICEF report, malnutrition kills 225,000 children annually in the five western Sahelian countries (those excluding former Sudan and the African Horn). But the problems of the Sahel go far beyond climate.

As Figures 6B-9 and 6B-1 show, the African Transition Zone is unlike the other four regions of this realm. This is where Subsaharan Africa's cultures intersect with those of the Muslim world. This Islamic incursion into Subsaharan Africa has engulfed some countries in their entirety (such as Senegal and Burkina Faso), and divides others into Muslim and non-Muslim sectors (Nigeria, Ivory Coast, and Chad). As elsewhere in the world where certain geographic realms meet and overlap, complications mark the African Transition Zone. In some areas, the transition from Muslim to non-Muslim society is spatially gradual, as is the case in Sierra Leone and Ivory Coast. In other places, the cultural divide is considerably sharper, as in eastern Ethiopia and former Sudan, where a traditional border between Christian/animist African cultures and Islamic African communities is more abrupt. The southern border of the African Transition Zone, the religious frontier occasionally referred to as Africa's **Islamic Front [7]**, is therefore neither static nor uniform.

Conflict marks much of the African Islamic Front (mapped in Fig. 6B-9): longstanding in former Sudan, where a decades-long war for independence by non-Islamic Africans in the south against the Arabized, Muslim north was successful but cost millions of lives; intermittent in Nigeria, where Islamic revivalism is currently clouding that country's prospects; and recent in Ivory Coast, where political rivalries have reawakened religious strife in a once stable country. Nonetheless, the most conflict-prone part of the African Transition Zone lies in the east, involving not only now-bifurcated Sudan but also the historic Christian state of Ethiopia, which is nearly encircled by predominantly Muslim societies.

The Horn of Africa

Africa's Horn (Fig. 6B-10) is an especially volatile subregion of the African Transition Zone, which contains Ethiopia, Eritrea, Djibouti, and Somalia. We will concentrate our attention here on the two largest of these countries. But note that the ministate of **Djibouti**, fully 97 percent Muslim, has taken on added importance since it lies directly across from Yemen and overlooks the narrow entry to the Red Sea, the Bab el Mandeb Strait, a **choke point [8]** in international commerce. **Eritrea** came into existence in 1993 when it split off from Ethiopia; still-ongoing boundary conflicts with Ethiopia have damaged the economies of both countries. Eritrea's early hopes for representative government and economic progress were dashed by war, dictatorial rule, and, more recently, involvement in Islamic sectarian strife.

The highlands of **Ethiopia**, where the country's core area, capital (Adis Abeba), and its historic Christian heartland lie, are virtually encircled by dominantly Muslim societies. As Figure 6B-9 reminds us, Ethiopia's eastern Ogaden is traditional Somali country and almost entirely Islamic. Today, one-third of Ethiopia's population of 91.2 million is Muslim, and here the Islamic Front is quite sharply defined. With the continent's second-largest population, Ethiopia is an important African cornerstone state but, landlocked and fragmented by the African Transition Zone, its economy and political system remain weak. However, given its substantial population and crucial relative location, Ethiopia's importance in this realm is certain to grow.

Somalia

The key component in the eastern sector of the African Transition Zone is Somalia, where 10.7 million people, virtually

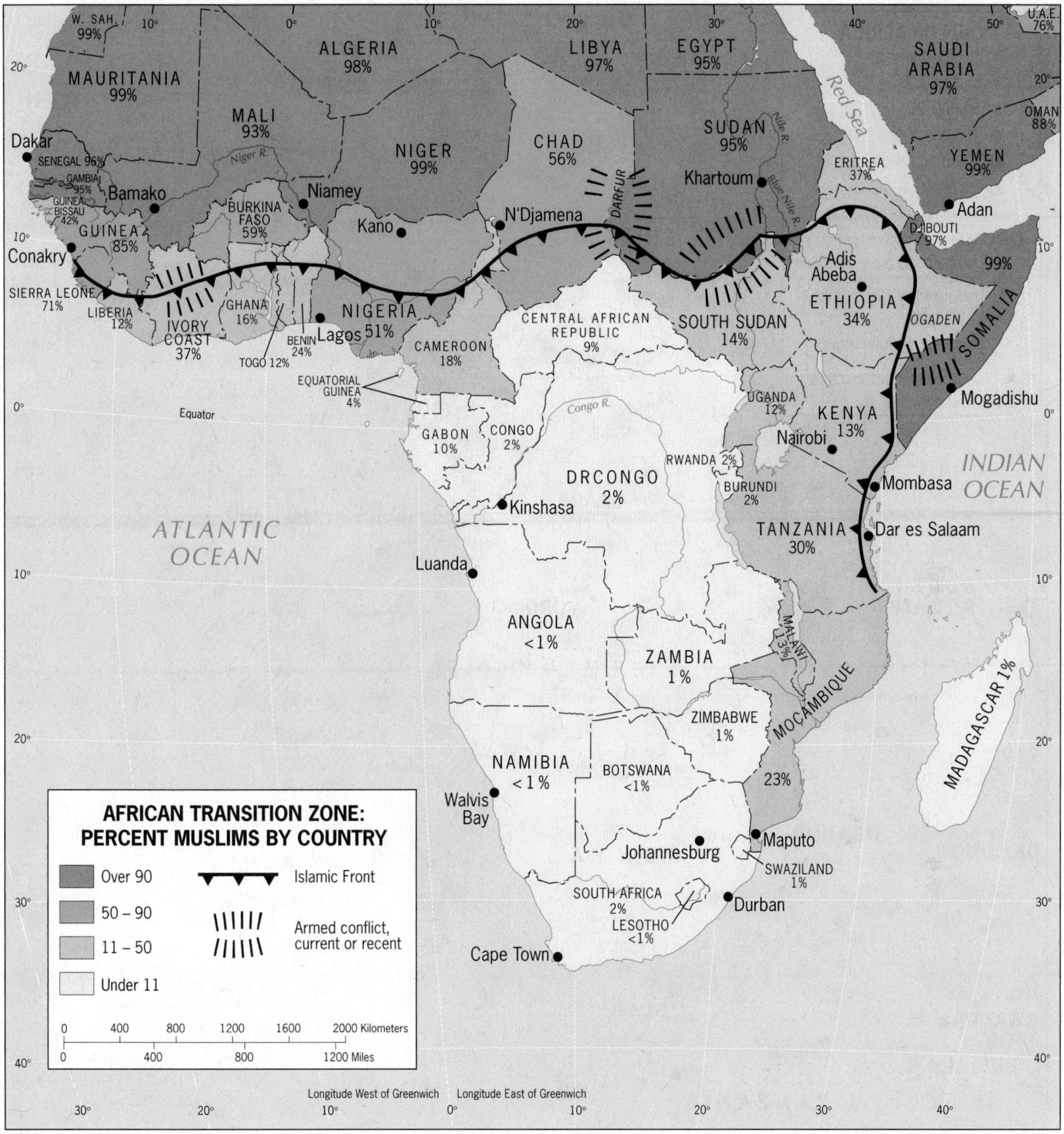

FIGURE 6B-9

all Muslim, live at the mercy of a desert-dominated climate that requires cross-border migration into Ethiopia's Ogaden zone in pursuit of seasonal pastures. As many as 3 to 5 million Somalis live permanently on the Ethiopian side of the border, and this is not the only division the Somali "nation" faces. The Somali people constitute an assemblage of five major ethnic groups fragmented into hundreds of clans, all engaged in an endless struggle for power as well as survival.

Early in this century's first decade, Somalia's condition as a **failed state [9]** led to the country's fragmentation into three subdivisions (Fig. 6B-10). In the north, the sector known as ***Somaliland***, which had proclaimed its independence in the 1990s and which remains by far the most stable of the three, functioned essentially as an African state, although the "international community" will not recognize it as such. In the east, a conclave of local chiefs declared their territory to be separate from the rest of Somalia, gave it the name of ***Puntland***, and asserted an unspecified degree of autonomy. In the south, where the official capital, Mogadishu, is located on the Indian Ocean coast, local secular warlords and Islamic militias continued their grim contest for

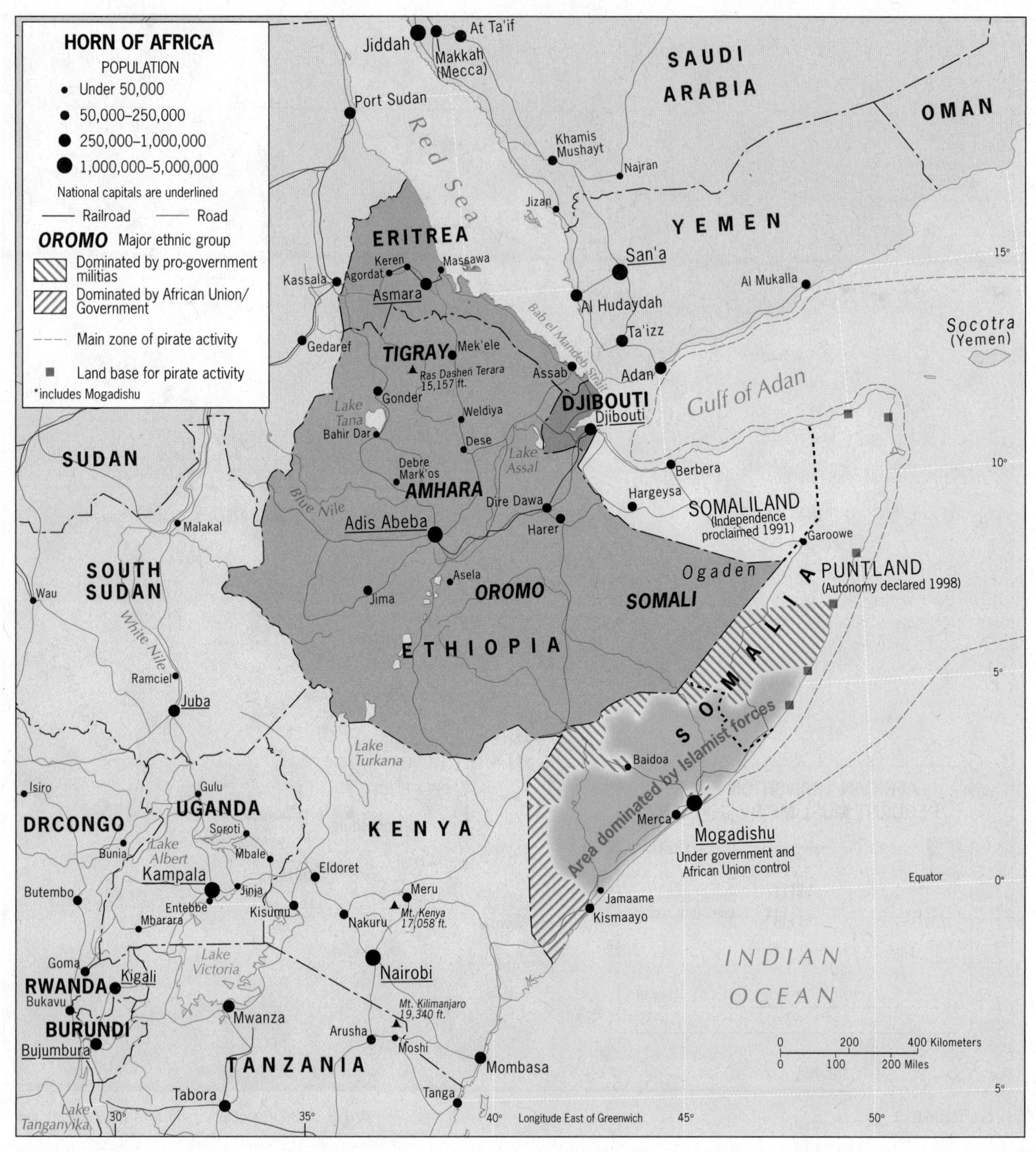

FIGURE 6B-10

supremacy, the warlords supported by U.S. funding. In 2006, the militias stormed into the capital and took control, ousting the warlords and proclaiming their determination to create an Islamic state. In the ensuing years, the al-Qaeda-linked al-Shabaab Islamic terrorist movement has used southern Somalia as a base from which it launched its attacks.

The absence of a functioning government also enabled organized and well-equipped pirate groups, operating from the Somali coast, to hijack dozens of ships including huge oil tankers, demanding ransom and sometimes killing their captives. Eventually, coordinated international responses involving protective measures at sea and improved surveillance began to reduce the pirates' success rate.

In late 2012, the Somali government was able, with the help of African Union troops, to recapture Mogadishu along with a number of other areas in the southern part of the country. Islamic rebels, however, continued their hold

Reuters/AU-UN 1st Photo/Stuart Price/Landov

In late 2012, Somali government troops supported by African Union forces took back the capital city of Mogadishu. The surrounding areas remained in control of Islamist groups and it was hard to know how stable this new political arrangement would be. At least the city's residents could enjoy the beach again after the withdrawal of the extremists who banned such social gatherings of men and women. But note in this photo that few if any females dared to venture here.

on large territorial expanses in 2013, and Somalia remained a failed state.

As this chapter has underscored, Africa is a continent of infinite social diversity, and Subsaharan Africa is a realm of matchless cultural history. Since the beginning of the twenty-first century, the realm has become more dynamic than ever. An encouraging number of countries are developing rapidly in response to the global commodity boom, although several others remain economic backwaters. Democracy is taking hold in an unprecedented number of states, but elsewhere war and dislocation persists. Within countries, differences are growing with respect to well-being, consumption patterns, and lifestyles. South Africa, long the most vibrant Subsaharan African economy, has slowed its growth and now faces major social and political challenges. Nigeria, with the second-biggest economy, aims to become a global player but its own territorial integrity seems to be unraveling in its northern tier of States. Oil, gas, and other natural resources are the cornerstones of this realm's progress in this decade, but building a prosperous future will require greater economic diversification and tightening cooperation throughout Subsaharan Africa.

POINTS TO PONDER

- South Africa was freed from Apartheid two decades ago, but it is now one of the most unequal societies on Earth.
- In 2010, Angola was China's top oil supplier; since then it has ranked second behind Saudi Arabia and ahead of third-place Iran.
- A number of low-income Subsaharan African countries have seen major revenue increases from raw material exports. How should they be spending those revenues?

7A THE NORTH AFRICAN/SOUTHWEST

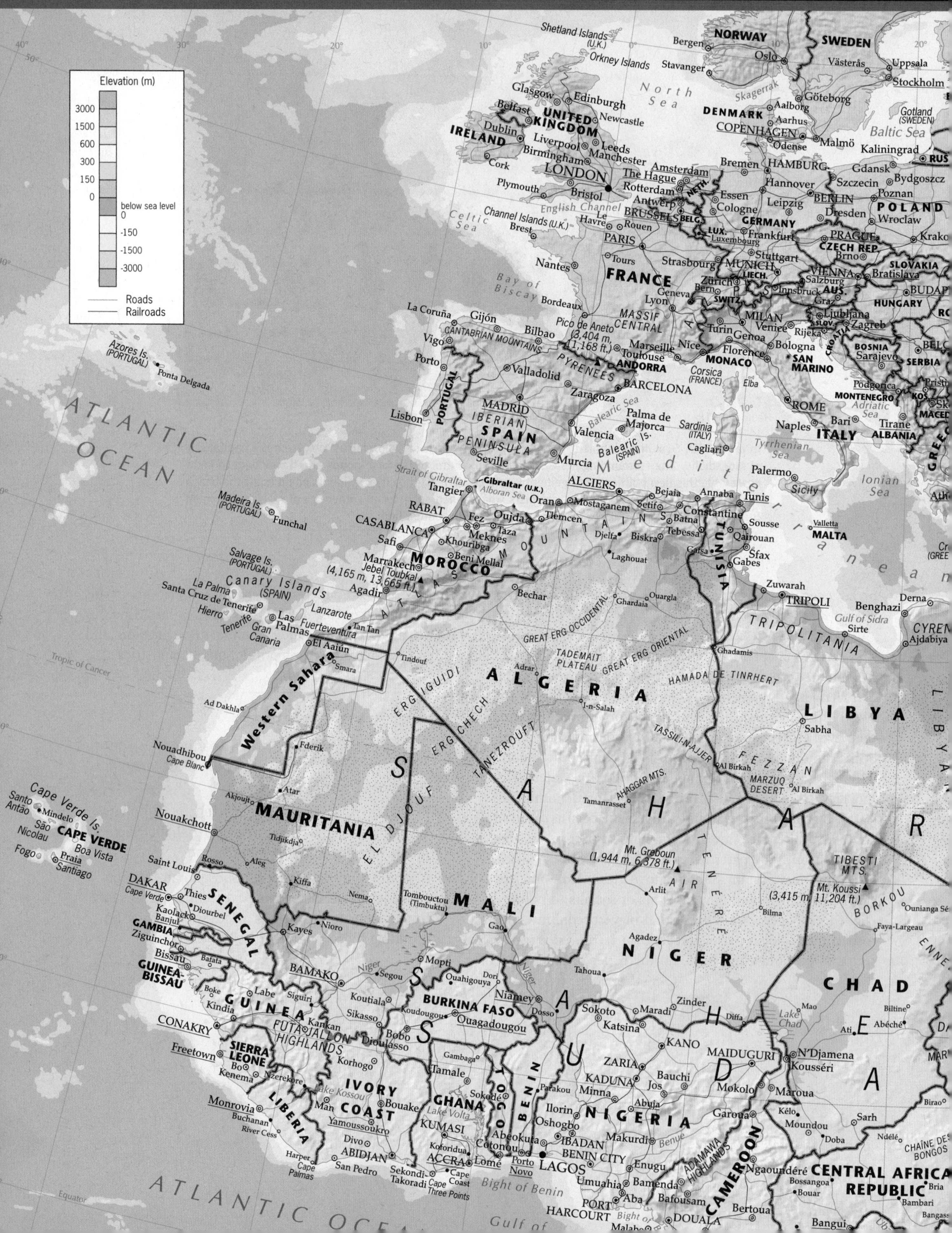

ST. PETERSBURG
PERM
URAL MOUNTAINS
YEKATERINBURG
OMSK
Ob
Gora Belukha (4,506 m, 14,783 ft.)
Volga
NIZHNIY NOVGOROD
KAZAN
CHELYABINSK
Petropavl
Pavlodar
Irtysh
Qostanay (Kustanay)
Kökshetav (Kokchetav)
Semey (Semipalatinsk)
Öskemen (Ust-Kamenogorsk)
LATVIA
MOSCOW
Kama
RUSSIA
Astana
Tobyl
KAZAKH UPLANDS
Qaraghandy
Vilnius
MENSK (MINSK)
SAMARA
Orenburg
BELARUS
Oral
Aqtöbe
JUNGGAR BASIN
Ural
Zhezqazghan
KAZAKHSTAN
Lake Balqash
SARYESIK-ATYRAU DESERT
Taldyqorghan
ÜRÜMQI
KIEV (KYYIV)
KHARKIV
Don
VOLGOGRAD
ALMATY
Vinnytsya
UKRAINE
DNIPROPETROVSK
DONETSK
Volga
CASPIAN DEPRESSION
Atyrau
Aral Sea
MOYYNQUM
Bishkek
Ysyk Köl
TIAN SHAN
Aksu
TARIM BASIN
Dnieper
Qyzylorda
Syr Darya
Zhambyl
MOLDOVA
Chisinau
ROSTOV
KYRGYZSTAN
ODESA
Sea of Azov
USTYURT PLATEAU
Shymkent
Namangan
Andijon
TAKLAMAKAN DESERT
CRIMEAN PEN.
QIZILQUM
TASHKENT
UZBEKISTAN
Ismoili Somoni Peak (7,495 m, 24,590 ft.)
BUCHAREST
Sevastopol
Groznyy
Aqtau
Caspian Sea
Nukus
Amu Darya
Jizzax
CHINA
Dashhowuz
Urgench
Samarqand
CAUCASUS MTS.
Black Sea
GEORGIA
Garabogaz Bay
Bukhoro
Qarshi
Dushanbe
TAJIKISTAN
KUNLUN MOUNTAINS
BULGARIA
TBILISI
Türkmenbashi
TURKMENISTAN
Kulob
K2 (8,611 m, 28,250 ft.)
AZERBAIJAN
Charjew
Qurghonteppa
KARAKORAM RANGE
ISTANBUL
Zonguldak
Samsun
Trabzon
ARMENIA
BAKI
GARAGUM
Termiz
Kunduz
Ordu
YEREVAN
Ashgabat
Mary
HINDU KUSH
BURSA
Kocaeli
ANKARA
Sivas
Erzurum
AZER.
Mazar-e-Sharif
Jalalabad
PESHAWAR
Islamabad
KASHMIR
Eskisehir
Lake Van
TABRIZ
Ardabil
IZMIR
Usak
TURKEY
Elazig
Van
Rasht
CENTRAL
KABOL (KABUL)
GUJRANWALA
Kayseri
Malatya
Diyarbakir
Orumiyeh
Zanjan
Sari
MASHHAD
Khyber Pass
RAWALPINDI
HIMALAYAS
Konya
Kahramanmaras
Lake Orumiyeh
ELBURZ MTS.
PAROPAMISUS RA.
Denizli
Sanliurfa
GAZIANTEP
MOSUL
KARAJ
TEHRAN
Herat
HIGHLANDS
LAHORE
Antalya
ADANA
Irbil
Sanandaj
Hamadan
SALT DESERT
AFGHANISTAN
LUDHIANA
ALEPPO
Tigris
ZAGROS
QOM
DASHT-E LUT
Crete
Al Ladhiqiyah
Kirkuk
Arak
FAISALABAD
Ganges
Nicosia
Homs
SYRIA
MESOPOTAMIA
Euphrates
CYPRUS
SYRIAN
Kermanshah
IRAN
Kandahar
MULTAN
DELHI
LEB.
BAGHDAD
ISFAHAN
IRANIAN PLATEAU
RIGESTAN
Quetta
New Delhi
BEIRUT
Karbala
Kut
Yazd
Sea
DAMASCUS
DESERT
IRAQ
Amarah
PAKISTAN
THAR DESERT
LUCKNOW
Haifa
KANPUR
ISRAEL
Jerusalem
Najaf
Ahvaz
Kerman
JAIPUR
Tel Aviv-Jaffa
Diwaniyah
CHAGAI HILLS
Port Said
AMMAN
Nasiriyah
Abadan
Zahedan
Indus
ALEXANDRIA
Ar'ar
BASRA
SHIRAZ
INDIA
JORDAN
Kuwait
Bushehr
CAIRO
Suez
Sakakah
KUWAIT
MOUNTAINS
BALUCHISTAN
El Faiyum
SINAI
Eilat
AN NAFUD DESERT
Al Ahmadi
Bandar-e-Abbas
HYDERABAD
BHOPAL
Abu Zenima
Hafir al Batin
Persian Gulf
INDORE
Qattara Depression (-133 m, -436 ft.)
El Tur
Tabuk
Jubail
Strait of Hormuz
VINDHYA RANGE
Minya
EASTERN DESERT
Nile
Ha'il
Ad Damman
BAHRAIN
Dubai
OMAN
KARACHI
AHMADABAD
WESTERN
HEJAZ
ARABIAN
Manama
Gulf of Oman
Tropic of Cancer
Gulf of Kutch
SATPURA RANGE
Asyut
Doha
Suhar
Mutrah
VADODARA
NAGPUR
DESERT
Qena
Buraydah
Al Mobarraz
QATAR
EGYPT
Al Hofuf
Abu Dhabi
As Sib
Muscat
Gulf of Khambhat
El Kharga
Thebes
RIYADH
Al Kharj
UNITED ARAB EMIRATES
Sur
Nizwa
Ra's al Hadd
PUNE
Al Madinah (Medina)
SAUDI
Aswan
Yanbu
(BOMBAY) MUMBAI
Aswan High Dam
Red Sea
ARABIA
Lake Nasser
WESTERN GHATS
Masira
A
PENINSULA
MAKKAH (MECCA)
RUB AL KHALI (EMPTY QUARTER)
Wadi Halfa
JIDDAH
NUBIAN DESERT
OMAN
Panaji
Port Sudan
TIHAMAT ASH SHAM
Arabian Sea
Dongola
Khamis Mushayt
Merowe
Abha
Najran
Salalah
Mengaluru (Mangalore)
Jizan
SUDAN
YEMEN
Saywan
SAN'A
Asmara
Ash Shihr
OMDURMAN
Kassala
ERITREA
Hudaydah
Dhamar
Al-Mukalla
KHARTOUM
Wad Madani
Ibb
'Abd al Kuri (YEMEN)
Socotra (YEMEN)
Gedaref
Mek'ele
Ta'izz
Adan (Aden)
Gulf of Adan
El Obeid
Kosti
Ras Dejen (4,620 m, 15,158 ft.)
Assab
Cape Gwardafuy
Fasher
Gonder
DJIBOUTI
Boosaaso
Lake Tana
Blue Nile
White Nile
Djibouti
Bahir Dar
Dese
Berbera
HORN OF AFRICA
N
Dire Dawa
Burao
ETHIOPIAN
Hargeysa
Harer
Garoowe
ADIS ABEBA
Malakal
ETHIOPIA
SOMALIA
Gore
OGADEN
HIGHLANDS
INDIAN OCEAN
Wau
Awasa
SOUTH SUDAN
Arba Minch
Obo
Juba
Luq
Lake
Equator

0 km 400 800 1000 1200
0 miles 200 400 600 800
Albers Equal-Area Projection
Scale 1:26,600,000

40° 50° 60° 80°

IN THIS CHAPTER

- The cradle of civilization
- The Arab Spring and its volatile aftermath
- Oil: A mixed blessing
- The shifting geographies of terrorism in the name of Islam

FIGURE 7A-1

CONCEPTS, IDEAS, AND TERMS

Cultural geography 1
Culture hearth 2
Cultural diffusion 3
Cultural landscape 4
Hydraulic civilization theory 5
Climate change 6
Spatial diffusion 7
Expansion diffusion 8
Relocation diffusion 9
Islamization 10
Stateless nation 11
Choke point 12
Fragmented modernization 13
League of Nations 14
Religious revivalism 15
Jihad 16
Wahhabism 17
Domino effect 18
Political regime 19

From Morocco on the shores of the Atlantic to the mountains of Afghanistan, and from the Horn of Africa to the steppes of inner Asia, lies a vast geographic realm of enormous cultural complexity. It stands at the crossroads where Europe, Asia, and Africa meet, and it is part of all three (Fig. 7A-1). Throughout history, its influences have radiated across Eurasia and Africa and reached practically every other part of the world as well. This is one of humankind's primary source areas. On the Mesopotamian Plain between the Tigris and Euphrates Rivers (in modern-day Iraq) and on the banks of the Egyptian Nile arose several of the world's earliest civilizations. In its soils, plants were domesticated that are now grown from the Americas to Australia. Along its paths walked prophets whose religious teachings are still followed by hundreds of millions. And in this second decade of the twenty-first century, the heart of this realm is plagued by political turbulence, religious strife, and some of the most dangerous conflicts in the world.

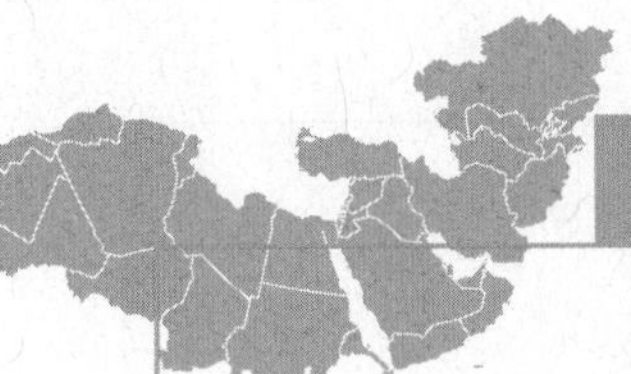

major geographic qualities of NORTH AFRICA/SOUTHWEST ASIA

1. North Africa and Southwest Asia were the scene of several of the world's great ancient civilizations, based in its river valleys and basins.
2. From this realm's culture hearths diffused ideas, innovations, and technologies that changed the world.
3. The North Africa/Southwest Asia realm is the source of three world religions: Judaism, Christianity, and Islam.
4. Islam, the most recent of the major religions to arise in this realm, transformed, unified, and energized a vast domain extending from Europe to Southeast Asia and from Russia to East Africa.
5. Drought and unreliable precipitation dominate natural environments in this realm. Population clusters occur where water supply is adequate to marginal.
6. Certain countries of this realm have enormous reserves of oil and natural gas, creating great wealth for some but doing little to raise the living standards of the majority.
7. The boundaries of the North Africa/Southwest Asia realm consist of volatile transition zones in several places in Africa and Asia.
8. Conflict over water sources and supplies is a constant threat in this realm, where population growth rates are high by current world standards.
9. The Middle East, as a region, lies at the heart of this realm; and Israel lies at the center of the Middle East conflict.
10. The broad commonalities and interrelationships across much of the realm were demonstrated in the recent spread of political uprisings against autocratic regimes. But the results of this movement remain quite unclear and vary from country to country.

DEFINING THE REALM

NAMING THIS PIVOTAL REALM

As you may conclude from its long and somewhat cumbersome name—North Africa/Southwest Asia—this is a sprawling, geographically complex realm. In our era of high-speed communication you will sometimes see it referred to as ***NASWA***, the first letters of its regional deployment. And it is tempting to refer to it in other kinds of geographic shorthand, based on some of its dominant features. But such generalizations can be misleading, and some examples follow.

© Shawn Baldwin/Corbis

Cairo, February 1, 2013: Egyptian protesters chant anti-government slogans during a rally in Tahrir Square. More than a year after the ouster of its longtime president, Hosni Mubarak, Egypt's 'democratic revolution' has stalled. The Arab Spring that began in late 2010 shook much of this realm, but has yet to produce any genuine, lasting progress toward democracy.

A "Dry World"?

This realm is, for instance, sometimes called the Dry World, containing as it does the vast Sahara as well as the Arabian Desert. But most of the realm's people live where there is water—along the Nile River, along the hilly Mediterranean coastal strip or *tell* (meaning mound in Arabic) of northwesternmost Africa, along the Asian eastern and northeastern shores of the Mediterranean Sea, in most of the Tigris-Euphrates Basin, in far-flung desert oases, and along the lower mountain slopes of Iran south of the Caspian Sea as well as Turkestan to the northeast. Figure 7A-2 reflects this co-varying relationship between population distribution and available water. Quite a few of this realm's prominent geographic features—river valleys, basins, and deltas; moist coastlines; adequately watered mountain basins are virtually defined by the clusters you see on this map.

So we know this world realm as a place where water is almost always at a premium; where consumers (such as Israelis and their Arab neighbors) compete for limited

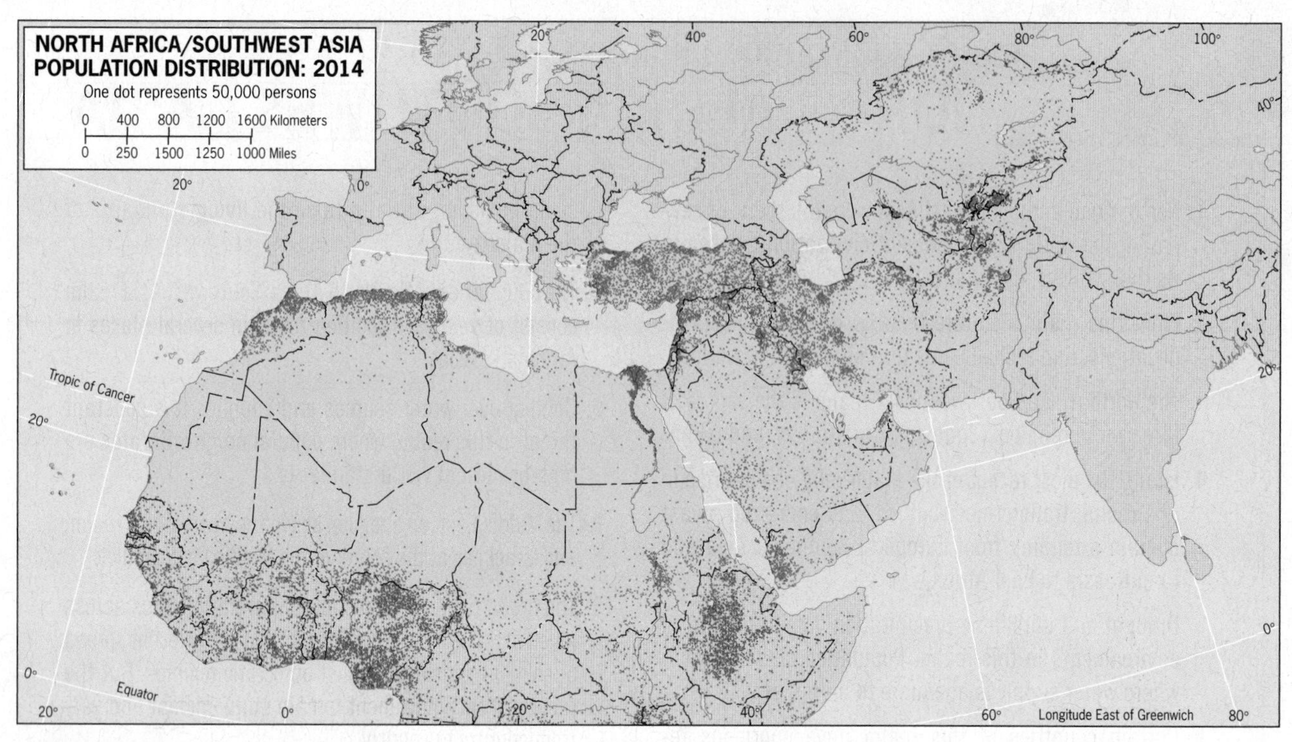

FIGURE 7A-2

supplies; where peasants often struggle to make soil and moisture yield a small harvest; where nomadic peoples and their livestock still migrate across dust-blown flatlands; where oases are green islands in the desert that sustain local farming, supply weary travelers, and support trade across vast expanses of aridity.

But this is also the land of the Nile, the lifeline of Egypt, the crop-covered *tell* of coastal Algeria, the verdant shores of western Turkey, and the meltwater-fed valleys of Central Asia. Compare Figure 7A-1 to Figure G-7, and the dominance of ***B*** climates becomes evident, underscoring just how water dependent this realm's clustered population is.

Is *This* the "Middle East"?

This realm is commonly referred to as the Middle East. That must sound odd to someone in, say, India, who might think of a Middle West rather than a Middle East! The name, of course, reflects the biases of its source: the Western world, which saw a Near East in Turkey; a Middle East in Egypt, Arabia, and Iraq; and a Far East in China and Japan. Still, the term has taken hold, and it can be seen and heard in everyday usage by scholars, journalists, and members of the United Nations. Even so, we feel that it should only be applied to one of the regions of this sprawling realm, not to NASWA as a whole.

An "Arab World"?

Another shorthand label often used for North Africa/Southwest Asia is the Arab World. This term implies a uniformity that does not actually exist. First, the name *Arab* is applied loosely to the peoples of this area who speak Arabic and related languages, but ethnologists normally restrict it to certain occupants of the Arabian Peninsula—the Arab "source." In any case, Turks are not Arabs, and neither are most Iranians or Israelis. Moreover, although the Arabic language prevails from Mauritania in the west across all of North Africa to the Arabian Peninsula, Syria, and Iraq in the east, it is not spoken in other parts of this realm. In Turkey, for example, Turkish is the major language, and it has Ural-Altaic rather than Arabic's Semitic or Hamitic roots; the Iranian language is a member of the Indo-European linguistic family (see Fig. G-9). Other Arab World languages that have separate ethnological identities are spoken by the Jews of Israel, the Tuareg people of the western Sahara, the Berbers of northwestern Africa, and the myriad peoples of the wide transition zone between North Africa and Subsaharan Africa to the south.

An "Islamic World"?

Finally, and perhaps most significantly, this realm is routinely referred to as the Islamic World. Indeed, the prophet Muhammad was born on the Arabian Peninsula in AD 571, and Islam, the religion to which he gave rise, spread across this realm and beyond after his death in 632. This great saga of Arab conquest saw Islamic armies, caravans, and fleets carry Muhammad's teachings across deserts, mountains, and oceans, penetrating Europe, converting the kings of West African states, threatening the Christian strongholds of Ethiopia, invading Central Asia, conquering most of what is now India, and even reaching the peninsulas and islands of

Southeast Asia. Today, the world's largest Muslim state is Indonesia, and the Islamic faith extends far outside the realm under discussion (Fig. G-10). In this context, the term Islamic World is also misleading in that it suggests that there is no Islam beyond NASWA's borders.

In any case, this World of Islam is not entirely Muslim either. Christian minorities continue to survive in all of the regions and many of the countries of the North Africa/Southwest Asia realm. Judaism has its base in the Middle East region; and smaller religious communities, such as Lebanon's Druze and Iran's Bahais, many of them under pressure from Islamic regimes, continue to diversify the religious mosaic.

Nonetheless, Islam's impact on this realm's cultural geography is pervasive. It ranges from dominant in such countries as Iran, Saudi Arabia, and Egypt to laminate (that is, more of a veneer) in several of the countries of Central Asia's Turkestan, where the former Soviet rulers discouraged the faith and where its comeback is still in progress—one of the criteria that allow us to identify Turkestan as a discrete region within this realm. The rise, impact, and continuing evolution of Islam lie at the heart of the story of this chapter.

States and Nations

Although Islam and its cultural expressions dominate this geographic realm, its political and social geographies are divided and often fractious. We will encounter states with strong internal divisions (Lebanon, Iraq), nations as yet without their own states (Palestinians, Kurds, Berbers), and territories still in the process of integration (Western Sahara, Palestine). No single state dominates this realm as Mexico eclipses the rest of Middle America or Brazil looms over South America. Of its nearly 30 states and territories, the three largest and in many ways the most important are Egypt in North Africa, Turkey on the threshold of Europe, and Iran at the margins of Turkestan. All three have populations between 77 and 86 million, modest by world standards. But at the other end of this continuum there are mini-states such as Bahrain, Qatar, and Djibouti, each with fewer than two million inhabitants. It is quite important to study Figure 7A-1 attentively because the boundary framework—much of it inherited from the colonial era—lies at the root of some of the realm's current problems.

It is also useful to take a careful look at Figure 7A-2 because it clearly shows how differently populations are distributed within the countries we will examine. Compare the three largest countries just mentioned: Egypt's population forms a river-valley ribbon and delta cluster; Turkey's, between the Black Sea and the Mediterranean Sea, is far more evenly dispersed, although its west is more populous than the east; and in Iran you can draw a line down its center from the Caspian Sea to the Persian Gulf that separates the comparatively crowded west, where the capital is located, from the drier and far more sparsely peopled east. In North Africa, almost everyone seems to live along or near the Mediterranean coast, but no such coastal concentrations mark the Arabian Peninsula except in its far south, where Yemen's inhabitants cluster near the Red Sea. Certainly the historic role of water can be seen in the population patterns displayed in Figure 7A-2.

HEARTHS OF CULTURES

This geographic realm occupies a pivotal part of the world: here Africa, the source of humankind, meets Eurasia, crucible of human cultures. Two million years ago, the ancestors of our species walked from East Africa into North Africa and Arabia and then spread all across Asia. Less than one hundred thousand years ago, *Homo sapiens* crossed these lands on their way to Europe, Australia, and, eventually, the Americas. Ten thousand years ago, human communities in what we now call the Middle East began to domesticate plants and animals, learned to irrigate their fields, enlarged some of their settlements into towns, and formed the earliest states. The world's dominant monotheistic religions originated here: first Judaism, then came Christianity, and most recently the heart of the realm was stirred and mobilized by the teachings of Muhammad and the Quran (Koran), and Islam arose. Today this realm is a cauldron of religious and political activity as well as turmoil—complicated by foreign interventions; weakened by conflict; empowered and enriched by oil in some places; and plagued by poverty in many others.

Dimensions of Culture

In the introductory chapter, we discussed the concept of culture and its regional expression in the cultural landscape. **Cultural geography [1]**, we observed, is a wide-ranging and comprehensive field that studies spatial aspects of human cultures, focusing not only on cultural landscapes but also on **culture hearths [2]**—the crucibles of civilization, the sources of ideas, innovations, and ideologies that transformed regions and realms. Those ideas and innovations spread far and wide through a set of processes that we study under the rubric of **cultural diffusion [3]**. Because we understand these processes better today, we can reconstruct many of the ancient routes by which the knowledge and achievements of culture hearths spread—that is, *diffused*—to other areas.

Another aspect of cultural geography, particularly noteworthy in the context of the North Africa/Southwest Asia realm, is the study of **cultural landscapes [4]** that a dominant culture creates. Human cultures exist in long-term accommodation with (and adaptation to) their natural environments, exploiting opportunities that these environments present and coping with the extremes they can impose. In the process, they fuse their physical landscape and cultural landscape into an interacting unity.

Rivers and Communities

In the basins of the major rivers of this realm (the Tigris-Euphrates system of modern-day Turkey, Syria, and Iraq as well as the Nile of Egypt) lay two of the world's earliest culture hearths (Fig. 7A-3). Mesopotamia (which means

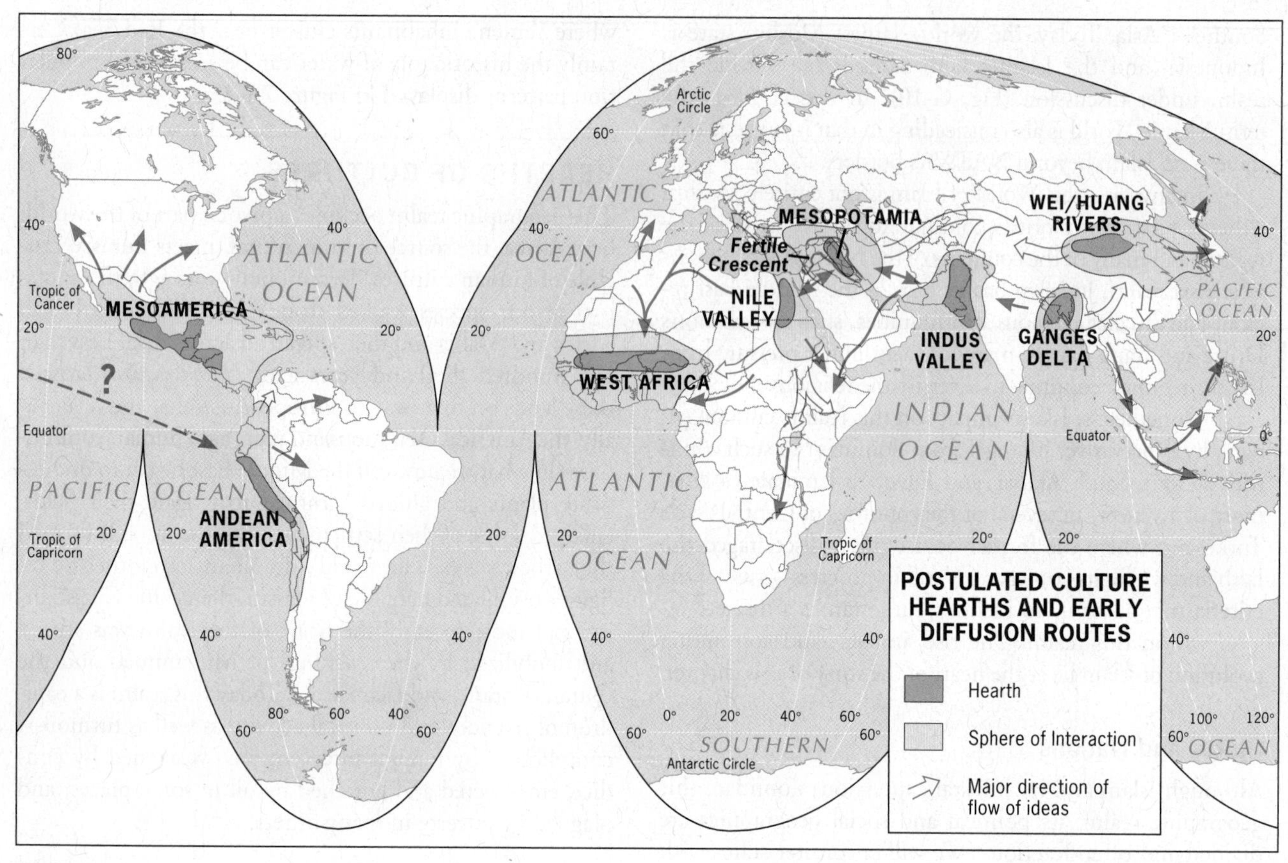

FIGURE 7A-3

land between the rivers) possessed fertile alluvial soils, abundant sunshine, ample water, and animals and plants that could be domesticated. Here, in the Tigris-Euphrates lowland between the Persian Gulf and the uplands of present-day Turkey, arose one of humanity's first culture hearths, a cluster of communities that grew into larger societies and, eventually, into the world's first states.

Mesopotamia

Mesopotamians were innovative farmers who domesticated cereal grains such as wheat and knew when to sow and harvest various other crops, water their fields, and store their surpluses. Their knowledge spread to villages near and far, and a ***Fertile Crescent***, a region of significant agricultural productivity, evolved, extending from Mesopotamia across southeastern Turkey into Syria and the eastern Mediterranean coast beyond (Fig. 7A-3).

Irrigation was the key to prosperity and power in Mesopotamia, and urbanization was its major reward. Among many settlements in the Fertile Crescent, some thrived, grew, enlarged their hinterlands, and diversified socially and occupationally; others failed. What determined success? One theory, the **hydraulic civilization theory [5]**, holds that cities that could control irrigated farming over large hinterlands held power over others, used food as a weapon, and prospered. One such city, Babylon on the Euphrates River, endured for nearly 4000 years (from 4100 BC). A busy port, its walled and fortified center endowed with temples, towers, and palaces, Babylon for a time was the world's largest city.

Egypt and the Nile

Egypt's cultural evolution may have started even earlier than Mesopotamia's, and its focus lay upstream from (south of) the Nile Delta and downstream from (north of) the first of the Nile's series of rapids, or cataracts. This part of the Nile Valley is surrounded by inhospitable desert, and unlike Mesopotamia (which lay open to all comers), the Nile provided a natural fortress here. The ancient Egyptians converted their security into progress. The Nile was their highway of trade and interaction; it also supported agriculture through irrigation (see photo). The Nile's cyclical ebb and flow was much more predictable than that of the Tigris-Euphrates river system. By the time Egypt fell victim to outside invaders (around 1700 BC), a full-scale urban civilization had emerged. Ancient Egypt's artist-engineers left a magnificent legacy of massive stone monuments, some of them containing treasure-filled crypts of god-kings called Pharaohs. These tombs have enabled archeologists to reconstruct the ancient history of this culture hearth.

Today, the world continues to benefit from the accomplishments of the ancient Mesopotamians and Egyptians. They domesticated cereals (wheat, rye, barley), vegetables (peas, beans), fruits (grapes, apples, peaches), and several animals (horses, pigs, sheep). They also advanced the study

From the Field Notes . . .

"Flying north over central Egypt from Luxor's ruins toward Cairo along the Nile's ribbon of green, I pondered the famous remark by the ancient Greek historian Herodotus that 'Egypt is the gift of the River Nile.' The Nile brings water from the interior of Africa that crosses the vast northern deserts and is crucially important to a country with as little rainfall as Egypt. About 95 percent of the Egyptian population lives within 20 kilometers (12 mi) of the river's banks. Almost all of Egypt's agriculture has always depended on irrigation, which is how this narrow strip of desert turned fertile. Notice the razor-sharp boundaries between irrigated and non-irrigated land. Without a doubt, the Nile was the lifeline that supported one of the ancient world's major civilizations."

© Jan Nijman

www.conceptcaching.com

of the calendar, mathematics, astronomy, government, engineering, metallurgy, and a host of other skills and technologies. In time, many of their innovations were adopted and then modified by other cultures in the Old World and eventually in the New World as well. Europe was the greatest beneficiary of these legacies of Mesopotamia and ancient Egypt, whose achievements constituted the foundations of Western civilization.

Decline and Decay

As Figure G-7 reminds us, many of the early cities of this culture hearth lay in what is today desert territory. Assuming that there were no good reasons to build large settlements in the middle of deserts, we may hypothesize that **climate change [6]** sweeping across this region, not a monopoly over irrigation techniques, gave certain cities in the ancient Fertile Crescent an advantage over others. Climate change, closely associated with shifting environmental zones following the final Pleistocene glacial retreat, may have destroyed the last of the old civilizations. Perhaps overpopulation and human destruction of the natural vegetation contributed to the process. Indeed, some cultural geographers suggest that the momentous innovations in agricultural planning and irrigation technology were not "taught" by the seasonal flooding of the rivers but were forced on the inhabitants as they tried to survive changing environmental conditions.

The scenario is not difficult to imagine. As outlying areas began to fall dry and farmlands were destroyed, people congregated in the already crowded river valleys—and made every effort to increase the productivity of the land that could still be watered. Eventually overpopulation, destruction of the watershed, and perhaps reduced rainfall in the rivers' headwater areas dealt the final blow. Towns were abandoned to the encroaching desert; irrigation channels filled with drifting sand; croplands dried up. Those who could migrated to areas that were still reputed to be productive. Others stayed, their numbers dwindling, increasingly reduced to subsistence.

As old societies disintegrated, power emerged elsewhere. First the Persians, then the Greeks, and later the Romans imposed their imperial designs on the tenuous lands and disconnected peoples of North Africa/Southwest Asia. Roman technicians converted North Africa's farmlands into irrigated plantations whose products went by the boatload to Roman Mediterranean shores. Thousands of people were carried off as slaves to the cities of the new conquerors. Egypt was quickly colonized, as was the area we now call the Middle East. One region that lay distant, and therefore remote from these invasions, was the Arabian Peninsula, where no major culture hearth or large cities had emerged and where the turmoil had not affected Arab settlements and nomadic routes.

STAGE FOR ISLAM

In a remote place on the Arabian Peninsula, where the foreign invasions of the Middle East had had little effect on the Arab communities, an event occurred early in the seventh century that was to change history and affect the destinies of people in many parts of the world. In a town called Mecca (Makkah), about 70 kilometers (45 mi) from the Red Sea coast in the Jabal Mountains, a man named Muhammad in the year AD 611 began to receive revelations from Allah (God). Muhammad (571–632) was then in his early forties. Arab society was in social and cultural disarray, but Muhammad forcefully taught Allah's lessons and began to transform his culture. His personal power soon attracted enemies, and in 622 he fled from Mecca to the safer haven of Medina (Al Madinah), where he continued his work until his death. This moment, the *hejira* (meaning migration), marks the starting date of the Muslim era, Year 1 on Islam's calendar. Mecca, of course, later became Islam's holiest place.

The Faith

The precepts of Islam in many ways constituted a revision and embellishment of Judaic and Christian beliefs and traditions. All of these faiths have but one god, who occasionally communicates with humankind through prophets; Islam acknowledges that Moses and Jesus were such prophets but considers Muhammad to be the final and greatest prophet.

Islam brought to the Arab World not only the unifying religious faith it had lacked but also a new set of values, a new way of life, a new individual and collective dignity. Islam dictated observance of the Five Pillars: (1) repeated expressions of the basic creed, (2) daily prayer, (3) a month each year of daytime fasting (Ramadan), (4) the giving of alms, and (5) at least one pilgrimage to Mecca in each Muslim's lifetime (the *hajj*). Islam prescribed and proscribed in other spheres of life as well. It forbade alcohol, smoking, and gambling. It tolerated polygamy, even though it acknowledged the virtues of monogamy. Mosques appeared in Arab settlements, not only for the (Friday) sabbath prayer, but also as social gathering places to knit communities closer together. Mecca became the spiritual center for a divided, widely dispersed people for whom a collective focus was something new.

The Arab-Islamic Empire

Muhammad provided such a powerful stimulus that Arab society was mobilized almost overnight. The prophet died

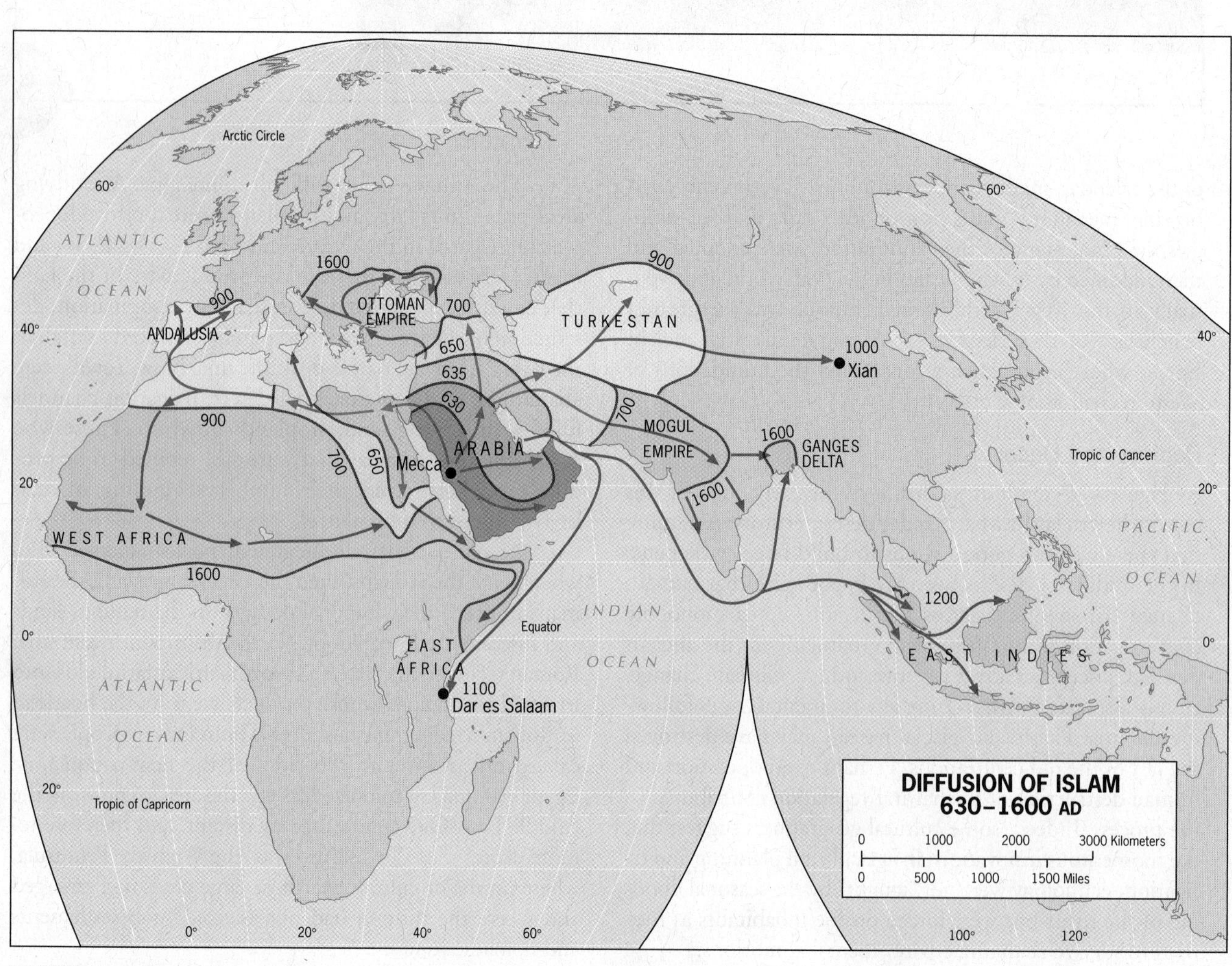

FIGURE 7A-4

in 632, but his faith and fame spread like wildfire. Arab armies carrying the banner of Islam formed, invaded, conquered, and converted wherever they went. As Figure 7A-4 shows, by AD 700 Islam had reached far into North Africa, into Transcaucasia, and into most of Southwest Asia. In the centuries that followed, it penetrated southern and eastern Europe, Central Asia's Turkestan, West Africa, East Africa, and South and Southeast Asia, even reaching China by AD 1000.

The spread of Islam provides a good illustration of a series of processes known as **spatial diffusion [7]** that focus on the way ideas, inventions, and cultural practices propagate through a population in space and time. Diffusion takes place in two forms: **expansion diffusion [8]**, when propagation waves originate in a strong and durable source area and spread outward, affecting an ever larger region and population; and **relocation diffusion [9]**, in which migrants carry an innovation or an idea (or belief). Islam spread through expansion as well as relocation diffusion, reaching such faraway places as the Ganges Delta in South Asia, present-day Indonesia, and East Africa.

The heart of Islam remained in Southwest Asia. There, Islam became the cornerstone of an Arab Empire with Medina as its first capital. As the empire expanded, its headquarters shifted from Medina to Damascus (in what is now Syria) and later to Baghdad on the Tigris River (in modern Iraq). And it prospered. In architecture, mathematics, and science, the Arabs overshadowed their European contemporaries. The Arabs established institutions of higher learning in many cities including Cairo, Baghdad, and Toledo (Spain), and their distinctive cultural landscapes (see photos) united their far-flung domain (see box titled "The Flowering of Islamic Culture"). Non-Arab societies in the path of the Muslim drive were not only Islamized, but also Arabized, adopting other Arab traditions as well. Islam had spawned a culture; it still lies at the heart of that culture today.

As we have noted, Islam's diffusion eventually was checked in Europe, Russia, and elsewhere. But a map showing the total area under Muslim sway in Eurasia and Africa reveals the enormous dimensions of the domain affected by **Islamization [10]** at one time or another (Fig. 7A-5). Islam continues to expand, now mainly by relocation diffusion. There are Islamic communities in cities as widely scattered as Vienna, Singapore, and Cape Town, South Africa; Islam is also growing rapidly in the United States. With about 1.7 billion adherents today (24 percent of humankind), Islam is a vigorous, burgeoning, and still-growing cultural force around the world.

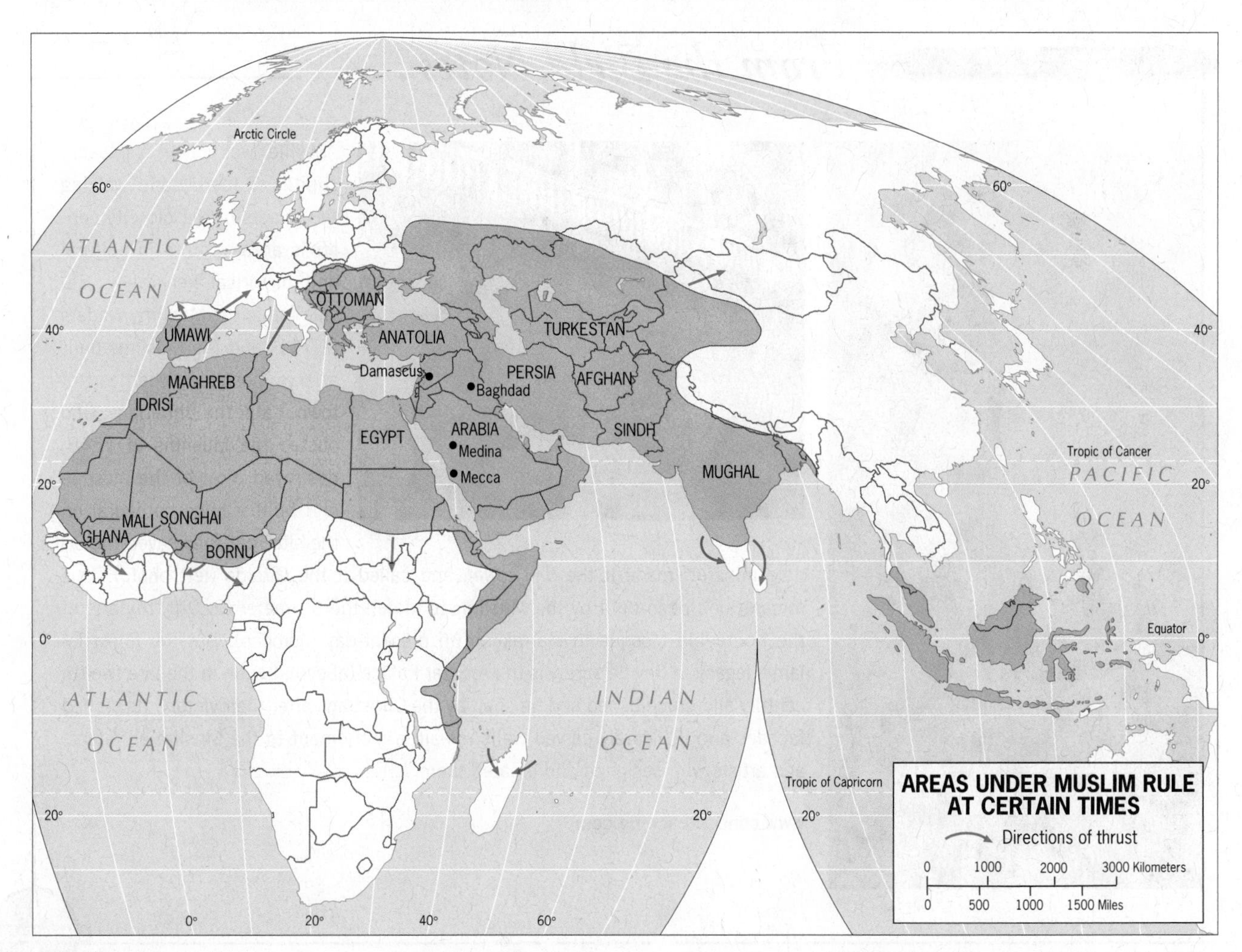

FIGURE 7A-5

The Flowering of Islamic Culture

THE CONVERSION OF a vast realm to Islam was not only a religious conquest: it was also accompanied by a glorious explosion of Arab culture. In science, the arts, architecture, and other fields, Arab society far outshone European society. While the western European remnants of the Roman Empire languished, Arab energies soared.

When the wave of Islamic diffusion reached the Maghreb (western North Africa), the Arabs saw, on the other side of the narrow Strait of Gibraltar, an Iberia ripe for conquest and ready for renewal. An Arab-Berber alliance, called the Moors, invaded Spain in AD 711 and controlled all but northern Castile and Catalonia before the end of the eighth century.

It took seven centuries for Catholic armies to recapture all of the Arabs' Iberian holdings, but by then the Muslims had made an indelible imprint on the Spanish and Portuguese cultural landscape. The Arabs brought unity and imposed the rule of Baghdad, and their works soon overshadowed what the Romans had wrought.

Al-Andalus, as this westernmost outpost of Islam was called, was endowed with thousands of magnificent castles, mosques, schools, gardens, and public buildings. The ultimately victorious Christians destroyed most of the less durable art (pottery, textiles, furniture, sculpture) and burned the contents of Islamic libraries, but the great Islamic structures survived, including the Alhambra in Granada, the Giralda in Seville, and the Great Mosque of Córdoba, three of the world's greatest architectural achievements. While Spanish and Portuguese culture became Hispanic-Islamic culture, the Muslims were transforming their cities from Turkestan to the Maghreb in the image of Baghdad. The lost greatness of a past era still graces those townscapes today.

From the Field Notes . . .

© H.J. de Blij

© H.J. de Blij

"Seville was once a Muslim capital of Iberia, and walking the streets of the old city center is an adventure in cultural and historical geography. Seville fell to the Muslim invaders in 711, and the Muslims built a large mosque in the heart of town. Later the Christians, who ousted the Muslims in 1248, destroyed most of the mosque and built a huge cathedral on the site, but they saved the ornate minaret, made it the bell tower, and called it the Giralda (left photo). This minaret had been built by the Muslims to match the 67-meter (220-ft) minaret of the Koutoubia Mosque in Marrakech (in present-day Morocco). But the finest Islamic legacy in Seville surely is the Alcazar Palace (above), begun in the late twelfth century and embellished and finished by the Christians after their victory. Its arched façades and intricately carved walls remain a monument to the Muslim architects and artists who designed and created them."

www.conceptcaching.com

Islam and Other Religions

Two additional major faiths also had their sources in the ***Levant*** (the area extending from Greece eastward along the Mediterranean coast to northern Egypt)—Judaism and Christianity—and both are older than Islam. Islam's rise submerged many smaller Jewish communities, but the Christians, not the Jews, waged centuries of holy war against the Muslims, seeking, through the Crusades, not only to drive Islam back but to reestablish Christian communities where they had dominated before the Islamic expansion. The aftermath of that campaign still marks the realm's cultural landscape today. A substantial Christian minority remains in Lebanon, and Christian minorities also survive in Israel, Egypt, Syria, and Jordan. By far the most intense conflict in modern times has pitted the Jewish state, Israel, against its Islamic neighbors near and far. Jerusalem—holy city for Judaism, Christianity, and Islam—lies in the crucible of this confrontation.

ISLAM DIVIDED

For all its vigor and success, Islam is still fragmented into sects. The earliest and most consequential division arose after Muhammad's death between those who believed that only a blood relative should follow the prophet as leader of Islam (***Shi'ites***) and others who felt that any devout follower of Muhammad was qualified to take over (***Sunnis***). In all, the Sunnis prevailed and, in the following centuries, the great expansion of Islam was largely propelled by Sunnis; the Shi'ites survived as minorities scattered throughout the realm. Today, more than 80 percent of all Muslims are Sunnis, but the Shia–Sunni schism is at the root of various conflicts in this realm.

The minority Shi'ites have always vigorously promoted their version of the faith. In the early sixteenth century their work paid off: the royal house of Persia (present-day Iran) made Shi'ism the only legal religion throughout its vast empire. That domain extended from Persia into lower Mesopotamia, into Azerbaijan, and into western Afghanistan and Pakistan. As the world map of religions (Fig. G-10) shows, this created for Shi'ism a large culture region and gave the faith unprecedented strength. Iran remains the bastion of Shi'ism in the realm today, and its appeal continues to radiate into neighboring countries and even farther afield.

During the late twentieth century, the schism was reignited when the *shah* (king) of Persia tried to secularize the country and to limit the power of the *imams* (mosque officials); he provoked a revolution that cost him the throne and made Iran a Shi'ite Islamic republic—the opposite of what he had intended. Before long, Iran was at war with neighboring Sunni-ruled but Shi'ite-majority Iraq, and Shi'ite parties and communities elsewhere were invigorated by the newfound power of Shi'ism. From Arabia to Africa's northwestern corner, Sunni-ruled countries warily monitored their Shi'ite minorities, newly imbued with religious fervor. Mecca, the sacred place for both Sunnis and Shi'ites, became a battleground during the week of the annual pilgrimage, and for a time the (Sunni) Saudi Arabian government denied entry to Shi'ite pilgrims. That rift has healed somewhat, but intra-Islamic sectarian differences run deep. And the regional upheaval accompanying the "Arab Spring" movement that arose in late 2010 has in some ways brought these differences back to the surface—from Shi'ites protesting minority Sunni rule in Bahrain to Syria's majority Sunnis rebelling against a regime whose leaders adhere to a branch of Shi'ism.

The Ottoman Empire and Its Aftermath

If we want to understand this sprawling realm today, we must appreciate the historical role of the Ottoman Empire, its enormous expansion, and the ways it was undone by other great powers. It is ironic that Islam's last great advance into Europe eventually led to European occupation of Islam's very heartland. The Ottomans (named after their leader, Osman I), based in what is today Turkey, conquered Constantinople (now Istanbul) in 1453 and then pushed into southeastern Europe. Soon Ottoman forces were on the doorstep of Vienna; they also invaded Persia, Mesopotamia, and North Africa (Fig. 7A-6). At its height, the Ottoman Empire under Suleyman the Magnificent (ruled 1522–1560) was the most powerful state in western Eurasia. As Figure 7A-6 shows, its armies even advanced toward Moscow, Kazan, and Krakow from a base at Azov (near present-day Rostov in southwestern Russia). The Turks also launched marine attacks on Sicily, Spain, and France.

The Ottoman Empire survived for more than four centuries (it ended in 1923), but it lost territory as time went on, first to the Hungarians, then to the Russians, and later to the Greeks and Serbs until, after World War I, the European powers took over its provinces and made them colonies—colonies we now know by the names of Syria, Iraq, Lebanon, and Yemen (Fig. 7A-7). As the map shows, the French and the British took large possessions; even the Italians annexed part of the Ottoman domain.

The boundary framework that the colonial powers created to delimit their holdings was not satisfactory. As Figure 7A-2 illustrates, this realm's population of just under 630 million is clustered, fragmented, and strung out in river valleys, coastal zones, and crowded oases. The colonial powers laid out long stretches of boundary as ruler-straight lines across uninhabited territory; they saw no need to adjust these boundaries to cultural or physical features in the landscape. Other boundaries, even some in desert zones, were poorly defined and never marked on the ground. Later, when the colonies had become independent states, such boundaries led to quarrels, even armed conflicts, among neighboring Muslim states. In some instances, they left entire nations, such as Kurds and Palestinians, without any territory of their own, so-called **stateless nations [11]** (see box titled "A Future Kurdistan?").

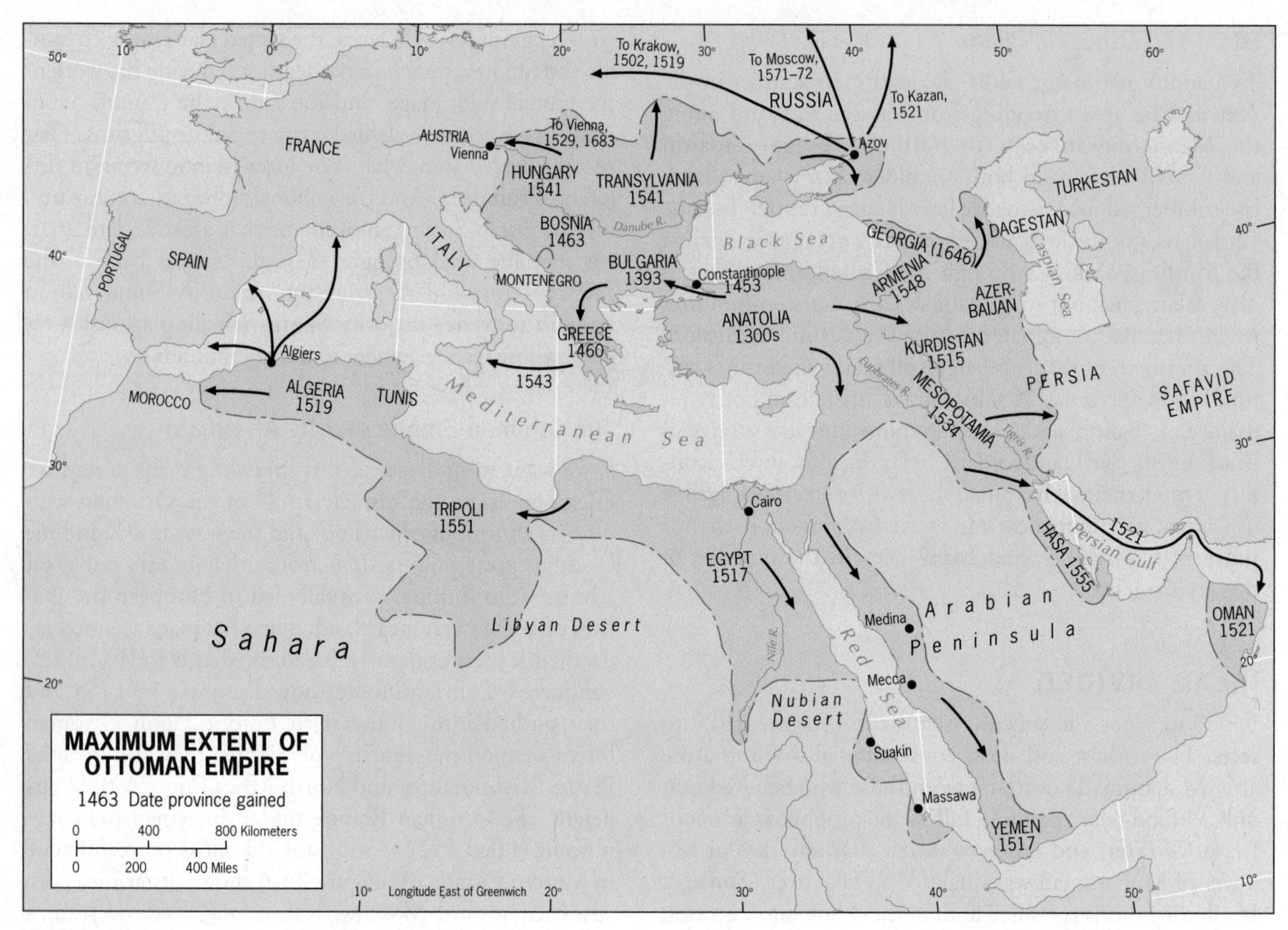

FIGURE 7A-6

A Future Kurdistan?

POLITICAL MAPS OF the heart of the North Africa/Southwest Asia realm do not show it, but where Turkey, Iraq, and Iran meet, the cultural landscape is not Turkish, Iraqi, or Iranian. Here live the Kurds, a fractious and fragmented nation of more than 35 million (their numbers are uncertain). More Kurds live in Turkey than in any other country (perhaps as many as 20 million); possibly as many as 6.5 million in Iraq; at least another 6 million in Iran; and smaller numbers in Syria, Armenia, and even Azerbaijan (see Fig. 7B-4).

The Kurds have occupied this isolated, mountainous, frontier zone for over 3000 years. They are a nation, but they have no state; nor do they enjoy the international attention that peoples of other *stateless nations* (such as the Palestinians) receive. Turkish as well as Iraqi repression of the Kurds, and Iranian betrayal of their aspirations, briefly make the news but are soon forgotten. Relative location has much to do with this: spatial remoteness and the obstacles created by the ruling regimes inhibit access to their landlocked domain.

Many Kurds dream of a day when their fractured homeland will become a nation-state. Most would agree that the city of Diyarbakir, now in southeastern Turkey, would become the capital. It is the closest any Kurdish town comes to a primate city, although the largest urban concentration of Kurds today is in the shantytowns of Istanbul, where well over 3 million have migrated. In their heartland, meanwhile, the Kurds, their shared goals notwithstanding, are a divided people whose intense disunity has thwarted their objectives; and, as in the case of Europe's Basques, a small minority of extremists has tended to use violence in pursuit of their aims.

The Kurds—without a seacoast, without a powerful patron, without a global public relations machine—are victims of a conspiracy between history and geography. But geography did provide them with one intriguing opportunity: Iraqi Kurdistan sits on major oil reserves and the Kurds are claiming this fuel is theirs to sell. Part of the intrigue is that the main customer is energy-hungry Turkey, which is also not on very good terms with the Iraqi government in Baghdad. Oil may provide the Kurds with a lifeline but it is fraught with risk and in the long run it may only dampen prospects of Iraqi recognition of independence. One of the world's largest stateless nations keeps hoping for a deliverance that is unlikely to come.

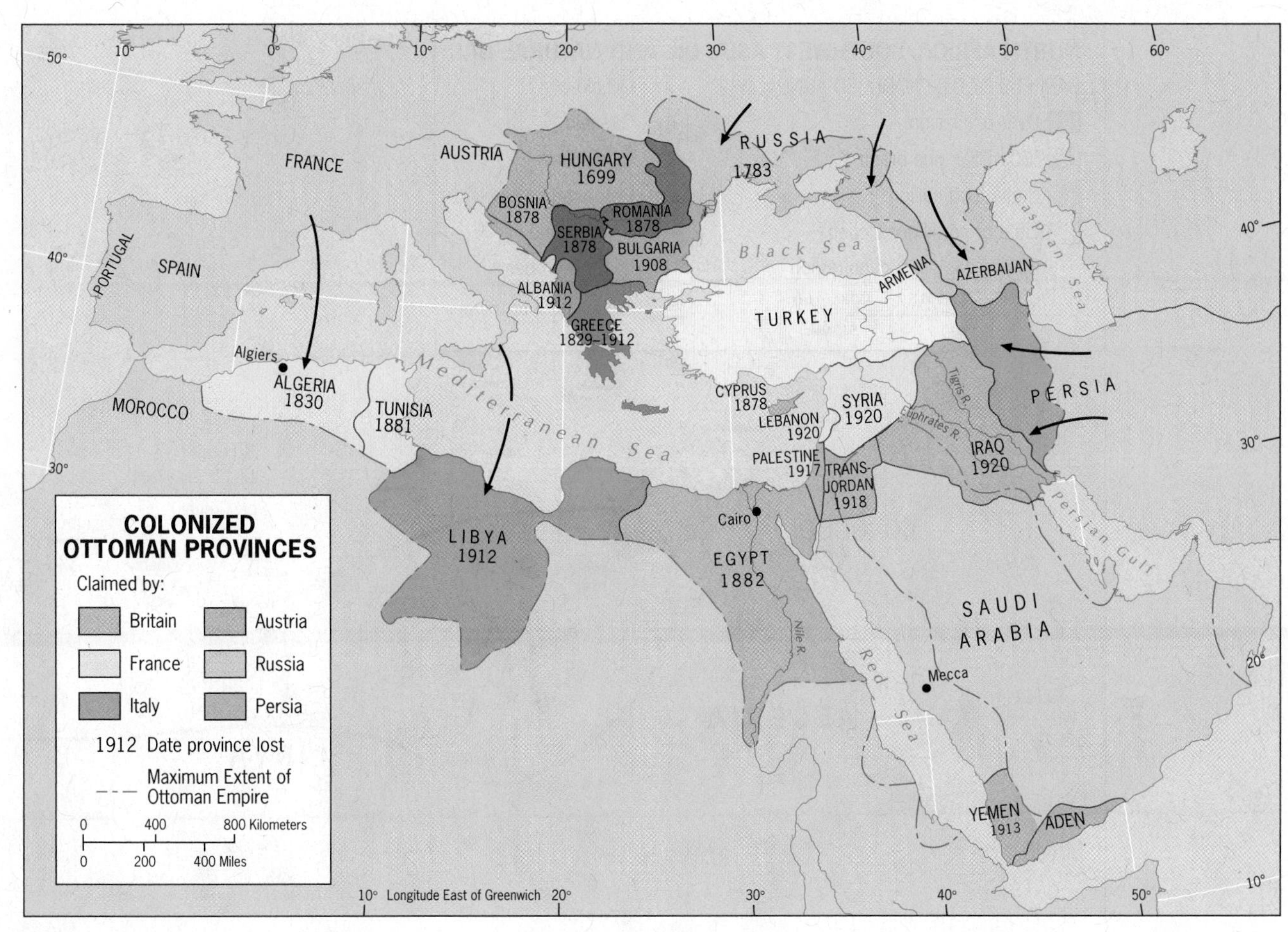

FIGURE 7A-7

THE POWER AND PERIL OF OIL

About 30 of the world's countries have significant oil reserves, with six of the nine leading countries located in this realm: Saudi Arabia, Iran, Iraq, Kuwait, the United Arab Emirates, and Libya. In general terms, oil (and associated natural gas) exists in this realm in three discontinuous zones (Fig. 7A-8). The most productive of these zones extends from the southern and southeastern part of the Arabian Peninsula northwestward around the rim of the Persian Gulf, reaching into Iran and continuing northward into Iraq, Syria, and southeastern Turkey, where it peters out. The second zone lies across North Africa and extends from north-central Algeria eastward across northern Libya to Egypt's Sinai Peninsula. The third zone begins on the margins of the realm in eastern Azerbaijan, continues eastward under the Caspian Sea into Turkmenistan and Kazakhstan, and also reaches into Uzbekistan, Tajikistan, Kyrgyzstan, and Afghanistan.

Producers and Consumers

Saudi Arabia is the world's largest oil exporter, but in recent years Russia has risen to second place. As we noted in Chapters 2A and 2B, oil and natural gas are by far Russia's most valuable commodities, and the Russian state urgently needs the foreign revenues they generate. Thus Russia, with modest (though expanding) known reserves, is vigorously exporting to international markets and is now a leading force in the global energy picture. In combination, however, oil production by the countries of the North Africa/Southwest Asia realm far exceeds that from all other sources. Throughout recent times, the United States has been the world's leading importer, but in 2012 it was surpassed by China (mainly due to an upsurge in American domestic petroleum production).

As Figure G-11 indicates, the production and export of oil and gas has vaulted a number of this realm's countries into the higher-income categories. But petroleum wealth also has entangled these Islamic societies and their governments in global strategic affairs. When regional conflicts create instability in producing countries that have the potential to disrupt international supply lines, powerful consumers are tempted to intervene and have done so.

Colonial Legacy

When the colonial powers laid down the boundaries that partitioned this realm among themselves, no one knew about the riches that lay beneath the ground. A few wells had been drilled, and production in Iran had begun as early as 1908 and in Egypt's Sinai Peninsula in 1913. But

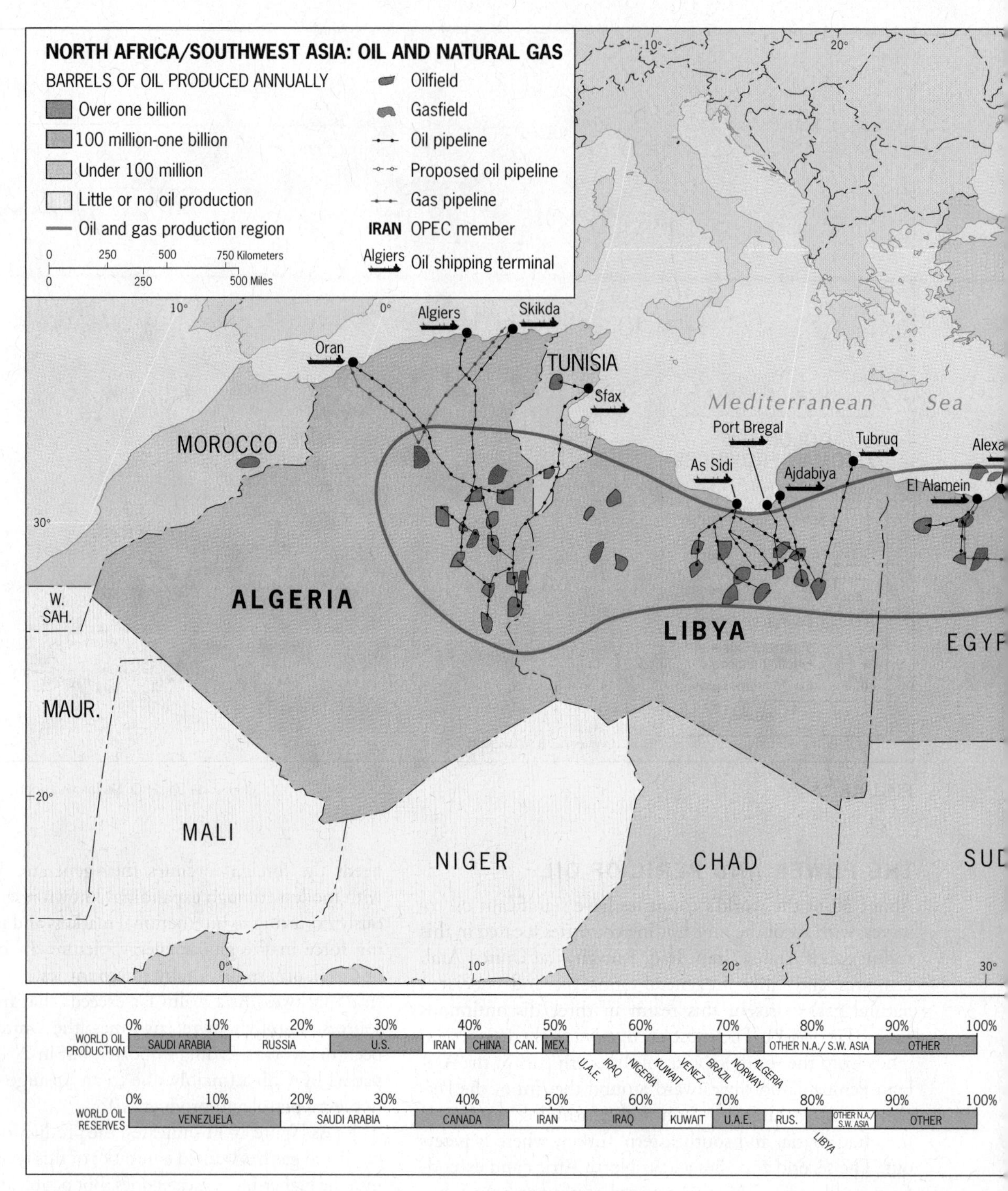

FIGURE 7A-8

the major discoveries came later, in some cases after the colonial powers had already withdrawn. Some of the newly independent countries, such as Libya, Iraq, and Kuwait, found themselves endowed with wealth undreamed of when the Turkish Ottoman Empire collapsed. As Figure 7A-8 shows, however, others were less fortunate. A few countries had (and still have) potential. The smaller, weaker emirates and sheikdoms on the Arabian Peninsula always feared that powerful neighbors would try to annex them (Kuwait faced this prospect in 1990 when Iraq invaded it). The unevenly distributed oil wealth, therefore, created another source of division and distrust among Islamic neighbors.

A Foreign Invasion

The oil-rich countries of the realm found themselves with a coveted energy source but lacked in the skills, capital, or equipment to exploit it. These had to come from the Western world and entailed what many tradition-bound Muslims feared most: a strong foreign presence on Islamic soil, outside intervention in political as well as economic affairs, and penetration of Islamic societies by the vulgarities of Western ways. The transport of oil from these countries to destinations in other geographic realms also raised the importance of strategic arteries that connect NASWA to the rest of the world. Examples of such *choke points*—discussed in the box titled "Choke Points: Danger on the Sea Lanes"—include Egypt's Suez Canal and the Strait of Hormuz at the outlet of the Persian Gulf (Fig. 7A-1).

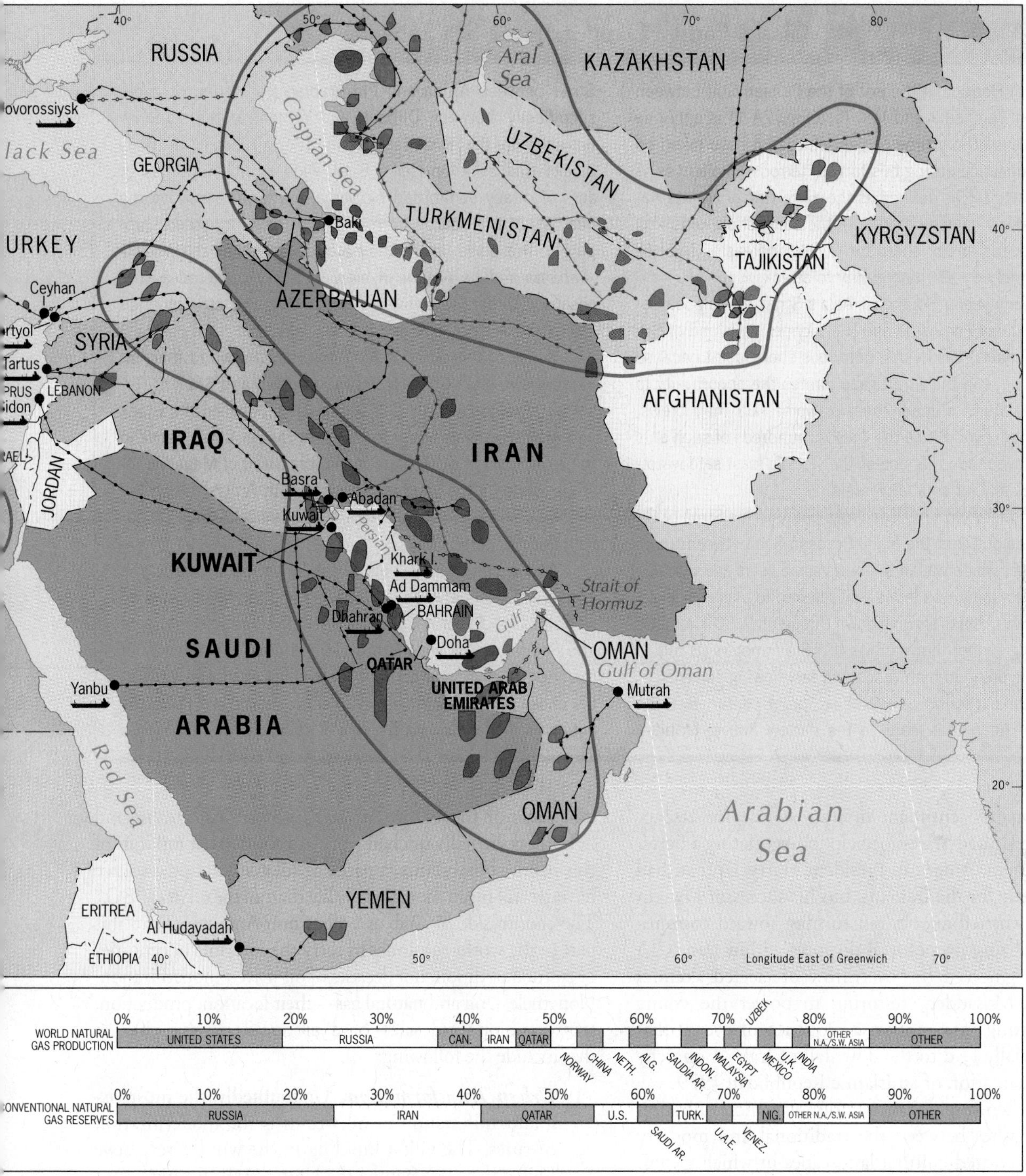

Many Muslims' worst fears came true in Iran during the 1950s, when its government tried to control British oil exploitation centered on Abadan and the Persian Gulf. The imperious ways of the British, the luxurious life of the European expatriates compared to the abject poverty of Iranian workers, and the unfair terms of the concession under which the oil was exported all contributed to the

Choke Points: Danger on the Sea Lanes

The Strait of Hormuz at the exit of the Persian Gulf between the Arabian Peninsula and Iran (see Fig. 7A-1) is only one among dozens of maritime choke points that have taken on added significance during this time of terrorist challenges. A **choke point [12]** is defined as the narrowing of an international waterway causing marine traffic congestion, requiring reduced speeds and/or sharp turns, and increasing the risk of collision as well as vulnerability to attack.

For many years, Southeast Asia's Strait of Malacca between the Malay Peninsula and the Indonesian island of Sumatera (Sumatra) has been a notorious choke point because reduced speeds in the Strait gave pirates the opportunity to board vessels and plunder them or worse, kill their crews, and take them over. Since the 1980s, hundreds of such acts of piracy have made this one of the world's least safe waterways, especially for smaller vessels.

One of the world's busiest choke points is the Strait of Gibraltar at the outlet of the Mediterranean Sea between Spain and Morocco, which was in the news some years ago when an al-Qaeda document was found that referred to plans for an attack on Western ships slowing down through this 60-kilometer (35-mi)-long funnel that narrows to 13 kilometers (8 mi) in width and is prone to high winds and fast-flowing currents.

Another particularly risky choke point continues to be the Gulf of Adan that leads to the narrow Bab el Mandeb Strait between Africa and the Arabian Peninsula, or, more specifically, between Djibouti and Yemen, at the southern entrance to the Red Sea. Barim (Perim) Island partially blocks this 30-kilometer (18-mi)-wide strait. Hundreds of acts of piracy committed by Somali outlaws, based on the beaches of the nearby Horn of Africa, have imperiled shipping in these sea lanes. Their attempt to hijack the *Maersk Alabama* and its American crew in 2009 triggered a successful counterattack by the U.S. Navy and strengthened international resolve to end this scourge.

Some of the other choke points on the world map include the English Channel between Britain and France, Turkey's narrow Bosphorus and Dardanelles between the Black and Mediterranean Seas, Indonesia's Sunda Strait between the Java Sea and the Indian Ocean, the Strait of Magellan for ships rounding the southern end of South America, and the Hainan Strait between southernmost China's Leizhou Peninsula and Hainan Island.

And certain choke points involve artificial waterways. The Suez Canal between the Red Sea and the Mediterranean as well as the Panama Canal connecting Atlantic (Caribbean) and Pacific waters are the busiest and most prominent, but many other route-shortening canals also constitute vulnerable choke points, especially in an era when the risk of hostile action looms larger than the risk of collision.

rise of nationalist sentiment in Iran, whose leaders appealed to the United States for help in negotiating a better deal with Britain. American President Harry Truman had some sympathy for the Iranians, but his successor Dwight Eisenhower worried about Iran turning toward communism. Capitalizing on political disputes within Iran, CIA operatives engineered the overthrow of elected Premier Mohammad Mosaddeq, restoring to power the young shah and setting into motion a sequence of events that would eventually lead to the downfall of Iran's monarchy and the proclamation of an Islamic Republic in 1979.

Even when foreign intervention was more subtle, it intensified clashes between the traditional and modern. Oil revenues created cultural landscapes in which gleaming skyscrapers towered over ornate mosques and historic communities. The social chasms between the rich, well-connected, and Westernized elites and less fortunate local citizens bred resentment. In Iran, these issues played a role in the alienation of the shah from his own people, and they fueled a religious revival that ultimately led to his ouster. Oil, foreign involvement, economics, and religion are all connected in many countries in the realm.

The Geography of Oil's Impact

As is true of every natural resource found across the planet, oil has completely transformed cultural landscapes in certain parts of the North Africa/Southwest Asia realm and left others virtually unchanged. For hundreds of millions of this realm's inhabitants, a patch of tillable soil and a source of water still mean more to daily life than all the oil in OPEC.* The countryside in Arab as well as non-Arab regions in this part of the world continues to carry the imprints of centuries of cultural tradition, not decades of oil-driven modernization. Nonetheless, oil and natural gas—their location, production, transportation, and sale—have produced massive changes that include the following:

1. ***Urban Transformation.*** Undoubtedly, the most visible manifestation of oil wealth is the modernization of cities. The tallest building in the world rises above Dubai in a state named the United Arab Emirates, but it is only one in a forest of glass-encased skyscrapers, many of which test the limits of design and engineering. Besides the capitals (such as Saudi Arabia's Riyadh) a number of other cities on the Arabian Peninsula also reflect the riches oil has brought—yes, even Mecca.

*OPEC, the Organization of Petroleum Exporting Countries, is the international oil cartel (syndicate) formed by 12 producing countries to promote their common economic interests through the setting of joint pricing policies and the limitation of market options for consumers. Its eight NASWA members are Algeria, Iran, Iraq, Kuwait, Libya, Qatar, Saudi Arabia, and the United Arab Emirates.

And, as we note in Chapter 7B, entirely new cities are springing up in the deserts and along the coasts.

2. ***Variable Incomes.*** Oil and natural gas prices fluctuate on world markets (in 2008 oil sold for more than [U.S.] $140 per barrel before plunging to less than $40, but by 2012 the price of a barrel was once again above $75 and continued rising to near $100 by press time in mid-2013). When energy prices are high, several states in this realm rank among the highest-income societies in the world. Even when they decline, many petroleum-exporting countries manage to remain at least in the upper-middle-income category (Fig. G-11).

3. ***Infrastructure.*** Massive spending on airports, seaports, bridges, tunnels, four-lane highways, public buildings, shopping malls, recreational facilities, and other components of national infrastructure creates an image of comfort and affluence quite unlike that prevailing in countries without significant oil or gas revenues. Saudi Arabia, for instance, is now engaged in a vast modernization project extending from coast to coast.

4. ***Industrialization.*** A number of far-sighted governments among those with oil wealth, realizing that reserves will not last forever, are investing some of their income in industries that will outlast the era of massive oil exporting. Petrochemical manufacturing using domestic supplies, aluminum, steel, and fertilizers are among those industries, although others potentially more promising, for example those in high-technology fields, are not yet a significant part of this important initiative.

5. ***Regional Disparities.*** Oil wealth, like other high-value resources, tends to create strong regional contrasts both within and among countries. Saudi Arabia's ultramodern east coast is a world apart from most of its interior, which remains a land of barren deserts, widely scattered oases, vast distances, isolated settlements, and—until the 2010s—slow change. This is not unique to countries in this realm: sharp regional disparities mark oil-rich countries from Algeria to Azerbaijan.

6. ***Foreign Investment.*** Governments and private entrepreneurs have invested enormous amounts of oil-generated wealth in foreign countries, buying financial securities as well as acquiring prestigious hotels, famous stores, and other high-cachet properties. These investments have created a network of international linkages that not only connects NASWA states and individuals to the national economies of other realms but also links them to the growing Islamic communities in those countries.

7. ***Foreign Involvement.*** To many inhabitants of this realm, especially those with strong Islamic-revivalist (fundamentalist) convictions, the inevitable presence of foreigners (including businesspeople, politicians, architects, engineers, and even armed forces) on Islamic soil is an unwelcome byproduct of the energy era. In fact, public opinion in Saudi Arabia forced its ruling regime to negotiate the 2003 departure of American troops from the country's territory.

8. ***Intra-Realm Migration.*** Oil wealth allows governments, industrialists, and private individuals to hire workers from less-favored parts of the realm to labor in the oilfields, ports, and numerous menial service occupations. This has brought many Shi'ites to the countries of the eastern zone of the Arabian Peninsula, where they now form significant sectors of national populations; hundreds of thousands of Palestinian Arabs also sought temporary employment in local industries here. In 2013, it was estimated that Saudi Arabia, with a population of just under 30 million, hosted about 5 million foreign workers from elsewhere in the realm.

9. ***Migration from Other Realms.*** The labor market in such places as Dubai and Abu Dhabi in the United Arab Emirates has also attracted workers from beyond the realm's borders during recent periods of rapid growth. Wages in such countries as Pakistan, India, Sri Lanka, and Bangladesh are even lower than those paid by the building industry or private employers in oil-boom-driven NASWA states. These workers serve mainly as domestics, gardeners, trash collectors, and the like. In 2013, more than 90 percent of the total population of the United Arab Emirates consisted of foreign employees! Over the past few years, substandard working conditions and wage issues have also led to protests that have exposed the harsh conditions faced by many guest workers throughout this realm.

10. ***Diffusion of Revivalism.*** Oil and gas revenues are utilized by Islamist regimes to support Muslim communities as well as their mosques and cultural centers throughout the world. No NASWA country has spent more money on such causes than Saudi Arabia, and thousands of mosques from England to Indonesia prosper as a result. This example of relocation diffusion creates myriad nodes of recruitment for the faith—and ensures the dissemination of revivalist principles.

FRAGMENTED MODERNIZATION

More than any other realm, North Africa/Southwest Asia is marked by a human spatial mosaic of **fragmented modernization [13]**: in certain respects it is highly modern and prosperous, but in many more ways it remains traditional, stagnant, and quite poor. These disparities that characterize so many of the realm's societies are best explained by the uneven impact of oil, the absence of democracy, and the tenuous role of religion. It explains why much of the

The geography of inequality is literally visible anywhere in this realm—even here just blocks away from some of the world's highest real-estate values along Dubai's main axis, Sheikh Zayed Road. In astonishing contrast to the city's glittering, ultramodern skyline, these ramshackle structures sitting atop the desert surface house poverty-stricken migrants whose meager shacks are all the protection they have against searing summer heat and nocturnal winter cold in this virtually rain-free climate.

© Horizons WWP/Alamy

realm often seems on edge, and it provides a vital backdrop to the recent revolutionary turmoil that has surged from Tunisia to Egypt to Syria to Yemen and engulfed several countries in between.

The Uneven Impact of Oil

Oil brought this realm into contact with the outside world in ways unforeseen just a century ago. Oil has strengthened and empowered some of its peoples, but has dislocated and imperiled others. It has truly been a double-edged sword.

We should remind ourselves that the great majority of NASWA's inhabitants are not, in their daily lives, directly affected by the changes the energy era has brought. Most Moroccans, Tunisians, Egyptians, Jordanians, Yemenis, Lebanese, and countless millions of others—Kurds, Palestinians, Berbers, Tuaregs—make ends meet by trading or farming or working at jobs their parents and grandparents performed. Take the case of Iran, which as a country earns about two-thirds of its income from oil and natural gas. Only one-half of one percent of that country's workforce (just over 100,000 out of more than 20 million workers) earn a salary from energy or energy-related work. By far the largest number engaged in any single occupation are the farmers (5 million). And for all their oil, Iranians in 2012 earned a mere (U.S.) $13,000 per person—less than in energy-poor Turkey and less than one-quarter of the per capita income in Singapore. As we shall discover, cultural-geographic rather than economic-geographic forces mainly shape the regionalization of this realm.

The nearly 30 states of the North African/Southwest Asia geographic realm display an often stunning degree of variety and diversity. Oil and natural gas have generated wealth, but also severe inequality and disparity—not only within the countries with large reserves but also between their have and have-not citizens. Ultramodern city skylines on the Arabian Peninsula stand in sharp contrast to Nile Valley villages virtually unchanged for millennia. Superhighways cross deserts still traversed by nomadic traders riding camels. Social geographies vary just as greatly: for instance, the female literacy rate in Lebanon is more than 85 percent, but only 48 percent in Yemen. And the demographic range is also enormous: the annual growth of oil-poor Tunisia's population about equals the world average (which is far below the realm's average), while in oil-rich Saudi Arabia it is 50 percent higher than the global rate.

The Absence of Democratic Traditions

Democracy has been a rarity across this realm; although specific reasons vary, the majority of countries share a history of autocratic, conservative regimes, usually backed strongly by their military forces. Most countries were part of the Ottoman Empire before World War I (see Fig. 7A-7), and only became independent after a period of (attempted) European rule that was, in some cases, sanctioned by the **League of Nations [14]** (the forerunner of the United Nations that functioned from 1919 to 1946). Accordingly, Egypt, Iraq, Palestine, Transjordan, and parts of Yemen were administered by Britain; Lebanon, Syria, Tunisia, Morocco, and Algeria were ruled by France; and Libya was controlled by Italy.

In most cases, the Europeans faced major and tenacious opposition from local populations but were reluctant to let go. Independence was generally preceded by intense conflict or even war, and none of these countries can be said to have been prepared to function as European-style democracies. Indeed, virtually everywhere the newly independent countries fell into the hands of the military (such as Egypt) or restored monarchs with strong military support (such as Yemen).

After World War II, these autocratic regimes were often cemented into place through foreign involvement. Mostly, this was about oil: foreign powers needed oil, showed little interest in domestic politics, and provided local governments with abundant revenues to build up their military power, secure their political position, and maintain order. Saudi Arabia, the world's richest petro-state, is the quintessential example. In other cases, it was Cold War politics that effectively supported the realm's non-democratic governments. For instance, the United States and the Soviet Union competed to incorporate Egypt into their spheres of influence. Egypt was first championed by the Soviet Union from the mid-1950s through the early 1970s, and then by the United States since 1973 (Egypt has for many years been the second-biggest recipient of U.S. foreign aid after Israel).

Political regimes in this realm vary but, at least until the Arab Spring movement emerged in 2010, most government leaders had been in power for many years, and their main concern was to preserve their control and the status quo. Their reigns have been generally conservative, controlled top-down with little concern for the "people on the street," and sometimes marked by repression and violence. Thus, across most of this realm, the common people have not only been deprived of economic advancement but political and religious freedoms as well. For how long, one might wonder, could the masses put up with that kind of situation?

Religious Revivalism

Another schism in this realm relates to the resurgence of religious fundamentalism, or as Muslims refer to it, **religious revivalism [15]**. During the 1970s, the imams in Shi'ite Iran sought to reverse the shah's moves toward liberalization and secularization: they wanted to (and soon did) recast society in a traditional, revivalist Islamic mold. An *ayatollah* (leader under Allah) replaced the shah in 1979; Islamic rules and punishments were instituted. Urban women, many of whom had been considerably liberated and educated during the shah's regime, were forced to return to more traditional roles. Vestiges of Westernization, encouraged by the shah, disappeared. Under these new conditions, even the inconclusive war against neighboring Iraq (1980–1990), which began as a conflict over territory, became a holy crusade that cost more than a million lives.

Islamic revivalist fundamentalism did not rise in Iran alone, nor was it only confined to Shi'ite communities. Many Muslims—Sunnis as well as Shi'ites—in all parts of the realm disapproved of the erosion of traditional Islamic values, the corruption of society by European colonialists and later by Western modernizers (which brought economic benefits only to the elites), and the declining power of the faith within the secular state. As noted above, most governments in this realm are both secular *and* repressive, and they do not provide the economic growth their citizens yearn for. In that sense, it is not so hard to see why many people would seek solace in religion (or even turn fundamentalist) as a way to regain hope and dignity.

But some have taken their return to the faith much farther and imbued it with a fanaticism that is alien to most Muslims—and this has set Muslim against Muslim in many parts of NASWA. Revivalists fired the faith with a new militancy, aggressively challenging the status quo from Afghanistan to Algeria. The militants forced their governments to ban "blasphemous" books, to re-segregate schools according to gender, to enforce traditional dress codes, to legitimize religion-based political parties, and to heed the wishes of the *mullahs* (teachers of Islamic ways). Militant Muslims proclaimed that Western ideas about modernization, inherited from colonialists and adopted by Arab nationalists, were incompatible with the dictates of the Quran.

Terrorism in the Name of Islam

The great majority of Muslims are not fundamentalists; not all fundamentalists are militants; and not all militants are terrorists. Most religious scholars point out that Islam in itself is no more predisposed toward militancy than other religions. We should also note that terrorism is a tool of war that has been used around the world, and indeed throughout modern history, by people of different beliefs and ethnic backgrounds.

But nowadays the religion of Islam, in particular, has proven to be a significant vehicle for political mobilization and a powerful source of inspiration for militants with terrorist agendas. To be sure, their extreme interpretation of the Quran is contested by many; but at the same time, there can be little doubt that their actions are in no small part motivated by their particular fundamentalist beliefs. Theirs is a "holy war"—a ***jihad*** **[16]**—and as such it is a deeply reactionary movement that looks to the past rather than the future.

In Afghanistan, the so-called ***Taliban*** ("seekers of religion"), a kind of Islamist militia, control sizeable parts of the country. They adhere to **Wahhabism [17]**, an orthodox form of Sunni Islam similar to what is practiced in Saudi Arabia. The Taliban impose an especially rigid interpretation of Islamic law and have decreed a reversion to what in many ways resembles a premodern society. For example, criminals are executed in public, men are jailed for not growing beards, and education for women is outlawed. The Taliban also exert a strong influence across the border in parts of northern Pakistan and are well known for harboring ***al-Qaeda*** terrorists.

If the Taliban are mainly concerned with regional territorial control (they emerged from rebel groups that opposed the Soviet occupation of Afghanistan during the 1980s), al-Qaeda has a much broader global agenda. *Al-Qaeda* means "base" or "foundation," and it is better understood as a multinational organization or network. It possesses a tightly knit core but is difficult to define along its edges, with participants and sympathizers in several countries. Al-Qaeda's stated aim is to establish Islamic rule across the realm and to banish all foreign influence. To them, all means are justified. Their actions have killed thousands of people, mostly innocent bystanders, inside and outside the realm, with the most infamous being the 9/11 attacks of 2001 in the United States. The organization was co-founded in Afghanistan in the early 1990s by Usama bin Laden, a wealthy Saudi who particularly resented the presence of U.S. military personnel in his home country. Bin Laden was a primary target of American retaliation for well over a decade before he was finally found and killed by U.S. Special Forces in Pakistan in 2011. Al-Qaeda was weakened at the core, and by 2013 U.S. anti-terrorist efforts had shifted, at least in part, from Afghanistan and Pakistan to Yemen, Somalia, and northern Africa.

Regional ISSUE Islam and Democracy

ISLAMIC REVIVAL IS THE ONLY WAY

"We Muslims must return to, and protect, the fundamental tenets of our faith. Too many of us have strayed, lured and seduced by the vulgarities of Western and other foreign ways. Consider our history. The Prophet set us on course, the Quran was and is the source of all truth, and the message took root from Mecca to Morocco to Malaya. Don't forget: Allah's revelations came many centuries after other religions—Hinduism, Buddhism, Christianity—had captured the minds of millions. But many saw that Islam is the only way, and today practicing Muslims (which means every believer) outnumber Christians, millions of whom never see the inside of a church and do not live pious lives.

"As a devout Muslim living in Egypt and having suffered the consequences of my devotion, let me say this: we Muslims have tried to live harmoniously with other faiths. We not only brought science and enlightenment to the world we converted, but we lived in peace with non-Arabs. In Islamic Andalusia, Jewish communities were safe and secure, just as they were in Marrakech. It was the Christian Catholics who invented the horrors of the Inquisition. From West Africa to Central Asia, we proved our willingness to live in peace with infidels. So what did we get in return? The 'three c's'—the crusades, colonialism, and commercial exploitation! From my American friends I understand that there's an expression, 'three strikes and you're out.' Well, you non-Muslims are out. And so are those Muslims who still cooperate with foreigners, including former President Mubarak of my own country (his reward for being an infidel!) and the entire Saudi royal family. I agree with Sheikh Hamoud: whoever supports the infidel against Muslims is himself an infidel. When you sell your soul for oil dollars or foreign aid, you're subject to a *fatwa*.

"Yes, I'm a revivalist, or what you people call a fundamentalist. Our salvation lies in a return to the strictest rules of Islam. Every Muslim must adhere to every tenet of the faith. Youngsters should learn the Quran by heart: it is the source of all truth beyond which nothing else matters. Women should know and keep their place as it is accorded to them by the Quran. Sharia law should be imposed by all Islamic countries, so that the punishment of transgressors will serve as an example to others. We should dress conservatively, resist the degenerate temptations of other cultures, and follow the teachings of our religious leaders. Let me ask you this: are we better off today than when Sheikh al-Wahhab and Ibn Saud were forging the Islamic state of Arabia? I would say not. We have to go back to our roots to undo the damage that has been done to us.

"Ayatollah Khomeini's Islamic revolution failed in Iran, and our Algerian brothers were prevented from making Algeria an Islamic state, but these were just harbingers of things to come. Our goal is Saudi Arabia, where it all started fourteen centuries ago—and where we will make the new start that will transform the world."

ISLAMIC COUNTRIES NEED DEMOCRATIC REFORM

"How is it that we Muslims, with our rich cultural and scientific heritage, wealth of oil, history of interaction with other societies, and central location in the world have been held back so long by autocratic, corrupt, and incompetent governments, interfering and medieval clerics, stultifying traditions, and internal discord? As a practicing Tunisian Muslim who took part in the Jasmine Revolution that ousted our corrupt former regime, I can tell you that we need democratic reforms, not the religious revivalism about which I keep hearing.

"Let's be objective about this. Look at the world's social and economic statistics, and you'll see that countries where Islam is the dominant religion tend to rank near the bottom—not always, but far too often for this to be a coincidence. The globe is in the grip of a revolution in information and communication technology, but not here. The planet has a mosaic of religions, each professing to be the real path to salvation, and each bedeviled by a fundamentalist fringe, but not here. Here the fundamentalists are a constituency, not a fringe. And they are using tactics ranging from intimidation to terrorism to maintain and increase their clout.

"The problem is we are caught between despotic and self-serving governments on one side and extremist revivalists on the other side. And they put us in a downward spiral. Repressive governments that persecute revivalists only make the problem worse, and Islamic extremism only invites the government to become even more repressive. This has to end.

"It all boils down to this: freedom. We need the freedom to debate the issues that confront us without fear of *fatwas* or imprisonment. So we need reform on both sides, political and religious. Here in Tunisia we have rid ourselves of a government that delivered nothing to the people while it persecuted revivalists. We need a government that allows people to practice the religion they want and practice it the way they want. And, most importantly, we need a government that is democratically elected and that works for the people, that cares about our youth and helps them get a proper education and jobs.

"At the same time, people must understand that religious revivalism is not a solution but is like stepping back in time. It will only bring more darkness. If anything, Islam needs a reformation, a reinterpretation of its inspired writings in light of modern times. And the mullahs need to get out of the business of controlling government, which is their constant aim. The separation of mosque and state is indispensable to our progress.

"And our reformation will have to address the plight of women. As a woman, it infuriates me that half of all women in the Arab World can neither read nor write, that they have unequal citizenship, and that they play virtually no role in government. Nowhere in the world is the domination of men more absolute, and I refuse to believe that the Prophet would approve of this. We need a democratic revolution, not a religious revival."

Vote your opinion at www.wiley.com/go/deblijpolling

THE ARAB SPRING AND ITS AFTERMATH

On December 17, 2010, a 26-year-old Tunis fruit vendor named Mohammed Bouazizi became so distraught when police in the Tunisian capital, after months of routine harassment, confiscated his unlicensed produce stand, that he set himself on fire. His act of desperation resonated with countless Tunisians whose frustration with their government had reached a boiling point. Bouazizi died soon afterward, but his self-immolation quickly came to symbolize Tunisia's "Jasmine Revolution." By mid-January of 2011, popular opposition to the government of President Zine El Abidine Ben Ali had grown so fierce that he was forced to step down and flee to Saudi Arabia after 23 years in power.

Ben Ali's departure was the climax to an unprecedented sequence of events in this realm, where autocratic regimes routinely clung to power for decades with the silent acquiescence of the masses. In the Arab World, conventional wisdom suggested, not just democratic government but even the *desire* for democracy was nonexistent. That latter myth appeared to explode as Tunisians of all backgrounds, including large numbers of women as well as youths, demanded an end to cronyism, corruption, repression, and economic mismanagement. The Tunisian people insisted on being heard—and their voices swiftly echoed throughout North Africa and Southwest Asia.

The Diffusion of Popular Revolts

What began in Tunisia, as 2010 turned into 2011, rapidly diffused across North Africa and into adjacent parts of Southwest Asia. Within a few months, unrest had spread to Egypt, Libya, Syria, Yemen, and Bahrain. The conditions in all these countries were similar: they were ruled by long-established autocratic regimes that failed to bring economic progress, repressed their own people, and had lost touch (most especially with the younger generation). Across the realm, people followed the events in Tunisia via television and the Internet, and revolutionary fervor spread like wildfire. It may well have been the most spectacular **domino effect [18]** in recent history.

Undoubtedly, the most important of these falling dominoes was pivotal Egypt. The most heavily populated, historically prominent, and centrally located country in the Arab World, Egypt had for decades suffered from stunted economic growth and an out-of-touch, repressive government. In Cairo and elsewhere, the masses took to the streets and demanded the ouster of President Mubarak, even though protesters were repeatedly confronted by harsh and bloody encounters with the police and military forces. In Egypt alone, an estimated 850 people died in these clashes and thousands were wounded; nonetheless, after nearly three weeks of demonstrations Mubarak, too, was compelled to resign. Tahrir Square (literally, "Liberation Square") in central Cairo had turned into a monumental site of rebellion and popular demonstration.

The events in Egypt generated shockwaves throughout the realm. Whereas Tunisia is a small and comparatively peripheral state and Ben Ali was not that well known outside of his country, Egypt lies at the epicenter of the Arab World and Mubarak was widely regarded as unassailable as a pharaoh. If the tide could be turned here, it could happen anywhere. At least that seemed to be the conviction of fearless protesters in surrounding countries.

A New Restless Generation

In most countries of the North Africa/Southwest Asia realm, more than half of the population is under 25 years of age. This enormous demographic cohort has grown up in a world in which information flows almost uncontrollably, so these young people know their home countries deny them the opportunities that exist elsewhere. Moreover, most have known no other national leader than the autocrat already in power at the time of their birth.

The youthfulness of these populations stands in stark contrast to the antiquated nature of their governments. Consider this list of despotic longevity at the outset of 2011: Tunisia's Ben Ali, 75 years old, 24 years in office; Egypt's Mubarak, 82/30; Yemen's Abdullah Saleh, 69/33; Libya's Muammar Qadhafi, 69/42; and Syria's al-Assad father-and-son rulers, in office for 41 years.

The 2011 uprisings were predominantly (although not exclusively) led by the younger generation, from teenagers to thirty-somethings. In Tunisia, Egypt, and Syria, students took the lead, using the Internet (Facebook, Twitter, blogging) to organize demonstrations. This emphasized one crucial aspect of the "Arab Spring": the disconnect between the realm's aging autocrats and their young, globally aware subjects.

The Arab Spring and Its Repercussions

If the very phrase *Arab Spring* expressed the hopefulness of a new dawn, such romantic notions were completely dispelled by mid-2013. First of all, the relatively peaceful ouster of the long-time leaders of Tunisia and Egypt was not repeated elsewhere except in Yemen. The more despotic regimes in Libya and Syria responded with stubborn resistance and iron fists. Libya quickly degenerated into civil war, where Muammar Qadhafi's unleashing of military forces against his unarmed people was so brutal that a makeshift alliance of France, the United Kingdom, and the United States (later subsumed by NATO) intervened with air attacks to protect the Libyans from annihilation by their own government. By the autumn of 2011, the tide had turned, the capital of Tripoli fell, and Qadhafi was cornered and killed in his home town on October 20. The situation in Syria has been far worse because government forces triggered a civil war by unleashing a particularly ruthless campaign of repression in April 2011 that showed no sign of letting up more than two years later; by the summer of 2013, the civilian death toll resulting from this horrible, unrelenting carnage surpassed 100,000.

And where it had looked like a transition of government had been achieved, as in Tunisia and Egypt, democratic prospects swiftly lost their glamour. Although the

Arab Spring was undoubtedly fueled in large part by the aspirations of masses of young people yearning for progress and democracy, the protesters faced an uphill battle. Earlier revolutionary movements in the world—such as the "Prague Spring" that led to reforms in communist former Czechoslovakia in 1968, the overthrow of dictatorships in Portugal (1975) and South Korea (1980), and the nonviolent democratic transformation of South Africa (1994)—occurred in countries and regions in which certain social and political conditions already prevailed and could be used and modified to channel the onrushing process of change. Tellingly, not one of the Arab Spring states produced an opposition leader who could articulate the demands of the reformers, organize their energies, and negotiate a transition of power—so no Arab Nelson Mandelas emerged. Not a single country permitted genuinely free speech to enable its broadcast and print media to carry the messages of the oppressed; no country possessed even a flawed judiciary capable of conversion into a dependable and impartial one; and no country had police to impartially enforce the law. Thus the odds against the success of the Arab Spring movement were daunting.

Religion and Geopolitics

The Arab Spring should also be viewed in the context of the Sunni-Shi'ite split discussed earlier. Superimposed on the conflicts in several Arab states is the role of Shi'ite minorities in propelling the uprisings—or of crushing them. This issue is explored further in the regional discussion in Chapter 7B, but it should be noted here that Shi'ites played major roles in the uprisings in Bahrain (where they are in the majority but ruled by a Sunni monarchy), in Yemen (where their rebellion began even before Tunisia fell apart), and, paradoxically, in Syria, where a Shi'ite sect (the Alawites) formed the ruling minority whose murderous response to the people's demands draws ever greater international condemnation.

The Arab Spring arose from a combination of populist grievances ranging from economic issues to religious repression. Obviously, religious revivalists saw their opportunity as part of the wider revolt that was aimed primarily at the removal of repressive regimes: those aging autocrats had relentlessly persecuted fundamentalists of every stripe. In several countries, particularly Egypt, the Muslim Brotherhood had a tenuous, conflict-ridden relationship with the ruling regime. But during the early stages of the Arab Spring, religious revivalists did not play leading roles in the popular uprisings. In fact, the young rebels and their notions of democracy ran counter to the fundamentalist views of reactionary Islamists, a divide that adds yet another layer to the complex social organization marking this fractious part of the world.

But at the regional level, the current upheaval is not just about democracy or religious revivalism because it has rapidly assumed important wider geopolitical dimensions. Much of this goes back to the conflict between Sunnis and Shi'ites and the rivalry between Saudi Arabia and Iran. Sectarian conflict was never far from the surface because many (autocratic) governments in this realm represent a minority of either Sunnis or Shi'ites. Thus Saudi Arabia supported Bahrain's government by sending military troops to confront (Shi'ite) protesters—yet at the same time it has taken a strong position against Syrian dictator Bashar al-Assad for his bloody repression of Sunni rebels. Iran, in the meantime, supports Assad as well as the Shi'ite government in Iraq (in 2011 it had declared support for Bahrain's protesters). And at an even broader scale, the reluctance of Russia and China to take a meaningful stand against Syria is explained by the enmity between Syria and Saudi Arabia as well as the close relations between the latter and the United States. The outcome of this conflict in Syria, therefore, is crucial to the geopolitical order of both the Middle East region and the realm as a whole.

The State of the Realm Today

As this book went to press in mid-2013, the geographical impact of the Arab Spring across this enormous realm was felt in varying degrees and in different ways—with each national outcome dependent on local circumstances and past experiences. Figure 7A-9 shows the state of the realm as expressed by the **political regimes [19]** in place in its different countries. This map may be a snapshot of a realm in flux, but it does provide a useful impression of NASWA's current political mosaic. It also reveals some important geographic patterns that help to put the Arab Spring in perspective. Chapter 7B presents more detailed profiles of individual countries and how they have fared in the aftermath of the Arab Spring; here we are mainly concerned with general, realmwide trends.

First, the Arab Spring was in essence an Arab-Islamic series of events that mainly impacted northern Africa and the Middle East. It is here that we find most of the countries that now confront a post-Arab Spring crisis—in particular, Egypt, Libya, and Tunisia. As for Syria, it experienced the most catastrophic conflict of this still-young century; even though the fate of the Assad regime was not yet clear, Syria by mid-2013 had already become a failed state.

Second, whereas the Arab popular revolts resonated and touched nerves across the Arabian Peninsula, they did not seriously challenge any of that region's governments except in Yemen. It was here—as well as in Jordan, Iran, and Morocco—that the established monarchies and/or theocracies prevailed with only modest responses to popular demands for reform. Nonetheless, there are some important differences among them: for instance, Saudi Arabia and Jordan are both monarchies, but the former is Islamic while the latter is secular. And even though these regimes endure, some continue to deal with substantial dissent. Bahrain, especially, is an interesting case because its Shi'ite majority has intermittently engaged in street protests since 2011.

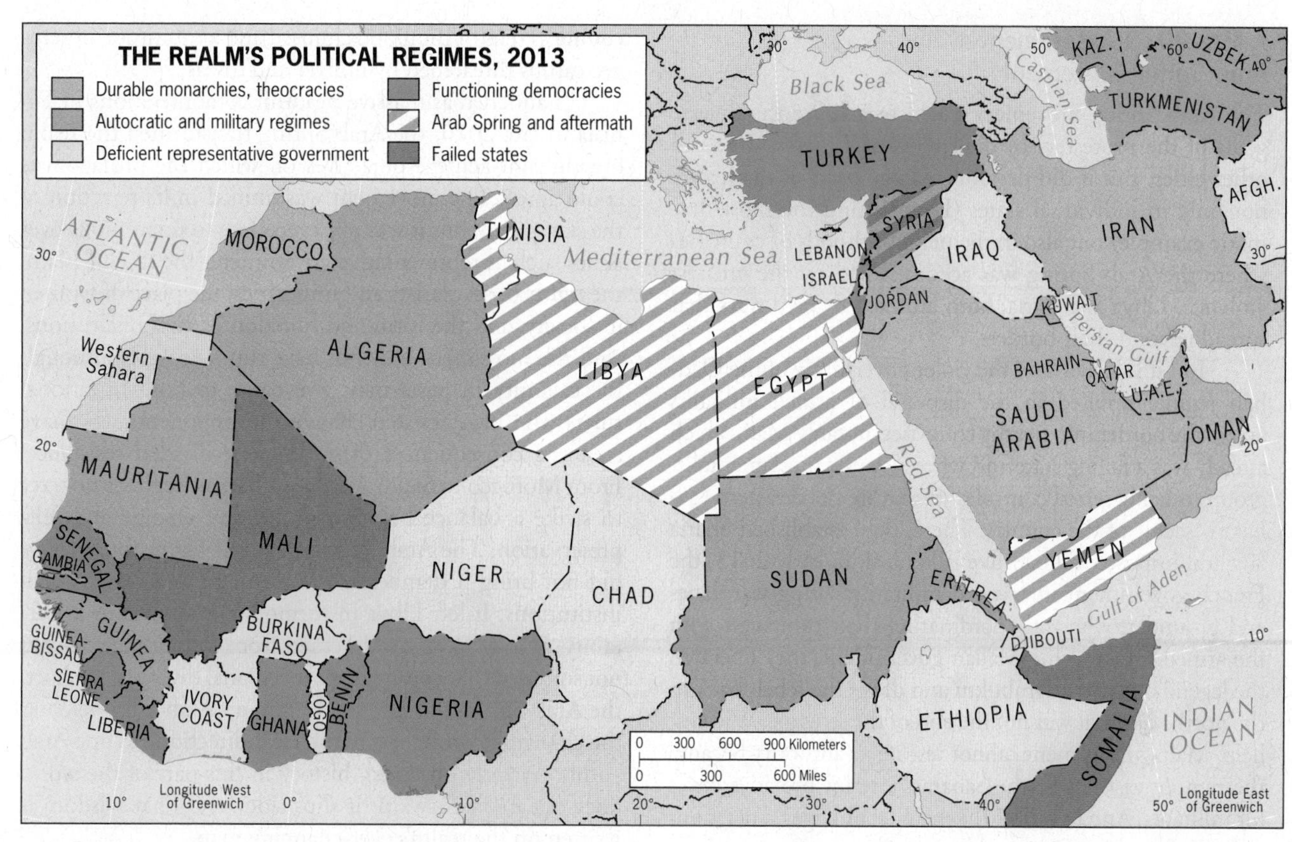

FIGURE 7A-9

© H. J. de Blij, P. O. Muller, J. Nijman, and John Wiley & Sons, Inc.

Elsewhere, the impact of the Arab Spring has been far less significant, notably in the African Transition Zone (elaborated in Chapter 6B). The ATZ's environmental, economic, ethnic, and religious conditions interact to form a major obstacle to efficient, democratic governance. Most countries within this wide east-west belt adjoining the southern Sahara have autocratic rulers or flawed representative governments, as well as instabilities that produce frequent internal conflict (and, as in Somalia and Mali, the occasional outcome is a failed state). Note, too, that several of these countries are only partially Arab or Islamic (see Fig. 6B-9).

© Arnaud Roine/ECPAD/Corbis

January 28, 2013: French troops are greeted by residents of historic Timbuktu in central Mali after wresting control of the city from Islamic extremists. The French then joined Malian military forces in the wider effort to drive out the rebels and restore government control of the central and northern parts of the country.

Senegal and Turkey (along with Israel) are the realm's only durable democracies, but the populations of both countries are more than 95 percent Islamic—a sure sign that Islam and democracy can coexist.

Finally, even though the countries that constitute Turkestan are dominantly Islamic, they were too far removed from the Arab core to be seriously affected by the Arab Spring. Here, the Soviet imperial legacy of autocracy, nepotism, and expropriation, combined with severe geographic isolation, presents an all but insurmountable challenge to those who long for democracy. Tajikistan is in the worst condition of all, and is now often labeled as a failed state as well.

A Mid-2013 Assessment of the Arab Spring Movement

The Arab Spring movement failed to achieve the diverse goals of the protesters in the streets of Tunis, Cairo, and other cities. But it did destabilize large parts of the realm, not only in individual states (Egypt being the most dramatic example) but also in the neighborhoods of countries where the Arab Spring was accompanied by the greatest violence, Libya and Syria, both afflicted by civil wars that spread beyond their borders.

In Libya's civil war, the violent overthrow of the Qadhafi regime resulted in the dispersal of armed militants across the border into nearby countries, notably Mali, which already was a failing state and where Islamic fundamentalist groups quickly seized control of the entire northeastern (Saharan) sector of the country. There, they established Sharia law and imposed Taliban-like rule until, in early 2013, the French government intervened militarily. Using warplanes and ground troops and coordinating their campaign with the armed forces of the Malian government, they liberated the legendary city of Timbuktu and drove the rebels back to the north. But that was not the end of the struggle. Without help, Mali's government cannot restore its authority because the remote vastness of the Saharan interior provides cover for militants. And in mid-2013 the state of Niger, a neighbor of Mali, found itself targeted by extremists even as northern Nigeria to its south became a battleground between *Boko Haram* Islamic jihadists and government forces.

But it was the terrible conflict in Syria that threatened to destabilize an entire region. What began as an Arab Spring uprising against the minority (Alawite) regime headed by Bashar al-Assad turned into a full-scale civil war whose forces—government and rebels alike—received outside support from as far afield as Iran and Russia. As Figure 7A-1 shows, Syria borders Lebanon, Israel (in the Golan Heights), Jordan, Iraq, and Turkey, and well over a million Syrian refugees flooded into Jordan and Turkey. When the Assad regime seemed on the verge of defeat, the Lebanon-based Shi'ite terrorist organization, *Hizbullah,* entered the war and altered the balance of power. Following deadly car bombings inside Turkey it seemed as though the Turks might enter the war on the rebel side; meanwhile the rebels did not achieve a unified front and got support from Sunnis of diverse religious and political persuasions. By the summer of 2013, the death toll from this Arab-Spring-triggered catastrophe was estimated by the UN to exceed 100,000, with countless tens of thousands injured and a multitude of refugee camps threatened by hunger and disease.

Launched as massive peaceful demonstrations in Tunisia in late 2010, the Arab Spring transformed the realm but did not achieve objectives on which the protagonists could agree. The movement was unified in its rejection of the status quo, but it was not even close to unanimous over issues such as representative government, the role of Islam, the rights of secularists and minorities, the place of women in society, and the form and function of civil institutions. And the movement may not have run its course, although the calamity in Syria may give pause to both traditional rulers who have resisted change and opponents who have seen the consequences of violent but divided resistance. From Morocco to Saudi Arabia, still-stable regimes now try to strike a balance between progressive change and self-preservation. The Arab Spring changed Egypt forever, but did not bring it democratic government and needed civil institutions. It left Libya in turmoil and Tunisia in uncertainty. It may have saved Yemen from imploding, but has not softened that country's tense regional division. In short, the Arab Spring did not evolve into an Arab Summer, and the changes it wrought have no clear direction. As one Arab author recently observed, history in this part of the world does not move forward: it slips sideways. That wisdom is written on the realm's ever-changing map.

POINTS TO PONDER

- Islamic fundamentalism has become an important force in the realm's politics, but at the same time the religious divide between Sunni and Shi'ite Muslims has deepened.
- Oil-rich countries tend to have non-diversified economies and exhibit a wide gap between rich and poor.
- In Afghanistan, the Taliban outlawed education for females; in Tunisia and Egypt, women played an important role in the Arab Spring movement that demanded democratic reforms.
- The Arab Spring did not produce major advances toward democracy but it did play a role in regional destabilization.

POLITICAL UNITS AND GEOGRAPHIC REGIONS OF NORTH AFRICA/SOUTHWEST ASIA

- Egypt and the Lower Nile Basin
- The Maghreb and Its Neighbors
- Middle East
- Arabian Peninsula
- The Empire States
- Turkestan
- African Transition Zone

POPULATION

- Under 50,000
- 50,000–250,000
- 250,000–1,000,000
- 1,000,000–5,000,000
- Over 5,000,000

National capitals are underlined

- Railroad
- Road
- Canal

0 400 800 1200 1600 Kilometers

0 200 400 600 800 1000 Miles

Longitude West of Greenwich

ATLANTIC OCEAN

North Sea

Mediterranean Sea

Black Sea

Red Sea

Volga R.

Don R.

Dnieper R.

Nile R.

Blue Nile R.

White Nile R.

Niger R.

Benue R.

Congo R.

Lake Nasser

Lake Victoria

Tropic of Cancer

Equator

0° 15° 30° 45°

NORWAY, Oslo, SWEDEN, Stockholm, FINLAND, Helsinki, St. Petersburg, ESTONIA, LATVIA, LITHUANIA, RUSSIA, Moscow, BELARUS, Mensk, DENMARK, UNITED KINGDOM, London, NETH., BELG., LUX., GERMANY, Berlin, POLAND, Warsaw, CZECH REPUBLIC, SLOVAKIA, UKRAINE, Kiev, MOLDOVA, Rostov, FRANCE, Paris, SWITZ., AUSTRIA, Vienna, HUNGARY, Budapest, ROMANIA, Bucharest, SLO., CRO., BOSNIA-HERZEGOVINA, SERBIA, Belgrade, MONTENEGRO, KOS., MAC., ALBANIA, BULGARIA, GREECE, Athens, ITALY, Milan, Rome, Naples, MALTA, PORTUGAL, SPAIN, Barcelona, Gibraltar (U.K.), GEORGIA, ARMENIA

TURKEY, Istanbul, Bursa, Ankara, Izmir, Adana, Gaziantep, EMPIRE STATES, CYPRUS, Nicosia, SYRIA, Aleppo, Euphrates, LEBANON, Beirut, Damascus, MIDDLE EAST, ISRAEL, Tel Aviv-Jaffa, Jerusalem, JORDAN, Amman, Port Said, ARABIAN PENINSULA, Jiddah

MAGHREB, Atlas Mountains, MOROCCO, Tangier, Rabat, Casablanca, Marrakech, Bechar, Canary Is. (Sp.), El Aaiún, WESTERN SAHARA, ALGERIA, Algiers, Oran, Constantine, Annaba, TUNISIA, Tunis, Sfax, LIBYA, Tripoli, Benghazi, Libyan Desert, Sahara, EGYPT, Alexandria, Cairo, Thebes, Aswan, EGYPT AND THE LOWER NILE BASIN

MAURITANIA, Nouakchott, MALI, NIGER, CHAD, SUDAN, Port Sudan, Khartoum, Wad Madani, ERITREA, SENEGAL, Dakar, GAMBIA, GUINEA-BISSAU, GUINEA, Conakry, SIERRA LEONE, Freetown, LIBERIA, Monrovia, IVORY COAST, Yamoussoukro, Abidjan, BURKINA FASO, Ségou, Mopti, Bamako, Niamey, GHANA, Accra, TOGO, BENIN, NIGERIA, Kano, Kaduna, Abuja, Ibadan, Lagos, N'Djamena, AFRICAN TRANSITION ZONE, SUDD

SOUTH SUDAN, Juba, ETHIOPIA, Adis Abeba, CENTRAL AFRICAN REPUBLIC, Bangui, CAMEROON, Yaoundé, EQ. GUINEA, SÃO TOME & PRINCIPE, GABON, CONGO, Brazzaville, DRCONGO, Kinshasa, UGANDA, Kampala, KENYA, Nairobi, TANZANIA

SIAN REGIONS

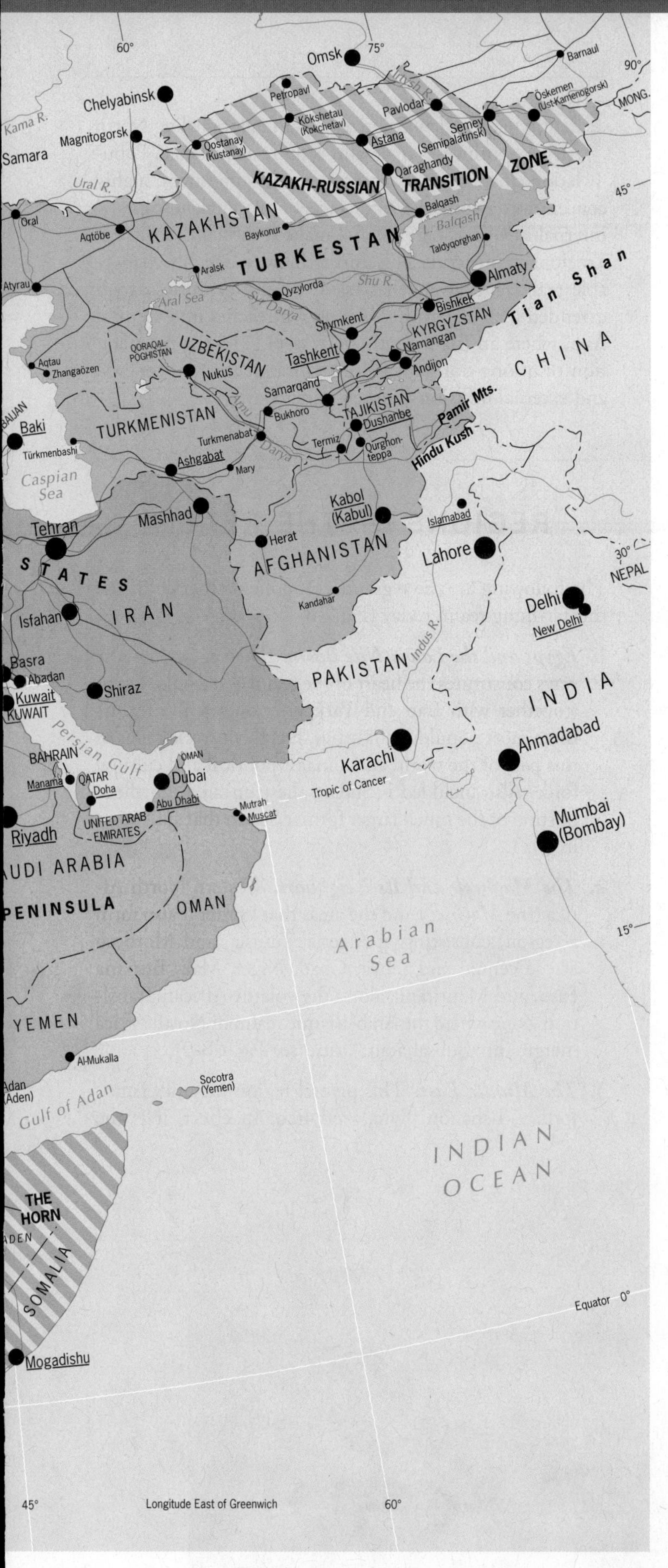

FIGURE 7B-1 © H. J. de Blij, P. O. Muller, and John Wiley & Sons, Inc.

REGIONS

Egypt and the Lower Nile Basin

The Maghreb and Its Neighbors

The Middle East

The Arabian Peninsula

The Empire States

Turkestan

IN THIS CHAPTER

- Syria's tragedy
- The near-breakup of Mali
- Turkey on the rise
- Israel and the Palestinians . . . again
- Egypt: Muddling through
- Planning for Afghanistan's future

CONCEPTS, IDEAS, AND TERMS

Cultural revival	1
Turkish Model	2
Rain shadow effect	3
Police state	4
Stateless nation	5
Choke point	6
Qanat	7
Forward capital	8
Buffer state	9
Taliban	10
al-Qaeda	11

© Jan Nijman

Where were these photos taken?
Find out at www.wiley.com/college/deblij

Identifying and delimiting regions in the sprawling North Africa/Southwest Asia realm is quite a challenge. Population clusters are widely scattered in some areas, highly concentrated in others. Cultural transitions and landscapes—internal as well as peripheral—make it difficult to discern a regional framework. This, furthermore, is a highly changeable realm and always has been. Several centuries ago it extended into eastern Europe; now it reaches into Central Asia, where an Islamic **cultural revival [1]**—the regeneration of a long-dormant culture through internal renewal and external infusion—is well under way.

REGIONS OF THE REALM

The following are the regional components that comprise this far-flung realm today (Fig. 7B-1):

1. ***Egypt and the Lower Nile Basin.*** This region in many ways constitutes the heart of the realm as a whole. Egypt (together with Iran and Turkey) is one of the realm's three most populous countries. It is the historic focus of this part of the world and a major political and cultural force. Also included is Sudan, the truncated, northern portion of the much larger former Sudan that split apart in 2011.

2. ***The Maghreb and Its Neighbors.*** Western North Africa (the *Maghreb*) and the areas that border it also form a region, consisting of Algeria, Tunisia, and Morocco at the center, and Libya, Chad, Niger, Mali, Burkina Faso, and Mauritania along the volatile African Transition Zone where the Arab-Islamic realm of North Africa merges into Subsaharan Africa (see Fig. 6B-9).

3. ***The Middle East.*** This pivotal region includes Israel, Jordan, Lebanon, Syria, and Iraq. In effect, it is the

© Jan Nijman

crescent-shaped zone of countries that extends from the eastern coast of the Mediterranean to the head of the Persian Gulf.

4. ***The Arabian Peninsula.*** Dominated by the enormous territory of Saudi Arabia, the Arabian Peninsula also encompasses the United Arab Emirates (UAE), Kuwait, Bahrain, Qatar, Oman, and Yemen. Here lies the source and focus of Islam, the holy city of Mecca; here, too, lie many of the world's greatest oil deposits.

5. ***The Empire States.*** We refer to this region as the Empire States because two of the realm's giants, both the centers of major historic empires, dominate its geography: Turkey and Iran.

6. ***Turkestan.*** Turkish influences mix with Islam and Soviet legacies in the five former Central Asian republics; fractious Afghanistan also forms part of this region. Not long ago considered isolated and remote, Turkestan is assuming an increasingly strategic position between Russia, China, Iran, and South Asia.

EGYPT AND THE LOWER NILE BASIN

Egypt occupies a pivotal location in the heart of a realm that extends over 9600 kilometers (6000 mi) longitudinally and some 6400 kilometers (4000 mi) latitudinally. At the northern end of the Nile and of the Red Sea, at the eastern end of the Mediterranean Sea, in the northeastern corner of Africa across from Turkey to the north and Saudi Arabia to the east, adjacent to Israel, to Sudan,* and to Libya, Egypt lies in the crucible of this realm. And because it owns the Sinai Peninsula, Egypt, alone among states on the African continent, has a foothold in Asia. This foothold gives it a coast overlooking the strategic Gulf of Aqaba (the northeasternmost arm of the Red Sea). Egypt also controls the Suez Canal, the vital link between the Indian and Atlantic oceans and a lifeline of Europe. The capital, Cairo, is the realm's second-largest city and a leading center of Islamic civilization.

Egypt has six subregions, mapped in Figure 7B-2. Most Egyptians live and work in Lower (i.e., northern) and Middle Egypt, subregions ① and ②. This is the country's core area anchored by Cairo and flanked by the leading port and second-largest manufacturing center, Alexandria. The other four regions are remote and sparsely populated, especially Upper Egypt ③, the Western Desert ④, and the Sinai ⑥. The northern stretches of the Eastern Desert ⑤ are now in the path of Cairo's encroaching urban sprawl (see *Field Note* photo).

MAJOR CITIES OF THE REALM

City	Population* (in millions)
Istanbul, Turkey	12.2
Cairo, Egypt	11.8
Tehran, Iran	7.6
Baghdad, Iraq	6.6
Riyadh, Saudi Arabia	6.1
Khartoum, Sudan	5.1
Tel Aviv-Jaffa, Israel	3.6
Kabol (Kabul), Afghanistan	3.3
Casablanca, Morocco	3.2
Algiers, Algeria	3.2
Damascus, Syria	2.9
Tashkent, Uzbekistan	2.3
Beirut, Lebanon	2.1
Almaty, Kazakhstan	1.5
Tripoli, Libya	1.2
Jerusalem, Israel	0.8

*Based on 2014 estimates.

Gift of the Nile

Egypt's Nile is the aggregate of the two great branches upstream: the White Nile, which originates in the streams that feed Lake Victoria in East Africa, and the Blue Nile, whose source lies in Lake Tana in Ethiopia's highlands. The two Niles converge at Khartoum, capital of present-day Sudan. About 95 percent of Egypt's 85.6 million people live within 20 kilometers (12 mi) of the great river's banks or in its delta (Figs. 7B-2, 7A-2).

Before dams were constructed on the Nile, the ancient Egyptians used *basin irrigation*, building fields with earthen ridges and trapping the annual floodwaters with their fertile silt, to grow their crops. That practice continued for thousands of years until, during the nineteenth century, the construction of permanent dams made it possible to irrigate Egypt's farmlands year-round. These dams, with locks for navigation, controlled the Nile's annual flood, expanded the country's cultivable area, and allowed farmers to harvest more than one crop per year on the same field. The largest of these dams, the Aswan High Dam completed in 1970, created Lake Nasser (the reservoir that extends southward 150 kilometers [100 mi] into northern Sudan) and increased Egypt's irrigable land by nearly 50 percent. Today it still provides about 15 percent of the country's electricity.

Much farther upstream, in the western highlands of Ethiopia where the Blue Nile originates, its government is constructing the Grand Ethiopian Renaissance Dam, with more than one-third of the funding coming from China. After completion in 2015, the dam is expected to boost

**Sudan* refers to the truncated, northern portion of the former country of Sudan that remained after South Sudan (discussed in Chapter 6B) became independent and split away on July 9, 2011. The much larger country that existed before that date is referred to as *former Sudan.*

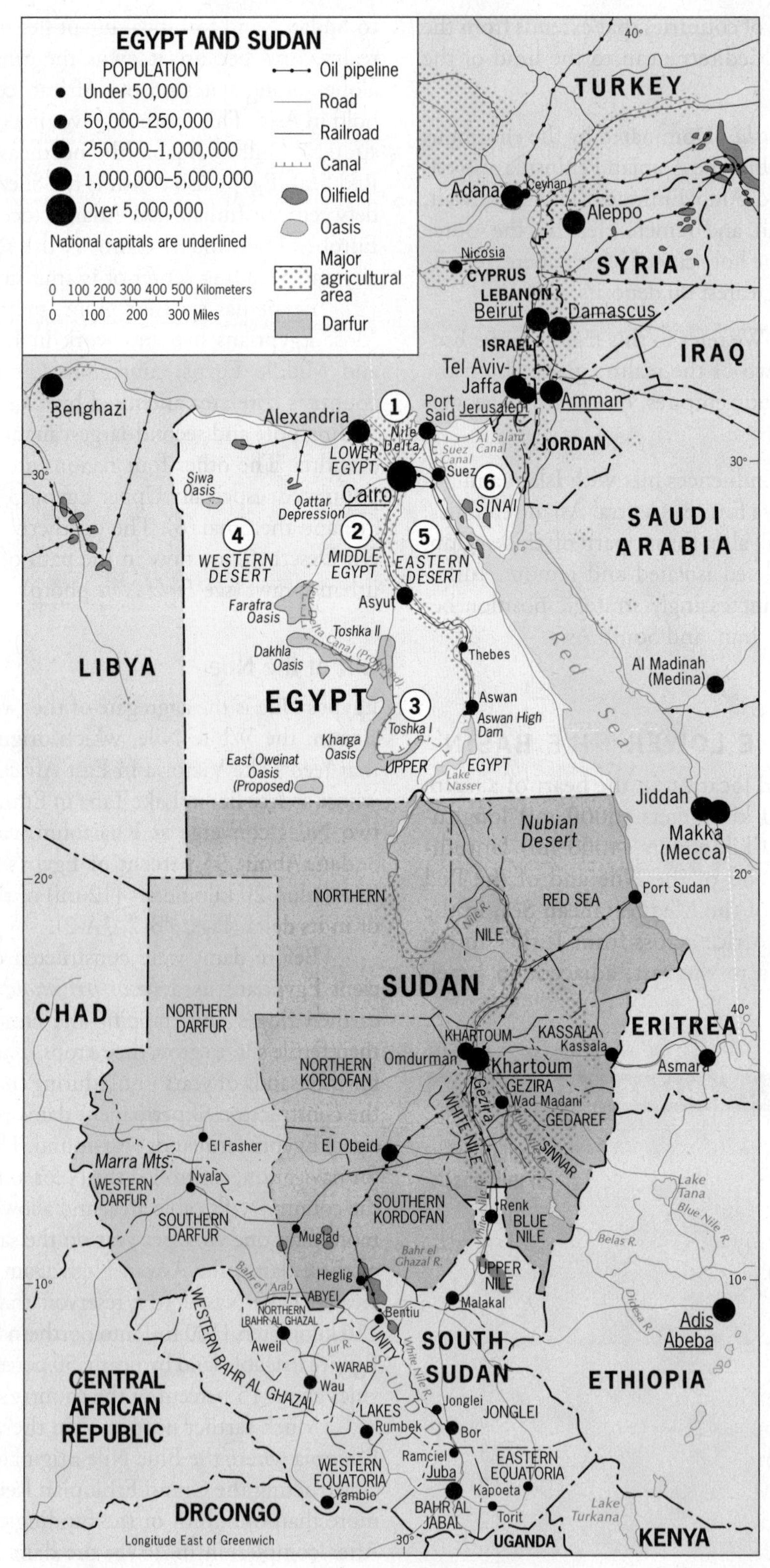

FIGURE 7B-2 © H. J. de Blij, P. O. Muller, and John Wiley & Sons, Inc.

From the Field Notes . . .

"Taking off from Cairo International Airport on the northeast edge of this sprawling megacity, we flew over one of Cairo's newest housing developments at the edge of the northern reaches of the Eastern Desert. Look at this residential density—in a desert! With a birth rate twice that of the United States, Egypt's critical problem is the provision of a sufficient water supply—which right now is steadily decreasing. Earlier on this day in November 2009, we took a field trip through this area led by a local geographer. These are middle- and upper-middle-class residential complexes. All those new apartment towers will need water, and so will anything green planted here. Where will these people work and how will they get there? Is this sustainable urban development?"

© Jan Nijman

www.conceptcaching.com

electrical power production to meet all domestic demands as well as earn new revenues from sales to Ethiopia's energy-deficient neighbors. But the project has also attracted growing opposition (and raised the potential for conflict) in downstream Sudan and Egypt, who fear a reduction in water supplies flowing into the lower Nile Basin.

Egypt's elongated oasis along the Nile, just 5 to 25 kilometers (3 to 15 mi) wide, broadens north of Cairo across a delta anchored in the west by the historic city of Alexandria and in the east by Port Said, gateway to the Suez Canal. The delta contains extensive farmlands, but it is a troubled area today. The ever more intensive use of the Nile's water and silt upstream is depriving the delta of desperately needed replenishment. And the low-lying delta is also at risk due to geological subsidence and sea-level rise, increasing the danger of salt-water intrusion from the Mediterranean that can severely damage soils.

Egypt's millions of subsistence farmers, the *fellaheen*, struggle daily to make their living off the land, as did the peasants of the Egypt of five millennia ago. Some rural landscapes seem barely to have changed; ancient farming methods are still used, and dwellings remain rudimentary. Persistent poverty, disease, and moderately high infant mortality rates prevail among Egypt's masses. The government has announced grandiose plans to expand irrigated acreage into the desert, but almost nothing has taken place so far.

Economic and Political Discontent

Until recently, Egypt had experienced steady economic growth for almost a decade at an annual rate of about 5 percent, and the economy diversified somewhat. At the same time, its annual rate of population growth has declined, while literacy and per capita consumption have increased markedly. Egypt now exports fruits, vegetables, rice, and textiles in addition to its longtime staple, cotton; it also has a sizeable tourist industry (which is frequently damaged by the actions of Islamic extremists and, since 2010, by widespread social unrest). Now that Egypt has begun to exploit its domestic resources, the country's expenditures on oil imports have substantially declined. These gains notwithstanding, the economy has slowed in this decade, and today economic growth (down to about 2 percent yearly) barely exceeds population growth. And even when growth rates were higher, the primary beneficiaries were the wealthy and the upper-middle class.

Egypt's poor today are as deprived and entrenched in poverty as ever. As the population continues to grow, the gap between food supply and demand widens, and Egypt must now import roughly U.S. $6 billion worth of food every year. Nearly 60 percent of the population is under 25 years of age, and the young, most of whom lack economic opportunities, are restless. After the Mubarak regime came into office following the assassination of President Anwar Sadat in 1981, it failed to connect with Egypt's new generation and in the end was unable to justify its three-decade lock on power. All along, this government repressed Islamic groups such as the Muslim Brotherhood, but discontent had also been rising among broader segments of the population, especially the secular middle class and the young.

AMONG THE REALM'S GREAT CITIES . . .

CAIRO

STAND ON THE roof of one of the high-rise hotels in the heart of Cairo, and in the distance you can see the great pyramids, monumental proof of the longevity of human settlement in this area. But the present city was not founded until Muslim Arabs chose the site as the center of their new empire in AD 969. Cairo (al Qahira) became and remains Egypt's primate city, situated where the Nile River opens onto its delta, home today to nearly one-seventh of the entire country's population.

Cairo's population of 11.8 million ranks 20th among the world's largest urban agglomerations, and it shares with other cities of the less-advantaged realms the staggering problems of crowding, inadequate sanitation, crumbling infrastructure, and substandard housing. But even among such cities, Cairo is noteworthy for its stunning social contrasts. Along the Nile waterfront, elegant skyscrapers rise above carefully manicured, Parisian-looking surroundings. But look toward the east and the metropolitan landscape extends gray, dusty, almost featureless as far as the eye can see. Not far away, more than a million squatters live in the sprawling cemetery known as the City of the Dead (see *Field Note* photo). On the urban outskirts, millions more survive in overcrowded shantytowns of mud huts and hovels, while exclusive new residential developments for the middle and upper classes sprawl well into the desert beyond.

Nevertheless, Cairo is not only the dominant city of Egypt but also serves as the cultural capital of the entire Arab World, with high-quality universities, splendid museums, world-class theater and music, as well as magnificent mosques and Islamic learning centers. Although Cairo has always primarily been a focus of government, administration, and religion, it also is a river port and an industrial complex, a commercial center, and, as it sometimes seems, one giant bazaar. Cairo is truly the heart and soul of the Arab World, a creation of its geography and a repository of its history.

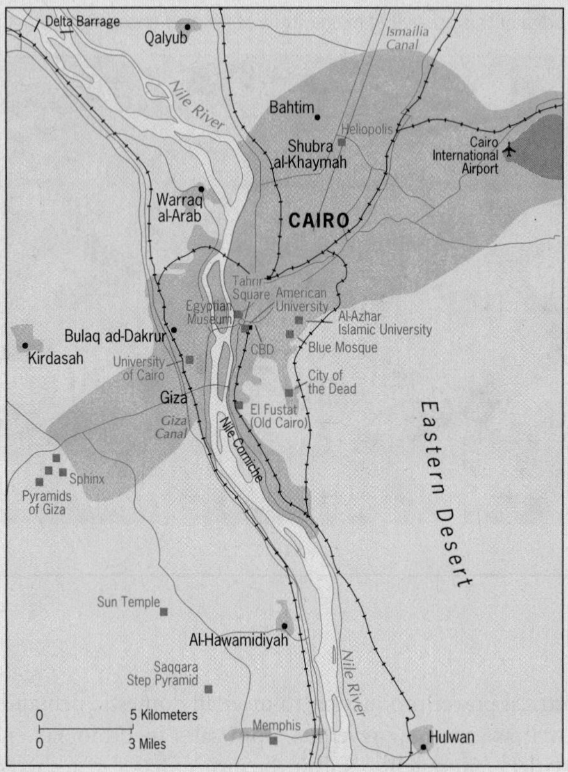

Tahrir Square, located in the center of the city, has become Cairo's most famous public space in this decade—a cauldron of protest, a crucible of solidarity for Egypt's anti-government activists, as well as an arena of repression. Literally 'Freedom Square,' it has served as the main stage for demonstrations since 2010, and images broadcast from here have become familiar on television and computer screens around the world (see opening photo in Chapter 7A).

Political Upheaval

As we noted in Chapter 7A, when Tunisia's longtime president was forced to resign in early 2011 after massive demonstrations in Tunis, it was the spark that ignited an "Arab Spring" of political upheaval all across the realm, most notably in Egypt. Hundreds of thousands of people quickly assembled in Cairo's Tahrir Square to demand the removal of President Mubarak. Following initial efforts by the authorities to crush the protests, and after more than a week of violent confrontations in which an estimated 850 people perished, the regime caved in and Mubarak resigned shortly thereafter. Such an outcome would have been unthinkable just a few months earlier, and many Egyptians—even those risking everything at Tahrir Square—could scarcely believe they had been victorious.

With Mubarak arrested and put on trial, military leaders took control of the situation and swiftly organized a referendum on constitutional change, as demanded by the protesters. New constitutional amendments approved overwhelmingly, included term limits for the president and judicial oversight of free elections. Following prolonged demonstrations urging more rapid change, parliamentary elections were held in early 2012 and Egypt's Muslim Brotherhood prevailed over the combined opposition by 52 to 48 percent of the vote. A few months later, the Muslim Brotherhood's leader, Mohamed Morsi, was elected president.

At first, the new leadership seemed to espouse governance in line with the **Turkish model [2]**, a multi-party democracy that has a place for, yet is not dominated by, Islamic parties. But soon President Morsi took a winner-take-all view of Egyptian democracy, announcing decrees that

From the Field Notes . . .

© H.J. de Blij

"On the eastern edge of central Cairo we saw what looked like a combination of miniature mosques and elaborate memorials. Here lie buried the rich and the prominent of times past in what locals call the 'City of the Dead.' But we found it to be anything but a dead part of the city. Many of the tombs here are so large and spacious that squatters have occupied them. Thus the City of the Dead is now an inhabited graveyard, home to well over a million people. The exact numbers are impossible to determine; indeed, whereas metropolitan Cairo in 2014 has an officially estimated population of just under 12 million, many knowledgeable local observers believe that 17 million is closer to the mark."

www.conceptcaching.com

gave his office sweeping new executive and legislative powers and ignoring minority rights and concerns. Once again the protesters took to the streets, now reflecting the president's shrinking popularity and fueled by a failing economy and growing unemployment. At first, the military cracked down on the protests, inflicting serious casualties, but by mid-2013 it was clear that the government had failed. In July the army, having issued an ultimatum requiring presidential compromise, intervened amid massive celebrations and installed a "transitional" team that would reset Egypt's course toward more competent administration. But a clear outcome was not in sight.

Sudan

As Figure 7B-2 shows, Egypt is bordered by two countries that have recently gone through even more turbulent times than Egypt itself: Sudan to the south and Libya to the west. Sudan, the northern remnant of the 2011 breakup of former Sudan, is almost twice the size of Egypt (but with barely 40 percent of the population), and lies centered on the confluence of the White Nile (from Uganda) and the Blue Nile (from Ethiopia). Here the twin capital, Khartoum–Omdurman, anchors a sizeable agricultural area where cotton was planted during colonial times.

Almost all of Sudan is desert, with irrigated agriculture along the banks of the While Nile and the Blue Nile. The country also has a 500-kilometer (300-mi) coastline on the Red Sea, where Port Sudan lies almost directly across from Jiddah and Makkah (Mecca) in Saudi Arabia. For several years after former Sudan became independent in 1956, the economy was typical of the energy-poor global periphery, exchanging sheep, cotton, and sugar for oil, with Saudi Arabia the main trading partner.

Moreover, former Sudan's decades of civil war between north and south following the end of colonial rule

VOICE FROM THE Region

CAIRO AS MIRROR

© Bahia Shehab

Bahia Shehab, Artist, Designer, Historian, Cairo

"Two years into the Egyptian revolution and we are still facing police brutality on the streets. My newsfeed is filled with images of constant and consistent violence against peaceful protesters. As an artist the only weapon I have is a spray can, so I decided to use it. I spray messages on the streets of Cairo. In my messages I say no. In Arabic to say no, we say "No, and a thousand times no". I decided to say no to military rule, no to emergency laws, no to postponing trials, no to military trials, no to stripping the people, no to blinding heroes, no to snipers, no to sectarian divisions, no to killing, no to killing men of religion, no to sexual harassment, no to burning books, no to violence, no to barrier walls, no to bullets, no to tear gas, no to infiltration, no to stealing the revolution, no to a new pharaoh . . . . no, and a thousand times no! My series of "no" is all over the streets of Cairo and I do not see myself as somebody who is vandalizing the city; on the contrary, I think I am making it more human by creating a mirror that reflects the dreams of the people living in it. I add color to a gray smoky city. I want to make people think about their dreams and to remind them every day, as they drive to work or go to get their groceries, that the revolution will continue and it will not stop until every street in Egypt is a clean and safe place for any human being to exist."

had impoverished the Islamic regime in Khartoum. One particularly violent chapter of Sudan's recent history was written in the western province of ***Darfur*** (Fig. 7B-2). The local people there, the Fur, have for centuries lived as Arabized pastoralists in the north and as settled farmers in the south. It is not clear how the conflict began, but a number of officials of the Khartoum regime suspected the southern Fur of sympathizing with rebels in the far south who were opposed to Islamic rule. In 2003, northern pastoralist militias, called the *Janjaweed* and encouraged by the regime, rode into the villages and fields of the southern Fur, burning homes, destroying crops, and killing thousands. An estimated 2 million people were driven from their homes, and a combination of violence and resultant disease took a toll that by now may exceed 400,000.

Oil in the Transition Zone

In the 1980s, it became clear that significant oil reserves lay in former Sudan, and during the 1990s discoveries were made in Kordofan Province and elsewhere in the central part of the country, including major deposits in the embattled south. The government then saw its income rise, allowing it to purchase more weapons. Local peoples living on top of or near the newly tapped deposits were forcibly relocated. A construction boom of ultramodern buildings in Khartoum reflected the newfound wealth for the elite few. But the discovery of oil in the south also fueled the drive for independence in that part of former Sudan and culminated in the breakup of 2011.

As a single political entity, former Sudan had been a fixture on the map of northeastern Africa for over a century. But on July 9, 2011, after 55 years of independence, that country split in two. Post-2011 Sudan constitutes the much bigger, Islamic, northern part; the new South Sudan is a typical Equatorial African country dominated by Christianity and African religions. The breakup followed more than five decades of bitter and bloody conflict with an estimated 1.5 million fatalities. Sudan is now a more homogeneously Islamic state, with a Muslim population of more than 90 percent (compared to the 68 percent of pre-breakup Sudan). Even though the unofficial international boundary puts the majority of the former Sudan's oil reserves in South Sudan, the pipelines for export run northward across post-2011 Sudan.

Both Sudan and South Sudan remain involved in costly military operations and internal as well as cross-border conflicts, while their populations remain in critical need of economic and political progress. One of their disputes concerns the ownership of Abyei Province—located astride the still-unresolved international border between the two—which contains a mixed population and sits atop an important oil reserve. Although South Sudanese troops crossed the border in early 2012 to briefly seize the key Heglig oilfield (see photo), a year later tensions had subsided and both sides were abiding by a new agreement to avoid open conflict and allow the oil pipelines to resume operations. By mid-2013, relations between the two countries were also improving in other spheres, but the final chapter in this still-unfolding geopolitical drama has yet to be written.

THE MAGHREB AND ITS NEIGHBORS

The countries of northwestern Africa are collectively called the ***Maghreb***, but the Arab name for them is more elaborate than that: *Jezira-al-Maghreb*, or "Isle of the West," in recognition of the great Atlas Mountain range rising like a huge island from the Mediterranean Sea to the north and the sandy flatlands of the immense Sahara to the south.

The countries of the Maghreb (sometimes spelled *Maghrib*) include Morocco, last of the North African kingdoms; Algeria, a secular republic beset by religious-political problems; and Tunisia, smallest and most Westernized of the three (Fig. 7B-3). Neighboring Libya, facing the Mediterranean between the Maghreb and Egypt, is unlike any other North African country: an

Sudanese technicians inspect the oil facilities damaged during the April 2012 conflict at the Heglig oilfield in Sudan. Border clashes between the two neighbors had been ongoing since South Sudan became independent in July 2011. In 2012 the two countries signed an agreement to honor a demilitarized zone along the border and allowing minorities on both sides to cross the border freely—but tension remains and hostilities could resume.

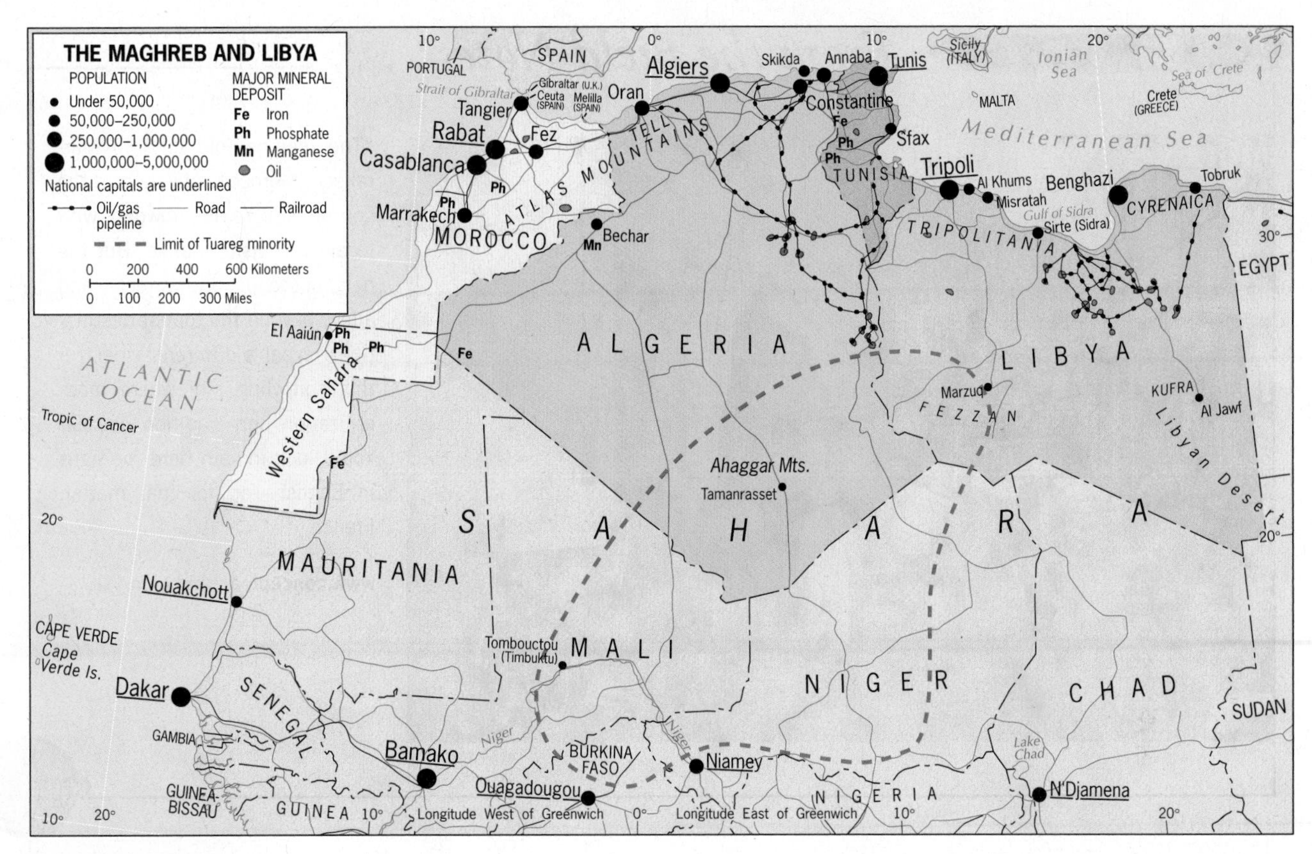

FIGURE 7B-3

oil-rich desert state whose population is almost entirely clustered in settlements along the coast—and continues to reel in the political aftermath of a brief but especially violent episode of civil war that overthrew the Qadhafi dictatorship.

Atlas Mountains

Whereas Egypt is the gift of the Nile, the Atlas Mountains facilitate the settled Maghreb. These high ranges wrest from the rising air enough orographic rainfall to sustain life in the intervening valleys, where good soils support productive farming. From the vicinity of Algiers eastward along the coast into Tunisia, annual rainfall averages greater than 75 centimeters (30 in), a total more than three times as high as that recorded for Alexandria in Egypt's delta. Even 240 kilometers (150 mi) inland, the slopes of the Atlas still receive over 25 centimeters (10 in) of rainfall. The effect of this topography can even be read on the world climate map (Fig. G-7): where the highlands of the Atlas terminate, aridity immediately takes over.

The Atlas Mountains are aligned southwest-northeast and begin inside Morocco as the High Atlas, with elevations close to 4000 meters (13,000 ft). Eastward, two major ranges dominate the landscapes of Algeria proper: the Tell Atlas to the north, facing the Mediterranean, and the Saharan Atlas to the south, overlooking the great desert. Between this pair of mountain chains, with each consisting of several parallel ranges and foothills, lies a series of intermontane basins markedly drier than the northward-facing slopes of the Tell Atlas. Within these valleys, the **rain shadow effect [3]** of the Tell Atlas is reflected not only in the steppe-like natural vegetation but also in land-use patterns: pastoralism replaces cultivation, and stands of short grass and shrubs blanket the countryside.

Colonial Impact

During the colonial era, which began in Algeria in 1830 and lasted until the early 1960s, well over a million Europeans came to settle in North Africa—most of them French, and a substantial majority bound for Algeria—and these immigrants soon dominated commercial life. They stimulated the renewed development of the region's towns, and Casablanca, Algiers, Oran, and Tunis became the urban foci of the colonized territories. Although the Europeans dominated trade and commerce and integrated the North African countries with France and the European Mediterranean world, they did not confine themselves to the cities and towns. They recognized the agricultural possibilities of the favored parts of the *tell* (the lower Tell Atlas slopes and narrow coastal plains that face the Mediterranean) and

From the Field Notes . . .

© H.J. de Blij

"The medina (old city) in Marrakech, Morocco is lively, and the central square is crowded with locals and visitors alike. But the neighborhoods and streets and alleys beyond the tourist destinations reveal a different aspect of this sprawling city where modernization and tradition coexist. You'll look in vain here for signs in English—or, for that matter, French . . ."

www.conceptcaching.com

established prosperous farms. Not surprisingly, agriculture here is Mediterranean. Algeria soon became known for its vineyards and wines, citrus groves, and dates; Tunisia has long been one of the world's leading exporters of olive oil; and Moroccan oranges went to many European markets.

The Maghreb Countries

Between desert and sea, the Maghreb states display considerable geographic diversity (Fig. 7B-3). **Morocco**, a conservative kingdom in a revolution-marked region, is tradition-bound and weak economically. Like some of the other Arab monarchies in the realm, Morocco's government introduced some modest reforms to preempt growing popular discontent. Its core area lies in the north, anchored by four major cities; but the Moroccans' political attention is focused on the south, where the government is seeking to absorb its neighbor, ***Western Sahara*** (a former Spanish dependency with almost 600,000 inhabitants, many of them immigrants from Morocco). Even if this campaign is successful, it will do little to improve the lives of most of Morocco's 33.5 million people, of whom hundreds of thousands have migrated to Europe.

Algeria, France's one-time colony whose agricultural potential drew more than one million European settlers, now has an economy based primarily on its substantial oil and gas reserves. Its bitter war of liberation (1954–1962) was followed by recurrent conflict between Islamists and military-backed secular authorities, costing over 100,000 lives and resulting in a long-delayed compromise still punctuated by occasional skirmishes. The country's primate city, Algiers, is centrally situated on the Mediterranean coast and contains nearly 10 percent of Algeria's 39 million people. But a large contingent of Algerians—officially estimated at between 1.7 and 2 million—has emigrated to France, where they form one of Europe's largest Muslim population sectors. Hardnosed military control in combination with generous social spending kept the Arab Spring at bay. But in early 2013, a major terrorist attack by Muslim extremists took place in the remote Sahara of far southeastern Algeria. Jihadists associated with insurgents in northern Mali overran a well-protected gasfield, occupied its main production complex, and captured dozens of hostages. A few days later, Algerian troops stormed the facility and killed all of the rebels . . . but also left 38 hostages dead.

The smallest of the Maghreb states, **Tunisia**, lies at the eastern end of the region. As Appendix B shows, Tunisia in many ways outranks surrounding countries: it has a higher urbanization level, higher social indicators, and a much lower growth rate (among its population of just over 11 million) than elsewhere in the Maghreb. Most of the country's productive capacity lies in the north in the hinterland of its historic capital, Tunis.

Nonetheless, in January 2011 Tunisia took center stage as the launching pad for a wave of revolt that swept through most of the Arab World and beyond. Despite its rather modern appearance, Tunisia's repressive and corrupt government had become intolerable to the people. Their response was the *Jasmine Revolution*, a sudden and explosive outburst of street protests that forced out the regime of President Ben Ali, who fled the country. It was an amazing feat: the dictator of a **police state [4]** entrenched for a generation overthrown in a matter of weeks by a spontaneous uprising.

All across the Arab World, the events in Tunisia were closely monitored by excited populations and nervous governments. But in Tunisia, as in Egypt shortly thereafter, the ouster of the regime gave way to a power vacuum, rendering the governance system dysfunctional. Still, free elections were held later in 2011 and the moderate Islamist party, Ennahda, emerged as the biggest winner, announcing that it would continue the country's secular tradition and not seek to impose Islamic law. But once in office, Ennahda increasingly showed signs of a conservative Islamist turn (the assassination of a secular opposition leader in early 2013 darkened the mood further). Nevertheless, of all the revolutions sparked by the Arab Spring movement, Tunisia's was by far the most successful, and democratic prospects remain alive.

Libya

Almost rectangular in shape, Libya (with a population of 6.7 million), faces the Mediterranean Sea between the Maghreb states and Egypt (Fig. 7B-3). What limited coastal-zone agricultural possibilities exist lie in the two districts known as Tripolitania in the northwest, centered on the capital of Tripoli, and in the northeast in Cyrenaica, where Benghazi is the urban focus. But it is oil, not farming, that propels the economy of this highly urbanized country. The oilfields are located well inland from the central Gulf of Sidra, linked by pipelines to coastal terminals. Libya's two interior corners, the desert Fezzan district in the southwest and the Kufra oasis in the southeast, are connected to the coast by two-lane roads subject to frequent sandstorms.

Muammar Qadhafi had been one the Arab World's most violent despots, clinging to power for more than four decades. His iron-fisted response to anti-government demonstrations in several Libyan cities that arose in early 2011 did not come as a surprise. Within a few months, however, advancing rebel forces captured a number of cities including their main base of Benghazi as well as Tobruk in the northeast; Misratah, about 200 kilometers (125 mi) east of Tripoli; and a few smaller towns in the far northwest near the Tunisian border. From there, they extended their reach and by the fall of 2011 had seized control of Tripoli. They could not have done so without the intervention of air power supplied by a coalition of NATO countries (including the United States), which enforced a no-fly zone for Qadhafi's air force throughout northern Libya and carried out a series of bombardments on key military targets. After nine months of fighting the rebels prevailed, and on October 20, 2011 Qadhafi was cornered and killed in Sidra, his home town located on the shore of the gulf of the same name, midway between Tripoli and Benghazi.

But even though the Libyan people had rid themselves of a despised dictator, their country remains in political chaos. The militias involved in the uprising have not disarmed, and elections have yet to reach the discussion stage. Local tribes and ethnic groups, long suppressed under Qadhafi, have reemerged and are contesting each other as Libya lingers in crisis.

Adjoining Saharan Africa

Between the string of North Africa's coastal states and the southern margin of the Sahara lies a tier of five states, all but one of them landlocked, dominated by the world's greatest desert, sustained by modest ribbons of water, and under the sway of Islam. Figure 7B-1 best displays this chain of countries whose populations are hardly insignificant by African standards, totaling some 70 million.

Only coastal **Mauritania** contains a minuscule population even in this company: less than 4 million in a territory the size of Texas and New Mexico combined, half of it concentrated in and around the Atlantic coastal capital, Nouakchott, base of a small fishing fleet. Overwhelmingly Muslim and experiencing its first democratic presidential election as recently as 2007, Mauritania has been infamous for tolerating slavery despite its official abolition in 1981; following that unprecedented election, its parliament voted to impose prison terms on slave-owners continuing the practice. Despite that breakthrough, democracy was short-lived, and in 2008 the military seized power again.

Neighboring **Mali** (17.0 million), a huge country the combined size of Texas and California, was until recently held up by many as a democratic model before it descended into chaos in 2012 after a military coup sidelined its president. The military was said to be frustrated with the government's handling of the Tuareg insurgency in the sparsely populated north (Fig. 7B-3). But the coup triggered a rebel surge, and within months the country was effectively split in two. The Islamic jihadists (espousing Taliban sympathies) took control of the fabled city of Timbuktu—located near the center of this hourglass-shaped country—and declared the north an independent state, naming it *Azawad*.

In early 2013, the military returned control to a civilian government in the capital, Bamako; shortly thereafter France sent in 4000 of its troops to help recapture Timbuktu. But once this key city was reclaimed the French began to withdraw their forces, turning over the effort to restore full government control of northern Mali to a UN-mandated all-African military force. Exacerbating this conflict was a natural crisis in the form of a severe drought that plagued the country's central Sahel zone, which by mid-2013 had already forced more than 100,000 Malians to flee their impoverished villages. At press time, it was clear that Mali's long post-independence era of stability was ending as it became a failed state, the latest victim of the turmoil endemic to the African Transition Zone.

Uranium is now the leading export of **Niger**, which is pronounced *nee-ZHAIR*. As in Mali, the Niger River is the lifeline of this nation of 17.5 million, as reflected by Figure 7A-2, and Niamey, the capital, lies on its banks in the far southwestern "tail" of this goldfish-shaped country. We should keep in mind that these landlocked and low-income countries emerged more than half a century ago as independent states from French imperial rule; when their boundaries were first laid out nobody seriously imagined these colonial dependencies would ever become sovereign states.

This explains why there is a country called **Burkina Faso**—lying south of Mali and southwest of Niger—where there is no Niger River, no coastline, and little economic opportunity (this is one of the world's poorest countries, with nearly half the population of 18.6 million subsisting on one U.S. dollar a day or less). Environmental variability here is marked by destructive floods alternating with severe droughts, the latter driving people and their livestock southward into Ivory Coast, resulting in deadly conflicts. Despite all this, a representative government in the capital of Ouagadougou manages to play a regional and international role in this part of the continent.

The fifth member of this tier of desert-margin states, **Chad**, lies east of Niger and, as Figure 7B-1 indicates, is also bisected by the wide African Transition Zone. Here, divisions between Chad's Islamized north and Christian/animist south are strong, and the map shows its capital, N'Djamena, lies directly on the Islamic Front. Since 2000, this country of 12.5 million has been in upheaval, the crisis in Sudan's Darfur provinces spilling across the border into its eastern territorial margin, the south experiencing an oil boom (see Fig. 6B-6), and the capital the focus of a struggle among rebel factions trying to overthrow the government.

Islamic militancy, diverse resources increasingly in demand, and restive minorities are now at long last drawing the international attention of global powers to this broad, remote, Subsaharan periphery of north-central Africa long in the shadow of more prominent coastal neighbors to the north and south.

■ THE MIDDLE EAST: CRUCIBLE OF CONFLICT

The regional term ***Middle East***, we noted earlier, is not satisfactory, but it is so common and generally used that avoiding it creates more problems than it solves. It originated when Europe was the world's dominant realm and when places were *near*, *middle*, and *far* from Europe: hence a Near East (Turkey), a Far East (China, Japan, and other countries of East Asia), and a Middle East (Egypt, Arabia, Iraq). If you check definitions used in the past, you will see that these terms were applied inconsistently: Syria, Lebanon, Palestine, and even Jordan sometimes were included in the Near East, and Persia and Afghanistan in the Middle East.

Today, the geographic designation *Middle East* has a more specific meaning. And at least half of it has merit: this region, more than any other, lies at the middle of the far-flung Islamic realm (Fig. 7B-1). To the north and east of it, respectively, lie Turkey and Iran, with Muslim Turkestan beyond the latter. To the south lies the Arabian Peninsula. And to the west lie the Mediterranean Sea and Egypt plus the rest of North Africa. This, then, is the key region of the realm, its very heart.

Five countries comprise the Middle East (Fig. 7B-4): Iraq, largest in population and territorial size, facing the Persian Gulf; internally-embattled Syria, next in both categories and fronting the Mediterranean; Jordan, linked by the narrow Gulf of Aqaba to the Red Sea; Lebanon, whose survival as a unified state has come into question; and Israel, Jewish nation in the crucible of the Muslim world. Because of the extraordinary importance of this region in world affairs, we focus in some detail on issues of cultural, economic, and political geography.

Iraq's Enduring Importance

Figure 7B-4 explains, even at a glance, why Iraq is pivotal among the states of the Middle East. With about 60 percent of the region's total area, more than 45 percent of its predominantly Arab population, and most of its valuable energy and agricultural resources, Iraq is key to the Middle East's fortunes. Iraq also is heir to the early Mesopotamian states and empires that emerged in the basin of its twin rivers, the Tigris-Euphrates, and the country is studded with matchless archeological sites and museum collections. Much of this heritage was disastrously damaged in 2003 when the United States led a military invasion of Iraq in the aftermath of the 9/11 terrorist attacks in America. The map indicates why this attack had (and continues to have) critical consequences for the region and beyond: Iraq has six neighbors. To the west lie Syria, Iraq's staunch ally during dictator Saddam Hussein's despotic rule, and Jordan, whose leaders chose Iraq's side during an early 1990s conflict with southern neighbor Kuwait. To the south, Iraq possesses a lengthy border with Saudi Arabia. And to the north, Iraq has borders with Iran, against which it fought a bitter war during the 1980s, and with Turkey, source of its crucial river lifelines.

With so many land neighbors, it is no surprise that Iraq is nearly landlocked. Figure 7B-4 shows how short and congested its single outlet to the Persian Gulf is; this was the main reason that Iraq tried to conquer and annex neighboring Kuwait in 1990 (Kuwait has significant oil reserves as well). As a result, a network of pipelines across Iraq's neighbors must carry much of Iraq's oil export volume to coastal terminals in other countries—another reason these countries have an interest in what happens to Iraq in the years ahead.

Discordant Cultural and Political Geographies

Let us now turn to the larger-scale map of Iraq (Fig. 7B-5), which shows how Iraq is divided into three cultural domains in which religion, tradition, and custom form the basis of its political geography. The largest population sector

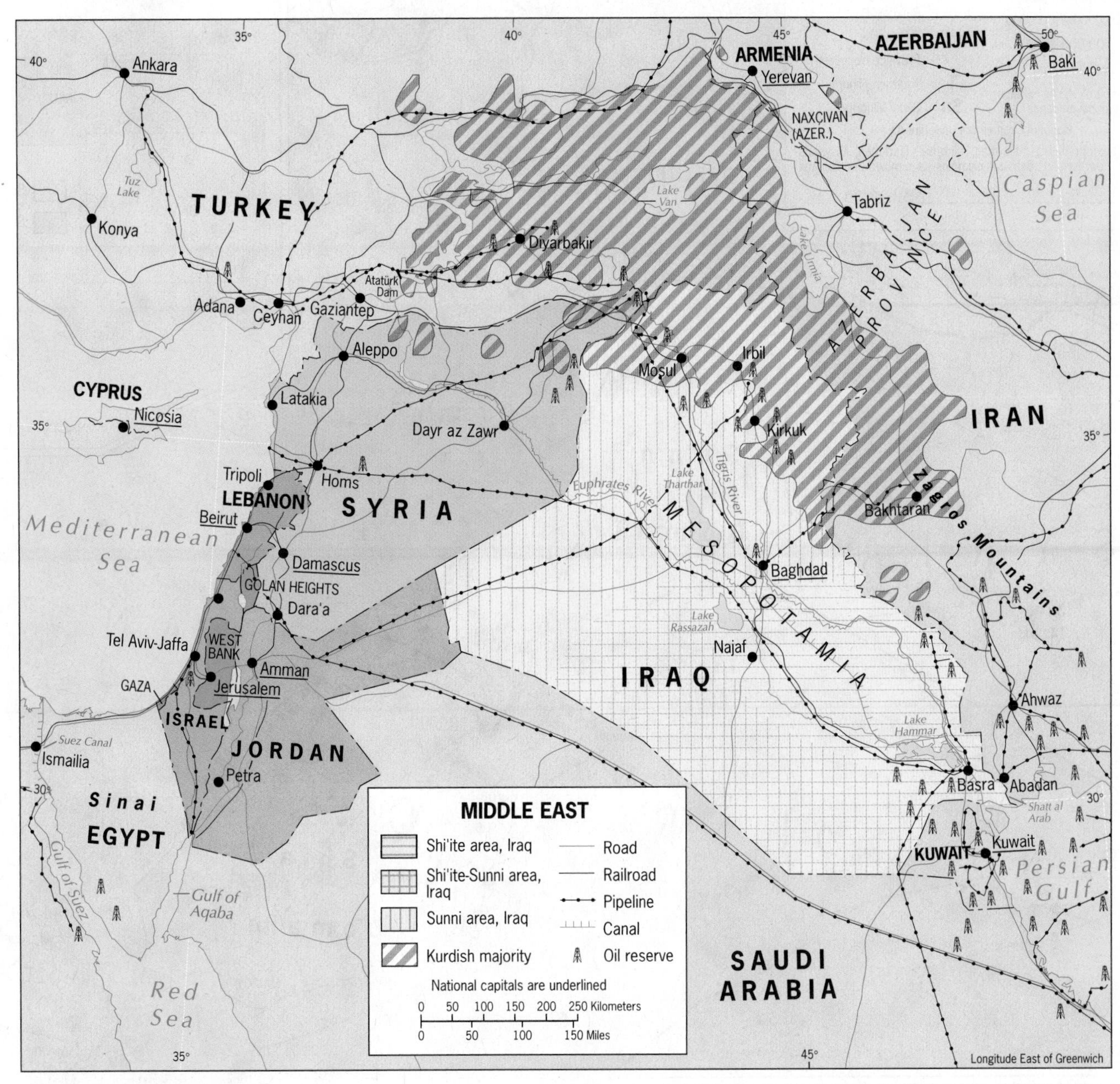

FIGURE 7B-4

(within the national total of 35.7 million) is that of the 19 million Shi'ites in the southeast, whose religious affinities are with neighboring Iran. The next largest is the Sunni minority of about 10 million in the north and west, which ruled the country before the 2003 U.S. invasion. Perhaps the most striking feature in Figure 7B-5 is the orange-colored zone that covers northeastern Iraq, large parts of Turkey and Iran, and smaller parts of other countries. This is the area where the majority of residents are not Arabs but Kurds, one of the world's largest **stateless nations [5]** (discussed further in Chapter 7A). More than 6 million Kurds inhabit the hilly north of Iraq; adjacent southeastern Turkey is the traditional home of about 20 million more, and another 6 million or so live in western Iran.

The U.S.-Led Invasion and Withdrawal

The United States invaded Iraq in 2003 on the grounds that it possessed weapons of mass destruction and that al-Qaeda had important bases there. Neither proved to be true, and the invasion was heavily criticized in the United States and around the world—even though Saddam Hussein's regime was widely considered to be one of the most brutal in this realm. Hussein was swiftly removed but the invasion unleashed horrendous sectarian violence and enabled al-Qaeda terrorists, who had not previously been active here, to worsen the security situation.

More recently, Iraq's people have repeatedly voted in elections that advanced the country toward representative government, and some middle-class Iraqis (among the hundreds of thousands who fled the country in the mid-2000s)

IRAQ
POPULATION
Under 50,000
50,000–250,000
250,000–1,000,000
1,000,000–5,000,000
Over 5,000,000
Limit of strong Kurdish regional influence
Oilfield
Road
Railroad
Oil pipeline
Province boundary
T Turkmen Minority
Su Sunni Minority
National capitals are underlined
Domains depict prevailing majorities; minorities (Turkmen, Assyrians) not shown. Borderlands between domains have mixed populations.
0 40 80 120 160 200 Kilometers
0 20 40 60 80 100 Miles
CARTOGRAM OF POPULATION GROUPS IN IRAQ
MOSUL
IRBIL
KIRKUK
BAGHDAD
BASRA
Shia
Sunni
Kur
Mix
Longitude East of Greenwich
TURKEY
SYRIA
IRAN
IRAQ
JORDAN
KUWAIT
SAUDI ARABIA
Kurdish Domain
Sunni Domain
Shia Domain
Desert
Zagros Mts.
KHUZESTAN PROVINCE
Persian Gulf
Rumaylah Oilfield
Lake Van
Lake Orumiyeh
Lake Asad
Tharthar Lake
Razzaza Lake
Qadisiyah Lake
Euphrates River
Tigris River
Naxcivan
Tatvan
Van
Salmas
Hakkari
Orumiyeh
Gaziantep
Sanliurfa
Qamishli
Kilis
Zakhu
Aleppo
Mosul
Irbil
Sulaymaniyah
Sanandaj
Kirkuk (Tamim)
Hamadan
Dayr az Zawr
Tikrit
Samarra
Kermanshah
Arak
Borujerd
Mandali
Baqubah
Khorramabad
Fallujah
Ramadi
Baghdad
Rutbah
Karbala
Hillah
Kut
Dezful
Najaf
Diwaniyah
'Amarah
Nukhayb
Ahvaz
Nasiriyah
Badanah
Basra
Abadan
Rafha
Kuwait
Subayhiyah
Hafar al Batin
PROVINCES OF IRAQ
0 100 200 Kilometers
0 50 100 Miles
Dahuk
Arbil
Ninawa (Nineveh)
Tamim (Kirkuk)
Sulaymaniyah
Salah ad Din (Saladin)
Diyala
Baghdad
Anbar
Babil
Wasit
Karbala
Qadisyah
Maysan
Najaf
Dhi Qar
Muthanna
Basra

have returned to their homeland. The allied presence also helped to create a more open economy. In 2010, the official combat role of U.S. troops in Iraq was declared to be at an end, following a protracted war that claimed an estimated 150,000 Iraqi lives and caused the death of some 5000 allied soldiers; the last U.S. troops withdrew from Iraq on December 18, 2011.

Today, the Kurdish and Shi'ite zones are relatively stable and prosperity has risen as the level of violence has subsided. But the Sunni domain, especially around the capital, Baghdad, remains conflicted and very poor. Tensions between the Shi'ite-dominated government and the Sunni minority show no signs of letting up. In a typical month of early 2013, a wave of car bombings in Baghdad's Shi'ite neighborhoods claimed more than 30 fatalities and dozens of severe injuries. It remains uncertain as to whether Iraq will be able to achieve anything resembling structural stability, let alone lasting political democracy.

Syria

Like Iraq before 2003, Syria was ruled with an iron fist by a minority (that was still brutally fighting for its political life as this book went to press). Although Syria's population of 23.4 million is about 75 percent Sunni Muslim, the ruling elite comes from a smaller Shi'ite Islamic sect based in the country—the Alawites—who account for only around 12 percent of the population. Leaders of this powerful minority have retained control over the state for decades, mostly by ruthless suppression of dissent. In 2000, president-for-life Hafez al-Assad died and was succeeded by his son, Bashar, signaling a continuation of the political status quo. For 25 years, part of this status quo was the occupation and control of neighboring Lebanon, but in 2005 the Syrians finally withdrew.

Like Lebanon and Israel, Syria has a Mediterranean coastline where crops can be raised without irrigation. Behind this densely populated coastal zone, Syria has a much larger interior than its neighbors, but its areas of productive capacity are widely dispersed (Fig. 7B-4). Damascus, in the southwestern corner of the country, was built on an oasis and is considered to be one of the world's oldest continuously inhabited cities. It is now the capital of Syria, with a population of 2.9 million.

The far northwest is anchored by Aleppo (ravaged in 2013 by the regime's vicious military attacks on Syria's defenseless civilians), the focus of cotton- and wheat-growing areas close to the Turkish border. Here the Orontes River is the chief source of irrigation water, but in the northeast the Euphrates Valley is the crucial lifeline. It was in Syria's interest to develop its northeastern provinces (which also contain recently discovered oil reserves), but all development plans came to a screeching halt when the Arab Spring movement arrived. As the latter engulfed the country and ignited a ghastly civil war, disintegrating Syria was swiftly transformed into a failed state.

Syria's dictatorship had first seemed immune to the wave of unrest that cascaded across much of the Arab World in early 2011. But demonstrations soon broke out in a number of cities, and the protest movement spread rapidly in response to the Assad regime's savage military crackdown, with the worst of the brutalities inflicted on the western cities of Aleppo, Latakia, Dara'a, and Homs. By mid-2013, at least 100,000 people had died in the violence while thousands more were reported to be in custody or missing. The conflict also created about two million refugees, many of whom crossed the border into Turkey, Jordan, Lebanon, and, to a lesser extent, Iraq; for those who stayed, widespread food shortages only added to their suffering. Internationally, even though the regime had been almost universally condemned, it continued to receive steady (if tacit) support from Russia, Iran, and China—as discussed in Chapter 7A.

Jordan

With a poverty-stricken capital (Amman), lacking in oil reserves, and possessing only a small and remote outlet to the Gulf of Aqaba, Jordan (population 6.7 million) has survived with U.S., British, and other aid. It lost its West Bank territory in the 1967 war with Israel, including its sector of Jerusalem (then the kingdom's second-largest city). No third country has a greater stake in a settlement between Israel and the Palestinians than does Jordan, especially because Palestinians are the majority here.

Jordan did not experience the kind of upheaval that shook many other Arab countries in 2011 because the monarchy had enjoyed widespread support and government policies lacked the level of repression found in Syria or Tunisia. Yet there were some demonstrations to air grievances, and King Abdullah II had to dismiss his cabinet to quell the protests. Jordan's long-term challenges are economic, and in 2012 deepening deprivation began to turn into political discontent; by mid-2013, this convergence had strengthened to a level that threatened the fundamental tribal alliance critical to the monarchy's continued control of the state.

Lebanon

The map suggests that Lebanon has significant geographic advantages in this region: a lengthy coastline on the Mediterranean Sea; a well-situated capital, Beirut, on its shoreline; oil terminals along its coast; and a major capital (Syria's Damascus) in its hinterland. The map at the scale of Figure 7B-4 cannot reveal yet another asset: the fertile, agriculturally productive Bekaa Valley in the eastern interior. But if Lebanon is well endowed from a physical geographical point of view, its social geography and geopolitical situation are problematic. As a result, the country has gone back and forth between good times and bad.

For so small a country, Lebanon is permeated with religious and ethnic factions and is prone to sectarian breakdown. The Lebanese were unhappy with what they felt to

FIGURE 7B-5

be a Syrian occupation, yet that situation persisted for well over two decades. In the meantime, an Iran-sponsored terrorist movement, Hizbullah, came into being and even became a political force. The Syrians finally departed in 2005 under United Nations auspices, and a stable new Lebanese government coalesced around a power-sharing agreement. Peace, however, was short-lived. In 2006, a Hizbullah kidnapping of Israeli soldiers provoked an attack from Israel, shattering the reconstructed physical and political infrastructure. By 2013, Beirut seemed to be regaining some of its luster, once again attracting international business as well as more tourists. Lebanon is sometimes aptly described as a "garden without a fence" because of its vulnerability to outside interference, its latest burden being the more than 600,000 refugees of the Syrian war (as of mid-2013). With so many external powers and influences involved in its affairs, Lebanon is unlikely to achieve stability and significant progress in the foreseeable future.

Israel and the Palestinian Territories

Israel, the Jewish state, lies in the center of the Arab World (Fig. 7B-1). Since 1948, when Israel was created as a homeland for the Jewish people on the recommendation of a United Nations commission, the Arab-Israeli conflict has affected all else in the region.

Figure 7B-6 helps us understand the complex issues involved here. In 1946, the British, who had administered

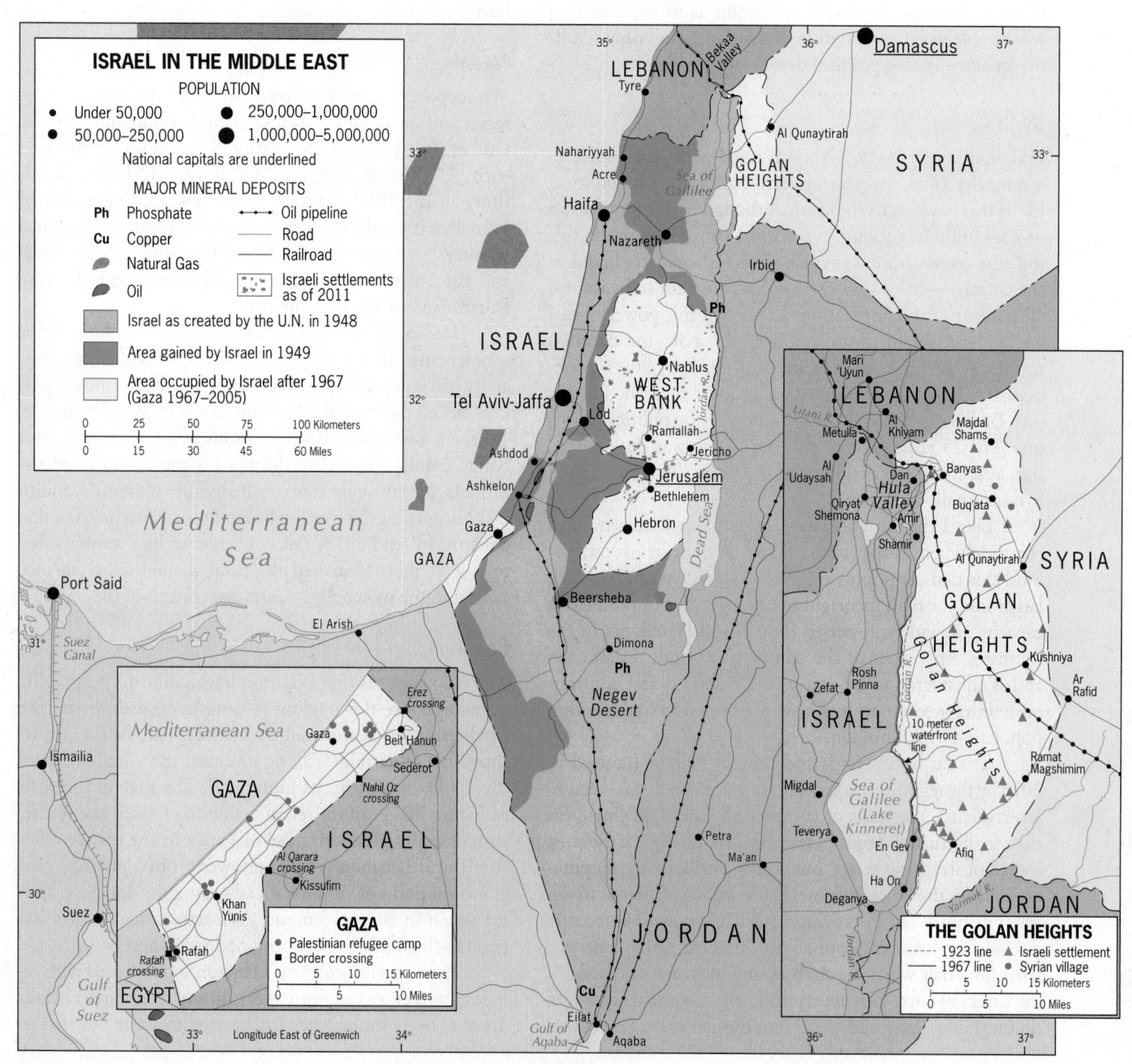

FIGURE 7B-6

this area in post-Ottoman times, granted independence to what was then called Transjordan, the kingdom east of the Jordan River. In 1948, the orange-colored area became the UN-sponsored state of Israel—including, of course, land that had long belonged to Arabs in this territory called Palestine.

As soon as Israel proclaimed its independence, neighboring Arab states attacked it. Israel, however, not only held its own but pushed the Arab forces back beyond its borders, gaining the green areas in 1949 (Fig. 7B-6). Meanwhile, Transjordanian armies crossed the Jordan River and annexed the yellow-colored area named the West Bank (of the Jordan River), including part of the city of Jerusalem. The king called his newly enlarged country Jordan.

Further conflict soon followed. In 1967, a six-day-long war produced a major Israeli victory: Israel took the Golan Heights from Syria, the West Bank from Jordan, and the Gaza Strip from Egypt; it also conquered the entire Sinai Peninsula all the way to Egypt's Suez Canal. In later peace agreements, Israel returned the Sinai to Egypt but not the Gaza Strip.

All this strife produced a huge outflow of Palestinian Arab refugees and displaced persons. The Palestinian Arabs constitute another of this realm's ***stateless nations***; about 1.7 million continue to live as Israeli citizens within the borders of Israel, but more than 2.3 million are in the West Bank and about another 1.7 million in the Gaza Strip. Many Palestinians live in neighboring and nearby countries, including Jordan (2.9 million), Syria (600,000), Lebanon (450,000), and Saudi Arabia (350,000); another 450,000 or so reside in Iraq, Egypt, and Kuwait; and some live in other countries around the world, including the United States (ca. 300,000). The estimated size of the total, stateless Palestinian population in 2014 was between 11 and 12 million.

Israel is approximately the size of Massachusetts and has a population of 8.2 million (including its 1.7 million Arab citizens), but because of its location, a powerful military, and its strong international links, these data do not reflect Israel's importance. Israel has been a recipient of massive U.S. financial aid, and American foreign policy has been to seek an accommodation between Jews and Palestinians, as well as between Israel and its Arab neighbors. Washington is strongly pushing toward a two-state solution, but the political will on either side does not appear to be forthcoming.

Geographic Obstacles to a Two-State Solution

1. ***The West Bank.*** Even after its capture by Israel in 1967, the West Bank might have become a Palestinian homeland, but steady and deliberate Jewish immigration into this territory made such a future problematic. In 1977, only 5000 Jews lived in the West Bank, including East Jerusalem. In 2013, there were nearly 400,000 living in the West Bank (plus at least another 300,000 in East Jerusalem)—constituting fully 15 percent of the population and creating a seemingly inextricable jigsaw of Jewish and Arab settlements (Fig. 7B-7) that makes a two-state solution increasingly difficult. Moreover, the West Bank is almost entirely bounded by the Israeli-built 'security fence' (see below).
2. ***The Golan Heights.*** The right inset map in Figure 7B-6 suggests how difficult the Golan Heights issue is. The Heights overlook a large area of northern Israel, and they flank the Jordan River as well as crucial Lake Kinneret (Sea of Galilee), the main water reservoir for Israel. Relations even with a post-Assad Syria are not likely to become normalized until the Golan Heights are returned, but in democratic Israel the political climate may make the ceding of this territory impossible.
3. ***Jerusalem.*** The United Nations intended Tel Aviv to be Israel's capital, and Jerusalem an international city. But the Arab attack and the 1948–1949 war allowed Israel to drive toward Jerusalem (see Fig. 7B-6, the green wedge into the West Bank). By the time a cease-fire was arranged, Israel held the western part of the city, and Arab forces the eastern sector. But in this eastern sector lay major Jewish historic sites, including the Western Wall. Still, in 1950 Israel declared the western sector of Jerusalem its capital, making this, in effect, a ***forward capital***. Figure 7B-8 shows the position of the (black) armistice line, leaving most of the Old City in Jordanian hands. Then, in the 1967 war, Israel conquered all

Jerusalem's urban landscape is rife with cultural contrasts—and is also home to the holy sites that mean so much to Jews, Christians, and Muslims

© H.J. deBlij

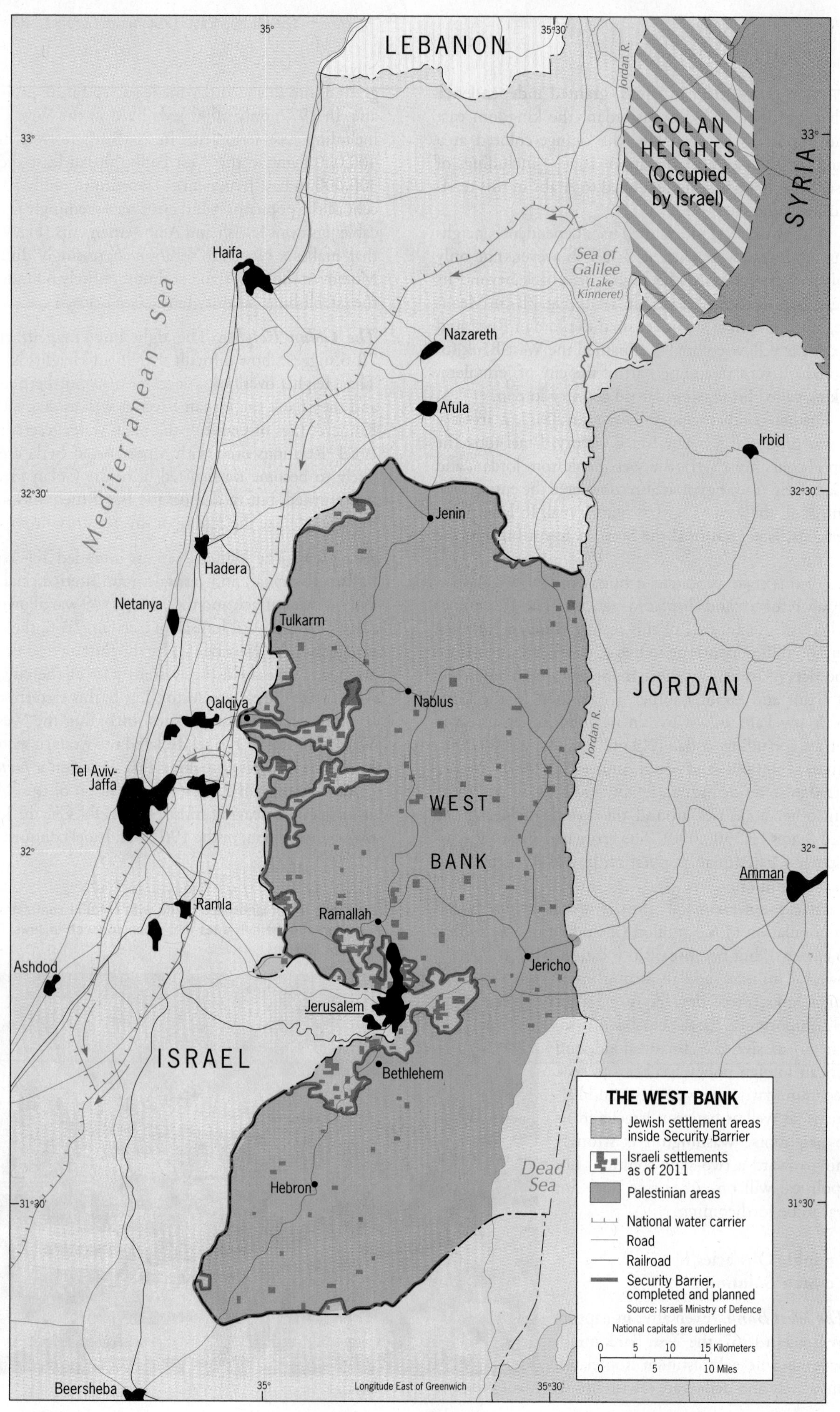
LEBANON
GOLAN HEIGHTS (Occupied by Israel)
SYRIA
Mediterranean Sea
Sea of Galilee (Lake Kinneret)
Jordan R.
Haifa
Nazareth
Afula
Irbid
Jenin
Hadera
Netanya
Tulkarm
JORDAN
Nablus
Qalqiya
Tel Aviv-Jaffa
WEST BANK
Amman
Ramla
Ramallah
Ashdod
Jericho
Jerusalem
ISRAEL
Bethlehem
Dead Sea
Hebron
Beersheba
35°
35°30'
33°
32°30'
32°
31°30'
Longitude East of Greenwich
THE WEST BANK
Jewish settlement areas inside Security Barrier
Israeli settlements as of 2011
Palestinian areas
National water carrier
Road
Railroad
Security Barrier, completed and planned
Source: Israeli Ministry of Defence
National capitals are underlined
0 5 10 15 Kilometers
0 5 10 Miles

FIGURE 7B-7

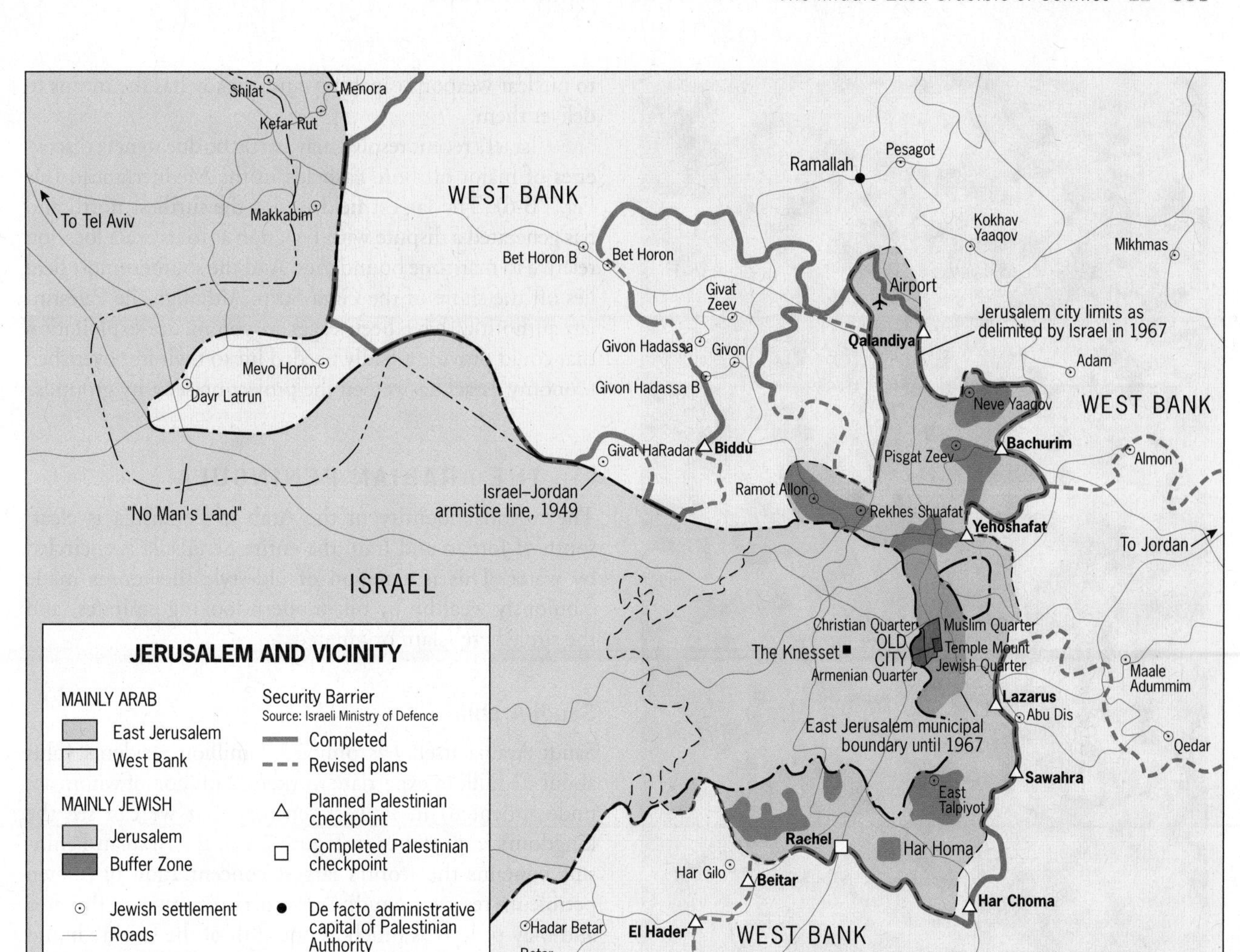

FIGURE 7B-8 © H. J. de Blij, P. O. Muller, and John Wiley & Sons, Inc.

of the West Bank, including East Jerusalem; in 1980, the Jewish state reaffirmed Jerusalem's status as capital, calling on all countries to relocate their embassies from Tel Aviv (a few did, but none are here today). Meanwhile, the government redrew the map of the ancient city, building Jewish settlements in a ring around East Jerusalem that would terminate the old distinction between Jewish west and Arab east (Fig. 7B-8). This enraged Palestinian leaders, who still view Jerusalem as the eventual headquarters of a hoped-for Palestinian state. Renewed U.S. efforts to persuade Israel to halt the construction of new settlements in East Jerusalem have thus far been futile, despite promises of increased American aid to Israel.

4. ***The Security Fence.*** In response to the infiltration of suicide terrorists, Israel has now walled off most of the West Bank along the Security Barrier border shown in Figure 7B-7 and the next photo. This reinforcement of the West Bank border imposes a hardship on Palestinians, and for some it runs between their homes and their farm fields. Palestinians demand that the wall be taken down; Israelis reply that the Palestinian National Authority (the interim self-government body) must control the terrorists.

5. ***The Gaza Strip.*** Israel in 2005 decided to withdraw from the Gaza Strip, removing all Jewish settlements and yielding control of the area to the (then) Palestinian Authority (Fig. 7B-6, left inset map). A power struggle ensued between the two main Palestinian political factions, Fatah and the more militant Hamas, and Hamas won a disputed 2006 election. As rockets smuggled in from Egypt through tunnels dug beneath the Gaza-Egyptian border (see photo) fell on Israel, causing some deaths, injuries, and damage, the Israelis warned the Palestinian (Hamas) regime of retaliation. Hamas authorities either would not or could not stop the rockets fired from Gaza, and in 2009 Israel launched a massive attack, killing about 1300 Gaza residents including many civilians and inflicting considerable infrastructural damage. Condemnation of Israel's assault as disproportionate was worldwide, but a majority of Israeli citizens approved of it. A cease-fire led to a relatively peaceful interlude during which Gaza received considerable investments from abroad (especially from oil-rich

© AP/Wide World Photos

Israel's Security Barrier takes several forms—a concrete wall, an iron fence—and its effect also has multiple dimensions. This section of the barrier separates the village of Abu Dis on the outskirts of Jerusalem (left) from the West Bank (right). While Israel defines this demarcation as a security issue, the Arabs, citing its location inside Palestinian territory, argue that it is motivated primarily by politics.

to nuclear weapons capability and already has the means to deliver them.

Israel's recent resolve may partly be due to new discoveries of major offshore gasfields in the Mediterranean (see Fig. 7B-6). The largest field is also the furthest north and has generated a dispute with Lebanon as to its exact location relative to maritime boundaries. And the southernmost field lies off the shore of the Gaza Strip. Although the Palestinian authorities have been eager to commence exploitation that could provide a badly needed lift to their impoverished economy, Israel has vetoed the project on security grounds.

THE ARABIAN PENINSULA

The regional identity of the Arabian Peninsula is clear: south of Jordan and Iraq, the entire peninsula is encircled by water. This is a region of old-style sheikdoms made fabulously wealthy by oil, modern-looking emirates, and the site where Islam originated.

Saudi Arabia

Saudi Arabia itself has only 29.7 million residents (plus about 11 million expatriate workers, 2 million of whom are undocumented) in its vast territory, but we can see the kingdom's importance in Figure 7A-8: the Arabian Peninsula contains the world's largest concentration of known petroleum reserves. Saudi Arabia occupies most of this area and may possess as much as one-fifth of the world's liquid

February 9, 2013: A Palestinian works inside a smuggling tunnel beneath the Egyptian-Gaza border in the town of Rafah (see inset map in Fig. 7B-6). As the Israelis systematically curtailed traffic between the Gaza Strip and Israel, Gaza has become highly dependent on Egypt for all kinds of goods. But these tunnels are used to smuggle arms and ammunition into the Strip for attacks against Israel. A few weeks after this photo was taken, Egypt, fearing the destabilization of the Sinai Peninsula, ordered its military to flood all cross-border tunnels, some of them with raw sewage.

© Abed Rahim Khatib/Demotix/Corbis

Persian Gulf states) and the economy entered an upswing, particularly its construction sector. But in late 2012, another episode of violent cross-border exchanges occurred, with rockets flying into southern Israel and the retaliating Israeli security forces killing more than 200 Palestinians in Gaza.

Israel lies in the path of a fast-moving geopolitical storm. The issues raised above are only part of the overall problem: others involve water rights, compensation for expropriated land, Arabs' "right of return" to pre-refugee abodes, and the exact form a Palestinian state would take. Meanwhile, an Iranian regime continues its aggressive declaratory stand toward Israel while that country appears to be on its way

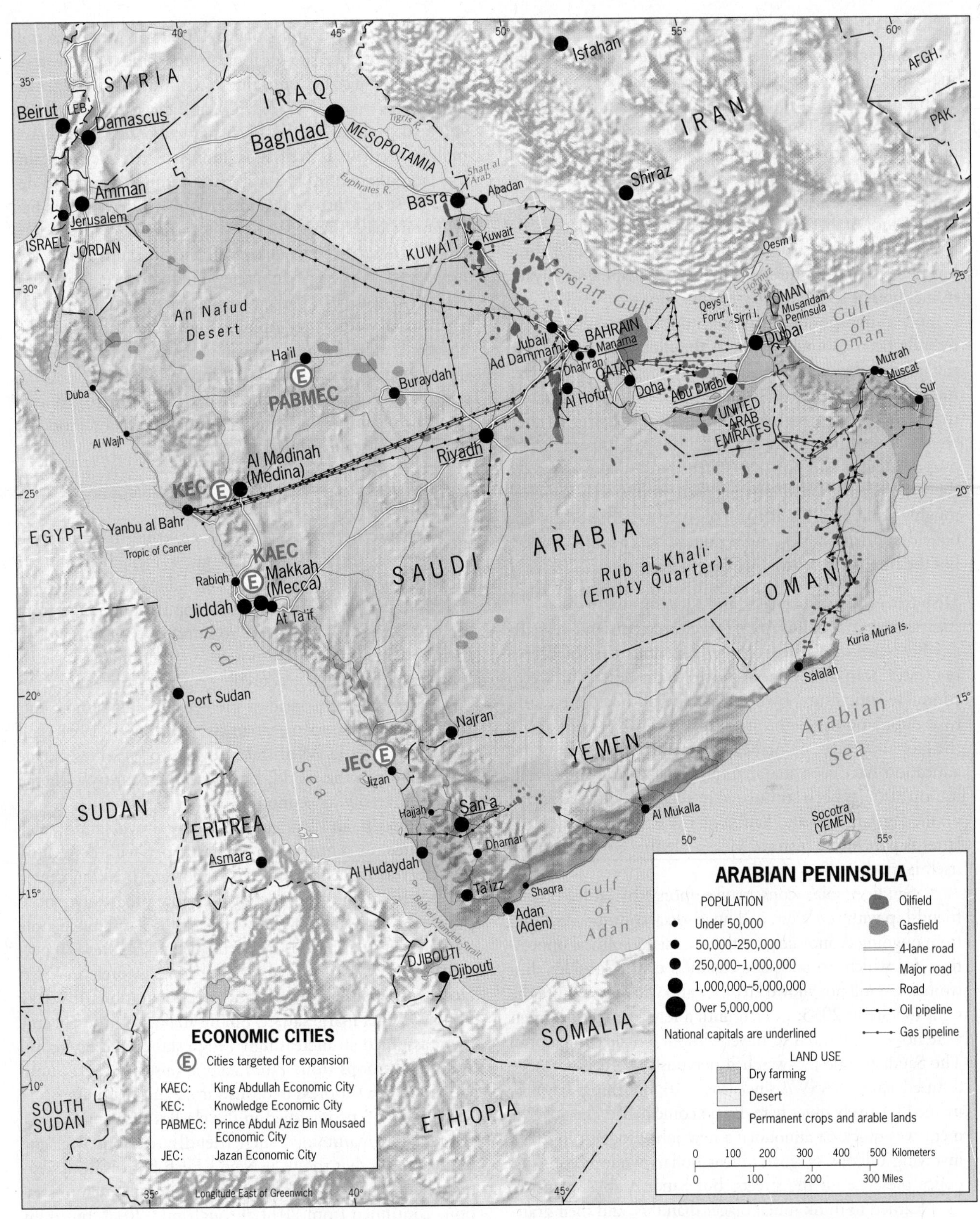

FIGURE 7B-9

oil deposits, second only to reserve leader Venezuela. As Figure 7B-9 shows, these reserves lie in the eastern part of the country, particularly in the vicinity of the Persian Gulf, and extend southward into the Rub al Khali (Empty Quarter).

As a region, the Arabian Peninsula is environmentally dominated by a desert habitat and politically dominated by the Kingdom of Saudi Arabia (Fig. 7B-9). Containing 2,150,000 square kilometers (830,000 sq mi), Saudi Arabia is the realm's third-biggest state; only Kazakhstan and Algeria cover more territory. On the peninsula, Saudi Arabia's neighbors (proceeding clockwise from the head of the Persian Gulf) are Kuwait, Bahrain, Qatar, the United Arab Emirates (UAE), the Sultanate of Oman, and the Republic of Yemen. Together, these countries on the eastern and southern fringes of the peninsula contain 45 million inhabitants; the largest by far is Yemen, with 27 million.

Figure 7B-9 reveals that most of the economic activities in Saudi Arabia are concentrated in a wide belt across the "waist" of the peninsula, from the oil boomtown of Dhahran on the Persian Gulf through the national capital of Riyadh in the interior to the Mecca–Medina area near the Red Sea. A modern transportation and communications network has recently been completed, but in the more remote zones of the interior Bedouin nomads still ply their ancient caravan routes across the vast deserts. For decades, Saudi Arabia's royal families were virtually the sole beneficiaries of their country's wealth, and there was hardly any impact on the lives of villagers and nomads. When the oil boom arrived in the 1950s, foreign laborers were brought in (today there are roughly 11 million, many of them Shi'ite) to work in the oilfields, ports, factories, and as domestics. The east boomed, but the rest of the country has lagged behind.

Disparities, Uncertainties, and Development Plans

Efforts to reduce Saudi Arabia's regional economic disparities have been impeded by the enormous cost of bringing water from deep-seated sources to the desert surface to stimulate agriculture, by the sheer size of the country, and by a population growth rate that is still 50 percent above the global average. Nonetheless, housing, health care, and education have seen major improvement, and industrialization also has been stimulated in such new cities as Jubail on the Persian Gulf and Yanbu on the Red Sea. In fact, the latter is part of the historic Mecca-Medina corridor which itself is modernizing rapidly.

Saudi Arabia's conservative monarchic rule, official friendship with the West, and huge social contrasts resulting from ongoing economic growth have raised political opposition, for which no adequate channels exist. Consider this: women are still not allowed to drive and will only acquire the right to vote in 2015; in fact, until a few years ago, women were not even allowed to travel without a male companion! The Saudi regime has watched nervously as uprisings have diffused across NASWA since late 2010, but things have remained relatively quiet here. By no coincidence, the government was quick to announce a new jobs program in 2011 involving billions of government dollars while beefing up police presence on the streets. But Saudi Arabia's planners have learned to think much bigger than this, and their grandiose plans call for nothing less than a complete economic transformation of the country.

The Saudis have good reason to invest in their economic future. Today, 70 percent of the population is under the age of 30, and unemployment among these young people may run as high as 40 percent. In order to accelerate the generation of the millions of jobs that will be needed, the Saudi royal rulers are spending their country's oil money on six new "economic cities" that by 2020 will create 1.3 million new jobs. The sites of four of these cities have been selected, and construction is under way (see Fig. 7B-9). *King Abdullah Economic City* (*KAEC*), at Rabiqh on the Red Sea, will be completed first by 2016 and will house more than 2 million residents as well as an industrial complex containing more than 2000 factories. According to the grand design, these new urban agglomerations will have diversified economic bases but specialize in particular combinations of industries: Medina will become the site of the *Knowledge Economic City (KEC)*, with a new graduate-level university; *PABMEC* near Ha'il in the desert north will concentrate on agricultural research, equipment, and innovation; and *JEC* next to Jizan in the far south will focus on environmentally-friendly energy and manufacturing industries. Overall, in anticipation of the end of the oil era, the ultimate goal is to transform Saudi Arabia from a petrodollar-based economy into a global-scale industrial power.

On the Peninsular Periphery

Five of Saudi Arabia's six neighbors on the Arabian Peninsula face the Persian and Oman gulfs (Fig. 7B-9) and are monarchies in the Islamic tradition. All five also derive substantial revenues from oil. Their populations range from 1.3 to 8.3 million in addition to hundreds of thousands of foreign workers; these are not strong or influential states, but they do display considerable geographic diversity.

Kuwait, at the head of the Persian Gulf, almost cuts Iraq off from the open sea, an issue the 1990–1991 Gulf War did not settle. Mini-state Kuwait's oil reserves are the sixth-largest in the world, but even this country feels the need to diversify its economy.

Bahrain, an island-state ruled by a royal family since the eighteenth century, is a tiny territory with dwindling oil reserves. It has become an important banking center for the region and has some of the more progressive social policies in this part of the world. Its 1.3 million people are split between Sunnis (30 percent) and Shi'ites (70 percent); nearly two-thirds of its labor force is foreign. When faced with popular unrest and demonstrations in 2011 (mainly by its marginalized Shi'ite majority), the government assumed an increasingly tough stance and invited in additional troops from Saudi Arabia (also run by a Sunni monarchy) to suppress demonstrators and maintain order (more than 50 people died in the clashes). Over the years, Bahrain has maintained such friendly relations with the United States that the U.S. Navy's Fifth Fleet is headquartered here. The country's repression of popular demonstrations continued from 2011 through mid-2013, but drew only minimal criticism from Washington.

Neighboring **Qatar** consists of a peninsula jutting out into the Persian Gulf that has capitalized on its oil reserves and its increasing natural gas reserves (this may be the richest country in the world in terms of per capita annual income). With a modest population of 1.9 million but a dominant Sunni majority, this emirate is trying to establish itself as not only an Arab World focus for

© Jose Fuste Raga/Corbis

The Burj Khalifa, centerpiece of the ultramodern city of Dubai in the United Arab Emirates. Completed in 2010, the 163-story Burj is the world's tallest building, topping out at 828 meters (2716 ft)—just over half a mile high, almost twice the height of New York's Empire State Building.

international economic, social, and political reform, but as a global center for financial deliberation. Its capital, Doha, is the site of annual "Doha Round" meetings of the world's leading economic powers, and it is home to global media giant Al Jazeera (which played a leading role in telecasting the Arab Spring across the realm). The political unrest that coursed through much of the realm during 2011 bypassed Qatar, which was the first Arab country to grant full recognition to Libya's anti-Qadhafi insurgents. Qatar will host soccer's World Cup tournament in 2022, its winning bid a surprise to most observers. This minuscule nation has no soccer tradition to speak of, and its summer temperatures can reach 50° Celsius (122° F). The local organizers, hardly lacking monetary resources, have grandiose plans to build several indoor, fully air-conditioned stadiums.

The **United Arab Emirates (UAE)**, a federation of seven emirates that is home to 8.3 million citizens, faces the Persian Gulf between Qatar and Oman. A reigning sheik is the absolute monarch in each of the emirates, and the seven sheiks together form the Supreme Council of Rulers. In terms of oil revenues, however, there is no equality: two emirates—Abu Dhabi and Dubai—have most of the reserves. As the photo reveals, Dubai (Dubayy) has converted its favorable geographic location together with its oil revenues into a booming economy (staggered by the recent global recession but now recovering) in which banking, trade, tourism, and international transit functions (led by a prestigious global airline) are symbolized by cutting-edge architecture, state-of-the-art engineering feats, and world-class sports, leisure, and entertainment complexes.

The eastern corner of the Arabian Peninsula is occupied by the Sultanate of **Oman**, another absolute monarchy, centered on the capital of Muscat. Figure 7B-9 shows that Oman is fragmented into two parts: the large

From the Field Notes . . .

"The port of Mutrah, Oman, like the capital of Muscat nearby, lies wedged between water and rock, the former encroaching by erosion, the latter crumbling as a result of tectonic plate movement. From across the bay one can see how limited Mutrah's living space is, and one of the dangers here is the frequent falling and downhill sliding of large pieces of rock. It took about five hours to walk from Mutrah to Muscat; it was extremely hot under the desert sun, but the cultural landscape was fascinating. Oil also drives Oman's economy, but here you do not find the total transformation seen in Kuwait or Dubai. Townscapes (as in Mutrah) retain their Arab-Islamic qualities; modern highways, hotels, and residential areas have been built, but not at the cost of the older and the traditional. Oman's authoritarian government is slowly opening the country to the outside world after long-term isolation."

© H.J. de Blij

www.conceptcaching.com

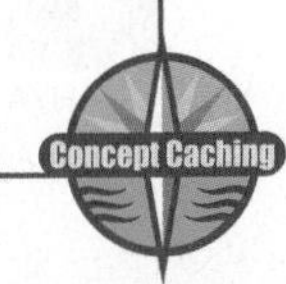

eastern corner of the peninsula, and a small but highly critical cape to the north—the Musandam Peninsula—which protrudes into the outlet of the Persian Gulf to form a narrow channel that is a classic **choke point [6]**, the Hormuz Strait (Iran lies on the opposite shore). Tankers leaving the other Gulf states must negotiate this slender, twisting channel at slow speed, and during politically tense times warships have needed to protect them. Iran's claim to several small islands near the Strait which are owned by the UAE remains the source of a dispute that has intensified since 2012.

This brings us to what is, in so many ways, Saudi Arabia's most substantial peninsular neighbor: **Yemen**. The boundary between the two has only recently been satisfactorily delimited in an area where there may be substantial oil reserves. Another boundary, between former North Yemen and South Yemen, was erased in 1990, when the two countries joined to form the present state.

A sea of Yemeni protesters on May 13, 2011 carry an enormous flag during a demonstration in San'a to demand the ouster of President Saleh. Here and in most other countries touched by the Arab Spring, demonstrators often defiantly displayed their national flag to confirm their nationalistic feelings—but also as a gesture of scorn to remind their government how badly it had failed the country and its people.

San'a, formerly the capital of North Yemen, retained that status; Adan (Aden), the only major port along a lengthy stretch of the peninsula's Arabian Sea coast, anchors the south.

Yemen, with a population of 27.3 million—approximately 53 percent Sunni and 45 percent Shi'ite—is faced with poverty, centrifugal political forces, and terrorist activity—all in abundance. A Shi'ite rebellion against the central government has destabilized the north even as a secessionist movement flares in the south, where oilfields encourage such aspirations.

Figure 7B-9 underscores the geographic attractions Yemen has for al-Qaeda and other Islamic terrorist groups. Not only does Yemen lie at the back door of Saudi Arabia: its southern tip overlooks another of the world's busiest choke points at the entrance to the Red Sea (the Bab el Mandeb ["Gate of Grief"] Strait), where ships converge and risk capture by pirates. Across the Gulf of Adan lies another terrorist haven (and leading pirate base), Somalia, and across the Red Sea lies Eritrea, a likely target for extremists of the al-Qaeda stripe. From Yemen, too, these militants can stoke turmoil all along Africa's Islamic Front (see Fig. 6B-9).

Yemen's precarious position became even more complicated in 2011 when the country was inundated by mass demonstrations against the 32-year-old regime of President Ali Abdullah Saleh. The government responded with a mix of conciliatory measures and violent confrontation, but courageous protesters, many students among them, defied the brutalities inflicted by police and returned to Yemen's streets again and again (see photo). After obtaining promises of immunity from prosecution, President Saleh finally resigned in early 2012. He was succeeded by his vice president, but many of Saleh's relatives and operatives continued to occupy key positions in government and the military. Although a slide into anarchy may have been averted, the current situation does not bode well for a more democratic future.

■ THE EMPIRE STATES

Two major states, both with imperial histories, dominate the region that lies immediately to the north of the Middle East and Persian Gulf (Fig. 7B-1), where Arab ethnicity gives way but Islamic culture endures—Turkey and Iran. Although they share a short border and are both Muslim countries, they display significant differences as well. Even their versions of Islam are different: Turkey is an officially secular but dominantly Sunni state, whereas Iran is the heartland of Shi'ism. Turkey's leaders have worked to establish satisfactory relations with Israel; not long ago, Iran's president promised to "wipe Israel off the map." In recent years, Turkey has been involved in negotiations aimed at entry into the European Union. Iran's goals have been quite different as that country has defied international efforts to constrain its nuclear ambitions. Two smaller countries are inextricably bound up with these two regional powers: island ***Cyprus*** to the southwest, still divided today between Turks and Greeks in a way that threatens Turkey's European ambitions (discussed in Chapter 1B); and oil-rich, Caspian Sea-bordering ***Azerbaijan*** to the northwest, with ethnic and religious ties to Iran but economic links elsewhere (covered in Chapter 2B).

Turkey

As Figure 7B-10 indicates, Turkey is a mountainous country with generally moderate relief; it also tends to exhibit considerable environmental diversity, ranging from steppe to highland (see Fig. G-7). On central Turkey's dry Anatolian Plateau, villages tend to be small and subsistence farmers grow cereals and raise livestock. Coastal plains are not large, but they are productive as well as densely populated. Textiles (from home-grown cotton) and farm products dominate the export economy, but Turkey also has substantial mineral reserves, some oil in the southeast as well as massive dam-building projects on the Tigris and Euphrates rivers, and a small steel industry based on domestic raw materials.

In Chapter 7A we chronicled the historical geography of the Ottoman Empire, its expansion, cultural domination, and collapse. By the beginning of the twentieth century, the country we now know as Turkey lay at the center of that disintegrating state, ripe for revolution and renewal. This occurred in the 1920s and thrust into prominence a leader who became known as the father of modern Turkey: Mustafa Kemal, known after 1933 as Atatürk, meaning "Father of the Turks."

Capitals New and Old

The ancient capital of Turkey was Constantinople (now Istanbul), located on the Bosphorus, part of the strategic Turkish Straits that connect the Black and Mediterranean seas. But the struggle for Turkey's survival had been waged from the heart of the country, the Anatolian Plateau, and it was here that Atatürk (who ruled the country from 1923 to 1938) decided to place his seat of government (Fig. 7B-10). Ankara, the new capital, possessed some unique advantages: it would remind the Turks that they were (as Atatürk always said) Anatolians, it lay much closer to the center of the country than Istanbul, and it could therefore act as a stronger unifier.

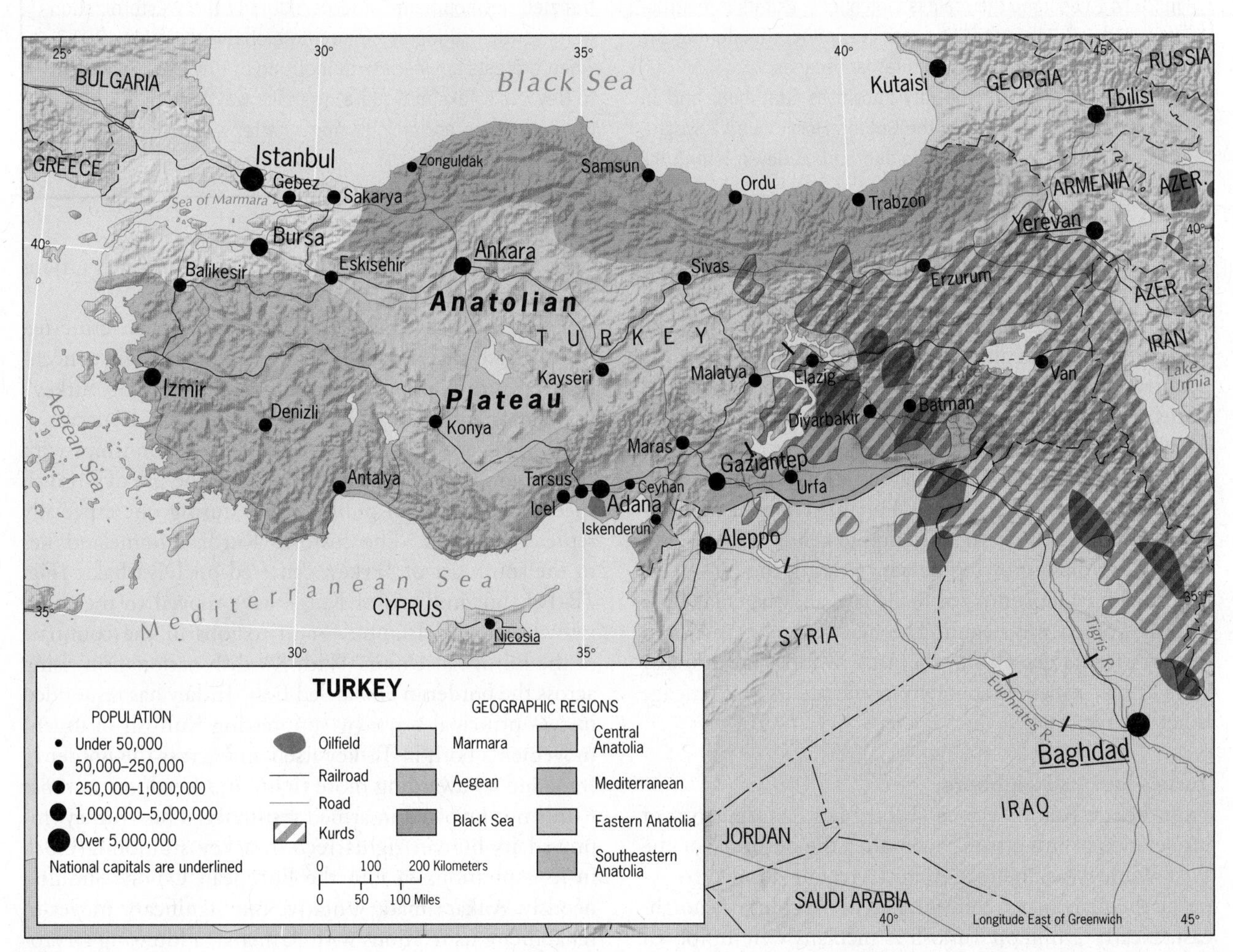

FIGURE 7B-10

AMONG THE REALM'S GREAT CITIES . . .

ISTANBUL

FROM BOTH SHORES of the Sea of Marmara, northward along the narrow Bosphorus waterway toward the Black Sea, sprawls the fabulous city of Istanbul, known for centuries as Constantinople, headquarters of the Byzantine Empire, capital of the Ottoman Empire, and, until Atatürk moved the seat of government to Ankara in 1923, capital of the modern Republic of Turkey as well.

Istanbul's site and situation are exceptional. Situated where Europe meets Asia and where the Black Sea joins the Mediterranean, the city was built on the requisite seven hills (as Rome's successor), rising over a deep inlet on the western side of the Bosphorus, the famous Golden Horn. A sequence of empires and religions endowed the city with a host of architectural marvels that give it, when approached from the water, an almost surreal appearance.

Turkey's political capital may have moved to Ankara, but Istanbul remains its cultural and commercial headquarters. It also is the country's leading urban magnet, luring millions from the countryside. Istanbul's populaltion is doubling every 15 years at current rates, having reached 12.2 million in 2014 to become the realm's largest city. But its infrastructure is crumbling under this massive influx, threatening the legacy of two millennia of cultural landscapes.

In the heart of the city known as Stamboul and in the "foreign" area north of the Golden Horn called Beyoğlu, modern buildings vie for space and harbor views. Istanbul's frenzied development—and particularly the Westernization of its youth—suddenly came into global view in May 2013, when protests against the demolition of tiny Gezi Park (next to Beyoğlu's Taksim Square) to make way for a mall quickly escalated into nationwide unrest after a harsh crackdown by police.

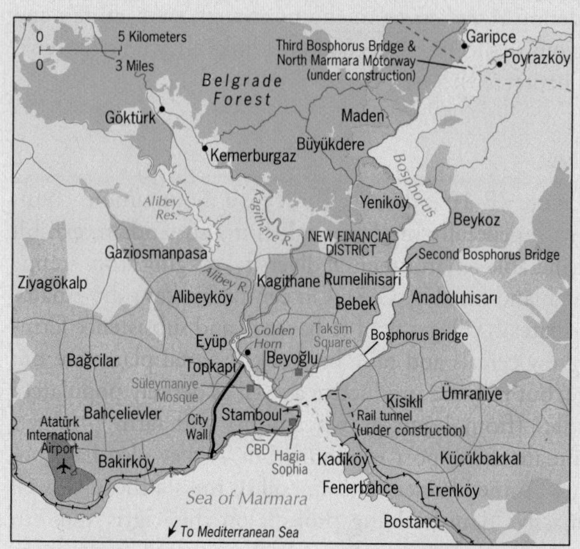

© H. J. de Blij, P. O. Muller, J. Nijman, and John Wiley & Sons, Inc.

Even though Atatürk moved the capital eastward and inward, his orientation was westward and outward. To implement his plans for Turkey's modernization, he initiated reforms in almost every sphere of life within the country. Islam, formerly the state religion, lost its official status, and Turkey became a secular state whose army ensured that the Islamists would not take over again. The state took over most of the religious schools that had controlled education. The Roman alphabet replaced the Arabic. A modified Western code supplemented Islamic law. Symbols of old—growing beards, wearing the fez—were prohibited. Monogamy was made law, and the emancipation of women was begun. The new regime emphasized Turkey's separateness from the Arab World, and in many ways it has remained aloof from the affairs that engage other Islamic states.

Turkey and Its Neighbors

From before Atatürk's time, Turkey has had a history of intolerance of minorities. Soon after the outbreak of the First World War, the (pre-Atatürk) regime decided to expel all the Armenians, concentrated in the country's northeast. Nearly 2 million Turkish Armenians were uprooted and brutally forced out; an estimated 600,000 died in a campaign that continues to arouse anti-Turkish emotions among Armenians today.

In modern times, Turkey also has been criticized for its harsh treatment of its large and regionally concentrated Kurdish population. About one-fifth of Turkey's population of 76.7 million is Kurdish, and successive Turkish governments have mishandled relationships with this minority nation, even prohibiting the use of Kurdish speech and music in public places during one especially repressive period. The historic Kurdish homeland lies in the southeast of Turkey, centered on Diyarbakir (Fig. 7B-10), but millions of Kurds have moved to the shantytowns around Istanbul—and to jobs in the countries of the European Union. With Kurdish nationalism rising across the border in embattled Iraq, Turkey has responded in two principal ways: by suppressing Kurdish insurgent movements both in Turkey itself and across the border in Iraq, and by awarding more rights and freedoms to those Kurds not involved in armed resistance. Thus Turkey improved its human rights record, a key stumbling block in its aspirations to join the European Union. Simultaneously, Ankara made tentative but significant moves to rehabilitate its relations with Armenia, addressing an erosive, longstanding problem.

From the Field Notes . . .

"Having spent the night in Istanbul's Beyoğlu District on the European side, I walked across the Atatürk (Unkapani) Bridge to the Eminönü District on the south side of the Golden Horn, intending to make it to the upscale Fatih District later. But getting a clear view of the Hagia Sophia 'Museum' was not easy from this angle, amid the congested buildup in this hilly area. In a line of taxis at a taxi stand one had a sign saying 'Speak English' so I asked the driver about getting to a good vantage point, and he showed me on my map how to find a lookout point on the terrace of a local restaurant. From there, I could see the bridge I had crossed to my left, the Hagia Sophia Church with its later minarets in the distance, the Sultan Haseki Hürrem Baths to the right, and in the foreground a simple, local mosque. 'You should know the rules about mosques,' he said. 'The number of minarets reveals the importance of any mosque. One minaret, and it's local, neighborhood. Two, and it's likely to be more prosperous, perhaps the work of a very successful citizen or community. Any more than two, and the builders have to have not only religious sanction but also state permission to build. And if they get that, they go for four.' Thousands of mosques, many of them architectural treasures, grace the urban landscape of Istanbul, whose population today, according to local geographers, is really 21 million [almost twice the official 12.2 million]."

© H.J. de Blij

www.conceptcaching.com

Turkey's main foreign policy issue at the present time, besides the Kurdish problem in as far as that extends beyond its borders, is Syria. By mid-2013, at least 500,000 Syrian refugees had settled on Turkish soil and the alleged cross-border movements of Syrian rebel forces have invited fire from government troops inside Syria. Turkey is a strong critic of Syria's violently repressive regime, particularly its treatment of majority Sunnis within that country. Growing tension between Turkey and Syria since 2011 has on more than one occasion brought them to the brink of war.

Turkey and the EU

With its diversified and robustly growing economy, secular government, and independent posture in international affairs, Turkey is the key predominantly Islamic state potentially eligible for admission into the European Union. However, major obstacles lie in the way of this becoming a reality, beginning with the Cyprus and Kurdish issues as well as Turkey's checkered human rights record. Nonetheless, Turkey has made considerable progress over the past decade. The state is now a functioning democracy, and its secular values are constitutionally protected so that even an Islamic party majority in government does not portend radicalization even as it may encourage symbols of piety.

Even though Turkey might be closing in on European Union targets and requirements, enthusiasm among Turks for joining the EU is diminishing. The economy has been doing so well in recent years that the need for EU membership is not as alluring as it once looked—especially as some established EU members (Greece in particular) effectively entered bankruptcy in the early 2010s. In addition, many Turks take offense at the unrestrained public debate in Europe concerning their candidacy, often including criticisms of Turkish customs and traditions not only in Turkey itself but also in Turkish communities already established in Europe. So we have now reached a point that if the EU should eventually extend an invitation to Turkey, it is by no means certain that a future Turkish government would accept it. Polls indicate that popular support for EU membership has dropped from 75 percent in 2000 to less than 50 percent in 2012, with a mere 17 percent believing that EU accession would ever take place.

Turkey is a crucial country in a unique geographical position that is trying to steer an independent and singular course in one volatile realm while accommodating the demands of another. Its importance was confirmed in a most dramatic fashion when protesters across the Arab World in 2011 repeatedly espoused the *Turkish Model* as the ideal future for their countries: a mildly Islamist government that does not alienate the religiously devout and that prioritizes democratic reform while at the same time pursuing vigorous economic growth. For the EU to pass up Turkey's admission, many observers say, looks increasingly imprudent.

Iran

Iran has its own history of conquest and empire, collapse and revival. Oil-rich and vulnerable, Iran already was just a remnant of a once-vast regional empire that waxed and waned, from the Danube to the Indus, over more than two millennia. Long known to the outside world as *Persia* (although the people had called themselves Iranians for centuries), the state was officially renamed *Iran* by the reigning shah (king) in 1935. His intention was to stress his country's Indo-European cultural heritage as opposed to its Arab and Mongol infusions.

But for all the rhetoric about modernization, the shahs failed to advance the lot of the common people (a failure so often repeated throughout this realm), and by 1979 a fundamentalist revolution had engulfed the country. The monarchy was replaced by an Islamic republic ruled by an ayatollah (leader under Allah), and Islamists continue to tightly control the state today. Most recently, the disputed 2009 presidential election—that resulted in a wave of street protests, violence, reprisals, and murder—dramatically exposed these divisions.

A Crucial Location and Dangerous Terrain

As Figure 7B-1 shows, Iran occupies a critical position in this turbulent realm. It controls the entire corridor between the Caspian Sea and the Persian Gulf. To the west it adjoins Turkey and Iraq, both historic enemies. To the north (west of the Caspian Sea) Iran borders both Azerbaijan and Armenia, where once again Islam confronts Christianity. To the east Iran meets Afghanistan and Pakistan, and to the northeast lies similarly volatile Turkmenistan.

Iran, as Figure 7B-11 reveals, is a country of mountains and deserts. The heart of it is a sizeable upland, the Iranian Plateau, that lies surrounded by even higher mountains, including the Zagros in the west, the Elburz in the north along the Caspian Sea coast, and the mountains of Khurasan to the northeast. This mountainous topography signals geologic danger: here the Eurasian and Arabian tectonic plates converge, causing major and often devastating earthquakes. The Iranian Plateau is actually a huge highland basin marked by salt flats and wide expanses of sand and rock. The highlands wrest some moisture from the air, but elsewhere only oases interrupt the arid monotony—oases that for countless centuries have been stops on the area's caravan routes.

In ancient times, Persepolis in southern Iran (located near the modern city of Shiraz) was the focus of a powerful Persian kingdom, a city dependent on **qanats [7]**, underground tunnels carrying water from moist mountain slopes to dry flatlands some distance away. Today, Iran's population of 81 million is 69 percent urban, and the capital, Tehran (population 7.6 million), lies far to the north on the southern, interior slopes of the Elburz Mountains, at the heart of modern Iran's core area.

The Ethnic-Cultural Map

Iranians dominate Iran's affairs, but the country is no more ethnically unified than Turkey. Figure 7B-11 shows that three corners of Iran are inhabited by non-Iranians, two in the west and one in the southeast. Iran's northwestern corner is part of greater Kurdistan, the multi-state region in which Kurds are in the majority and of which southeastern Turkey also is a part (Fig. 7B-4). Between the Kurdish area and the Caspian Sea the majority population is Azeri, who number more than 13 million and also form the majority in the neighboring state of Azerbaijan to the north.

In the southwest, the minority is Arab and is concentrated in Khuzestan Province. During the Iran-Iraq War of the 1980s, Saddam Hussein had his eye on this part of Iran, whose Arab minority often complained of mistreatment by the Iranian regime, and where a large part of Iran's huge oil reserve is located (note the situation of the export terminal of Abadan in Fig. 7B-11). The Arab minority here is among Iran's poorest and most restive, in a province whose stability is key to the country's economy.

The third non-Iranian corner of the country lies in the southeast, where a coastal Arab minority focused on the key port of Bandar-e-Abbas on the Strait of Hormuz gives way eastward to a scattered and still-nomadic population of Baluchis who have ties to fellow ethnics in neighboring Pakistan (see Fig. 7B-11). On the Iranian side of the border, this is a remote and currently inconsequential area, but in Pakistani Baluchistan the government faces a persistent rebellion by tribes who feel slighted by official neglect. Iran's ethnic-cultural map, therefore, is anything but simple.

Energy and Conflict

As Figure 7A-8 shows, Iran possesses the realm's second-largest concentration of oil reserves, which amount to just over half of those in Saudi Arabia. Petroleum and natural gas production account for about 90 percent of the country's income. The reserves lie in a zone along the southwestern periphery of Iran's territory, and Abadan became its petroleum capital near the head of the Persian Gulf. But Iran is a large and populous country, and the wealth oil generated could not transform it in the way the shah, or the ayatollahs who succeeded him, intended. Modernization remained but a veneer: in the villages distant from Tehran's polluted air, the holy men continued to dominate the lives of ordinary Iranians. As elsewhere in the Islamic world, urbanites, villagers, and nomads remained enmeshed in a web of production and profiteering, serfdom and indebtedness that has always characterized traditional society here.

The Islamic revolution swept this system away, but it did not improve the lot of Iran's millions: the economic failure of the shah was followed by the economic failure of the ayatollahs. A devastating war with Iraq (1980–1990), into which Iran ruthlessly poured hundreds of thousands of its young men, sapped both the coffers and energies of the state. When it was over, Iran was left poorer, weaker, and aimless, its revolution spent on unproductive pursuits.

In the early years of the twenty-first century, evidence abounded that the people of Iran remained divided between conservatives determined to protect the power of the mullahs (teachers of Islamic ways) and reformers seeking to modernize and liberalize Iranian society. Economically, the

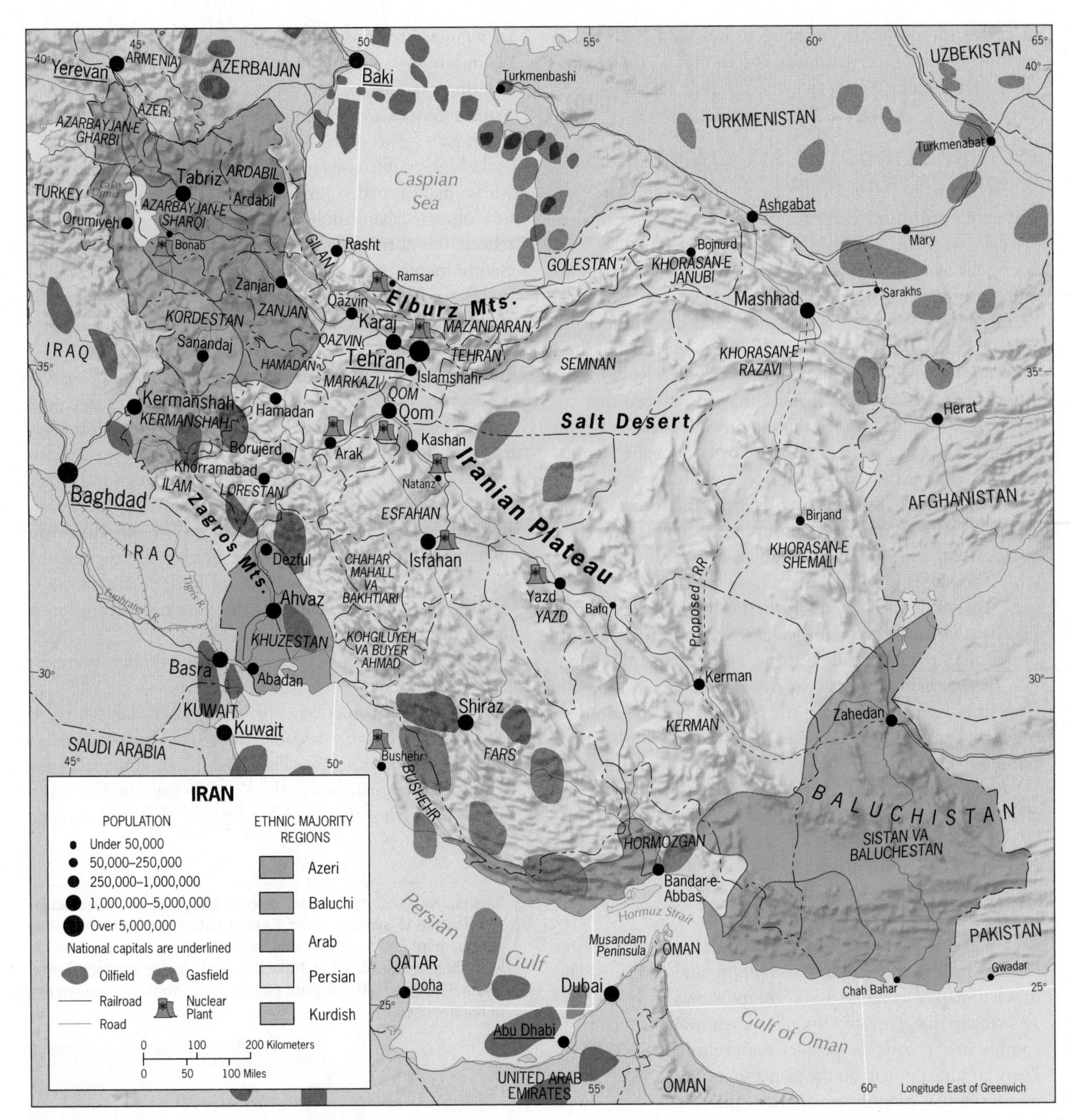

FIGURE 7B-11

© H. J. de Blij, P. O. Muller, and John Wiley & Sons, Inc.

country is at a low ebb, with prices rising about 25 percent per year, in no small part due to international embargoes related to Iran's efforts to acquire nuclear weapons capabilities.

Iran's rise as a (near) nuclear power and its regional ambitions are not surprising. Although revolutionary Iran disavowed the imperial designs of its Persian predecessors, this does not mean that its national interests now end at its borders. Tehran has a major stake in developments in Iraq and Afghanistan, both neighbors; it has an already-nuclear and unstable Pakistan on its southeastern border; it has a strong interest in the fate of Shi'ite majorities and minorities elsewhere in the realm; it has long been an avowed enemy of Israel and a strong supporter of Palestinian causes; and it has funded organizations labeled terrorist whose actions have even reached across the Atlantic.

Since the beginning of this decade, there has been talk of possible attacks by the United States and/or Israel on Iran to destroy its escalating nuclear capabilities. Given the regime's extremely hostile position toward Israel, that country is especially nervous about Iran's intentions. However, it would be difficult to actually destroy key (increasingly subterranean) facilities, in part because intelligence may be lacking. Iran's worsening economic condition, accentuated by international sanctions, might make the regime more receptive to voluntarily agreeing to abandon its nuclear-weapon ambitions in exchange for an end to those sanctions, but nothing was in the works at press time.

TURKESTAN: THE SIX STATES OF CENTRAL ASIA

For many centuries, Turkish (Turkic) peoples spread outward across a vast central Asian domain that extended from Mongolia and Siberia to the Black Sea. Propelled by population growth and energized by Islam, they penetrated Iran, defeated the Byzantine Empire in present-day Turkey, and colonized much of southeastern Europe. Eventually, their power declined as Mongols, Chinese, and then Russians invaded their strongholds. During the twentieth century, the Soviet Union ruled most of this region and, between China and the Caspian Sea, created five Soviet Socialist Republics named after the majority peoples within their borders. Thus the Kazakhs, Turkmen, Kyrgyz, and other Turkic peoples retained some geographic identity in what was then Soviet Central Asia. These Republics became independent when the USSR imploded in 1991 (Fig. 7B-12).

As we define it here, Turkestan includes six states: (1) ***Kazakhstan***, territorially larger than the other five combined but situated astride a major ethnic transition zone; (2) ***Turkmenistan***, relatively closed and dictatorial, but with important frontage on the Caspian Sea and bordering both Iran and Afghanistan; (3) ***Uzbekistan***, the most populous state and situated at the heart of the region; (4) ***Kyrgyzstan***, wedged between powerful neighbors and chronically unstable; (5) ***Tajikistan***, regionally and culturally divided as well as strife-torn; and (6) ***Afghanistan*** (which was never part of the USSR), engulfed in almost continuous warfare since it was invaded by Soviet forces in 1979.

Central Asia—***Turkestan***—is a still-evolving region marked by enormous ethnic complexity and diversity. The detailed map of ethnolinguistic groups shown in Figure 7B-13 actually is a generalization of an even more complicated mosaic of peoples and cultures. Every cultural domain on the map also includes minorities that cannot be mapped at this scale, so that people of different faiths, languages, and ways of life rub shoulders everywhere. Often this results in friction, and at times such friction escalates into ethnic conflict. It should not come as a surprise, in this geographic context and given the Soviet legacy, that democracy is hard to come by: these countries are either autocratic or have flawed representative governments. Tajikistan, in fact, may already be labeled a failed state.

The States of Former Soviet Central Asia

During their hegemony over Central Asia, the Soviets tried to suppress Islam and install secular regimes (this was their objective when they invaded Afghanistan as well), but today Islam's revival is one of the defining qualities of this region. From Almaty to Samarqand, mosques are being repaired and revived, and Islamic dress again is part of the cultural landscape. National leaders make high-profile visits to Mecca; most are sworn into office with a hand on the Quran. All of Central Asia's countries now observe Islamic holidays. In other ways, too, Turkestan reflects the norms of the NASWA realm: in its dry-world environments and the clustering of its population; its mountain-fed streams irrigating farms and fields; its sectarian conflicts; and its oil-based economies. It also is a region where democratic government remains an elusive goal.

Turkestan's intricate ethnolinguistic mosaic is vividly represented in Figure 7B-13. Although states in this region tend to be named after their largest ethnic population sectors, political boundaries do not conform to the distribution of cultures. Perhaps the most complex cultural fabric is that of Afghanistan, populated by Pushtuns (occasionally spelled Pashtuns), Tajiks, Hazaras, Uzbeks, Turkmen, Baluchis, and several lesser groups. Add to this the region's rugged relief, extensive deserts, and limited surface communications, and you can see why national integration is such a difficult challenge here.

Kazakhstan is the region's giant and borders two even greater territorial titans: Russia and China (Fig. 7B-12). During the Soviet period, northernmost Kazakhstan was heavily Russified and become, in effect, part of Russia's Southeastern Frontier (see Fig. 2B-1). Rail and road links crossed the area mapped as the Kazakh–Russian Transition Zone, connecting the north to Russia. The Soviets made

Astana, Kazakhstan's capital since 1998, exhibits spectacular architecture to symbolize the Kazakh nation, its rapid economic development and modernization, and its aspiration to become the key urban center for all of Turkestan. This is the Bayterek Tower, which depicts a mythical giant poplar tree containing a golden egg in its branches.

© Jane Sweeney/JAI/Corbis

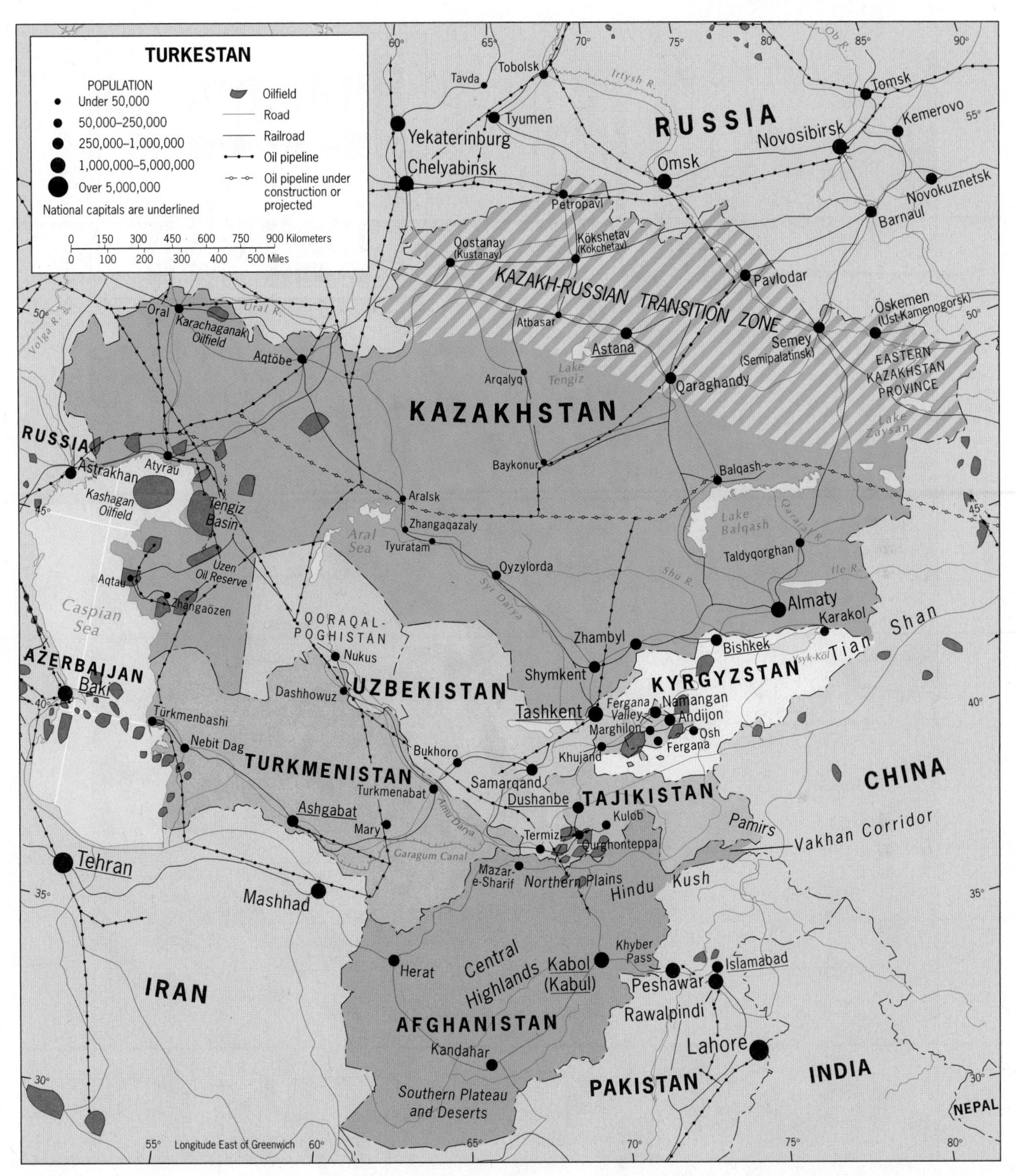

FIGURE 7B-12

Almaty, deep inside the Kazakh domain, the territory's capital. Today the Kazakhs are in control, and they in turn have moved the capital to Astana (see photo), right in the heart of the Transition Zone, where over 4 million Russians (who constitute 24 percent of the total national population of 17.3 million) still live. Clearly, Astana is another **forward capital [8]**.

Kazakhstan is now leading an energy boom in Central Asia: it already ranks among the world's leading producers of uranium as well as the top 20 oil-producing countries. Economic growth has averaged 8 percent per year over the past decade. Figure 7B-12 further reveals Kazakhstan's position as a bridge between the huge oil reserves of the northern Caspian Basin and China. New oil and gas

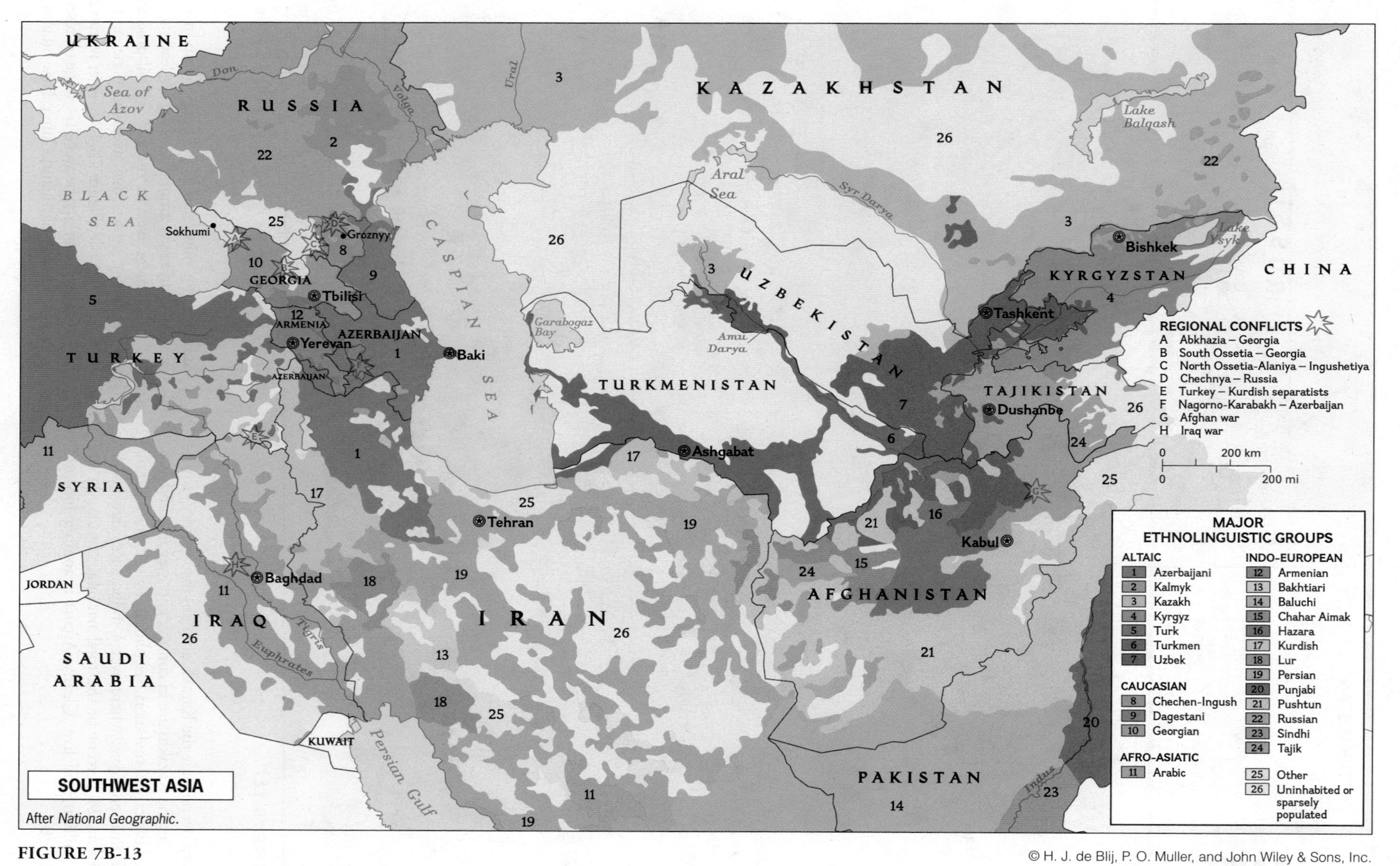
SOUTHWEST ASIA
After National Geographic.
UKRAINE
RUSSIA
KAZAKHSTAN
CHINA
KYRGYZSTAN
UZBEKISTAN
TURKMENISTAN
TAJIKISTAN
AFGHANISTAN
PAKISTAN
IRAN
IRAQ
TURKEY
SYRIA
JORDAN
SAUDI ARABIA
KUWAIT
GEORGIA
ARMENIA
AZERBAIJAN
BLACK SEA
CASPIAN SEA
Sea of Azov
Aral Sea
Lake Balqash
Lake Ysyk
Garabogaz Bay
Persian Gulf
Don
Volga
Ural
Syr Darya
Amu Darya
Tigris
Euphrates
Indus
Sokhumi
Groznyy
Tbilisi
Yerevan
Baki
Tehran
Baghdad
Ashgabat
Tashkent
Bishkek
Dushanbe
Kabul
REGIONAL CONFLICTS
A Abkhazia – Georgia
B South Ossetia – Georgia
C North Ossetia-Alaniya – Ingushetiya
D Chechnya – Russia
E Turkey – Kurdish separatists
F Nagorno-Karabakh – Azerbaijan
G Afghan war
H Iraq war
0 200 km
0 200 mi
MAJOR ETHNOLINGUISTIC GROUPS
ALTAIC
1 Azerbaijani
2 Kalmyk
3 Kazakh
4 Kyrgyz
5 Turk
6 Turkmen
7 Uzbek
CAUCASIAN
8 Chechen-Ingush
9 Dagestani
10 Georgian
AFRO-ASIATIC
11 Arabic
INDO-EUROPEAN
12 Armenian
13 Bakhtiari
14 Baluchi
15 Chahar Aimak
16 Hazara
17 Kurdish
18 Lur
19 Persian
20 Punjabi
21 Pushtun
22 Russian
23 Sindhi
24 Tajik
25 Other
26 Uninhabited or sparsely populated

FIGURE 7B-13

© Diana K. Ter-Ghazaryan

Kyrgyzstan's dry scrublands offer few agricultural opportunities besides pastoralism, and these quickly peter out with rising elevation here in the foothills of the Tian Shan mountain corridor that marks the Chinese border. The semi-nomadic herders of these marginal highlands must constantly move their animals in search of pasture grass, and reside in portable tent-like dwellings known as *yurts*. The yurt is native to Turkestan and its distinctive crown is the central symbol on the post-Soviet flag of Kyrgyzstan.

pipelines operating across Kazakhstan are serving to reduce China's dependence on oil imports carried by tankers along distant international sea lanes.

Uzbekistan occupies the heart of Turkestan and borders every other state in it. Uzbeks not only make up more than 80 percent of the population of 31 million, but also form substantial minorities in a number of neighboring states. The capital, Tashkent, lies in the eastern core area of the country, where most of the people live in towns and farm villages, and the crowded Fergana Valley is the focus. In the far west lies the shrunken Aral Sea, whose feeder rivers were diverted into cotton fields and croplands during the Soviet occupation; heavy use of pesticides contaminated the groundwater, and countless thousands of local people continue to suffer from cancers and other catastrophic medical problems as a result.

Turkmenistan, the autocratic desert republic that extends all the way from the Caspian Sea to the borders of Afghanistan, has a population of 5.3 million, of which nearly 90 percent are ethnic Turkmen. During the Soviet era, communist planners initiated work on a massive project—the Garagum (Kara Kum) Canal designed to transfer water from Turkestan's eastern mountains into the heart of the desert. Today the canal is 1100 kilometers (700 mi) long, and it has enabled the cultivation of more than one million hectares (ca. 3 million acres) of cotton, vegetables, and fruits. The plan is to extend the canal all the way to the Caspian Sea, but meanwhile Turkmenistan has hopes of greatly expanding its oil and gas output from Caspian coastal and offshore reserves. But take another look at Figure 7B-12: this country's relative location is clearly disadvantageous for export routes.

Kyrgyzstan's topography and political geography are reminiscent of the Caucasus. The agricultural economy is generally weak, consisting of pastoralism in the highlands and farming in the valleys. Mapped in yellow in Figure 7B-12, Kyrgyzstan lies intertwined with Uzbekistan and Tajikistan to the point of having exclaves and enclaves along their common borders. Ethnic Kyrgyz constitute about 70 percent of the population of 5.9 million. More than three-fourths of the people profess allegiance to Islam, and (orthodox-Sunni) Wahhabism has achieved a strong foothold here as well; the town of Osh is often referred to as the headquarters of this movement in Turkestan. In 2010, ethnic rioting erupted in Osh between the Kyrgyz majority and the Uzbek minority, killing hundreds and placing the country in jeopardy of breaking apart. Tens of thousands of Uzbeks, who make up roughly 14 percent of the total population but form a prosperous merchant class, fled their homes.

Tajikistan's mountainous scenery is even more spectacular than Kyrgyzstan's, and here as well topography is a barrier to the integration of a multicultural society. The Tajiks, who constitute more than 80 percent of the population of 7.4 million, are ethnically Persian (Iranian), not Turkic, and speak an Indo-European language related to Persian. Most Tajiks, despite their Persian affinities, are Sunni Muslims, not Shi'ites. Small though it is, regionalism to the point of state failure plagues Tajikistan: the government in Dushanbe is repeatedly at odds with the barely connected northern part of the country (Fig. 7B-12), a boiling cauldron not only of Islamic revivalism but also of anti-Tajik, Uzbek activism.

Fractious Afghanistan

Afghanistan, the southernmost country in this region, exists because the British and Russians, competing for hegemony in this area during the nineteenth century, agreed to tolerate it as a cushion, or **buffer state [9]**, between them. This is how Afghanistan acquired the narrow extension leading from its main territory eastward to the Chinese border—the Vakhan Corridor (Fig. 7B-14). As the colonialists delimited it, Afghanistan adjoined the domains of the Turkmen, Uzbeks, and Tajiks to the north, Persia (now Iran) to the west, and the western flank of British India (now Pakistan) to the east.

Landlocked and Fragmented

Geography and history seem to have conspired to divide Afghanistan. As Figure 7B-14 shows, the towering Hindu Kush range dominates the center of the country, creating three broad environmental zones: the relatively well-watered, fertile northern plains and basins; the rugged, earthquake-prone

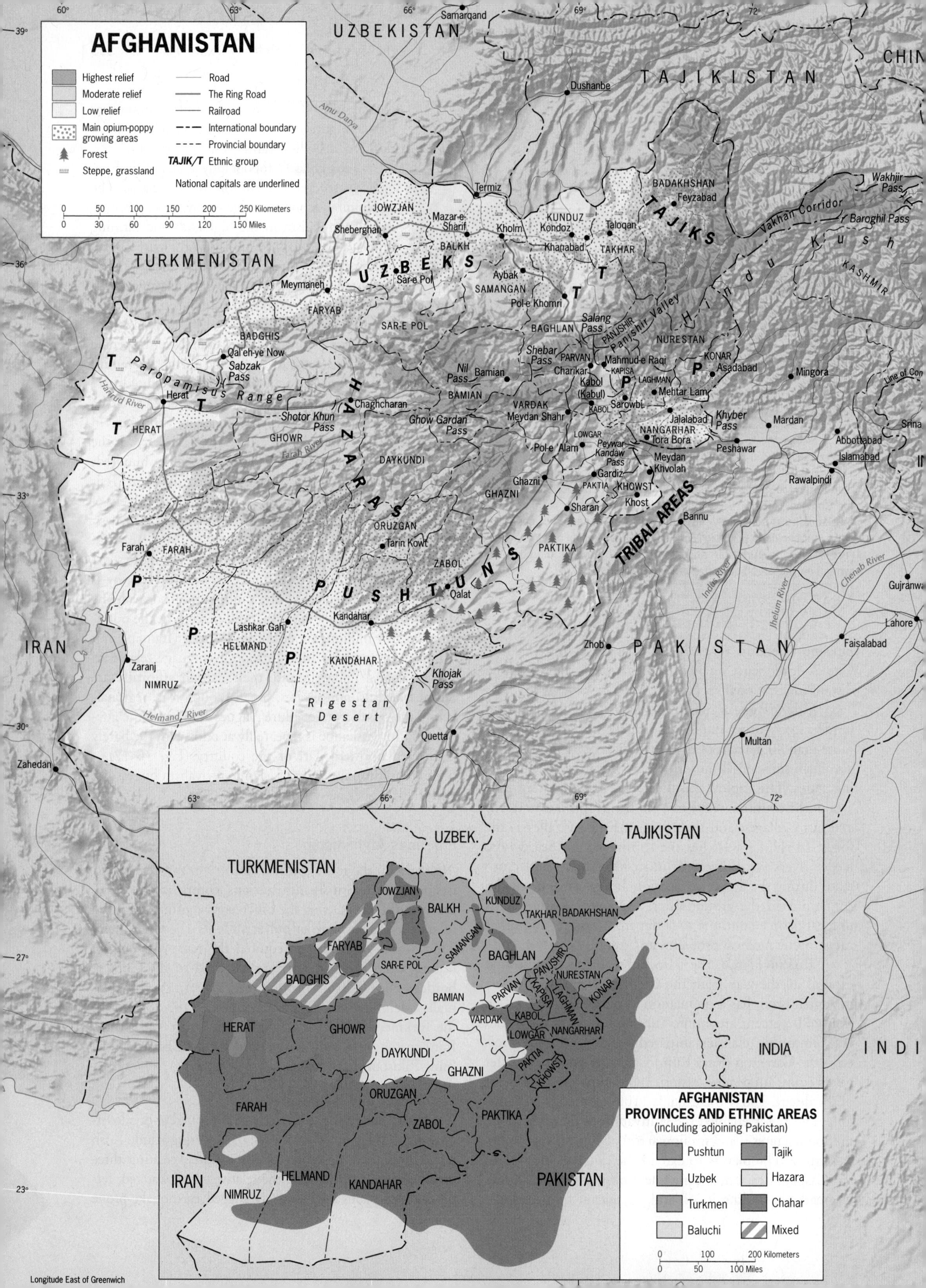

AFGHANISTAN
Highest relief
Moderate relief
Low relief
Main opium-poppy growing areas
Forest
Steppe, grassland
Road
The Ring Road
Railroad
International boundary
Provincial boundary
TAJIK/T Ethnic group
National capitals are underlined
0 50 100 150 200 250 Kilometers
0 30 60 90 120 150 Miles
UZBEKISTAN
TAJIKISTAN
TURKMENISTAN
IRAN
PAKISTAN
UZBEKS
TAJIKS
HAZARAS
PUSHTUNS
TRIBAL AREAS
Hindu Kush
Vakhan Corridor
Paropamisus Range
Rigestan Desert
Samarqand
Dushanbe
Termiz
Mazar-e-Sharif
Sheberghan
Kholm
Kunduz
Kondoz
Taloqan
Feyzabad
Khanabad
Aybak
Pol-e Khomri
Meymaneh
Sar-e Pol
Qal'eh-ye Now
Herat
Chaghcharan
Bamian
Charikar
Kabol (Kabul)
Mahmud-e Raqi
Asadabad
Mehtar Lam
Sarowbi
Jalalabad
Tora Bora
Meydan Shahr
Pol-e Alam
Gardiz
Ghazni
Khost
Sharan
Tarin Kowt
Qalat
Kandahar
Lashkar Gah
Zaranj
Farah
Peshawar
Islamabad
Rawalpindi
Mardan
Mingora
Abbottabad
Bannu
Zhob
Quetta
Zahedan
Multan
Lahore
Faisalabad
Gujranwa
Salang Pass
Shebar Pass
Nil Pass
Sabzak Pass
Shotor Khun Pass
Ghow Gardan Pass
Khyber Pass
Peywar-Kandaw Pass
Khojak Pass
Wakhjir Pass
Baroghil Pass
Panjshir Valley
Amu Darya
Harirud River
Farah River
Helmand River
Indus River
Jhelum River
Chenab River
Longitude East of Greenwich
AFGHANISTAN PROVINCES AND ETHNIC AREAS
(including adjoining Pakistan)
Pushtun
Tajik
Uzbek
Hazara
Turkmen
Chahar
Baluchi
Mixed
0 100 200 Kilometers
0 50 100 Miles
UZBEK.
INDIA
JOWZJAN
BALKH
KUNDUZ
TAKHAR
BADAKHSHAN
FARYAB
SAR-E POL
SAMANGAN
BAGHLAN
PANJSHIR
NURESTAN
KONAR
BADGHIS
BAMIAN
PARVAN
KAPISA
LAGHMAN
VARDAK
KABOL
LOWGAR
NANGARHAR
HERAT
GHOWR
DAYKUNDI
GHAZNI
PAKTIA
KHOWST
ORUZGAN
FARAH
ZABOL
PAKTIKA
NIMRUZ
HELMAND
KANDAHAR

central highlands; and the desert-dominated southern plateaus. Kabol (Kabul), the capital, lies on the southeastern slopes of the Hindu Kush, linked by narrow passes to the northern plains and by the famous Khyber Pass to Pakistan.

Across this rugged, variegated landscape moved countless peoples: Greeks, Turks, Arabs, Mongols, and others. Some settled here, their descendants today speaking Persian, Turkic, and other languages. Others left archeological remains or no trace at all. The present population of Afghanistan (just over 35 million) has no ethnic majority. This is a country of minorities in which the Pushtuns of the east are the most numerous but make up only 42 percent of the total. The second-largest minority are the Tajiks (27 percent), a world away across the massive Hindu Kush mountain range, concentrated in the zone near Afghanistan's border with Tajikistan. The Hazaras (9 percent) of the central highlands and the south, the Uzbeks (9 percent) and Turkmen (3 percent) in the northern border areas, the Baluchis (2 percent) of the southern deserts, and a handful of smaller groups scattered across this country create one of the world's most complicated cultural mosaics (see Figs. 7B-14 inset; 7B-13). Two major languages, Pushtun and Dari (the local variant of Persian), plus several others create a veritable Tower of Babel here.

The Emergence of the Taliban and al-Qaeda

Episodes of conflict have marked the history of Afghanistan, but none was as costly as its involvement in the Cold War (1945–1990). Following the Soviet invasion of 1979, the United States supported the Muslim opposition, the *mujahideen* (strugglers), with up-to-date weapons and money, and in 1989 the Soviets were forced to withdraw. Soon the factions that had been united during the anti-Soviet campaign were in conflict, delaying the return of some 4 million refugees who had fled to Pakistan and Iran. The situation resembled the pre-Soviet past: a feudal country with a weak and ineffectual government in Kabol.

In 1994, what at first seemed to be just another warring faction appeared on the scene: the **Taliban [10]** (seekers of religion), from religious schools (*madrassas*) in Pakistan. Their avowed aim was to end Afghanistan's chronic factionalism and endemic corruption by instituting strict Islamic law. Popular support in the war-weary country, especially among the Pushtuns, led to a series of successes, and by 1996 the Taliban had taken Kabol.

The Taliban's imposition of strict Islamic (*Sharia*) law was so uncompromising and severe that Islamic as well as non-Islamic countries objected. Restrictions on the activities of women ended their professional education, employment, and freedom of movement, and had a devastating impact on children as well. Public amputations and stonings enforced the Taliban's code.

In the process, Afghanistan became a haven for groups of revolutionaries whose agendas and goals went far beyond those of the Taliban: they plotted attacks on Western interests throughout the realm and threatened Arab regimes they deemed to be compliant with Western priorities. Taliban-ruled, cave-riddled, remote and isolated Afghanistan was an ideal refuge for these outlaws. Already in possession of arms and ammunition (Soviet as well as American) left over from the Cold War, they also took advantage of Afghanistan's huge, thriving, illicit opium trade. Afghanistan has long been the world's foremost producer of opium, and in 2012 accounted for fully 75 percent of the global heroin supply; not surprisingly, much of this drug revenue found its way into the coffers of the conspirators.

In 1996, a terrorist organization named **al-Qaeda [11]** took root in the country, an expanding global network that would further the aims of the once-loosely-allied revolutionaries. Afghanistan now became al-Qaeda's headquarters, and Usama bin Laden, a notorious Saudi renegade, its director. Al-Qaeda's most fateful operation came with the suicide attacks of September 11, 2001 that destroyed New York's Twin Towers, killed thousands, caused billions of dollars in physical destruction, and inflicted an untold amount of psychological damage. Several weeks later, United States and British forces, with the acquiescence of Pakistan, attacked both the Taliban regime and the al-Qaeda infrastructure in Afghanistan.

The U.S.-Led Invasion of Afghanistan and Its Aftermath

In the immediate aftermath of al-Qaeda's 9/11 attacks on New York and Washington, the American response provided

The Nowruz Festival, marking the first day of the Persian New Year, is celebrated in many parts of Southwest Asia, as here on March 21, 2013 at the Kart-e Sakhi Mosque in Kabol, Afghanistan. Despite the massive challenges this city faces, daily life, most of the time, finds a way to go on.

© Ahmad Massoud/Xinhua Press/Corbis

FIGURE 7B-14 © H. J. de Blij, P. O. Muller, and John Wiley & Sons, Inc.

reasons for optimism: the Taliban were swiftly defeated, a more representative government was forming, warlords were co-opted or sidelined, Pushtun and other refugees were streaming back into Afghanistan, and life in Kabol and other urban centers returned to a semblance of normal. In 2004, Afghanistan held elections that produced a representative government headed by President Hamid Karzai.

Proof of bin Laden's and al-Qaeda's complicity in the September 11 assault was discovered in videotape and documentary form after the invasion, and successive U.S. governments declared it a priority to bring him to justice. After nearly a decade, he was found not in Afghanistan but in the city of Abbottabad in the central flatlands of northern Pakistan. On May 2, 2011, a helicopter-supported unit of U.S. Navy SEALs stormed a house inside a fortified compound and fatally shot bin Laden. They took his body and, after retaining DNA samples, immediately buried him at sea in the nearby Indian Ocean. The role of Pakistan's military and intelligence services in the hiding of bin Laden remains unclear, but people in the United States overwhelmingly received the news of his death as justified retribution.

Overall, however, the U.S. occupation of Afghanistan seemed to suffer from a lack of purpose (the Americans were distracted by the war in Iraq between 2003 and 2011) as well as from adverse circumstances as it proved extremely difficult to foster a workable consensus among various factions and build democracy. The Taliban regrouped into smaller militias and resumed their attacks on foreign forces and local facilities (such as reopened girls' schools). By 2009, Afghanistan seemed to have regained its prominence among American priorities as the U.S. government approved a significant increase in combat troops and support personnel to augment its inadequate forces.

In 2011, the governments of the United States and Afghanistan negotiated a transition that would involve the withdrawal of most American and all Allied forces by the end of 2014 as well as the transfer of their responsibilities to the Afghan military. It was agreed that a small contingent of American troops would remain for training and security purposes. Taliban attacks continued during this phased retreat, and political as well as military leaders voiced doubts concerning the ability of Afghan forces to prevent the Taliban from destabilizing or fracturing the country. As the 2014 presidential election loomed, the fragile political, social, and economic order of Afghanistan hung in the balance. Ominously, the names of some of the warlords who had seized opportunities presented by the Taliban's earlier success resurfaced in a replay of recent history. Were this to happen, a return to chaos would have regional consequences extending far beyond Afghanistan's borders.

This chapter opened by emphasizing the diversity and variety among the cultures, countries, and regions of the North Africa/Southwest Asia realm. Obviously, its common geographic attribute is the predominance of Islam and that faith's varying tendencies toward modernization, politicization, and militancy. The Arab Spring that began in late 2010 and the rebellions it spawned in Tunisia, Libya, Egypt, Yemen, Syria, and elsewhere exposed yet another dimension: widespread, simmering discontent with repressive regimes that failed to bring economic progress. Powerful, initially peaceful demonstrations demanding democratic reforms affected many countries and brought different responses from autocratic rulers. Geography had much to do with these differences: the Sunni-Shi'ite schism, the distribution and relative location of oilfields, proximity to the African Transition Zone. The initially romanticized hopes of the Arab Spring were dashed on the anvil of harsh reality: effective democracy requires mature social institutions, which throughout this realm, to (again) varying degrees, had failed to develop. Even democratic Turkey, often cited as the model for Islamic states in this realm, confronted social disorder of unanticipated intensity that began as a quarrel over a building plan for a historic Istanbul city square and spread throughout much of the country as anti-government protests. The tensions between Islam and governance defy solution.

POINTS TO PONDER

- In Saudi Arabia, one of America's leading trade partners and military allies, women cannot drive and will be given only limited voting privileges in 2015.
- Natural gas deposits have been discovered off the coast of the Gaza Strip. Should the Palestinian National Authority have the right to exploit this resource? If not, why not?
- Turkey, for all its democratic achievements and economic progress, remains the world's leading jailer of journalists.
- The United Arab Emirates, home to ultramodern Dubai and burgeoning Abu Dhabi, and where women may drive as needed, in mid-2013 promulgated a series of laws to preclude Arab Spring-type demonstrations.
- Russia and Iran gave crucial support to the murderous Assad regime ruling Syria. What might be their strategic reasons?

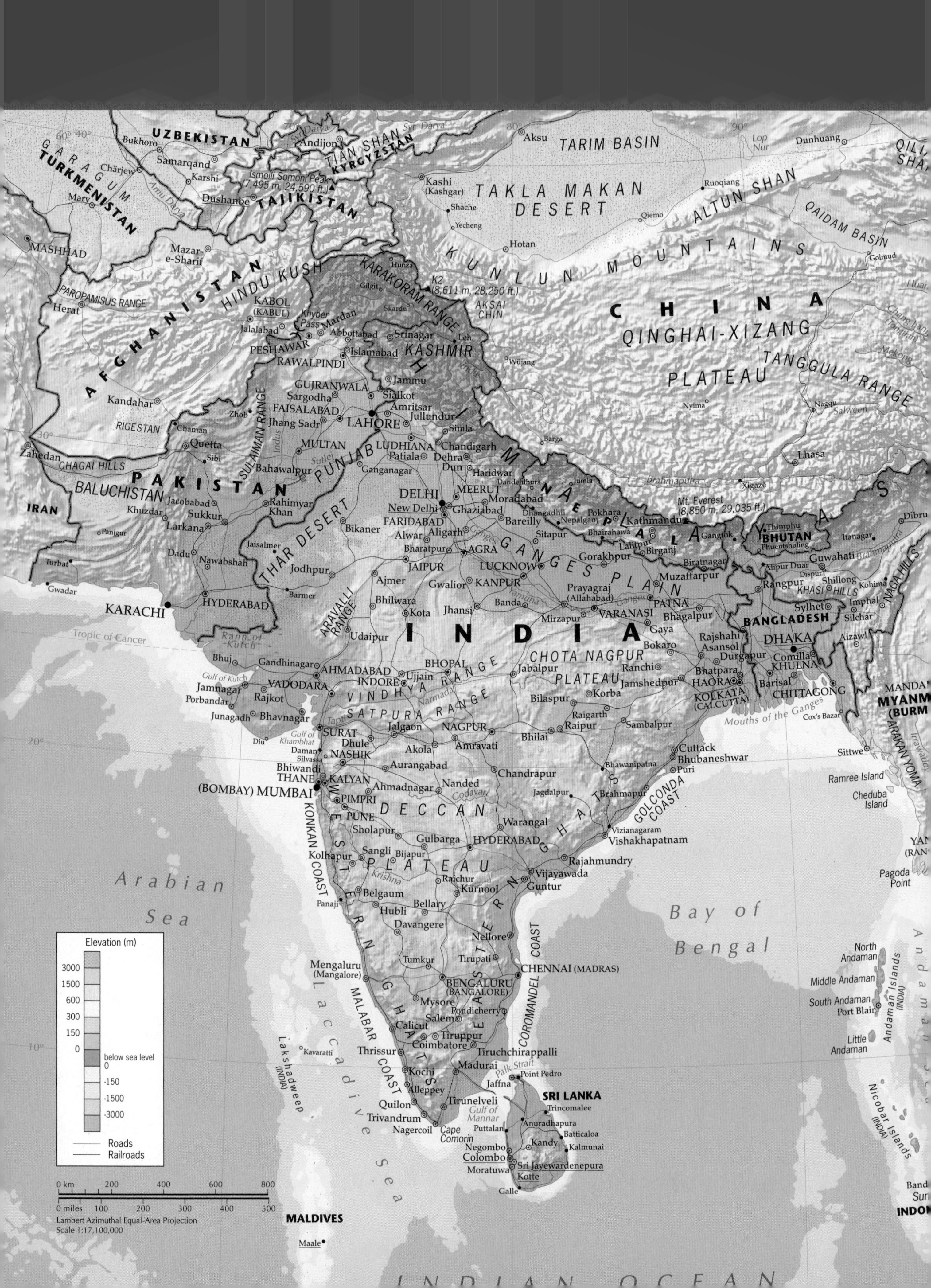

UZBEKISTAN
TURKMENISTAN
GARAGUM
TAJIKISTAN
KYRGYZSTAN
TIAN SHAN
Ismoili Somoni Peak (7,495 m, 24,590 ft.)
Bukhoro
Samarqand
Chärjew
Karshi
Mary
Dushanbe
Andijon
Syr Darya
Amu Darya
Aksu
TARIM BASIN
Lop Nur
Dunhuang
Kashi (Kashgar)
Shache
Yecheng
TAKLA MAKAN DESERT
Ruoqiang
Qiemo
ALTUN SHAN
QAIDAM BASIN
Golmud
Hotan
KUNLUN MOUNTAINS
MASHHAD
Mazar-e-Sharif
AFGHANISTAN
HINDU KUSH
PAROPAMISUS RANGE
Herat
KABOL (KABUL)
Jalalabad
Khyber Pass
Mardan
Abbottabad
PESHAWAR
RAWALPINDI
Islamabad
Hunza
Gilgit
Skardu
KARAKORAM RANGE
K2 (8,611 m, 28,250 ft.)
AKSAI CHIN
Srinagar
Leh
KASHMIR
Indus
Wujiang
CHINA
QINGHAI-XIZANG PLATEAU
TANGGULA RANGE
Mekong
Salween
Nagqu
Nyima
Kandahar
Zhob
SULAIMAN RANGE
GUJRANWALA
Sargodha
FAISALABAD
Jhang Sadr
LAHORE
Jammu
Sialkot
Amritsar
Jullundur
Simla
RIGESTAN
Chaman
Quetta
Sibi
MULTAN
LUDHIANA
Patiala
Chandigarh
Dehra Dun
Haridwar
Zahedan
CHAGAI HILLS
PAKISTAN
BALUCHISTAN
PUNJAB
Sutlej
Bahawalpur
Ganganagar
Barga
Lhasa
Brahmaputra
Xigazê
Dandeldhura
Jumla
HIMALAYAS
IRAN
Jacobabad
Khuzdar
Sukkur
Larkana
Rahimyar Khan
THAR DESERT
DELHI
New Delhi
MEERUT
Moradabad
Ghaziabad
FARIDABAD
Bareilly
Dhangadhi
Nepalganj
Pokhara
Kathmandu
NEPAL
Mt. Everest (8,850 m, 29,035 ft.)
Thimphu
BHUTAN
Phuentsholing
Dibru
Itanagar
Panjgur
Bikaner
Alwar
Aligarh
Ganges
Sitapur
Bhairahawa
Lalitpur
Birganj
Gangtok
Turbat
Dadu
Nawabshah
Jaisalmer
Bharatpur
AGRA
GANGES PLAIN
Gorakhpur
Biratnagar
Guwahati
Gwadar
Jodhpur
JAIPUR
LUCKNOW
Muzaffarpur
Alipur Duar
Dispur
Shillong
KHASI HILLS
Kohima
NAGA HILLS
Ajmer
Gwalior
KANPUR
Prayagraj (Allahabad)
PATNA
Rangpur
Imphal
KARACHI
HYDERABAD
Barmer
ARAVALLI RANGE
Bhilwara
Kota
Jhansi
Banda
Yamuna
Mirzapur
VARANASI
Bhagalpur
Sylhet
Silchar
BANGLADESH
Rann of Kutch
Tropic of Cancer
Udaipur
INDIA
Gaya
Rajshahi
Asansol
DHAKA
Aizawl
CHOTA NAGPUR PLATEAU
Bokaro
Durgapur
Comilla
KHULNA
Bhuj
Gandhinagar
AHMADABAD
BHOPAL
Jabalpur
Ranchi
Bhatpara
HAORA
Barisal
CHITTAGONG
MANDA
MYANM (BURM
Gulf of Kutch
Jamnagar
VADODARA
INDORE
Ujjain
VINDHYA RANGE
Narmada
Jamshedpur
Korba
Bilaspur
KOLKATA (CALCUTTA)
Porbandar
Rajkot
Junagadh
Bhavnagar
SATPURA RANGE
Tapti
Raigarth
Raipur
Sambalpur
Mouths of the Ganges
Cox's Bazar
ARAKAN YOMA
Irrawaddy
SURAT
Jalgaon
NAGPUR
Gulf of Khambhat
Diu
Dhule
Akola
Amravati
Bhilai
Daman
Silvassa
NASHIK
Cuttack
Bhubaneshwar
Sittwe
Bhiwandi
THANE
KALYAN
Aurangabad
Chandrapur
Bhawanipatna
Puri
Ramree Island
(BOMBAY) MUMBAI
Ahmadnagar
Nanded
Godavari
Jagdalpur
Brahmapur
GOLCONDA COAST
Cheduba Island
PIMPRI
PUNE
DECCAN
WESTERN GHATS
EASTERN GHATS
KONKAN COAST
Sholapur
Gulbarga
HYDERABAD
Warangal
Vizianagaram
Vishakhapatnam
YAN (RAN
Kolhapur
Sangli
Bijapur
PLATEAU
Rajahmundry
Vijayawada
Guntur
Krishna
Raichur
Kurnool
Pagoda Point
Arabian Sea
Panaji
Belgaum
Bellary
Hubli
Davangere
Nellore
Bay of Bengal
Elevation (m)
3000
1500
600
300
150
0
below sea level
0
-150
-1500
-3000
Roads
Railroads
Mengaluru (Mangalore)
Tumkur
Tirupati
CHENNAI (MADRAS)
COROMANDEL COAST
North Andaman
Middle Andaman
South Andaman
Port Blair
Andaman Islands (INDIA)
Andaman Sea
BENGALURU (BANGALORE)
Mysore
Salem
Pondicherry
MALABAR COAST
Calicut
Tiruppur
Coimbatore
Laccadive Sea
Lakshadweep (INDIA)
Kavaratti
Thrissur
Tiruchchirappalli
Madurai
Little Andaman
Palk Strait
Point Pedro
Jaffna
SRI LANKA
Kochi
Alleppey
Trincomalee
Quilon
Trivandrum
Tirunelveli
Gulf of Mannar
Anuradhapura
Nagercoil
Cape Comorin
Puttalan
Batticaloa
Kandy
Kalmunai
Negombo
Colombo
Sri Jayewardenepura Kotte
Moratuwa
Galle
Nicobar Islands (INDIA)
Band Sur
INDO
0 km 200 400 600 800
0 miles 100 200 300 400 500
Lambert Azimuthal Equal-Area Projection
Scale 1:17,100,000
MALDIVES
Maale
INDIAN OCEAN

IN THIS CHAPTER

- South Asia as a birthplace of religions
- Cutting-edge IT, backward agriculture
- Two nuclear powers quarrel over Kashmir
- The Indian Ocean: A crucial geopolitical arena
- South Asia's missing girls

CONCEPTS, IDEAS, AND TERMS

Monsoon	1
Social stratification	2
Indo-European languages	3
Dravidian languages	4
Partition	5
Refugees	6
Forward capital	7
Terms of trade	8
Neoliberalism	9
Population geography	10
Population density	11
Physiologic density	12
Demographic transition	13
Fertility rate	14
Demographic burden	15
Population pyramid	16
Sex ratio	17

FIGURE 8A-1

South Asia is a realm of almost magical geographic names: Mount Everest, Kashmir, the Khyber Pass, the Ganges River. There was a time when this realm was legendary and prized. Remember that it was "India" and its fabled wealth that the European explorers were after, from Vasco da Gama to Columbus to Magellan. Before them, the fourteenth-century North African geographer Ibn Battuta had traveled overland to South Asia, and his writings about its riches were met with astonishment and even disbelief. From the sixteenth century onward, European trading companies derived enormous profits from commerce in this realm.

However, by the late nineteenth century South Asia seemed to have become remote from the affairs of the world—hungry, weak, exploited, the prototype of the global periphery. Even after independence in 1947, India as well as the other countries of this realm long remained among the world's poorest. For decades, population growth outstripped economic expansion.

Today, for a number of reasons, South Asia commands the world's attention once again. It became the most populous geographic realm on Earth in 2011 (see the Data Table in Appendix B). Two of its states, India and Pakistan, often find themselves in conflict and both are nuclear powers. In the remote mountain hideaways of Pakistan, a terrorist organization's leaders planned attacks that changed the skyline of New York and presaged the battleground of Iraq. In the ports of India, a growing navy reflects the emergence of the Indian Ocean as a new global geopolitical arena in which China, too, is asserting itself. Meanwhile, outsourcing by U.S. companies to India has become a hot topic, and India's spectacular rise in information technology has changed that industry. Our daily lives will increasingly be affected by what happens in this crowded and restive part of the world.

DEFINING THE REALM

THE GEOGRAPHIC PANORAMA

The Eurasian landmass incorporates all or part of 6 of the world's 12 geographic realms, and of these half-dozen none is more clearly defined by nature than the one we call South Asia. Figure 8A-1 shows us why: the huge triangular subcontinent that divides the northern Indian Ocean between the Arabian Sea and the Bay of Bengal is so sharply demarcated by mountain walls and desert wastes that you could take a pen and mark its boundary, from the Naga Hills in the far east through the Great Himalaya and Karakoram in the north to the Hindu Kush and the Iran-bordering wastelands of Baluchistan in the west. Note how short the distances are over which the green of habitable lowlands turns to the dark brown of massive, snowcapped mountain ranges.

South Asia's kaleidoscope of cultures may be the most diverse in the world, proving that neither formidable mountains nor forbidding deserts could prevent foreign influences from further diversifying an already variegated realm. We will encounter many of these influences in this chapter and the next, but South Asia also possessed one unifying force of sorts: the British Empire, which in its late-nineteenth-century heyday came to hold

Entering the beautiful medieval walled city of Jaisalmer in India's Thar Desert, not far from the border with Pakistan.

© Jan Nijman

major geographic qualities of SOUTH ASIA

1. South Asia is clearly defined physiographically, and much of the realm's boundary is marked by mountains, deserts, and the Indian Ocean.
2. South Asia's great rivers, especially the Ganges, have for tens of thousands of years supported huge population clusters.
3. South Asia, and especially northern India, was the birthplace of major religions that include Hinduism and Buddhism.
4. Due to the realm's natural boundaries, foreign influences in premodern South Asia came mainly via a narrow passage in the northwest (the Khyber Pass).
5. South Asia covers just over 3 percent of the Earth's land area but contains nearly 24 percent of the world's human population.
6. South Asia's annual monsoon continues to dominate life for hundreds of millions of subsistence and commercial farmers. Failure of the monsoon cycle spells economic crisis.
7. Certain remote areas in the realm's northern mountain perimeter are a dangerous source of friction between India and both Pakistan and China.
8. South Asia is still predominantly rural with hundreds of thousands of small villages; but it also contains some of the biggest cities in the world.

sway over all of it in a ***raj*** (period of rule) that endured through the mid-twentieth. When, in the immediate aftermath of World War II, the British wanted to transfer their authority to a single regional government, local objections swiftly nullified this notion. That regional government would have been Hindu-dominated, but Muslims concentrated in the realm's eastern and western flanks refused, as did a pair of small kingdoms in the mountainous north as well as the Buddhist-dominated southern island then called Ceylon (now Sri Lanka). Negotiations and compromises produced partition and the political boundaries seen in Figure 8A-1. As a result, India, the realm's giant, is flanked by six countries (in clockwise order, Pakistan, Nepal, Bhutan, Bangladesh, Sri Lanka, and the Maldives) as well as a remaining disputed territory in the far north, Kashmir.

Since Islam is Pakistan's official religion (India has none) and that faith is a key criterion in defining the realm we designated as North Africa/Southwest Asia, should Pakistan be included within the latter? The answer lies in several aspects of Pakistan's historical geography. One criterion is ethnic continuity, which links Pakistan to India rather than to Afghanistan or Iran. Another factor involves language: although Urdu is Pakistan's official language, English is the *lingua franca*, as it is in India. Still another factor, of course, is Pakistan's evolution as part of the British Indian Empire. Furthermore, the boundary between Pakistan and India does not signify the eastern frontier of Islam in Asia. As we shall see, more than 200 million of India's 1.3 billion citizens are Muslims (which is just about as many as there are in all of Pakistan), and millions live very close to the Indian side of the border whose creation cost so many lives in 1947. And not only are Pakistan and India linked in the cultural-historical arena: they are locked in a deadly and dangerous embrace in embattled Kashmir.

The tight integration of Pakistan with South Asia will not surprise you after studying the realm's physiography in Figure 8A-1: the natural boundary in this part of the realm lies west of the Indus River, not east of it. Pakistan today remains part of a realm that changes not in the Punjab, but at the Khyber Pass, the highland gateway to Afghanistan.

SOUTH ASIA'S PHYSIOGRAPHY

From snowcapped peaks to tropical forests and from bone-dry deserts to lush farmlands, this part of the world presents a virtually endless array of environments and ecologies, a diversity that is matched by its cultural mosaic. The broad outlines of this realm's physiography are best understood against the backdrop of its fascinating geologic past.

A Tectonic Encounter

As Figure 8A-2 shows, the spectacular relief in the north of this realm resulted initially from the collision of two of the Earth's great tectonic plates (see Fig. G-4). About 10 million years ago, after a lengthy geologic journey following the breakup of the supercontinent Pangaea (see Chapter 6A), the Indian Plate encountered Eurasia. In this huge, slow-motion, accordion-like collision, parts of the crust were pushed upward, thereby creating the mighty Himalaya mountain range. That process is still going on—at the rate of 5 millimeters (0.2 in) per year—and this is one of the most earthquake-prone areas in the world. One major outcome of the tectonic collision was that the northern margins of the South Asian realm were thrust upward to elevations where permanent snow and ice make the landscape appear polar. The march of the seasons melts enough

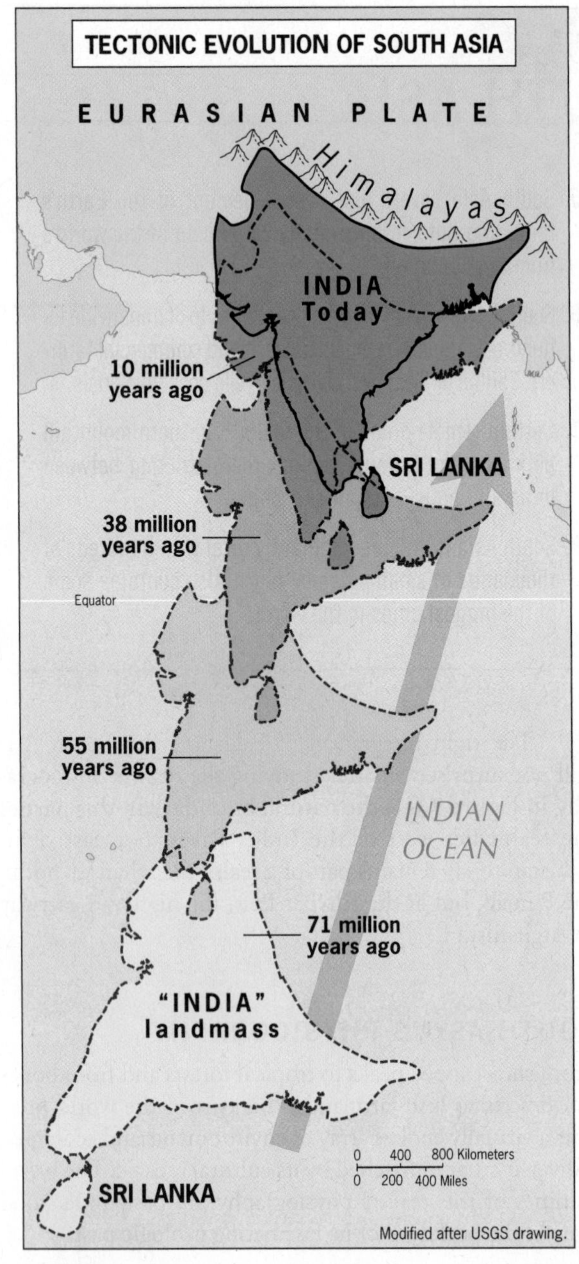

FIGURE 8A-2 © H. J. de Blij, P. O. Muller, J. Nijman, and John Wiley & Sons, Inc.

of this snow in spring and summer to sustain the great rivers below, providing water for farmlands that support hundreds of millions of people. The Ganges, Indus, and Brahmaputra all have their origins in the Himalaya. Only south of the Ganges Basin does the massive plateau begin that marks the much older geologic core of the Indian Plate as it drifted northeastward toward Eurasia.

The Monsoon

Physical geography, therefore, is crucial here in South Asia—but not just on and below the surface. What happens in the atmosphere is critical as well. The name "South Asia" is almost synonymous with the term **monsoon [1]** because the annual rains that come with its onset, usually in June, are indispensable to subsistence as well as commercial agriculture in the realm's key country, India.

Figure 8A-3 shows how the monsoon works. As the South Asian landmass heats up during the spring, a huge low-pressure system forms above it. This low-pressure system begins to draw in vast volumes of air from over the ocean onto the subcontinent. When the inflow of moist oceanic air reaches critical mass in early June, the ***wet monsoon*** has arrived. It may rain for 60 days or more. The countryside turns green, the paddies fill, and another dry season's dust and dirt are washed away. The region is reborn (see photo pair). The moisture-laden air flowing onshore from the Arabian Sea is forced upward against the Western Ghats, cooling as it rises and condensing large amounts of rainfall. The other branch of the wet monsoon originates in the Bay of Bengal and gets caught up in the convection (rising hot air) over northeastern India and Bangladesh. Seemingly endless rain now inundates a much larger area, including the entire North Indian Plain. The Himalaya mountain wall blocks the onshore airstream from spreading into the Asian interior and the rain from dissipating. Thus the moist airflow is steered westward, drying out as it advances toward Pakistan. After persisting for weeks, this pattern finally breaks down and the wet monsoon gives way to periodic rains and, eventually, another dry season. Then the anxious wait begins for the next year's summer monsoon, for without it India would face disaster. In much of rural India, life can hang by a meteorological thread.

Physiographic Regions

Figure 8A-3 underscores South Asia's overall division into three physiographic zones: northern mountains, southern plateaus, and, in between, a wide crescent of river lowlands.

The arrival of the annual rains of the wet monsoon transforms the Indian countryside. By the end of May, the paddies lie parched and brown, dust chokes the air, and it seems that nothing will revive the land. Then the rains begin, and blankets of dust turn into layers of mud. Soon the first patches of green appear on the soil, and by the time the monsoon ends all is green. The photo on the left, taken just before the onset of the wet monsoon in the west-coast State of Goa, shows the paddies before the rains begin; three months later this same countryside looks as on the right.

© Steve McCurry/Magnum Photos, Inc.

© Steve McCurry/Magnum Photos, Inc.

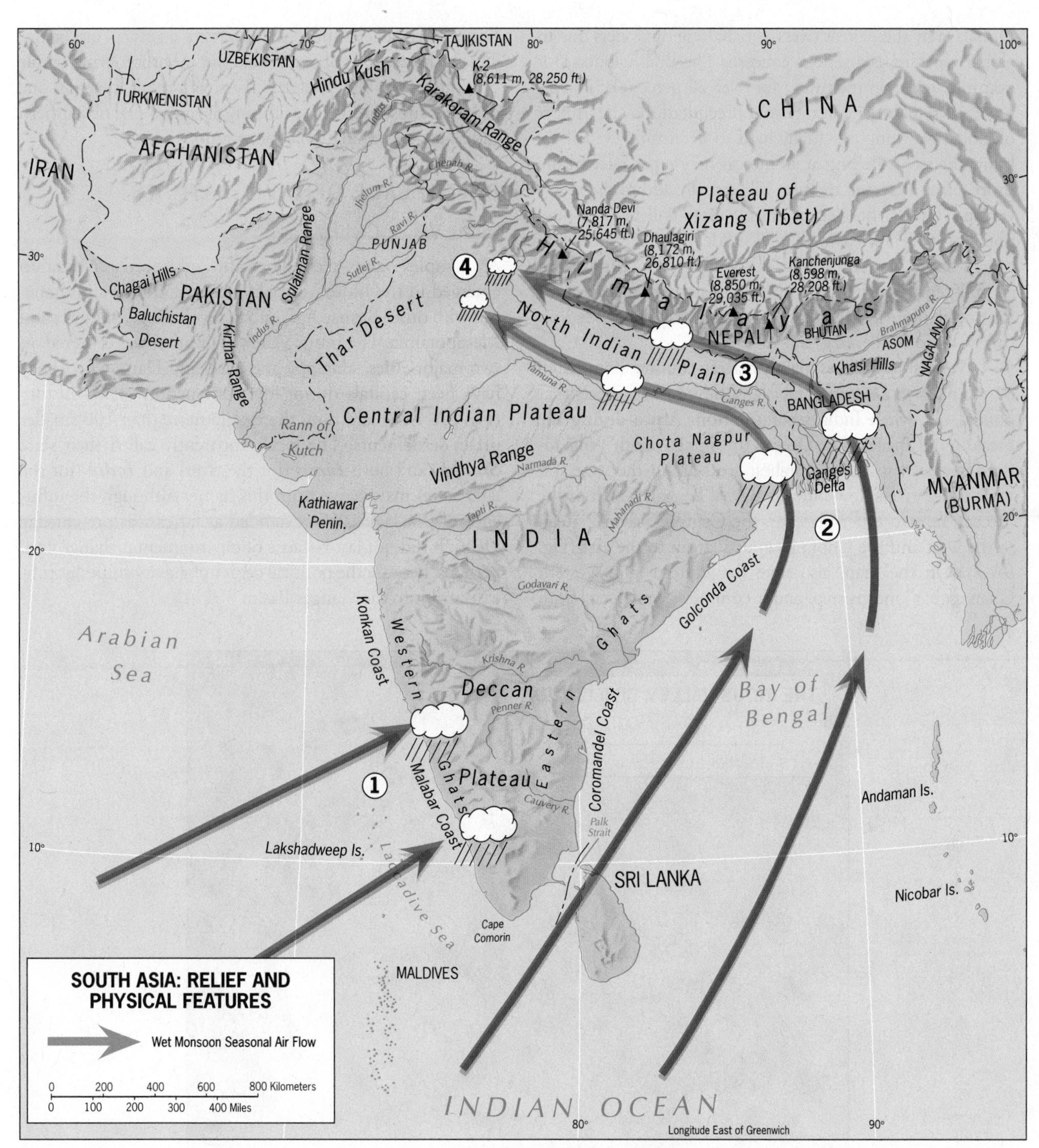

FIGURE 8A-3

The ***northern mountains*** extend from the Hindu Kush and Karakoram ranges in the northwest through the Himalaya in the center (Everest, the world's tallest peak, lies on the crestline that forms the Nepal-China border) to the ranges of Bhutan and the Indian State of Arunachal Pradesh in the east. Dry and barren in the west on the Afghanistan border, the ranges become green and tree-studded in Kashmir, forested in the lower-lying sections of Nepal, and even more densely vegetated in Arunachal Pradesh. Transitional foothills, with many deeply eroded valleys cut by rushing meltwater, lead to the river basins below.

The belt of ***river lowlands*** extends eastward from Pakistan's lower Indus Valley (the area known as Sindh) through India's wide Gangetic Plain and then on across the great double delta of the Ganges and Brahmaputra in Bangladesh (Fig. 8A-3). In the east, this physiographic region is often called the North Indian Plain. To the west lies the lowland of the Indus River, which rises in Tibet, crosses Kashmir, and then bends southward to receive its major tributaries from the Punjab ("Land of Five Rivers") to the east.

With so much of the realm, and so many of its people, depending on the water transported down from the High

Himalaya by these great rivers, the melting of glaciers due to global warming is a serious concern. The Brahmaputra, for example, relies on meltwater for over 20 percent of its volume, the remainder coming from precipitation. The impact of global warming is actually hard to determine even if we know the ice is melting at an increasingly rapid rate. In the short term, increased melting will actually add to the river's volume, but as the glaciers shrink there will be a tipping point at which meltwater will suddenly and rapidly decrease. Some models indicate this is likely to occur as soon as mid-century, with potentially devastating consequences for populations at lower elevations that have depended on this water for millennia.

Peninsular India is mainly plateau country, dominated by the massive ***Deccan***, a tableland built of lava sheets that poured out when India separated from Africa during the breakup of Pangaea. The Deccan (meaning "South") tilts toward the east, so that its highest areas are in the west and most of the rivers flow into the Bay of Bengal. North of the Deccan lie two other plateaus, the Central Indian Plateau to the west and the Chota Nagpur Plateau to the east (Fig. 8A-3). On the map, also note the Eastern and Western Ghats: "ghat" means step, and it connotes the descent from Deccan Plateau elevations to the narrow coastal plains below. The onshore winds of the annual wet monsoon bring ample precipitation to the Western Ghats; as a result, here lies one of India's most productive farming areas and one of southern India's largest population concentrations.

BIRTHPLACE OF CIVILIZATIONS

Indus Valley Civilization

A complex and technologically advanced civilization emerged in the Indus Valley by about 2500 BC, simultaneous with other Bronze Age "urban revolutions" in Egypt and Mesopotamia. The Indus Valley civilization was centered on two major cities, Harappa and Mohenjo-Daro, which may have been capitals during different periods of its history (Fig. 8A-4); in addition, there were more than 100 smaller urban settlements. The locals apparently called their state ***Sindhu***, and both ***Indus*** (for the river) and ***India*** (for the later state) may derive from this name. Although the influence of this civilization extended as far east as present-day Delhi, it did not last because of environmental change and, perhaps, because the political center of gravity shifted southeastward into the Ganges Basin.

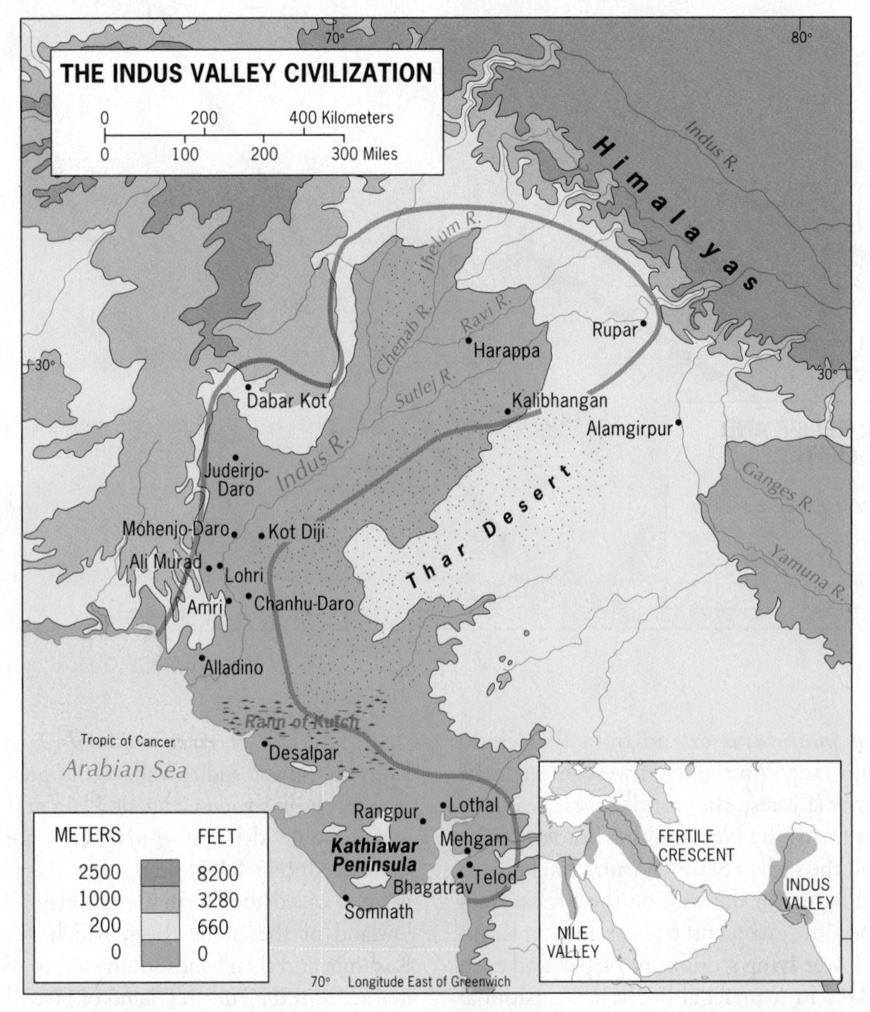

FIGURE 8A-4

Aryans and the Origins of Hinduism

Around 1500 BC, northern India was invaded by the ***Aryans*** (peoples speaking Indo-European languages based in what is today Iran). As the Iron Age dawned in India, acculturated Aryans began the process of integrating the Ganges Basin's isolated tribes and villages into a new organized system, and urbanization made a comeback. The Aryans brought their language (Sanskrit, related to Old Persian) and a new social order to the vast riverine flatlands of northern India. Their settlement here was also accompanied by the emergence of a religious belief system, ***Vedism***. Out of the texts of Vedism and local creeds there arose a new religion—***Hinduism***—and with it a new way of life.

It is thought that the arrival and accommodation of Aryans in this new society forged a system of **social stratification [2]** that would solidify the powerful position of the Aryans and be legitimized through religion. Starting about 3500 years ago, a combination of regional integration, the organization of villages into controlled networks, and the emergence of numerous small city-states produced a hierarchy of power among the people, a ranking from the very powerful (Brahmins—highest-order priests) to the weakest. This class-based ***caste system*** of Hinduism is highly controversial in the West (and among groups of Indians too) because of its rigidity and the ways in which it justifies structural inequality. Those in the lowest castes, deemed to be there because they deserved it given their past lives, are worst off, without hope of advancement and at the mercy of those ranking higher on the social ladder. In recent years, the caste system appears to be eroding from the combined effects of globalization, economic growth, and urbanization; this is true for India's bigger cities, but much less so for the majority of people in the country who still reside in rural areas.

Although Hinduism spread across South Asia and even reached the Southeast Asian realm (especially Cambodia and Indonesia), Indo-European languages never took hold in the southern portion of the subcontinent. As Figure 8A-5 shows, **Indo-European languages [3]** (several of which are rooted in Sanskrit) predominate in the western

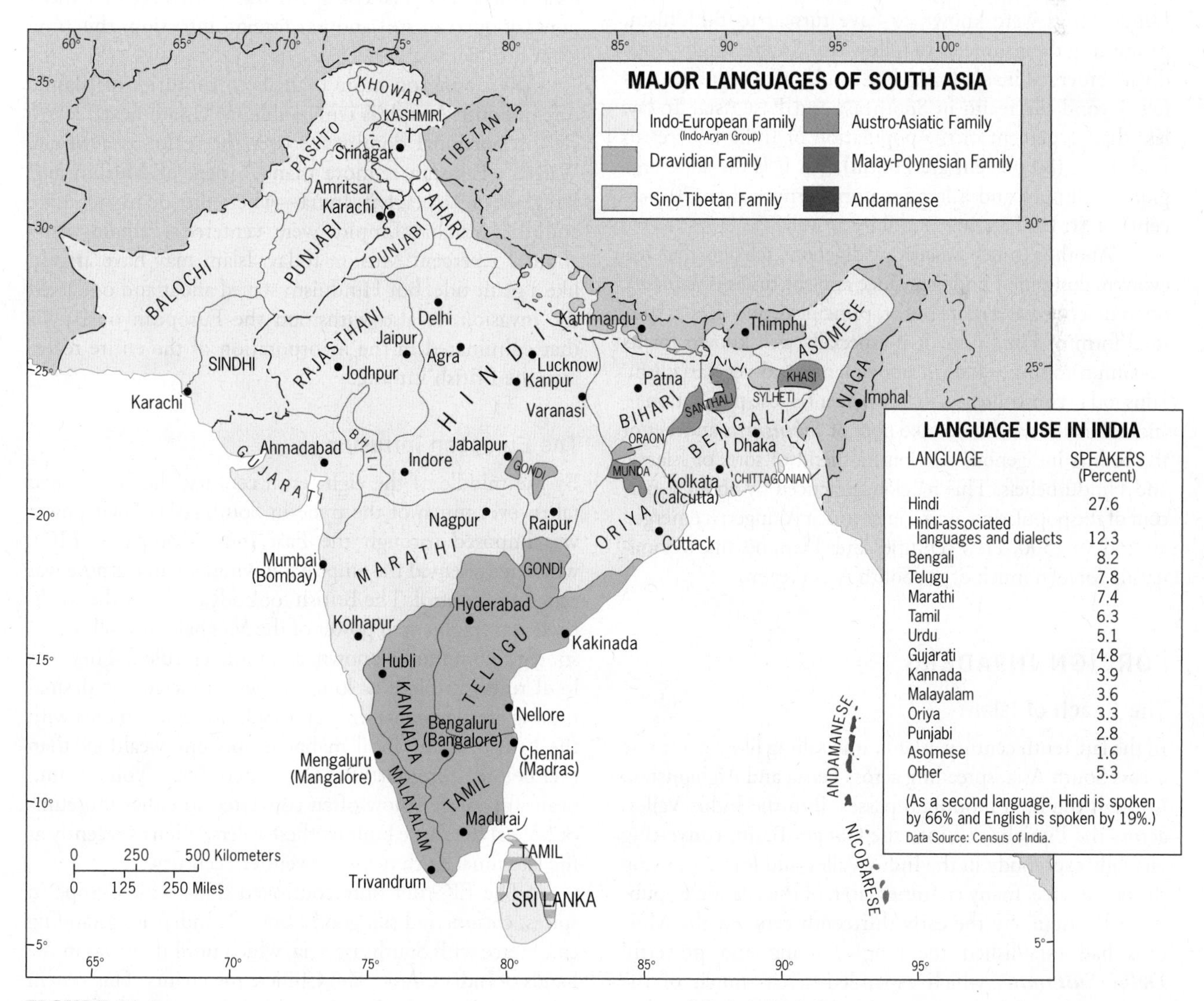

FIGURE 8A-5

and northern parts of the realm, whereas the southern languages belong to the **Dravidian [4]** family—languages that were indigenous to the realm even before the arrival of the Aryans. But these are not fossil languages: they remain vibrant today and have long literary histories. Telugu, Tamil, Kanarese (Kannada), and Malayalam are spoken by some 275 million people. In India's northern and northeastern fringes, Sino-Tibetan languages predominate, and smaller pockets of Austro-Asiatic speakers can be found in eastern India and neighboring Bangladesh.

Buddhism and Other Indigenous Religions

Hinduism is not the only religion that emerged in this realm. Around 500 BC, ***Buddhism*** arose in the eastern Ganges Basin in what is today the Indian State of Bihar. The famous story of the "enlightenment" of the Prince Siddhartha (the Buddha) took place in the town of Bodh Gaya, and his following soon expanded in all directions. The appeal of Buddhism was (and is) especially strong among lower-caste Hindus, and substantial numbers have converted through the ages. Even some prominent ancient Hindu kings were known to have turned to Buddhism, prompting their subjects to follow suit. Interestingly, Buddhism emerged inside India, but its ultimate influence was felt beyond the realm in Southeast and East Asia. Today, less than 1 percent of the population of India adheres to Buddhism (80 percent are Hindu), but it is the state religion in Bhutan and a large majority (more than 70 percent) of Sri Lankans are Buddhist as well.

Another (much smaller) indigenous religion that has evolved alongside Hinduism since ancient times is ***Jainism***, often described as a more purist, principled, and deeply spiritual form of Hinduism. It is especially well known for its uncompromising stand on nonviolence and vegetarianism. Jains today constitute less than 1 percent of the population in India. Finally, we should take note of ***Sikhism*** as another of the realm's indigenous religions, a blend of sorts of Islamic and Hindu beliefs. This religion, practiced by about 2 percent of the population, is of course much younger; it emerged around AD 1500, a few centuries after Islam became a dominating force in much of the South Asian realm.

FOREIGN INVADERS

The Reach of Islam

In the late tenth century, Islam came rolling like a giant tide across South Asia, spreading across Persia and Afghanistan, through the high mountain passes, into the Indus Valley, across the Punjab, and into the Ganges Basin, converting virtually everybody in the Indus Valley and foreshadowing the emergence, many centuries later, of the Islamic Republic of Pakistan. By the early thirteenth century, the Muslims had established the long-surviving and powerful ***Delhi Sultanate***, which expanded across much of the northern tier of the peninsula. The Muslims also came by sea, arriving at the Ganges-Brahmaputra Delta and spreading their faith from the east as well as the west, in the process laying the foundation of today's predominantly Islamic state of Bangladesh.

In the early 1500s, a descendant of Genghis Khan named Babur placed his forces in control of Kabol (Kabul) in Afghanistan, and from that base he penetrated the Punjab and challenged the Delhi Sultanate. In the 1520s, his Islamicized Mongol armies ousted the Delhi rulers and established the ***Mughal (Mogul) Empire***.

By most accounts, Mughal rule was at times remarkably enlightened, especially under the leadership of Babur's grandson Akbar, who expanded the empire by force but adopted tolerant policies toward Hindus under his sway; Akbar's grandson, Shah Jahan, made his enduring mark on India's cultural landscape through such magnificent architectural creations as the Taj Mahal in the city of Agra.

Nonetheless, by the early eighteenth century the Mughal Empire was in decline. Maratha, a Hindu state in the west, expanded not only into the peninsular south but also northward toward Delhi, capturing the allegiance of local rulers and weakening Islam's hold. Fractured India now lay open to still another foreign intrusion, this time from Europe.

Reflecting on more than seven centuries of Islamic rule in South Asia, it is remarkable that Islam never achieved proportional dominance over the realm as a whole. Whereas Pakistan is more than 96 percent Muslim and Bangladesh 91 percent, India—where the Delhi Sultanate and the Mughal Empire were centered—remains only about 15 percent Muslim today. Islam may have arrived like a giant tide, but Hinduism stayed afloat and outlasted the invasion. It also withstood the European onslaught that culminated in the incorporation of the entire realm into the British Empire.

The European Intrusion

By the middle of the eighteenth century, the British had taken over much of the trade in South Asia. Their power was imposed through the East India Company (EIC), which represented the empire but whose main purpose was economic control. The British took advantage of the weakened and fragmented power of the Mughals and followed a strategy commonly known as "indirect rule." They left local rulers in place as long as they extracted the desired trading arrangements. In fact, thanks to arrangements with the British, many local maharajas became wealthier than ever before: from northern to southern India, you can find beautiful palaces (now often converted to either museums or hotels) that were built by these rulers, often as recently as the late nineteenth or early twentieth century.

The EIC not only controlled trade with Europe in spices, cotton, and silk goods, but also India's longstanding commerce with Southeast Asia, which until then was in the hands of Indian, Arab, and Chinese merchants. This system worked well (for the British and their Indian trade partners) for almost a century, but by then political developments and heightened tensions were making it inevitable that the British government itself would take over from the EIC and assume responsibility. Thus "East India" became part of the British colonial empire in 1857—a ***raj*** that would endure for the next 90 years—and Queen Victoria officially became its empress 20 years later.

Colonial Transformation

British colonialism in South Asia coincided with the Industrial Revolution in Europe, and the impact of Britain on the realm must be understood in that context. South Asia became, in large part, a supplier of raw materials needed to keep the factories going in Manchester, Birmingham, and other industrial centers in Britain. For instance, when the supply of cotton from the American South came to a halt during the U.S. Civil War in the early 1860s, the British quickly encouraged (and indeed enforced) cotton production in what is today western India.

When the British took power in South Asia, this was a realm with already considerable industrial development (notably in metal goods and textiles) and an active trade with both Southeast and Southwest Asia. The colonialists saw this as competition, and soon India was exporting raw materials and importing manufactured products—from Europe, of course. Local industries declined, and Indian merchants lost their markets.

Colonialism did produce some assets for India. The country was bequeathed one of the most extensive transport networks of the colonial era, particularly the railroad system—even though the network focused on interior-to-seaport linkages rather than fully interconnecting the various parts of the country. British engineers laid out irrigation canals through which millions of hectares of land were brought into cultivation. Coastal settlements that had been founded by Britain developed into major cities and bustling ports, led by Bombay (now renamed Mumbai), Calcutta (now Kolkata), and Madras (now Chennai). These three cities still rank among India's largest urban centers, and their cityscapes bear the unmistakable imprint of colonialism (see photo in *From the Field Notes*).

British rule also produced a new elite among the South Asian natives. They had access to education and schools that combined English and Indian traditions, and their Westernization was reinforced through university education in Britain. This elite drew from Hindu and Muslim communities, and it was to play a major part in the rising demands for self-rule and independence. These demands started to gather momentum in the early twentieth century and could no longer be denied when World War II came to an end in the mid-1940s.

From the Field Notes . . .

© H.J. de Blij

"More than a half-century after the end of British rule, the centers of India's great cities continued to be dominated by the Victorian-Gothic buildings the colonizers constructed here. This also is evidence of a previous era of globalization, when European imprints transformed urban landscapes. Walking the streets of some parts of Mumbai (the British called it Bombay) you can turn a corner and be forgiven for mistaking the scene for London, double-decker buses and all. One of the British planners' major achievements was the construction of a nationwide railroad system, and railway stations were given great prominence in the urban architecture. I had walked up Naoroji Road, having learned to dodge the wild traffic around the circles in the Fort area, and watched the throngs passing through Victoria (now Chhatrapati Shivaji) Station. Inside, the facility is badly worn, but the trains continue to run, bulging with passengers hanging out of doors and windows."

www.conceptcaching.com

THE GEOPOLITICS OF MODERN SOUTH ASIA

Partition and Independence

Even before the British government decided to yield to demands for independence, it was clear that British India would not survive the coming of self-rule as a single political entity. As early as the 1930s, Muslim activists were promoting the idea of a separate state. As the colony moved toward independence, a major political crisis developed that eventually resulted in the separation of India and Pakistan. But **partition [5]** was no simple matter. True, Muslims were in the majority in the western and eastern sectors of British India, but smaller Islamic clusters were scattered throughout the realm. Furthermore, the new boundaries between Hindu and Muslim communities had to be drawn right through areas where both sides coexisted—thereby displacing millions (see photo).

The consequences of this migration for the social geography of India were especially far reaching. Comparing the country's 1931 and 1951 distributions of Muslims in Figure 8A-6, you can see the impact on the Indian Punjab and in what is today the State of Rajasthan. (Since Kashmir was mapped as three entities before the partition and as one afterward, the change there represents an administrative, not a major numerical, alteration.) Even in the east a Muslim exodus occurred, as reflected on the map by the State of West Bengal, adjacent to Bangladesh, where the Islamic component in the population declined substantially.

The world has seen many **refugee [6]** migrations but none involving so many people in so short a time as the one resulting from British India's partition (which occurred on Independence Day, August 15, 1947). Scholars who study the refugee phenomenon differentiate between "forced" and "voluntary" migrations, but as this case underscores, it is not always possible to separate the two. Many Muslims, believing they had no choice, feared for their future in the new India and joined the stampede. Others had the means and the ability to make a decision to stay or leave, but even these better-off migrants undoubtedly sensed a threat.

The great majority of Hindus who lived on the "wrong" side of the border moved as well. The Hindu component of present-day Pakistan may have been as high as 16 percent in 1947 but is barely more than 1.5 percent today; in Bangladesh, which was named East Pakistan at the time of partition, it declined from 30 percent to around 9 percent today. Partition therefore created an entirely new cultural and geopolitical landscape in South Asia.

Flight was one response to the 1947 partition of what had been British India, resulting in one of the greatest mass population transfers in human history. Here, two trainloads of eastbound Hindu refugees, fleeing (then) West Pakistan arrive at the station in Amritsar, the first city inside India."

India–Pakistan

From the moment of their separate creation, India and Pakistan have had a tenuous relationship. Upon independence, present-day Pakistan was united with present-day Bangladesh, and the two countries were respectively called West Pakistan and East Pakistan. As we have noted, the basis for this scheme was Islam: in both Pakistan and Bangladesh, Islam is the state religion. Between the two Islamic wings of Pakistan lay Hindu India. But there was little else to unify the Muslim easterners and westerners, and their union lasted less than 25 years. In 1971, a costly war of secession, in which India supported East Pakistan, led to the collapse of this unusual arrangement. East Pakistan, upon its "second independence" in 1971, took the name Bangladesh; and since there was no longer any need for a "West" Pakistan, that qualifier was dropped and the name Pakistan remained on the map.

India's encouragement of independence for Bangladesh emphasized the continuing tension between Pakistan and India, which had already led to war in 1965, to further conflict during the 1970s over Jammu and Kashmir, and to periodic flare-ups over other issues. During the Cold War, India tilted toward Moscow, while Pakistan found favor in Washington because of its strategic location adjacent to Afghanistan. Armed conflict between the two South Asian countries seemed to be a regional matter—until the early 1990s, when their arms race took on ominous nuclear proportions.

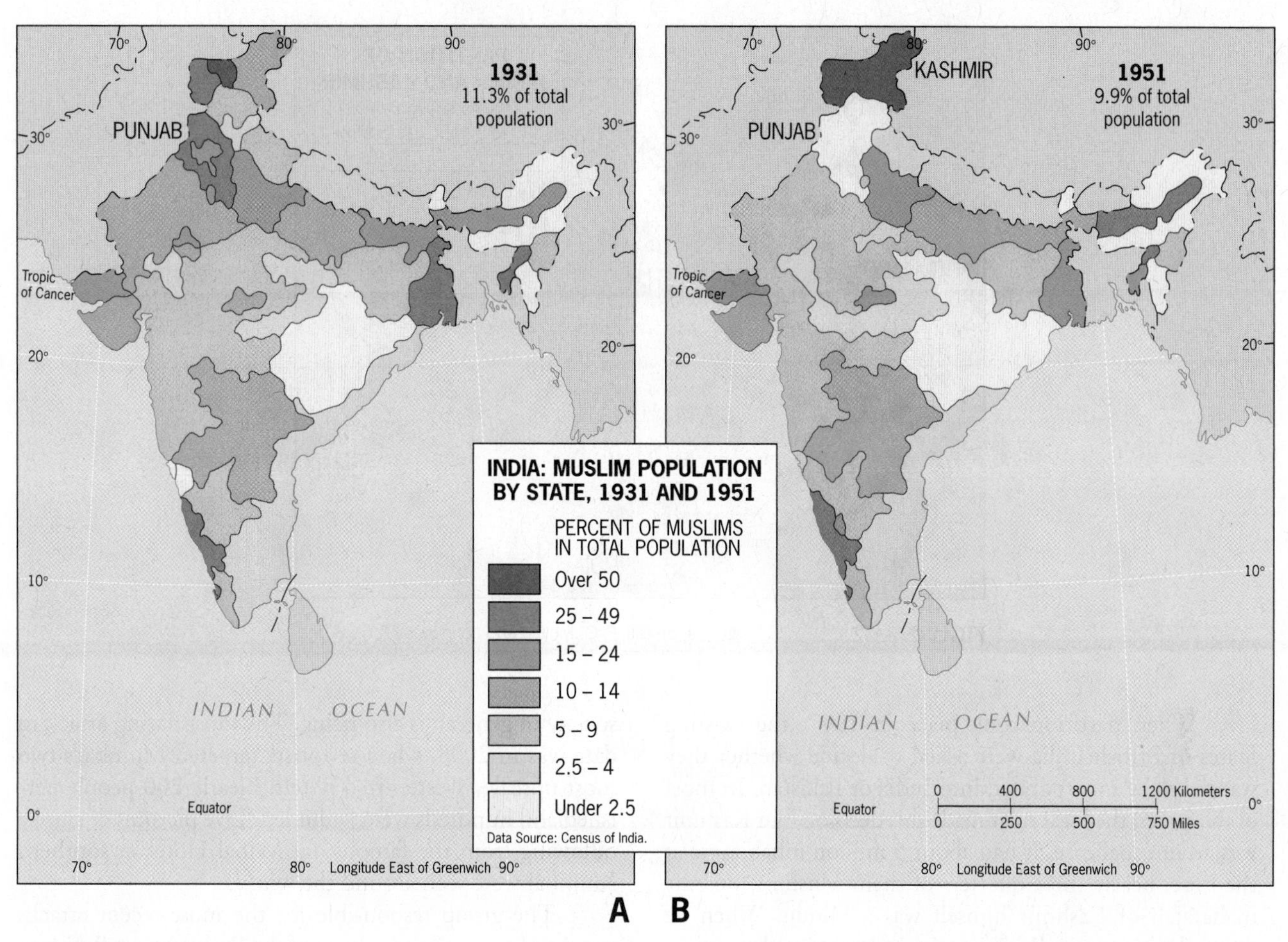

FIGURE 8A-6

Since then, the specter of nuclear war has hung over the conflicts that continue to embroil Pakistan and India, a concern not just for the South Asian realm but for the world as a whole. No longer merely a decolonized, divided, and disadvantaged country trying to survive, Pakistan has taken a crucial place in the political geography of a geographic realm in turbulent transition.

The relationship between India and Pakistan is especially sensitive because there are still so many Muslims in India. Massive as the 1947 refugee movement was, it left far more Muslims in India than those who had departed. The number of Muslims in India declined sharply, but it remained a huge minority, one that was growing rapidly to boot. By 2013, it surpassed 200 million, just over 15 percent of the total population—the largest cultural minority in the world and almost 10 percent larger than Pakistan's entire population (188 million).

What this means is that a sizeable portion of India's population tends to have more or less "natural" sympathies vis-à-vis Pakistan. Their presence works at times as a brake on hawkish Indian policies toward Islamabad. On the other hand, conflict with Pakistan can have detrimental effects on Hindu-Muslim relations inside India, and over the years has led to communal violence and deadly clashes. This issue is further complicated today by the Indian Muslims' alleged role in terrorist activities in India, orchestrated from Pakistan.

Contested Kashmir

When Pakistan became an independent state following the partition of British India in 1947, its capital was Karachi on the south coast, near the western end of the Indus Delta. As the map shows, however, the present capital is Islamabad. By moving the capital from the "safe" coast to the embattled interior, and by placing it on the doorstep of the contested territory of Kashmir, Pakistan announced its intent to stake a claim to its northern frontiers. And by naming the city Islamabad, Pakistan proclaimed its Muslim foundation here in the face of the Hindu challenge. This politico-geographical usage of a national capital can be assertive, and as such Islamabad exemplifies the principle of the **forward capital [7]**.

Kashmir is a territory of high mountains surrounded by Pakistan, India, China, and, along more than 50 kilometers (30 mi) in the far north, Afghanistan (Fig. 8A-7). Although known simply as Kashmir, the area actually consists of several political divisions, including the Indian State properly referred to as Jammu and Kashmir, a major bone of contention between India and Pakistan.

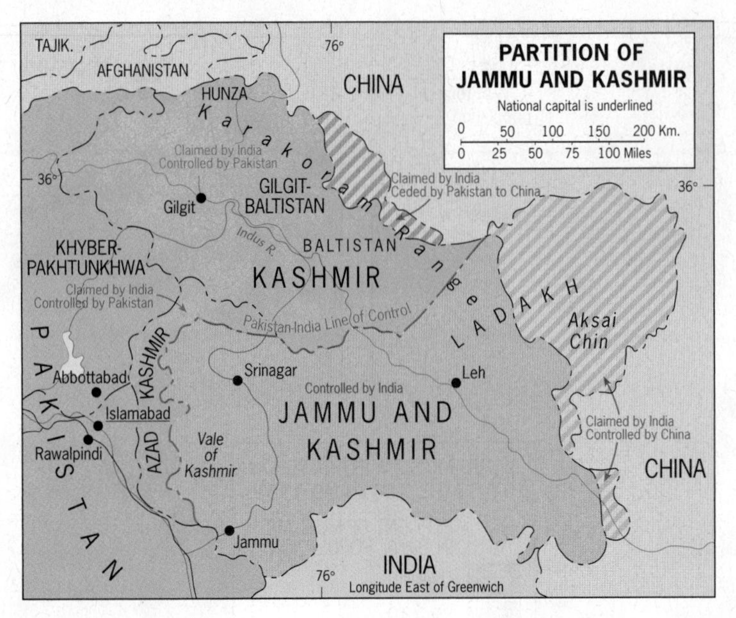

FIGURE 8A-7 © H. J. de Blij, P. O. Muller, and John Wiley & Sons, Inc.

When partition took place in 1947, the existing States of British India were asked to decide whether they wanted to be incorporated into India or Pakistan. In most of the States, the local ruler made this decision, but Kashmir was an unusual case. It had about 5 million inhabitants at the time, nearly three-quarters of them Muslims, but the maharajah of Kashmir himself was a Hindu. When he decided not to join Pakistan and instead aimed to retain autonomous status, this was answered with a Muslim uprising supported by Pakistan. The maharajah, in turn, called for help from India. After more than a year's fighting and through the intervention of the United Nations, a cease-fire line left most of Jammu and Kashmir (including nearly four-fifths of the territory's population) in Indian hands. Eventually, this line—now known as the Line of Control—began to appear on maps as the final boundary settlement, and Indian governments have proposed that it be so recognized.

With about two-thirds of the Kashmiris still Muslim, Pakistan has for decades demanded a referendum in Jammu and Kashmir in which the people can decide for themselves to remain with India or become part of Pakistan. India has refused, arguing that there is a place for Muslims in secular India but not for Hindus in the Islamic Republic of Pakistan. Given the specter of terrorism in India and the dangerous precedent of a concession in light of India's enormous ethnic and regional diversity, the Kashmir conflict has been left to smolder and is unlikely to be resolved for years to come.

The Specter of Terrorism

Comparatively successful as the integration of India's Muslim communities into the fabric of the Indian state has been, the risk of Islamic violence, directed against Indian society in general, is also rising. The most daring attack to date was in 2008, when terrorists targeted Mumbai's two most upscale, Westernized hotels. Nearly 200 people perished and hundreds were wounded. Live pictures of smoke billowing from the famous Taj Mahal Hotel in southern Mumbai were seen around the world.

The group responsible for the most recent attacks was *Lashkar-e-Taiba* (the Party of the Righteous), a Pakistan-based organization that among other things aims to return Kashmir to Islamic rule. These events have significant implications for India, where the overwhelming majority of Muslim citizens have remained uninvolved in extremist causes. It seems that a small number of Indian Muslims are joining local terrorist cells with links to Pakistan and perhaps other Islamic countries. Their terrorist acts lead to investigations and a reactionary climate that offend ordinary and peaceful Indian Muslims, radicalizing a number of them and expanding the market for Islamic militancy. It is still too early to gauge the potential impact of this development on a country that has long and justly prided itself on its multicultural democracy, but the portents for India's political, social, and economic geography are obviously serious.

In the meantime, Pakistan's northwestern frontier is effectively managed by the Taliban, the Afghan Islamic extremists who also have a history of collaboration with al-Qaeda. This border zone with Afghanistan is well beyond the control of the Pakistani government. U.S. efforts to defeat the Taliban in Afghanistan continued in 2013, but were thwarted by the Taliban's ability to move back and forth across this border. Thus the United States exerts increasing pressure on Islamabad to confront the Taliban on the Pakistani side of the border in this remote mountain refuge.

It is a delicate geopolitical chess game. Pakistan is careful not to alienate its Islamic base even if it despises the

Regional ISSUE Who Should Govern Kashmir?

KASHMIR SHOULD BE PART OF PAKISTAN!

"I don't know why we're even debating this. Kashmir should and would have been made part of Pakistan in 1947 if that colonial commission hadn't stopped mapping the Pakistan–India boundary before they got to the Chinese border. And the reason they stopped was clear to everybody then and there: instead of carrying on according to their own rules, separating Muslims from Hindus, they reverted to that old colonial habit of recognizing "traditional" States. And what was more traditional than some Hindu potentate and his minority clique ruling over a powerless majority of Muslims? It happened all over India, and when they saw it here in the mountains they couldn't bring themselves to do the right thing. So India gets Jammu and Kashmir and its several million Muslims, and Pakistan loses again. The whole boundary scheme was rigged in favor of the Hindus anyway, so what do you expect?

Here's the key question the Indians won't answer. Why not have a referendum to test the will of all the people in Kashmir? India claims to be such a democratic example to the world. Doesn't that mean that the will of the majority prevails? But India has never allowed the will of the majority even to be expressed in Kashmir. We all know why. About two-thirds of the voters would favor union with Pakistan. Muslims want to live under an Islamic government. So people like me, a Muslim carpenter here in Srinagar, can vote for a Muslim collaborator in the Kashmir government, but we can't vote against the whole idea of Indian occupation.

Life isn't easy here in Srinagar. It used to be a peaceful place with boats full of tourists floating on beautiful lakes. But now it's a violent place with shootings and bombings. Of course we Muslims get the blame, but what do you expect when the wishes of a religious majority are ignored? So don't be surprised at the support our cause gets from Pakistan across the border. The Indians call them terrorists and they accuse them of causing the 60,000 deaths this dispute has already cost, but here's a question: why does it take an Indian army of 600,000 to keep control of a territory in which they claim the people prefer Indian rule?

Now this so-called War on Terror has made things even worse for us. Pakistan has been forced into the American camp, and of course you can't be against 'terrorism' in Afghanistan while supporting it in Kashmir. So our compatriots on the Pakistani side of the Line of Control have to stay quiet and bide their time. But don't underestimate the power of Islam. The people of Pakistan will free themselves of collaborators and infidels, and then they will be back to defend our cause in Kashmir."

KASHMIR BELONGS TO INDIA!

"Let's get something straight. This stuff about that British boundary commission giving up and yielding to a maharajah is nonsense. Kashmir (all of it, the Pakistani as well as the Indian side) had been governed by a maharajah for a century prior to partition. What the maharajah in 1947 wanted was to be ruled by neither India nor Pakistan. He wanted independence, and he might have gotten it if Pakistanis hadn't invaded and forced him to join India in return for military help. As a matter of fact, our Prime Minister Nehru prevailed on the United Nations to call on Pakistan to withdraw its forces, which of course it never did. As to a referendum, let me remind you that a Kashmir-wide referendum was (and still is) contingent on Pakistan's withdrawal from the area of Kashmir it grabbed. And as for Muslim 'collaborators', in the 1950s the preeminent leader on the Indian side of Kashmir was Sheikh Muhammad Abdullah (get it?), a Muslim who disliked Pakistan's Muslim extremism even more than he disliked the maharajah's rule. What he wanted, and many on the Indian side still do, is autonomy for Kashmir, not incorporation.

In any case, Muslim states do not do well by their minorities, and we in India generally do. As far as I am concerned, Pakistan is disqualified from ruling Kashmir by the failure of its democracy and the extremism of its Islamic ideology. Let me remind you that Indian Kashmir is not just a population of Hindus and Muslims. There are other minorities—for example, the Ladakh Buddhists—who are very satisfied with India's administration but who are terrified at the prospect of incorporation into Islamic Pakistan. You already know what Sunnis do to Shi'ites in Pakistan. You are aware of what happened to ancient Buddhist monuments in Taliban Afghanistan (and let's not forget where the Taliban came from). Can you imagine the takeover of multicultural Kashmir by Islamabad?

To the Muslim citizens of Indian Kashmir, I, as a civil servant in the Srinagar government, say this: look around you, look at the country of which you are a subject. Muslims in India are more free, have more opportunities to participate in all spheres of life, are better educated, have more political power and influence than Muslims do in Islamic states. Traditional law in India accommodates Muslim needs. Women in Muslim-Indian society are far better off than they are in many Islamic states. Is it worth three wars, 60,000 lives, and a possible nuclear conflict to reject participation in one of the world's greatest democratic experiments?

Kashmir belongs to India. All inhabitants of Kashmir benefit from Indian governance. What is good for all of India is good for Kashmir."

Vote your opinion at www.wiley.com/go/deblijpolling

AP/Wide World Photos

Smoke billows from the landmark Taj Mahal Hotel in Mumbai on November 29, 2008. The Taj was one of several sites in southern Mumbai that were simultaneously targeted for attack by Islamic militants. The siege lasted four days and almost 200 people died in the violence. Indian commandos killed all of the terorists but one, who was captured, tried, sentenced to death, and executed in 2012. It was not the first time that such terrorism had struck India and, unfortunately, not the last: in early 2013, bombings killed a dozen people in the southern city of Hyderabad.

northern extremists, and it is fearful of an economically stronger India and its growing ties to the United States. India is deeply concerned about Pakistan's role in terrorism on Indian soil and, even worse, the possibility of fundamentalists taking control of Pakistan's nuclear arsenal; at the same time, it must also guard against increased tensions between Hindus and Muslims inside India. India is also impatient with American reluctance to choose its side in the Kashmir conflict. The United States, in turn, is sympathetic to the world's largest democracy but needs Pakistan to be an ally in the global counterterrorism campaign. Each of these parties is walking a tightrope, where the slightest mistake could have deadly consequences.

Chinese Border Claims

An overview of this realm's geopolitical framework would not be complete without noting the powerful and sometimes invasive presence of China. As Figure 8A-7 shows, China claims the northeastern extension of Jammu and Kashmir State. This issue has been quiet in recent years, but officially neither China nor India shows any sign of conceding

China's control over Tibet (called Xizang by the Chinese) and its efforts to influence the lives of Tibetans both within and outside Tibet create additional issues. To the dismay of Beijig, Tibet's exiled Dalai Lama calls India his second home. In recent years China has been pressuring the government of Nepal, wedged between India and Tibet, to discourage Tibetan immigration and to constrain the activities of Tibetans already in the country. Farther east, China claims the bulk of India's State of Arunachal Pradesh ("land of the dawn-lit mountains"), based on the assertion that the boundary, established in 1914, was never ratified by Beijing—even though it was approved by the then-independent Tibetans themselves. China has also claimed rights to a small area in northern India located between Sikkim and Bhutan on the basis that the people there are Tibetans and therefore belong under Chinese jurisdiction.

China's power is felt in other ways. In recent years the Chinese have announced plans to construct dams on the Tibetan headwaters of the upper Brahmaputra River, potentially jeopardizing water supply in both northeastern India and Bangladesh. Kashmir is by no means the only place along South Asia's northern frontier where trouble can erupt at any time.

Indian Ocean Geopolitics

China needs access to markets for its products and supplies of raw materials to sustain its rapidly growing industrial production, and a major part of this access runs through the Indian Ocean. Along the way, it is extending Chinese political and military power through the expansion of its navy in the Indian Ocean and building bases in Pakistan, Myanmar, and Bangladesh. India, increasingly concerned about Chinese intentions, has responded by building its own new alliances with such Southeast Asian states as Indonesia and Vietnam.

From a broader pan-Asian perspective (South and East), geopolitical developments are increasingly a matter of U.S.-China-India relations in which China is asserting itself in the Indian Ocean Basin as well as along the northern Indian border; in which India's economic rise has given it a new assertiveness in the political arena; and in which the United States is ideologically inclined to sympathize with India while the imperatives of political reality motivate it to steer toward maintaining the traditional political balance of power between the two Asian giants. The future could be a U.S.-Chinese-Indian condominium or, more

AMONG THE REALM'S GREAT CITIES . . .

DELHI NEW AND OLD

FLY DIRECTLY OVER the Delhi–New Delhi conurbation into its new international airport (opened in 2010), and you may not see the place at all. A combination of smog and dust creates an atmospheric soup that can limit visibility to a few hundred meters for weeks on end. Relief comes when the rains arrive, but Delhi's climate is mostly dry. The tail-end of the wet monsoon reaches here during late June or July, but altogether the city only gets about 60 centimeters (25 in) of rain a year. When the British colonial government decided to leave Calcutta more than a century ago and build a new capital city adjacent to Delhi, conditions were different. South of the old city lay a hill about 15 meters (50 ft) above the surrounding countryside, on the right bank of the southward-flowing Yamuna River, a tributary of the Ganges. Compared to Calcutta's hot, swampy environment, Delhi's was agreeable. In 1912 it was not yet a megacity. Skies were mostly clear. Raisina Hill became the site of a New Delhi.

This was not the first time rulers chose Delhi as the seat of empire. Ruins of numerous palaces mark the passing of powerful kingdoms. But none brought to the Delhi area the transformation the British did. In 1947, the Indian government decided to keep its headquarters here. In 1970, the metropolitan-area population exceeded 4 million. By 2014, it was a staggering 24.8 million, India's largest urban region.

Delhi is popular as a seat of government for the same reason as its ongoing expansion: the city has a fortuitous relative location. The regional topography creates a narrow corridor through which all land routes from northwestern India to the North Indian Plain must pass, and Delhi lies in this gateway. Thus the twin cities not only contain the government functions; they also anchor the core area of this massively populated country.

Old Delhi was once a small, traditional, homogeneous town. Today Old and New Delhi form a multicultural, multifunctional urban giant. From above, the Delhi conurbation looks like an inkblot, a nearly concentric region that has steadily expanded in all directions. The fastest growth has been to the south, where formerly separate towns and satellite cities such as Faridabad are now part of this sprawling conurbation. Gurgaon is especially well-known, a leading activity hub south of the airport that has witnessed explosive growth over the past 15 years or so, an agglomeration of IT companies, international call centers, and new middle-class residential developments. Another sign of Delhi's modernization is the construction of a new, heavy-rail transit system, aimed at providing some relief to the metropolitan area's massive congestion and pollution problems.

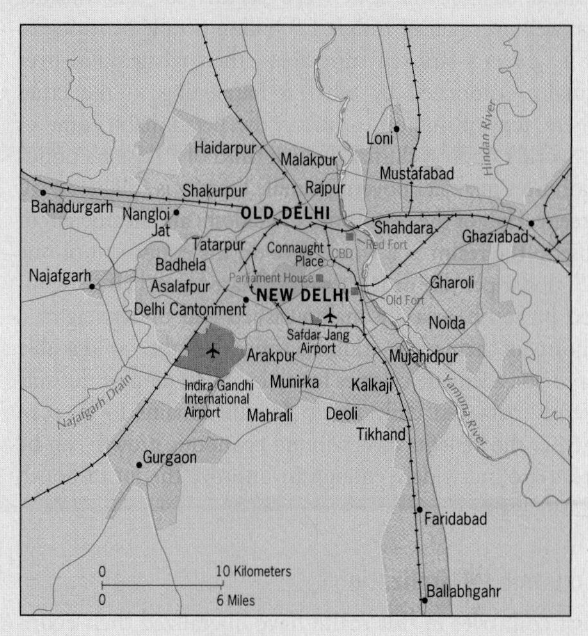

likely, a continued informal alliance of the United States and India seeking to put the brakes on China's assertiveness. U.S. support for Indian membership in the United Nations Security Council is but one manifestation of this evolving U.S.-India relationship.

There is some indication of a rapprochement between the two Asian giants. In a notable turnaround, a Chinese spokesperson in early 2013 indicated that China would not oppose Indian membership in the Security Council. Perhaps most importantly, growing economic interdependence will help to keep things in check. China wants to penetrate India's vast and growing consumer markets. China-India trade is growing rapidly, surpassing U.S. $75 billion in 2012 ($58 billion from China to India and less than one-third of that in the other direction). China runs a significant surplus, and this reflects, up to now, the fact that the **terms of trade [8]** are to China's advantage: India supplies mainly raw materials while China sells finished goods with a higher added value. In 2010, the two pledged to increase their trade to $100 billion by 2015, with India (so far unsuccessfully) seeking to reduce its trade deficit.

EMERGING MARKETS AND FRAGMENTED MODERNIZATION

In recent years, optimistic reports in the media have been proclaiming a new era for South Asia, marked by rising growth rates for the realm's national economies, rewards from globalization and modernization, and increasing integration into the global economy. India, obviously the key to the realm, has even been described as "India Shining" during this wave of enthusiasm.

And indeed, a combination of circumstances, ranging from America's involvement with Pakistan in the campaign against terrorism to the real estate and stock market booms in India, suggest that a new era has arrived. But consider this: well over half of India's 1.3 billion people continue to live in poverty-stricken rural areas, their villages and lives virtually untouched by what is happening in the cities (where, tens of millions of urban dwellers inhabit some of the world's poorest slums). Fully a third of Pakistan's population lives in abject poverty; female literacy is still below 50 percent. Half of the people of Bangladesh, and nearly half of those in the realm as a whole, live on the equivalent of one U.S. dollar per day or less. It is estimated that half the children in South Asia are malnourished and underweight, a majority of them girls—this at a time when the world is able to provide adequate calories for all its inhabitants, if not sufficiently balanced daily meals. It still remains to be seen whether the benefits of newfound economic growth can be spread around widely enough to improve the lot of South Asia's masses.

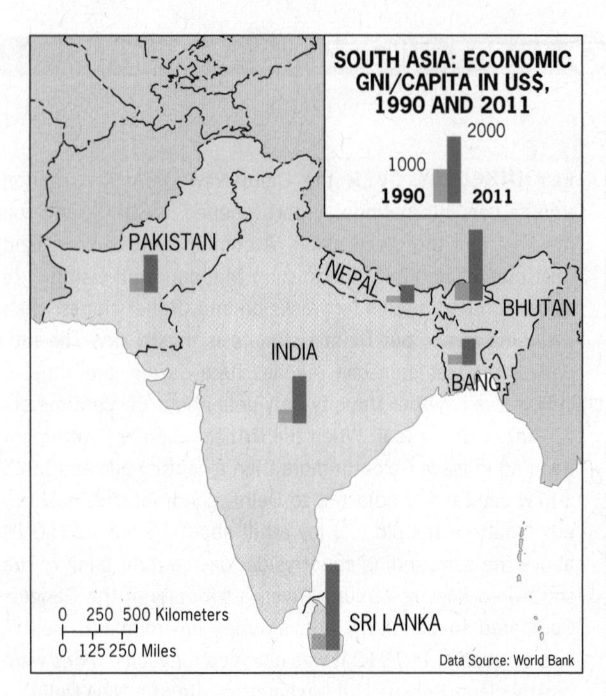

FIGURE 8A-8 © H. J. de Blij, P. O. Muller, J. Nijman, and John Wiley & Sons, Inc.

Economic Liberalization

Most countries in this realm have liberalized their economies since the late 1980s as part of a worldwide turn toward **neoliberalism [9]**. This involves privatization of state-run companies, lowering of international trade tariffs, reduction of government subsidies, cutting of corporate taxes, and overall deregulation to spur business activity. It was an important change from previous times in which markets were tightly controlled by unyielding central governments that since independence had espoused ideologies opposed to unrestrained capitalism. A change of direction was unavoidable. The ineffective policies of the past, continuing grinding poverty, and near fiscal bankruptcy demanded support from the International Monetary Fund; the IMF, in turn, insisted on structural reforms.

The results of these reforms have been especially noticeable in India, Bangladesh, and Pakistan, where economic growth rates climbed to levels never seen before (Fig. 8A-8). Most of this growth is in manufacturing, services, finance, and, in India, information technology (IT). A more open economy has attracted increased foreign investment, and during the past two decades a new (urban) middle class has emerged. This steadily expanding new class may account for only 25 percent or so of the population, but in South Asia that translates into more than 400 million people—a huge new consumer market for an array of products ranging from cars to smartphones. Nevertheless, that still leaves well over a billion South Asians who have *not* attained middle-class status, for whom almost nothing has changed, who remain overwhelmingly dependent on agriculture, and who are not likely to log on to the new information economy anytime soon.

One striking feature of the South Asian realm, then, is that even if the majority of the people live traditional lives in rural villages, there are also a fair number of megacities in which social and economic change is the order of the day. The urban regions of Mumbai, Delhi–New Delhi, Kolkata, Dhaka, and Karachi all contain populations larger than 15 million that have grown rapidly over the past two decades. Their density is often overwhelming, environmental conditions are poor, and the contrast between rich and poor is usually staggering. But almost always it is better to be poor in the city than in the countryside, simply because cities offer opportunities that do not exist in the villages—which helps to explain the burgeoning flow of rural-to-urban migrants that drives the growth of towns and cities all across South Asia.

The Significance of Agriculture

More than half of the entire workforce of South Asia is employed in agriculture, ranging from about 43 percent in Pakistan to about 75 percent in Nepal. But overall productivity is low, and the contribution of agriculture to the national economy is only around 20 percent. Incomes in rural areas are much lower than in major cities, and the same is true for the standard of living. Almost 70 percent of South Asia's population is rural, and even those who do not work in agriculture tend to rely on it indirectly.

Millions of lives every year depend on a good harvest. As Figure 8A-3 shows, the wet monsoon brings life-giving rains to the southwestern (Malabar) coast, and a second branch from the Bay of Bengal sweeps across north-central India toward Pakistan, losing strength (and moisture) as it proceeds. This means that amply watered eastern India and Bangladesh as well as the southwestern

From the Field Notes . . .

© Jan Nijman

"During my travels in northeastern India in early 2011, I visited some agricultural areas in the Brahmaputra Valley that benefit from plentiful irrigation. Tea and rice are two major crops. Because big companies run the large tea plantations, many of the rice farmers have to make do with small plots of land. Here a farmer in Asom State shows off part of his rice harvest. I encountered him and several others at a small, machine-operated mill shop. Note that the rice is brown—milling it removes the bran layer and makes the rice white. 'We have the best rice in all of India,' he proudly said, as he slowly poured it from his hand back into the bag. Having enjoyed some excellent meals in the area, I could not disagree."

www.conceptcaching.com

coastal strip grow rice as their staple crop; but drier northwestern India and Pakistan raise wheat.

Farmers' fortunes tend to vary geographically, as can be illustrated in the situation on either side of India's Western Ghats upland. The monsoon rains are generally plentiful along the coast-paralleling, western slopes of this linear highland, from the southern tip of the subcontinent as far north as the vicinity of Mumbai. Here you can see the hillside vegetation assume all shades of fresh green come the month of June, a sure sign that the harvest will be bountiful. But on the eastern (rain shadow) side and deeper into the interior of the Deccan Plateau, it is a different story. The rains come less often and do not last as long. Farming becomes a gamble with nature, and life becomes precarious. Many of the farmers here are members of lower castes, landless and indebted, and have the hardest time making ends meet. Almost every year, Maharashtra's inland districts report several thousand (!) farmer suicides as desperately poor peasants end their lives because they can no longer provide for their families.

It is clear that the majority of people in this realm depend on agriculture and that governments must aim their economic policies at improving agricultural productivity to raise the standard of living in rural areas. But they have a long way to go. The demands on governments are many, and they often seem distracted by economic sectors that can make a faster and greater contribution to the tax base, such as manufacturing, financial services, and IT—economic activities that almost always are centered on the big cities far from the impoverished countryside.

SOUTH ASIA'S POPULATION GEOGRAPHY

Given its enormous human content, the South Asian realm's areal size is relatively quite small. It totals less than 40 percent of the size of similarly populous East Asia. Comparing the world's two giants shows that China's territory is almost three times as large as India's. The total population of Subsaharan Africa is less than half of South Asia's, in an area almost five times as large. Adjectives such as "teeming," "overcrowded," and "crammed" are often used to describe the realm's habitable living space, and with good reason. South Asia's intricate cultural mosaic is tightly packed, with only the deserts in the west and the mountain

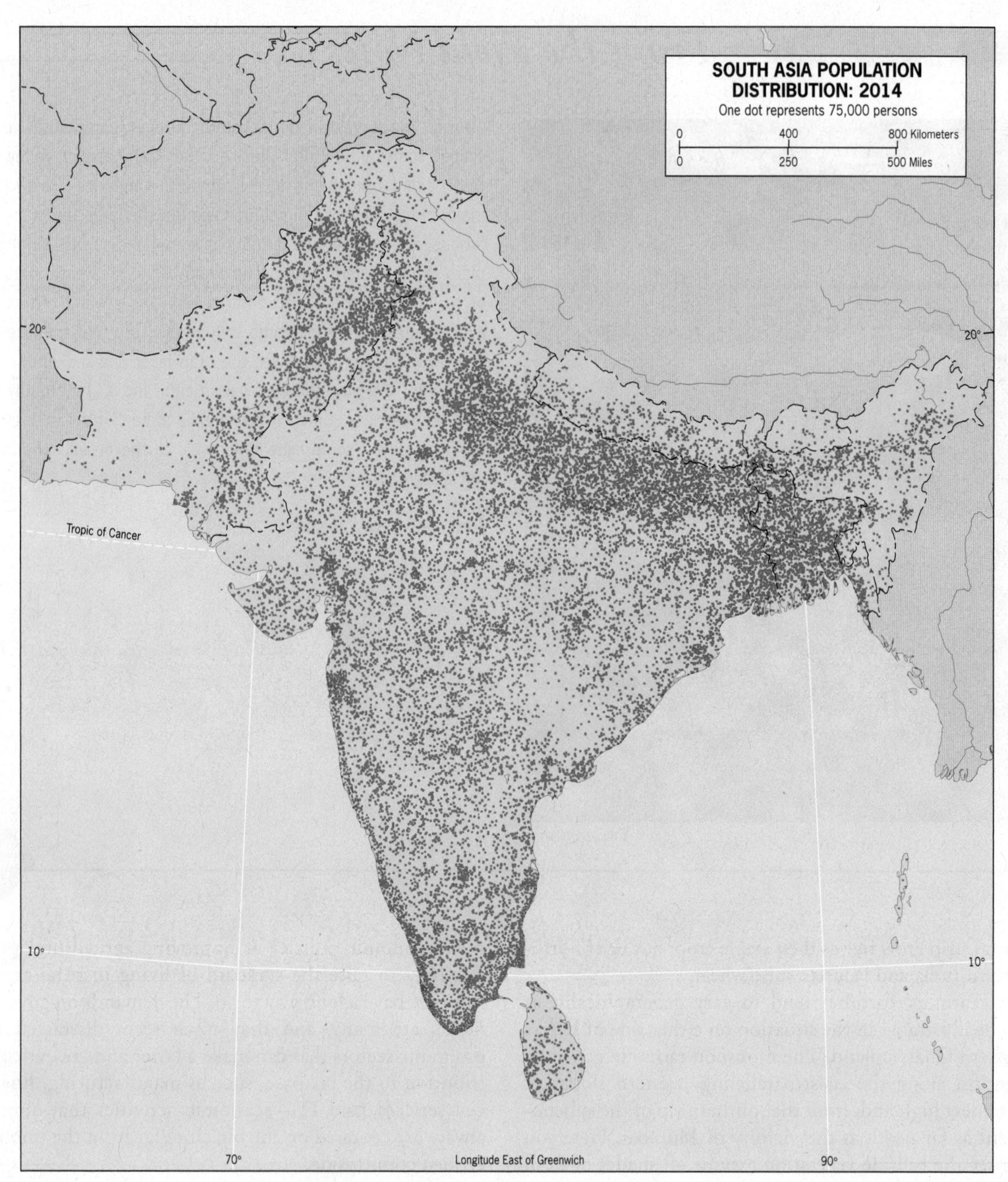

FIGURE 8A-9

fringe in the north displaying extensive empty spaces. The outlines of the densely populated river basins are clearly visible in the dot patterns of the population distribution map (Fig. 8A-9).

The field of **population geography [10]** focuses on the characteristics, distribution, growth, and other aspects of spatial demography in a country, region, or realm as this relates to soils, climates, land ownership, social conditions, economic development, and other factors. In the South Asian context, it is useful to concentrate on four demographic dimensions: the role of density; the demographic transition; demographic burdens; and the gender bias in birth rates. As we shall see, population issues are often more complicated than they first appear.

Population Density and the Question of Overpopulation

Population density [11] measures the number of people per unit area (such as a square kilometer or square mile) in a country, province, or an entire realm. We distinguish

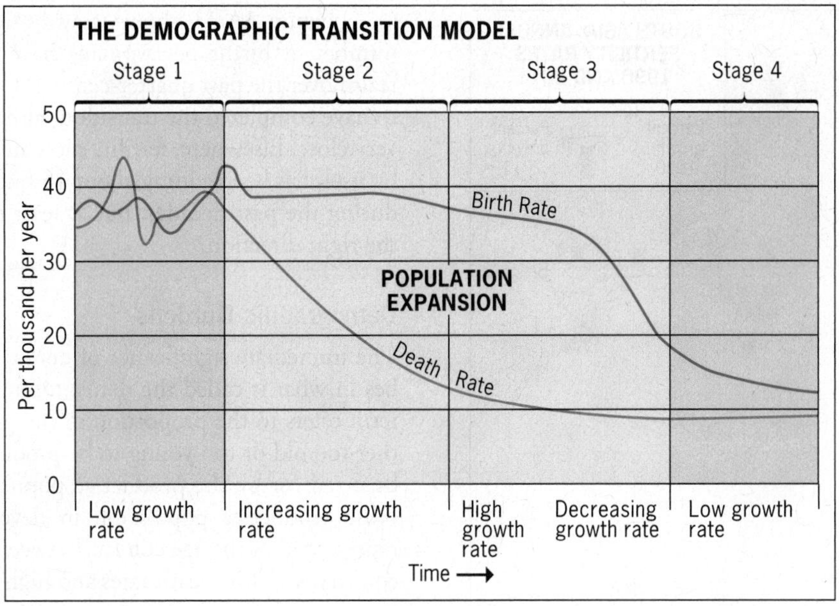

FIGURE 8A-10

between two types of measures. ***Arithmetic density*** is simply the number of people per area, usually a country. **Physiologic density [12]** is a more meaningful measure because it takes into account only land that is arable and can be used for food production. Please take a careful look at the data displayed for South Asia in Appendix B, and you will see that, for example in Pakistan, the two measures are quite different because of that country's large deserts and inhospitable mountain ranges.

Until recently, South Asia's persistent poverty was often related to its enormous and rapidly growing population and its high population densities. The idea was that there were simply "too many mouths to feed"—the realm was "overpopulated." The notion of ***overpopulation*** can be compelling and seems to make sense at an intuitive level since every country or region can be thought of as having a limited "carrying capacity."

But things are more complex than that. If you look again at Appendix B, you will find that some countries with high densities, such as the Netherlands or Japan, are doing very well, and it is not necessarily because they have an impressive natural resource base (neither does). The point is that high density in itself is not always a problem and that, in certain circumstances, population can be considered as a ***human resource***. If productivity is high, there does not appear to be a problem, but if productivity is low, then large populations can be a drain on the economy. Countries with high education levels, institutional effectiveness, and technological know-how are able to use their natural resources more efficiently.

Thus in South Asia, with large numbers of people still illiterate and undereducated, population tends to function as a *burden* rather than a resource. The problem is not so much that there are too many, but that too many are not sufficiently productive. The good news is that, as a result of higher economic growth rates stemming from economic reforms, there is more money to invest in education. The bad news so far is that not enough of this money is actually being spent that way.

The Demographic Transition

The relevance of population issues to development goes far beyond density, which is really just a snapshot of the population pattern at a moment in time. It gets more interesting—and more complicated—when we relate population change to economic trends. For instance, for a considerable time South Asia's population grew faster than the realm's economy. Clearly that was a problem because more and more people had to survive on less. Today, fortunately, it is the other way around: in most of the realm, the economy is growing faster than the population.

The term **demographic transition [13]** refers to a structural change in birth and death rates resulting, first, in rapid population increase and, subsequently, in declining growth rates and a stable population (Fig. 8A-10). The United States and other highly developed countries had already passed through this transition by the mid-twentieth century, and most countries in the South Asian realm are in the third stage today. Note that Stage 2 and part of Stage 3, with high birth rates and low death rates (due to medical advances), entail a population expansion. In South Asia, this expansion occurred from the 1950s through the 1970s. The key issue, of course, is for birth rates to come down so that overall growth rates will drop and the population will stabilize. This is happening today, but the process is not yet complete.

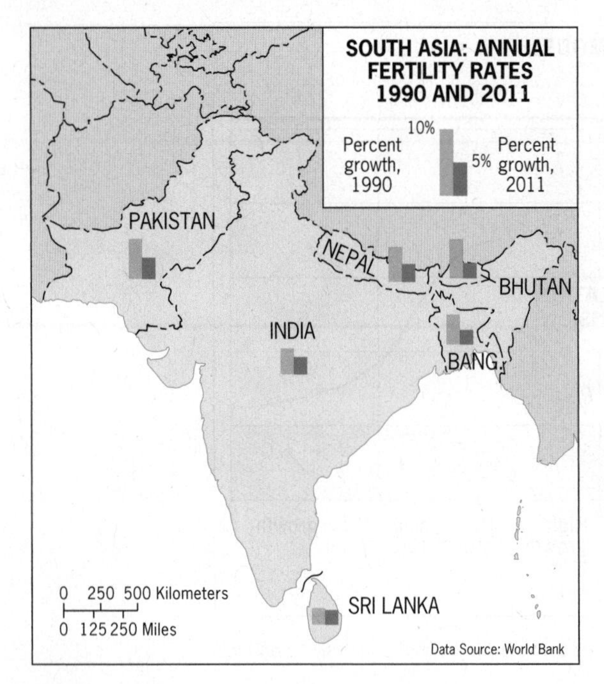

FIGURE 8A-11

Figure 8A-11 shows how **fertility rates [14]** (the number of births per woman) have dropped across the realm over the past quarter-century. Only Sri Lanka seems to have completed the transition, although Bhutan is now very close. Elsewhere, fertility rates are still too high (India by itself has been adding about 15 million people per year during the past decade), but at least they are trending in the right direction.

Demographic Burdens

The immediate significance of demography to economics lies in what is called the **demographic burden [15]**. This term refers to the proportion of the population that is either too old or too young to be productive and that must be cared for by the productive population. Typically, the most productive population in developing countries is represented in the age cohorts between 20 and 50 years. A country with low death rates and high birth rates will have a relatively large share of old and young people, and thus a large demographic burden. Obviously, the way to reduce this burden is to lower birth rates.

Let us now examine Figure 8A-12 and compare today's **population pyramids [16]** (diagrams showing the

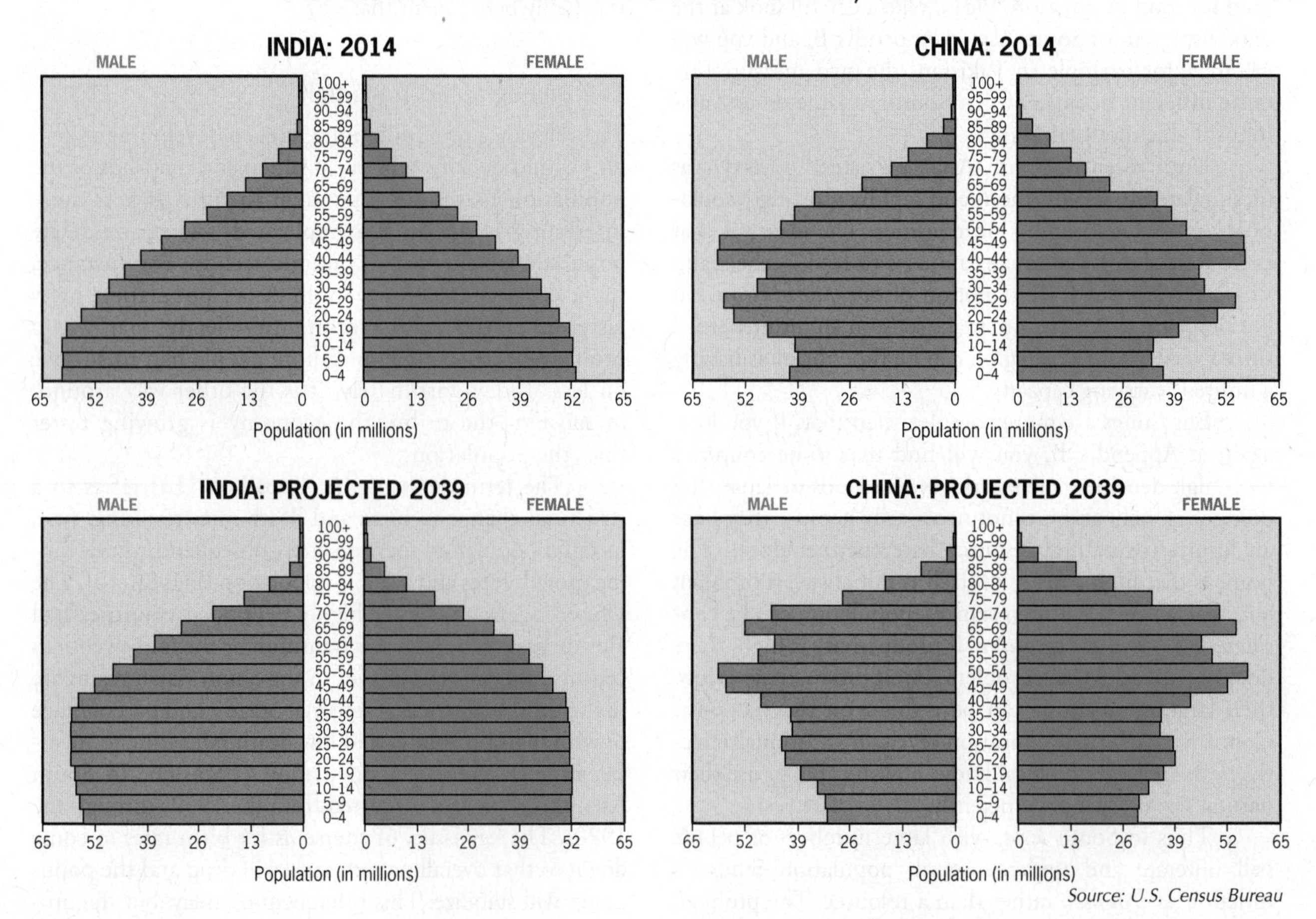

FIGURE 8A-12

age–sex structure) for India and China. The latter has been more successful in curtailing births since 1980, so China now faces a lower demographic burden than India. But, interestingly, what is advantageous today can become a disadvantage tomorrow. Look at Figure 8A-12 again and see what the population profiles are projected to be 25 years from now. Assuming that India will be able to further reduce its birth rate in the coming years, its demographic burden will be less than China's one generation from now. In China, today's productive cohort will have moved on to old age, thereby adding to its national demographic burden. It is another reason for India boosters to be optimistic about the future. Yet some other major hurdles remain, and certain forms of birth control are as morally reprehensible as they are economically counterproductive.

© Jan Nijman

A group of girls in a park in Delhi showing off their henna tattoos. "Mehndi" is a centuries-old festive tradition involving decorative painting on the hands and wrists. The designs usually wear off in a few weeks' time.

The Missing Girls

Issues of family planning and birth control shed some fascinating light on the "fragmented modernization" of South Asia. As we saw, the realm finds itself in an advanced stage of the demographic transition wherein birth rates have begun to decrease. But take a good look at India's population pyramid for 2014 and note that among young children boys far outnumber girls. In fact, males outnumber females well into middle age.

Traditionally, boys are valued more than girls because they are thought to be more productive income-earners, because they are entitled to land and inheritance, and because they do not require a dowry at the time of marriage. When a couple gets married (often arranged, and at a young age), the bride comes into the care of the groom's family, where she also contributes her work in and around the house. For this, the bride's family must provide a dowry that can impose a major expense on her parents. For these reasons, the birth of a boy is a greater cause for celebration than that of a girl. "Raising a daughter," as one saying goes, "is like watering your neighbor's garden."

One reason for the high fertility rates in the past was that families would continue to have children until there were enough sons to take care of the parents in their old age (the girls, after all, would be taking care of their future husband's parents). Hence, this gender bias is in itself a major factor in South Asia's population growth—but that is not the only problem. When a poor couple repeatedly produces girls and not boys, in some instances the family decides to end the life of the newborn daughter. It is this ***female infanticide*** or *gendercide* that causes the unnatural gender bias in South Asia's population profiles (the same applies to parts of China as well).

But why—as the economic situation has improved, as birth rates have receded, and as modernization has begun to set in—do we still observe this skewed **sex ratio [17]**? The answer is that with fewer children, the importance of having at least one boy has for many families become even more pressing. And here is where "modernization" throws another curve ball: newly available technologies of ultrasound scanning plus rising incomes (i.e., the growing affordability of a scan) have induced many families to determine the gender of the unborn child and decide on abortion if the child is female. Thus in recent years the sex ratio has become more, not less, skewed, and the most extreme ratios are now found in some of the most developed parts of the realm, such as the Indian States of Punjab and Haryana.

In the long run, of course, this leads to a shortage of females, which becomes particularly apparent at marriage age. In some areas, families now face a problem in finding brides for their sons, and this "bachelor angst," in turn, is leading to a change in attitudes. Look at India's population pyramid for 2039 and note that over the next quarter-century the sex ratio is expected to become less skewed.

Discrimination against women and girls in South Asia is expressed in a variety of ways in daily life. One matter that is now receiving more attention is the lack of sanitation for females. In many workplaces, schools, slum neighborhoods, and public spaces, women still have no (or insufficient) access to toilets. For safety and

From the Field Notes . . .

"Strolling the grounds of one of India's leading high-tech companies, Wipro, in the southern city of Bengaluru (formerly Bangalore), I was struck by the thought that their use of the word 'campus' was right on target: the great majority of the 33,000 employees on this site (120,000 in total) are in their mid-20s and almost all have a college degree; a large number of them are involved in research and product development; and the layout of the premises could easily be mistaken for a (well-funded) university. This Silicon Valley-like campus is the heart of the company's booming outsourcing and consulting business. The company is active in almost all major Indian cities and has a sizeable global presence. I was shown around by Rohit, a 26-year-old with a bachelor's degree in engineering from an Indian college and an MBA from Singapore National University. He reveled in the opportunity: 'I joined the company just four months ago in their marketing division and feel extremely fortunate. Everybody wants this job! It's a great professional opportunity and the salary is very good.' And Bengaluru is a pleasant place to live: at 900 meters (3000 ft) above sea level, even the summer weather is tolerable. Bengaluru, however, is growing so rapidly that its infrastructure cannot keep up, so that congestion, traffic jams, and commuting times are all growing as well."

© Jan Nijman

www.conceptcaching.com

Concept Caching

personal reasons women need more privacy, but often that is not available. In India, an estimated 330 million females lack proper access to toilets, with major implications for health and social functioning. Imagine a small rural school with one toilet (if indeed there is one): it will be used by boys at the exclusion of girls. Girls will have no access at all the entire day. This simple fact leads to girls missing school and even dropping out. Recent efforts by development organizations to install toilets for girls have so far only led to significant decreases in female-student dropout rates.

Still, it is difficult to make generalizations about gender relations across this populous realm, in part because of religious and regional diversity as well as rural-urban differences. To be sure, these are in many respects male-dominated societies, especially at a young age. But it is useful to remember that Pakistan, India, Sri Lanka, and Bangladesh have all had female prime ministers who held their countries' most powerful political office. That has yet to happen in the United States.

FUTURE PROSPECTS

South Asia is a realm in transition—politically, economically, and demographically. It is a realm that seems clearly bounded by nature, yet it is vitally linked to Southwest Asia and, increasingly, the entire world. It is also a realm that at times is difficult to read. India–Pakistan tensions continue to be a cause for concern, and the specter of terrorism haunts those who wish only to live in peace. This is not just in the hands of the governments of the two biggest states in the realm. Religious movements (Muslim and Hindu) and the way they engage in politics are crucially important, and the United States and China have major roles to play as well.

Economically, there is no question that India's rise will increasingly command the world's attention. Indian transnational corporations will continue to penetrate the global economy, and the growing Indian middle class with its appetite for consumption will increasingly draw interest from producers around the world. That English is the subcontinent's *lingua franca,* and that IT is a leading economic sector, give it an enormous advantage into the future. And the fact that India, the realm's giant, can claim to be the world's largest democracy gives it tremendous credibility.

When, during the next several decades, South Asia passes through the demographic transition; if it keeps the peace; if it continues its leading role in the global IT sector; if its economic growth is used to educate and empower the masses; and if the reorganization of agriculture allows more productive and prosperous lives—and these are all real possibilities—then this populous and wondrous South Asian realm may yet turn out to be the biggest story of the twenty-first century.

POINTS TO PONDER

- The South Asian realm contains three of the world's mightiest rivers and the world's largest human concentration.
- After two-thirds of a century, the conflict over Kashmir is still unresolved.
- The most skewed sex ratios in the population occur in the most prosperous parts of this realm.
- The Indian Ocean may be the world's most crucial geopolitical arena of the twenty-first century.
- Hundreds of millions of females in South Asia suffer from a lack of access to basic human needs, including sanitary facilities.

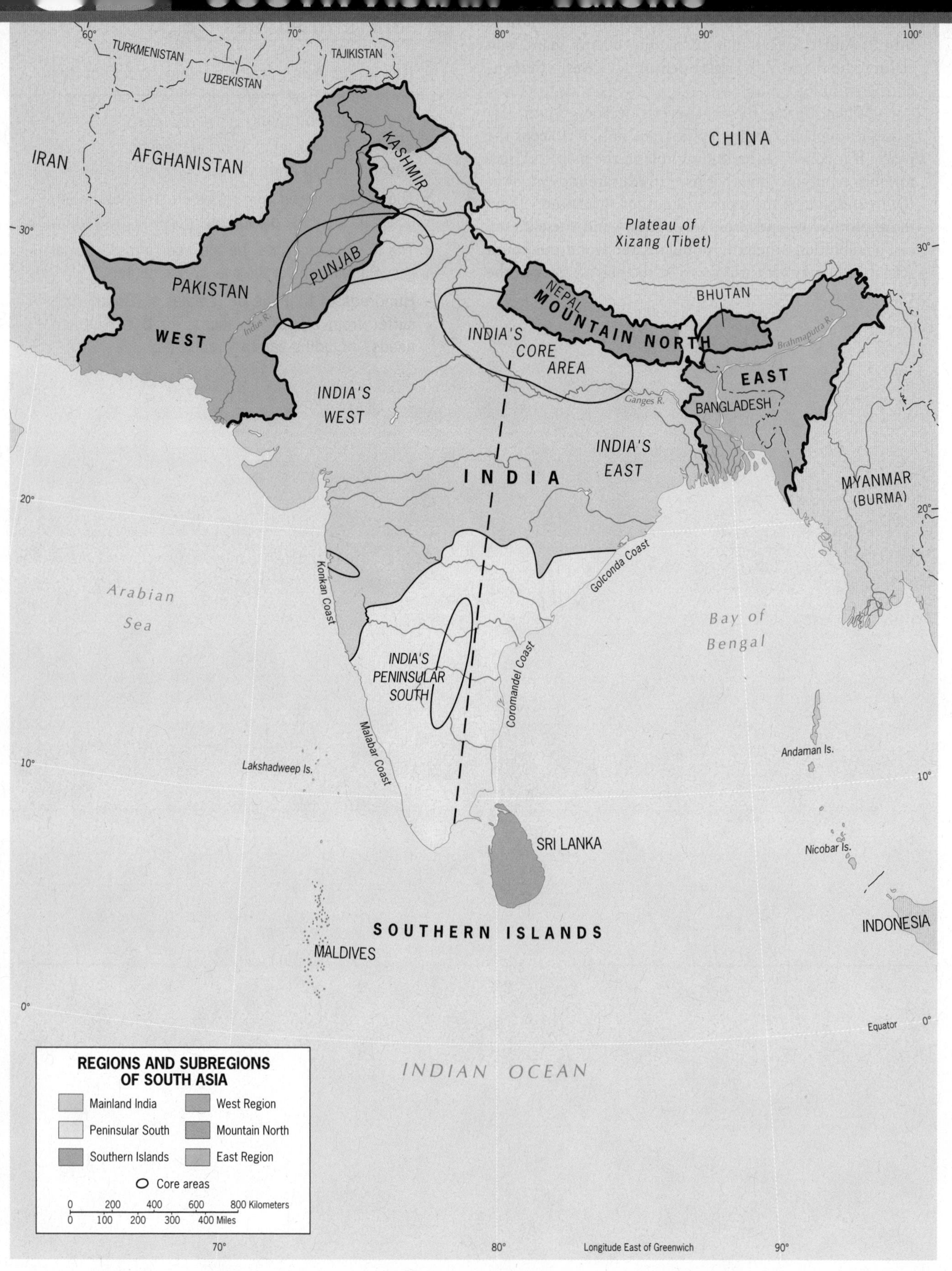

TURKMENISTAN
TAJIKISTAN
UZBEKISTAN
IRAN
AFGHANISTAN
CHINA
KASHMIR
Plateau of
Xizang (Tibet)
PUNJAB
PAKISTAN
WEST
Indus R.
NEPAL
MOUNTAIN NORTH
BHUTAN
Brahmaputra R.
INDIA'S
CORE
AREA
EAST
BANGLADESH
Ganges R.
INDIA'S
WEST
INDIA'S
EAST
INDIA
MYANMAR
(BURMA)
Golconda Coast
Konkan Coast
Arabian
Sea
Bay of
Bengal
INDIA'S
PENINSULAR
SOUTH
Coromandel Coast
Malabar Coast
Lakshadweep Is.
Andaman Is.
SRI LANKA
Nicobar Is.
SOUTHERN ISLANDS
INDONESIA
MALDIVES
Equator
INDIAN OCEAN
REGIONS AND SUBREGIONS
OF SOUTH ASIA
Mainland India
West Region
Peninsular South
Mountain North
Southern Islands
East Region
Core areas
0 200 400 600 800 Kilometers
0 100 200 300 400 Miles
Longitude East of Greenwich

FIGURE 8B-1

REGIONS

Pakistan

India

Bangladesh

Mountainous North

Southern Islands

IN THIS CHAPTER

- Pakistan: A hard place to govern
- Caste in modern India
- India's new consumers
- Bangladesh: Survival in the double delta
- Sri Lanka: Stability at last?

CONCEPTS, IDEAS, AND TERMS

Communal tension	1
Hindutva	2
Caste system	3
Outsourcing	4
Informal sector	5
Double delta	6
Physiologic density	7
Non-governmental organization (NGO)	8
Micro-credit	9
Buffer state	10
Failed state	11
Global warming	12
Insurgent state	13

Developments in the South Asian realm, discussed in Chapter 8A, require analysis at multiple scales, Local, regional, and realmwide geographic issues, from gender to geopolitics and from refugees to religion, are closely tied to place, whether it is the Taliban in Pakistan, the Tibetans in Nepal, or the Tamils in Sri Lanka. From Kashmir to the Indian Ocean and from the textile factories of Bangladesh to the economic rise of India, this chapter provides a better understanding of the endless diversity of this realm, all in geographic context.

REGIONS OF THE REALM

REGIONS AND STATES

India, the cornerstone of South Asia, has a core area centered on the broad basin of the Ganges River, the historic heart of this realm (Fig. 8B-1). India is a region as well as a state, with key subregions based on cultural and other criteria to be discussed later in this chapter. To the west lies Islamic Pakistan, whose lifeline, the Indus River and its tributaries, creates a core area in the Punjab. The boundary between India and Pakistan originated with the partition of the former British Indian Empire, the tragic separation that occurred on the eve of independence in 1947 (see Chapter 8A).

In South Asia's east, the British had earlier drawn a border between (Hindu) Bengal and (dominantly Muslim) East Bengal, and that border became the boundary of what became known as East Pakistan, the political connection based on shared Islamic values. In 1971, however, East Pakistan severed its link with West Pakistan (which thereupon renamed itself Pakistan) and became independent Bangladesh.

As Figure 8B-1 shows, three separate entities make up the region defined as the Mountainous North. From east to west, these are the traditional kingdom of Bhutan, as isolated a country as the world has today; the troubled

© Jan Nijman

Where were these pictures taken?
Find out at www.wiley.com/college/deblij

© Jan Nijman

MAJOR CITIES OF THE REALM

City	Population* (in millions)
Delhi–New Delhi, India	24.8
Mumbai (Bombay), India	20.8
Dhaka, Bangladesh	16.9
Karachi, Pakistan	15.1
Kolkata (Calcutta), India	15.0
Bengaluru (Bangalore), India	9.6
Chennai (Madras), India	9.6
Hyderabad, India	8.6
Lahore, Pakistan	8.2
Ahmadabad, India	7.1
Varanasi, India	1.5
Kathmandu, Nepal	1.1
Colombo, Sri Lanka	0.8

*Based on 2014 estimates.

state of Nepal, where royal rule failed but the struggle to establish representative government continues; and, at the western end, the disputed territory of Kashmir, where the British boundary-making effort failed to resolve a complicated cultural and political situation, and where India and Pakistan have repeatedly clashed in armed conflict.

Finally, the region mapped as the Southern Islands consists of two very different countries: dominantly Buddhist Sri Lanka, where a quarter-century-long civil war ended in 2009; and the ministate Republic of the Maldives, whose official religion is Islam.

Physiographically, India mainly consists of plateau-and-basin country, but its neighbors to the north, Nepal and Bhutan, adhere to the southern slopes of the mighty Himalaya range and its rugged foothills. The waters of the Indian Ocean separate India from neighbors Sri Lanka and the Maldives in the south. And with Pakistan separated from its western and northern neighbors by the soaring ranges of the Hindu Kush and Karakoram, there are compelling geographic reasons for using the political framework to designate the regions of this realm. Culture and nature constitute formidable regionalizing factors here, even within India itself. The cultural forces that broke British India apart at the time of independence continue to exert enormous power in still-evolving South Asia.

PAKISTAN: ON SOUTH ASIA'S WESTERN FLANK

Gift of the Indus

If, as is so often said, Egypt is the gift of the Nile, then Pakistan is the gift of the Indus. The Indus River and its principal tributary, the Sutlej, nourish the ribbons of life that form the heart of this populous country (Fig. 8B-2). Territorially, Pakistan is not large by Asian standards; its area is about the same as that of Texas plus Louisiana. But Pakistan's population of 187.7 million makes it one of the world's ten most populous states. Among Muslim countries (its official name is the Islamic Republic of Pakistan), only Southeast Asia's Indonesia is larger.

Pakistan lies like a gigantic wedge between Iran and Afghanistan to the west and India to the east. Here in Pakistan lay South Asia's earliest urban civilizations, whose innovations radiated southeastward into the massive triangular peninsula. Here lies South Asia's Muslim frontier, contiguous to the great Islamic realm to the west and irrevocably linked to the enormous Muslim minority to its east.

Pakistan's cultural landscapes bear witness to its transitional location. Teeming, disorderly Karachi is the typical South Asian city; as in India, the largest urban center lies on the coast. Historic, architecturally Islamic Lahore is reminiscent of the scholarly centers of Muslim Southwest Asia. In Pakistan's east, the boundary of the 1947 partition divides a Punjab subregion that is otherwise continuous—a land of villages, wheatfields, and irrigation canals. In the northwest, Pakistan resembles Afghanistan in its huge migrant populations and its mountainous frontier. And in the far north, Pakistan and India are locked in a deadly and intractable conflict over Kashmir (discussed in Chapter 8A). A legacy of the time of partition, this territory is claimed by Pakistan because the majority of the inhabitants are Muslim, while India refuses to give up control because it claims that the Hindu minority would have no future inside Pakistan. This issue has plagued relations between the two countries for decades and is not likely to be resolved in the foreseeable future.

A Hard Place to Govern

At independence (West) Pakistan had a bounded national territory, a capital, a cultural core, and a population—but few centripetal forces to bind state and nation. The disparate subregions of Pakistan shared the Islamic faith and an aversion for Hindu India, but little else. Karachi and the coastal south, the southwestern desert of Baluchistan, the city of Lahore and the Punjab, the rugged northwest along Afghanistan's border, and the mountainous far north remain worlds apart. Urdu is the official language, yet English remains the *lingua franca* of the elite. Several other major languages, however, prevail in different areas (see Fig. 8A-5), and ways of life vary enormously.

Successive Pakistani governments, civilian as well as military, turned to Islam to provide the common bond that history and geography had denied this nation. In the process, Pakistan became one of the world's most theocratic states. But even Islam itself is not unified in restive Pakistan. Almost 80 percent of the people are Sunni Muslims, and the Shia minority numbers approximately 20 percent. Sunni fanatics intermittently attack Shi'ites, leading to retaliation and establishing grounds for subsequent revenge.

To govern so diverse and fractious a country would challenge any system, and so far Pakistan has failed the test. Democratically elected governments have repeatedly

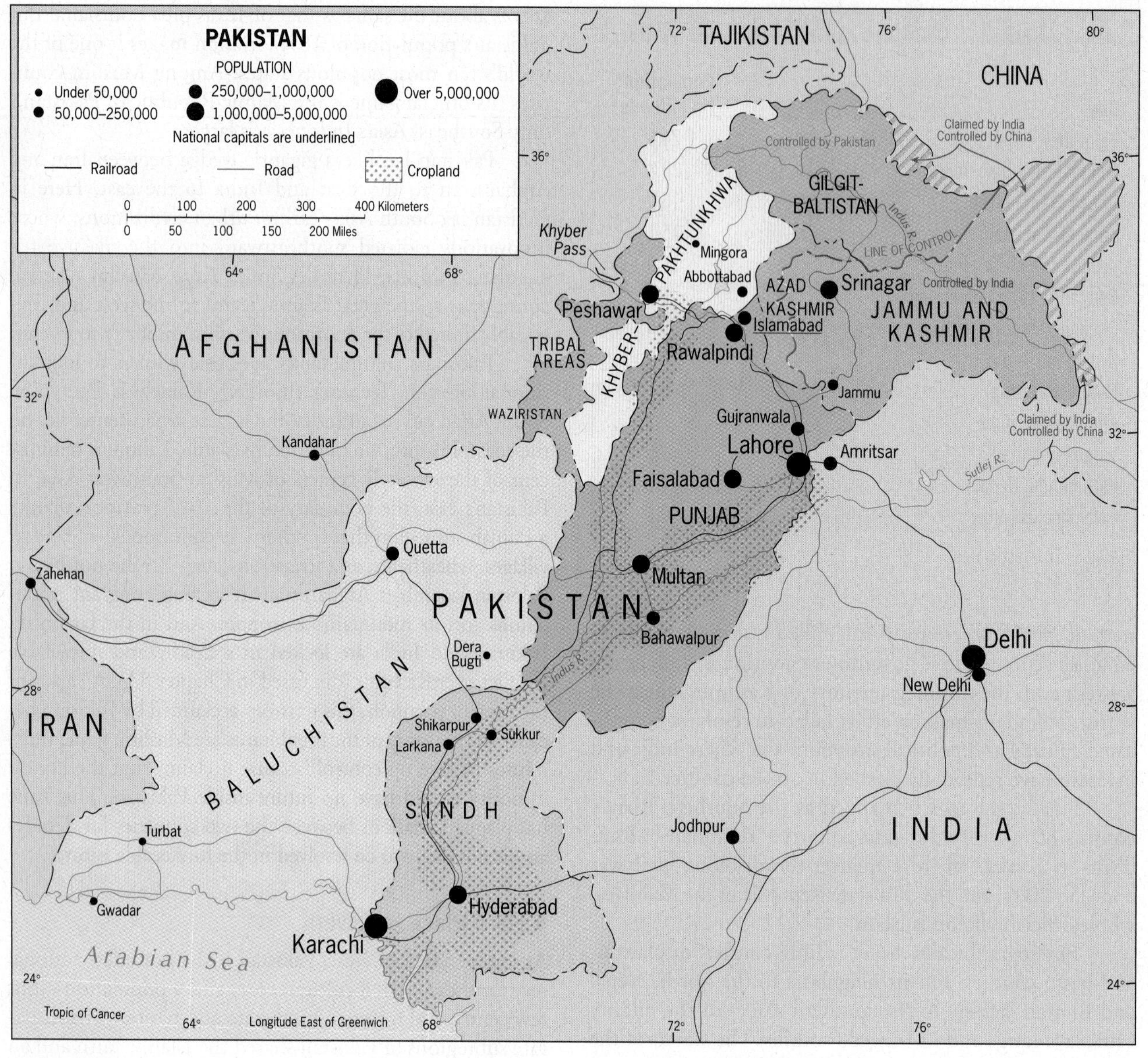

FIGURE 8B-2

squandered their opportunities, only to be overthrown by military coups. Pakistan's recent economic boom has not filtered down to the poor; literacy rates are not rising; health conditions are not improving significantly; national institutions are weak (for instance, there are only about 2 million registered taxpayers in a country of over 180 million); and one consequence of the global antiterrorism campaign is that Pakistanis, who used to be overwhelmingly secular in their political choices, are now increasingly joining Islamic parties.

Meanwhile, too little is being done to confront a growing water-supply crisis, an insurgency festers in Baluchistan, the army is incapable of establishing control over mountainous Waziristan (where al-Qaeda and Taliban groups maintain hideouts), the issue of Kashmir costs Pakistan dearly, and relations with neighboring India (which if satisfactory would bring enormous benefits) remain conflicted.

Adding to Pakistan's woes, the second half of 2010 witnessed the worst floods in the country's history, a result of the wettest monsoon in decades. The Indus River, the lifeline that traverses the entire country from north to south, swelled beyond capacity, backed up many of its tributaries, and flooded fully one-quarter of the country. This disaster seemed to push Pakistan to the brink as the government, already facing so many major challenges, was not in a position to provide adequate relief. Asked if the crisis would threaten the current government, President Zardari commented that "I don't think anybody in their right mind would want to take over Pakistan right now." A hard country to govern indeed!

Subregions of Pakistan

Punjab

Pakistan's core area is the Punjab (Fig. 8B-2), the Muslim heartland across which the post-independence boundary

ASIF HASSAN/AFP/Getty Images, Inc.

The floods that plagued Pakistan in the late summer and fall of 2010 occurred almost everywhere along the mighty Indus River, from Khyber Pakhtunkhwa Province in the north to Sindh in the south. This photo, taken on October 30th, shows the extent of the flooded area around Jacobabad (located about 50 kilometers [30 miles] from the river's main channel on the wide Indus floodplain), near the border between Sindh and Baluchistan just north of Shikarpur (see Fig. 8B-2). From this medium-sized city alone, some 20,000 people were forced to flee for their lives. The massive flood also destroyed much of that year's harvest, further adding to Pakistan's mounting miseries.

between Pakistan and India was superimposed. (As a result, India also has a State named Punjab, sometimes spelled Panjab there.) Pakistan's Punjab is home to just over half of the country's population; in the triangle formed by the Indus River and its tributary, the Sutlej, live more than 100 million people. Punjabi is the language here, and wheat farming is the mainstay.

Three cities anchor this core area: Lahore, the outstanding center of Islamic culture in the realm; Faisalabad; and Multan. Lahore, now home to 8.2 million people, lies close to the India–Pakistan border. Founded around 2000 years ago, Lahore was situated favorably to become a great Muslim center during the Mughal period, when the Punjab was the main corridor into India. After partition in 1947, the city received hundreds of thousands of refugees and grew rapidly. Lahore did lose its eastern hinterland, but its new role in independent Pakistan sustained its growth. Punjab's relationship with Pakistan's other three provinces is one of the country's weak points. Both the governments and residents of those provinces feel uneasy about the dominance of Punjab, the populous, powerful core of the country from which most of the army is drawn.

Sindh

The lower Indus River is the key to life in Sindh (Fig. 8B-2), but the Punjab controls the waters upstream, which is one of the issues dividing Pakistan. When the Punjab-dominated regime proposes to build dams across the Indus and its major tributaries, Sindhis (who make up almost one-fifth of the national population) are reminded of their underrepresentation in government and talk of greater autonomy. Nationalist, anti-Pakistan rage swept Sindh following the 2007 assassination of Sindh's presidential candidate, Benazir Bhutto.

The ribbon of fertile, irrigated, alluvial land along the lower Indus, where the British laid out irrigation systems, makes Sindh a Pakistani breadbasket for wheat and rice. Commercially, cotton is king here, supplying textile factories in the cities and towns (textiles account for more than half of Pakistan's exports by value).

But the dominant presence in southern Sindh is the chaotic, crime-ridden megacity of Karachi, with its stock market, dangerous streets, crowded beaches, and poverty-stricken shantytowns, a place of searing contrasts under a broiling sun. Karachi grew explosively during and after partition in 1947, when refugee Muhajirs from across the new border with India streamed into this urban area, setting off riots and gang warfare whose aftermath still simmers. With little effective law enforcement, Karachi has become a hotbed of terrorist activity, but somehow the city still functions as Pakistan's (and Afghanistan's) major maritime outlet and the seat of Sindh's provincial government.

Khyber Pakhtunkhwa

As Figure 8B-2 shows, this province lies wedged between the powerful Punjab to the east and troubled Afghanistan to the west, with the territory long known as the ***Tribal Areas*** intervening in the south. The name "Pakhtunkhwa" connotes "belonging to the Pushtuns," the Afghan-associated tribes that inhabit this subregion.

AMONG THE REALM'S GREAT CITIES . . .

KARACHI

LOCATED NEXT TO the Indus Delta and on the Arabian Sea, Karachi is Pakistan's biggest city as well as its economic and financial center. In former days, it was the capital too, but that function was shifted far inland to Islamabad in the 1960s. Until independence in 1947, Karachi was a typical colonial port city, designed to serve British interests. This function can still be seen in the urban landscape: the city was organized around the port, and this is where you can still find the main railroad terminals. Raw materials and agricultural goods were brought in from the hinterland to be loaded onto ships headed for England's industrial complexes. The old city center lies just north of the port, and today's main business district is located just to the east of it.

Karachi has grown enormously over the years, from around 100,000 at the beginning of the twentieth century to about half a million in 1950 to 15.1 million today. This undoubtedly is Pakistan's "world city" because it dominates the country's linkages to the global economy (even more so than Mumbai does for India). Karachi is home to most of the foreign multinational firms, accounts for about three-quarters of international trade, and produces about 20 percent of Pakistan's GDP. It has the busiest airport and two seaports that together handle more than 90 percent of all transoceanic trade (the newer port of Qasim was built east of the metropolis in the 1970s). The city's economic importance is manifest in a major central business district (Saddar) with an imposing skyline.

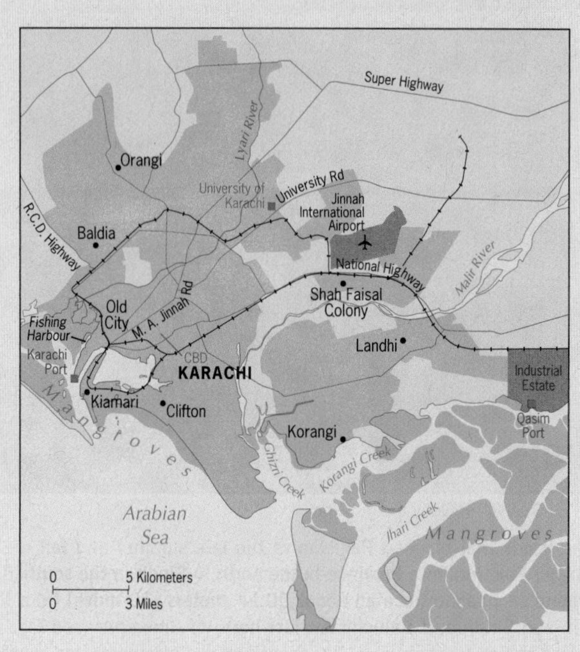

But this is a city besieged with serious problems. Rampant and unplanned growth has resulted in congestion and pollution. Many people live in grinding poverty, and the contrast with the small but very wealthy urban elite is stark. Compared to most other cities in this realm, street crime is rampant. Karachi sits near the bottom in rankings of the world's "most liveable" cities. The great majority of its residents are first- or second-generation immigrants from a wide variety of ethnic backgrounds (a large number of them originally hailing from northern India prior to the 1947 partition), and recent years have witnessed growing communal tension and violent conflict (more than 2500 people were killed in such violence during 2012 alone). In addition, the city is widely described as a nurturing ground for Islamic fundamentalists and increasingly serves as a base for Taliban leaders and followers. Karachi's image as Pakistan's most cosmopolitan center is being fatally threatened.

Why should there be a zone designated "Tribal Areas" in a country otherwise subdivided into provinces? The designation began during colonial times when the British assigned a special status to the obstinate Pushtun villages and their chiefs living in the remote hills and mountains along the Afghan border. The "Tribal Agencies" were the responsibility of an official "agent" who exercised control over the local chiefs through ample reward and harsh punishment—and little or no accountability. There were seven of them, some very small, several large enough to appear on a relatively small-scale map such as Figure 8B-2. North and South Waziristan were among the largest.

The Tribal Areas have had a certain degree of autonomy ever since British times, and the Pakistani government's reach into these parts is quite limited to this day. The province's mountainous physical geography reflects its remoteness and the isolation of many of its people. Mountain passes lead to Afghanistan; the Khyber Pass, already noted as the historic route of invaders, is legendary (see photo). Coming from Afghanistan, the Khyber's road leads directly to the provincial capital, Peshawar, which lies in a broad, alluvium-filled, fertile valley where wheat and corn drape the countryside.

During the Soviet occupation of Afghanistan in the 1980s and later during the Taliban regime, several million Pushtun refugees streamed through the Khyber and other passes into refugee camps in this area. Following the defeat of the Taliban in Afghanistan in late 2001, the great majority of Pushtuns returned home. Khyber Pakhtunkhwa remains a conservative, deeply religious, militant province, where Islamic political parties and movements are proportionately stronger than in any other part of the country and where the obstruction of national policies (including antiterrorist operations) is a common goal.

© Barbara A. Weightman

The Khyber Pass that links Afghanistan and Pakistan across the Hindu Kush. One of the most strategic passes in the world, it was no easy passage for invading armies in times past. Nowadays, the roads and tunnels facilitate the movement of refugees, drugs, and weapons, and they are used by militant separatist forces from both sides of the border. Pakistan's North West Frontier Province [now renamed Khyber Pakhtunkhwa], and especially the city of Peshawar, are hotbeds of these activities.

Pakistan's northwest frontier is critical to the U.S. war in Afghanistan as well as the wider campaign against Islamic terrorism, because it is extremely difficult to monitor the cross-border movements of Taliban and al-Qaeda forces. The area is a hotbed of terrorist activity and training camps. The United States constantly prods Pakistan to try to secure the province militarily, but it is not clear to what extent Pakistan is willing—or able—to comply.

Baluchistan

As Figure 8B-2 indicates, Baluchistan is by far the biggest of Pakistan's four provinces, accounting for (not including Kashmir) nearly half of the national territory but inhabited only by an estimated 13.5 million people or barely seven percent of the country's population. For a sense of its terrain, take a look at Figure 8A-1; much of this vast territory is desert, with mostly barren mountains that wrest some moisture from the air only in the northeast. Sheep raising is the leading livelihood here, and wool the primary export. In its northern extremity, Baluchistan abuts the Tribal Areas and Afghanistan. The provincial capital, Quetta, lies in this zone. This province, in fact, could easily be called the "South West Frontier Province."

Baluchistan is of considerable economic-geographic importance. Beneath its parched surface lie possibly substantial reserves of oil as well as coal, and already the province produces most of Pakistan's natural gas. The major new seaport of Gwadar on the southwestern coast not only handles raw materials from Baluchistan but is also a transshipment point for oil and gas from Iran and the Caspian Basin, destined for markets in East Asia (China was a major investor in the construction of this port).

But Baluchis are dissatisfied with the central government and complain that Pakistan is dominated by the Punjabis. Ninety percent of Baluchis have no energy supply and eight out of ten do not have access to clean water. For decades, short-lived local rebellions have signaled dissatisfaction with the government, but more recently the Baluchistan Liberation Army (BLA) has instigated a more serious and durable insurgency. Since 2006, an estimated 600 Punjabi settlers in Baluchistan have been killed. And local Sunnis are also venting their frustration on 'immigrant' Shi'ite Muslims who originally hail from Central Asia. A bombing attack in early 2013 killed 86 Shi'ites in the city of Quetta, and the president of Pakistan flew in to express his sympathy to the victims' families. The regional problem of Baluchistan is compounded by the presence of Afghan Taliban leaders in their hideout in Quetta (Fig. 8B-2) and by the fact that the Baluchi population spills over the Pakistani border into Afghanistan and Iran. Hence, Pakistan's government must also be wary of irredentist support for the Baluchis from abroad.

Pakistan's Prospects

Pakistan is a country of enormous cultural contrasts, where the modern and medieval exist side by side, where you hear residents of one province call those of another (but not themselves) "Pakistanis," and where a sense of nationhood is still elusive among many people whose loyalties to their family, clan, and village are stronger.

Pakistan has experienced so many cycles of progress and failure that confidence is in short supply. And yet, against all odds, there are areas of progress: a combination of expanded irrigation and Green Revolution farming techniques has, during the past decade, enabled Pakistan to export some rice (although wheat imports continue), and the country's manufacturing and services sectors have shown substantial growth. Exports include not only cotton-based textiles but also carpets, tapestries, and leather goods. Domestic manufacturing continues to be quite limited, but Pakistan does have its own steel mill near Karachi. Meanwhile, the authorities struggle to control Pakistan's growing production and trade in opium and hashish. Neighboring Afghanistan remains the world's dominant source of this illegal commerce, and the spillover effect continues; but Pakistan has many remote corners where poppy fields yield high returns and trade routes are well established. In this as in so many other spheres, Pakistan displays the contradictory symptoms of a state in transition.

This western flank of South Asia is the realm's most critical region, today more so than ever before. Islamic Pakistan's coherence and stability have now become crucial at a time when the global struggle against terrorism is entangling its leaders with Western power and priorities. The role of Pakistan's government in this struggle is disputed and resented by many of its own people, who express their distaste by voting for militant Islamic parties and voicing support for the resurgent Taliban movement. Militancy and instability are no longer confined to the northwest or the far north: the cities of Lahore and Karachi have been targeted repeatedly in recent years by terrorists. The government blames the Taliban.

Islamic militants are far from a majority but their recourse to violence and the politics of intimidation are casting a deepening shadow across Pakistan. In 2011, the former governor of Punjab, Salman Taseer, who had returned to his legal profession, was shot dead on a street in Islamabad—because he had spoken out against the country's pernicious blasphemy laws. He had done so in connection with the case against his client, a poor Christian woman whom a court had condemned to death for blasphemy (swearing). The killer, one of Mr. Taseer's own security guards, was later showered with rose petals by groups of lawyers (!) outside the courthouse in Islamabad. However, liberal critics in Pakistan were quick to voice their fears of a wholesale revival of religious intolerance across the country.

Geographically complex Pakistan confronts many difficult challenges, and it is easy to envision a number of ways in which the country could descend into disastrous turmoil. Pakistan's future hangs in the balance and with it the stability of South (as well as Southwest) Asia.

INDIA: GIANT OF THE REALM

If you have been reading the press and watching television over the past few years, you have seen the increasing attention being paid to India—not just in North America but around the world. Examples of news coverage include the purportedly growing "strategic partnership" between the United States and India, the outsourcing of American jobs to India, the rapid rise of a new Indian middle class, and the emergence of large Indian companies that are making their presence felt around the world. Undoubtedly, India is on the move—but will it really become the next economic superpower?

Certainly India has the dimensions to make the world take notice. Not only does it occupy three-quarters of the great land triangle of South Asia: India also is poised to overtake China to become the most populous country on Earth before the middle of this century. According to some estimates, India will possess the fourth-largest economy in the world by 2020 (after the United States, China, and Japan). Already, India is the world's biggest democracy, a federation of 28 States and several additional Territories with a population of nearly 1.3 billion.

That India has endured as a unified state is a politico-geographical miracle. The country contains a cultural mosaic of immense ethnic, religious, linguistic, and economic diversity and contrast; it is truly a state of many nations. Upon independence in 1947, India adopted a democratic, secular, federal system of government, giving regions and peoples some autonomy and identity, and allowing others to aspire to such status.

India's pluralist as well as democratic complexion cannot be fully understood without considering the huge influence of Mahatma Gandhi, the great spiritual and political leader who played a central role in the achievement of independence from the British. More than anyone else, he symbolized tolerance, reason, nonviolence, and perseverance—the basic principles upon which the Indian freedom struggle and the new state were founded. Jawaharlal Nehru, India's founding prime minister, also exerted enormous influence through his strong, unshakable beliefs in democracy and secular government. It is also worth noting that those beliefs and principles can be said to be firmly embedded in Hindu culture. After all, Hinduism is in some ways an extremely open, diverse, and introspective religion and way of life.

Political Geography

A Federation of States and Peoples

The map of India's political geography shows a federation of 28 States, 6 Union Territories (UTs), and 1 National Capital Territory (NCT) (Fig. 8B-3; Table 8B-1). The federal government retains direct authority over the UTs, all of which are small in both territory and population. The NCT, however, includes most of the Delhi/New Delhi urban region and now contains more than 17 million inhabitants.

This form of political spatial organization is mainly the product of India's restructuring following independence from Britain. Its State boundaries reflect the broad outlines of the country's cultural mosaic: insofar as possible, the system recognizes languages, religions, and cultural traditions. Indians speak 14 major and numerous minor languages, and while Hindi is the official language, it is by no means universal (see Fig. 8A-5). At the time of independence, Hindi was the most widely spoken native language, but it was the mother tongue for only about one-third of the population. It is telling that the famous Midnight Speech of Prime Minister Nehru at the moment of independence in 1947 was given in English—Hindi or any other language would have been very divisive and politically impossible. Thereafter, English swiftly became the new country's *lingua franca*.

As Figure 8B-3 shows, the territorially largest States lie in the heart of the country as well as on the massive southward-pointing peninsula. Uttar Pradesh (200 million, according to the 2011 census) and Bihar (104 million) together cover much of the Ganges Basin and also form the core area of modern India. Maharashtra (112 million), anchored by the great coastal city of Mumbai (known as Bombay before 1996), also has a population larger than

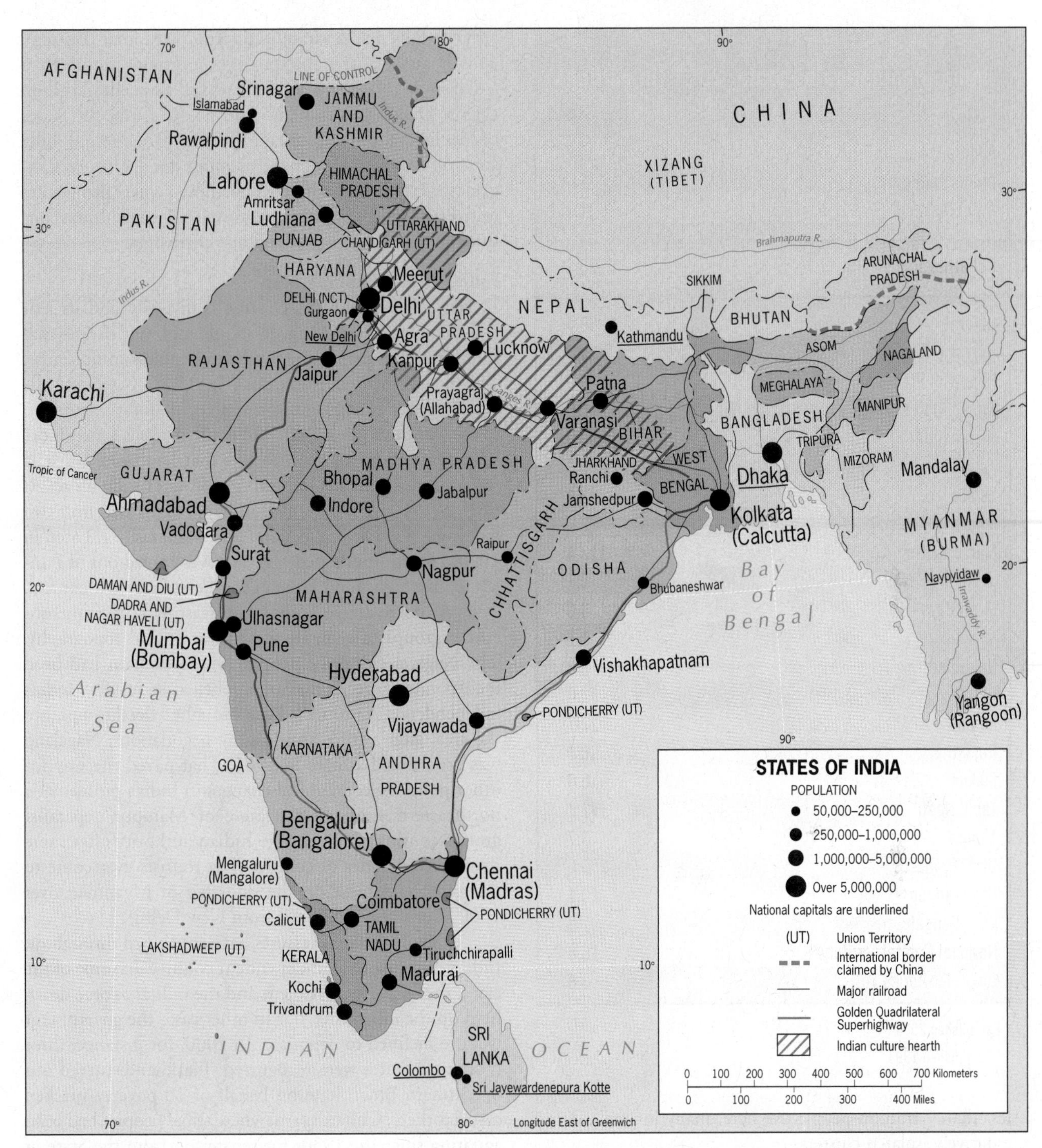

FIGURE 8B-3

that of most countries. West Bengal, the State that adjoins Bangladesh, contains more than 91 million residents, 15 million of whom live in its urban focus, Kolkata (known as Calcutta before 2000).

These are staggering numbers, and they do not decline much toward the south. Southern India consists of four States linked by a discrete history and by their distinct Dravidian languages. Facing the Bay of Bengal are Andhra Pradesh (85 million) and Tamil Nadu (72 million), both part of the hinterland of the city of Chennai (formerly Madras) and located on the coast near their joint border. Facing the Arabian Sea are Karnataka (61 million) and Kerala (33 million).

As Figure 8B-3 indicates, India's smaller States lie mainly in the northeast, on the far side of Bangladesh, and in the northwest, toward Jammu and Kashmir. North of Delhi, physical and cultural landscapes change from the flatlands of the Ganges to the hills and mountains of spurs of the Himalaya. In the State of Himachal Pradesh, forests blanket the hillslopes and high relief reduces living space;

TABLE 8B-1
India: Population by State, 2011 (in millions)

State	Population
Andhra Pradesh	84.7
Arunachal Pradesh	1.4
Asom (Assam)	31.2
Bihar	103.8
Chhattisgarh*	25.6
Goa	1.5
Gujarat	60.4
Haryana	25.4
Himachal Pradesh	6.9
Jammu and Kashmir	12.6
Jharkhand*	33.0
Karnataka	61.1
Kerala	33.4
Madhya Pradesh	72.6
Maharashtra	112.4
Manipur	2.7
Meghalaya	3.0
Mizoram	1.1
Nagaland	2.0
Odisha	42.0
Punjab	27.7
Rajasthan	68.6
Sikkim	0.6
Tamil Nadu	72.1
Tripura	3.7
Uttar Pradesh	200.0
Uttarakhand*	10.1
West Bengal	91.3
National Capital Territory**	16.8
Union Territories	3.8

* Established 2000
** Established 1993

less than 7 million people live here, many in small, comparatively isolated clusters.

The map becomes even more complicated in the distant northeast, beyond the narrow corridor between Bhutan and Bangladesh. The dominant State here is Asom (Assam), home to just over 31 million, famed for its tea plantations and important because its petroleum and gas production amounts to more than 40 percent of India's total. In the Brahmaputra Valley, Asom resembles the India of the Ganges. But in almost all directions from Asom, things change.

North of Asom, in sparsely populated Arunachal Pradesh (1.4 million), we are in the Himalayan offshoots once again. To the east, in Nagaland (2.0 million), Manipur (2.7 million), and Mizoram (1.1 million), lie the forested and terraced hills that separate India from Myanmar (Burma). This is an area of numerous ethnic groups (more than a dozen in Nagaland alone) and of frequent rebellion against Delhi's government. And to the south, the States of Meghalaya (3.0 million) and Tripura (3.7 million), hilly and still wooded, border the teeming floodplains of Bangladesh. Here in the country's northeast, where peoples are restive and where population growth still soars, India confronts one of its strongest regional challenges.

India's Ever-Changing Map

In view of the country's enormous diversity and its federal democratic structure, it should not surprise us that India's political map is the product of endless compromise. The newly devised framework in 1947, based on the major regional languages, proved to be unsatisfactory to many communities in India. In the first place, many more languages are in use than the 14 that had been officially recognized. Demands for additional States soon arose. As early as 1960, the State of Bombay was divided into two language-based States, Gujarat and Maharashtra. Later, in 1966, Hindu-dominated Haryana was carved out of Punjab to give Sikhs (and Punjabi speakers) more self-control.

As noted above, India's northeast harbors numerous ethnic groups in a highly varied, forest-clad topography. The Naga, a cluster of peoples whose domain had been incorporated into Asom State, rebelled soon after India's independence. A protracted war brought federal troops into the area; after a truce and lengthy negotiations, Nagaland was proclaimed a State in 1961. That paved the way for other politico-geographical changes in India's problematic northeastern wing. In the State of Manipur, separatist groups continue to challenge Indian authority after more than three decades of conflict (few tourists ever come to this remote outpost on the doorstep of Myanmar, over 1600 kilometers [1000 mi] from New Delhi).

Devolutionary pressures have continued throughout India's existence as an independent country. In some of the cases, the federal government and the military come down hard on the insurgents, but in other cases the government is more inclined to negotiate. In 2000, for instance, three more new States were recognized: Jharkhand, carved out of southern Bihar State on behalf of 18 poverty-stricken districts there; Chhattisgarh, where tribal peoples had been agitating since the 1930s for separation from the State of Madhya Pradesh; and Uttarakhand (originally named Uttaranchal), which split from India's most populous Ganges Basin, core-area State of Uttar Pradesh on the basis of its highland character and lifeways (Fig. 8B-3).

At the time of writing, India's central government is in protracted negotiations about carving a new State of Telangana out of the northwestern sector of Andhra Pradesh. Telangana is an inland area that sees itself as marginalized by coastal elites. The proposed State would contain more than 35 million people, cover approximately one-third of Andhra Pradesh, and its capital would be Hyderabad—one of India's leading technopoles and its sixth-largest city. Other demands for independent Statehood persist, particularly the proposal to subdivide almost unmanageable Uttar Pradesh (2011 population: 200 million) into four new States.

During the past several years, there have been troubling signs of yet another challenge to India's federal government: communist- (avowedly Maoist-) inspired rebellions that seem to be transforming into a coordinated revolutionary campaign. It is known in India as the ***Naxalite*** movement, named after a village in the State of West Bengal where it was founded in the 1960s. Mainly active in India's poorest and most disaffected States—such as Bihar, Jharkhand, and Andhra Pradesh—the Naxalites appeal to the poor and other minorities (especially tribal people), whose plight is blamed on India's elites and neoliberal economic policies. According to Indian observers, they remain active in one-third of India's more than 600 districts (the administrative unit below the State) and maintain a violent presence in about 90 districts (Fig. 8B-4). They blow up railroad tracks, attack police stations, and intimidate villagers (see photo). Stability is key to India's economic prospects, and this is what the Naxalites try to undermine. The Indian government has declared the Naxalites to be the country's greatest internal security threat and has embarked on a major counteroffensive. This appears to have had some success: the 2010 death toll associated with Naxalite violence surpassed 1100, but in 2011-2012 it fell to about half that number.

Communal Tensions

The term **communal tension [1]** (communal disharmony) refers to the several different categories of conflicts that recur among India's highly diverse sociocultural groups. Most commonly, these conflicts have a base in (politicized) religion, but they can also be caste-based (castes will be discussed shortly).

The Sikhs

The Sikhs (the word means "disciples") adhere to a religion that was created about five centuries ago to unite warring Hindus and Muslims into a single faith. They rejected what they considered to be the negative aspects of Hinduism and Islam, and Sikhism gained millions of followers in the Punjab and surrounding areas. During the colonial period, many Sikhs supported the British administration of India, and by doing so they won the respect and trust of the colonialists, who employed tens of thousands of Sikhs as soldiers and police officers. By 1947, there was a large Sikh middle class in the Punjab. Today they still exert a strong influence

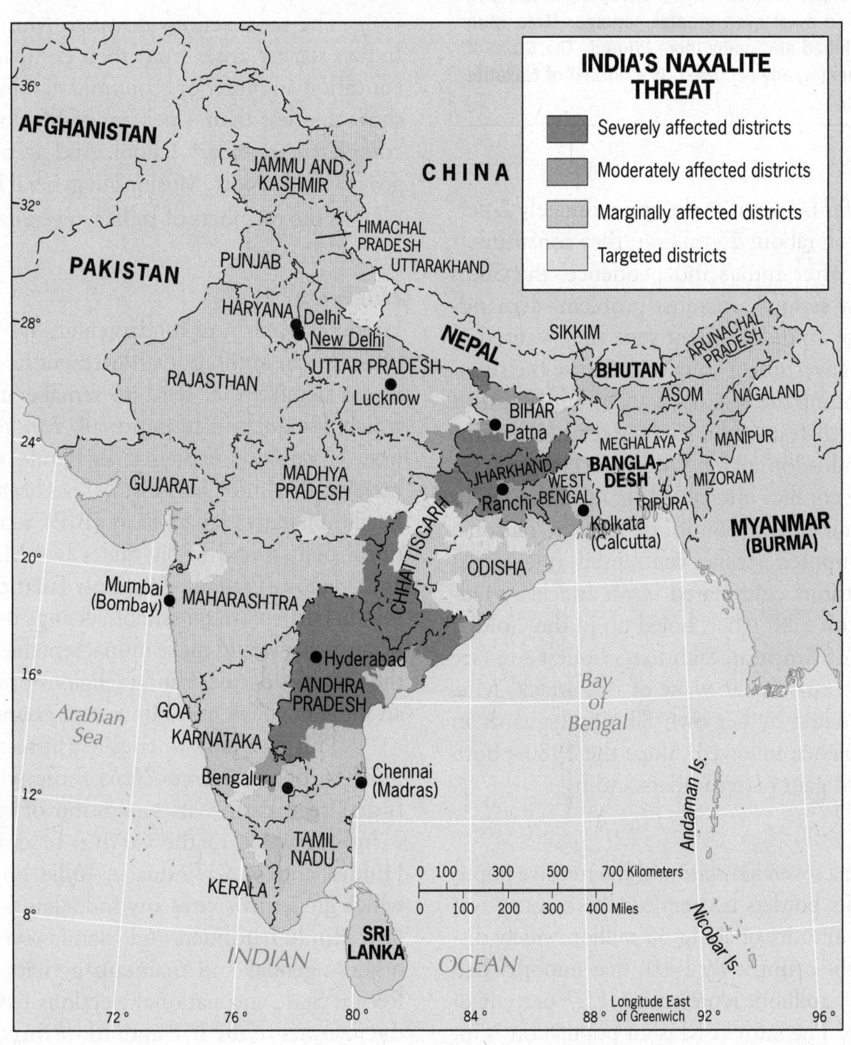

FIGURE 8B-4

AP/Wide World Photos

Although the Maoist Naxalite rebels denied involvement, Indian authorities blamed their leadership for planning and executing the carefully timed destruction of two trains in May 2010, their most audacious terrorist attack through mid-2013. In this dramatic photo, the cars of the rust-colored train, struck by an explosion on the track about 150 kilometers (90 mi) west of Kolkata on the way to Mumbai, lie crushed against those of the oncoming blue train, which could not stop to avert even greater carnage. More than 150 passengers were killed and many more injured. The State of West Bengal, where this tragedy occurred, is a hotbed of Naxalite antigovernment activity.

over Indian affairs, far in excess of the approximately 2 percent of the population (about 25 million) they constitute.

For a period after India's independence, the Sikhs created India's most serious separatist problem, demanding the formation of an independent state they wanted to call *Khalistan*. The government sought to defuse the situation in 1966 by dividing the original State of Punjab into a Sikh-dominated northwest—which would retain the name Punjab—and a Hindu-dominated southeast (Haryana; see Fig. 8B-3). But the conflict intensified again in the 1970s when Prime Minister Indira Gandhi declared a "national emergency" and imposed strong centralized rule across the federation. Tensions culminated in an attack by federal military forces on Sikh rebels holed up in the Golden Temple in the city of Amritsar, Sikhdom's holiest site (see chapter-opening photo). In the wake of this attack, Mrs. Gandhi was assassinated by her own Sikh bodyguards in 1984 and more violence followed. Since the 1980s, both sides have worked to gain better understanding.

The Muslims

When India became a sovereign state and the massive population shifts across its borders had ended, the country was left with a Muslim minority of about 35 million widely dispersed throughout the country. By 1991, that minority had grown to nearly 100 million, representing 11.7 percent of the total population. The current Muslim population is estimated to exceed 200 million (roughly 15.4 percent), and it continues to grow faster than the overall Indian population. Muslims are in the majority in Jammu and Kashmir, Asom, and West Bengal, while their largest absolute numbers are in the big States of Uttar Pradesh, Bihar, and Maharashtra. To India's great geographic advantage, the Muslim minority is not regionally concentrated, avoiding what otherwise could lead to a dangerous secessionist movement.

The position of Muslims in Indian society is not easy to separate from India's relations with Pakistan, from the issue of Kashmir, and from acts of Islamic terrorism that have bedeviled India in recent years (particularly the attacks on Parliament in 2001 and Mumbai's upscale hotels in 2008). At present, India holds the unenviable distinction of being the most frequently targeted country in the world for terrorist acts. As a result, relations between Muslims and Hindus are deteriorating. Concern over linkages between foreign (read Pakistani) terrorists and local Muslim subversives is changing Hindu attitudes and tactics, a topic elaborated in the next section.

The most serious threat to Muslim integration into Indian society arises from their comparatively low level of education and inferior economic standing. Official statistics show that less than 4 percent of Muslims nationwide have completed secondary school. And according to one recent government report, Muslims in general have now become as poor as the members of India's lowest-ranked Hindu castes.

Hindutva

Hindu extremism or fundamentalism may seem like a contradiction in terms, but a movement has come to the fore in recent decades that seeks to remake India as a society in which Hindu principles prevail. ***Hindutva*** **[2]**, or Hinduness, is variously expressed as Hindu nationalism, Hindu heritage, or Hindu patriotism. It has been the guiding agenda for the Bharatiya Janata Party (BJP), a powerful force in national politics and in big States like Maharashtra, Gujarat, and Madhya Pradesh. *Hindutva* fanatics want to impose a Hindu curriculum on schools, change the flexible family law in ways that would make it unacceptable to Muslims, inhibit the activities of non-Hindu religious proselytizers, and forge an India in which non-Hindus are essentially outsiders.

This naturally worries Muslims as well as other minorities, but it also concerns those who understand that India's secularism—its separation of religion and state—is indispensable to the survival of democracy. Moderate Hindus and non-Hindus in India oppose such notions, which are as divisive as any India has faced. It is easy to see how Hindu hardliners and Islamic revivalists can fuel each other's agendas and maintain a vicious cycle of conflict. Recent State and national elections have seen a decline in the fortunes of the BJP and, more importantly, an internal

From the Field Notes . . .

© Jan Nijman; photo by Zach Woodward

"It is a Friday afternoon as we make our way through a main thoroughfare in one of Mumbai's biggest slums. Dharavi 'houses' approximately half a million people on less than 2 square kilometers! About 20 percent are Muslims and the overwhelming majority of all others are dalits. Increasingly, the Muslims live spatially segregated in their own tightly knit neighborhoods. There are several mosques but little space inside them. So the men, taking a break from work nearby, place their mats alongside the road and prepare for prayer outdoors. Having a faith is important to people living and working in Dharavi's filthy and impoverished environs, especially for the Muslim minority."

www.conceptcaching.com

party struggle between moderates and *Hindutva* hardliners in which the moderates have seemed to prevail. Overall, Indian voters have not rushed to embrace this radicalization of Hinduism, a confirmation of the continuing robustness and vitality of India's democracy.

The Persistence of Caste

Hinduism's benign and admirable properties combine with a system of social stratification that is generally derided in the West (and by a large number of Indians as well). Under Hindu dogma, ***castes*** are fixed layers in society whose ranks

From the Field Notes . . .

© Jan Nijman

"Varanasi, on the banks of the holy Ganges River. I was able to get up close to the Hindu priests performing the *aarti* (fire) ritual, just after sunset. The ritual follows ancient conventions and choreography, drawing thousands of devotees and spectators. Varanasi is India's holiest city, said to have been created by the god Shiva. Archeological evidence indicates that Varanasi (also known as Benares) has existed for more than 3000 years, making it one of the oldest living cities in the world. Consider this: for more than three millennia, every day of every year, Hindus have descended the stepped banks of the holy Ganges River here to pray and worship, to find solace, to wash away their sins and purify their souls, and to cremate their dead. Daily life on the banks of the Ganges is a moving spectacle, for Hindus and non-Hindus alike."

www.conceptcaching.com

are based on ancestries, family ties, and occupations. The **caste system [3]** has its origins in the early social divisions into priests and warriors, merchants and farmers, and it is also thought to have had a racial basis (the Sanskrit term for caste, *varna*, means color). More specifically, caste became associated with specific professions and, over the centuries, its complexity increased until India had thousands of castes, some of them with a few hundred members and others containing millions.

Hindus believe in reincarnation, and a person is thought to be born into a particular caste based on his or her actions in a previous existence. Hence it would not be appropriate to counter such ordained caste assignments by permitting movement (or even contact) from a lower caste to a higher one. Persons of a particular caste could perform only certain jobs, wear only certain clothes, and worship only in prescribed ways at particular places. They or their children could not eat, play, or even walk with people of a higher social status.

The *untouchables* occupying the lowest tier of all were the most debased, wretched members of this rigidly structured social system. Indeed, the term "untouchable" acquired such negative connotations that it was replaced several times. Mahatma Gandhi, a powerful critic of the system, introduced the term *harijans*, meaning children of God. But more recently this label was thought to have a condescending connotation, and it gave way to the term ***dalits*** (the oppressed), indicating a greater sense of awareness and assertiveness among them.

In the isolated villages deep in the countryside, dalits still suffer from severe discrimination and harsh treatment by higher castes. Children often are made to sit on the floor of their classroom (if they go to school at all); dalits are not allowed to draw water from the village well because they might pollute it; they must take off their shoes, if they own any, when they pass higher-caste houses; and they cannot take jobs in professions other than the one they were born into (sweeping, cleaning, and the like).

Today, dalits are estimated to constitute more than 15 percent of all Hindus while Brahmins, the highest caste, account for a comparable share. The rest of the population occupies a wide range of in-between castes. The Indian government officially abolished castes at independence, but the system has proven difficult to dismantle. Many dalits have chosen to convert to other religions, most notably Buddhism. It is estimated that nearly 90 percent of Buddhists in India were originally dalits, but it seems that even after converting they do not lose this stigma.

Successive Indian governments have now introduced an elaborate system of affirmative action on behalf of *Scheduled Castes* (the official government label for dalits). This effort has had more effect in urban than in rural areas of India. In any case, dalits now have reserved for them places in the schools, a fixed percentage of State and federal government jobs, and a quota of seats in national and State legislatures. These jobs are often highly desirable because they tend to be white-collar and much better paid than those generally available to dalits.

Because of the huge number of minorities in India, affirmative action schemes have become increasingly complicated over time and are at the heart of seemingly never-ending debates. In addition to the existing quota of 15 percent of federal jobs for dalits, the central government is considering whether it should reserve public-sector jobs for Muslims and other religious minorities such as Sikhs. In another proposal now under discussion, one-third of all Parliament seats would go to women. This was protested by some Muslim organizations that feared there would not be enough eligible Muslim women, and they subsequently demanded a subquota for them. In the world's largest and most diverse democracy, politics can be a highly complicated affair!

Just how far India's political pendulum can swing was shown in the 2007 provincial legislative elections in Uttar Pradesh State, where a dalit party won an absolute majority and where a woman named Mayawati Kumari was the first dalit to become a chief minister of one of India's major States. It was a stunning victory that made headlines throughout the country and shook up the political establishment, revealing the growing power of the lowest castes and marking a turning point for India's representative government. But most lower-caste Indians are still faced with very limited opportunities, widespread discrimination, and abject poverty. In traditional India, caste provided stability and continuity; in contemporary India, it constitutes an often painful and difficult legacy.

Economic Geography

The most commonly cited, and most clearly evident, regional division of India is between north and south. The north is India's heartland, the south its Dravidian appendage; the north speaks Hindi as its *lingua franca*, the south prefers English over Hindi; the north is bustling and testy, life in the south seems more measured and less agitated.

East and West

But there is another, as yet less obvious, but potentially more significant divide across India. In Figure 8B-5 draw a line from Lucknow, on the Ganges River, south to Chennai, near the northern tip of Tamil Nadu State (see Fig. 8B-1). To the west of this line, India is showing signs of economic progress, the kind of productive activity that has brought Pacific Rim countries such as Thailand and Indonesia a new life. To the east, India has more in common with the less-promising countries that also face the Bay of Bengal: Bangladesh and Myanmar (Burma). The manufacturing map may seem to suggest that much of India's industrial strength lies near Kolkata, but the heavy industries built here by the state in the 1950s are now outdated, uncompetitive, and in steady decline. The State of Bihar represents the stagnation that afflicts much of India east of our line: by several measures it ranks among the poorest of the 28 States.

Compare this to western India. The State of Maharashtra, the hinterland of Mumbai, leads India in many

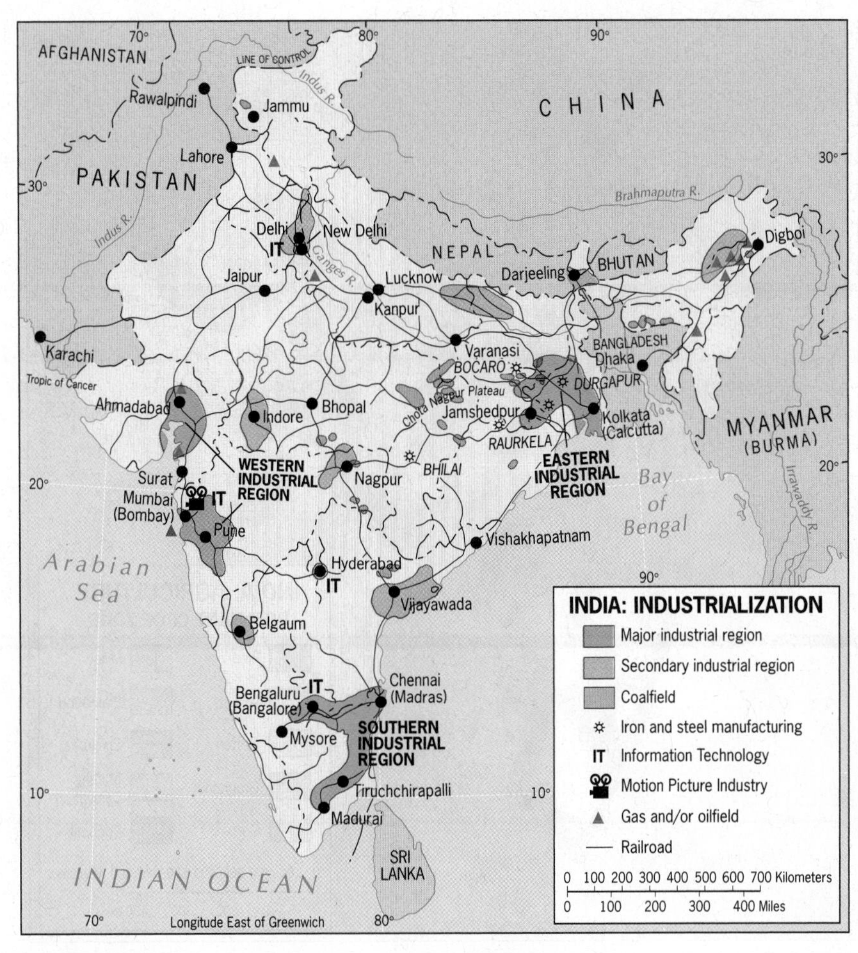

FIGURE 8B-5

categories, and Mumbai leads Maharashtra. Many smaller, private industries have taken root here, manufacturing goods that range from umbrellas to satellite dishes and from toys to textiles. Across the Arabian Sea lie the oil-rich economies of the Persian Gulf countries. Hundreds of thousands of workers from western India have found jobs there, sending remittances back to families from Punjab to Kerala. More importantly, many have used their foreign incomes to establish service industries back home. Outward-looking western India, in sharp contrast to the inward-looking east, has begun to establish additional ties to the outside world. Satellite and fiber-optic cable links have propelled Bengaluru (formerly Bangalore) to become the center of a burgeoning software-producing complex that reaches world markets. Maharashtra's economic success has also spilled over into Gujarat, its northern neighbor, and even landlocked Rajasthan (the next State to the north beyond Gujarat) is experiencing the beginnings of what, by Indian standards, is a boom.

Life Is in the Harvest

Agriculture provides more jobs in India than any other employment sector, and India's fortunes (and misfortunes) remain strongly tied to farming. Fully two-thirds of the population still lives on (and from) the land, spread in and around the country's hundreds of thousands of villages. There, traditional farming methods persist, yields per unit area remain among the world's lowest, and hunger and malnutrition still afflict millions even as grain surpluses accrue. The relatively few areas of modernization, as in the wheatlands of the Punjab, are islands in a sea of stagnation. Land reform has essentially failed; roughly one-quarter of India's entire cultivated area, including much of the best land, is still owned by less than 5 percent of the country's landholders, who have great political influence and obstruct redistribution. Perhaps half of all rural families own less than 1 hectare (2.5 acres) or no land at all; and an estimated 135 million live and work as tenants, always uncertain of their fate.

As everywhere, agriculture depends very strongly on physiography and climate, and agricultural fortunes vary significantly across the country. In India, the monsoon is absolutely critical, its abundance and timing a reliable predictor of the harvest. Figure 8B-6 displays India's agricultural geography and the map reflects the rainfall patterns and monsoonal cycle depicted in Figure 8A-3. Rice dominates all along the Arabian Sea-facing southwestern coast (on the rainy side of the Western Ghats) and in the monsoon-drenched peninsular northeast; where drier conditions develop, wheat and other grains prevail.

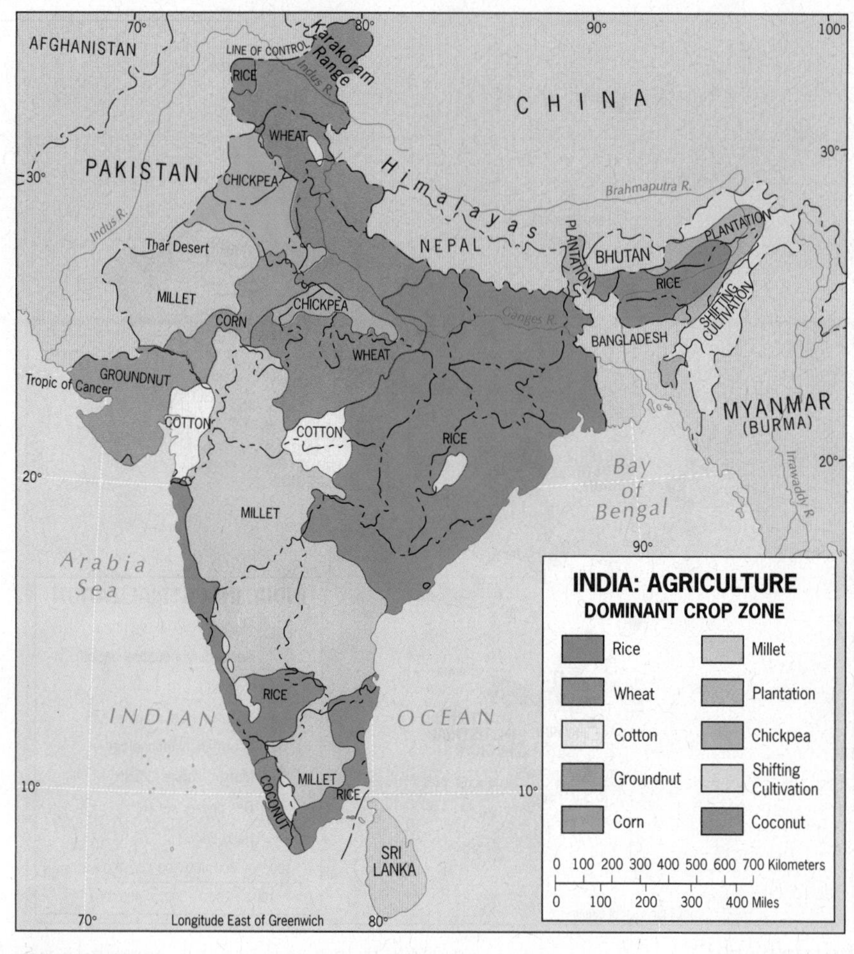

FIGURE 8B-6

From the Field Notes . . .

"In late 2012, I visited a small village in Rajasthan, a couple of hours away from the city of Jodhpur. A farmer took me to the small plot where he grows wheat, then showed me his home, made of mud with a roof of wood and straw. In this semiarid environment, people eat mainly wheat and other hardy grains, and far less rice. The farmer's wife and their children sit around the fire making thin round breads called *chapati*. The hospitality of these Rajasthanis is matched by the colorfulness of their dress."

www.conceptcaching.com

Even though the country produces an ample variety of crops, yields are often too low as a result of inefficient land ownership and small farmers' lack of access to inputs such as fertilizer, irrigation equipment, and machinery. If there is any to sell, getting produce to market is yet another struggle for millions of farmers. In 2010, almost half of India's 600,000-plus villages could not be accessed by truck let alone automobile, and in this era of modern transportation animal-drawn carts still far outnumber motor vehicles nationwide.

Manufacturing and Information Technology

India's industrial sectors have significantly improved over the past two decades, but the pace of change is still too slow to meet the country's needs. The map of manufacturing geography (Fig. 8B-5) is a legacy of colonial times, with

AMONG THE REALM'S GREAT CITIES . . .

MUMBAI (BOMBAY)

IN SOME WAYS, Mumbai is a microcosm of India, a burgeoning, crowded, chaotic, fast-moving agglomeration of humanity. Shrines, mosques, temples, and churches evince the pervasive power of religion in this multicultural society. Street signs come in a bewildering variety of scripts and alphabets. The Victorian-Gothic architecture of the city center is a legacy of the British colonial period (see photo in Chapter 8A). Creaking double-decker buses compete with ox-wagons and handcarts on the congested roadways. The throngs on the sidewalks—businesspeople, holy men, sari-clad women, beggars, clerks, homeless wanderers—spill over into the streets.

In precolonial times, fishing folk living on the seven small islands at the entrance to this harbor named the place after their local Hindu goddess, Mumbai Devi. The Portuguese, first to colonize it, called it Bom Bahia, "Beautiful Bay." The British, who came next, corrupted both names to form *Bombay*, and so it remained for more than three centuries. In 1996, the government of the State of Maharashtra, with support from the federal authorities, changed the city's name back to Mumbai.

With 20.8 million people spread across the metropolis, this is India's second-largest urban region. Think of Mumbai as a combination of New York and Los Angeles—India-style. This is the country's commercial and financial center with most of the banks, the leading stock exchange, and most foreign multinationals. And it is also India's most glamorous city, owing in no small part to the presence of Bollywood, center of the Hindi film industry that produces more motion pictures (for more viewers) than anywhere else on Earth.

The 2009 movie *Slumdog Millionaire*, a box office hit around the world, focused on the darker side of this megacity, the approximately 8 million slum dwellers who struggle to make a living alongside the movie stars and the burgeoning middle classes. In part because its peninsular geography imposes strict limits, Mumbai is an extremely crowded city; its overall density is about seven times that of major U.S. central cities, and its slums exhibit much higher densities than that. Anyone visiting the city is bound to be overwhelmed by the sheer scale of its poverty, which stands in sharp contrast to the lavish, hugely expensive homes of the rich.

Mumbai is very much a product of colonial times. During the Industrial Revolution, England needed raw materials at low prices and India was a leading supplier, especially of

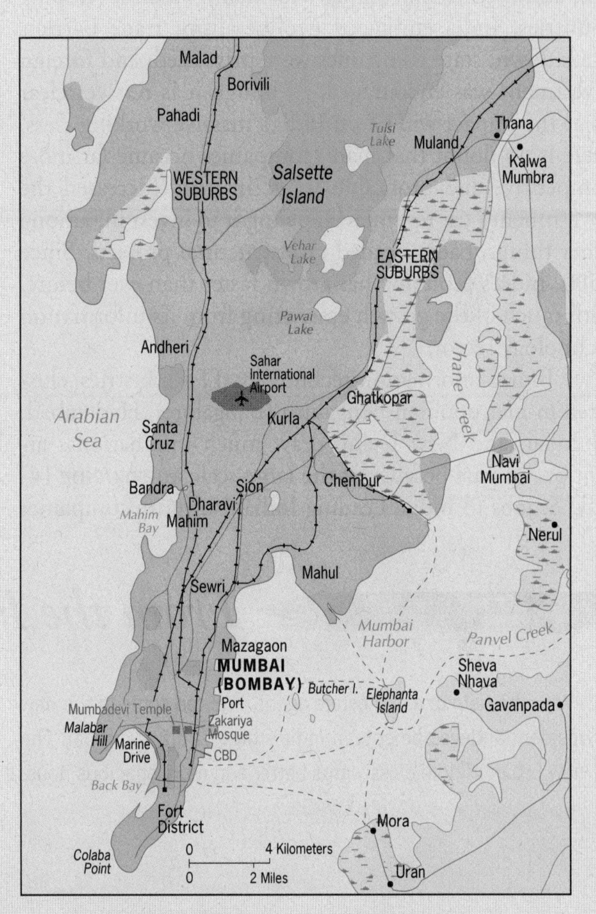

cotton. Much of the raw cotton was processed in the textile mills of Bombay and then shipped to London, Manchester, or Liverpool to be converted into finished products. From the 1860s on, Bombay became one of the most important colonial cities and a vital node in the far-flung British Empire. The imprint of that era is still visible in the architecture of the Fort District in the far south, the port, the railways, and the textile mills lining the rail corridors farther north.

Today not much is left of the textile industry in the urban economy. Instead, Mumbai has become the country's most important world-city with its growing sectors of finance and producer services that include accounting, advertising, and consulting.

coastal Mumbai, Kolkata, and Chennai anchoring major industrial zones, and textiles—the entry-level industry of disadvantaged countries—dominating the manufacturing scene. Other industries, such as steel, machinery, and building materials, have become much stronger, and some of the companies involved have even become major global players. For example, Mittal Steel (now officially called ArcelorMittal) and Tata Steel rank among the dozen largest steel producers in the world.

These changes were made possible through major policy changes during the 1990s, when India liberalized its economy following four decades of stifling overregulation and inefficiency. Along with many other developing countries, India embraced neoliberalism: trade barriers came down, state companies were privatized, and foreign investment was encouraged. Although it is not yet clear what the impact will be on India's massive working class, there is no doubt that many companies became far more competitive and profitable. This, in turn, increased the government's tax revenues, allowing it to invest in, among other things, badly needed infrastructural projects. Since 2000, India's economy has grown faster than ever before, with much of the growth emanating from its information technology sector.

India's information-technology (IT) industries, clustered in metropolitan Bengaluru (Bangalore), Hyderabad, Mumbai, and New Delhi, draw much international attention, in part because of the large-scale **outsourcing [4]** of U.S. jobs to India. Leading Indian software companies such as Infosys, Wipro, and Cognizant are now household names all across urban India. The growth of software and IT services has been spectacular and now accounts for about 8 percent of GDP and no less than one-quarter of merchandise exports. But this sector employs only around 2 percent of the workforce, with about ten times as many people holding government jobs.

What India needs far more, in terms of employment growth, are manufacturing industries competitively selling goods on world markets, putting tens of millions to work, and transforming the economy. Clearly, India is very different from China. The two may have comparably-sized populations, but the Chinese economy employs 12 times as many workers in manufacturing and it has a much larger urban middle class. India, on the other hand, has shown spectacular growth; but too much of it is confined to IT and benefits only a small, highly educated urban elite.

Urbanization

India is well known for its enormous and teeming cities, but the country is not yet an urbanized society. Only about 31 percent of the population currently lives in cities and towns—compared to an average of roughly 80 percent across the developed world. But whereas rural-to-urban migration has long stabilized at a low level in the West, the rate of urbanization in India is much higher. People by the hundreds of thousands are arriving in the already overcrowded cities, swelling urban India by about

From the Field Notes . . .

"With the help of a real estate agent, I toured some of the new residential developments in India's biggest city, Mumbai. The new urban middle class wants better and more spacious housing, but given the lack of space in the city it has to be a high-rise far away from the center. From a new construction site I looked out onto the northeastern edge of the city, marked by an already completed condominium complex. These homes are of a much better quality than the average dwelling in Mumbai, containing most of the luxuries taken for granted in the United States. Two-bedroom apartments sell for about U.S. $75,000 and up—a huge sum of money even for India's upwardly-mobile, affluent classes and totally beyond reach for the city's masses. Half of Mumbai's 21 million residents live in sub-standard housing, and many in the kind of slums you see in the foreground, nestled up against the coveted middle-class residences. As the new middle class takes off, the contrasts are getting sharper."

© Jan Nijman

www.conceptcaching.com

3 percent annually—twice as fast as the country's overall population growth. Not only do the cities attract as they do everywhere, but many villagers are also driven off the land by desperate conditions in the countryside. As these migrants manage to establish themselves in Mumbai or Kolkata or Chennai, they help their relatives and friends to join them in squatter settlements that often are populated by newcomers from the same locality, bringing their language and customs with them and cushioning the stress of relocation.

As a result, India's cities display staggering social contrasts. Squatter shacks without any amenities crowd cheek-by-jowl against the walls of modern high-rise apartment buildings and condominiums (see photo). Hundreds of thousands of homeless people roam the streets and sleep in parks, under bridges, and even on sidewalks. As crowding intensifies, social stresses multiply.

Figure 8B-3 displays the distribution of major urban centers in India. Except for Delhi–New Delhi (24.8 million), the biggest cities have coastal locations: Kolkata (15.0 million) dominates the east, Mumbai (20.8 million) the west, and Chennai (9.6 million) the south. The overriding influence of these coastal cities is a colonial legacy. But urbanization also has expanded in the interior, notably within the core area. In 2014, even though only 31 percent of its citizens were urbanites, India had more than 50 metropolitan areas containing populations of at least one million.

When you arrive in any Indian city, you are struck by the abundance of small shops everywhere—tiny businesses wedged into every available space in virtually every nonpublic building along every street. Even the upper, walk-up floors are occupied by shops, their advertising signs suspended from windows and balconies. According to a study by the Indian government's Department of Consumer Affairs, India has the highest density of retail outlets of any country in the world: a total in excess of 15 million shops (compared to well below 1 million in the United States, where the marketplace is 13 times richer). After farming, the retail sector is India's largest provider of jobs.

What keeps all these small stores in business? Most of them earn very little and can afford to stay open only because they are part of what economic geographers call the **informal sector [5]**—they are essentially unregistered, pay no rent and probably no taxes, utilize family labor, have been handed down through generations, and keep going because India's economic geography, bound by longstanding protective government regulation, has been slow to change. When you are in India you may wonder how so many shops can survive, but the answer is in the throngs of people on the sidewalks (spilling over into the clogged traffic in the streets). Not many of these people are well off, but all of them need basic goods and some can afford small luxuries, and so the shops tend to be busy all day.

This bustling retail scene notwithstanding, the newest statistics confirm that change is under way, especially in the cities. India's middle class, now estimated to include over 300 million people (equal to the entire U.S. population), is expanding rapidly, and something quite new is beginning to appear in the urban landscape: shopping malls (see photo below). As recently as 2000, this vast country containing more than one-sixth of humankind did not have a single shopping center. But only five years later the 100th had opened its doors, and today more than 1300 are in operation. You will see American fast-food restaurants among the establishments in these malls as well as the brand names of numerous other foreign companies, proving that globalization has already breached India's walls.

From the Field Notes . . .

© Jan Nijman

"As I set foot into this spacious (and partially air-conditioned) mall from the steamy and crowded Andheri Link Road in Mumbai, I thought of the implications—is this the future in a land of small family-owned shops where 'retail' has a different and historic meaning? Infinity Mall, an upscale shopping center opened in 2005 in suburban Mumbai, caters to the megacity's new middle class. My Indian friend Pankaj thought it was all for the better. 'On the weekends my wife and I often come here. There is a nice food court and the shopping is great. These are the things you Americans want and these are the things we Indians want.' Malls are now springing up across India's major cities, transforming urban landscapes as well as consumer behavior."

www.conceptcaching.com

AMONG THE REALM'S GREAT CITIES . . .

KOLKATA (CALCUTTA)

THE NAME "CALCUTTA" seems to invoke all that can go wrong in large cities: poverty, disease, pollution, congestion, and lack of planning. The city was renamed Kolkata in 2001, one in a series of cities in India whose names were "decolonized" and returned to vernacular pronunciation. But that may have done little to change its reputation. It wasn't always so.

Calcutta's heyday began in 1772 when the British made it their colonial capital. The British chose the site, 130 kilometers (80 mi) up the Hooghly River from the Bay of Bengal and less than 9 meters (30 ft) above sea level, not far from some unhealthy marshes but well placed for commerce and defense as well as accessibility to the famous tea plantations in the northeast. When the British East India Company was granted freedom of trade in this populous hinterland, the British sector of the city was drained and raised, and so much wealth accumulated here that Calcutta became known as the "city of palaces." Outside the British town, rich Indian merchants built magnificent mansions.

But even in those times the city knew incredible poverty and deprivation. Beyond the "palaces" lay neighborhoods that were often based on occupational caste, whose names are still on the map today (such as Kumartuli, the potters' district). Almost everywhere, on both banks of the Hooghly, lay the huts and hovels of the poorest of the poor. Searing social contrasts characterized Calcutta and the city was notorious for its lack of hygiene and frequency of epidemics.

The twentieth century was not kind to Calcutta. In 1912, the British moved their colonial capital to New Delhi. The 1947 partition that created Pakistan also created then-East Pakistan (now Bangladesh), cutting off a large part of Calcutta's hinterland and burdening the city with a flood of refugees. The Indian part of the city had arisen virtually without any urban planning, and the influx created almost unimaginable conditions.

Kolkata today counts 15.0 million residents, and the city is in some ways left behind in India's modernization. It was surpassed by Bombay (now Mumbai) in the industrial era, and the current information age belongs to Bengaluru (formerly Bangalore) and Hyderabad. But despite its poverty and urban problems, Kolkata still has a reputation as India's true cultural capital, home of the poet Rabindranath Tagore (who in 1913 became the first non-European Nobel laureate), the well-known 1950s art-film maker Satyajit Ray, and the late great musician Ravi Shankar (legendary maestro of the stringed instrument known as the sitar).

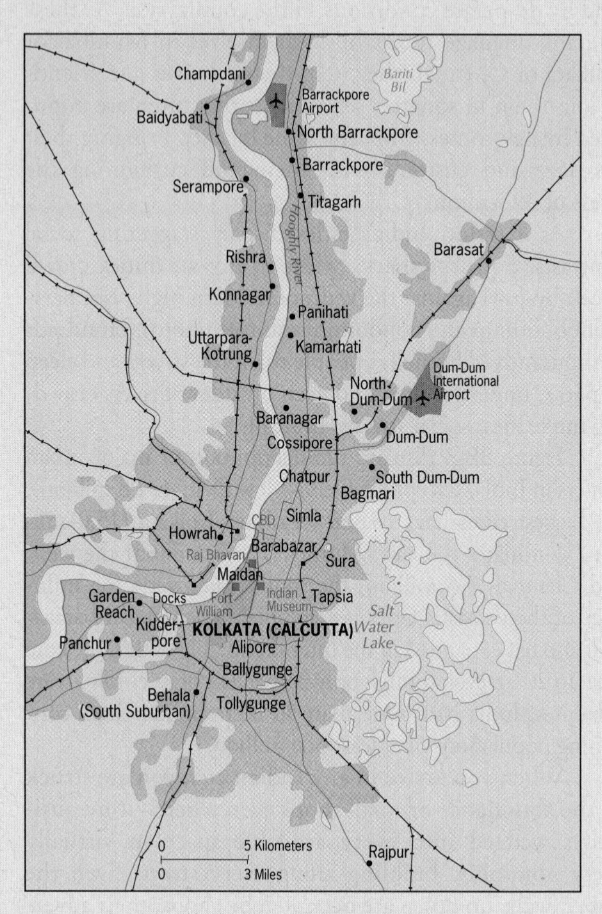

The Indian government forced another breakthrough reform in late 2011: foreign supermarkets such as Walmart were for the first time permitted to enter India, albeit under strict limitations, such as mandatory joint ventures and with access limited to cities of more than one million population. But this move provoked major protests across the country where it is (rightly) perceived as a direct threat to the myriad small-scale, mom-and-pop stores. Yet most economists argue this incursion is for the best because the big chains will bring greater efficiency and lower prices.

Nonetheless, if 300-plus million middle-class Indians can afford to visit a mall, more than 900 million Indians cannot—and the gap widens by the day. Can India's ongoing economic transformation be accomplished without severe social dislocation?

Infrastructural Challenges

India is now in the midst of realizing an unprecedented series of infrastructural projects, ranging from superhighways to state-of-the-art airports. One of these megaprojects is the construction of a nationwide four-lane superhighway—the Golden Quadrilateral—linking the four anchors of the urban system (Delhi, Mumbai, Chennai, and Kolkata), and in the process connecting 15 other major cities along this all-important route (Fig. 8B-3). The impacts of this project are multiple: it is expanding urban hinterlands; commuters are using it to travel farther to work than ever before; several once-remote rural areas now have a link to markets; and it is accelerating the rural-to-urban migration flow that will continue to transform India in the decades ahead.

Improving India's infrastructure, however, is hardly enough to overcome all of the serious impediments to the country's economic advancement. Take another look at Figure 8B-3 and note that the Golden Quadrilateral crosses a number of State boundaries. In the United States, we are used to seeing thousands of trucks on the interstate highways, crossing from one State into another without slowing down; there are truck-weighing stations along these expressways, but in the newest ones all the truck has to do is slow down enough for electronic surveillance to record its passing.

In India, truck drivers face a very different experience. It can take as long as nine days, including more than 30 hours waiting at State-border checkpoints and tollbooths, for a loaded truck to travel from Kolkata to Mumbai via Chennai, or less than the distance from Los Angeles to New Orleans on Interstate–10. Drivers are subject to daunting piles of paperwork as well as incessant demands for bribes, and when mechanical problems occur it may take days to get a truck back on the road. It will take more than expanding India's highway network to improve the overall circulation the country so desperately needs.

The Energy Problem

If you have a friend in (or from) India, you know someone who is familiar with power outages. They are a way of life in India, where electricity demand routinely exceeds the available supply, where governments cannot bring themselves to require customers to pay for the actual cost of the power they consume, and where power grids, generating equipment, and other associated infrastructure are in bad shape. On top of all this, keep in mind that hundreds of millions of villagers still have no electrical supply at all.

And yet electrical power is crucial to India's modernization. Already, some foreign companies doing business in India are importing their own generators, and others are discouraged from investing in factories and other facilities because the power supply is so unreliable. Most of India's electricity is generated in thermal plants burning coal, oil, or natural gas; about 20 percent comes from hydroelectric sources, and about 2 percent from nuclear plants. The problems are many: India has substantial coal deposits, but the railroads cannot handle the transport to power plants. So India must import coal, but the ports do not have the required capacity. And India possesses only limited oil and natural gas reserves. Add to this an increasingly inadequate national power-supply grid in a country with a still-exploding population and even faster-growing demand, and you have trouble.

A major remedy, of course, lies in increased oil and gas imports, but here India runs into geopolitical problems. Although the U.S. government was motivated to give India leeway in the nuclear arena, the United States made it clear that it does not like India's plan to buy Iranian natural gas via a pipeline across Pakistan. India's other options lie in interior Asia, but those sources are more distant and pipeline construction would involve additional diplomatic complications. Once again, the other alternative—importing oil and gas via tankers and ports—is constrained by inadequate infrastructure. Energy is going to remain a problem for some time to come.

India's Prospects

So what lies in store for India? Some economic geographers suggest that India might leapfrog China and move quickly from an "underdeveloped" to a "postindustrial" information-based economy. Certainly India has the requisite intellectual clout. But to reach India's hundreds of millions of potential wage-earners, India also needs a vigorous expansion of its secondary industries, those that make goods (other than textiles) and sell them at home and abroad. Here's how it may happen: when an economy churns along the way China's has over the past three decades, labor and production costs tend to escalate. That causes manufacturers to look for places where cheaper labor will reduce such costs. India, with its long history of local manufacturing, huge domestic market, and vast reservoir of capable labor, would then take its turn on the world stage.

India's prospects are brightening, if not yet in the dramatic terms used in popular-media hype. A middle class 300-million strong demands goods that range from cell phones (in 2012, India had over 900 million subscribers encompassing 73 percent of the population) to motor bikes (more than 30,000 are now sold every day). During the past few years, Chennai has already become a leading Indian automobile center: Korean and German cars roll off its assembly lines not only to be marketed in India but also to be shipped to foreign consumers.

India's economy today ranks as the world's fourth-largest and is expected to move up to third place by 2020. Over the past several years, the Indian economy has grown by an average of 6 percent annually—less than China's, but far ahead of growth rates in most other parts of the world. China's dramatic growth resulted from decisions at the top, a transformation both planned and implemented in controlled detail. In India the economy is growing from the bottom up, accompanied by all the traditional chaos that makes it a country like no other. As with China's provinces, some of India's States will advance ahead of others, and the country's already incredible socioeconomic

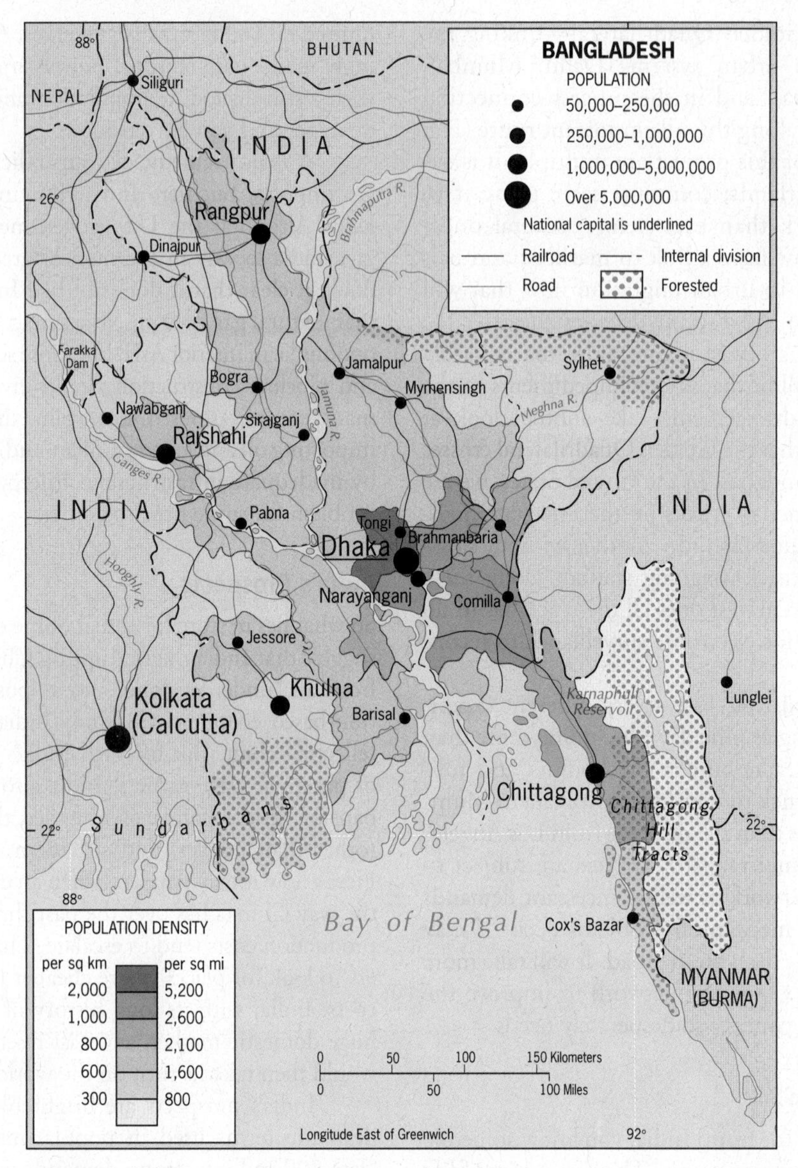

FIGURE 8B-7 © H. J. de Blij, P. O. Muller, and John Wiley & Sons, Inc.

contrasts will intensify. But over time, India could achieve what China hitherto has not: an economic and cultural geography of consensus.

BANGLADESH: CHALLENGES OLD AND NEW

On the map of South Asia, Bangladesh looks like another State of India: the country occupies the area of the **double delta [6]** of India's great Ganges and Brahmaputra rivers, and India almost completely surrounds it on its landward side (Fig. 8B-7). But Bangladesh is an independent country, born in 1971 after its brief war for independence against Pakistan, with a territory about the size of Wisconsin. Today it remains one of the poorest and least developed countries on Earth, with a population of 158.1 million that is growing at an annual rate of 1.6 percent.

Bangladesh remains largely a nation of subsistence farmers. Barely one-fourth of the population lives in urban settlements, and more than half of the workforce is engaged in agriculture. Dhaka, the megacity capital, and the cities of Chittagong, Rangpur, Khulna, and Rajshahi are the only urban centers of consequence. Moreover, Bangladesh has one of the highest **physiologic densities [7]** (people per unit area of arable land) in the world: 1743 people per square kilometer/4514 per square mile.

A Vulnerable Territory

Not only is Bangladesh a poor country; it also is highly susceptible to damage from natural hazards. During the twentieth century, two of the deadliest weather disasters in the world struck this small country. The most recent of these occurred in 1991 when a cyclone (as hurricanes are called in this corner of the world) killed more than 150,000 people. Smaller cyclones strike several times every year and almost routinely kill anywhere between dozens and thousands of people.

AMONG THE REALM'S GREAT CITIES . . .

DHAKA

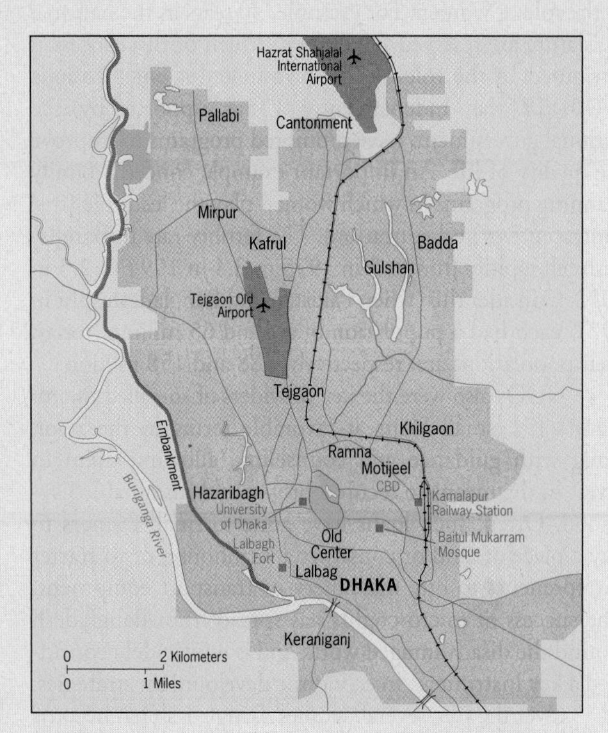

DHAKA (formerly spelled Dacca) is situated in the heart of Bangladesh in the northern sector of the double delta formed by the Ganges and Brahmaputra. It is the economic, political, and cultural focus of the country, and with a population of 16.9 million it is by the far the biggest city.

Known as the "city of mosques and muslin" (a type of cotton), the modern history of Dhaka began in the early seventeenth century when the area came under the rule of the Islamic Mughals, along with what is today northern India. Dhaka occupied an advantageous position on the main waterways, and to this day the local Buriganga River is at the center of the city's activities. The oldest section of Dhaka lines the north bank of the waterfront. The British turned the place into a provincial capital of sorts, and their legacy can still be seen in some of the colonial architecture near the old center alongside the remaining Mughal structures. At independence in 1947 the city became the administrative capital of East Pakistan, and in 1971 it became the seat of government for the newly formed state of Bangladesh.

As part of the great double delta, Dhaka lies close to sea level and is prone to flooding during the summer monsoon. Every so often, major cyclones barrel in from the Bay of Bengal and cause massive human and material devastation. The greatest disaster in modern times was triggered by Cyclone Bhola in 1970 that killed as many as 500,000 of the country's citizens, destroyed hundreds of thousands of homes, and flooded most of the city.

Dhaka has grown rapidly in recent decades as it attracted large numbers of migrants from rural areas, and the city gradually expanded northward. At the time of partition, many Hindus left and Muslims arrived, and now more than 90 percent of the population adheres to Islam. Religious fanaticism is not as widespread as it is in Pakistan. The main schism here is between the tiny rich minority and the enormous poverty-stricken majority. Dhaka is also known as the bicycle-riksha capital of the world: some 400,000 colorfully painted rikshas criss-cross the thoroughfares of this city on a daily basis—surely the cheapest mode of urban transportation on the planet.

The reasons for Bangladesh's vulnerability can be deduced from Figures 8B-7 and 8A-1. Southern Bangladesh consists of the deltaic plain of the combined Ganges–Brahmaputra river system, integrating extremely fertile alluvial (river-deposited silt) soils that attract farmers with the low elevations that endanger them when the water rises. Moreover, the shape of the Bay of Bengal forms a funnel that steers cyclones and their storm surges of wind-driven water to barrel into the double delta's coast. Without money to build seawalls, floodgates, elevated shelters in sufficient numbers, or adequate escape routes, hundreds of thousands of people are at continuous risk, with deadly consequences.

The country's relations with neighboring India have at times been strained over water resources (India controls the Ganges River, which is the lifeline of Bangladesh), cross-border migration (8 percent of the population is Hindu), and transit between parts of India across Bangladesh's north (refer to Fig. 8B-5 to see the reason).

Limited Opportunities—Creative Development Strategies

Even though geography seems to offer Bangladesh limited options and the country overall remains very poor, it should be noted that in recent years this country has achieved some remarkable successes. Since 1990, life expectancy in Bangladesh increased by ten years and now exceeds that of India, despite Indian incomes being considerably higher. Second, primary school enrollment among girls today is double that of 2010 and the female literacy rate is steadily improving. And third, since 1990 infant (<1 year) mortality rates have been halved and child (<5 years) mortality rates as well as maternal mortality rates have dropped by three-fourths. According to these indicators, Bangladesh outperforms India and also fares much better than Pakistan.

Interestingly, these improvements were achieved in the absence of notable economic growth. Indeed, incomes in Bangladesh are considerably lower than in India and even Pakistan does better in this regard. Bangladesh is an overwhelmingly Muslim society but in some ways it has also been a very progressive one, particularly when it comes to the role of women. For example, 30 seats in the national legislature are reserved for women. Much of this success is attributed to the role of **non-governmental organizations (NGOs) [8]** that, independently of (but supported by) the national government, have promoted programs to improve the quality of life. An important example concerns family planning programs in which women play the lead role (free contraceptives and education). The fertility rate in Bangladesh fell rapidly from 6.5 in 1975 to 3.4 in 1994 to 2.3 in 2012. Consider this: when Pakistan and Bangladesh split in 1971, each had a population of around 65 million; today, their populations are, respectively, 188 and 158 million.

NGOs also were the key providers of so-called **micro-credits [9]**, small loans at favorable terms to the poor, along with guidance and counseling, allowing them to invest in the means to secure a proper and sustainable livelihood. Often, such loans have gone to small farmers to buy a piece of land or construct a farmhouse, or to starter entrepreneurs to buy machinery or transport equipment. The success of micro-credits has spread from Bangladesh around the disadvantaged world and is now widely considered a key instrument in economic development strategies.

Over the past several decades, Bangladesh has become known as the "textile capital of the world." Many of the cheapest clothing items you see sold under popular labels from big-name marketers are made in factories in Bangladesh that are unsafe and where the minimum wage is little more than a U.S. dollar a day. Western companies impose strict production quotas and deadlines, and factory owners and managers force their laborers to work long hours under often dangerous conditions. There are no unions to protect workers; terrible fires and building failures are not uncommon. In April 2013, a factory building in the Dhaka suburb of Savar known to be unsafe collapsed, killing more than 1,000; terrified workers had alerted police and other officials about widening cracks in the structure but were ordered to stay on the job. Globalization may have made Bangladesh's garment industry a valuable contributor to the nation's commercial economy, but those low prices you see advertised in the Western media come at a high cost to the poor and the powerless in the countries of the global periphery.

■ THE MOUNTAINOUS NORTH

As Figure 8B-1 shows, a tier of landlocked countries and territories lies across the mountainous zone that walls India off from China. One of them, Kashmir, is in a condition of near-war (see Chapter 8A). Another, Sikkim, was absorbed by India in 1975 and made into one of its federal States. But Nepal and Bhutan retain their independence.

Nepal

A **buffer state [10]** between India and China, Nepal exhibits an internal politics that is increasingly a reflection of

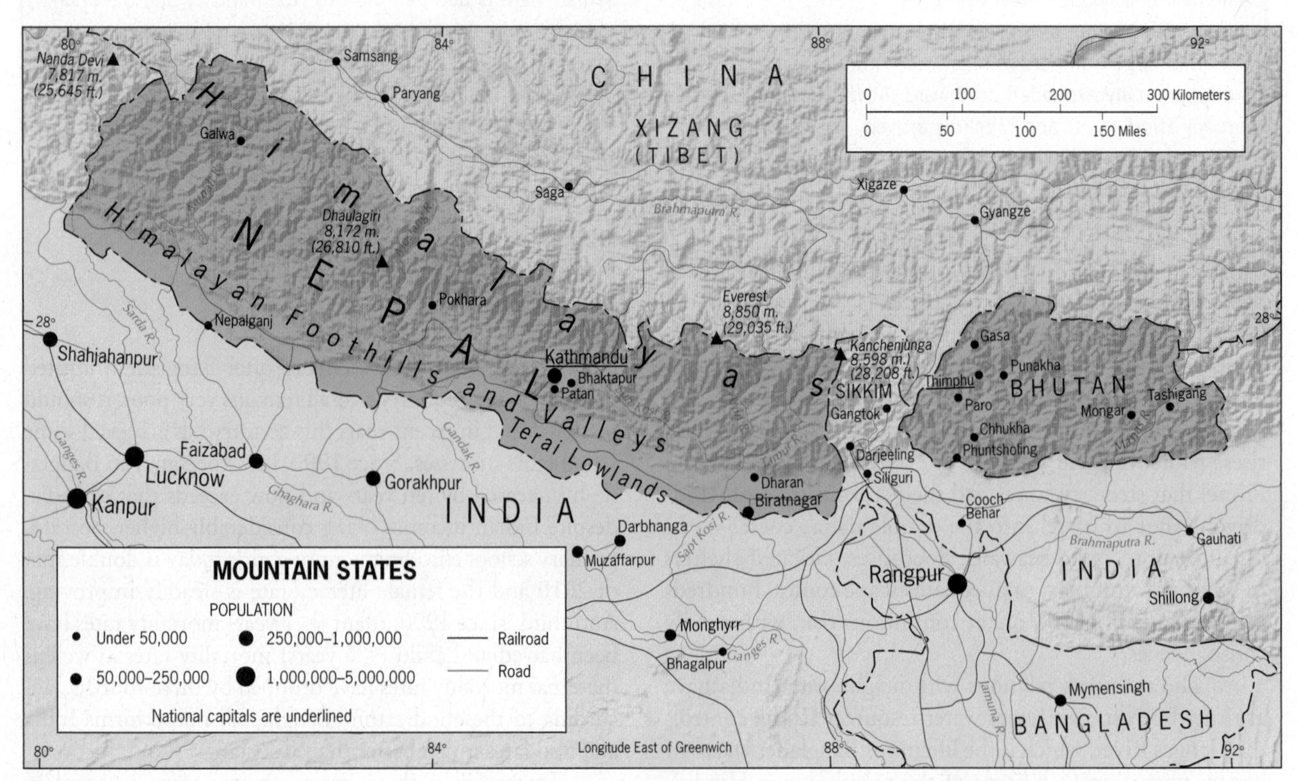

FIGURE 8B-8

the competitive influences of these two giant neighbors. Culturally and, until recently, economically, Nepal lies within India's sphere of influence, but China is making its presence felt ever more strongly. Maoist groups are supported by China but loathed by India, resulting in fractious and contentious governance. China, as it does just about everywhere in the world nowadays, is using its economic clout as leverage to gain heightened political influence.

Nepal, located just northeast of India's Hindu core, has a population of 32 million and is the size of Illinois. It has three geographic zones (Fig. 8B-8): a southern, subtropical, fertile lowland called the Terai; a central belt of Himalayan foothills with swiftly flowing streams and deep canyons; and the spectacular high Himalaya itself (topped by Mount Everest) in the north. The capital, Kathmandu, lies in the east-central part of the country in an open valley of the central hill zone.

Nepal, birthplace of Buddha, is materially poor but culturally rich. The Nepalese are a people of many origins, including India, Tibet, and interior Asia. About 82 percent are Hindu, and Hinduism is the country's official religion; but Nepal's Hinduism is a unique blend of Hindu and Buddhist ideals. Thousands of temples and pagodas, ranging from the simple to the ornate, grace the cultural landscape, especially in the Vale of Kathmandu, the country's core area. Although well over a dozen languages are spoken, 85 percent of the people also speak Nepali, a language related to Indian Hindi.

This is a troubled country that suffers from severe underdevelopment, with the lowest GNI in the realm. It is the only country in the realm with a declining income per capita over the past two decades. It also faces strong centrifugal social and political forces. Environmental degradation, crowded farmlands, and soil erosion, as well as deforestation, scar the highly corrugated countryside. The soaring Himalayan peaks form a world-renowned tourist attraction, but expenditures by tourists in Nepal, always relatively modest, have been cut back because of the recurrent disorder associated with a Maoist-communist insurrection and the collapse of the country's monarchy.

Nepal's cultural and political geographies have long been in turmoil. Strong regionalism divides the country both north-south and east-west. The southern Terai with its tropical lowlands is a world apart from the foothills of the Himalaya in the central interior, and the peoples of the west have origins and traditions different from those in the east. An absolute (Hindu) monarchy held Nepal together until 1990, when antigovernment demonstrations and resulting casualties persuaded the king to lift the ban on political parties. But chaos soon resumed, the king was assassinated, and by 2002 Maoist rebels controlled nearly half of the country, turning Nepal into a **failed state [11]**. The ruling king (the brother of the former king) was effectively forced from his throne in 2006, the insurgents were brought into the political system, and in the following year a new constitution formalized the abolition of the monarchy and the reinvention of Nepal as a secular federal republic. Peaceful elections in 2008 confirmed the capacity of Nepal's new system to accommodate the still-powerful centrifugal forces in the country, however it remains to be seen if the new government can deliver on the demands of the country's highly diverse communities and constituencies.

Bhutan

Mountainous Bhutan, with a population of about 750,000, lies wedged between India and China's Tibet (Fig. 8B-8), the only other buffer between Asia's giants. In landlocked, fortress-like Bhutan, time seems to have stood still. Bhutan long was a constitutional monarchy ruled by a king whose absolute power was unquestioned by his subjects. But in 2007 the reigning king, who had recently succeeded his father, and perhaps with an eye on what had happened to the monarchy in nearby Nepal, decided to order his subjects to vote for a political party in a newly created democracy. Thus in 2008 Bhutan went from an absolute monarchy to a multi-party democracy on the orders of its monarch, and Thimphu, the capital, became the seat of a new national assembly.

In the mountainous countryside, the symbols of Buddhism, the state religion, dominate the cultural landscape, and the government's development policies stress the importance of spiritual fulfillment alongside the satisfaction of material needs. Still, there is social tension driven by the presence of a Nepalese (Hindu) minority, and a Bhutanese refugee population remains housed in camps in eastern Nepal. Add to this a still-unresolved boundary issue with China and the possibility of rising material expectations, and newly democratic Bhutan faces some pressing challenges.

Regarding economic geography, forestry, hydroelectric power, and tourism all have much potential here, and Bhutan possesses considerable mineral resources. But isolation and inaccessibility continue to preserve traditional ways of life in this mountainous nation.

THE SOUTHERN ISLANDS

As Figure 8B-1 shows, South Asia's subcontinental landmass is flanked by several sets of islands: Sri Lanka just off the southern tip of India, the Maldives in the Indian Ocean to the southwest, and the Andaman Islands (which belong to India) marking the eastern edge of the Bay of Bengal.

The Maldives

The Maldives consists of more than 1000 tiny islands whose combined area is just 300 square kilometers (115 sq mi) and whose highest elevation is less than 2 meters (6 ft) above sea level. Its population of 340,000 from Dravidian and Sri Lankan sources is now virtually 100 percent Muslim, with one-quarter of it concentrated on the island of the capital named Maale. The Maldives might be unremarkable, except that, as the table in Appendix B shows, this country has the realm's highest GNI per capita. The locals have translated

VOICE FROM THE Region

© Jan Nijman

Ahmed Rasheed, 28 years of Age, The Maldives

"I live with my wife and 1-year old daughter in Hithadhoo, one of the southernmost islands of the Maldives. I was born in the north in Maale, the capital, and went to school there through the 10th grade; that is how far education goes for most of us. I moved down here to work at one of the nearby resorts as a water sports instructor, mainly surfing, boating, and parasailing. Every morning I take the ferry to the resort and in the evening I return home. I enjoy my work and I love the islands, the water. The rest of the world looks at the Maldives like a paradise and it is my home! There is much talk about how global warming and rising sea levels will affect us, about the vulnerability of the coral reefs, and other environmental issues. I sometimes hear the fishermen say that the catch is not as good as it used to be. For me personally it is actually hard to see the impact; perhaps these things just go so slowly.... What I read in the papers or see on television scares me more than what I see with my own eyes. There is even talk of all of us being forced to move sometime in the future. I have never been out of my country and cannot imagine having to leave. I don't think it will happen in my lifetime and I hope my daughter will live all of her life here too."

their palm-studded, beach-fringed islands into a tourist mecca that attracts tens of thousands of mainly European visitors annually. The tourism sector, accounting for nearly one-third of national income, took a hit during the recent global recession; nonetheless, by 2011 national economic growth had rebounded to a level near 4 percent.

The Maldives' low elevation (80 percent of the country lies less than one meter [3 ft] above sea level) has been singled out repeatedly in assessments of the future impact of **global warming [12]**, which will likely result in rising sea levels. Scientific research indicates sea levels worldwide have been rising at a rate of 3.5 millimeters (0.14 in) per year since the early 1990s. Those fears became sudden reality on December 26, 2004 when the devastating Indian Ocean tsunami generated near Indonesia (discussed in Chapters 10A and 10B) swept over the islands, killing more than 100 residents as well as tourists and destroying resort facilities along the shore as well as inland.

This country faces social and political challenges, too. Its nascent democracy was dealt a blow in 2012 when the first democratically elected President, Mohamed Nasheed, was deposed in a *coup d'état* involving the country's former dictator and radical Islamists.

Sri Lanka: Paradise Lost and Regained?

Sri Lanka (known as Ceylon before 1972), the compact, pear-shaped island located just 35 kilometers (22 mi) across the Palk Strait from southernmost India, became independent from Britain in 1948 (Fig. 8B-9). There were good reasons to create a separate sovereignty for Sri Lanka. This is neither a Hindu nor a Muslim country: the large majority of its population of 21.7 million people—some 70 percent—are Buddhists. Furthermore, unlike Pakistan or India, plantations dominate Sri Lanka, and commercial farming still is the mainstay of the agricultural economy.

The great majority of Sri Lanka's population is descended from migrants who came to this island from northwestern India beginning about 2500 years ago. Most of them probably walked there, as the island was connected to southeast India with a land bridge (now submerged as Palk Strait) that was also known as *Rama's Bridge*. Look closely at Figure 8B-9 and you can easily see the Sri Lanka-end remains of that shallow bridge, part of which was destroyed by a cyclone in the fifteenth century. The migrants brought with them the advanced culture of their source area,

Sinhalese Buddhist monks visit the famous ruins of the twelfth century Hatadag Temple in the ancient city of Polonnaruwa in north-central Sri Lanka. The temple that once stood here was also known as the "tooth relic temple" because it was said to have treasured an actual tooth of the Buddha.

© Jan Nijman

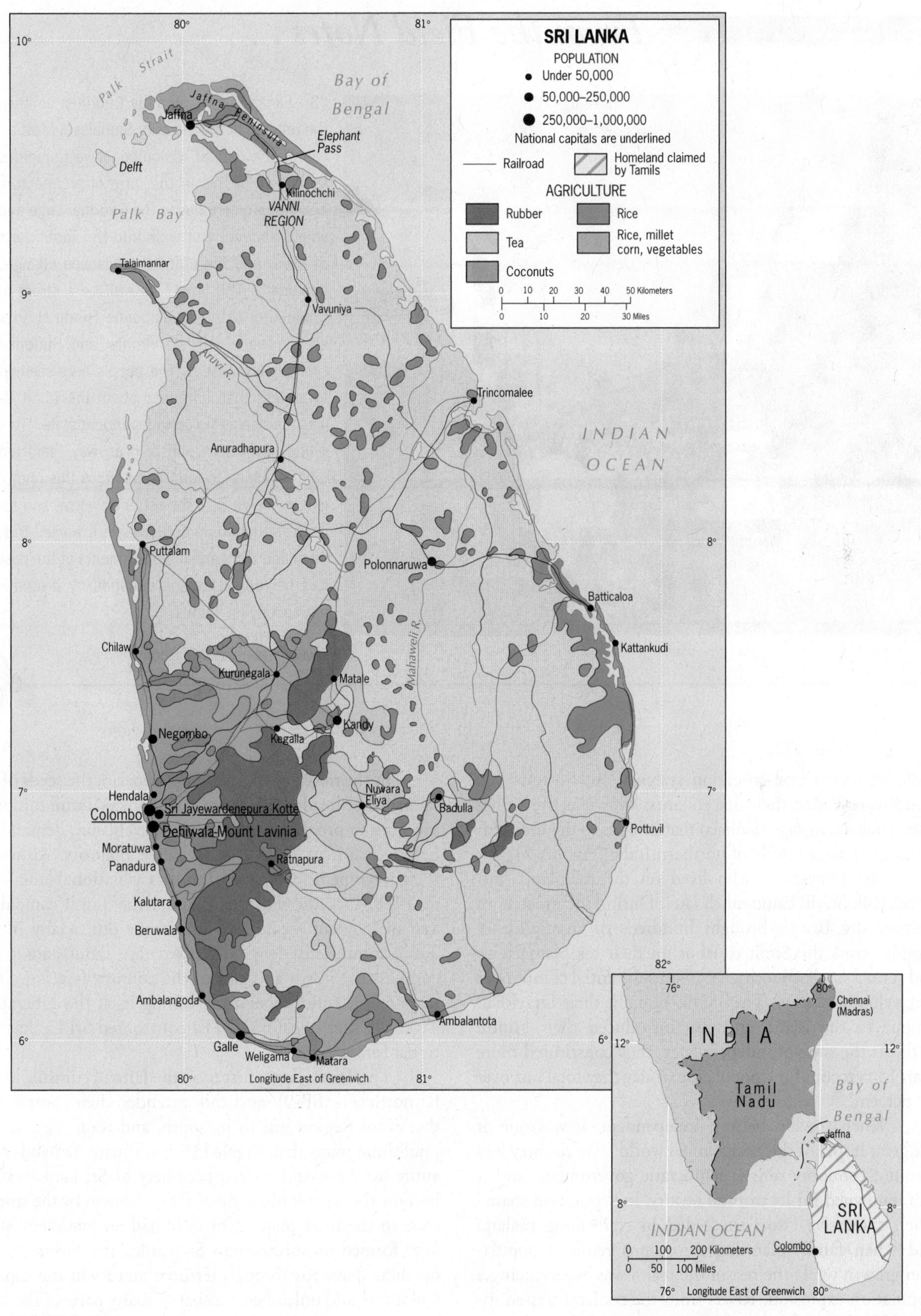
SRI LANKA
POPULATION
Under 50,000
50,000–250,000
250,000–1,000,000
National capitals are underlined
Railroad
Homeland claimed by Tamils
AGRICULTURE
Rubber
Rice
Tea
Rice, millet corn, vegetables
Coconuts
0 10 20 30 40 50 Kilometers
0 10 20 30 Miles
Palk Strait
Bay of Bengal
Jaffna
Jaffna Peninsula
Elephant Pass
Delft
Kilinochchi
VANNI REGION
Palk Bay
Talaimannar
Vavuniya
Aruvi R.
Trincomalee
INDIAN OCEAN
Anuradhapura
Puttalam
Polonnaruwa
Batticaloa
Mahaweli R.
Chilaw
Kattankudi
Kurunegala
Matale
Kandy
Negombo
Kegalla
Nuwara Eliya
Hendala
Badulla
Colombo
Sri Jayewardenepura Kotte
Dehiwala-Mount Lavinia
Pottuvil
Moratuwa
Panadura
Ratnapura
Kalutara
Beruwala
Ambalangoda
Ambalantota
Galle
Weligama
Matara
Longitude East of Greenwich
INDIA
Chennai (Madras)
Tamil Nadu
Bay of Bengal
Jaffna
SRI LANKA
INDIAN OCEAN
Colombo
0 100 200 Kilometers
0 50 100 Miles

FIGURE 8B-9

From the Field Notes . . .

© H.J. de Blij

"Sri Lanka is a dominantly Buddhist country, the religion of the majority Sinhalese. Most areas of the capital, Colombo, have numerous reminders of this in the form of architecture and statuary: shrines to the Buddha, large and small, abound. But walk into the Tamil parts of town, and the cultural landscape changes drastically. This might as well be a street in Chennai or Madurai: elaborate Hindu shrines vie for space with storefronts, and Buddhist symbols are absent. The people here seemed to be less than enthused about the Tamil Tigers' campaign for an independent state. 'This would never be a part of it anyway,' said the fellow walking toward me as I took this photograph. 'We're here for better or worse, and for us the situation up north makes it worse.' But, he added, Sri Lankan governments of the past had helped create the situation by discriminating against Tamils."

www.conceptcaching.com

Concept Caching

building towns and irrigation systems, and introducing Buddhism. Today, their descendants, known as the Sinhalese, speak a language (Sinhala) that belongs to the Indo-European language family of northern India (Fig. 8A-5).

The Dravidians who lived on the mainland, just across Palk Strait, came much later. During the nineteenth century the British brought hundreds of thousands of Tamils across the Strait to labor on their tea plantations, and soon a small minority became a substantial component of Ceylonese society. The Tamils brought their Dravidian tongue to the island and also introduced their Hindu faith. At the time of independence, they constituted more than 15 percent of the population; today they total just over 11 percent.

When Ceylon became independent, it was one of the great hopes of the postcolonial world. The country had a sound economy plus a democratic government, and it was renowned for its tropical beauty. Its reputation soared when a massive effort succeeded in eradicating malaria and when family-planning campaigns reduced population growth while the rest of the realm was experiencing a population explosion. Rivers from the cooler, forested interior highlands fed the paddies that provided ample rice; crops from the moist southwest paid most of the bills, and the capital, Colombo, grew to reflect the optimism that prevailed.

In the midst of this glowing scenario, the seeds of disaster were already being sown. Sri Lanka's Tamil minority soon began proclaiming its sense of exclusion, demanding better treatment from the Sinhalese majority. Although the government recognized Tamil as a "national language" in 1978, sporadic violence marked the Tamil campaign, and in 1983 full-scale civil war broke out. Many in the Tamil community demanded a separate Tamil state to encompass the north and east of the country (see Fig. 8B-9 inset map), and a rebel army calling itself the Liberation Tigers of Tamil Eelam (LTTE) confronted Sri Lanka's national forces.

The Tamil Tigers first took the Jaffna Peninsula in the far north (Fig. 8B-9), next they extended their control over the Vanni Region just to its south, and soon they started publishing maps that revealed their ultimate demand—the entire northern and eastern periphery of Sri Lanka would become the secessionist state of *Eelam* (shown by the striped zone in the inset map). Not only had an **insurgent state [13]** formed in northeastern Sri Lanka: the Tigers backed up these demands through terrorist attacks in the capital, Colombo, and unleashed combat in many parts of the state they were seeking to establish.

From 2002 until 2007, the government and the LTTE engaged in a repeatedly interrupted cease-fire as well as what appeared to be half-hearted negotiations about the country's possible partitioning. But in the end the Colombo government decided on a powerful, iron-fisted offensive that broke the Tamil resistance. On January 2, 2009, President Mahinda Rajapaksa announced the capture of the strategic town of Kilinochchi and proclaimed that the threat of partition in Sri Lanka had ended. An international outcry over the tactics of the Sri Lankan military was followed by expressions of concern over the future of Sinhalese-Tamil relations in this obviously still-fractured country. The United Nations in 2011 concluded that there had been major human rights violations by both sides. Today, the Rajapaksa government continues to be accused by the opposition of nepotism and a turn to dictatorship. The early years of this decade have seen peace but reconciliation is difficult. Sri Lanka's promise, based in natural endowment and geography, could still be fulfilled but political challenges must first be overcome.

POINTS TO PONDER

- Pakistan is crucially important to the stability of both Southwest and South Asia, yet it is becoming increasingly difficult to govern and faces growing religious fanaticism and political turmoil.
- India's IT sector accounts for one-quarter of all exports, but employs only 2 percent of the national workforce.
- If India proceeds with plans to reserve 33 percent of its Parliament seats for women, it would have more than two and a half times the number of female legislators than the 113th (2013–2015) U.S. Congress.
- Nepal, struggling to emerge from failed-state status, is the only country in South Asia in which per capita incomes have declined in recent years.

Novokuznetsk
KAZAKH UPLANDS
Qaraghandy
KAZAKHSTAN
Semey
Rubtsovsk
Irtysh
Öskemen
SAYAN MOUNTAINS
Kyzyl
Yenisey
Mt. Belukha (4,506 m, 14,783 ft.)
Mt. Munku Sardyk (3,491 m, 11,453 ft.)
Angarsk
Irkutsk
Angara
Ulaangom
Ölgiy
Lake Balqash
SARYESIK-ATYRAU DESERT
MOYYNQUM
Syr Darya
Altay
Dund-Us
ALTAY MOUNTAINS
HANGAYN MTS.
Uliastay
Erdenet
Darhan
Ulaanbaatar
MONGOL
TASHKENT
Bishkek
KYRGYZ RANGE
ALMATY
Karamay
JUNGGAR BASIN
UZBEKISTAN
KYRGYZSTAN
Ysyk Köl
Yining
Shihezi
Changji
ÜRÜMQI
Ulaan-Uul
Arvayheer
MONGOLIAN PLATEAU
TIAN SHAN
ALAY MTS.
TAJIKISTAN
Baicheng
Aksu
Kuqa
Yanqi
Korla
Turpan
Turpan Depression (-154 m, -505 ft.)
Hami
Dalandzadgad
Ismoil Somoni Peak (7,495 m, 24,590 ft.)
Kashi (Kashgar)
TARIM BASIN
TAKLA MAKAN DESERT
Shache
Yecheng
Pamir
AFGHANISTAN
HINDU KUSH
KARAKORAM RANGE
K2 (8,611 m, 28,250 ft.)
Hotan
Qiemo
Ruoqiang
Lop Nur
Dunhuang
Ejin Qi
Yumen
ALTUN SHAN
KUNLUN MOUNTAINS
QILIAN SHAN
Khyber Pass
PAKISTAN
Indus
KASHMIR
Islamabad
RAWALPINDI
Lenghu
Zhangye
Wuda
Shizuishan
Da Qaidam
QAIDAM BASIN
Jinchang
Wuwei
Qingtongxia
Yinchuan
Wuzhong
LAHORE
FAISALABAD
LUDHIANA
Chandigarh
Wujiang
Tianjun
Qinghai Hu
Golmud
Dulan
Qabqa
Xining
Nangdo
LANZHOU
Guyuan
Huang He
QINGHAI-XIZANG PLATEAU
Shiquanhe
Qagcaka
CHINA
Tianshui
Wei
QIN L
DELHI
New Delhi
MEERUT
Ghaziabad
FARIDABAD
Barga
Wenquan
Nyima
TANGGULA RANGE
Gyêgu
Wudu
JAIPUR
AGRA
Zhongba
Nagqu
Salween
Riwoqê
Mekong
Guangyuan
NYAINQÊNTANGLHA SHAN
HIMALAYAS
NEPAL
Mianyang
LUCKNOW
KANPUR
Brahmaputra
Xigazê
Gyangzê
Lhasa
Nyingchi
CHENGDU
Qianning
SICHUAN (RED) BASIN
Nanchong
Kathmandu
Mt. Everest (8,850 m, 29,035 ft.)
Markâm
DAXUE MOUNTAINS
Yaan
GANGES PLAIN
Yamuna
Ganges
Thimphu
BHUTAN
FULIN
Neijiang
Zigong
Chang Jiang
Yibin
Prayagraj (Allahabad)
VARANASI
PATNA
Xichang
BHOPAL
VINDHYA RANGE
Narmada
SAPTURA RANGE
CHOTA-NAGPUR PLATEAU
Tropic of Cancer
Guwahati
KHASI HILLS
NAGA HILLS
Zhaotong
Zunyi
Panzhihua
Liupanshui
Anshun
GUIYANG
INDIA
Rajshahi
BANGLADESH
DHAKA
Myitkyina
Dali
Qujing
YUNNAN PLATEAU
NAGPUR
KOLKATA (CALCUTTA)
KHULNA
Baoshan
KUNMING
Yuxi
CHITTAGONG
MYANMAR (BURMA)
Gejiu
Mouths of the Ganges
Bhubaneshwar
ARAKAN MTNS.
MANDALAY
Irrawaddy
Nanning
Qinzhou
VIETNAM
HANOI
Haiphong
Bay of Bengal
Vishakhapatnam
Ramree Island
Cheduba Island
Nay Pyi Daw
Salween
LAOS
Gulf of Tonkin
Chiang Mai

Sakhalin Island (RUSSIA)
Sea of Okhotsk
Kurile Islands (RUSSIA)
Tatar Strait
SIKHOTE-ALIN RANGE
GREATER KHINGAN RANGE
NORTHEAST CHINA PLAIN
YABLONOVYY RANGE
NORTH KOREA
SOUTH KOREA
PYONGYANG
SEOUL
BEIJING
TOKYO
East Sea (Sea of Japan)
Yellow Sea
East China Sea
Korea Bay
Bo Hai Gulf
Hokkaido
Honshu
Shikoku
Kyushu
JAPAN
PACIFIC OCEAN
Philippine Sea
Ryukyu Islands (JAPAN)
Nampo Shoto (JAPAN)
Izu Islands
Bonin Is.
Volcano Is.
TAIWAN
TAIPEI
Taiwan Strait
Luzon Strait
South China Sea
Babuyan Islands (PHILIPPINES)
Hainan
XIANGGANG (HONG KONG)
SHANGHAI
NORTH CHINA PLAIN
DABIE SHAN
WUYI SHAN
LULIANG SHAN
Tsushima (JAPAN)
Korea Strait
Cheju I. (S. KOREA)
(3,776 m, 12,388 ft.) Mt. Fuji
Tropic of Cancer
Elevation (m)
3000
1500
600
300
150
0
below sea level
0
-150
-1500
-3000
Roads
Railroads
0 km 200 400 600 800 1000 1200
0 miles 200 400 600 800
Albers Equal-Area Projection
Scale 1:20,000,000

IN THIS CHAPTER

- East Asia: The world's economic powerhouse
- China's massive water diversion projects
- Geopolitics in the seas adjoining China
- North Korean provocations
- The Taiwan conundrum

FIGURE 9A-1

CONCEPTS, IDEAS, AND TERMS

Tsunami	1
State formation	2
Neolithic	3
Dynasty	4
Confucianism	5
Sinicization	6
Hanification	7
Extraterritoriality	8
Isolationism	9
Modernization	10
Westernization	11
Asian Tigers	12
High-value-added goods	13

East Asia is a geographic realm like no other. At its heart lies the world's most populous country, the product of what may be the world's oldest continuous civilization. On its Pacific mainland shores an economic transformation has taken shape with no parallel in world history. Its offshore islands witnessed the first use of atomic weapons on civilian populations and the postwar emergence of one of the world's most powerful economies. Few lives in this world were left unaffected, directly or indirectly, by the momentous events that occurred in East Asia over the past two generations. Just look around you. Chinese-made televisions (and so much more), Japanese-designed automobiles, South Korean smartphones, Taiwanese computers—from toys to textiles and from hardware to software—East Asian products fill streets and stores, homes and hotels.

DEFINING THE REALM

China is the world's biggest exporter and home to three of the five busiest container ports in the world: Shanghai, Shenzhen, and Hong Kong (shown here). The other two—Singapore and South Korea's Busan—are also located in the Asian Pacific Rim.

© Jan Nijman

It has all happened with astonishing speed. Some of us can remember the time when you were no more likely to find anything useful made in China than you were to buy anything from Russia. But after end of the 1960s, things changed rapidly. Japan led the way, turning its World War II defeat into postwar economic triumph. By the mid-1970s, Japan's economic growth—compared to China's seemingly total stagnation—appeared to justify Japan's recognition as a discrete geographic realm, an economic engine for the world

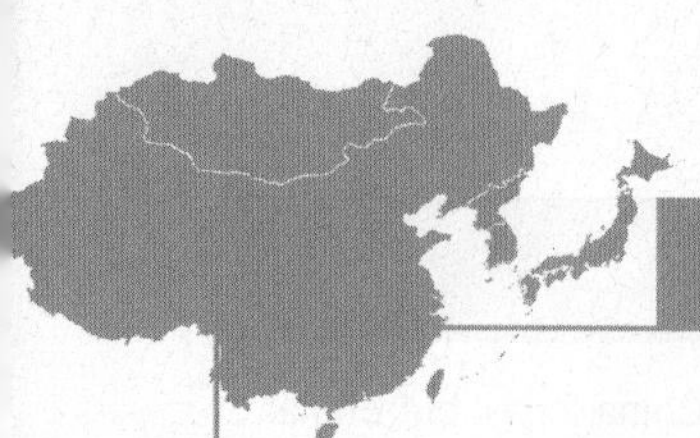

major geographic qualities of EAST ASIA

1. East Asia is encircled by snowcapped mountains, vast deserts, cold climates, and Pacific waters.
2. East Asia was one of the world's earliest culture hearths, and China is one of the world's oldest continuous civilizations.
3. East Asia is the second most populous geographic realm after South Asia; its population remains heavily concentrated in its eastern regions.
4. China, the world's largest state demographically, is the current rendition of an empire that has expanded and contracted, fragmented and unified many times during its long existence.
5. China's sparsely peopled western regions are strategically important to the state, but they lie exposed to minority pressures and Islamic influences.
6. Along China's east coast, an economic transformation launched more than thirty-five years ago is now rapidly expanding westward.
7. Increasing regional disparities and fast-changing cultural landscapes are straining East Asian societies.
8. Japan, one of the economic giants of the East Asian realm, has a history of colonial expansion and wartime conduct that still affects international relations here.
9. East Asia is home to the world's newest superpower as China's economic and political influence is increasingly projected around the globe.
10. The political geography of East Asia contains a number of flashpoints capable of generating conflict, including North Korea, Taiwan, and several island groups in the seas adjoining the realm.

and a strong competitor for the top-ranked U.S. economy. But then Hong Kong, South Korea, and autonomous Taiwan began to show the world what other peoples in East Asia were capable of. When the ruler of communist China, the realm's giant, opened the door to an American president in 1972, a pivotal moment in global history (and geography) had arrived. Before the end of the century, China had taken its place as modern East Asia's core. That event in 1972 signified a new direction in Chinese political thinking, and it would facilitate major revisions in economic policy. Today, East Asia is the most dynamic realm in the global economy: led by China, it is changing the world—and in the process is itself undergoing a profound transformation.

THE GEOGRAPHIC PANORAMA

As Figure 9A-1 shows, the East Asian geographic realm forms a roughly triangular wedge between the vast expanses of eastern Russia to the north and the populous countries of South and Southeast Asia to the south, its edges often marked by high mountain ranges or remote deserts. The darker brown on the map designates the highest mountains and plateaus, which create a vast arc north of the Himalayas before bending southward and becoming lower (tan shading) toward Myanmar, Laos, and Vietnam in Southeast Asia. Here in the southwest, where Tibet is located, mountains and plateaus alike are covered by permanent ice and snow, the soaring ranges crumpled up like the folds of an accordion. Three major rivers, their valleys parallel for hundreds of kilometers, disclose the orientation of this high-relief topography. Northward, note how rapidly the mountains give way to broad, flat deserts whose names appear prominently: the Takla Makan in the far west, the Gobi where China meets Mongolia, the Ordos in the embrace of what looks like a huge meander of a river we will learn more about shortly, the Huang He (Yellow River).

One basin, lodged between high mountains to its west and lower ones elsewhere, and of special interest, is China's Sichuan (Red) Basin. It looks rather small on the map, but is home to more than 120 million people and, we will discover, is important for other reasons as well. In this country of mountains and deserts, living space is at a premium. And speaking of living space, the green areas on the map, which have the lowest relief and (often) the most fertile soils, are home to the vast majority of this realm's population. Here the great rivers that come from the melting ice and snow in the interior highlands have been depositing their sediment load for eons, and when humans domesticated plants and started to grow crops, this was the place to be. That was thousands of years ago—perhaps as long as 10,000 years—and ever since, this has been the largest human cluster on Earth.

But the East Asian realm is not confined to the mainland of mountains and river basins. You can imagine how its offshore islands were populated: the Korean Peninsula seems to form part of a bridge pointing toward the southernmost island of what is today Japan, and from there it seems likely that the early migrants moved farther

Places and Names

AS IN OTHER parts of the world, European colonists in China wrote down the names of places and people as they heard them—and often got them wrong. The Wade-Giles system put such place names as Peking, Canton, and Tientsin on the map, but that's not how the Chinese knew them. In 1958, the communist regime replaced the foreign version with the ***pinyin*** system, based on the pronunciation of Chinese characters in Northern Mandarin, which was to become the standard form of the Chinese language throughout China. The world now became familiar with these same three cities written as Beijing, Guangzhou, and Tianjin. Among prominent Chinese geographic names, only one has not gained universal recognition: Tibet, called Xizang by China but referred to by its old name on many maps for reasons that will become obvious in Chapter 9B.

Pinyin usage may be standard in China today, but China remains a country of many languages and dialects. Not only do minorities speak many different languages, but Mandarin-speakers use numerous dialects—so that when you sit down with some Chinese friends who have just met, it may take just one sentence for one to say to the other "ah, you're from Shanghai!" Migration is changing this, of course. China's population is on the move, and a growing number of residents of Shanghai today were not born in that city. But China's language map is anything but simple. And not just place names, but personal names, too, were revised under pinyin rules. The communist leader who used to be called Mao Tse-tung in the Western media became Mao Zedong. Also remember that the Chinese write their last name first: President Xi Jinping is Jinping to his friends and Mr. Xi to others. To the Chinese, therefore, it is Obama Barack, not the other way around.

north until they reached Hokkaido. In warmer times, they may even have ventured beyond, onto the Kurile Islands. And in the south, Taiwan lies even closer to mainland China than Japan does to Korea, whereas tropical Hainan Island, the realm's southernmost extremity, is almost—but not quite—connected to the small peninsula that reaches toward it.

In terms of total land area, though, East Asia is mostly mainland—but the islands and their peoples have played leading roles in forging this realm's regional geography. The waters between mainland and islands (the Taiwan Strait, the South China Sea, the East China Sea, the Yellow Sea, the Korea Strait, and others) also figure prominently in the geographic saga of this realm. Today, the Japanese and the Chinese are arguing over the ownership of small islands in these waters, specks of land with large oil reserves and fishing grounds claimed by both sides. So this realm's map is considerably more complicated than that wedge-shaped triangle in Figure 9A-1 initially suggests—even the spellings of many of its contents (see box titled "Places and Names").

POLITICAL GEOGRAPHY

It is all too easy to refer to China when you mean East Asia, because China is the latter's dominant country, contains more than 85 percent of the realm's population, and has taken an increasingly prominent role on the world stage. But there are five other political entities on East Asia's map: Japan, South Korea, North Korea, Mongolia, and Taiwan. Note that we refer here to *political entities* rather than *states*. In this realm, the distinction is important. Taiwan refers to itself as the Republic of China (ROC), but it is not recognized as a sovereign state by most members of the international community; the communist administration in Beijing, capital of the People's Republic of China (PRC), regards Taiwan as part of China and as a temporarily wayward province. And North Korea is widely viewed as a rogue state, a brutal and archaic dictatorship that has failed its people terribly. Having compiled one of the world's most dreadful human rights records, North Korea is not even a fully functional member of the United Nations.

Nevertheless, China is the realm's dominant entity: demographically, economically, and politically. It is important to keep in mind, as you read Chapters 9A and 9B, that portions of what we map today as regional components of China were not part of the country in the past, and that other areas now lying outside China are regarded by many Chinese as Chinese property (for example, a large sector of the Indian State of Arunachal Pradesh and portions of the Russian Far East). As we will see, China's imperial past saw the state expand and contract and expand again, leaving unfinished business on land as well as at sea. On such issues, Chinese emotions can run quite deeply.

Six hundred years ago China already was the largest nation on the planet, its history encompassing thousands of years, its unmatched fleets exploring lands and peoples in South Asia and East Africa, its technologies unequaled. But even the Chinese could not curb the growing presence of Europe's colonial powers, and eventually the British, French, Russians, and Germans accomplished the unthinkable—taking control over most of the Chinese state. When the Japanese emulated the Europeans and forged their own colonial empire, much of it at China's expense, humiliation was complete—and never forgotten. To this day, the sense that China's borders (including those with Russia and India) were imposed by outsiders is never far from the surface.

Gambling With Nature

THE CATASTROPHIC NATURAL disaster that hit Japan in March 2011 overwhelmed the northeastern coast of its main island (Honshu) as well as the country as a whole—but it did not come as a surprise. The Earth's crustal dynamics in this part of the realm guarantees earthquakes at a relentless frequency (see Fig. G-5).

As Figure 9A-2 shows, it all stems from the configuration of the underlying tectonic plates that converge on Japan. Consider the distribution of all those smaller quakes before and after the big one. Any one of them could have been the epicenter, and it is obvious that the entire east coast of Japan lies in the danger zone, including all of the gigantic conurbation centered on Tokyo. So even the misfortunes that befell Japan on March 11, 2011 could have been far worse.

All Japanese, and especially the nearly 30 million living in and around Tokyo, know about the "70-year" rule: over the past three and a half centuries, the Tokyo area has been struck by major earthquakes roughly every 70 years—in 1633, 1703, 1782, 1853, and 1923. The Great Kanto Earthquake of 1923 set off a firestorm that swept over the city and took the lives of an estimated 143,000 people. Tokyo Bay virtually emptied of water; then a killer tsunami roared back in, sweeping everything before it.

Today, metropolitan Tokyo is decidedly more vulnerable than it was a century ago. True, building regulations are stricter and civilian preparedness is better. But entire expanses of industries have been built on landfill that will liquefy; the city is honeycombed by underground gas lines that will rupture and stoke countless fires; congestion in the area's maze of narrow streets will hamper rescue operations; and many older high-rise buildings do not have the structural integrity that has lately emboldened builders to erect skyscrapers in excess of 50 stories. Add to this the burgeoning population of the greater Kanto Plain—which surpasses 40 million on what may well be the most dangerous 4 percent of Japan's territory—and we realize that the next major earthquake along northern Japan's Pacific coast could produce an ever bigger disaster than the historic Tohoku quake of 2011.

ENVIRONMENT AND POPULATION

To understand the complex physical geography of East Asia shown in Figure 9A-1, it is useful to refer back to Figures G-4 and G-5 in the introductory chapter. The high, snowcapped mountain ranges of the realm's southwestern interior result from the gigantic collision of the Indian and Eurasian tectonic plates (see Fig. 8A-2), pushing the Earth's crust upward and creating not only the mountain ranges of which the Himalaya is the most famous, but also popping up the enormous, domelike Qinghai-Xizang (Tibetan) Plateau. In Figure G-5, note the high incidence of earthquakes associated with this collision, converging on a narrow zone of instability that crosses southwestern China and stretches into Southeast Asia. The calamitous 2008 earthquake (magnitude 7.9) in Sichuan Province that killed almost 90,000 originated in this danger zone but was only one in an endless series that will continue to take its toll (the latest occurring here in April 2013—a 7.0 temblor that killed at least 200 and injured more than 10,000).

From Figure G-5 it is also obvious that the Pacific Ring of Fire, with its lethal combination of volcanism and earthquakes, endangers Japan far more than it does China. This threat became a horrific reality on March 11, 2011 when a monstrous 9.0-magnitude earthquake struck off Japan's northeastern coast near the city of Sendai. Figure 9A-2 shows the configuration of the underlying tectonic plates that converge on Japan. The Pacific Plate moves westward at an average of 7–10 centimeters (3–4 in) per year, and the only way it can do so is by pressing forward (or subducting) *beneath* the North American Plate (Japan sits atop the western tip of the latter). It was the most powerful Japanese quake ever recorded, and one of the five biggest in the world since 1900. Even worse, this massive temblor and its torrent of violent aftershocks triggered a series of devastating **tsunamis [1]** (seismic sea waves) that swept across the narrow, densely populated coastal plains

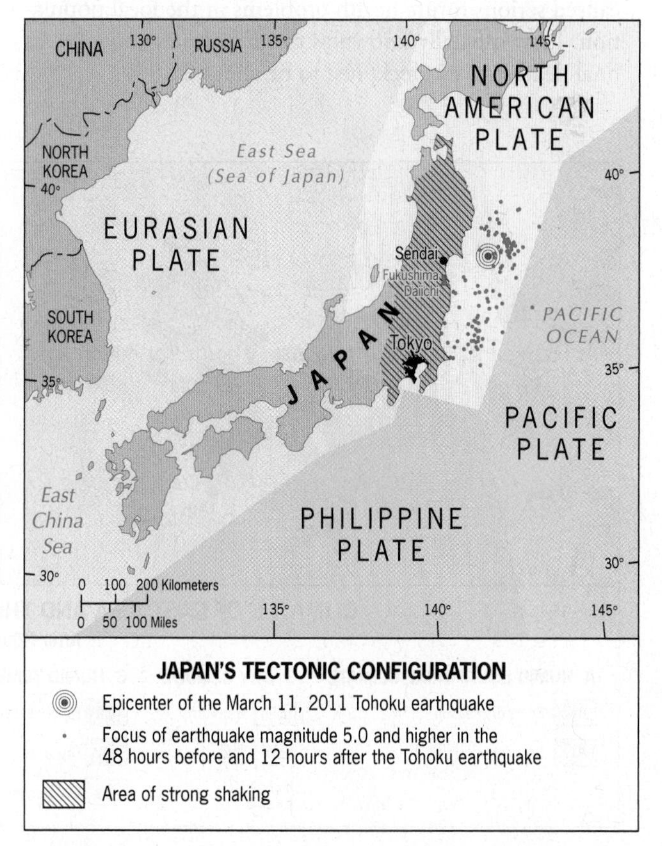

FIGURE 9A-2

© Kyodo/XinHua/Xinhua Press/Corbis

Houses and other buildings being washed away by massive tsunami waves minutes after the gigantic Tohoku earthquake of March 11, 2011. Virtually the entire area around the city of Sendai was destroyed, including this satellite town of Natori. It was a truly apocalyptic event: first the magnitude-9.0 quake, then the tsunamis, and finally the radioactive pollution emitted by the damaged reactors of the Fukushima Daiichi nuclear complex 95 kilometers (60 mi) south of Sendai (see Fig. 9A-2).

that hug the shoreline of Honshu (Japan's largest island) north of the Tokyo area (see photo). And as if the destruction and death left behind was not enough, the leakage of radioactivity from a heavily damaged nuclear-power complex near Fukushima south of Sendai (Fig. 9A-2) may have caused serious future health problems in the local population. Now officially known as the Tohoku earthquake, its final death toll was reckoned to be about 21,000.

The vulnerability of Japan to such disasters results from a dangerous combination of circumstances: it is located in a particularly active tectonic-plate collision zone, and many of the country's habitable (and most densely populated) areas are confined, low-lying plains on the islands' east coast open to flooding by Pacific tsunamis. Not surprisingly, the location of most Japanese nuclear power plants along these susceptible shorelines has now become a hotly debated issue.

Looking at the map of East Asian climates (Fig. 9A-3, left map), we should not be surprised that the western and northern sectors of this realm are dominated by conditions that do not favor substantial population clusters. Permanent snow and ice cover much of the area mapped as ***H*** (highland climes) including Tibet (Xizang) and Qinghai. Northward, the ***B*** climates (desert and steppe) prevail because this vast expanse lies about as far from maritime influences (and moist air masses) as you can get in Asia. Mongolia is one of the driest countries in the world, but even here—and also in frosty Tibet—there are places where people manage to eke out a living. But the map leaves no doubt as to why the great majority of East Asians inhabit the eastern part of this realm.

When we compare the climates prevailing in the East Asian realm to those familiar to us in North America (Fig. 9A-3, right map), it is immediately obvious that the ***C*** or humid-temperate climates are more extensive in the United States than in East Asia. Note especially the comparative location of the milder ***Cfa*** climate, which in the United States extends beyond 40° North latitude up to New England, but which in China yields to colder ***D*** climates at a latitude equivalent to Virginia's. Thus the capital, Beijing,

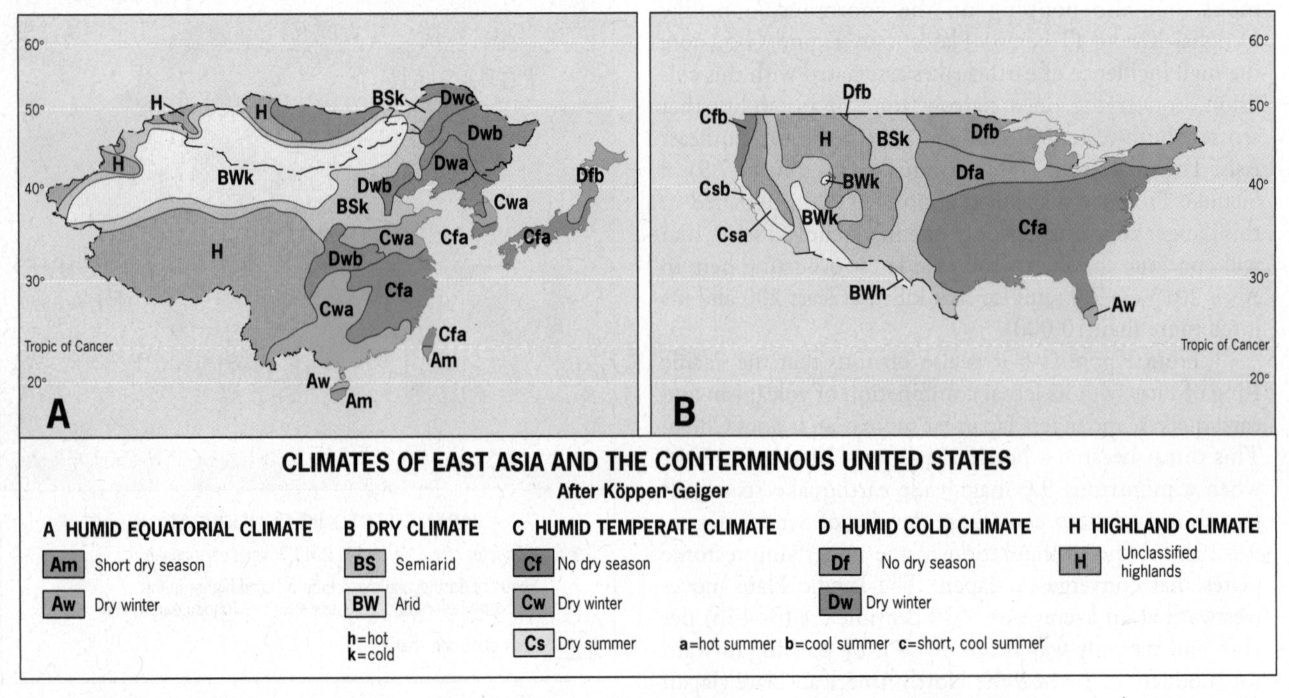

FIGURE 9A-3

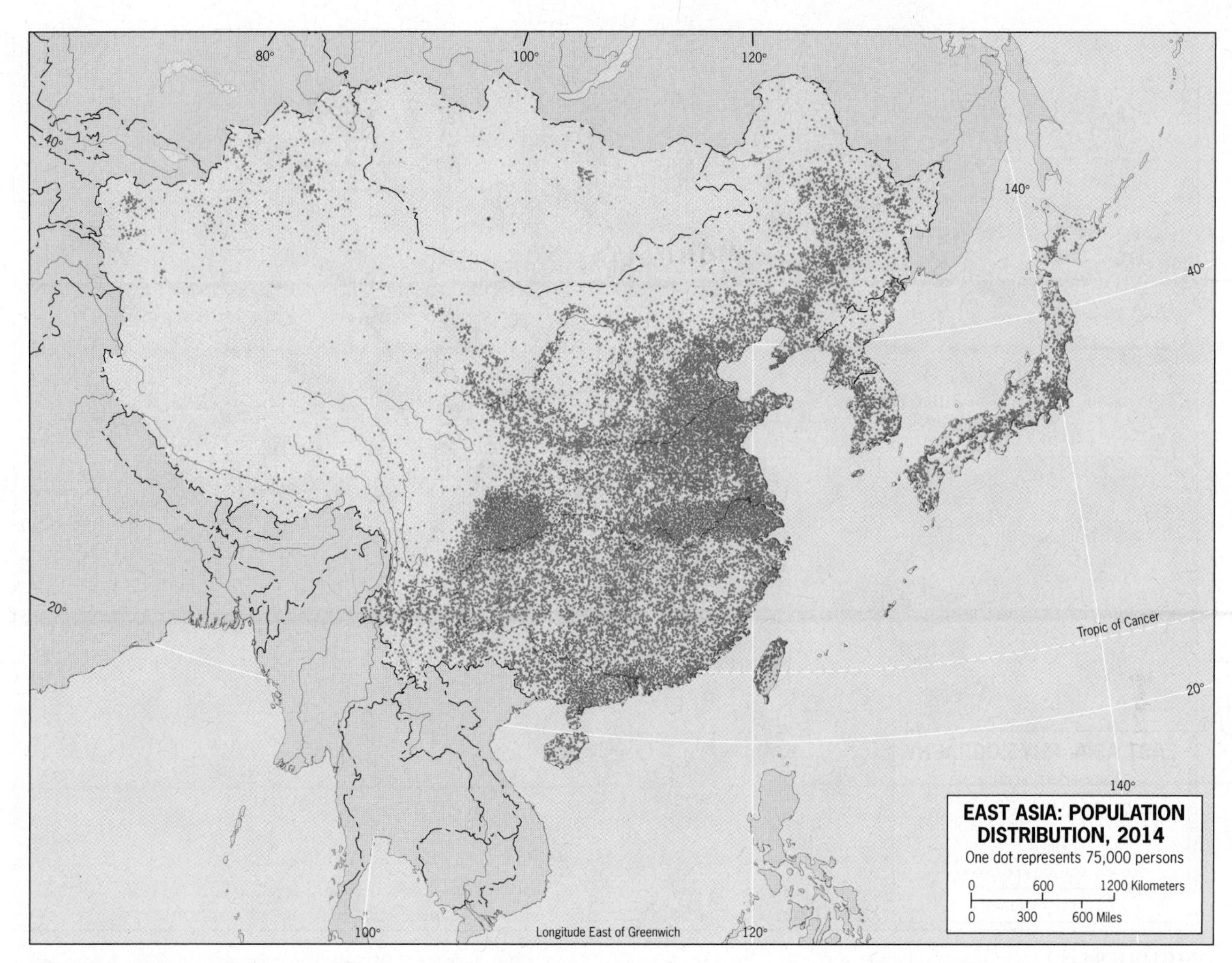

FIGURE 9A-4

has the warm summers but long, bitterly cold winters characteristic of ***D*** climates. Take a closer look at Figure 9A-3 and it is clear that both the Korean Peninsula and the Japanese islands lie astride this transition from ***C*** to ***D*** climates. South Korea is significantly milder and moister than North Korea, and Japan's largest island (Honshu) is temperate in the south but cold in the north. Not surprisingly, the least densely populated major island of Japan is northernmost Hokkaido, where the climate is rather like that of northern Wisconsin.

Comparing just the United States and China, note that whereas ***C*** or ***D*** climates prevail over more than half of the U.S., these climates predominate over less than one-third of China—even though the United States has only 317 million people compared to China's nearly 1.4 billion. That is what makes the population distribution map (Fig. 9A-4) so noteworthy: the overwhelming majority of East Asia's people are located in the easternmost one-third of the realm's territory, creating the biggest, most densely settled population cluster in the world, which mainly depends on the limited green-colored area in Figure 9A-1.

Several times in earlier chapters of this book we have noted the capacity of humans to live under virtually all environmental circumstances; technological developments enable the survival of year-round communities in Antarctica, on oil platforms at sea, in the driest of deserts, and on the highest of plateaus. But the world population distribution map (Fig. G-8) still reminds us how we got our modern start—through crop-raising and herding. The fertile river basins and coastal plains of East Asia supported ever-larger farming populations, whose descendants still live on that same land: for all its current industrialization and urbanization, just under half the population of China remains rural to this day. Environment, in the form of elevation, relief, water supply, soil fertility, and climate played the crucial role in the evolution of the population distribution displayed in Figure 9A-4—and will continue to do so for centuries to come.

The Great Rivers

China is in some respects the product of four great river systems and their basins, valleys, deltas, and estuaries. These rivers and their tributaries are visible, but do not stand out as clearly in Figure 9A-1 as they do on the physiographic map (Fig. 9A-5). So important are they as

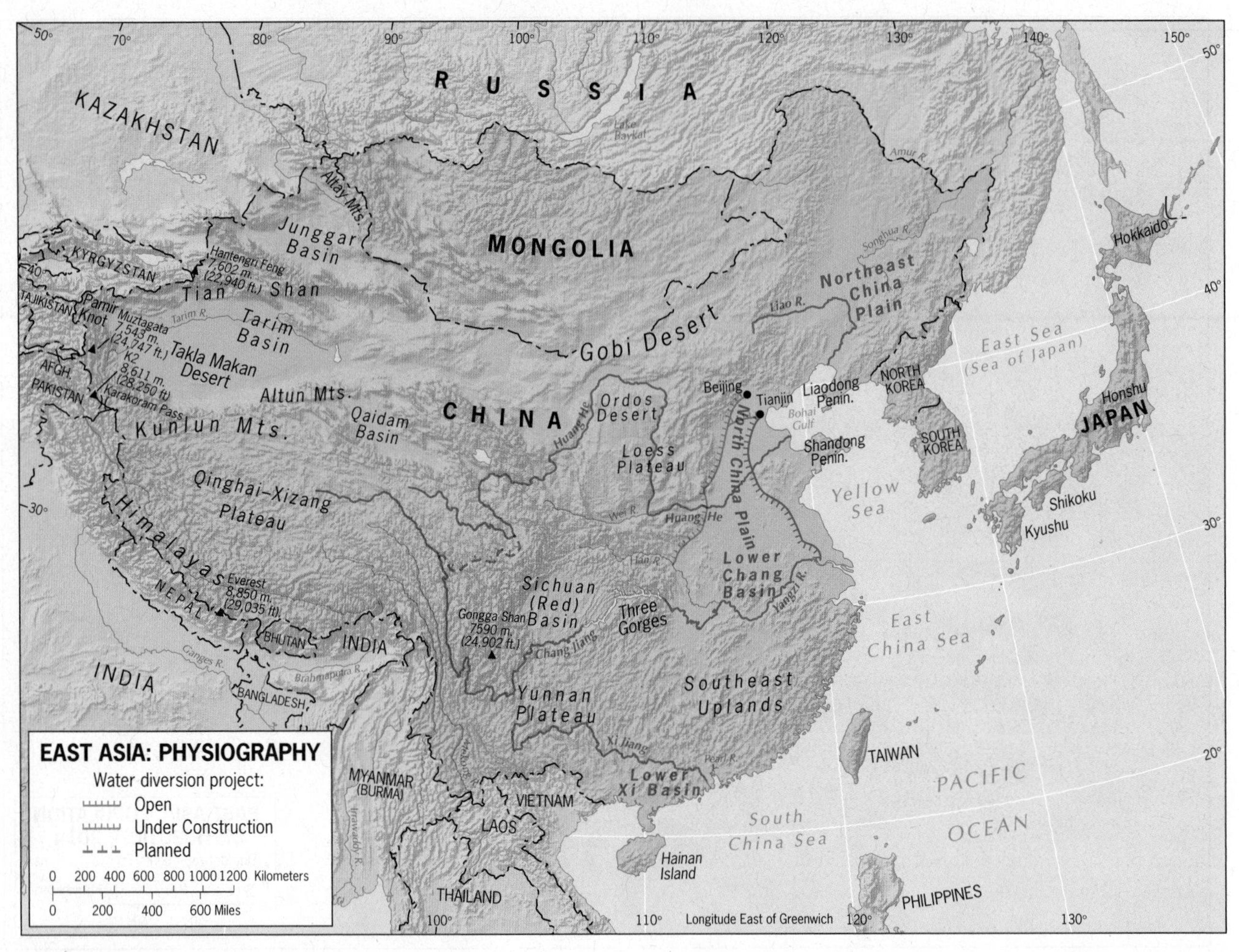

FIGURE 9A-5 © H. J. de Blij, P. O. Muller, J. Nijman, and John Wiley & Sons, Inc.

designators of the realm's regional geography that we should acquaint ourselves with them here. Of the four, the two in the middle are in many ways the most important: the ***Huang He*** (Yellow River) that makes a huge loop around the Ordos Desert and then flows across the North China Plain into the Bohai Gulf, and the ***Yangzi*** River, probably the most famous river in China's historical geography, called the ***Chang Jiang*** (Long River) upstream.

As the map shows, the Yellow River and its tributaries form the sources of water vital to the historic core area of China, the North China Plain, where the capital, Beijing, is located. The Yangzi River is the major artery of the Lower Chang Basin; by its mouth lies China's largest city, Shanghai, and in its middle course the water flow is controlled by the world's biggest dam (see photo). Both the Huang and the Yangzi rivers originate in the snowy mountains of the Qinghai-Xizang Plateau, a reminder that these remote environments are critical to hundreds of millions of people who live thousands of kilometers away.

The other two rivers have much shorter courses, but the ***Pearl*** River outlet of the one in the south, the ***Xi Jiang*** (West River), forms an estuary that has become China's (and East Asia's) greatest hub of globalization—an ongoing saga discussed in Chapter 9B. This is where you find Hong Kong and, right next to it, the fastest-growing major city in the history of humankind—Shenzhen.

Finally, the northernmost of China's four major rivers, the ***Liao*** River, originates near the margin of the Gobi Desert and then forms an elbow as it crosses the Northeast China Plain to reach the Bohai Gulf flowing southward. Here the climate is colder, low relief scarcer, agriculture lagging, and population smaller than in the more southerly river basins, but mineral resources create opportunities for mining and industry. Each of East Asia's river-based clusters has its own combination of potentials and problems.

But even these major river systems, together with China's lesser ones, are increasingly pressed to satisfy the country's ever-growing demands for water. In order to meet these urgent needs, China has embarked on the massive South-North Water Transfer Project to bring new supplies to its thirsty northeastern core area, especially metropolitan Beijing and Tianjin (Fig. 9A-5). When completed, the entire project will annually divert more than 40 billion cubic meters of water from the Chang River to the north. The Eastern Route (following China's Grand Canal) opened in 2013, and the Central Route is expected to begin operating in 2014; the Western Route in the rugged highlands of the upper Chang and Huang basins is still in the planning stage. This

Three Gorges Dam is emblematic of China's contemporary "era of the megaproject." The dam wall rises 180 meters (600 ft) above the inundated valley floor of the Chang/Yangzi River. It is 2 kilometers (1.2 mi) wide and creates a reservoir—largest of its kind in the world—that extends more than 600 kilometers (400 mi) upstream.

gargantuan project, however, comes with a price that is not only monetary: the reservoirs and canals along the Central Route will submerge more than 300 square kilometers of land, forcing well over 300,000 people from their homes in Henan and Hubei provinces.

Along the Coast

East Asia's Pacific margin is a jumble of peninsulas and islands. The Korean Peninsula looks like a near-bridge from Asia to Japan, and indeed it has served as such in the past. The Liaodong and Shandong peninsulas protrude into the Yellow Sea, which continues to silt up from the sediments of the Huang and Liao rivers. Off the mainland lie the islands that have played such a crucial role in the modern human geography of Asia and, indeed, the world: Japan, Taiwan, and Hainan. Japan's environmental range is expressed by cold northern Hokkaido and warm southern Kyushu, but Japan's core area lies on its main island, Honshu. As Figure 9A-1 shows, myriad smaller islands flank the mainland and dot the East and South China seas. As we will discover, some of these smaller islands significantly affect the political geography of this realm.

NATURAL RESOURCES

Given that the East Asian realm contains nearly one-fourth of the world's population, it is not difficult to imagine the magnitude of the demand for natural resources here. The world first received notice of this about a hundred years ago when Japan's imperial expansion was in part driven by the need for raw materials to feed its rapidly expanding industrialization. Today, East Asia's demand for natural resources is expressed not in imperialism but in the global marketplace: it is driving a commodity boom all over the world, from Russia to Australia to Brazil to Subsaharan Africa.

While China was moribund under its early communist administration, and before post–World War II Japan embarked on its headlong rush to become a world economic power, East Asia's requirements remained modest by global standards. But then Japan's economic success, followed by China's swift adoption of market economics, created unimagined and unprecedented needs.

Japan, about the size of Montana but with a population of more than 100 million, showed what lay ahead. With limited domestic resources to support large-scale manufacturing, the Japanese set up global networks through which flowed commodities ranging from oil and natural gas to iron ore and chemicals. Urbanizing and modernizing populations demand ever more consumer goods, and Japanese products poured onto domestic as well as foreign markets. Japanese-owned fleets of freighters plied the oceans, and for a while Japan even became remote Australia's top-ranked customer for commodities.

When China took off in the 1980s, economic geographers cast a wary eye on the geologic map. Until then, China's biggest resource had been its fertile, river-deposited alluvium (silt): despite the communist regime's best efforts, state-run industries planned to satisfy domestic needs were no match for globalizing Japan. Most Chinese were farmers. Staving off famine was a never-ending preoccupation. But when China opened its doors to the world, and its cities burgeoned even as its industries proliferated, its needs—for oil, natural gas, metals, food, electricity, water—multiplied. Before long, China had replaced Japan as Australia's number-one customer. And Chinese manufacturers and suppliers searched for commodities from Indonesia to Iraq and from Tanzania to Brazil. If China holds an advantage, it is in the so-called *rare earth* elements not commonly known, such as thulium (used in lasers), praseodymium (aircraft parts), lanthanum (electric automobiles), and promethium (X-ray equipment). By some measures, East Asia contains almost 95 percent of the known deposits of these minerals, which are also increasingly utilized in missile technology and "green" energy applications.

But China clearly needs the world because East Asia's storehouse of other known resources is not encouraging (Fig. 9A-6). Coal reserves are widespread and can satisfy the expanding coal-fired energy system (but only at the cost of thousands of miners' lives every year). Although oil reserves, are also widespread, they tend to be rather modest and diminishing. Deposits in northeastern and far western China are the largest; moreover, exploration is proceeding offshore. Yet nothing in China compares with the massive iron ore deposits and plentiful ferroalloys available to Russia when it

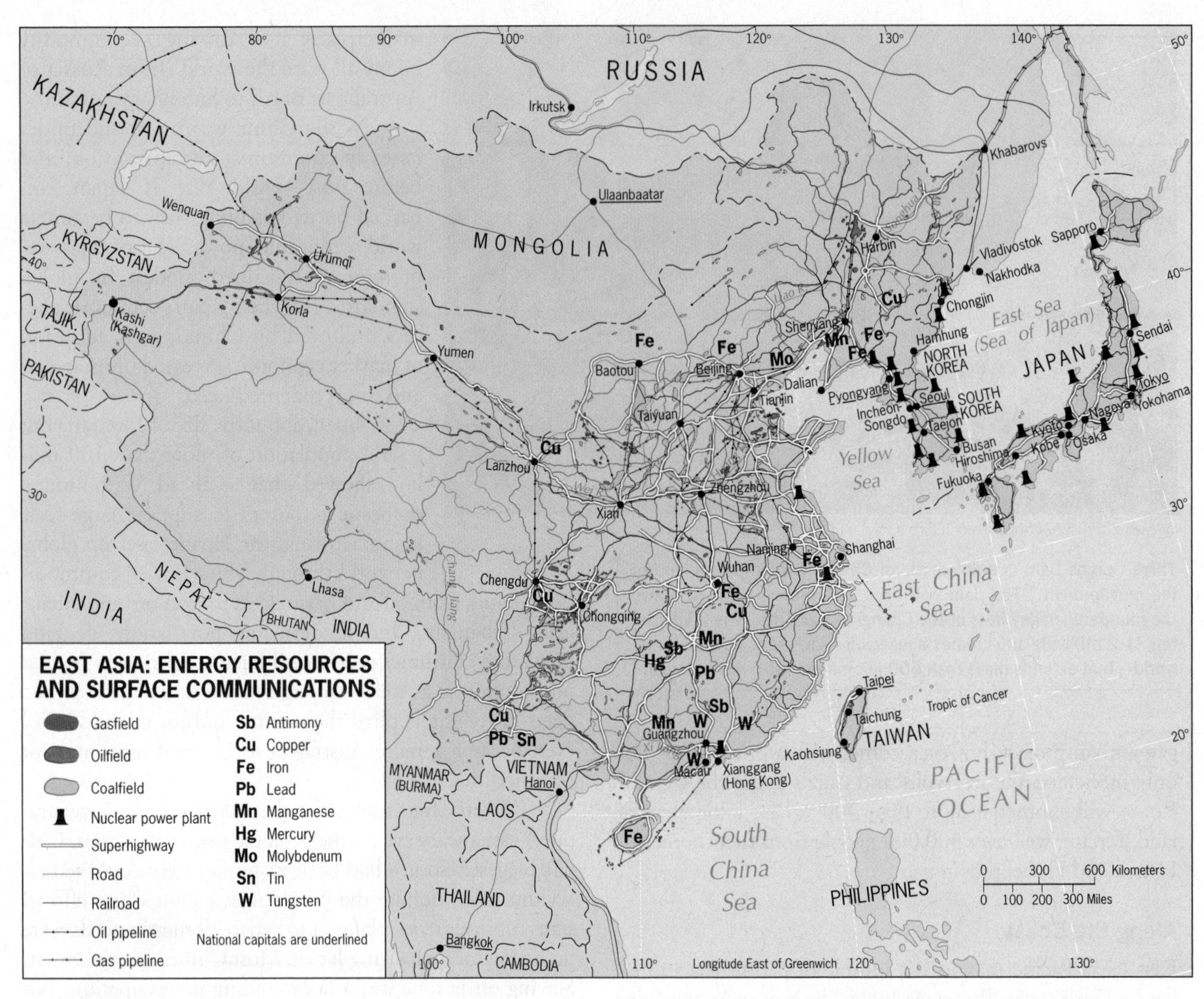

FIGURE 9A-6

industrialized to meet the Nazi challenge; and nothing in China can match the abundant gas and oil reserves of contemporary Russia, let alone the massive deposits of Southwest Asia's Persian Gulf region. Hence China competes with Japan for energy pipelines from Russia and with other sources for its industrial raw materials. So in just a few short years, China has become both the world's biggest consumer as well as its leading exporter.

East Asia's ascent to the globalizing world's principal stage has put it in the forefront of regional development, but at a high environmental cost in terms of air and water pollution (see photo). The East Asian geographic realm is now in the midst of an industrial, social,

In China's headlong rush to become a global economic power, environmental concerns have always taken a back seat. Thus it is hardly surprising that a recent study reported atmospheric pollution in 2010 was a factor in well over one million premature deaths countrywide. This scene along Beijing's Guangha Road during rush hour in February 2013 particularly underscores the daily challenge facing urban residents. In fact, days before this picture was taken, air pollution measurements here broke the global record for smog by a wide margin—registering *30 times* higher than the World Health Organization's uppermost "safe" level! The unusually-shaped structure to the right, one of Beijing's newest landmarks, is CCTV Tower, headquarters of state-run China Central Television, the PRC's broadcasting behemoth that controls a vast network of 22 channels that reach more than a billion viewers.

AMONG THE REALM'S GREAT CITIES . . .

XIAN, ANCIENT AND MODERN

THE CITY KNOWN today as Xian in Shaanxi Province is the site of one of the world's oldest urban centers. It may have been a settlement during the Shang-Yin Dynasty more than 3000 years ago; it was a town during the Zhou Dynasty, and the Qin emperor was buried here along with 6000 life-sized terracotta soldiers and horses, reflecting the city's importance. During the Han Dynasty (ca. 200 BC to AD 200) the city, then called Chang'an, was one of the greatest centers of the ancient world, the Rome of ancient China. Chang'an formed the eastern terminus of the Silk Route, a storehouse of enormous wealth. Its architecture was unrivaled, from its ornamental defensive wall with elaborately sculpted gates to the magnificent public buildings and gardens at its center.

Situated on the fertile loess plain of the upper Wei River, Chang'an was the focus of ancient China during crucial formative periods. After two centuries of Han rule, political strife led to a period of decline, but the Sui emperors rebuilt and expanded Chang'an when they made it their capital. During the Tang Dynasty, Chang'an again became a magnificent city with three districts: the ornate Palace City; the impressive Imperial City, which housed the national administration; and the busy Outer City containing the homes and markets of artisans and merchants.

After its Tang heyday the city again declined, although it remained a bustling trade center. During the Ming Dynasty it was endowed with some of its architectural landmarks, including the Great Mosque marking the arrival of Islam; the older Big Wild Goose Pagoda dates from the influx of Buddhism. After the Ming period, Chang'an's name was changed to Xian (meaning "Western Peace"), then to Siking, and in 1943 back to Xian again.

Having been a gateway for Buddhism and Islam, Xian in the 1920s became a center of Soviet communist ideology.

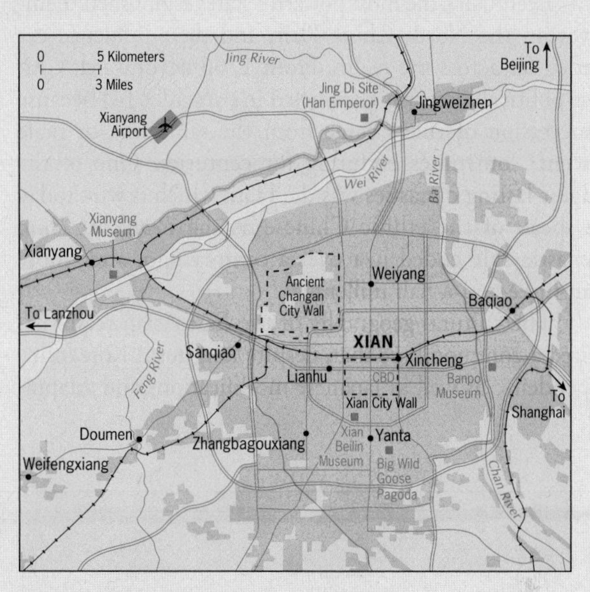

© H. J. de Blij, P. O. Muller, J. Nijman, and John Wiley & Sons, Inc.

The Nationalists, during the struggle against the Japanese, moved industries from the vulnerable east coast to Xian, and when the communists took power they enlarged Xian's industrial base still further. The present city (population: 5.4 million) lies southwest of the famed tombs, its cultural landscape now dominated by a large industrial complex that includes a steel mill, textile factories, chemical plants, and machine-making facilities. Little remains (other than some prominent historic landmarks) of the splendor of times past, but Xian's location on the main railroad line to the vast western frontier of China sustains its long-term role as one of the country's key gateways.

and political revolution the outcome of which is far from clear.

HISTORICAL GEOGRAPHY

We in the Western world often take it for granted that the pivotal events that led to the domestication of animals and the selective farming of plants, the herding of livestock and the harvesting of crops, the storage of produce and the growth of villages into towns, all began in what we now call the Middle East. The story of rivers ebbing and flowing, the "lessons" of natural irrigation, the planned harvesting of grains, the rise of cities, and the organization of the earliest states is the story of "Western" civilization—the saga of Mesopotamia and the Tigris-Euphrates Basin, and ancient Egypt and the Nile. From the Fertile Crescent, we were always taught, these revolutionary changes diffused outward to other parts of Eurasia and then to the rest of the world.

The Chinese have long taken a different view, and we now know from extensive archeological research that East Asia was one of the few places in the world where the process of **state formation [2]** occurred independently thousands of years ago, and that the modern Chinese state could trace its roots to that ancient time, long before there was a Greece or a Rome.

Ancient China

We now know that the East Asian realm was forged from numerous cultures that flourished around 6000 BC during the early **Neolithic [3]** (New Stone Age) in several areas, including the Lower Chang Basin to the south and even beyond. Crops like millet, rice, and wheat were domesticated here, independently of developments in the Middle East, and urbanization and state formation followed suit.

Eventually, the most powerful states established themselves on the North China Plain and their influence extended far across the realm. From 1766 BC onward, Chinese political history is chronicled in **dynasties [4]** because a succession of rulers came from the same line of male descent, sometimes enduring for centuries. One of the longest-lasting dynasties was the Han, which is why today we speak of the ethnic Chinese as the *People of Han*. Dynastic rule ended just over a century ago, after more than three-and-a-half millennia.

The cultural geography of the East Asian realm acquired a common base that, despite long-term Korean independence as well as Japanese modifications and adaptations to the Chinese norms they had borrowed, acquired an overall Chinese denomination. Buddhism matured in China and diffused to Japan; *Confucianism* infused Korean kingdoms. In Japan, where local tradition places the country's beginnings in 660 BC, the ruling elites borrowed heavily from Chinese culture, including town plans, building styles, legal models, and even writing systems.

Peoples of the East Asian Realm

If we were studying the East Asian realm six centuries ago, during China's Ming Dynasty, we would have had no doubt as to the location of core and periphery. China was the core. Encircling the Chinese state in the periphery were not only the Koreans and the Japanese but also the Mongols and Tatars to the north and northwest; the Kazakhs, Kyrgyz, Tajiks,

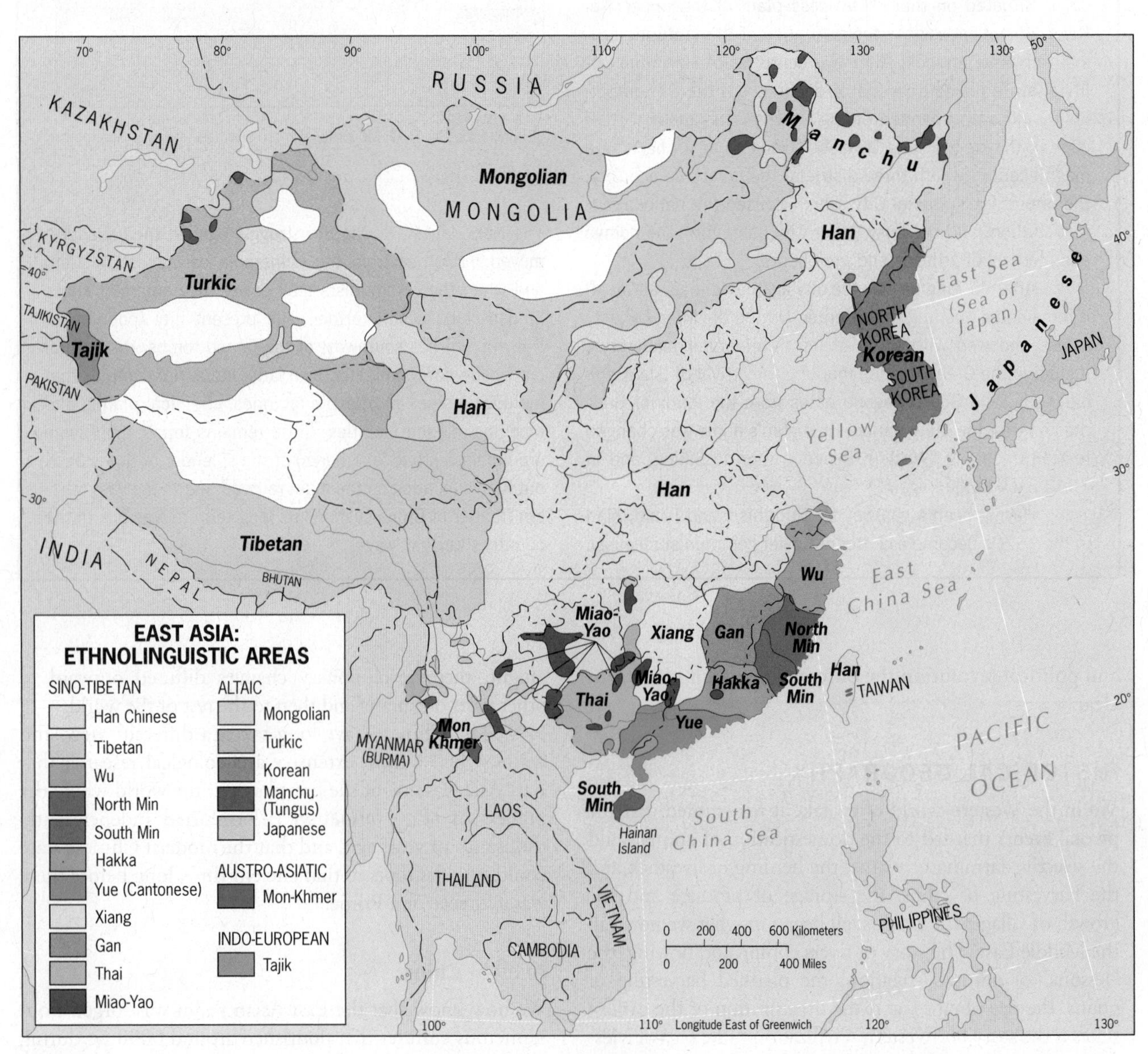

FIGURE 9A-7

and Uyghurs to the west; the Tibetans, Nepalese, and others to the southwest; and peoples too numerous to identify individually to the south, including both majorities (Burmans, Thais, Vietnamese) and minorities in states that were still forming in South and Southeast Asia.

Like all empires, the China of the Ming emperors expanded and contracted over time, but during the next—and final—dynasty, that of the Qing (whose rule began in AD 1644), many of the states and peoples just mentioned fell under Chinese rule. But it was the emperors' last hurrah: European and Japanese imperialists challenged Qing rule, took control over most of the state's core area, ousted the Chinese from much of the periphery, and left the country in chaos, bringing about the end of nearly 3700 years of dynastic rule in 1911.

But by that time, the ethnic geography of East Asia had become highly complicated. Even after losing control over peoples from Korea to Vietnam and from Mongolia to Burma (now called Myanmar), the Chinese state of the twentieth century still governed numerous minorities. East Asia today, therefore, remains an especially complex mosaic of ethnicities and languages.

Figure 9A-7 highlights the cultural diversity of a realm in which China dominates numerically, but where infusions from elsewhere are evident. We already know of the Mongols, the Muslim Uyghurs, and the Buddhist Tibetans; but the most varied and most numerous minority groups inhabit the southeastern corner of this realm, from the island of Hainan to the mouth of the Yangzi River. For example, the Yue language, middle green on the map (it used to be called Cantonese), is the common language in the pivotal Pearl River Estuary. Many of the other minority tongues shown here have links to Southeast Asian languages.

And do not be misled by what the map seems to show about Han Chinese. Officialdom, the elites, and the well-educated speak Standard Chinese (also called "Mandarin"), with those in the south somewhat less true to the Beijing-area version than those in the north. But ordinary people in villages a few kilometers apart may not be able to understand each other at all. What they are able to do is read the characters in which Standard Chinese (locally called *Putonghua*) is written. And so, wherever you are in China, when you watch an anchorperson read the news on television, you will see a moving ticker below showing in characters what is being said on-screen. Viewers in Sichuan may not understand the newscaster in Beijing, but still will be able to get the news by reading the crawl.

The ethnolinguistic mosaic of the East Asian realm is paralleled by regional variations in belief systems that are discussed in Chapter 9B, because beliefs—their origins, diffusion, and current regional expression—all reflect the diverse ways that cultural and political geographies are interconnected in this populous realm.

The Great Wall of China, portions of it built and rebuilt from the seventh century BC to the sixteenth century AD, primarily aimed at fending off Mongol invaders (see Fig. 9A-8). Some parts are completely gone but other sections have been renovated in recent times, such as this one about 65 kilometers (40 mi) north of Beijing.

CHINA'S HISTORICAL ROLE WITHIN EAST ASIA

Chinese Empires and Dynasties

China's great antiquity, as noted above, is chronicled in dynasties, the earliest of which are still shrouded in mystery. But, as with any long-existing state, some of China's dynastic rulers proved more productive than others. Some dynasties bequeathed long-term geographic legacies to the state. Others molded the Chinese nation. The Han Dynasty did both, which again is why even today we speak of the Chinese as the ***People of Han***.

But the Zhou Dynasty, centuries before the Han, witnessed the arrival of Buddhism in China, saw Confucius walk the pathways of the north, started the building of the Great Wall (see photo), and, in another context, spread the use of chopsticks for eating. **Confucianism [5]** was to become China's guiding philosophy for two millennia (see box titled "Confucius").

More than a thousand years after the Han Dynasty, something happened that the Russians would understand. The Mongols drove the local rulers from power and took over the state. But instead of imprinting

Confucius

CONFUCIUS (*Kongfuzi* in pinyin) was China's most influential philosopher and teacher. His ideas dominated Chinese life and thought for over 20 centuries. Confucius was born in 551 BC and died in 479 BC. Appalled at the suffering of ordinary people during the Zhou Dynasty, he urged the poor to assert themselves and demand explanations for their harsh treatment by the feudal lords. He tutored the indigent as well as the privileged, giving the poor an education that had hitherto been denied them and ending the aristocracy's exclusive access to the knowledge that constituted power.

Confucius's revolutionary ideas extended to the rulers as well as the ruled. He abhorred supernatural mysticism and cast doubt on the divine ancestries of China's aristocratic rulers. Human virtues, not godly connections, should determine a person's place in society, he taught. Accordingly, he proposed that the dynastic rulers turn over the reins of state to ministers chosen for their competence and merit.

His earthly philosophies notwithstanding, Confucius took on the mantle of a spiritual leader after his death. His thoughts, distilled from the mass of philosophical writing (including Daoism) that poured forth during his lifetime, became the guiding principles of the formative Han Dynasty. The state, he said, should not exist just for the power and pleasure of the elite; it should be a cooperative system for the well-being and happiness of the people.

With time, a mass of writings evolved, much of which Confucius never wrote. At the heart of this body of literature lay the *Confucian Classics*, 13 texts that became the basis for education in China for 2000 years. From government to morality and from law to religion, the *Classics* were Chinese civilization's guide. The entire national system of education (including the state examinations through which everyone, poor or privileged, could enter the civil service and achieve political power) was based on the *Classics*. Confucius championed the family as the foundation of Chinese culture, and the *Classics* prescribe a respect for the aged that was a hallmark of Chinese society.

But his philosophies also were conservative and rigid, and when the colonial powers penetrated China, his ideas came face to face with practical Western education. For the first time, some Chinese leaders began to call for reform and modernization, especially of teaching. Confucian principles, they said, could guide an isolated China, but not China in the new age of competition.

The communists who took power in 1949 attacked Confucian thought on all fronts. The *Classics* were abandoned, indoctrination pervaded education, and, for a time, even the family was viewed as an institution of the past. Here the communists miscalculated. It proved impossible to eradicate two millennia of cultural conditioning in a few decades. As soon as China entered its post-Mao period in 1976, public interest in Confucius surged, and the shelves of bookstores again sagged under the weight of his writings. The spirit of Confucius will pervade physical and mental landscapes in China for generations to come.

Mongol rules on their Chinese subjects, the Mongols underwent **Sinicization [6]** or, as it is sometimes said, **Hanification [7]**—that is, they adopted many of the ways of the People of Han. Hostility between the Mongols and the Chinese never diminished, however. Marco Polo saw it first-hand during his early medieval travels to "Cathay" (the name used for Northern China) and reported on it when he returned to Europe.

Mongol (Yuan Dynasty) rule ended in 1368, and it is amazing that, in the five and a half centuries dynastic China was subsequently to endure, only two dynasties were to rule: the Ming (1368–1644) and the Qing (1644–1911). The Ming, ruling their empire from the Forbidden City in Beijing (where ordinary mortals were forbidden entry), started out triumphantly: they annexed northern Korea, Mongolia, even Myanmar (Burma). They dispatched huge fleets into Pacific and Indian Ocean waters to explore the wider world. But it all came crashing down under conditions we talk a great deal about today: climate change. A major environmental shift called the *Little Ice Age* that had struck Europe in the previous century now afflicted Ming China. Wheatfields lay barren. With 100 million mouths to feed, the emperors worried about revolution. They burned the fleets, built barges, and extended the Grand Canal from the Lower Chang Basin to transport rice to the hungry north. But the Ming Dynasty never recovered its initial verve.

By the mid-1600s, the Manchus, a northern people with historic Mongol-Tatar links, saw their opportunity and seized control in Beijing. The People of Han are still shaking their heads about it—a people numbering about a million managed to grab the reins of power over a nation of several hundred million. The invaders adopted Chinese ways, kept the Ming systems of administration and education, called themselves Qing, and set about expanding the empire. As noted earlier, it was the emperors' last hurrah—although a long-drawn-out one. The Qing Empire became the largest Chinese empire ever (Fig. 9A-8), but it was stretched thin, failed to modernize, and could not withstand the nineteenth-century onslaught of foreign powers on its shores. Our map shows the maximum dimensions the Qing Empire acquired as well as the colonial domains the Europeans, Russians, and Japanese appropriated. Thus 36 centuries of dynastic rule over China soon ended in war, revolution, chaos, and the final collapse in 1911.

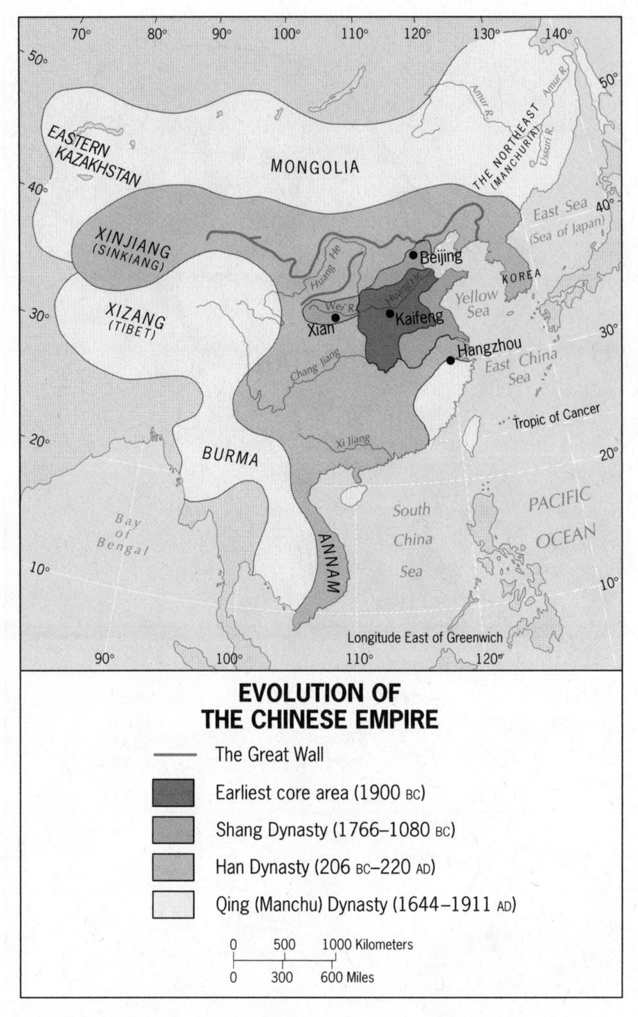

FIGURE 9A-8 © H. J. de Blij, P.O. Muller, and John Wiley & Sons, Inc.

China in Disarray

Figure 9A-9 provides only a mere hint of China's disarray. Chinese emperors (and their subjects) had come to regard the state as invincible, but the colonial powers demonstrated otherwise. Economically, the colonialists imported cheap manufactures into China, whose handicraft industries succumbed in the face of unbeatable competition. In addition, British merchants imported large quantities of opium from India into China, and soon this addictive intoxicant was destroying the very fabric of Chinese cultural life. When the Qing government tried to resist this influx, the First Opium War (1839–1842) proved disastrous. The victorious British forced China's rulers to acquiesce to opium importation, and the breakdown of Chinese sovereignty was underway. When the Chinese tried that approach again, the Second Opium War 15 years later resulted in the Chinese being forced to allow the cultivation of the opium poppy in China itself. Very soon thereafter, Chinese society was disintegrating beneath a narcotic tide that proved uncontrollable.

Meanwhile, the modern ethnic complex of Figure 9A-7 was in the making. The Beijing court was forced to grant concessions and leases to foreign merchants; China ceded Hong Kong to Britain even as British ships sailed up the Yangzi to consolidate a huge sphere of influence in China's crucial midsection. Portugal took Macau, Germany established itself on the Shandong Peninsula, the French encroached on China from their Asian colonies farther south, and the Russians entered the Northeast, then still known as *Manchuria.* The Japanese subsequently invaded Korea, annexed the Ryukyu Islands (they still possess them, including Okinawa, today), and colonized Formosa (now called Taiwan) in 1895.

All over China, under the doctrine of *extraterritoriality*, the British and other colonial powers carved out ***concessions*** in which Europeans, as well as Russians and Japanese, were immune from local Chinese law (see box titled "Extraterritoriality"). Some of these urban enclaves, business as well as residential, were even off-limits to Chinese citizens. In many other buildings, parks, and additional facilities, Chinese found themselves unable to enter without permission from the entrenched foreigners. This contributed to the loss of face that helped provoke the so-called Boxer Rebellion of 1900, when bands of revolutionaries roamed cities and countryside, attacking and killing not only foreigners but also their Chinese collaborators. It was

Extraterritoriality

DURING THE NINETEENTH century, as China weakened and European colonial invaders entered China's coastal cities and sailed up its rivers, the Europeans forced China to accept a European doctrine of international law—**extraterritoriality [8]**. Under this doctrine, foreign states and their representatives are immune from the jurisdiction of the country in which they are based. Today this applies to embassies and diplomatic personnel. But in Qing (Manchu) China, it went far beyond that.

The European, Russian, and Japanese invaders established as many as 90 ***treaty ports***—extraterritorial enclaves in China's cities under unequal treaties enforced by gunboat diplomacy. In their "concessions," diplomats and traders were exempt from Chinese law. Not only port areas but also the best residential suburbs of large cities were declared to be "extraterritorial" and made inaccessible to Chinese citizens. In the city of Guangzhou (Canton of colonial times), Sha Mian Island in the Pearl River was a favorite extraterritorial enclave. A sign at the only bridge to the island stated, in English and Cantonese, "No Dogs or Chinese."

FIGURE 9A-9

a portentous uprising, put down with great loss of life by a multinational force consisting of British, Russian, French, Italian, German, Japanese, and American soldiers. China had no friends to call upon.

Revolutionary China

Even as China's public order continued to disintegrate, a better-organized revolutionary movement arose: the so-called Nationalist movement under a prominent leader named Sun Yat-sen. Blaming the sclerotic dynastic regime for China's troubles, these Nationalists in 1911 mounted a fierce attack on the emperor's garrisons all over China. Within a few months the 267-year-old Qing Dynasty was overthrown, and with it the system that had survived for thousands of years at China's helm.

The Nationalists, however, faced insurmountable problems in their own efforts to impose a new order on chaotic China. They did negotiate an end to the extraterritorial treaties, and, because of their well-orchestrated military campaign, they did acquire greater legitimacy than the previous regime had when confronting colonial interests. But the Nationalist government set up its base in the southern city of Canton (now Guangzhou), leaving another would-be government to attempt to rule from Peking (Beijing), the old imperial headquarters. Meanwhile, in 1921, a group of intellectuals in Shanghai founded the Chinese Communist Party; one of its prominent co-founders was a young man named Mao Zedong.

During the chaotic 1920s, the Nationalists and the Communist Party at first cooperated, with the remaining foreign presence their joint target. After Sun Yat-sen's death in 1925, Chiang Kai-shek became the Nationalists' leader, and by 1927 the foreigners were on the run, escaping by boat and train or falling victim to rampaging Nationalist forces. But soon the Nationalists began purging communists even as they pursued foreigners, and in 1928, when Chiang established his Nationalist capital in the city of Nanjing on the banks of the lower Yangzi, it appeared that the Nationalists would emerge victorious from their campaigns. They had driven the communists ever deeper into the interior, and by 1933 the Nationalist armies were on the verge of encircling the last communist stronghold in the area around Ruijin in Jiangxi Province. This led to a momentous event in Chinese history: the ***Long March***. Nearly 100,000 people—soldiers, peasants, leaders—marched westward from Ruijin in 1934, a communist column that included Mao Zedong and Zhou Enlai. Nationalist forces rained down attack after attack on the marchers, and of the original 100,000, about three-quarters were eliminated. But new sympathizers joined along the way (see the route marked in Fig. 9A-9), and the 20,000 survivors found a refuge in the remote mountainous interior of Shaanxi Province, 3200 kilometers (2000 miles) away. There, they prepared for a renewed campaign that would eventually bring them to power.

JAPAN'S HISTORICAL ROLE IN EAST ASIA

From Isolationism to Imperialism

Since the seventeenth century, Japan had cultivated and enforced a strict policy of **isolationism [9]**. Foreign influences were shunned, the Japanese people were not allowed to travel outside of Japan, and foreigners were not tolerated on Japanese soil. Contact with the outside world was extremely limited, therefore, and even trade was heavily restricted. For some time during the seventeenth and eighteenth centuries, the Dutch East India Company obtained the exclusive right to trade with Japan, but Company personnel were prohibited from setting foot on Japanese soil. So the Japanese built a small artificial island in Nagasaki Bay named Deshima where Dutch ships could dock and sailors stretch their legs. Japanese workers would cross the bridge to the island to haul the cargo ashore, but the foreigners were strictly confined to Deshima.

But as Western imperial powers (especially the British and the French) imposed their designs on Asia, isolationism became an increasingly unrealistic option for Japan because it did not want to end up itself one day as a colonial prize of the Europeans. The Industrial Revolution had yet to reach Japan, and its ability to wage war was premodern compared to the more advanced military technology (e.g., steamships) of the Western powers. Prodded by a local show of force by the U.S. Navy in the 1850s, the so-called ***Meiji Restoration*** (the return of "enlightened rule" focused on the Emperor Meiji) in 1868 introduced a wholesale change of Japanese foreign policy: away from isolationism and aimed at rapid modernization. Japan had decided to emulate the West.

After these modernizers took control of Japan, they turned to Britain for guidance in reforming their nation and its economy. During the decades that followed, the British advised the Japanese on the layout of cities and the construction of a railroad network, the location of industrial plants, and the organization of education. The British influence is still visible in the Japanese cultural landscape: the Japanese, like the British, drive on the left side of the road. (Consider how this affects the effort to open the Japanese market to U.S.-made automobiles!)

The Japanese reformers of the late nineteenth century undoubtedly saw many geographic similarities between Britain and Japan. At that time, most of what mattered in Japan was concentrated on the country's largest island, Honshu (literally, *mainland*). The ancient capital, Kyoto, lay in the interior, but the modernizers wanted a coastal, outward-looking headquarters. So they chose the town of Edo, situated on a large bay where Honshu's eastern coastline bends sharply (Fig. 9A-1). They renamed the place ***Tokyo*** (meaning *eastern capital*), and little more than a century later it was the largest urban agglomeration on Earth. Honshu's coasts were also close to mainland Asia, where raw materials and potential markets for Japanese products could be found. Importantly, the notion of a greater Japanese empire, too, was inspired in part by British example.

But the Japanese reformers who oversaw this process of **modernization [10]** managed to build on, not replace, Japanese cultural traditions. We in the Western world tend to equate modernization with **Westernization [11]**: urbanization, the spread of transport and communications facilities, the establishment of a market (money) economy, the breakdown of local traditional communities, the proliferation of formal schooling, and the acceptance and adoption of foreign innovations. In Japan, and later in other East and Southeast Asian countries, modernization was viewed as progress mainly confined to the introduction of new technologies and better methods of production. Modernization served to increase efficiency, from producing goods to running railroads to fighting war. But modernization need not affect culture, especially cultural norms and values. In their view, a society can be modernized without being Westernized.

In this context, Japan's modernization was a novelty. Having long resisted foreign intrusion, the Japanese did not achieve the transformation of their society by importing a Trojan horse; it was guided by Japanese planners, building on the existing Japanese infrastructure, to fulfill Japanese objectives. Certainly, the Japanese imported foreign technologies and adopted innovations from the British and others, but the Japan that was built, a unique combination of modern and traditional elements, was essentially a Japanese achievement.

Japan in China

By the end of the 1880s, Japan had already built the strongest military force in East Asia and was prepared to use it. The so-called First Sino-Japanese War of 1894–1895 resulted in a Japanese victory: a long-term Japanese presence on the mainland in the Northeast, Japanese control of Korea, and the effective Japanese annexation of Taiwan (renamed Formosa). If it was not yet clear that power in East Asia had shifted to Japan, this would become obvious when the Japanese defeated Russia in their brief war of 1904–1905, both powers having had imperialist designs on northeastern China (the Russians were especially interested in Port Arthur, the Yellow Sea port now known as Dalian). Along the way, Japan became ever more aggressive in its foreign policy: in 1910, it annexed Korea; in 1931, it took firm control of Manchuria (Northeast China); in 1940, it invaded French Indochina, the U.S.-controlled Philippines, the Dutch East Indies (now Indonesia), and the British colonies of Burma and modern-day Malaysia; and finally, on December 7, 1941, it ignited World War II in the Pacific by attacking the United States at Pearl Harbor in Hawai'i (Fig. 9A-10).

The Japanese-Chinese conflict during this era had exposed the weakness of China, and that in turn fueled a drive for change (and rehabilitation) among the Chinese. Although many foreigners fled China during the 1920s and 1930s, others took advantage of the opportunities presented by the contest between the Nationalists and

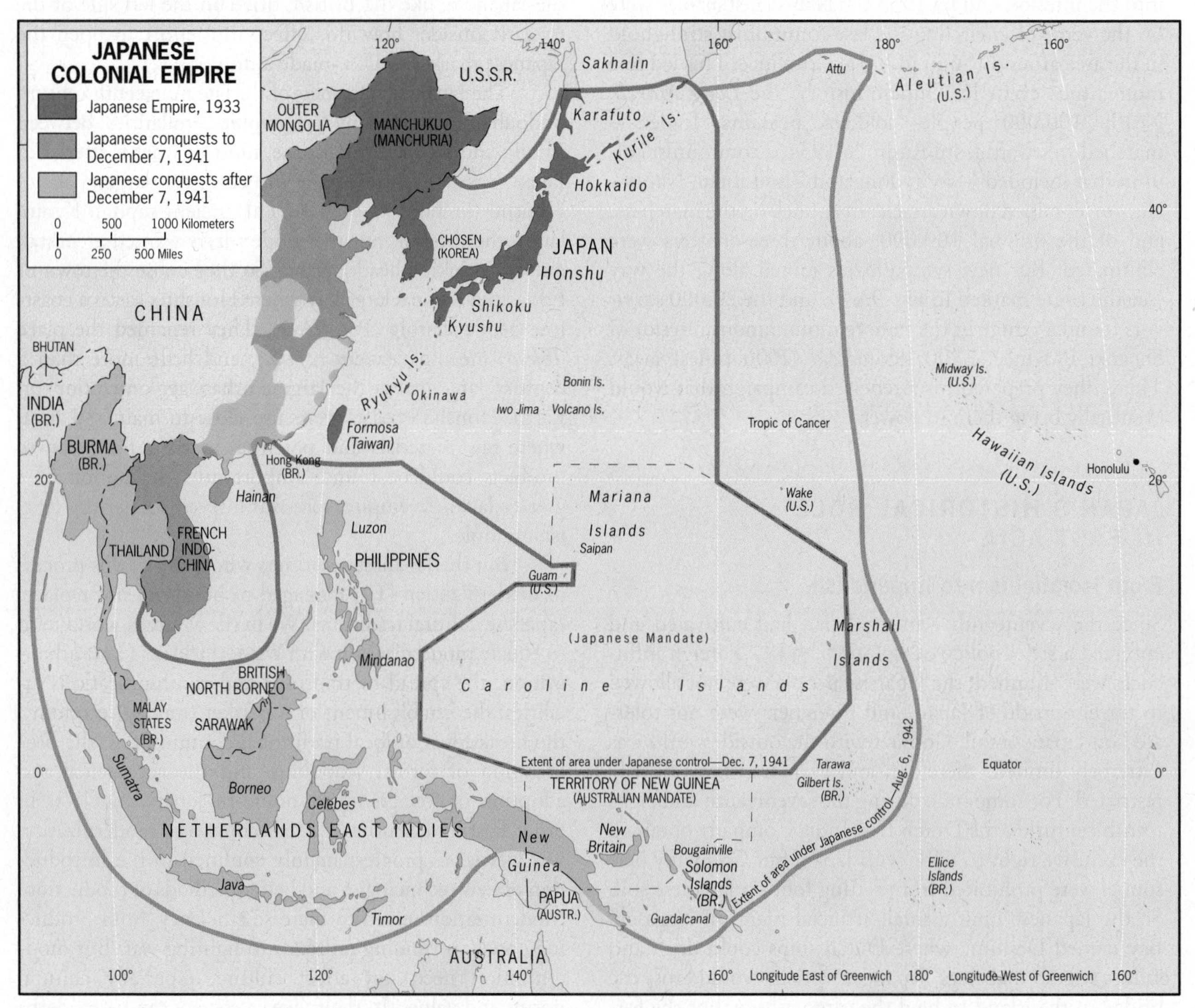

FIGURE 9A-10

communists. The Japanese took control over the Northeast, and when the Nationalists proved unable to dislodge them, they established a puppet state there, appointed a Manchu ruler to represent them, and named their possession *Manchukuo*.

The inevitable full-scale war between the Chinese and the Japanese broke out in 1937, with the Nationalists bearing the brunt of it (providing the communists with even more time and room to regroup). Figures 9A-9 and 9A-10 show how much of China the Japanese conquered. The Nationalists moved their capital inland to Chongqing, and the communists controlled the territory centered on Yanan to the north. China had been effectively broken into three pieces.

POST–WORLD WAR II EAST ASIA

Communist China

After the U.S.-led Western allies defeated Japan in 1945, the civil war in China quickly resumed. The United States, hoping for a stable and friendly government in China, attempted to mediate the conflict but at the same time recognized the Nationalists as the legitimate government. The United States also aided the Nationalists militarily, destroying any chance of genuine and impartial mediation. By 1948, it was quite clear that Mao Zedong's well-organized militias would defeat Chiang Kai-shek's forces. The remnants of Chiang's faction gathered Chinese treasures and valuables and then fled to the nearby island of Taiwan. There they swiftly acquired control of the government and proclaimed their own Republic of China. Meanwhile, on October 1, 1949, standing in front of the assembled masses at the Gate of Heavenly Peace in Beijing's Tiananmen Square, Mao Zedong proclaimed the birth of the People's Republic of China.

Under communism, Chinese society was completely overhauled and the dynasties soon seemed part of an ancient past. Benevolent or otherwise, the dynastic rulers of old China headed a country in which—for all its splendor, strength, and cultural richness—the fate of landless people and of serfs often was indescribably miserable; in which floods, famines, and diseases could decimate the populations of entire regions without any help from the state; in which local lords could (and often did) repress the people with impunity; in which children were sold and brides were purchased. The European intrusion made things even worse, bringing slums, starvation, and deprivation to the millions who had moved to the cities.

The communist regime, dictatorial and brutal though it was, attacked China's weaknesses on several fronts, mobilizing virtually every able-bodied citizen in the process. Land was confiscated from the wealthy; farms were collectivized; dams and levees were built with the hands of thousands; the threat of hunger for millions receded; health conditions improved; child labor was reduced; literacy was encouraged.

At the same time, the regime committed colossal errors and engaged in systematic repression of its own people at an unheard of scale. The so-called ***Great Leap Forward*** (the propaganda term for what this was supposed to be) became what was perhaps the worst human-engineered catastrophe in the history of the world. A combination of delusional attempts at revolutionary glory, economic incompetence, and extreme ruthlessness is estimated to have caused between 30 and 45 million deaths—mostly as a direct result of starvation, but with many others falling victim to murder, torture, and exhaustion due to forced labor. The idea was to enforce labor-intensive industrialization through the compulsory enlistment of large rural populations. But the results were inferior industrial products and a horrific disruption of agriculture.

Mao ruled China from 1949 to 1976, long enough to leave lasting imprints on the state. Another aspect of his communist ideology had to do with population. Like the Soviets (and influenced by a horde of Soviet advisors and planners), Mao refused to impose or even recommend any population policy, arguing that such a policy would represent a capitalist plot to constrain China's human resources. As a result, China's population mushroomed explosively during his rule.

Yet another disastrous episode of Mao's rule was the so-called ***Great Proletarian Cultural Revolution***, launched during his final decade in power (1966–1976). Fearful that Maoist communism was becoming contaminated by Soviet "deviationism" and concerned about his own stature as its revolutionary architect, Mao unleashed a vicious campaign against what he viewed as emerging elitism in society. He mobilized young people living in cities and towns into cadres known as Red Guards and commanded them to attack "bourgeois" elements throughout China, criticize Communist Party officials, and root out "opponents" of the system. The results were truly staggering: thousands of China's leading intellectuals died; moderate leaders were purged; and teachers, elderly citizens, and older revolutionaries were tortured to make them confess to crimes they did not commit. As the economy suffered, food and industrial production declined steadily. Violence and famine killed as many as 30 million people as the Cultural Revolution spun out of control. One of those who survived was a Communist Party leader who had himself been purged and thereafter been reinstated—Deng Xiaoping. Deng was destined to emerge as the country's leader in the post-Mao period of economic transformation.

Japan's Defeat and Recovery

During World War II, Japan had expanded its domain farther than the architects of the 1868 modernization could ever have imagined. But by 1945, when American nuclear bombs devastated two Japanese cities and produced surrender shortly thereafter, the expansionist era was over and the country lay in ruins. But once again, aided this time by an enlightened U.S. postwar administration, Japan surmounted

From the Field Notes . . .

© H.J. de Blij

© H.J. de Blij

"Visiting the Peace Memorial Park in Hiroshima is a difficult experience. Over this site on August 6, 1945 began the era of nuclear weapons use, and the horror arising from that moment in history, displayed searingly in the museum, is an object lesson in this time of nuclear proliferation. In the museum is a model of the city immediately after the explosion (the red ball marks where the detonation occurred), showing the total annihilation of the entire area with an immediate loss of more than 80,000 people and the death from radiation of many more subsequently. In the park outside, the Atomic Bomb Memorial Dome, the only building to partially survive the blast, has become the symbol of Hiroshima's devastation and of the dread of nuclear war."

www.conceptcaching.com

Concept Caching

The Wages of War

JAPAN AND THE former Soviet Union never signed a peace treaty to end their World War II conflict. Why? There are four reasons, and all of them are on the map just to the northeast of Japan's northernmost large island, Hokkaido (see Fig. 9A-1). Their names are Habomai, Shikotan, Kunashiri, and Etorofu. The Japanese call these rocky specks at the southern end of the Kurile Island chain their "Northern Territories." The Soviets occupied them just before the end of the war and never gave them back to Japan. Now they are part of Russia, and the Russians have not given them back either.

The islands themselves are no great prize. During World War II, the Japanese brought 40,000 forced laborers, most of them Koreans, to mine the minerals there. When Russia's Red Army overran them in 1945, the Japanese were ordered out, and most of the Koreans fled. Today the population of about 19,000 is overwhelmingly Russian, most of them members of the military based on the islands and their families. At their closest point the islands are only 5 kilometers (3 mi) from Japanese soil, a constant and visible reminder of Japan's defeat and loss of land. Moreover, territorial waters bring Russia even closer, so that the islands' geostrategic importance far exceeds their economic potential.

All attempts to resolve this issue have failed. In 1956, Moscow offered to return the two smallest, Shikotan and Habomai, but the Japanese declined, demanding all four islands back. In 1989, then-Soviet President Mikhail Gorbachev visited Tokyo in the hope of securing an agreement. The Japanese, it was widely reported, offered an aid-and-development package worth U.S. $26 billion to develop Russia's eastern zone—its Pacific Rim and the vast resources of the eastern Siberian interior. This would have begun the transformation of Russia's Far East, stimulated the ports of Nakhodka and Vladivostok, and made Russia a participant in the spectacular growth boom of the western Pacific Rim.

But it was not to be. Subsequently, Russian presidents Yeltsin, Putin, and Medvedev also were unable to come to terms with Japan on this issue, facing opposition from the islands' inhabitants and from their own governments in Moscow. And so the Second World War, nearly 70 years after its conclusion, continues to cast a shadow over this northernmost segment of the Asian Pacific Rim.

disaster. The Japanese were forced to accept a new constitution, and territorial adjustments were imposed (see the box titled "The Wages of War"). The new constitution stipulated that the country could not spend more than 1 percent of its GDP on the military and it had to accept the permanent stationing of American troops on its soil. This was meant to constrain Japan in terms of any possible future expansionist urges—and it probably did. It also induced Japan to shift its focus from military might to economic prowess. And once again everything turned out to be extraordinarily successful.

EAST ASIA'S ECONOMIC STATURE

Japan's Postwar Transformation

Japan's accelerated economic recovery and rise to the status of world economic superpower was one of the greatest success stories of the second half of the twentieth century. Japan had lost a war and an empire, but now was scoring many economic victories in a new global arena. Japan became an industrial giant, a technological pacesetter, a fully urbanized society, a political power, and one of the most affluent nations on Earth.

Japan once again proved adept at emulating the West: by 1980, the U.S. automobile industry had effectively been competed out of the market by the likes of Toyota and Honda. The very same thing happened in the domain of consumer electronics and other high-technology products. Japan did it better and cheaper than anyone else. Cities everywhere have reliable Japanese cars on their streets; people across the globe listen to Japanese portable audio players; laboratories the world over use Japanese optical equipment. From microwave ovens to smartphones, from oceangoing ships to high-definition TVs, Japanese-designed products flood global markets—even if many of these Japanese companies now have their products made in China.

For the past two decades, however, it should be noted that Japan's economy has been virtually stalled. Even though it is still the third-biggest economy in the world, it has lost much of its dynamism and momentum. Japan's current challenges are elaborated in Chapter 9B.

The Asian Tigers

The Japanese economic miracle was replicated, in turn, by the four **Asian Tigers [12]**: Hong Kong, South Korea, Taiwan, and Singapore. In the 1960s and 1970s, they embarked on similar strategies that resulted in rapid industrialization propelled by the attraction of foreign investment and the creation of export processing zones for the manufacturing of **high-value-added goods [13]**, including computers, mobile phones, kitchen appliances, and a plethora of electronic devices. The Tigers quickly became trading nations, and they particularly oriented themselves to the most affluent Western markets. Today, it is no coincidence that most of the world's largest ports are located on the shores of East Asia (see photo on the opening page of this chapter), with their exports shipped mainly to North America and Europe.

Note that the Asian Tigers (as well as Japan) share yet another characteristic: while they all implemented significant market liberalization measures (e.g., reducing tariffs or limits on foreign investment), they also maintained strong central governments that exerted a major, sometimes authoritarian influence on their economies. It might even be argued that this is a reflection of "Asian values" reminiscent of Confucius—especially loyalty to the family, the corporation, and the nation combined with the suppression of individual desires.

China's Economic Miracle

After Mao Zedong's death in 1976, China began a historic metamorphosis that was to have the widest global impact. The essence of China's transformation was comparable to what the Asian Tigers had achieved earlier: the creation of a favorable environment for foreign investment to support the growth of a manufacturing sector mostly geared toward exports. Chinese wages, at least initially, were kept low, and training programs were aimed at constantly upgrading the skills of the local workforce. At the same time, political conditions remained stable because in China, more than anywhere else, the government maintains tight controls. This was (and is) a communist state—but one that proved to be extremely adept at understanding how global capitalism operates and how to put it

Hong Kong Island (left photo), long the focus of a British Crown Colony (until 1997) but always closely linked to adjacent mainland China, is one of the most densely built-up areas in the world, with its high-rise towers crammed together cheek by jowl. This former city-state has a major, globally-oriented central business district and offers a high-end residential lifestyle for those who can afford it. English is still the chief language and many multinational corporations have major offices here. And just a short ferry ride across the water to the north lies Hong Kong's port and Kowloon (right photo), a decidedly less internationalized sector of the metropolis—still vibrant but with a lower cost of living and catering primarily to ethnic Chinese residents.

© Jan Nijman

© Jan Nijman

From the Field Notes . . .

"I could not quite believe my eyes when I came upon this Starbucks in the middle of Beijing's Forbidden City: this icon of American consumerism, wrapped in ancient Chinese architecture, in the heart of this revered complex, in a country that calls itself communist? It was back in 2004 that the Chinese authorities had agreed to allow the Starbucks café on the premises in return for a hefty financial contribution to ongoing restoration efforts. But not everybody agreed this was a good idea. Several Chinese groups protested against this "'erosion" of Chinese culture, and by 2007 the government felt compelled to reverse its decision. The deal was rescinded and Starbucks was evicted from the Forbidden City. It was a splendid illustration of the pragmatism that marks the culture of the Chinese and permeates their economic policies. Call it capitalism or authoritarianism, but the Chinese prefer to do whatever works given the circumstances, and they have little time for ideological principles."

© Jan Nijman

www.conceptcaching.com

to use. These days, pragmatism has unmistakably become the hallmark of Chinese policies.

It was as if Japan and its Tiger emulators had merely been warm-ups for East Asia's main economic act: the colossal rise of China. In less than a quarter of a century, China emerged as the most dynamic and fastest-growing component of the world economy. To be sure, Japan and the Tigers had experienced double-digit growth rates as well, but it is a different story altogether when the leading character is a country of more than a billion people. In 2010, China surpassed Japan to become the second-largest economy in the world, and its ascent steadily proceeds in this decade.

GEOPOLITICS IN EAST ASIA

Sino-Japanese Relations

It should not come as a surprise that Chinese-Japanese relations are problematic. A hundred years ago, China was in disarray and Japan was taking East Asia by storm—as well as by force. The contrast was stark: one was a wounded civilization, the other was beating the Western powers at their own game (at least for a time). Japan proudly proclaimed to stand for pan-Asian ideals, and that is how it legitimized its invasion and occupation of China. But the Japanese committed unspeakable atrocities during their campaign in China. Millions of Chinese citizens were shot, burned, drowned, subjected to gruesome chemical and biological experiments, and otherwise wantonly victimized. When you ask the Japanese, they say World War II began in 1941 when the United States declared war right after Pearl Harbor. When you ask the Chinese, they say World War II began in 1931 when the Japanese invaded Manchuria.

Decades later, when China's economic reforms of the 1980s and 1990s led to a renewed Japanese presence in China, the Chinese public and its leaders called for Japan to acknowledge and apologize for these wartime crimes against humanity. The unqualified apology the Chinese desire, in word and deed, has not been forthcoming. Some Japanese history textbooks still avoid acknowledging what happened to the satisfaction of the Chinese, and surveys indicate strong public sentiment on this still-sensitive issue.

But now that China has surpassed Japan in terms of economic prowess; now that China is the biggest exporter in the world; and now that Japan's economy has been stagnating for nearly a generation—this time it is China that is full of confidence and on its front foot. Although the two countries have close economic ties, their diplomatic relations are strained by clashing interests, historical memory, and cultural friction.

One of the most recent flashpoints to arise involves the Sino-Japanese-Taiwanese dispute over the Senkaku Islands in the East China Sea (Fig. 9A-11). These tiny uninhabited islands were seized by Japan in 1895 but are claimed by the Chinese, who call them the Diaoyu Islands. And they are also claimed by Taiwan, which refers to them as the Diaoyutai Islands. It is not so much the intrinsic

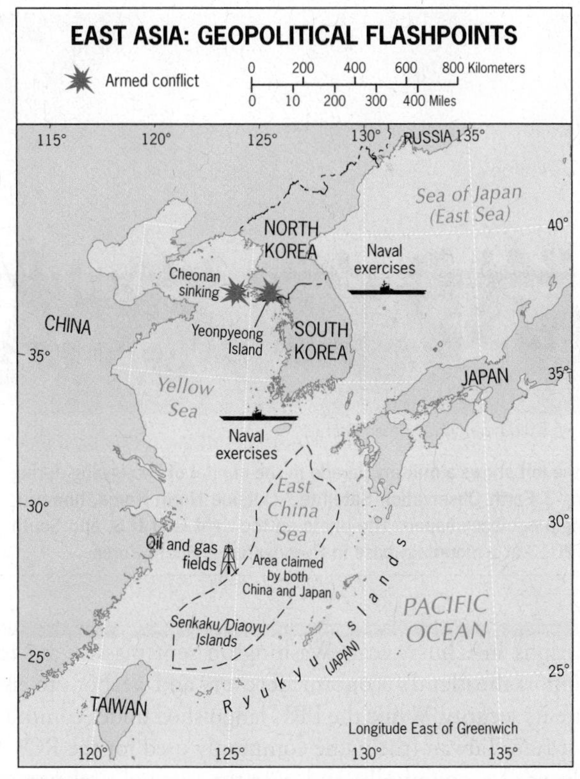

FIGURE 9A-11 © H. J. de Blij, P. O. Muller, J. Nijman, and John Wiley & Sons, Inc.

value of the islands that is at stake here: it is far more a matter of national pride and entitlement. Moreover, some oil and gas deposits have recently been discovered in this area, and ownership of the islands grants rights to their territorial waters and what lies beneath them. In any event, by 2013 China's heightened belligerence over these 'rocks' had brought Chinese-Japanese relations to their lowest level in decades. It may well be that China has the most compelling claims to these islands, because eighteenth- and nineteenth-century maps (from both China and Japan) seem to corroborate that they were in Chinese hands before the Japanese takeover.

The Korea Factor

Throughout history, the Koreans have repeatedly been divided, partitioned, colonized, and occupied. And even when outsiders were not involved, their indigenous kingdoms struggled for supremacy. During the Qing Dynasty, Chinese emperors intervened at will. As China fell apart after 1900, the Japanese conquered the Korean Peninsula and annexed all of it as their colony in 1910.

When Japan was defeated in 1945, the Allied powers divided Korea for "administrative" purposes. The territory north of the 38th parallel was placed under the control of Soviet forces; south of this latitude, the United States was in control. In 1950, communist forces from North Korea invaded the South in a forced-unification drive, unleashing a devastating conflict that first swept southward across the peninsula, then back northward across the 38th parallel, and finally drew in Chinese armed forces who pushed the front southward again. By 1953, a military stalemate halted the hostilities, and the Korean War ended at the Cease-Fire Line not far from where the 38th parallel had marked the original 1945 boundary. Ever since, a heavily armed demilitarized zone (DMZ) has more or less hermetically sealed North from South—with the two Koreas, having grown apart, still in danger of renewed conflict.

But this is not just a bilateral issue. The Korean conflict has long had realmwide and even global implications. One reason is that North Korea's nuclear capability has been in the hands of perhaps the most repressive and archaic regime on Earth—unreliable, unpredictable, and without scruples. North Korea's nuclear missiles may be intended primarily for South Korea, but (theoretically at least) can also reach China, Russia, Japan and perhaps even North America. Moreover, the North Koreans have carried out three underground nuclear weapons tests since 2006, the last one in February 2013, thereby increasing the possibility of a nerve-wracking East Asian arms race.

The North Korea issue also deepens divisiveness in the realm. South Korea and Japan are diametrically opposed to the regime based in Pyongyang, whereas China takes a far more neutral position, at times even seeming to lend the North Korean regime support. The Chinese appear to use North Korea in their dealings with Japan and the United States, since it is widely thought that China is crucial to containing the North Korean threat.

Ever since the end of the Second World War in 1945, Japan has adhered to a constitution that essentially forbids its rearmament and commits it to a relationship with the United States involving the stationing of tens of thousands of American armed forces on Japanese soil. But Japan's 2009 election campaign brought to the fore an unusually forceful reappraisal of these issues, and public opinion has been shifting toward a stronger military posture and the ouster of U.S. troops. One result of the international community's failure to constrain North Korea's nuclear aspirations may be Japan's military revival. The country has gradually increased its military posture and billions have been invested in a sophisticated missile defense system to protect against potential North Korean (or Chinese) attacks.

The North Koreans, in turn, are aware of latent Chinese support, and this seems to encourage them to test the waters. In 2010, a North Korean submarine torpedoed the *Cheonan*, a South Korean naval vessel patrolling off the west coast of South Korean near the boundary with the North, killing 46 (Fig. 9A-11). South Korea, the United States, Japan, and much of the world strongly condemned this act of violence—yet China pointedly did not.

The death of the North's leader, Kim Jong-il, in December 2011 and the accession of his son, Kim Jong-un, coincided with another critical food shortage among the North Korean people, more desperately in need of outside aid than ever. For a moment, in early 2012, it looked like an agreement had been achieved for U.S. food aid in

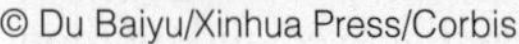
© Du Baiyu/Xinhua Press/Corbis

© Park Jin-hee/Xinhua Press/Corbis

North Korea continues to be a major regional security concern. The photo on the left shows a military parade in the capital of Pyongyang during December 2012, celebrating the successful launch of the "Kwangmyongsong-3 Earth Observation Satellite." Outside North Korea, however, it was widely interpreted as a test of a long-range missile capable of delivering a nuclear weapon. The photo on the right is of U.S. and South Korean marines involved in a joint cold-weather training exercise in February 2013 at a mountain base in Pyeongchang, South Korea.

return for a promise to end nuclear (missile) testing—but the deal collapsed a few weeks later following North Korea's (failed) launch of a rocket capable of delivering a ballistic missile. The North Koreans finally managed a successful launch several months later and U.S.-North Korean talks were suspended.

In the spring of 2013, North Korea's blustering reached new heights with repeated threats of "thermonuclear war" against the South and even against the United States. Such threats have been part of a standard playbook in recent years, and tend to coincide with the annual joint military exercises of South Korean and U.S. troops. But this time they were louder than ever, and the unpredictability of the young, untested Kim Jong-un raised alarms throughout the region and well beyond.

Taiwan: The Other China

Mention the island of Taiwan in China, and you are likely to be greeted with a frown and a headshake. Taiwan, your host may tell you, is a problem foreigners do not understand. Virtually all of the 23 million people of Taiwan are Chinese. Taiwan was part of China during the Qing Dynasty. Taiwan was stolen from China by Japanese imperialists in 1895, when it was known as Formosa. Then, when communists and Nationalists were fighting each other for control of mainland China right after World War II, and the communists were about to win, the Nationalists in 1949 fled by plane and boat to Taiwan, where they overpowered the locals. Even as Mao Zedong was proclaiming the birth of the People's Republic of China (PRC) in Beijing, the loser, Chiang Kai-shek, named his regime in Taiwan the Republic of China (ROC)—and told the world that he headed China's "legitimate" government.

The PRC, of course, ridiculed this assertion, but the ROC had powerful friends, especially the United States. Chiang Kai-shek's regime was soon installed at the United Nations in China's seat. Washington sent massive aid to support the island's economic recovery and weapons to ensure its security. While the PRC languished under communist rule, Taiwan (the name commonly used for the ROC) advanced economically, and over time its political system matured into a functioning (if turbulent) democracy. What the Taiwanese sought was economic prosperity, and they got it.

But to the PRC, Taiwan is regarded as a "wayward province" that must be reunited with the motherland. When U.S. President Nixon arrived in Beijing in 1972 for a historic visit that was to change the world, Taiwan was a bargaining chip. Soon, the ROC's UN delegation was dismissed and representatives of the Beijing government were seated in its place. Many countries around the world that had recognized Taiwan as the legitimate heir to China's leadership now suddenly changed sides. Meanwhile, Beijing's leaders set about trying to isolate the ROC, and to a large extent they succeeded.

Geography, however, was to intervene. With billions of U.S. dollars in reserves, fruitful connections to Overseas Chinese in Southeast Asia, and its emergence as an Asian Tiger economic powerhouse, Taiwan had some significant cards to play—and the Beijing regime discovered it could not afford to deny Taiwanese companies permission to exploit opportunities in the PRC's development zones. And so, via the "back door" of Hong Kong, Taiwanese entrepreneurs built thousands of factories in mainland China, many of them located directly across the Taiwan Strait. Taiwanese businesspeople now pumped hundreds of millions of dollars into China's development boom, and the economies of Taiwan and the PRC found themselves on a path toward ever tighter integration. In 2012, an estimated 27 percent of Taiwan's exports went to China along with substantial Taiwanese foreign direct investment, and at least 800,000 Taiwanese were living and working (temporarily) on the mainland.

Today, even in difficult economic times, per capita annual income in Taiwan exceeds U.S. $20,000—more than triple that of China and on a par with South Korea. Even though a sizeable majority of Taiwanese oppose re-unification with the PRC, few would want to see their economic well-being imperiled by political adventures.

China Goes Global

Napoleon once famously described China as a sleeping giant that would shake the world when it finally awoke. China today is indeed awake and changing the world—but it is changing itself and the surrounding East Asian realm even more. A closer look at this transformation within the framework of East Asia's regions is the objective of Chapter 9B.

POINTS TO PONDER

- The fortunes of China and Japan have always seemed to be inversely related: when one was rising, the other was in decline.
- East Asia constitutes the biggest manufacturing complex in the world, but the realm's natural resources are limited.
- In the past decade, China's defense spending has increased by about 12 percent per year, much of it intended to deter the United States from intervening in a possible crisis over Taiwan.
- The impoverished North Korean people continue to suffer horrendous famines and mass imprisonment while their government dedicates massive resources to building nuclear weapons.

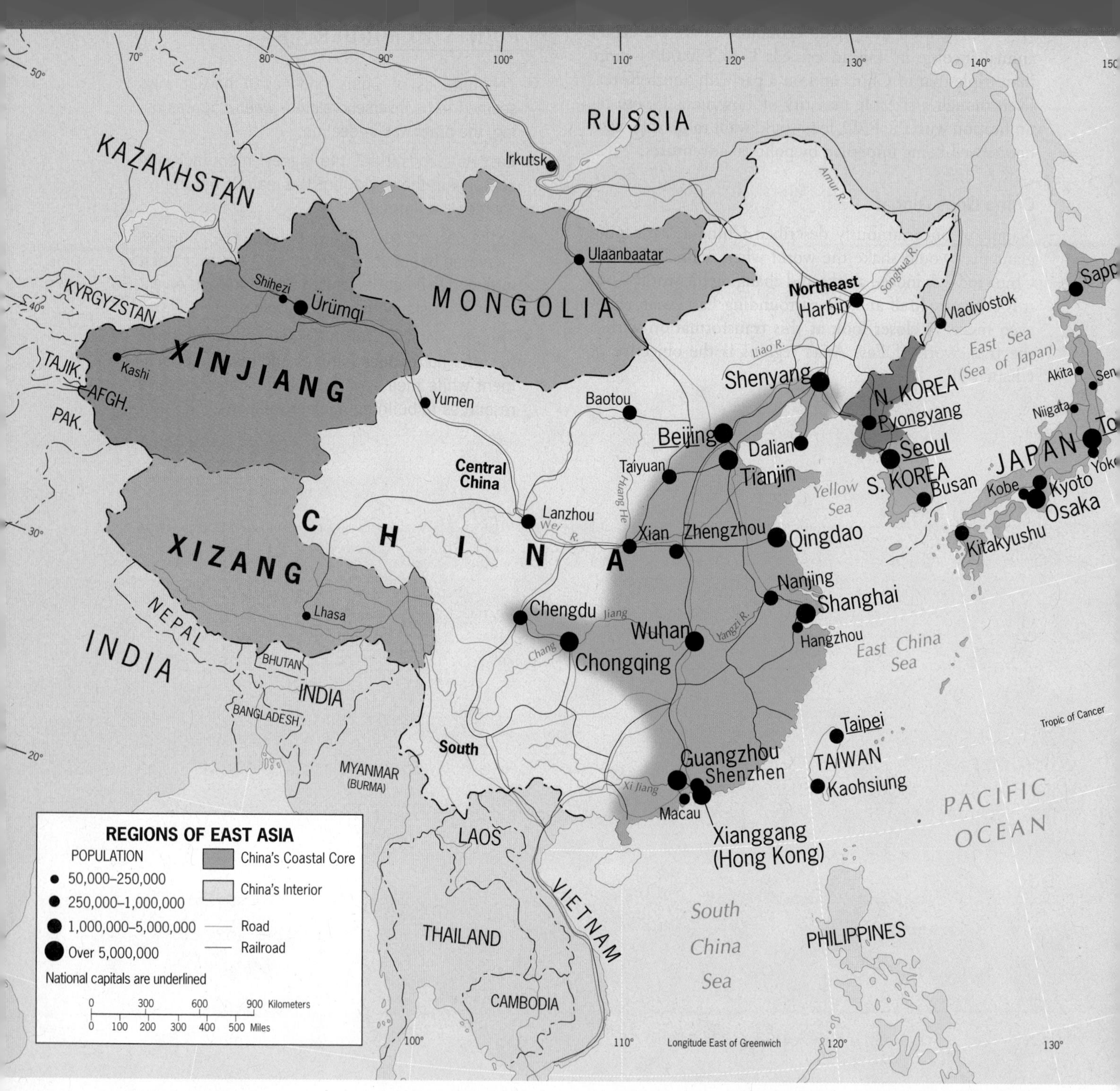
REGIONS OF EAST ASIA
POPULATION
50,000–250,000
250,000–1,000,000
1,000,000–5,000,000
Over 5,000,000
China's Coastal Core
China's Interior
Road
Railroad
National capitals are underlined
0 300 600 900 Kilometers
0 100 200 300 400 500 Miles
RUSSIA
KAZAKHSTAN
MONGOLIA
XINJIANG
XIZANG
CHINA
KYRGYZSTAN
TAJIK.
AFGH.
PAK.
NEPAL
INDIA
BHUTAN
BANGLADESH
MYANMAR (BURMA)
LAOS
THAILAND
CAMBODIA
VIETNAM
N. KOREA
S. KOREA
JAPAN
TAIWAN
PHILIPPINES
Irkutsk
Ulaanbaatar
Shihezi
Ürümqi
Kashi
Yumen
Baotou
Central China
Northeast
Harbin
Vladivostok
Shenyang
Beijing
Tianjin
Dalian
Taiyuan
Lanzhou
Xian
Zhengzhou
Qingdao
Nanjing
Shanghai
Hangzhou
Chengdu
Chongqing
Wuhan
Lhasa
South
Guangzhou
Shenzhen
Macau
Xianggang (Hong Kong)
Taipei
Kaohsiung
Pyongyang
Seoul
Busan
Kobe
Kyoto
Osaka
Kitakyushu
Niigata
Akita
Amur R.
Songhua R.
Liao R.
Huang He
Wei R.
Chang Jiang
Yangzi R.
Xi Jiang
Yellow Sea
East Sea (Sea of Japan)
East China Sea
South China Sea
PACIFIC OCEAN
Tropic of Cancer
Longitude East of Greenwich

FIGURE 9B-1

REGIONS

China's Coastal Core

China's Interior

China's Western Periphery

Mongolia

The Korean Peninsula

Japan

Taiwan

IN THIS CHAPTER

- North Korea's bluster
- The miracle of Shenzhen
- China's persistent regional disparities
- One-child policy under fire
- The smothering of Tibetan culture
- Japan's energy dilemma
- Taiwan's future

CONCEPTS, IDEAS, AND TERMS

Floating population 1
Hukou system 2
Special Economic Zone (SEZ) 3
Economic geography 4
Overseas Chinese 5
Foreign direct investment (FDI) 6
Buffer state 7
Regional complementarity 8
State capitalism 9
Conurbation 10
Demographic burden 11
Geothermal power 12
Technopole 13

In Chapter 9A we defined the boundaries of the East Asian geographic realm, using its politico-geographical layout as a frame of reference. But when it comes to the regions that constitute this physiographically and culturally diverse realm, things get rather more complicated. One problem is the speed with which East Asia is changing. In the introductory chapter we emphasized that regions are always subject to change, but East Asia over the past half-century has been a special case. The spectacular takeoff and subsequent faltering of Japan; the meteoric rise of eastern China; the transformation of South Korea from moribund dictatorship to burgeoning, high-income democracy—these are just some of the changes affecting East Asia. So what follows reflects the situation in 2013, and we can only imagine what the regional geography of this realm will look like a decade from now.

© Jan Nijman

Where were these pictures taken?
Find out at www.wiley.com/college/deblij

REGIONS OF THE REALM

Half a century ago, some geographers described East Asia as a realm with an "empty heart"—not because China, its once and future core area, was depopulated, but because China lay isolated, with very limited contact with the outside world and less influence over its Eurasian neighbors. Imagine this: after the Chinese communists quarreled with their Soviet advisors in the 1950s and threw them out, just a few *dozen* foreigners were left in all of China. What little contact there was with the world beyond its borders went via Hong Kong, then still a British crown colony. China's cities were huge but showed little evidence of modernization. The old, colonial-era banks and other corporate buildings were still the newest structures (other than the gray, faceless tenements the Soviet advisors introduced) in their CBDs. Today, hundreds of thousands of foreigners transact business, teach, study, and live long-term in Chinese cities that now have skylines to match those of any city in the world, and more. And China is again—as it

© Jan Nijman

MAJOR CITIES OF THE REALM

City	Population* (in millions)
Tokyo, Japan	26.6
Shanghai (Shi), China	22.2
Beijing (Shi), China	17.4
Guangzhou, China	12.0
Shenzhen, China	11.8
Osaka, Japan	11.7
Chongqing (Shi), China	10.8
Wuhan, China	9.9
Seoul, South Korea	9.8
Tianjin (Shi), China	9.3
Chengdu, China	7.5
Hong Kong SAR, China	7.3
Nanjing, China	6.5
Harbin, China	6.3
Shenyang, China	5.9
Xian, China	5.4
Busan, South Korea	3.3
Taipei, Taiwan	2.7
Macau SAR, China	0.6

*Based on 2014 estimates

was during its dynastic period—the core of the East Asian realm. Indeed, this time around China is asserting itself as nothing less than a full-fledged global power.

The colonial powers employed the name China Proper to denote the "real" China, the China of those who call themselves the ***people of Han***. They do so because the Han Dynasty (206 BC–AD 220) is seen as the time when China became the real China, when the state adopted Confucian principles, culture flourished as never before, cities burgeoned (present-day Xian served as the "Rome" of China), armies conquered distant lands, trading goods moved across the heart of Eurasia via the Silk Road to and from the Roman Empire, and population grew steadily. Ever since, the ethnic Chinese majority has referred to itself and the country as *Han China*. Later, in the eyes of colonial-era foreigners, Han China was China Proper.

Han China approximately corresponds to the extensive ethnolinguistic area mapped as Han Chinese in Figure 9A-7 and, as Figure 9B-1 shows, it consists of two major regions. The core of Han China (and of China overall) lies in the eastern coastal provinces, anchored by the great cities of Beijing, Shanghai, Hong Kong, and Guangzhou. Today, this Coastal Core is not only East Asia's most influential region but also the most dynamic spatial component of the global economic system. The development success story that began here less than 40 years ago in a few designated port cities on the coast has steadily expanded to encompass a zone of rapid economic growth that now reaches as far west as Chengdu in Sichuan Province. But still farther west, in that vast crescent of territory centered on Lanzhou, lies quite another China—the region we call the Interior. Here the Han are rural, poor, and far removed—geographically as well as functionally—from the booming Pacific seaboard. The contrasts between this increasingly affluent Coastal Core and the much poorer, still-rural Interior have now become a matter of grave concern to the rulers of China, and policies to counter this unevenness have risen to the top of their agenda. Will the Core fracture? Is there still time for central government to bridge the gap between Coast and Interior?

Beyond the Interior, as Figure 9B-1 shows, lies the Western Periphery that includes blue-colored Tibet (which the Chinese call Xizang) and Xinjiang (colored brown) under Chinese control, and independent Mongolia (mapped in green) outside of it. And in the east, seaward of the mainland coast, is a part of the realm that could not be more different from the landlocked Western Periphery: here lie North Korea, where China has significant influence; South Korea, now a strong trading partner of China; Japan, a longtime rival of China; and Taiwan, which is claimed by mainland China and viewed as a temporarily wayward province.

To summarize, we will discuss East Asia's regions in the following sequence:

1. China's Coastal Core
2. China's Interior
3. China's Western Periphery
4. Mongolia
5. The Korean Peninsula
6. Japan
7. Taiwan

But in order to set the stage, we need to begin with a general overview of China as well as a profile of its administrative structure, because the latter is critical to the way in which this huge country is governed.

THE PEOPLE'S REPUBLIC OF CHINA

Citizens of the People's Republic of China (PRC) have no doubt: not only does their country still have the largest population of any state in the world ("still" because India is closing in), but theirs is the oldest continuous civilization on Earth. Yet China's territorial size does not match its demography: even with its 1 billion, 364 million people, China is only slightly larger than the conterminous United States (Fig. 9B-2). Distances in China are comparable to those in the Lower-48. It is about the same distance from Beijing to Shanghai as it is from New York to Chicago. And as the map shows, east-west distances also are similar. Latitudinally, however, China extends farther north as well as south of the conterminous United States, creating a greater range of natural environments. In the far northeast, China comes close to Siberian cold. The extreme south exhibits Caribbean

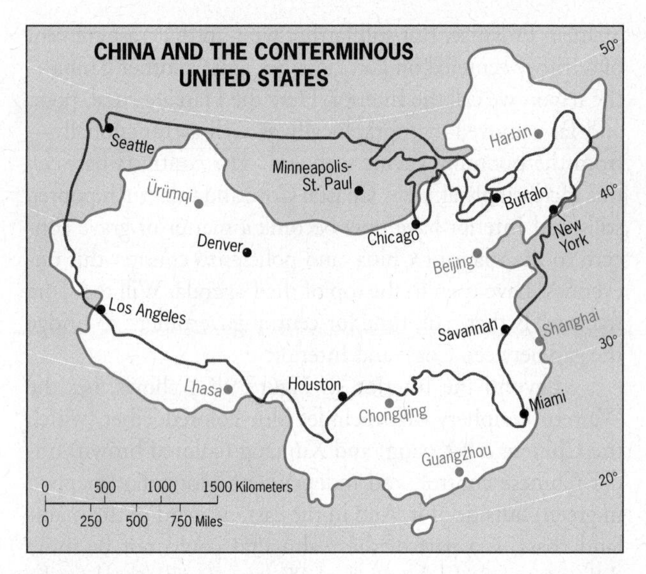

FIGURE 9B-2 © H. J. de Blij, P. O. Muller, and John Wiley & Sons, Inc.

warmth—with typhoons (as hurricanes are called in Pacific Asia) to match.

Political and Administrative Divisions

China and the United States may be about the same size, but they have very different governance structures: the United States is a federation whereas China is a very highly centralized unitary state. For administrative purposes, China is divided into the following units (Fig. 9B-3):

4 Central-government-controlled Municipalities (each is known as a *Shi*)

22 Provinces

2 Special Administrative Regions (SARs)

5 Autonomous Regions (ARs)

The four central-government-controlled ***Municipalities*** (or ***Shi's***) are the capital, Beijing; the nearby port

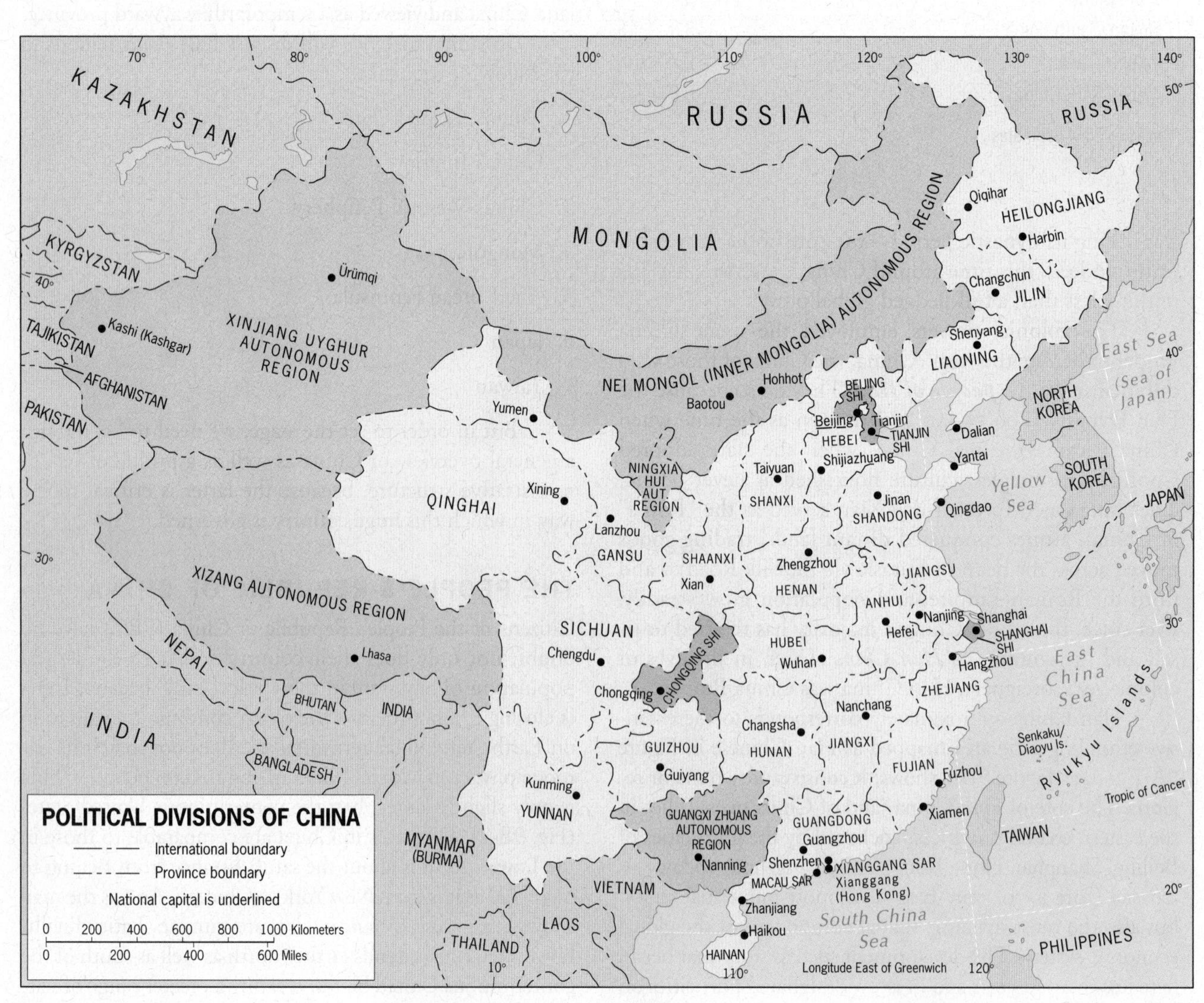

FIGURE 9B-3 © H. J. de Blij, P. O. Muller, and John Wiley & Sons, Inc.

AMONG THE REALM'S GREAT CITIES . . .

BEIJING

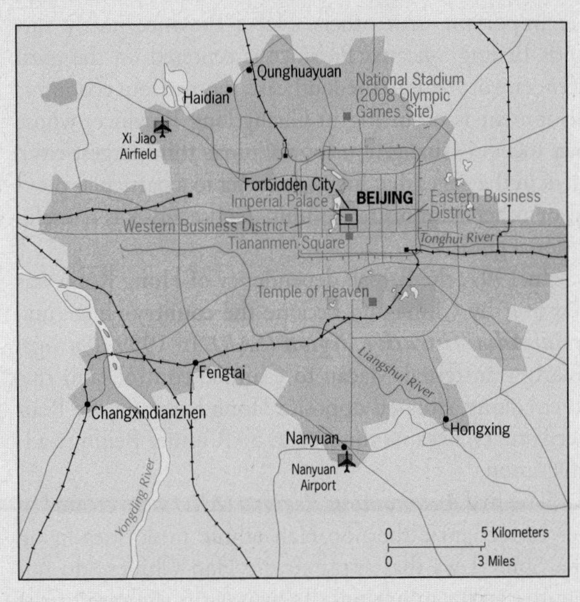

© H. J. de Blij, P. O. Muller, and John Wiley & Sons, Inc.

BEIJING (population 17.4 million), capital of China, lies at the northern apex of the roughly triangular North China Plain, just over 160 kilometers (100 mi) from its port, Tianjin, on the Bohai Gulf. Urban sprawl has now reached the hills and mountains that bound the Plain on the north, a defensible natural barrier fortified by the builders of the Great Wall. Today you can drive to the Great Wall from central Beijing in about an hour.

Although settlement on the site of Beijing began thousands of years ago, the city's rise to national prominence began during the Mongol (Yuan) Dynasty, more than seven centuries ago. The Mongols, preferring a capital close to their historic heartland, endowed the city with walls and palaces. Following the Mongols, later rulers at times moved their capital southward, but the government always returned to Beijing (whose name means "northern capital"). From the third Ming emperor onward, Beijing was China's focus; it was ideally situated for the Manchus of the northeast when they took over in 1644. During the twentieth century, China's Nationalists again chose a southern capital, but when the communists prevailed in 1949 they immediately reestablished Beijing as their (and China's) headquarters.

Ruthless destruction of historic monuments, carried out from the time of the Mongols to the communists, has diminished but not totally destroyed Beijing's heritage. With fifteenth-century monuments such as the Forbidden City of the Qing emperors (left photo) and the Temple of Heaven, Beijing remains an outdoor museum and the cultural focus of China. Successive emperors and aristocrats, moreover, bequeathed the city numerous parks and additional recreational spaces that other cities lack. Over the past two decades, Beijing has been transformed by China's new economic policies. A forest of ultramodern high-rises now towers above the retreating traditional cityscape, avenues have been widened, and expressways have been built. Unmistakably, a new era has dawned in this time-honored capital (right photo).

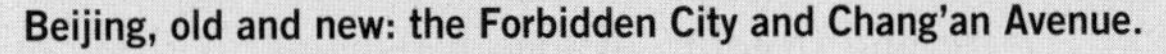

Beijing, old and new: the Forbidden City and Chang'an Avenue.

© Jan Nijman

© Jan Nijman

city of Tianjin; China's largest metropolis, Shanghai; and the Chang River port of Chongqing in the Sichuan Basin. These Shi's form the cores of China's most populous and important subregions, and direct control over them from the capital entrenches the central government's power.

We should keep in mind that the administrative map of China continues to change—and to pose problems for geographers. The city of Chongqing was made a Shi in 1996, and its municipal territory was enlarged to incorporate not only the central urban area but an enormous hinterland covering all of eastern Sichuan Province. As a result, the urban population of Chongqing is now officially 30 million, making this the world's "largest metropolis"—but in truth, the central urban area contains just under 11 million inhabitants. And because Chongqing's population is officially not part of the province that borders it to the west (Fig. 9B-3), the official population of Sichuan dropped by 30 million when the Chongqing Shi was created.

China's 22 ***Provinces***, like U.S. States, tend to be smallest in the east and largest toward the west. The territorially smallest are the three easternmost provinces on China's coastal bulge: Zhejiang, Jiangsu, and Fujian. The two largest are Qinghai, flanked by Xizang (Tibet), and Sichuan, China's Midwest. As with all large countries, some provinces are more important than others. Hebei Province nearly surrounds Beijing. Shaanxi Province is centered on the great ancient city of Xian. In the southeast, momentous economic development is occurring in Guangdong Province, whose urban focus is Guangzhou (now China's third-largest city). When, in the pages that follow, we refer to a particular province or other administrative unit, Figure 9B-3 is a useful locational guide.

In 1997, the British dependency of Hong Kong was taken over by China and became the country's inaugural ***Special Administrative Region (SAR)***. In 1999, Portugal similarly transferred Macau to Chinese control, and this former colony, situated opposite Hong Kong on the Pearl River Estuary, became the second SAR under Beijing's administration.

The five ***Autonomous Regions (ARs)*** were created in order to recognize the non-Han ethnic minorities living there. Some laws that apply to the Han Chinese do not apply to certain minorities. As we saw in the case of the former Soviet Union, however, demographic changes and population movements affect such regions, and the policies of the 1940s may not work in the twenty-first century. Han Chinese immigrants now outnumber several minorities in their own ARs. The label "autonomous" should not be taken literally here, and more than anything else it is a political gesture because Beijing maintains firm control in every corner of China. The five Autonomous Regions are: (1) Nei Mongol AR (Inner Mongolia); (2) Ningxia Hui AR (adjacent to Inner Mongolia); (3) Xinjiang Uyghur AR (China's broad northwestern corner); (4) Guangxi Zhuang AR (far south, bordering Vietnam); and (5) Xizang AR (Tibet, anchoring the southwest).

China's "Capitalist" Turn

It was not hard to see why China needed to change course in the late 1970s: agriculture had rebounded from the disastrous effects of the "Great Leap Forward" two decades earlier, but meaningful industrial progress continued to elude Beijing's economic planners. China was now falling further behind in terms of technology, a liability made abundantly clear in any comparison to the miraculous ascent of postwar Japan. And even that "wayward province" of Taiwan was outperforming the mainland—and at a Tigerish pace to boot. Something had to change.

Taking their cues from the accelerated development of Japan and the four trailing Asian Tigers, the new leadership saw opportunities that could not even be mentioned during Mao's long reign (1949–1976). For example, the Lower Chang Basin—the subregion anchored by Shanghai and interconnected by the Yangzi and its tributary waterways—had retained some of its historic identity and energy. Drab, teeming, and decrepit though it had become, Shanghai had not completely lost its intellectual vigor, artistic individuality, risk-taking entrepreneurs, or even its opponents of communist dogma. And Shanghai's geography still offered immense opportunities. From day one, the new regime in Beijing led by Deng Xiaoping and his pragmatists saw in Shanghai what they could not yet foresee in their own backyard. Here was a vibrant hub on the Pacific coast right at the mouth of the realm's greatest river, loaded with talent and accustomed to taking chances. And in its vast hinterland lay much more than a farmscape of flood-prone wheatfields: the Sichuan Basin alone had a population (at the time) of more than 100 million, growing everything from rice to tea and from fruits to spices.

If China's planners needed encouragement, all they had to do was look to the southernmost subregion of eastern China: the Pearl River Estuary, the primary outlet of the Xi River. There was Hong Kong, about as vivid a contrast between success and failure as China could present. When Deng took charge, Hong Kong was a thriving port city importing raw materials by the shipload and disgorging manufactured products that sold on markets all around the world. Hong Kong may have been a British crown colony, but it was Chinese managers and Chinese workers who propelled its economy. If you were a visitor to Hong Kong in the 1970s, they took you to a place on a nearby hillside from where you could peer across the fortified border into "Red China"—and observe some villages with duck ponds and rice-producing paddies as well as wooden fishing boats. The contrast was crystal clear.

China's newly formulated policies of the late 1970s, however, did not entail an official departure from communism. This is still the People's Republic of China: the Communist Party remains firmly in control, the Politburo still calls the shots, Mao is still officially revered like a deity, and freedom of speech remains anything but. Yet the Chinese leadership fundamentally reorganized the way the economy operates in China—at least in certain vital parts of it. China created the "perfect" conditions for foreign investors from around the globe with a docile, skilled, hardworking labor force; efficient facilities; a good infrastructure; fiscal advantages; and, not least, superb accessibility vis-à-vis the rest of East Asia, already the fastest-growing regional component in the world economy.

Population Issues

China has been the world's most populous state for many centuries, and during the population explosion of the twentieth century it was also one of the fastest-growing. Throughout Mao's rule, when China still was a dominantly agricultural society, families were encouraged to produce numerous children, and China grew at a rate approaching 3 percent per year. That was the policy inherited by Deng and his reformers, who immediately grasped that slowing down China's rate of growth was critical to its economic future. Accordingly, the new regime embarked on a vigorous population-control program that imposed on Han Chinese

Regional ISSUE The One-Child Policy

POPULATION CONTROL IS KEY TO OUR FUTURE

"I was born here in Shanghai in 1963, the youngest in a family of five children. My father and mother worked very hard all their lives but there was very little they could give us. We lived in a small tenement, sharing sanitation facilities with everyone else on our block. My siblings and I all got a basic education and basic health care but food was scarce and luxury did not exist. Of course, in the countryside things were much worse and the stories about the terrible famines and grinding poverty there are well-known.

"There can be no doubt that, if our government had not enforced family planning, China would never have taken off as a great economic power. And, more importantly, it would have been impossible for the Chinese people to improve their lives as they have done. Why do you think we are so far ahead of India in terms of life expectancy, literacy, income, child mortality rates, and so on? Indeed, within ten years or something like that, they say that India will surpass China as the biggest country in the world in terms of population. My condolences to them!

"Our policy works with incentives meant to discourage couples from having more than one child. Taxes can go up for those having more than one, or their youngest children will not get certain benefits from the state such as a free education. Don't forget, our state provides lots of things to the people, so it seems to me that it is only reasonable that they prevent people from abusing the system.

"Now let's look to the future. I run a factory here on the outskirts of the city and we employ 600 people, but because of rising wages we get more and more into automation, using robots. I know there have been temporary shortages in past years but most likely we will need less labor in the future. Agriculture is more and more mechanized; our economy moves from labor-intensive manufacturing to high-tech, IT, and services; and China is increasingly producing things abroad. Will we really need all that labor in the future? I don't think so.

"And I have not even said anything about the huge environmental burden of such an enormous population: the air in Beijing and Shanghai is hard to breathe and there aren't enough parking spaces for all the cars. It is hard to imagine the environmental cost of every Chinese family driving a car, isn't it? We must stick with tight family planning, for the good of China and its prosperity."

THE ONE-CHILD POLICY IS WRONGHEADED AND A DISGRACE

"Rather than getting into these abstract debates about what is good for China's future—and nobody can tell the future anyway—let me tell you a real-life story that recently circulated on the Internet all over China. And I think you will understand that I would rather remain anonymous when expounding my views. There was a young couple in a small village in the north-central province of Shaanxi. They had one child and the woman was 7 months pregnant with their second. The local authorities found out and slapped them with a huge fine of 40,000 yuan (about U.S. $5,500). If she paid the fine, she could keep the child and obtain *hukou* privileges. Well the husband earned just 4,000 yuan a month working at a local hydroelectric power station and so they did not have the money. On May 30, 2012, the husband set out for the coal mines of Inner Mongolia for a month or so to earn extra money. Then the local officials (thugs, better put) go to their home, basically abduct the woman, and force her to sign a consent form to abort the baby. They restrained her and gave her an injection in her belly. The next day she gave birth to a dead 7-month old baby.

"This is the daily reality of the one-child policy for lots of people, especially the poor. The reason that more and more Chinese are protesting the policy is that the stories get out on the Internet and people are now daring to raise their voices. It was the family of the woman who posted the story online. You see, this is about basic human rights! How can such practices ever be approved on the basis of abstract economic planning? The so-called 'incentives' are measures of force, that is what they are, and they deprive people of fundamental human choices.

"Moreover, this policy is now completely outdated. First, China's population will begin to shrink by 2026 and the working population is already declining as we speak. We need more, not fewer, people who can provide for the old and the young. When will our government understand that we actually have, as a result of this one-child policy, a shortage of working-age people as well as a shortage of women?

"Finally, most people in China are already deciding for themselves that they don't want a second child, let alone a third. The cost of living is too high and living quarters are too small. Today in China, people want to improve their material circumstances and quality of life. If a smaller number of couples, mostly living in rural areas, desire more than one child, let it be so because, overall, our population is not going to keep growing anyway.

"The one-child policy was immoral and unethical from the start. Today it has become a counterproductive approach as well. The sooner it is suspended, the better."

Vote your opinion at www.wiley.com/go/deblijpolling

families (but not minorities) a one-child limit. Enforced by sometimes draconian methods, this policy had the desired effect: by the mid-1980s China's annual population growth was down to 1.2 percent, and by 2012 the rate had dipped to just 0.47 percent (less than half the worldwide average of 1.2 percent).

The one-child policy had its desired economic impact, but its other results were less salutary. China's is a dominantly patriarchal society, where male children are much preferred; rates of female fetus abortion, infanticide, and abandonment skyrocketed, resulting in a gender imbalance that raises concern for the future. China's administration estimates that, if the policy is not modified, the society will be short some 30 million brides (the reported gender imbalance in births for 2012 was 117 boys for every 100 girls). This is already increasing the trafficking of women both within China and from neighboring countries, where widespread resentment of Chinese males pursuing (and sometimes abducting) local females is growing.

The one-child policy had other outcomes as well. An important one was that China's became an aging population as its proportion of youngsters shrank while the older age cohorts mushroomed. That raised the specter of a population implosion (discussed in the chapters on Europe and Russia): would there be a sufficient number of workers in the younger age groups to support the ever-expanding older population? For China, this concern was magnified by the prospect that, unlike in European countries and Japan, China might grow old before it grew rich.

In 2009, a rare event occurred in the PRC: official toleration of public debate regarding Politburo policy. In the English-language newspaper *China Daily* as well as other media, brief articles mentioned that a national debate over the one-child policy was "tolerable," signaling open discussion of it at high levels in the Communist Party. Should China abandon the policy, demographers will need to revise their projected world population totals for the twenty-first century.

A crowd of Chinese passengers lined up at an entrance to Beijing Railway Station, one of the capital's two main terminals, to return home for the Spring Festival (Chinese New Year). For many migrants, this official week-long holiday is the only one long enough for such a visit; in recent years, this event has become internationally known as the biggest single movement of people on Earth. In 2013, for many this Festival (centered on the lunar-new-year date of February 10th) lasted as long as 40 days, and produced an astounding 3.4 billion trips, an increase of more than 8 percent from the year before. Of that trip total, 3.1 billion were by road, 225 million by rail, and some 35 million by air. In no small way, this army of annual migrants serves to maintain what is left of the ties between China's rapidly developing cities and the far-flung rural hinterlands.

China's Urban Transformation

As recently as 1980, barely one-fifth of the Chinese people were urban dwellers; today, just over half live in towns and cities, that milestone recorded at the beginning of 2012. China's urban transformation is without precedent and on a scale the world has never seen before. The urban population increased from 18 percent in 1978 to 26 percent in 1990 to 31 percent in 2000. Once into the twenty-first century, however, that rate has skyrocketed—surpassing the 50-percent mark only a dozen years later! Much of this growth has been planned and controlled by the government, and that by itself is an incredible feat. But it has brought profound changes to Chinese society, uprooting tens of millions, infusing awareness even in remote rural areas, creating inequality both within cities and between city and countryside, and spawning a huge **floating population [1]** that consists of temporary urban dwellers with restricted residency rights.

The so-called ***hukou* system [2]** is based on residency permits that indicate where individuals are from and where they may exercise particular rights that include education, health care, and housing. The *hukou* tradition dates back to ancient China and is not uncommon in Japan, Vietnam, and other parts of Asia. During Mao's rule, this residency-permit system became far more rigid and was used as a tool of central planning to restrict the movement of people. More recently, *hukou* has been employed to manage the needed migration to China's cities—supplying enough migrants to meet local demands for factory workers (but not too many), yet always with the possibility of a mandated return to their rural points of origin should the need for their services decline.

Shanghai, China's largest metropolis, is the prototype, having attracted millions of such migrants in recent decades. Over the past ten years, however, their numbers have tripled and now account for as many as 9 million

VOICE FROM THE Region

© Zheng Yue

Zheng Yue, Beijing

HUKOU AND LIFE IN CHINA

"*Hukou* is the term we use in China for a person's residential status: it tells you where you have residency rights to access health care or housing or education and how many social benefits you can enjoy. In the past, during the time of the Planned Economy, *hukou* was very important because it basically determined your entire course of life, even your access to food. When I was a little kid, my family's *hukou* entitled us to 250 grams of pork per month for all three of us—my parents usually let me eat most of it. It was the same with sugar, rice, and other things. Lots of people always want to move to the cities because everything is better there, but it was impossible for most people to get their *hukou* changed. Today, *hukou* no longer determines access to food but it still decides where you can go to school, if you are allowed to buy a car, a house, and other important things. It is still difficult to change your *hukou*. The government has made moving to selected cities easier because more workers are needed there. They will adjust your *hukou* so that you can get a house. Sometimes the government gives certain big state-owned companies the right to provide a local-city *hukou* to attract new talent from other parts of the country. I was born in Xian, and that's where I had my first *hukou*. When I was 9 years old, my parents moved to Shenzhen to work, and so I had my second *hukou* from there. When I was 27, I settled in Beijing after getting my Masters degree in the USA. Beijing provides a *hukou* for those with degrees from good overseas universities, so I got my third *hukou* from there; otherwise, it would be very difficult to obtain a Beijing *hukou*. Now, because of my status, my son has a Beijing *hukou* as well. I won't say *hukou* is a headache for everyone, but it still affects the lives of a lot of people—you'd better plan ahead if you can! And *hukou* changes from time to time. When my baby son grows up, I wonder how it will affect him and his generation."

of the 22-plus million people who inhabit this burgeoning urban region. Among Shanghai's 20–34 age cohort, nearly two-thirds are migrants. This is the life-cycle stage during which many couples start a family, and in 2013 the metropolis had an estimated 400,000 children below the age of six who were officially categorized as migrants without a Shanghai *hukou*. This is clearly a growing social problem—elaborated in the *Voice From the Region* box—that the Chinese government will have to resolve sooner or later.

China now counts more than 160 cities containing at least one million people, three times as many as the United States. By 2025, there could be at least 60 more. But the future is uncertain. Will there continue to be enough jobs for all these rural-urban migrants? What will be the consequences of China's accelerated program to expand its urban housing stock by millions of new units every year (see the box titled "A Chinese Real Estate Bubble?")? Can cities that grow at such a hectic pace be managed in an environmentally sustainable way? Will the Communist Party be able to maintain control over the new agglomerations of urbanites with their changing lifestyles, identities, and expectations? For how long can the government justify the widening income inequalities within the major cities (this, after all, is communist China), and can they meet the steadily increasing demands of industrial workers? Could these enormous economic and social changes at some point force a political transformation—and would that be for the better? Clearly, there is no blueprint for China as it tries to chart its course into the next stage of its remarkable odyssey.

CHINA'S COASTAL CORE

China's economic transformation was first introduced in 1980 at a few carefully chosen places. By establishing new economic rules that would initially apply only to certain urban areas on the Pacific coast, most of the rest of the country and the majority of the population would remain comparatively unaffected, at first. Figure 9B-4 shows how it was planned to work.

Special Economic Zones

The government introduced a three-tier system of **Special Economic Zones (SEZs) [3]**, so-called ***Open Cities***, and ***Open Coastal Areas***, which would attract technologies and investments from abroad and reshape the economic geography of eastern China. In these economic zones, investors were offered numerous incentives that included low taxes, eased import-export regulations, and simplified land leases. The hiring of labor under contract was permitted. Products made in the economic zones could be sold on foreign markets and, under some restrictions, in China as well. Even Taiwanese enterprises could operate here. And profits earned were allowed to be sent back to the investors' home countries.

When the government made the decisions that would reorient China's economic geography, location was a leading consideration. Beijing wanted China to participate in the global market economy, but it also wanted to produce as little impact on interior China as possible—at least in the initial stages. The obvious answer was to position the Special Economic Zones along the coast. Initially, four SEZs were

AMONG THE REALM'S GREAT CITIES . . .

SHANGHAI

SAIL INTO THE mouth of the great Yangzi River, and you see little to prepare you for your encounter with China's largest city (population: 22.2 million). For that, you turn left into the narrow Huangpu River, and for the next several hours you will be spellbound. To starboard lies a fleet of Chinese warships. On the port side you pass oil refineries, factories, and crowded neighborhoods. Next you see an ultramodern container facility, white buildings, and rust-free cranes, built by Singapore. Soon rusty tankers and freighters line both sides of the stream. High-rise tenements tower behind the cluttered, chaotic waterfront where cranes, sheds, boatyards, piles of rusting scrap iron, mounds of coal, and stacks of cargo vie for space. In the river, your boat competes with ferries, barge trains, and cargo ships. Large vessels, anchored in midstream, are being offloaded by dozens of lighters tied up to them in rows. The air is acrid with pollution. The noise—bells, horns, whistles—is deafening.

What strikes you is the vastness and sameness of Shanghai's cityscape, until you pass beneath the first of two gigantic suspension bridges. Suddenly, everything changes. To the left, or east, lies *Pudong*, an ultramodern district with the space-age Oriental Pearl Television Tower rising like a rocket on its launchpad (see photo) above a forest of modern, glass-and-chrome skyscrapers that make the Huangpu look like Hong Kong's Victoria Harbor. To the right, along the waterfront Bund (Embankment), stand the remnants of Victorian buildings, monuments to the British colonialists who made Shanghai a treaty port and started the city on its way to greatness. Everywhere, construction cranes rise above the cityscape, and Shanghai now boasts more skyscrapers over 400 meters (1300 ft) than any other city except Chicago. The biggest of all opens in 2014—Shanghai Tower, at 824 meters (2703 ft) the world's second-tallest building after Dubai's Burj Khalifa.

The Chinese spent heavily to improve infrastructure in metropolitan Shanghai, including the new Pudong International Airport, connected to the city by the world's first maglev (magnetic levitation) train—the fastest anywhere with a top speed of 430 kph (267 mph). Pudong itself has become a magnet for foreign investment, attracting as much as 10 percent of the country's annual total. Shanghai's income is rising much faster than China's, and the Yangzi River Delta is becoming a counterweight to the massive Pearl River Estuary development in China's South. Among other things, Shanghai is becoming China's "motor city," complete with a Formula One racetrack seating 200,000 at the center of an automobile complex where all components of the industry, from manufacturing to sales, are being concentrated.

In 2010, Shanghai hosted World Expo on the banks of the Huangpu River (the maglev route was extended to the site), and the city's planners, who have already transformed the place into a quintessential symbol of the New China, took advantage of the opportunity to show the world what has been accomplished here in just a single generation.

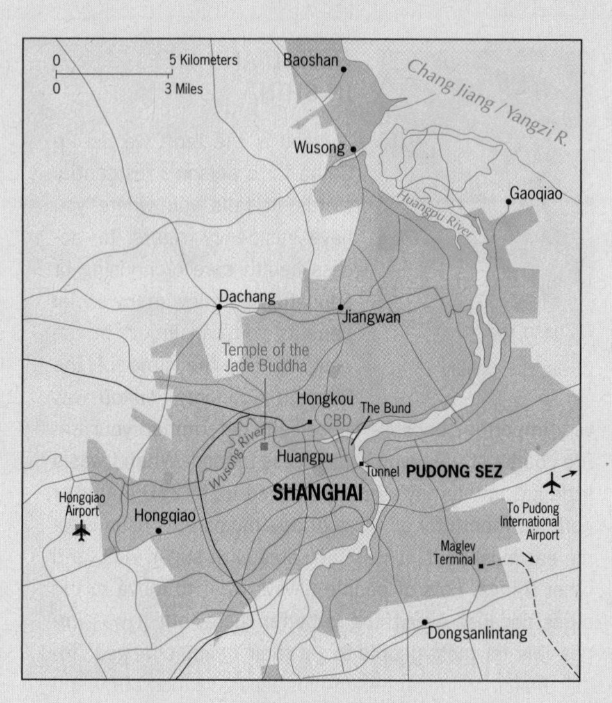

established in 1980, all with particular geographical properties (Fig. 9B-4):

1. ***Shenzhen***, adjacent to then-booming British Hong Kong on the Pearl River Estuary in Guangdong Province
2. ***Zhuhai***, across from (then still Portuguese) Macau, also on Guangdong's Pearl Estuary
3. ***Shantou***, opposite southern Taiwan, a colonial treaty port, also in Guangdong Province, source of many expatriate Chinese now residing in Thailand
4. ***Xiamen***, on the Taiwan Strait, another colonial treaty port, in Fujian Province, source of many Chinese now based in Singapore, Indonesia, and Malaysia

In 1988 and 1990, respectively, two additional SEZs were proclaimed:

5. ***Hainan Island***, declared an SEZ in its entirety, its potential success linked to its forward location vis-à-vis Southeast Asia
6. ***Pudong***, on the east bank of the Huangpu River across from central Shanghai, China's largest city, different

From the Field Notes . . .

© H.J. de Blij

"Shanghai shows no signs of economic slowdown, even if the environmental consequences are dire. From my vantage point on the top floor of the Jin Mao Tower in Pudong District, buildings beyond the Huangpu River disappeared in a dense smog on a hot spring day that locals described to me as 'pretty clear.' China's prodigious economic growth has made the People's Republic the world's largest polluter of the atmosphere, surpassing the United States. Some projections suggest that, by 2020, China may be emitting twice as much pollution as America, but since China has more than four times the population, the U.S. would still lead in terms of pollutants per capita. That's just one reason the United States shouldn't invoke the 'China excuse' to weaken its own efforts to reduce pollution from motor vehicles, factories, and other sources. 'Enjoy this view of the Oriental Pearl Television Tower,' Li Sheng told me. 'We're building the Shanghai Tower [China's tallest building as of 2014] right down there where you see the construction site, and it'll be in the way. And it'll be almost twice the height of your Empire State Building.'" [Well, let's hope the view from the top will be clearer.]

www.conceptcaching.com

A Chinese Real Estate Bubble?

CONCERNS ARE GROWING that China's construction and real estate sectors, which play an enormous role in the country's overall economic growth, have overheated and perhaps escalated out of control. China's megacities, most notably Shanghai, Beijing, and Shenzhen, have expanded at astonishing rates; at the same time, entirely new cities to house more than a million apiece have been planned and built from scratch. Governments at various levels are closely involved and the market is far from transparent. Since 2012, it has become clear that tens of millions of new high-rise apartment buildings—from the suburbs of Zhengzhou on the North China Plain to the brand new city of Ordos in Inner Mongolia—are sitting empty along with the huge shopping malls that are part of the same developments. The price range of most apartments today is U.S. \$60,000–\$120,000, way out of reach for the average Chinese resident. A substantial number of these apartments have actually been sold—to affluent members of the new urban middle class who have no intention of ever living in those units! With limited private investment opportunities in the PRC, these owners of course are speculators who are betting that prices will continue to rise as they did over the past several years. Real estate bubbles are common to economies all over the world, and certainly the United States has experienced its share. In China, however, everything occurs at a bigger scale, so if this bubble does burst it is likely to produce the greatest real estate meltdown in history.

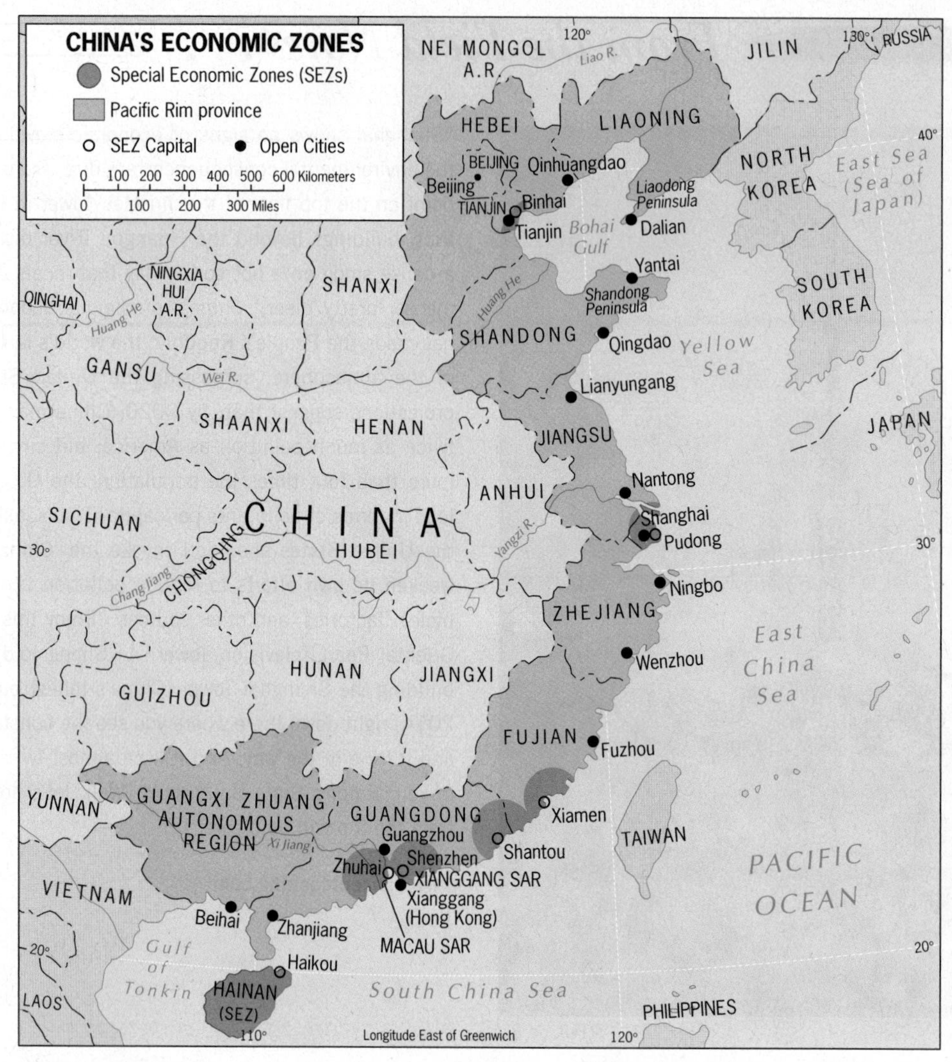

FIGURE 9B-4

from other SEZs because it was a gigantic state-financed project designed to attract large multinational corporations

And the process continues, with the newest SEZ authorized in 2006:

7. ***Binhai New Area***, the coastal zone of the northern port city of Tianjin (just over 100 kilometers [70 mi] southeast of Beijing), a long-established Open City with considerable foreign investment, now elevated to SEZ status, projected to outperform even Shanghai–Pudong and Shenzhen itself. On the outskirts of Tianjin the Chinese are building what is planned to be the biggest financial district in the world, *Yujiapu* (see photo).

The grand design of China's economic planners, therefore, was to stimulate economic growth in the coastal provinces and to capitalize on the exchange opportunities created by location; the availability of funding; the proximity of foreign investors in Southeast Asia, Taiwan, Japan, and, importantly, still-British Hong Kong; the presence of abundant cheap labor; and the promise of world markets eager for low-cost Chinese products.

To date, the poster-child by far of China's economic planners' grand design is Shenzhen. In the 1970s the name Shenzhen was hardly known, but during the following decade those rice paddies and duck ponds gave way to the fastest-growing city in the history of humanity—from 80,000 (mainly fishermen and farmers) in the late 1970s to nearly 12 million today (see photo). Virtually everyone who lives in Shenzhen comes from somewhere else, and because all are really outsiders they speak to each other in Standard ("Mandarin") Chinese rather than the Cantonese dialect that prevails all around them.

Meanwhile, the entire Pearl River Estuary was rapidly evolving into a world-class industrial powerhouse. What caused that to happen? When China opened the Shenzhen SEZ, hundreds of companies moved their factories here from neighboring Hong Kong, capitalizing on lower wages and taxes, less environmental regulation, and weaker official oversight (Fig. 9B-5). At the same time, the Chinese government also built business-friendly infrastructure; expanded seaport

© Inmaginechina/Corbis

Have you ever seen a construction project of this size? China is studded with these big bets that the country's headlong economic growth will continue apace, but few are more grandiose than the *Yujiapu* financial district (shown here in the fall of 2012). Already nicknamed "China's new Manhattan," this massive complex aspires to be a world-class, highest-order activity hub. It encompasses more than 50 skyscrapers, and is being built within a tight bend of the Hai River on the coastal flats 50 kilometers (30 miles) from central Tianjin. Financed by huge loans from state-owned banks, this immense undertaking is a pet public works project closely associated not only with powerful regional politicians but also Hu Jintao, the PRC's president from 2003 to 2013. Even though Yujiapu is on track to be completed in 2019, it is not at all certain that it will be able to live up to expectations. In fact, recent reports have warned about substantial overcapacity and that many financial firms may well choose to stay in such established hubs as Hong Kong, Shanghai, or nearby Beijing.

facilities, roads, railroads, and airports; and constructed countless new apartment buildings to accommodate hundreds of thousands of workers. Soon Shenzhen was the shining example of Deng's Open China policy, to be emulated (inasmuch as possible) by the other SEZs from Zhuhai (next to Macau) to Pudong (Shanghai) to Binhai (Tianjin).

Think about it: Shenzhen, a city barely more than three decades old, now contains 45 percent more residents than New York City, and Guangdong Province has an industrial workforce larger than that of the entire United States. The Pearl River Estuary is the most productive regional economy of its size in the world, accounting for just about half of all of China's exports. If you have a car, television, computer, smartphone, or any similar device, it is certain to contain parts that came via the Pearl Estuary's massive manufacturing complex (see the box titled "A Corporate Giant in China").

Regional Engines of Growth

None of this would hurt Hong Kong's economy. In fact, it ushered in a new and even more prosperous era for the then-still crown colony. Hong Kong now engaged less in manufacturing and far more in banking, finance, fiscal management, and business services, taking advantage of the heightened skills already known to exist among its workforce during earlier economic transitions. Such matters are the concern of the field of **economic geography [4]**, which focuses on raw-material distributions, historical-cultural factors (such as the lingering effects of colonialism), environmental issues, and particularly the role of spatial economic networks in regional development.

Chinese strategies were based on the understanding that rapid economic growth can be achieved and managed if it is spatially concentrated, because local conditions can be manipulated and controlled. Once growth takes off, it can have a stimulating effect on surrounding places, and the benefits can be spread around by a strong central government. China devised a truly remarkable strategy combining pure capitalist principles with powerful government involvement.

This strategy includes competition among these regional growth poles and cities using tax breaks and other incentives that lure domestic and foreign investors. Thus

© Photolibrary, Royalty-Free

All of us who witnessed the incredible conversion of a fishing village with duck ponds and paddies into a metropolis of millions in a few years of frenzied economic and demographic growth wondered, inevitably, what kind of city Shenzhen would become. After all, shouldn't cities grow slowly, have histories etched in their architecture, develop traditional neighborhoods, and negotiate by consensus the pitfalls of planning and zoning? But look at Shenzhen only a third of a century later, and it seems far more orderly, established, and accommodating than could have been imagined. Certainly Shenzhen has its richer and poorer neighborhoods—not everyone can afford to live in the CBD shown here. And when China was buffeted by the global economic downturn, Shenzhen was among the SEZs in which factories closed and workers had to leave. But this full-fledged megacity is here to stay, a prominent icon of the new China.

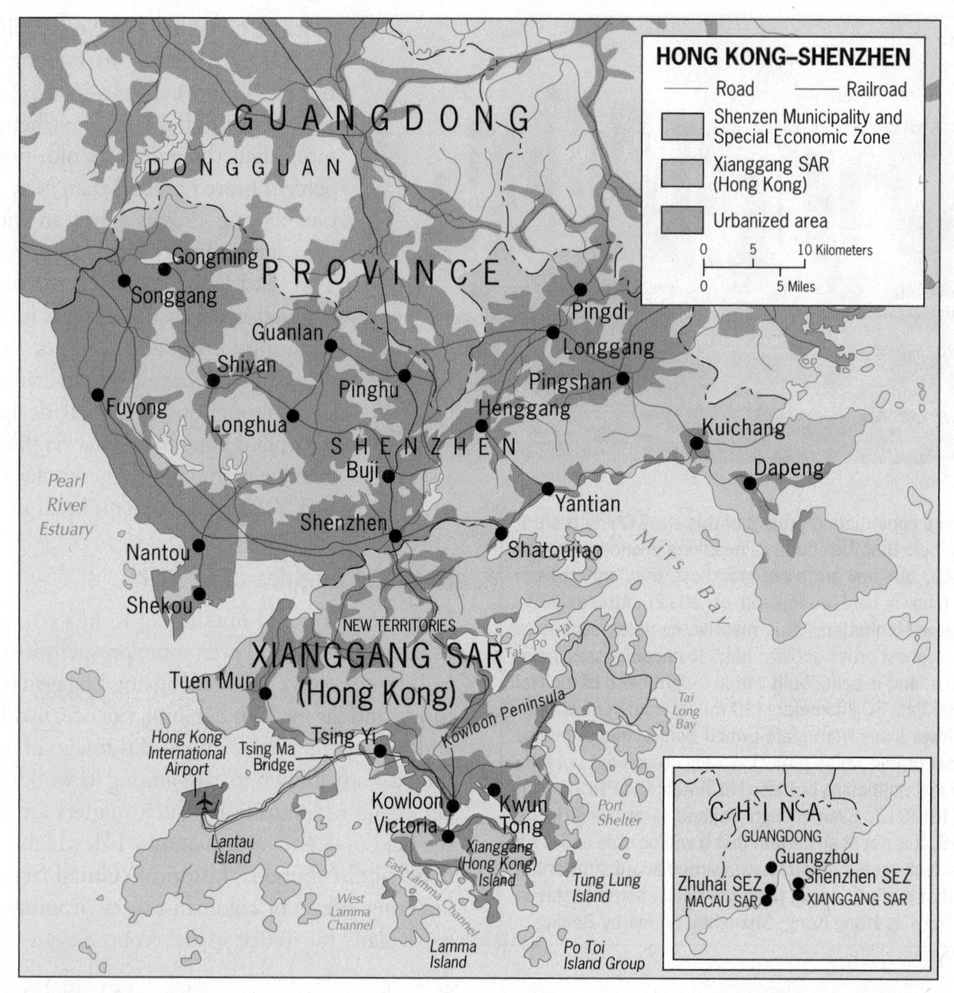

FIGURE 9B-5

© H. J. de Blij, P. O. Muller, and John Wiley & Sons, Inc.

A Corporate Giant in China

IF THE UNITED States has long represented "big" to the rest of the world—whether in terms of houses, automobiles, skyscrapers, or the size of soft drinks—China today increasingly stands for "huge." ***Foxconn***, also known as *Hon Hai*, is a giant electronics contract manufacturer. Think of Foxconn as a huge manufacturing subcontractor for corporations such as Apple, Dell, Hewlett-Packard, and PlayStation. This company was established in 1978 in Taiwan and opened its first gigantic factory on the mainland ten years later in the Longhua Subdistrict of Shenzhen. In 2013, Foxconn was operating 28 plants across the PRC. At Longhua by then, the company employed 240,000 workers on a sprawling "campus" complete with dormitories, stores, banks, and sport facilities as well as a food court that daily dispenses three tons of meat. You are quite likely to own something made by Foxconn, the leading maker of iPhones, iPads, Kindles, myriad video games, and so much more. Worldwide, Foxconn now employs about 1.5 million workers (ten times as many as a decade ago) and, beyond China, operates factories in Mexico, India, Brazil, the Czech Republic, Hungary, and Malaysia. The company also has a contentious record of labor relations, and protests, demonstrations, and even worker suicides are not unknown. Foxconn's leadership says it is addressing labor issues and working conditions, and is also planning to vigorously invest in the company's growth through the foreseeable future, perhaps even expanding into the United States.

Employees on the assembly line at Hon Hai Group's massive Foxconn plant in Shenzhen.

© Qilai Shen/In Pictures/Corbis

urban municipalities are allowed important freedoms to stimulate local economic growth. At the same time, the central government encouraged millions of **Overseas Chinese [5]** (most of whom had left China from those very coastal provinces to settle—and often thrive—in other countries) to invest their money in their ancestral homeland—as well as attracting ever more substantial **foreign direct investment [6]** from Japan, the United States, and other key countries.

The Expanding Core

China's economic miracle began with the Special Economic Zones, and the government continues to use this strategy to launch new regional engines of growth through massive, concentrated investment plans. But economic development has also spilled over into adjacent areas and diffused to other provinces, so an increasing number and variety of localities now carry a "special" designation of one sort or another.

We should not get too fixated on the SEZs: some have received much less attention and investment than others. Hainan, for instance, has only experienced modest increases in manufacturing, which for the most part involves the processing of agricultural products. Furthermore, it does not necessarily require the official status of SEZ to obtain inflows of investment (particularly from domestic sources) and to achieve economic growth. Thus the entire corridor of seaboard provinces from Guangdong Province in the south to Liaoning in the north now forms the centerpiece of a highly productive, Pacific Rim-based economy, and in recent years the Core region itself has expanded as far west as Chengdu in the Sichuan Basin (see Fig. 9B-1).

Export industries continue to be clustered along the coast, as Figure 9B-6 shows, but much economic growth has decidedly been channeled toward the interior around mushrooming manufacturing centers near such major cities as Zhengzhou, Xian, Shenyang, Wuhan, Chongqing, and Chengdu (Fig. 9B-7). And then there are places like the Inner Mongolia AR that have seen rapid economic growth (mainly through mining) but their exporting activity is minimal because most of the commodities they produce are consumed by industries elsewhere in China.

In all, the Chinese government has made substantial efforts to spread economic growth from the Special Economic Zones to other parts of the country, mainly along

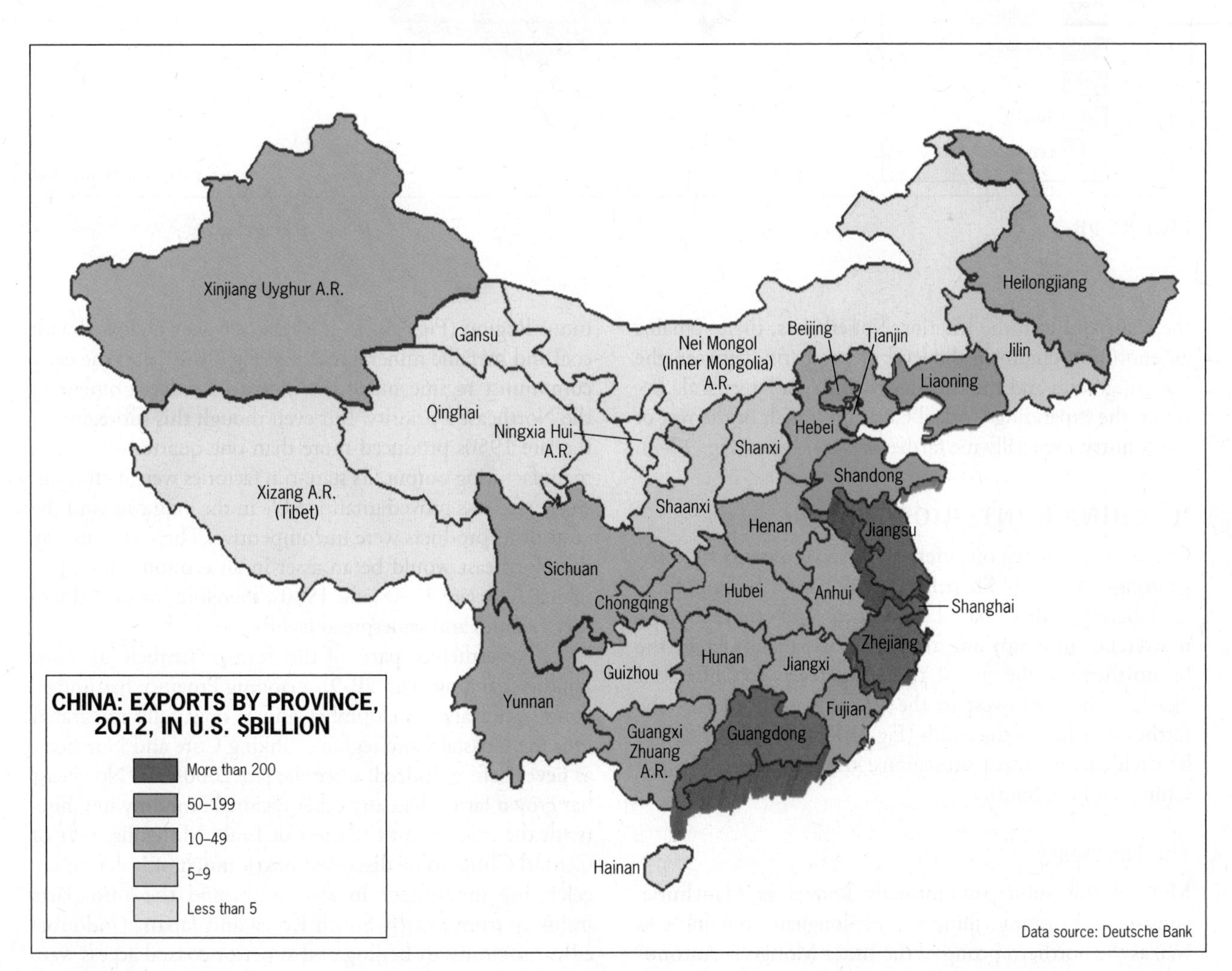

FIGURE 9B-6

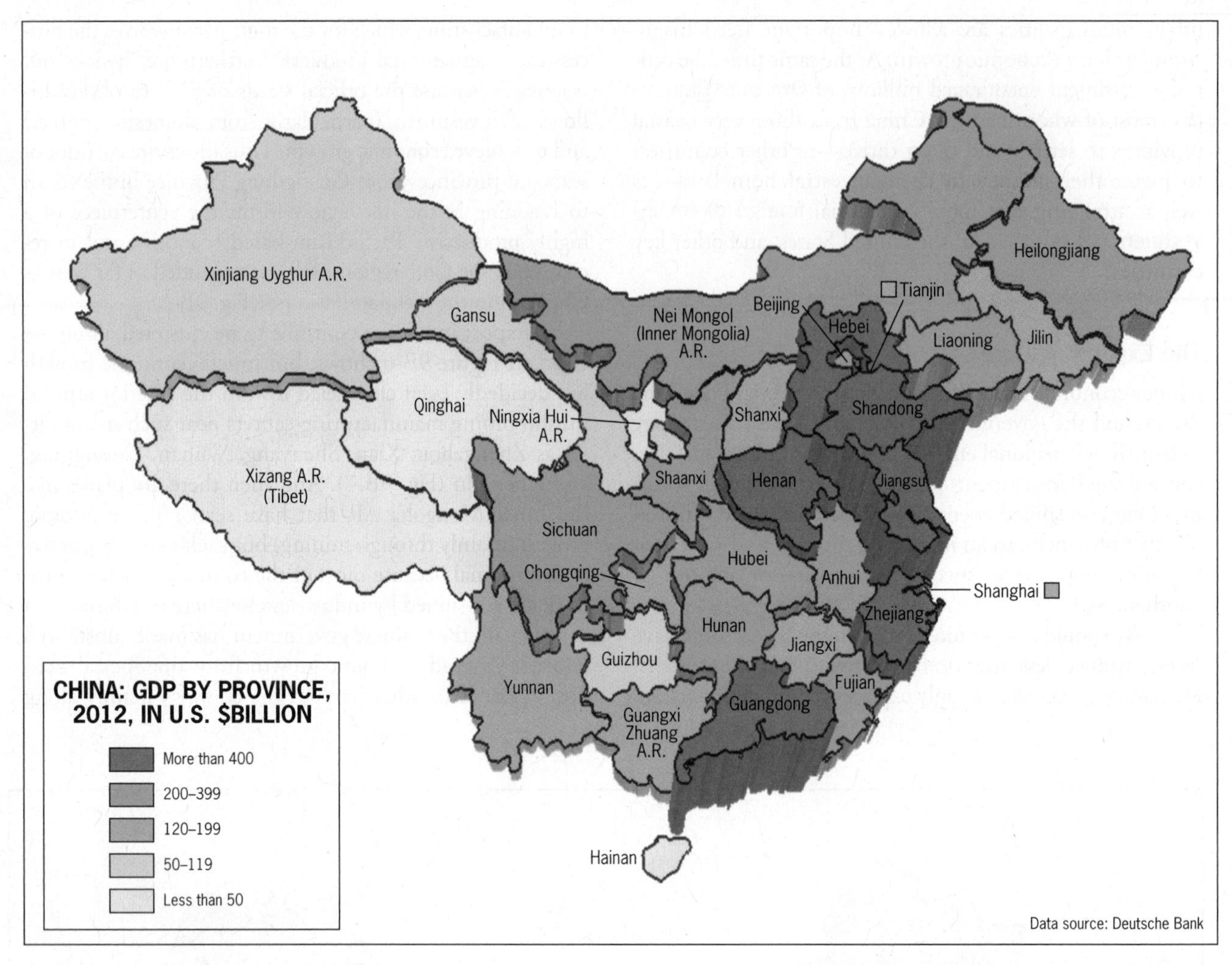

FIGURE 9B-7

the coast and into the Interior. Nonetheless, there remains an enormous chasm in the level of prosperity between the booming cities and the rural areas, and at a larger scale between the expanding Coastal Core and much of the rest of the country, especially the farther west one goes (Fig. 9B-7).

CHINA'S INTERIOR

China's Interior region offers a stark contrast to the burgeoning environs of Shenzhen, Shanghai, or Beijing. Situated between the Coastal Core and Western Periphery, it stretches in a vast arc from the Russian border in the far northeast to the city of Yumen (gateway to China's far west) in the northwest to the Vietnamese border in the farthest reaches of the south (Fig. 9B-1). The Interior can be divided into three subregions: the Northeast, Central China, and the South.

The Northeast

Most of this subregion, formerly known as Manchuria, consists of Liaoning, Jilin, and Heilongjiang provinces as well as the northern prong of the Inner Mongolia Autonomous Region (Fig. 9B-3). It is relatively well endowed with coal and metallic mineral reserves (Fig. 9B-8), and the early communist regime made the industrial redevelopment of the Northeast a priority. But even though this subregion in the late 1950s produced more than one-quarter of China's manufacturing output, its state-run factories were inefficient, workers' perks proved unaffordable in the long run, and the costs of its products were uncompetitive. There was no way the Northeast would be an asset in an economically open China. The late 1970s and 1980s, therefore, witnessed factory closures and widespread layoffs.

Nevertheless, parts of this former 'rustbelt' are now rapidly changing (Fig. 9B-9). Liaoning Province has undergone significant development and is now fully integrated into the Coastal Core region—linking Core and Northeast as never before. Indeed, since the late 2000s, the Northeast has grown faster than any other regional economy in China (with the exception of the rest of Inner Mongolia, part of Central China to be discussed next), much of it due to accelerating investment in electronics and the auto parts industry from nearby South Korea and Japan. Undoubtedly, proximity to Beijing and superior coastal access were

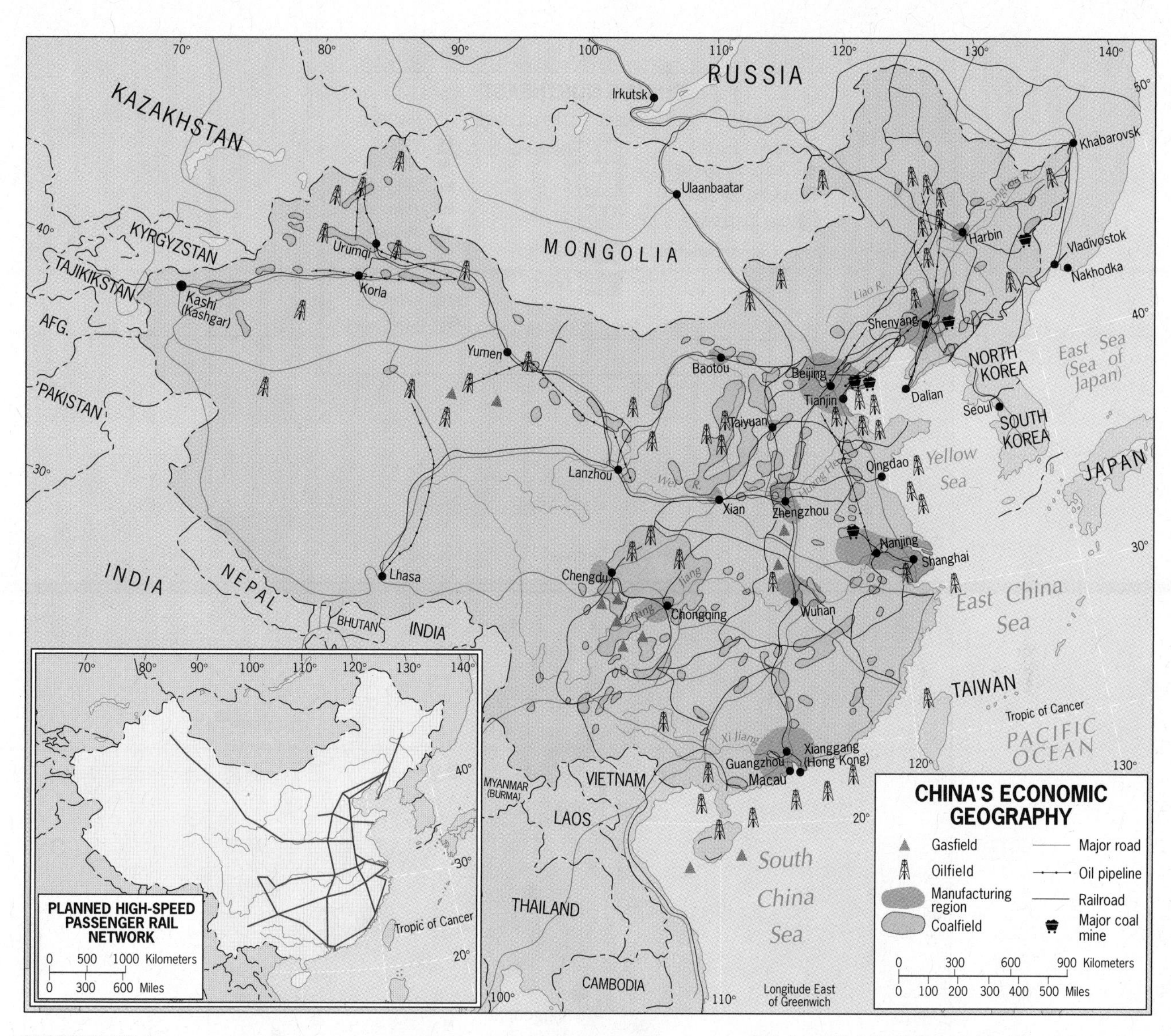

FIGURE 9B-8

The city of Chengdu, capital of Sichuan Province, has grown and modernized at a dizzying pace since 2000, part of China's unbelievably rapid urbanization. Not long ago, this was part of the Interior; now the city has not only been drawn into China's westward-advancing Core region but also forms its spearhead. Traditionally known as one of the country's most laid-back and liveable cities, Chengdu has quickly acquired a new reputation as a pacesetting economic center that specializes in information technology, finance, and biopharmaceutical industries.

© Jan Nijman

major factors in this transformation. Liaoning's leading city, Shenyang (population: 5.9 million), now showcases a diversified economy that, besides heavy industry, includes aerospace, electronics, and banking. The Northeast's primary outlet port—the city of Dalian, located where the Bohai Gulf meets the Yellow Sea—has become a leading export focus for trade with Japan and the Koreas.

Central China

The Central subregion of the Interior borders Mongolia in the north, Xinjiang and Tibet in the west, and the spearheading westward protrusion of the Coastal Core around Chengdu in the south (Fig. 9B-1). The sparsely populated western margins of this subregion are dominated by Tibetan pastoralists. Lanzhou, centrally located in the upper basin of the Huang He, is the subregion's leading urban center. With a population of 2.8 million, this key crossroads city is best known for its oil refineries and petrochemical industries. Much of the

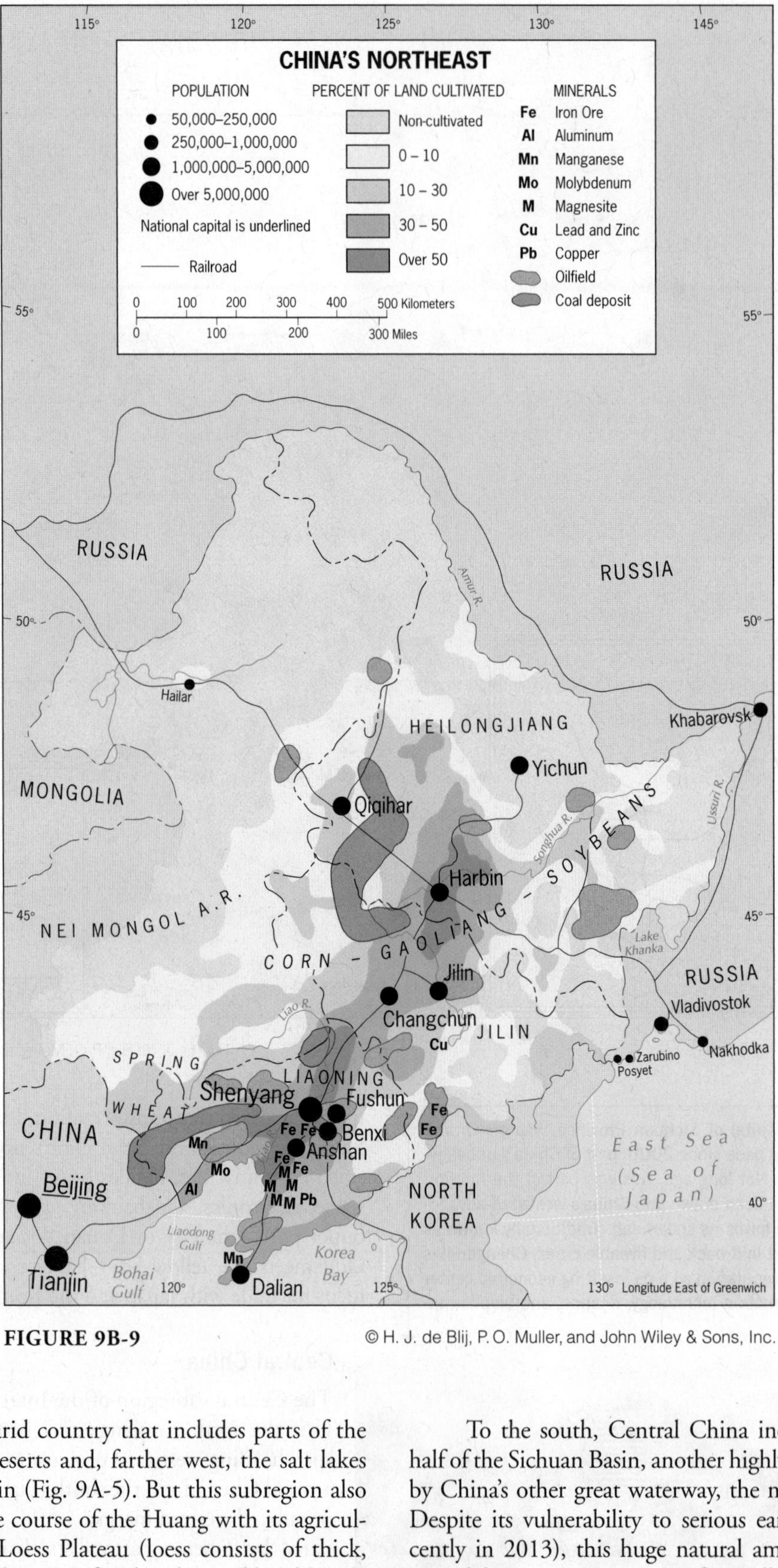

FIGURE 9B-9 © H. J. de Blij, P. O. Muller, and John Wiley & Sons, Inc.

central Interior is arid country that includes parts of the Gobi and Ordos deserts and, farther west, the salt lakes of the Qaidam Basin (Fig. 9A-5). But this subregion also includes the middle course of the Huang with its agriculturally productive Loess Plateau (loess consists of thick, fertile, windblown deposits of rock pulverized by glaciers). Add the water of this great river together with an adequate growing season, and a sizeable population has materialized here.

To the south, Central China includes the northern half of the Sichuan Basin, another highly fertile area, crossed by China's other great waterway, the mighty Chang Jiang. Despite its vulnerability to serious earthquakes (most recently in 2013), this huge natural amphitheater has supported human communities for millennia, and you can readily discern its population cluster in Figure 9A-4. The Sichuan Basin, encircled as it is by mountains, is also one of the world's most sharply delineated physiographic regions,

From the Field Notes . . .

"It is about eight in the morning when, after a long hike through the cloud forest, we reach the first major monastery on Mount Emei in Sichuan Province. The Shaolin temple appears magically from the mist as we get closer. A UNESCO World Heritage Site, Mount Emei is the highest (elevation 3099 meters/10,168 ft) of China's Four Sacred Buddhist Mountains. This is where the oldest Buddhist temple in China was built during the first century AD and where Shaolin martial arts originated in the sixteenth century. Most of China's famous old temples are located well away from the coast and major cities, undisturbed by the atheist ideologies of the communist state. Mount Emei is a major destination for Buddhist pilgrims from all over China. It is hard to imagine a more serene venue, a more perfect place for spiritual dedication."

www.conceptcaching.com

and the concentration of its roughly 125 million inhabitants reflects that configuration.

Although there are minorities in the Interior, the great majority of the people are Han Chinese. Understandably, they feel entitled to participate in China's remarkable achievements, but hundreds of millions of them still continue to subsist on less than 5 U.S. dollars a day. China's Pacific Rim transformation drew many millions more to the east-coast factories as well as urban and regional construction sites in what came to be the largest internal national migration in human history. Nonetheless, those who stayed behind found themselves remote from the opportunities introduced by globalization—and often exploited by local mayors, party bosses, and others when it came to land, property, or produce markets. It does not make the international news, but China's inner periphery is rife with public protests in the streets and villages. It is an unstable, even dangerous situation, and the Chinese communist regime is showing an increasing awareness of the need for mitigation.

One way Beijing's planners are responding to this uneven development is by spreading the privileges enjoyed by the coastal SEZs into the urban centers of the Interior. With workers' hourly salaries rising in the seaboard provinces, some corporations are finding it profitable to move their production facilities into this transitional periphery. To encourage them to do so, new economic zones have been established in Lanzhou, Kunming, and other inland cities. Another incentive is the construction of massive public works in the central Interior, such as the ongoing diversion of waters from the Chang to the increasingly thirsty northern cities of the Core. Moreover, to improve linkages between the Coastal Core and Interior, China is constructing tens of thousands of kilometers of four-lane highways; a far-flung, high-speed passenger rail network (Fig. 9B-8, inset map); and dozens of new or expanded airports.

The South

The Southern subregion of the Interior, like its two counterparts, is dominated by Han Chinese, but also includes large clusters of minorities, especially near the borders with Myanmar, Laos, and Cambodia (see Fig. 9A-7). Indeed, adjoining Vietnam in the southeast we find the Guangxi Zhuang Autonomous Region, which was specially created with these minorities in mind (Fig. 9B-3).

The South is mainly covered by the Yunnan Plateau, source of the streams that feed the Xi Basin and its Pearl Estuary outlet. Much of southeastern China has comparatively high relief; it is hilly and in places even mountainous. Among other things, this challenging terrain has inhibited overland contact between China and Southeast Asia. Yunnan Province itself, in addition to its geopolitical significance, possesses important mineral deposits. And thanks to long-established terrace systems, farming fares quite well here despite those steep slopes.

CHINA'S WESTERN PERIPHERY

Historically, empires have expanded and contracted, acquiring and losing territory as well as subjects, and leaving their imprints where they once ruled. Russians in Central Asia, British in East Africa, French in Indochina, Japanese

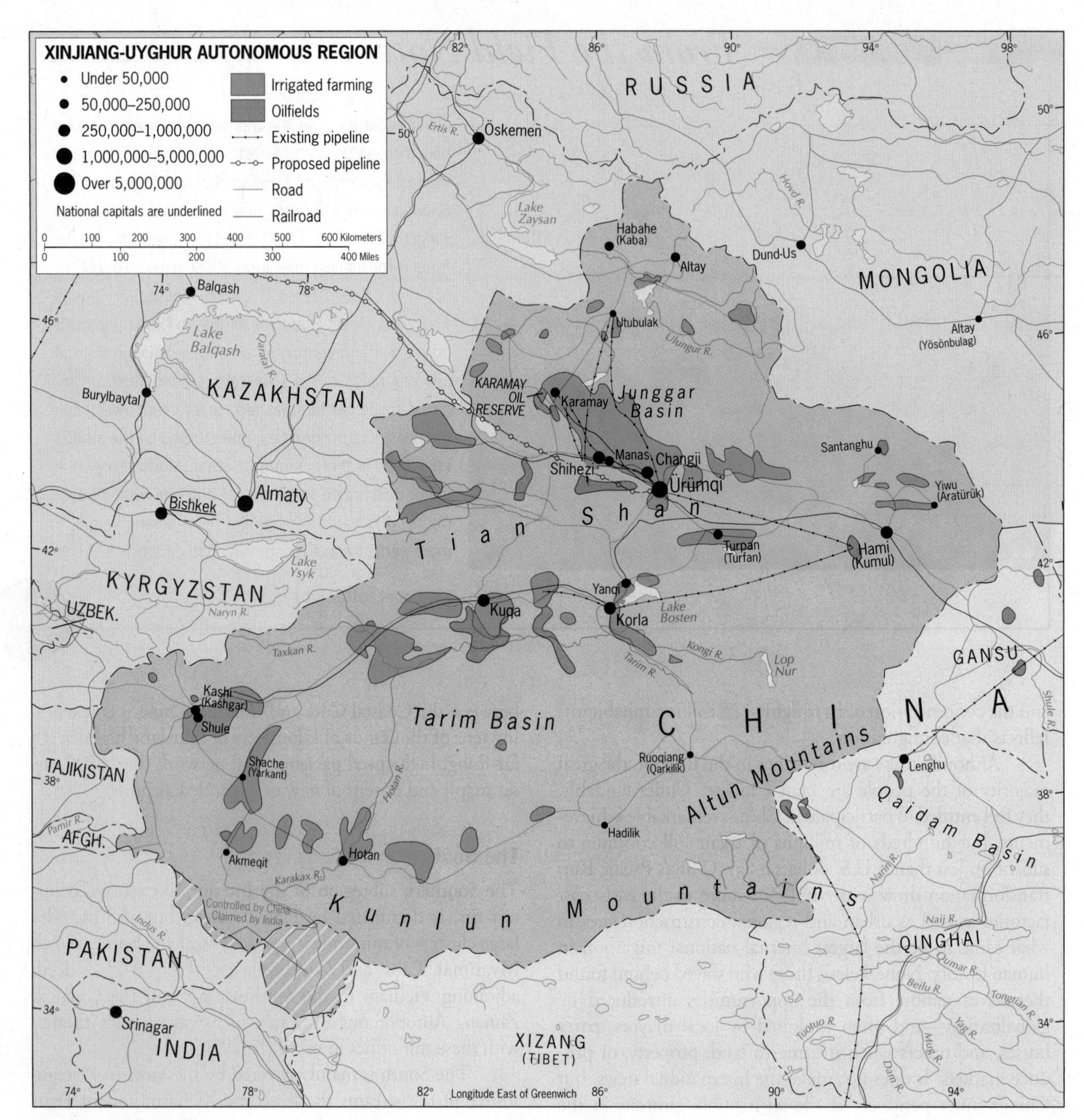

FIGURE 9B-10

in Taiwan all came and went, leaving behind languages, religions, infrastructures, and traditions.

And so it was with China. Figure 9A-8 reminds us just how vast China's Qing Dynasty empire was and how much it lost—but Figure 9B-1 reveals that China is still an empire today. Its beyond-the-Han domain includes a pair of large Autonomous Regions and a third one rapidly being integrated into Han China itself. Two other, smaller Autonomous Regions located in the inner periphery are similarly misnamed (see Fig. 9B-3). There is nothing "autonomous" about China's far western minority entities: just ask the Tibetans of Xizang or the Uyghurs of Xinjiang.

Xizang (Tibet)

Tibet (called *Xizang* by the Han) is the icebound heartland of Tibetan-Buddhist culture that extends northeastward into Qinghai Province and, importantly, into a corner of far northeastern India in the Indian State of Arunachal Pradesh. Tibet shook off Chinese domination at the end of the nineteenth century, but in 1950, almost immediately after the communist regime took control in Beijing, it ordered the Red Army into Xizang to recapture the territory. Tibetan society had been organized around the fortress-like monasteries of Buddhist monks who paid allegiance

© Jan Nijman

Tibetan Buddhist pilgrims at Jokhang Temple, Lhasa.

to their supreme leader, the Dalai Lama (see chapter-opening photos).

The Chinese wanted to modernize this feudal system, but the resistant Tibetans clung to their traditions, and in 1959 the army had to crush an uprising. With the Dalai Lama now exiled and the people powerless, the Chinese destroyed much of Tibet's cultural heritage, looting its religious treasures and works of art.

Under Deng's administration part of this heritage was returned to Xizang, but Beijing also encouraged Han Chinese to move to Tibet and built "the world's highest railroad" (reaching an altitude of 5072 meters/16,640 feet) across the Tibetan Plateau to the capital of Lhasa (Fig. 9B-8). While the Dalai Lama travels the world making the 3-million-plus Tibetans' case and asking not for independence but for genuine autonomy, Hanification continues. Most Tibetans alive today were born after the Chinese seized control; they have gone through Chinese-mandated schools and, for the young especially, the Chinese presence has become a fact of life. A leading casualty is Tibetan culture, which is steadily being extinguished.

Tibet is now fiercely controlled by the Chinese, not only for geopolitical reasons (e.g., the border with India) but also because many critical parts of China depend on the water provided by the great rivers that rise in this soaring highland. In fact, the Chinese government refers to the Tibetan Plateau as "China's water tower" and has launched canal projects to divert some of Tibet's river waters to the populated areas of the Western Periphery's arid northern interior.

Economic growth in Tibet is strong and incomes are climbing. But popular resistance continues, accompanied by violence and drama. Between 2010 and mid-2013, over 100 self-immolation suicides took place, mainly by Buddhist monks to protest the Chinese occupation, the marginalization of Tibetan culture in schools, and the destruction

From the Field Notes . . .

© Jan Nijman

"Tibet is one of the most colorful and spiritual places on Earth, high in the Himalaya with its steely blue skies and dry, crisp air. Since 1950, Tibet has been controlled by the People's Republic of China. The Chinese military seem omnipresent, and it is hard to escape a sense of forceful occupation, especially in the capital city, Lhasa. I explored the busy streets around the famous Jokhang Temple, a major pilgrimage destination for Tibetan Buddhists. There, too, the Chinese were conspicuously present, with observation posts perched atop buildings as if to remind ordinary Tibetans of the political order of the day."

www.conceptcaching.com

of the Tibetan landscape through mining. A landslide not far from Lhasa in the spring of 2013 near a major copper and gold mine claimed at least 80 lives and morbidly underscored many Tibetan grievances, despite the economic development that comes from the exploitation of local natural resources. In China's Periphery, therefore, rising prosperity neither guarantees stability nor compensates for the lack of political freedom.

Xinjiang

Xinjiang constitutes the westernmost margin of China's modern empire and is even larger than Xizang (Fig. 9B-10). With just over 22 million people, about 40 percent of them Han Chinese, the Xinjiang-Uyghur AR is even more important than its Buddhist, Tibetan neighbor. Here China meets the peoples of Turkestan and the faith of Islam; here China has significant energy reserves; and here, in the remote, cloudless, desert-dominated far west, China built its original space program. Moreover, only a single country lies between this inner Asian outpost and the much larger oil and gas reserves of the Caspian Sea Basin: Kazakhstan. As noted in Chapter 7B, connecting pipelines are already in place and more are under construction.

The modern, highly controlled, and urbanized component of Xinjiang is focused on the revitalized capital of Ürümqi and the nearby model city of Shihezi, located in the northern Junggar Basin. The traditional, religious, and rural component is anchored by the historic town of Kashgar (Kashi on Chinese maps) in the extreme western corner of the southern Tarim Basin (Fig. 9B-10).

Xinjiang's Autonomous Region (AR) is no more autonomous than Tibet. During the Qing Dynasty, the Muslim peoples here—including Uyghurs, Kazakhs, Tajiks, Kyrgyz, and others—fell under Chinese control, a numerical dominance that persisted for half a century after the communist regime came to power in Beijing. But in the twenty-first century, Hanification has shifted into high gear: today at least 40 percent of the population are Han, and Xinjiang has become a two-tiered society. Led by the Uyghurs, who still constitute 46 percent of the AR's population, the peoples of the lower, Islamic tier disdain "Xinjiang," preferring the appellation "East Turkestan" to signify their inter-realm connection to the Turkic nations of neighboring Turkestan.

Where the Uyghurs meet the Han, as occurs in the factories of Ürümqi and the streets of Kashgar, the results are sometimes tragic. In 2009, a skirmish between Uyghur workers and Han managers at a plant outside Ürümqi spun out of control and led to rioting in which some 200 Han Chinese citizens were killed and hundreds more wounded. Despite the growing risk of government retaliation, Uyghur militants continue to pursue an intermittent—and fruitless—campaign of violence against Han domination.

In the meantime, Beijing is tightening its grip on the Western Periphery for economic as well as geopolitical reasons. In 2010, the city of Kashgar was designated an "economic development zone," the first in this subregion. Consider the geographic significance of Kashgar: it has a long history as an important waystation on the ancient Silk Road; it is close to the borders with Pakistan, Afghanistan, Kyrgyzstan, Uzbekistan, and Tajikistan; and it is already the home of Central Asia's largest marketplace. The city is by far the most effective link between China and this sector of Central Asia, and is to be swiftly developed into a major trading and logistical bridge. Clearly, China today is also looking closely at its far west as it continues to focus on its westward-expanding Core region.

Both Xizang and Xinjiang exhibit the properties of peripheries: local cultures and traditional economies are overpowered by national interests and global systems, widening disparities not only between core and periphery but also between and among societies within the periphery itself. When, as in China's case, the political system affords inadequate opportunity for the expression of grievances and the representation of local interests, the symptoms of marginalization become entrenched.

Even though China has made enormous strides and now holds a position of considerable power on the global stage, it faces some formidable internal challenges. First, rising (regional) inequality will be very difficult to legitimize by a regime that still pays homage—at least on paper—to socialist principles. "Let some people get rich first," were the famous words of the great reformer Deng Xiaoping more than 3 decades ago. But now it seems that getting rich has become an obsession for many yet is attainable for relatively few. China's Gini coefficient in 2012 measured a comparatively high 0.47—behind South Africa and Brazil but ahead of the United States. Second, there has been a major shift in the constituencies of the Communist Party: peasants and workers, the traditional pillars of support for the regime, have grown increasingly critical as they feel ever more left behind. With the Party today most strongly supported by the urban middle class and the *nouveau riche*, how much longer can it maintain the ideological foundation of its policies? Third, as noted earlier, rapid urbanization and migration are testing the longevity of government control through the *hukou* system while dissent steadily rises. And fourth, the proliferation of the Internet and social media is providing an ever more powerful outlet for government critics. More generally, we have to wonder for how long a country as massive as China can experiment with capitalism while maintaining a centrally-planned system and suppressing demands for democracy. As the new regime led by Xi Jinping entered office in 2013, the world was watching intently for signs of new approaches to confronting China's challenges.

MONGOLIA

We now turn to the East Asian regions beyond China and first we look to the north, to Mongolia, where the effects of the realm's economic transformation have only recently arrived. This immense, landlocked, isolated country wedged between China and Russia, with an area larger than Alaska but a sparse population of just 3 million, suggests a steppe-and-desert-dominated vacuum between two of the world's most powerful countries (Fig. 9B-1).

This used to be the domain of a powerful people who, centuries ago, swept westward to challenge the Russians and southeastward to rule China. But in more modern times, Mongolia became a weak and vulnerable country whose 800,000 herders and their millions of sheep today follow nomadic tracks along the fenceless fringes of the immense Gobi Desert, where Siberian cold periodically causes severe human and livestock losses. During the Soviet era, the location of the capital of Ulaanbaatar symbolized the country's security against Chinese encroachment, and Mongolia functioned as a typical **buffer state [7]**.

In the 2010s, Chinese involvement and investment are growing, and much of this revolves around Mongolia's enormous storehouse of raw materials. Major deposits of gold, copper, and coal have been found, and exploration by large foreign companies has begun. Mongolia was the fastest growing economy in the world in 2011 (at a rate of almost 18 percent), and the International Monetary Fund is predicting double-digit annual growth for years to come. But the country has now become overly dependent on a small number of large foreign mining companies, and negotiations concerning taxes and regulations have proven to be a cause of economic volatility. The main buyer of Mongolia's commodities is China, and Chinese investments have been pouring in, mainly for infrastructural projects. It is not clear whether ordinary Mongolians, among the poorest people in Asia, will benefit from this bonanza anytime soon, but some economists speculate that Mongolia may soon become another Qatar—where a small population grew rich from its enormous natural endowments.

THE KOREAN PENINSULA

Take another look at the map of East Asian regions and you will see one prominent peninsula that seems to reach out from the East Asian mainland toward the islands of Japan. As Koreans will tell you, this isn't just another peninsula. This is a place where human geography took momentous turns as a corridor of migration, an incubator of culture, a cauldron of warfare, a cradle of economic miracles. Today this Idaho-sized tongue of land has a population of 75 million (Idaho: 1.6 million) that is divided between two states so unbelievably different that north and south seem to belong to different worlds. Unbelievable, because the Koreans are a single nation. It is ideology that divides them.

Given its shape and situation, it is no surprise that Korea has been a turbulent stage. Thousands of years ago, even before there was a China, lower sea levels made the crossing from Korea to Japan possible and migrants from the mainland began to challenge the even earlier inhabitants of the islands. Later, long-stable states, always influenced by rising China, forged a durable cultural geography with distinctly Korean attributes including ethnicity, language, and customs. Then, more than a century ago, disaster struck as Japan embarked on a campaign of colonial conquest that eventually overran not only Korea, but also large parts of neighboring China. Later, in the aftermath of Japan's World War II defeat, the Korean Peninsula became the scene of a terrible great-power conflict between China-supported communist armies and American-led anticommunist forces. For more than three years (1950–1953) this devastating conflict swept back and forth across the peninsula, claiming more than 3 million lives and ending in an armistice whose ceasefire line can be seen in Figure 9B-11. Ever since, that line has separated—almost hermetically sealed—a ruthless communist dictatorship in North Korea from a South Korea that started as a brutal capitalist dictatorship but transformed itself into one of the world's most free-wheeling democracies and most successful economies. This partition remains as the centerpiece of what has been called the world's saddest political-geographical tragedy. Outsiders, not the Koreans themselves, caused it.

North Korea

As the map shows, North Korea is territorially larger than South Korea, but its population of 25 million is half the size of South Korea's. Seven decades of the harshest variety of communist rule have turned North Korea into one of the poorest, hungriest, and most regimented countries on Earth, but with one of the largest standing armies as well as advanced nuclear weapons and missile technologies. A regime that imprisons its citizens for the slightest offenses operates a gulag of prison camps reputedly worse than that of the Stalin-era Soviets, starves its people as punishment, and isolates its subjects from the rest of the world—yet invests heavily in weaponry with which to blackmail its neighbors and infect the world. Refugees' and escapees' stories tell of the most extreme forms of poverty and misery, but China, North Korea's ideological ally, supports the regime based in the capital of Pyongyang in the interest of regional "stability" rather than pressuring it to ameliorate its policies.

North Korean dictators rule with powers unimagined even by the Arab rulers overthrown in the "Arab Spring," and the North Korean people have become a nation of informers—on each other. Any sign of disaffection risks imprisonment or worse, and mass displays of regimented solidarity reminiscent of Soviet times attend dynastic transfers of power. So it was when the ostensibly "affable" Kim Jong-un, son of the previous dictator Kim Jong-il, took control of the state in 2011. Shortly thereafter he threatened to resume nuclear testing, he and his ruling elite kept comfortably in power by China's provision of food and energy.

South Korea

The tragedy of the Korean Peninsula is especially painful when put in geographic context. As environmental maps reveal, North Korea and South Korea actually need each

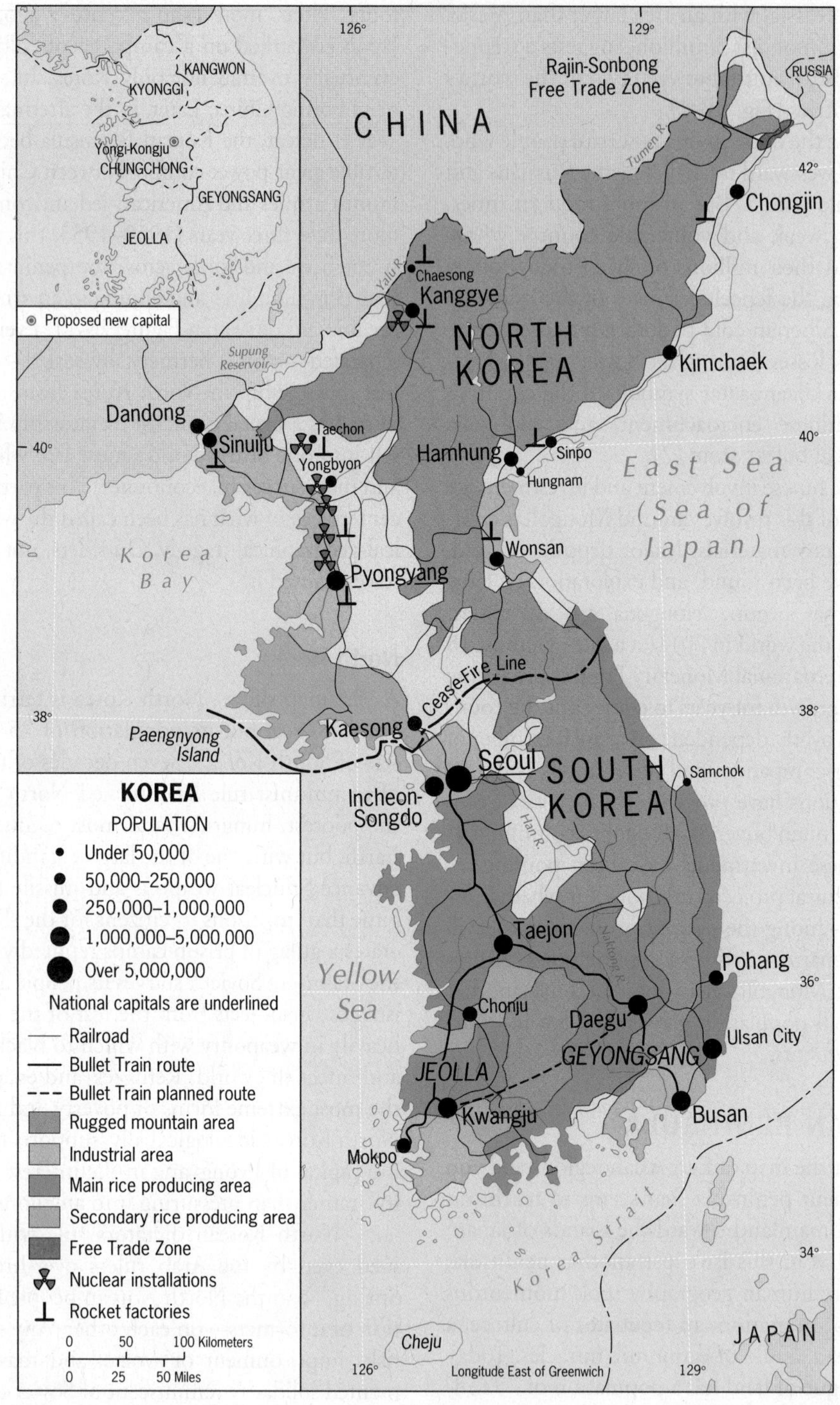

FIGURE 9B-11

other. The North has raw materials wanted by South Korean industries. The South produces food the North could use. This is a condition referred to as **regional complementarity [8]**, but in Korea it is destroyed by political segregation. This makes the narrow demilitarized zone (DMZ) that separates the two Koreas an ultimate expression of core-periphery demarcation, because even as the North stagnated, the South became one of the Asia Pacific's famed ***economic tigers***.

As we just noted, South Korea (49.4 million) emerged from the Korean War as an unstable, dictatorial, politically and economically corrupt nation where, as in the North, you could pay with your life for dissent. But South Korea's rulers also encouraged industrial growth, propelled by powerful industrial conglomerates in cahoots with the politicians. Such **state capitalism [9]** generated rapid development, and the politicians, corrupt as they were, knew that the workers would demand better treatment and that the economic and

AMONG THE REALM'S GREAT CITIES . . .

SEOUL

SEOUL (population: 9.8 million), located on the Han River, is ideally situated to be the capital of all of Korea, North and South (its name means "capital" in the Korean language). Indeed, it served as such from the late fourteenth century until the early twentieth, but events in that century changed its role. Today the city lies in the northwest corner of South Korea, for which it serves as its capital; not far to the north lies the tense DMZ (demilitarized zone) that contains the cease-fire line with North Korea. That line cuts across the mouth of the Han River, thereby depriving Seoul of its inland waterborne traffic. Its bustling ocean port, Incheon, has emerged as a result (Fig. 9B-11).

Seoul's undisciplined growth, attended by a series of major accidents including the failure of a key bridge over the Han and the collapse of a six-story department store, reflects the unbridled expansion of the South Korean economy as a whole, as well as the political struggles that carried the country from autocracy to democracy. Central Seoul lies in a basin surrounded by hills to an elevation of about 300 meters (1000 ft), and this near-megacity has sprawled outward in all directions, even toward the DMZ. An urban plan designed in the 1960s was subsequently overwhelmed by the steady inflow of immigrants.

During the era of Japanese colonial control, Seoul's surface links to other parts of the Korean Peninsula were improved, and this infrastructure played a role in the city's postwar success. Seoul is not only the capital but also the leading industrial center of South Korea, exporting sizeable quantities of textiles, clothing, footwear, and (increasingly) electronic goods. South Korea continues to thrive as an Asian-Pacific economic tiger, and Seoul is its heart.

political landscape would have to change. And so it was: dictatorial rule ended, representative government took hold, corruption was confronted if not altogether tamed, and the economy, once described by early postwar observers as never-to-be-developed, took off like a rocket. South Korea in short order became a boisterous democracy, an economic giant, a nation in universal pursuit of education, the world's leading shipbuilder, a manufacturer of iron and steel as well as chemicals and electronics plus countless other industrial products in addition to producing ample food supplies. South Korean social institutions became ever stronger, the media are as free as any on Earth, freedom of religion prevails, Korean art and culture find a global market, and Korean athletes excel in the Olympic Games and other international sports venues. As a key member of the international community, South Korea participates in peacekeeping and security operations worldwide.

The map of South Korea (Fig. 9B-11) reveals several of these achievements, most notably the metropolitan agglomeration formed by Seoul (the capital) and neighboring Incheon, which today constitutes one of Asia's greatest urban complexes and contains about one-quarter of the country's population. Even before democracy took hold, South Korea's rulers decided to further enhance the country's infrastructure by introducing a high-speed, bullet-train network to link Seoul to the traditionally separated subregions of Geyongsang and Jeolla. For a long time, workers in parts of the country not favored by politicians and manufacturers had trouble finding employment in fast-growing industrial areas, an economic-geographic pattern still not completely erased. As regional integration progressed, the southeastern subregion of Geyongsang, centered on the city of Busan, was always ahead of the southwest, where Jeolla was historically marginalized by the powers in Seoul. Today Geyongsang remains in the lead: here 300,000-ton tankers are built along with smaller freighters and warships; here are the famous Hyundai and Kia automobile plants; and here is the home base of POSCO, headquartered in Pohang, one of the world's largest steelmaking corporations.

Given all this success, it is surprising to learn that while the world is quite pleased with these achievements, the South Koreans themselves are not especially happy as a society. In recent surveys, only 9 percent of them declared themselves to be happy—a lower percentage than in China, Indonesia, Russia, and the great majority of the approximately 100 surveyed countries. Why should this be so? Interpretations of these data tend to focus on the danger from the North with which South Koreans live daily, frequently reminded of the

SJ. Kim/Getty Images, Inc.

Songdo New City, South Korea's answer to China's Pacific Rim hub cities. Songdo is the centerpiece of the ultramodern urban complex springing up around the country's second-largest port of Incheon at the southwestern corner of the Seoul-centered conurbation. Modeled after Singapore, Songdo is not only designed to be "the world's gateway to Northeast Asia" (more than 25 percent of the world's population lives within four hour's flying time) but also to be a prototype "Smart City". The gateway function is based on the global accessibility enabled by adjacent, state-of-the-art Incheon International Airport—making Songdo an international magnet or "aerotropolis" serving globalized businesses as well as offering a Las Vegas-style "pleasure carnival" featuring mega-casino/resorts, theme parks, and a Jack Nicklaus-built golf course. At the same time, still-building Songdo's smart-city technology is regarded as a cutting-edge example of sustainable urbanization, one that so highly impressed the UN's Green Climate Fund that it located its headquarters here in 2012.

risk when North Korea inflicts lethal attacks and launches ballistic missiles. But there is another dimension not reflected by the North-South contrasts displayed in Appendix B. In states that achieve great economic success, women tend to participate in the action even if they are not paid as well as men. But South Korea's success story is almost entirely male. The percentage of women gainfully employed in South Korea is not representative of a developed economy. The number of women in senior positions is far lower. Late in 2012, the South Koreans elected as their president Park Geun-hye, the daughter of one of the modern country's early dictators. Korean women hope that one of her priorities is to start correcting a gender deficit that contradicts South Korea's success story.

JAPAN

Japan consists of four main islands—Honshu, the largest; Hokkaido to the north; and Shikoku and Kyushu to the south—in addition to numerous small islands and islets (Fig. 9B-12). Most of the country is mountainous and steep-sloped, geologically young, earthquake-prone, and studded with volcanoes. A mere 18 percent of the country is considered to be habitable. Japan's high-relief topography has constrained its economic development. Except for the ancient capital of Kyoto, all of Japan's major cities are perched along the coast, and virtually all lie partly on artificial land reclaimed from the sea.

From the Field Notes . . .

"The city of Kyoto, chronologically Japan's second capital (after Nara; before Tokyo), is the country's principal center of culture and religion, education, and the arts. Tree-lined streets lead past hundreds of Buddhist temples; tranquil gardens provide solace from the bustle of the city. I rode the bullet train from Tokyo and spent my first day following a walking route recommended by a colleague, but got only part of the way because I felt compelled to enter so many of the temple grounds and gardens. And not only Buddhism, but also Shinto makes its mark on the cultural landscape. I passed under a *torii*—a gateway usually formed by two wooden posts topped by two horizontal beams turned up at their ends—which signals that you have left the secular and entered the sacred, and found this beautiful Shinto shrine with its orange trim and olive-green glazed tiles."

© H.J. de Blij

www.conceptcaching.com

Sailing into Kobe harbor near Osaka (the country's second-largest city), one passes artificial islands designed for high-volume shipping and connected to the mainland by automatic, space-age trains. Enter Tokyo Bay, and the refineries and factories to the east and west stand on huge expanses of landfill that have pushed the bay's shoreline outward.

With just over 127 million people, seven-eighths of whom reside in towns and cities, Japan uses its habitable living space very intensively—and expands it wherever possible. As Figure 9B-12 shows, farmland in Japan is both limited and quite regionally fragmented. Urban sprawl has invaded much of the cultivable land. In the hinterland of Tokyo, around Osaka, and surrounding Nagoya, major farming zones are under relentless urban pressure. All three of these lowlands lie within Japan's fragmented but well-defined core area (delimited by the red line on the map), the heart of Japan's prodigious manufacturing complex.

Coastal Development

Figure 9B-12 also tells us a great deal about the nature of Japan's external orientation and its dependence on foreign trade. All of the country's primary and secondary regions are found on the coast. Dominant among these regions is the ***Kanto Plain***, the heart of Japan's core area, which is focused on metropolitan Tokyo and contains about one-third of the country's population. Among its advantages are an unusually extensive area of low relief, a fine natural harbor at Yokohama, a relatively mild and moist climate, and a central location with respect to the country as a whole. (Its principal disadvantage lies in its vulnerability to major earthquakes [see Fig. 9A-2].)

The second-ranking primary economic region within Japan is named the ***Kansai District*** (Fig. 9B-12); it contains the Osaka-Kobe-Kyoto triangle and is located at the eastern end of the Seto Inland Sea between Honshu and Shikoku. Osaka and Kobe are leading industrial centers and busy ports, but the Kansai District also yields large harvests of rice, Japan's staple food. Between the Kanto Plain and the Kansai District lies the ***Nobi Plain*** (Fig. 9B-12), centered on the city of Nagoya. And, as the map shows, the Japanese core area is anchored in the west by the **conurbation [10]** centered on ***Kitakyushu***, situated not on Honshu Island but in the northwestern corner of Kyushu, Japan's southernmost major island. This still-expanding edge of the core area is particularly favored by its location relative to South Korea and China.

A Trading Nation

Japan is a remarkable country that despite its poor natural endowments (except for its advantageous location) has done extremely well. As noted in Chapter 9A, it became the dominant power of East and Southeast Asia through military force prior to World War II, and its postwar economic miracle astounded the world. In the 1970s and 1980s, people referred to this corner of the world as the "Yen Bloc," one that was dominated by the Japanese economy.

The country defeated the odds by calling on old Japanese virtues: organizational efficiency, massive productivity, dedication to quality, and tight adherence to common goals. Note that this has always been an exceptionally homogeneous, collectively oriented, and consensus-minded nation with a strong work ethic and a deep-seated conviction that common interests override those of the individual. It is hard to overestimate this cultural context in Japan's outstanding achievements even if it can turn into a liability, as when collective rigidity impedes adaptation to new circumstances.

More concretely, Japan dealt with its own limited geographic opportunities by engaging the alternatives. Look at Figures 9A-6 as well as 9B-12 and note Japan's substantial reliance on nuclear energy—a strategy not without major risks in this earthquake-prone region, as became so painfully clear in the aftermath of the massive Tohoku quake/tsunami of 2011 (see Fig. 9A-2). Most important of all, Japan had no choice but to engage other countries, near and far, in order to compensate for what it does not have itself. Japan, therefore, became wealthy through trade, imported raw materials, the accomplished production and processing of high-value-added goods, and the ability to turn the last into lucrative exports.

Tokyo, at the center of one of the largest metropolises in the world, continues to change. Land filling and bridge building in Tokyo Bay continue; skyscrapers sprout amid low-rise neighborhoods in this earthquake-prone area; traffic congestion steadily worsens. The red-painted Tokyo Tower, a beacon in this section of the city, was modeled after the Eiffel Tower in Paris but, as a billboard at its base announces, is an improvement over the original: lighter steel, greater strength, and less weight.

Yann Arthus-Bertrand/Science Source

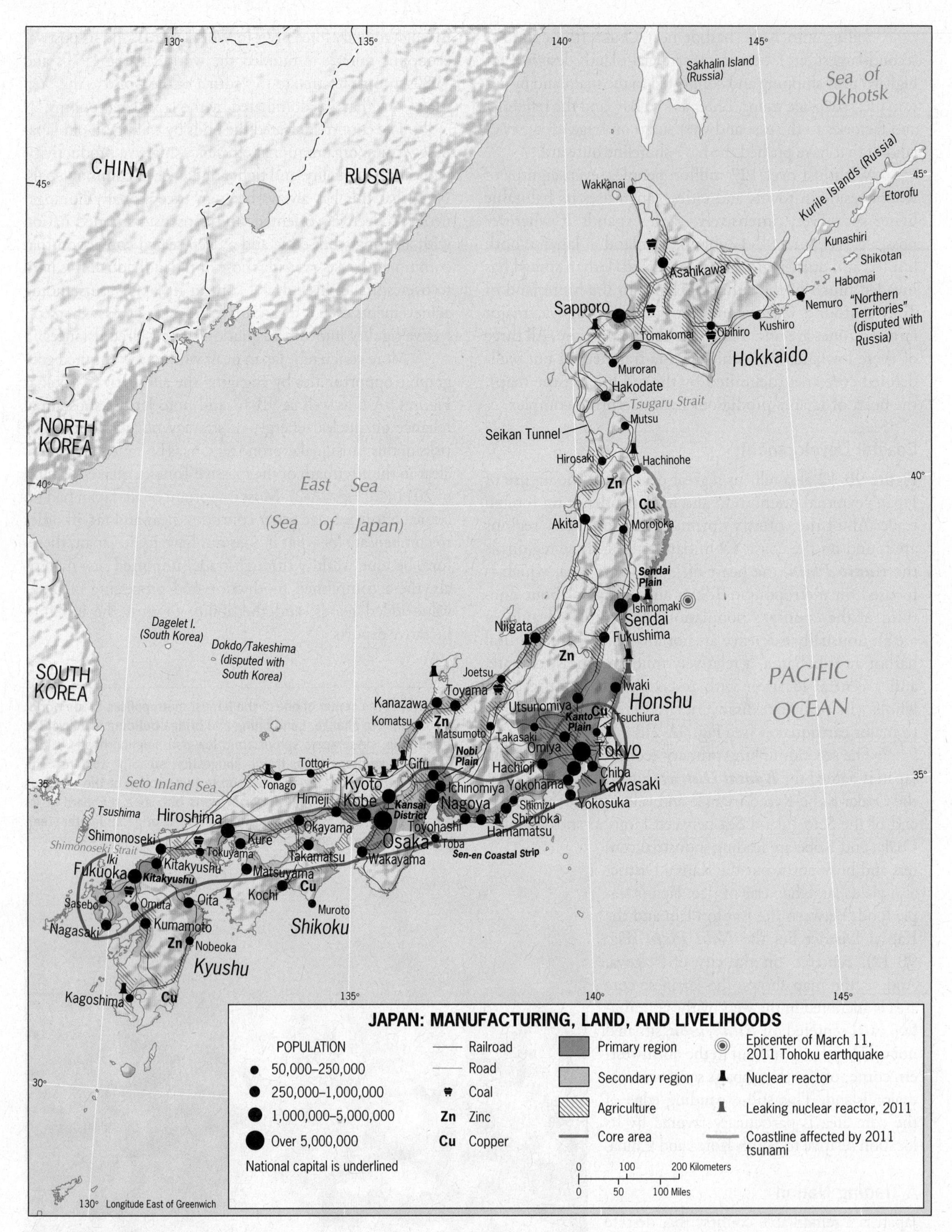

JAPAN: MANUFACTURING, LAND, AND LIVELIHOODS
CHINA
RUSSIA
NORTH KOREA
SOUTH KOREA
Sakhalin Island (Russia)
Sea of Okhotsk
Kurile Islands (Russia)
Etorofu
Kunashiri
Shikotan
Habomai
"Northern Territories" (disputed with Russia)
Wakkanai
Asahikawa
Sapporo
Tomakomai
Obihiro
Kushiro
Nemuro
Hokkaido
Muroran
Hakodate
Tsugaru Strait
Seikan Tunnel
Mutsu
Hirosaki
Hachinohe
Akita
Morojoka
Sendai Plain
Ishinomaki
Sendai
Fukushima
Niigata
Joetsu
Toyama
Kanazawa
Komatsu
Matsumoto
Takasaki
Utsunomiya
Kanto Plain
Tsuchiura
Omiya
Tokyo
Chiba
Kawasaki
Yokohama
Yokosuka
Hachioji
Iwaki
Honshu
PACIFIC OCEAN
East Sea
(Sea of Japan)
Dagelet I. (South Korea)
Dokdo/Takeshima (disputed with South Korea)
Gifu
Nobi Plain
Ichinomiya
Nagoya
Shimizu
Shizuoka
Hamamatsu
Toyohashi
Toba
Sen-en Coastal Strip
Tottori
Yonago
Seto Inland Sea
Kyoto
Kobe
Himeji
Kansai District
Osaka
Wakayama
Okayama
Takamatsu
Hiroshima
Kure
Tokuyama
Shimonoseki
Shimonoseki Strait
Tsushima
Iki
Kitakyushu
Fukuoka
Matsuyama
Kochi
Muroto
Shikoku
Oita
Omuta
Sasebo
Nagasaki
Kumamoto
Nobeoka
Kyushu
Kagoshima
Zn
Cu
130°
135°
140°
145°
45°
40°
35°
30°
POPULATION
50,000–250,000
250,000–1,000,000
1,000,000–5,000,000
Over 5,000,000
National capital is underlined
Railroad
Road
Coal
Zn Zinc
Cu Copper
Primary region
Secondary region
Agriculture
Core area
Epicenter of March 11, 2011 Tohoku earthquake
Nuclear reactor
Leaking nuclear reactor, 2011
Coastline affected by 2011 tsunami
0 100 200 Kilometers
0 50 100 Miles
130° Longitude East of Greenwich

FIGURE 9B-12

AMONG THE REALM'S GREAT CITIES . . .

TOKYO

MANY URBAN AGGLOMERATIONS are named after the city that lies at their heart, and so it is with the second-largest of all: Tokyo (26.6 million). Even its longer name—the Tokyo–Yokohama–Kawasaki conurbation—does not begin to describe the congregation of cities and towns that form this massive, crowded metropolis that encircles the head of Tokyo Bay and continues to grow, outward and upward.

Near the waterfront, some of Tokyo's neighborhoods are laid out in a grid pattern. But the urban area has sprawled over hills and valleys, and much of it is a maze of narrow winding streets and alleys. Circulation is slow, and traffic jams are legendary. The train and subway systems, however, are models of efficiency—although during rush hours you must get used to the *shirioshi* pushing you into the cars to get the doors closed.

At the heart of Tokyo lies the Imperial Palace with its moats and private parks. Across the street, buildings retain a respectful low profile, but farther away Tokyo's skyscrapers seem to ignore the peril of earthquakes. Nearby lies one of the world's most famous avenues, the Ginza, lined by department stores and luxury shops. In the distance you can see an edifice that looks like the Eiffel Tower, only taller: this is the Tokyo Tower (see photo), a multipurpose structure designed to test lighter Japanese steel, transmit television and cellphone signals, detect Earth tremors, monitor air pollution, and attract tourists.

Tokyo is the epitome of modernization, but Buddhist temples, Shinto shrines, historic bridges, and serene gardens still grace this burgeoning urban behemoth, a cultural landscape that reflects Japan's successful marriage of the modern and the traditional.

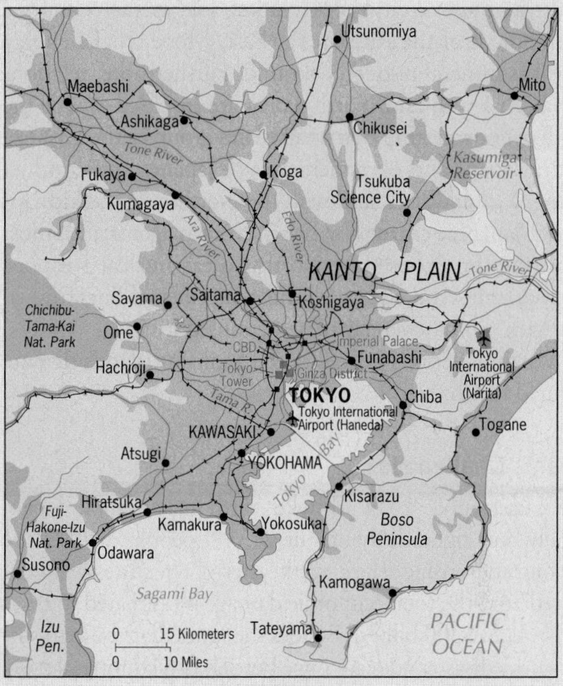

© H. J. de Blij, P. O. Muller, J. Nijman, and John Wiley & Sons, Inc.

Japan's Two Lost Decades

At the end of the 1980s, it seemed that Japan's meteoric rise had been so prodigious that it would only be a matter of time before it would surpass even the United States' economy. But everything came to a rather sudden halt during the early 1990s, and Japan has unexpectedly been stuck in a rut ever since. This now more than 20-year-long economic slowdown had many causes, ranging from government mismanagement and inefficiency to intensifying international competition (such as from South Korean cars and Taiwanese electronic products).

One of the main problems was that Japan is deficient in adaptability, so its culturally embedded economic system could not cope with rapidly changing circumstances. When the economy turned sour, there was no vigorous response as one might expect in a true free market. With bankruptcy a near-taboo, too many weak firms were kept alive, and this kind of economic sclerosis greatly inhibits the creation of new companies (the United States in any given year has three times the number of start-ups as Japan and twice the number of bankruptcies).

Japan's conservative culture became a liability. Whereas women constitute an ever greater share of Japan's highly educated workforce, only 7 percent of all senior management executives are female (compared to 20 percent in the U.S. and 31 percent in China). For the first time in decades, Japan suffered a crisis of confidence and, more tangibly, something unfamiliar to the burgeoning Japan of the late twentieth century—rising unemployment. Faced with much leaner economic times, men especially seemed to seek refuge in the past as growing numbers preferred to stay with the same dormant big companies instead of starting their own businesses.

Japan is a rapidly aging society as well, projected to decline from 127.3 million today to barely 95 million in 2050 and less than 65 million by the end of this century. It faces an enormous and increasing **demographic burden [11]**: by 2025, it is projected there will be only two workers for every retiree. Ethnically homogeneous Japan has historically resisted immigration, so when the government tried to recruit ethnic Japanese living in Brazil (and elsewhere) to return home, the experiment was not very successful. Thus a shrinking base of qualified workers threatens the government's capacity to sustain social programs for all.

Still another set of problems has to do with Japan's international relations. Japan never signed a peace treaty with the (then) Soviet Union after World War II because the Russians had occupied and refused to return four small island groups in the Kurile chain northeast of Hokkaido (Fig. 9B-12). Failed negotiations for the return of these "Northern Territories," as the Japanese call them, have cost Japan the opportunity to play a key role in the economic development of the Russian Far East, where crucial energy as well as mineral resources abound. Furthermore, relations with both South and North Korea are troubled—with the South over ownership of a small island group in the East Sea (Sea of Japan) and over memories of Japanese misconduct during World War II, and with the North over the kidnapping of Japanese citizens by North Korean agents and, most of all, the threat posed by North Korea's nuclear weapons development. And add to all this a number of lingering issues with China over Japanese actions during the colonial and wartime periods as well as the Senkaku Islands dispute outlined in Chapter 9A.

Japan's Challenges

The catastrophic 2011 Tohoku earthquake shook Japan, literally and figuratively: about 21,000 people died, entire communities in Honshu's northeastern prefectures were destroyed, and the total economic damage is estimated to be as high as U.S. $300 billion. Besides the immediate human suffering, the disaster added a huge burden to a country already straining to return its economy to a healthier state. Before the quake, Japan had already amassed a debt exceeding twice its GDP, bigger than that of any other high-income country (the U.S., in comparison, has a national debt that about equals its GDP and many consider that hugely problematic). The staggering costs of recovery are certain to push that deficit even deeper into financial distress.

Interestingly, now that the country's long-term plans for reliance on nuclear power have been called into question, some Japanese power companies in 2013 were turning their attention to **geothermal power [12]**. The potential is excellent because the latter is derived from underground heat sources, generated here by the incessant tectonic activity that dominates Japan's geology. Cutting-edge geothermal technologies do offer promising future opportunities, but the country's needs are far more immediate: energy demand today is rising steadily and, for years to come, will be impossible to satisfy without nuclear power.

Japan has fared poorly since 1990 in adjusting to new economic realities, and the situation today has become even more acute. Quite apart from marshaling the resources to complete its earthquake recovery, to get its house back in order Japan also has to deal with: (1) its demographic burden, by increasing births as well as immigration; (2) its moribund economy, by allowing inadequately performing companies to fail and letting the creative destruction of capitalism do its work; (3) the issue of fostering business start-ups, by rewarding both creativity and risk; (4) abandoning its self-defeating attitudes in the sphere of gender equality; and (5) the resolution of major geopolitical disputes that serve to impede potentially fruitful rewarding new economic relationships, most importantly with Russia and China (see Chapter 9A).

■ TAIWAN

When we see where Taiwan is located, it would seem to be inextricably bound up with China because it lies a mere 200 kilometers (125 mi) from the mainland. Not surprisingly, Taiwan has a history of immigration from coastal Chinese provinces. As discussed in Chapter 9A, China's rulers view Taiwan as a "wayward province" that must one day return to the People's Republic. Taiwan lost its seat in the United Nations in 1971 and was replaced by the PRC. The United States at the time also rescinded its official recognition of Taiwan in favor of improved relations with mainland China.

Taiwan's Island Geography

Taiwan, as Figure 9B-13 shows, is not a large island. It is smaller than Switzerland but has a much larger population (23.4 million), most of it concentrated in a crescent lining the western and northern coasts. The Chungyang Mountains, an area of high elevations (some more than 3000 meters [10,000 ft]), steep slopes, and dense forests dominate the eastern half of the island. Westward, these mountains yield to a zone of hilly topography and, facing the Taiwan Strait, a wide, substantial coastal plain. Streams from the mountains irrigate the paddyfields, and farm production has more than doubled since 1950 even though hundreds of thousands of farmers left the land for employment in Taiwan's expanding industries.

Today the lowland urban-industrial corridor lining western Taiwan is anchored by the capital, Taipei, at the island's northern end and rapidly growing Kaohsiung in the far south. The Japanese developed Chilung, Taipei's outport, to export nearby coal, but now the raw materials flow the other way. Taiwan imports raw cotton for its textile industry, bauxite (for aluminum) from Indonesia, oil from Brunei, and iron ore from Africa. Taiwan has a developing iron and steel industry, nuclear power plants, shipyards, a substantial chemical industry, and modern transport networks. Increasingly, however, it has been exporting high-technology products, especially personal computers, telecommunications equipment, and precision electronic instruments. Taiwan offers enormous brainpower resources, and many foreign firms join in the research and development carried on in such places as Hsinchu in the north, where the government helped to establish a **technopole [13]** centered on the microelectronics and personal computer industries. In the south, the "science city" of Tainan specializes in microsystems and information technology.

The establishment of China's Special Economic Zones went a long way toward mitigating longstanding problems between Beijing and Taipei. China needed investment capital and Taiwan had it, and in the almost-anything-goes environment of the SEZs, Taiwanese entrepreneurs could buy or build factories just as "real" foreigners were able to do.

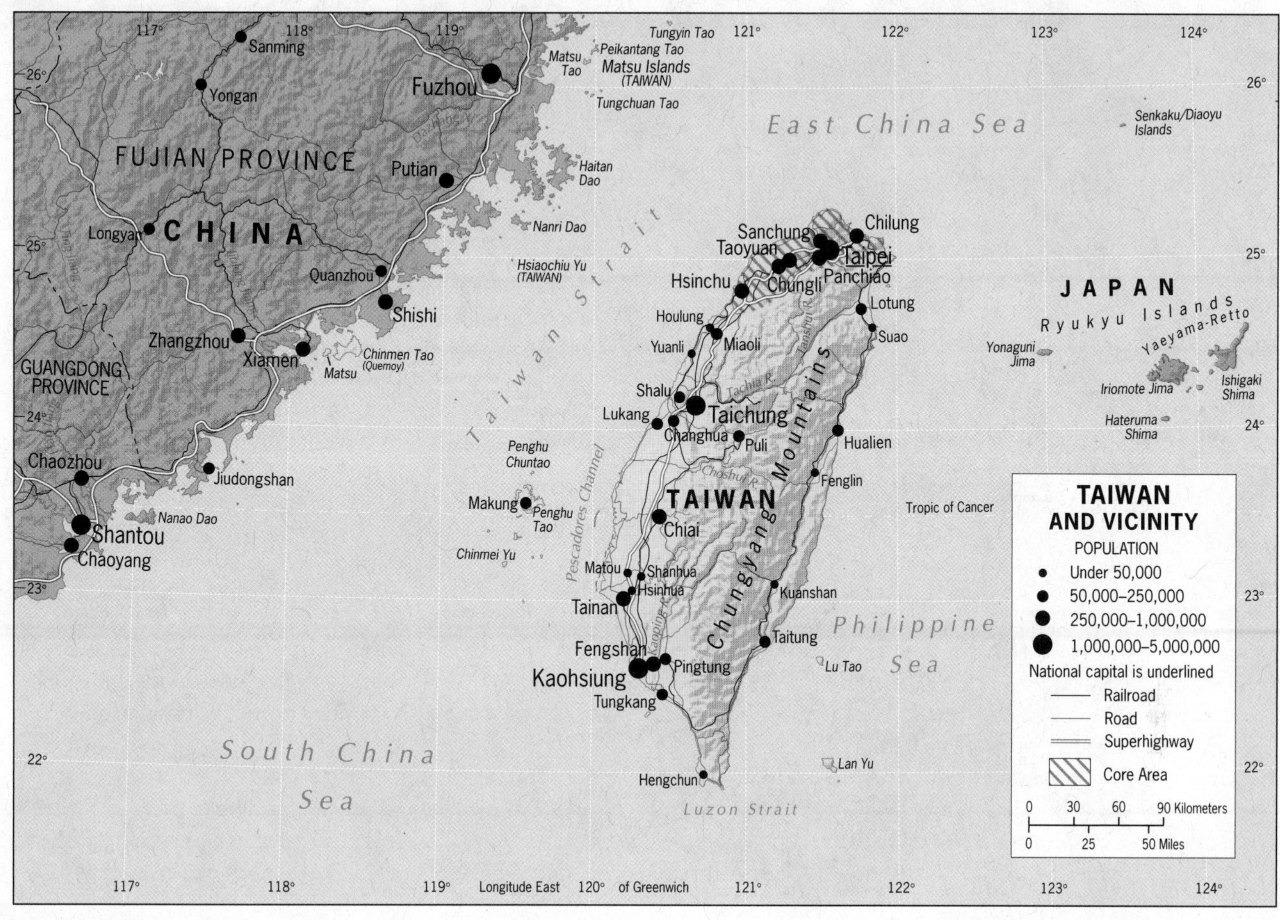

FIGURE 9B-13

Importantly, as we noted earlier, the Xiamen SEZ was laid out directly across from Taiwan for this purpose (people there will tell you that 80 percent of locals have relatives across the Taiwan Strait). As a result, thousands of Taiwanese-owned factories began operating in China, their owners and investors having a strong interest in avoiding violent confrontation over political issues.

Taiwan's Future

Taiwan's unresolved status entails risks, but there are signs that both parties want to find a solution (in 2012 the Taiwanese reelected President Ma Ying-jeou, effectively voting for another four years of rapprochement with the PRC). Given that the PRC will never allow Taiwan to attain independence, and the growing prospect that the United States, Taiwan's chief guarantor, could not secure Taiwan against Chinese military intervention, many Taiwanese as well as their allies and adversaries are seeking a long-range, negotiated solution. Among the options is one that appears to be working in Hong Kong, where the principle of ***One Nation, Two Systems*** has functioned more successfully than many observers anticipated (even though Hong Kong occasionally experienced political protests and demonstrations against Chinese rule). As a Special Administrative Region (SAR), Hong Kong's economy has thrived, and freedoms not available in the Shi's, provinces, or Autonomous Regions continue to prevail. In time, the Taiwanese may negotiate a comparable status, internationally as well as domestically, that would benefit all concerned.

POINTS TO PONDER

- In 2013, the world's three biggest construction companies, in terms of revenues generated, were Chinese (a decade earlier none was in the top 10).
- The urban complex focused on China's Pearl River Estuary has a manufacturing workforce larger than that of the entire United States.
- Increasing social inequalities resulting from China's rapid economic growth are becoming ever more difficult to justify by China's self-proclaimed communist government.
- After nearly 70 years of dictatorship, malnutrition, and famine, North Koreans at the age of 21 are an average 6 centimeters (2.4 in) shorter than their South Korean neighbors.

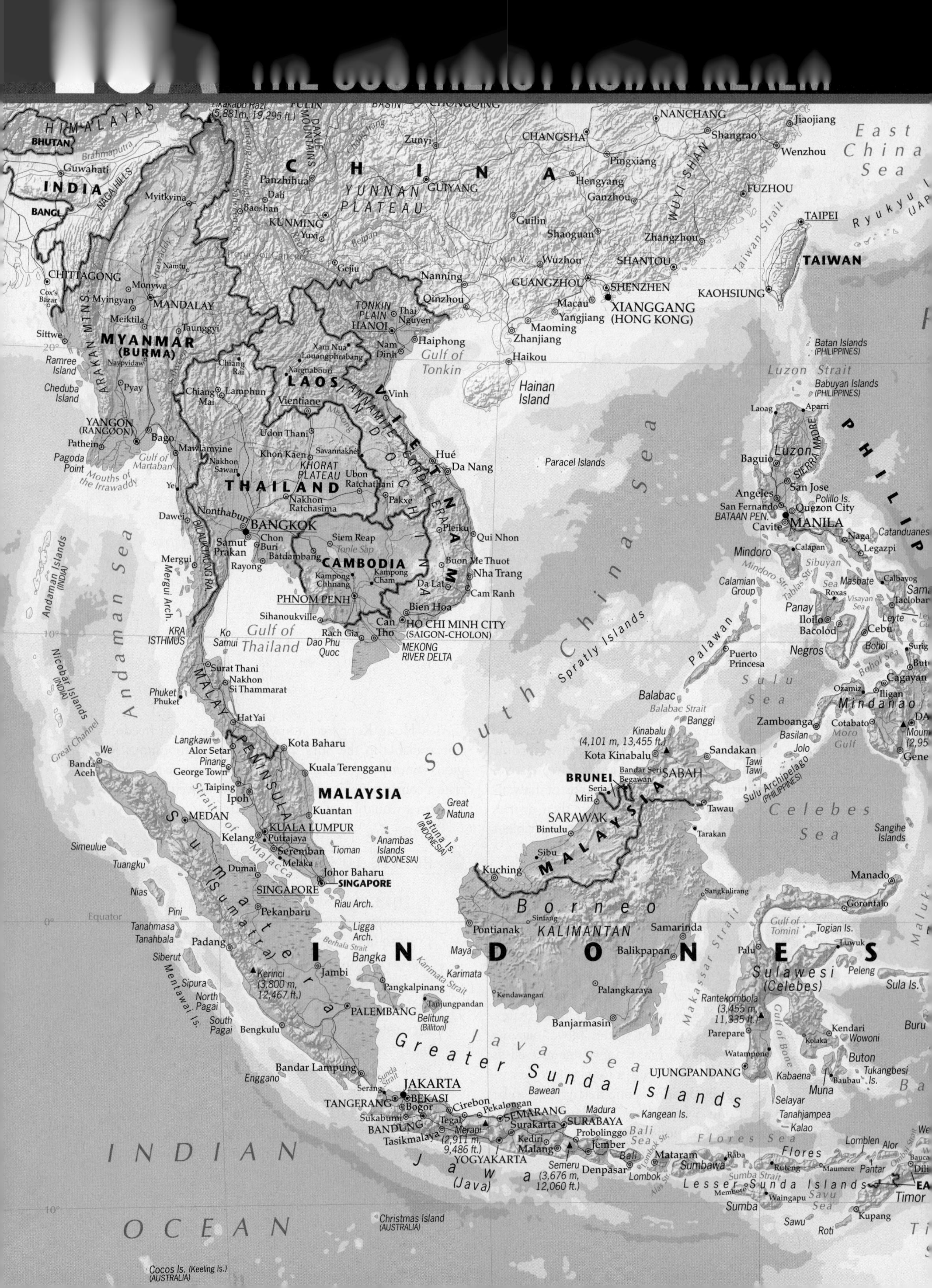

HIMALAYAS
Hkakabo Razi (5,881 m, 19,295 ft.)
BHUTAN
Brahmaputra
Guwahati
INDIA
NAGA HILLS
BANGL.
Myitkyina
CHITTAGONG
Cox's Bazar
Namtu
Monywa
Myingyan
MANDALAY
Meiktila
Taunggyi
Sittwe
ARAKAN MTNS.
MYANMAR (BURMA)
Naypyidaw
Ramree Island
Cheduba Island
Pyay
YANGON (RANGOON)
Pathein
Bago
Mawlamyine
Gulf of Martaban
Pagoda Point
Mouths of the Irrawaddy
Ye
Dawei
Mergui
Mergui Arch.
Andaman Sea
Andaman Islands (INDIA)
Nicobar Islands (INDIA)
Great Channel
BILAUKTAUNG RA.
KRA ISTHMUS
Phuket
We
Banda Aceh
Irrawaddy
Salween
CHINA
FULIN
DAXUE MOUNTAINS
BASIN
CHONGQING
Zunyi
Panzhihua
Dali
Baoshan
KUNMING
Yuxi
YUNNAN PLATEAU
GUIYANG
Gejiu
Tropic of Cancer
CHANGSHA
NANCHANG
Jiaojiang
Shangrao
Wenzhou
Pingxiang
Hengyang
Ganzhou
WUYI SHAN
FUZHOU
Guilin
Shaoguan
Zhangzhou
Wuzhou
SHANTOU
GUANGZHOU
SHENZHEN
XIANGGANG (HONG KONG)
Macau
Yangjiang
Maoming
Zhanjiang
Haikou
Hainan Island
Nanning
Qinzhou
TAIPEI
TAIWAN
KAOHSIUNG
Taiwan Strait
East China Sea
Ryukyu
TONKIN PLAIN
HANOI
Thai Nguyen
Haiphong
Nam Dinh
Gulf of Tonkin
Xam Nua
Louangphrabang
Xaignabouri
Chiang Rai
Chiang Mai
Lamphun
LAOS
Vinh
Vientiane
ANNAMITE
Udon Thani
Khon Kaen
Savannakhet
KHORAT PLATEAU
Nakhon Sawan
THAILAND
Ubon Ratchathani
Nakhon Ratchasima
Pakxe
Hué
Da Nang
CORDILLERA
VIETNAM
Nonthaburi
BANGKOK
Samut Prakan
Chon Buri
Rayong
Siem Reap
Tonle Sap
Batdambang
CAMBODIA
Kampong Chhnang
Kampong Cham
PHNOM PENH
Pleiku
Qui Nhon
Buon Me Thuot
Nha Trang
Da Lat
Cam Ranh
Bien Hoa
HO CHI MINH CITY (SAIGON-CHOLON)
Sihanoukville
Rach Gia
Can Tho
Dao Phu Quoc
MEKONG RIVER DELTA
Ko Samui
Gulf of Thailand
Surat Thani
Nakhon Si Thammarat
Hat Yai
MALAY PENINSULA
Langkawi
Alor Setar
Pinang
George Town
Kota Baharu
Kuala Terengganu
Taiping
Ipoh
MALAYSIA
Kuantan
KUALA LUMPUR
Putrajaya
Kelang
Seremban
Melaka
Tioman
Strait of Malacca
Johor Baharu
SINGAPORE
Riau Arch.
Anambas Islands (INDONESIA)
Great Natuna
Natuna Is. (INDONESIA)
Paracel Islands
South China Sea
Spratly Islands
Luzon Strait
Batan Islands (PHILIPPINES)
Babuyan Islands (PHILIPPINES)
Laoag
Aparri
Luzon
SIERRA MADRE
Baguio
PHILIPPINES
Angeles
San Jose
San Fernando
Polillo Is.
Quezon City
BATAAN PEN.
MANILA
Cavite
Naga
Catanduanes
Legazpi
Calapan
Mindoro
Mindoro Str.
Sibuyan Sea
Tablas Str.
Masbate
Calbayog
Samar
Tacloban
Calamian Group
Roxas
Visayan Sea
Panay
Iloilo
Bacolod
Leyte
Cebu
Negros
Bohol
Surigao
Bohol Sea
Palawan
Puerto Princesa
Sulu Sea
Balabac
Balabac Strait
Banggi
Ozamiz
Iligan
Cagayan
Mindanao
Zamboanga
Cotabato
Basilan
Moro Gulf
Jolo
Tawi Tawi
Sulu Archipelago (PHILIPPINES)
Kinabalu (4,101 m, 13,455 ft.)
Kota Kinabalu
Sandakan
SABAH
BRUNEI
Bandar Seri Begawan
Seria
Miri
Tawau
SARAWAK
Bintulu
Tarakan
Sibu
MALAYSIA
Kuching
Celebes Sea
Sangihe Islands
Manado
Gorontalo
Gulf of Tomini
Togian Is.
Sangkulirang
Borneo
Sintang
Pontianak
KALIMANTAN
Samarinda
Balikpapan
Makassar Strait
Palu
Luwuk
Peleng
Sula Is.
Maluku
Sulawesi (Celebes)
INDONESIA
Palangkaraya
Kendawangan
Banjarmasin
Rantekombola (3,455 m, 11,335 ft.)
Parepare
Gulf of Bone
Kendari
Kolaka
Wowoni
Buru
Buton
Watampone
Kabaena
Baubau
Muna
Tukangbesi Is.
UJUNGPANDANG
Selayar
Tanahjampea
Kalao
Simeulue
MEDAN
Tuangku
Nias
Dumai
Pini
Equator
Pekanbaru
Sumatera (Sumatra)
Tanahmasa
Tanahbala
Padang
Siberut
Mentawai Is.
Sipura
North Pagai
South Pagai
Kerinci (3,800 m, 12,467 ft.)
Jambi
Bengkulu
PALEMBANG
Lingga Arch.
Berhala Strait
Bangka
Maya
Karimata
Karimata Strait
Pangkalpinang
Tanjungpandan
Belitung (Billiton)
Java Sea
Greater Sunda Islands
Bandar Lampung
Enggano
Sunda Strait
Serang
JAKARTA
BEKASI
TANGERANG
Bogor
Sukabumi
BANDUNG
Cirebon
Pekalongan
Tegal
SEMARANG
Merapi (2,911 m, 9,486 ft.)
Tasikmalaya
Surakarta
YOGYAKARTA
Kediri
Malang
Madura
SURABAYA
Probolinggo
Bawean
Kangean Is.
Semeru (3,676 m, 12,060 ft.)
Jember
Jawa (Java)
Bali
Bali Sea
Denpasar
Lombok
Lombok Str.
Alas Str.
Mataram
Sumbawa
Raba
Flores Sea
Flores
Ruteng
Maumere
Lomblen
Alor
Pantar
Dili
Sumba Strait
Lesser Sunda Islands
Sumba
Memboro
Waingapu
Savu Sea
Sawu
Roti
Kupang
Timor
INDIAN OCEAN
Christmas Island (AUSTRALIA)
Cocos Is. (Keeling Is.) (AUSTRALIA)

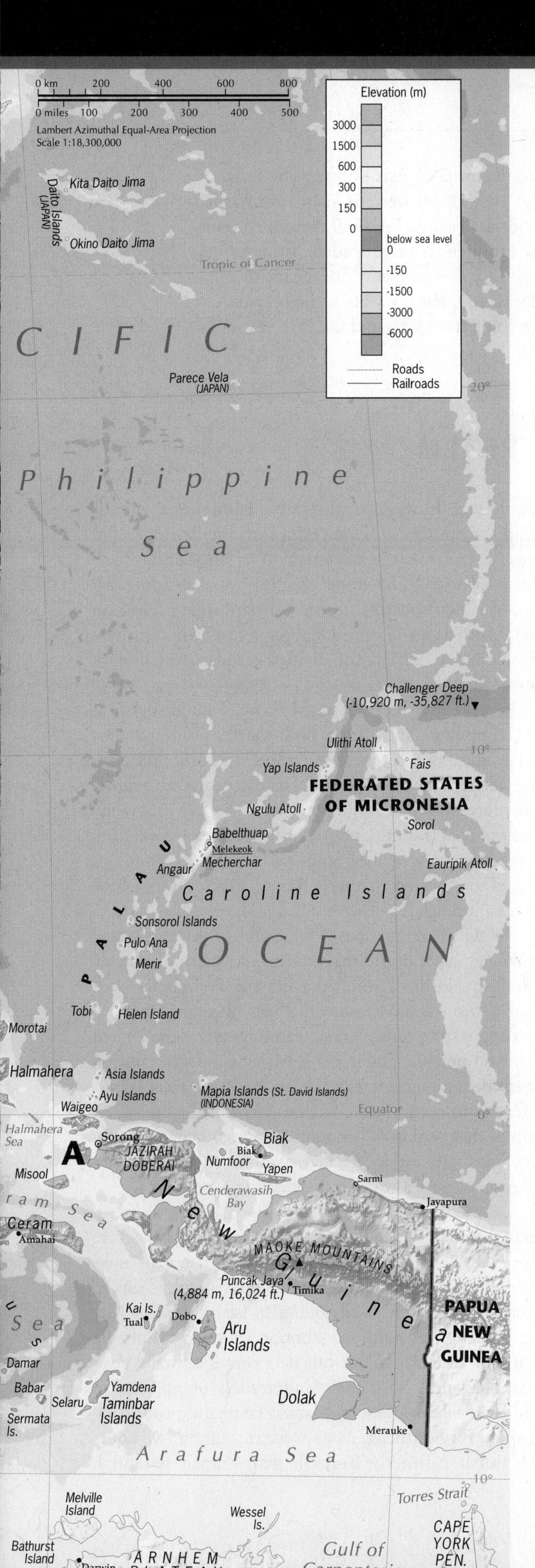

IN THIS CHAPTER

- Earthquakes, tsunamis, and volcanoes
- The mixed blessings of palm oil
- The Overseas Chinese
- Singapore's centrality
- ASEAN's impotence
- Geopolitics in the South China Sea
- Why the shape of a country matters

CONCEPTS, IDEAS, AND TERMS

Buffer zone 1
Shatter belt 2
Tsunami 3
Biodiversity 4
Overseas Chinese 5
Emerging market 6
Node 7
ASEAN 8
AFTA 9
Antecedent boundary 10
Subsequent boundary 11
Superimposed boundary 12
Relict boundary 13
Compact state 14
Protruded state 15
Elongated state 16
Fragmented state 17
Perforated state 18

FIGURE 10A-1

Southeast Asia is a realm of peninsulas and islands, a corner of Asia bounded by India on the northwest and China on the northeast (Fig. 10A-1). Its western coasts are washed by the Indian Ocean, and to the east stretches the vast Pacific. From all these directions, Southeast Asia has been penetrated by outside forces. From India came traders; from China, settlers; from across the Indian Ocean, Arabs to engage in commerce and Europeans to build empires; and from across the Pacific, Americans. Southeast Asia has been the scene of countless contests for power and primacy—the competitors have come from near and far.

DEFINING THE REALM

Southeast Asia's geography in some ways resembles that of eastern Europe, even if the physiography is very different. It is a mosaic of smaller countries on the periphery of two of the world's largest states. It has been a **buffer zone [1]** between powerful adversaries. It is a **shatter belt [2]** in which stresses and pressures from without and within have fractured the political geography. Like eastern Europe, Southeast Asia exhibits great cultural diversity. This is a realm of hundreds of cultures and ethnicities, numerous languages and dialects, global as well as local religions, and diverse national economies ranging from high- to low-income.

A GEOGRAPHIC OVERVIEW

Figure 10A-1 displays the relative location and dimensions of the Southeast Asian geographic realm, an assemblage of 11 countries situated on the Asian mainland as well as on thousands of islands, large and small, extending from Sumatera* in the west to New Guinea in the east, and from Luzon in the north to Timor in the south. We will become familiar with only the largest and most populous of these islands, but sailing the seas of this part of the world you would pass dozens of smaller ones every day. And if you could stop, you would find even the tiniest populated islands to have their own character resulting from the cultural sources of their residents, the environmental challenges they face and the opportunities they exploit, their modes of dress, and the structure of their dwellings, even the vivid colors with which they often decorate their boats.

The giant of this realm, in terms of both territory and population, is the far-flung ***archipelago*** (island chain) of Indonesia, labeled appropriately on the map by the largest letters. In the east, the state of Indonesia extends beyond the Southeast Asian realm into the Pacific Realm because it controls the western half of an island—New Guinea—whose indigenous peoples are not Southeast Asian. We focus on this unusual situation later (and on New Guinea as a whole in Chapter 12), but note it here because unusual borders and divided islands are a hallmark of this realm.

In mainland Asia, as noted above, Southeast Asia is bordered by India and China, both sources of immigrants, cultural infusions, economic initiatives, and other relationships evident in the realm's cultural landscapes. Also originating there are rivers that play a central role in the lives of many millions of people south of the realm border. As we will find, the core areas of several of Southeast Asia's most populous countries are located in the basins of major rivers whose sources lie inside China.

Before we get started, it is helpful to get acquainted with the key states of this realm, some of which will already be familiar. North of Indonesia lies the Philippines, well known to Americans because this was once an American colony and is still a source of many immigrants to the United States. Also north of Indonesia is Malaysia, easy to find on the map because its core area lies on the long Malay Peninsula that almost connects mainland Asia with Indonesia. At the tip of that peninsula lies another famous geographic locality: Singapore, the spectacular economic success story of Southeast Asia and a world-city by every measure.

On the Asian mainland, the name Vietnam still resonates in America, which fought a bitter and costly war there in the 1960s and 1970s. As the map shows, Vietnam looks like a sliver of land extending from its border with China to the delta of the greatest of all Southeast Asian rivers, the Mekong. Looking westward, neighboring Laos and Cambodia may not be all that familiar, but next comes centrally positioned Thailand, the dream (and reality) of millions of tourists and one of the world's most fascinating countries. And then, on Southeast Asia's western margin, we come to Myanmar (formerly Burma): here human potential,

---

*As in Africa and South Asia, names and spellings have changed with independence. In this chapter we will use contemporary spellings, except when we refer to the colonial era. Thus Indonesia's four major islands are Jawa, Sumatera, Kalimantan (the Indonesian portion of Borneo), and Sulawesi. Prior to independence they were respectively known as Java, Sumatra, Dutch Borneo, and Celebes.

major geographic qualities of SOUTHEAST ASIA

1. Southeast Asia extends from the peninsular mainland to the archipelagos offshore. Because Indonesia controls part of New Guinea, its functional region reaches into the neighboring Pacific geographic realm.
2. Southeast Asia, like eastern Europe, has been a shatter belt between powerful adversaries and has a fractured cultural and political geography shaped by foreign intervention.
3. Southeast Asia's physiography is dominated by high relief, crustal instability marked by volcanic activity and earthquakes, and tropical climates.
4. A majority of Southeast Asia's 625 million people live on the islands of just two countries: Indonesia, with the world's fourth-largest population, and the Philippines. The rate of population increase in the Insular region of Southeast Asia exceeds that of the Mainland region.
5. Although the great majority of Southeast Asians have the same ancestry, cultural divisions and local traditions abound, which the realm's divisive physiography sustains.
6. Southeast Asia's political geography exhibits a variety of boundary types and several categories of state territorial morphology.
7. The Mekong River, Southeast Asia's Danube, has its source in China and borders or crosses five Southeast Asian countries, sustaining tens of millions of farmers, fishing people, and boat owners.
8. Singapore is the leading world-city in Southeast Asia and lies at the realm's center of trade and business relations.
9. Southeast Asia contains a number of rapidly emerging markets and fast-growing economies that in some respects follow in the tracks of the neighboring realm of East Asia.
10. China's influence in this realm has increased markedly in recent years and in some instances triggered a local backlash.

A section of a vast palm oil plantation in Sarawak State in the southern part of Malaysian Borneo. These plantations have proliferated in Malaysia and Indonesia, bringing in export revenues but at a mounting environmental cost.

natural endowment, and opportunity were thwarted by a half-century of extremely harsh military rule (that plunged the country into devastating poverty)—which in 2012 quite suddenly relaxed its vise-like grip and reopened Myanmar to the world.

SOUTHEAST ASIA'S PHYSICAL GEOGRAPHY

Physiographically, Southeast Asia is in some ways reminiscent of Middle America, a fractured realm of islands and peninsulas flanking a populous mainland studded with high mountains and deep valleys. It is useful to look back at Figures G-4 and G-5 to see why: both Southeast Asia and Middle America are dangerous places, where the Earth's crust is unstable as tectonic plates are in collision, earthquakes are a constant threat, volcanic eruptions take their toll, tropical cyclones lash sea and land, and floods, landslides, and other natural hazards make life riskier than in most other parts of the world.

In terms of human geography, Southeast Asia is part of the Pacific Rim. But in physiographic terms, it forms part of the Pacific Ring of Fire and all the hazards this designation brings with it. As recently as 2004, an undersea earthquake off westernmost Indonesia caused a **tsunami [3]** (seismic sea wave) in the Indian Ocean that killed more than 300,000 people along coastlines from Sumatera to Somalia. This was only the latest in an endless string of natural disasters originating in Southeast Asia whose effects were felt far beyond the realm's borders. In 1883, the Krakatau volcano lying between Sumatera and Jawa exploded, resulting in a death toll estimated at more than 36,000. In 1815, the Tambora volcano in the chain of islands east of Jawa known as the Lesser Sunda Islands blew up, darkening skies throughout the world and affecting climates around the planet (the year that followed is still known as the "year without a summer" when crops failed, economies faltered, and people went hungry as far away as Egypt, New England, and France). Research on what may have been the most calamitous of all such eruptions suggests that, about 73,000 years ago, the Toba volcano on Sumatera exploded with such force that its ash and soot not only darkened skies and affected weather, but changed global climate for perhaps as long as 20 years and threatened the very survival of the human population, then still small in number and widely dispersed. In fact, a number of scientists postulate that this eruption caused such widespread casualties that human genetic diversity was significantly diminished.

It therefore goes without saying that high relief dominates Southeast Asia, from the Arakan Mountains in western Myanmar to the glaciers (yes, glaciers!) of New Guinea. Take a close look at Figure 10A-1 and you can see how many elevations approach or exceed 3000 meters (10,000 ft), and, using the elevation guide in the upper right-hand corner, how mountainous and hilly much of the realm's topography is. Lengthy ranges form the backbones not only of islands such as Sulawesi and Sumatera, but also of the Malay Peninsula, most of Vietnam, and the border zone between Thailand and Myanmar. In Figure G-5 it is possible to trace these volcano-studded mountain ranges by their earthquake epicenters and volcanic records.

Exceptional Borneo

But among the islands there is one significant exception. The bulky island named Borneo in Figure 10A-1 has high elevations (Mount Kinabalu in the north reaches 4101 meters [13,455 ft]), yet has no volcanoes and negligible Earth tremors. This island has been called a stable "mini-continent" amid a mass of volcanic activity, a slab of ancient crust that long ago was pushed high above sea level by tectonic forces and was subsequently eroded into its present landscapes. Borneo's soils are not nearly as fertile as those of the volcanic islands, so that an equatorial rainforest developed here that long survived the human population explosion, giving sanctuary to countless plant and animal species including Southeast Asia's great ape, the orangutan. That era is now ending as human encroachment on Borneo's tropical habitat is accelerating, logging is destroying the remaining forest, and roads and farms are penetrating its interior. Borneo, along with other parts of Indonesia and Malaysia, has also experienced the rapid expansion of palm oil plantations at the expense of tropical woodlands (see box titled "Palm Oil Plantations and Deforestation").

Borneo's tropical forests are a remnant of a much larger stand of equatorial rainforest that once covered most of this realm. Besides Borneo, eastern Sumatera still contains limited expanses of rainforest, including a few orangutan sanctuaries. So does less populous and more remote New Guinea, which was never reached by these great apes. As we noted in the introductory chapter, equatorial rainforests can still be found today in three low-latitude areas of the world: the Amazon Basin of South America, west-equatorial Africa, and here in Southeast Asia. The combination of climatic conditions that sustains these forests—consistently warm temperatures and year-round rainfall—produces the biologically richest and ecologically most complex vegetation regions on Earth. Huge numbers and varieties of trees and other plants grow in very close proximity, vying for space and sunlight both horizontally and vertically. Yet despite all this luxuriant growth, the soils beneath rainforests are nutrient poor. Most of the surface nutrients are derived from the decaying vegetation on the forest floor, which nurture the growth of the next generation of plants. Long-term indigenous inhabitants have developed ways to use the plants, animals, and soils to eke out a living—but when migrant farmers from elsewhere remove the trees to utilize the soil, mistakenly assuming it will be able to support their crops, failure swiftly results. This reality notwithstanding, intensifying population pressure throughout the tropics continues to shrink what remains of the world's rainforests.

Relative Location and Biodiversity

Look again at Figure 10A-1, but in a more general way. The Malay Peninsula, adjacent Sumatera, Jawa, and the Lesser Sunda Islands east of Jawa seem to form a series of stepping stones toward New Guinea, and in the extreme lower-right corner of the map you can see Australia's Cape York Peninsula appearing to reach toward New Guinea. What would happen, you might ask, if sea level were to drop and those narrow bodies of water between the peninsulas and the islands dried up?

That is precisely what occurred—not just once, but repeatedly in the geologic history of Southeast Asia. Long ago, the orangutans whose descendants remain in Indonesia today were able to migrate from the mainland into warmer equatorial latitudes because global sea level dropped during periods of glaciation, and islands separated by water today were temporarily connected by land bridges. More recently, early human arrivals here also got assistance from

Palm Oil Plantations and Deforestation

IN THE MID-NINETEENTH century, it was discovered that the reddish fruit of a species of palm tree in West Africa contained rich oils that could be used for everything from making soap to lubricating steam engines. The European colonial powers, who liked to experiment with different crops throughout their empires to obtain the highest and most profitable yields, eventually learned that these palm trees grew especially well in Southeast Asia. Beginning in the 1930s, oil-palm plantations were established across this realm, particularly in what is now Malaysia and Indonesia. Today, palm oil is used in an enormous range of products; in fact, it is so widely utilized in processed foods that up to half of all products sold in supermarkets contain it.

This form of agriculture is also quite lucrative, because the yields of oil palms per hectare (2.5 acres) are much higher than those of such competing oil crops as soybeans and sunflowers. Moreover, the price of this commodity on global markets has risen steadily over the past few years, triggering successive rounds of production increases. Malaysia and Indonesia have benefited the most, and in 2012 together supplied no less than 85 percent of all the world's palm oil.

The achievement of this success, however, has come at a high price. According to UN agencies, the expansion of oil-palm plantations now constitutes the single biggest threat to tropical forest preservation in the two countries. Most at risk for survival are such key wildlife species as orangutans, pygmy elephants, Sumatran rhinos, and tigers. In 2010, the swiftly expanding closed-canopy palm oil plantations were conservatively estimated to already cover an area the size of Austria, with about half located in Sumatera and the rest divided about evenly between Borneo and peninsular Malaysia (Fig. 10A-2 below; see also the photo at the beginning of this chapter).

The palm oil industry is now being pressured by various groups to adhere to more sustainable practices, mainly by avoiding start-ups of new plantations in areas of high conservation value, and by introducing a "certified" label for oil produced in a sustainable manner. But progress on this front is slow, and as long as global demand remains high, the temptation to expand production in suitable areas will be hard to resist.

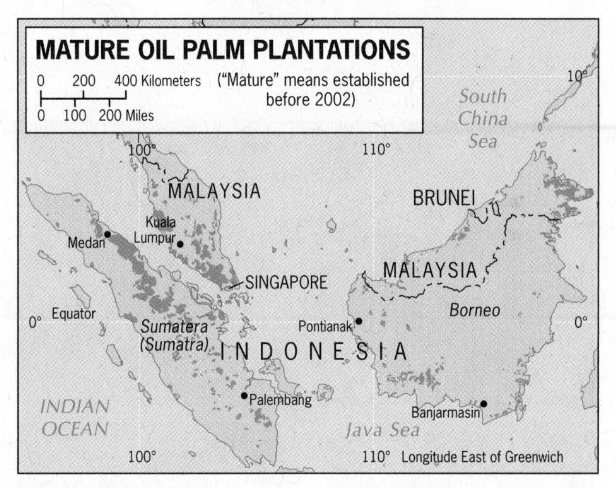

FIGURE 10A-2 © H. J. de Blij, P. O. Muller, J. Nijman, and John Wiley & Sons, Inc.

nature: Australia's Aborigines probably managed to cross what remained of the deeper trenches between islands by building rafts, but their epic journey was facilitated by land bridges as well.

Time and again, therefore, Southeast Asia was a receptacle for migrating species. Combine this with the realm's tropical environments, and it is no surprise that it is known for its **biodiversity [4]**. Scientists estimate that fully 10 percent of the Earth's plant and animal species are found in this comparatively small realm. Biogeographers can trace the progress of many of these species from the mainland across the archipelago toward Australia, but some were halted by deeper trenches that remained filled with water even when sea levels dropped. An especially deep trench lies in the Lombok Strait between the islands of Bali (next to Jawa) and Lombok (Fig. 10A-1), a key biogeographical boundary (discussed in Chapter 11) first recognized by the naturalist Alfred Russel Wallace, a contemporary of Charles Darwin.

As we shall presently observe, Southeast Asia's biodiversity had a fateful impact on its historical geography. Among the realm's specialized plants, it was the spices that attracted outsiders from India, China, and Europe, with consequences still visible on the map today.

Four Major Rivers

Water is the essence of life, and among world realms Southeast Asia is comparatively well endowed with moisture (Fig. G-7). Ample, occasionally even excessive, rainfall fills the rice-growing paddies of Indonesia and the Philippines. On the realm's mainland, where annual rainfall averages are somewhat lower and the precipitation is more seasonal, major rivers and their tributaries fill irrigation channels and form fertile deltas. The map of population distribution clearly highlights this spatial relationship between rivers and people, expressed in distinct coastal clustering (Fig. 10A-3).

As Figure 10A-1 reminds us, rivers that are crucial to life in certain realms sometimes have their sources in neighboring realms, a geographic issue that can lead to serious regional discord. What right does an "upstream" state in one realm have to dam, or otherwise interfere with, a river whose flow is vital to a state (or states) downstream?

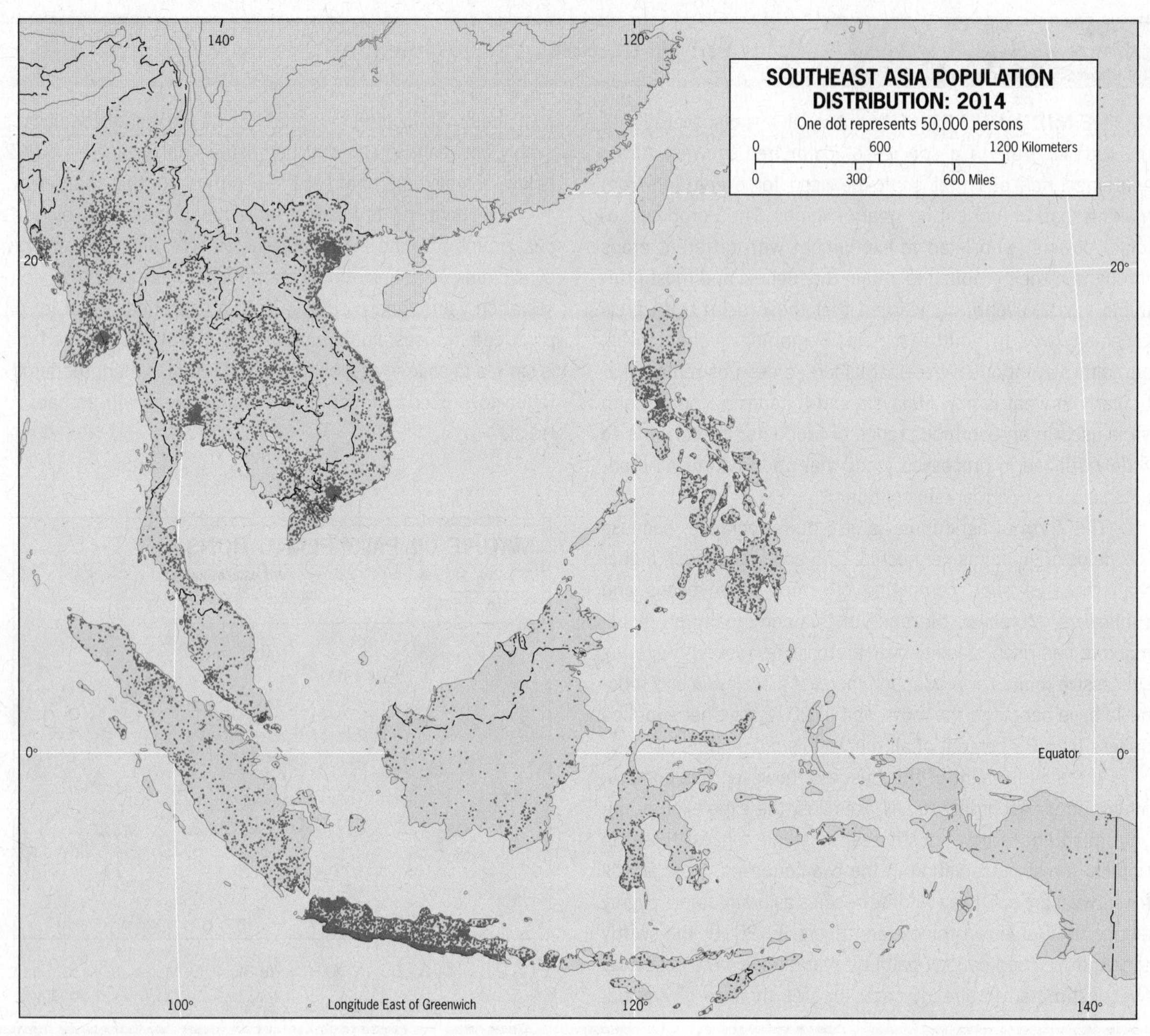

FIGURE 10A-3

In Southeast Asia's case, three of its four major rivers originate in China: (1) the ***Mekong***, which crosses the mainland from north to south; (2) the ***Red*** River that reaches the sea via its course across northern Vietnam; and (3) the ***Irrawaddy***, which forms the lifeline of Myanmar. The fourth is an intra-realm river, the ***Chao Phraya***, that functions as the key artery of Thailand.

The Mighty Mekong

From its headwaters high in the snowy uplands of China's Qinghai-Xizang (Tibetan) Plateau, the Mekong River rushes and flows some 4200 kilometers (2600 mi) to its delta in southernmost Vietnam. This "Danube of Southeast Asia" traverses or borders five of the realm's countries, supporting rice farmers and fishing people, forming a transportation route where roads are few, and providing electricity from dams upstream. Tens of millions of people depend on the waters of the Mekong, from subsistence farmers in Cambodia to apartment dwellers in China. The Mekong Delta in southern Vietnam is one of the realm's most densely populated areas and produces enormous harvests of rice.

But problems loom. China is building a series of dams across the Lancang (as the Mekong is called there) to supply surrounding Yunnan Province with electricity. Although such hydroelectric dams should not interfere with water flow, countries downstream worry that a severe dry spell in the interior would impel the Chinese to slow the river's flow to keep their reservoirs full. Cambodia is especially concerned over the future of the Tonlé Sap, a large natural lake whose water is supplied by the Mekong (see Fig. 10B-2). In Vietnam, farmers worry about salt water invading the Delta's paddies should the Mekong's level drop. And the Chinese may not be the only dam builders in the future: Thailand has expressed an interest in building a dam on the Thai-Laos border where it is defined by the Mekong.

In such situations, the upstream states have an advantage over those downstream. Several international organizations have been formed to coordinate development in the Mekong Basin, most notably the Mekong River Commission (MRC) founded more than 50 years ago. China has offered to sell electricity generated by its dams to Thailand, Laos, and Myanmar. Coordinated efforts to reduce

From the Field Notes . . .

"After a three-hour drive by bus southward from Hanoi, we reached the Van Long Nature Reserve in northeastern Vietnam's Red River Delta. It lies in a limestone terrain known as karst, part of the same geology that continues to the north through Halong Bay and into southern China beyond. Karst landscapes are formed through the slow but steady dissolution of limestone by (naturally) acidic water; the harder rock formations are more resistant to erosion and remain as towers, resulting in unique and sometimes whimsical landforms (left photo). The Van Long Nature Reserve is a biologically rich ecotourism site that was established in 2001 and involves local communities. Women from nearby villages guide visitors around in their basket boats, and they welcome the extra income in an area barely touched by Vietnam's emerging economy (right photo)."

© Jan Nijman

© Jan Nijman

www.conceptcaching.com

deforestation in the Mekong's drainage basin have had some effect. After consultations with the MRC, Australia built a bridge linking Laos and Thailand. There is even a plan to make the Mekong navigable from Yunnan to the coast, creating an alternative outlet for interior China. Sail the Mekong today, however, and you are struck by the slowness of development along this key artery. Wooden boats, thatch-roofed villages, and teeming paddies mark a river still crossed by antiquated ferries and flanked by few towns. Of modern infrastructure, there is very little to observe (see final photo of this chapter). And yet the Mekong and its basin form the lifeline of mainland Southeast Asia's predominantly rural societies.

Rivers and States

Whereas the Mekong impacts the lives of peoples in several Southeast Asian states, the effects of the other three major Southeast Asian rivers are for the most part internal. As Figure 10A-1 shows, the Red River forms the focus for the heavily populated Tonkin Plain in northern Vietnam, where the capital, Hanoi, lies on its banks. In Thailand, the relatively short but crucial Chao Phraya, on which Bangkok is situated, is just one of a series of channels in that river's delta, formed by numerous streams rising in the country's interior. And Myanmar's Irrawaddy River (some of whose headwaters arise in China) crosses that country from north to south, its wide valley one of the world's leading rice-producing areas and its largest city, Yangon, at the corner of its delta.

POPULATION GEOGRAPHY

Examine the map of Southeast Asia's population distribution (Fig. 10A-3), and you are immediately struck by the huge concentration of people on a relatively small island in Indonesia—a cluster larger than any other in the realm, the four mainland river deltas included. This is Jawa, and its population of close to 150 million not only accounts for well over half of Indonesia's national total but also exceeds that of every other country in the realm.

This population concentration is particularly noteworthy because Indonesia is not yet a highly urbanized country. Today more than half of the Indonesians still live in rural areas, and although Jawa, as shown in Chapter 10B, is the country's most urbanized island, over 60 million of its inhabitants still live off the land. What makes all this possible is a combination of fertile volcanic soil, ample water, and extremely warm temperatures that enable Jawa's farmers to raise three crops of rice in a single paddy during a single year, helping feed a national population still growing faster than the global average. But we should also be aware that the thickest red clusters on Jawa denote fast-growing urban areas, cornerstones in the building of a new economy with increasingly global linkages.

Within Indonesia, the contrasts between Jawa and the four other major islands—Sumatera, Borneo (Kalimantan), Sulawesi, and most especially Indonesian New Guinea—reflect the core-periphery relationship between these two sectors of the country. As the map suggests, such contrasts, less sharply defined, also mark other Southeast Asian countries, and the primate cities here (Bangkok, Manila, Yangon, Kuala Lumpur) are particularly dominant. Vietnam even contains two such anchors, which respectively represent its historic northern (Hanoi) and southern (Ho Chi Minh City) core areas.

As the table in Appendix B indicates and as the map confirms, Southeast Asia's states are not, by world standards, especially populous. Indonesia today is the world's fourth-ranking country in terms of population (just under 250 million), but no other Southeast Asian country contains even half that total. Three mainland countries contain between

50 and 100 million people. But take Laos, quite a large country territorially (about the size of the United Kingdom) and note that its population is less than 7 million; similarly, Cambodia, half the size of Germany, has only 15.5 million. In part, such modest numbers on the mainland reflect natural conditions less favorable to farming than those prevailing on volcanic soils or in fertile river basins, but more generally this realm did not grow as explosively during the past century as did neighboring realms. Indeed, Appendix B indicates that several countries in Southeast Asia today are growing at or below the global average of 1.2 percent annually.

The Ethnic Mosaic

Southeast Asia's peoples come from a common stock just as (Caucasian) Europeans do, but this has not prevented the emergence of regionally or locally discrete ethnic or cultural groups. Figure 10A-4 displays the broad distribution

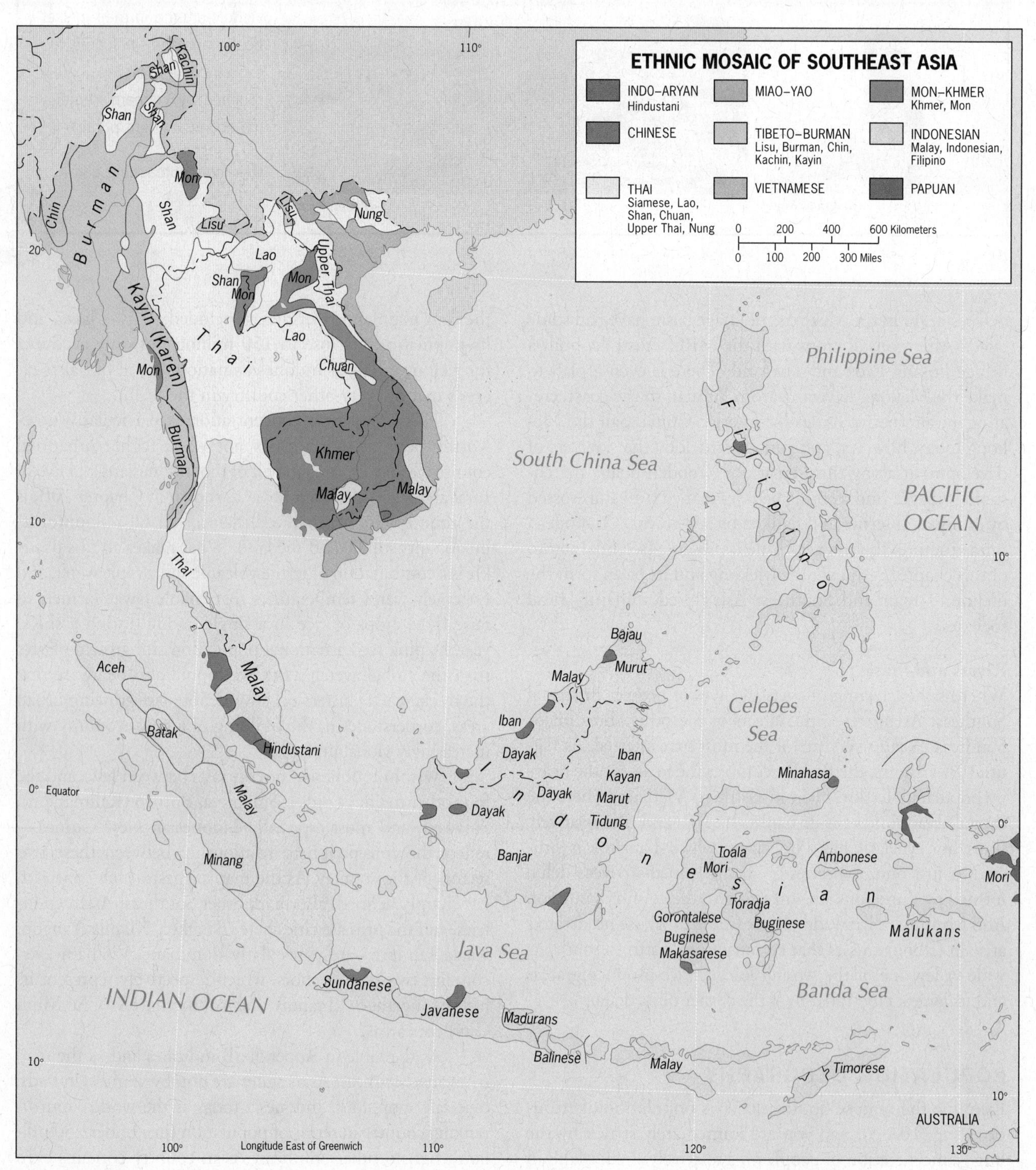

FIGURE 10A-4

of ethnolinguistic groups in the realm, but be aware that this is a generalization. At the scale of this map, myriad smaller groups cannot be represented.

Figure 10A-4 shows the rough spatial coincidence, on the mainland, between major ethnic group and contemporary political state. The Burman dominate in the country formerly called Burma (now Myanmar); the Thai occupy the state once known as Siam (now Thailand); the Khmer form the nation of Cambodia and extend northward into Laos; and the Vietnamese inhabit the long strip of territory facing the South China Sea.

Territorially, by far the largest population shown in Figure 10A-4 is classified as Indonesian, the inhabitants of the great island chain that extends from Sumatera west of the Malay Peninsula to the Malukus (Moluccas) in the east and from the Lesser Sunda Islands in the south to the Philippines in the north. Collectively, all these peoples shown on the map—the Filipinos, Malays, and Indonesians—are known as Indonesians, but they have been divided by history and politics. And note as well that the Indonesians in Indonesia itself include Javanese, Madurese, Sundanese, Balinese, and other major groups; again, hundreds of smaller ones cannot be mapped at this scale. In the Philippines, too, island isolation and contrasting ways of life are reflected in the cultural mosaic. Also part of this Indonesian ethnic-cultural complex are the Malays, whose heartland lies on the Malay Peninsula but who form minorities in other areas as well. Like most Indonesians, the Malays are Muslims, although Islam is a more powerful force within Malay society than it generally is in Indonesian culture.

In the northern part of the mainland region, numerous minorities inhabit remote corners of the countries in which the Burman (Burmese), Thai, and Vietnamese dominate. Those minorities tend to occupy areas on the peripheries of their countries, where the terrain is mountainous and the forest is dense, and where the governments of their national states do not exert complete control. This remoteness and sense of detachment give rise to aspirations of secession, or at least resistance to government efforts to establish full authority, often resulting in bitter ethnic conflict.

Immigrants

Figure 10A-4 further reminds us that, again like eastern Europe, Southeast Asia is home to major ethnic minorities from outside the realm. On the Malay Peninsula, note the South Asian (Hindustani) cluster. Such Hindu communities with Indian ancestries exist in many parts of the peninsula, but in the southwest they form the majority in a small area. In Singapore, too, South Asians form a significant minority. These communities emerged during the European colonial period, but South Asians had arrived in this realm many centuries earlier, propagating Buddhism and leaving their architectural and cultural imprints on places as far removed as Jawa and Bali.

From the Field Notes . . .

"Like most major Southeast Asian cities, Bangkok's urban area includes a large and prosperous Chinese sector. No less than 14 percent of Thailand's population of 71 million is of Chinese ancestry, and the great majority of Chinese live in the cities. In Thailand, this large non-Thai population is well integrated into local society, and intermarriage is common. Still, Bangkok's 'Chinatown' is a distinct and discrete part of the great city. There is no mistaking Chinatown's limits: Thai commercial signs change to Chinese, goods offered for sale also change (Chinatown contains a large cluster of shops selling gold, for example), and the urban atmosphere, from street markets to bookshops, is dominantly Chinese. This is a boisterous, noisy, energetic part of multicultural Bangkok, a vivid reminder of the Chinese commercial success in Southeast Asia."

www.conceptcaching.com

The Chinese

By far the largest immigrant minority in Southeast Asia, however, is Chinese. The Chinese began arriving here during the Ming and early Qing (Manchu) dynasties, and the largest exodus occurred during the late colonial period (1870–1940), when as many as 20 million immigrated. The European powers at first encouraged this influx, using the Chinese in administration and trade. But soon these **Overseas Chinese [5]** began to congregate in the major cities, where they established Chinatowns and gained control over much of the commerce. By the time the Europeans tried to limit Chinese immigration, World War II was about to break out and the colonial era would end soon thereafter.

Figure 10A-5 shows the migration routes and current concentrations of Chinese in Southeast Asia. Most migrants originated in southern China's Fujian and Guangdong provinces, and a large number invested much of their wealth back in China when it opened up to foreign businesses three decades ago. Clearly, the Overseas Chinese of Southeast Asia have played a significant role in shaping the economic miracle in this sector of the Pacific Rim.

Southeast Asia is now home to about 33 million Overseas Chinese, more than two-thirds of the world total. Their lives have often been difficult. The Japanese relentlessly persecuted Chinese living in Malaya during World War II. Later, during the 1960s, Chinese in Indonesia were accused of communist sympathies, and hundreds of thousands were killed. In the late 1990s, Indonesian mobs again attacked Chinese and their property because of their relative wealth and because many Chinese had become Christians during the colonial era and were now targeted by Islamic throngs. Resentment still continues, expressed by episodic flare-ups against Chinese in various parts of Southeast Asia.

During this decade, Singapore is experiencing its own unique challenges with the latest wave of Overseas Chinese. Unique, because the overwhelming majority of native Singaporeans are themselves of Chinese descent—even though they are third-, fourth-, or fifth-generation and have, to varying degrees, blended in with surrounding Malay culture (Singapore and Malaya were part of the same British colonial entity for more than 130 years before independence in

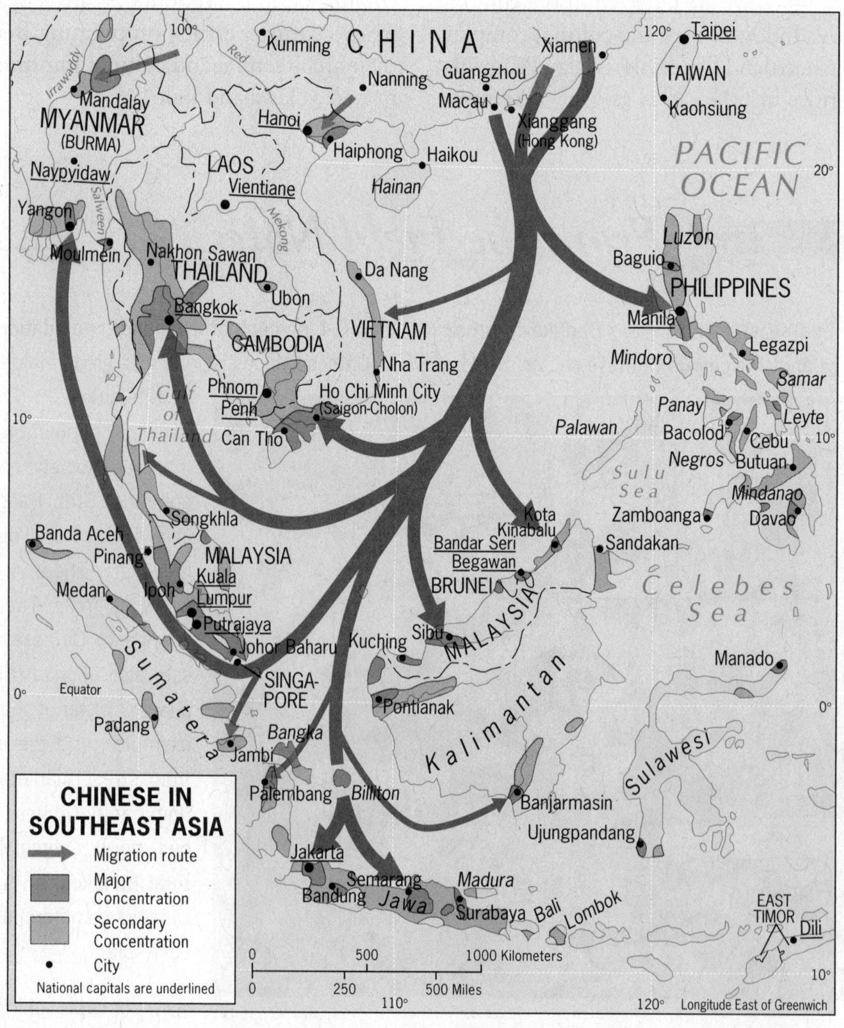

FIGURE 10A-5

Regional ISSUE The Chinese Presence in Southeast Asia

THE CHINESE ARE TOO INFLUENTIAL!

"It's hard to imagine that there was a time when we didn't have Chinese minorities in our midst. I think I understand how Latin Americans [*sic*] feel about their 'giant to the north.' We've got one too, but the difference is that there are a lot more Chinese in Southeast Asia than there are Americans in Latin America. The Chinese even run one country because they are the majority there, Singapore. And if you want to go to a place where you can see what the Chinese would have in mind for this whole region, go there. They're knocking down all the old Malay and Hindu quarters, and they've got more rules and laws than we here in Indonesia could even think of. I'm a doctor here in Bandung and I admire their modernity, but I don't like their philosophy.

"We've had our problems with the Chinese here. The Dutch colonists brought them in to work for them, they got privileges that made them rich, they joined the Christian churches the Europeans built. I'm not sure that a single one of them ever joined a mosque. And then shortly after Sukarno led us to independence they even tried to collaborate with Mao's communists to take over the country. They failed and many of them were killed, but look at our towns and villages today. The richest merchants and the money lenders tend to be Chinese. And they stay aloof from the rest of us.

"We're not the only ones who had trouble with the Chinese. Ask them about it in Malaysia. There the Chinese started a full-scale revolution in the 1950s that took the combined efforts of the British and the Malays to put down. Or Vietnam, where the Chinese weren't any help against the enemy when the war happened there. Meanwhile they get richer and richer, but you'll see that they never forget where they came from. The Chinese in China boast about their coastal economy, but it's coastal because that's where our Chinese came from and where they sent their money when the opportunity arose.

"As a matter of fact, I don't think that China itself cares much for or about Southeast Asia. Have you heard what's happening in Yunnan Province? They're building a series of dams on the Mekong River, our major river, just across the border from Laos and Burma. They talk about the benefits and they offer to destroy the gorges downstream to facilitate navigation, but they won't join the Mekong River Commission. When the annual floods cease, what will happen to Tonlé Sap, to fish migration, to seasonal mudflat farming? The Chinese do what they want—they're the ones with the power and the money."

THE CHINESE ARE INDISPENSABLE!

"Minorities, especially successful minorities, have a hard time these days. The media portray them as exploiters who take advantage of the less fortunate in society, and blame them when things go wrong, even when it's clear that the government representing the majority is at fault. We Chinese arrived here long before the Europeans did and well before some other 'indigenous' groups showed up. True, our ancestors seized the opportunity when the European colonizers introduced their commercial economy, but we weren't the only ones to whom that opportunity was available. We banded together, helped each other, saved and shared our money, and established stable and productive communities. Which, by the way, employed millions of locals. I know: my family has owned this shop in Bangkok for five generations. It started as a shed. Now it's a six-floor department store. My family came from Fujian, and if you added it all up we've probably sent more than a million U.S. dollars back home. We're still in touch with our extended family and we've invested in the Xiamen SEZ.

"Here in Thailand we Chinese have done very well. It is sometimes said that we Chinese remain aloof from local society, but that depends on the nature of that society. Thailand has a distinguished history and a rich culture, and we see the Thais as equals. Forgive me, but you can't expect the same relationship in East Malaysia. Or, for that matter, in Indonesia, where we are resented because the 3 percent Chinese run about 60 percent of commerce and trade. But we're always prepared to accommodate and adjust. Look at Malaysia, where some of our misguided ancestors started a rebellion but where we're now appreciated as developers of high-tech industries, professionals, and ordinary workers. This in a Muslim country where Chinese are Christians or followers of their traditional religions, and where Chinese regularly vote for the Malay-dominated majority party. So when it comes to aloofness, it's not us.

"The truth is that the Chinese have made great contributions here, and that without the Chinese this place would resemble parts of South or Southwest Asia. Certainly there would be no Singapore, the richest and most stable of all Southeast Asian countries, where minorities live in peace and security and where incomes are higher than anywhere else. In fact, mainland Chinese officials come to Singapore to learn how so much was achieved there. Imagine a Southeast Asia without Singapore! It's no coincidence that the most Chinese country in Southeast Asia also has the highest standard of living."

Vote your opinion at www.wiley.com/go/deblijpolling

the 1960s). That shared ethnicity has not stopped many Singaporeans from becoming ever more critical of the rapid recent influx of 'newly rich' mainland Chinese, who tend to come with a sense of entitlement coupled with a lack of interest in the regional culture. Tensions peaked in 2012 when an intoxicated Chinese immigrant ran a red light, crashed his Ferrari (!) into a taxicab, and killed two natives. More generally, resentment focuses on the role of arrogant wealthy immigrants in driving up real-estate prices, taking the best jobs, and showing disrespect for the city-state's strict codes of public conduct. Overall, this increasingly erodes social relations because 'old' Chinese and 'new' Chinese now refer to distinctly different identities within Singapore's multicultural mosaic.

RELIGIONS OF SOUTHEAST ASIA

Southeast Asia, more than any other realm, is a historic crossroads of religions. With the migrants from the Indian subcontinent came their faiths: first Hinduism and Buddhism, later Islam. The Muslim religion, promoted by the growing number of Arab traders who appeared on the scene, became the predominant religion in Indonesia (where nearly 90 percent of the population adheres to Islam today). But in Myanmar, Thailand, and Cambodia, Buddhism remained supreme, and in all three countries today more than 90 percent of the people are adherents. In culturally diverse Malaysia, the Malays are Muslims (to be a Malay *is* to be a Muslim), and almost all Chinese are Buddhists; but most Malaysians of Indian ancestry remain Hindus.

Although Southeast Asia has generated its own local cultural expressions, most of what remains in tangible form has resulted from the infusion of foreign elements. Take, for instance, Angkor Wat, the enormous complex of religious structures built in Cambodia during the twelfth century AD and today one of the world's most famous monuments (see photos). It was originally constructed as a Hindu temple, dedicated to the god Vishnu. The carvings on the walls of Angkor Wat tell the stories of the Hindu epics, and the temple's designs are closely associated with Hindu cosmology. But during the fourteenth century Buddhism took over this area and the temple complex became a place of worship for Buddhists.

Vietnam is perhaps the ultimate religious crossroads of the realm, where for many centuries there has been an almost casual blending of early Hinduism with Buddhism, Daoism, and Confucianism—all mixing comfortably with age-old traditions of ancestor worship.

Cambodia's renowned Angkor Wat offers a blend of Hindu and Buddhist iconographies: a statue of Vishnu (left) and a nearby one of the Buddha (right).

© Jan Nijman

© Jan Nijman

COLONIALISM'S HERITAGE: HOW THE POLITICAL MAP EVOLVED

When the European colonizers arrived in Southeast Asia, they encountered a patchwork of kingdoms, principalities, sultanates, and other traditional political entities whose leaders they tried to co-opt, overpower, or otherwise fold into their imperial schemes. There was no single powerful center of indigenous culture as had developed in Han-dominated China. In the river basins and on the plains of the mainland, as well as on the islands offshore, a flowering of cultures had produced a diversity of societies whose languages, religions, arts, music, foods, and other achievements formed an almost infinitely varied mosaic—but none of those cultures had risen to imperial power when the Europeans arrived. Those European colonizers forged empires here, often by playing one state off against another; the Europeans divided and ruled. Out of this foreign intervention came the modern map of Southeast Asia, and only Thailand (formerly Siam) survived the colonial era as an independent entity. Thailand was useful to two competing powers, the French to the east and the British to the west: it served as a convenient buffer, and although the colonists carved pieces off Thailand's domain, the kingdom endured.

Indeed, the Europeans accomplished what local powers could not: the formation of comparatively large, multicultural states that encompassed diverse peoples and societies and welded them together. Were it not for the colonial intervention, it is unlikely that the 17,000 islands of far-flung Indonesia would today constitute the world's fourth-largest country in terms of population. Nor would the nine sultanates of Malaysia have been united, let

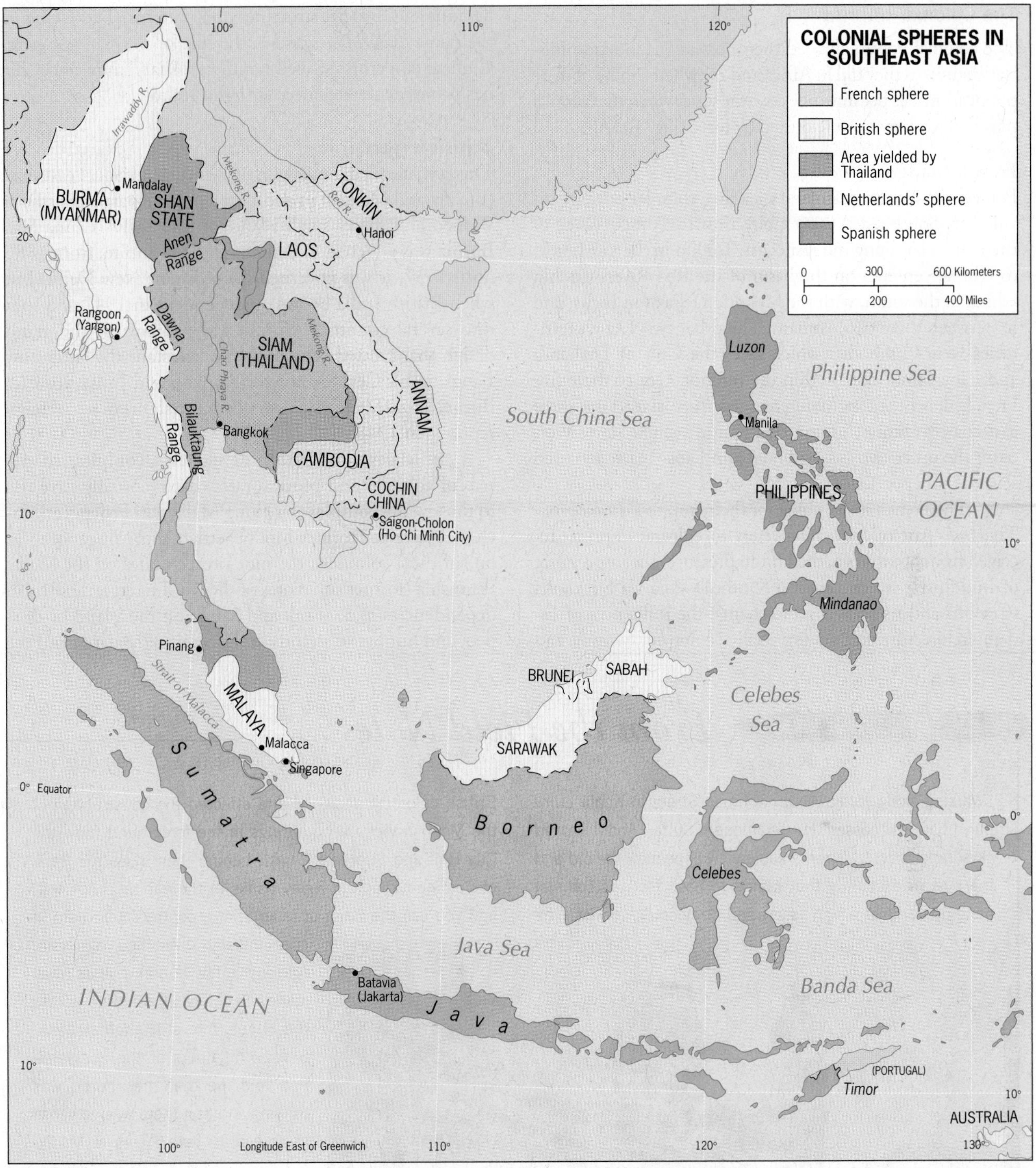

FIGURE 10A-6

alone with the peoples of northern Borneo across the South China Sea. For good or ill, the colonial intrusion consolidated a realm of few culture cores and numerous ministates into less than a dozen countries. The leading colonial competitors here in Southeast Asia were the Dutch, British, French, and Spanish (with the Spanish later replaced by the Americans in their stronghold, the Philippines). The Japanese had colonial objectives here as well, but those aspirations came and went during the course of World War II.

Figure 10A-6 shows the colonial framework in the late nineteenth century, before the United States assumed control over the Philippines in 1898. Note again that only Thailand survived as an independent state, but it was compelled to yield territory to the British in Malaya and Burma and to the French in Cambodia and Laos.

The Colonial Imprint

The colonial powers divided their possessions into administrative units as they did in Africa and elsewhere. Some of these political entities became independent states when the colonial powers withdrew or were ousted by force (Fig. 10A-6).

French Indochina

France, one of the mainland's leading colonial powers, divided its Southeast Asian empire into five units. Three of these units lay along the east coast: Tonkin in the north next to China, centered on the basin of the Red River; Cochin China in the south, with the Mekong Delta as its focus; and in between these two, Annam. The other two French territories were Cambodia, which faces the Gulf of Thailand; and Laos, landlocked within the interior. Out of these five French dependencies there emerged three states: the three east-coast territories ultimately became a single state, Vietnam; the other two—Cambodia and Laos—each achieved separate independence.

The French had a name for their empire—***Indochina***. The *Indo* part of Indochina refers to cultural imprints received from South Asia: the Hindu presence; the importance of Buddhism, which came to Southeast Asia via Sri Lanka (Ceylon) and its seafaring merchants; the influences of Indian architecture and art (especially sculpture), writing and literature, and social structures and patterns. The *China* in the name *Indochina* signifies the role of the Chinese here. Chinese emperors coveted Southeast Asian lands, and China's power penetrated deep into this realm.

British Imperialism

The British ruled a pair of major entities in Southeast Asia (Burma and Malaya) in addition to a large part of northern Borneo and many small islands in the South China Sea. Burma was attached to Britain's Indian Empire; from 1886 until 1937, it was governed from distant New Delhi. But when British India became independent in 1947 and split into several countries, Burma was not part of the grand design that created West and East Pakistan (the latter now Bangladesh), Ceylon (now Sri Lanka), and India. Instead, Burma (now Myanmar) was given the status of a sovereign republic in 1948.

In Malaya, the British developed a complicated system of colonies and protectorates that eventually gave rise to the equally complex, far-flung Malaysian Federation. Included were the former Straits Settlements (Singapore was one of these colonies), the nine protectorates on the Malay Peninsula (former sultanates of the Muslim era), the British dependencies of Sarawak and Sabah on the island of Borneo, and numerous islands in the Strait of Malacca and the

From the Field Notes . . .

"Walking along Tunku Abdul Rahman Street in Kuala Lumpur, I had just passed the ultramodern Sultan Abdul Samad skyscraper when this remarkable view appeared: the old and the new in a country that seems to have few postcolonial hang-ups and in which Islam and democracy coexist. The British colonists designed and effected the construction of the Moorish-Victorian buildings in the foreground (now the City Hall and Supreme Court); behind them rises the Bank of Commerce, one of many banks in the capital. Look left, and you see the Bank of Islam, not a contradiction here in economically diversified Malaysia. And just a few hundred yards away stands St. Mary's Cathedral, across the street from still another bank. Several members of the congregation told me that the church was thriving and that there was no sense of insecurity here. 'This is Malaysia, sir,' I was told. 'We're Muslims, Buddhists, Christians. We're Malays, Chinese, Indians. We have to live together. By the way, don't miss the action at the Hard Rock Café on Sultan Ismail Street.' Now there, I thought, was a contradiction as remarkable as this scene."

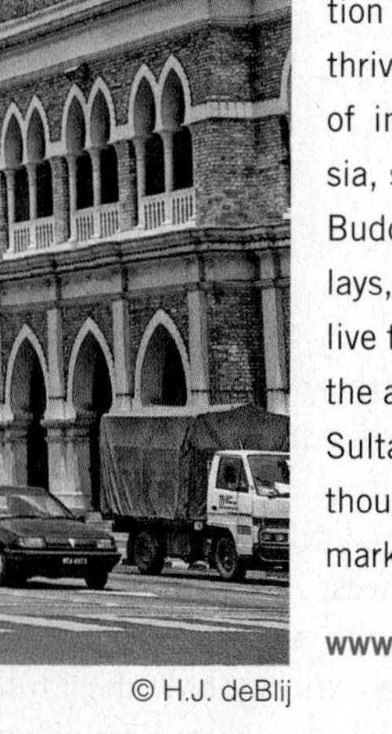

© H.J. deBlij

www.conceptcaching.com

Concept Caching

South China Sea. The original Federation of Malaysia was created in 1963 by the political unification of recently independent mainland Malaya, Singapore, and the former British dependencies on the largely Indonesian island of Borneo. Singapore, however, left the Federation in 1965 to become a sovereign city-state, and the remaining units were later restructured into peninsular Malaysia and, on Borneo, Sarawak and Sabah. Thus the term *Malaya* properly refers to the geographic territory of the Malay Peninsula, including Singapore and other nearby islands; the term *Malaysia* identifies the politico-geographical entity of which Kuala Lumpur is the capital city.

Netherlands "East Indies"

Following in the wake of the Portuguese, the first European colonizers in this realm, the Dutch, came in search of what Southeast Asia had to offer and now one aspect of this realm's biodiversity had fateful consequences. Among the plants domesticated by the local people on the islands of present-day Indonesia was a group collectively known as the *spices*. We know them today as black pepper, cloves, cinnamon, nutmeg, ginger, turmeric, and other condiments essential to flavorful meals. In Figure 10A-1 you can detect, in eastern Indonesia between Sulawesi and New Guinea, a group of small islands called the Maluku Islands (formerly Moluccas). The Dutch colonizers called these the ***Spice Islands*** because of the lucrative commerce in spices long carried on by Arab, Indian, and Chinese traders. What the Europeans wanted was control over this trade, and they were willing to go to war for it.

It may seem odd in today's world that spices could be important enough to be fought over, but at that time, in the prerefrigeration era, spices not only conserved food but also added flavor to otherwise bland diets. Spices commanded sky-high prices on European markets. The Dutch East India Company, the commercial arm of the Dutch government, managed to take control of the Spice Islands, bringing untold wealth to the Netherlands and ushering in an era of enrichment through colonial exploitation known as the country's Golden Age.

Java (Jawa), the most populous and productive island, became the focus of Dutch administration; from its capital at Batavia (now Jakarta), the Dutch East India Company extended its sphere of influence into Sumatra (Sumatera), Celebes (Sulawesi), and much of Borneo (Kalimantan) as well as the smaller islands of the East Indies. This was not accomplished overnight, and the struggle for territorial control was carried on long after the Company had yielded its administration to the Netherlands government in 1800. Dutch colonialism therefore threw a girdle around Indonesia's more than 17,000 islands, paving the way for the creation of the realm's largest and most populous state that is now home to a quarter-billion people.

From Spain to the United States

In the colonial tutelage of Southeast Asia, the Philippines, long under Spanish domination, had a unique experience. As early as 1571, the islands north of Indonesia were under Spain's control (they were named for Spain's King Philip II). There was much profit to be made, but the indigenous peoples shared little in it. Great landholdings were awarded to loyal Spanish civil servants and to men of the church. Oppression eventually yielded revolution, and Spain was confronted with a major uprising in the Philippines just as the Spanish-American War broke out elsewhere in 1898.

As part of the settlement of that brief war, the United States replaced Spain in Manila as colonial proprietor. That was not the end of the revolution, however. The Filipinos now took up arms against their new foreign ruler, and not until 1905, after terrible losses of life, did American forces manage to pacify their new dominion. Subsequently, U.S. administration in the Philippines was more progressive than Spain's had been and eventually (after a temporary occupation by Japan during World War II) led the Philippines to independence in 1946.

Today all of Southeast Asia's states are independent, but centuries of imperial rule have left strong cultural imprints. In their urban landscapes, their education systems, and countless other ways, this realm still carries the legacy of its colonial past.

SOUTHEAST ASIA'S EMERGING MARKETS

If Japan and the Asian Tigers set the early example in the 1960s and 1970s, with post-Mao China following suit since the mid-1980s, it now seems to be the turn of several Southeast Asian countries to join the ranks of the world's rapidly **emerging markets [6]**. Vietnam, Indonesia, and Malaysia, in particular, have in recent years attracted substantial foreign investment and exhibited decidedly robust economic growth rates.

In the table in Appendix B, look at the column that shows per-capita income and note the very considerable disparities that mark this realm. The tiny oil state of Brunei has an income higher than that of the United States (although this figure is not indicative of the earnings of its relatively large foreign workforce). However, it is comprehensively developed Singapore that records Southeast Asia's highest income—at a level halfway between those of wealthy Luxembourg and Switzerland (and 18 percent higher than that of the U.S.). Malaysia comes in a distant third and is followed by Thailand. Populous Indonesia and the Philippines exhibit per-capita incomes less than one-twelfth of Singapore's. Finally, Vietnam, Laos, Cambodia, and Myanmar, in that order, rank at the bottom of the list. This table, however, only provides a 2010 snapshot and does not reflect the newest economic trends. According to many current observers, Vietnam has been growing the fastest in this decade. Of course, it remains a very poor country (whose per-capita income is only 5.6 percent of Singapore's) in urgent need of high growth rates to lift its masses to a better standard of living. Vietnam also remains a communist state, and in certain ways its government is more conservative than China's. But a stock exchange did open in Ho Chi Minh City (formerly Saigon) in 2000, and seven years later the country joined the World Trade Organization. If Vietnam takes further steps to liberalize its economy, observers say, it could become Asia's next economic miracle.

Singapore's Leadership

With its soaring per-capita income, there can be no doubt that Singapore is the economic heart of Southeast Asia. With a mere 5.4 million citizens and only 619 square kilometers (240 sq mi) of territory, we are obviously not referring to size—this is all about the connections and centrality that enable Singapore to function as the leading **node [7]** in a realmwide economic network. At the same time, geographers also rank Singapore as a top-tier world-city because it has major international linkages and exerts global influence. Singapore's container port is not only the largest in the realm but in the entire world, underscoring its key role within and far beyond Southeast Asia (see photo). Moreover, ultramodern Changi Airport Singapore is rated among the world's best international airports.

Singapore's exceptional regional position is based on its superb relative location. In the seventeenth century, it was Malacca farther up the Malay Peninsula that was the leading hub for trade and shipping in Southeast Asia. The Strait of Malacca (Fig. 10A-6), named after that town, was already the most important sea route providing access to the realm's waters for ships coming from the west. When the British sought to displace Dutch dominance in the region during the eighteenth century, they discovered an even better local base of operations: Singapore Island. Because it possessed a larger and deeper natural harbor than Malacca to accommodate the larger steamships of the time, Singapore swiftly rose to prominence as the British consolidated their power over this key corner of the world.

Singapore, Southeast Asia's unmatched world-city, has the biggest container port on Earth. This burgeoning city-state simultaneously functions as the heart of the realm's spatial economic system and a vitally important gateway to the global economy.

Most importantly, Singapore today is a symbol of modernity, a model for Southeast Asia's future. Throughout the realm, those who can afford it go there to shop, to connect to international flights, to transact business, to invest in real estate, or to send their children to one of the city-state's highly ranked universities. Imagine major parts of Southeast Asia following in Singapore's footsteps over the next two decades and how that would affect the realm and the world!

Prospects of Realmwide Integration: ASEAN

The overall development of Southeast Asia still has a long way to go. Political stability and increased regional integration will facilitate the process, and that is the long-term goal of **ASEAN [8]**, the Association of Southeast Asian Nations. Founded in the late 1960s, this supranational organization has primarily been concerned with security. But that has been a constantly challenging effort because a wide range of conditions mark its ten member-states. With the lone exception of minuscule East Timor, ASEAN encompasses all of the realm's countries. These include one influential city-state; an Islamic oil state (Brunei); two impoverished communist regimes (Vietnam and Laos); and a reforming military dictatorship whose population is rising from the ranks of the most deprived on Earth (Myanmar).

One conflict in which ASEAN has been conspicuously absent is a long-running border dispute between Cambodia and Thailand, directly north of the Cambodian city of Siem Reap (Fig. 10A-1). At issue is the location of the ancient Preah Vihear Temple, which according to a 1962 ruling by the International Court of Justice belongs to Cambodia but continues to be aggressively claimed by Thailand. The temple was declared a UNESCO World Heritage Site in 2008, and that only exacerbated matters. Gunfire across the border in 2010 and 2011 killed a dozen soldiers before a bilateral agreement was reached to withdraw troops from the area. In the spring of 2013, the Thai and Cambodian defense ministers met to strengthen peace and cooperation in the border area (ASEAN could have played a significant role in this dispute, but did not).

Another problem that ASEAN has failed to resolve, one that literally affects health across much of this realm, is the recurrent air pollution caused by Indonesia's massive, human-ignited forest fires. Depending on weather conditions and prevailing winds, thick plumes of smoke emanating from Indonesia's Sumatera (where most of the burning occurs, often to enlarge oil palm plantations) stream out toward Jawa, Singapore, Malaysia, and countries farther afield. Repeatedly, Singapore's government has had to advise residents to stay indoors (a new record high pollution level was set in June 2013). Indonesia, the realm's most powerful country, has been painfully slow to address this environmental crisis and even refused to ratify the 2002 ASEAN Agreement on Transboundary Haze Pollution.

In 1992, 25 years after its founding, ASEAN was able to expand into the economic domain through **AFTA [9]**, the ASEAN Free Trade Agreement, and here the payoff has been more substantial. AFTA has both triggered the lowering of tariffs and encouraged an upsurge in trade within Southeast Asia. With lower wages than China, certain foreign investments (e.g., in the garment industry) have shifted to Southeast Asia. Moreover, intra-realm trade has surged in recent years—an important development to avoid being completely overshadowed by China.

CHINA IN SOUTHEAST ASIA TODAY

This is hardly the first time that Southeast Asia finds itself entangled in the spheres of influence of external powers. Over the past century alone, this realm was first dominated by European powers; next it was absorbed into Japan's short-lived empire; following postwar independence, it was later integrated within the Japanese-led *Yen bloc*; and since 2000, it is increasingly drawn into the economic (and political) orbit of China. In 2010, ASEAN and China concluded a free-trade agreement, with the Chinese especially interested in acquiring raw materials from mainland Southeast Asia as well as accessing the region's growing export markets.

Boosting Economic Development

The results of this agreement have been significant. China now provides the biggest share of imports flowing into Cambodia, Indonesia, Malaysia, Myanmar, the Philippines, Singapore, and Vietnam; at the same time, China has become the leading export destination for Thailand, Vietnam, Singapore, and Malaysia (Fig. 10A-7). China's growing trade dominance in Southeast Asia has been achieved largely at the expense of Japan and the United States (the latter's trade with this realm now totals less than half of China's).

In addition to boosting trade, the Chinese are investing heavily across Southeast Asia in the exploitation of raw materials, construction, and massive infrastructure projects that facilitate the shipment of these commodities to China. They include a state-of-the-art rail connection from Kunming, capital of China's southern Yunnan Province, to Vientiane, Laos, fanning out from there in stages to Cambodia, Vietnam, Thailand, Malaysia, and Singapore; and if obstacles can be overcome, another key link will be a direct rail line from Kunming to Myanmar's Indian Ocean coast.

Although China is directly and indirectly contributing to the realm's economic development, its assertiveness and drive for dominance is also cause for growing unease. Some of this stems from China's formidable economic prowess, which produces rather asymmetrical relationships; for instance, many infrastructure projects are huge joint ventures that require comparatively larger investments from these much smaller Southeast Asian countries. And, more often than not, the benefits are clearly more favorable for China than for the host nation. Moreover, it is not all about economics: for countries such as Vietnam or Japan, historic sensitivities and/or longstanding rivalries come into play. Another issue is a perceived lack of Chinese

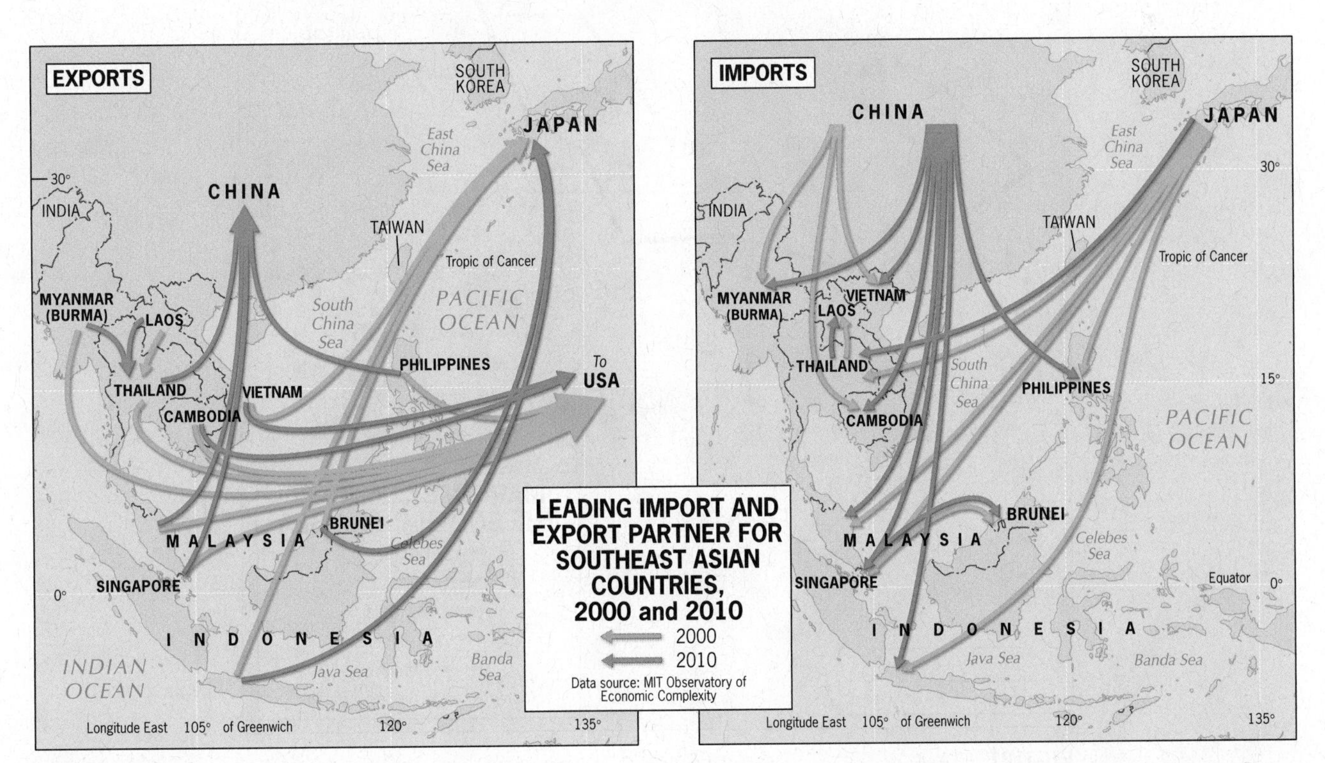

FIGURE 10A-7

diplomacy and consideration for regional and national interests. Consider this 2010 remark by the Chinese foreign minister that "China is a big country and other countries are small countries, and that is just a fact." In response, the increasingly prevalent view in Southeast Asia is that, in order to avoid excessive dependence on China, its emerging markets must develop rapidly, diversify their trade partners, and strengthen intra-realm economic connections. And, as we are about to discover, this mindset also has a consequential geopolitical dimension, with a growing number of countries (most notably Vietnam, the Philippines, and Indonesia) wanting the United States to play a more prominent regional role.

Geopolitics in the South China Sea

If China's economic role in Southeast Asia is often viewed with mixed feelings, its recent geopolitical and military forays are increasingly met with indignation and opposition. Much of this revolves around China's maritime ambitions and its claims in the South China Sea. Since 2009, the Chinese government has circulated the so-called *nine-dash*

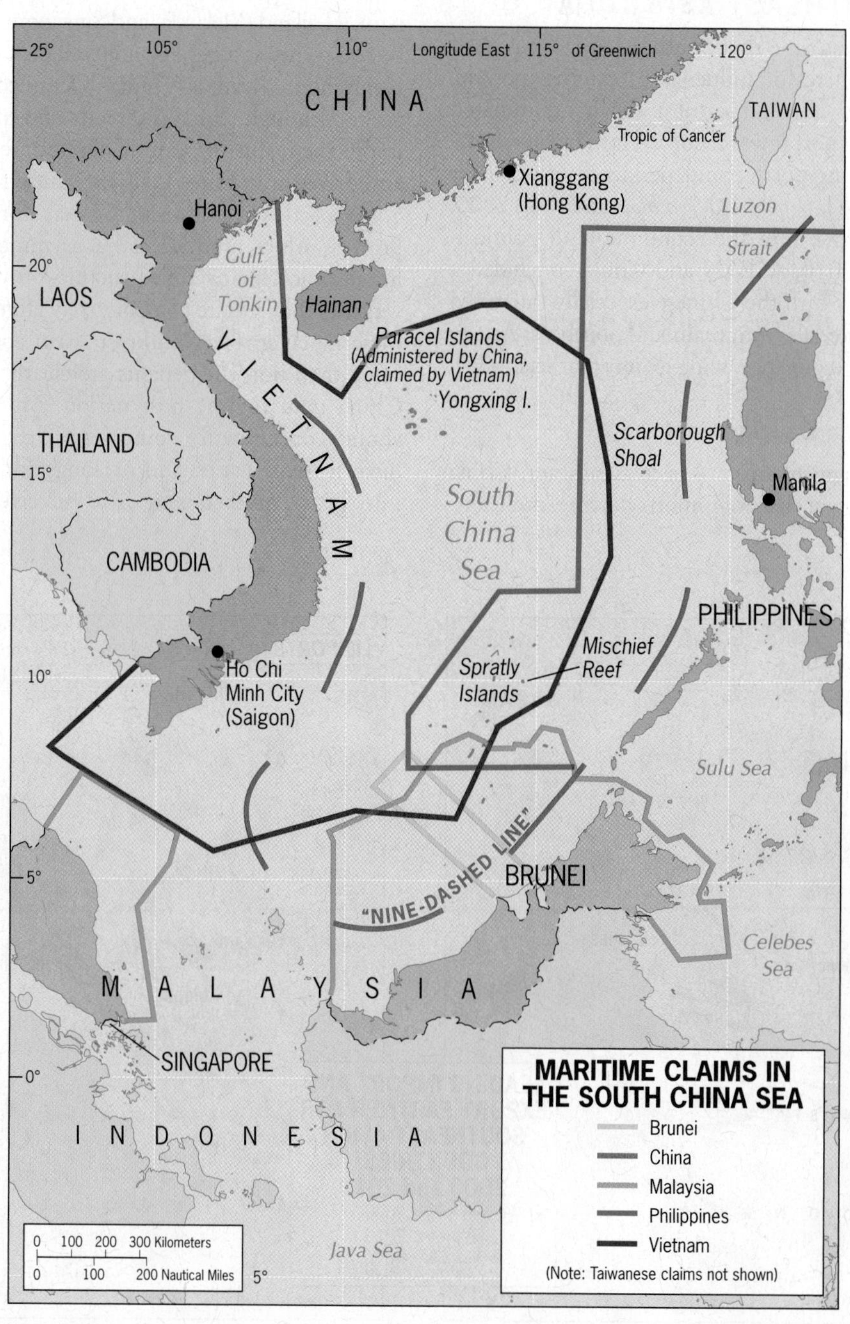

FIGURE 10A-8 © H. J. de Blij, P. O. Muller, J. Nijman, and John Wiley & Sons, Inc.

map showing Chinese claims in these waters—a map said to date back to 1949, the year the PRC was founded. The 'nine dashes' refer to the delimitation of Chinese claims that effectively cover most of the Sea (Fig. 10A-8). In the words of one high-ranking Chinese official in 2012, "China does not want all of the South China Sea, it just wants 80 percent." That 80 percent includes some 40 islands that China says are illegally occupied by other countries as well as seafloor zones believed to be rich in deposits of oil and natural gas.

The South China Sea, according to official government communications, "is crucial to the future of China as a growing maritime nation." Indeed, the PRC already controls many islands in this body of water, and in 2012 it expanded its military facilities on Yongxing (largest of the Paracel Islands), more than 200 miles southeast of Hainan Island, with the declared purpose of 'exercising sovereignty' across the South China Sea. The Sea contains about 250 islands, most of them lifeless rocks inundated at high tide, as well as numerous shoals that are permanently submerged. Even though the Chinese-controlled islands are inhabited by only a few hundred people, the PRC's assertion is mainly about sovereignty over surface waters and automatic ownership of oil and gas reserves that may be buried in the seabed below them.

The disputes and conflicts concerning territorial waters are numerous and complicated, but the most important concern: (1) the Paracel Islands, claimed by China, Taiwan, and Vietnam; (2) the Spratly Islands, claimed by China, Vietnam, Malaysia, and Brunei; and (3) the Scarborough Shoal, claimed by China, Taiwan, and the Philippines (Fig. 10A-8). Not surprisingly, ASEAN has been hopelessly divided on these geopolitical controversies, with Vietnam, Malaysia, and the Philippines as well as Malaysia strongly countering Chinese claims; Cambodia and Laos (tacitly) supporting China; and Singapore and Indonesia trying to steer a more neutral, diplomatic course. Recent ASEAN summit meetings have ended in open disagreement. Interestingly, Vietnam and the Philippines have sought a closer relationship with the United States, clearly in search of a counterweight to Chinese leverage.

Whereas the United States has declared its neutrality on the matter of who owns the islands and shoals, it has emphasized the paramount importance of free and open international access to the waters of the South China Sea. Why? Because this is one of the most essential oceanic trade corridors in the world. Think not only of the major shipping routes within the Southeast Asian realm that interconnect its chief ports of Manila, Singapore, Saigon, Bangkok, and Jakarta, but also of all the intercontinental trade coming from the west—from Europe and Africa as well as South and Southwest Asia—that passes through the Malacca and Singapore straits and then across the South China Sea on the way to China, Taiwan, South Korea, and Japan (not to mention all the seaborne cargoes that are transported along this route in the opposite direction).

In a recent joint military exercise, Filipino and American troops enacted the recapture of a small Philippine island from 'hostile forces.' The United States already had a mutual defense agreement with the Philippines; in 2012, it initiated joint military training with Vietnamese forces as well as signing a new military pact with Thailand. China is never mentioned by name as the hypothetical enemy in these exercises, but nobody is under any delusions as to where the threat to free access to these waters comes from. The South China Sea has quickly become a global geopolitical hotspot in this decade, and is certain to be closely monitored for the foreseeable future.

STATES AND BOUNDARIES

Although we tend to think of boundaries as lines on the map or fences on the ground, the legal definition of a boundary goes much farther than that. In fact, boundaries are actually invisible vertical planes extending above the ground into the air and below the ground into soil and rock (or water). Where these planes intersect with the ground, they form lines—the lines on the map.

A useful way to think about boundaries is to regard them as contracts between states. Such contracts take the form of treaties that contain the ***definition*** of every segment of the boundaries between them. These written definitions refer to actual landforms of the terrain through which the boundary lies—streams, larger water bodies, hills, ridges. Next, surveyors translate these descriptions into lines on large-scale maps that show every detail in the landscape. This process of ***delimitation*** creates the official boundary agreed to by the parties, and we see the results in generalized form on atlas maps. Sometimes neighboring states start arguing over the treaty language or the outcome of delimitation, which can result in armed conflict (as occurred between Sudan and South Sudan in 2012). To avoid such problems, states mark certain stretches of their borders with fences, walls, or other barriers on the cultural landscape, a process referred to as ***demarcation.*** Even though this clearly expresses a state's claims, it does not always resolve the dispute.

Classifying Boundaries

Boundaries have many functions, and to understand this (and to learn why some boundaries tend to produce more trouble than others) it is helpful to view them in a categorical perspective. Some boundaries conform to elongated features in the natural landscape (mountain ranges, rivers) and have ***physiographic*** origins. Others coincide with historic breaks or transitions in the cultural landscape and are sometimes referred to as *anthropogeographic,* or more recently as ***ethnocultural*** boundaries. And as any world political map shows, many boundaries are simply straight lines, defined by endpoint coordinates with no reference to physical or human landscape features; these ***geometric***

From the Field Notes . . .

"I stood on the Laotian side of the great Mekong River which, during the dry season, did not look so great! On the opposite side was Thailand, and it was rather easy for people to cross here at this time of the year. But, the locals told me, it is quite another story in the wet season. Then the river inundates the rocks and banks you see here, it rushes past, and makes crossing difficult and even dangerous. The buildings where the canoes are docked are built on floats, and rise and fall with the seasons. The physiographic-political boundary between Thailand and Laos lies in the middle of the valley we see here."

 www.conceptcaching.com

Concept Caching

boundaries often lead to problems when valuable natural resources are found to lie across or beneath them.

In general, the colonial powers and their successor governments defined the boundaries of Southeast Asia more judiciously than was the case in several other now-postcolonial parts of the world that lay in remote and/or sparsely peopled areas (for instance, across interior Borneo). Nonetheless, a number of Southeast Asian boundaries have triggered disputes, among them the geometric boundary between Papua, the portion of New Guinea ruled by Indonesia, and the country of Papua New Guinea, which occupies the eastern part of that island. In fact, the artificiality of this boundary continues to intensify secessionist feelings among the population of Indonesian Papua.

Even on a small-scale map of the kind we use in this chapter, we can categorize the boundaries in this realm. A comparison between Figures 10A-1 and 10A-4 demonstrates that the boundary between Thailand and Myanmar over long stretches is ethnocultural, most notably where the name ***Kayin (Karen)***, the Myanmar minority, appears in Figure 10A-4. And Figure 10A-1 shows that a lengthy segment of the Vietnam–Laos boundary is physiographic-political, in that it coincides with the crest of the Annamite Cordillera (Highlands).

Boundaries in Changing Times

A number of the world's boundaries are centuries old, whereas others are of more recent origin. Hence, another way of interpreting their functions is to examine their evolution as part of the cultural landscape they partition. We distinguish four types of these ***genetic*** (evolutionary) boundaries; as it happens, examples of all four can be found in Southeast Asia.

Certain boundaries were defined and delimited before the present-day human landscape materialized. In Figure 10A-9 (upper-left map), the border between Malaysia and Indonesia across the island of Borneo is an example of the first boundary type, an **antecedent boundary [10]**. Most of this border passes through very sparsely inhabited tropical rainforest, and the break in the settlement pattern can even be detected in the realm's population map (Fig. 10A-3).

A second category of boundaries evolved as the cultural landscape of an area took shape and became part of the ongoing process of accommodation between neighboring states. These **subsequent boundaries [11]** are represented in Southeast Asia by the map in the upper right of Figure 10A-9, which shows in some detail the border between Vietnam and China. This border is the result of a long process of adjustment and modification, the end of which may not yet have occurred.

The third category involves boundaries drawn forcibly across a unified or at least homogeneous cultural landscape. The colonial powers did this when they divided the island of New Guinea by delimiting a boundary in a nearly straight line (curved in only one place to allow for a bend in the Fly River), as shown in the lower-left map of Figure 10A-9. The **superimposed boundary [12]** they delimited gave the Netherlands the western half of New Guinea. When Indonesia became independent in 1949, the Dutch did not yield their part of New Guinea, which is peopled

GENETIC POLITICAL BOUNDARY TYPES

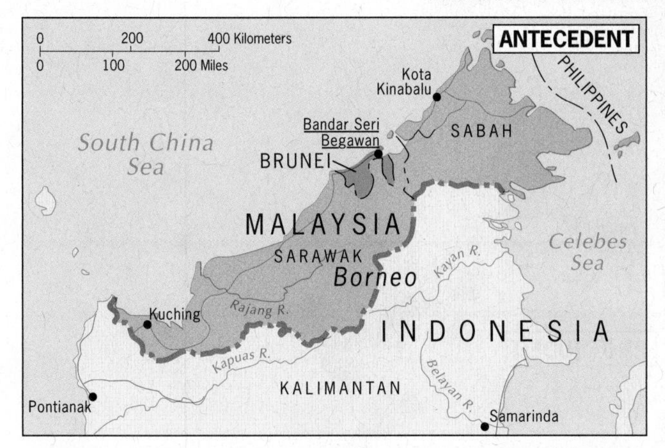

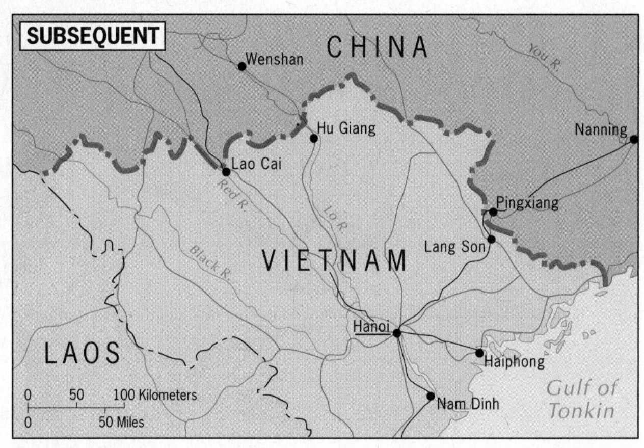

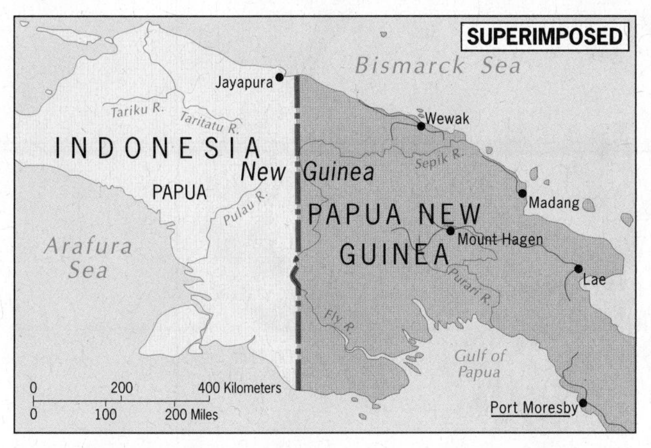

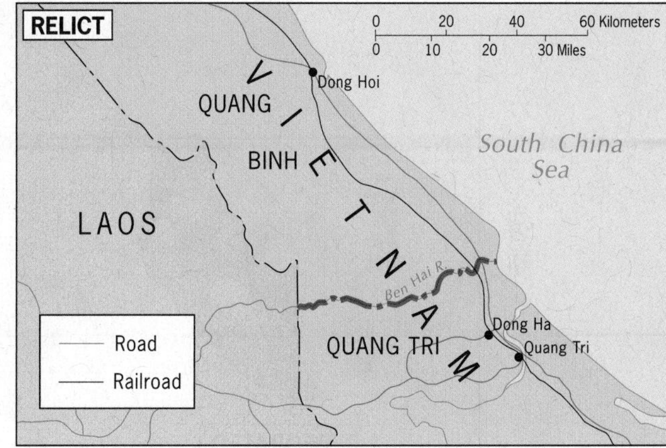

FIGURE 10A-9

mostly by ethnic Papuans, not Indonesians. In 1962, the Indonesians invaded the territory by force of arms, and in 1969 the United Nations recognized its authority there. This made the colonial, superimposed boundary the eastern border of Indonesia and had the effect of projecting Indonesia from the Southeast Asian realm into the adjoining Pacific Realm. Geographically, all of New Guinea lies within the Pacific Realm.

The fourth genetic boundary type is the so-called **relict boundary [13]** a border that has ceased to function but whose imprints (and sometimes influence) are still evident in the cultural landscape. The boundary between former North and South Vietnam (Fig. 10A-9, lower-right map) is a classic example: once demarcated militarily, it has held relict status since 1976 following the reunification of Vietnam in the aftermath of the Indochina War (1964–1975).

Southeast Asia's boundaries have colonial origins, but they have continued to influence the course of events in postcolonial times. Take one instance: the physiographic boundary that separates the main island of Singapore from the rest of the Malay Peninsula, the Johor Strait (see Fig. 10B-6). That physiographic-political boundary facilitated, perhaps crucially, Singapore's secession from the state of Malaysia in 1965. Without it, Malaysia might have been persuaded to halt the separation process; at the very least, territorial issues would have arisen to slow the sequence of events. As it was, no land boundary needed to be defined: the Johor Strait demarcated Singapore and left no question as to its limits.*

State Territorial Morphology

Boundaries define and delimit states; they also create the mosaic of often interlocking territories that give individual countries their shape. This shape or ***territorial morphology*** can affect a state's condition, even its survival. Vietnam's extreme elongation has influenced its existence since time immemorial. And, as is noted in Chapter 10B, Indonesia has tried to redress its fragmented nature (thousands of islands) by promoting unity through the "transmigration" of residents of Jawa from the most populous island to many of the others.

*Except one: a tiny island at the eastern entrance to the Strait named Pedra Blanca (as Singapore calls it) or Pulau Batu Putih (the Malaysian version), which is still disputed today.

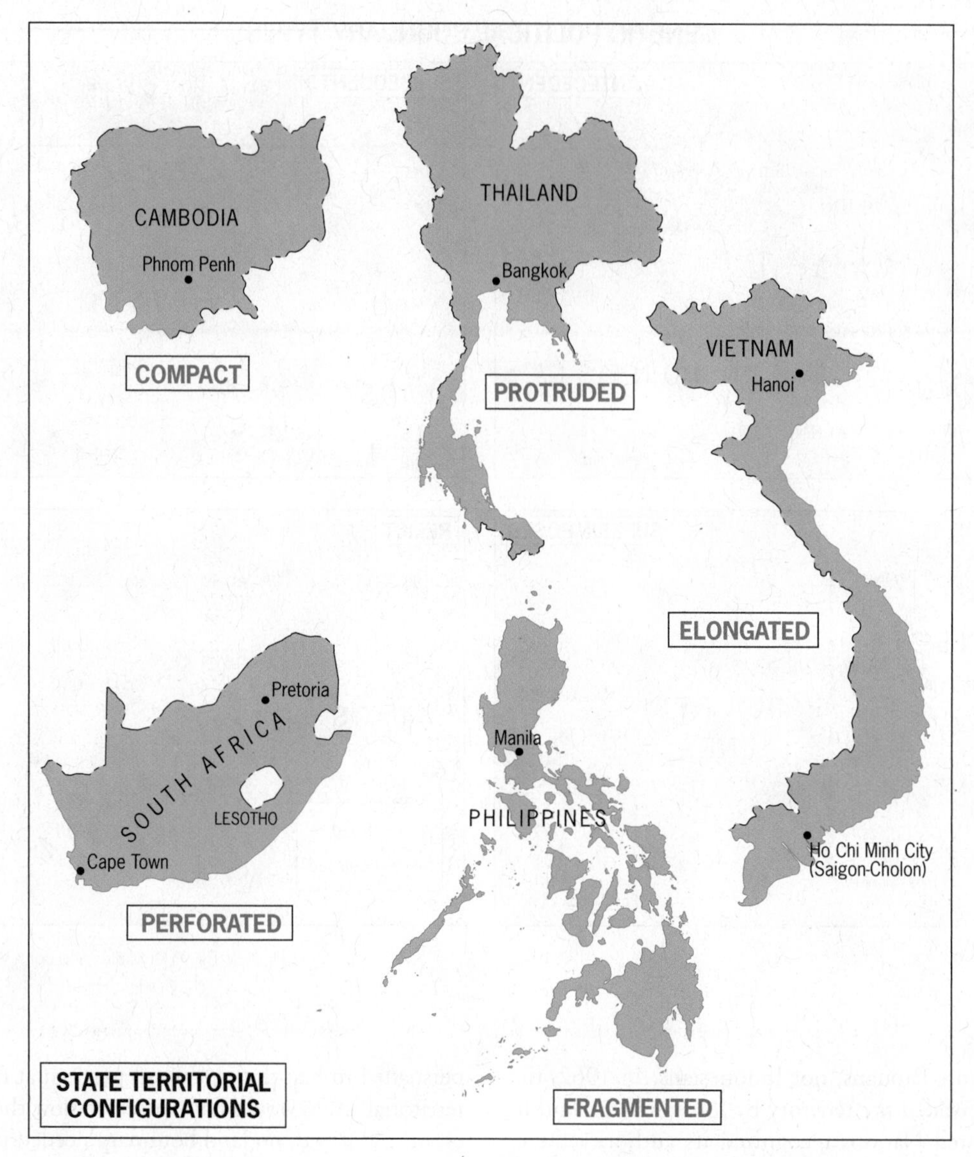

FIGURE 10A-10

Political geographers identify five dominant state territorial configurations, all of which we have encountered in our world regional survey but which we have not categorized until now. All but one of these shapes is represented in Southeast Asia, and Figure 10A-10 provides the terminology and examples:

- **Compact states [14]** have territories shaped somewhere between round and rectangular, without major indentations. This encloses a maximum amount of territory within a minimum length of boundary. Southeast Asian example: Cambodia.
- **Protruded states [15]** (sometimes called ***extended***) have a substantial, usually compact territory from which extends a peninsular or other corridor that may be landlocked or coastal. Southeast Asian examples: Thailand and Myanmar.
- **Elongated states [16]** (also known as ***attenuated***) have territorial dimensions in which the length is at least six times the average width, creating a state that lies astride environmental or cultural transitions. Southeast Asian example: Vietnam.
- **Fragmented states [17]** consist of two or more territorial units separated by foreign territory or a substantial body of water. Subtypes include mainland-mainland, mainland-island, and island-island. Southeast Asian examples: Malaysia, Indonesia, the Philippines, and East Timor.
- **Perforated states [18]** completely surround the territory of other states, so that they have a "hole" in them. No Southeast Asian example; the most illustrative case on the world political map is South Africa, perforated by Maryland-sized Lesotho.

In Chapter 10B, we will have frequent occasion to refer to the shapes of Southeast Asia's states. For so comparatively small a realm with so few countries, Southeast Asia displays a considerable variety of state morphologies. But one point of caution: states' territorial morphologies do not determine their viability, cohesion, unity, or lack thereof; they can, however, influence these qualities. Cambodia's compactness has not ameliorated its divisive political geography, for instance. But as we will find in our survey of the realm's regional geography, shape does play a key role in the still-evolving political and economic geography of Southeast Asia.

REALM BETWEEN THE GIANTS

Southeast Asia's 11 states create a geographic panorama of diverse cultures, economic contrasts, and political options, but this realm's looming reality today is the rise of China on land and rea. Take another look at Figure 10A-1: it is somehow symbolic that the realm's mainland and offshore regions encircle a large body of water named the South China Sea even though, by the time you sail the water between Vietnam and Malaysia, you are well over 1500 kilometers (930 mi) from the nearest coast of China. The Chinese colossus is everywhere in Southeast Asia today—from the bustling Chinatowns in the cities to the commercially productive minorities on remote islands, from the dams and pipelines being laid for local governments to the farmlands the Chinese are leasing for the future. Diversity is Southeast Asia's hallmark, and certainly the realm's governments are reacting differently to China's growing influence. On the mainland, the government of Cambodia has become a strong ally of China while neighboring Vietnam takes a very different view. Offshore, the Philippines two decades ago sought to distance itself from greater American involvement, but today the Philippine government sees the United States as a counterweight to China's potential dominance.

If all eyes are on China in Southeast Asia these days, it is useful to remember that another colossus adjoins this realm to the west: India. Southeast Asia is where the Indian Ocean meets the Pacific, and although India is no China in terms of its economic strength or political power, India's geopolitical view of the Indian Ocean is not totally unlike China's perspective on the western Pacific: India, too, takes a proprietary view of "its" maritime region. India, like China but with fewer resources, is expanding and modernizing its navy and proclaiming its presence among Asia's naval forces. India eyed warily China's long-term relationship with the generals who ran Myanmar, its neighbor, until Myanmar's current transition began. Now India takes an active interest in the fate of its neighbor's reformist efforts, but with a different attitude. And let us not forget that while Chinese minorities are numerous and economically prominent within Southeast Asia, there also are substantial Indian minorities (and cultural legacies) in Malaysia, Singapore, Indonesia, and elsewhere. Might India someday become the counterweight to China in this realm? At present it appears unlikely, but political geography can take unexpected turns.

As we traverse the physical and cultural landscapes of the countries that make up this fascinating geographic realm, be prepared for some surprises. In one of the most modern, you cannot say uncomplimentary things about the king or risk going to jail (Thailand). In another, you would believe that you had arrived in an Arab petro-sheikdom (Brunei). In still another, you are at the center of globalization—the airport by itself will make you think you've come from the Third World (Singapore). But drive into some of the countrysides, and it will seem as though you are back in an earlier century, and time has stood still. This truly is an entire world in a single, diminutive realm.

POINTS TO PONDER

- Certain Southeast Asian states have achieved much in their efforts to accommodate diverse cultural traditions.
- The two smallest countries of the realm are also by far the wealthiest.
- China's influence in Southeast Asia is growing. Is the Mainland region becoming a Chinese periphery?
- In your daily life, how many products do you use that contain palm oil? How can you find out? As a consumer, do you contribute to deforestation?

10B SOUTHEAST ASIAN REGIONS

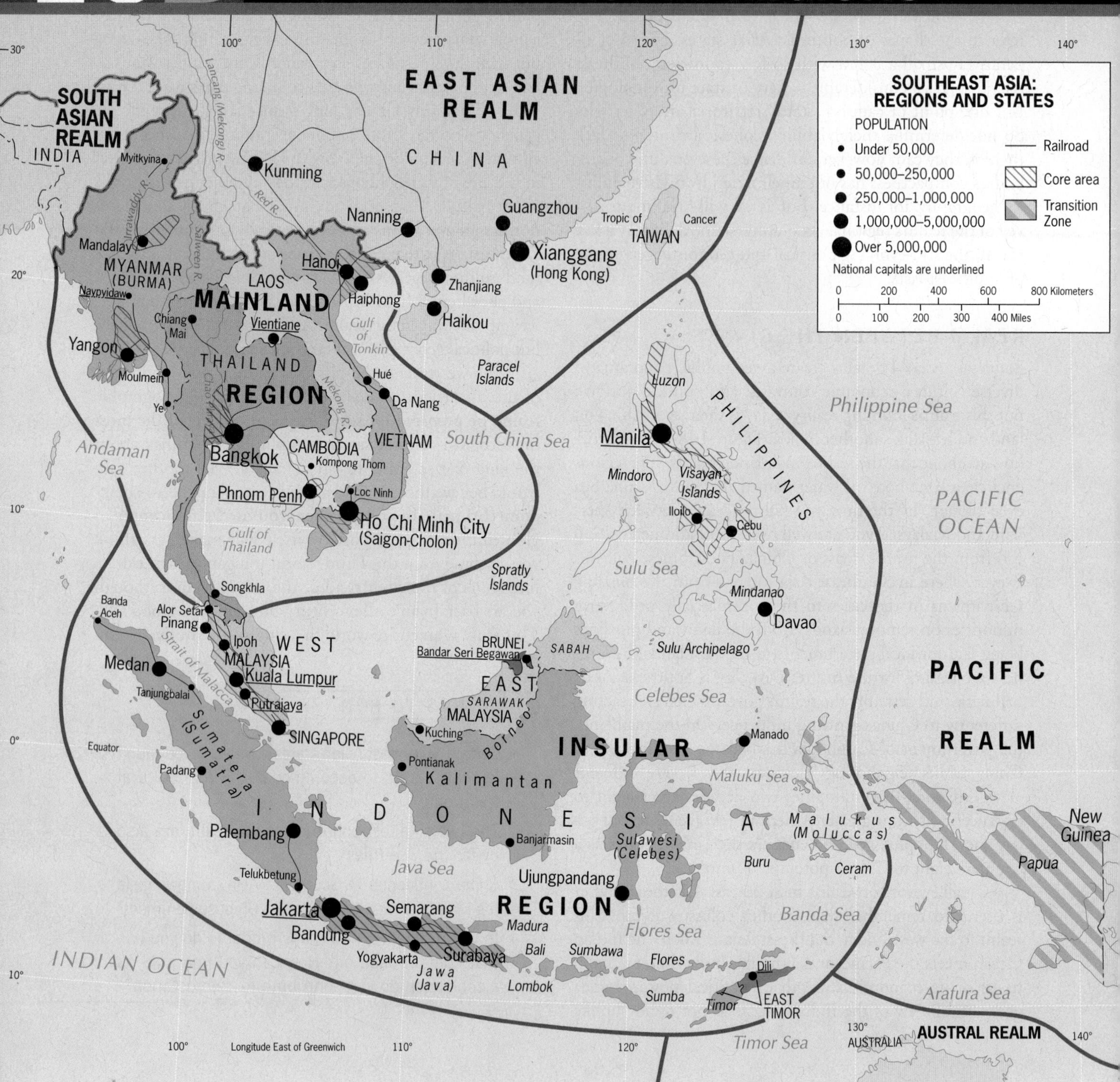

FIGURE 10B-1

REGIONS

Mainland Southeast Asia

Insular Southeast Asia

IN THIS CHAPTER

- Myanmar: Two steps forward, one step back
- Chinese railroads in Laos
- Oil wealth in Brunei and East Timor
- Vietnam allied with the United States
- Indonesia takes the lead
- The Philippines: Wary of China

CONCEPTS, IDEAS, AND TERMS

Domino theory 1

Diffusion 2

Protrusion 3

Choke point 4

Entrepôt 5

Growth Triangle 6

Archipelago 7

Transmigration 8

Exclave 9

Remittances 10

The regional geography of Southeast Asia evolved from the former empires of competing colonial powers whose domination ensured compliance with their imperial designs. The postcolonial states that emerged in this realm confronted strong centrifugal forces but generally succeeded in maintaining the stability of this regional framework. Neither outsiders (U.S. intervention in the 1960s aimed at dividing Vietnam into a communist North and a non-communist South) nor locals (such as insurgents in Malaysia and secessionists in Indonesia) could fracture that framework, whose national components did not begin their independent existence with much in common, either culturally or economically. There were times when it seemed that Malaysia might break up, but only Singapore left that federation in a negotiated secession without violence. Sprawling Indonesia dealt with separatists by force as well as through persuasion and accommodation. The troubled birth of East Timor in 2002, the realm's youngest state, is therefore the exception rather than the rule. Meanwhile, Southeast Asia's governments have become more representative, its economies are maturing, and the realm's key statistical indicators (see Appendix B) reflect the internal development that has taken place.

© Jan Nijman

Where were these pictures taken?
Find out at www.wiley.com/college/deblij

REGIONS OF THE REALM

So much of the Southeast Asian realm consists of the waters between and among its land areas that, on the globe, it seems to be much larger territorially than it really is. Indeed, the realm's mainland component is actually smaller than its island entities combined: the mainland (including the peninsular part of Malaysia) consists of about 2.1 million square kilometers (800,000 sq mi) whereas the islands, even without including Indonesian Papua (which belongs to the Pacific Realm), account for some 2.4 million square kilometers (930,000 sq mi)—a ratio of 46 to 54 percent. But, as we shall see, Malaysia overall exhibits stronger insular

© Jan Nijman

MAJOR CITIES OF THE REALM

City	Population* (in millions)
Manila, Philippines	12.7
Jakarta, Indonesia	10.3
Bangkok, Thailand	9.1
Ho Chi Minh City (Saigon), Vietnam	7.1
Singapore, Singapore	5.4
Yangon, Myanmar (Burma)	4.8
Hanoi, Vietnam	3.4
Kuala Lumpur, Malaysia	1.7
Phnom Penh, Cambodia	1.7
Vientiane, Laos	1.0

*Based on 2014 estimates.

than mainland properties, tilting the balance even further toward offshore Southeast Asia.

With 11 countries ranging from mini-states Singapore and Brunei to regional giants Indonesia and Myanmar, Southeast Asia's states fall into every World Bank income category (see Fig. G-11). Singapore got rich from trade, Brunei from oil. Myanmar remained extremely poor when it did not need to because despotic mismanagement prevented it from realizing its potential. Environmental, locational, and administrative problems plagued, and then restrained, Laos and Cambodia. But Malaysia forged ahead, Thailand is again upward bound after recent political setbacks, and conservatively communist Vietnam continues to loosen its economic shackles. There was a time not long ago when this realm as a whole was classified as part of the 'underdeveloped world', but those days are over.

Southeast Asia's first-order regionalization recognizes the mainland–island dichotomy that is obvious from any map of this realm. But things are somewhat more complicated than that because the southern part of the Malay Peninsula, occupied by the states of Malaysia and Singapore, exhibits physiographic, historical, and cultural characteristics to justify its inclusion in the insular (island) rather than the mainland region. Using the political framework as our grid, we can map the broadest regional division of Southeast Asia in Figure 10B-1:

- **Mainland region:** Vietnam, Cambodia, Laos, Thailand, and Myanmar (Burma)
- **Insular region:** Malaysia, Brunei, Singapore, Indonesia, East Timor (Timor-Leste), and the Philippines

Special note should also be taken of the discordance between the political matrix and the cultural-geographic reality along the eastern periphery of the Insular region. Here, Indonesia controls the western half of the island of New Guinea—but all of New Guinea belongs to the neighboring Pacific Realm.

MAINLAND SOUTHEAST ASIA

Five countries form the Mainland region of Southeast Asia (Fig. 10B-1): the three remnants of Indochina—Vietnam, Cambodia, and Laos—in the east; Thailand in the center; and Myanmar (formerly Burma) in the west. Even though one religion, Buddhism, dominates cultural landscapes, this is a multicultural, multiethnic region (see Fig. 10A-4). Although still one of the less urbanized regions of the world, the Mainland contains several major cities and the pace of urbanization is accelerating. And as Figure 10B-1 shows, two countries (Vietnam and Myanmar) possess dual core areas.

We approach this region from the east, beginning our survey in Indochina where the United States fought and lost a disastrous war that ended in 1975, but whose impact on America continues to be felt today. After the Indochina War formally began in 1964 (U.S. involvement in Vietnam actually began earlier), some scholars warned that the conflict might spill over from Vietnam into Laos and Cambodia, and from there into Thailand, Malaysia, and even then-Burma. This view was based on the **domino theory [1]**, which holds that destabilization and conflict from any cause in one country can result in the collapse of order in one or more neighboring countries, triggering a falling-domino chain of events that can affect a series of contiguous states in a region.

That did not happen, and some of the rhetoric of the time was infused with alarmist views from the West. Cambodia and Laos were indeed disastrously impacted as the original Vietnam War became the wider Indochina War, but not the other "dominoes" (in fact, the last Laotian Hmong refugees who had been driven from their homes and who had found refuge in Thailand were evicted as recently as 2009).

Still, that does not necessarily invalidate the idea of the domino effect. Powerful states that intervene in weaker countries, as the United States did in Iraq in 2003 and Russia in Georgia in 2008, do so at the risk of destabilizing those countries' neighbors by setting in motion arms races and sowing the seeds of future intraregional conflict. Ethnic-cultural strife in one state can also engulf neighboring countries, as happened in central Africa during the 1990s when Rwanda's civil wars spread into the neighboring Congo (where a government fell as a result), and also drew in Burundi and Uganda.

The domino theory remains an intriguing model framework, and in our crowded world potential dominoes abound. Its most recent, dramatic manifestation, as shown in Chapters 7A and 7B, was the **diffusion [2]** of the popular, Arab Spring uprisings across North Africa and parts of Southwest Asia in 2011.

Vietnam

Vietnam (population: 90.8 million) still carries the deep scars of the Indochina War (1964–1975), although the vast majority of Vietnamese have no personal memory of that terrible conflict. The more immediate concerns in Vietnam

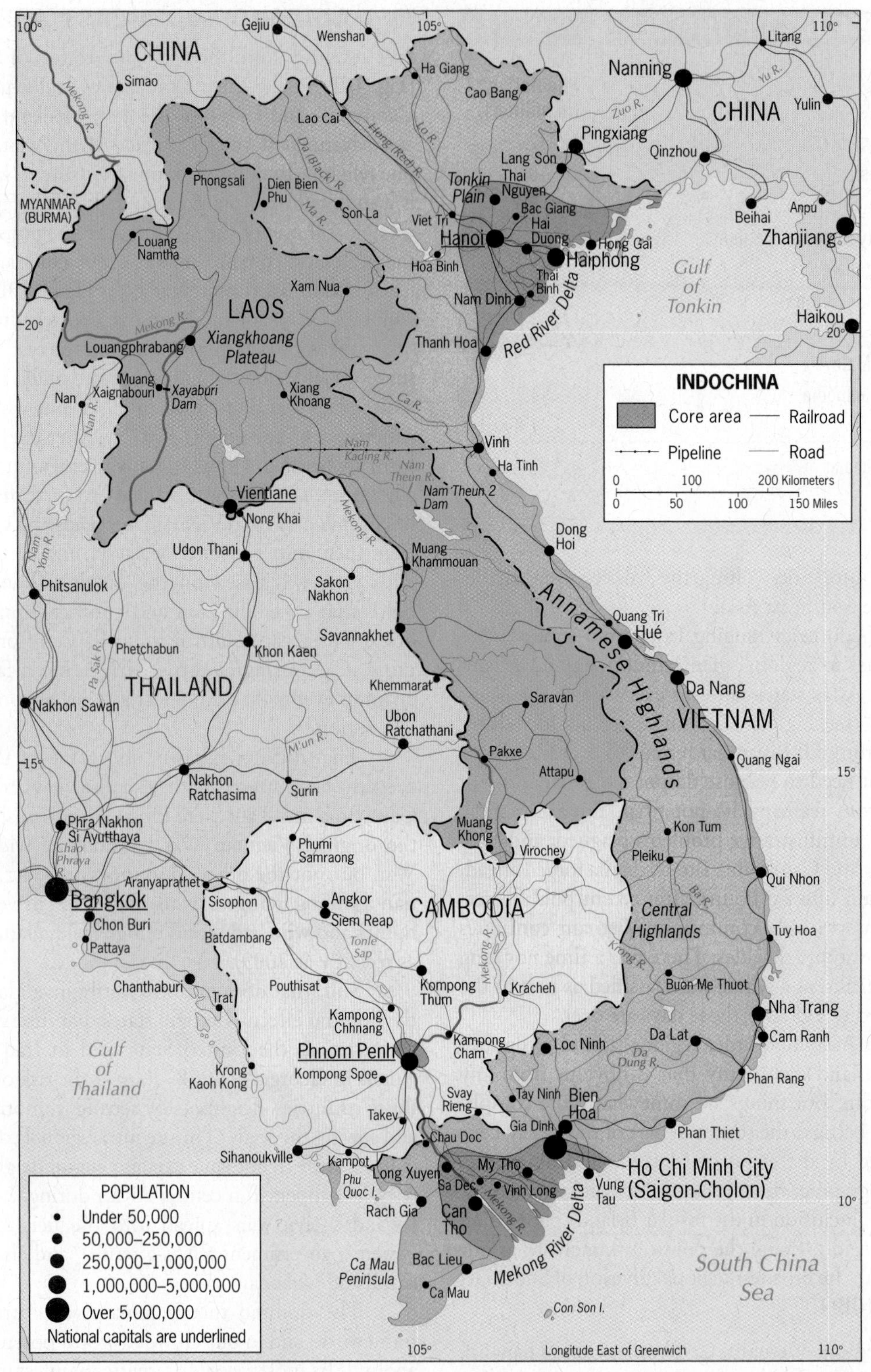

FIGURE 10B-2

today are to reconnect the country with the outside world and to integrate its 2000-kilometer (1200-mi) strip of highly elongated territory through upgraded infrastructure (Fig. 10B-2). This also is an emerging economy that is now making considerable progress. Poverty has declined dramatically over the past decade as Vietnam became a player in the global economy. In the mid-2010s, it is regarded as a "hot" new location for assembly plants, which are attracted by wages lower than China's and the incentives offered by more than a dozen economic development zones.

North, South, and Unification

Japan had invaded Vietnam during World War II, after which the French tried to regain control, fighting Vietnamese

© Jan Nijman

© Jan Nijman

Two faces of changing Vietnam, three blocks apart in Ho Chi Minh City (Saigon): a statue of the revered former communist leader Ho Chi Minh in front of City Hall; and the first Louis Vuitton store, catering to the expanding base of higher-income consumers.

nationalists in a war that lasted nine years before they were decisively ousted in 1954. But newly independent Vietnam did not become a unified state. Separate regimes took control: a communist one in the north based in Hanoi and a non-communist counterpart in the south headquartered at Saigon. Vietnam's pronounced elongation had made things difficult for the French, and now it played the same role during the postcolonial era. Note, in Figure 10B-2, that Vietnam is widest in the north and south, with a slender "waist" in its middle zone—ensuring that North and South Vietnam were worlds apart in a number of ways. For more than a decade the United States tried to prop up the Saigon regime that controlled the south, but the communists eventually prevailed; like China, Vietnam still has a communist government today. After hostilities terminated in 1975, as many as 2 million Vietnamese refugees set out in often-flimsy watercraft onto the South China Sea; of those "boat people" who survived, a majority settled in the United States.

Vietnam officially became a unified state in 1976, and since then the contrasts between north and south have slowly diminished. The capital, Hanoi, has long lagged behind bustling Saigon (now renamed Ho Chi Minh City after North Vietnam's revolutionary leader), but today its growing skyline reflects modernization and the overland links to its outport of Haiphong have been upgraded. With 3.4 million residents, Hanoi anchors the northern (Tonkin Plain) core area of Vietnam, the lower basin of the Red River (its agricultural hinterland). On rural roadways, goods are still moved by human- or animal-drawn cart, but the roads themselves are being improved. The south, however, is experiencing even more significant change and faster economic growth (see photo pair) as the government slowly, and on its own terms, opens up the country to the global economy.

Vietnam in Transition

Since the late 1980s, Vietnam's self-declared communist government has cautiously followed China's pathbreaking approach to economic development. By the mid-1990s, free enterprise was encouraged; farmers were finally allowed to cultivate for profit; and foreign investment was welcomed, although under far more restrictive terms than in China. The government also prioritized the provision of beneficial social services, resulting in steadily rising attendance in secondary schools and health indicators that now surpass China's.

A leading development focus in Vietnam since 2000 has been the expansion of agricultural production. One result is that the country now ranks among the five leading coffee producers in the world. However, because Vietnam's mass-produced bulk coffee is raised plantation-style in the sun, the diffusion of this crop into the interior highlands has triggered substantial deforestation, worsening the annual monsoon-related floods that plague this now-unstable countryside. Other major Vietnamese crops include rice, rubber, tea, sugar, and spices—all with increasingly favorable export opportunities.

Vietnam has been able to maintain a relatively fast-growing national economy, and was only marginally affected by the economic difficulties that have plagued much of the world since 2008. Today, Ho Chi Minh City is thriving, sparked by the rapid development of the east bank of the Saigon River, which includes a Chinese-style Special Economic Zone as well as the booming New Saigon business/residential district. In fact, the Saigon area now accounts for more than 25 percent of Vietnam's industrial output and about one-third of its tax revenues. Nonetheless, economists argue that the country is overly dependent on its state-run companies and urgently needs to enhance opportunities for private-sector initiatives. That will require relaxation of the tight economic controls that Hanoi's communist planners have imposed in their determination to prevent runaway capitalism of the Chinese variety, particularly the income disparities that come with it.

Surprisingly, Vietnam these days has become one of the most pro-American countries in Southeast Asia despite its persistence as a communist state. Because three-quarters

From the Field Notes . . .

© H.J. de Blij

© H.J. de Blij

"Hanoi and most of the country's other urban areas look and function much as they have for generations. Main arteries throb with commerce and traffic (still mostly bicycles and mopeds); side streets are quieter and more residential. No high-rises here, but the problems of dense urban populations are nevertheless evident. Every time I turned into a side street, the need for street cleaning and refuse collection was obvious, and drains were clogged more often than not. 'We haven't allowed our older neighborhoods to be destroyed like they have in China,' said my colleague from Hanoi University, 'but we also don't have an economy growing so fast that we can afford to provide the services these people need.' Certainly the contrast between burgeoning coastal China and slower-growing Vietnam—a matter of communist-government policy—is evident in their cities. 'As you see, nothing much has changed here,' she said. 'Not much evidence of globalization. But we're basically self-sufficient, nobody goes hungry, and the gap between the richest and the poorest here in Vietnam is a fraction of what it is now in China. We decided to keep control.' You would find more agreement on these points here in Hanoi than in Saigon, but the Vietnamese state is stable and progressing."

www.conceptcaching.com

of all Vietnamese alive today were born after 1975, what they call their (winning) "American War" is history not memory. Undoubtedly, these good feelings toward their former enemy at least partially reflect the perceived threat of China, against whom the U.S. can provide a valuable counterweight. American-Vietnamese relations have warmed considerably in recent years, and we noted in Chapter 10A that joint military exercises were initiated in 2012. In fact, Vietnam has now opened its strategically located port at Cam Ranh Bay to U.S. naval forces—who significantly modernized and expanded this superb deepwater port a half-century ago to make it their primary base of operations during the Indochina War (Fig. 10B-2).

Cambodia

Compact Cambodia is heir to the ancient Khmer Empire whose capital was Angkor and whose legacy is a vast landscape of imposing monuments including the great Buddhist-inspired temple complex, Angkor Wat. Today, more than 85 percent of Cambodia's 15.5 million inhabitants are ethnic Khmers, with most of the remainder divided between Vietnamese and Chinese. The present capital, Phnom Penh, lies on the Mekong River, which crosses Cambodia before it enters and forms its massive delta in Vietnam (Fig. 10B-2).

Geographically, Cambodia enjoys a number of advantages. Compact states (see Fig. 10A-10) enclose a maximum amount of territory within a minimum of boundary, and cultural homogeneity tends to diminish centrifugal forces. But sometimes the most obvious geographic opportunities are undone by destructive politics. Neither spatial morphology nor ethnic uniformity could withstand the impact of the Indochina War and the subsequent murderous exploits of the Khmer Rouge communist regime in the late 1970s.

Postwar Problems

Cambodia's present social geography continues to suffer from the aftereffects of that war. Its internal spatial advantages notwithstanding, this country could not overcome the liability of relative location adjacent to a conflict that was bound to spill across its borders (Fig. 10B-2). In 1970, domestic discord fueled by foreign interference led to the ouster of the last king by the military; five years later, that regime was itself overthrown by communist revolutionaries, the so-called Khmer Rouge. These new rulers embarked on a vicious course of terror and destruction in order to reconstruct *Kampuchea* (as they called Cambodia) as a rural society. They drove townspeople into the countryside where they had no place to live or work, emptied hospitals and sent the sick and dying into the streets, outlawed religion and family, and in the process killed as many as 2 million Cambodians (out of a population of 8 million). In the late 1970s, Vietnam, having won its own war, invaded Cambodia to drive the Khmer Rouge away. But this action sparked another reign of terror, and a stream of refugees crossed the border into Thailand. Eventually, remnants of the Khmer Rouge managed to establish a base in northwestern Cambodia, where their murderous leader, Pol Pot, who was never brought to justice for the genocide he masterminded, committed suicide in 1998.

The country's postwar trauma still lingers. Once self-sufficient and able to feed others, it now must import food. Rice and beans are the subsistence crops, but the dislocation in the farmlands set production back severely, and continuing strife in the countryside has disrupted supply routes for years. If there is hope for future economic growth, it may lie in recently discovered offshore oil reserves near Cambodia's relatively short coastline. Income from oil exports began in 2012 and could approach U.S. $2 billion annually by the end of this decade, a significant boost for an otherwise weak economy that relies heavily on its garment-industry exports. There are some bright spots as well. Investments from abroad have increased in this decade, mostly by producers seeking to shift their operations away from China (where wages are steadily rising) to capitalize on the much lower cost of labor in Cambodia and other disadvantaged Southeast Asian countries. One result is that employment in the Phnom Penh Special Economic Zone, which strongly relies on foreign revenues, impressively doubled its workforce from 10,000 in 2012 to 20,000 in 2013. Cambodians are also optimistic that growing tourism, largely focused on the Angkor temple complex, is now on the way to becoming a mainstay of the economy.

Laos

Landlocked Laos has no fewer than five neighbors, one of which is East Asia's giant, China (Fig. 10B-2). The Mekong River forms a long stretch of its western boundary, and the important sensitive border with Vietnam to the east lies in mountainous terrain. With only 6.8 million people (over half of them ethnic Lao, related to the Thai of Thailand), Laos lies surrounded by comparatively powerful states. The country has a woefully inadequate infrastructure, hardly any industry, and less than 5 percent of its land is suitable for agriculture. Laos is barely more than 25 percent urbanized; the capital, Vientiane, lies on the Mekong and has an oil pipeline to Vietnam's coast.

Laos has long included one small corner of the ***Golden Triangle*** of opium poppy-cultivation fame, but under international (especially U.S. and European) pressure, the communist regime has forced the mainly hill-tribe people who produce most of it to abandon those crops. Because this was the only way they could make a living, these farmers were sent to resettlement villages in the lowlands, where they contracted malaria and other diseases not prevalent in their upland domain and were subject to cultural disintegration. The reward for the authorities was the continuation of foreign aid; the cost to the powerless hill people (especially the women) is incalculable. Here is an outstanding example of the power of the globalizing core reaching into the weakest of peripheries.

Communist Laos has been very slow to follow the examples from elsewhere in the region to open up its economy. A much belated move in 2011 was the opening of the first stock market in Vientiane. A major new hydropower project on the central Laotian Nam Theun River, a tributary of the Mekong, began operating in 2010 and now provides electricity to much of the middle part of the country as well as selling surplus power to Thailand. And a high-speed rail line now under construction from Vientiane to Kunming in China is certain to pull the country more tightly into the Chinese orbit. For now, though, Laos is more closely tied to neighboring Thailand, still its number-one trading partner.

In 2012, the country launched its biggest hydroelectric dam project on the Mekong at Xayaburi in northern Laos (Fig. 10B-2), but almost immediately drew criticism from downstream Cambodia and Vietnam. Many knowledgeable observers claim this dam will have disastrous consequences in those countries, especially for the tens of millions who depend on fishing for their livelihood. This contested project has already been suspended and restarted in 2013, making it unlikely that a clear path can be forged toward completing construction by the 2019 target date.

Thailand

In virtually every way, Thailand is the leading Mainland state. In contrast to its neighbors, Thailand has been a strong participant in the realm's economic development. Its capital, Bangkok (population: 9.1 million), is by far the largest urban center in the Mainland region and one of the world's most prominent primate cities. The country's population of nearly 71 million is growing at the slowest rate (equal to Singapore's) in this geographic realm.

As Figure 10B-1 reveals, Thailand occupies the heart of Southeast Asia's Mainland region. Even though Thailand has no Red, Mekong, or Irrawaddy Delta, its central lowland is watered by a set of streams that flow off the northern highlands and the Khorat Plateau to the east. One of these waterways, the Chao Phraya, is the Rhine of Thailand and forms the axis of the country's core area (Fig. 10B-3). From the head of the Gulf of Thailand to Nakhon Sawan, this river is a highway for boat traffic. Barge trains loaded with rice head toward the coast, ferries move upstream, and freighters transport tin and tungsten (of which Thailand is among the world's leading

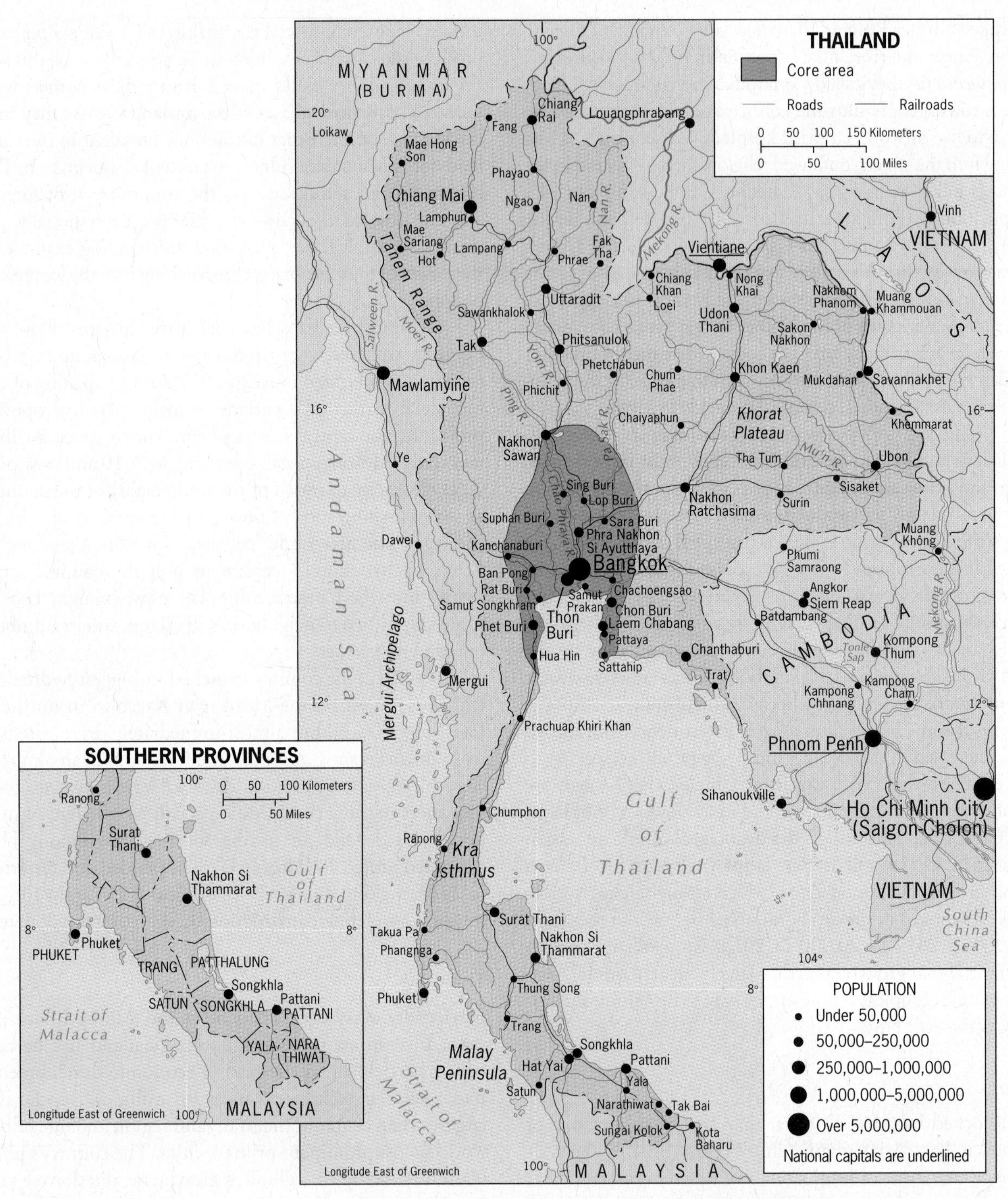

FIGURE 10B-3

producers). Bangkok sprawls along both sides of the lower Chao Phraya floodplain, here flanked by skyscrapers, pagodas, modern factories, boatsheds, ferry landings, luxury hotels, and myriad modest dwellings, all built in crowded confusion on swampy terrain. During the second half of 2011, Thailand suffered its worst flooding since 1940, most of it occurring in the densely populated Chao Phraya Basin. This months-long disaster displaced well over a million people, drowning more than 400 of them; temporarily inundated almost two-thirds of the country, including the low-lying area around Bangkok (see photo); and caused such widespread damage to both agriculture and industry that at least 700,000 people were put out of work, many permanently.

Over the past few decades, only political instability and uncertainty have inhibited economic progress. Thailand is a constitutional monarchy with an elected parliament, but its progress toward stable democracy has experienced a number of setbacks. In 2006, the armed forces ousted a controversial but popular prime minister, and uncertainty reigned until the country's first female prime minister was elected in

AMONG THE REALM'S GREAT CITIES . . .

BANGKOK

WITH A POPULATION of just over 9 million, Bangkok is mainland Southeast Asia's largest city, a sprawling metropolis on the banks of the Chao Phraya River, an urban agglomeration without a true center, an aggregation of neighborhoods ranging from the immaculate to the impoverished. In this city of great distances and sweltering tropical temperatures, getting around is often difficult because roadways are choked with traffic. A (diminishing) network of waterways affords the easiest way to travel, and life focuses on the busiest waterway of all, the Chao Phraya. Ferries and water taxis carry tens of thousands of commuters and shoppers across and along this bustling riverine artery, flanked by a growing number of high-rise office buildings, luxury hotels, and ultramodern condominiums. Many of these contemporary structures reflect the Thais' fondness for domes, columns, and small-paned windows, creating a skyline unique in Asia.

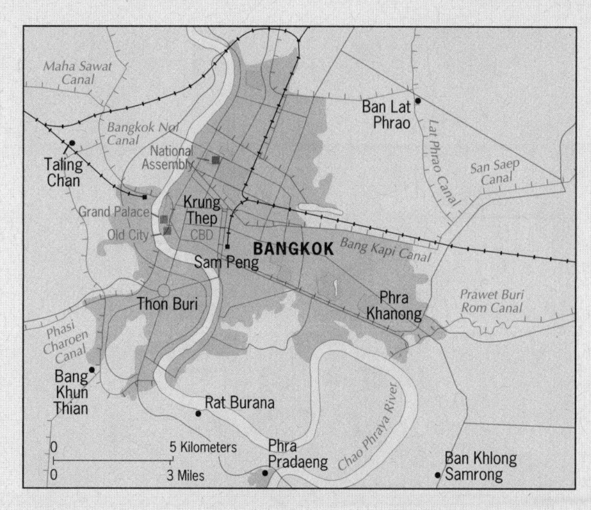

Gold is Buddhist Thailand's symbol, adorning religious and nonreligious architecture alike. From a high vantage point in the city, you can see hundreds of golden spires, pagodas, and façades rising above the townscape. The Grand Palace, where royal, religious, and public buildings are crowded inside a white, crenellated wall 2 kilometers (more than 1 mi) long, is a gold-laminated city within a city embellished by ornate gateways, dragons, and statuary. Across the mall in front of the Grand Palace are government buildings sometimes targeted by rioters in Bangkok's volatile political environment. And not far away are Chinatown (see photo in Chapter 10A) and myriad markets, all components of Bangkok's throbbing commercial life.

2011. But Yingluck Shinawatra is the sister of that ousted predecessor, which casts a shadow over her rule as the country tries to calm down, refocus its energies on resuming economic progress, and deal with the continuing challenges of flood recovery as well as Islamic violence in the far south.

Crowds awaiting transportation swarm onto one of the few buses available during the height of Bangkok's historic, months-long flood in November 2011. This is the upscale Lat Phrao District, home to one of the city's largest shopping complexes, located northeast of the CBD and well away from the banks of the overflowing Chao Phraya River.

The Restive Peninsular South

Thailand has a compact heartland that contains the core area, capital, and leading zones of productive capacity, whereas a 1000-kilometer (600-mi) **protrusion [3]** (a corridor of land extending away from the rest of the state), in places less than 32 kilometers (20 mi) in width, extends southward to the border with Malaysia (Fig. 10B-3). The boundary that defines this protrusion runs down the length of the upper Malay Peninsula to the narrow Kra Isthmus, where neighboring Myanmar peters out and what's left of Thailand to the south fronts both the Andaman Sea (an arm of the Indian Ocean) and the Gulf of Thailand.

From the Field Notes . . .

© H.J. de Blij

"Thailand exhibits one of the world's most distinctive cultural landscapes, both in its urban and its rural areas. Graceful pagodas, stupas, and spirit houses of Buddhism adorn city and countryside alike. Gold-layered spires rise above, and beautify cities and towns that would otherwise be drab and featureless. The architecture of Buddhism has diffused into public architecture, so that many secular buildings, from businesses to skyscrapers, are embellished by something approaching a national style. The grandest expression of the form undoubtedly is a magnificent assemblage of structures within the walls of the Grand Palace in the capital, Bangkok, of which a small sample is shown here. Climb onto the roof of any tall building in the city's center, however, and the urban scene will display hundreds of such graceful structures, interspersed with the modern and the traditional townscape."

www.conceptcaching.com

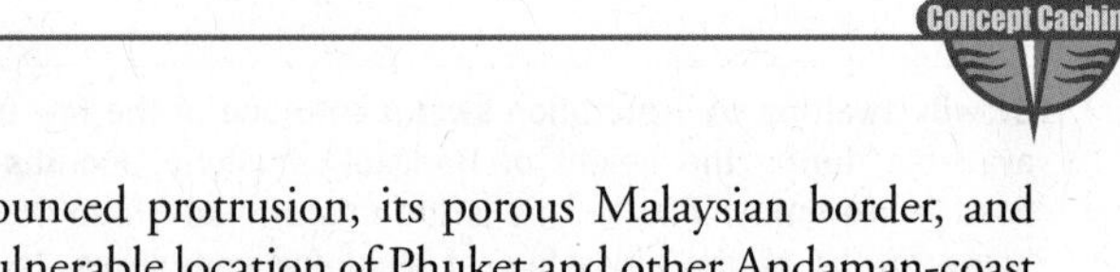

In the entire country, no place lies farther from the capital than the southern end of this tenuous protrusion. Consider this spatial situation in the cultural context of Figure 10A-4. The Malay ethnic population group extends northward from Malaysia more than 300 kilometers (200 mi) into Thai territory. In Thailand's five southernmost provinces (Fig. 10B-3, inset map), close to 90 percent of the inhabitants are Muslims (the figure for Thailand as a whole is less than 6 percent). Moreover, the international border between Thailand and Malaysia is porous; in fact, you can cross it at will almost anywhere along the Kolok River by canoe, and inland it is a matter of walking a forest trail.

For more than a century, the southern provinces have had closer ties with Malaysia across the border than with the Thai capital 1000 kilometers (600 mi) to the north. Rising Islamic militancy and violence pose a growing challenge to the Bangkok government (more than 5000 have been killed and thousands more maimed by terrorist attacks here over the past dozen years). This schism was also reflected in the most recent elections, where support for the national party in power (the Pheu Thai) was concentrated in northern Thailand and almost nonexistent in the far south. Most recently, terrorist bombs in 2012 claimed more than 100 victims in Yala (Fig. 10B-3) and were also aimed at the town's large ethnic-Chinese community. Overall, Thailand's pronounced protrusion, its porous Malaysian border, and the vulnerable location of Phuket and other Andaman-coast resorts (key components of the important Thai tourist industry), combine to increasingly raise concern over this distant outpost of the national territory.

Myanmar

Thailand's western neighbor, Myanmar (unofficially also referred to as Burma, its older name), is one of Asia's poorest countries. Long oppressed by one of the world's most corrupt and brutal military dictatorships, its government turned a corner in 2011 and soon surprised the world by taking bold steps forward. Grinding poverty, crippling isolation along with sanctions (imposed by much of the world because of human rights violations), and growing dependence on China must finally have played a part in pushing Myanmar's leadership in a new direction.

In December 2011, then Secretary of State Hillary Clinton visited Myanmar (and very publicly met with the revered, former dissident Daw Aung San Suu Kyi), the first official U.S. visit to the country in half a century. In the wake of this opening, foreign investments and an upsurge in tourism followed, and the United States, Norway, and Australia were among the first countries to lift sanctions and restart economic and diplomatic relations. But Myanmar's ties with China are still strong and the PRC has vested interests in keeping this country inside its orbit. When the West embargoed and worked to isolate Myanmar's military regime in years past, China took the opportunity to invest an estimated U.S. $27 billion here; the Chinese are likely to retain their influence in Myanmar for some time to come even as the country opens up to the rest of the world.

Myanmar's geographic challenges are complicated by a shift during colonial times of the Burmese core area from north to south. Prior to the colonial period, the core of embryonic Burma lay in the so-called dry zone between the Arakan Mountains in the far west and the Shan Plateau, which covers the country's triangular easternmost extension (Fig. 10B-4). The urban hub of the state was Mandalay, with its central situation and relative proximity to the surrounding

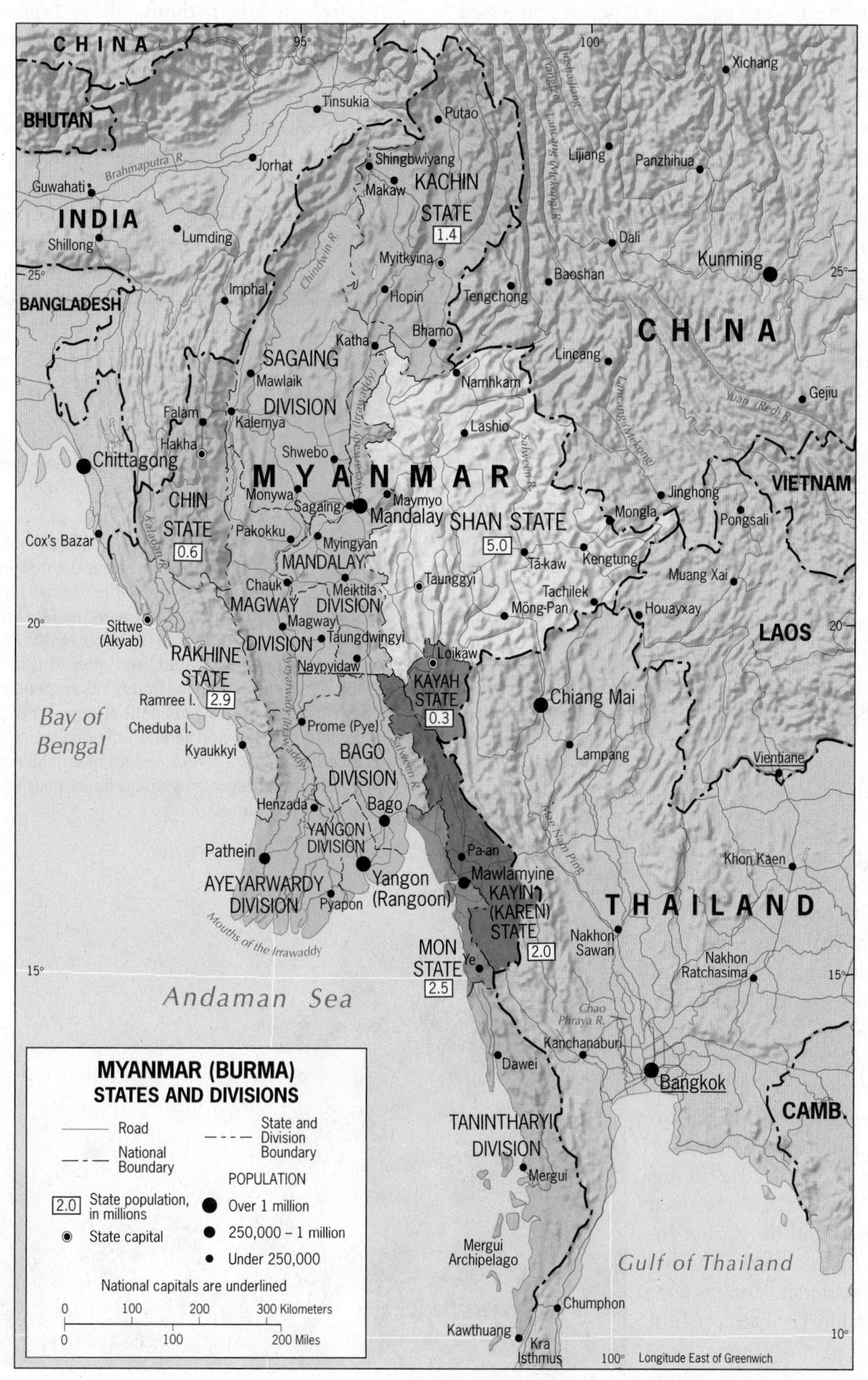

FIGURE 10B-4

non-Burmese highlands. Then the British developed the agricultural potential of the Irrawaddy Delta farther south, and Rangoon—now renamed Yangon—became the capital and leading focus of the colony. The Irrawaddy River links the old and new core areas, but the center of gravity has been located in the south since the late nineteenth century.

In 2006, the military junta moved the capital 300 kilometers (200 mi) northward from Yangon to a newly constructed national-government center—ostensibly a "city", but it barely qualifies as such because it is poorly connected to the rest of the country and has few urban amenities. Situated next to the interior town of Pyinmana, the new capital was named Naypyidaw, with the government asserting that its more central location would yield greater administrative efficiency as well as better protection against (unspecified) enemies.

As Figure 10B-4 indicates, the peripheral peoples of Myanmar occupy significant parts of the country that are designated as seven different States. The majority Burman-dominated areas (68 percent of a national population of just under 56 million) are designated as Divisions. Among the leading minorities, the Shan of the northeast and far north, who are related to the neighboring Thai, account for about 9 percent of the population, or 5 million. The Kayin (Karen), who constitute more than 7 percent (4 million), live in the neck of Myanmar's protrusion and have proclaimed their desire to create an autonomous territory within a federal Myanmar. The Rakhine (3.5 percent, roughly 2 million) of the far western coast facing the Bay of Bengal have long struggled with their Muslim minority and have talked about secession since the time of British of rule. The Mon (2.2 percent, or 1.2 million) were in what is today Myanmar long before the Burmans and introduced Buddhism to the Irrawaddy Basin; they want the return of ancestral lands from which they were ousted.

When Myanmar finally shook off the excesses of dictatorship in 2011 and embraced long awaited reforms, it was hoped this would usher in a period of national progress and democratic advance. In some respects, this has occurred as political prisoners were freed, press censorship was largely ended, and long overdue elections took place. But the end of that lengthy era of repression was accompanied by the resurfacing of long-simmering ethnic tensions which have already spawned two serious outbreaks of sectarian violence. The first was a bloody conflict over autonomy that flared up in 2011–2012 in the northernmost State of Kachin between government forces and the Kachin Independence Army. This strife also has potential international ramifications because the Kachin (1.5 percent [just under one million] of Myanmar's population) are part of a larger ethnic community that spills across both the Chinese and Indian borders. The wider second conflict emerged in late 2012 in the form of violent attacks on the country's Muslim minority, who constitute about 2 million or just under 4 percent of the national population. The worst violence broke out in Rakhine State, home to almost a million Muslims known as Rohingyas, whose ancestors hailed from what is now neighboring Bangladesh. By early 2013, the communal unrest in selected places had evolved into ethnic cleansing, with hundreds killed, thousands of homes destroyed or taken over, and tens of thousands driven away to become internal refugees. Anti-Islamic rioting was also spreading to other parts of the country: the town of Meiktila south of Mandalay in central Myanmar experienced an especially bloody wave of fighting between Buddhists and Muslims (see photo).

Myanmar's long coastline on the waters of the Indian Ocean offers promising opportunities now that the country is opening up to international trade. In the north at Sittwe, a joint venture with India has facilitated major port renovations. India's chief interest is to increase the accessibility of its landlocked easternmost States, an impoverished highland periphery difficult to reach via overland routes from the Brahmaputra Valley to the north. A fully

Smoke rises from a burning Muslim neighborhood torched by Buddhist mobs during the height of a four-day riot in Meiktila in late March 2013. Communal rage in this central Myanmar town was triggered by a heated quarrel between the Muslim owner of a gold shop and a pair of Buddhist customers over the value of a gold hair clip. Events rapidly escalated, and after Muslim youths attacked and killed a Buddhist monk, Buddhists responded in an extended rampage against Muslim residential areas and several mosques, killing more than 40 and temporarily displacing 13,000. Despite a presidential declaration of a state of emergency in this locality, police authorities reportedly stood by without intervention as the violence ran its course.

developed port at Sittwe would allow the opening of a shorter and easier southern routeway. Sittwe's enhanced port is designed so that ocean-going ships can offload their goods onto smaller vessels that are able to sail up the Kaladan River to the Indian border; from there, trucks using improved roads will cover the final leg (Fig. 10B-4).

In the south, a far more significant enterprise from Myanmar's point of view is the so-called Dawei Development Project, a joint initiative with neighboring Thailand to convert the small southern city of Dawei on the Andaman Sea into a world-class port facility. A key first step is the construction of a 160-kilometer (100-mi) highway eastward from Dawei to Thailand's Kanchanaburi (a town with good surface connections to the lower Chao Phraya Basin), located halfway between Dawei and Bangkok (Fig. 10B-4). Plans also call for Dawei to become Myanmar's primary industrial node, a newly-built complex modeled after the special economic zones of the Pacific Rim. For Myanmar, this could spark a noteworthy entry into the global economy, while far more advanced Thailand would gain valuable direct access for its booming core area to the nearby Andaman Sea and India beyond it.

INSULAR SOUTHEAST ASIA

On the peninsulas and islands of Southeast Asia's southern and eastern periphery lie six of the realm's 11 states (Fig. 10B-1). Few regions in the world contain so diverse a set of countries. Malaysia, the former British colony, consists of two major territories separated by hundreds of kilometers of South China Sea. The realm's southernmost state, Indonesia, sprawls across thousands of islands from Sumatera in the west to New Guinea in the east. North of the Indonesian archipelago lies the Philippines, another island-chain country that once was a U.S. colony. These are three of the most severely fragmented states on Earth, and each has faced the challenges that such politico-geographical division brings. This Insular region of Southeast Asia also contains three small but significant sovereign entities: a city-state, a sultanate, and a mini-state. The city-state is Singapore, once a part of Malaysia (and one instance in which internal centrifugal forces were too powerful to be overcome). The sultanate is Brunei, an oil-rich Muslim territory on the island of Borneo that seems to be transplanted from the Persian Gulf. The mini-state is impoverished East Timor, which only achieved independence in 2002. Few parts of the world are more varied or interesting geographically.

Mainland–Island Malaysia

The state of Malaysia represents one of the three categories of fragmented states discussed in Chapter 10A: the mainland–island type, in which one portion of the national territory lies on a continent and the other on one or more islands. Malaysia is a colonial political artifice that combines a pair of decidedly disparate components into a single state: the southern end of the Malay Peninsula and the northern sector of the island of Borneo. These are known, respectively, as ***West Malaysia*** and ***East Malaysia*** (Fig. 10B-1). The appellation "Malaysia" came into use in 1963, when the original Federation of Malaya, on the Malay Peninsula, was expanded to incorporate the territories of Sarawak and Sabah in Borneo. When the name Malaya is used, it refers to the peninsular part of the Federation, whereas Malaysia refers to the total political entity.

Ethnic Components

The Malays of the peninsula, traditionally a rural people, displaced older aboriginal communities there and now constitute just over 50 percent of the country's population of 29.9 million. They possess a compelling cultural identity expressed in adherence to the Muslim faith, a common language, and a sense of territoriality that arises from their perceived Malayan origins and their collective view of Chinese, Indian, European, and other foreign intruders.

The Chinese came to the Malay Peninsula as well as to northern Borneo in substantial numbers during the colonial period, and today they constitute just under 25 percent of Malaysia's population (they are the largest ethnic group in Sarawak). Hindu South Asians were in this area long before the Europeans, and for that matter before the Arabs and Islam arrived on these shores; today they still form a substantial minority of 7 percent of the population, clustered, like the Chinese, on the western side of the peninsula.

Malaysia's ethnic and racial groups have long coexisted in relative harmony but relations today are marked by rising social tensions. As Malays increasingly make claims based on 'majority rights', the Chinese find themselves in much the same minority position that characterize the Overseas Chinese across the realm. The Indian population is much smaller and feels ever more crowded out. Add religious tensions to the mix (for example, Muslim demands for a separate Sharia legal system) and the notion of a common Malaysian identity now looks more fragile than in the past. The ruling ethnic Malay government extended its hold on power in the 2013 elections, but did so by the narrowest of margins—a sign of the deepening political polarization in this country and the increasingly vocal dissent among its minorities.

West Malaysia: The Dominant Peninsula

The populous peninsular part of Malaysia remains the country's dominant sector with 11 of its 13 States and about 80 percent of its population. Here, the Malay-dominated government has very tightly controlled economic and social policies while pushing the country's modernization. During the Asian economic boom of the 1990s, Malaysia's government embraced the notion of symbols: the capital, Kuala Lumpur, was endowed with (at that time) the world's tallest building; a space-age airport that outpaced Malaysia's needs; a high-tech administrative capital was built at Putrajaya; and a nearby development was called Cyberjaya—all part of a so-called ***Multimedia Supercorridor*** to anchor Malaysia's core area (Fig. 10B-5).

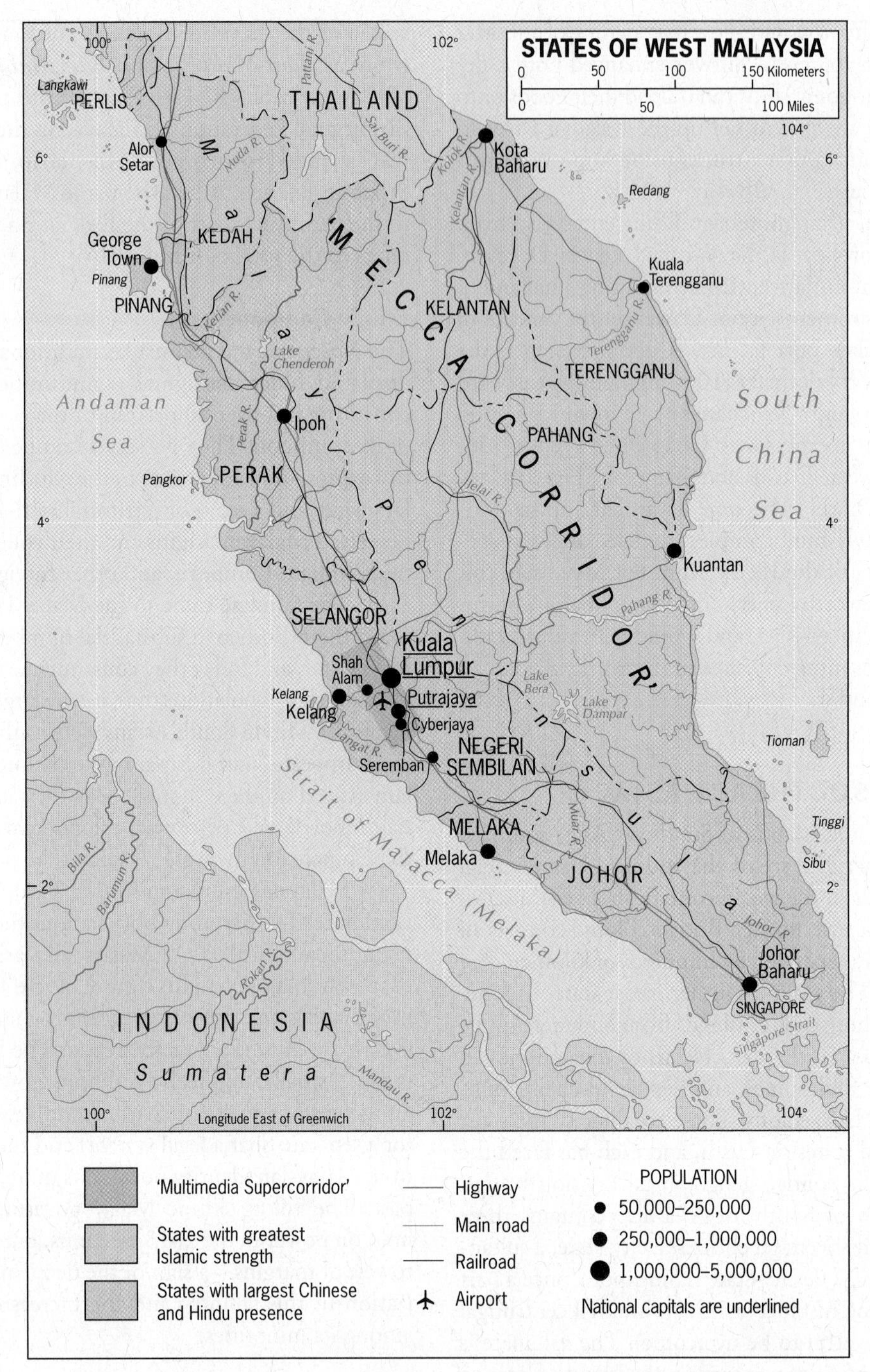

FIGURE 10B-5 © H. J. de Blij, P.O. Muller, and John Wiley & Sons, Inc.

Malaysia's headlong rush to modernize caused a backlash among more conservative Muslims, which in 2001 led to Islamist victories in two States, tin-producing Kelantan and energy-rich but socially poor Terengganu. As the fundamentalist governments in those eastern States imposed strict religious laws, Malaysians talked of two corridors marking their country: the Multimedia Supercorridor in the west and the Mecca Corridor in the east (Fig. 10B-5). But over the past decade, Islamist fervor in the Mecca Corridor, which had featured calls for a *jihad* in Malaysia, has steadily subsided.

The Strait of Malacca (Melaka) to the west continues to be one of the world's busiest and most strategic waterways, a crucial **choke point [4]** (a narrow waterway that constrains navigation) in the flow of resources and goods between major world realms. Malaysia, despite the loss of Singapore and notwithstanding its recurrent ethnic troubles, has become a leading player in Southeast Asia. The strong skills and modest wages of the local workforce have attracted many companies, and the government has capitalized on its opportunities, such as encouraging the creation of a high-technology manufacturing complex on the far northwestern island of Pinang, where Chinese outnumber Malays by two to one.

East Malaysian Borneo

The decision to combine the 11 Sultanates of Malaya with the States of Sabah and Sarawak on Borneo (Fig. 10B-1), creating the country we now call Malaysia, had far-reaching consequences. Those two States may constitute 60 percent of Malaysia's territory, but they are home to barely 20 percent of the population. They endowed Malaysia with major energy resources, huge stands of timber, and vast tracts of land suitable for palm oil plantations (the country is now the world's leading exporter, and these two States produce 55 percent of the national crop [see box in Chapter 10A]). But they also complicated Malaysia's ethnic complexion because each State contains more than two dozen indigenous groups (as noted above, the immigrant Chinese form Sarawak's largest ethnic group). These locals complain that the federal government in Kuala Lumpur treats Malaysian Borneo as a colony, and politics here are contentious and fractious. It is therefore likely that Malaysia will eventually confront devolutionary forces in East Malaysia.

Brunei

Also located on Borneo—along the coast near where Sarawak and Sabah meet—is Brunei (Fig. 10B-1), a rich, oil-exporting Islamic sultanate far from the Persian Gulf. Brunei, the remnant of a former Islamic kingdom that once controlled all of Borneo and areas beyond, came under British control and was granted independence in 1984. Just slightly larger than Delaware and with a population of about 450,000, the sultanate is a mere mini-state—except for the discovery of oil in 1929 and natural gas in 1965, which made it Southeast Asia's richest country. And there are indications that further discoveries will be made in the offshore zone owned by Brunei. The Sultan of Brunei rules as an absolute monarch; his palace in the capital, Bandar Seri Begawan, is reputed to be the world's largest. He will have no difficulty finding customers for his oil in energy-poor eastern Asia.

Brunei's territory is not only small but is also fragmented (see the inset map in Fig. 10B-7). A tiny sliver of Malaysian land separates the larger west (where the capital lies) from the east, and Brunei Bay provides a water link between the country's two parts. It has been proposed that Brunei purchase this separating corridor from the Malaysians, but no action has yet been taken.

Brunei's steadily growing population—in which immigration plays a larger role than internal natural increase—is ethnically two-thirds Malay and 12 percent Chinese. Most inhabitants live and work near the offshore oilfields in the northern district of Brunei Muara that includes the capital city. Evidence of profligate spending is everywhere, ranging from sumptuous palaces to magnificent mosques to the most luxurious of hotels. In this respect Brunei offers a stark contrast to surrounding East Malaysia, but even within Brunei there are significant differences. The country's interior, where a small minority of indigenous groups survive, remains an area of subsistence agriculture and rural isolation, its villages a world apart from the modern splendor of Bandar Seri Begawan.

Singapore

In 1965, a fateful event occurred in Southeast Asia. Singapore, the crown jewel of British colonialism in this realm,

VOICE FROM THE Region

SINGAPORE

Teresa Tan Sze Kai, student, 28 years old

"Like every Singapore citizen, I own a pink identity card, which features my mugshot, thumbprint, name, date and country of birth, gender, and—contentiously—race. I was born in Singapore and learned to be proud of the multicultural, multiracial, and multireligious society that is inscribed in our Constitution. Traditionally, we are divided into Chinese, Malay, Indian, and Other Groups (CMIO), but many of us don't fit these labels anymore. I am half-Peranakan (Peranakan is a mixed race of Chinese and Malay) but my identity card indicates that I am Chinese. I have friends who are of mixed parentage but they are registered only under their father's race (which is already mixed). Race matters because public policies are formulated according to these supposed racial lines. For example, at school each race is assigned a mother tongue: those labeled as Chinese are all taught Mandarin; those designated as Indians must learn Tamil; and the wide range of people who are classified as Malay are taught that language, and they are also closely linked to the Muslim religion. This leads to some ethnic stereotyping that can cause problems. Our government is trying to maintain the race ratio by orchestrating an influx of Mainland Chinese to replace the majority Singaporean Chinese (whose numbers are shrinking due to low fertility rates). But the Singaporean Chinese believe they are culturally different from the Mainlanders. Some of the new immigrants haven't really assimilated and there have been incidents that have heightened tensions. We are a global city and we have thrived in a world that is increasingly interconnected, yet we are always grappling with our sense of identity and degree of "Singaporeanness." Recently, the government started giving people a choice to register under dual races, and that is a step in the right direction. We are evolving rapidly, and I am confident we will overcome these challenges."

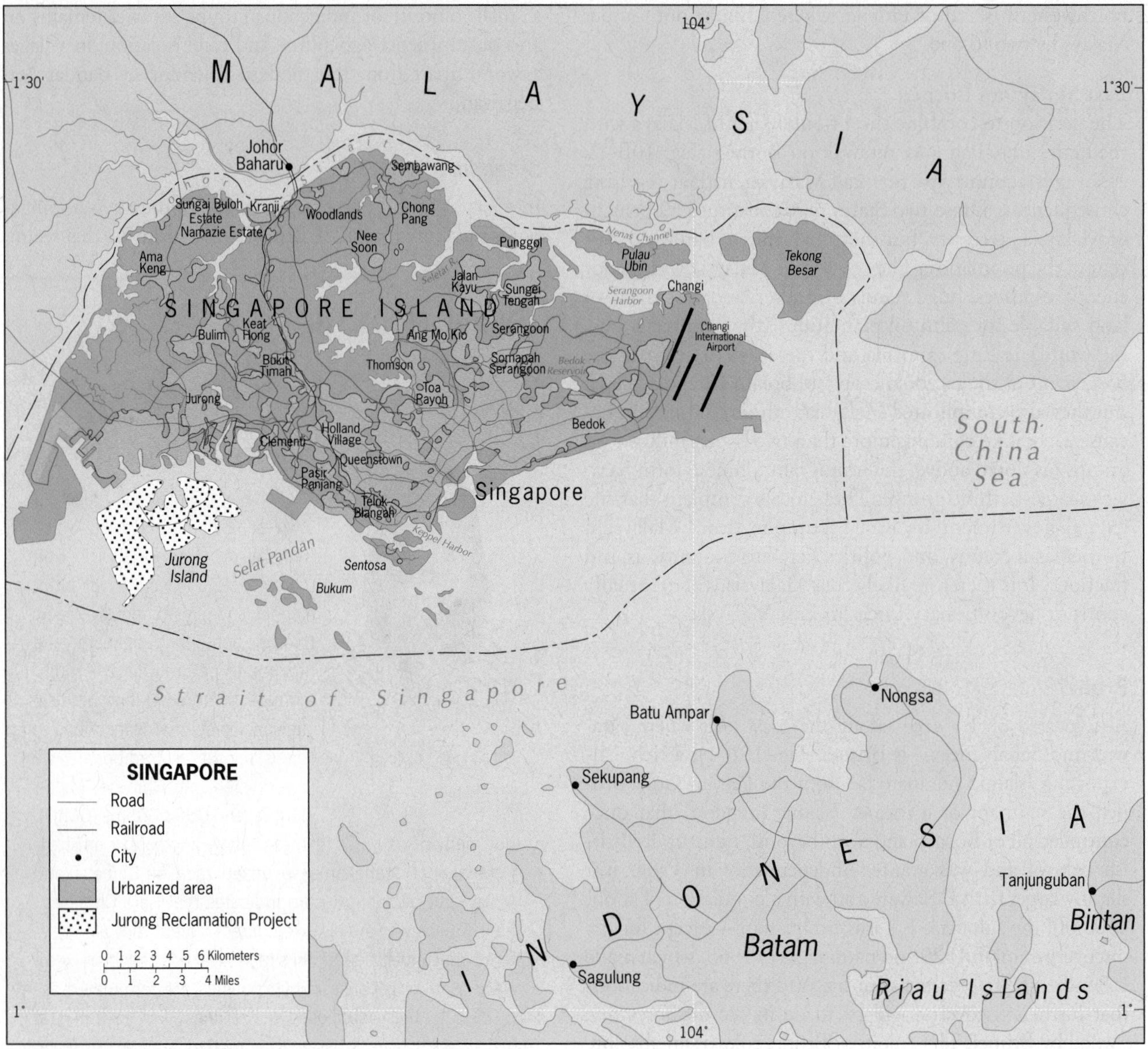

FIGURE 10B-6

seceded from the recently independent (1963) Malaysian Federation and became a sovereign state, albeit a mini-state (Fig. 10B-6). With its magnificent relative location, its multiethnic and well-educated population, and its no-nonsense government, Singapore then overcame the severe limitations of space and the absence of raw materials to become one of the leading economic tigers of East and Southeast Asia. While GDP per capita skyrocketed from U.S.$511 in 1965 to $51,162 in 2012, that of neighboring Malaysia advanced from $333 to only $10,304 during the same time span.

With a mere 619 square kilometers (239 sq mi) of territory, space is at a premium in Singapore, and this is a constant challenge for the government. Singapore's only local spatial advantage over Hong Kong was that its small territory is less fragmented (there are just a few small islands in addition to the compact main island). With a population of 5.4 million and a robustly expanding economy, Singapore's highest priority is to develop ultramodern but space-conserving manufacturing and service industries. Benefiting from its relative location, the old port of Singapore had become one of the world's busiest (by number of ships served) even before its independence. It thrived as an **entrepôt [5]** between the Malay Peninsula, the rest of Southeast Asia, Japan, and other emerging economic powers on the western Pacific Rim. Crude oil from Southeast Asia still is unloaded and refined in Singapore, then shipped to East Asian destinations. Raw rubber from the adjacent peninsula and from Indonesia's nearby island of Sumatera is shipped to China, Japan, the United States, and many other countries. Timber from Malaysia, rice, spices, and other foodstuffs are processed and forwarded via Singapore. In return, automobiles, machinery, and equipment are imported into Southeast Asia and distributed almost exclusively through Singapore.

Singapore's current development strategy stresses two primary objectives. The first is to focus its industries more tightly on cutting-edge information technology, automation, and biotechnology. The second is the forging of an ever stronger **Growth Triangle [6]** involving Singapore's developing neighbors, Malaysia and Indonesia; accordingly, the latter two countries would supply the necessary raw materials and relatively inexpensive labor, and Singapore would supply the capital and technical expertise.

The ethnic composition of Singapore's population today is 74 percent Chinese, 13 percent Malay, and 9 percent South Asian. The government is Chinese-dominated, and its policies have served to sustain ethnic-Chinese control. Indeed, the Singapore government has encouraged immigration from the PRC to stabilize the ethnic mosaic of the city-state, where natural-increase rates, especially among its native Chinese, have for some time been well below the replacement level. However, as the *Voice From the Region* box illustrates, ethnic as well as racial harmony in Singapore is still a work in progress.

The skyline of Jakarta's CBD as seen from atop the towering Monas (National Monument) that rises at the center of the Indonesian capital's main ceremonial space, Merdeka (Freedom) Square. Air quality on this day is as good as it gets here in this megacity usually shrouded in smog, produced by local factories and motor vehicles, and intensified by the region's frequent large-scale forest fires and eruptions of volcanic ash.

Indonesia

The fourth-largest country in the world in terms of human numbers is also the globe's most expansive **archipelago [7]**. Spread across a chain of more than 17,000 mostly volcanic islands, Indonesia's 247 million people live both separated and clustered—separated by water and clustered on islands large and small.

The complicated map of Indonesia requires some close study (Fig. 10B-7). Five large islands dominate the archipelago territorially, but one of these, easternmost New Guinea, is not part of the Indonesian culture sphere, although its western half is under Indonesian control. The other four major islands are collectively known as the Greater Sunda Islands: ***Jawa*** (Java),* the smallest but by far the most heavily populated and important; ***Sumatera*** (Sumatra) in the west, directly across the Strait of Malacca from Malaysia; ***Kalimantan***, the southern, Indonesian sector of sizeable, compact, mini-continent Borneo; and wishbone-shaped ***Sulawesi*** (Celebes) to the east. Extending eastward from Jawa are the Lesser Sunda Islands, including Bali and, towards the eastern end, Timor. Another important island chain that lies within Indonesia is the Maluku (Molucca) Islands, between Sulawesi and New Guinea. The central water body of Indonesia is the Java Sea.

The Major Islands

Indonesia is a Dutch colonial creation they named the East Indies, and Jawa was chosen as its colonial headquarters with Batavia (now Jakarta) as the capital. Today, **Jawa** remains the core of Indonesia. With almost 150 million inhabitants, Jawa is one of the world's most densely peopled places (see Figs. G-8 and 10A-3) as well as one of the most agriculturally productive, with its terraced rice paddies rising up the highly fertile flanks of dozens of active volcanoes. This combination of extreme population density atop a tectonically active crustal zone has the potential for disaster. As recently as late 2010, a ferocious two-week-long eruption of Mount Merapi near the south-coast city of Yogyakarta killed nearly 200 people and disrupted the lives of thousands.

Jawa also is the most highly urbanized part of a country in which 57 percent of the people still live off the land. The primate city of Jakarta has today become the heart of the much larger Greater Jakarta conurbation that also encompasses the large cities of Bogor, Depok, Bekasi, and Tangerang/South Tangerang. Since 1990, the population of this megalopolis has nearly doubled from less than 15 to more than 28 million today—and is already home to 11 percent of Indonesia's entire population as well as 27 percent of its urban population. Thousands of factories, their owners taking advantage of low prevailing wages, have been

*As in Africa and South Asia, names and spellings have changed with independence. In this chapter we use contemporary spellings, except when we refer to the colonial era. Thus Indonesia's four major islands are Jawa, Sumatera, Kalimantan (the Indonesian portion of Borneo), and Sulawesi. Prior to independence they were respectively known as Java, Sumatra, Dutch Borneo, and Celebes.

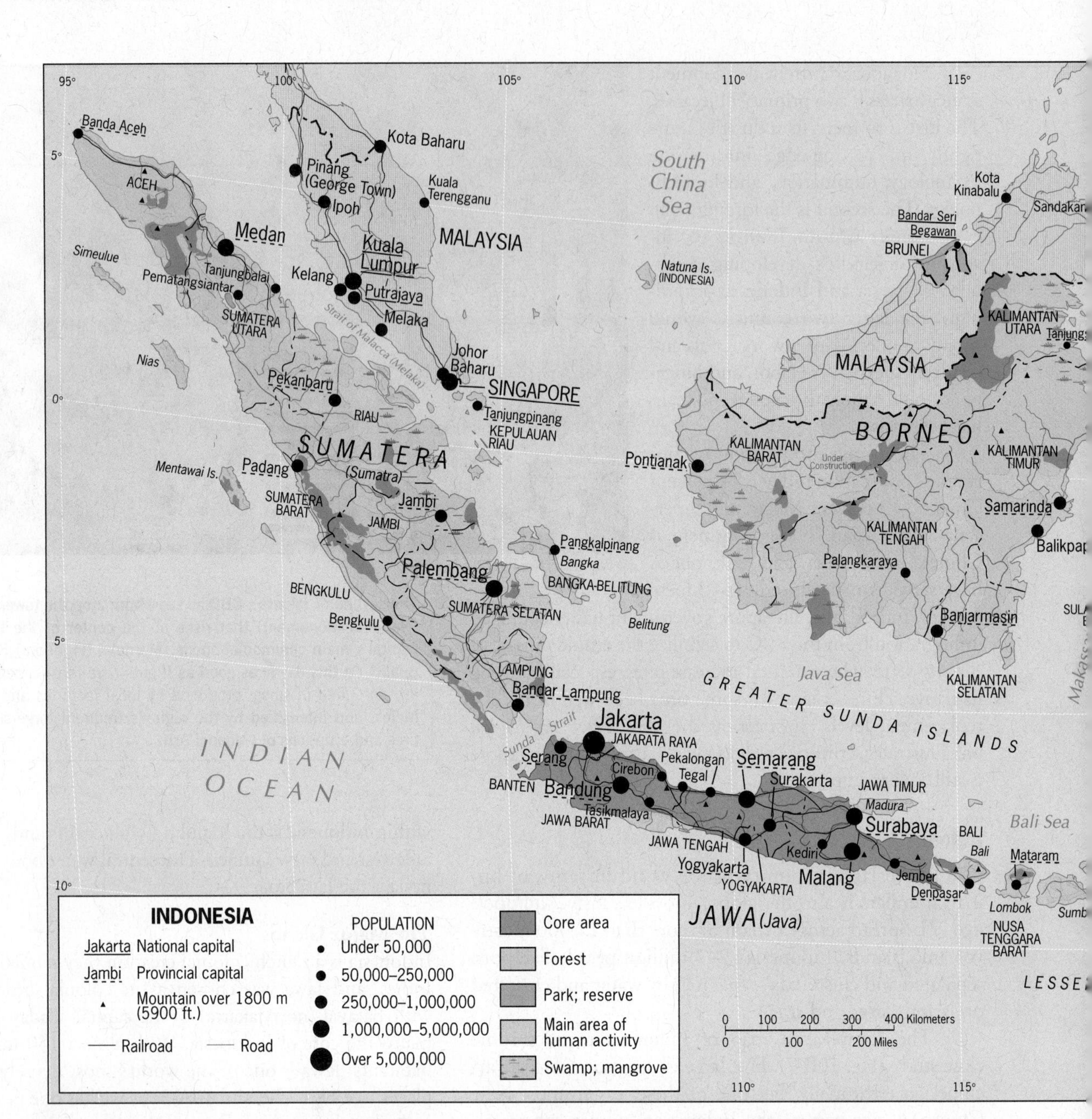

FIGURE 10B-7

built in this sprawling conurbation, badly straining its infrastructure and overburdening the port of Jakarta.

As has always been the case in Indonesia, Jawa is where the power lies. As a cultural group (though itself heterogeneous), the ethnic Jawanese constitute about 45 percent of Indonesia's population. And with Indonesia containing the biggest Muslim population of any country, Jawa also is the main base for two broad national Islamic movements; one is comparatively moderate, but the other is increasingly fundamentalist and militant (its most radical group being *Jemaah Islamiyah (JI)*, perpetrator of several deadly terrorist bombings).

Sumatera, Indonesia's westernmost island, forms the western shore of the busy Strait of Malacca; Singapore lies across the Strait from approximately the middle of the island. Although much larger than Jawa, Sumatera has only about one-third as many people (almost 50 million). In colonial times the island became a base for rubber and palm oil production; its high relief makes possible the cultivation of a wide variety of crops, and neighboring Bangka and Belitung yield petroleum and natural gas. Palembang is the key urban center in the south. In recent years, the northeastern coastlands of Sumatera, along with parts of mainland Malaysia and southern Kalimantan, have been blanketed by large, expanding palm oil plantations that drive up export revenues (but at huge cost to the environment, as discussed in the box in Chapter 10A).

In the island's far north, the Acehnese fought the Dutch well into the twentieth century and, after Indonesia became a sovereign state, demanded autonomy and even outright independence (yet all they gained was recognition as a "Special Territory" rather than a province). Rebels fought the Indonesian military to a costly stalemate here, and thousands died in the conflict—which would probably still be going on but for a dramatic turn of events. The seafloor epicenter of the great 2004 Indian Ocean tsunami lay near the far northern coast of Sumatera, and Aceh was directly in the path of the most powerful ocean waves. Entire towns and roads were swept away by tsunamis, and tens of thousands died; Banda Aceh, the Special Territory's capital city, was devastated. The international relief effort opened Aceh to foreigners in ways the Indonesians had long prevented, and the Indonesian army as well as the rebels were engaged in rescue missions rather than warfare. This combination of circumstances facilitated a truce and a negotiated peace agreement under which the rebels agreed to drop their demand for independence and the Indonesian military withdrew.

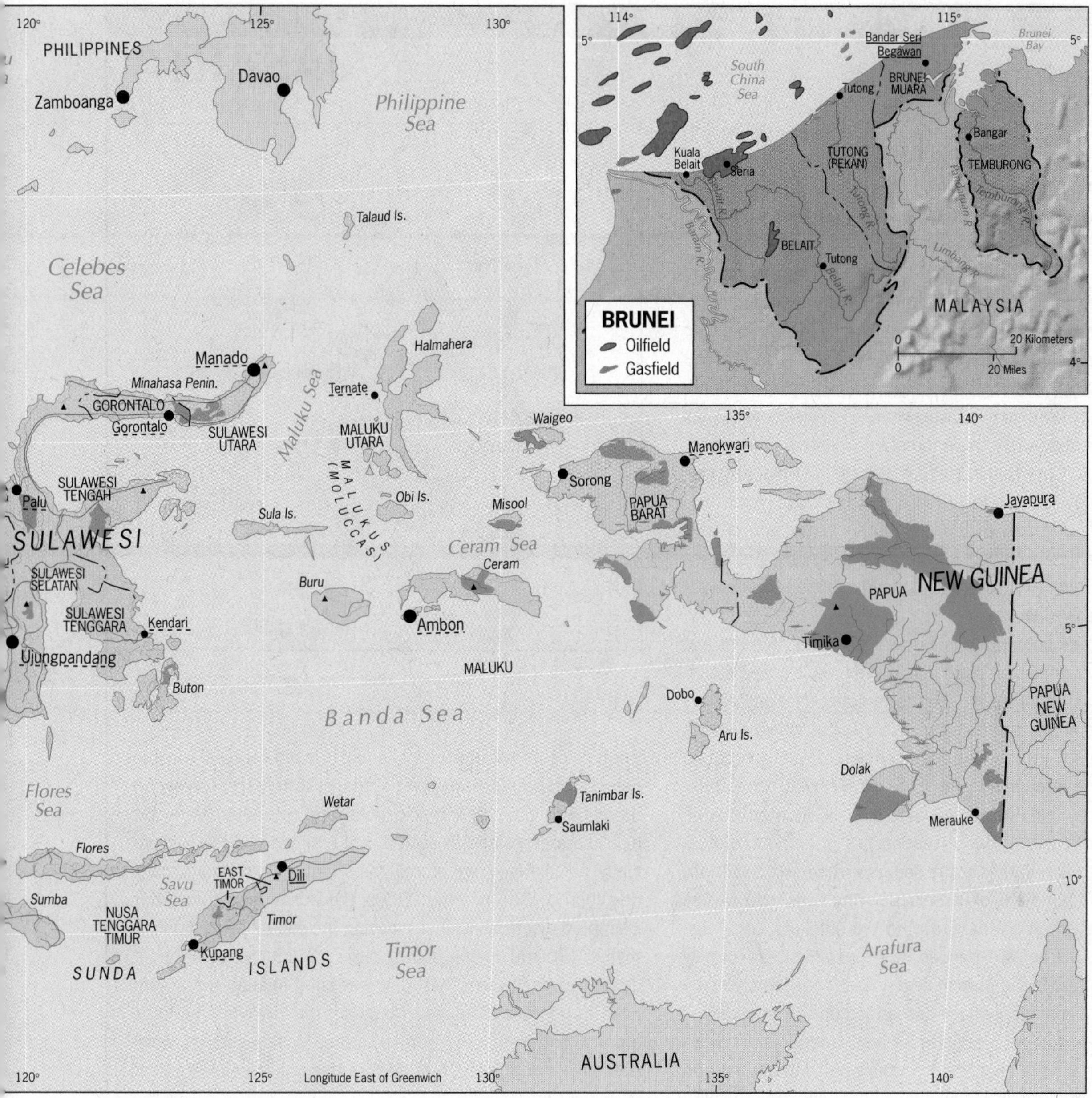

Kalimantan is the Indonesian sector of the island of Borneo, a slab of the Earth's crystalline crust whose backbone of tall mountains is of erosional, not volcanic, origin. Larger than Texas, Borneo has a deep, densely rainforested interior that is a last refuge for some 35,000 orangutans as well as dwindling numbers of Asian elephants, rhinoceroses, and tigers. Along with a number of indigenous peoples, these wildlife species survive even as loggers and farmers relentlessly penetrate their shrinking habitat. Borneo's Pleistocene heritage sustains a comparatively small human population (about 15 million on the Indonesian side, just 6 percent of the country's population) on poor tropical soils. Indigenous

AMONG THE REALM'S GREAT CITIES . . .

JAKARTA

JAKARTA, CAPITAL OF Indonesia and the realm's second-largest city, sometimes is called the Kolkata (Calcutta) of Southeast Asia. Stand on the elevated highway linking the port to the city center and see the villages built on top of garbage dumps by scavengers using what they can find in the refuse, and the metaphor seems to fit. There is poverty here unlike that in any other Southeast Asian metropolis.

But there are other sides to Jakarta, which has just attained megacity status with its 2014 population of 10.3 million. Indonesia's economic progress has made its mark here, and the evidence is everywhere. Television antennas and satellite dishes rise like a forest from rusted, corrugated-iron rooftops. Cars (almost all, it seems, late-model), mopeds, and bicycles clog the streets, day and night. A meticulously manicured part of the city center contains a cluster of high-rise hotels, office buildings, and luxury-apartment complexes. Billboards advertise planned communities on well-located, freshly cleared land.

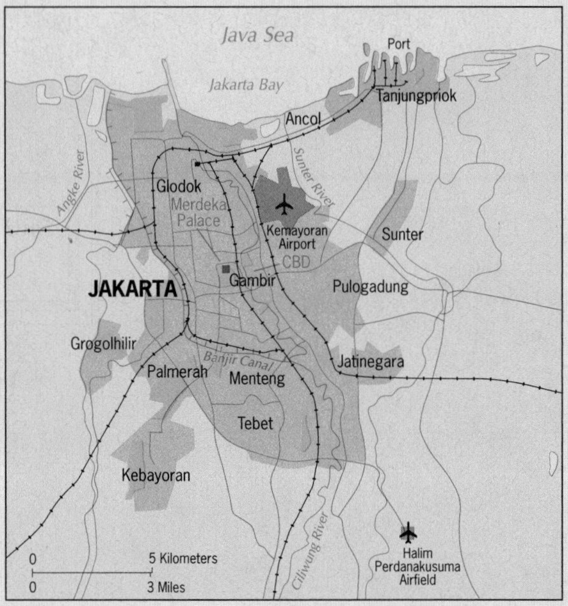

Jakarta's population is a cross-section of Indonesia's, and the silver domes of Islam rise above the cityscape alongside Christian churches and Hindu temples. The city always was cosmopolitan, beginning as a cluster of villages at the mouth of the Ciliwung River under Islamic rule, becoming a Portuguese stronghold and later, as Batavia, the capital of the Dutch East Indies. Advantageously situated on the northwestern coast of Jawa, Indonesia's most populous island, Jakarta is bursting at the seams with growth. Sail into the port, and hundreds of vessels, carrying flags from Russia to Argentina, await berths. Travel to the outskirts, and huge shantytowns are being expanded by a constant stream of new arrivals. So vast is the human agglomeration—nobody really knows how many people have descended on this megacity—that most live without adequate (or any) amenities.

Jakarta is the biggest city in the world without a metro transit system, but one is now finally under construction with the first of its two routes expected to open in 2017. In the meantime, public transport continues to rely exclusively on buses, and the city's traffic jams are notorious (transport delays are estimated to cost U.S. $3 billion per year), especially when heavy equatorial rains cause flooding in many neighborhoods. Since the 2000s, the worsening gridlock has prompted discussions about moving the capital some 50 kilometers (30 mi) to the south; but in 2013 the government distanced itself from that idea, stressing instead the urgent need to improve drainage throughout the city while building a much better highway infrastructure. Most Jakartans, however, are unimpressed, and none of this talk has made a dent in alleviating their frustrations.

peoples, principally the Dayak clans, have traditionally had less impact on the natural environment than the Indonesian and Chinese immigrants as well as the multinational corporations that log the forests and clear woodland for farms. As Figure 10B-7 shows, the only towns of any size in Kalimantan lie on or near the coast; routes into the interior are still few and far between.

Sulawesi consists of a set of intersecting, volcanic mountain ranges rising above sea level; the 800-kilometer (500-mi) Minahasa Peninsula, propelled by volcanic activity, continues to build itself into the Celebes Sea. This northern peninsula, a favorite of the Dutch colonizers, remains the most developed part of an otherwise rugged and remote island, with Manado its relatively prosperous urban focus. Seven major ethnic groups inhabit the valleys and basins between the mountains, but the population of about 18 million also includes many immigrants from Jawa, especially in and around the southern center of Ujungpandang. Subsistence farming is the leading mode of life, although logging, some mining, and fishing augment the economy. Clashes between Muslims and Christians occur intermittently in remote areas.

Papua, the Indonesian name for the western part of the island of New Guinea, has become an issue in Indonesian politics. Bordered on the east by a classic superimposed geometric boundary (Figs. 10B-7; 10A-9, lower-left map), it was taken over by Indonesia from the Dutch in 1969. Papua contains about 22 percent of Indonesia's territory, but

From the Field Notes . . .

"I drove from Manado on the Minahasa Peninsula in northeastern Sulawesi to see the ecological crisis at Lake Tondano, where a fast-growing water hyacinth is clogging the water and endangering the local fishing industry. On the way, in the town of Tomolon, I noticed this side street lined with prefabricated stilt houses in various stages of completion. These, I was told, were not primarily for local sale. They were assembled from wood taken from the forests of Sulawesi's northern peninsula, then taken apart again and shipped from Manado to Japan. 'It's a very profitable business for us,' the foreman told me. 'The wood is nearby, the labor is cheap, and the market in Japan is insatiable. We sell as many as we can build, and we haven't even begun to try marketing these houses in Taiwan or China.' At least, I thought, this wood was being converted into a finished product, unlike the mounds of logs and planks I had seen piled up in the ports of Borneo awaiting shipment to East Asia."

www.conceptcaching.com

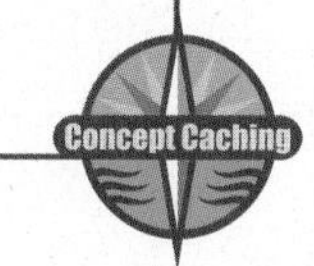

its (fast-growing) population is only 3.8 million—just 1.5 percent of the national total. The indigenous inhabitants of this territory, which is in effect a colony, are Papuan, most living in the remote reaches of this mountainous and densely forested island. Papua is economically important to Indonesia, for it contains what is reputed to be the world's richest gold mine and its third-largest open-pit copper mine. But political consciousness has now reached the Papuans: the Free Papua Movement has become increasingly active, holding small demonstrations in the capital (Jayapura), displaying a Papuan flag, and demanding recognition.

Diversity in Unity

Indonesia's survival as a unified state is as remarkable as India's or Nigeria's. With more than 300 discrete ethnic clusters, over 250 languages, and just about every religion practiced on Earth (although Islam dominates), actual and potential centrifugal forces are quite powerful here. Wide waters and high mountains perpetuate both cultural distinctions and differences. Indeed, Indonesia's national motto is *bhinneka tunggal ika*: diversity in unity.

It is not just the numerous Jawanese who have their own ethnic identity; there are also the Sundanese (who constitute roughly 16 percent of Indonesia's population), the Madurese (around 3.5 percent), and others. Perhaps the best impression of this cultural diversity comes from the string of islands that extends eastward from Jawa to Timor, the Lesser Sunda Islands (Fig. 10B-7). The rice growers of Bali adhere to a modified version of Hinduism, giving the island a unique cultural atmosphere; the population of Lombok is mainly Muslim, with some Balinese Hinduism; next comes Sumbawa, a Muslim community; Flores is mostly Roman Catholic. In western Timor, Protestant groups dominate; in independent East Timor, where the Portuguese ruled, Roman Catholicism still prevails.

Nevertheless, Indonesia nominally is the largest Muslim country in the world: overall, 88 percent of the people adhere to Islam, and in the cities the silver domes of neighborhood mosques rise densely above the townscape. Although until recently Indonesian Islam has been relatively moderate, as noted earlier more overt Islamization has been on the rise, with new laws banning public displays of affection and limiting the kinds of clothing women may wear in public.

Transmigration and the Outer Islands

Jawa contains no less than 58 percent of the total population of nearly 250 million on barely 7 percent of Indonesia's territory; with close to 150 million people on an island the size of Louisiana, population pressure here is staggering. Since colonial times, Jawa leaned toward overpopulation while many of the outer islands were very sparsely inhabited. From 1974 to 2001, the Indonesian government embarked on a policy known as **transmigration [8]** (*transmigrasi* in the Bahasa Indonesian language), inducing many from the densely populated inner islands (especially Jawa, Bali, and Madura) to relocate to such sparsely inhabited outer islands as Kalimantan and Sulawesi. During the last quarter of the twentieth century, millions moved to these peripheral locales as part of this government-sponsored program.

As many as 8 million Jawanese were relocated to other islands, but surveys showed that half of these migrants experienced a decline in their standard of living; many were reduced to bare subsistence on tropical-forest land confiscated from its indigenous inhabitants that turned out to be unsuitable for the farming methods used by the settlers. Cultural conflict, ecological havoc, and rampant deforestation finally led the Indonesian government to cancel the program in 2001, but the damage it did will long outlive this ill-fated initiative.

Today, Indonesia's stature within Southeast Asia is on the rise and some observers argue it is about ready to join the ranks of the BRIC countries (Brazil, Russia, India, China). That may be a stretch, especially since Indonesia's growth has been disproportionately dependent on the global commodity boom of recent years (e.g., coal, natural gas, palm oil, rubber) and the economy is not sufficiently diversified. But this is already by far the biggest country in the realm and it is now making steady progress toward achieving a commensurate level of economic development and prosperity. But centrifugal forces are intensifying as well (growing Islamic militancy in Jawa is of particular concern), and ongoing decentralization and devolution throughout this immense archipelago will continue to challenge the central government.

East Timor

As Figure 10B-8 shows, Timor is the easternmost of the sizeable Lesser Sunda Islands, and throughout the Dutch colonial period the Portuguese maintained a colony on the eastern half of it. In 1975, Indonesian forces overran that colony and annexed it formally in 1976. Indonesian rule, however, was even less benign than the Portuguese had been, and soon

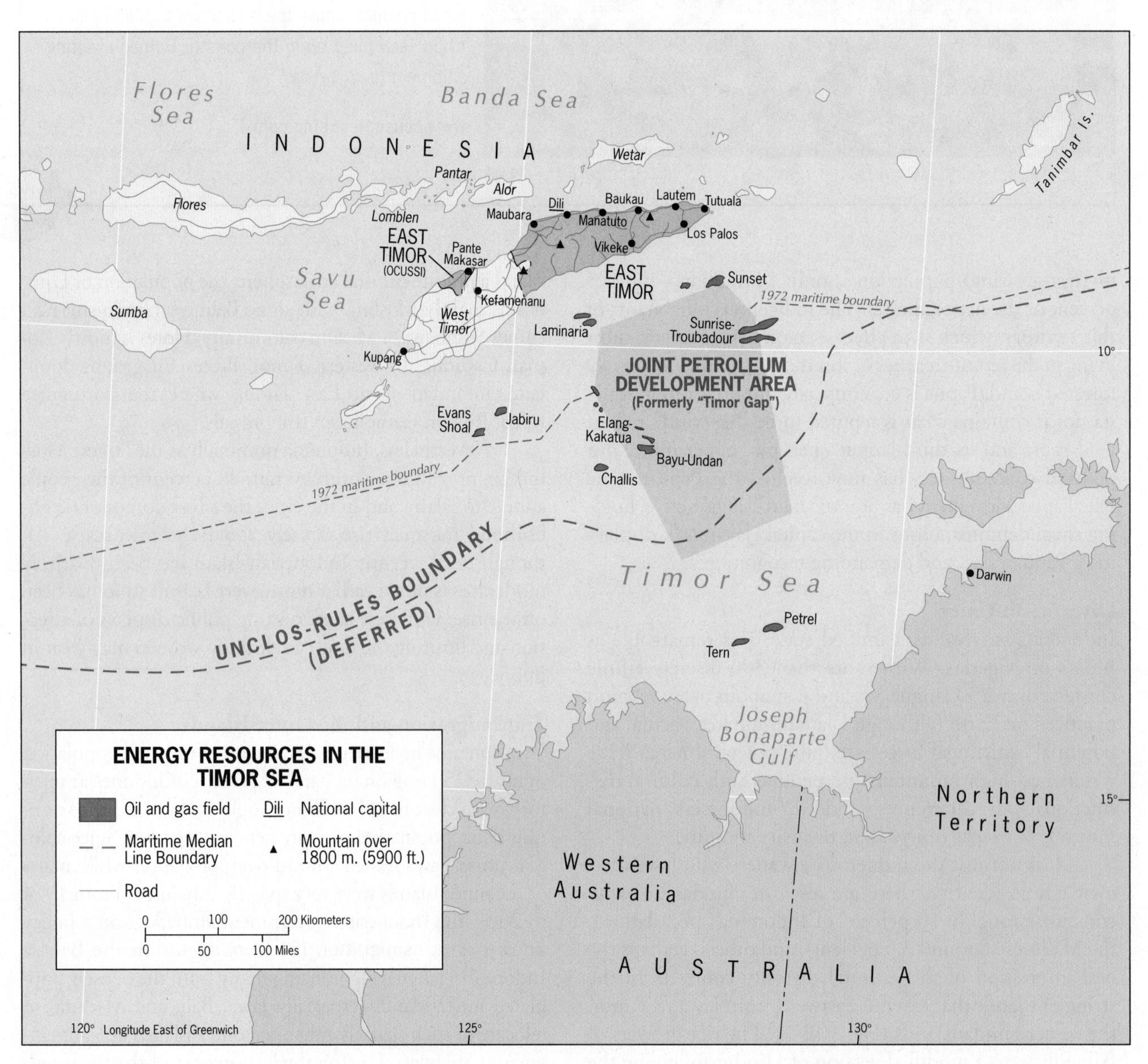

FIGURE 10B-8

a bitter struggle for independence was under way. Eventually, under UN supervision, the people of East Timor voted overwhelmingly for independence in 1999; but Indonesia refused to let go, and unleashed a brutal military occupation. It took another three years of armed conflict and foreign intervention (led by Australia, which had supported Indonesia before 1999) until independence was finally realized in 2002. By then, the violence had devastated much of this mini-state and its infrastructure.

East Timor's leaders proclaimed the official name of their country to be ***Timor-Leste***, the Portuguese version of "Timor East." Nation-building in this Connecticut-sized country continues to be a formidable proposition. And Figure 10B-8 helps explain why Australia first supported Indonesia in its effort to absorb East Timor: by doing so, Australia would benefit from Indonesia's compliance in the delimitation of maritime boundaries in the Timor Sea, giving Australia a larger share of the known oil and gas reserves than an independent East Timor would be likely to relinquish. In the end, difficult negotiations between powerful and technologically capable Australia and weak and incapable East Timor produced a settlement that gave East Timor an acceptable share of future revenues—but not before Australia's international reputation for fairness was further damaged. On the map, the red maritime boundary is the one delimited under United Nations regulations; the blue boundary represents the bilateral agreement between Indonesia and Australia. At issue is the "gap" between them, most of it on East Timor's side but now a "joint petroleum development area."

As Figure 10B-8 shows, East Timor is a fragmented country, with a dominant east (where the coastal capital, Dili, is located) and a tiny **exclave [9]** on the northern coast of Indonesian West Timor named Ocussi (sometimes mapped as "Ocussi-Ambeno Province"). Even though East Timor is an overwhelmingly agricultural country, farming has been neglected during the turmoil of recent decades. To make matters worse, its population of 1.2 million continues to grow explosively, exhibiting a yearly natural-increase rate of 2.6 percent (30 percent higher than that of Laos, the realm's second-fastest-growing country); not surprisingly, East Timor's fertility rate of 5.7 children per woman is one of the world's highest. Despite encouraging annual economic growth rates since the late 2000s—and the prospect of billions of dollars of future oil revenues—the nation-building project here has barely begun to deal with the country's severe poverty as well as the lingering devastation of a prolonged conflict in which some 100,000 civilians were killed by the Indonesian military.

The Philippines

North of Indonesia, across the South China Sea from Vietnam, and south of Taiwan lies a lengthy archipelago of more than 7000 islands (less than 500 of them larger than 1 square kilometer [0.4 sq mi] in area) inhabited by 100 million people. These islands of the Philippines can be viewed as three groups: (1) Luzon, largest of all, and Mindoro in the north; (2) the Visayan group in the center; and (3) Mindanao, second-largest, located in the south (Fig. 10B-9). Southwest from Mindanao lies a small group of islands, the Sulu Archipelago, nearest to Malaysian Borneo, where Muslim-based insurgencies have kept this area in turmoil.

Few of the generalizations we have been able to make for Southeast Asia could apply to the Philippines without qualification. The country's location relative to the mainstream of change in this part of the world has had much to do with this situation. The islands, inhabited by peoples of Malay ancestry with Indonesian strains, shared with much of the rest of Southeast Asia an early period of Hindu cultural influence, which was strongest in the south and southwest and diminished northward. Next came the Chinese invasion, felt more strongly on the largest island of Luzon in the northern part of the Philippine archipelago. Islam's arrival was delayed somewhat by the position of the Philippines well to the east of the mainland and to the north of the Indonesian islands. The few southern Muslim beachheads were soon overwhelmed by the Spanish invasion during the sixteenth century. Today the Philippines, adjacent to the world's largest Muslim state (Indonesia), is 81 percent Roman Catholic, 9 percent Protestant, and only 5 percent Muslim.

Out of the Philippines melting pot, where Malay, Arab, Chinese, Japanese, Spanish, and American elements have met and mixed, has emerged the distinctive Filipino culture. It is not a homogeneous or unified culture, as is reflected by the dozens of Malay languages in use in the islands, but it is in many ways unique. At independence in 1946, the largest of the Malay languages, Tagalog (also called Pilipino), became the country's official language. But English is widely learned as a second language, and a Tagalog-English hybrid, "Taglish," is increasingly heard today. The Chinese component of the population is small (barely 2 percent) but dominant in local business.

The Philippines' small Muslim population, concentrated in the southernmost flank of the archipelago, and especially on densely forested Basilan Island (Fig. 10B-9), has long decried its marginalization in this predominantly Christian country. Over the past generation, a half-dozen Muslim organizations have promoted the Muslim cause through tactics ranging from peaceful negotiation with the government to violent insurgency.

The Philippines' population, concentrated where the good farmlands lie, is densest in three general areas: (1) the northwestern and south-central part of Luzon; (2) the southeastern extension of Luzon; and (3) the islands of the Visayan Sea between Luzon and Mindanao (see Fig. 10A-3). Luzon is the site of the capital, Manila-Quezon City (12.7 million—one-eighth of the entire population of the country), a sprawling megacity facing the South China Sea.

Prospects

The Philippines seems to get little mention in discussions of Asia-Pacific developments, even though it would seem to be well positioned to share in the realm's economic growth. Agriculture continues to dominate the Philippines'

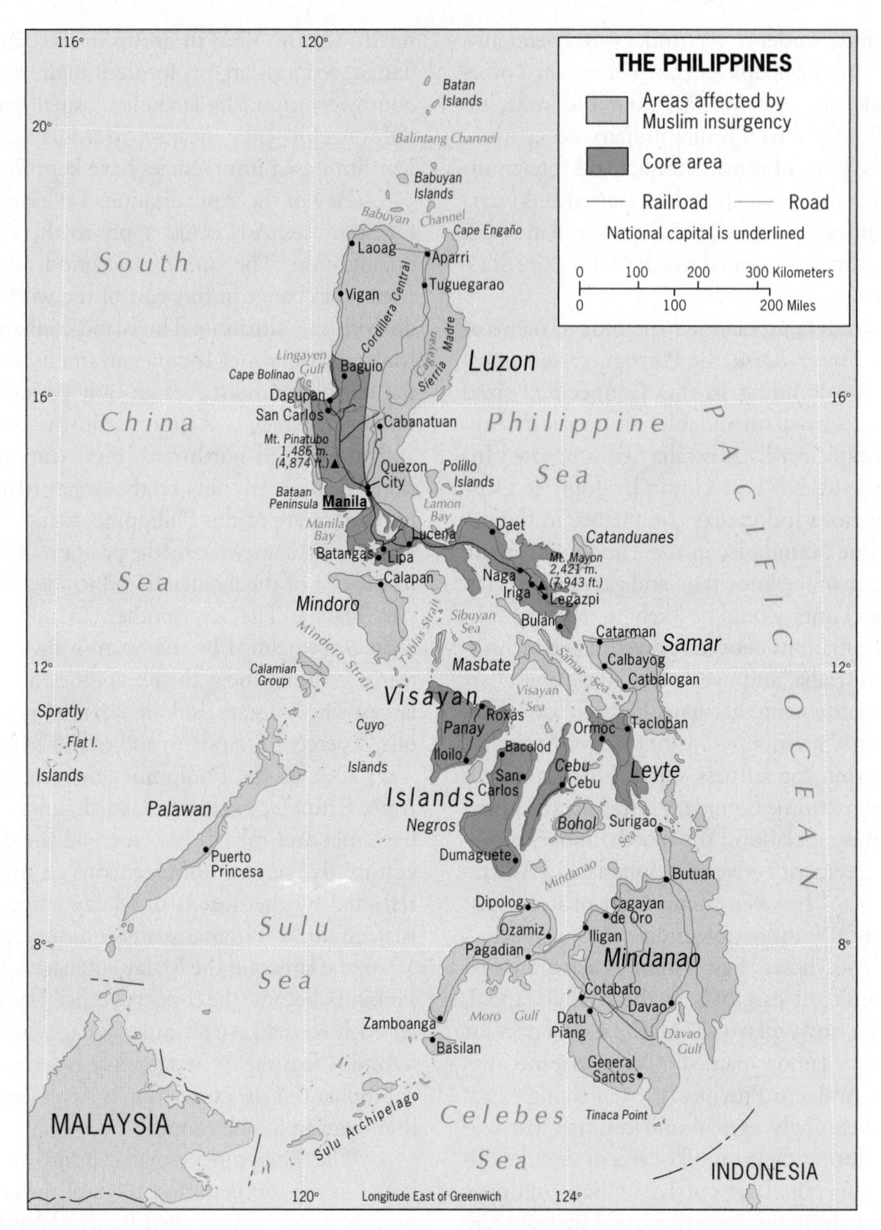

FIGURE 10B-9

economy. Alluvial as well as volcanic soils, together with ample moisture in this tropical environment, enable self-sufficiency in rice and other staples and make the Philippines a net exporter of farm products despite a high population growth rate of 1.9 percent. In the industrial sphere, the country has participated in offshore manufacturing in its numerous Export Processing Zones, which are similar to the Mexican maquiladoras discussed in Chapters 4A and 4B; electronics and textile production in particular are expanding continuously (mostly in metropolitan Manila), propelled by growing foreign investment.

Nonetheless, unemployment remains high, trade linkages are insufficient, further land reform is badly needed, and social restructuring (reducing the controlling influence of a comparatively small group of families over national affairs) must proceed. Perhaps more than any other people, Filipinos take jobs in foreign countries in massive numbers, proving their capacity to succeed in jobs they cannot find at home. More than 10 million Filipinos, about a quarter of the total workforce, are now employed abroad. The global merchant marine would not exist without Filipino sailors, and Filipina nurses and domestic workers can be found from Dubai to Dublin to Dubuque. Funds sent home to family members by the emigrants make the Philippines a world leader in monetary inflow known as **remittances [10]**, constituting about 10 percent of the country's GDP.

Overall, the Philippines now sustains a lower-middle-income economy, and given a longer period of political stability and success in mitigating the problems outlined above, it should be able to rise to the next level and finally take its place among the Pacific Rim's growth poles. Currently, the Philippines is pursuing a pair of niches in the

AMONG THE REALM'S GREAT CITIES . . .

MANILA

MANILA, CAPITAL OF the Philippines and the realm's largest city, was founded by the Spanish invaders of Luzon more than four centuries ago. The colonial conquerors made a good choice in terms of site and situation. Manila sprawls at the mouth of the Pasig River where it enters one of Asia's finest natural harbors. To the north, east, and south a crescent of mountains encircles the city, which lies just 1000 kilometers (600 mi) southeast of China's Hong Kong.

Manila, named after a flowering shrub in the local marshlands, is bisected by the Pasig, which is bridged at numerous points. The old walled city, Intramuros, lies to the south. Despite heavy bombardment during World War II, some of the colonial heritage survives in the form of churches, monasteries, and convents. St. Augustine Church, completed in 1599, is one of the city's landmarks.

The CBD of Manila lies on the north side of the river. Although Manila has a well-defined commercial center with several avenues of luxury shops and modern buildings, the skyline does not reflect the high level of energy and activity common in Pacific Rim cities on the opposite side of the South China Sea. Neither is Manila a city of notable architectural achievements. Wide, long, and straight avenues flanked by palm, banyan, and acacia trees give it a look similar to San Juan, Puerto Rico.

In 1948, a newly built city immediately to the northeast of Manila was inaugurated as the *de jure* capital of the Philippines and named Quezon City. The new facilities were eventually to house all government offices, but many functional components of the national government never made the move. In the meantime, Manila's growth overtook Quezon's, so that it became part of the Greater Manila metropolis (which is now home to 12.7 million). Although the proclamation of Quezon City as the Philippines' official capital was never rescinded, Manila remains the *de facto* capital today.

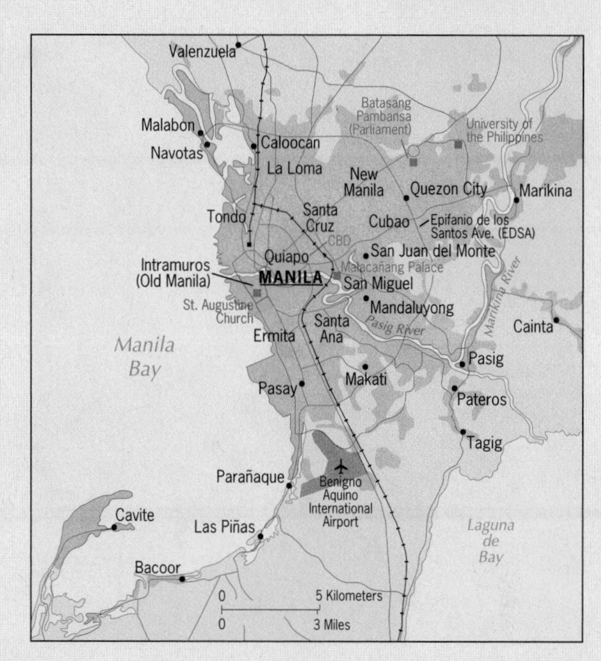

global economy: international call centers and the outsourcing of digital services through such Internet websites as *oDesk*—one of several that serves as a global marketplace for digital services by freelancers performing data entry and other basic back-office work. However, the latter is highly unpredictable as a growth industry, and the pay is certainly low by worldwide standards.

In its foreign relations, this country is changing course. In the 1990s it terminated the lease for a U.S. naval base, but now the Philippines government is once again seeking closer relations with the United States. This is in no small part driven by the growing assertiveness of China in the South China Sea and the persistent disputes between China and the Philippines over the Spratly Islands and the Scarborough Shoal (discussed in Chapter 10A). In response, American and Filipino forces now annually conduct joint naval exercises in the South China Sea, which are widely interpreted as a show of solidarity in contested waters that have long attracted Chinese attention and (still-ambiguous) claims.

POINTS TO PONDER

- China's growing regional dominance affects not only lands and peoples, but also waters and unpopulated islands. What do Vietnam and the Philippines have in common in this respect?
- Reforming Myanmar is witnessing the eruption of long-suppressed religious strife as Buddhists and Muslims clash. How are other states in the realm dealing with religious tensions?
- Under Malaysia's constitution, a person of Malay ancestry (unlike someone with Chinese or South Asian roots) is automatically registered as a Muslim.
- Singapore in 2013 experienced the worst air pollution in the city-state's history. The cause lay in Indonesia. What does the growth of palm oil cultivation have to do with this?

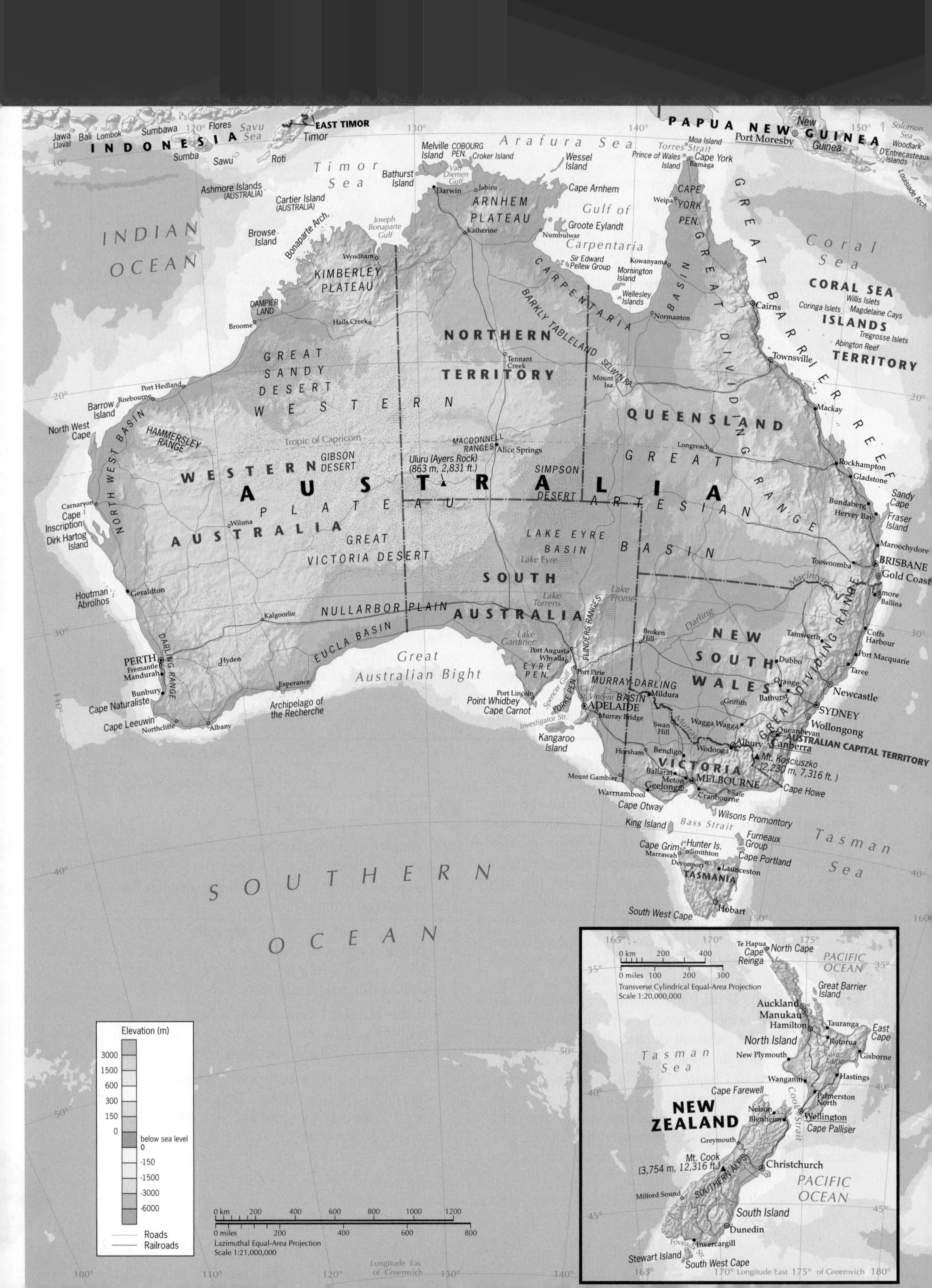

INDIAN OCEAN
SOUTHERN OCEAN
AUSTRALIA
WESTERN AUSTRALIA
NORTHERN TERRITORY
QUEENSLAND
SOUTH AUSTRALIA
NEW SOUTH WALES
VICTORIA
TASMANIA
AUSTRALIAN CAPITAL TERRITORY
CORAL SEA ISLANDS TERRITORY
PAPUA NEW GUINEA
INDONESIA
EAST TIMOR
Timor Sea
Arafura Sea
Coral Sea
Tasman Sea
Great Australian Bight
Gulf of Carpentaria
Bass Strait
Torres Strait
PERTH
ADELAIDE
MELBOURNE
SYDNEY
BRISBANE
Canberra
Darwin
Hobart
Alice Springs
Uluru (Ayers Rock) (863 m, 2,831 ft.)
Mt. Kosciuszko (2,230 m, 7,316 ft.)
GREAT SANDY DESERT
GIBSON DESERT
GREAT VICTORIA DESERT
SIMPSON DESERT
NULLARBOR PLAIN
GREAT ARTESIAN BASIN
LAKE EYRE BASIN
MURRAY-DARLING BASIN
GREAT DIVIDING RANGE
GREAT BARRIER REEF
Tropic of Capricorn
NEW ZEALAND
North Island
South Island
Auckland
Wellington
Christchurch
Dunedin
Mt. Cook (3,754 m, 12,316 ft.)
PACIFIC OCEAN
Transverse Cylindrical Equal-Area Projection
Scale 1:20,000,000
Elevation (m)
3000
1500
600
300
150
0
below sea level
0
-150
-1500
-3000
-6000
Roads
Railroads
Lazimuthal Equal-Area Projection
Scale 1:21,000,000
Longitude East 175° of Greenwich 180°

DEFINING

REGIONS

Australia

New Zealand

IN THIS CHAPTER

- Australia's amazing biogeography
- Australia's Asian turn
- China covets Australian commodities
- Aboriginal claims to land and resources
- Foreign policy dilemmas Down Under
- New Zealand's matchless physiography

CONCEPTS, IDEAS, AND TERMS

Austral	1
Southern Ocean	2
Subtropical Convergence	3
West Wind Drift	4
Biogeography	5
Wallace's Line	6
Aboriginal population	7
Federation	8
Unitary state	9
Outback	10
Import-substitution industries	11
Primary sector	12
Aboriginal land issue	13
Environmental degradation	14
Peripheral development	15

FIGURE 11-1

© John Hicks/Corbis

Where were these pictures taken?
Find out at www.wiley.com/college/deblij

The Austral Realm is geographically unique (Fig. 11-1). It is the only realm that lies entirely in the Southern Hemisphere. It is also the only one that has no land link of any kind to a neighboring realm and is thus completely surrounded by ocean and sea. It is second only to the Pacific as the world's least populous realm. Appropriately, its name refers to its location (the word **austral [1]** comes from the Latin for "south")—a location far from the sources of its dominant cultural heritage, but close to its newfound economic partners on the Asian Pacific Rim.

DEFINING THE REALM

Two countries constitute the Austral Realm: Australia, in every way the dominant one, and New Zealand, physiographically more varied but demographically much smaller than its giant partner (Fig. 11-2). Between them lies the Tasman Sea. To the west lies the Indian Ocean, to the east the Pacific, and to the south the frigid Southern Ocean.

This southern realm is at a crossroads. On the doorstep of populous eastern Asia, its Anglo-European legacies are now infused by many other cultural strains. Polynesian Maori in New Zealand and Aboriginal communities in Australia are demanding greater rights and more acknowledgment of their cultural heritage. Pacific Rim markets are buying growing quantities of raw materials. Chinese and other Asian tourists fill hotels and resorts. The streets of Sydney and Melbourne display a multicultural panorama unimagined just two generations ago. All these changes have stirred political debate. Issues ranging from immigration quotas to indigenous land rights dominate, exposing social fault lines (city versus Outback in Australia; North Island versus South Island in New Zealand). Aborigines and Maori first settled

© H. J. de Blij

major geographic qualities of THE AUSTRAL REALM

1. Australia and New Zealand constitute a geographic realm by virtue of territorial dimensions, relative location, and dominant cultural landscape.
2. Despite their inclusion in a single geographic realm, Australia and New Zealand differ physiographically. Australia has a vast, dry, low-relief interior; New Zealand is mountainous and has a temperate climate.
3. Australia and New Zealand are marked by peripheral development—Australia because of its aridity, New Zealand because of its topography.
4. The populations of Australia and New Zealand are not only peripherally distributed but also highly clustered in urban centers.
5. The economic geography of Australia and New Zealand is dominated by the export of livestock products and specialty goods such as wine. Australia also has significant wheat production and a rich, diverse base of mineral resources.
6. Australia and New Zealand are now integrated into the economic framework of the Asian Pacific Rim, principally as suppliers of raw materials. Mineral-rich Australia's newly discovered and exploited gas reserves are adding to its export base.
7. As a result of heightened immigration with respect to the western Pacific Rim, Australia in a cultural sense is "returning" to Asia.

this realm, then the Europeans arrived, and now Asians are an increasingly significant economic and cultural element.

LAND AND ENVIRONMENT

Physiographic contrasts between massive, compact Australia and elongated, fragmented New Zealand are related to their locations with respect to the Earth's tectonic plates (see Fig. G-4). Australia, with some of the geologically most ancient rocks on the planet, lies at the center of its own plate, the Australian Plate. New Zealand, younger and less stable, lies at the convulsive convergence of the Australian and Pacific plates. Earthquakes are rare in Australia, and volcanic eruptions are unknown; New Zealand has plenty of both (the big Christchurch quakes of 2010–2011 have only been the latest reminder). This locational contrast is also reflected by differences in relief (Fig. 11-1). Australia's highest relief occurs in what Australians call the Great Dividing Range, the mountains that line the east coast from the Cape York Peninsula in the north to southern

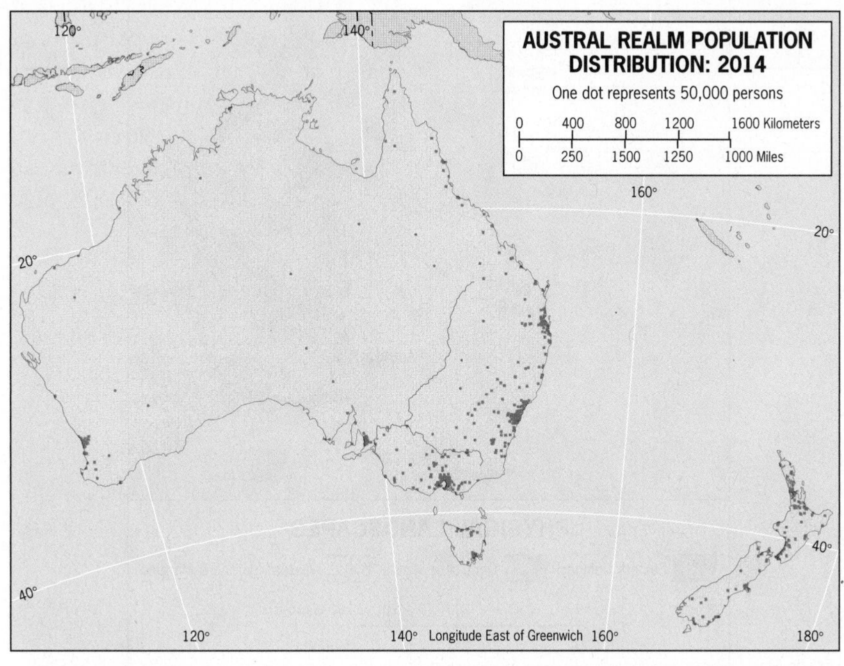

FIGURE 11-2

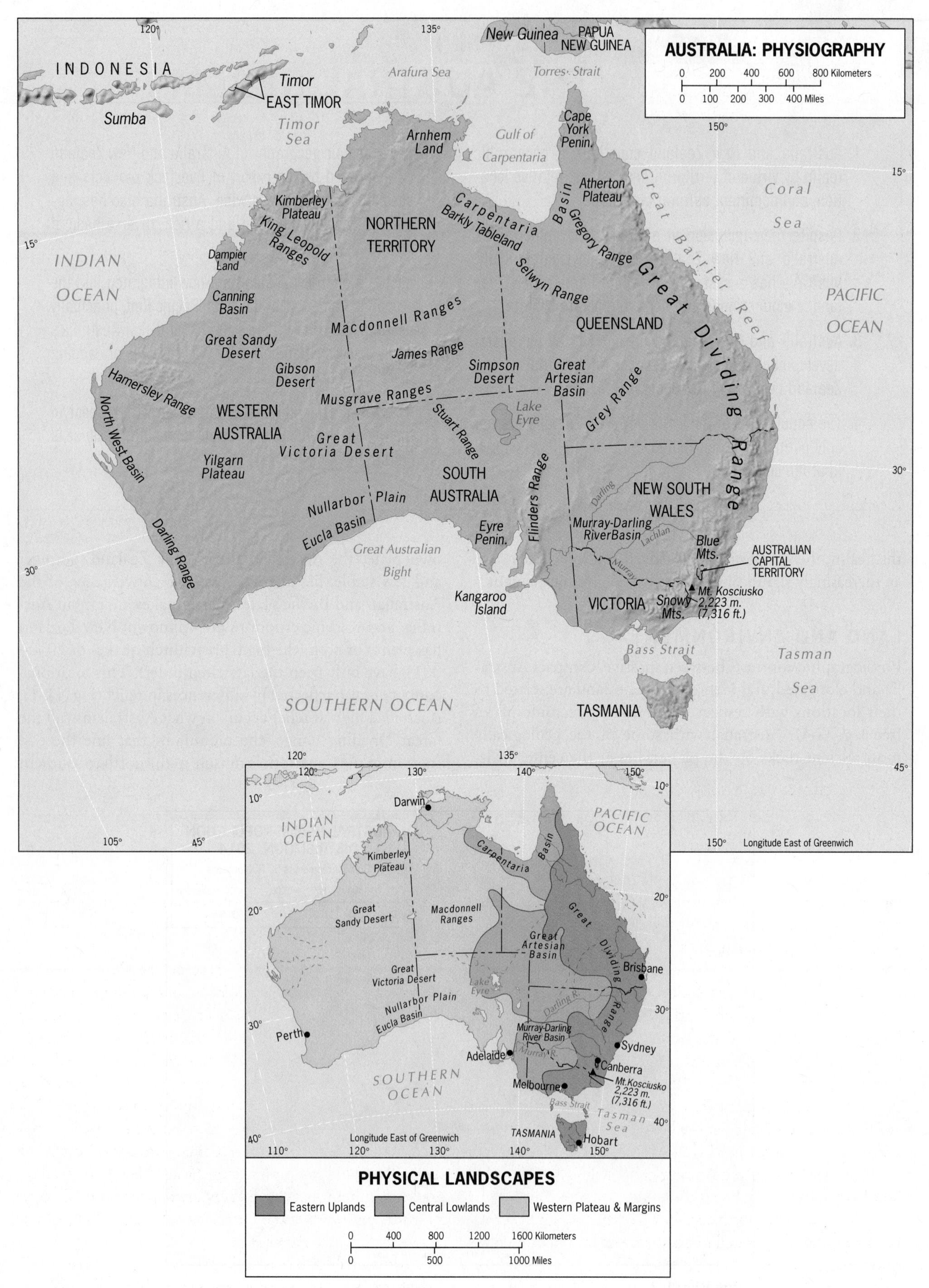
AUSTRALIA: PHYSIOGRAPHY
INDONESIA
Timor
EAST TIMOR
Sumba
New Guinea
PAPUA NEW GUINEA
Arafura Sea
Torres Strait
Timor Sea
Arnhem Land
Gulf of Carpentaria
Cape York Penin.
Kimberley Plateau
King Leopold Ranges
Dampier Land
NORTHERN TERRITORY
Carpentaria Basin
Barkly Tableland
Atherton Plateau
Gregory Range
Coral Sea
Great Barrier Reef
INDIAN OCEAN
Canning Basin
Selwyn Range
Great Dividing Range
PACIFIC OCEAN
Macdonnell Ranges
QUEENSLAND
Great Sandy Desert
James Range
Gibson Desert
Simpson Desert
Great Artesian Basin
Hamersley Range
Musgrave Ranges
WESTERN AUSTRALIA
Lake Eyre
Grey Range
North West Basin
Stuart Range
Great Victoria Desert
Yilgarn Plateau
SOUTH AUSTRALIA
NEW SOUTH WALES
Flinders Range
Darling
Nullarbor Plain
Eucla Basin
Eyre Penin.
Murray-Darling River Basin
Lachlan
Blue Mts.
AUSTRALIAN CAPITAL TERRITORY
Darling Range
Great Australian Bight
Murray
Mt. Kosciusko 2,223 m. (7,316 ft.)
Kangaroo Island
VICTORIA
Snowy Mts.
Bass Strait
Tasman Sea
SOUTHERN OCEAN
TASMANIA
Longitude East of Greenwich
Darwin
Brisbane
Perth
Adelaide
Sydney
Canberra
Melbourne
Hobart
PHYSICAL LANDSCAPES
Eastern Uplands
Central Lowlands
Western Plateau & Margins

FIGURE 11-3

Victoria State, with an outlier in Tasmania. The highest point along these old, now eroding mountains is Mount Kosciusko, 2230 meters (7316 ft) tall. In New Zealand, entire ranges are higher—Mount Cook in the Southern Alps, for example, reaches 3754 meters (12,316 ft). West of Australia's Great Dividing Range, the physical landscape mostly exhibits low relief, with some local exceptions such as the Macdonnell Ranges near the continental center; plateaus and plains dominate (Figs. 11-1; 11-3, top map). The Great Artesian Basin is a key physiographic region, providing underground water sources in what is otherwise desert country. To its south lies the continent's predominant Murray-Darling river system. The area mapped as *Western Plateau and Margins* in the lower map of Figure 11-3 contains much of Australia's mineral wealth.

Climates

Figure G-7 reveals the effects of latitudinal position and interior isolation on Australia's climatology. In this respect, Australia is far more varied than New Zealand, its climates ranging from tropical in the far north, where rainforests flourish, to Mediterranean in two corners of the south. The interior is dominated by desert and steppe conditions, the semiarid steppes providing the grasslands that sustain tens of millions of livestock. Only in the extreme east does Australia possess a zone of humid temperate climate, and here lies most of the country's core area. New Zealand, by contrast, is totally under the influence of the Southern and Pacific oceans, creating moderate, moist conditions, temperate in the north and progressively colder in the south.

The Southern Ocean

Twice now we have referred to the **Southern Ocean [2]**, but try to find this ocean on maps and globes produced by such well-known cartographic organizations as the National Geographic Society. From their maps you would conclude that the Atlantic, Pacific, and Indian oceans reach all the way to the shores of Antarctica. Australians and New Zealanders know better. They experience the frigid waters and persistent winds of this great weather-maker on a daily basis.

For us geographers, it is a good exercise to turn the globe upside down now and then. After all, the usual orientation is quite arbitrary. Modern mapmaking started in the Northern Hemisphere, and the cartographers put their hemisphere on top and the other at the bottom. That is now the norm, and it can distort our view of the world. In bookstores in the Southern Hemisphere, you see upside-down maps showing Australia and Argentina at the top, and Europe and Canada at the bottom. But this matter has a serious side. An inverted view of the globe shows us how vast the ocean encircling Antarctica is (see Fig. 12-4). The Southern Ocean may be remote, but its existence is real.

Where do the northward limits of the Southern Ocean lie? This ocean is bounded not by land but by a marine transition called the **Subtropical Convergence [3]**. Here the very cold, extremely dense waters of the Southern Ocean meet the warmer waters of the Atlantic, Pacific, and Indian oceans. It is quite sharply defined by changes in temperature, chemistry, salinity, and marine fauna. Flying over it, you can actually observe it in the changing colors of the water: the Antarctic side is a deep gray, the northern side a greenish blue.

Although the Subtropical Convergence shifts seasonally, its position does not vary far from latitude 40° South, which also is the approximate northern limit of Antarctic icebergs. Defined this way, the great Southern Ocean is a huge body of water that circulates clockwise (from west to east) around Antarctica, which is why we also call it the **West Wind Drift [4]**.

Biogeography

One of this realm's defining characteristics is its wildlife. Australia is the land of kangaroos and koalas, wallabies and wombats, possums and platypuses. These and numerous other *marsupials* (animals whose young are born very early in their development and then are carried in an abdominal pouch) owe their survival to Australia's early isolation during the breakup of Gondwana (see Fig. 6A-3). Before more advanced mammals could enter Australia and replace the marsupials, as happened in every other part of the world, this landmass was separated from Antarctica and India, and today it contains the world's largest assemblage of marsupial fauna. Australia's vegetation has distinctive qualities as well, notably the hundreds of species of eucalyptus trees native to this geographic realm. The study of fauna and flora in spatial perspective integrates the disciplines of biology and geography in a field known as **biogeography [5]**, and Australia is a gigantic laboratory for biogeographers.

Biogeographers are especially interested in the distribution of plant and animal species, as well as in the relationships between plant and animal communities and their natural environments. (The study of plant life is called *phytogeography*; the study of animal life is called *zoogeography*.) In 1876, one of the founders of biogeography, Alfred Russel Wallace, posited that the zoogeographic boundary of Australia's fauna was located beyond Australia in the Sunda island chain to the northwest, between Borneo and Sulawesi, and just east of Bali (Fig. 11-4).

Wallace's Line [6] soon was challenged by other researchers, who found species Wallace had missed and who visited islands Wallace had not. There was no question that Australia's zoogeographic realm terminated somewhere in Indonesia's Sunda archipelago, but where? Western Indonesia was the habitat of nonmarsupial animals such as tigers, rhinoceroses, and elephants in addition to primates; New Guinea clearly was part of the realm of the marsupials. How far had the more advanced mammals progressed eastward along the island stepping stones toward New Guinea? The zoogeographer Max Weber found evidence that led him to postulate his own ***Weber's Line***, which, as Figure 11-4 shows, lay very close to New Guinea.

In Australia, the arrival of the **Aboriginal population [7]** about 50,000 years ago appears to have triggered an

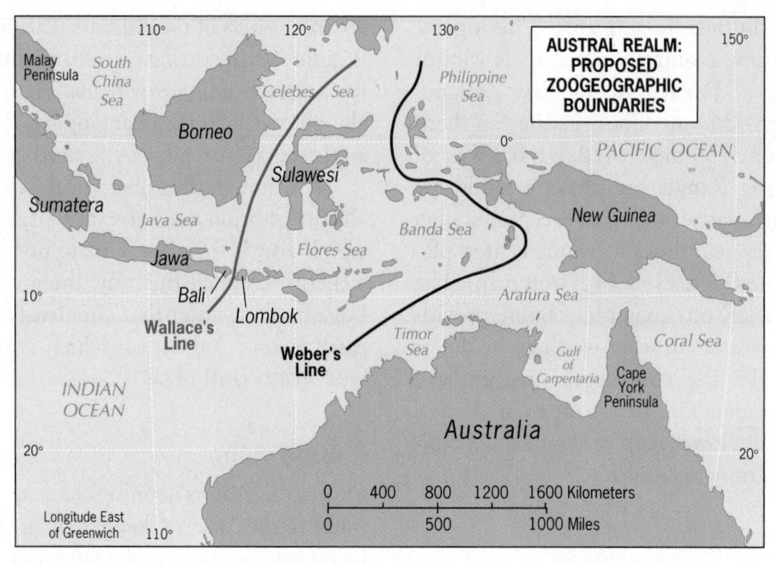

FIGURE 11-4

ecosystem collapse. Widespread burning of the existing forest, shrub, and grassland vegetation all across Australia probably led to the spread of desert scrub and to the rapid extinction of most of the continent's large mammals soon after the human invasion occurred. The species that survived faced a second crisis much later when European colonizers introduced their livestock, leading to the further destruction of remaining wildlife habitats. Survivors include marsupials such as the koala bear and the wombat, but the list of extinctions is much longer.

MAJOR CITIES OF THE REALM

City	Population* (in millions)
Sydney, Australia	4.8
Melbourne, Australia	4.2
Brisbane, Australia	2.2
Perth, Australia	1.8
Auckland, New Zealand	1.6
Adelaide, Australia	1.3
Canberra, Australia	0.4
Wellington, New Zealand	0.4

*Based on 2014 estimates.

REGIONS OF THE REALM

Australia is the predominant component of the Austral Realm, a continent-scale country in a size category that also includes China, Canada, the United States, and Brazil. For two reasons, however, Australia has fewer regional divisions than do the aforementioned countries: the comparatively uncomplicated physiography of Australia and its diminutive human numbers. Our discussion, therefore, uses the core-periphery concept as a basis for investigating Australia and focuses on New Zealand as a region by itself.

AUSTRALIA

Positioned on the Pacific Rim, almost as large as the 48 contiguous U.S. States, well endowed with farmlands and vast pastures, rivers, groundwater supplies, mineral deposits, and energy resources, served by good natural harbors,

The breathtakingly beautiful Great Barrier Reef, off the coast of Queensland in northeastern Australia (Fig. 11-3, top map). This is the largest coral reef on the planet, sometimes described as the only living thing on Earth that is visible from space. It was designated a World Heritage Site in 1981, and most of it now lies within Great Barrier Reef Marine Park. Not surprisingly, it ranks as one of the world's most prominent ecotourism destinations.

and populated by 23.1 million mostly well-educated people, Australia is one of the most geographically fortunate countries in the world.

Historical Geography

From the late eighteenth century onward, the Europeanization of Australia doomed the continent's Aboriginal societies. The first to suffer were those situated in the path of British settlement on the coasts, where penal colonies and free towns were founded. Distance protected the Aboriginal communities of the northern interior longer than elsewhere; but in Tasmania, the indigenous Australians died off in just decades after having lived there for perhaps 45,000 years.

The Seven Colonies

Eventually, the major coastal settlements became the centers of seven different colonies, each with its own hinterland; by 1861, Australia was delimited by its now-familiar pattern of straight-line boundaries (Fig. 11-5). Sydney was

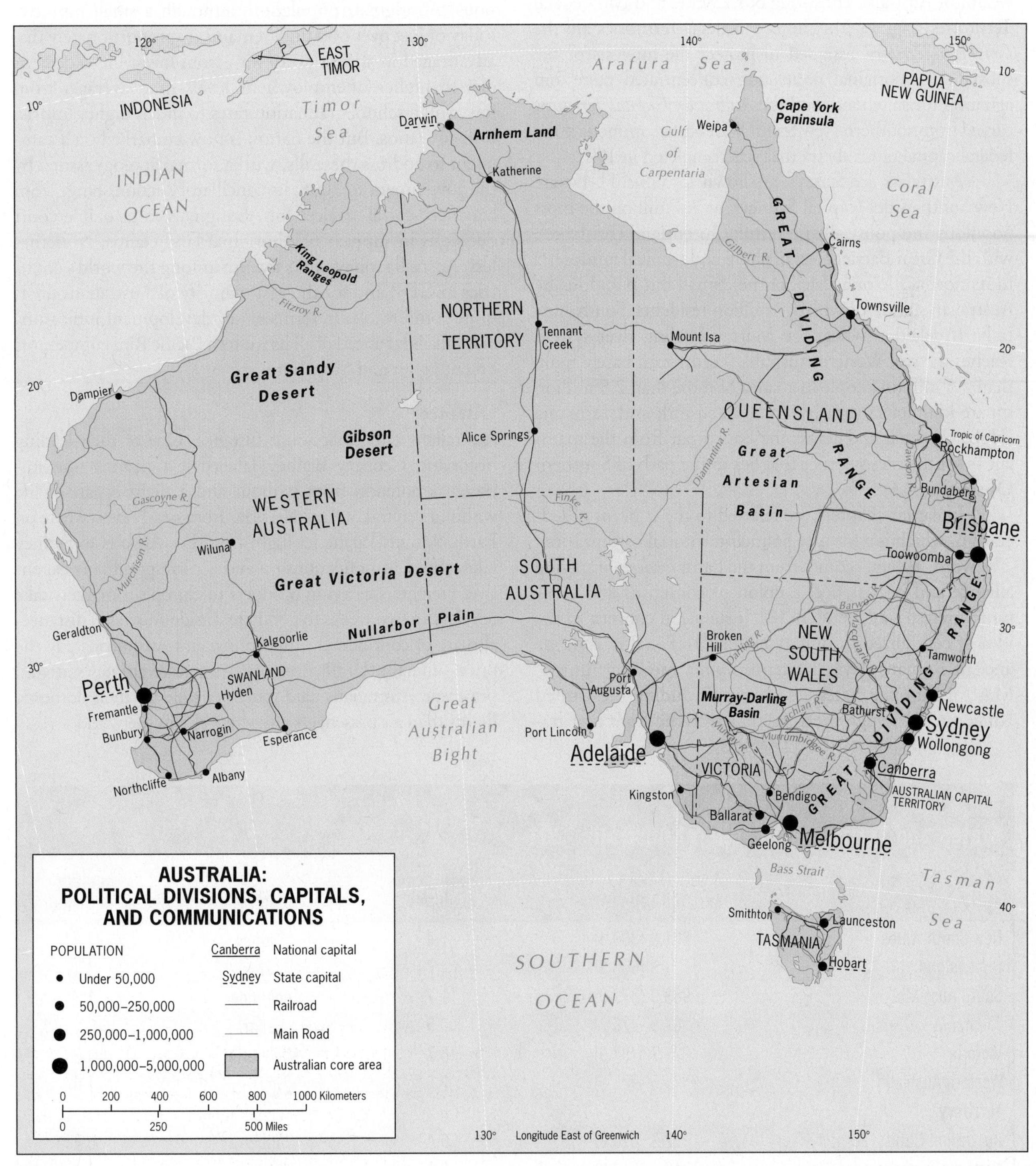

FIGURE 11-5

the focus for New South Wales; Melbourne, Sydney's rival, anchored Victoria. Adelaide was the heart of South Australia, and Perth lay at the core of Western Australia. The largest clusters of surviving Aboriginal people were located in the so-called Northern Territory, with Darwin, on Australia's tropical north coast, its colonial city.

Successful Federation

On January 1, 1901, following years of difficult negotiations, the Australia we know today finally emerged: the Commonwealth of Australia, consisting of six States and two Federal Territories (Fig. 11-5). The two Federal Territories are the *Northern Territory*, assigned to protect the interests of the substantial Aboriginal population concentrated there and agitating for statehood, and the *Australian Capital Territory*, carved from southern New South Wales to accommodate the federal capital of Canberra that was completed in 1927.

Australia's six States—as shown in Table 11-1—are New South Wales (capital Sydney), at 7.4 million the most populous and politically powerful; Queensland (Brisbane), with the Great Barrier Reef offshore and tropical rainforests in its north; Victoria (Melbourne), small but populous by Australian standards with 5.7 million residents; South Australia (Adelaide), where the Murray-Darling river system reaches the sea; Western Australia (Perth) with barely more than 2.5 million people in an area of more than 2.5 million square kilometers (nearly 1 million sq mi); and Tasmania (Hobart), the island across the Bass Strait from the mainland's southeastern corner that lies in the path of Southern Ocean storms.

In earlier chapters, we referred to the concept of federalism, which is a form of politico-territorial organization. The word "federal" comes from the Latin *foederis*, implying alliance and coexistence, a union of consensus and common interest—a **federation [8]**. It stands in contrast to the idea of centralized or **unitary states [9]**. For this, too, the ancient Romans devised a term: *unitas*, meaning "unity." Most of Europe's states are unitary, including the United Kingdom. Although most Australians came from that tradition, they managed to overcome their differences and forge a Commonwealth that was, in effect, a federation of States with different viewpoints, economies, and objectives, separated by enormous distances along the rim of a remote island continent.

Sharing the Bounty

Despite the country's good fortunes, not everyone in Australia adequately shares in the national wealth. The indigenous (Aboriginal) population, although a small minority today of just over 600,000, remains disproportionately disadvantaged in almost every way, from lower life expectancies to higher unemployment levels than average, from lower high school graduation rates to much higher imprisonment ratios. But the nation is now embarked on a campaign to address these ills, with a formal apology issued by the government in 2008; its conciliatory actions range from enhanced social services for Aboriginals to favorable court decisions in support of Aboriginal land claims. Nonetheless, Australia today ranks twelfth among the world's countries in GDP, and for the vast majority of Australians life is quite comfortable. In terms of key development indicators, Australia is far ahead of all its western Pacific Rim competitors except Japan and Singapore.

Distance

Australians often talk about distance. One of their leading historians, Geoffrey Blainey, labeled it a tyranny—an imposed remoteness from without and a divisive part of life within. Even today, Australia is far from nearly everywhere on Earth. A trans-Pacific jet flight from Los Angeles to Sydney takes about 14 hours nonstop and is correspondingly expensive. Freighters carrying products to European markets take ten days to two weeks to get there. Inside Australia, distances also are of continental proportions, and Australians pay the price—literally. Until some upstart private airlines started a price war, Australians paid more per mile for their domestic flights than air passengers anywhere else in the world.

TABLE 11-1

States and Territories of Federal Australia, 2014

State	Area (1000 sq km [1000 sq mi])	Population (millions)	Capital	Population (millions)
New South Wales	801.6 (309.5)	7.4	Sydney	4.8
Queensland	1727.3 (666.9)	4.7	Brisbane	2.2
South Australia	983.9 (379.9)	1.7	Adelaide	1.3
Tasmania	67.9 (26.2)	0.5	Hobart	0.2
Victoria	227.7 (87.9)	5.7	Melbourne	4.2
Western Australia	2525.5 (975.1)	2.5	Perth	1.8
Territory				
Australian Capital Territory	2.3 (0.9)	0.4	Canberra	0.4
Northern Territory	1346.3 (519.8)	0.2	Darwin	0.1

But distance also was an ally, permitting Australians to ignore the obvious. Australia was a British progeny, a European outpost. Once you had arrived as an immigrant from Britain or Ireland, there were a wide range of environments, magnificent scenery, vast open spaces, and seemingly limitless opportunities. When the Japanese Empire expanded, Australia's remoteness saved the day. When boat people by the hundreds of thousands fled Vietnam in the mid-1970s aftermath of the Indochina War, almost none reached Australian shores. When immigration became an issue, its self-perceived comforts of isolation led Australia to adopt an all-white admission policy (officially terminated only as recently as 1976) that was out of step with most of the rest of the world.

Immigrants

Today Australia is changing and rapidly so. Immigration policy now focuses on the would-be immigrants' qualifications, skills, financial status, age, and facility with the English language. With regard to skills, high-technology specialists, financial experts, and medical personnel are particularly welcome. Relatives of earlier immigrants, as well as a quota of genuine asylum-seekers, are admitted as well. In recent years, immigration has averaged between 140,000 and 200,000 annually, which keeps Australia's population growing because its (declining) natural rate of increase is only 0.7 percent. Indeed, the country is fast becoming a truly multicultural society. In Sydney, one in five residents is now of Asian ancestry. Overall, no less than a quarter of Australia's population is foreign-born, and another quarter consists of first-generation Australians.

Core and Periphery

As Figure 11-2 shows, Australia is a large landmass, but its population is almost entirely concentrated in a (discontinuous) core area that lies in the east and southeast, most of which faces the Pacific Ocean (locally named the Tasman Sea between Australia and New Zealand). Figure 11-5 shows that this crescent-shaped Australian heartland extends from north of the city of Brisbane to the vicinity of Adelaide and includes the largest city, Sydney; the capital, Canberra; and the second-largest city, Melbourne. A secondary core area has developed in the far southwest, centered on Perth. Beyond lies the vast periphery, which the Australians call the **Outback [10]** (see *Field Note*).

To better understand the evolution of this spatial arrangement, it helps to refer again to the map of world climates (Fig. G-7). Environmentally, Australia's most favored strips face the Pacific and Southern oceans, and they are not large. We can describe the country as a coastal rimland with cities, towns, farms, and forested slopes giving way to the vast, arid, interior Outback. On the western flanks of the

From the Field Notes . . .

"My most vivid memory from my first visit to Alice Springs in the heart of the Outback is spotting vineyards and a winery in this parched, desert environment as the plane approached the airport. I asked a taxi driver to take me there, and got a lesson in economic geography. Drip irrigation from an underground water supply made viticulture possible; the tourist industry made it profitable. None of this, however, is evident from the view seen here: a spur of the Macdonnell Ranges overlooks a town of bare essentials under the hot sun of the Australian desert. What Alice Springs has is centrality: it is the largest settlement in a vast area Australians often call "the centre." Not far from the midpoint on the nearly 3200-kilometer (2000-mi) Stuart Highway from Darwin on the Northern Territory's north coast to Adelaide on the Southern Ocean, Alice Springs also was the northern terminus of the Central Australian Railway (before it was extended north to Darwin in 2003), seen in the middle distance. The shipping of cattle and minerals is a major industry here. You need a sense of humor to live here, and the locals have it: the town actually lies on a river, the intermittent Todd River. An annual boat race is held, and in the absence of water the racers carry their boats along the dry river bed. No exploration of Alice Springs would be complete without a visit to the base of the Royal Flying Doctor Service, which brings medical help to outlying villages and homesteads."

© H.J. de Blij

www.conceptcaching.com

AMONG THE REALM'S GREAT CITIES . . .

SYDNEY

MORE THAN TWO centuries ago, Sydney was founded by Captain Arthur Phillip as a British outpost on one of the world's most magnificent natural harbors. The free town and penal colony that struggled to survive evolved into Australia's largest city. Today, metropolitan Sydney (4.8 million) is home to more than one-fifth of the country's total population. An early start, the safe harbor, fertile nearby farmlands, and productive pastures in its hinterland combined to propel Sydney's growth. Later, as road and railroad links made Sydney the focus of Australia's growing core area, industrial development and political power augmented its primacy.

With its incomparable setting and mild sunny climate, its many urban beaches, and its easy reach to the cool Blue Mountains of the Great Dividing Range, Sydney is one of the world's most liveable cities. Good public transportation (including an extensive cross-harbor ferry system from the doorstep of the waterfront CBD), fine cultural facilities headed by the multi-theatre Opera House complex, and many public parks and other recreational facilities make Sydney highly attractive to visitors as well (see chapter-opening photos). A healthy tourist trade, much of it from East Asian countries, bolsters the city's economy. Sydney's hosting of the 2000 Olympic Games was further testimony to its rising global visibility.

Increasingly, Sydney is evolving as a multicultural city. Its small Aboriginal sector is being overwhelmed by the arrival of large numbers of Asian immigrants. The Sydney suburb of Cabramatta symbolizes the impact: nearly 75 percent of its more than 23,000 residents were born elsewhere, at least a third of them in Vietnam. Unemployment is high, drug use is a problem, and crime and gang violence persist. Yet, despite the deviant behavior of a small minority, tens of thousands of Asian immigrants have established themselves in some profession (as the photo and caption suggest).

These developments herald Sydney's coming of age. The end of Australia's isolation has brought Asia across the country's threshold, and again the leading metropolis is showing the way.

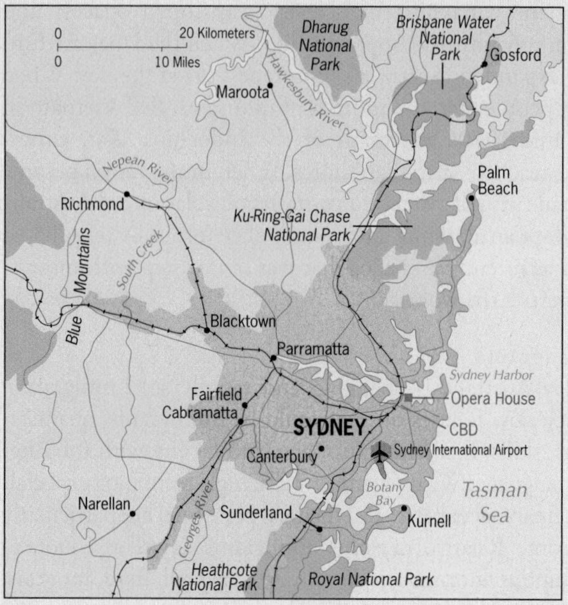

Great Dividing Range lie the extensive grassland pastures that catapulted Australia into its first commercial age—and on which still graze one of the largest sheep herds in the world (about 76 million sheep, producing over 40 percent of all the apparel wool sold in the world). Where it is moister, to the north and east, cattle by the millions graze on ranchlands. This is frontier Australia, over which livestock have ranged for nearly two centuries.

An Urban Culture

Despite the vast open spaces and romantic notions of frontier and Outback, Australia is an urban country, with 82 percent of all Australians living in cities and towns. On the map, Australia's economic spatial organization is similar to Japan's: large cities lie along the coast, the centers of manufacturing complexes as well as the foci of agricultural areas. Contributing to this situation in Japan was mountainous topography; in Australia it was the arid interior. There, however, the similarity ends. Australia's territory is 20 times larger than Japan's, and Japan's population is more than five times that of Australia's. Japanese port cities are built to receive raw materials and to export finished products. Australian cities forward minerals and farm products from the Outback to foreign markets and import manufactures from overseas.

Distances in Australia are much greater, and spatial interaction (which tends to decline with increasing distance) is reduced. In comparatively small, tightly organized Japan, you can travel from one end of the country to the other on superhighways, through tunnels, and over bridges with the utmost speed and efficiency. In Australia, the overland trip from Sydney to Perth, or from Darwin to Adelaide (using the railroad link via Alice Springs), is time consuming and slow. Nothing in Australia compares to Japan's high-speed bullet trains.

The Cities

For all its vastness and youth, Australia has developed a remarkable cultural identity, a sameness of urban and rural landscapes that persists from one end of the continent to the other. Sydney, often called the New York of Australia, lies on a spectacular estuarine site, its compact, high-rise central

© Robert Francis/Alamy

The name Cabramatta conjures up varied reactions among Australians. During the 1950s and 1960s, many immigrants from southern Europe settled in this western suburb of Sydney, attracted by affordable housing. In the 1970s and 1980s, Southeast Asians arrived here in large numbers, and during this period Cabramatta, in the eyes of many, became synonymous with gang violence and drug dealing. More recently, however, Cabramatta's ethnic diversity has come to be viewed in a more favorable light, and it is now seen as the "multicultural capital of Australia," a tourist attraction and proof of Australia's capacity to accommodate non-Europeans. Meanwhile, Cabramatta has been spruced up with Oriental motifs of various kinds. This "Freedom Gate" in the Vietnamese community is flanked by a Ming horse and a replica of a Forbidden City lion—all reflecting better times for an old gateway for immigrants as well as refugees.

business district overlooking a port bustling with ferry and freighter traffic. Sydney is a vast, sprawling metropolis of 4.8 million, with multiple outlying centers dominating its far-flung suburban ring; brash modernity and reserved British ways somehow blend here. Melbourne (4.2 million), sometimes regarded as the Boston of Australia, prides itself on its more interesting architecture and more cultured ways. Brisbane, the capital of Queensland, which also anchors Australia's Gold Coast and adjoins the Great Barrier Reef, is the Miami of Australia; unlike Miami, however, its residents can find nearby relief from the summer heat in the mountains of its immediate hinterland (as well as at its beaches). Perth (*Field Note* photo), Australia's San Diego, is separated from its nearest Australian neighbor by two-thirds of a continent and from Southeast Asia and Africa by thousands of kilometers of ocean—but due to the ever-expanding mining activities of Western Australia it is increasingly drawn into the global economy.

And yet, each of these cities—as well as the capitals of South Australia (Adelaide), Tasmania (Hobart), and, to a lesser extent, the Northern Territory (Darwin)—exhibits an Australian character of unmistakable quality. Life is both orderly and unhurried. Streets are clean, slums are uncommon, graffiti rarely seen. By American and even European standards, violent crime (although rising) is uncommon. Standards of public transport, city schools, and health-care provision are high. Spacious parks, pleasing waterfronts, and plentiful sunshine make Australia's urban life more acceptable than just about anywhere else on Earth.

Economic Geography

Agricultural Abundance

Australia's initial prosperity was achieved in the mines and on the farms, not in the cities. Australia has material assets

From the Field Notes . . .

© H.J. de Blij; inset photo © Jim Winkley/Andaulucia Plus Image Bank/Alamy

"Throughout Australia, summer weather in 2012 has been abnormal, and so it is today here in Perth on February 16th. People are wearing sweaters and jackets as a pall of smoke from nearby forest fires obscures the sun. Here on a hillside in King's Park, from where the view over the city is spectacular on a clear day (inset photo), are some historic markers, and this one gives pause: a simple but moving monument (second structure from the right) to commemorate the citizens of Perth who died in the Bali terrorist attack on October 12, 2002, including one who died of his injuries three years later. Look at Figure 11-1, and you can see why Bali is such a popular tourist destination for locals: Perth lies closer to this Indonesian island than any other major Australian city."

www.conceptcaching.com

of which other countries on the Pacific Rim can only dream. In agriculture, sheep raising was the earliest commercial venture, but it was the technology of refrigeration that brought world markets within reach of Australian beef producers. Wool, meat, and wheat have long been the country's big three income earners; Figure 11-6 displays the immense pastures in the east, north, and west that constitute the ranges of Australia's huge herds. The zone of commercial grain farming forms a broad crescent extending from northeastern New South Wales through Victoria into South Australia, with a major outlier covering much of the hinterland of Perth.

And keep in mind the scale of this map: Australia is only slightly smaller than the 48 contiguous States of the United States. Commercial grain farming in Australia is big

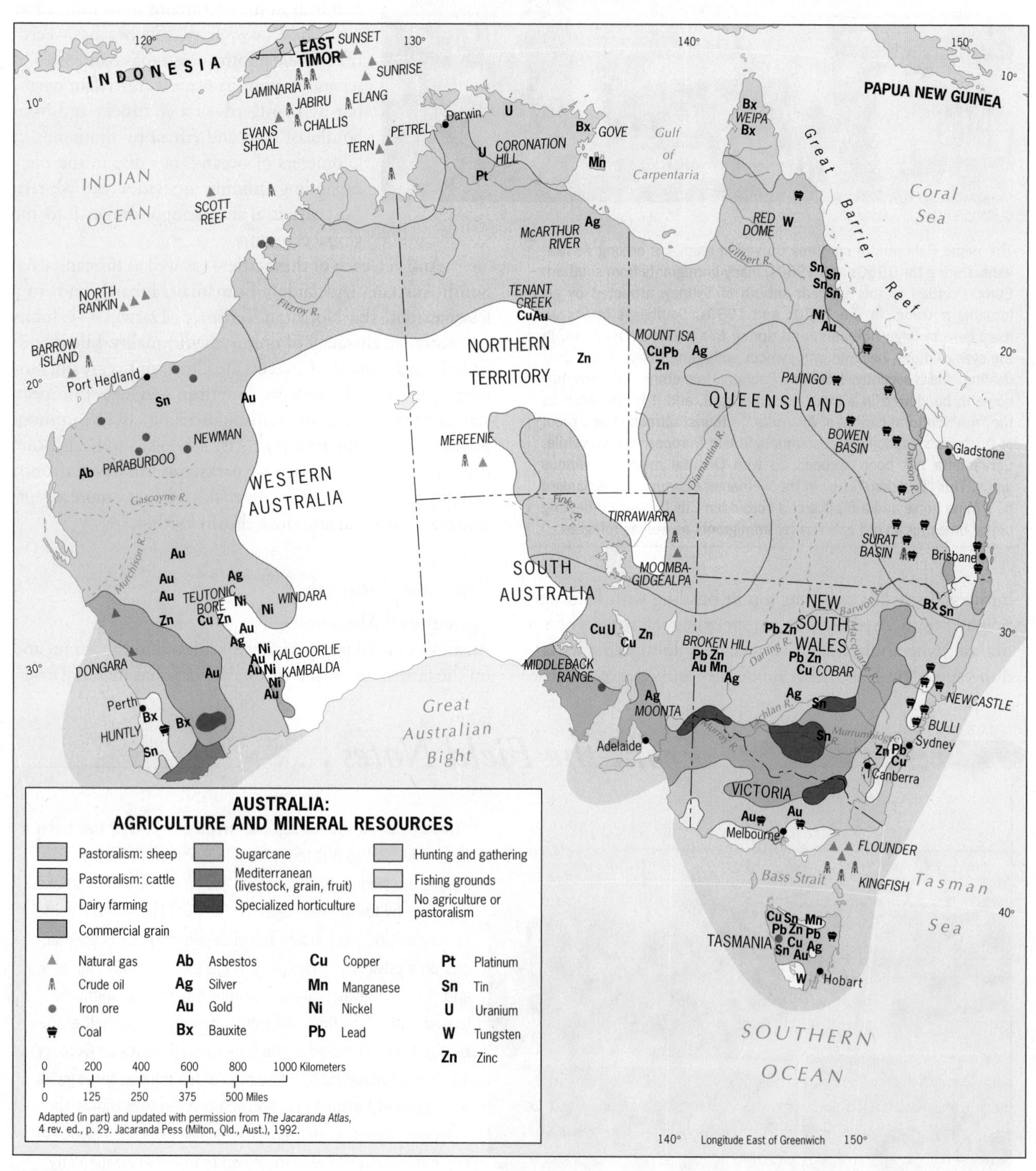

FIGURE 11-6

© Photoshot Holdings Ltd/Alamy

About 320 kilometers (200 mi) south-southeast of Port Hedland in the Pilbara region of Western Australia, Mount Whaleback near the town of Newman is one of the world's leading sources of high-grade iron ore. China's insatiable demand for this ore has generated the planet's largest "open-cut" iron mine, a huge and growing gash in the natural landscape—just one such impact resulting from Australia's role as raw-material supplier to industrializing economies.

business. As the climate map (Fig. G-7) would suggest, sugarcane grows along most of the humid, subtropical, coastal-lowland strip of Queensland, and Mediterranean crops (including grapes for Australia's highly successful wine industry) cluster in the hinterlands of Adelaide and Perth. Mixed horticulture concentrates in the Murray River Basin, including rice, grapes, and citrus fruits, all under irrigation. In addition, as elsewhere in the world, dairying has developed near the large metropolitan areas. With its considerable range of environments, Australia yields a great diversity of crops.

Mineral Wealth

Australia's mineral resources, as Figure 11-6 shows, also are diverse. Major gold discoveries in Victoria and New South Wales produced a ten-year gold rush starting in 1851 and ushered in a new economic era. By the middle years of that decade, Australia was producing 40 percent of the world's gold. Subsequently, the search for more gold led to the discoveries of additional minerals. New finds are still being made today, and even oil and natural gas have been found both inland and offshore (see the symbols in Fig. 11-6 in the Bass Strait between Tasmania and the mainland, and off the northwestern coast of the continent). The energy boom is transforming Australia's northernmost city of Darwin, which faces the fossil-fuel riches of the Timor Sea; offshore natural gas is processed here, converting it into liquid natural gas (LNG) for export to East and Southeast Asia. Similar developments are occurring in the central Queensland port town of Gladstone on Australia's east coast; there an expanding processing-plant complex produces LNG from piped-in coal-seam gas, which is extracted with new technologies from coal deposits in interior Queensland's Bowen and Surat Basins (Fig. 11-6). But environmental groups are increasingly opposed because the new hydraulic fracturing techniques threaten the State's agricultural resources; moreover, Gladstone's coastal facilities lie within the supposedly protected confines of the Great Barrier Reef.

Coal itself is mined at numerous locations, notably in the east near Sydney and Brisbane, in Western Australia, and in Tasmania (before coal prices fell this was an especially valuable export). Substantial deposits of metallic and nonmetallic minerals abound—from the complex at Broken Hill and the mix of minerals at Mount Isa to the huge nickel deposits at Kalgoorlie and Kambalda, the copper of Tasmania, the tungsten and bauxite of northern Queensland, and the asbestos of Western Australia. A closer look at the map reveals the wide distribution of iron ore, and for this raw material as for many others, Japan was Australia's best customer for many years until it was surpassed by China during the 2000s. By 2011, in fact, the enormous demand for raw materials from China as well as emerging India marked the longest-running commodity boom in recent Australian history—and was still going in mid-2013.

Manufacturing's Limits

When Australia became established as a state, however, it needed goods from overseas, and here the "tyranny" of distance played a key role. Imports from Britain (and later the United States) were expensive mainly because of transport costs. This encouraged local entrepreneurs to establish their own industries to produce these goods more cheaply. Economic geographers call such industries **import-substitution industries [11]**, and this is how local industrialization got its start.

Australian manufacturing remains oriented to domestic markets, and its automobiles, electronic equipment, and cameras are decidedly not challenging the Pacific Rim's economic tigers for a place on world markets. Australian manufacturing is diversified, producing some machinery and equipment made of locally produced steel as well as textiles, chemicals, paper, and many other items. These industries cluster in and near the major urban areas where the markets are located. The domestic market in Australia is not large, but it remains relatively affluent. This makes it attractive to foreign producers, and Australia's shops overflow with high-priced goods from Japan, South Korea, Taiwan, the United States, and Europe. Indeed, despite its long-term protectionist practices, Australia still does not produce many goods that could be manufactured at home. Overall, the continuing prominence of the **primary sector [12]** indicates that further economic development is still needed.

Because of its early history as a treasure trove of raw materials, Australia to this day is still seen worldwide as a country whose economy depends primarily on its exportable natural resources. In fact, however, Australia's economy depends mostly on services, not commodity exports, just as is the case in all highly developed economies. Indeed, tourism by itself contributes around 5 percent—about the same as the value of mineral exports. Australia's natural resources are vitally important, but most of the money is earned in those bustling coastal metropolitan areas, not in the Outback.

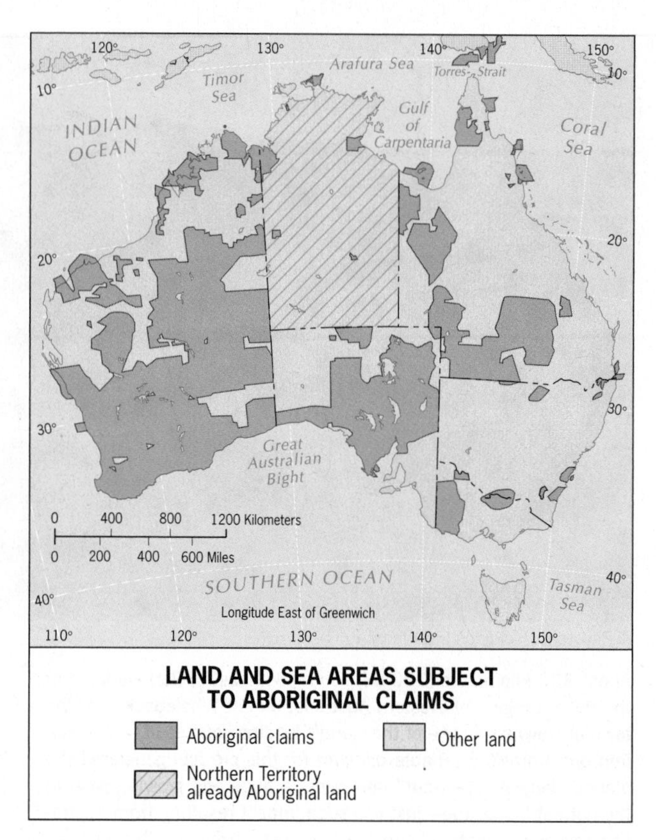

FIGURE 11-7 © H. J. de Blij, P. O. Muller, and John Wiley & Sons, Inc.

Australia's Challenges

The Commonwealth of Australia may be changing, but its neighbors in Southeast and South Asia are changing even faster. Australia's European bonds are weakening as its Asian ties strengthen, and it plays a growing role in the western Pacific Rim. Australia does face certain challenges at home as well. These include: (1) Aboriginal claims; (2) concerns involving immigration; (3) environmental degradation; and (4) issues related to Australia's status and regional role.

Aboriginal Issues

For several decades, the Aboriginal issues focused on two questions: formal acknowledgment, by the government and majority, of mistreatment of the Aboriginal minority with official apologies and reparations; and land ownership. The first question was resolved in 2008 when Prime Minister Kevin Rudd offered a formal apology for the historic mistreatment of the Aborigines. The second question has major geographic implications. Although making up less than 3 percent of the total population, the Aboriginal population of just over 600,000 (including many of mixed ancestry) has been gaining influence in national affairs, and in the 1980s Aboriginal leaders began a campaign to obstruct exploration on what they designated as ancestral and sacred lands.

Until 1992, Australians had taken it for granted that Aboriginals had no right to land ownership, but in that year the Australian High Court made the first of a series of rulings in favor of Aboriginal claimants. A subsequent court decision implied that vast areas (probably as much as 78 percent of the entire continent) could potentially be subject to Aboriginal claims (Fig. 11-7). Today the **Aboriginal land issue [13]** remains mostly (though not exclusively) an Outback issue, but it has the potential to overwhelm Australia's court system and to inhibit economic growth.

Lately, Australians have initiated debate on ways to bring so-called market-driven incentives to Aboriginal areas. At the moment, much Aboriginal land is administered communally by Aboriginal Land Councils, which makes the people land-rich but keeps them dirt-poor because the system in effect prevents private enterprise, including even the construction of family-owned housing. Some Aboriginal leaders are arguing that the system perpetuates an undesirable dependence on federal handouts, prevents jobs from becoming available, and restrains economic incentive. But not all Aboriginal leaders agree, worrying that privatization of any kind will undermine the Land Councils and create a small minority of better-off businesspeople, leaving the great majority of tradition-bound Aborigines even worse off. Still, the fact that the idea has some Aboriginal support is a sign of changing times in the relationships among Australian communities.

In 2007, Australia was shaken by a report on conditions in Aboriginal desert camps in the Northern Territory that chronicled rampant alcoholism, domestic violence, unemployment, school truancy, and a general breakdown of Aboriginal culture. The prime minister at the time, calling this a national emergency, ordered the federal government to take control of 60 Aboriginal communities in the Northern Territory through the deployment of police and army patrols, health checks, and other measures. Even though there were accusations of political grandstanding and questions about sincerity, there could be no doubt that Australia's conscience was once again troubled by the fate of the country's first inhabitants left behind in the era of modernization.

Regional ISSUE Aborigines, Immigrants, Rights and Wrongs

FIRST AUSTRALIANS FIRST

"Australia is becoming a multiracial society. Immigrants from Tajikistan to Thailand are changing the (mostly urban) cultural landscape. Long debates over their rights and privileges roil the political scene. Meanwhile, the Australians who got here first, most of whom live far beyond the burgeoning cities and out of sight, remain the most disadvantaged of minorities. We should remind ourselves of what they went through, and why Australia owes them—big time.

"When my European ancestors arrived on these shores there were more than a million Aboriginal people here, organized into numerous clans and subcultures. Like all peoples, they had their vices and virtues. Among the latter was that they didn't appropriate land. Land was assigned to them by the creator (in what they call The Dreaming), and their relationship to it was spiritual, not commercial. They didn't build fences or walls. Neither had they adopted some bureaucratic religion. Just imagine: a world where land was open and free, and religion was local and personal.

"Then the British showed up and started claiming and fencing off land that, under their European rules, was there for the taking. If the Aboriginal clans got in the way, they were pushed out, and if they resisted, they got killed. You don't even want to *think* about what happened in Tasmania: a campaign of calculated extermination. Between 1800 and 1900, the Aboriginal population dropped from 1 million to about 50,000. When we became a 'nation' in 1901, they weren't even accorded citizenship, They didn't get the vote until 1962.

"None of this stopped certain white Australian men from getting Aboriginal women pregnant. And from 1910 on, church and state managed to make things even worse. They took these young children and put them in institutions, where they would be 'Europeanized' and then married off to white partners, so that they would lose their Aboriginal inheritance. This, if you'll believe it, went on into the 1960s! Think of the scenes, these kids being kidnapped from their mothers by armed officials never to be seen again.

"It's hard to believe that it took nearly another half-century before the Australian government, following a contentious and divisive debate, finally offered a formal apology for this and other misdeeds of the past. But now the question is, do Aboriginal Australians benefit from the country's growing ethnic complexity? I work in a State government office here in Sydney that assists Asian immigrants, and all I can say is that I wish that we'd done for the First Australians what we're doing for the stream of immigrants we admit today."

NO MORE SPECIAL TREATMENT

"Australia is a nation of immigrants, and we've all gone through rough times. I'm not complaining to the British government for what happened to my ancestors when they were shipped out here as prisoners. Like my father, I was born on this Outback sheep station about 40 years ago, and we employ a dozen Aboriginal workers, most of whom were also born in this area. I had nothing to do with what happened more than a century ago. What I would or might have done is irrelevant, and I can't be blamed for what my great-great-grandparents may have done. All over the world people are born into situations not of their making.

"And I believe that this country has bent over backwards, in my time at least, to undo the alleged wrongs of the past. Look, the Aboriginal minority counts a bit over 600,000 or 2.8 percent of the population. Take a look at this on the map: they've got the whole Northern Territory and other parts of the country too, and that's 15 percent of Australia. And now we're required to give them even more? The Australian High Court keeps awarding Aboriginal claimants more and more land. That affects all of us. Pretty soon all of Australia will be targeted by these Aboriginals and their lawyers. And, by the way, in the old days those people moved around all the time. Who is to say what clans owned what land when it comes to claiming 'native title'? These court cases are going to tie us up in legal knots for generations to come. So, if you're a Japanese or Chinese buyer in search of commodities, are you going to sign contracts when you're not sure who will own the land?

"We should take a lesson from the Kiwis, who agreed to a land deal with the Maori minority in New Zealand, and now look what's happening over there. The place is overrun with Chinese who are making deals for supposedly 'native' land and are converting the whole place to milk-powder production. And the easy residence rules are making New Zealand a stepping stone for entering Australia.

"Look, I like the fellows working here, but you've got to realize that no laws or treaties are going to solve all of the problems they have. They're getting all kinds of preferential access to government employment, remedial help in many areas, but still they wind up leaving school, abandoning jobs, winding up in jail. They have to grab the opportunities they now have rather than ask for more. In a lot of countries they would have never gotten them: Aussies are a pretty decent people. It's up to them to make the best of it."

Vote your opinion at www.wiley.com/go/deblijpolling

VOICE FROM THE Region

Courtesy of Julie Tregale

Julie Tregale, Sydney, Australia

BEING ASIAN IN AUSTRALIA

"I live in a suburb on the north side of Sydney, Australia's number-one city. Born and raised in Indonesia on the island of Jawa, I am one of many Asians who have migrated to Australia's cities in recent times. Most immigrants come here to look for work or to get an education. At the workplace and in the universities, there are always Asian faces. The government encourages immigration because Australia has an aging population and they need foreign workers. As a result, the country is changing fast. It's good that Australia has opened up to the rest of Asia and that it has become a much more diverse culture. But I can also see that people are worried about everybody getting along, about integration. Some ethnics turn their area into their little country where no one speaks English. That makes it difficult for Australians to go there. Also, it seems that gang fights between different ethnicities are happening more and more. Being a mother of two, I also see some issues at schools. The students who excel academically are usually Asians. Sometimes, Australian parents complain that Asian students have been coached and how unfair it is, etc. But I think it's about the different priorities that each family has on how they want to prepare their child and that comes from values they grew up with. I see that more Asians are geared into high academic achievement whereas the Australians are more into sports (rugby, soccer, swimming). A friend of mine who is Vietnamese-born but has lived here nearly all her life said that it's not always an advantage to have high academic achievements because in reality, in the workplace, you also need to be accepted, have the social skills, and fit in. In all, I am not surprised that the growing number of immigrants has caused some discrimination and tension. Government should take an active role to balance things out and make Australia a great place to live for all."

Immigration Issues

The immigration issue is older than Australia itself. Fifty years ago, when Australia had less than half the population it has today, 95 percent of the people were of European ancestry, and more than three-quarters of them came from Britain and Ireland. Eugenic (race-specific) immigration policies maintained this situation until the mid-1970s. Today, the picture is dramatically different: of 23 million Australians, only about one-third are of British-Irish origin, and Asian immigrants outnumber both European immigrants and the natural increase each year. During the early 1990s, about 150,000 legal immigrants arrived in Australia annually, most from Hong Kong, Vietnam, China, the Philippines, India, and Sri Lanka. Annual immigration quotas have since been reduced, but were then allowed to rise again—reaching 190,000 in 2012–2013—with Asian immigrants continuing to outnumber those from Western sources. Sydney, the leading recipient of the Asian influx, has become a mosaic of ethnic neighborhoods, some of which have gone through periods of gang violence and drug dealing, but have stabilized and even prospered over time. Still, as the economy has grown, particularly in the mining sector, the country will have to rely on immigration to meet growing skilled-labor demands. Multiculturalism will undoubtedly remain a long-term challenge for Australia.

Environmental Issues

Environmental degradation [14], unfortunately, is practically synonymous with Australia. First the Aborigines, then the Europeans and their livestock, inflicted catastrophic damage on Australia's natural environments and ecologies. Great stands of magnificent forest were destroyed. In Western Australia, centuries-old trees were simply ringed and left to die so that the sun could penetrate through their leafless crowns to nurture the grass below for pasture for introduced livestock. In island Tasmania, where Australia's native eucalyptus tree reaches its greatest concentrations (comparable to California's redwood stands), tens of thousands of hectares of this irreplaceable treasure have been lost to chain saws and pulp mills. Many of Australia's unique marsupial species have been driven to extinction, and many more are endangered or threatened. "Never have so few people wreaked so much havoc on the ecology of so large an area in so short a time," observed a geographer in Australia not long ago. But awareness of this environmental degradation is growing. In Tasmania, the Green environmentalist political party has now become a force in State affairs, and its activism has slowed deforestation, dam building, and other development projects. Still, many Australians fear the environmentalist movement as an obstacle to economic growth, and this too is an issue for the future.

Another environmental problem involves Australia's wide and long-term climatic variability. In a dominantly arid continent, droughts in the moister fringes are the worst enemy, and Australia's history is replete with devastating dry spells. Australia is vulnerable to El Niño events (see Chapter 5B), but recent global warming may also be playing a role in the process. One such serious drought (see photo pair), regarded as the worst in living memory, lasted throughout the 2000s and imperiled the entire region watered by the Murray-Darling river system, Australia's breadbasket (Fig. 11-6). Part of this is caused by nature, although other factors have to do with the increasing demand

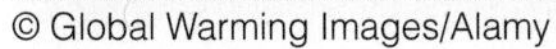

© Global Warming Images/Alamy

Photo by James Croucher/Newspix/Getty Images, Inc.

During the first decade of the twenty-first century Australia lay in the grip of its worst drought on record, with calamitous consequences ranging from deadly forest fires to parched farmlands. One of Australia's most profitable agricultural industries, winemaking, suffered severely as many winegrowers were driven off their land. The left photo shows a once-mature vineyard near Yarrawonga, on the New South Wales-Victoria border north of Melbourne along the Murray River, desiccated and abandoned during the summer of 2009. When the rains finally came, they caused disastrous floods, washing away topsoil and eroding the heat-baked countryside. The right photo, taken in January 2010, shows the results at Coonamble, New South Wales along a tributary of the Darling River.

for water in Australia's burgeoning urban areas as well as by excessive damming, well-drilling, and water diversion in the upstream tributaries of these two vital rivers. The Australian government has now embarked on a coordinated drainage-basin-control program in this region that involves both State governments and local farmers. On the world's driest continent, water management now finally seems to be catching up.

Australia's Place in the World

Several issues involving Australia's status at home, relations with neighbors, and position in the world are also stirring up national debate. A persistent domestic question is whether Australia should become a republic, ending the status of the British monarch serving as head of state, or whether it should continue to participate in the British Commonwealth.

Relations with neighboring *Indonesia* and *East Timor* have become more complex. For many years, Australia had what may be called a special relationship with Indonesia, whose help it needs in curbing illegal seaborne immigration. It was also profitable for Australia to counter international (UN) opinion and recognize Indonesia's 1976 annexation of Portuguese East Timor, for in doing so Australia could deal directly with Jakarta for the oil and gas reserves beneath the Timor Sea (see Fig. 10B-8). Thus Australia gave neither recognition nor support to the rebel movement that fought for independence in East Timor. But this story had a surprisingly happy ending. When the East Timorese campaign for independence succeeded in 1999 and Indonesian troops began an orgy of murder and destruction, Australia sent an effective peacekeeping force and spearheaded the United Nations effort to stabilize the situation (for more see Chapter 10B). Today, a new era has opened in Australia's relations with these northern neighbors.

Australia also maintains a long-term relationship with *Papua New Guinea (PNG)* (see Chapter 12). This association, too, has gone through difficult times. In recent years, the inhabitants of Papua New Guinea have strongly resisted privatization, World Bank involvement, and globalization generally. Australia assists PNG in several spheres, but its motives are sometimes questioned. Not long ago, the construction of a projected gas pipeline from PNG to the Australian State of Queensland precipitated fighting among tribes over land rights, resulting in dozens of casualties, and Australian public opinion reflected doubts concerning the appropriateness of this venture. When political violence and chaos overtook the *Solomon Islands* lying east of Papua New Guinea in 2003, Australian forces intervened: a failing state in Australia's neighborhood could become a base for terrorist activity.

Along with some other Southeast Asian countries, Australia is seeking closer relations with the *United States* as a counterweight to China's rapidly increasing influence across these realms; not surprisingly, the U.S. is pleased to reciprocate. Starting in 2012, several thousand rotating U.S. troops have been stationed at Australian military bases; moreover, the U.S. Air Force will now have access to Australian airfields in the Northern Territory that are within easy flying distance of the South China Sea.

Overall, Australia's growing stature in the world and its closer ties with neighboring realms has prompted a reconsideration of the country's very identity and where it best fits. Australia has both assumed a wider global presence and in certain ways (re-)turned to Asia. The exclusive relationship with Britain now belongs to the past; trade and investment relationships with nearby countries are intensifying; political sensitivity to issues in Australia's corner of the world continues to grow; and the proliferation of Asian faces on the streets of Sydney and other cities underscores the changing, twenty-first-century story.

Territorial dimensions, relative location, and raw-material wealth have helped determine Australia's place in the world and, more specifically, on the western Pacific Rim. Australia's population is still less than 25 million but the country's importance in the international community far exceeds its human numbers.

NEW ZEALAND

Twenty-four hundred kilometers (1500 mi) east-southeast of Australia, in the Pacific Ocean across the Tasman Sea, lies New Zealand, also known as *Aotearoa* in Maori (meaning "land of the long white cloud"). In an earlier age, New Zealand would have been part of the Pacific geographic realm because its population was entirely Maori, a people with Polynesian roots. But New Zealand, like Australia, was invaded and occupied by Europeans. Today, its population of 4.5 million is about 70 percent European, and the Maori form a substantial minority of about 700,000, with many of mixed Euro-Polynesian ancestry (including Pacific Islanders).

New Zealand consists of two large mountainous islands and many scattered smaller islands (Fig. 11-8). The two main islands, with the South Island somewhat larger than the North Island, look diminutive in the great Pacific

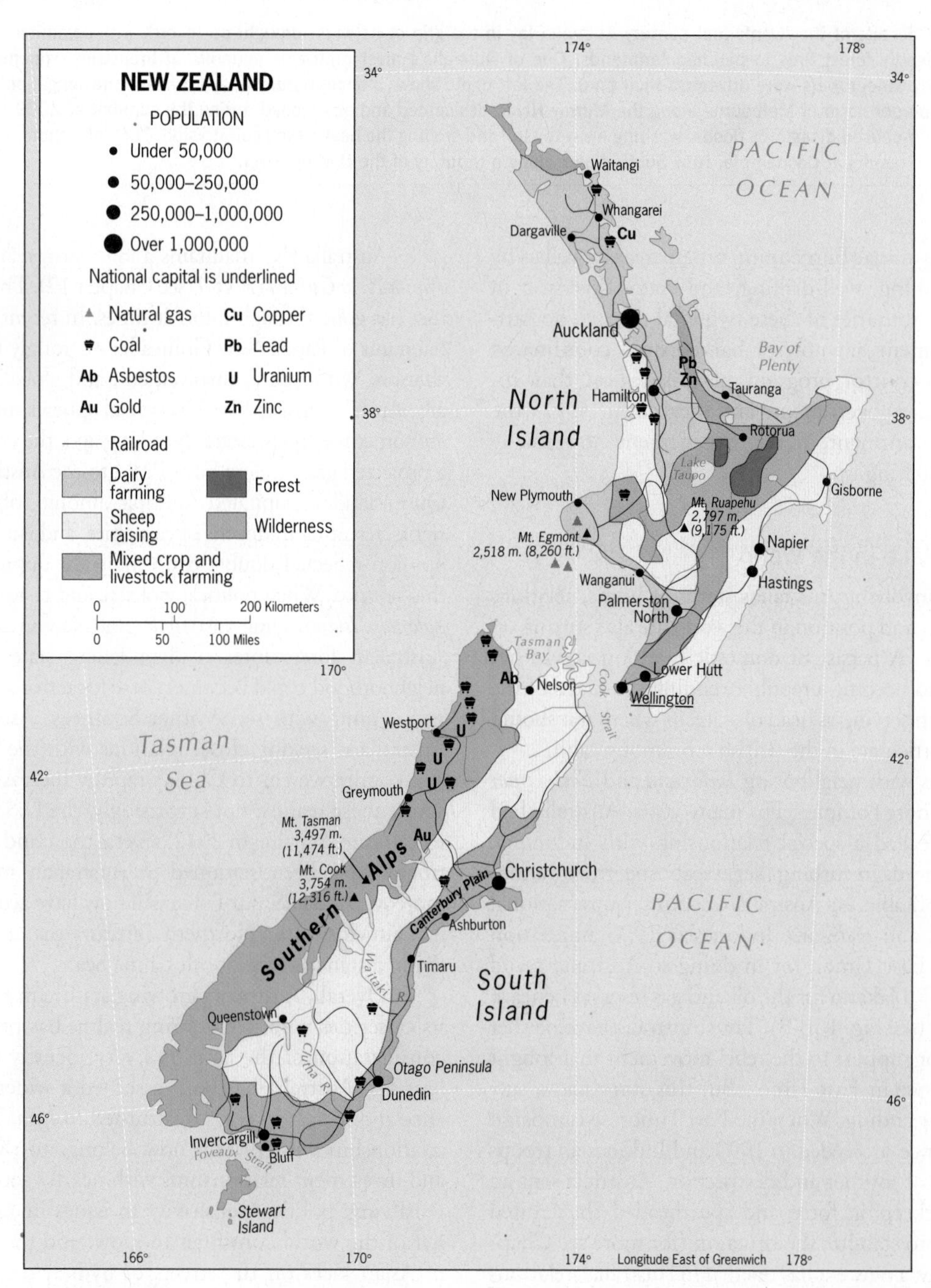

FIGURE 11-8

From the Field Notes . . .

© H.J. de Blij

"The drive from Christchurch to Arthur's Pass on the South Island of New Zealand was a lesson in physiography and biogeography. Here, on the east side of the Southern Alps, you leave the Canterbury Plain and its agriculture and climb into the rugged topography of the glacier-cut, snowcapped mountains. A last pasture lies on a patch of flatland in the foreground; in the background is the unmistakable wall of a U-shaped valley sculpted by ice. Natural vegetation ranges from pines to ferns, becoming even more luxuriant as you approach the moister western side of the island."

www.conceptcaching.com

Basin, but together they are larger than Britain. In contrast to generally low-relief Australia, the more rugged terrain of the two large islands contains several peaks rising far higher than any on the Australian landmass. The South Island has a spectacular snowcapped range appropriately called the Southern Alps, with numerous summits reaching beyond 3300 meters (10,000 ft). The smaller North Island has proportionately more land under low relief, but it also has an area of central highlands along whose lower slopes lie the pastures of New Zealand's chief dairying district.

As we noted at the beginning of this chapter, the convergence of the Australian and Pacific tectonic plates underlies much of New Zealand, and it renders the country prone to volcanic eruptions and earthquakes. On February 22, 2011, a magnitude 6.3 quake struck very close to the center of Christchurch, the South Island's biggest city (essentially a delayed aftershock of an even stronger earthquake five months earlier, whose epicenter was located at the edge of this urban region). Structural damage to the CBD and its immediate surroundings was severe, and the death toll reached 182, making it New Zealand's deadliest quake in 80 years.

Human Spatial Organization

The most promising areas for habitation, therefore, are the lower-lying slopes and lowland fringes on both islands. On the North Island, the largest urban area, Auckland, occupies a comparatively low-lying peninsula. On the South Island, the largest lowland is the agriculture-dominated Canterbury Plain, focused on Christchurch. What makes these lower-elevation zones so attractive, apart from their availability as cropland, is their magnificent pastures. The range of soils and pasture plants allows both summer and winter grazing. Moreover, the Canterbury Plain, the chief farming region, also produces a wide variety of vegetables, cereals, and fruits. About half of all New Zealand is pasture land, and much of the farming provides fodder for the pastoral industry. About 31 million sheep, 6 million dairy cattle, and 4 million beef cattle dominate these livestock-raising activities, with wool, milk products, and meat providing about two-thirds of the islands' export revenues.

Despite their contrasts in size, shape, physiography, and history, New Zealand and Australia share a number of characteristics. Apart from their joint British heritage, they share a sizeable pastoral economy with growth in specialty goods such as wines, a small local market, the problem of great distances to world markets, and a desire to stimulate (through protection) domestic manufacturing. The very high degree of urbanization in New Zealand (86 percent of the total population) once again resembles Australia: substantial employment in city-based industries—mostly the processing and packing of livestock and farm products—as well as government jobs.

Spatially, New Zealand further shares with Australia its pattern of **peripheral development [15]** (Fig. 11-2), imposed not by deserts but by high rugged mountains and the fragmented layout of the country. The country's major cities—Auckland and the capital of Wellington on the North Island; Christchurch and Dunedin on the South Island—are all located on the coast, and the rail and highway

From the Field Notes . . .

© H.J. de Blij

"It was Sunday morning in Christchurch on New Zealand's South Island, and the city center was quiet. As I walked along Linwood Street I heard a familiar sound, but on an unfamiliar instrument: the Bach sonata for unaccompanied violin in G minor—played magnificently on a guitar. I followed the sound to the artist, a Maori musician of such technical and interpretive capacity that there was something new in every phrase, every line, every tempo. I was his only listener; there were a few coins in his open guitar case. Shouldn't he be playing before thousands, in schools, maybe abroad? No, he said, he was happy here, he did all right. A world-class talent, a street musician playing Bach on a Christchurch side street, where tourists from around the world were his main source of income. Talk about globalization."

www.conceptcaching.com

networks are therefore entirely peripheral in their configuration. Moreover, the two main islands are separated by Cook Strait, a windswept waterway that can only be crossed by ferry or air (Fig. 11-8). On the South Island, the Southern Alps are New Zealand's most formidable barrier to surface communications.

The Maori Factor and New Zealand's Future

Like Australia, New Zealand has had a history of difficult relations with its indigenous population. The Maori, who account for 15.5 percent of the country's population today, appear to have reached the islands during the tenth century AD. By the time the European colonists arrived, the Maori had had a tremendous impact on the islands' ecosystems, especially on the North Island where most of them lived. In 1840, the Maori and the British signed a treaty at Waitangi that granted the colonizers sovereignty over New Zealand but guaranteed the Maori rights covering established tribal lands. Although the British abrogated parts of the treaty in 1862, the Maori had reason to believe that vast reaches of New Zealand, as well as offshore waters, were theirs in perpetuity.

As in Australia, judicial rulings during the 1990s supported the Maori position, which led to expanded claims and growing demands. Culturally, the declaration of Maori as an official language in New Zealand and its teaching throughout the school system are seen as significant progress toward an acceptance of the Maori cultural heritage. But the most persistent Maori complaint concerns the slow pace of integration of this minority into modern New Zealand society. Although Maori claims encompass much of rural New Zealand, they also cover prominent sites in the major cities. Today, the Maori question is the leading domestic issue.

The Green Factor

New Zealand is well known for its progressive politics and superior quality of life. Among the factors that contribute to the country's environmental progressiveness is its status as one of the leading "green" societies in the world, with a long-active Green Party and an established program of environmental conservation. Although the Maori and then the European colonists degraded New Zealand's landscapes, their descendants have been exemplary in observing environmentally friendly and sustainable policies.

Environmental scientists recently ranked New Zealand first in the world (the United States was 28th) in a report that examined a range of environmental indices such as clean water, air pollution, renewable energy, and biodiversity conservation. With approximately 30 percent of its land area now protected from development, New Zealand in 2007 declared that it would become the world's first carbon-neutral country by 2020. More than 75 percent of its energy is already derived from renewable sources (hydro and geothermal), compared to 10 percent in the United States. New Zealand is also a nuclear-free country that does not permit even visiting naval vessels with nuclear capabilities to anchor at its ports. And the country has even established Environmental Courts to hear cases involving environmental management decisions. With its pioneering green initiatives, New Zealand demonstrates that a country, albeit one with a small population, can successfully work to improve its natural environment if its leaders possess the necessary political will.

Dominant cultural heritage and prevailing cultural landscape form two criteria on which the delimitation of the Austral Realm is based. But in both Australia and New Zealand, the cultural continues to change and the convergence with neighboring realms is well underway.

POINTS TO PONDER

- Australia's population growth is marked by an annual immigration rate that exceeds the natural increase rate.
- Britain was Australia's most important trading partner until the 1980s, when it was displaced by Japan. China took over that role in 2007, and since then has been pulling away as the country's leading trading partner.
- New Zealand experiences more than 10,000 earthquakes each year; as many as 150 can be felt, and about 20 are strong enough to cause damage on the human landscape.
- Australia may soon be on the road to energy independence thanks to the 2012–2013 discoveries of massive, fracking-friendly shale oil reserves. These deposits lie beneath South Australia's Arckaringa Basin northwest of Adelaide, and may be nearly equal in size to the estimated reserve total for all of Saudi Arabia.

12 PACIFIC REALM AND POLAR FUTURES

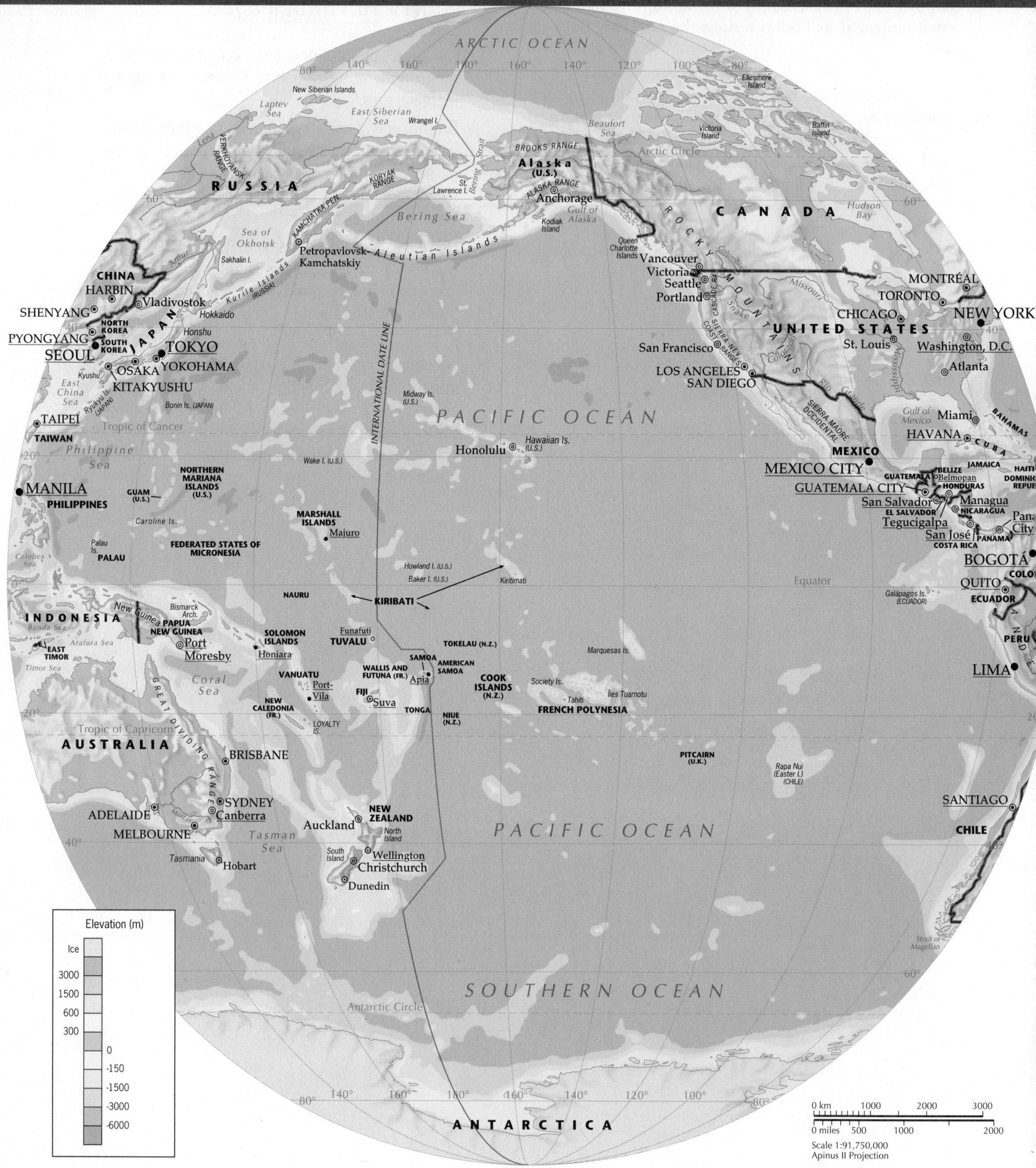

FIGURE 12-1

REGIONS

Melanesia

Micronesia

Polynesia

IN THIS CHAPTER

- Water, water everywhere—but who owns it?
- Islands high and low
- Exploring the deep seas
- The conundrum of divided Papua
- Colonists and foreigners: Still there, still dominant
- The partitioning of Antarctica
- Geopolitics in the warming Arctic Basin

CONCEPTS, IDEAS, AND TERMS

Marine geography 1
Territorial sea 2
High seas 3
Continental shelf 4
Exclusive Economic Zone (EEZ) 5
Maritime boundary 6
Median-line boundary 7
British Commonwealth 8
High-island cultures 9
Low-island cultures 10
Antarctic Treaty 11
Northeast Passage 12

© H. J. de Blij

Where were these pictures taken?
Find out at www.wiley.com/college/deblij

In this final chapter, we focus on three of the world's vast expanses so different from those discussed in previous chapters that we need an entirely new perspective. Water and ice, not land and soil, dominate the physiography. Importantly, environmental changes occurring here have consequences for the entire planet.

The largest of these expanses, the ***Pacific Realm***, covers almost half the globe from the Bering Sea to the Southern Ocean and from Central America to Indonesia. The Pacific Ocean, larger than all the world's land areas combined, is studded with tens of thousands of islands of which New Guinea is by far the largest. The two polar zones to which it is linked could hardly be more different: the ***Arctic*** is water, frozen at the surface, but you can reach the North Pole in a submarine; the ***Antarctic*** is mainly land, weighed down by the planet's largest accumulation of permanent ice. The surface journey to the South Pole, first accomplished a bit more than a century ago, has cost many an explorer's life.

Climate change is altering geography everywhere, as is the enhanced technology of mineral (especially oil and gas) exploration and extraction. Rising sea levels would imperil thousands of populated low-elevation islands. Areas of ocean floor once beyond the scope of human intervention are now coming within reach. Melting ice may be clearing new maritime shipping routes. But, as we will see, international rules to govern who owns what in these remote frontiers are subject to dispute. When land boundaries jut outward into the sea, it depends on who draws the maps. We should be aware of the potential for competition, even conflict.

© H. J. de Blij

major geographic qualities of THE PACIFIC REALM

1. The Pacific Realm's total area is the largest of all geographic realms. Its land area, however, is the smallest, as is its population.
2. The island of New Guinea, with 11.1 million people, alone contains more than three-quarters of the Pacific Realm's population.
3. The Pacific Realm, with its huge expanses of water and numerous islands, has been strongly affected by United Nations Law of the Sea provisions regarding states' rights over economic assets in their adjacent waters.
4. The highly fragmented Pacific Realm consists of three island regions: Melanesia (including New Guinea), Micronesia, and Polynesia.
5. The Pacific Realm's islands and cultures may be divided into volcanic high-island cultures and coral-based low-island cultures.
6. In Micronesia, U.S. influence has been particularly strong and continues to affect local societies.
7. In Polynesia, indigenous culture exhibits remarkable consistency and uniformity throughout the region, its enormous dimensions and dispersal notwithstanding. Yet at the same time, local cultures are nearly everywhere severely strained by external influences. In Hawai'i, as in New Zealand, indigenous culture has been largely submerged by Westernization.

DEFINING THE REALM

Our survey begins with the Pacific geographic realm that covers just about an entire hemisphere, the one commonly called the Sea Hemisphere (Fig. 12-1). This Sea Hemisphere meets the Russian and North American realms in the far north and merges into the Southern Ocean in the south. Despite the preponderance of water, this fragmented, culturally complex realm does possess regional identities. It includes the Hawaiian Islands, Tahiti, Tonga, and Samoa—fabled names in a world apart.

In terms of modern cultural and political geography, Indonesia and the Philippines are not part of the Pacific Realm, though Indonesia's political system reaches into it; nor do Australia and New Zealand belong to it. Before the European invasion and colonization, Australia would have been included (because of its Aboriginal population) as well as New Zealand (its Maori population has Polynesian affinities). But the Europeanization of their countries has engulfed indigenous Australians and Maori New Zealanders, and the regional geography of Australia and New Zealand today is decidedly not Pacific. In New Guinea, on the other hand, Pacific peoples, numerically and culturally, remain the dominant element.

In examining the Pacific Realm, it is important to keep in mind the dimensional contrasts this part of the world presents. The realm is enormous, but its total land area is a mere 975,000 square kilometers (377,000 sq mi), about the size of Texas plus New Mexico, and over 90 percent of this lies in New Guinea.* The population is so widely scattered that a distribution map would be impractical, and it totals only around 14 million (about as many people as there are in Buenos Aires, Argentina).

COLONIZATION AND INDEPENDENCE

The Pacific islands were colonized by the French, British, and Americans; an indigenous Polynesian kingdom in the Hawaiian Islands was annexed by the United States and since 1959 has functioned as the fiftieth State. Yet today's map is still an assemblage of independent and colonial territories (Fig. 12-1). France controls New Caledonia and French Polynesia. The United States administers Guam as well as American Samoa, the Line Islands, Wake Island, Midway Islands, and several smaller island groups; the U.S. also has special relationships with other territories, former dependencies that are now nominally independent. The British, through New Zealand, have responsibility for the Pitcairn group of islands, and New Zealand administers and supports the Cook, Tokelau, and Niue Islands. Easter Island, that storied speck of land in the southeastern Pacific, is part

*The numbers in the Data Table in Appendix B do not match these totals because only the political entity of Papua New Guinea is listed, not the Indonesian province of Papua that occupies the western part of the island. Here, as Figure 10B-1 shows, the political and the realm boundaries do not coincide.

of Chile. Indonesia rules Papua, the western half of the island of New Guinea.

Other island groups have now become independent states. The largest are Fiji, once a British dependency, the Solomon Islands (also formerly British), and Vanuatu (until 1980 ruled jointly by France and Britain). Also on the current map, however, are such microstates as Tuvalu, Kiribati, Nauru, and Palau. Foreign aid is crucial to the survival of most of these countries. Tuvalu, for instance, has a total area of 26 square kilometers (10 sq mi), a population of barely 10,000, and a per-capita GNI of about U.S. $5000, derived from fishing, sales of copra (coconut meat used to make oil), and some tourism. But what really keeps Tuvalu going is an international trust fund set up by Australia, New Zealand, the United Kingdom, Japan, and South Korea. Annual grants from that fund, as well as remittances sent back to families by workers who have left for New Zealand and elsewhere, allow Tuvalu to survive.

THE PACIFIC REALM AND ITS MARINE GEOGRAPHY

Certain land areas may not be part of the Pacific Realm (we mentioned the Philippines and New Zealand), but the Pacific Ocean extends from the shores of North and South America to mainland East and Southeast Asia and from the Bering Sea off Alaska southward to the Subtropical Convergence around latitude 40°S. This means that several seas, including the Sea of Japan (East Sea), the East China Sea, and the South China Sea, are part of the Pacific Ocean. As we shall see below, this relationship matters. Pacific coastal countries, large and small, mainland and island, compete for jurisdiction over the waters that bound them.

The Pacific Realm and its ocean, therefore, form an ideal place to focus on **marine geography [1]**. This field encompasses a variety of approaches to the study of oceans and seas; some marine geographers focus on the biogeography of coral formations, others on the geomorphology of beaches, and still others on the movement of currents. A particularly interesting branch of marine geography has to do with the definition and delimitation of political boundaries at sea. Here geography intersects political science and maritime law.

The State at Sea

Littoral (coastal) states do not end where atlas maps suggest they do. States have claimed various forms of jurisdiction over coastal waters for centuries, closing off bays and estuaries and ordering foreign fishing fleets to stay away from nearby fishing grounds. Thereby arose the notion of the **territorial sea [2]**, where all the rights of a coastal state would prevail. Beyond lay the **high seas [3]**, free, open, and unfettered by national interests.

It was in the interest of colonizing, mercantile states to keep territorial seas narrow and high seas wide, thereby interfering as little as possible with their commercial fleets. In the seventeenth and eighteenth centuries the territorial sea was 3, 4, or at most 6 nautical miles wide, and the colonizing powers claimed the identical widths for their colonies (1 nautical mile = 1.85 kilometers or 1.15 statute miles).*

During the twentieth century, these constraints weakened. States without trading fleets saw no reason to limit their territorial seas. States with nearby fishing grounds traditionally exploited by their own fleets wanted to keep the increasing number of foreign trawlers away. States with shallow **continental shelves [4]** (offshore continuations of coastal plains) wished to control their resources on and below the seafloor, made more accessible by improved technology. States also disagreed on the methods by which offshore boundaries, whatever their width, should be defined. Early efforts by the League of Nations (the precursor to the United Nations) during the 1920s to resolve these issues met with only partial success, mainly in the technical sphere of boundary delimitation.

Scramble for the Oceans

In 1945, the United States helped precipitate what has become known as the scramble for the oceans. President Truman issued a proclamation that claimed U.S. jurisdiction and control over all the resources "in and on" the continental shelf down to its margin, around 100 fathoms (183 meters/600 ft) deep. In some areas, the shallow continental shelf of the United States extends more than 300 kilometers (200 mi) offshore, and Washington did not want foreign countries drilling for oil just beyond the 3-mile territorial sea.

Few observers foresaw the impact the Truman Proclamation would have, not only on U.S. waters but on the oceans everywhere, including the Pacific. It set off a rush of additional claims. In 1952, a group of South American countries, some with little continental shelf to claim, issued the Declaration of Santiago, claiming exclusive fishing rights up to a distance of 200 nautical miles off their coasts. Meanwhile, as part of the Cold War competition, the Soviet Union claimed a 12-mile territorial sea and urged its allies to do likewise.

UNCLOS Intervention

At this point the United Nations intervened, and a series of UNCLOS (United Nations Conference on the Law of the Sea) meetings began. These meetings addressed issues ranging

*Here and in the rest of this section all distances involving maritime boundaries are given in nautical miles only. Throughout the modern history of maritime law, the latter unit has been the only one used for this kind of boundary making. Any distance stated in nautical miles can be converted to its metric equivalent by multiplying it by 1.85. The conversion table in Appendix A will aid in this computation.

from the closure of bays to the width and delimitation of the territorial sea, and after three decades of negotiations they achieved a convention that changed the political and economic geography of the world's oceans forever. Among its key provisions were the authorization of a 12-mile territorial sea for all countries as well as the establishment of a 200-mile (230-statute-mi/370-km)-wide **Exclusive Economic Zone (EEZ) [5]** over which a coastal state would have total economic rights. Resources in and under this EEZ (fish, minerals, oil, and gas) belong to the coastal state, which could either exploit them or lease, sell, or share them as it saw fit. These zones were noted in Chapter 2A in regard to the new competition for the Arctic Ocean, but as we will see EEZs take on even greater importance in this Pacific Realm.

These new provisions had a far-reaching impact on the world's oceans and seas (Fig. 12-2), especially the Pacific. Unlike the Atlantic Ocean, the Pacific is studded with islands large and small, and a microstate consisting of a single small island suddenly acquired an EEZ covering 166,000 square nautical miles. European colonial powers that still controlled minor Pacific possessions (most notably France) saw their maritime jurisdictions vastly expanded. Small, low-income archipelagos could now bargain with large, rich fishing nations over fishing rights in their EEZs. And for all the UNCLOS Convention's provisions for the "right of innocent passage" of shipping through EEZs and via narrow straits, the extent of the world's high seas has been diminished accordingly.

Maritime Boundaries

The extension of the territorial sea to 12 nautical miles and the EEZ to an additional 188 nautical miles created new **maritime boundary [6]** problems. Waters less than 24 nautical miles wide separate many countries all over the world, so that **median lines [7]**, equidistant from opposite shores, have been delimited to establish their territorial seas. And even more countries lie closer than 400 nautical miles apart, requiring further maritime-boundary delimitation to determine their EEZs. In such maritime arenas as the North Sea, the Caribbean Sea, and the East and South China seas (and more recently the Arctic Ocean), a maze of maritime boundaries emerged, some of them subject to dispute. Furthermore, political changes on land can also lead to significant modifications at sea.

A case in point involves recently independent East Timor and its neighbor across the Timor Sea, Australia (see Fig. 10B-8). Australia had divided the waters and seafloor of the Timor Sea with Indonesia while recognizing Indonesia's 1976 annexation of East Timor. Thus when East Timor achieved independence in 2002, the so-called *Timor Gap* became an issue: where was the median line that would divide Timorese and Australian claims to the oil and gas reserves in this energy-rich zone? The Australians argued that since the line defined with Indonesia predated East Timor's independence, it should continue in effect, giving Australia the bulk of the energy resources. But UNCLOS regulations, to which Australia subscribed, required a redelimitation, giving East Timor a much larger share. This led Australia to withdraw from UNCLOS in 2002 so that it would not be bound by its rules. After difficult negotiations, the Australians in 2005 offered East Timor a half share of the energy revenues from the natural gas reserve that was in dispute, in addition to income already flowing from another major gasfield in the so-called Joint Petroleum Development Area (Fig. 10B-8). In return, East Timor's government agreed to defer the maritime-boundary issue for 50 years, leaving a future generation to resolve this question. It was estimated that, over the next 30 years, East Timor might receive U.S. $13 billion from these sources, and, depending on the life and capacity of the reserves, perhaps even more. This will go a long way toward meeting that impoverished fledgling state's most urgent needs.

EEZ Implications

The UNCLOS provisions created new opportunities for some states to expand their spheres of influence. Wider territorial-sea and EEZ allocations raised the stakes: claiming an island now entailed potential control over a huge maritime area. In Chapter 2A, we mentioned the need to carve up the Arctic Basin as the ice melts and resources beckon. In Chapters 9B and 10B, we referred to several island disputes off mainland East and Southeast Asia, which involve Japan and Russia, Japan and South Korea, Japan and China, China and Vietnam, and China and the Philippines. Ownership of many islands there is uncertain, and small specks of island territory have become large stakes in the scramble for the oceans. For instance, in the case of the Spratly Islands (see Fig. 10A-8), six countries claim ownership, including both Taiwan and China. China's island claims in the South China Sea support Beijing's contention that this body of water is part and parcel of the Chinese state—a position that increasingly worries other states with coasts facing it.

Figure 12-2 reveals what EEZ regulations have meant to Pacific Realm countries such as Tuvalu, Kiribati, and Fiji, with nearly circular EEZs now surrounding the clusters of islands in this vast ocean space. Japan, Taiwan, and other nations with fisheries have purchased fishing rights in these EEZs from the island governments. Nonetheless, violations of EEZ rights do occur; not long ago, Vanuatu and the Philippines were at odds over unauthorized Filipino fishing in Vanuatu's EEZ. The process of boundary delimitation continues, and the Pacific (and world) map of maritime boundaries remains very much a work in progress.

That point is underscored by a development that occurred at the end of the 1990s, when the implications of provisions in Article 76 of the 1982 UNCLOS Convention were reexamined. Whenever a continental shelf extends beyond the 200-mile limit of the EEZ, the Convention

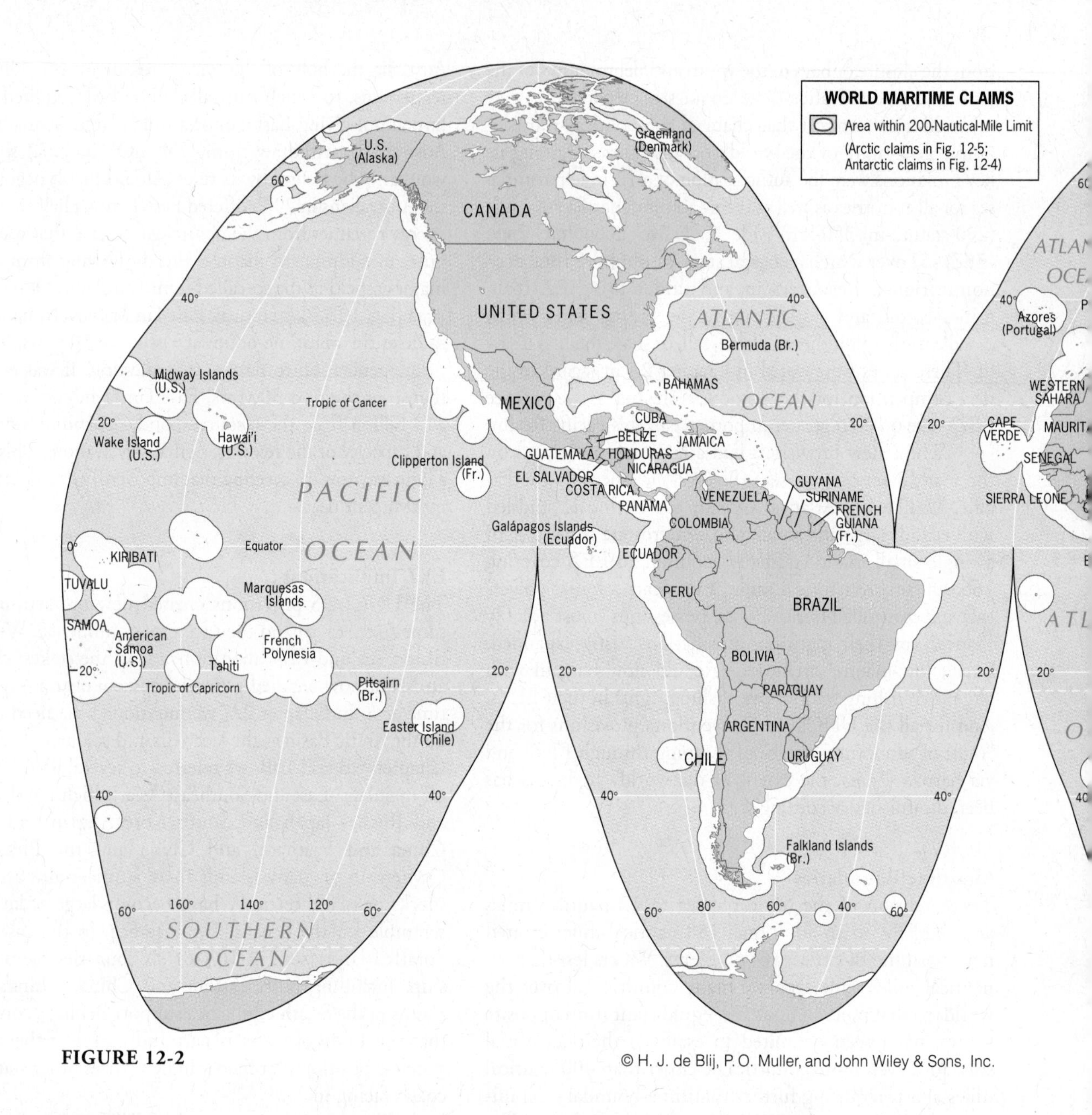

FIGURE 12-2

apparently permits a coastal state to claim that extension as a natural prolongation of its landmass. Although there is as yet no regulation to also allow the state to extend its EEZ to the edge of the continental shelf, it is not difficult to foresee such an amendment to the Convention. As it stands, states are now delimiting their proposed natural prolongations under a current deadline, which puts poorer states at a disadvantage since the required marine surveys are quite expensive. Today, this issue has become quite heated, especially in the Arctic Ocean as countries scramble to claim prolongations (most notably Russia, as we shall see at the end of this chapter). The big winners are likely to include Russia, the United States, Canada, Australia, New Zealand, and India—although a total of 60 coastal countries could ultimately benefit to some extent from the provision.

Another complicating factor is the role of new technologies that are accelerating deep-sea exploration and the exploitation of seabed resources. Today, the quest to discover additional mineral deposits is increasingly an underwater endeavor, led by the search for so-called 'massive sulfides'—enormous concentrations of gold, silver, copper, and other metallic minerals formed in sulfide-rich, subseafloor, volcanic environments. The technological breakthroughs entail sensors, advanced robotic devices, and other state-of-the-art gear (such as "sea gliders") developed by the offshore oil and gas industry. The growing use of sea gliders is of special interest because these small, slow-moving, unmanned vessels can track enormous distances as they collect detailed information on everything below—from marine life to seabed resources to the movements of submarines. And so the scramble for the oceans marches

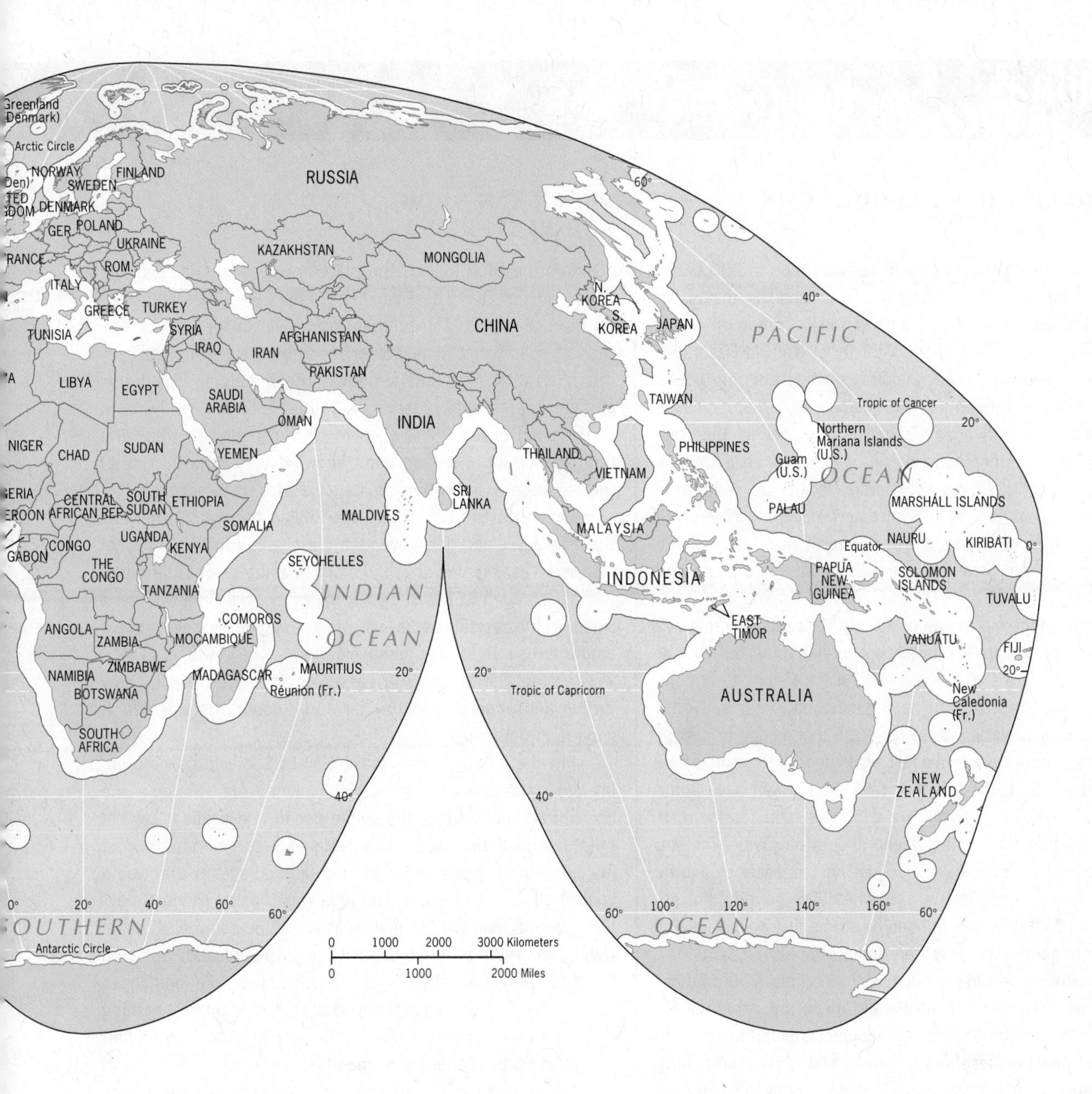

ever onward, and with it the constriction of the world's high seas and open waters.

● REGIONS OF THE REALM

Sail across the Pacific Ocean, and one spectacular vista follows another. Dormant and extinct volcanoes, sculpted by erosion into basalt spires draped by luxuriant tropical vegetation and encircled by reefs and lagoons, tower over azure waters. Low atolls with nearly snow-white beaches, crowned by stands of palm trees, seem to float in the water. Pacific islanders, where foreign influences have not overtaken them, appear to take life with enviable ease.

More intensive investigations, however, reveal that such Pacific sameness is more apparent than real. Even the Pacific Realm, with its centuries-old sailing traditions, its still-diffusing populations, and its historic migrations, has a durable regional framework. Figure 12-3 outlines the three regions that constitute this realm:

Melanesia: Papua (province of Indonesia), Papua New Guinea, Solomon Islands, Vanuatu, New Caledonia (France), Fiji

Micronesia: Palau, Federated States of Micronesia, Northern Mariana Islands, Republic of the Marshall Islands, Nauru, western Kiribati, Guam (United States)

Polynesia: Hawaiian Islands (U.S.), Samoa, American Samoa, Tuvalu, Tonga, eastern Kiribati, Cook and other New Zealand-administered islands, French Polynesia, Easter Island (Chile)

Regional ISSUE Who Should Own the Oceans?

WE WHO LIVE HERE SHOULD OWN THE WATERS

"Without exclusive fishing rights in our waters up to 200 nautical miles offshore, our economy would be dead in the water, no pun intended, sir. As it is, it's bad enough. I'm an official in the Fisheries Department of the Ministry of Natural Resources Development here in Tuvalu, and this modest building you're visiting is the most modern in all of the capital, Funafuti. That's because we're the only ones making some money for the state other than our Internet domain (*tv*), which has been the big story around here since we commercialized it. We don't have much in the way of natural resources and we export some products from our coconut and breadfruit trees, but otherwise we need donors to help us and major fishing nations to pay us for the use of our 750,000 square kilometers of ocean.

"Which brings me to one of our big concerns. Three-quarters of a million square kilometers sound like a lot, but look at the map. It's a mere speck in the vast Pacific Ocean. Enormous stretches of our ocean seem to belong to nobody. That is, they're open to fishing by nations that have the technology to exploit them, but we don't have those fast boats and factory ships they use nowadays. Still, the fish they take there might have come to our exclusive grounds, but we don't get paid for them. It seems to us that the Pacific Ocean should be divided among the countries that exist here, not just on the basis of nearby Exclusive Economic Zones but on some other grounds. For example, islands that are surrounded by others, like we are, ought to get additional maritime territory someplace else. Islands that aren't surrounded should get much more than just 200 nautical miles. To us the waters of the Pacific mean much more than they do to other countries, not only economically but historically too. I don't like Japanese and American and Norwegian fishing fleets on these nearby 'high seas.' They don't obey international regulations on fishing, and they overfish wherever they go. When there's nothing left, they'll simply end their agreement with us, and we'll be left without income from a natural resource over which we never had control.

"I don't expect anything to change, but we feel that the era of the 'high seas' should end and all waters and subsoil should be assigned to the nearest states. The world doesn't pay much attention to our small countries, I realize. But we've put up with colonialism, world wars, nuclear weapons testing, pollution. It's time we got a break."

THE OCEAN SHOULD BE OPEN AND RESTRICTED

"Ours is the Blue Planet, the planet of oceans and seas, and the world has fared best when its waters were free and open for the use of all. Encroachment from land in the form of 'territorial' seas (what a misnomer!) and other jurisdictions only interfered with international trade and commercial exploitation and caused far more problems than they solved. Already much of the open ocean is assigned to coastal countries that have no maritime tradition and don't know what to do with it other than to lease it. And when coastal countries lie closer than 400 nautical miles to each other, they divide the waters between them for EEZs and don't even leave a channel for international passage. Yes, I realize that there are guarantees of uninhibited passage through EEZs. But mark my words: the time will come when small coastal states will start charging vessels, for example cruise ships, for passage through 'their' waters. That's why, as the first officer of a passenger ship, I am against any further allocation of maritime territory to coastal states. We've gone too far already.

"In the 1960s, when those UNCLOS meetings were going on, a representative from Malta named Arvid Pardo came up with the notion that the whole ocean and seafloor beyond existing jurisdictions should be declared a 'common heritage of mankind' to be governed by an international agency that would control all activities there. The idea was that all economic activity on and beneath the high seas would be controlled by the UN, which would ensure that a portion of the proceeds would be given to 'Geographically Disadvantaged States' like landlocked countries, which can't claim any seas at all. Fortunately nothing came of this—can you imagine a UN agency administering half the world for the benefit of humanity?

"So you can expect us mariners to oppose any further extensions of land-based jurisdiction over the seas. The maritime world is complicated enough already, what with choke points, straits, narrows, and various kinds of prohibitions. To assign additional maritime territory to states on historic or economic grounds is to complicate the future unnecessarily and to invite further claims. From time to time I see references to something called the 'World Lake Concept,' the preliminary assignment of all the waters of the world to the coastal states through the delimitation of median lines. Such an idea may be the natural progression from the EEZ concept, but it gives me nightmares."

Vote your opinion at www.wiley.com/go/deblijpolling

MAJOR CITIES OF THE REALM

City	Population* (in millions)
Honolulu, Hawai'i (U.S.)	1.0
Port Moresby, Papua New Guinea	0.3
Nouméa, New Caledonia	0.2
Suva, Fiji	0.2

*Based on 2014 estimates.

Ethnic, linguistic, and physiographic criteria are the major foundations for this regionalization of the Pacific Realm, but we should not lose sight of the dimensions. Not only is the land area small, but as we also noted above the current population of this entire realm (including Indonesia's Papua Province) totals just about 14 million—barely 10 million without Papua—about the same as one very large city. Even fewer people live in this realm than in another vast area of far-flung settlements—the oases of North Africa's Sahara.

MELANESIA

The large island of New Guinea lies at the western end of a Pacific region that extends eastward to Fiji and includes the Solomon Islands, Vanuatu, and New Caledonia (Fig. 12-3). The human mosaic here is complex, both ethnically and culturally. Most of the 11.1 million people of New Guinea (divided between the Indonesian province of Papua and the independent state of Papua New Guinea) are Papuans, and a large minority is Melanesian. Altogether there are more than 800 communities speaking different languages; the Papuans are most numerous in the densely forested highland interior and in the lowland south, whereas the Melanesians inhabit the north and east. Minus Papua, Melanesia contains over 9 million inhabitants, making this the most populous Pacific region by far.

With 7.3 million people today, **Papua New Guinea (PNG)** became a sovereign state in 1975 after nearly a century of British and Australian administration. Almost all of PNG's limited development is taking place along its coasts, whereas most of the interior remains hardly touched by the changes that transformed neighboring Australia. Perhaps four-fifths of the population is part of the self-sufficient subsistence economy, growing root crops, raising pigs, hunting wildlife, and gathering forest products. Old traditions of the kind lost in Australia continue on here, protected by remoteness and the rugged terrain.

Welding this disparate population into a nation is a task hardly begun, and PNG faces numerous obstacles in addition to its cultural complexity. Not only are hundreds of languages in use, but more than 40 percent of the population is estimated to be illiterate. English, the official language, is used by the educated minority but is of little use beyond the coastal zone and its towns. The capital, Port Moresby, contains about 330,000 residents, reflecting the

From the Field Notes . . .

"Arriving in the capital of New Caledonia, Nouméa, in 1996, was an experience reminiscent of French Africa 40 years earlier. The French tricolor was much in evidence, as were uniformed French soldiers. European French [Caldoche] residents occupied hillside villas overlooking palm-lined beaches, giving the place a Mediterranean cultural landscape. And New Caledonia, like Africa, is a source of valuable minerals. It is one of the world's largest nickel producers, and from this vantage point you could see the huge treatment plants, complete with concentrate ready to be shipped (left, under conveyor). What you cannot see here is how southern New Caledonia has been ravaged by the mining operations, which have denuded whole mountainsides. Working in the mines and in this facility are the local Kanaks, Melanesians who make up around 45 percent of the population of roughly 275,000. Violent clashes between Kanaks and Caldoche French have obstructed government efforts to change New Caledonia's political status in such a way as to accommodate pressures for independence as well as continued French administration."

© H.J. de Blij

www.conceptcaching.com

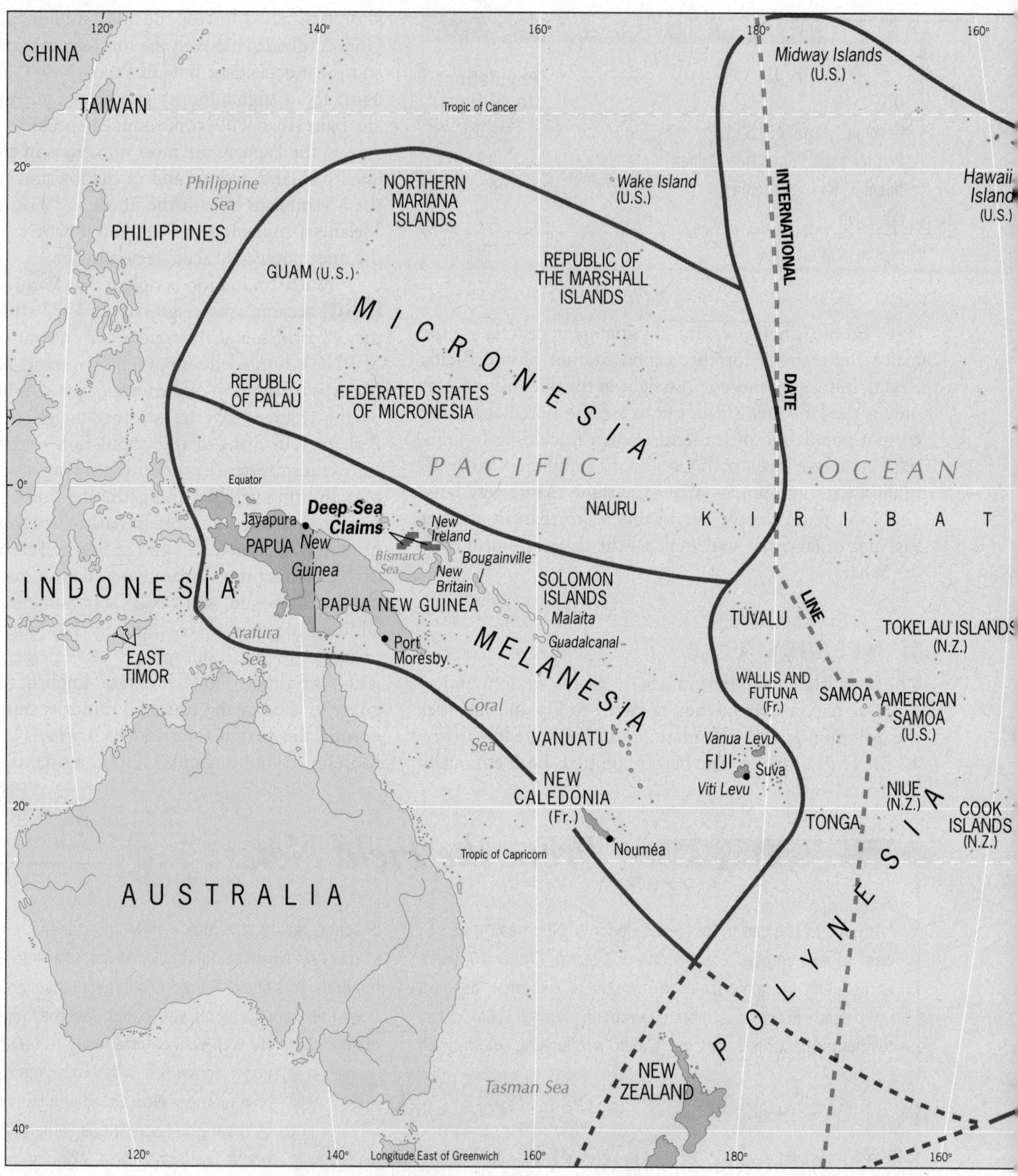

FIGURE 12-3

very low level of urbanization (13 percent) in this developing economy.

Yet Papua New Guinea is not without economic opportunities. Oil was discovered in the 1980s, and by 2000 crude oil was PNG's largest export by value. More recently, newly found gas deposits have propelled the growth of the energy sector, and a major liquid natural gas development project will be completed by 2015. Other important exports include gold, copper, silver, timber, and several agricultural products that include coffee and cocoa, reflecting the country's diversity of environments and resources. And PNG has not been immune to Pacific Rim influences: although the biggest share of exports still go to its nearest neighbor, Australia, Japan ranks second, and fast-rising China is close behind.

Turning eastward, it is a measure of Melanesia's cultural fragmentation that as many as 120 languages are spoken in the approximately 1000 islands that make up the

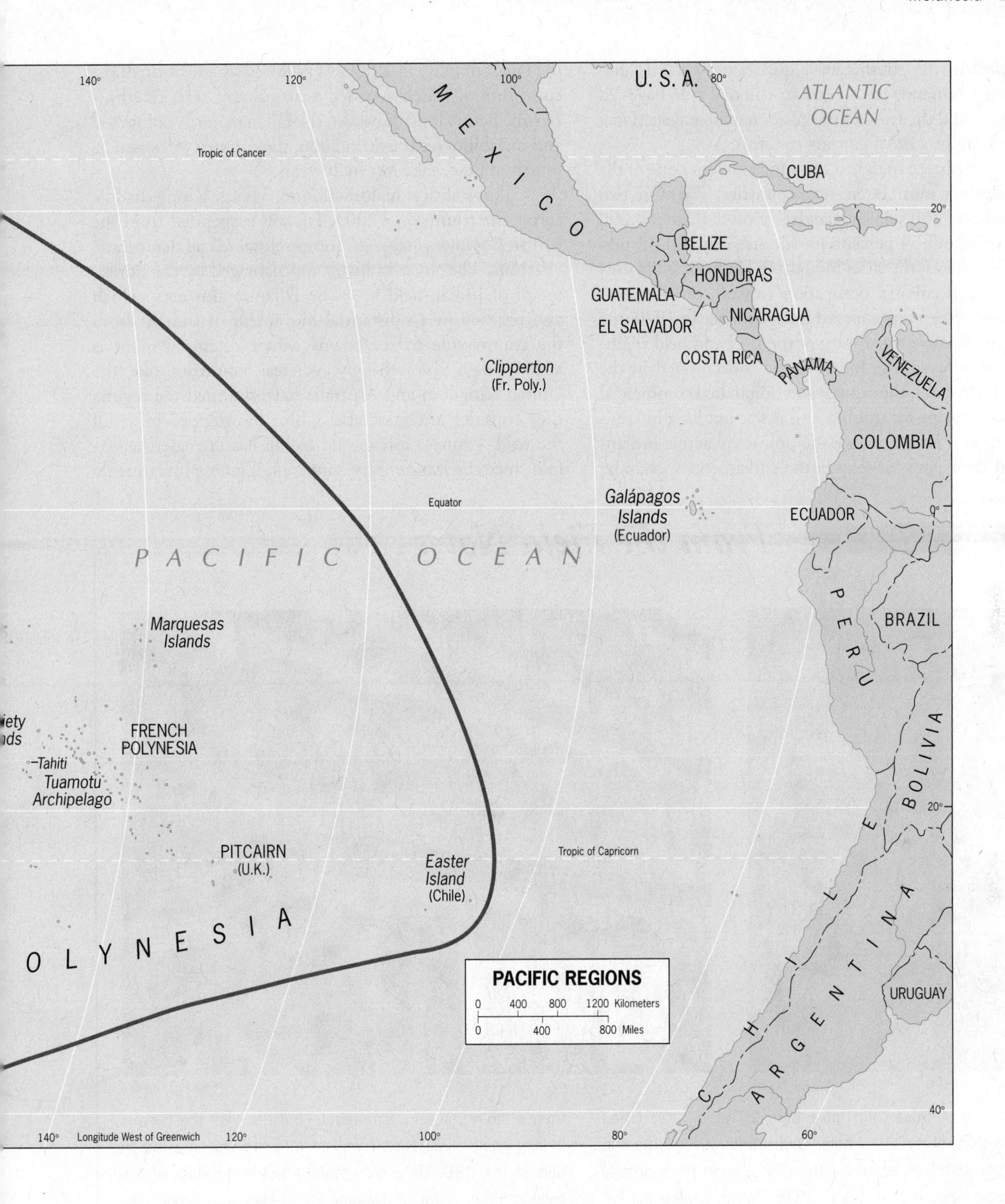

Solomon Islands (about 80 of them support almost all the people, who number around 600,000). Inter-island, historic animosities among the islanders were worsened by the events of World War II, when U.S. forces moved thousands of Malaitans to Guadalcanal. This started a postwar cycle of violence that finally led Australia to intervene in 2003.

New Caledonia, still under French rule, is in a very different situation. Only around 45 percent of its population of about 300,000 are Melanesian; not that far behind are the 33 percent of European ancestry (known as the Caldoche), many of them descended from the inhabitants of the penal colony France established here during the nineteenth century. Nickel mines, based on reserves that rank among the world's largest, dominate New Caledonia's export economy (see *Field Note*). The mining industry attracted additional French settlers, and social problems soon arose. Most of the French population lives in or near the capital city of Nouméa, steeped in French cultural landscapes, in the southeastern quadrant of the island. Melanesian demands for an end to colonial rule have led to violence, and the two communities remain engaged in a long-running process of coming to terms.

On its eastern margins, Melanesia includes one of the Pacific Realm's most interesting countries, **Fiji**. On two larger and over 100 smaller islands live more than 800,000 Fijians, of whom 54 percent are Melanesians and 38 percent South Asians, the latter brought to Fiji from India during the British colonial occupation to work on the sugar plantations. When Fiji achieved independence in 1970, the indigenous Fijians owned most of the land and held political control, whereas the Indians were concentrated in the towns (chiefly Suva, the capital) and dominated commercial life. It was a recipe for trouble, and it was not long in coming when, in a later election, the politically active Indians outvoted the Fijians for seats in the parliament. A coup by the Fijian military in 1987 was followed by a revision of the constitution, which awarded a majority of seats to ethnic Fijians. Before long, however, the Fijian majority splintered and instability returned. In 2006, the country witnessed its fourth military takeover in 20 years.

This political malfunctioning has cast a long shadow across the country. In 2009, Fiji was suspended from the **British Commonwealth [8]** for its refusal to call democratic elections. The sugar industry was damaged by the nonrenewal of Indian-held leases by Fijian landowners, which also resulted in a substantial movement of Indians from the countryside to the towns, where unemployment is already high. Interestingly, whereas countries like the United Kingdom and Australia have shunned the regime they consider undemocratic, China has stepped in to fill the void. Chinese foreign aid to Fiji has increased sevenfold since the last military coup, and Chinese investments

From the Field Notes . . .

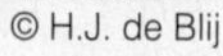

© H.J. de Blij

© H.J. de Blij

"Returning in Suva (Fiji) more than 20 years after I had done a study of its CBD, I noted that comparatively little had changed, although somehow the city seemed more orderly and prosperous than it was in 1978. At the Central Market I watched a Fijian woman bargain with a seller over a batch of taro, the staple starch provider in local diets. I asked her how she would serve it. 'Well, it's like your potato,' she said. 'I can make a porridge, I can cut it up and put it in a stew, and I can even fry pieces of it and make them look like the French fries they give you at the McDonald's down the street.' Next I asked the seller where his taro came from. 'It grows over all the islands,' he said. 'Sometimes, when there's too much rain, it may rot—but this year the harvest is very good.' . . . Next I walked into the crowded Indian part of the CBD. On a side street I got a reminder of the geographic concept of agglomeration. Here was a colonial-period building, a former hotel converted into a business center. I counted 15 enterprises, ranging from a shoe store to a photographer and including a shop where rubber stamps were made and another selling diverse tobacco products. Of course (this being the Indian sector of downtown Suva) tailors outnumbered all other establishments."

www.conceptcaching.com

in infrastructural projects have increased as well. This is but another sign that China's influence is now being projected on a global scale, and not always in a manner that existing powers find reassuring.

Melanesia, the most populous region in the Pacific Realm, is bedeviled by centrifugal forces of many kinds. No two countries present the same form of multiculturalism; each has its own challenges to confront, and some of these spill over into neighboring (or more distant foreign) islands.

MICRONESIA

North of Melanesia and to the east of the Philippines lie the islands that constitute the region known as Micronesia (Fig. 12-3). The name refers to the size of the islands: the 2000-plus islands of Micronesia are not only tiny (many of them no larger than one square kilometer [half a square mile]), but they are also much lower-lying, on average, than those of Melanesia. Some are volcanic islands—**high islands [9]**, as the people call them—but they are outnumbered by islands composed of coral, the **low islands [10]** that barely protrude above sea level (see photo). Guam, with 550 square kilometers (210 sq mi), is Micronesia's largest island, and no elevation anywhere in Micronesia reaches 1000 meters (3300 ft).

Take, for example, the country of **Kiribati**, consisting of three main island groups that lie across Micronesia and Polynesia (Fig. 12-3). The average elevation of Kiribati's 33 islands is 2 meters (6.5 ft) and, like all other low islands in this realm, they are highly vulnerable to rising sea level at a time of global warming.

A classic example of a *low island* in the archipelago that forms Kiribati, which lies near where the equator and International Date Line intersect—as well as the realm's three regions.

© H.J. de Blij

The high-island/low-island dichotomy is useful not only in Micronesia, but also throughout the realm. Both the physiographies of these islands and the economies they support differ in crucial ways. High islands wrest substantial orographic rainfall from the moist maritime air and tend to be well watered with good volcanic soils. As a result, agricultural products show some diversity, and life is reasonably secure. Populations tend to be larger on these high islands than on the low islands, where drought is the rule and fishing and the coconut palm are the mainstays of life. Small communities cluster on the low islands, and over time many have died out. The major migrations, which sent fleets to populate islands from Hawai'i to New Zealand, tended to originate in the high islands.

Until the mid-1980s, for the most part, Micronesia was a United States Trust Territory (the last of the post–World War II trusteeships supervised by the United Nations). But, as Figure 12-3 shows, Micronesia is now divided into countries bearing the names of independent states. Today the **Marshall Islands**, where the United States tested nuclear weapons (giving prominence to the name *Bikini*), is a republic in "free association" with the United States, having the same status as Palau and the Federated States of Micronesia. The **Northern Mariana Islands** are a commonwealth "in political union" with the United States, which in effect provides billions of dollars in assistance to these countries, in return for which they commit themselves to avoid foreign policy actions that are contrary to American interests.

The Republic of **Palau** is another fascinating example of how some of these small Pacific nations make a living. This group of islands some 800 kilometers (500 mi) east of the Philippines became officially independent in 1994 and has a total population of roughly 25,000. Palau has no military but entered into an agreement with the United States to provide security in return for a 50-year lease for a base. It is highly dependent upon American foreign assistance. Palau also gets funds and investments from Taiwan, a reward for Palau's official recognition of Taiwan's sovereign status as the Republic of China (and for withholding recognition of the People's Republic of China).

Also part of Micronesia is the U.S. territory of **Guam**, where independence is not in sight and where U.S. military installations and tourism provide most of its income, and the remarkable Republic of **Nauru**. With a population of only about 10,000 and just 20 square kilometers (8 sq mi) of land, Nauru got rich by selling its phosphate deposits to Australia as well as New Zealand, where they are used as fertilizer. Per capita incomes had risen above U.S. $12,000, making Nauru one of the realm's high-income societies. But the main phosphate deposits have run out; now uncertainty prevails as mining of a secondary layer proceeds.

In this region of tiny islands, most people subsist on farming or fishing, and virtually all the countries need infusions of foreign aid to survive. The natural economic complementarity between the high-island farming cultures and the low-island fishing communities all too often is negated by distance, spatial as well as cultural. Life here may seem idyllic to the casual visitor, but for the Micronesians it is usually a major, daily challenge.

POLYNESIA

To the east of Micronesia and Melanesia lies the heart of the Pacific, enclosed by a great triangle stretching from the Hawaiian Islands to Chile's Easter Island to New Zealand. This is Polynesia (Fig. 12-3), a region of numerous islands (*poly* means many), ranging from volcanic mountains rising above the Pacific's waters (Mauna Kea on Hawai'i reaches over 4200 meters [nearly 13,800 ft]), clothed by luxuriant tropical forests and drenched by well over 250 centimeters (100 in) of rainfall each year, to low coral atolls where a few palm trees form the only vegetation and where drought is a persistent problem.

Its vastness and the diversity of its natural environments notwithstanding, Polynesia clearly constitutes a geographic region within the Pacific Realm. Polynesian culture, though spatially fragmented, exhibits a remarkable consistency and uniformity from one island to the next, from one end of this widely dispersed region to the other. This consistency is particularly expressed in vocabularies, technologies, housing, and art forms. The Polynesians are uniquely adapted to their marine environment, and long before European sailing ships began to arrive in their waters, Polynesian seafarers had learned to navigate their wide expanses of ocean in huge double canoes as long as 45 meters (150 ft). They traveled hundreds of kilometers to favorite fishing zones and engaged in inter-island barter trade, using maps constructed from bamboo sticks and cowrie shells and navigating by the stars. However, modern descriptions of a Pacific Polynesian paradise of emerald seas, lush landscapes, and gentle people distort harsh realities. Polynesian society was forced to accept much loss of life at sea when storms claimed their boats; families were ripped apart by accident as well as migration; hunger and starvation afflicted the inhabitants of smaller islands; and the island communities were often embroiled in violent conflicts and cruel retributions.

The political geography of Polynesia is complex. In 1959, the **Hawaiian Islands** became the fiftieth State to join the United States. The State's population is now 1.4 million, with over 70 percent living on the island of Oahu. There, the superimposition of cultures is symbolized by the panorama of Honolulu's seaside skyscrapers framed against the famous extinct volcano at nearby Diamond Head. The Kingdom of **Tonga** became an independent country in 1970 after seven decades as a British protectorate; the British-administered Ellice Islands were renamed **Tuvalu** and received independence from Britain in 1978. Other islands continue under French control (including the Marquesas Islands and Tahiti), under New Zealand's administration (Cook Islands), and under British, U.S., and Chilean flags.

VOICE FROM THE Region

Courtesy of Fatima Sauafea-Leau

Fatima Sauafea-Le'au, fisheries biologist, American Samoa

FA'A SAMOA—THE SAMOAN WAY

"I was born and raised in American Samoa in the small coastal village of Paepaeulupo'o. Legends and stories are passed down from several generations and form the foundation of our cultural and traditional values, known as the *Fa'a Samoa*. One cultural tradition that has been carried on for many years is the management of the reef areas next to village communities. For centuries we Samoans have lived off the ocean reefs that support a rich fish fauna. Management of the reefs depended greatly on how well organized the village was in regard to its different institutions: the *matai* (village chief) system; the women's group; and the men's group. These are the different systems that make up a village community, and they each utilize the reefs and resources in different ways. But in modern times American Samoa has gone through many changes that have altered our social fabric and that have resulted in poor management practices in many places. As a fisheries biologist, I think modern scientific approaches are very useful and effective in demonstrating key processes to understand the status and health of our resources and environment, and how human activities damage them. At the same time, as a Samoan I truly believe that the incorporation of traditional knowledge with modern science will enhance effective management of our resources. Traditional ways are important in sustaining the reefs in the long run, and they also provide our people with a sense of ownership. They possess knowledge about local resources and their problems, as well as information based on their daily observations of changes occurring within their habitat. We can be modern and at the same time live in harmony with nature. As a scientist and as a Samoan, my hope is to see more village communities on my island engaged in managing our reefs and resources, and keeping *Fa'a Samoa* alive."

In the process of politico-geographical fragmentation, Polynesian culture has suffered severe blows. Land developers, hotel builders, and tourist dollars have set **Tahiti** on a course along which Hawai'i has already traveled far. The Americanization of eastern **Samoa** has created a new society different from the old. Polynesia has lost much of its ancient cultural consistency; today, the region is a patchwork of new and old—the new often bleak and barren, with the old under mounting pressure.

The countries and cultures of the Pacific Realm lie in an enormous ocean basin along whose rim a great drama of economic and political transformation will play itself out during the twenty-first century. Already, the realm's own former margins—in Hawai'i in the north and in New Zealand in the south—have been so recast by foreign intervention that little remains of the kingdoms and cultures that once prevailed. Today, the world's largest country and next superpower (China) faces the richest and most powerful (the United States) across the Pacific. How will the microstates of the Pacific Realm fare?

POLAR FUTURES

PARTITIONING THE ANTARCTIC

South of the Pacific Realm lies Antarctica and its encircling Southern Ocean. The combined area of these two gargantuan geographic expanses constitutes 40 percent of the entire planet—two-fifths of the Earth's surface containing a mere one one-thousandth of the world's population.

Is Antarctica a geographic realm? In physiographic terms, yes—but not on the basis of the criteria we use in this book. Antarctica is a continent, practically twice the size of Australia, but virtually all of it is covered by a dome-shaped ice sheet nearly 3.2 kilometers (2 mi) thick near its center. The continent frequently is referred to as "the white desert" because, despite all of its ice and snow, annual precipitation is extremely low—averaging less than 15 centimeters (6 in) per year. Temperatures are frigid, with winds so powerful that Antarctica also is called the home of the blizzard. For all its size, no functional regions have developed here, nor have any towns or transport networks except the supply lines of research stations. Antarctica still is a frontier, even a scientific frontier only grudgingly giving up its secrets. Underneath all that ice lie some 70 lakes of which Lake Vostok is the largest at over 14,000 square kilometers (5400 sq mi). It may be as much as 600 meters (2000 ft) deep. The first samples of its water were brought to the surface by Russian scientists after their decades-long drilling effort through 4000 meters (13,000 ft) of ice finally reached the lake in 2012.

Like virtually all frontiers, Antarctica has always attracted explorers and pioneers. Whale and seal hunters destroyed huge populations of Southern Ocean fauna during the eighteenth and nineteenth centuries, and explorers planted the flags of their countries on Antarctic shores. By the first decade of the twentieth century, the quest for the South Pole had become an international obsession; Roald

From the Field Notes . . .

"The Antarctic Peninsula is geologically an extension of South America's Andes Mountains, and this vantage point in the Gerlach Strait leaves you in no doubt: the mountains rise straight out of the frigid waters of the Southern Ocean. We have been passing large, flat-topped icebergs, but these do not come from the peninsula's shores; rather, they form on the leading edges of ice shelves or on the margins of the mainland where continental glaciers slide into the sea. Here along the peninsula, the high-relief topography tends to produce jagged, irregular icebergs. The mountain range that forms the Antarctic Peninsula continues across Antarctica under thousands of feet of ice and is known as the Transantarctic Mountains. Looking at this place you are reminded that beneath all this ice lies an entire continent, still little-known, with fossils, minerals, even lakes yet to be discovered and studied."

© H.J de Blij

www.conceptcaching.com

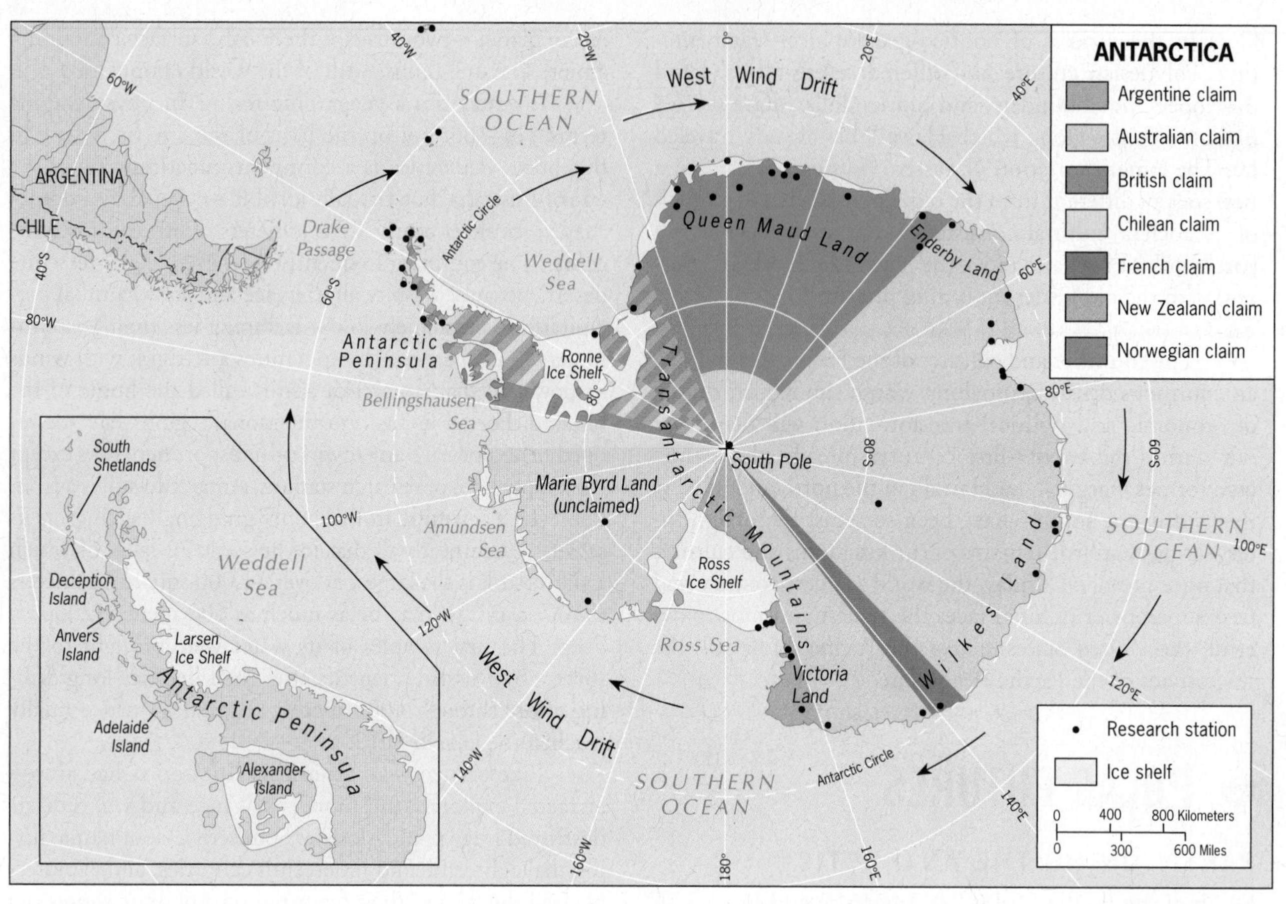

FIGURE 12-4

Amundsen, the Norwegian, reached it first in 1911. All this activity soon led to national claims in Antarctica during the interwar era (1918–1939).

The geographic effect was the partitioning of Antarctica into pie-shaped sectors centered on the South Pole (Fig. 12-4). In the least frigid area of the continent, the Antarctic Peninsula, British, Argentinian, and Chilean claims overlapped—and they still do. One sector, Marie Byrd Land (shown in neutral beige on the main map), was never formally claimed by any country.

Why should states be interested in territorial claims in so remote and difficult an area? Because these remote areas near the poles contain substantial resources and are attractive for possible future use, despite the difficulties of access. Antarctica is no different as both land and sea contain raw materials that may some day become crucial: proteins in the waters, and fossil-fuel and mineral deposits beneath the ice. As noted above, Antarctica at 14.2 million square kilometers (5.5 million sq mi) is almost twice the size of Australia, and the Southern Ocean is nearly as large as the North and South Atlantic combined. However distant actual future exploitation may be, countries want to hang onto their stakes here.

But the claimant states (those with territorial claims) recognize the need for cooperation. During the late 1950s, they participated in the International Geophysical Year (IGY) that launched major scientific programs and established a number of permanent research stations throughout the continent. This spirit of cooperation led to the 1961 signing of the **Antarctic Treaty [11]**, which ensures continued scientific collaboration, prohibits military activities, safeguards the environment, and holds national claims in abeyance. In 1991, when the treaty was extended under the terms of the Wellington Agreement, concerns were raised that it does not do enough to control future resource exploitation.

Notice, in Figure 12-4, that the map shows no maritime claims off the pie-shaped sectors that blanket most of Antarctica. In fact, some of the claimant states did draw maps that extended their claims into the Southern Ocean, but the 1991 treaty extension terminated that initiative and restricted the existing claims to the landmass. It is, of course, a possibility that those claims could be reinstated should the Wellington Agreement disintegrate: national self-interest has abrogated many an international treaty in the past. But delimiting maritime claims off Antarctica is particularly challenging for practical reasons. As Figure 12-4 shows, Antarctica in a number of places is flanked by ***ice shelves*** attached to its coast, permanent slabs of floating ice (such as the Ronne and the Ross). These ice shelves are replenished on the landward side and break off (calve) on the seaward side, where huge icebergs float northward into the Southern Ocean to become part of the wide belt of *pack ice* that encircles the continent. From where would any territorial sea or EEZ be measured? From the inner edge of the ice shelf? That would put the territorial "sea" boundary on the ice shelf! From the ever-changing outer edge of the ice shelf? Measuring a territorial sea or EEZ from such an unstable perimeter would not work either. No UNCLOS regulations would be applicable in situations like this, so no matter what the maps showed, such claims are legally unsupportable. That is just as well. The last thing the world needs is a scramble for Antarctica.

In an age of growing national self-interest and increasing raw material consumption, the possibility nonetheless exists that Antarctica and its offshore waters may yet become an arena for international rivalry. Until now, its remoteness and its forbidding environments have saved it from that fate. The entire world benefits from this because evidence is mounting that Antarctica plays a critical role in the global environmental system, so that human modifications are likely to have worldwide (and unpredictable) consequences.

GEOPOLITICS IN THE ARCTIC BASIN

To observe how different the physiographic as well as the political situation is in the Arctic, consider the implications of Figure 12-5. Not only does the North Pole lie on the floor of a relatively small body of water grandiosely labeled the Arctic Ocean, but the entire Arctic is ringed by countries whose EEZs, delimited under UNCLOS rules, would allocate much of the ocean floor (the *subsoil*, to use its legal designation) to those states. Moreover, again under UNCLOS regulations, states can expand their rights to the seafloor even farther than the 200-mile EEZ if they are able to prove their continental shelves continue beyond that limit—in fact, up to 350 nautical miles offshore.

As the main map shows, the Siberian Continental Shelf is by far the largest in the Arctic; indeed, it is the largest in the world. It extends from Russia's north coast beneath the waters of the Arctic Oacean and under the floating ice cap at the ocean's center. And because several island groups off the mainland (such as Franz Josef Land, North Land, and the New Siberian Islands) belong to Russia, the Russians can claim virtually all of Eurasia's northern continental shelf under existing regulations. But even that is not enough: the Russians want to extend their claim all the way to the North Pole itself, where in 2007 they sent a submarine to plant a Russian flag on the seafloor nearly 4000 meters (13,000 ft) below the Pole (see photo).

To bolster their claim, the Russians assert that the Lomonosov Ridge, a submerged mountain range, is a "natural extension" of their Siberian Continental Shelf. Few if any neutral observers would agree that this ridge, or several others rising from the deep floor of the Arctic Ocean, is a continental-shelf landform. But this is fodder for international disputes over maritime boundaries, and the legal wrangling is certain to go on for decades.

Another look at Figure 12-5 reveals the other states with frontage on the Arctic Ocean: Norway, through its Svalbard Islands (including Spitsbergen); Denmark, because it still has ultimate authority over now-autonomous Greenland; Canada, with tens of thousands of kilometers of Arctic coastline but proportionately less continental shelf; and the United States, with Alaska's limited but important continental shelves to its north and west. The United States stands alone among these candidates by not having ratified the UNCLOS treaties, so that the Americans do not feel bound by the regulations to which everyone else adheres. In any case, it is obvious that an Antarctic-type, pie-shaped-sector partitioning would not work here; but it is clear as well that the evolving contest among claimant states is fraught with risk. And the stakes are high: estimates of the quantity of oil and natural gas reserves to be found below Arctic waters and ice run as high as 25 percent of the world's remaining total. No matter what method is used to calculate which country gets how much, the biggest winner will be Russia, regardless of the final outcome of its Lomonosov Ridge claim.

On August 3, 2007, tens of millions of Russian television viewers witnessed live the scene pictured here: the operator of a Russian mini-submarine planting a metal Russian flag at the North Pole, on the seafloor of the Arctic Ocean beneath the polar icecap. As Figure 12-5 shows, the North Pole lies far beyond the Russian continental shelf, but in the upcoming international negotiations relating to national claims in the Arctic Basin, the Russians will claim that the Lomonosov Ridge actually constitutes an extension of their continental shelf, thereby entitling Russia to draw international boundaries up to, and even beyond, the North Pole. Other countries with actual and pending Arctic claims make less dramatic but equally assertive moves in the run-up to negotiations that have the potential to transform the map of the Arctic region. With new sea routes and newly exploited oil and gas reserves in prospect, diplomacy will be difficult but crucial.

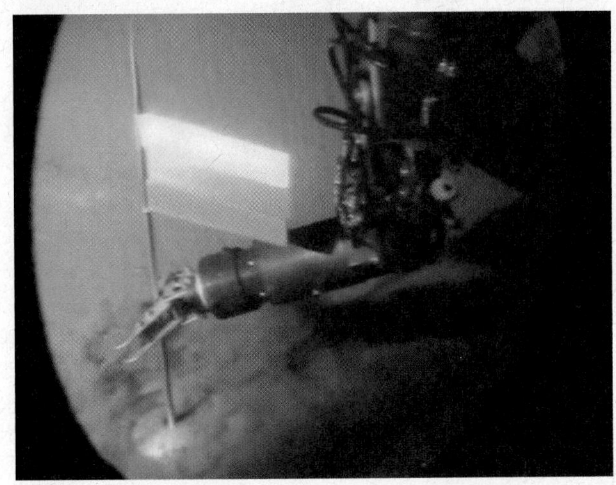

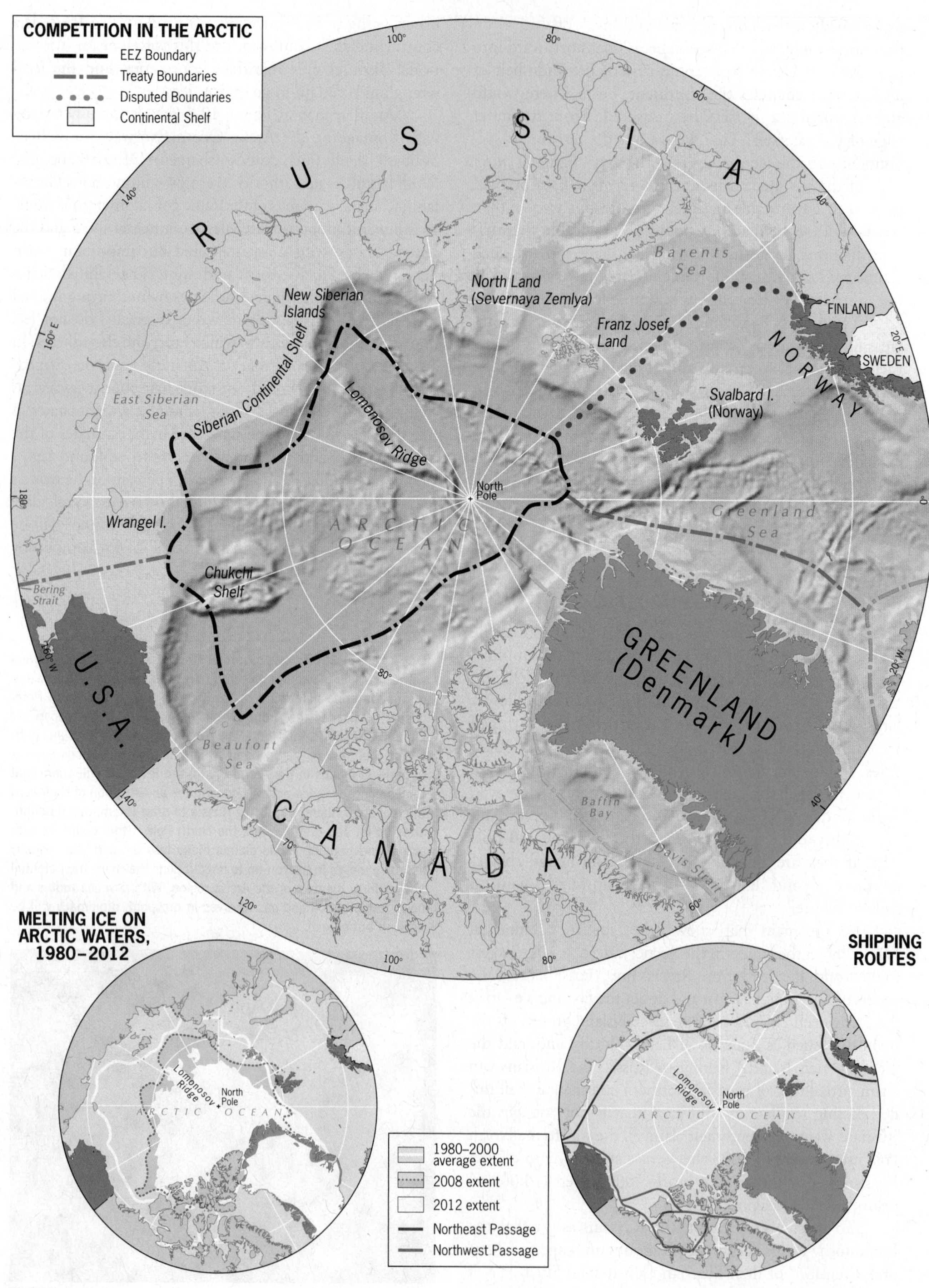
COMPETITION IN THE ARCTIC
EEZ Boundary
Treaty Boundaries
Disputed Boundaries
Continental Shelf
R U S S I A
New Siberian Islands
North Land (Severnaya Zemlya)
Franz Josef Land
Barents Sea
FINLAND
SWEDEN
NORWAY
Svalbard I. (Norway)
East Siberian Sea
Siberian Continental Shelf
Lomonosov Ridge
North Pole
Wrangel I.
ARCTIC OCEAN
Greenland Sea
Chukchi Shelf
Bering Strait
U.S.A.
GREENLAND (Denmark)
Beaufort Sea
CANADA
Baffin Bay
Davis Strait
MELTING ICE ON ARCTIC WATERS, 1980–2012
SHIPPING ROUTES
1980–2000 average extent
2008 extent
2012 extent
Northeast Passage
Northwest Passage

FIGURE 12-5

Disputation and Navigation

Even as a dispute over Arctic maritime boundaries looms, another major issue that has emerged in recent years involves the effects of sustained global warming. The Greenland Ice Sheet, in many ways a smaller version of Antarctica's, is now in a melting phase, and the zone of seasonally expanding and contracting Arctic sea ice has also shown significant contraction in both surface extent and thickness. Researchers with the Greenland Ice Core Project have reported in *Science* that during a previous warming episode, more than 450,000 years ago, the southernmost part of Greenland was covered by boreal forest—suggesting that both the ice sheet and the sea ice may disappear entirely if the current global warming trend continues. Indeed, this process seems well underway: the polar ice cap is now less than half of what it was in 1979, and the lower-left subsidiary map in Figure 12-5 highlights the most recent phase of ongoing shrinkage. The ecological consequences are far-reaching, threatening the habitats of polar bears, whales, walruses, seals, and other species. But the economic- and politico-geographical implications would be significant as well.

When the sea ice melts in summer and does not fully recover in winter, once-blocked waterways begin to open up and accessibilities can be transformed. The **Northeast Passage [12]**, that legendary high-latitude route between the Far East and Europe, has been the object of hope and despair for centuries, causing much loss of human life in the quest to forge a viable sea lane. But in 2007, for the first time in recorded history, it was completely ice-free for a brief period during the late-summer peak of seasonal melting. Since then, the Northeast Passage has been ice-free every year during that September window; by 2012, a record 46 vessels—carrying 1.2 million tons of cargo—were able to successfully navigate the route (mapped in the lower-right subsidiary map of Fig. 12-5). That map also shows the presence of a similar route around the northern perimeter of North America, the *Northwest Passage*. This other icon of (unsuccessful) pre-twentieth-century polar exploration would connect the Pacific and Atlantic oceans, following a watercourse along Alaska's northern coast, through Canada's massive Arctic Archipelago, and then a final southward traverse via the Davis Strait between Canada and Greenland.

If either or both of these waterways can be activated, the impact on global shipping would be momentous. Consider what it would mean if ships bound for Asia's Pacific Rim from the U.S. eastern seaboard, and too large to fit even through the upgraded Panama Canal, could shorten their voyages by several days and thousands of kilometers by rounding North America instead of South America! No wonder countries like China, India, Japan, and even Singapore have applied for 'observer status' at the meetings of the Arctic Council, whose regular members include the United States, Russia, Canada, Norway, Sweden, Denmark, Finland, and Iceland.

But who really owns the waterways vacated by the ice? Canada regards the Northwest Passage as a domestic waterway, giving it the right to control all shipping. The United States holds that the Northwest Passage is an international waterway, and American and Russian ships have sailed through it without Canada's permission. The Canadian government has affirmed its position by starting work on a deepwater port at Nanisivik on Nunavut's Baffin Island at the eastern entrance to the Northwest Passage, and by building a military base at Resolute Bay to the west; it has also announced the construction of eight Arctic patrol vessels to police the route. In the meantime, the United States is opening a new Coast Guard station to patrol the narrow Bering Strait. The lines, if not yet battle lines, are already being drawn in the warming Arctic.

Humanity's growing numbers, escalating demands, environmental impacts, and technological capacities are transforming even the most remote recesses of our resilient planet. No international resolution is more critical than the avoidance of destructive conflict in fragile polar environments.

POINTS TO PONDER

- Governments of low-lying island-states in both the Pacific and Indian oceans are demanding international assistance as they are confronted by the threat of the world ocean's rising sea level.
- Fiji receives support from Beijing for officially recognizing the People's Republic of China (PRC) but not Taiwan; Palau gets support from the Taiwanese for officially recognizing Taiwan but not the PRC.
- As national claims in Antarctica lie dormant, states with coastlines bordering the Arctic Ocean—including Russia, Canada, and the United States—are sketching maritime boundaries as well as charting claims in the far north.
- Singapore, located almost directly on the equator, has been granted 'observer status' to monitor the meetings of the Arctic Council because of the possible impact of the opening of ice-free, far northern passages on global shipping.

APPENDIX A

NOTE: Appendices C, D, and E may be found on the book's website at www.wiley.com/college/deblij

Note that Appendix B is found on pages DT-1 to DT-7 that run backward from the inside cover

METRIC (STANDARD INTERNATIONAL [SI]) AND CUSTOMARY UNITS AND THEIR CONVERSIONS

Length

Metric Measure	
1 kilometer (km)	= 1000 meters (m)
1 meter (m)	= 100 centimeters (cm)
1 centimeter (cm)	= 10 millimeters (mm)
Nonmetric Measure	
1 mile (mi)	= 5280 feet (ft)
	= 1760 yards (yd)
1 yard (yd)	= 3 feet (ft)
1 foot (ft)	= 12 inches (in)
1 fathom (fath)	= 6 feet (ft)
Conversions	
1 kilometer (km)	= 0.6214 mile (mi)
1 meter (m)	= 3.281 feet (ft)
	= 1.094 yards (yd)
1 centimeter (cm)	= 0.3937 inch (in)
1 millimeter (mm)	= 0.0394 inch (in)
1 mile (mi)	= 1.609 kilometers (km)
1 foot (ft)	= 0.3048 meter (m)
1 inch (in)	= 2.54 centimeters (cm)
	= 25.4 millimeters (mm)

Temperature

To change from Fahrenheit (F) to Celsius (C)

$$°C = \frac{°F - 32}{1.8}$$

To change from Celsius (C) to Fahrenheit (F)

$$°F = °C \times 1.8 + 32$$

Area

Metric Measure	
1 square kilometer (km^2)	= 1,000,000 square meters (m^2)
	= 100 hectares (ha)
1 square meter (m^2)	= 10,000 square centimeters (cm^2)
1 hectare (ha)	= 10,000 square meters (m^2)
Nonmetric Measure	
1 square mile (mi^2)	= 640 acres (ac)
1 acre (ac)	= 4840 square yards (yd^2)
1 square foot (ft^2)	= 144 square inches (in^2)
Conversions	
1 square kilometer (km^2)	= 0.386 square mile (mi^2)
1 hectare (ha)	= 2.471 acres (ac)
1 square meter (m^2)	= 10.764 square feet (ft^2)
	= 1.196 square yards (yd^2)
1 square centimeter (cm^2)	= 0.155 square inch (in^2)
1 square mile (mi^2)	= 2.59 square kilometers (km^2)
1 acre (ac)	= 0.4047 hectare (ha)
1 square foot (ft^2)	= 0.0929 square meter (m^2)
1 square inch (in^2)	= 6.4516 square centimeters (cm^2)

GLOSSARY*

Aboriginal land issue The legal campaign in which Australia's **indigenous peoples** have claimed title to traditional land in several parts of that country. The courts have upheld certain claims, fueling Aboriginal activism that has raised broader issues of indigenous rights.

Aboriginal population Native or *aboriginal* peoples; often used to designate the inhabitants of areas that were conquered and subsequently colonized by the **imperial** powers of Europe.

Absolute location The position or place of a certain item on the surface of the Earth as expressed in degrees, minutes, and seconds of **latitude** and **longitude**.

Accessibility The degree of ease with which it is possible to reach a certain location from other locations. *Inaccessibility* is the opposite of this concept.

Acculturation Cultural modification resulting from intercultural borrowing. In **cultural geography**, the term refers to the change that occurs in the **culture** of **indigenous peoples** when contact is made with a society that is technologically superior.

Advantage The most meaningful distinction that can now be made to classify a country's level of economic **development**. Takes into account geographic **location, natural resources**, government, political stability, productive skills, and much more.

AFTA The ASEAN Free Trade Agreement that since 1992, through lowered tariffs and other incentives, has fostered increased trade within Southeast Asia. This is the economic centerpiece of the **Association of Southeast Asian Nations (ASEAN)**, a **supranational** organization whose members include 10 of that realm's 11 states (only East Timor does not participate).

Agglomeration **Process** involving the clustering or concentrating of people or activities.

Agrarian Relating to the use of land in rural communities or to agricultural societies in general.

Agriculture The purposeful tending of crops and livestock in order to produce food and fiber.

Al-Qaeda The terrorist organization that evolved into an expanding global network under the directorship of Usama bin Laden between the mid-1990s and his elimination by the U.S. in 2011. It sought to coordinate the efforts of once loosely allied Muslim revolutionary movements, and unleash a ***jihad*** aimed at what it perceived to be Islam's enemies in the West.

Alluvial Referring to the mud, silt, and sand (collectively *alluvium*) deposited by rivers and streams. *Alluvial plains* adjoin many larger rivers; they consist of the renewable deposits that are laid down during floods, creating fertile and productive soils. Alluvial **deltas** mark the mouths of rivers such as the Nile (Egypt) and the Ganges (Bangladesh).

Altiplano High-elevation plateau, basin, or valley between even higher mountain ranges, especially in the Andes of South America.

Altitudinal zonation Vertical regions defined by physical-environmental zones at various elevations (see Fig. 4A-4), particularly in the highlands of South and Middle America.

American Manufacturing Belt North America's near-rectangular Core Region, whose corners are Boston, Milwaukee, St. Louis, and Baltimore.

Animistic religion The belief that inanimate objects, such as hills, trees, rocks, rivers, and other elements of the natural landscape, possess souls and can help or hinder human efforts on Earth.

Antarctic Treaty International cooperative agreement on the use of Antarctic territory.

Antecedent boundary A political boundary that existed before the **cultural landscape** emerged and stayed in place while people moved in to occupy the surrounding area.

Anthracite coal Hardest and highest carbon-content coal, and therefore of the highest quality.

Apartheid Literally, *apartness*. The Afrikaans term for South Africa's pre-1994 policies of racial separation, a system that produced highly segregated socio-geographical patterns.

Aquaculture The use of a river segment or an artificial pond for the raising and harvesting of food products, including fish, shellfish, and even seaweed. The Japanese pioneered the practice, which is now spreading globally, and is already the dominant source of seafood as the oceans become fished out.

Aquifer An underground reservoir of water contained within a porous, water-bearing rock layer.

Arable Land fit for cultivation by one farming method or another. See also **physiologic density**.

Archipelago A set of islands grouped closely together, usually elongated into a *chain*.

Area A term that refers to a part of the Earth's surface with less specificity than **region**. For example, *urban area* alludes generally to a place where urban development has occurred, whereas *urban region* requires certain specific criteria on which such a designation is based (e.g., the spatial extent of commuting or the built townscape).

Areal interdependence A term related to **functional specialization**. When one area produces certain goods or has certain raw materials and another area has a different set of raw materials and produces different goods, their needs may be *complementary*; by exchanging raw materials and products, they can satisfy each other's requirements.

Arithmetic density A country's population, expressed as an average per unit area, without regard for its **distribution** or the limits of **arable** land. See also **physiologic density**.

Aryan From the Sanskrit *Arya* (meaning "noble"), a name applied to an ancient people who spoke an **Indo-European language** and who moved into northern India from the northwest.

ASEAN The **Association of Southeast Asian Nations (ASEAN)**, a **supranational** organization whose members include 10 of that realm's 11 states: Brunei, Cambodia, Indonesia, Laos, Malaysia, Myanmar (Burma), the Philippines, Singapore, Thailand, and Vietnam (only minuscule East Timor does not participate).

Asian Tiger See **economic tiger**.

Atmosphere The Earth's envelope of gases that rests on the oceans and land surface and penetrates open spaces within soils. This layer of nitrogen (78 percent), oxygen (21 percent), and traces of other gases is densest at the Earth's surface and thins with altitude.

Atoll A ring-like coral reef surrounding an empty lagoon that probably formed around the rim of a now-completely-eroded volcanic cone standing on the seafloor. They are common in certain tropical areas of the Pacific Ocean where they are classified among that realm's **low islands**.

Austral South.

Autocratic A government that holds absolute power, often ruled by one person or a small group of persons who control the country by despotic means.

Balkanization The fragmentation of a **region** into smaller, often hostile political units. Named after the historically contentious Balkan Peninsula of southeastern Europe.

Barrio Term meaning "neighborhood" in Spanish. Usually refers to an urban community in a Middle or South American city.

Biodiversity Shorthand for *biological diversity*; the total variety of plant and animal species that exists in a given area.

*Words in **boldface type** within an entry are defined elsewhere in this Glossary

Biodiversity hotspot A much higher than usual, world-class geographic concentration of natural plant and/or animal species. Tropical rainforest environments have dominated, but their recent ravaging by **deforestation** has had catastrophic results.

Biogeography The study of *flora* (plant life) and *fauna* (animal life) in spatial perspective.

Biome One of the broadest justifiable (geographic) subdivisions of the plant and animal world (less than a dozen exist overall). An assemblage and association of plants and animals that forms a regional ecological unit of subcontinental dimensions.

Birth rate The *crude birth rate* is expressed as the annual number of births per 1000 individuals within a given population.

Bituminous coal Softer coal of lesser quality than **anthracite**, but of higher grade than **lignite**. When heated and converted to coking coal or *coke*, it is used to make steel.

Boreal forest The subarctic, mostly **coniferous** snowforest that blankets Canada south of the **tundra** that lines the Arctic shore; known as the **taiga** in Russia.

Break-of-bulk point A location along a transport route where goods must be transferred from one carrier to another. In a port, the cargoes of oceangoing ships are unloaded and put on trains, trucks, or perhaps smaller river boats for inland distribution. An ***entrepôt***.

BRIC Acronym for the four biggest emerging national markets in the world today—**B**razil, **R**ussia, **I**ndia, and **C**hina.

British Commonwealth Now renamed *The Commonwealth of Nations* (and known simply as *The Commonwealth*), this nonpolitical inter-governmental organization is constituted by 54 member-states. All but two (Rwanda and Moçambique) formerly belonged to the British Empire, from which the organization originated. Its members cooperate in free association to promote, among other things, democracy, human rights, free trade, and world peace.

Buffer state or Buffer zone A country or set of countries separating ideological or political adversaries. In southern Asia, Afghanistan, Nepal, and Bhutan were parts of a buffer zone set up between British and Russian-Chinese imperial spheres. Thailand was a *buffer state* between British and French colonial domains in mainland Southeast Asia.

Caliente See ***tierra caliente***.

Cartogram A specially transformed map not based on traditional representations of **scale** or area.

Cartography The art and science of making maps, including data compilation, layout, and design. Also concerned with the interpretation of mapped patterns.

Caste system The strict **social stratification** and segregation of people—specifically in India's Hindu society—on the basis of ancestry and occupation.

Cay A low-lying small island usually composed of coral and sand. Pronounced *kee* and often spelled "key."

Cellphone revolution Applied specifically to much of Subsaharan Africa and certain other parts of the developing world, the linking of farmers with centers of information for weather and market conditions via cellphone text messaging—allowing them to negotiate higher prices for their products.

Central business district (CBD) The downtown heart of a central city; marked by high land values, a concentration of business and commerce, and the clustering of the tallest buildings.

Centrality The strength of an urban center in its capacity to attract producers and consumers to its facilities; a city's "reach" into the surrounding region.

Centrifugal forces A term employed to designate forces that tend to divide a country—such as internal religious, linguistic, ethnic, or ideological differences.

Centripetal forces Forces that unite and bind a country together—such as a strong national culture, shared ideological objectives, and a common faith.

Cerrado Regional term referring to the fertile savannas of Brazil's interior Central-West that make it one of the world's most promising agricultural frontiers. Soybeans are the leading crop, and other grains and cotton are expanding. Inadequate transport links to the outside world remain a problem.

Choke point A narrowing of an international waterway causing marine-traffic congestion, requiring reduced speeds and/or sharp turns, and increasing the risk of collision as well as vulnerability to attack. When the waterway narrows to a distance of less than 38 kilometers (24 mi), this necessitates the drawing of a **median line (maritime) boundary**.

City-state An independent political entity consisting of a single city with (and sometimes without) an immediate **hinterland**.

Climate The long-term conditions (over at least 30 years) of aggregate **weather** over a region, summarized by averages and measures of variability; a synthesis of the succession of weather events we have learned to expect at any given location.

Climate change theory An alternative to the **hydraulic civilization theory**; holds that changing **climate** (rather than a monopoly over **irrigation** methods) could have provided certain cities within the ancient Fertile Crescent with advantages over other cities.

Climate region A **formal region** characterized by the uniformity of the **climate** type within it. Figure G-7 maps the global distribution of such regions.

Climatology The geographic study of **climates**. Includes not only the classification of climates and the analysis of their regional distribution, but also broader environmental questions that concern climate change, interrelationships with soil and vegetation, and human–climate interaction.

Coal See **anthracite coal, bituminous coal, fossil fuels**, and **lignite**.

Collectivization The reorganization of a country's **agriculture** under communism that involves the expropriation of private holdings and their incorporation into relatively large-scale units, which are farmed and administered cooperatively by those who live there.

Colonialism Rule by an autonomous power over a subordinate and an alien people and place. Though often established and maintained through political structures, colonialism also creates unequal cultural and economic relations. Because of the magnitude and impact of the European colonial thrust of the last few centuries, the term is generally understood to refer to that particular colonial endeavor.

Command economy The tightly controlled economic system of the former Soviet Union, whereby central planners in Moscow assigned the production of particular goods to particular places, often guided more by socialist ideology than the principles of **economic geography**.

Commercial agriculture For-profit **agriculture**.

Common market A **free-trade area** that not only has created a **customs union** (a set of common tariffs on all imports from outside the area) but also has eliminated restrictions on the movement of capital, labor, and enterprise among its member countries.

Communal tension Persistent stress among a country's sociocultural groups that can often erupt into communal violence.

Compact state A politico-geographical term to describe a state that possesses a roughly circular, oval, or rectangular territory in which the distance from the geometric center to any point on the boundary exhibits little variance.

Complementarity Exists when two regions, through an exchange of raw materials and/or finished products, can specifically satisfy each other's demands.

Confucianism A philosophy of ethics, education, and public service based on the writings of Confucius (*Kongfuzi*); traditionally regarded as one of the cornerstones of Chinese **culture**.

Congo Two countries in Africa have the same short-form name, *Congo*. In this book, we use ***DRCongo*** for the larger Democratic Republic of the Congo, and ***Congo*** for the smaller Republic of Congo.

Coniferous forest A forest of cone-bearing, needleleaf evergreen trees with straight trunks and short branches, including spruce, fir, and pine. See also **taiga** and **boreal forest**.

Connectivity The degree of direct linkage between a particular location and other locations within a regional, national, or global transportation **network**.

Contagious diffusion The distance-controlled spreading of an idea, innovation, or some other item through a local population by contact from person to person—analogous to the communication of a contagious illness.

Conterminous United States The 48 **contiguous** or adjacent States that occupy the southern half of the North American realm. Alaska is not contiguous to these States because western Canada lies in between; neither is Hawai'i, separated from the mainland by over 3000 kilometers (2000 mi) of ocean.

Contiguous Adjoining; adjacent.

Continental drift The slow movement of continents controlled by the processes associated with **plate tectonics**.

Continental shelf Beyond the coastlines of many landmasses, the ocean floor declines very gently until the depth of about 660 feet (200 m). Beyond the 660-foot line the sea bottom usually drops off sharply, along the *continental slope*, toward the much deeper mid-oceanic basin. The submerged continental margin is called the continental shelf, and it extends from the shoreline to the upper edge of the continental slope.

Continentality The variation of the continental effect on air temperatures in the interior portions of the world's landmasses. The greater the distance from the moderating influence of an ocean, the greater the extreme in summer and winter temperatures. Continental interiors also tend to be dry when the distance from oceanic moisture sources becomes considerable.

Conurbation General term used to identify a large multimetropolitan complex formed by the coalescence of two or more major **urban areas**.

Copra The dried-out, fleshy interior of a coconut that is used to produce coconut oil.

Cordillera Mountain chain consisting of sets of parallel ranges, especially the Andes in northwestern South America.

Core See **core area**; **core-periphery relationships**.

Core area In geography, a term with several connotations. *Core* refers to the center, heart, or focus. The core area of a **nation-state** is constituted by the national heartland, the largest population cluster, the most productive region, and the part of the country with the greatest **centrality** and **accessibility**—probably containing the capital city as well.

Core-periphery relationships The contrasting spatial characteristics of, and linkages between, the *have* (core) and *have-not* (periphery) components of a national, regional, or the global **system**.

Corridor In general, refers to a spatial entity in which human activity is organized in a linear manner, as along a major transport route or in a valley confined by highlands. More specifically, the politico-geographical term for a land extension that connects an otherwise **landlocked state** to the sea.

Cross-border linkages The ties between two closely-connected localities or regions that face each other across an international boundary. These relationships are often longstanding, and intensify further as **supranationalism** proceeds (especially among the EU countries of western Europe).

Cultural diffusion The **process** of spreading and adopting a cultural element, from its place of origin across a wider area.

Cultural diversity A society marked by a variety of cultures, especially in its ancestral backgrounds.

Cultural ecology or cultural environment The myriad interactions and relationships between a **culture** and its **natural environment**.

Cultural geography The wide-ranging and comprehensive field of geography that studies spatial aspects of human **cultures**.

Cultural landscape The forms and artifacts sequentially placed on the **natural landscape** by the activities of various human occupants. By this progressive imprinting of the human presence, the physical (natural) landscape is modified into the cultural landscape, forming an interacting unity between the two.

Cultural pluralism A society in which two or more population groups, each practicing its own **culture**, live adjacent to one another without mixing inside a single **state**.

Cultural revival The regeneration of a long-dormant **culture** through internal renewal and external infusion.

Culture hearth (also called Cultural hearth) Heartland, source area, or innovation center; place of origin of a major **culture.**

Culture The sum total of the knowledge, attitudes, and habitual behavior patterns shared and transmitted by the members of a society. This is anthropologist Ralph Linton's definition; hundreds of others exist.

Culture hearth Heartland, source area, innovation center; place of origin of a major **culture.**

Culture region A distinct, culturally discrete spatial unit; a **region** within which certain cultural norms prevail.

Customs union A **free-trade area** in which member countries set common tariff rates on imports from outside the area.

Dalit The term now used for members of the lowest-ranking social layer in the **caste system** of Hindu India.

Death rate The *crude death rate* is expressed as the annual number of deaths per 1000 individuals within a given population.

Deciduous A deciduous tree loses its leaves at the beginning of winter or the onset of the dry season.

Definition In **political geography**, the written legal description (in a treaty-like document) of a boundary between two countries or territories. See also **delimitation.**

Deforestation The clearing and destruction of forests (especially tropical rainforests) to make way for expanding settlement frontiers and the exploitation of new economic opportunities.

Deglomeration Deconcentration.

Deindustrialization **Process** by which companies relocate manufacturing jobs to other regions or countries with cheaper labor, leaving the newly-deindustrialized region to convert to a service economy while struggling with the accompanying effects of increased unemployment and meeting the retraining needs of its workforce.

Delimitation In **political geography**, the translation of the written terms of a boundary treaty (the **definition**) into an official cartographic representation (map).

Delta **Alluvial** lowland at the mouth of a river, formed when the river deposits its alluvial load on reaching the sea. Often triangular in shape, hence the use of the Greek letter whose symbol is D.

Demarcation In **political geography**, the actual placing of a political boundary on the **cultural landscape** by means of barriers, fences, walls, or other markers.

Demographic burden The proportion of a national population that is either too old or too young to be productive and that must be cared for by the productive population.

Demographic transition Multi-stage **model**, based on western Europe's experience, of changes in population growth exhibited by countries undergoing industrialization. High **birth rates** and **death rates** are followed by plunging death rates, producing a huge net population gain; birth and death rates then converge at a low overall level.

Demography The interdisciplinary study of population—especially **birth rates** and **death rates**, growth patterns, longevity, **migration**, and related characteristics.

***Dependencia* theory** Originating in South America during the 1960s, it was a new way of thinking about economic development and underdevelopment that explained the persistent poverty of certain countries in terms of their unequal relations with other (i.e., rich) countries.

Desert An arid expanse supporting sparse vegetation, receiving less than 25 centimeters (10 in) of precipitation per year. Usually exhibits extremes of heat and cold because the moderating influence of moisture is absent.

Desertification **Process** of **desert** expansion into neighboring **steppelands** as a result of human degradation of fragile semiarid environments.

Development The economic, social, and institutional growth of national **states.**

Devolution The **process** whereby regions within a **state** demand and gain political strength and growing autonomy at the expense of the central government.

Dhows Wooden boats with characteristic triangular sails, plying the seas between Arabian and East African coasts.

Dialect Regional or local variation in the use of a major language, such as the distinctive accents of many residents of the U.S. South.

Diffusion The spatial spreading or dissemination of a **culture** element (such as a technological innovation) or some other phenomenon (e.g., a disease outbreak). For the various channels of outward geographic spread from a source area, see **contagious, expansion, hierarchical**, and **relocation diffusion**.

Distance decay The various degenerative effects of distance on human spatial structures and interactions.

Diurnal Daily.

Divided capital In **political geography**, a country whose central administrative functions are carried on in more than one city is said to have divided capitals. The Netherlands and South Africa are examples.

Domestication The transformation of a wild animal or wild plant into a domesticated animal or a cultivated crop to gain control over food production. A necessary evolutionary step in the development of humankind: the invention of **agriculture**.

Domino effect or theory The belief that political destabilization in one **state** can result in the collapse of order in a neighboring state, triggering a chain of events that, in turn, can affect a series of **contiguous** states.

Double complementarity Complementarity exists when two regions, through an exchange of raw materials and/or finished products, can specifically satisfy each other's demands; *double complementarity* exists when that interaction occurs in both directions simultaneously.

Double cropping The planting, cultivation, and harvesting of two crops successively within a single year on the same plot of farmland.

Double Delta South Asia's combined **delta** formed by the Ganges and Brahmaputra rivers. All of Bangladesh lies on this enormous deltaic plain, which also encompasses surrounding parts of eastern India. Well over 200 million people live here, attracted by the fertility of its soils that are constantly replenished by the **alluvium** transported and deposited by these two of Asia's largest river systems. Natural hazards abound here as well, ranging from the flooding caused by excessive **monsoonal** rains to the intermittent storm surges of powerful cyclones (**hurricanes**) that come from the Bay of Bengal to the south.

Dravidian languages The language family, indigenous to the South Asian realm, that dominates southern India today; as opposed to the Indo-European languages, whose tongues dominate northern India.

Dry canal An overland rail and/or road **corridor** across an **isthmus** dedicated to performing the transit functions of a canalized waterway. Best adapted to the movement of containerized cargo, there must be a port at each end to handle the necessary **break-of-bulk** unloading and reloading.

Dynasty A succession of Chinese rulers that came from the same line of male descent, sometimes enduring for centuries. Dynastic rule in China lasted for thousands of years, only coming to an end just a bit more than a century ago in 1911.

Ecology The study of the many interrelationships between all forms of life and the natural environments in which they have evolved and continue to develop. The study of *ecosystems* focuses on the interactions between specific organisms and their environments. See also **cultural ecology**, **biome**, and **biodiversity**.

Economic geography The field of geography that focuses on the diverse ways in which people earn a living and on how the goods and services they produce are expressed and organized spatially.

Economic restructuring The transformation of China into a market-driven economy in the post-Mao era, beginning in the late 1970s.

Economic tiger One of the burgeoning beehive countries of the western **Pacific Rim**. Following Japan's route since 1945, these countries have experienced significant modernization, industrialization, and Western-style economic growth since 1980. Three leading economic tigers are South Korea, Taiwan, and Singapore. The term is increasingly used more generally to describe any fast-developing economy.

Economies of scale The savings that accrue from large-scale production wherein the unit cost of manufacturing decreases as the level of operation enlarges. Supermarkets operate on this principle and are able to charge lower prices than small grocery stores.

Ecosystem See **ecology**.

Ecumene The habitable portions of the Earth's surface where permanent human settlements have arisen.

Ejidos Mexican farmlands redistributed to peasant communities after the Revolution of 1910–1917. The government holds title to the land, but user rights are parceled out to village communities and then to individuals for cultivation.

Elite A small but influential upper-echelon social class whose power and privilege give it control over a country's political, economic, and cultural life.

El Niño-Southern Oscillation (ENSO) A periodic, large-scale, abnormal warming of the sea surface in the tropical latitudes of the eastern Pacific Ocean that has global implications, disturbing normal weather patterns in many parts of the world, especially South America.

Elongated state A **state** whose territory is decidedly long and narrow in that its length is at least six times greater than its average width.

Emerging market The world's fastest growing national market economies as measured by economic growth rates, attraction of **foreign direct investment**, and other key indicators. Led by the **BRICs** (Brazil, Russia, India, and China), but this club is now expanding to include many other countries.

Emigrant A person **migrating** away from a country or area; an out-migrant.

Empirical Relating to the real world, as opposed to theoretical abstraction.

Enclave A piece of territory that is surrounded by another political unit of which it is not a part.

Endemism Referring to a disease in a host population that affects many people in a kind of equilibrium without causing rapid and widespread deaths.

Entrepôt A place, usually a port city, where goods are imported, stored, and transshipped; a **break-of-bulk point**.

Environmental degradation The accumulated human abuse of a region's **natural landscape** that, among other things, can involve air and water pollution, threats to plant and animal **ecosystems**, misuse of **natural resources**, and generally upsetting the balance between people and their habitat.

Epidemic A local or regional outbreak of a disease.

Escarpment A cliff or very steep slope; often marks the edge of a plateau. The Great Escarpment that lines much of Africa's east coast is a classic example.

Estuary The widening mouth of a river as it reaches the sea; land subsidence or a rise in sea level has overcome the tendency to form a **delta**.

Ethanol The leading U.S. biofuel that is essentially alcohol distilled from corn mash. Much of it is produced in the historic Corn Belt centered on Iowa and Illinois. Many risks and problems accompany this energy source, which overall is not an efficient replacement for **fossil fuels**.

Ethnic cleansing The slaughter and/or forced removal of one **ethnic** group from its homes and lands by another, more powerful ethnic group bent on taking that territory.

Ethnicity The combination of a people's **culture** (traditions, customs, language, and religion) and racial ancestry.

European state model A **state** consisting of a legally defined territory inhabited by a population governed from a capital city by a representative government.

European Union (EU) **Supranational** organization constituted by 28 European countries to further their common economic interests. In alphabetical order, these countries are: Austria, Belgium, Bulgaria, Croatia, Cyprus, the Czech Republic, Denmark, Estonia, Finland, France, Germany, Greece, Hungary, Ireland, Italy, Latvia, Lithuania, Luxembourg, Malta, the Netherlands, Poland, Portugal, Romania, Slovakia, Slovenia, Spain, Sweden, and the United Kingdom.

Exclave A bounded (non-island) piece of territory that is part of a particular **state** but lies separated from it by the territory of another state.

Exclusive Economic Zone (EEZ) An oceanic zone extending up to 200 **nautical miles** (370 km) from a shoreline, within which the coastal **state** can control fishing, mineral exploitation, and additional activities by all other countries.

Expansion diffusion The spreading of an innovation or an idea through a fixed population in such a way that the number of those adopting it grows continuously larger, resulting in an expanding area of dissemination.

Extraterritoriality The politico-geographical concept suggesting that the property of one **state** lying within the boundaries of another actually forms an extension of the first state.

Failed state A country whose institutions have collapsed and in which anarchy prevails.

Fatwa Literally, a legal opinion or proclamation issued by an Islamic cleric, based on the holy texts of Islam, long applicable only in the *Umma*, the realm ruled by the laws of Islam. In 1989, the Iranian Ayatollah Khomeini extended the reach of the *fatwa* by condemning to death author Salman Rushdie, a British citizen living in the United Kingdom.

Favela Shantytown on the outskirts or even well within an urban area in Brazil.

Fazenda Coffee plantation in Brazil.

Federal state A political framework wherein a central government represents the various subnational entities within a **nation-state** where they have common interests—defense, foreign affairs, and the like—yet allows these various entities to retain their own identities and to have their own laws, policies, and customs in certain spheres.

Federal system A political framework wherein a central government represents the various subnational entities within a country where they have common interests—defense, foreign affairs, and the like—yet allows these various entities to retain their own identities and to have their own laws, policies, and customs in certain spheres.

Federation A country adhering to a political framework wherein a central government represents the various subnational entities within a **nation-state** where they have common interests—defense, foreign affairs, and the like—yet allows these various entities to retain their own identities and to have their own laws, policies, and customs in certain spheres.

Fertile Crescent Crescent-shaped zone of productive lands extending from near the southeastern Mediterranean coast through Lebanon and Syria to the **alluvial** lowlands of Mesopotamia (in Iraq). Once more fertile than today, this is one of the world's great source areas of **agricultural** and other innovations.

Fertility rate More technically the Total Fertility Rate, it is the average number of children born to women of childbearing age in a given population.

First Nations Name given Canada's **indigenous peoples** of American descent, whose U.S. counterparts are called Native Americans.

Fjord Narrow, steep-sided, elongated, and inundated coastal valley deepened by glacier ice that has since melted away, leaving the sea to penetrate.

Floating population China's huge mass of mobile workers who respond to shifting employment needs within the country. Most are temporary urban dwellers with restricted residency rights, whose movements are controlled by the ***hukou*** system.

Floodplain Low-lying area adjacent to a mature river, often covered by **alluvial** deposits and subject to the river's floods.

Forced migration Human **migration** flows in which the movers have no choice but to relocate.

Foreign direct investment (FDI) A key indicator of the success of an **emerging market** economy, whose growth is accelerated by the infusion of foreign funds to supplement domestic sources of investment capital.

Formal region A type of **region** marked by a certain degree of homogeneity in one or more phenomena; also called *uniform region* or *homogeneous region.*

Formal sector The total activities of a country's legal economy that is taxed and monitored by the government, whose **gross domestic product (GDP)** and **gross national product (GNP)** are based on it; as opposed to an **informal economy**.

Forward capital Capital city positioned in actually or potentially contested territory, usually near an international border; it confirms the **state's** determination to maintain its presence in the region in contention.

Fossil fuels The energy resources of **coal**, natural gas, and petroleum (oil), so named collectively because they were formed by the geologic compression and transformation of tiny plant and animal organisms.

Four Motors of Europe *Rhône-Alpes* (France), *Baden-Württemberg* (Germany), *Catalonia* (Spain), and *Lombardy* (Italy). Each is a high-technology-driven region marked by exceptional industrial vitality and economic success not only within Europe but on the global scene as well.

Fragmented modernization A checkerboard-like spatial pattern of **modernization** in an **emerging-market** economy wherein a few localized regions of a country experience most of the development while the rest are largely unaffected.

Fragmented state A **state** whose territory consists of several separated parts, not a **contiguous** whole. The individual parts may be isolated from each other by the land area of other states or by international waters.

Francophone French-speaking. Quebec constitutes the heart of Francophone Canada.

Free-trade area A form of economic integration, usually consisting of two or more **states**, in which members agree to remove tariffs on trade among themselves. Frequently accompanied by a **customs union** that establishes common tariffs on imports from outside the trade area, and sometimes by a **common market** that also removes internal restrictions on the movement of capital, labor, and enterprise.

Free Trade Area of the Americas (FTAA) The ultimate goal of **supranational** economic integration in North, Middle, and South America: the creation of a single-market trading bloc that would involve every country in the Western Hemisphere between the Arctic shore of Canada and Cape Horn at the southern tip of South America.

Fría See ***tierra fría***.

Frontier Zone of advance penetration, usually of contention; an area not yet fully integrated into a national **state**.

Functional region A **region** marked less by its sameness than by its dynamic internal structure; because it usually focuses on a central **node**, also called *nodal region* or *focal region*.

Functional specialization The production of particular goods or services as a dominant activity in a particular location. See also **local functional specialization**.

Fundamentalism See **revivalism (religious)**.

Gentrification The upgrading of an older residential area through private reinvestment, usually in the downtown area of a central city. Frequently, this involves the displacement of established lower-income residents, who cannot afford the heightened costs of living, and conflicts are not uncommon as such neighborhood change takes place.

Geographic change Evolution of **spatial** patterns over time.

Geographic realm The basic spatial unit in our world regionalization scheme. Each realm is defined in terms of a synthesis of its total human geography—a composite of its leading cultural, economic, historical, political, and appropriate environmental features.

Geography of development The subfield of economic geography concerned with spatial aspects and regional expressions of **development**.

Geometric boundaries Political boundaries **defined** and **delimited** (and occasionally **demarcated**) as straight lines or arcs.

Geomorphology The geographic study of the configuration of the Earth's solid surface—the world's landscapes and their constituent landforms.

Ghetto An intraurban region marked by a particular **ethnic** character. Often an inner-city poverty zone, such as the black ghetto in U.S. central cities. Ghetto residents are involuntarily segregated from other income and racial groups.

Glaciation Period of global cooling during which continental ice sheets and mountain glaciers expand.

Globalization The gradual reduction of regional differences at the world **scale**, resulting from increasing international cultural, economic, and political exchanges.

Global warming A general term referring to the temperature increase of the Earth's atmosphere over the past century and a half, how humans may be contributing to this warming, and scenarios of future environmental change that could result if this trend continues.

Green Revolution The successful recent development of higher-yield, fast-growing varieties of rice and other cereals in certain developing countries.

Gross domestic product (GDP) The total value of all goods and services produced in a country by that state's economy during a given year.

Gross national product (GNP) The total value of all goods and services produced in a country by that state's economy during a given year, plus all citizens' income from foreign investment and other external sources.

Growth pole An urban center with a number of attributes that, if augmented by investment support, will stimulate regional economic **development** in its **hinterland**.

Growth triangle An increasingly popular economic **development** concept along the western **Pacific Rim**, especially in Southeast Asia. It involves the linking of production in growth centers of three countries to achieve benefits for all.

Hacienda Literally, a large estate in a Spanish-speaking country. Sometimes equated with the **plantation**, but there are important differences between these two types of agricultural enterprise.

Hanification Imparting a cultural imprint by the ethnic Chinese (the "people of Han"). Within China often refers to the steadily increasing migration of Han Chinese into the country's **periphery**, especially Xinjiang and Xizang (Tibet). **Overseas Chinese** imprints, more generally referred to as **Sinicization**, have been significant as well, most importantly in the Southeast Asian realm.

Heartland theory The hypothesis, proposed by British geographer Halford Mackinder during the early twentieth century, that any political power based in the heart of Eurasia could gain sufficient strength to eventually dominate the world. Furthermore, since Eastern Europe controlled access to the Eurasian interior, its ruler would command the vast "heartland" to the east.

Hegemony The political dominance of a country (or even a region) by another country.

Helada See ***tierra helada***.

Hierarchical diffusion A form of **diffusion** in which an idea or innovation spreads by trickling down from larger to smaller adoption units. An urban **hierarchy** is usually involved, encouraging the leapfrogging of innovations over wide areas, with geographic distance a less important influence.

Hierarchy An order or gradation of phenomena, with each level or rank subordinate to the one above it and superior to the one below. The levels in a national urban hierarchy—or **urban** system—are constituted by hamlets, villages, towns, cities, and (frequently) the **primate city**.

High–island cultures Cultures associated with volcanic islands of the Pacific Realm that are high enough in elevation to wrest substantial moisture from the tropical ocean air (see **orographic precipitation**). They tend to be well watered, their volcanic soils enable productive agriculture, and they support larger populations than **low islands.**

High seas Areas of the oceans away from land, beyond national jurisdiction, open and free for all to use.

High value-added goods Products of improved net worth.

Highveld A term used in South Africa to identify the high, grass-covered plateau that dominates much of the country. The lowest-lying areas (mainly along the narrow coastlands) in South Africa are called *lowveld*; areas that lie at intermediate elevations form the *middleveld*.

Hindutva "Hinduness" as expressed through Hindu nationalism, Hindu heritage, and/or Hindu patriotism. The cornerstone of a fundamentalist movement that has been gaining strength since the late twentieth century that seeks to remake India as a society dominated by Hindu principles prevail. It has been the guiding agenda of the Bharatiya Janata Party (BJP), which has emerged a powerful force in national politics and in big States like Maharashtra, Gujarat, and Madhya Pradesh.

Hinterland Literally, "country behind," a term that applies to a surrounding area served by an urban center. That center is the focus of goods and services produced for its hinterland and is its dominant urban influence as well.

Historical inertia A term from manufacturing geography that refers to the need to continue using the factories, machinery, and equipment of heavy industries for their full, multiple-decade lifetimes to cover major initial investments—even though these facilities may be increasingly obsolete.

Holocene The current *interglacial* epoch (the warm period of glacial contraction between the glacial expansions of an **ice age**); extends from 10,000 years ago to the present. Also known as the *Recent Epoch*.

Human evolution Long-term biological maturation of the human species. Geographically, all evidence points toward East Africa as the source of humankind. Our species, *Homo sapiens*, emigrated from this hearth to eventually populate the rest of the **ecumene**.

***Hukou* system** A longstanding Chinese system whereby all inhabitants must obtain and carry with them residency permits that indicate where an individual is from and where they may exercise particular rights such as education, health care, housing, and the like.

Hurricane A tightly-wound, tropical cyclonic storm capable of inflicting great wind and water damage in low-lying coastal zones. Originates at sea in the hot, moist atmosphere of the lower latitudes, and can reach Category-5 wind speeds in excess of 249 kph (155 mph). Minimal hurricane wind speed is 119 kph (74 mph); but even below that intensity, tropical storms (63-119 kph [39-74 mph]) can wreak significant damage. The name *hurricane* is confined to North and Middle America; in the western North Pacific Ocean, such storms are called *typhoons*; in the Indian Ocean Basin, they are known as *(tropical) cyclones*.

Hurricane Alley The most frequent pathway followed by tropical storms and **hurricanes** over the past 150 years in their generally westward movement across the Caribbean Basin. Historically, hurricane tracks have bundled most tightly in the center of this route, most often affecting the Lesser Antilles between Antigua and the Virgin Islands, Puerto Rico, Hispaniola (Haiti/Dominican Republic), Jamaica, Cuba, southernmost Florida, Mexico's Yucatán, and the Gulf of Mexico.

Hydraulic civilization theory The theory that cities which managed to control **irrigated** farming over large **hinterlands** held political power over other cities. Particularly applies to early Asian civilizations based in such river valleys as the Chang (Yangzi), the Indus, and those of Mesopotamia.

Hydrologic cycle The **system** of exchange involving water in its various forms as it continually circulates between the **atmosphere**, the oceans, and above and below the land surface.

Ice age A stretch of geologic time during which the Earth's average atmospheric temperature is lowered; causes the equatorward expansion of continental ice sheets in the higher latitudes and the growth of mountain glaciers in and around the highlands of the lower latitudes.

Immigrant A person **migrating** into a particular country or area; an in-migrant.

Imperialism The drive toward the creation and expansion of a **colonial** empire and, once established, its perpetuation.

Import-substitution industries The industries local entrepreneurs establish to serve populations of remote areas when transport costs from distant sources make these goods too expensive to import.

Inaccessibility See **accessibility**.

Indentured workers Contract laborers who sell their services for a stipulated period of time.

Indigenous Aboriginal or native; an example would be the pre-Columbian inhabitants of the Americas.

Indigenous peoples Native or *aboriginal* peoples; often used to designate the inhabitants of areas that were conquered and subsequently colonized by the **imperial** powers of Europe.

Indo-European languages The major world language family that dominates the European **geographic realm**. This language family is also the most widely dispersed globally (Fig. G-9), and about half of humankind speaks one of its languages.

Industrial Revolution The term applied to the social and economic changes in agriculture, commerce, and especially manufacturing and urbanization that resulted from technological innovations and greater specialization in late-eighteenth-century Europe.

Informal sector Dominated by unlicensed sellers of homemade goods and services, the primitive form of capitalism found in many developing countries that takes place beyond the control of government. The complement to a country's **formal sector**.

Infrastructure The foundations of a society: urban centers, transport networks, communications, energy distribution systems, farms, factories, mines, and such facilities as schools, hospitals, postal services, and police and armed forces.

Insular Having the qualities and properties of an island. Real islands are not alone in possessing such properties of **isolation**: an **oasis** in the middle of a **desert** also has qualities of insularity.

Insurgent state Territorial embodiment of a successful guerrilla movement. The establishment by antigovernment insurgents of a territorial base in which they exercise full control; thus a **state** within a state.

Intercropping The planting of several types of crops in the same field; commonly used by **shifting cultivators**.

Interglacial Period of warmer global temperatures between the end of the previous glaciation and the onset of the next one.

Intermontane Literally, between the mountains. Such a location can bestow certain qualities of natural protection or **isolation** to a community.

Internal migration **Migration** flow within a country, such as ongoing westward and southward movements toward the **Sunbelt** in the United States.

International migration **Migration** flow involving movement across an international boundary.

Intervening opportunity In trade or **migration** flows, the presence of a nearer opportunity that greatly diminishes the attractiveness of sites farther away.

Inuit **Indigenous** peoples of North America's Arctic zone, formerly known as Eskimos.

Irredentism A policy of cultural extension and potential political expansion by a **state** aimed at a community of its nationals living in a neighboring state.

Irrigation The artificial watering of croplands.

Islamic Front The southern border of the African Transition Zone that marks the religious **frontier** of the **Muslim** faith in its southward penetration of Subsaharan Africa (see Fig. 6B-9).

Islamization Introduction and establishment of the **Muslim** religion. A **process** still under way, most notably along the **Islamic Front**, that marks the southern border of the African Transition Zone.

Isohyet A line connecting points of equal rainfall total.

Isolation The condition of being geographically cut off or far removed from mainstreams of thought and action. It also denotes a lack of receptivity to outside influences, caused at least partially by poor **accessibility**.

Isoline A line connecting points of equal value; see **isohyet**, **isotherm**.

Isotherm A line connecting points of equal temperature.

Isthmus A **land bridge**; a comparatively narrow link between larger bodies of land. Central America forms such a link between Mexico and South America.

Jihad A doctrine within Islam. Commonly translated as *holy war*, it entails a personal or collective struggle on the part of **Muslims** to live up to the religious standards prescribed by the *Quran* (Koran).

Juxtaposition Contrasting places in close proximity to one another.

Karst The distinctive natural landscape associated with the chemical erosion of soluble limestone rock.

Land alienation One society or culture group taking land from another.

Land bridge A narrow **isthmian** link between two large landmasses. They are temporary features—at least when measured in geologic time—subject to appearance and disappearance as the land or sea level rises and falls.

Land hemisphere The half of the globe containing the greatest amount of land surface, centered on western Europe.

Land reform The spatial reorganization of **agriculture** through the allocation of farmland (often expropriated from landlords) to **peasants** and tenants who never owned land.

Land tenure The way people own, occupy, and use land.

Landlocked location/state An interior **state** surrounded by land. Without coasts, such a country is disadvantaged in terms of **accessibility** to international trade routes, and in the scramble for possession of areas of the **continental shelf** and control of the **exclusive economic zone** beyond.

Language family Group of languages with a shared but usually distant origin.

"Latin" American city model The Griffin-Ford model of intraurban spatial structure in the Middle American and South American realms.

Latitude Lines of latitude are **parallels** that are aligned east-west across the globe, from 0° latitude at the equator to 90° North and South latitude at the poles.

Leached soil Infertile, reddish-appearing, tropical soil whose surface consists of oxides of iron and aluminum; all other soil nutrients have been dissolved and transported downward into the subsoil by percolating water associated with the heavy rainfall of moist, low-latitude climates.

League of Nations The international organization that emerged after the First World War (1914–1918) whose purpose was to maintain international peace and promote cooperation in solving international economic, social, and humanitarian problems. It consisted of as many as 58 member-countries by the mid-1930s, but it failed to prevent World War II (1939–1945) and is best remembered as the flawed predecessor of the United Nations.

Leeward The protected or downwind side of a **topographic** barrier with respect to the winds that flow across it.

Liberation Theology A powerful religious movement that arose in South America during the 1950s, and subsequently gained followers all over the global **periphery**. At its heart is a belief system, based on a blend of Christian faith and socialist thinking, that interprets the teachings of Christ as a quest to liberate the impoverished masses from oppression.

Lignite Low-grade, brown-colored variety of **coal**.

Lingua franca A "common language" prevalent in a given area; a second language that can be spoken and understood by many people, although they speak other languages at home.

Littoral Coastal or coastland.

Llanos The interspersed **savanna** grasslands and scrub woodlands of the Orinoco River's wide basin that covers most of interior Venezuela and Colombia.

Local functional specialization A hallmark of Europe's **economic geography** that later spread to many other parts of the world, whereby particular people in particular places concentrate on the production of particular goods and services.

Location Position on the Earth's surface; see also **absolute location** and **relative location**.

Location theory A logical attempt to explain the locational pattern of an economic activity and the manner in which its producing areas are interrelated.

Loess Deposit of very fine silt or dust that is laid down after having been windborne for a considerable distance. Notable for its fertility under **irrigation** and its ability to stand in steep vertical walls.

Longitude Angular distance (0° to 180°) east or west as measured from the *prime meridian* (0°) that passes through the Greenwich Observatory in suburban London, England. For much of its length across the mid-Pacific Ocean, the 180th meridian functions as the *international date line*.

Low island cultures Cultures associated with low-lying coral islands of the Pacific Realm that cannot wrest sufficient moisture from the tropical maritime air to avoid chronic drought. Thus productive agriculture is impossible, and their modest populations must rely on fishing and the coconut palm for survival.

Lusitanian The Portuguese sphere, which by extension includes Brazil.

Madrassa **Revivalist** (**fundamentalist**) religious school in which the curriculum focuses on Islamic religion and law and requires rote memorization of the *Quran* (Koran), Islam's holy book. Founded in former British India, these schools were most numerous in present-day Pakistan but have **diffused** as far as Turkey in the west and Indonesia in the east. Afghanistan's **Taliban** emerged from these institutions in Pakistan

Maghreb The region occupying the northwestern corner of Africa, consisting of Morocco, Algeria, and Tunisia.

Main Street Canada's dominant **conurbation** that is home to nearly two-thirds of the country's inhabitants; extends southwestward from Quebec City in the mid-St. Lawrence Valley to Windsor on the Detroit River.

Mainland-Rimland framework The twofold regionalization of the Middle American realm based on its modern cultural history. The Euro-Indigenous *Mainland*, stretching from Mexico to Panama (minus the Caribbean coastal strip), was a self-sufficient zone dominated by **hacienda land tenure**. The Euro-African *Rimland*, consisting of that Caribbean coastal zone plus all of the Caribbean islands to the east, was the zone of the **plantation** that relied heavily on trade with Europe. See Figure 4A-7.

Maquiladora The term given to modern industrial plants in Mexico's U.S. border zone. These foreign-owned factories assemble imported components and/or raw materials, and then export finished manufactures, mainly to the United States. Import duties are disappearing under **NAFTA**, bringing jobs to Mexico and the advantages of low wage rates to the foreign entrepreneurs.

Marchland An area or **frontier** of uncertain boundaries that is subject to various national claims and an unstable political history. Refers specifically to the movement of various armies, refugees, and migrants across such zones.

Marine geography The geographic study of oceans and seas. Its practitioners investigate both the physical (e.g., coral-reef **biogeography**, ocean–**atmosphere** interactions, coastal **geomorphology**) as well as human (e.g., **maritime boundary**-making, fisheries, beachside development) aspects of oceanic environments.

Maritime boundary An international boundary that lies in the ocean. Like all boundaries, it is a vertical plane, extending from the seafloor to the upper limit of the air space in the atmosphere above the water.

Median-line boundary An international **maritime boundary** drawn where the width of a sea is less than 400 **nautical miles**. Because the **states** on either side of that sea claim **exclusive economic zones** of 200 nautical miles, it is necessary to reduce those claims to a (median) distance equidistant from each shoreline. **Delimitation** on the map almost always appears as a set of straight-line segments that reflect the configurations of the coastlines involved.

Medical geography The study of health and disease within a geographic context and from a spatial perspective. Among other things, this geographic field examines the sources, **diffusion** routes, and distributions of diseases.

Megacity Informal term referring to the world's most heavily populated cities; in this book, the term refers to a **metropolis** containing a population of greater than 10 million.

Megalopolis When spelled with a lower-case *m*, a synonym for **conurbation**, one of the large coalescing supercities forming in diverse parts of the world. When capitalized, refers specifically to the multimetropolitan (*Bosnywash*) corridor that extends along the northeastern U.S. seaboard from north of Boston to south of Washington, D.C.

Melting pot Traditional characterization of American society as a blend of numerous **immigrant ethnic** groups that over time were assimilated into a single societal mainstream. This notion always had its challengers among social scientists, and is now increasingly difficult to sustain given the increasing complexity and sheer scale of the U.S. ethnic mosaic in the twenty-first century.

Mental maps Maps that individuals carry around in their minds that reflect their constantly evolving perception of how geographic space (ranging from their everyday activity space to the entire world) is organized around them.

Mercantilism Protectionist policy of European **states** during the sixteenth to the eighteenth centuries that promoted a **state**'s economic position in the contest with rival powers. Acquiring gold and silver and maintaining a favorable trade balance (more exports than imports) were central to the policy.

Meridian Line of **longitude**, aligned north-south across the globe, that together with **parallels** of **latitude** forms the global grid system. All meridians converge at both poles and are at their maximum distances from each other at the equator.

Mestizo Derived from the Latin word for *mixed*, refers to a person of mixed European (white) and Amerindian ancestry.

Métis **Indigenous** Canadian people of mixed native (**First Nations**) and European ancestry.

Metropolis Urban **agglomeration** consisting of a (central) city and its suburban ring. See also **urban** (**metropolitan**) **area**.

Metropolitan area See **urban (metropolitan) area**.

Micro-credit Small loans extended to poverty-stricken borrowers who would not otherwise qualify for them. The aim is to help combat poverty, encourage entrepreneurship, and to empower poor communities—especially their women.

Microstate A sovereign **state** that contains a minuscule land area and population. They do not have the attributes of "complete" states, but are on the map as tiny yet independent entities nonetheless.

Migration A change in residence intended to be permanent.

Migratory movement Human relocation movement from a source to a destination without a return journey, as opposed to cyclical movement (see also **nomadism**).

Model An idealized representation of reality built to demonstrate its most important properties. A **spatial** model focuses on a geographical dimension of the real world.

Modernization In the eyes of the Western world, the **Westernization process** that involves the establishment of **urbanization**, a market (money) economy, improved circulation, formal schooling, the adoption of foreign innovations, and the breakdown of traditional society. Non-Westerners mostly see "modernization" as an outgrowth of **colonialism** and often argue that traditional societies can be modernized without being Westernized.

Monsoon Refers to the seasonal reversal of wind and moisture flows in certain parts of the subtropics and lower-middle latitudes. The *dry monsoon* occurs during the cool season when dry offshore winds prevail. The *wet monsoon* occurs in the hot summer months, which produce onshore winds that bring large amounts of rainfall. The air-pressure differential over land and sea is the triggering mechanism, with windflows always moving from areas of relatively higher pressure toward areas of relatively lower pressure. Monsoons make their greatest regional impact in the coastal and near-coastal zones of South Asia, Southeast Asia, and East Asia.

Mosaic culture Emerging cultural-geographic framework of the United States, dominated by the fragmentation of specialized social groups into homogeneous communities of interest marked not only by income, race, and **ethnicity** but also by age, occupational status, and lifestyle. The result is an increasingly heterogeneous sociospatial complex, which resembles an intricate mosaic composed of myriad uniform—but separate—tiles.

Mulatto A person of mixed African (black) and European (white) ancestry.

Multilingualism A society marked by a mosaic of local languages. Constitutes a **centrifugal force** because it impedes communication within the larger population. Often a ***lingua franca*** is used as a "common language," as in many countries of Subsaharan Africa.

Multinationals Internationally active corporations capable of strongly influencing the economic and political affairs of many countries in which they operate.

Muslim An adherent of the Islamic faith.

NAFTA (North American Free Trade Agreement) The **free-trade area** launched in 1994 involving the United States, Canada, and Mexico.

Nation Legally a term encompassing all the citizens of a **state**, it also has other connotations. Most definitions now tend to refer to a group of tightly knit people possessing bonds of language, **ethnicity**, religion, and other shared **cultural** attributes. Such homogeneity actually prevails within very few states.

Nation-state A country whose population possesses a substantial degree of **cultural** homogeneity and unity. The ideal form to which most **nations** and **states** aspire—a political unit wherein the territorial state coincides with the area settled by a certain national group or people.

NATO (North Atlantic Treaty Organization) Established in 1950 at the height of the Cold War as a U.S.-led **supranational** defense pact to shield postwar Europe against the Soviet military threat. NATO is now in transition, expanding its membership while modifying its objectives in the post-Soviet era. Its 28 member-states (as of mid-2013) are: Albania, Belgium, Bulgaria, Canada, Croatia, Czech Republic, Denmark, Estonia, France, Germany, Greece, Hungary, Iceland, Italy, Latvia, Lithuania, Luxembourg, the Netherlands, Norway, Poland, Portugal, Romania, Slovakia, Slovenia, Spain, Turkey, the United Kingdom, and the United States.

Natural hazard A natural event that endangers human life and/or the contents of a **cultural landscape**.

Natural increase rate Population growth measured as the excess of live births over deaths per 1000 individuals per year. Natural increase of a population does not reflect either **emigrant** or **immigrant** movements.

Natural landscape The array of landforms that constitutes the Earth's surface (mountains, hills, plains, and plateaus) and the physical features that mark them (such as water bodies, soils, and vegetation). Each **geographic realm** has its distinctive combination of natural landscapes.

Natural resource Any valued element of (or means to an end using) the environment; includes minerals, water, vegetation, and soil.

Nautical mile By international agreement, the nautical mile—the standard measure at sea—is 6076.12 feet in length, equivalent to approximately 1.15 statute miles (1.85 km).

Near Abroad The 14 former Soviet republics that, in combination with the dominant Russian Republic, constituted the USSR. Since the 1991 breakup of the Soviet Union, Russia has asserted a sphere of influence in these now-independent countries, based on its proclaimed right to protect the interests of ethnic Russians who were settled there in substantial numbers during Soviet times. These 14 countries include Armenia, Azerbaijan, Belarus, Estonia, Georgia, Kazakhstan, Kyrgyzstan, Latvia, Lithuania, Moldova, Tajikistan, Turkmenistan, Ukraine, and Uzbekistan.

Neocolonialism The term used by developing countries to underscore that the entrenched **colonial** system of international exchange and capital flow has not changed in the postcolonial era—thereby perpetuating the huge economic advantages of the developed world.

Neoliberalism A national or regional development strategy based on the privatization of state-run companies, lowering of international trade tariffs, reduction of government subsidies, cutting of corporate taxes, and overall deregulation to business activity.

Neolithic The "New Stone Age" (ca. 9500-4500 BC), which was the final stage of cultural evolution and technological development among prehistoric humans. Among the numerous accomplishments that took place, two of the most important for human geography were: (1) the rise of farming through the domestication of plants and animals, and (2) the first permanent village settlements that mark the origin of urbanization.

Network (transport) The entire regional **system** of transportation connections and **nodes** through which movement can occur.

Nevada See ***tierra nevada***.

New World Order A description of the international system resulting from the 1991 collapse of the Soviet Union whereby the balance of nuclear power theoretically no longer determines the destinies of **states**.

Node A center that functions as a point of **connectivity** within a regional **network** or **system**. All urban settlements possess this function, and the higher the position of a settlement in its **urban system** or **hierarchy**, the greater its nodality.

Nomadism Cyclical movement among a definite set of places. Nomadic peoples mostly are **pastoralists**.

Non-governmental organization (NGO) Legitimate organizations that operate independently from any form of government and do not function as for-profit businesses. They mostly seek to improve social conditions, but are not affiliated with political organizations.

North American Free Trade Agreement See **NAFTA**.

Northeast Passage The high-latitude sea route of the Arctic Ocean that follows the entire north coast of Eurasia from northern Norway in the west to the northeasternmost corner of Russia where it meets the Bering Strait. Increased seasonal melting of the Arctic ice cap in recent years has begun to open up this waterway as a summer route for shipping between Europe and East Asia.

Northwest Passage The high-latitude, Arctic Ocean sea route around North America extending from Alaska's Bering Strait in the west to the Davis Strait between Canada and Greenland in the east. Heightened summertime melting of the Arctic ice cap in recent years has increased the likelihood of opening up this waterway for shipping between Asia's Pacific Rim and the Atlantic seaboard of the Americas.

Nucleation Cluster; **agglomeration**.

Oasis An area, small or large, where the supply of water (from an **aquifer** or a major river such as the Nile) permits the transformation of the immediately surrounding **desert** into productive cropland.

Occidental Western. Also see ***Oriental***.

Offshore banking Term referring to financial havens for foreign companies and individuals, who channel their earnings to accounts in such a country (usually an "offshore" island-state) to avoid paying taxes in their home countries.

Oligarchs Opportunists in post-Soviet Russia who used their ties to government to enrich themselves.

OPEC (Organization of Petroleum Exporting Countries) The international oil *cartel* or syndicate formed by a number of producing countries to promote their common economic interests through the formulation of joint pricing policies and the limitation of market options for consumers. The 12 member-states (as of mid-2013) are: Algeria, Angola, Ecuador, Iran, Iraq, Kuwait, Libya, Nigeria, Qatar, Saudi Arabia, United Arab Emirates (UAE), and Venezuela.

Oriental The root of the word "oriental" is from the Latin for *rise*. Thus it has to do with the direction in which one sees the sun "rise"—the east; *oriental* therefore means Eastern. ***Occidental*** originates from the Latin for fall, or the "setting" of the sun in the west; *occidental* therefore means Western.

Orographic precipitation Mountain-induced precipitation, especially when air masses are forced to cross **topographic** barriers. Downwind areas beyond such a mountain range experience the relative dryness known as the **rain shadow effect**.

Outback The name given by Australians to the vast, peripheral, sparsely settled interior of their country.

Outer city The non-central-city portion of the American **metropolis**; no longer "sub" to the "urb," this outer ring was transformed into a full-fledged city during the late twentieth century.

Outsourcing Turning over the partial or complete production of a good or service to another party. In **economic geography**, this usually refers to a company arranging to have its products manufactured in a foreign country where labor and other costs are significantly lower than in the home country (which experiences a commensurate loss in jobs).

Overseas Chinese The approximately 50 million ethnic Chinese who live outside China. About two-thirds live in Southeast Asia, and many have become quite successful. A large number maintain links to China and as investors played a major economic role in stimulating the growth of **SEZs** and Open Cities in China's **Pacific Rim**.

Pacific Rim A far-flung group of countries and components of countries (extending clockwise on the map from New Zealand to Chile) sharing the following criteria: they face the Pacific Ocean; they exhibit relatively high levels of economic development, industrialization, and urbanization; their imports and exports mainly move across Pacific waters.

Pacific Ring of Fire Zone of crustal instability along **tectonic plate** boundaries, marked by earthquakes and volcanic activity, that ring the Pacific Ocean Basin.

Paddies (paddyfields) Ricefields.

Pandemic An outbreak of a disease that spreads worldwide.

Pangaea A vast, singular landmass consisting of most of the areas of the present-day continents. This *supercontinent* began to break up more than 200 million years ago when still-ongoing **plate** divergence and **continental drift** became dominant **processes** (see Fig. 6A-3).

Parallel An east-west line of **latitude** that is intersected at right angles by **meridians** of **longitude**.

Partition The subdivision of the British Indian Empire into India and Pakistan at the end of colonial rule on August 15, 1947. Shortly before their departure from what is now the South Asian realm, the British were persuaded to create separate countries for South Asia's massive Hindu (India) and Muslim (West and East Pakistan) populations. Boundaries were drawn hurriedly and ineffectively through thousands of kilometers of highly complicated cultural terrain, triggering the biggest mass migration in human history as millions fled their homes to be sure they were not on the "wrong" side of the new border. Both India and Pakistan (the successor to West Pakistan; East Pakistan seceded in 1971 to become independent Bangladesh) continue to suffer the consequences of the Partition that occurred 65 years ago, including the scars of four subsequent armed conflicts, the dangerous unresolved impasse in Kashmir, and persistent tensions magnified by the nuclear capabilities of both.

Pastoralism A form of **agricultural** activity that involves the raising of livestock.

Peasants In a **stratified** society, peasants are the lowest class of people who depend on **agriculture** for a living. But they often own no land at all and must survive as tenants or day workers.

Peninsula A comparatively narrow, finger-like stretch of land extending from the main landmass into the sea. Florida and Korea are examples.

Peon *(peone)* Term used in Middle and South America to identify people who often live in serfdom to a wealthy landowner; landless **peasants** in continuous indebtedness.

Per capita Capita means *individual.* Income, production, or some other measure is often given per individual.

Perforated state A **state** whose territory completely surrounds that of another state.

Periodic market Village market that is open every third day or at some other regular interval. Part of a regional network of similar markets in a preindustrial, rural setting where goods are brought to market on foot and barter remains a leading mode of exchange.

Peripheral development Spatial pattern in which a country's or region's development (and population) is most heavily concentrated along its outer edges rather than in its interior.

Periphery Used in conjunction with core areas at many geographic scales. At the scale of the city-region, it is the outlying tributary area served by a city that produces food and raw materials for the urban core in exchange for receiving goods and services. At the world scale, it is all those less advantaged parts of the Earth's surface lying outside the affluent global core.

Permafrost Permanently frozen water in the near-surface soil and bedrock of cold environments, producing the effect of completely frozen ground. Surface can thaw during brief warm season.

Physical geography The study of the geography of the physical (natural) world. Its subfields encompass **climatology, geomorphology, biogeography, soil geography, marine geography**, and water **resources**.

Physical landscape Synonym for **natural landscape**.

Physiographic political boundaries Political boundaries that coincide with prominent physical features in the **natural landscape**—such as rivers or the crest ridges of mountain ranges.

Physiographic region (province) A **region** within which there prevails substantial **natural-landscape** homogeneity, expressed by a certain degree of uniformity in surface **relief, climate**, vegetation, and soils.

Physiography Literally means *landscape description*, but commonly refers to the total **physical geography** of a place; includes all of the natural features on the Earth's surface, including landforms, **climate**, soils, vegetation, and water bodies.

Physiologic density The number of people per unit area of **arable** land.

Pilgrimage A journey to a place of great religious significance by an individual or by a group of people (such as a pilgrimage [*hajj*] to Mecca for **Muslims**).

Plantation A large estate owned by an individual, family, or corporation and organized to produce a cash crop. Almost all plantations were established within the tropics; in recent decades, many have been divided into smaller holdings or reorganized as cooperatives.

Plate tectonics Plates are bonded portions of the Earth's mantle and crust, averaging 100 kilometers (60 mi) in thickness. More than a dozen such plates exist (see Fig. G-4), most of continental proportions, and they are in motion. Where they meet one slides under the other **(subduction)**, crumpling the surface crust and producing significant volcanic and earthquake activity; a major mountain-building force.

Pleistocene Epoch Recent period of geologic time that spans the rise of humankind, beginning about 2 million years ago. Marked by *glaciations* (repeated advances of continental ice sheets) and more moderate *interglacials* (ice sheet contractions). Although the last 10,000 years are known as the **Holocene** Epoch, Pleistocene-like conditions seem to be continuing and we are most probably now living through another Pleistocene interglacial; thus the glaciers likely will return.

Polder Land reclaimed from the sea adjacent to the shore of the Netherlands by constructing dikes and then pumping out the water trapped behind them.

Police state A **state** in which the government (usually marked by dictatorial leadership) exercises totalitarian control over the political, social, and economic life of its citizens. Repression of the people includes rigid restrictions on their movements, freedom to communicate and express their views, and aggressive monitoring and enforcement by means of a secret-police force.

Political geography The study of the interaction of geographic space and political **process**; the spatial analysis of political phenomena and processes.

Political regime An assemblage of political structures that constitute a state; the form of government administered by those in power.

Pollution The release of a substance, through human activity, which chemically, physically, or biologically alters the air or water it is discharged into. Such a discharge negatively impacts the environment, with possible harmful effects on living organisms—including humans.

Population decline A decreasing national population. Russia, which now loses about half a million people per year, is the best example. Also see **population implosion**.

Population density The number of people per unit area. Also see **arithmetic density** and **physiologic density** measures.

Population distribution The way people have arranged themselves in geographic space. One of human geography's most essential expressions because it represents the sum total of the adjustments that a population has made to its natural, cultural, and economic environments. A population distribution map is included in every chapter in this book.

Population expansion (explosion) The rapid growth of the world's human population over the past century, attended by accelerating *rates* of increase.

Population geography The field of geography that focuses on the spatial aspects of **demography** and the influences of demographic change on particular countries and regions.

Population implosion The opposite of **population explosion**; refers to the declining populations of many European countries and Russia in which the **death rate** exceeds the **birth rate** and **immigration** rate.

Population movement See **migration** and **migratory movement**.

Population projection The future population total that demographers forecast for a particular country. For example, in the Data Table in Appendix B such projections are given for all the world's countries for 2025.

Population pyramid Graphic representation or *profile* of a national population according to age and gender. Such a diagram of age-sex structure typically displays the percentage of each age group (commonly in five-year increments) as a horizontal bar, whose length represents its relationship to the total population.

Primary sector/economic activity Activities engaged in the direct extraction of natural resources from the environment such as mining, fishing, lumbering, and especially agriculture.

Postindustrial economy Emerging economy, in the United States and a number of other highly advanced countries, as traditional industry is increasingly eclipsed by a higher-technology productive complex dominated by services, information-related, and managerial activities.

Primary economic activity Activities engaged in the direct extraction of **natural resources** from the environment such as mining, fishing, lumbering, and especially **agriculture**.

Primate city A country's largest city—ranking atop its urban **hierarchy**—most expressive of the national culture and usually (but not in every case) the capital city as well.

Process Causal force that shapes a spatial pattern as it unfolds over time.

Productive activities The major components of the spatial economy. For individual components see: **primary economic activity, secondary economic activity, tertiary economic activity, quaternary economic activity**, and **quinary economic activity**.

Protruded state Territorial shape of a **state** that exhibits a narrow, elongated land extension (or *protrusion*) leading away from the main body of territory.

Protrusion A pronounced extension of national territory that leads away from an otherwise compact state either as a long peninsula or land corridor.

Push-pull concept The idea that **migration** flows are simultaneously stimulated by conditions in the source area, which tend to drive people away, and by the perceived attractiveness of the destination.

Qanat In **desert** zones, particularly in Iran and western China, an underground tunnel built to carry **irrigation** water by gravity flow from surrounding mountains (where **orographic precipitation** occurs) to the arid flatlands below.

Quaternary economic activity Activities engaged in the collection, processing, and manipulation of *information*.

Quinary economic activity Managerial or control-function activity associated with decision-making in large organizations.

Racial profiling The use, by police and other security personnel, of an individual's race or ethnicity in the decision to engage in law enforcement.

Rain shadow effect The relative dryness in areas downwind of mountain ranges resulting from **orographic precipitation**, wherein moist air masses are forced to deposit most of their water content as they cross the highlands.

Rate of natural population increase See **natural increase rate**.

Realm See **geographic realm**.

Refugees People who have been dislocated involuntarily from their original place of settlement.

Region A commonly used term and a geographic concept of paramount importance. An **area** on the Earth's surface marked by specific criteria, which are discussed in the Introduction.

Regional boundary In theory, the line that circumscribes a **region**. But razor-sharp lines are seldom encountered, even in nature (e.g., a coastline constantly changes depending on the tide). In the **cultural landscape**, not only are regional boundaries rarely self-evident, but when they are ascertained by geographers they most often turn out to be **transitional** borderlands.

Regional complementarity Exists when two regions, through an exchange of raw materials and/or finished products, can specifically satisfy each other's demands.

Regional concept The geographic study of **regions** and regional distinctions.

Regional disparity The spatial unevenness in standard of living that occurs within a country, whose "average" overall income statistics invariably mask the differences that exist between the extremes of the affluent **core** and the poorer, disadvantaged **periphery**.

Regional geography Approach to geographic study based on the spatial unit of the **region**. Allows for an all-encompassing view of the world, because it utilizes and integrates information from geography's topical (**systematic**) fields, which are diagrammed in Figure G-14.

Regional state A "natural economic zone" that defies political boundaries and is shaped by the global economy of which it is a part; its leaders deal directly with foreign partners and negotiate the best terms they can with the national governments under which they operate.

Regionalism The consciousness of and loyalty to a **region** considered distinct and different from the **state** as a whole by those who occupy it.

Relative location The regional position or **situation** of a place relative to the position of other places. Distance, **accessibility**, and **connectivity** affect relative location.

Relict boundary A political boundary that has ceased to function, but the imprint of which can still be detected on the **cultural landscape**.

Relief Vertical difference between the highest and lowest elevations within a particular area.

Religious revivalism Religious movement whose objectives are to return to the foundations of that faith and to influence state policy. Often called *religious fundamentalism*; but in the case of Islam, **Muslims** prefer the term *revivalism*.

Relocation diffusion Sequential **diffusion process** in which the items being diffused are transmitted by their carrier agents as they relocate to new areas. The most common form of relocation diffusion involves the spreading of innovations by a **migrating** population.

Remittances Money earned by **emigrants** that is sent back to family and friends in their home country, mostly in cash; forms an important part of the economy in poorer countries.

Restrictive population policies Government policy designed to reduce the **rate of natural population increase**. China's one-child policy, instituted in 1979 after Mao's death, is a classic example.

Revivalism (religious) Religious movement whose objectives are to return to the foundations of that faith and to influence state policy. Often called *religious fundamentalism*; but in the case of Islam, **Muslims** prefer the term *revivalism*.

Rift valley The trough or trench that forms when a thinning strip of the Earth's crust sinks between two parallel faults (surface fractures).

Rural-to-urban migration The dominant **migration** flow from countryside to city that continues to transform the world's population, most notably in the less advantaged geographic realms.

Russification Demographic resettlement policies pursued by the central planners of the Soviet Empire (1924–1991), whereby ethnic Russians were encouraged to **emigrate** from the Russian Republic to the 14 non-Russian republics of the USSR.

Sahel Semiarid **steppeland** zone extending across most of Africa between the southern margins of the arid Sahara and the moister tropical **savanna** and forest zone to the south. Chronic drought, **desertification**, and overgrazing have contributed to severe famines in this area since 1970.

Satellite states The countries of eastern Europe under Soviet **hegemony** between 1945 and 1989. This tier of countries—the "satellites" captured in Moscow's "orbit" following World War II—was bordered on the west by the Iron Curtain and on the east by the USSR. Using the names then in force, they included Bulgaria, Czechoslovakia, East Germany, Hungary, Poland, and Romania.

Savanna Tropical grassland containing widely spaced trees; also the name given to the tropical wet-and-dry climate type (***Aw***).

Scale Representation of a real-world phenomenon at a certain level of reduction or generalization. In **cartography**, the ratio of map distance to ground distance; indicated on a map as a bar graph, representative fraction, and/or verbal statement. *Macroscale* refers to a large area of national proportions; *microscale* refers to a local area no bigger than a county.

Secession The act of withdrawing from a political entity, usually a **state,** as when the U.S. South tried unsuccessfully to secede from the United States in 1861 and sparked the Civil War.

Secondary economic activity Activities that process raw materials and transform them into finished industrial products; the *manufacturing* sector.

Sedentary Permanently attached to a particular area; a population fixed in its location; the opposite of **nomadic**.

Separate development The spatial expression of South Africa's "grand" **apartheid** scheme, whereby nonwhite groups were required to settle in segregated "homelands." The policy was dismantled when white-minority rule collapsed in the early 1990s.

Sex ratio A **demographic** indicator showing the ratio of males to females in a given population.

Shantytown Unplanned slum development on the margins of cities in disadvantaged countries, dominated by crude dwellings and shelters mostly made of scrap wood and iron, and even pieces of cardboard.

Sharecropping Relationship between a large landowner and farmers on the land wherein the farmers pay rent for the land they farm by giving the landlord a share of the annual harvest.

***Sharia* (law)** The strict criminal code based in Islamic law that prescribes corporal punishment, amputations, stonings, and lashing for both major and minor offenses. Its occurrence today is associated with the spread of **religious revivalism** in **Muslim** societies.

Shatter belt **Region** caught between stronger, colliding external cultural-political forces, under persistent stress, and often fragmented by aggressive rivals.

Shifting agriculture Cultivation of crops in recently cut-down and burned tropical-forest clearings, soon to be abandoned in favor of newly cleared nearby forest land. Also known as *slash-and-burn agriculture*.

Sinicization Giving a Chinese cultural imprint; Chinese **acculturation**. See also **Hanification**.

Site The internal locational attributes of an urban center, including its local spatial organization and physical setting.

Situation The external locational attributes of an urban center; its **relative location** or regional position with reference to other non-local places.

Small-island developing economies The additional disadvantages faced by lower-income island-states because of their often small territorial size and populations as well as overland **inaccessibility**. Limited resources require expensive importing of many goods and services; the cost of government operations per capita are higher; and local production is unable to benefit from **economies of scale**.

Social stratification In a layered or stratified society, the population is divided into a **hierarchy** of social classes. In an industrialized society, the working class is at the lower end; **elites** that possess capital and control the means of production are at the upper level.

Southern Ocean The ocean that surrounds Antarctica (discussed in Chapter 11).

Sovereignty Controlling power and influence over a territory, especially by the government of an autonomous state over the people it rules.

Spatial Pertaining to space on the Earth's surface. Synonym for *geographic(al)*.

Spatial diffusion The spatial spreading or dissemination of a **culture** element (such as a technological innovation) or some other phenomenon (e.g., a disease outbreak).

Spatial perspective: Broadly, the geographic dimension or expression of any phenomenon; more specifically, anything related to the organization of space on the Earth' surface.

Spatial system The components and interactions of a **functional region**, which is defined by the areal extent of those interactions.

Special Administrative Region (SAR) Status accorded the former dependencies of Hong Kong and Macau that were taken over by China, respectively, from the United Kingdom in 1997 and Portugal in 1999. Both SARs received guarantees that their existing social and economic systems could continue unchanged for 50 years following their return to China.

Special Economic Zone (SEZ) Manufacturing and export center in China, created since 1980 to attract foreign investment and technology transfers. Seven SEZs—all located on China's Pacific coast—currently operate: Shenzhen, adjacent to Hong Kong; Zhuhai; Shantou; Xiamen; Hainan Island, in the far south; Pudong, across the river from Shanghai; and Binhai New Area, next to the port of Tianjin.

State A politically organized territory that is administered by a sovereign government and is recognized by a significant portion of the international community. A state must also contain a permanent resident population, an organized economy, and a functioning internal circulation system.

State boundaries The borders that surround **states** which, in effect, are derived through contracts with neighboring states negotiated by treaty.

State capitalism Government-controlled corporations competing under free-market conditions, usually in a tightly regimented society.

State formation The creation of a **state**, exemplifying traditions of human **territoriality** that go back thousands of years.

State planning Involves highly centralized control of the national planning process, a hallmark of communist economic systems. Soviet central planners mainly pursued a grand political design in assigning production to particular places; their frequent disregard of the principles of **economic geography** contributed to the eventual collapse of the USSR.

State territorial morphology A **state's** geographical shape, which can have a decisive impact on its spatial cohesion and political viability. A **compact** shape is most desirable; among the less efficient shapes are those exhibited by **elongated, fragmented, perforated**, and **protruded** states.

Stateless nation A national group that aspires to become a **nation-state** but lacks the territorial means to do so.

Steppe Semiarid grassland; short-grass prairie. Also the name given to the semiarid climate type (***BS***).

Stratification (social) In a layered or stratified society, the population is divided into a **hierarchy** of social classes. In an industrialized society, the working class is at the lower end; **elites** that possess capital and control the means of production are at the upper level. In the traditional **caste system** of Hindu India, the "untouchables" form the lowest class or caste, whereas the still-wealthy remnants of the princely class are at the top.

Subduction In **plate tectonics**, the **process** that occurs when an oceanic plate converges head-on with a plate carrying a continental landmass at its leading edge. The lighter continental plate overrides the denser oceanic plate and pushes it downward.

Subsequent boundary A political boundary that developed contemporaneously with the evolution of the major elements of the **cultural landscape** through which it passes.

Subsistence Existing on the minimum necessities to sustain life; spending most of one's time in pursuit of survival.

Subsistence agriculture Farmers who eke out a living on a small plot of land on which they are only able to grow enough food to support their families or at best a small community.

Subtropical Convergence A narrow marine **transition zone**, girdling the globe at approximately latitude 40°S, that marks the equatorward limit of the frigid **Southern Ocean** and the poleward limits of the warmer Atlantic, Pacific, and Indian oceans to the north.

Suburban downtown In the United States (and increasingly in other high-income countries), a significant concentration of major urban activities around a highly accessible suburban location, including retailing, light industry, and a variety of leading corporate and commercial operations. The largest are now coequal to the American central city's **central business district (CBD)**. A leading feature of the **outer city**.

Sunbelt The popular name given to the southern tier of the United States, which is anchored by the mega-States of California, Texas, and Florida. Its warmer climate, superior recreational opportunities, and other amenities have been attracting large numbers of relocating people and activities since the 1960s; broader definitions of the Sunbelt also include much of the western United States, even Colorado and the coastal Pacific Northwest.

Superimposed boundary A political boundary emplaced by powerful outsiders on a developed human landscape. Frequently ignores preexisting cultural-spatial patterns, such as the border that still divides North and South Korea.

Supranational A venture involving three or more **states**—political, economic, and/or cultural cooperation to promote shared objectives.

System Any group of objects or institutions and their mutual interactions. Geography treats systems that are expressed spatially, such as in **functional regions**.

Systematic geography Topical geography: **cultural, political, economic geography**, and the like.

Taiga The subarctic, mostly **coniferous** snowforest that blankets northern Russia and Canada south of the **tundra** that lines the Arctic shore. Known as the **boreal forest** in North America.

Takeoff Economic concept to identify a stage in a country's **development** when conditions are set for a domestic Industrial Revolution.

Taliban The term means "students" or "seekers of religion." Specifically, refers to the Islamist militia group that emerged from ***madrassas*** in Pakistan and ruled neighboring Afghanistan between 1996 and 2001; it has been trying to regain control of that country in its continuing conflict with U.S.-led NATO troops. Taliban rule, in adherence with an extremist interpretation of ***Sharia*** law, was marked by one of the most virulent forms of militant Islam ever seen.

Tar sands The main source of oil from non-liquid petroleum reserves. The oil is mixed with sand and requires massive open-pit mining as well as a costly, complicated process to extract it. The largest known deposits are located in the northeast of Canada's province of Alberta, and by most estimates these Athabasca Tar Sands constitute one of the largest oil reserves in the world. The high oil prices of recent years have led to greatly expanded production here, but the accompanying environmental degradation caused by strip-mining and waste disposal has triggered a widening protest movement that may limit the exploitation of this resource.

Technopole A planned techno-industrial complex (such as California's Silicon Valley) that innovates, promotes, and manufactures the products of the **postindustrial** informational economy.

Tectonic plates The slabs of heavier rock on which the lighter rocks of the continents rest. The plates are in motion, propelled by gigantic circulation cells in the red-hot, molten rock below. Most earthquakes and volcanic eruptions are associated with collisions of the mobile plates, as is the building of mountain chains.

Terms of trade In international economics refers to an agreement between trading partners that stipulates the quantity of imports that can be purchased by one country through the sale of a fixed quantity of exports to the other.

Terracing The transformation of a hillside or mountain slope into a step-like sequence of horizontal fields for intensive cultivation.

Territoriality A country's or more local community's sense of property and attachment toward its territory, as expressed by its determination to keep it inviolable and strongly defended.

Territorial sea Zone of seawater adjacent to a country's coast, held to be part of the national territory and treated as a component of the sovereign **state**.

Tertiary economic activity Activities that engage in *services*—such as transportation, banking, retailing, education, and routine office-based jobs.

Tierra caliente The lowest of the **altitudinal zones** into which the human settlement of Middle and South America is classified according to elevation. The ***caliente*** is the hot humid coastal plain and adjacent slopes up to 750 meters (2500 ft) above sea level. The natural vegetation is the dense and luxuriant tropical rainforest; the crops include sugar and bananas in the lower areas, and coffee, tobacco, and corn along the higher slopes.

Tierra fría Cold, high-lying **altitudinal zone** of settlement in Andean South America, extending from about 1800 meters (6000 ft) in elevation up to nearly 3600 meters (12,000 ft). **Coniferous** trees stand here; upward they change into scrub and grassland. There are also important pastures within the *fría*, and wheat can be cultivated.

Tierra helada In Andean South America, the highest-lying habitable **altitudinal zone**—ca. 3600 to 4500 meters (12,000 to 15,000 ft)—between the tree line (upper limit of the ***tierra fría***) and the snow line (lower limit of the ***tierra nevada***). Too cold and barren to support anything but the grazing of sheep and other hardy livestock.

Tierra nevada The highest and coldest **altitudinal zone** in Andean South America (lying above 4500 meters [15,000 ft]), an uninhabitable environment of permanent snow and ice that extends upward to the Andes' highest peaks of more than 6000 meters (20,000 ft).

Tierra templada The intermediate **altitudinal zone** of settlement in Middle and South America, lying between 750 meters (2500 ft) and 1800 meters (6000 ft) in elevation. This is the "temperate" zone, with moderate temperatures compared to the ***tierra caliente*** below. Crops include coffee, tobacco, corn, and some wheat.

Topography The surface configuration of any segment of **natural landscape**.

Toponym Place name.

Transculturation Cultural borrowing and two-way exchanges that occur when different **cultures** of approximately equal complexity and technological level come into close contact.

Transferability The capacity to move a good from one place to another at a bearable cost; the ease with which a commodity may be transported.

Transition zone An area of spatial change where the **peripheries** of two adjacent **realms** or **regions** join; marked by a gradual shift (rather than a sharp break) in the characteristics that distinguish these neighboring geographic entities from one another.

Transmigration The now-ended policy of the Indonesian government to induce residents of the overcrowded, **core-area** island of Jawa to move to the country's other islands.

Treaty ports **Extraterritorial enclaves** in China's coastal cities, established by European colonial invaders under unequal treaties enforced by gunboat diplomacy.

Triple Frontier The turbulent and chaotic area in southern South America that surrounds the convergence of Brazil, Argentina, and Paraguay. Lawlessness pervades this haven for criminal elements, which is notorious for money laundering, arms and other smuggling, drug trafficking, and links to terrorist organizations, including money flows to the Middle East.

Tropical deforestation The clearing and destruction of tropical rainforests in order to make way for expanding settlement frontiers and the exploitation of new economic opportunities.

Tsunami A seismic (earthquake-generated) sea wave that can attain gigantic proportions and cause coastal devastation. The tsunami of December 26, 2004, centered in the Indian Ocean near the Indonesian island of Sumatera (Sumatra), produced the first great natural disaster of the twenty-first century. Our new century's second major tsunami disaster occurred along the coast of Japan's northeastern Honshu Island on March 11, 2011.

Tundra The treeless plain that lies along the Arctic shore in northernmost Russia and Canada, whose vegetation consists of mosses, lichens, and certain hardy grasses.

Turkestan Northeasternmost region of the North Africa/Southwest Asia realm. Known as Soviet Central Asia before 1992, its five (dominantly Islamic) former Soviet Socialist Republics have become the independent countries of Kazakhstan, Uzbekistan, Turkmenistan, Kyrgyzstan, and Tajikistan. Today Turkestan has expanded to include a sixth state, Afghanistan.

Turkish model In the wake of the regime changes in the North Africa/Southwest Asia realm brought about by the "Arab Spring" of 2011, moderates have cited Turkey as the best model of democratic governance for this part of the world. Specifically, this involves a multi-party democracy that has a place for, but is not dominated by, Islamic political parties.

Uneven development The notion that economic development varies spatially, a central tenet of **core-periphery relationships** in realms, regions, and lesser geographic entities.

Unitary state A **nation-state** that has a centralized government and administration that exercises power equally over all parts of the **state.**

United Nations See discussion under **League of Nations.**

Unity of place The great German natural scientist Alexander von Humboldt's notion that in a particular locale or region intricate connections exist among climate, geology, biology, and human cultures. This laid the foundation for modern geography as an *integrative discipline* marked by a spatial perspective.

Urbanization A term with a variety of connotations. The proportion of a country's population living in urban places is its level of urbanization. The **process** of urbanization involves the movement to, and the clustering of, people in towns and cities—a major force in every geographic realm today. Another kind of urbanization occurs when an expanding city absorbs rural countryside and transforms it into suburbs; in the case of cities in disadvantaged countries, this also generates peripheral **shantytowns.**

Urban (metropolitan) area The entire built-up, non-rural area and its population, including the most recently constructed suburban appendages. Provides a better picture of the dimensions and population of such an area than the delimited municipality (central city) that forms its heart.

Urban realms model A spatial generalization of the contemporary large American city. It is shown to be a widely dispersed, multinodal **metropolis** consisting of increasingly independent zones or *urban realms*, each focused on its own **suburban downtown**; the only exception is the shrunken central realm, which is focused on the central city's **central business district** (see Fig. 3A-11).

Urban system A **hierarchical** network or grouping of urban areas within a finite geographic area, such as a country.

Veld See **highveld.**

Voluntary migration Population movement in which people relocate in response to perceived opportunity, not because they are forced to **migrate.**

Wahhabism A particularly virulent form of (Sunni) **Muslim revivalism** that was made the official faith when the modern **state** of Saudi Arabia was founded in 1932. Adherents call themselves "Unitarians" to signify the strict fundamentalist nature of their beliefs.

Wallace's Line The zoogeographical boundary proposed by Alfred Russel Wallace that separates the marsupial fauna of Australia and New Guinea from the non-marsupial fauna of Indonesia (see Figure 11-4).

Weather The immediate and short-term conditions of the **atmosphere** that impinge on daily human activities.

West Wind Drift The clockwise movement of water as a current that circles around Antarctica in the **Southern Ocean.**

Westernization The Western view of the **process** of **modernization** that involves the establishment of **urbanization**, a market (money) economy, improved circulation, formal schooling, adoption of foreign innovations, and the breakdown of traditional society. Non-Westerners mostly see "modernization" as an outgrowth of **colonialism** and often argue that traditional societies can be modernized without being Westernized.

Windward The exposed, upwind side of a **topographic** barrier that faces the winds that flow across it.

World-city Either London, New York, or Tokyo. The highest-ranking urban centers of **globalization** with financial, high-technology, communications, engineering, and related industries reflecting the momentum of their long-term growth and **agglomeration.**

INDEX

A page number followed by "f" indicates the entry is within a figure; a page number followed by "t" indicates the entry is within a table.

C

D

E

F

G

N

Q

R

S